CHI 2000 Conference Proceedings
Conference on Human Factors in Computing Systems

Editors
..er, *Conference and Technical Program Co-Chair*
..lus, *Conference and Technical Program Co-Chair*
Mary Czerwinski, *Papers Co-Chair*
Fabio Paternò, *Papers Co-Chair*

Associate Editor
Steven Pemberton

 Sponsored by ACM's Special Interest Group on Computer-Human Interaction (ACM SIGCHI)

ACM Order Info

The Association for Computing Machinery, Inc.
1515 Broadway
New York, NY 10036

Copyright © 2000 by the Association for Computing Machinery, Inc. (ACM). Permission to make digital or hard copies of part or all of this work for personal or classroom use is granted without fee provided that the copies are not made or distributed for profit or commercial advantage, and that copies bear this notice and the full citation on the first page. Copyrights for components of this work owned by others than ACM must be honored. Abstracting with credit is permitted. To copy otherwise, to republish, to post on servers, or to redistribute to lists, requires prior specific permission and/or a fee. Request permission to republish from: Publications Department, ACM, Inc. FAX to +1 212 869 0481 or email to permissions@acm.org.

For other copying of articles that carry a code at the bottom of the first or last page or screen display, copying is permitted provided that the per copy fee indicated in the code is paid through the Copyright Clearance Center, 222 Rosewood Drive, Danvers, MA 01923.

. . . In Proceedings of the CHI 2000, ACM, New York, pp. 00,
The Hague, Netherlands April 1-5, 2000.

Ordering Information

ACM Members
A limited number of copies are available at the ACM member discount. Send order with payment in US dollars to:

ACM Order Department
P.O. Box 11414
New York, NY 10286-1414

Credit Card Orders
From the U.S. and Canada call:
+1 800 342 6626
From New York metropolitan area and outside of the U.S. call:
Tel: +1 212 626 0500 or
Fax: +1 212 944 1318

Please include your ACM Member Number and the ACM Order Number with your order:

ACM Order Number: 608001
ACM ISBN Number: 1-58113-216-6

Nonmembers
Nonmembers orders placed within the U.S. should be directed to:

Pearson Education Order Department
200 Old Tappan road
Old Tappan, NJ 07675

Call TOLL FREE: +1 800 922 0579
FAX: +1 800 445 6991
PROMPT + PLUS: +1 800 742 5599

Addison-Wesley will pay postage and handling on orders accompanied by a check. Credit card orders may be placed by mail or by calling the Addison-Wesley Order Department at the number above. Follow-up inquiries should be directed to the Customer Service Department at the same number.

Please include the Addison-Wesley ISBN number with your order:

A-W ISBN 0-201-48563-X

Nonmember orders from outside the U.S. should be addressed as noted below:

Europe/Middle East/Africa
Pearson Education
Edinburgh Gate Harlow
Essex CM20 2JE
UNITED KINGDOM
Tel: +44 1279 623 925
Fax: +44 1279 623 627

Asia
Pearson Education
317 Alexandra Road #04-01
IKEA Building
Singapore 159965
Tel: +65 476 4688
Fax: +65 378 0370

Japan
Pearson Education
Nishi-Shinjuku, KF Building
8-14-24 Nishi-Shinjuku, Shinjuku, ku
Tokyo 160-0023
Japan
Tel: +813 3365 9224
Fax: +813 3365 9225

Australia/New Zealand
Pearson Education Australia
Level 2 Unit 4
14 Aquatic Drive
Frenchs Forest NSW 2086
Australia
Tel: +612 9454 2200
Fax: +612 9453 0089

Latin America
Pearson Education
Calle 4, #25 2nd Piso
Fracc. Industrial Alce Blanco
Naucalpan de Juarez
Estado de Mexico 53330
Tel: +525 358 8400
Fax: +525 358 8400

Canada
Pearson Education
26 Prince Andrew Place
P.O. Box 580
Don Mills, Ontario M3C 2T8
Tel: +416 447 5101
Fax: +416 443 0948

vii	Society Welcome
viii	Co-Chairs' Welcome
ix	Acknowledgements
x	Committees
xvii	Technical Program Overview

Papers: Beyond the Workplace
Tuesday, 4 April 11:00 – 12:30

1 **Unleashed: Web Tablet Integration into the Home**
Anne McClard, Patricia Somers, *MediaOne Labs*

9 **Predicting Text Entry Speed On Mobile Phones**
Miika Silfverberg, *Nokia Research Center*
I. Scott MacKenzie, *York University*
Panu Korhonen, *Nokia Research Center*

17 **Developing a Context-aware Electronic Tourist Guide: Some Issues and Experiences**
Keith Cheverst, Nigel Davies, Keith Mitchell, Adrian Friday, Christos Efstratiou, *Lancaster University*

Papers: Multi-Hand + Multi-DOF
Tuesday, 4 April 11:00 – 12:30

25 **Measuring the Allocation of Control in a 6 Degree-of-Freedom Docking Experiment**
Maurice Masliah, Paul Milgram, *University of Toronto*

33 **Symmetric Bimanual Interaction**
Ravin Balakrishnan, *University of Toronto & Alias/Wavefront*
Ken Hinckley, *Microsoft Research*

41 **Two-Handed Input Using a PDA and a Mouse**
Brad A. Myers, Kin Pou Lie, Bo-Chieh Yang: *Carnegie Mellon University*

Papers: Agents
Tuesday, 4 April 14:00 – 15:30

49 **The Effects of Animated Characters on Anxiety, Task Performance, and Evaluations of User Interfaces**
Byron Reeves, Raoul Rickenberg, *Stanford University*

57 **Helper Agent: Designing an Assistant for Human-Human Interaction in a Virtual Meeting Space**
Katherine Isbister, *NTT Open Lab*
Hideyuki Nakanishi, Toru Ishida, *Kyoto University*
Clifford Nass, *Stanford University*

65 **Agents to Assist in Finding Help**
Henry Lieberman, Adriana Vivacqua, *MIT Media Lab*

Papers: Communication Environments
Tuesday, 4 April 14:00 – 15:30

73 **Lurker Demographics: Counting the Silent**
Blair Nonnecke, Jenny Preece, *University of Maryland Baltimore County*

81 **Talking In Circles: Designing A Spatially-Grounded Audioconferencing Environment**
Roy Rodenstein, Judith Donath, *MIT Media Lab*

89 **Jotmail: A Voicemail Interface That Enables You To See What Was Said**
Steve Whittaker, *ATT-Labs Research*
Julia Hirschberg, *AT&T*
Richard Davis, *Virtual Ink*
Urs Muller, *AT&T*

Papers: Models
Tuesday, 4 April 16:00 – 17:30

97 **Instructional Interventions in Computer-Based Tutoring: Differential Impact on Learning Time and Accuracy**
Albert Corbett, *Carnegie Mellon University*
Holly Trask, *American Management Systems*

105 **Keystroke Level Analysis of Email Message Organization**
Olle Bälter, *Interaction and Presentation Laboratory*

113 **Using Naming Time to Evaluate Quality Predictors for Model Simplification**
Benjamin Watson, Alinda Friedman, Aaron McGaffey, *University of Alberta*

Papers: Tangible UI Systems
Tuesday, 4 April 16:00 – 17:30

121 **Interactive Textbook and Interactive Venn Diagram: Natural and Intuitive Interfaces**
Hideki Koike, Yoshinori Kobayashi, Hiroaki Tobita, Motoki Kobayashi, *University of Electro-Communications*
Yoichi Sato, *University of Tokyo*

129 **curlybot: Designing a New Class of Computational Toys**
Phil Frei, Victor Su, Bakhtiar Mikhak, Hiroshi Ishii, *MIT Media Lab*

137 **HandSCAPE: A Vectorizing Tape Measure for On-Site Measuring Applications**
Jay Lee, Victor Su, Sandia Ren, Hiroshi Ishii, *MIT Media Lab*

Papers: Bringing Order Out of Chaos
Tuesday, 4 April 16:00 – 17:30

145 **Bringing Order to the Web: Automatically Categorizing Search Results**
Susan Dumais, *Microsoft Research*
Hao Chen, *University of California at Berkeley*

Table of Contents

153 Enhancing a Digital Book with a Reading Recommender
Allison Woodruff, Rich Gossweiler, James Pitkow, Ed H. Chi, Stuart K. Card, *Xerox PARC*

161 The Scent of a Site: A System for Analyzing and Predicting Information Scent, Usage, and Usability of a Web Site
Ed H. Chi, Peter Pirolli, James Pitkow, *Xerox PARC*

Papers: Video Summarization
Wednesday, 5 April 8:30 – 10:00

169 Browsing Digital Video
Francis Li, *Group for User Interface Research*
Anoop Gupta, Elizabeth Sanocki, Liwei He, Yong Rui, *Microsoft Research*

177 Comparing Presentation Summaries: Slides vs. Reading vs. Listening
Liwei He, Elizabeth Sanocki, Anoop Gupta, Jonathan Grudin, *Microsoft Research*

185 An Interactive Comic Book Presentation for Exploring Video
John Boreczky, Andreas Girgensohn, Gene Golovchinsky, Shingo Uchihashi, *FX Palo Alto Laboratory*

Papers: Emotions and Values
Wednesday, 5 April 8:30 – 10:00

193 Face to InterFace: Facial Affect in (Hu)Man and Machine
Diane J. Schiano, Sheryl M. Ehrlich, Krishnawan Rahardja, Kyle Sheridan, *Interval Research*

201 Hedonic and Ergonomic Quality Aspects Determine a Software's Appeal
Marc Hassenzahl, Axel Platz, Michael Burmester, Katrin Lehner, *Siemens AG*

209 Alternatives: Exploring Information Appliances Through Conceptual Design Proposals
Bill Gaver, Heather Martin, *Royal College of Art*

Papers: Tangible UI Design Issues
Wednesday, 5 April 8:30 – 10:00

217 An Observational Study of How Objects Support Engineering Design Thinking and Communication: Implications for the Design of Tangible Media
Margot Brereton, Ben McGarry, *University of Queensland*

225 Tagged Handles: Merging Discrete and Continuous Manual Control
Karon MacLean, Scott Snibbe, *Interval Research*
Golan Levin, *MIT Media Lab*

233 Traversable Interfaces Between Real and Virtual Worlds
Boriana Koleva, Holger Schnädelbach, Steve Benford, Chris Greenhalgh, *University of Nottingham*

Papers: WWW Navigation Aids
Wednesday, 5 April 11:00 –12:30

241 Tradeoffs In Displaying Peripheral Information
Paul Maglio, Christopher S. Campbell, *IBM Almaden Research Center*

249 The Impact of Fluid Documents on Reading and Browsing: An Observational Study
Polle Zellweger, Jock Mackinlay, Bay-Wei Chang, *Xerox PARC*
Susan Harkness Regli, *Carnegie Mellon University*

257 Effects of Contextual Navigation Aids on Browsing Diverse Web Systems
Joonah Park, Jinwoo Kim, *Yonsei University*

Papers: Eye Gaze
Wednesday, 5 April 11:00 –12:30

265 Interacting with Eye Movements in Virtual Environments
Vildan Tanriverdi, Robert Jacob, *Tufts University*

273 Intelligent Gaze-Added Interfaces
Dario Salvucci, *Cambridge Basic Research*
John Anderson, *Carnegie Mellon University*

281 Evaluation of Eye Gaze Interaction
Linda Sibert, *Naval Research Laboratory*
Robert Jacob, *Tufts University*

Papers: User Experience in E-Commerce
Wednesday, 5 April 11:00 –12:30

289 Enriching Buyers' Experiences: The Smartclient Approach
Pearl Pu, Boi Faltings, *EPFL*

297 Quality is in the Eye of the Beholder: Meeting Users' Requirements for Internet Quality of Service
Anna Bouch, *University College London*
Allan Kuchinsky, Nina Bhatti, *Hewlett Packard Labs*

305 What Makes Internet Users Visit Cyber Stores Again? Key Design Factors for Customer Loyalty
Jungwon Lee, Jinwoo Kim, *Yonsei University*
Jae Yun Moon, *New York University*

Papers: Speech
Wednesday, 5 April 14:00 – 15:30

313 Speak Out and Annoy Someone: Experiences with Intelligent Kiosks
Andrew Christian, Brian Avery, *Compaq*

321 The Effect of Task Conditions on the Comprehensibility of Synthetic Speech
Jennifer Lai, *IBM T.J. Watson Research Center*
David Wood, *IBM Research*
Michael Considine, *Rice University*

329 Does Computer-Generated Speech Manifest Personality? An Experimental Test of Similarity-Attraction
Clifford Nass, Kwan Min Lee, *Stanford University*

Papers: Usability
Wednesday, 5 April 14:00 – 15:30

337 A Toolkit for Strategic Usability: Results from Workshops, Panels, and Surveys
Stephanie Rosenbaum, *Tec-Ed, Inc.*
Janice Anne Rohn, *Siebel Systems, Inc.*
Judee Humburg, *JL Humburg Associates*

345 Measuring Usability: Are Effectiveness, Efficiency, and Satisfaction Really Correlated?
Erik Frøkjær, *University of Copenhagen*
Morten Hertzum, *Centre for Human-Machine Interaction*
Kasper Hornbæk, *University of Copenhagen*

353 The Streamlined Cognitive Walkthrough Method: Working Around Social Constraints Encountered in a Software Development Company
Richard Spencer, *Microsoft*

Papers: Novel Input
Wednesday, 5 April 16:00 – 17:30

360 Visual Similarity of Pen Gestures
A. Chris Long, James Landay, Lawrence Rowe, Joseph Michiels, *University of California at Berkeley*

368 Providing Integrated Toolkit-Level Support for Ambiguity in Recognition-Based Interfaces
Jennifer Mankoff, *Georgia Institute of Technology*
Scott Hudson, *Carnegie Mellon University*
Gregory D. Abowd, *Georgia Institute of Technology*

376 Programming and Enjoying Music with Your Eyes Closed
Steffen Pauws, Don Bouwhuis,
IPO, Center for User-System Interaction
Berry Eggen, *Philips Research Laboratories Eindhoven*

Papers: Awareness and Gaze in Group Communication
Wednesday, 5 April 16:00 – 17:30

384 Presenting to Local and Remote Audiences: Design and Use of the TELEP System
Gavin Jancke, *Microsoft*
Jonathan Grudin, Anoop Gupta, *Microsoft Research*

392 Coming to the Wrong Decision Quickly: Why Awareness Tools Must be Matched with Appropriate Tasks
Jonathan Cadiz, *Microsoft Research*
Alberto Espinosa, Luis Rico-Gutierrez, Robert Kraut,
William Scherlis, *Carnegie Mellon University*
Glenn Lautenbacher, *University of Pittsburgh*

400 Gaze Communication Using Semantically Consistent Spaces
Michael Taylor, Simon Rowe,
Canon Research Centre Europe

Papers: Haptic Force Feedback
Thursday, 6 April 8:30 – 10:00

408 Eye-Hand Coordination with Force Feedback
Roland Arsenault, Colin Ware,
University of New Brunswick

415 Putting the Feel in 'Look and Feel'
Ian Oakley, Marilyn Rose McGee, Stephen Brewster,
Philip Gray, *University of Glasgow*

423 Force-Feedback Improves Performance For Steering And Combined Steering-Targeting Tasks
Jack Tigh Dennerlein, *Harvard University*
David Martin, *Dartmouth College*
Christopher Hasser, *Stanford University*

Papers: Glimpses of the Future
Thursday, 6 April 8:30 – 10:00

430 Power Browser: Efficient Web Browsing for PDAs
Orkut Buyukkokten, Hector Garcia-Molina,
Andreas Paepcke, Terry Winograd, *Stanford University*

438 A Diary Study of Information Capture in Working Life
Barry Brown, Abigail Sellen, Kenton O'Hara,
Hewlett Packard Labs

446 Instrumental Interaction: An Interaction Model for Designing Post-WIMP User Interfaces
Michel Beaudouin-Lafon, *University of Aarhus*

Papers: Chat
Thursday, 6 April 11:00 – 12:30

454 Anchored Conversations: Chatting in the Context of a Document
Elizabeth Churchill, Jonathan Trevor, Lester Nelson,
FX Palo Alto Laboratory
Sara A. Bly, *Sara A. Bly Consulting*
Davor Cubranic, *University of British Columbia*

462 The Social Life of Small Graphical Chat Spaces
Marc Smith, Shelly Farnham, Steven Drucker,
Microsoft Research

470 The Effect of Communication Modality on Cooperation in Online Environments
Carlos Jensen, *Georgia Institute of Technology*
Shelly Farnham, Steven Drucker, *Microsoft Research*
Peter Kollock, *University of California at Los Angeles*

Table of Contents

Papers: 3D Environments
Thursday, 6 April 11:00 – 12:30

478 Using a Large Projection Screen as an Alternative to Head-Mounted Displays for Virtual Environments
Emilee Patrick, *Motorola Labs*
Dennis Cosgrove, Aleksandra Slavkovic,
Jennifer Ann Rode, Thom Verratti, Greg Chiselko,
Carnegie Mellon University

486 Alice: Lessons Learned from Building a 3D System for Novices
Matthew Conway, Steve Audia, Tommy Burnette,
Jim Durbin, Rich Gossweiler, Shuichi Koga, Chris Long,
Beth Mallory, Steve Miale, Kristen Monkaitis, James Patten,
Joe Shochet, David Staack, Richard Stoakley, John Viega,
Jeff White, George Williams, *University of Virginia*
Dennis Cosgrove, Kevin Christiansen, Rob Deline,
Jeff Pierce, Brian Stearns, Chris Sturgill, Randy Pausch,
Carnegie Mellon University

494 The Task Gallery: A 3D Window Manager
George Robertson, Maarten van Dantzich, Mary Czerwinski,
Ken Hinckley, David Thiel, *Microsoft Research*
Daniel Robbins, Kirsten Risden, Vadim Gorokhovsky,
Microsoft

Papers: Tools for Design
Thursday, 6 April 14:00 – 15:30

502 A Comparison of Tools for Building GOMS Models
Lynn Baumeister, Bonnie John, *Carnegie Mellon University*
Mike Byrne, *Rice University*

510 DENIM: Finding a Tighter Fit Between Tools and Practice for Web Site Design
James Lin, Mark Newman, Jason Hong, James Landay,
University of California at Berkeley

518 Tool Support for Cooperative Object-Oriented Design: Gesture Based Modeling on an Electronic Whiteboard
Christian Heide Damm, Klaus Marius Hansen,
Michael Thomsen, *University of Aarhus*

Papers: 3D Input
Thursday, 6 April 14:00 – 15:30

526 The Cubic Mouse:
A New Device for Three-Dimensional Input
Bernd Fröhlich, John Plate, *GMD*

532 The Role of Contextual Haptic and Visual Constraints on Object Manipulation in Virtual Environments
Yanqing Wang, Christine L. MacKenzie,
Simon Fraser University

540 Non-Isomorphic 3D Rotational Techniques
Ivan Poupyrev, *ATR MIC Labs*
Suzanne Weghorst, *University of Washington*
Sidney Fels, *University of British Columbia*

Papers: Story Telling
Thursday, 6 April 14:00 – 15:30

548 Joking, Storytelling, Artsharing, Expressing Affection: A Field Trial of How Children and Their Social Network Communicate with Digital Images in Leisure Time
Anu Mäkelä, *Helsinki University of Technology*
Verena Giller, Manfred Tscheligi, Reinhard Sefelin,
CURE & University of Vienna

556 Designing Storytelling Technologies to Encourage Collaboration Between Young Children
Steve Benford, Victor Bayon, Rob Ingram, Helen Neale,
Claire O'Malley, Danaë Stanton, *University of Nottingham*
Benjamin Bederson, Allison Druin, Juan Pablo Hourcade,
University of Maryland
Karl-Petter Åkesson, Pär Hansson, Kristian Simsarian, *SICS*
Yngve Sundblad, Gustav Taxén, *Royal Institute of Technology*

564 Storytelling with Digital Photographs
Marko Balabanović, Gregory Wolff, *Ricoh Silicon Valley*
Lonny Chu, *Stanford University*

572 Video Figures

573 Author Index

574 Keyword Index

579 Color Plates

Welcome
From the ACM SIGCHI Chair and Executive Vice-Chair

We wish you a warm welcome to CHI 2000! The CHI conference provides a forum for people to meet both formally and informally, to share and to learn. We trust that you will find here the intellectually exciting and personally rewarding experiences that bring people back to this conference year after year.

The CHI conference is sponsored by ACM's Special Interest Group on Computer-Human Interaction (SIGCHI). SIGCHI is an international group of researchers, practitioners, educators, designers, students and others who share an interest in human-computer interaction. SIGCHI is committed to advancing the field of human-computer interaction and supporting the exchange of information within the HCI community. This conference will provide you with an opportunity to learn more about SIGCHI activities and to explore taking an active role in one or more such activities. You can do this by attending the Newcomer's Orientation at the conference, stopping by the SIGCHI booth, attending the SIGCHI Business Meeting or talking to any one of the Executive Committee members who are identified by the EC ribbon on their badges.

While the CHI conference is the largest and most visible activity of SIGCHI, we also support numerous other conferences. They include conferences on computer-supported cooperative work (CSCW), virtual reality software and technology (VRST), user interface software and technology (UIST), design of interactive systems (DIS), intelligent user interfaces (IUI), design of augmented reality environments (DARE), creativity and cognition (C&C) and universal usability (CUU). SIGCHI also distributes the quarterly *SIGCHI Bulletin*, a research journal, *Transactions on Information Systems* (TOCHI), and a magazine for practitioners called *Interactions*.

The SIGCHI Development Fund has been supporting HCI activities proposed by SIGCHI members for many years. If you have ideas for advancing our field or for communicating about our field to others, we encourage you to submit a proposal. Details about how to submit can be found on our SIGCHI website. The SIGCHI website also has many other useful pieces of information from descriptions of the many conferences and workshops we run to access to the HCI Bibliography. We urge you to visit it regularly to find out what is happening in the world of HCI - http://www.acm.org/sigchi/.

On behalf of the SIGCHI Executive Committee and all SIGCHI members we thank and congratulate all those volunteers who have participated in creating and shaping the CHI 2000 conference into a valuable experience for each and every one of us.

Marilyn Tremaine, SIGCHI Chair
Wendy Mackay, SIGCHI Executive Vice Chair

Conference Partners

ACM SIGCAPH
ACM SIGDOC
ACM SIGGRAPH
ACM SIGGROUP
ACM SIGWEB
Association Francophone d'Interaction Homme-Machine (AFIHM)
Austrian Computer Society (OCG)
British HCI Group (BCS)
CHISIG New Zealand
European Association of Cognitive Ergonomics (EACE)
German Society for Informatics (GI), Special Committee on "Ergonomics in Informatics"
Human Communication Group [HCG] of the IEICE of Japan
Human Factors and Egonomics Society (HFES)
Human Interface Society
Information Processing Society of Japan SIGGroupware
Information Processing Society of Japan, SIG Human Interface (SIGHI)
ISS (Interactive System and Software) of Japan Software Society
Italian Association for Artificial Intelligence (AI*IA)
Japan Ergonomics Society
Swedish Interdisciplinary Interest Group for Human-Computer Interaction (STIMDI)

Local SIGCHI Chapters
Asia / Australia
New Zealand: SIGCHI_NZ
Europe
Bulgaria: BulSIGCHI
Czech Republic: Czech SIGCHI
France: Toulouse SIGCHI
Italy: SIGCHI Italy
Netherlands: SIGCHI.NL
Portugal: Universidade Fernando Pessoa SIGCHI in Portugal
Russia: Russia SIGCHI
Switzerland: SwissCHI
The Americas
Mexico: CHI-MEXICO
Ottawa: CapCHI
Toronto: ToRCHI
Vancouver: VanCHI
Waterloo: WatCHI, University of Waterloo
California: BayCHI, LA-SIGCHI
Georgia: CHI-Atlanta
Illinois: Chi-Squared
Massachusetts: GB/SIGCHI, LowellCHI
Michigan/Ohio: MOCHI
Minnesota: TwinCHI
Missouri: KC-CHI, Gateway CHI
New York: NYC-CHI
North Carolina: TriCHI
Ohio: BuckCHI
Oregon: CHIFOO
Texas: CHI-Austin, Lone Star SIG-CHI
Utah: NUCHI
Washington: Puget Sound SIGCHI
Wisconsin: Milwau-CHI

THE FUTURE IS HERE

Welcome
From the Conference Co-Chairs

The year 2000 marks the end of the 20th and the beginning of the 21st century. For years, it was a synonym for "The Future." Now, suddenly, The Future is Here! The onset of the Digital Age at the end of the 20th century has led to substantial changes in all aspects of peoples lives and issues related to Human-Computer Interaction are expanding in importance. Computers are increasingly integrated into most aspects of our lives. While most computers were on the desktop or in the computer room twenty years ago, now they are portable, held in the hand, stored in the pocket, worn, embedded in offices, homes, cars, in seemingly everything. The emerging link between television and computers through set top boxes, the relative simplicity of Internet boxes and the rapid growth in Internet access are changing entertainment habits, information seeking strategies and communication. Thus we see Human-Computer Interaction moving "beyond the desktop" and into nearly every environment. In doing so, computers affect a growing community of people, more diverse in their background, skills and computer training. CHI 2000 highlights the HCI of these varied new situations of computer use while the conference explores the visions, challenges, and opportunities of the 21st century. As a European CHI, taking place in The Hague, The Netherlands, CHI 2000 will also provide a stage for the European HCI scene – making it better known within the HCI community. See the *Technical Program Overview* on page xvii for an idea of how CHI 2000 addresses these issues and topics throughout the technical program through its special area themes, plenary speakers, and invited sessions.

The CHI Conference is a global forum, supporting and encouraging our growing global community and this is reflected in the CHI 2000 conference committee. CHI 2000 has nine Regional Liaisons from six continents, as well as a highly international conference committee. The review committees appointed by our Area Chairs are also globally distributed. Not only were there record numbers of submissions in most categories, but the accepted submissions are an excellent representation of the international flavor of the field of HCI.

The Future Is Here, at CHI, to consider, create, and to take hold of – and we invite you to make your own contribution to the future.

Thea Turner and Gerd Szwillus
CHI 2000 General Conference Co-Chairs

Special Thanks

CHI depends heavily on the efforts of volunteers. The CHI conference would not be possible without the contributions made by the conference committee. We would like to thank the committee members and their organizations, listed below, whose support has made their individual participation possible.

Argus Associates
AT&T Labs - Research
CNUCE-C.N.R.
DARPA
Darmstadt University of Technology
Delft University of Technology
Dray and Associates
Drexel University
FX Palo Alto Laboratory
GMD
Genentech
The Hiser Group
IBM Research
IDEO Product Development
INRIA
Jacques Hugo Associates
Kings College
KPN Research
Linköping University
Microsoft Research
Miramontes Computing
The MITRE Corporation
Motion Container
Motorola Labs
Nielsen Norman Group
Nokia Research Center
Philips Consumer Electronics
Practical Reasoning
Reactivity
Rensselaer Polytechnic Institute
Rivendel Consulting
Simon Fraser University
Shizuoka University
Stanford University
Sun Microsystems
UERJ/PUC-Rio
UMIST
UNISYS
University College London
University of Aarhus
University of Hagen
University of Limerck
University of Maryland
University of Massachusetts Lowell
University of Paderborn
University of Sunderland
University of Toulouse 1
University of Vienna
US WEST Advanced Technologies
Xerox PARC
Xerox Research Centre Europe

CHI 2000 • 1-6 APRIL 2000 — Acknowledgements

Sponsor Program

CHI 2000 gratefully acknowledges its Sponsor Program participants. The generosity of these organizations enables the conference to provide technical content and operational services that otherwise might not be possible.

Champion Sponsors

Contributing Sponsors

CHI 2000 thanks Eastman Kodak, Monkeymedia, Morgan Kaufmann, KPN Royal Dutch Telecom, KTH - The Royal Institute of Technology Stockholm and User Interface Engineering for their support of CHIkids!

THE FUTURE IS HERE

Committees

CHI 2000 • 1-6 APRIL 2000

Management Team

Conference Co-Chairs
Gerd Szwillus
University of Paderborn, Germany

Thea Turner
Motorola Labs, USA

SIGCHI CMC Liaison
Michael Tauber
University of Paderborn, Germany

ACM SIGCHI Program Director
Heather Levell
ACM, USA

Conference Manager
Paul Henning
Conference & Logistics Consultants, USA

Process Advisor
Carol Klyver
Foundations of Excellence, USA

Technical Program

Technical Program Co-Chairs
Gerd Szwillus
University of Paderborn, Germany

Thea Turner
Motorola Labs, USA

Special Area Chairs

Beyond the Desktop
Panu Korhonen
Nokia Research Center, Finland

European HCI
Manfred Tscheligi
Center for Usability Research & Engineering, University of Vienna, Austria

Future HCI
Jim Miller
Miramontes Computing, USA

Interaction
Robin Jeffries
Sun Microsystems, USA

CHIkids
Angela Boltman
University of Maryland, USA

Demonstrations
David Crow
Reactivity, USA

Hans de Graaff
KPN Research, The Netherlands

Development Consortium
Ian McClelland
Philips Consumer Electronics, The Netherlands

Doctoral Consortium
Gilbert Cockton
University of Sunderland, UK

Mentoring Liaisons
Gene Golovchinsky
FX Palo Alto Laboratory, USA

John Tang
Sun Microsystems, USA

Organization Overviews
Birgit Bomsdorf
University of Paderborn, Germany

Elizabeth Churchill
FX Palo Alto Laboratory, USA

Panels
David Gilmore
IDEO Product Development, USA

Jean Scholtz
DARPA, USA

Papers
Mary Czerwinski
Microsoft Research, USA

Fabio Paternò
CNUCE-C.N.R., Italy

Short Talks and Interactive Posters
Mike Atwood
Drexel University, USA

Françoise Détienne
INRIA, France

Special Interest Groups
Philippe Palanque
University of Toulouse 1, France

Alistair Sutcliffe
UMIST, UK

Student Posters
Charles van der Mast
Delft University of Technology, The Netherlands

Willemien Visser
INRIA, France

Tutorials
Liam Bannon
University of Limerick, Ireland

Alan Edwards
UNISYS, USA

Video Papers
Ben Bederson
University of Maryland, USA

M. Angela Sasse
University College London, UK

Workshops
Hans-Jürgen Hoffmann
Darmstadt University of Technology, Germany

Dianne Murray
Independent Consultant and Kings College, London, UK

Regional Liaisons

Africa
Jacques Hugo
Jacques Hugo Associates, South Africa

Asia Pacific
Masaaki Kurosu
Shizuoka University, Japan

Australia
Sarah Bloomer
The Hiser Group, Australia

Eastern Europe
Claus Unger
University of Hagen (Fernuniversität), Germany

North America
Susan Dray
Dray and Associates, USA

Scandinavia
Yvonne Wærn
Linköping University, Sweden

South and Central America
Raquel Oliveira Prates
UERJ/PUC-Rio, Brazil

United Kingdom
Allan MacLean
Xerox Research Centre Europe, UK

Western Europe
Michel Beaudouin-Lafon
University of Aarhus, Denmark

Operations

Doctoral Consortium Liaison
Marian G. Williams
University of Massachusetts Lowell, USA

Local Arrangements
Boyd de Groot
Motion Container, The Netherlands

Reviewer Liaison
Keith Instone
Argus Associates, USA

Special Conferences Liaison
Austin Henderson
Rivendel Consulting & Design, USA

Sponsorship Liaison
Jakob Nielsen
Nielsen Norman Group, USA

Student Volunteers
Garett Dworman
Practical Reasoning, USA

Tom Gross
GMD, Germany

Technology Support Liaisons
John "Scooter" Morris
Genentech, USA

Marilyn Salzman
US WEST Advanced Technologies, USA

Staff

Conference Administration
Vicki Currie
Conference & Logistics Consultants

European Logistics
Iris Allebrandi
Lidy Groot Congress Events

Exhibits
Tim Fitzgerald
Outbound Direct

Publications
Jean Tullier
Tullier Marketing Communications

Publicity
Rosemary Wick Stevens
Ace Public Relations

Registration
Carole Mann
Registration Systems Lab

Sponsorship
Carol Klyver
Foundations of Excellence

Web Design
Tom Brinck
Diamond Bullet Design

Papers Associate Chairs

Sara A. Bly
Sara Bly Consulting, USA

Jack Carroll
Virginia Tech, USA

Gilbert Cockton
University of Sunderland, UK

Francesca Costabile
Università di Bari, Italy

George Fitzmaurice
Alias|Wavefront, Canada

Beverly Harrison
SoftBook Press, USA

Michael Harrison
University of York, UK

Marti Hearst
Univeristy of California at Berkeley, USA

Ken Hinckley
Microsoft Research, USA

Robin Jeffries
SUN Microsystems, USA

Jürgen Koenemann
humanIT, Germany

Panu Korhonen
Nokia Research Center, Finland

Alison Lee
IBM T.J. Watson Research Center, USA

Arnold M. Lund
US WEST Advanced Technologies, USA

Jock Mackinlay
Xerox PARC, USA

Allan MacLean
Xerox Research Centre Europe, UK

Ian McClelland
Philips Consumer Electronics, The Netherlands

Jim Miller
Miramontes Computing, USA

Laurence Nigay
CLIPS-IMAG, France

Judith Olson
University of Michigan, USA

Philippe Palanque
University Toulouse 1, France

Angel Puerta
RedWhale Software, USA

George Robertson
Microsoft Research, USA

Chris Schmandt
MIT Media Lab, USA

Alistair Sutcliffe
UMIST, UK

Loren Terveen
AT&T Labs, USA

Manfred Tscheligi
Center for Usability Research & Engineering, University of Vienna, Austria

Gerrit C. van der Veer
Vrije Universiteit Amsterdam, The Netherlands

Shumin Zhai
IBM Almaden Research Center, USA

Jürgen E. Ziegler
Fraunhofer Institute IAO, Germany

Papers Reviewers

Gregory D. Abowd
Georgia Institute of Technology, USA

Mark Ackerman
University of California at Irvine, USA

Eleonora Acosta
Universidad Central de Venezuela, Venezuela

Seffah Ahmed
Computer Research Institute of Montreal, Canada

Motoyuki Akamatsu
National Institute of Bioscience and Human-Technology, Japan

Ghassan Al-Qaimari
Royal Melbourne Institute of Technology, Australia

Robert B. Allen
University of Maryland, USA

Mark W. Altom
Lucent Technologies, Bell Laboratories, USA

Keith Andrews
Graz University of Technology, Austria

Mark Apperley
University of Waikato, New Zealand

Francine Arble
Capital One, USA

Wendy Ark
IBM Almaden Research Center, USA

Albert G. Arnold
Delft University of Technology, The Netherlands

Jonathan Arnowitz
Informaat, The Netherlands

Michael Atyeo
Nortel Networks, Canada

Irene Au
Yahoo! Inc., USA

Maribeth Back
Xerox PARC, USA

Ravin Balakrishnan
University of Toronto & Alias|Wavefront, Canada

Sandrine Balbo
CSIRO-MIS, Australia

Michelle Baldonado
Xerox PARC, USA

Olle Bälter
Interaction and Presentation Laboratory, Sweden

Todd Barlow
SAS Institute Inc, USA

Remi Bastide
Université Toulouse 1, France

J. M. Christian Bastien
INRIA, France

Thomas Baudel
Ilog, France

Michel Beaudouin-Lafon
University of Aarhus, Denmark

Ben Bederson
University of Maryland, USA

Dave Bell
Philips Research Laboratories, UK

Alexandr Belyshkin
Moscow State University, Russia

Peter Benda
Pacific Access, Australia

Steve Benford
University of Nottingham, UK

David Benyon
Napier University, UK

Olav W Bertelsen
Aarhus University, Denmark

Nigel Bevan
Serco Usability Services, UK

Mark Billinghurst
University of Washington, USA

Helen Blakey
Potomac Research International, USA

Harry Blanchard
AT&T, USA

Ann Blanford
Middlesex University, UK

Jeanette Blomberg
Xerox PARC, USA

Sara A. Bly
Sara Bly Consulting, USA

Susanne Bodker
University of Aarhus, Denmark

Heinz-Dieter Boecker
German National Research Center for Information Technology, Germany

Deborah Boehm-Davis
George Mason University, USA

Katy Boerner
Indiana University, USA

Richard A. Bolt
MIT Media Laboratory, USA

James Boritz
ISG Technologies Inc., Canada

James H Bradford
Western Illinois University, USA

Dr Stephen Brewster
University of Glasgow, UK

Amy Bruckman
Georgia Institute of Technology, USA

Arnout Bruins
Telematica Instituut, The Netherlands

Hans Brunner
IBM Global Services, USA

Margaret Burnett
Oregon State University, USA

Robert Burns
The Boeing Company, USA

Myra Bussemakers
Catholic University of Nijmegen, The Netherlands

Mike Byrne
Rice University, USA

Judy Cantor
AT&T Labs, USA

Andy Cargile
drugstore.com, USA

Sheelagh Carpendale
Simon Fraser University, Canada

David A. Carr
Linköping University, Sweden

Jack Carroll
Virginia Tech, USA

Chad Carson
University of California at Berkeley, USA

Chaomei Chen
Brunel University, UK

Ed H. Chi
Xerox PARC, USA

Luca Chittaro
University of Udine, Italy

Yee-Yin Choong
GE Information Services, USA

Michael Christel
Carnegie Mellon University, USA

Elizabeth Churchill
FX Palo Alto Laboratory, USA

Steven Clarke
Independent, UK

Janette Coble
Washington University, USA

William Cockayne
Stanford University, USA

Tom Cocklin
Hewlett-Packard, USA

Gilbert Cockton
University of Sunderland, UK

Andrew L Cohen
Lotus Research, USA

Maxine Cohen
Nova Southeastern University, USA

Martin Colbert
University of Kingston, UK

Penny Collings
University of Canberra, Australia

Albert Corbett
Carnegie-Mellon University, USA

Jose Coronado
Hyperion Solutions Corporation, USA

Nuno Correia
New University of Lisbon, Portugal

Francesca Costabile
Università di Bari, Italy

Joelle Coutaz
CLIPS (IMAG), France

Ed Cutrell
Microsoft Research, USA

Allen Cypher
Stagecast Software, USA

Per Dahlberg
Viktoria Institute, Sweden

Berardina Nadja De Carolis
University of Bari, Italy

Hans de Vries
Cap Gemini Nederland, The Netherlands

Fernando Diaz-Prado
Univ. Cd. Juarez, Mexico

Andrew Dillon
Indiana University, USA

Miwako Doi
Toshiba, Japan

Jianming Dong
IBM Corporation, USA

Paul Dourish
Xerox Palo Alto Research Center, USA

Laurie Dringus
Nova Southeastern University, USA

Allison Druin
University of Maryland, USA

Professor D A Duce
Rutherford Appleton Laboratory, UK

Tim Dudley
Nortel Networks, Canada

Martin Dulberg
North Carolina State University, USA

Susan T. Dumais
Microsoft Research, USA

Elizabeth Dykstra-Erickson
Apple Computer, USA

Alistair D N Edwards
University of York, UK

Committees

Kate Ehrlich
Lotus Development, USA

Michael Eisenberg
University of Colorado, USA

Don Elman
Microsoft, USA

L. Miguel Encarnação
Fraunhofer CRCG, USA

George Engelbeck
Active Voice, USA

David England
Liverpool John Moores University, UK

Tom Erickson
IBM T.J. Watson Research Center, USA

Stefania Errore
Tivoli, Italy

Robert Fein
Trilogy Development Group, USA

Steven Feiner
Columbia University, USA

Jean-Daniel Fekete
Ecole des Mines de Nantes, France

Evan Feldman
Microsoft, USA

Daniel Felix
ETH Zürich, Switzerland

Sidney Fels
University of British Columbia, Canada

Thomas Finholt
University of Michigan, USA

Robert S. Fish
Panasonic, USA

Brian Fisher
University of British Columbia, Canada

George Fitzmaurice
Alias|Wavefront, Canada

Peter Forbrig
University of Rostock, Germany

Paola Forcheri
CNR-IMA, Italy

Jean Fox
Bureau of Labor Statistics, USA

Björn N. Freeman-Benson
Rational Software Corporation, USA

Nancy Frishberg
New Media Centers, USA

Wai-Tat Fu
George Mason University, USA

Robert Fulk
Strategic Interactive, US

George W. Furnas
University of Michigan, USA

Susan Fussell
Carnegie Mellon University, USA

Steve Gershik
Nuance Communications, USA

Douglas Gillan
New Mexico State University, USA

Patrick Girard
ENSMA, France

Gene Golovchinsky
FX Palo Alto Laboratory, USA

Cleotilde Gonzalez
Carnegie Mellon University, USA

Narciso Gonzalez
Research Psychologist, Spain

Michael Good
SAP Labs, USA

Peter Gorny
University of Oldenburg, Germany

Thomas Graefe
Compaq Corporation, USA

Nicholas Graham
Queen's University, Canada

Saul Greenberg
University of Calgary, Canada

Irene Greif
Lotus Development, USA

Tom Gross
GMD, Germany

Chris Grounds
Schafer Corporation, USA

Jonathan Grudin
Microsoft Research, USA

Steve Guest
Groupworks, USA

Ashok Gupta
University College London, UK

Nils-Erik Gustafsson
Ericsson Utvecklings AB, Sweden

Carl Gutwin
University of Saskatchewan, Canada

Lynne Hall
University of Northumbria at Newcastle, UK

David Hamilton
IBM Santa Teresa Laboratory, USA

Judy Hammond
University of Technology at Sydney, Australia

Libby Hanna
Independant, USA

Morten Borup Harning
Copenhagen Business School, Denmark

Beverly Harrison
SoftBook Press, USA

Michael Harrison
University of York, UK

Marti Hearst
University of California at Berkeley, USA

Frans Heeman
Elsevier Science, The Netherlands

Giorgios Heliadis
Loughborough University, UK

Austin Henderson
Rivendel Consulting & Design, USA

Scott Henninger
University of Nebraska-Lincoln, USA

James D. Herbsleb
Bell Labs, Lucent Technologies, USA

Harry M. Hersh
Fidelity Investments, USA

Stacie Hibino
Bell Labs, Lucent Technologies, USA

David R. Hill
University of Calgary, Canada

Will Hill
AT&T Research, USA

Ken Hinckley
Microsoft Research, USA

Debby Hindus
Interval Research, USA

Stephen Hirtle
University of Pittsburgh, USA

Challis Hodge
HannaHodge, USA

James D. Hollan
UCSD, USA

Stefan Holmlid
University of Linköping, Sweden

H. Ulrich Hoppe
University of Duisburg, Germany

Steve Howard
Swinburne University of Technology, Australia

Roland Hubscher
Auburn University, USA

Scott Hudson
Carnegie Mellon University, USA

Jacques Hugo
Jacques Hugo Associates, South Africa

Kori Inkpen
Simon Fraser University, Canada

Keith Instone
Argus Associates, USA

Hiroshi Ishii
MIT Media Laboratory, USA

Ismail Ismail
University College London, UK

Julie Jacko
University of Wisconsin-Madison, USA

Kai Jakobs
Technical University of Aachen, Germany

Robin Jeffries
Sun Microsystems, USA

Pamela Jennings
SRI International, USA

Ljubomir Jerinic
University of Novi Sad, Yugoslavia

Andrew Johnson
Electronic Visualization Laboratory, USA

Chris Johnson
University of Glasgow, UK

Peter Johnson
University of Bath, UK

Mark Jones
Andersen Consulting, USA

Matt Jones
Middlesex University, UK

Joaquim Armando Pires Jorge
INESC, Portugal

Anker Helms Jorgensen
Copenhagen University, Denmark

Luc Julia
CHIC! SRI International, USA

Tomas Kalen
University of Skövde, Sweden

Tomonari Kamba
NEC, Japan

Lisa Kamm
IBM, USA

Eser Kandogan
SGI, USA

Panagiotis Kappos
University of Patras, Greece

John Karat
IBM T.J. Watson Research Center, USA

Demetrios Karis
GTE Laboratories, USA

Judy Kay
University of Sydney, Australia

Susan Keenan
DAVOX Corporation, USA

Wendy A. Kellogg
IBM T.J. Watson Research Center, USA

Dr. Kinshuk
Massey University, New Zealand

Mark Kirby
University of Huddersfield, UK

Arthur Kirkpatrick
University of Oregon, USA

Muneo Kitajima
National Institute of Bioscience and Human-Technology, Japan

Jonathan Klein
IS Robotics, Inc., USA

Jayne Klenner
Pennsylvania State University, USA

Chris Knowles
University of Waikato, New Zealand

Jürgen Koenemann
humanIT, Germany

Raghu Kolli
Meru Research, The Netherlands

Piet Kommers
University of Twente, The Netherlands

Shinichi Konomi
University of Colorado at Boulder, USA

Joseph A. Konstan
University of Minnesota, USA

Panu Korhonen
Nokia Research Center, Finland

Committees

Peter Koss-Nobel
Microsoft, USA

Axel Kramer
Fragment Art & Research, USA

Kari Kuutti
Helsinki University of Technology, Finland

Jennifer Lai
IBM T.J. Watson Research Center, USA

Mary LaLomia
Intel Corporation, USA

James Landay
University of California at Berkeley, USA

Kevin Larson
University of Texas, USA

Sharon Laskowski
NIST, USA

Alison Lee
IBM T.J. Watson Research Center, USA

Adrienne Lee
New Mexico State University, USA

Mark Lee
Old Dominion University, USA

Laura Leventhal
Bowling Green State University, USA

Michael Levi
US Bureau of Labor Statistics, USA

Henry Lieberman
MIT Media Lab, USA

Heui-Seok Lim
Chonan University, Korea

Gitte Lindgaard
Swinburne University of Technology at Melbourne and Gitte Lindgaard & Associates, Australia

Fredrik Ljungberg
Viktoria Research Institute, Sweden

Lucy Lockwood
Constantine & Lockwood, USA

Ben Loh
Northwestern University, USA

A. Chris Long
University of California at Berkeley, USA

Christopher Lueg
University of Zürich, Switzerland

Paul Luff
King's College, UK

Arnold M. Lund
US WEST Advanced Technologies, USA

Jay Lundell
Intel, USA

Clemens Lutsch
Icon Medialab, Germany

Kipp Lynch
Cambridge Technology Partners, USA

Blair MacIntyre
Georgia Institute of Technology, USA

Wendy Mackay
University of Aarhus, Denmark

Christine MacKenzie
Simon Fraser University, Canada

Jock Mackinlay
Xerox PARC, USA

Allan MacLean
Xerox Research Centre Europe, UK

Paul Maglio
IBM Almaden Research Center, USA

Scott D. Mainwaring
Interval Research, USA

Meera Manahan
Compaq Computer, USA

Thomas Mandl
University of Hildesheim, Germany

Andrea Mankoski
Sun Microsystems, USA

Frank Marchak
Veridical Research and Design, USA

Gary Marsden
University of Cape Town, South Africa

Steve Marsh
National Research Council, Canada

Masood Masoodian
Natural Interactive Systems Laboratory, Denmark

John McCarthy
University College Cork, Ireland

Ian McClelland
Philips Consumer Electronics, The Netherlands

Marshall R. McClintock
Microsoft, USA

Richard McDaniel
Siemens Technology-To-Business Center, USA

Sharon McDonald
University of Sunderland, UK

Joanna McGrenere
University of Toronto, Canada

Jean McKendree
University of Edinburgh, UK

Hans van der Meij
University of Twente, The Netherlands

Luciano Krob Meneghetti
USP, Brazil

David R. Miller
IBM Global Services, USA

Jim Miller
Miramontes Computing, USA

Shailey Minocha
The Open University, UK

Anant Kartik Mithal
Sun Microsystems, USA

Naomi Miyake
Chukyo University, Japan

Tom Moher
University of Illinois at Chicago, USA

Andrew Monk
University of York, UK

Anders Morch
University of Bergen, Norway

Hirohiko Mori
Musashi Institute of Technology, Japan

Osamu Morikawa
National Institute of Bioscience and Human-Technology, Japan

John "Scooter" Morris
Genentech, USA

Michael Muller
Lotus Development, USA

Alice Mulvehill
BBN Technologies, USA

Dianne Murray
Consultant, UK

Pardo Mustillo
Media Renaissance, Canada

Brad A. Myers
Carnegie Mellon University, USA

Elizabeth Mynatt
Georgia Institute of Technology, USA

Frank Nack
GMD-IPSI, Germany

Jocelyne Nanard
University of Montpellier, France

N. Hari Narayanan
Auburn University, USA

Lester Nelson
FX Palo Alto Laboratory, USA

Todd Nelson
Air Force Research Laboratory, USA

Christine Neuwirth
Carnegie Mellon University, USA

Stephanie Newland
Homestead Technologies & Center for the Study of Language and Information, USA

Laurence Nigay
CLIPS-IMAG, France

Alexander Nikov
Technical University of Sofia, Bulgaria

Lorraine F. Normore
OCLC, USA

David G. Novick
University of Texas at El Paso, USA

Lucy Nowell
Battelle/Pacific Northwest National Lab, USA

Dan R. Olsen Jr.
Brigham Young University, USA

Gary M. Olson
University of Michigan, USA

Judith Olson
University of Michigan, USA

Scott P. Overmyer
Drexel University, USA

Sharon Oviatt
Oregon Graduate Institute of Science & Technology, USA

Nadine Ozkan
CSIRO Mathematical and Information Sciences, Australia

Sherrill Packebush
SBC Technology Resources, USA

Colleen Page
Microsoft, USA

Philippe Palanque
University Toulouse 1, France

Leysia Palen
University of Colorado at Boulder, USA

Antonia Palmer
University of Waterloo, Canada

Marilyn Panayi
Natural Interactive Systems Laboratory, Denmark

Erin Panttaja
Comverse Network Systems, USA

Holly Patterson-McNeill
Texas A&M University - Corpus Christi, USA

Lynn Pausic
Reactivity, USA

Caroline Pawlowsky
Federal University of Parana and Choose Technologies, Brazil

Gaynor Peacock
Simba Technologies, Canada

Matt Pearcey
Intergral, UK

Elin Ronby Pedersen
FX Palo Alto Laboratory, USA

Ted Pedersen
University of Minnesota at Duluth, USA

Samuli Pekkola
University of Jyvaskyla, Finland

Manuel A. Perez-Quinones
Universidad de Puerto Rico-Mayaguez, USA

Kara Pernice Coyne
Iris Associates, USA

Daniela Petrelli
IRST, Italy

Richard W. Pew
BBN Technologies, USA

Marios Pittas
Kent Ridge Digital Labs, Singapore

Catherine Plaisant
University of Maryland, USA

Lydia Plowman
Scottish Council for Research in Education, UK

Andrea Polli
Columbia College Chicago, USA

Peter G. Polson
University of Colorado, USA

Steven E. Poltrock
The Boeing Company, USA

Gokul Prabhakar
Bell Labs, USA

Committees

Scott Preece
Motorola, USA

Rob Procter
Edinburgh University, UK

Angel Puerta
RedWhale Software, USA

Chris Quintana
University of Michigan, USA

Kari-Jouko Raiha
University of Tampere, Finland

Roope Raisamo
University of Tampere, Finland

Arvind Ramakrishnan
iNautix Technologies, USA

Rossen Rashev
GMD-FIT, Germany

Michael J Rees
Bond University, Australia

Susan Harkness Regli
Carnegie Mellon University, USA

Brian Reilly
UC Riverside, USA

Mary Beth Rettger
The MathWorks, USA

John Rheinfrank
seespace llc, USA

Luc Richard
STAR, France

John T. Richards
IBM T.J. Watson Research Center, USA

Kirsten Risden
Microsoft, USA

Chris R. Roast
Sheffield Hallam University, UK

Daniel Robbins
Microsoft, USA

Dave Roberts
IBM, UK

Teresa L. Roberts
Sun Microsystems, USA

George Robertson
Microsoft Research, USA

Kerry Rodden
University of Cambridge, UK

Anne Rose
University of Maryland, USA

Tony Rose
Canon Research Centre Europe, UK

Elizabeth Rosenzweig
Eastman Kodak, USA

Dmitri Roussinov
Syracuse University, USA

David Roy
University of Southern Denmark, Denmark

Richard Rubinstein
Perot Systems, USA

Arnold Rudorfer
Meta4 S.A., Spain

Daniel Russell
Xerox PARC, USA

Dario Salvucci
Cambridge Basic Research, USA

Alfredo Sanchez
Universidad de las Americas-Puebla, Mexico

Carmen Santoro
CNUCE - CNR, Italy

Martina Angela Sasse
University College London, UK

Mark Schlager
SRI International, USA

Egbert Schlungbaum
CiS GmbH & University Rostock, Germany

Chris Schmandt
MIT Media Lab, USA

Kurt Schmucker
Apple Computer, USA

Kevin Schofield
Microsoft, USA

Doree Duncan Seligmann
Bell Labs, USA

John Seton
BT Laboratories, UK

Brian Shackel
HUSAT Research Institute Loughborough University, UK

Chris Shaw
University of Regina, Canada

Donald P. Sheridan
Auckland University, New Zealand

Hal Shubin
Interaction Design, USA

Stéphane Sire
CENA, France

Bruce Sklar
Enterworks, USA

Brian Smith
MIT Media Laboratory, USA

Jeff Sokolov
Open Market, USA

Benjamin Somberg
AT&T Labs, USA

Kirsty Spence
Telstra Research Laboratories, Australia

Jared M. Spool
User Interface Engineering, USA

Jan Stage
Aalborg University, Denmark

Jeffrey Staw
The MITRE Corporation, USA

Harald Stegavik
Telenor Mobil, Norway

Patrick Steiger
PricewaterhouseCoopers, Switzerland

Arnd Steinmetz
IBM T. J. Watson Research Center, USA

Constantine Stephanidis
Foundation for Research and Technology-Hellas, Greece

Markus Stolze
IBM Research, Switzerland

Hank Strub
Interval Research, USA

Bernhard Suhm
BBN Technologies, USA

Piyawadee Sukaviriya
IBM T.J. Watson Research Center, USA

Yasuyuki Sumi
ATR Media Integration & Communications Research Labs, Japan

Tamara Sumner
University of Colorado, USA

Alistair Sutcliffe
UMIST, UK

Merryanna Swartz
Georgetown Visitation Prep School, USA

Desiree Sy
Alias|Wavefront, Canada

Peter Szmyt
JetForm, Canada

John C. Tang
Sun Microsystems, USA

Michael Tauber
University of Paderborn, Germany

Said Tazi
University Toulouse 1, France

Ross Teague
Intel, USA

Barbee Teasley
SBC Technology Resources, Inc.

Unalome Pim Techamuanvivit
Cisco Systems, University of California at San Diego, USA

Loren Terveen
AT&T Labs - Research, USA

Victor Theoktisto
Universidad Simon Bolivar, Venezuela

Jennifer Thomas
Pace University, USA

John C. Thomas
IBM Research, USA

Richard Thomas
University of Glasgow, UK

Sandra Thwaites
Corel, Canada

Elaine Toms
Dalhousie University, Canada

Jozsef A. Toth
Pacific Sierra Research - An Operating Company of Veridian, USA

Michael D. Travers
Afferent Systems, USA

Marilyn Tremaine
Drexel University, USA

Susan Trickett
George Mason University, USA

Randall Trigg
Xerox Palo Alto Research Center, USA

Spiros Tsaltas
Compuware Europe Uniface Lab, University of Utrecht, The Netherlands

Manfred Tscheligi
Center for Usability Research & Engineering & University of Vienna, Austria

Susan Turner
University of Northumbria at Newcastle, UK

Lisa Tweedie
The Oracle Corporation, UK

David Ungar
Sun Microsystems, USA

Zita Vale
Polytechnic Institute of Porto, Portugal

Maarten van Dantzich
Microsoft Research, USA

Richard van de Sluis
Philips Research, The Netherlands

Charles van der Mast
Delft University of Technology, The Netherlands

Gerrit C. van der Veer
Vrije Universiteit, The Netherlands

Susanne van Mulken
DFKI, Germany

Gerard van Os
Ericsson Business Mobile Networks, The Netherlands

Bradley Vander Zanden
University of Tennessee, USA

Jean Vanderdonckt
Université Catholique de Louvain, Belgium

Dan Venolia
Microsoft Research, USA

Harry Vertelney
Sun Microsystems, USA

Frank Vetere
Swinburne Universtity of Technology, Australia

Robert Virzi
GTE Laboratories Incorporated, USA

Bruno von Niman
Ericsson Business Networks AB, Sweden

Pawan Vora
NexTag.com, USA

Yvonne Wærn
Linköping University, Sweden

Andrew Walker
Square Design, USA

Colin Ware
University of New Brunswick, Canada

Clive P. Warren
*British Aerospace plc,
Sowerby Research Centre,* UK

Benjamin Watson
University of Alberta, Canada

Gerhard Weber
Harz University, Germany

Karon Weber
Pixar Animation Studios, USA

Hans Wegener
Credit Suisse, Switzerland

Steven Weintraub
Logical Design Solutions, USA

Janet Wesson
University of Port Elizabeth,
South Africa

Joyce Westerink
Philips Research, The Netherlands

Alan Wexelblat
Mainpring, USA

Steve Whittaker
ATT-Labs Research, USA

Susan Wiedenbeck
University of Nebraska at Omaha, USA

Rob Willems
Scope Principal Solutions,
The Netherlands

Gayna Williams
Microsoft, USA

Marian G. Williams
University of Massachusetts Lowell,
USA

Michael Wilson
CLRC Rutherford Appleton Laboratory,
UK

Terry Winograd
Stanford University, USA

G. Bowden Wise
Rensselaer Polytechnic Institute, USA

Dennis Wixon
Microsoft Corporation, USA

Christa Womser-Hacker
University of Hildesheim, Germany

Allison Woodruff
Xerox PARC, USA

Volker Wulf
University of Bonn, Germany

Ming-Hsuan Yang
*University of Illinois at Urbana-
Champaign,* USA

Nicole Yankelovich
Sun Microsystems Laboratories, USA

Polle Zellweger
Xerox PARC, USA

Ronald J. Zeno
Interaction Architects, USA

Shumin Zhai
IBM Almaden Research Center, USA

Qiping Zhang
University of Michigan, USA

Michelle Zhou
IBM T. J. Watson, USA

Jürgen E. Ziegler
Fraunhofer Institute IAO, Germany

CHI 2000 The Future Is Here

Technical Program
The CHI 2000 Technical Program represents the leading edge of work in the theory and practice of HCI. This year's conference theme, *The Future is Here*, explores the visions, challenges, and opportunities of the twenty-first century. In keeping with this focus, the opening plenary speaker, John Thackara of Doors of Perception, examines two cutting-edge developments that are transforming our work: pervasive computing and social agendas for innovation. Almost everything people make will soon combine hardware and software. Within ten years there will be millions of computers, everywhere–all talking to each other. The interaction of pervasive computing with social agendas for innovation represents a revolution in the way our products and our systems are designed in the way we use them–and how they relate to us. Kim Binsted, of SONY CSL's Interaction Lab, the closing plenary speaker, considers characteristics that advanced technology and magic have in common. We use specialized software and hardware to communicate ideas, transport objects and predict future events. Therefore, Binsted argues that technology will increasingly come to resemble imagined magic.

Special Areas of the Theme
Beyond the Desktop, a special area which highlights an aspect of the conference theme, is addressed in the CHI 2000 *Development Consortium* and an Invited Session. The consortium considers the growth of computer-based devices and services worldwide that are becoming embedded in the way we live. It also highlights emerging technologies and their use, interaction techniques, and contexts of use. The invited session presents interviews with three leading-edge experts from the design and communication industry, discussing the "Art of Beyond the Desktop" and includes a musical conductor's interactive experience.

In celebration of the European location, CHI 2000 spotlights European HCI through a *European HCI Village* and invited presentations from eminent Europeans in the field. The *Interaction* special area is concerned with encouraging participants to have provocative discussions. A special session called *Interactionary*, features teams of designers tackling a design problem in real-time, allowing the audience to contrast design approaches and activities. The Interactive Posters venue offers attendees opportunities to have intimate discussions with poster authors about their work. Finally, *Future HCI* is discussed in the many forward-looking presentations such as Organization Overviews, which offer attendees a view of how several organizations are approaching the HCI issues of the future.

Global HCI
Just as the world of the future offers increasingly global communication and collaboration, CHI 2000 is truly a global conference. Explore the immense potential presented here during the conference days, and enjoy the presented submissions, selected from record numbers of submissions in most categories, providing an excellent representation of the international flavor of the field of HCI.

Gerd Szwillus and Thea Turner
CHI 2000 Technical Program Co-Chairs

Pre-Conference Events

Tutorials
Tutorials are courses designed for novices to experienced participants. Courses cover a wide range of topics from theory to practice.

Development Consortium
The Development Consortium will examine issues and directions that the HCI community and SIGCHI should develop in the coming years.

Doctoral Consortium
The Doctoral Consortium provides an opportunity for a group of invited doctoral students to explore their research interests in an interdisciplinary workshop with other students as well as a group of experienced researchers.

Workshops
Workshops provide a valuable opportunity for small communities of people with diverse perspectives to engage in rich one- to two-day discussions on a topic of common interest. Workshop participants are pre-selected and workshops offer an opportunity to explore and develop work collaboratively.

Conference Events

Demonstrations (Live and Video)
Demonstrations offer an opportunity to show an innovative interface concept, HCI system, technique or methodology. Attendees are able to view systems in action and discuss them with the people who created them. Demonstrations may be either live or video presentations. CHI demonstrations include both refereed demonstrations and demonstrations informally organized by participants on site.

Organization Overviews
Organization Overviews describe the work of leading organizations engaged in HCI research and practice. The emphasis is on the circumstances under which the work is done, as well as the underlying goals, policies and organizational background and perspectives of the group's past, present and future HCI efforts.

Panels
Panels stimulate thought and discussion about ideas and issues of interest to the human-computer interaction community. Panels typically focus on controversial or emerging issues, allowing speakers and the audience to explore, debate, and reflect on these issues.

Papers
Papers present significant contributions by researchers and practitioners to the HCI field, capable of influencing the design life-cycle of current and future interactive systems. Papers are highly refereed and are published in the archival *CHI Conference Proceedings* and as an issue of *CHI Letters*.

Plenary Sessions
Plenary sessions are general sessions that open and close the conference. The key event of the session is usually an invited present-ation by a prominent person that supports the conference theme and offers a challenge to people interested in HCI.

Short Talks and Interactive Posters
Short Talks and Interactive Posters are particularly suitable for exciting new findings, ongoing work that has demonstrated special promise, preliminary results, timely work still in a state to be influenced, or tightly argued essays or opinion pieces. Posters are visual presentations of work that are displayed throughout the conference. Short Talks are presented in traditional technical sessions.

Special Interest Groups (SIGs)
SIGs enable conference attendees who share a common interest to meet informally for 90 minutes of discussion at the conference. They differ from workshops in that there is no pre-event selection of partici-pants and all attendees may participate.

Student Posters
Student Posters offer a unique opportunity for students to present their work at CHI and to receive encouragement in their development as HCI professionals. Student posters are displayed during the conference and provide an excellent opportunity to discuss late-breaking and on-going work in an informal setting.

The *CHI 2000 Conference Proceedings* is the archival print publication of the conference. It contains the full text of papers presented at the conference. These papers have undergone one of the most rigorous peer-review processes in the field.

The *CHI 2000 Extended Abstracts* contains summaries of technical presentations other than papers. It includes extended abstracts of Demonstrations (live and video), the Doctoral Consortium, Panels, Plenary Addresses, the Development Consortium, Special Interest Groups, Student Posters and Workshops.

The *CHI 2000 Video Program* contains the Demonstrations presented in video form. It also includes video figures belonging to the traditional papers, which show ideas that cannot readily be conveyed in print figures.

Unleashed: Web Tablet Integration into the Home

Anne McClard
MediaOne Labs
10355 Westmoor Drive, Suite 100
Westminster, CO 80021 USA
+1 303 404 8058
amcclard@mediaone.com

Patricia Somers
MediaOne Labs
10355 Westmoor Drive, Suite 100
Westminster, CO 80021 USA
+1 303 404 8088
psomers@mediaone.com

ABSTRACT
To understand how web access from a portable tablet appliance changes the way people use the Internet, MediaOne gave families pen-based tablet computers with a wireless connection to our high-speed data network. We used ethnographic and usability methods to understand how tablets would be integrated into household activities and to define user requirements for such devices. Participants viewed the tablet as conceptually different from a PC. The tablet enabled a high degree of multitasking with household activities, yet flaws in form and function affected use. Results suggest that correctly designed portable Internet appliances will fill a special role in peoples' daily lives, particularly if these devices share information with each other. They will allow spontaneous access to information and communication anywhere.

Keywords
Internet appliances, pen-based computing, hand-held computers, ergonomics, ethnography

INTRODUCTION
One might call the 1990s the decade of the Internet. In 1992 [2] when the first graphical browser became available to the public, few people accurately predicted the rapid growth of the World Wide Web. The second half of the decade has been full of predictions about the rise of Internet Appliances [6], small wireless devices dedicated to one or a few Internet-based services. As the century turns over, the first of these devices have made it to market (e.g., Internet cell phones, Cidco's MailStation) with the promise of many more to come (e.g., pizza magnets, tablet appliances, etc.). In an effort to better understand how Internet appliances will be integrated into the home, MediaOne Labs has conducted several studies. This paper focuses on one of these, an in-home study of a wireless tablet appliance. The goals of the study were to understand how a web tablet appliance would be integrated into household activities, to understand how using the tablet to access the Internet would differ from using a personal computer, and to define the functional and design requirements for a web tablet.

METHODS
The results of our study are based on findings from thirteen households to whom we gave tablets appliances and wireless network connections. The households were customers of MediaOne's high-speed data (HSD) service, a service providing a constant "always on" connection of 1.5 Mbps downstream and 356 Kbps upstream.

Equipment: What is a Tablet?
Tablet computers have been used for a variety of business applications over the last ten years. Tablets that are available on the market are smaller and lighter than laptops, but bigger and heavier than palm-type devices. They are designed to support highly mobile occupations (e.g., delivery, hospital workers, etc). Although many tablets have a detachable keyboard accessory, they are designed as pen-based; tapping and handwriting recognition are the primary input methods.

The tablet appliance we used had a Windows 95 operating system with a Pentium 133mHz processor. The screen was a 7.5 inch passive matrix 256 color touch-sensitive display. Input methods included an active stylus for tapping and writing, a detachable slim-line keyboard, and an on-screen keyboard. We opted to use a full Windows 95 tablet (a vertical market device) rather than a more consumer-oriented device with Windows CE because it offered more in the way of a broadband experience; Windows CE did not support the use of browser plug-ins, chat clients, or streaming media.

We installed a base set of applications on the tablets, including browsing software, email, instant messaging/chat client, and a variety of plug-ins supporting animation and streaming media. Participants could also load or download software of their choosing to the tablet from the Internet or by using file sharing with their home PC.

Demographics
Fifteen households were recruited from a random sample of 100 MediaOne HSD subscribers located in five suburbs of Detroit. We screened households by family composition, level of experience with PCs, and type of dwelling. By the

time the trial began, however, two households had dropped out because of concerns about the tracking methods that we planned to use; they were worried that we would have access to too much personal information.

Participating households varied demographically. Three households had no children, two had adult children living at home, five had children between the ages of 6 and 18, and three had children under the age of 6. Household incomes were distributed from $25,000 to more than $100,000. Nine participating households lived in multi-level single-family dwellings, three in apartments, and one in a townhouse.

Initially we estimated that we had a pool of 40 individuals of an age to use the Internet and therefore the tablet appliance (6-years-old and above). At end-of-trial however, we found that no one under the age of 9 had used the tablet, and that a few adults did not use it either, leaving us with a sample of 28 users: 23 adults, 4 teenagers, and 1 pre-teen. There were 13 females and 15 males, and in 12 households males were the primary users, even in cases where females were the owners of the primary PC in the household.

Data Collection

In an effort to characterize the tablet experience more precisely, we used a variety of methods to collect and analyze several types of data.

We used an ethnographic approach to understand the context of use and how the tablet appliance was integrated into the daily lives of participating households. This included two in-depth semi-structured formal interviews, both pre- and post-trial (3-4 hours total). Interviews covered topics about family dynamics, household schedules, activity mapping of the home, as well as specific questions related to use of technology in the home and the Internet. This data was augmented with spontaneous informal email interviews and interactions throughout the trial, as well as photographic documentation.

To calculate actual use of both the primary PC in the household and the tablet appliance we used a computer tracking program that saved information in a local database on the primary PC and the tablet appliance. It recorded the captions and start/stop times of active windows, as well as the executable running in each window. Participants started tracking usage on their primary PC for two weeks before they received the tablets and continued throughout the trial. Mid-trial, we asked participants to send us their data files so we could use them to provide content for our post-trial interviews. We also collected this tracking data at the end of the study.

While some usability issues were disclosed naturally in the course of our post-trial interviews, we also made an effort to observe participants using the tablets, in both the initial and final interview. When we first delivered the tablets, we gave participants a brief demonstration including use of handwriting recognition and the on-screen keyboard. Participants then used the tablet for about fifteen minutes while we observed their actions. We asked them to go to a web site and select links there, to use the Start menu, to close windows, and to type. We also showed them a different tablet design (one with an integrated keyboard and screen) and asked for their initial reaction to the form of each device. Participants also had an opportunity to type on the second tablet, which had a larger keyboard. During the trial, we asked participants about their problems with the tablets via e-mail, and when we returned for the final interview they showed us their biggest pet peeves.

Duration and Design of Study

Families had their tablets for approximately seven weeks. Two weeks before beginning the trial (mid-March 1999), we installed the tracking software on participants' primary PCs. In all cases, this was the PC with the HSD connection. We wanted to get baseline data about participants' use of general applications and the Internet with which we could compare our trial data. In the beginning of April we installed the tablet appliances and a wireless network. At this time, we conducted pre-trial interviews. Each household had their tablet appliance until the end of May when we returned to uninstall the networks and tablets, conduct post-trial interviews, and collect tracking data files.

LANGUAGE: THE CULTURAL CONSTRUCTION OF A TABLET APPLIANCE

Technological devices have no significance apart from the meaning, use and context that human beings construct around them. When trying to understand how a technological intervention (like a tablet appliance) "fits in" it is useful to ferret out native categories related to the objects in question. Native categories are spontaneous recurrent categorical references made by multiple individuals in relation to something else (in this case to the PC and tablet appliance). These categories become evident through a linguistic analysis of what people say, both in an informal conversational setting and in more formal interview situations.

In an effort to understand the semantic positioning of the tablet appliance we provided, we did a linguistic analysis of our interviews and informal interactions with participants. We found a striking contrast in the way people talked about their tablet in relation to their PCs (Table 1 below).

PC	TABLET
less comfortable	comfortable
work, serious	play, toy, fun, relaxing, casual
confined, chained	unchained, portable, mobile, take it anywhere, go anywhere
isolating	be with family
less convenient	convenient, handy, saved time

Table 1 Linguistic differences between tablet and PC

At our first interview, participants overwhelmingly used positive language with reference to their PCs taken in isolation, but at the end of the trial they talked about their PC experiences in contrast to their experience with the tablet appliance, often in more negative terms. PCs were referred to as being "less comfortable" to use because one had to sit in an upright chair at a desk, while the tablets were thought of as "comfortable" because one could sit, recline, or lie in any number of positions to use it. PCs were considered better than the tablet for "work" and "serious" activities, while the tablets were construed as "relaxing," "fun," and good for more "casual" use. PCs were frequently referred to as "confining," and "isolating," while the tablets engendered a sense of freedom, giving people the ability to go where they wanted, and to be with whom they wanted, typically family members. Finally, in contrast to PCs, the tablet appliances were considered "convenient," "handy," and "time-saving," primarily because they enabled a level of multi-tasking that was unobtainable with their PCs.

This linguistic juxtaposition between the PC and the tablet appliance forms a conceptual framework into which the following findings fit.

PORTABILITY: UNCHAINED, GO ANYWHERE

Portability was cited unanimously as the tablet appliance's best attribute. As one participant said, "It's [the tablet is] nice, it unchains you . . .you are not chained to a workstation." Consistent with findings of previous studies done by MediaOne [3], we found that households with children between the ages of 6 and 18 tended to place their primary PC in a common area of the house—in the living/dining areas. Households with adult children, children under 6, or no children in them tended to have their primary computer located away from common use areas—in a den or bedroom. In Figure 1, the locations of primary PCs are indicated with stars.

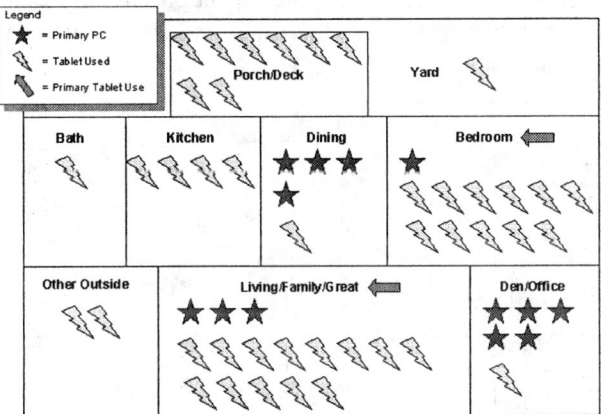

Figure 1 Locations in which people use tablet and PC

The tablet, represented by the lightening bolts in Figure 1, was used in every room of the house. All households reported using it in the living/family rooms[1], and eleven households reported using it in bedrooms. Eleven households also reported using it outside of their homes: on the porch or deck, in the yard, or at a neighbor's. The only constraint was the 150 foot limitation imposed by the wireless network solution we used.

Eight households reported that a household member had taken it somewhere away from home to show someone. Many participants lamented that their Internet connectivity ended when they got into their cars and drove away from the house and wireless bridge.

MULTI-TASKING: CONVENIENT, SAVES TIME, SOCIAL

While portability was explicitly mentioned as the most valued attribute of the tablet appliance, we argue that it was so highly valued because it enabled multi-tasking—doing an activity on the tablet appliance while engaging in other household activities. Some households reported a limited subset of multi-tasking activities with their stationary "always on" PCs, but all households reported extensive multi-tasking with their tablet appliances. The types of activities people reported engaging in simply would not have been possible with a stationary PC. (See Figure 2 below).

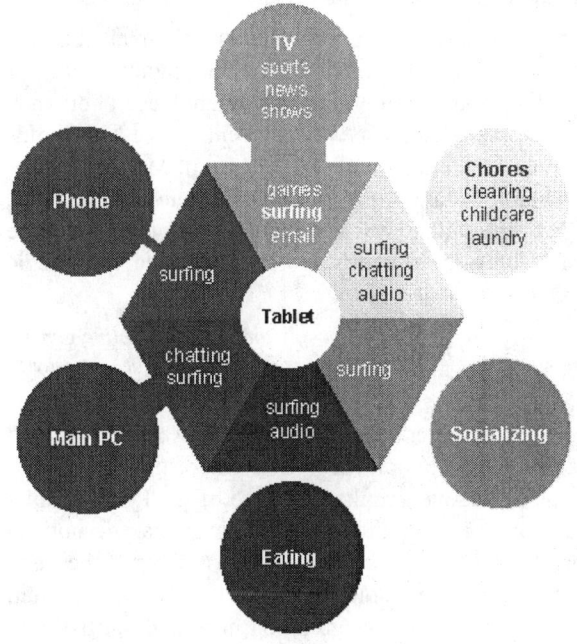

Figure 2 Schematic of multi-tasking

Figure 2 illustrates household activities (watching TV, doing chores, socializing, eating, using the main PC, and

---

[1] In the U. S., the family room is an informal living area that often adjoins the kitchen, while the living room is reserved for special occasions. In homes that do not have a family room, the living room serves the same informal purposes as the family room.

talking on the phone) in relation to activities that were done simultaneously on the tablet (surfing, playing stand-alone games, reading/sending email, chatting, and listening to audio programs). The bars linking the household activity to the tablet activities represent the strength of relationship between the corresponding activities. For example, when people were doing something on the tablet and doing chores, socializing, or eating, what they were doing on their tablet had no relation to the corresponding household activity. In contrast, there was often a relationship between simultaneous PC-tablet, phone-tablet, and TV-tablet activities.

The relationship between household-tablet activities was strongest for TV-tablet. All 13 households reported extensive television viewing while using the tablet. Eight of the households reported that their tablet activities were sometimes related to what they were watching on TV. The most commonly reported activity was looking up a web site based on a URL announced during programming or commercials. The second most frequent activity reported was watching sports events on TV and also following them on the web. Other simultaneous related activities included sending e-mail to news shows and looking up company information during investment-related programming.

Simultaneous unrelated (parallel) use of the main PC and the tablet was the norm for most households, but four households reported simultaneous related use. For each household, the related use was different. Three households reported using the tablet to do ICQ/Instant Messenger chat on their tablet while doing something related on their PC. One participant liked to use the tablet to chat while he played "his" online game; he felt he gained a strategic advantage.

> Talking [chatting] to other people and playing [online games] gave me a competitive advantage...I even thought of a way to cheat!

Three households reported using chat between the PC and the tablet. One example of this is interesting because it illustrates a creative simultaneous use of the PC and tablet to accomplish a task. From the tracking data, we noticed that one household was simultaneously looking at the same web site on both the tablet and the PC, while using chat. When we asked what was going on, it turned out that two household members were looking at *www.realtor.com* together to find a house to buy. Even though their PC was located in the dining area adjoining their living room and they were only feet away from each other, they were using chat to send each other URLs to look at. At the same time, they were vocally discussing the items they were viewing.

In sum, the tablet experience was a qualitatively different experience from the PC. This is evidenced in what people said about the two experiences, and also by how and where they used the tablet appliance. The tablet's portability gave families freedom to use the tablet anywhere in the home, and enabled a high degree of multi-tasking.

A TABLET IS NOT A PC: THE TRACKING DATA

In the previous sections we showed that users had differing conceptual frameworks for the tablet and the PC. In this part of the paper, we focus more specifically on how tablets were used compared to the PC. Tablet use and PC use differed in some striking ways, but they both followed a basic pattern in the way people used their on-line time.

Spending Time on the Tablet

Our first question was whether introduction of the tablet would change the way people used their desktop PCs. Would the availability of the tablet increase the total time spent on-line, or is there a limit to the amount of time people can spend on the Internet, so that they simply switch time from PC to tablet? And would the activities carried out on the tablet be the same as those on the PC?

Overall, families used their tablets less than their PCs. This was not surprising, since the tablets lacked some of the capabilities of the PCs, such as productivity software and the specialized hardware that supported favorite networked games. Before receiving their tablets, households spent an average of 3.4 hours per day on the PC, 56 percent of that on-line. The percentage of on-line time is similar for the tablet, 55 percent. This was a surprise to us; we had expected more on-line time on the tablet, since it didn't have many standalone applications. It turned out that participants were playing games like solitaire and FreeCell, and they reported to us that they often played these games while doing something else like watching TV or sitting with their families.

Households did not switch their on-line activities from the PC to the tablet but rather added tablet on-line activities to their day. As shown in Figure 3, they spent 1.85 hours per day on-line on their PCs during the trial (not significantly different from the 1.92 hours they spent on-line before the trial), and they spent an additional 32 minutes per day on-line on the tablet.

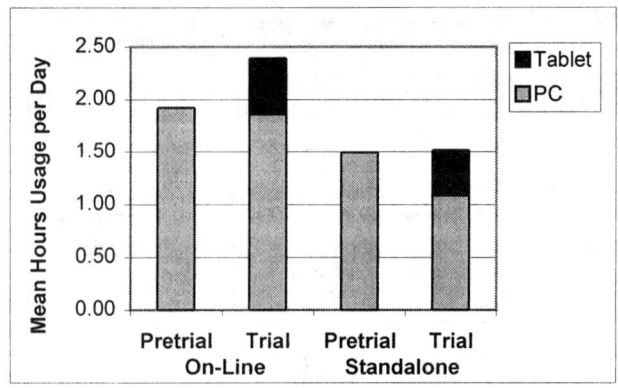

Figure 3 Time spent in on-line and standalone activities

Participants told us that one advantage of the tablet was that more than one person in the family could surf the web at the same time: the tablet reduced contention for the household's high-speed data connection. Thus, at least some of the increased overall on-line time may be due to additional family members being able to access the Internet. This relationship did not hold for standalone activities such as word processing and graphics: it appears that while households had the tablets, they spent less time on standalone activities on the PC.

As shown in Figure 4, the most time-consuming on-line activity for both tablet and PC was web browsing, followed by chat and e-mail. By browsing, we mean not only web surfing but listening to and viewing multimedia accessed from web sites, for example streaming audio. And "browsing" is probably not an accurate term, since participants reported more targeted access of specific web sites than wandering on the web. The dominance of web use over e-mail differs from patterns found in the HomeNet trial [5]. Our participants' high-speed, always-on Internet connection may facilitate web browsing.

Chat on both tablet and PC consisted primarily of using Instant Messenger or ICQ with friends, relatives, and game-player communities, not meeting strangers in a chat room. Chat and e-mail were not as common on the tablet as they were on the PC. Many participants reported that they were frustrated by the tablet's inadequate text input methods, which may account for the relatively low time spent in text-intensive activities. For example, one person told us, "Since the handwriting [recognition] feature doesn't work, and the keyboard is hard to use, I am not able to use the tablet for ICQ like I was hoping." We discuss this further in the usability section, below.

Averaged over households, the time spent on chat was not significantly higher than e-mail. (The difference was not statistically significant on either device). Interestingly, while only six of our households used chat on the tablet (eleven used e-mail), those households used chat more than twice as much as e-mail. When chat is used, it may to be taking over a position previously occupied by e-mail.

Web Destinations

On both the tablet and PC, there was a great diversity of web sites visited. On the tablet, 270 sites accounted for 75 percent of the time spent browsing, and 85 percent of the top 100 sites were visited by only one person. Certain categories of web sites were popular across households (for example, six of the thirteen households visited financial sites regularly), but each household had its own favorite site within those categories.

People surfed to different types of sites on their tablets than they did on their PCs. Just as participants often used their tablets while doing other things, the types of web sites they visited seemed, in general, to be more suited to multi-tasking, while sites visited on the PC required greater focus. Figure 5 categorizes the top 25 web sites participants visited using each device. (These are the top 25 sites based on total time spent on each site. These 25 sites accounted for 50% of all time spent browsing on each device.)

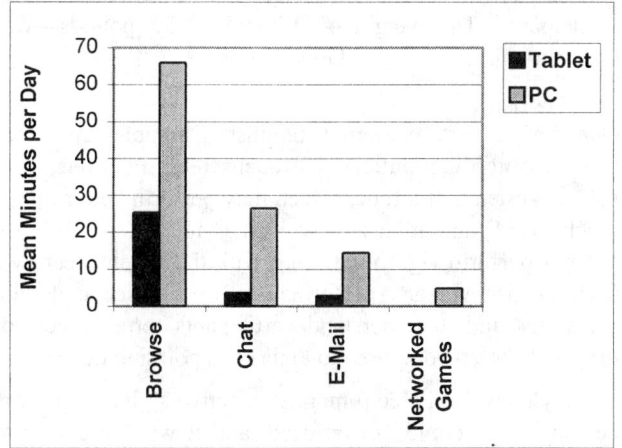

Figure 4 Time spent in various on-line activities

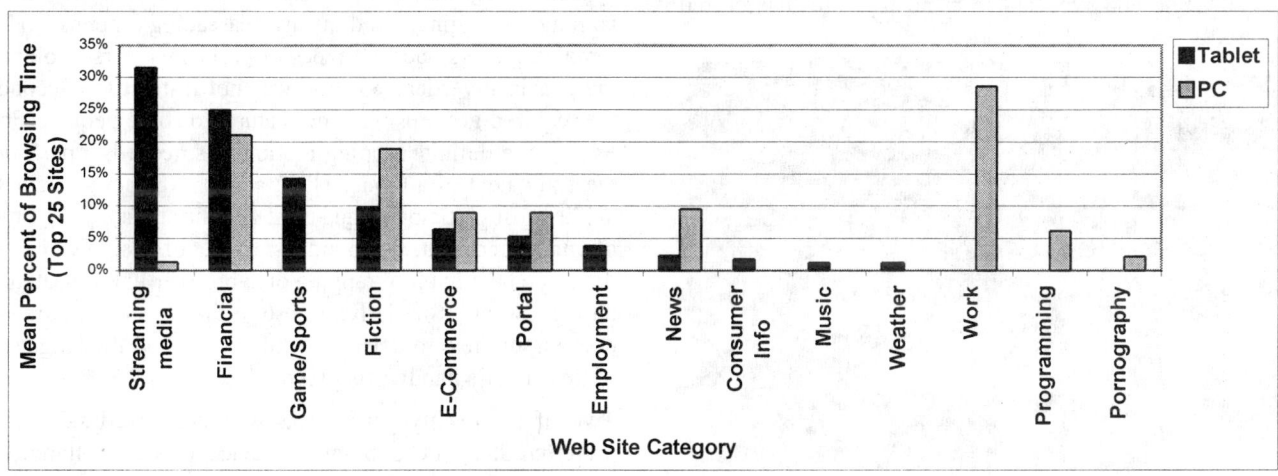

Figure 5 Time spent on web sites by category

The tablet and PC share some favorite web categories such as financial sites. Several of the top sites, however, differ on the two devices. On the tablet, participants spent a lot of time listening to streaming audio, primarily archived radio programs. One participant reported listening to audio while folding laundry and supervising his child in the bath. On the PC, using streaming media was an infrequent activity. Similarly, game and sports-related sites were frequently visited on the tablet but not on the PC.

To summarize, households used the tablets to reduce contention for access to browsing. There is evidence that they would have used e-mail and chat on the tablet as well, but for usability problems. They spent their time browsing different sites than they typically did on the PC: frequently visited sites facilitated time-sharing the tablet with other household activities and people.

ERGONOMICS AND USABILITY

The tablet that we used in this study was designed for vertical markets, for use in settings such as warehouses, oil fields, and beverage delivery. In these settings, users would often be standing and would not do extensive text entry. While our participants reported that it was easy to perform the basic activities with the tablet—selecting links in the browser—in many ways the design of the tablet was not well suited to home use. The following sections discuss the form of the tablet, its input methods, and the display.

Form

One of the biggest advantages of the tablet, according to participants, is that they can use it in places they can't use their PC, often in comfortable positions. We asked people to show us the most common locations and positions in which they used the tablet. Of 21 locations/positions, 18 were on a couch, easy chair, or in bed. Only two people reported using their tablet often at a desk or table.

The desire to use the tablet in non-traditional positions was its downfall. Figure 6 shows a typical user position, illustrating that it is difficult to hold the screen at the optimal angle and type at the same time in this relaxed posture.

Figure 6 A typical user position with the tablet

The most common complaint about the tablet (half of the users cited it as their biggest pet peeve) was that the screen and keyboard were not integrated. This was a problem for several reasons. First, the lack of a structural unit made it impossible for users to prop up the screen at the desired viewing angle without holding the screen or leaning it against an object like a pillow or the knees. As we discuss below, achieving the correct screen angle was important to achieve a clear display with this passive matrix technology. Second, it was inconvenient to carry two pieces—the screen and the keyboard—with the result that the keyboard was often not there when it was needed. Finally, the keyboard was difficult to position, especially while lounging. The user in Figure 6 kept his keyboard out of the way, and pulled it up onto his stomach only when he really needed to type. Some participants' favorite positions—lying on one's side propped up on an elbow, for example—would have made it difficult to type at all, and participants often used on-screen input methods in these positions. Thus, the ideal form must allow for adjustable screen tilt even with the keyboard tucked out of the way.

One aspect of the form-factor was acceptable to all participants. The weight of the tablet—2.2 pounds—was light enough to be comfortably carried.

The Stylus

One design feature than distinguishes a tablet appliance from a laptop computer is its touchscreen. Previous work has suggested that a touchscreen may perform better than a touchpad [4] and at least as well as a mouse [7]. While we did not perform rigorous testing with the touchscreen, we tried to discover any performance issues associated with the stylus and to understand participants' preferences for the touchscreen compared to their PC's pointing device.

The stylus was an electromagnetic, active stylus. A normal pen or finger could not be used, and it was necessary to hold the stylus perpendicular to the screen for the touch to register. In their first encounter with the stylus, approximately half the participants had no problems with tapping, dragging, and using cascading menus. The remaining users had two types of problems. First, holding the pen in a vertical position was not instinctive. Second, the required gestures were not natural to some people. For example, a definite tapping motion was needed to perform the equivalent of a mouse click; a press was not sufficient. While most participants had mastered the stylus by the time of our second visit, we found a small set of people who had developed the habit of tapping on a browser link repeatedly until it "went." Similarly, a gesture that seemed natural—crossing out text to delete it—did not work in the language of this tablet's handwriting recognizer.

Two of the twenty participants we interviewed said they preferred their PC's pointing device (a trackball and a mouse) to the touchscreen. Most participants, however, liked the convenience of being able to select an item directly on-screen, and it appears that this direct manipulation method was well suited to the positions in which they used the tablet. Participants adapted to the peculiarities of the touchscreen, and by the concluding visit, we observed few usability problems.

Text Entry

The tablets had three methods for entering text: the external keyboard, an on-screen keyboard, and handwriting recognition. We wondered whether all these methods were necessary, particularly given the slowness of selecting letters on the on-screen keyboard.

Participants' complaints suggested that none of the available text input methods was ideal. Their preferred input method depended on the task. For e-mail and chat, 71% of users preferred the external keyboard. For entering URLs, 67% preferred an on-screen method. Several users said they tried handwriting recognition first and then turned to the on-screen keyboard when the handwriting recognition was too inaccurate. It seems that the speed-accuracy tradeoff between handwriting and the on-screen keyboard necessitates both methods, given the current state of handwriting technology.

Browsing was by far the most common task, and two reasons for preferring an on-screen input method for that task were that the keyboard was awkward to use in a lounging position and that the keyboard frequently had been left behind in another room. Users seemed to turn to the keyboard only for text-intensive activities, and at least four users indicated they had limited their use of chat or e-mail because of input problems. Keyboard size was too small; some participants could not fit their fingers on the home row and abandoned the keyboard for this reason. Keys were 12 mm in diameter, which is within the ANSI standard [1], but key spacing (center-to center distance) was 15 mm, which is below standard.

We don't know the extent to which the keyboard would have been used if it had been integrated with the screen, but since the stylus is needed to select links in the browser, an on-screen input method for entering URLs would seem to be most convenient. Given users' postures, the ideal tablet would have an integrated keyboard available for text-intensive activities but would allow the keyboard to be tucked away for transport or while browsing, with an accurate handwriting recognition method available.

The Display

Most participants found the color, brightness, and contrast of the passive matrix display acceptable. The narrow viewing angle necessitated by the passive matrix technology and the small screen size (7.5 inch diagonal) received the majority of display-related complaints. While a few participants were annoyed with the need to scroll most web sites (resolution was 640 x 480 pixels), this was less important than physical size. That is, participants wanted larger pixels, not more of them. Informal observation revealed that several users hunched over the screen in order to see it better. We showed participants a different tablet that included a 9.4 inch diagonal display. While their experience with this tablet was limited, most participants said they preferred the larger size.

THE IDEAL TABLET

Our participants' behavior and comments suggest that an Internet appliance intended for general web access and text-based communication must have three characteristics. First, the software must be sufficiently full-featured that people find it useful. Second, the device must be highly portable and comfortable to use in relaxed positions, with an integrated keyboard and screen and accurate on-screen input method. And third, it must have a large enough screen and keyboard to be usable. Tablets available for the general market today tend to lack one or another of these characteristics. For example, some tablets have better form-factors than the one used in this study, but they run limited browsers that preclude activities that we found to be popular, such as streaming media and chat. By contrast, other tablets have full operating systems but inappropriate form-factors for the type of leisure home use we saw here. While a full operating system is not required (in fact, may be a detriment for less experienced PC users, as we found with some of our participants), our results suggest that the browser must be able to handle standard plug-ins, and that the tablet must support standard chat software.

To pin down critical features of an ideal tablet, we asked participants to define their ideal tablet, first in free-form discussion and then by rating a set of candidate features. Table 2 lists the mean importance ratings for the set of candidate features we supplied.

	Mean Importance Rating
Surf the web	5.91
Internet anywhere in house	5.68
E-Mail	5.24
File/print share	4.59
ICQ/Instant Messenger	4.23
Streaming Audio	4.18
Internet away from home	4.14
Internet in yard	3.95
Streaming Video	3.77
Create/save Notes	3.76
Internet Games	3.57
Chat (other)	3.32
Shockwave	3.27
Internet Phone Call	2.82
Videoconference	1.81

Table 2 Mean importance ratings for tablet features

The applications people are looking for are consistent with their usage: surfing, e-mail, Instant Messenger, ICQ, and streaming audio. The top capabilities included portability within the home and file/print sharing with other PCs in the home. Participants also dreamed of a device that they could use to access the Internet from anywhere; this was a top request in the free-form "ideal tablet" interviews as well. The other feature that came up frequently in those interviews was that users wanted a faster processor, to improve the quality of streaming video, download pages faster, and play networked games. It appears, then, that at least experienced PC users with broadband connections are looking for a tablet that combines portability with some of the power of a desktop PC, to take advantage of the broadband connection.

We should say a word about home networking, which was also mentioned frequently in ideal tablet discussions. Six households set up file and print sharing with the household PC. They transferred files from the PC to the tablet, primarily bookmarks and software, and they told us it was important to be able to print web pages from the tablet to the PC's printer. Usage data indicated that a primary benefit of home networking is that household members can surf the web from different devices at the same time. But home networking and multiple Internet access can lead to a problem: different devices will contain different subsets of information needed by the same person. For example, five of our participants ended up reconfiguring their e-mail program or forwarding e-mail to themselves so that they didn't end up with downloaded e-mail split between tablet and PC. As Internet appliances proliferate in the home, it will be important for them to be able to share information, to share e-mail, bookmarks, files, and perhaps software, without extra work on the part of the user.

CONCLUSIONS: THE FUTURE

This study suggests that Internet appliances will eventually fill a special role in people's daily lives. These small, portable appliances can make information and entertainment accessible without demanding focused attention. To optimize this type of use, the functional and design characteristics of a tablet-type appliance must conform to the way people conceptualize and use it. That is, a web tablet is not, and should not strive to be or replace the PC. Nonetheless, as PC users grow more sophisticated in their use of the Internet, they will come to expect appliances to be as powerful in some ways as a PC. We found that people expected the tablet to be able to access the broadband plug-ins and personal chat applications they used on their PC. It is likely that as broadband connections become more prevalent and people make greater use of video communication, they will expect the tablet to support this functionality as well. But they will expect it in a portable package that both conforms to relaxed usage positions and connects to the other devices in the home. In the future Internet appliances may really be fingers of a larger Internet access capability that enables true portability by providing access to information and tools anywhere.

REFERENCES

1. *American National Standard for Human Factors Engineering of Visual Display Terminal Workstations.* Human Factors Society, Santa Monica, CA, 1988.

2. Berners-Lee, T. "The World Wide Web: A very short personal history." Available at http://www.w3.org/People/Berners-Lee-Bio.html/ShortHistory.html.

3. Brekke, D. "High-Speed Habits," *Wired*, June, 1999, p. 90

4. Cohen, O. , Meyer, S., and Nilsen, E. Studying the Movement of High-Tech. Rodentia: Pointing and Dragging. *Proceedings of CHI '93, ACM Press, 135-136.*

5. Kraut, R., Tridas, M., Janusz, S., Kiesler, S., and Scherlis, W. Communication and Information: Alternative Uses of the Internet in Households, in *Proceedings of CHI '98* (Los Angeles, CA, April 1998), CAN Press, 368-375.

6. Miller, M. The Internet Appliance, *PC Magazine,* January 23, 1996. Available at http://www.zdnet.com/pcmag/issues/1502/pcm00012.htm

7. Sears, A., and Shneiderman, B. High precision touchscreens: design strategies and comparisons with a mouse. *International Journal of Man-Machine Studies 34, 4* (1991), p. 593-613.

Predicting Text Entry Speed on Mobile Phones

Miika Silfverberg
Nokia Research Center
P.O. Box 407
FIN-00045 Nokia Group
Finland
+358 40 528 7759
miika.silfverberg@nokia.com

I. Scott MacKenzie
Dept. of Mathematics & Statistics
York University
Toronto, Ontario
Canada M3J 1P3
+1 416 736 2100
smackenzie@acm.org

Panu Korhonen
Nokia Research Center
P.O. Box 407
FIN-00045 Nokia Group
Finland
+358 40 504 7123
panu.korhonen@nokia.com

ABSTRACT
We present a model for predicting expert text entry rates for several input methods on a 12-key mobile phone keypad. The model includes a movement component based on Fitts' law and a linguistic component based on digraph, or letter-pair, probabilities. Predictions are provided for one-handed thumb and two-handed index finger input. For the traditional multi-press method or the lesser-used two-key method, predicted expert rates vary from about 21 to 27 words per minute (wpm). The relatively new T9 method works with a disambiguating algorithm and inputs each character with a single key press. Predicted expert rates vary from 41 wpm for one-handed thumb input to 46 wpm for two-handed index finger input. These figures are degraded somewhat depending on the user's strategy in coping with less-than-perfect disambiguation. Analyses of these strategies are presented.

Keywords
Text entry, mobile systems, mobile phones, keypad input, human performance modeling, Fitts' law, digraph frequencies

INTRODUCTION
Designing new text entry methods for computing systems is labour intensive. It is also expensive, since a working prototype must be built, and then tested with real users. Because most text entry methods take time to learn, the testing should preferably take place in longitudinal settings. However, longitudinal user studies are tedious [13]. A pragmatic approach is to develop a predictive model to "test" new text entry methods a priori — without building prototypes or training users. Models, at their best, can be valuable and informative tools for designers of new text entry methods [1, 13].

Permission to make digital or hard copies of all or part of this work for personal or classroom use is granted without fee provided that copies are not made or distributed for profit or commercial advantage and that copies bear this notice and the full citation on the first page. To copy otherwise, to republish, to post on servers or to redistribute to lists, requires prior specific permission and/or a fee.
CHI '2000 The Hague, Amsterdam
Copyright ACM 2000 1-58113-216-6/00/04...$5.00

This research is concerned with the problem of text entry on mobile phones. Although we usually think of phones as devices for speech input and output, the transmission and reception of text messages on mobile phones is increasing rapidly. For example, Finland's largest teleoperator, Sonera, reports a six-fold increase of text messages during 1998 (http://www.sonera.fi/investor_en/publications/annualreports/sonera98_english.pdf).

Text entry on contemporary mobile phones is mainly based on the 12-key keypad (Figure 1). This paper describes a method for predicting potential expert user text entry speed for input methods that utilize the 12-key keypad. The model provides individual predictions for one-handed thumb and two-handed index finger use.

Figure 1. The 12-key keypad

State of the Art of Text Entry on Mobile Phones
The 12-key keypad consists of number keys 0-9 and two additional keys (# and *). Characters A-Z are spread over keys 2-9 in alphabetic order. The placement of characters is similar in most mobile phones, as it is based on an international standard [9]. The placement of the SPACE character varies among phones. In this paper, we assume the 0-key serves as the SPACE character.

Since there are fewer keys than the 26 needed for the characters A-Z, three or four characters are grouped on each key. Thus, ambiguity arises. For example, if the user presses key 2, the system must determine which of the characters A, B, or C the user intends. There are several approaches to this problem. We present three: the multi-press, the two-key, and the T9 methods.

Multi-press Input Method

The multi-press method is currently the main text input method for mobile phones. In this approach, the user presses each key one or more times to specify the input character. For example, the number key 2 is pressed once for the character 'A', twice for 'B', and three times for 'C'.

The multi-press approach brings out the problem of *segmentation*. When a character is placed in the same key as the previously entered character (e.g., the word *on*), the system must determine whether the new key press still "belongs to" the previous character or represents a new character. Therefore, a mechanism is required to specify the start of a new character.

There are two main solutions to this. One is to use a timeout period within which key presses belong to same character. Most phones have a timeout, typically between 1 and 2 seconds. The other solution is to have a special key to skip the timeout ("timeout kill") thus allowing the next character — on the same key — to be entered directly. Some phone models use a combination of the two solutions. For example, Nokia phones include both a 1.5-second timeout and the provision for a timeout kill using the arrow keys. The user may decide which strategy to use. We provide predictions for both.

Two-key Input Method

In the two-key method, the user presses two keys successively to specify a character. The first key press, as in the multi-press method, selects the "group" of characters (e.g., key 5 for 'JKL'). The second press is for disambiguation: one of the number keys, 1, 2, 3, or 4, is pressed to specify the position of the character within the group. For example to enter the character 'K', the user presses 5-2 ('K' is second character in 'JKL').

The two-key method is very simple. There are no timeouts or such. Each character A-Z is entered with exactly two key presses. SPACE is entered with a single press of the 0-key.

The two-key method is not in common use for entering Roman characters, however. In Japan, a similar method (often called the "pager" input method) is very common for entering Katakana characters.

T9 Input Method

A third way to overcome the problem of ambiguity is to add linguistic knowledge to the system. The T9 input method, patented by Tegic Communications, Inc. (Seattle, WA) [8], uses a dictionary as the basis for disambiguation. The method is based on the same key layout as the multi-press method, but each key is pressed only once. For example, to enter "the", the user enters the key sequence 8-4-3-0. The 0-key, for SPACE, delimits words and terminates disambiguation of the preceding keys. T9 compares the word possibilities to its linguistic database to "guess" the intended word.

Naturally, linguistic disambiguation is not perfect, since multiple words may have the same key sequence. In these cases, T9 gives the most common word as a default. To select an alternate word, the user presses a special NEXT function key. For example, the key sequence 6-6 gives "on" by default. If another word was intended, the user presses * to view the next possible word. In this case, "no" appears. If there are more alternatives, NEXT (*) is pressed repeatedly until the intended word appears. Pressing 0 accepts the word and inserts a SPACE character.

Based on our informal analyses, disambiguating works quite well. In a sample of the 9025 most common words in English (ftp://ftp.itri.bton.ac.uk/) produced from the British National Corpus, the user must press NEXT only after about 3% of the words. Naturally, the whole vocabulary is larger than 9025 words, so this estimate may be optimistic. However, 5% is a reasonable approximation, and will be used throughout this paper.

Most major mobile phone manufacturers have licensed the T9 input method, and, as of 1999, it has surfaced in commercial products (e.g., the Mitsubishi *MA125*, the Motorola *i1000Plus*, the Nokia *7110*). There is also a touch-screen version of T9 that is available for PDAs (e.g., the Palm Computing *Palm III*, the Philips *Nino*). Bohan et al. [2], describe an evaluation of the touch screen version; however, to our knowledge, there are no published evaluations of T9 with physical keys.

MODEL FOR MOBILE PHONE TEXT ENTRY

Our model is similar to that of Soukoreff and MacKenzie [15]. It is based on two components: a movement model (Fitts' law) and a linguistic model (digraph probabilities).

Movement Model (Fitts' Law)

The core of this paper is the application of Fitts' law to the mobile phone keypad. Fitts' law [6] is a quantitative model for rapid, aimed movements. It can be used to calculate the potential text entry speed of an expert user, assuming that the text entry performance of an expert is highly over-learned, and thus is limited only by motor performance. We will elaborate more on this assumption later.

Fitts' law has been applied with success to pointing devices [5, 12] and on-screen keyboards [13, 14]. There are only a few studies, however, that apply Fitts' law to physical keyboards. Card et al. [3] suggested using Fitts' law for keying times on a calculator. Drury and Hoffman [4] used Fitts' law to evaluate the performance tradeoffs of various inter-key gaps for data entry keyboards.

Fitts' law is expressed as

$$MT = a + b \log_2(A/W + 1) \quad (1)$$

where A is the length (amplitude) of a movement, and W is target size (width), in this case, the size of the pressed key [10]. Fitts' law is inherently one-dimensional, as evidenced by a single "width" term. Physical keys on a mobile phone keypad, however, are laid out in a two-dimensional array, and each key has both width and height. Therefore, we need to extend the model to two dimensions. For this purpose we substitute for W in Equation 1 the smaller of the width and height, as suggested by MacKenzie and Buxton [11]. In most cases, height is less than the width for keys of a mobile phone. Therefore, we used the height of the keys as W.

The log term in Equation 1 is called the index of difficulty (ID):

$$MT = a + b \times ID \quad (2)$$

The two constants, a and b, are determined empirically by regressing observed movement times on the index of difficulty. For this purpose, we collected empirical data from both one-handed thumb use (Experiment 1, see below) and two-handed index finger use (Experiment 2).

In mobile phone text entry, each character is entered with one or more key presses, i.e., movements. The first of these, the *initial movement*, M_0, consists of moving the finger over the desired key (e.g., key 'ABC' for character 'a') and pressing the key. Depending on the input method, there may be none, one or several *additional movements* (M_1, M_2, etc.).

For each movement (M_0, M_1, M_2, etc.), Fitts' law is used to predict the movement time (MT_0, MT_1, MT_2, etc.). The total time to enter a character, CT, is calculated as the sum of all the required movements:

$$CT_{ij} = \Sigma \, MT_k \quad (3)$$

The details, of course, depend on the text entry method. Below, we explain the models for each of the three text entry methods.

Movement Model for Multi-press Input Method
In the multi-press method, the user presses each key one or several times. There are two strategies, varying in their treatment of the timeout. We model these separately.

If the user allows the built-in timeout to segment consecutive characters on the same key, the character entry time is calculated as follows:

$$CT = MT_0 + N \times MT_{repeat} + T_{timeout} \quad (4)$$

MT_0 is the initial movement time, i.e., the time to move one's finger or thumb from the first key of the digraph to the second key. N is the number of key repetitions, which is an integer from 0 to 3 depending on character (e.g., character 'C' requires two extra presses of key 'ABC', $N = 2$). MT_{repeat} is the key-repetition time, which equals the intercept a in the Fitts' law equation ($ID = 0$).

For $T_{timeout}$, we used 1.5 seconds. This is the time used in Nokia phones, although it may vary among manufacturers. $T_{timeout}$ is required only if the second character, j, is on the same key as the first character, i.

Alternatively, the user may explicitly override the timeout by pressing a timeout kill key (the down-arrow key in Nokia phones). In the latter case, the character entry time is

$$CT = MT_0 + N \times MT_{repeat} + MT_{kill} \quad (5)$$

where MT_{kill} is the time to move to the arrow key.

Movement Model for Two-Key Input Method
In the two-key method, each character except SPACE requires two key presses. Therefore, character entry time is simply calculated as a sum of two movement times:

$$CT = MT_0 + MT_1 \quad (6)$$

With the SPACE character, the second movement time is zero.

Movement Model for T9 Input Method
In the T9 input method, each key is pressed only once. Also, there is no timeout. Therefore, the character entry time is simply calculated as:

$$CT = MT_0 \quad (7)$$

This model for T9 is for perfect disambiguation, and assumes the NEXT function is not needed. We will discuss the implications of this in detail later.

Linguistic Model (Digraph Probabilities)
Our linguistic model uses a 27×27 matrix of letter-pair (digraph) frequencies in common English [15]. The 27 characters include the letters A-Z and the SPACE character.

Each letter-pair, i-j, is given a probability P_{ij} based on an analysis of representative sample of common English. The sum of all probabilities is one:

$$\Sigma\Sigma \, P_{ij} = 1 \quad (8)$$

Since our predictions are based on a linguistic model, they are inherently language-specific, applying only to common English. However, because the language model is simple, the results are easy to adapt to another language by changing the digraph probabilities according to that language.

Combining the Models

To develop predictions we need to combine the motor and linguistic models. An average character entry time for the language (CT_L) is calculated as a weighted average of character entry times for all digraphs:

$$CT_L = \Sigma\Sigma \, (P_{ij} \times CT_{ij}) \qquad (9)$$

Taking the reciprocal of CT_L gives us the average number of characters per second, which can be transformed into words per minute by multiplying by 60 seconds per minutes and dividing by 5 characters per word:

$$WPM = (1 / CT_L) \times (60 / 5) \qquad (10)$$

METHOD

Our model is still incomplete, since the coefficients a and b in the Fitts' law equations are unknown for finger input on a 12-key phone keypad. Two experiments were carried out to determine these coefficients. Experiments 1 and 2 described below sought to determine these for one-handed thumb input and two-handed index finger input, respectively.

Experiment 1: One-handed Thumb Input

In this experiment participants held the phone in a preferred hand and pressed the keys with the thumb of the same hand. The other (non-preferred) hand was held idle.

Participants

Twelve volunteers (7 male, 5 female) participated in the study. Most participants were employees of the Nokia Research Center in Helsinki. Their age ranged from 24 to 47 years, with an average of 32.6 years.

Five of the participants were left-handed; however, two choose to hold the phone with their right hand. The right-handed participants held the phone in their right hand. All participants had prior experience in using a 12-key phone keypad. Ten participants were regular mobile phone users. The average mobile phone experience of all participants was 3.9 years.

Apparatus

The number keypad of a Nokia 5110 mobile phone was used as the model 12-key keypad (Figure 3). Only the number keys (1-9) and * and # keys were used in the experiment. Number keys are slightly larger than * and # keys. The dimensions are shown in Figure 2. As mentioned previously, the height of keys was used in calculating ID. Key dimensions and distances between keys were measured using a slide gauge.

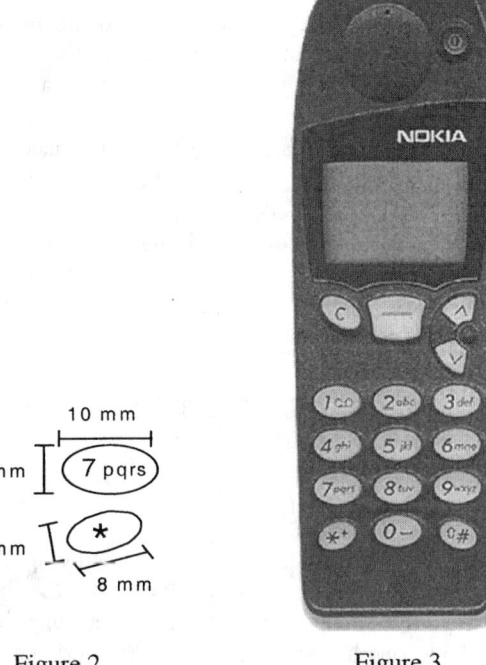

Figure 2   Figure 3

Test Tasks

The participant's task was to press specified keys on the phone keypad for a period of 10 seconds at a time. Participants were instructed to press the keys as fast as they could but to avoid errors.

There were two types of tasks:

(i) Single keys: In these tasks the participants pressed only a single key. There were four key repeat tasks altogether (keys 1, 3, 7 and 9).

(ii) Key pairs: In these tasks the participants pressed two keys alternately for 10 seconds. A subset of all possible pairs of keys was chosen to cover a range of movements, for example, from very short (e.g., key 1 - key 4) to very long (e.g., key 3 - key *). The inter-key distances ranged from 9 to 38 mm, with an average of 20.6 mm. The key pairs were selected to create similar movements for left- and right-handed participants.

There were 26 key-pair tasks per participant. This made up a total of 30 tasks per participant, 360 tasks altogether.

Procedure

The tasks were presented to participants in random order. The "write messages" function of the phone (with number mode selected) was enabled during the tasks; thus, the phone automatically showed the number of written characters.

A 10-second countdown timer controlled the task time (except for the first three participants, whose time was controlled manually with a stopwatch). The test moderator signaled the start of each task by saying "1-2-3-go" and pressing the start key on the countdown timer. After 10 seconds, the countdown timer gave a clearly audible sound. The participants were instructed to stop pressing the keys when they heard this sound. Key presses entered after the stop signal were ignored. The test moderator checked the number of key presses from the phone's display and recorded it in a spreadsheet file. The average movement time (in milliseconds) between successive key presses was then calculated using formula:

$$MT = 10000 / (N - 1) \qquad (11)$$

where N is the number of key presses during a 10 second interval. Error data were not collected.

Experiment 2: Two-handed Index Finger Input
Experiment 2 was similar to Experiment 1 except the index finger was used instead of the thumb. Users held the phone in one hand and entered key presses with the index finger of the other hand.

Participants
Twelve volunteers (7 male, 5 female) participated in the study. Seven had also participated in Experiment 1. Ages ranged from 23 to 41 years, with an average of 29.8 years.

Five of the participants were left-handed. One, however, choose to press keys with the right hand. The right-handed participants used their right hand. All participants had some experience using a standard phone keypad. Eleven were regular mobile phone users. The average mobile phone experience of all participants was 4.0 years.

RESULTS
Figure 4 shows the results from Experiment 1 and 2. In both experiments, the movement time (*MT*) grows linearly with index of difficulty (*ID*), as predicted by Fitts' law.

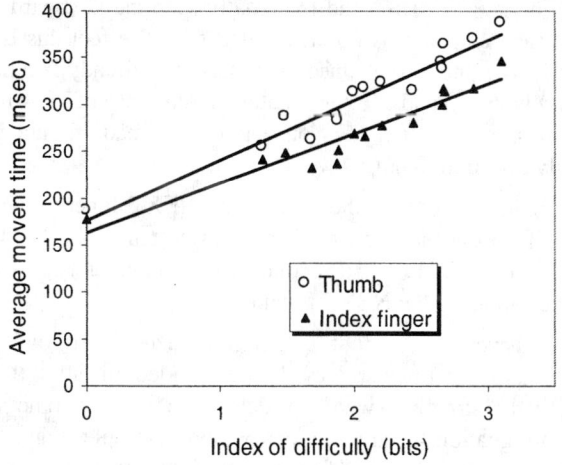

Figure 4. Results from Experiments 1 and 2

A linear regression of *MT* on *ID* was performed. The results are given in Figure 5.

	Intercept, *a* (ms)	Slope, *b* (ms/bit)	Correlation
Index finger	165	52	0.960
Thumb	176	64	0.970

Figure 5. Results from the linear regression

The correlations in the linear regression are high, indicating that Fitts' law predicts the movement time with high accuracy both with the index finger and thumb.

Overall, the index finger was faster than the thumb. The average movement time between successive key presses in all conditions was 273 ms for the index finger, and 309 ms for the thumb.

The analysis of variance of movement time showed clear main effects for both index of difficulty ($F_{13,692} = 85.9$, $p < .0001$) and input finger ($F_{1,692} = 114.1$, $p < .0001$). The *ID*-by-finger interaction was not significant ($F_{13,692} = 1.7$, $p > .05$).

The two points with $ID = 0$ are substantially to the left of the other points, and this is a concern. Although it has been suggested that Fitts' law does not apply when *ID* is small [7], a more legitimate explanation lies in the treatment of spatial variability in building the model. Fitts' law is predicated on the assumptions that (a) the spatial distribution of end-points is normal and (b) 4% of the distribution falls outside the target region. Where possible, it is desirable to use ID_e computed from W_e and A_e — the actual, or "effective", amplitude and width of the distributions. We could not do so in this simple experiment because there was no means to capture end-points. However, if W_e and A_e could have been used, then clearly the task with $ID = 0$ would have $W_e > 0$ and $A_e > 0$, and, hence, $ID_e > 0$. This would tend to shift the points at $ID = 0$ to the right [10].

Model Predictions for Mobile Phone Text Input
Based on Experiments 1 and 2, the movement time on mobile phone keypad can be reliably predicted using Fitts' law equations:

$$MT_{\text{index finger}} = 165 + 52 \log_2(A/W + 1) \text{ ms} \qquad (12)$$

and

$$MT_{\text{thumb}} = 176 + 64 \log_2(A/W + 1) \text{ ms} \qquad (13)$$

Incorporating this information, our model gives predictions of potential expert user text entry speeds. The results are shown in Figure 6.

Method	Index finger	Thumb
Multi-press		
- wait for timeout	22.5	20.8
- timeout kill	27.2	24.5
Two-key	25.0	22.2
T9	45.7	40.6

Figure 6. Results of model predictions (wpm)

Two predictions are given for the multi-press method corresponding to the two possible interaction strategies. If consecutive characters are on the same key, the user may either wait for the timeout to pass, or end it manually. In our model, the timeout was 1.5 seconds. Our model predicts that with this timeout, "timeout kill" is clearly the faster strategy (21% faster when using the index finger, 18% with the thumb). It is assumed that expert users adopt the faster strategy. This is supported by our observations of users at the Nokia Research Center: a majority of experienced multi-press users employ the timeout kill strategy.

Predictions are clearly higher for the T9 method than for the multi-press and two-key methods. These differences, and other interaction issues for T9, are discussed in detail later.

In comparing the multi-press and two-key methods, the multi-press method is slower if the user employs the timeout strategy — waiting for the timeout between consecutive characters on the same key. However, as expertise develops, users will invoke the timeout kill function. With an optimal use of timeout kill, the multi-press method is faster than the two-key method.

Input via the index finger is also consistently faster than with the thumb. The difference is largest with T9 and the two-key input methods where the index finger is 13% faster than with the thumb. The difference is smaller with the multi-press method. This is due to the steeper slope of the Fitts' law model for the thumb in Figure 4. With small *IDs* the difference between the index finger and thumb is quite small; the multi-press method involves many key-repetitions ($ID = 0$), which diminishes the difference between the index finger and thumb.

Within the multi-press method, the difference for the index finger is larger with the "timeout kill" strategy (11%) than with "wait for timeout" (8%). This is due to the constant length of the timeout, which diminishes the differences between input fingers in our "wait for timeout" results.

Comparisons With Empirical Data

There are, at present, no published data on text entry rates with mobile phones. As this mode of interaction evolves and enters the mainstream of mobile computing, however, this should change. Furthermore, there are few people who may be deemed "experts" in mobile phone text entry. The investigation described herein is in anticipation of an ever-increasing demand for this mode of text entry.

While formal user studies are preferred, we commonly perform quick and simple checks of novel text entry techniques using the well-known phrase, "*the quick brown fox jumps over the lazy dog*". This 43-character phrase includes every letter of the alphabet, and therefore ensures that each key, or key combination, is visited during entry.

Within our lab, one user performs a substantial amount of mobile phone text entry via the multi-press method, and, in our view, approaches the status of expert. For example, this person routinely uses the timeout kill feature where applicable. We asked this person to perform timed input of the quick-brown-fox phrase. In three repetitions using thumb input, the times were 27 s, 23 s, and 24 s, with 0 errors in each case. The mean entry time was 24.6 s. For a 43-character phrase, this translates into a text entry rate of 21.0 wpm. This is surprisingly close to our predicted expert entry rate of 24.5 wpm.

We asked the same user to perform the same test with a T9-enabled cell phone. We asked the user to ignore the possible need for the NEXT function, and to enter the phrase directly. The entry times were 15 s, 15 s, and 16 s. The only error was for the key sequence 5-2-9-9-0, which T9 incorrectly disambiguated to "jazz" instead of "lazy". The mean entry time of 15.7 s translates into an entry rate of 32.9 wpm. Keeping in mind that this user does not use T9 on a daily basis, the observed rate is reasonably close to our predicted expert entry rate of 40.6 wpm.

Interaction and Linguistic Issues for T9

The very generous predictions for T9 in Figure 6 should be viewed in the narrow context of our model. For example, our model is for experts and ignores the novice or learning behaviour that most users of this emerging input technique will experience.

As well, our model attends only to the motor component of the interaction. The need to visually scan the keyboard to find each character is not accounted for. We feel this is a relatively minor issue, since most users are already familiar with phone keypads. Expert status, in the sense of knowing the position of characters on a phone keypad, should be easily acquired in our view.

Of greater concern is the role of the NEXT function with T9. Two questions arise. First, how often is the NEXT function required? And second, what behaviour will users exhibit in using the NEXT function?

The answer to the first question is determined by the dictionary, or linguistic model, embedded in T9. It is relatively straightforward to determine the outcome of disambiguation for any dictionary. For example, Figure 7 provides an analysis using the word sample discussed earlier. The results are quite impressive. Of the 9025 words in the sample, 8437 (95%) can be entered and uniquely disambiguated.

The number of words requiring 1, 2, 3, 4, or 5 presses of the NEXT function is 476, 83, 23, 5, and 1, respectively.

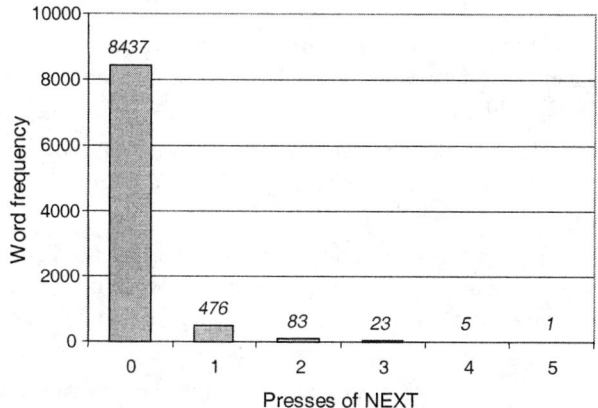

Figure 7. Use of NEXT for a sample of 9025 words

Figure 8 illustrates some ambiguous words – those requiring 4 or 5 presses of the NEXT function.

Key sequence	Initial word	Presses of NEXT	Subsequent words
2-2-7-3	case	5	care, base, card, bare, cape
2-6-9	any	4	boy, box, cow, box
7-2-4-3	said	4	page, paid, raid, rage
7-2-6	ran	4	sam, san, pan, ram
7-2-9	say	4	saw, pay, raw, ray

Figure 8. Examples of ambiguous words for T9

The initial word for any key sequence is the one with the highest probability in the linguistic model, while subsequent words are produced in decreasing order of their probability. Note that our word sample, as well as that in the T9 dictionary, includes proper nouns (e.g., Sam).

Although the T9 dictionary and the disambiguation process are considered proprietary by Tegic, we tested a T9-enabled mobile phone with the key sequences in Figure 8. All the words in Figure 8 were produced, although there were a few minor differences in the sequences.

Answering the second question above is much more difficult since it involves user strategies. Although as many as 95% of words entered will correctly appear by default, users will not necessarily anticipate this. Thus, there is a need for the user to visually verify input. This behaviour is outside the scope of our model, as noted earlier. It is also a behaviour that is difficult to empirically model, since there are both perceptual and cognitive processes at work.

Figure 9 presents a parametric analysis of the use of the NEXT function for two components of the behaviour for thumb input. First, the percentage of words for which visual inspection is performed is included: 0%, 25%, 50%, 75%, and 100%. For the 0% condition, the user never visually verifies input. For the 100% condition, the user visually verifies input after each word entered.

Second, the perceptual-plus-cognitive time associated with visual inspection is shown along the horizontal axis as a continuum from 0 to 1000 ms. Note that the movement time for multiple invocations of the NEXT function is quite small because it requires multiple presses of the same key (*).

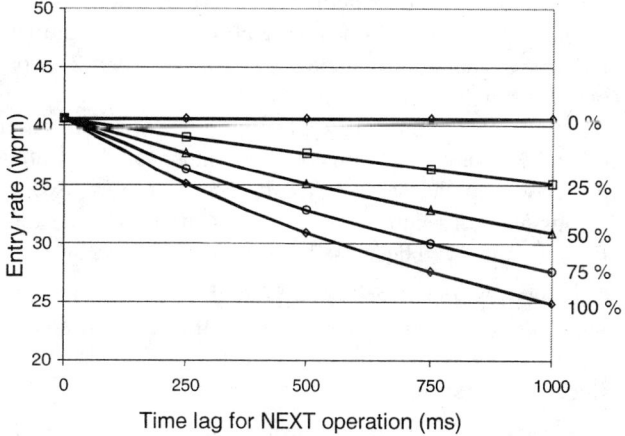

Figure 9. Parametric analysis for T9 user behaviour (see text for discussion)

Expert usage appears along the top line in Figure 9. That is, the user always knows when the next function is required and never visually verifies input. Although this behaviour will never fully occur, it may occur by degree. For example, a user may quickly learn that the word "on" requires the key sequence 6-6-0, whereas the word "no" requires the NEXT function: 6-6-*-0.

If the user visually verifies input 50% of the time (far more often than necessary, in fact), and each inspection takes 500 ms, then the T9 prediction falls to 35 wpm (see Figure 9). Bear in mind that this prediction is still predicated upon expert behaviour with respect to the keypad layout (i.e., no visual scan time to find the correct key). So, the prediction is still overly generous, perhaps.

Exploring hypothetical scenarios such as this, although important in characterizing user behaviour, is very weak in its ability to generate accurate predictions. Modeling expert performance is a luxury that affords a simplified view of user behaviour. Once we step off this ideal and attempt to accommodate more natural components of the interaction, there is an explosion in the sources and extent of variations. And so, the preceding exploration of T9 interaction will not be developed further. Suffice it to say that we expect T9 text entry rates to be slower than those cited above, consistent with a user's position on the learning curve and on the interaction strategy employed.

There are many other interactions issues, as well, such as the need to input numeric and punctuation symbols, or words not in the dictionary. Implementations of T9 we have tested include modes to insert words using the multi-press technique or to insert symbols from a displayed list. These important properties of the interaction are not accounted for in our current model.

CONCLUSIONS

We have provided predictions for expert text entry rates for several input schemes on mobile phones. The traditional multi-press method can support rates up to about 25 wpm or 27 wpm for one-handed thumb input or two-handed index finger input, respectively, provided the user effectively employs the timeout kill feature for consecutive characters on the same key. If the timeout is used to distinguish consecutive characters on the same key, then the entry rates will decrease by about 4 wpm in each case.

The two-key input technique is slightly slower than the multi-press method (using timeout kill): 22 wpm and 25 wpm for one-handed thumb input and two-handed index finger input, respectively.

The relatively new T9 technique requires only one key press per character, and relies on a built-in linguistic model to disambiguate input on a word-by-word basis. Text entry rates of 41 wpm and 46 wpm are predicted for one-handed thumb input and two-handed index finger input, respectively. These figures are for expert behaviour and a "perfect" disambiguation algorithm. Our analyses suggest that word-level disambiguation for English text with the traditional character layout on phone keypad is achievable with about 95% accuracy. The overhead of interacting with less-than-perfect disambiguation degrades performance, but the cost is difficult to quantify because of the complex and varied strategies that users may employ.

REFERENCES

[1] Bellman, T., and MacKenzie, I. S. A probabilistic character layout strategy for mobile text entry, *Proc of Graphics Interface '98*. Toronto: CIPS, 1998, 168-176.

[2] Bohan, M., Phipps, C. A., Chaparro, A., and Halcomb, C. G. A psychophysical comparison of two stylus-driven soft keyboards, *Proc of Graphics Interface '99*. Toronto: CIPS, 1999.

[3] Card, S. K., Moran, T. P., and Newell, A. *The psychology of human-computer interaction*, (Hillsdale, NJ: Lawrence Erlbaum, 1983).

[4] Drury, C. G., and Hoffmann, E. R. A model for movement time on data-entry keyboards, *Ergonomics 35* (1992), 129-147.

[5] Epps, B. W. Comparison of six cursor control devices based on Fitts' law models, *Proc Human Factors Society*. Santa Monica, CA: HFS, 1986, 327-331.

[6] Fitts, P. M. The information capacity of the human motor system in controlling the amplitude of movement, *Journal of Experimental Psychology 47* (1954), 381-391.

[7] Gan, K.-C., and Hoffmann, E. R. Geometrical conditions for ballistic and visually controlled movements, *Ergonomics 31* (1988), 829-839.

[8] Grover, D. L., King, M. T., and Kuschler, C. A. Patent No. US5818437, Reduced keyboard disambiguating computer. Tegic Communications, Inc., Seattle, WA (1998).

[9] ISO/IEC 9995-8. *Information systems - Keyboard layouts for text and office systems - Part 8: Allocation of letters to the keys of a numeric keypad*, International Organisation for Standardisation, 1994.

[10] MacKenzie, I. S. Fitts' law as a research and design tool in human-computer interaction, *Human-Computer Interaction 7* (1992), 91-139.

[11] MacKenzie, I. S., and Buxton, W. Extending Fitts' law to two-dimensional tasks, *Proc of CHI92*. New York: ACM, 1992, 219-226.

[12] MacKenzie, I. S., Sellen, A., and Buxton, W. A comparison of input devices in elemental pointing and dragging tasks, *Proc of CHI91*. New York: ACM, 1991, 161-166.

[13] MacKenzie, I. S., and Zhang, S. The design and evaluation of a high-performance soft keyboard, *Proc of CHI99*. New York: ACM, 1999, 25-31.

[14] Martin, G. L. Configuring a numeric keypad for a touch screen, *Ergonomics 31* (1988), 945-953.

[15] Soukoreff, W., and MacKenzie, I. S. Theoretical upper and lower bounds on typing speeds using a stylus and keyboard, *Behaviour & Information Technology 14* (1995), 370-379.

Developing a Context-aware Electronic Tourist Guide: Some Issues and Experiences

Keith Cheverst, Nigel Davies, Keith Mitchell, Adrian Friday, Christos Efstratiou

Distributed Multimedia Research Group
Department of Computing
Lancaster University
Lancaster, LA14YR, U.K.
+44 (0)1524 594539
{kc, nigel, mitchelk, adrian, efstrati}@comp.lancs.ac.uk

ABSTRACT

In this paper, we describe our experiences of developing and evaluating GUIDE, an intelligent electronic tourist guide. The GUIDE system has been built to overcome many of the limitations of the traditional information and navigation tools available to city visitors. For example, group-based tours are inherently inflexible with fixed starting times and fixed durations and (like most guidebooks) are constrained by the need to satisfy the interests of the majority rather than the specific interests of individuals. Following a period of requirements capture, involving experts in the field of tourism, we developed and installed a system for use by visitors to Lancaster. The system combines mobile computing technologies with a wireless infrastructure to present city visitors with information tailored to both their personal and environmental contexts. In this paper we present an evaluation of GUIDE, focusing on the quality of the visitor's experience when using the system.

Keywords
Mobile computing, context-awareness, adaptive hypermedia, user interface design, evaluation.

INTRODUCTION
The rapidly evolving field of mobile computing has massive potential for providing dynamic multimedia information to people on the move. Indeed, it has been predicted that in a few years time a large proportion of web browsing will be carried out via mobile devices. However, restricting the use of mobile devices to such tasks greatly underestimates their potential.

One area of research that is concerned with exploring the ways in which mobile devices can be used to provide more sophisticated services is that of context-aware computing [15]. Context-aware applications utilise contextual information, such as location, display medium and user profile, in order to provide tailored functionality.

This paper describes some of the issues and experiences gained while developing and evaluating GUIDE, a prototype context-aware tourist guide.

The GUIDE system [4,6] integrates the use of personal computing technologies, wireless communications, context-awareness and adaptive hypermedia [2] in order to support the information and navigation needs of visitors to the city of Lancaster. In more detail, GUIDE utilizes a cell-based wireless communications infrastructure in order to broadcast dynamic information and positioning information to portable GUIDE units that run a customized web-browser.

This paper focuses on three main parts of the development of GUIDE, namely:

- The requirements for supporting the information and navigation needs of city visitors.
- The design of a customized web-browser application to meet these requirements.
- An evaluation of GUIDE focusing on the quality of the visitor's experience.

GUIDE REQUIREMENTS
General Approach
We gathered an initial set of requirements for GUIDE from a series of semi-structured, one-to-one interviews with members of staff at Lancaster's Tourist Information Centre (TIC). In addition, several days were spent at the TIC observing the information needs of visitors.

Permission to make digital or hard copies of all or part of this work for personal or classroom use is granted without fee provided that copies are not made or distributed for profit or commercial advantage and that copies bear this notice and the full citation on the first page. To copy otherwise, to republish, to post on servers or to redistribute to lists, requires prior specific permission and/or a fee.
CHI '2000 The Hague, Amsterdam
Copyright ACM 2000 1-58113-216-6/00/04...$5.00

Identified Requirements

Flexibility

One of the key requirements for GUIDE was the need to provide sufficient flexibility to enable visitors to explore, and learn about, a city in their own way. For example, some visitors prefer to follow a guided tour while others may choose to explore on their own, following one or more guidebooks or street maps. So, therefore, the system should be capable of acting as an intelligent tour guide or as a richly featured guidebook depending on visitor's needs.

It is also important that the system enables visitors to control their pace of interaction with the system. For example, visitors should be able to interrupt a tour in order to take a coffee break whenever they desire. In addition, a visitor should not feel overly pressured by the system to leave an attraction prematurely.

Context-Sensitive Information

A further requirement was that the information presented to visitors should be tailored to their context. There are two classes of context that should be used, namely personal and environmental. Perhaps the most significant piece of personal context is the visitor's interests, e.g. history or architecture. Other examples of personal context that should be used include: the visitor's current location and any refreshment preferences they might have. Examples of environmental context to be used include: the time of day, and the opening times of attractions. When creating a tour of the city, GUIDE should use both personal and environmental context to create a suitably tailored tour.

Context should also be used when presenting information to the city visitor. For example, information should be presented in a way that is suitable given the age and technical background of the visitor and their preferred reading language. Context should also be used to adapt the presentation of information depending upon the information that the visitor has already seen. For example, if a visitor makes a return visit to a landmark then the information presented should reflect this fact, e.g. by welcoming the visitor back. Oberlander [12] uses the term *coherence* to describe the notion of tailoring the presentation of information based on what the user has already seen.

Support For Dynamic Information

During our study we found there to be a significant requirement for the support of dynamic information. Such information should be made available to visitors whenever their context deems this to be appropriate. For example, consider the hypothetical scenario in which a visitor touring the city has expressed a particular interest in Lancaster castle. When starting their tour, the castle was closed to the public because the courtroom, situated within the castle, was in session. However, because the court session finishes early the visitor should be notified that the castle is now open to the public.

Support for Interactive Services

Studying tourist activities in Lancaster revealed that a surprising number of visitors make repeat visits to the TIC, often during the course of a single day. In most cases this is because they either wish to ask a member of staff a specific question or they need to make use of a service offered by the TIC, most commonly the booking of accommodation. In order to help alleviate the need for visitors to walk back to the TIC to ask a question the system is required to support some form of electronic messaging service. In addition, the system should also enable visitors to make accommodation bookings without having to return to the TIC.

THE DEVELOPMENT OF GUIDE

The GUIDE system is based on a distributed and dynamic information model that is disseminated to hand-held GUIDE units using a cell-based wireless communications infrastructure.

Selection of the Hand-held GUIDE Unit

We considered a wide range of end-systems for use in GUIDE, including pen-based tablet PCs and PDAs, and finally selected the grayscale transflective version of the Fujitsu TeamPad 7600 [8] as illustrated in figure 1.

Figure 1: The GUIDE end-system.

The unit measures 213x153x15mm, weighs 850g and is based on a Pentium 166 MMX processor. It has a battery life of approximately two hours (driving the wireless networking card) and is readable even in direct sunlight.

Wireless Communications Infrastructure

The cell-based wireless communications infrastructure used to broadcast both location and dynamic information to mobile GUIDE units is shown in figure 2.

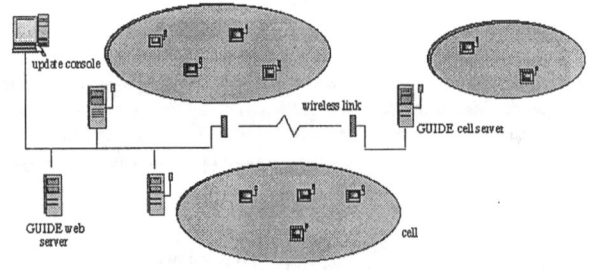

Figure 2: The GUIDE communications infrastructure.

In more detail, the city contains a number of WaveLAN cells, which conform to the IEEE 802.11 standard. Each cell provides a shared bandwidth of 2 Mbit/s and is supported by a GUIDE server. The fact that WaveLAN cells can be relatively large (up to 300m in diameter depending on the layout of buildings) means that GUIDE servers may have to support a potentially large number of GUIDE units. It was, therefore, decided that some form of broadcast based approach to data dissemination should be used for transferring information to the portables units.

Obtaining Positioning Information

In the current system, portable GUIDE units obtain positioning information by receiving location messages that are transmitted from strategically positioned base stations. We adopted this approach rather than one based on Differential Global Positioning System (DGPS) techniques for two reasons. Firstly, the approach requires no additional hardware and secondly because in a built up area it is often not possible to 'see' a sufficient number of satellites to obtain accurate positioning. However, using this approach does result in a lower resolution of positioning information.

The GUIDE Information Model

The GUIDE system required some form of information model in order to represent the following types of information:
- Geographic information.
- Hypertext information.
- Active components that can react to events.

Existing models are inadequate for representing all of the aforementioned information types [6] and so we designed a purpose built information model (figure 3).

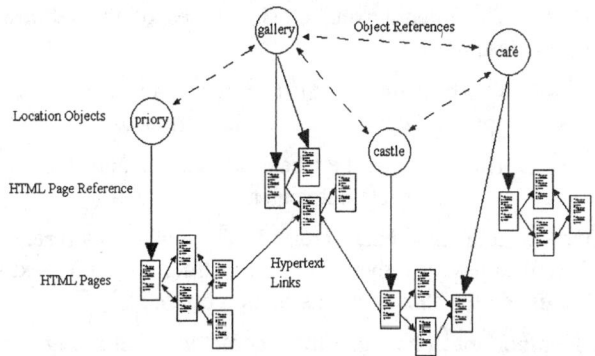

Figure 3: The GUIDE information model.

The information model manages the requirement for representing geographic information by including special navigation point objects. These can be used in conjunction with location objects for determining the best route between a source and destination location. One example of a location object is the city's castle. This object contains state representing various attributes, e.g opening times, and also contains hypertext links to related information.

Each GUIDE unit is able to locally cache parts of the information model and is therefore able to operate even when disconnected from the network. However, during periods of disconnection the cached information model can become stale which could result in out of date information being presented to the visitor.

APPLICATION AND USER INTERFACE DESIGN

The user interface to GUIDE is based around a modified browser metaphor. This decision was made on the basis of the growing acceptance of the web and the increasing familiarity of the browser metaphor as a tool for interaction. We hoped that positive transfer from the use of common web browsers would help make the system both easy to use and easy to learn for users with previous web experience. However, we also wanted to ascertain the extent to which the basic metaphor would be appropriate for the task of supporting the additional functionality required by GUIDE. In addition, we wanted to investigate the extent to which differences and inconsistencies with the standard would prove confusing to users.

In order to use GUIDE a visitor must first enter some personal details, such as their name, interests and preferred reading language. Having entered these details they are presented with the screen shown in figure 4.

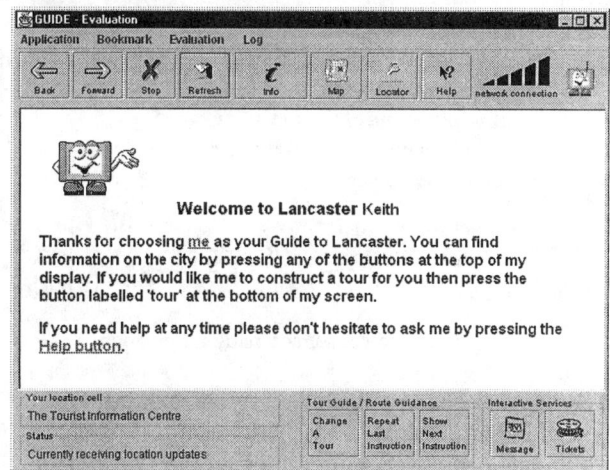

Figure 4: Welcoming the visitor to GUIDE.

In order to help the system appear more approachable to visitors we have attempted to give GUIDE a friendly personality. This decision was based on the observation [14] that, in general, novice users will find a computer-based interactive system more approachable if it is perceived as having a polite and friendly personality.

At this point, visitors have the flexibility to explore and retrieve information about the city using their own preferred methods (a requirement described in the 'Identified Requirements' section).

In more detail, the visitor can touch an appropriate button in order to perform one of the following tasks:

- Information retrieval.
- Navigation of the city using a map.
- Creating and then following a tour of the city.
- Communicating with other visitors or the TIC by sending a text message.
- Booking accommodation.

Alternatively, the visitor could simply head off to explore the city and resort to using the facilities provided by GUIDE as and when required. The ways in which the visitor can request information, navigate the city using a map or create and follow a tour of the city are described in the following three subsections.

Information Retrieval

Touching the info button enables the visitor to ask their GUIDE for information. More specifically, the visitor is presented with six choices for obtaining information as shown in figure 5.

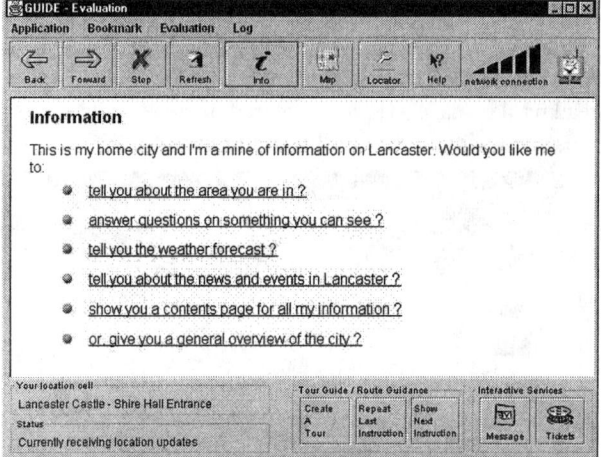

Figure 5: Choices for accessing information.

The first two options are context-sensitive in that they both lead to the presentation of information based on the visitor's current location. In particular, the second option is available in order to allow the visitor to query GUIDE in much the same way as they might query a person with local knowledge of the area. When choosing this option the visitor is shown thumbnail type pictures of things nearby with associated textual descriptions and links.

The latter three options allow the visitor to request information that is not connected with the current location. An earlier version of the GUIDE system did not support these three options but instead constrained the visitor's search for information by trying to pre-empt those specific pieces of information that we believed would be of interest to a visitor at each and every location. This was achieved by providing only a limited collection of hypertext links on every page. A series of initial trials revealed that this method for enabling users to access information was unsuitable. During the trials, visitors would, on occasion, became frustrated when the system did not provide the appropriate hypertext link for accessing specific information.

On a more general point, our experience with this aspect of the GUIDE system has taught us that designers of this kind of context-aware system should be careful not to be over zealous when deciding how to constrain information provided by the system based on a certain context.

Navigation Using a Map

The GUIDE system supports visitors wishing to navigate the city by enabling them to choose between viewing an overview map of Lancaster or a map of the local area.

At an early stage in the project we discussed whether or not the system should present maps because of the apparent sufficiency of providing succinct location-aware directions. However, from early trials with the system it soon became clear that a significant portion of visitors want to view a map at some point in their visit.

Creating and Following a Tour of the City

On touching the 'Create A Tour' button the visitor is asked to select those attractions that they wish to visit on their city tour. In more detail, the visitor is presented with various categories, such as 'Historic' and 'Recreation', from which to choose attractions. However, one of the problems with asking the visitor to choose attractions is that he or she does not necessarily appreciate what is special in a given town. For this reason, GUIDE provides a 'Popular Attractions' category that contains such special attractions.

When creating a tailored tour the system currently takes into account the following factors:

- The opening and closing times of the requested attractions.
- The best time to visit an attraction, e.g. avoiding opening time if there is often a queue.
- The distance between attractions and the most aesthetic route between them.

Once a tour has been created, the visitor can request GUIDE to navigate them from one attraction to the next by clicking on the show next instruction button.

It is important to note that the recommended ordering of the tour can change dynamically. This can occur when a visitor stays at a location longer than anticipated or if one of the attractions announces that it will close early. The system regularly calculates whether or not the current order for visiting the remaining attractions is appropriate given current time constraints.

The visitor can either agree to be taken to the next attraction recommended by GUIDE or override this recommendation by selecting a different attraction to be the next destination. The system provides this choice in order to prevent the system behaving in an overly authoritarian manner. It does, after all, seem reasonable to allow a visitor to delay their current tour and get directions to the nearest café instead.

However, providing this flexibility involved a significant increase in interface complexity and this part proved most difficult to visitors (see evaluation results).

Having asked the system to 'take me there' the visitor is presented with some directional information.

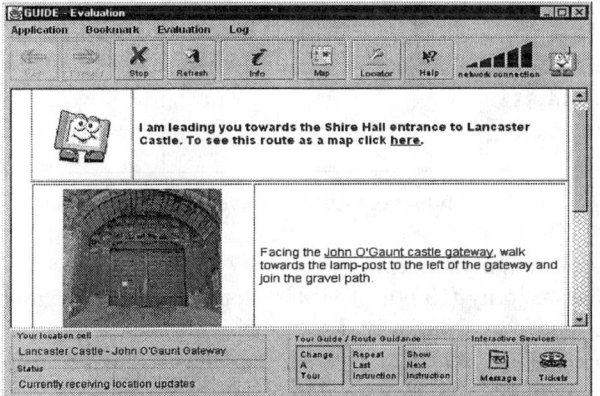

Figure 6: The presentation of navigation information.

The screen-shot (figure 6) illustrates the visitor being presented with succinct but detailed directions to their next location in their tour. In addition, the visitor is shown information on their current location, i.e. the gateway to the castle, and a hypertext link is available should the visitor wish to find out more information.

Providing an Awareness of Disconnected Operation

The current way in which the GUIDE system is engineered, i.e. using a cell-based communications infrastructure, results in situations where the mobile GUIDE unit does not have network connectivity. The fact that GUIDE units can cache large parts of the information model locally enables much of the system's functionality to remain available throughout periods of disconnection. However, disconnected operation clearly affects other aspects of GUIDE functionality such as location information, the messaging service, access to interactive services, e.g. ticket booking, and the reception of dynamic information.

Our key concern was that the system could appear unpredictable to visitors during periods of disconnection and that this would adversely affect their trust of the system. To help alleviate this problem, the user interface to GUIDE has been designed to encourage the user to form a suitable mental model of the system, i.e. one in which the functionality of the system is not static but dependant on whether or not wireless connectivity is currently available. This is achieved by providing the user with an appropriate level of mobile-awareness [3] to enable them to appreciate the affect of changes in connectivity on the system.

In more detail, we decided to incorporate a metaphor into the GUIDE user interface that would provide visitors with feedback regarding the current state of connectivity and also encourage them to associate this with available functionality. To choose a suitable metaphor, we considered how connectivity feedback is provided on mobile phones. The user of a mobile phone is given feedback of their current connectivity in the form of 'bars of connectivity' and when a user receives no bars of connectivity they expect limited functionality, i.e. the inability to send or receive calls.

In addition to the 'bars of connectivity' icon, the user interface also provides visitors with an awareness of the state of location updates. This is achieved using two text message boxes (positioned at the bottom-left of the display) one of which is used to state the visitor's current (or last known) location whilst the other provides feedback regarding the reception of location information.

The fact that the user interface is based on the direct manipulation paradigm implies that only buttons that actually do something should appear active. For this reason, we chose to 'grey-out' the ticket-booking icon when the facility is unavailable due to disconnection. We had also considered disabling the messaging icon when operating in disconnected mode, but instead chose to modify the messaging dialogue box to state that the message being composed would not be sent until on-line operation was resumed. We chose this approach to enable visitors to compose messages when out of communications coverage (a facility also common on mobile phones).

EVALUATION BY EXPERT WALKTHROUGH
Approach

The reason for evaluating the GUIDE system by expert walkthrough was to provide a crude first pass evaluation of the system's usability prior to its use by visitors.

Four experts, with backgrounds spanning user-centered design and computer supported learning, were asked to test the full range of GUIDE functionality for a period of approximately one hour. Experts were asked to use a talk-aloud protocol while using the system and were then interviewed and asked to criticize the system.

Findings

The expert walkthroughs revealed a number of problems with the system as described below.

- The button layout should be consistent with that of other browsers. This was fixed for the prototype used in the field trial.
- Animated feedback should be given to signify when a page is downloading. This was included for the prototype used in the field trial.
- The information button should be increased in size in order to encourage its use when the user may otherwise feel under encompassed. This adjustment was made for the field trial prototype.
- The system should learn the walking pace of the visitor and adjust the tour times appropriately.

- The presentation of lists of attractions, e.g. nearby attractions, should be adapted such that attractions already visited are moved further down the list.
- The visitor should be given some notion of how much information is still to be viewed on a particular topic and how much remains unseen.
- It can be difficult to select hypertext links using the touch sensitive screen. This was partially remedied by increasing the font size used.
- The existence of back and forward buttons and buttons for requesting to view the next or previous navigation instruction can be confusing because of the apparent semantic overlap. This was partially solved by graying-out the back and forward buttons when following a tour as opposed to a hypertext link.

Time constraints meant that only some of the suggested improvements could be made to the prototype before proceeding with the field trial evaluation. Another constraining factor was the mobile unit itself, i.e. its limited processing power and restricted screen size.

EVALUATION BY FIELD TRIAL

The main objective of our evaluation at this stage of the project was to validate and refine our initial set of requirements against a set of end-users. In addition, we wanted to know whether or not people were prepared to accept the use of a computer-based context-aware tourist guide. Consequently, we wanted to measure the quality of the visitors experience [9] as opposed to performance times for getting from A to B or accessing information X.

Approach

The evaluation of the GUIDE prototype by field trial was subject to a number of constraints. In particular, we felt acutely aware of the fact that we would be impinging on the leisure time of tourists. For this reason, we asked visitors to use the system as they would wish to use it and for only as long as they felt happy, rather than asking them to perform some predefined series of tasks.

Our method for evaluation was based on direct observation, with visitors encouraged to use a talk-aloud protocol for audio recording. In addition, we maintained a time-stamped log of their interaction with the system in order to gather a record of the number of links followed. Following each test, a semi-structured interview was performed in order to obtain the visitor's subjective opinion of the system.

We felt that this approach was suitable given the main objective of the evaluation. By shadowing users we could observe those parts of the interface causing problems. The semi-structured interview enabled us to follow up on any problems that were encountered during the trial and also enabled us to tailor the duration of the interview to match the time constraints of the visitor.

Findings

Over a period of approximately four weeks we had 60 people volunteer to use the system. The breakdown of these people in terms of age etc. is shown in table 1.

Age Profile	Number	Gender		Web Experience
		Male	Female	
10-20	6	4	2	6
21-35	15	7	8	7
36-55	26	12	14	8
56-70	13	6	7	1

Table 1. Profiles of visitors involved in the evaluation.

Validation of requirements

The majority (53/60) of visitors appreciated the flexibility provided by the system, i.e. the ability to use the system as a tour guide, a map or a guidebook. However, seven visitors thought that the system had too many choices available and expressed a desire for a 'less is more' system that could be easier to use.

All visitors expressed the opinion that the location-aware navigation and information retrieval mechanisms provided by the system were both useful and reassuring. In addition all visitors said that the ability to receive dynamic information, e.g. the 'specials' menu of a café, was a worthwhile feature.

However, the provision of access to interactive services, such as booking accommodation, had a more mixed response from visitors. Indeed, (5/60) of visitors would much rather speak to someone when booking accommodation (even if this meant queuing) and (48/60) of visitors said that they would want some form of confirmation that the booking had taken place. Suggestions for this included: a phone call back to the visitor's mobile phone or confirmation from the TIC.

Visitor's Subjective Opinion on Information Presentation.

All visitors appreciated the idea of being allowed to follow links to receive greater levels of detail (or related details) on an information topic. However, seven of the visitors expressed some concern that they might have missed information on a particular topic.

Despite the antagonism expressed by some expert users towards certain friendly interface features (e.g. the Microsoft paperclip), none of the visitors made negative comments regarding GUIDE's friendly personality.

The vast majority (59/60) of visitors stated that they enjoyed using GUIDE to explore the city. However, one person became frustrated when using the system because information was not available on a particular attraction.

The vast majority (59/60) of visitors said that they were prepared to trust the information presented by the system, including the navigation instructions. Interestingly, all visitors said they would be more inclined to trust such a system when provided by a reliable source, e.g. the TIC. A number of visitors suggested that their level of trust varied with the apparent accuracy of the information presented.

Visitor's Subjective Opinion on the GUIDE unit.
A reasonable majority (45/60) of visitors were basically happy with the dimensions and weight of the portable GUIDE unit. Of those that were not, only two stated that they would have preferred a smaller (PDA) sized device while 13 said they would have preferred a thinner device.

Interesting Results Based On Visitor Profile
All visitors in the 10 to 20 age profile seemed to revel in the technology and visited approximately twice as many links (per minute of usage) as those from other age profiles. This does not necessarily mean that visitors from this age group were learning more, but does suggest that they were more eager to explore the information available.

The vast majority (21/22) of visitors without previous web experience felt comfortable using the system to follow a tour and retrieve information by navigating hypertext links after a brief five minute training session.

Visitors Acceptance of Awareness Information
A large majority (54/60) of visitors said that they were aware that their GUIDE unit utilized wireless communications in a similar way to a mobile phone and that when no bars of connectivity were shown on the interface then reduced functionality would be available.

A reasonable majority (47/60) of visitors said that they appreciated that the system knew of their location to within a certain area by receiving location updates.

RELATED WORK

The earliest work on developing a location-aware tourist guide was Cyberguide [11]. An extended version of the system [13] was developed that utilised wireless connectivity in order to enable visitors on demonstration days to observe the location of other visitors.

Closely related work in the area of intelligent context-aware electronic tourist guides is currently being conducted as part of the HIPS (Hyper-Interaction within Physical Space) project [1].

Work on presenting 'intelligent labels', i.e. tailored information, to museum visitors is being carried out under the auspicious of the ILEX project [5]. Information is based on the visitor's profile and what they have seen previously.

FUTURE WORK

For future work, we intend to investigate the potential benefits of supplementing the existing GUDIE infrastructure with the latest low-power, micro-cellular, wireless communications technologies, such as Bluetooth. In particular, we hope to extend GUIDE services to within buildings and investigate the potential for developing additional context-aware interactive services. With this extended communications infrastructure in place we intend to assess the potential for performing highly computational tasks, such as the calculation of a tour, remotely. A number of the issues that can arise from performing remote computation in a mobile interactive system, e.g. the affect on interactive feedback, are investigated in [7].

Another future direction for GUIDE will be to utilize the growing acceptance of connected personal computing devices, e.g. WAP phones. It should be possible, in the near future, to enable visitors to download software onto their own device (with built-in Bluetooth support) in order to enable access to context-aware information and services.

A further avenue to explore is the potential for making the visitor's profile persistent. This raises some interesting possibilities, for example, if a visitor has shown an interest in castles on a previous city visit then this could be stored in their profile and used to tailor the presentation of information on future visits.

DISCUSSION AND CONCLUSION

This paper has described our development and evaluation of GUIDE, a prototype system for providing city visitors with context-aware information.

Through our evaluation of GUIDE, we found a surprisingly high level of acceptability across a wide range of users. However, for some visitors the flexibility provided by the system was a little bewildering and this illustrates the need to enable visitors to choose the level of functionality that they require. In addition, visitors should be able to choose GUIDE units based on different form factors and input devices. For example, use of the NaviPoint [10] input device could enable a system that supports one-handed operation.

A number of implications arise should systems like GUIDE become popular. For example, some form of agent will be required to enter dynamic information into the system and maintain/monitoring the accuracy of information. In Lancaster, the TIC is requesting additional council funding in order to employ a member of staff to act in this role.

Another implication is the potential effect of a system like GUIDE on the local business model. It will be interesting to discover the critical mass needed, i.e. the number of visitors using GUIDE, before local businesses consider GUIDE an important avenue for marketing their products.

The following conclusions could be used by others working on designing interactive systems based around mobile computing and/or context-aware systems.

- Interaction with a context-aware/location-aware system is not affected by the design of the user interface alone. In fact, interaction with GUIDE is, to a large extent, governed by the design of the infrastructure, i.e. the strategic placement of cells in order to provide appropriate areas of location resolution and network connectivity.

- Our experience with evaluating the presentation of context-aware information has taught us that designers need to be careful when deciding to pre-empt the information requirements of users based on current context. For example, when we restricted the information available to visitors, such that they could only access information on the attractions at their current location, some visitors became frustrated because they could not query the system on things visible in the distance.

- It is important to consider the potential advantages and disadvantages of borrowing or modifying familiar metaphors for use in different scenarios. For example, the modified browser metaphor used by GUIDE caused some confusion because of the semantic overlap between the standard back and forward buttons and the buttons for requesting to view the next or previous navigation instruction.

- In the leisure industry there appears to be a growing acceptance of the use of technology. Indeed, the uptake of personal technology by members of the public, such as mobile phones, digital cameras and personal organizers, suggests that more and more members of the public are prepared to make use of technology if it provides tangible benefits.

Following on from this last point, for a system like GUIDE to be accepted by the public at large it needs to show clear benefits over the traditional facilities available to tourists, such as paper-based guidebooks. Based on our initial evaluation, we believe that members of the public do appreciate the system's benefits.

ACKNOWLEDGMENTS

This work was carried out as part of the EPSRC funded GUIDE project (GR/L05280) with considerable cooperation from Lancaster City Council. The project has received support from HP Labs, Bristol and Lucent Technologies.

REFERENCES

1. Broadbent, J., and Marti, P. Location Aware Mobile Interactive Guides: usability issues, in *Proceedings of the Fourth International Conference on Hypermedia and Interactivity in Museums (ICHIM97)* (Paris, September 1997).

2. Brusilovsky, P. Methods and Techniques of Adaptive Hypermedia, in *User Modeling and User-Adapted Interaction* (6) (1996), Kluwer, 87-129.

3. Cheverst, K., Davies, N., Friday, A., and Blair, G. Supporting Collaboration in Mobile-aware Groupware, in *Proceedings of the Workshop on Handheld CSCW: ACM CSCW'98 Conference on Computer Supported Cooperative Work*, (Seattle, 1998), ACM Press, 59-6.

4. Cheverst, K., Davies, N., Mitchell, K., and Friday, A. The Role of Connectivity in Supporting Context-Sensitive Applications, in *Lecture Notes in Computer Science No. 1707*, Springer-Verlag, (1999), 193-207.

5. Cox, R., O'Donnell, M., and Oberlander, J. Dynamic versus static hypermedia in museum education: an evaluation of ILEX, the intelligent labelling explorer, in *Proceedings of the Artificial Intelligence in Education conference* (Le Mans, July 1999).

6. Davies, N., Mitchell, K., Cheverst, K., and Friday, A. Caches in the Air: Disseminating Tourist Information in the Guide System, in *Proceedings of the 2nd IEEE Workshop on Mobile Computing Systems and Applications* (New Orleans, 1999), 11-19.

7. Dix, A., Ramduny, D., Rodden, T., and Davies, N. Places to stay on the move: software architectures for mobile user interfaces, in *Proceedings of the 2nd Workshop on Human-Computer Interaction with Mobile Devices* (Edinburgh, August 1999), 65-71.

8. Fujitsu TeamPad 7600 Technical Page. Available at: http://www.fjicl.com/TeamPad/teampad76.htm.

9. Hook, K., and Svensson, R.E. Evaluating Adaptive Navigation Support, in *Proceedings of the Workshop on Personalised and Social Navigation in Information Space* (Stockholm, March 1998), ACM Press, 119-128.

10. Kawachiya, K., and Ishikawa, H. NaviPoint: an input device for mobile information browsing, in *Proceedings of CHI'98* (Los Angeles, 1998), ACM Press, 1-8.

11. Long, S., Kooper, R., Abowd, G.D., and Atkeson C.G. Rapid Prototyping of Mobile Context-Aware Applications: The Cyberguide Case Study, in *Proceedings of 2nd ACM International Conference on Mobile Computing* (Rye NY, 1996), ACM Press.

12. Oberlander, J., Mellish C., and O'Donnell, M. Exploring a gallery with intelligent labels, in *Proceedings of the Fourth International Conference on Hypermedia and Interactivity in Museums (ICHIM97)* (Paris, 1997).

13. Pinkerton, M. D. Ubiquitous Computing: Extending Access To Mobile Data, Master's Thesis, GVU Technical Report GIT-GVU-97-09 (1997).

14. Reeves, B., and Nass, C. The Media Equation: How People Treat Computers, Television, and New Media Like Real People and Places, Cambridge University Press; ISBN: 1575860538.

15. Schilit, B., Adams N., and Want R. Context-Aware Computing Applications, *in Proceedings of the Workshop on Mobile Computing Systems and Applications* (Santa Cruz, CA, 1994).

Measuring the Allocation of Control in a 6 Degree-of-Freedom Docking Experiment

Maurice R. Masliah and **Paul Milgram**
Ergonomics in Teleoperation and Control (ETC) Lab
Department of Mechanical and Industrial Engineering
University of Toronto, Toronto, Ontario
Canada, M5S 3G8
+1 416 978-3776
{moman, milgram}@sakura.rose.utoronto.ca

ABSTRACT
Coordination definitions and metrics are reviewed from the motor control, biomedical, and human factors literature. This paper presents an alternative measurement called the m-metric, the product of the simultaneity and efficiency of a trajectory, as a means of quantifying allocation of control within a docking task. A 6 degree-of-freedom (DOF) longitudinal virtual docking task experiment was conducted to address how control is allocated across six DOFs, how allocation of control changes with extended practice, and if differences in the allocation of control are input device dependent. The results show that operators, rather than controlling all 6 DOFs equally, allocate their control to the rotational and translational DOFs separately, and switch control between the two groups. With practice, allocation of control *within* the translational and rotational subsets increases at a faster rate than across all 6 DOFs together.

Keywords
Coordination, interaction techniques, allocation of control, virtual docking task, the m-metric, evaluation methods, motor control, input devices, 6 degree-of-freedom control

INTRODUCTION
As technologies for interacting with systems such as CAD workstations, process control plants and remotely controlled robots become more sophisticated, humans are frequently faced with the necessity of controlling multiple variables simultaneously. The question of how operators actually allocate their control across many degrees-of-freedom (DOFs) is important for the design of effective input devices and appropriate displays. Before this question can be addressed, however, it is essential first to develop a *metric* for evaluating allocation of control. Traditional performance measures such as task completion time and root-mean-square error do not tell us very much about how operators actually allocate their control. An allocation of control measurement tool is expeced to be useful not only for the design of systems such as those mentioned above, but also for applications involving neurological assessment and modeling of human motor control.

Applications which Motivate the Metric's Development
Teleoperation & Input device design
For high DOF systems, deciding whether the "optimal" distribution of DOFs should be across only one or two hands is less than clear. For example, Zhai [21] has pointed out that while the Space Shuttle Remote Manipulator, one of the most prominent instances of 6 DOF control, requires two-handed operation (due to zero gravity considerations), the literature supporting such design decisions is not unanimous.

During the past decade, many new input devices offering a large number (>2) of control DOFs have been introduced. Some of these devices, such as the IBM ScrollPoint® mouse and the PadMouse [3], are multi-channel or "mixed resistance mode" devices, for which different types of interaction devices, (such as isotonic mouse, touchpad, and isometric joystick) are merged together to form a single device. It has yet to be determined whether operators of mixed resistance devices are capable of using all available DOFs in a coordinated fashion, which will presumably be required as high-end computer applications such as computer-aided design, scientific data visualization, computer graphics animation, and 3D video games become increasingly common.

Process control
Measuring performance in a complex environment can help researchers understand the type of strategies operators may be using [19]. In process control environments, for example, operators are required to manipulate a multitude of different variables such as temperature, volume, pressure, and flow rates towards a goal state in real time. This type of manipulation is a coordination problem.

Human motor control
Understanding human coordination is one the fundamental issues in the motor control literature [16, 5]. A well known

coordination problem, known as "Bernstein's Problem", or the "degrees of freedom problem", is how a high number of DOFs in the body (individual muscles and joints) behave as if they were actually a much smaller number of DOFs. Studying and quantifying how a large number of DOFs work together in a coordinated fashion is one approach to understanding human motor control.

Neurologic assessment
Clinical evaluation of neurologic motor function often is "dependent on the skilled but subjective judgement of a physician"[11]. Such evaluations usually consist of having patients move their arms in certain prescribed motions while the physician assesses performance on an ordinal scale. Tests such as these are used, for example, to track the progression of Parkinson's disease, which is characterized by increased movement latency, slowing of movement execution, and difficulties in execution of multicomponent movements [8]. With the advent of relatively inexpensive computers, some researchers have focused on using computerized tracking tasks as a quantitative and objective means of assessing neurologic damage [18, 8, 1, 11]

Measures/Definitions of Coordination

Coordination is an all encompassing term meant to convey information about a trajectory that is usually lost in simple performance measures, such as task completion time or root-mean-square error. It is not obvious how to capture this trajectory information. Therefore, the literature contains many measures of coordination, each reflecting individual researchers' personal biases about what a definition of coordination should comprise. We believe that coordinated behavior, should result in *simultaneous and efficient* control of multiple variables. However, none of the existing measures address this allocation of control question. A selection of some of those measures/ definitions, with commentary, is presented here:

Time-on-target
Historically in the human factors literature, interest in human coordination [6] arose chiefly from military interests (and funding). Research in the late 1940's and early 1950's centered on anti-aircraft gunners' ability to hit targets, and time-on-target for each DOF was used mainly because it was feasible to compute. Poulton [15] has criticized time-on-target as a "not a very suitable measure", because errors which are slightly off-target are penalized just as much as errors which are far off.

Accuracy × speed
The "coordination index", proposed by Behbehani et al. [4] is a measure of accuracy multiplied by a measure of velocity, which is based on Fitts' speed/accuracy tradeoff law. This coordination index has been used in biomedical research as a means of quantifying upper extremity performance in Parkinson patients.

Spatial or temporal invariance
A common theme in the motor control literature is to use the amount of invariance in a repeated movement as a measure of coordination. Morrison & Newell wrote: "…coordination refers to the degree of spatial or temporal invariance, or both, in the motion of the respective limb effective units." [13] Measurement in the time domain is done by computing cross-correlations, while measurement in the frequency domain is through coherency and phase analysis. [13] This means of analysis is generally useful only for repeated rhythmic motions, such as walking, running and jumping.

Cross-correlations
Computing the cross-correlations [23] among error terms or joint angles [17] is another method for quantifying coordination. Unfortunately, this method usually restricts analysis to only two variables at a time.

Zhai [23] conducted a 6 DOF tracking study and analyzed the results by cross-correlating all pairs of DOF error terms. Interestingly, the correlation distributions showed that subjects were able to control all the DOFs equally well. However, simply correlating the error terms does not take into account task related performance, in that two errors which are decreasing, or even *increasing*, simultaneously will both result in high correlation coefficients.

Inefficiency
Addressing this problem, Zhai [21] has recently proposed a definition of coordination based on inefficiency. Applicable to docking tasks only, coordination is computed as the ratio of the length of the actual path followed divided by the length of the shortest path. With his unified metric, all the DOFs are combined to produce a single length, so it is not possible to make any conclusions as to relative allocation of subjects' control across different DOFs.

Integrality
Integrality [7] is not strictly a coordination measure, though it has been used in that way. (Balakrishman et al., for example, used integrality to demonstrate that subjects could control three degrees of freedom simultaneously with a two translational + one rotational DOF device, the Rockin'Mouse [2]). Two stimulus dimensions are considered integral if they are perceived as a single dimension, or separable if the dimensions seem unrelated [7]. Jacob, et al. [10] have proposed a means of quantifying integrality in the action domain, based upon whether or not movement exists simultaneously in all DOFs. Integrality, as defined by Jacob et al. is a task *independent* measure; that is, movement of any kind is considered integral regardless of whether or not the movement is contributing towards reaching the goal.

Allocation of control
The ability to measure the allocation of control is a new and powerful tool for the analysis of trajectory information in complex multi-DOF tasks. We propose the $\mathcal{M}$-metric,

introduced in the following section, as a means for quantifying allocation of control in a docking task. The $\mathcal{M}$-metric is a task dependent measure which may be used to analyze any number (≥2) of DOFs.

THE $\mathcal{M}$-METRIC

The $\mathcal{M}$-metric is based on the supposition that assessment of allocation of control across an n (where n ≥ 2) DOF docking task must take into account both the simultaneity and the efficiency of control across the DOFs. A docking task is defined here as a task for which an object, such as a cursor or a graphic object, must be moved from an initial position to a goal position, with no constraints on either the trajectories that may chosen or the time allowed to complete the movement from initial state to goal. Such tasks can thus be considered self-paced, or time-minimizing [12]. As justified in the following, it is the *product* of simultaneity and efficiency which defines the $\mathcal{M}$-metric.

Simultaneity of Control

Simultaneity of control is calculated by first computing the *normalized error reduction function* for each DOF separately. Error for each DOF is defined here as the difference between the goal position and the current position. Error reduction is the instantaneous amount by which the difference between the goal position and the current position is *reduced* (i.e. the error term moves closer to zero). Error reduction is a function of time and is set equal to zero during time periods in which the error may have actually increased (i.e. movement away from the goal). In other words, *error reduction represents the instantaneous value of the derivative of the error term*, but only for positive values.

The error reduction function for each DOF is normalized by dividing it by the total distance moved towards the goal over the entire docking task for that DOF. Thus, when all the normalized error reduction functions are graphed against time, the areas under each curve are all equal to each other. More formally, the Normalized Error Reduction Function, $NERF_i(t)$ (where i = 1,2, … n are the degrees of freedom being analyzed, and t is time), is defined as:

$$NERF_i(t) = \frac{-dE_i(t)}{dt} * \frac{1}{ACT_i} \quad , \text{for } \dot{E}_i < 0$$
$$= 0 \quad , \text{for } \dot{E}_i \geq 0$$

where E_i = error, (goal position - cursor position)

ACT_i = total actual error reduced for the *i*th DOF; in other words, ACT_i is equal to the area under the non-normalized error reduction function and can only be computed when the task is over.[1]

---

[1] Note that the $\mathcal{M}$-metric, as presently defined, can not be computed for docking tasks which are not successfully completed.

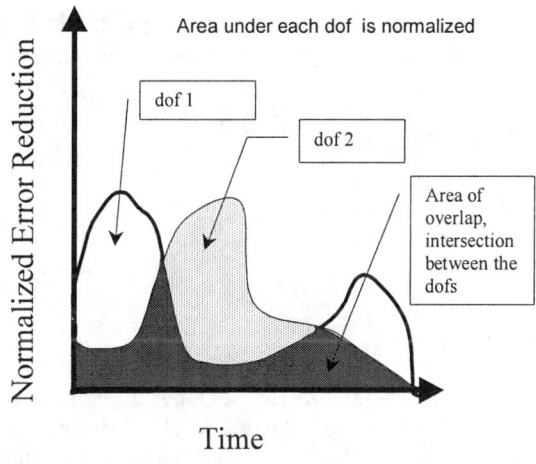

Figure 1. A normalized error reduction graph.

Figure 1 shows the normalized error reduction curves for two DOFs. The area of overlap between the curves depicts the simultaneity of control. Any number (n ≥ 2) or subset of DOFs may be analyzed by computing the overlaps between the normalized error reduction curves. Simultaneity of control (SOC) is therefore defined as:

$$SOC = \int_0^T Min(NERF_1(t), NERF_2(t), ... NERF_n(t)) dt$$

where *Min* where returns minimum value over all $NERF_i(t)$'s as a function of *t*, and T=total task completion time. The minimum function (*Min*) defines the contour of the curve to be integrated, for computation of the area of overlap, as illustrated in Figure 1.

Efficiency of Control

The *efficiency* component of the $\mathcal{M}$-metric is a weighted average of the ratios of the length of the optimal trajectory for each DOF (OPT = length of the optimal error reduction function trajectory) divided by ACT for that DOF. Efficiency (EFF) is thus defined here as:

$$EFF = \sum_{i=1}^n \left(\frac{OPT_i}{ACT_i} * W_i \right)$$

where the weights, W_i, are equal to: $W_i = \frac{k}{n} * \frac{OPT_i}{\sum_{j=1}^k OPT_j}$

and k = number of members in the same subset as the *i*th DOF. A *subset* of the total DOFs is a grouping of (k) DOFs that are similar in nature and measured in the same units. The purpose of the weights, W_i, is two-fold: 1) to weigh DOFs of the same units by their optimal trajectory magnitude, and 2) to deal with DOFs which might be measured in different units. Note that the sum of the n weights must be unity. Thus, for example, the W_i values for X-Y-RZ (where X, Y are translational DOFs, and RZ is a rotational DOF about the Z axis)[2], for a case in which the $OPT_{X,Y,RZ}$ values are 4 cm, 5 cm, 60° respectively, are:

$$W_X = \frac{2}{3} * \frac{4}{4+5} = 0.296$$

$$W_Y = \frac{2}{3} * \frac{5}{4+5} = 0.370$$

$$W_{RZ} = \frac{1}{3} * \frac{60}{60} = 0.333$$

To summarize, the object of the *M*-metric is to measure how control has been allocated among different DOFs during a task and to express this via a value between 0 and 1. A value closer to 1 indicates efficient and essentially synchronous control across DOFs, while a value closer to 0 indicates a switching of control between the DOFs in a relatively inefficient manner. The two components of the *M*-metric have been defined with this in mind, such that we can now define it as being equal to:

$$\mathcal{M}\text{ metric} = SOC \times EFF$$

A Note on the Time Dimension

The final *M*-metric score is not a function of the total length of time taken to complete a docking task, even though the metric explicitly takes into account the time dimension. What the "time dimension" refers to here rather is the *timing of events*. In other words, what the *M*-metric measures is the *degree of simultaneous error reduction occurring in multiple DOFs*, as opposed to measuring whether the error reduction took a particular amount of time to complete.

HYPOTHESIS

Similar to what has been observed in the motor control literature [17], we predict that novice operators attempting to control a large number of DOFs will not allocate their control equally across all the DOFs. Instead, subjects will control certain subsets of the total number of DOFs at a time and switch control between those subsets. Furthermore, the subsets controlled will not be arbitrary; rather, it is expected that rotation and translation DOFs will be treated separately. Imai and Garner [9] have identified a perceptual preference in discriminability between translation and rotation dimensions. We believe this perceptual preference extends into an action preference. Specifically, for a 6 DOF virtual docking task, operators will tend to allocate their control globally between the three translational and three rotational DOFs, and switch back and forth between them.

We further predict that input devices which support more "natural" modes of interactions (i.e. closer to real-world interactions) should show a more uniform allocation of control across all 6 DOFs.

Finally, we predict that, as expertise develops, and concurrently task completion time performance improves, one of two behavioral tendencies is likely to emerge:

1. Operators will continue to allocate their control between the translation and rotation DOFs, and only their control will improve, or

2. Uniform allocation of control across all 6 DOFs will continue to develop over time.

EXPERIMENT

Goals

Corresponding to the above hypotheses, the experiment presented here was designed to address three explicit questions:

1. How do people allocate their control across six DOFs in a virtual docking task?
2. Are differences in the allocation of control device dependent? Specifically are there differences between isometric (pressure or force sensing without movement) and isotonic (displacement sensing or free moving without resistance) devices? Isometric and isotonic represent the two extremes of possible controller resistance.
3. How does the allocation of control change over an extended period of practice?

Method

Subjects

12 right handed volunteers from the University of Toronto community were recruited as subjects. Three subjects were rejected for failing to discern a binocular disparity of at least 50 seconds of arc at 40 cm, as tested using the Randot® Stereotest (Stereo Optical Co., Inc., Chicago, IL). A fourth subject was rejected for being unable to complete the docking tasks. The remaining 4 male and 4 female subjects ranged in age from 25 to 32, and were paid $55 CND upon completion of the experiment. None of the subjects had previous experience with any 6 DOF computer input device, or any stereoscopic virtual displays.

Experimental Platform

The experiment was conducted on a Silicon Graphics Indigo™ workstation with a 20 inch color monitor, running MITS (Manipulation in Three Space) software, developed by Zhai [20], to create a through-the-window virtual environment. IMAX® liquid crystal glasses (IMAX Ltd, Toronto, Canada) operating at 120 Hz were used to allow stereoscopic viewing. For the isometric condition, subjects used a Spaceball® (Model #2003) manufactured by Labtec Inc. (Vancouver, WA), operating in rate control mode. For the isotonic condition, subjects used the Fingerball [22, 20] powered by a Flock of Birds™ (Ascension Technology Corp., Burlington, VT), operating in position control mode. The MITS software sampled subjects' data at 15 Hz.

---

[2] The symbols RX, RY, and RZ are used to represent the Euler angles ϕ, θ, and ψ respectively.

Task

A three-dimensional virtual docking task was used in this experiment (see [21, 20] for a more detailed description of the task). Subjects were told to align a tetrahedrally shaped cursor onto an identically shaped target tetrahedron as quickly as possible. Whenever a corner of the cursor was successfully matched to its corresponding target, the corner changed colour, indicating a correct docking. All four corners had to stay docked for 0.7 seconds to complete the trial. The sides of the tetrahedrons were colour coded and drawn in wire frame mode. For all trials, the initial position of the cursor was the center of the screen, while the target appeared in one of eight off-center locations. Target locations were selected so that a similar difference existed in all translational and rotational DOFs between the target and the cursor.

Procedure

Subjects were first tested for binocular disparity using the Randot® Stereotest. A short questionnaire was administered to collect subject information and experience with stereoscopic viewing and 6 DOF control devices. The questionnaire also included questions from the Edinburgh inventory [14] to assess handedness. Subjects were then introduced to their control device and asked to manipulate the cursor up/down, left/right, in/out, and then to rotate about those corresponding axes in a targetless environment. This introduction to the control device took less than two minutes. Subjects were then given a single docking trial as a training/explanation of the task. For all trials, subjects used only their dominant hand.

Design

The experiment was a $2 \times 5 \times 216$ between subjects design. The independent variables were as follows:

Input Device	isometric rate, isotonic position
Session	1, 2, … 5
Trial	1, 2, … 216

The above conditions with 8 subjects represent a total of 8640 trials, collected over 40 hours of experimentation. A stratified random method was used to assign subjects to either the isometric rate or isotonic position conditions, such that each group was composed of 2 males and 2 females. The number of trials, 216 per session, was chosen so that each session would last about an hour. Subjects completed one session per day on five consecutive days. A mandatory rest period of 90 seconds after every 24 trials was enforced by the software. Each block of 24 trials consisted of 8 randomly shuffled target locations selected three times.

RESULTS

The number of trials, 216, was chosen so that each session would take about an hour. In reality, the subjects' first session took between 90 and 100 minutes to complete, while the fifth (final) session took only 40-50 minutes to complete.

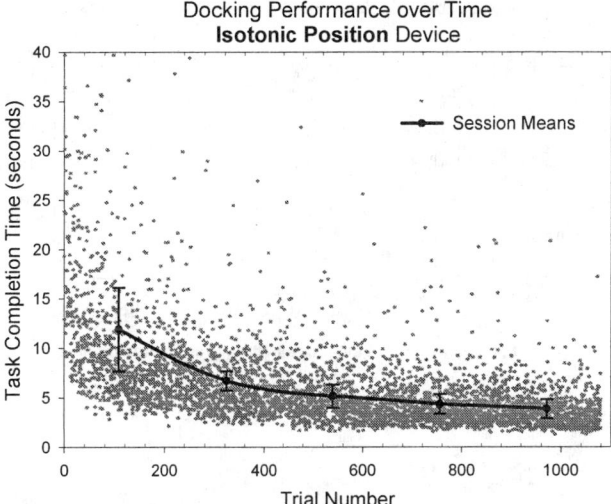

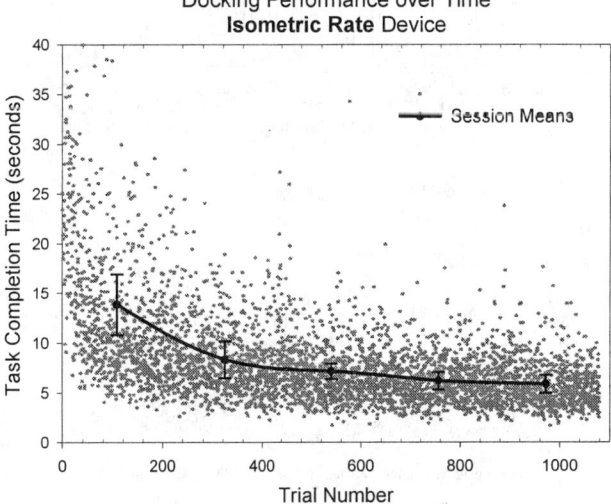

Figure 2. Task completion times by input device: Isotonic above and Isometric below. Session means, standard deviations (216 trials per session), and raw data times (4320 trials across 4 subjects per device) are shown.

Figure 2 shows task completions times over the five sessions, broken down by input device. Task completion times for the isotonic condition were an average 1.8 seconds faster per session than then isometric condition, however, this difference in mean times was not statistically significant.

m-metric scores were computed for all DOF groups (15 two-way comparisons, 20 three-way comparisons, 14 four-way comparisons, 6 five-way comparisons, and 1 six-way comparison) for a total of 56 different groupings. A *within* grouping subset is defined here as a set where all the DOFs in the set are either of the translation or rotation type. For example, for the 20 three-way comparisons, there are only 2 groups which are considered within groups, X-Y-Z and RX-RY-RZ. An *across* group subset is defined here as a set where at least one DOF is of the translation type and at least one is of the rotation type. So for the example of the 20

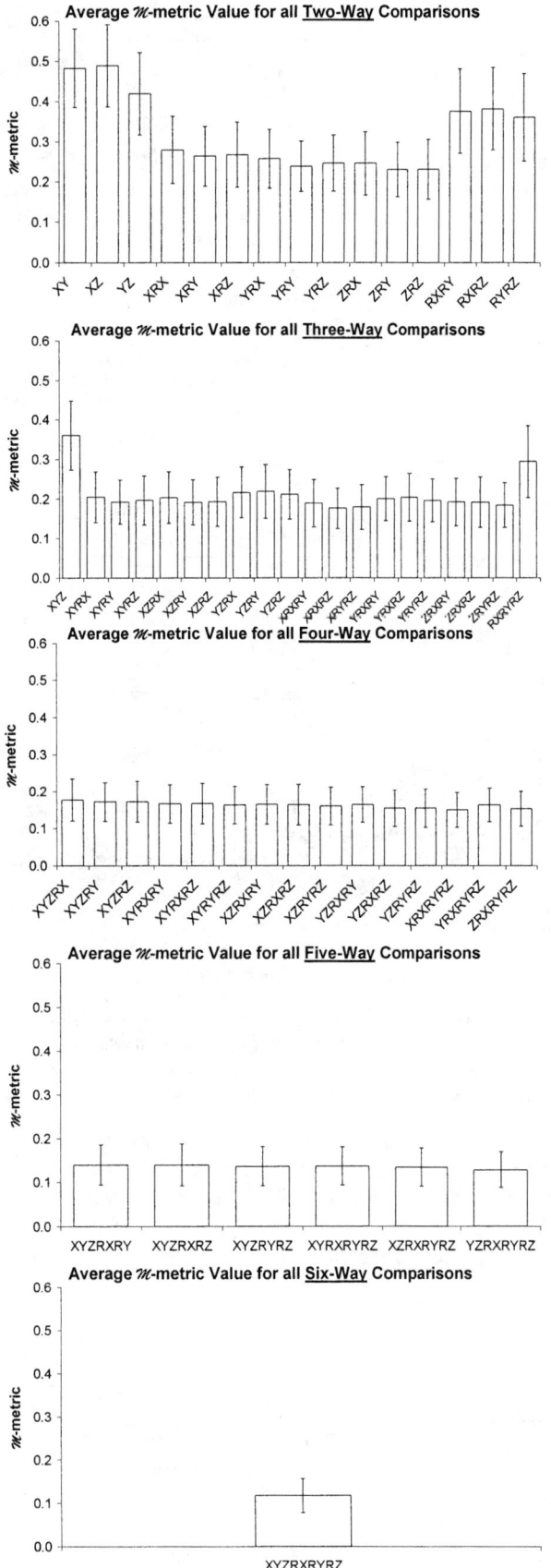

Figure 3. *M*-metric means and standard deviations across all trials.

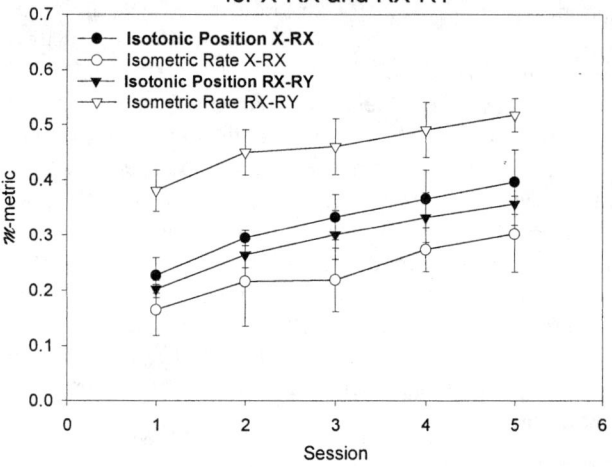

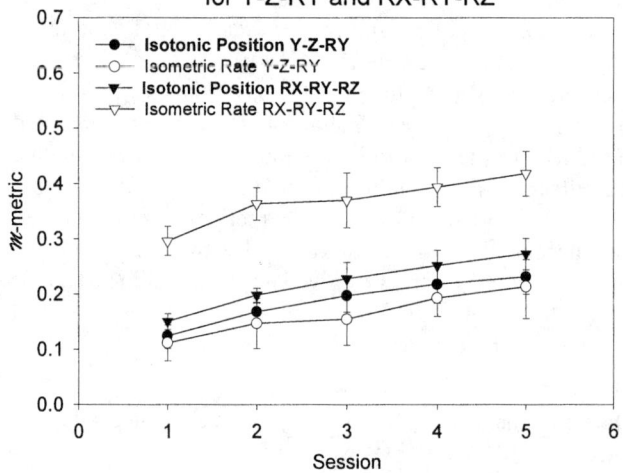

Figure 4. Changes in *M*-metric scores over sessions (216 trials) broken down by input device and number of compared DOFs. Only a representative sample of comparisons are shown.

three-way comparisons, the eighteen remaining groups are considered as *across* groups.

Mean *M*-metric scores broken down by number of DOFs compared are presented in Figure 3. For the two-way comparisons (X-Y, X-Z, ...RY-RZ), *within* rotation and *within* translation pairings had significantly higher (paired t test, $p < 0.0001$, using the Bonferroni method for multiple comparisons) *M*-metric scores than pair-wise comparisons *across* a rotation and translation DOF. The three-way comparisons showed similar results, with the within translation condition (X-Y-Z) and the within rotation condition (RX-RY-RZ) showing significantly higher (paired t test, $p < 0.0001$, Bonferroni) *M*-metric scores than their *across* rotation and translation counterparts (X-Y-RX, X-Y-RY,..., Z-RY-RZ). However, for the four-way and five-way comparisons, no significant differences in *M*-metric scores existed for any subset.

Figure 4 shows changes in $\mathcal{M}$-metric scores over time/session, broken down by input device and number of comparisons. Only a subset of $\mathcal{M}$-metric scores are shown, to save space. All $\mathcal{M}$-metric scores increase over session. For the two-way comparison case, X-RX and RX-RY have been selected as representative of an across translation-rotation pairing and a within rotation pairing respectively. Across all the two-way comparisons, the highest $\mathcal{M}$-metric scores always belonged to the within rotation isometric rate conditions. The isotonic scores for both the within and across conditions were always lower. The lowest $\mathcal{M}$-metric scores, for the two-way comparisons, always belonged to isometric rate across translation and rotation pairings.

The three-way pairings, depicted in Figure 4 by Y-Z-RY and RX-RY-RZ show this same pattern. Isometric within scores were the highest, followed by isotonic $\mathcal{M}$-metric scores, with the lowest scores belonging to the isometric across groups. For the four, five, and six-way comparisons, isotonic position $\mathcal{M}$-metric scores were in fact larger than their isometric rate counterparts, though in all cases the values were small.

DISCUSSION

Regardless of which input device was used, subjects tended to allocate control within rotation and translation groups separately. Previous research from the motor control literature has observed that novices control subsets of their total available degrees of freedom [17]. In addition, research from the psychology literature [9], has identified a perceptual discriminability preference to categorize stimuli into rotational and translational groups. However, this is the first time, to the authors' knowledge, that quantitative evidence has demonstrated an action preference to alternate between rotational and translational manipulations. This is exactly the type of analysis that is extremely difficult to do without the $\mathcal{M}$-metric.

An isotonic position control imposes fewer restrictions on an operator's allowed movement; muscle groups are allowed to move in a manner similar to perhaps how they are used in real word manipulation. This may mean that an isotonic position controller is a more "natural" means of interacting with a virtual environment. We expect that a more natural means of input should thus result in a more even distribution of control across available DOFs. Therefore, the higher $\mathcal{M}$-metric scores in the isotonic position condition for the across rotation and translation conditions, as compared to the isometric rate across conditions, are reasonable.

If an isometric rate controller is a less natural interaction method, and this could mean that the isometric rate device is comparably more difficult. A more complicated interaction method therefore should thus result in an uneven distribution of control. In Figure 4, in the 2-way comparison graph, the difference between the X-RX condition (an *across* translation and rotation group) and the RX-RY condition (a *within* rotation group) for the isometric rate is very large. Thus, for more complicated interaction devices, it is arguably more important for subjects to reduce the complexity of the task by controlling only a subset of the total 6 DOFs at a time. Switching control between subsets of the total available DOFs appears to be the method subjects used even after 1000 trials.

Task completions times across all subjects from session 4 to session 5 dropped an average of only 0.40 seconds (from 5.27 to 4.87 seconds), compared to a mean 5.38 second drop from the 1st to the 2nd session. Therefore, the task completion times in Figure 2 show evidence of subjects approaching a minimum time floor. With additional sessions, any further time reductions would probably be very small in magnitude.

Conversely, however, the $\mathcal{M}$-metric scores depicted in Figure 4 show much less evidence of approaching a limit. If $\mathcal{M}$-metric scores continue to change at a faster rate than task completion times, it may be possible to use $\mathcal{M}$-metric scores as a more sensitive measure of manual control expertise.

One of the goals of this experiment was to try to understand what happens to the allocation of control with extended practice. The hypothesized model was that one of two possibilities may occur: subjects will either continue to allocate control within subsets or allocate control across all DOFs. Rather than an either/or hypothesis, it now appears that both cases occur. That is, subjects continue to allocate control between rotation and translation subsets but at the same time show some improvement in their ability to simultaneously control all the DOFs.

While there is a temptation to claim that equal allocation of control across all DOFs is "better" than unequal control, this may or may not be the case, depending on the task, the environment, or the users. For example, if one imagines a task where the trajectory taken is irrelevant to performance, and the cognitive load of simultaneously manipulating multiple variables is high, then unequal allocation of control would probably be the best strategy. Either way, in order to make such judgments, a framework for measuring the allocation of control, such as the $\mathcal{M}$-metric, is a necessary tool.

CONCLUSIONS

In a 6 DOF virtual docking task, operators do not allocate their control equally across all available DOFs. Instead, operators allocate control among subsets, by controlling rotational and translational DOFs separately and switching control between those subsets. In addition, the type of input device used has an effect on the strategy used by operators to allocate control. A more "natural" type of input device should allow operators to exercise a more even distribution of control across the available DOFs. On the other hand, an "unnatural" input device might force operators to control only subsets of the total number of DOFs at a time. Some simultaneous allocation of control does exist across all 6

DOFs, and the amount of this allocation appears to be a function of the type of device used.

It is important to note that, in a docking task the operator does not have to follow a required trajectory; any trajectory which accomplishes the docking goal is acceptable. It remains to be seen whether, in a dynamic tracking task, where simultaneous control of all 6 DOFs may be required, operators continue to choose to allocate control to subsets of DOFs or instead change strategies to control all DOFs together.

ACKNOWLEDGEMENTS

This research is funded by the IRIS NCE project "Effective display and tele-control technology integration for real and virtual environments", with further contributions from DCIEM.

REFERENCES

[1] Andersen, O. T. A system for quantitative assessment of dyscoordination and tremor, *Acta Neurol Scand 73*, (1986), 291-294.

[2] Balakrishnan, R., Baudel, T., Kurtenbach, G., and Fitzmaurice, G. The Rockin'Mouse: Integral 3D Manipulation on a Plane, In *CHI '97 Conference on Human Factors in Computing Systems*. Addison Wesley, 1997, pp. 311-318.

[3] Balakrishnan, R., and Patel, P. The PadMouse: Facilitating Selection and Spatial Positioning for the Non-Dominant Hand, In *Proceedings of CHI '98 Conference on Human Factors in Computing Systems*. ACM, 1998, pp. 9-16.

[4] Behbehani, K., Kondraske, G. V., and Richmond, J. R. Investigation of Upper Extremity Visuomotor Control Performance Measures, *IEEE Transactions on Biomedical Engineering 35*, 7 (1988), 518-525.

[5] Bernstein, N. A. *The Co-ordination and Regulation of Movements*, (Oxford: Pergamon Press, 1967).

[6] Ellson, D. C. *The independence of tracking in two and three dimensions with the B-29 pedastal sight.*, TSEAA-694-2G. Aero Medical Laboratory, 1947.

[7] Garner, W. *The Processing of Information and Structure*, (New York: John Wiley & Sons, 1974).

[8] Hocherman, S., and Aharon-Peretz, J. Two dimensional tracing and tracking patients with Parkinson's disease, *Neurology 44*, (1994), 111-116.

[9] Imai, S., and Garner, W. R. Discriminalibility and preference for attributes in free and constrained classification, *Journal of Experimental Psychology 69*, 6 (1965), 596-608.

[10] Jacob, R. J. K., Sibert, L. E., McFarlane, D. C., and M. Preston Mullen, J. Integrality and Separability of Input Devices, *ACM Transactions on Computer-Human Interaction 1*, 1 (1994), 3-26.

[11] Kondraske, G. V., Potvin, A. R., Tourtellotte, W. W., and Syndulko, K. A computer-based system for automated quantitation of neurologic function, *IEEE Transactions on Biomedical Engineering 31*, (1984), 401-414.

[12] Meyer, D. E., Smith, J. E. K., Kornblum, S., Abrams, R. A., and Wright, C. E. Speed-Accuracy Tradeoffs in Aimed Movements: Toward a Theory of Rapid Voluntary Action, *Attention and Performance XIII*, ed. M. Jeannerod. ((Hillsdale, New Jersey: Lawrence Erlbaum Associates, 1990) 173-226.

[13] Morrison, S., and Newell, K. M. Interlimb Coordination as a Function of Isometric Force Output, *Journal of Motor Behavior 30*, 4 (1998), 323-342.

[14] Oldfield, R. C. The assessment and analysis of handedness: The Edinburgh inventory., *Neuropsychologia 9*, (1971), 97-113.

[15] Poulton, E. C. *Tracking Skill and Manual Control*, (New York: Academic Press, Inc., 1974).

[16] Turvey, M. T. Coordination, *American Psychologist 45*, 8 (1990), 938-953.

[17] Vereijken, B., Whiting, H. T. A., Newell, K. M., and Emmerik, R. E. A. v. Free(z)ing Degrees of Freedom in Skill Acquisition, *Journal of Motor Behavior 24*, 1 (1992), 133-142.

[18] Watson, R. W., Jones, R. D., and Sharman, N. B. Two dimensional tracking tasks for quantification of sensory-motor dysfunction and their application to Parkinson's disease, *Medical & Biological Engineering & Computing 35*, (1997), 141-145.

[19] Yu, X., Lau, E., Vicente, K. J., and Carter, M. W. Advancing Performance Measurement in Cognitive Engineering: The Abstraction Hierarchy as a Framework for Dynamical Systems Analysis, In *Proceedings of the Human Factors and Ergonomics Society 42nd Annual Meeting*. HFES, 1, 1998, pp. 359-363.

[20] Zhai, S. *Human Performance in Six Degree of Freedom Input Control*, Ph.D. University of Toronto, 1995.

[21] Zhai, S., and Milgram, P. Quantifying Coordination in Multiple DOF Movement and Its Application to Evaluating 6 DOF Input Devices, In *Proceedings of the Conference on Human Factors in Computing Systems CHI '98*. ACM, 1998, pp. 320-327.

[22] Zhai, S., Milgram, P., and Buxton, W. The Influence of Muscle Groups on Performance of Multiple Degree-of-Freedom Input, In *Proceedings of CHI96: ACM Conference on Human Factors in Computing Systems*. ACM, 1996, pp. 308-315.

[23] Zhai, S., and Senders, J. W. Investigating coordination in multidegree of freedom control II: time-on-target analysis of 6 DOF tracking, In *41st Annual Meeting of Human Factors and Ergonomics Society*. 2, 1997, pp. 1254-1258.

Symmetric Bimanual Interaction

Ravin Balakrishnan [1,2]
[1]Alias|wavefront
210 King Street East
Toronto, Ontario
Canada M5A 1J7
ravin@aw.sgi.com

[2]Dept. of Computer Science
University of Toronto
Toronto, Ontario
Canada M5S 3G4
ravin@dgp.toronto.edu

Ken Hinckley [3]
[3]Microsoft Research
One Microsoft Way
Redmond, WA
USA 98052
kenh@microsoft.com

ABSTRACT

We present experimental work that explores the factors governing symmetric bimanual interaction in a two-handed task that requires the user to track a pair of targets, one target with each hand. A symmetric bimanual task is a two-handed task in which each hand is assigned an identical role. In this context, we explore three main experimental factors. We vary the *distance* between the pair of targets to track: as the targets become further apart, visual diversion increases, forcing the user to divide attention between the two targets. We also vary the demands of the task by using both a slow and a fast *tracking speed*. Finally, we explore *visual integration* of sub-tasks: in one condition, the two targets to track are connected by a line segment which visually links the targets, while in the other condition there is no connecting line. Our results indicate that all three experimental factors affect the degree of parallelism, which we quantify using a new metric of bimanual parallelism. However, differences in tracking error between the two hands are affected only by the visual integration factor.

Keywords
two-handed input, symmetric interaction, Guiard theory, input, interaction techniques,

INTRODUCTION

Several promising two-handed interaction techniques have been described in the interface design literature [2, 3, 4, 10, 27, 28]. A solid theoretical basis for the design of such systems exists in the form of Guiard's Kinematic Chain theory [7, 8] and experimental studies in the human-computer interaction literature [1, 10, 11, 14] that have explored Guiard's theory as well as additional factors influencing cooperation of the hands when each hand is assigned a different, *asymmetric* role.

However, the literature suggests that a number of tasks that can be facilitated by two-handed input, such as two-handed line drawing, positioning and sizing a rectangle [5, 17], and 2D or 3D navigation [9, 16, 28] can be performed effectively with a *symmetric* assignment of roles to the hands. Unlike asymmetric two-handed interaction, which is well explained by the KC model, factors governing this second class of symmetric bimanual tasks have not been articulated as well in the research literature. Without better empirical data, there is little scientific knowledge to guide the design of interfaces that incorporate symmetric interaction techniques.

In this paper, we investigate how factors such as attention, task difficulty, and visual integration affect performance in a symmetric bimanual task. Of particular interest is whether symmetric bimanual tasks are fundamentally different from asymmetric bimanual tasks. At this point, it is important to note the difference between *task assignment* and *task performance*. Even if the task *assigned* to each hand is identical (i.e., symmetric), it is plausible that the combined task will not be *performed* in a symmetric and/or parallel manner. Under some conditions, it may be natural to perform a symmetric bimanual task in a sequential manner, moving one hand followed by the other, rather than moving both at the same time. The task could also be performed asymmetrically in the sense that one hand's performance could result in greater errors or poorer temporal performance than the other.

Note that we distinguish between symmetric and parallel performance. It is possible for bimanual performance to be sequential in nature, but nonetheless symmetric in the terms of error rate and/or time taken to perform each hand's subtask. Conversely, performance could be parallel (occur simultaneously) and yet asymmetric in terms of error and time measures. This raises the question of whether humans always perform symmetric tasks in a symmetric, parallel manner regardless of task difficulty, attentional demands, or visual integration of the sub-tasks assigned to each hand. Do users switch to a more sequential and/or asymmetric interaction style as these factors change?

Our results suggest that even when users are given a task with identical, symmetric role assignments for each hand, they do not always perform the task in a parallel, symmetric manner. We show that the lack of visual integration causes performance to become asymmetric in that root-mean square (RMS) error increases at a greater extent for the left hand[1]. Also, divided attention, task difficulty, and the lack of visual integration can all affect the degree of parallelism exhibited when performing the symmetric bimanual task. These results suggest that under some conditions, existing models of bimanual interaction [7, 21] may apply to tasks with a symmetric assignment of roles to the hands.

PREVIOUS WORK

There are several examples of symmetric two-handed interaction techniques in the literature. These include two-handed map manipulation [9], a two-handed "bulldozer" metaphor for 3D navigation [28], and symmetric rectangle and line editing [5, 17]. Furthermore, in the workflow of some two-handed input systems (e.g. Kurtenbach et. al. [16]) one can observe fluid transitions between asymmetric and symmetric two-handed actions, such as using a ToolGlass [3].

Leganchuk et. al. [17] used a rectangle editing task to reason about cognitive benefits of bimanual interaction. They showed that two different bimanual rectangle editing techniques resulted in superior performance to a unimanual technique. However, they found no difference between the bimanual technique that consistently assigned identical tasks to each hand (i.e., symmetric task assignment) and another technique that fluidly switched between asymmetric and symmetric task assignment.

Casalta and Guiard [5] found that in a rectangle editing task, symmetric task assignment resulted in better performance, as well as increased bimanual parallelism, than an asymmetric task assignment. This result suggests that for some tasks, a symmetric assignment of roles to the hands can result in better performance than an asymmetric role assignment.

Hinckley et. al. [9] describe a technique for two-handed manipulation (panning, zooming, and rotation) of maps. Their mapping of the degrees-of-freedom results in a technique that supports both symmetric and asymmetric use of the hands. For example, the user may zoom on a particular location by "pinning down" that location with one hand and "stretching" the map with the other hand; or conversely, the user may perform a more coarse zooming operation by moving both hands in opposite directions.

Balakrishnan and Kurtenbach [2] explore bimanual camera control and object manipulation. They report that in a 3D object docking task, subjects invariably adopt a symmetric style of interaction even though they could have adopted a asymmetric style of interaction to reduce the number of degrees-of-freedom that need to be controlled at once.

A number of bimanual tasks with a symmetric assignment of roles to the hands have been studied in the psychology and motor behavior literatures, including bimanual pointing to separate targets [15, 18, 25], bimanual tapping of rhythms [21, 26], circle drawing [24], and bimanual steering [22, 23].

Kelso, Southard, and Goodman [15] explore a two handed tapping task with targets of disparate difficulty for each hand (i.e., the task assignment is symmetric in that each hand performs a tapping task, but asymmetric in that the difficulty of each hand's task is different). They find that while the hands move at different speeds to different points in space, times to peak velocity and acceleration are highly synchronized. Thus, in a sense, performance is symmetric and parallel even though the task assignment is not completely symmetric.

Marteniuk, MacKenzie, and Baba [18] describe a similar experiment to Kelso et. al. [15]. From both their own data and a reanalysis of Kelso et. al.'s [15] data, they report some evidence for a left-right asymmetry between the two hands. In a more recent study, Jackson, Jackson, and Kritikos [12] find that in more complicated "reach and grasp" bimanual task, kinematic measures of performance are unaffected when each hand performs movements of identical or different levels of difficulty. They find that movements of both hands are scaled to a common time duration, whereas movement velocity and grip aperture are scaled independently. Hence, their data seems to support the findings of Kelso et. al. [15].

In a symmetric circle drawing task, Swinnen, Jardin, and Meulenbroek [24] report a distinct asymmetry in performance. Interestingly, they find that the dominant hand leads the non-dominant hand during the task. This is in contrast with Guiard's KC model, which postulates that the non-dominant hand precedes the dominant hand in the performance of asymmetric tasks. They also report that attentional cueing affects the size of the asymmetry: the amount of asymmetry (phase offset between the limbs) increases when subjects are told to monitor the dominant hand, and decreases when subjects are told to monitor the non-dominant hand.

Preilowski [22, 23] explored a two-handed steering task using hand cranks, each of which controls one degree-of-freedom of a cursor. After practice, normal subjects can steer the cursor (i.e., both hands are performing somewhat symmetrically and in parallel) without visual feedback, whereas patients with damage to the anterior commissure cannot. His focus however, was not on the symmetry/asymmetry and parallel/sequential issues per se.

In short, there appear to be many unresolved issues regarding symmetric bimanual tasks and exactly how these differ from, or when they may be preferable to, asymmetric assignments of roles to the hands. Prior studies have not quantified potential factors that may drive symmetric bimanual performance. The psychology and motor control literature are also inconclusive as to how bimanual tasks that assign essentially symmetric roles to each hand are performed. Some evidence [12, 15, 22, 23] suggests that

---

[1] For convenience, since the current experiment used only right-handed participants, we always refer to the preferred hand as the right hand and the nonpreferred hand as the left hand. For left-handers, these hand roles would be reversed.

performance is mostly symmetric, whereas others [18, 24] indicate asymmetric performance with attention being a contributing factor. The literature therefore suggests that this is an area in need of further experimental study.

EXPERIMENT
Task and Stimuli

We chose a bimanual target tracking task for two main reasons. First, the standard target docking or selection task that is widely used in motor behavior studies is unsuitable for our purposes because the only way to vary the difficulty of the task is to change the size of the target and its distance from the starting point. A large part of the task is therefore simply getting to the vicinity of the target; only at the last phase of the task does the size of the target affect performance. Hence, task difficulty does not apply uniformly throughout the task. In contrast, the task difficulty in a tracking task can be made to apply uniformly throughout the task (since the user must always attempt to stay on target), providing us with a rich set of data. Second, to the best of our knowledge, apart from Preilowski [22, 23], bimanual target tracking has not been studied in the literature. Thus, the present study contributes to the literature in the task aspect as well. Note that this tracking task is not intended to necessarily be representative of any particular symmetric bimanual user interface. Rather, we use this task as an experimental instrument to explore factors that can influence bimanual performance.

Participants tracked targets with both hands. There were two main conditions that varied the level of integration of the visual stimuli:

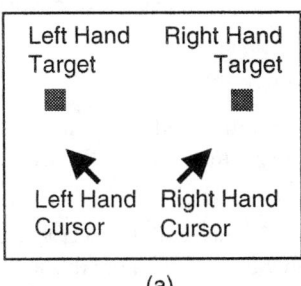 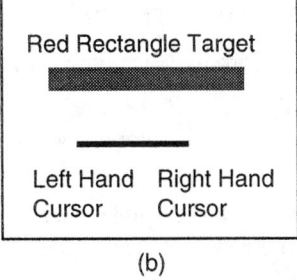

Figure 1. Experiment Stimuli. (a) Stimuli for the *Separated* condition. The Left and Right hand cursors are used to track the Left and Right hand targets, respectively. The distance between the centers of the targets are kept constant for a trial at either 100 or 840 pixels. (b) Stimuli for the *Integrated* condition. The Left and Right cursors control the position, orientation, and length of the line. The cursors themselves are not shown. The user tracks the red rectangle with the line. The length of the red rectangle is kept constant for a trial at either 100 or 840 pixels. None of the text in this diagram is displayed during the experiment.

Separated targets - Two red square (20x20 pixel) targets appeared at a given distance to the left and right of the center of the screen (Figure 1a). Participants controlled a white colored cursor with each hand. The left hand cursor always pointed towards the left side of the screen, the right hand cursor pointed towards the right. Participants were told to track the left square with the left cursor, and the right square with the right cursor. The two targets moved around the screen in a pseudorandom fashion, with the constraint that the movements of both targets were symmetric in the sense that they each moved the same amount in a given direction at a given time. The distance between the targets, and amount of movement at each time step (i.e., speed), were kept constant for a given trial (distance and speed were manipulated as experimental conditions). The background color of the screen was black throughout the experiment.

Integrated target - A single red rectangular (size: 20 pixels wide x *distance* pixels long) target appeared centered on the screen (Figure 1b). Instead of two cursors, a straight white line was drawn between the positions of the left hand and right hand cursors (the cursors were not shown). Participants were told to match the position, orientation, and length of the white line with that of the red rectangle. The rectangle moved around the screen in the same pseudorandom manner as the targets in the *Separated* condition. Essentially, the end points of the red rectangle were the same as the center points of the two targets in the *Separated* condition; henceforth we will refer to these as the "target points".

From the motor domain perspective, both *Separated* and *Integrated* conditions are identical in that the same motor actions are required to track the target(s). In the visual domain, however, they differ in that the *Separated* condition could be perceived as being two separate tasks whereas the *Integrated* condition could be perceived as being a single, integrated task [6].

The attentional demands of the task were manipulated by varying the distance between the target points. Two distances were used: 100 and 840 pixels. In the 100 pixel or *Singular Attention* condition, both target points (i.e., both targets in the *Separated* condition and the entire target in the *Integrated* condition) were visible in the participant's focal visual field. Thus, the participant only had to attend to a single area of on the screen at any one time. In the 840 pixel or *Divided Attention* condition, it was impossible to attend to both target points at the same time. This resulted in the participant having to divide attention between two areas of the screen.

The difficulty of the task was manipulated by varying the speed at which the target(s) moved. Two speeds were used: *Slow* (1 pixel/frame – the target moved 1 pixel in each of the x and y directions per frame update), and *Fast* (2 pixels/frame). The frame rate was kept constant at 60Hz

Apparatus

The experiment was conducted on a graphics accelerated two-processor workstation running Windows NT, with a 21-inch, 1280x1024 resolution, color display. Two pens on a Wacom Intuos 12x18 inch digitizing tablet were used as the input devices. The tablet was sampled at a constant rate of 60Hz, and the graphics update rate was also kept constant at 60Hz.

Participants
Eight right-handed volunteers participated in the experiment.

Design
A within-subjects repeated measures design was used. All participants performed the experiment for both the *Separated* and *Integrated* conditions. The presentation order of these two conditions was counterbalanced across the participants (Participants #1,3,5,7 did the *Separated* condition followed by the *Integrated* condition. Participants #2,4,6,8 did the *Integrated* condition followed by the *Separated* condition). For each condition, participants performed 7 blocks of trials. The first block of trials was considered to be practice trials and was excluded from the data analysis. Therefore, a total of 6 blocks of trials were used in the analysis. Each block consisted of 1 trial for each of the four combinations of attention and speed conditions. The presentation of these four trials within each block was randomized. Each trial lasted for 45 seconds. Participants were allowed breaks between trials. The experiment consisted of 384 total non-practice trials, as follows:

> 8 participants x
> 2 visual integration conditions (*Separated*, *Integrated*) x
> 6 blocks of trials for each integration condition x
> 2 attention conditions (*Singular*, *Divided*) per block x
> 2 speed conditions (*Slow*, *Fast*) per block
> = 384 total trials of 45 seconds each.

For each participant, the experiment was conducted in one sitting and lasted about one hour.

Participants initiated a trial by positioning the two cursors over the two targets (in the *Separated* condition. In the *Integrated* condition, they matched the position, orientation, and length of the white line with the red rectangle). No button presses were required. The target(s) then begin to move in a pseudorandom fashion for 45 seconds at the speed fixed for that trial. At the end of 45 seconds, the screen went blank for 2 seconds, and the next trial's stimuli were presented. The movement trajectories were precomputed and the same set of four trajectories (one for each attention x speed condition) was used for all the blocks in both the *Separated* and *Integrated* conditions. The use of a fixed set of trajectories allowed for a fair comparison between the conditions.

Hypotheses
We expect to find the following effects in our experimental data:

H1. The *Integrated* visual stimuli conditions will result in more accurate tracking than the *Separated* conditions.

H2. The *Singular Attention* conditions will result in more accurate tracking than the *Divided Attention* conditions.

H3. The *Slow* speed conditions will result in more accurate tracking than the *Fast* speed.

While accuracy is an important measure of performance in tracking tasks, the primary goal of this study is not to evaluate tracking performance per se. Rather, we are interested in how the experimental manipulations of visual integration, attentional demands, and task difficulty affect the level of parallelism and symmetry exhibited by the user when performing a symmetric bimanual task where each hand is assigned identical functional roles.

Two-handed performance can be considered to occur symmetrically, or in parallel, or possibly both (or neither). In the present discussion, we say that the two hands exhibit *symmetric* performance if the average root mean square (RMS) tracking error exhibited by the hands over the course of a trial have equal values – that is, if the difference in tracking error between the left hand and the right hand is statistically indistinguishable. Note, however, that this measure of symmetry ignores bimanual performance in the *time* dimension: the user might exhibit performance which, for example, adjusts only the right hand, and then only the left hand.

By contrast, our measure of *parallel* bimanual performance does consider time, by quantifying the *simultaneous magnitude and direction of movement* of each hand, using a new metric that is discussed later in this paper. By distinguishing symmetrical performance from parallel performance, our analyses take into account two different interpretations of bimanual performance, allowing us to produce a more complete characterization of our experimental results.

Accordingly, we further hypothesize that:

H4. The *Integrated* visual stimuli conditions will be performed more symmetrically than the *Separated* conditions.

H5. The *Singular Attention* conditions will be performed more symmetrically than the *Divided Attention* conditions.

H6. The *Slow* speed conditions will be performed more symmetrically than the *Fast* speed conditions.

H7. The *Integrated* visual stimuli conditions will be performed with greater parallelism than the *Separated* conditions.

H8. The *Singular Attention* conditions will be performed with greater parallelism than the *Divided Attention* conditions.

H9. The *Slow* speed conditions will be performed with greater parallelism than the *Fast* speed conditions.

Results
Overall Tracking Performance
Our first measure of tracking performance was the root mean square (RMS) error between each cursor position and the corresponding target point at each time step ($1/60^{th}$ of a

second) during the trial. The average RMS error for each hand per trial was computed, resulting in two RMS error metrics: RMS_{rh} for the right hand average RMS error, and RMS_{lh} for the left hand average RMS error. In addition a compound metric, $RMS_{tot} = RMS_{lh} + RMS_{rh}$, was computed to represent the total RMS error per trial.

The overall mean RMS_{tot} for our experimental conditions is shown in Figure 2. Repeated measures analysis of variance with RMS_{tot} as the dependent variable was conducted on the data. Overall, there was no significant difference between the two visual integration (*Separated*, *Integrated*) techniques ($F_{1,6} = 4.8$, $p=0.06$). Thus, using RMS_{tot} as the performance measure, hypothesis H1 is not confirmed. There was a significant effect for the attentional (*Singular* vs. *Divided Attention*) factors ($F_{1,6} = 109$, $p<0.01$), with *Singular Attention* resulting in superior performance, thus confirming hypothesis H2. A significant effect was found for the speed (*Slow* vs. *Fast*) factors ($F_{1,6} = 87$, $p<0.01$), with *Slow* speed resulting in superior performance, thus confirming hypothesis H3. The only other significant effect was an *Attention* x *Speed* interaction ($F_{1,6} = 6.62$, $p<0.05$), indicating that when tracking at the faster speed, divided attention has a greater effect.

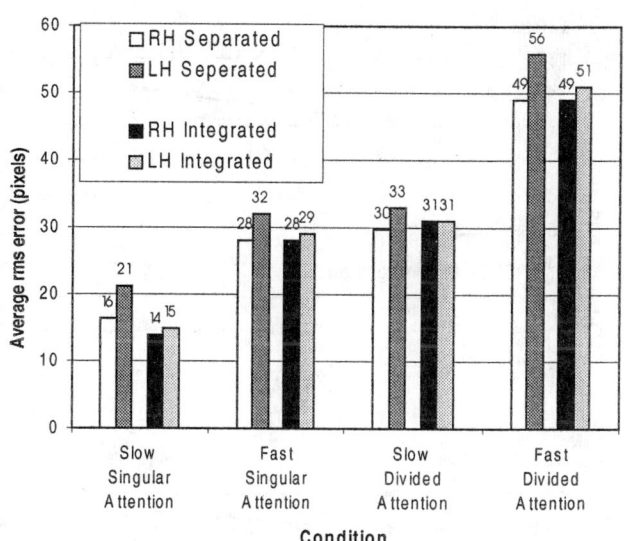

Figure 3. Tracking performance for each hand, broken down by experimental factors. Data for all trials from all 8 participants.

RMS_{lh} as the dependent variable showed a significant difference between the two visual integration conditions ($F_{1,6} = 7.6$, $p<0.05$). As the slopes in Figure 4(a) show, the RMS_{lh} measure was significantly higher than the RMS_{rh} measure for the *Separated* conditions, but did not differ significantly for the *Integrated* conditions. This result indicates that poor visual integration causes performance to become asymmetric, confirming hypothesis H4. There was no significant effect for the attention factor ($F_{1,6} = 3.14$, $p>0.05$) or the speed factor ($F_{1,6} = 0.94$, $p>0.05$), as illustrated by the identical slopes in Figures 4b and 4c). Thus, hypotheses H5 and H6 were not confirmed.

Parallelism Analysis

In order to analyze the level of parallelism exhibited by the two hands, we need an appropriate measure of parallelism. One such measure is the "Integrality" metric introduced by Jacob et. al. [13]. They proposed a means of quantifying parallelism (we use the term "parallel" instead of "integral" as originally proposed) in the time domain, based on whether movements in the dimensions of interest occurred simultaneously at each time step. This measure, however, classifies a set of movements as parallel as long as they moved by any amount during a time period. The relative magnitude and direction of movement in each dimension of interest is not taken into account.

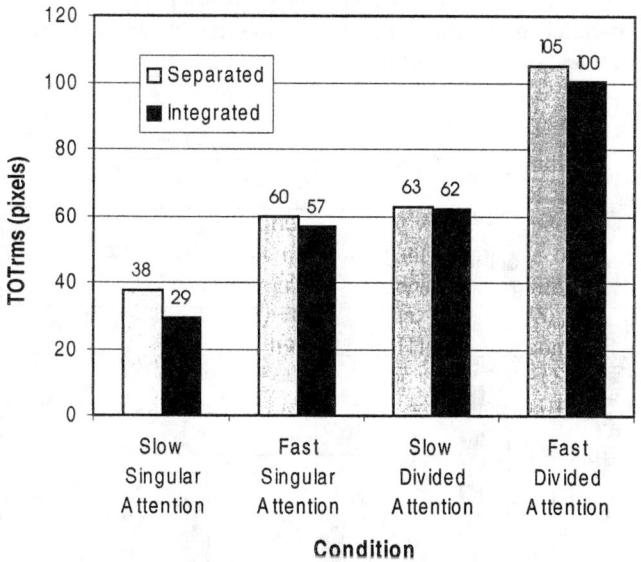

Figure 2. Overall tracking performance as measured by $RMStot$, broken down by the experimental factors. Data for all trials from all 8 participants.

Symmetry Analysis
Looking at the differences in performance between the two hands (Fig. 3), we find that the overall difference between performance of the right hand (RMS_{rh}) and the left hand (RMS_{lh}) was 8%, indicating that there was a slight asymmetry between the hands overall, although this result was not significant ($p=0.07$). Repeated measures analysis of variance conducted with the difference between RMS_{rh} and

Masliah [19] has proposed the m-metric to quantify coordination in multi-degree-of-freedom docking tasks. The m-metric takes into account the magnitude and direction of movement of each dimension of interest when computing simultaneity. The metric as originally proposed is only applicable to docking tasks. Here, we adapt it to measure parallelism in a tracking task. The basic idea behind this measure is to first compute how much the error between the

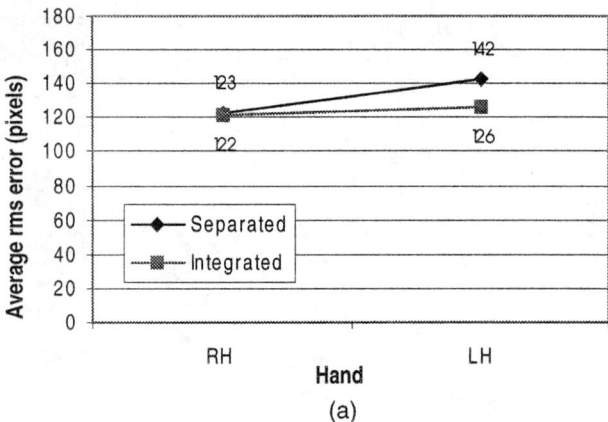

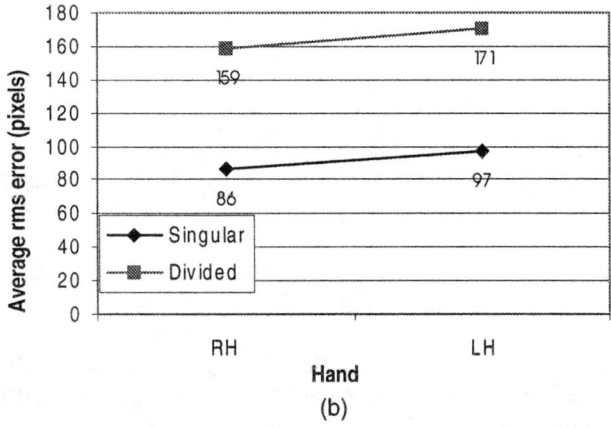

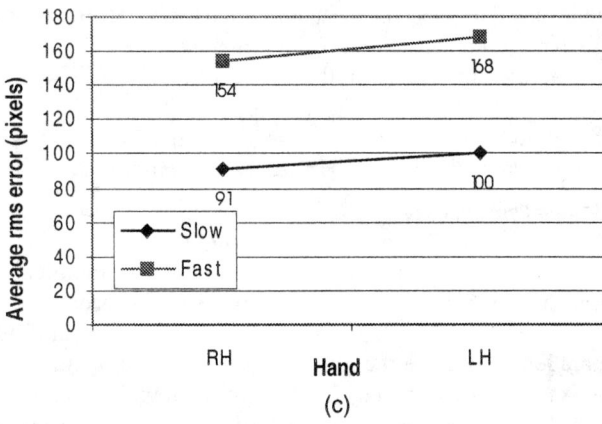

Figure 4. Tracking performance for each hand for (a) the two visual integration conditions, (b) the two attention factors, (c) the two speed factors.

current position and the target position is reduced at each time step. This percentage error reduction per time step is computed for each hand as follows:

$$\%ER = \frac{\text{actual magnitude of movement towards target}}{\text{movement required to reduce error to 0}}$$

This results in a number between 0 and 1, where 1 means the cursor is perfectly tracking the target and 0 means the cursor is not following the target at all.

The amount of parallelism at each time step is then calculated by taking the ratio of the two hand's %ER values, with the larger value taken as the denominator:

$$\text{Parallelism} = \frac{\text{Right Hand's \%ER}}{\text{Left Hand's \%ER}}$$

The average of all Parallelism measures over the duration of a trial thus results in a bounded measure between 0 and 1. Values closer to 1 indicate that both hands are simultaneously reducing their errors by the same amount (i.e., highly parallel, identical, movements), whereas values closer to 0 indicate that the hands are working in a sequential manner.

This metric not only considers if motion of the two hands is simultaneous, but also takes into account the magnitude and direction of any simultaneous motion. Thus, movements that occur at the same time but which do not contribute towards the accurate completion of the task are given much less weight in the metric. We feel that this results in a more meaningful measure of bimanual parallelism.

We analyzed our experimental data using this new parallelism metric. Figure 5 shows the mean parallelism values for each condition.

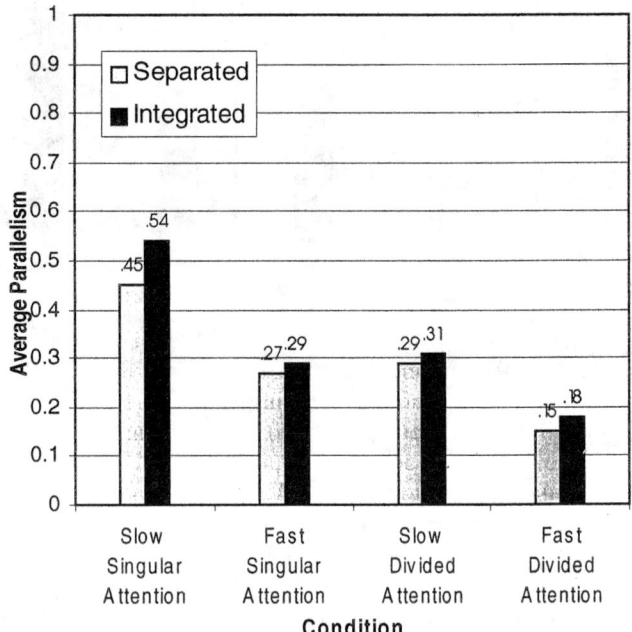

Figure 5. Parallelism between the two hands, broken down by experimental factors. Values close to zero indicate little parallelism, values close to 1 indicate a high degree of parallelism. Data for all trials from all 8 participants.

Overall, parallelism was not very high, at 0.31 units. There was a significant effect for the two visual integration conditions ($F_{1,6} = 7.28$, $p<0.05$), with the *Integrated* conditions exhibiting 12% more parallelism than the *Separated* conditions, thus confirming hypothesis H7.

Hypothesis H8 was also confirmed by a strong significant effect for the two attentional factors ($F_{1,6} = 108$, $p<0.01$), with *Singular Attention* conditions showing more parallelism than the *Divided Attention* conditions.

Hypothesis H9 was confirmed by a significant effect for the two speed factors ($F_{1,6} = 46$, $p<0.01$), with *Slow* conditions showing more parallelism than the *Fast* conditions.

CONCLUSIONS AND FUTURE WORK

We have presented experimental work that explores issues surrounding symmetric bimanual action. We also introduced a new metric, adapted from the coordination metric of Masliah [19], which quantifies the extent to which movements of the hands occur in parallel. The analysis of our data using this parallelism metric showed that increasing task difficulty, divided attention, and lack of visual integration can all cause the user to adopt a more sequential style of interaction.

Overall, our data showed a slight asymmetry (albeit not statistically significant at the 5% confidence level) with respect to RMS tracking error, with the left hand having 8% higher error than the right hand. We also found that a lack of visual integration results in significant asymmetry between the hands. Attentional demands and task difficulty, however, did not affect the level of symmetry in performance (i.e., both hands exhibited similar RMS tracking error rates).

Taking the symmetry and parallelism analyses as a whole, we see that decreased parallelism does not (except when visual integration is lacking) cause performance as measured by RMS tracking error to become more asymmetric. In other words, parallelism is not a requirement for performance to be symmetric.

From a practical viewpoint, although we used a bimanual tracking task as an experimental instrument to explore issues that can affect bimanual performance, and not necessarily to be representative of any particular symmetric bimanual user interface, the results can nonetheless yield design insights for symmetric bimanual interfaces. For example, our finding that lack of visual integration does not lend itself to symmetric interaction suggests that for a symmetric task like two-handed rectangle editing [5, 17] it would be not be good design to merely display the corners of the rectangle (as is sometimes done in the interest of not obscuring underlying geometry).

Also, our finding that dividing attention results in highly sequential performance suggests that symmetric tasks where the two hands are not operating nearby in the focal visual field should be avoided. This may be one reason that symmetric bimanual interaction lends itself to navigation tasks [9, 16, 28]. In a navigation task such as steering through a 3D environment [28], visual flow occurs across the entire display window in response to two-handed movements, so the focal visual field can provide sufficient feedback. A problem might arise in a bimanual interface using two cursors that may become widely separated, unless some secondary feedback in the focal visual field can be provided. For example, the map navigation example of [9] employs separate cursors for each hand, but the continuous visual flow of real-time feedback from the map moving, expanding, or shrinking provides sufficient feedback. If only two separate cursors were provided, our results suggest that the user's ability to control symmetric bimanual actions could be compromised.

From a theoretical perspective, given that our results show a slight general asymmetry in the performance of symmetric bimanual tasks, it is possible that existing theoretical models of asymmetric bimanual interaction [7, 21] could apply to symmetric bimanual tasks as well. However, since we also found that the level of symmetry does not easily degrade when task difficulty is increased or attention is divided, it is likely that performance in symmetric tasks also differ fundamentally in some aspects from asymmetric tasks. For example, our data clearly indicates that for symmetric tasks there is no tendency for the human motor system to devote more resources to the dominant hand when attention is divided.

By contrast, previous work by Peters [20] shows that when independent, asymmetric tasks are assigned to each hand, there is a tendency to devote more resources to the dominant hand. To the best of our understanding, the effect of task difficulty and visual integration on the performance of asymmetric bimanual tasks has not been explored. As such, we cannot draw any conclusions as to whether symmetric and asymmetric tasks differ along these factors. Clearly, more research is needed to quantify these differences and thus build better models that account for both symmetric and asymmetric bimanual tasks. The work presented in this paper is a step towards a more comprehensive understanding of symmetric (as well as asymmetric) two-handed interaction, including a better understanding of under what conditions symmetric, parallel action of the hands is possible.

ACKNOWLEDGEMENTS

We thank Bill Buxton, George Fitzmaurice, Gordon Kurtenbach, Russell Owen, and Jade Rubick for advice and assistance in various forms throughout the course of this work. We also thank all those who participated in our experiment, and Alias|wavefront and Microsoft for supporting this collaborative research study.

REFERENCES

1. Balakrishnan, R., and Hinckley, K. (1999). The role of kinesthetic reference frames in two-handed input performance. *ACM UIST'99 Symposium*, pp. 171-178.

2. Balakrishnan, R., and Kurtenbach, G. (1999). Exploring bimanual camera control and object manipulation in 3D graphics interfaces. *ACM CHI'99 Conference*, pp. 56-63.

3. Bier, E., Stone, M., Pier, K., Buxton, W., and DeRose, T. (1993). Toolglass and Magic Lenses: The see-through interface. *ACM Siggraph'93 Conference*, pp. 73-80.

4. Buxton, W., and Myers, B. (1986). A study in two-handed input. *ACM CHI'86 Conference*, pp. 321-326.

5. Casalta, D., Guiard, Y., and Beaudouin-Lafon, M. (1999). Evaluating two-handed input techniques: Rectangle editing and navigation. *ACM CHI'99 Conference (Extended Abstracts)*, pp. 236-237.

6. Garner, W.R. (1974). *The processing of information and structure*. Lawrence Erlbaum.

7. Guiard, Y. (1987). Asymmetric division of labor in human skilled bimanual action: The kinematic chain as a model. *Journal of Motor Behavior*, 19(4), pp. 486-517.

8. Guiard, Y., and Ferrand, T. (1995). Asymmetry in bimanual skills, in *Manual asymmetries in motor performance*. Elliot and Roy, Editors. CRC Press.

9. Hinckley, K., Czerwinski, M., and Sinclair, M. (1998). Interaction and modeling techniques for desktop two-handed input. *ACM UIST'98 Symposium*, pp. 49-58.

10. Hinckley, K., Pausch, R., Proffitt, D., and Kassell, N. (1998). Two-handed virtual manipulation. *ACM Transactions on Computer-Human Interaction*, 5(3), pp. 260-302.

11. Hinckley, K., Pausch, R., Proffitt, D., Patten, J., and Kassell, N. (1997). Cooperative bimanual action. *ACM CHI'97 Conference*, pp. 27-34.

12. Jackson, G.M., Jackson, S.R., and Kritikos, A. (1999). Attention for action: Coordinating bimanual reach-to-grasp movements. *British Journal of Psychology*, 90, pp. 247-270.

13. Jacob, R., Sibert, L., McFarlane, D., and Mullen, M. (1994). Integrality and separability of input devices. *ACM Transactions on Computer-Human Interaction*, 1(1), pp. 3-26.

14. Kabbash, P., Buxton, W., and Sellen, A. (1994). Two-handed input in a compound task. *ACM CHI'94 Conference*, pp. 417-423.

15. Kelso, J., Southard, D., and Goodman, D. (1979). On the coordination of two-handed movements. *Journal of Experimental Psychology: Human Perception and Performance*, 5(2), pp. 229-238.

16. Kurtenbach, G., Fitzmaurice, G., Baudel, T., and Buxton, W. (1997). The design of a GUI paradigm based on tablets, two-hands, and transparency. *ACM CHI'97 Conference*, pp. 35-42.

17. Leganchuk, A., Zhai, S., and Buxton, W. (1999). Manual and cognitive benefits of two-handed input. *ACM Transactions on Computer-Human Interaction*, 5(4), pp. 326-359.

18. Marteniuk, R., MacKenzie, C., and Baba, D. (1984). Bimanual movement control: Information processsing and interaction effects. *Quarterly J. of Experimental Psychology*, 36A, pp. 335-365.

19. Masliah, M., and Milgram, P. (1999). Measuring the allocation of control across degrees-of-freedom. *Graphics Interface'99*, pp. .

20. Peters, M. (1981). Attentional asymmetries during concurrent bimanual performance. *Quarterly Journal of Experimental Psychology*, 33A, pp. 95-103.

21. Peters, M. (1985). Constraints in the performance of bimanual tasks and their expression in unskilled and skilled subjects. *Quarterly J. of Experimental Psychology*, 37A, pp. 171-196.

22. Preilowski, B. (1972). Possible contribution of the anterior forebrain commissures to bilateral motor coordination. *Neuropsychologia*, 10, pp. 267-277.

23. Preilowski, B. (1990). Intermanual transfer, interhemispheric interaction, and handedness in man and monkeys, in *Brain Circuits & Functions of the Mind: Essays in Honor of Roger W. Sperry*. C. Trevarther, Editor. Cambridge University Press.

24. Swinnen, S.P., Jardin, K., and Meulenbroek, R. (1996). Between limb asynchronies during bimanual coordination: effects of manual dominance and attentional cueing. *Neuropsychologia*, 34(12), pp. 1203-1213.

25. Wing, A. (1982). Timing and coordination of repetitive bimanual movements. *Quarterly J. of Experimental Psychology*, 34A, pp. 339-348.

26. Wolff, P., Hurwitz, I., and Moss, H. (1977). Serial organization of motor skills in left- and right-handed adults. *Neuropsychologia*, 15, pp. 539-546.

27. Zeleznik, R., Forsberg, A., and Strauss, P. (1997). Two pointer input for 3D interaction. *ACM Symposium on Interactive 3D Graphics*, pp. 115-120.

28. Zhai, S., Kandogan, E., Smith, B., and Selker, T. (1999). In search of the "magic carpet": Design and experimentation of a bimanual 3D navigation interface. *Journal of Visual Languages and Computing*, February.

Two-Handed Input Using a PDA And a Mouse

Brad A. Myers, Kin Pou ("Leo") Lie, and Bo-Chieh ("Jerry") Yang

Human Computer Interaction Institute
School of Computer Science
Carnegie Mellon University
Pittsburgh, PA 15213
bam@cs.cmu.edu
http://www.cs.cmu.edu/~pebbles

ABSTRACT

We performed several experiments using a Personal Digital Assistant (PDA) as an input device in the non-dominant hand along with a mouse in the dominant hand. A PDA is a small hand-held palm-size computer like a 3Com Palm Pilot or a Windows CE device. These are becoming widely available and are easily connected to a PC. Results of our experiments indicate that people can accurately and quickly select among a small numbers of buttons on the PDA using the left hand without looking, and that, as predicted, performance does decrease as the number of buttons increases. Homing times to move both hands between the keyboard and devices are only about 10% to 15% slower than times to move a single hand to the mouse, suggesting that acquiring two devices does not cause a large penalty. In an application task, we found that scrolling web pages using buttons or a scroller on the PDA matched the speed of using a mouse with a conventional scroll bar, and beat the best two-handed times reported in an earlier experiment. These results will help make two-handed interactions with computers more widely available and more effective.

Keywords: Personal Digital Assistants (PDAs), Hand-held computers, Palm Pilot, Windows CE, Two-Handed Input, Smart Environments, Ubiquitous Computing, Pebbles.

INTRODUCTION

Many studies of two-handed input for computers have shown advantages for various tasks [1, 4, 8, 10, 15]. However, people rarely have the option of using more than just a mouse and keyboard because other input devices are relatively expensive, awkward to set up, and few applications can take advantage of them. However, increasing numbers of people now do have a device that they carry around that could serve as an extra input device for the computer: their Personal Digital Assistant (PDA). PDAs, such as 3Com's Palm Pilots and Microsoft's Windows CE devices, are designed to be easily connected to PCs and have a touch-sensitive screen which can be used for input and output. Furthermore, newer PDAs, such as the Palm V and the HP Jornada 420, are rechargeable, so they are *supposed* to be put in their cradles next to a PC when the user is in the office. Therefore, if using a PDA in the non-dominant hand proves useful and effective, it should be increasingly easy and sensible to use PDAs for two-handed input.

Another advantage of PDAs over the input devices studied in previous experiments is that they are much more flexible. PDAs have a display on which virtual buttons, knobs and sliders can be displayed, and they can be programmed to respond to a wide variety of behaviors that can be well-matched to particular tasks. However, a disadvantage is that the controls on the PDA screen are virtual, so users cannot find them by feel. Research is therefore needed to assess how well the PDA screen can work as a replacement for other input devices that have been studied for the left hand.

This paper reports on several experiments that measure various aspects of using a PDA as an input device in the non-dominant hand. Two experiments are new and are designed to measure the parameters of using a PDA. One experiment repeats an earlier study [15] but uses a PDA in the non-dominant hand. Since the actual pragmatics of input devices can have a large impact on their effectiveness [3, 9], we wanted to determine whether the results seen in prior experiments would also apply to using PDAs.

In summary, the results are:

- People can quickly and reliably hit large buttons drawn on the PDA with the fingers on their left hands without looking. 99% of the button taps were correct on buttons that are 1-inch square in a 2x2 arrangement. With a larger number of smaller buttons, the accuracy significantly decreases: 95% were correct for 16 buttons that are ½ inch on a side arranged 4x4. The time from stimulus to button tap was about 0.7 sec for the large buttons and 0.9 seconds for the smaller buttons.

- In a task where the subjects had to move both hands from the keyboard to the PDA and the mouse and then back, we found that it took an average of 0.791 seconds to move both hands from the devices to the keyboard. This was about 13% longer than moving one hand from the mouse to the keyboard (which took 0.701 sec). Moving to a PDA and mouse from the keyboard took an average of 0.838 seconds, which is about 15% longer than moving one hand to the mouse (0.728 seconds).

- In a repeat of the experiment reported in [15], subjects were able to scroll large web pages and then select a link on the page at about the same speed using buttons or a scroller on a PDA compared to using the mouse with a conventional scroll bar. The times we found for scrolling with buttons on the PDA were faster than any of the times in the earlier study, including the 2-handed ones.

RELATED WORK

There have been many studies of using input devices for computers in both hands, but none have tested PDAs in the left hand, and we were unable to find measurements of homing times from the keyboard to devices for two-handed use.

One of the earliest experiments measured the use of two hands in a positioning and scaling task and for scrolling to known parts of a document [4]. This study found that people naturally adopted parallel use of both hands and could scroll faster with the left hand. Theoretical studies show that people naturally assign different tasks to each hand, and that the non-dominant hand can support the task of the dominant hand [7]. This has motivated two-handed interfaces where the non-dominant hand plays a supporting role, such as controlling other drawing tools [10] and adjusting translation and scaling [4, 15]. Another study investigated how accurately gestures could be drawn with the non-dominant hand on a small touchpad mounted on top of a mouse [2]. Two-handed input for 3D interaction [1, 8] has also been found to be useful.

There has been prior work on using PDAs at the same time as regular computers for various tasks including meeting support [11], sharing information [12], and to help individuals at their desks, but we found no prior work on measuring performance of non-dominant hand use of PDAs.

EXPERIMENTAL DESIGN

Two new studies were performed. In the first, the subjects did five tasks in a row. The first task was a typing test to see how fast the subjects could type. Next, they performed a button size task to measure the error rates and speeds when tapping on different size buttons on the PDA. Next, the subjects performed a homing speed task where we measured the how quickly the subjects moved among the keyboard and the devices. Finally, they performed a scrolling task using a variety of devices, which is a repeat of an earlier experiment [15]. The subjects reported a number of problems with the scrolling devices on the PDA in the last task, so we redesigned the scrolling devices, and in a second study with new subjects, we evaluated the performance of the new scrollers on the same task.

Apparatus

Figure 1 shows the experimental set up. Subjects sat at a normal PC desktop computer that was running Windows NT. On the right of the keyboard was a mouse on a mouse pad. On the left was an IBM WorkPad 8602-30X PDA (which is the same as a Palm IIIx). In the first study, we put the WorkPad in its cradle. Subjects complained that the WorkPad was wobbly it its cradle, so for the second study, the new subjects used a WorkPad resting on a book and connected by a serial cable to the PC. There were no further comments about the positioning.

The WorkPad has a 3¼-inch diagonal LCD display screen (about 2 ¼ inches on a side) which is touch sensitive. It is 160x160 pixels.

Figure 2 shows a picture of the full WorkPad. Subjects used their fingers to touch the screen (and not the supplied stylus) as shown in Figure 1.

The software running on the WorkPad was our Pebbles Shortcutter program that allows panels of controls to be created so that each control sends specified events to the PC. The software on the PC consisted of various applications specifically created for this experiment (except in the scrolling task, which used the Netscape browser running a custom JavaScript program to collect the data).

Figure 1. Experimental setup.

Typing Test

We used a computerized typing test called "Speed Typing Test 1.0" [14]. The subjects were asked to type a paragraph displayed on the screen as quickly as possible.

Button Size Task

In this task, the PDA displayed between 4 and 16 buttons in eight different arrangements: 2 rows by 2 columns, 2x3, 3x2, 2x4, 4x2, 3x4, 4x3, and 4x4 (see Figure 2). Half of the subjects used the order shown above, and the other half used the reverse order (4x4 first down to 2x2 last). In the 2x2 condition, the buttons were about one inch square, and in the 4x4, they were about ½ inch square.

At the beginning of each condition, a picture was displayed on the PC screen showing the corresponding layout of the buttons (with the same size as the PDA). Then one of the buttons was shaded black (see Figure 2, bottom right). The subjects were asked to tap on the corresponding button on the WorkPad as quickly and accurately as possible with a finger on their left hand. The stimulus button was then cleared on the PC and the next stimulus button was shaded

500 milliseconds later. The stimuli appeared in random order. A total of 48 stimuli were used in each condition. Every button appeared the same number of times. For example, for the layout of 2 rows by 2 columns, each button appeared 12 times, while for the layout of 3 rows by 4 columns, each button appeared 4 times. There was a break after each condition. Our hypotheses were that people could more accurately select among fewer, larger buttons, and that people could make selections without looking at the WorkPad.

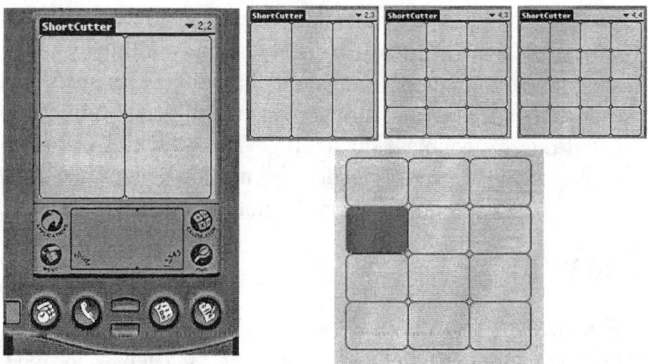

Figure 2. Left: a picture of a Palm Pilot (the WorkPad is similar) showing the 2x2 layout of buttons. Top: the screens for 2x3, 4x3 and 4x4. The other layouts are similar. Bottom: Part of the PC screen showing the stimulus during the 4x3 condition of the button task.

Homing Speed Task

The purpose of this task was to measure the times to move the hands back and forth from the keyboard to the mouse and WorkPad as the subjects switch between two-handed selection operation and two-handed typing. We compared moving a single hand to and from the keyboard to moving both hands.

There were three conditions with three trials in each. In each trial, 14 textboxes were shown on the screen with a label in front of each. The conditions were that the subjects had to first select a text box by either clicking in the field with the mouse in the usual way, tapping on a full-screen button on the WorkPad (which therefore worked like a "TAB" key and caused the cursor to jump to the next field), or tap on the WorkPad and click the mouse at the same time. In other words, the selection operation in this last condition was like a "Shift-Click" operation in which the button on the WorkPad was treated as a Shift key. After the textbox was selected, the subjects typed the word indicated on the left of the textbox. The word was either "apple" or "peach" (in alternating order). These words were chosen because they are easy to type and remember, and they start and end with keys that are under the opposite hands. The user could not exit the field until the word was typed correctly. After typing the word correctly into the textbox, the subject then continued to perform the same selection-typing operation in the next textbox. The trial ended when all 14 textboxes on the screen were filled in. There was a break after each trial. We measured the time from the mouse and WorkPad click to the first character typed, and from the last character typed to the first movement of the mouse or tap on the WorkPad. We did not count the time spent actually typing, and we eliminated the times for the first and last words, because they were biased by start-up and transients.

We hypothesized that moving to the WorkPad and the mouse would not take much longer than moving one hand since people would move both hands at the same time. We were also interested in the actual numbers for the time measurements. These might be used with future models of human performance for two-handed homing tasks.

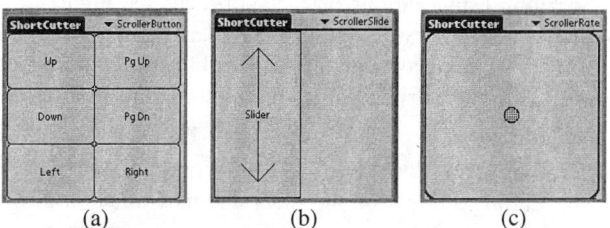

Figure 3. (a) The button scroller on the WorkPad used in the first experiment. (b) The slide scroller and (c) Rate scroller used in both experiments.

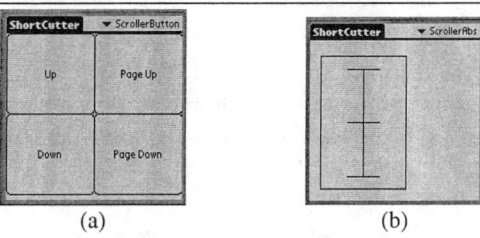

Figure 4. (a) The revised button scroller on the WorkPad used in the second experiment. (b) The "absolute scroller."

Scrolling Task

For this task, we were able to replicate the conditions of a previous experiment [15] exactly.[1] The purpose of this task was to evaluate and compare subjects' performance in scrolling web pages in a standard browser using different scrolling methods. The web pages contain text from an IBM computing terminology dictionary, and each page is about 12 screen-fulls of text. In each web page a hyperlink with the word "Next" is embedded at an unpredictable location. The subjects were asked to find the target hyperlink by scrolling the web page using the different scrolling mechanisms. Once the link was visible, they used the mouse in the conventional way to click on it. Clicking on the hyperlink brought the subject to the next web page. For each condition, the subjects first performed a practice run of 10 pages, during which they were asked to try out the scrolling method without being timed. Then, the subjects did two consecutive trials of 10 pages each as fast as they could.

The condition with the fastest time in the previous experiment used a "pointing stick" joystick to scroll, but we were

---

[1] Thanks very much to Shumin Zhai of IBM for supplying the experimental material from the earlier study.

not able to reproduce this condition.[2] The conditions we used in our first experiment were:

- Scrolling using the mouse and the regular scroll bar.
- Scrolling using a "scroll wheel" mounted in the center of the mouse (a Microsoft "IntelliMouse"). We were careful to explain to the subjects the three different ways the wheel can be used, including rolling the wheel, or tapping or holding the wheel down to go into "scroll mode" where the further you move the mouse from the tap point, the faster the text scrolls. The subjects could choose which methods to use.
- Scrolling using buttons on the WorkPad (see Figure 3a). There were 6 buttons that scrolled up and down a line, up and down a page, and left and right (which were not needed for this experiment). The buttons auto-repeated if held down.
- Scrolling using a "slider" on the WorkPad (see Figure 3b). Putting a finger on the slider and moving up or down moved the text the corresponding amount. If you reach the edge of the slider, then you need to lift your finger and re-stroke. Tapping on the slider has no effect since only relative movements are used. This is similar one of the sliders used in the early Buxton and Myers study [4].
- Scrolling using a "rate scroller," which acted like a rate-controlled joystick with three speeds (see Figure 3c). Putting a finger anywhere on the WorkPad and moving up or down started the text moving in that direction, and moving the finger further from the start point scrolled faster.

The order of the conditions was varied across subjects using a Latin square.

Revised Scrolling Task

We received a number of complaints and suggestions about the scrollers on the WorkPad in the first session, so we redesigned some of them and repeated the scrolling task in a second study with new subjects. In this study, we only used four buttons for the button scroller (since the left and right buttons were not needed—see Figure 4a). We also tried to improve the rate scroller, by adjusting the scroll speeds and the areas where each speed was in affect. Finally, we added a new (sixth) condition:

- Scrolling using an "absolute scroller," where the length of the scroller represented the entire document, so putting a finger at the top jumped to the top of the document, and the bottom represented the bottom (see Figure 4b). The user could also drag up and down to scroll continuously. Therefore, it was as if the scroll bar's indicator was attached to the finger. The motivation for this scroller was that we noticed that most people in the mouse condition

of the first session dragged the indicator of the scroll bar up and down, and we wanted to provide an equivalent WorkPad scroller. Also, this relates to the second kind of scroller used in the early Buxton and Myers study [4].

Subjects

There were 12 subjects in the first study, which took about an hour and they were paid $15 for participating. 12 different subjects did the second study, which took about ½ hour and they were paid $10. All subjects from both studies were Carnegie Mellon University students, faculty, or staff. 25% (6 out of 24) were female, and the age ranged from 19 to 46 with a median of 26. All were moderately to highly proficient with computers, and about half of the subjects had used PDAs. The data from some extra subjects were eliminated due to technical difficulties. The measures from two subjects who were left-handed are not included in the data, but informally, their numbers did not look different.

RESULTS

General

Pearson product-moment correlation coefficient between typing speed and tap speed in the button size task (namely the mean tap speed across all 8 layouts) was .60, which means the faster typists were somewhat faster at tapping. The correlation coefficient between typing speed and the speed for moving one hand from the PDA to the keyboard in the homing task was .79 which means, as expected, subjects who were better typists could put their hands in the home position more quickly. There was little correlation of typing speed to the other measures in the homing task. The correlation coefficient between typing speed and scrolling speed (in the revised scrolling task) across all 6 conditions and both trials was 0.32, which means there was little correlation for the scrolling task.

Age and gender did not affect the measures.

Button Size Task

Figure 5 shows the times to tap on the button measured from the time the stimulus appeared on the PC monitor. These numbers only include correct taps. There were two orders for the trials, so each condition was seen by some subjects early in the experiment, and by other subjects later. The chart presents the data for the early and late cases along with the average of both.

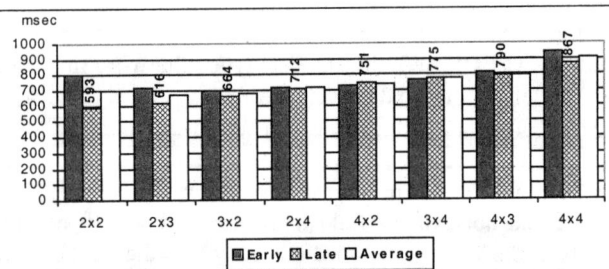

Figure 5. Times to tap each button depend on the size. The times are shown for the subjects who saw each condition later.

---

[2] We did not have a pointing stick to test, and anyway, it would have been difficult to connect one to the computers we had, which illustrates one of the claims of this paper—it can be difficult to connect multiple conventional input devices to today's computers. Since the experimental set up was identical to the original experiment [15], it should be valid to compare our times with the times reported there.

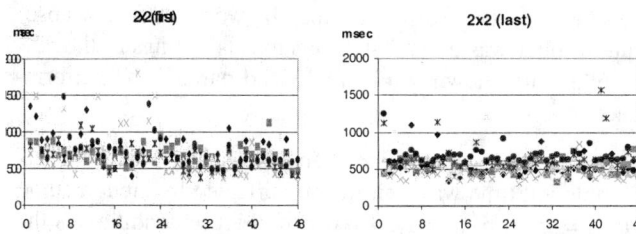

Figure 6. Plot of all times for the 2x2 layout shows (on the left) learning happening for those subjects who saw this condition first, but not (on the right) for those who saw it last.

Figure 6 shows the times to tap on a button in the 2x2 trial for each of the buttons for each of the subjects. The left graph is of those subjects who saw the 2x2 condition first, and roughly matches the power law of practice. However, for those subjects who did the 2x2 condition last, there was no apparent learning during that trial, and the times are flat. Therefore, we feel it is more valid to use the times from only the subjects who saw the condition later. The average time for just the second set is 593msec.

As shown in Figure 5, and predicted by Fitts's law [6, p. 55], the time to tap on a button is inversely proportional to the size of the button, ranging from 593 msec in the 2x2 condition to 867 msec in the 4x4 (for those the subjects who saw each condition later).

The times to tap differ significantly among different numbers of buttons ($F_{4,40}=16.8$, $p<.001$). There is significant interaction between button number and order of conditions (2x2->4x4 or 4x4->2x2) ($F_{4,40}=6.0$, $p=.001$), but the learning effect is most prominent among layouts with small number of buttons. The Tukey Test at .05 significant level indicates that there is no significant difference between the 4-button condition and the 6-button condition, between the 6-button and 8-button, or between the 8-button and 12-button. However, the 12-button condition is faster than the 16-button condition by a statistically significant margin.

The times for different layouts of the same number of buttons is not statistically significant, however: the Tukey Test at .05 significant level indicates that times for the 2×3 are not statistically different from 3×2, 2×4 compares to 4×2, and 3×4 compares to 4×3.

Figure 7 shows the error rates for the various configurations, which varies from 1.04% to 4.17% for the subjects who saw each condition later. The error rates do not differ significantly among different layouts ($F_{7,70}=1.6$, $p=.14$) nor among different numbers of buttons ($F_{4,40}=2.1$, $p=.07$). For the 4×4 layout, 45% of the errors were in the wrong row, 48% were in the wrong column, and 7% were wrong in both (on the diagonal from the correct button). There was no consistent pattern of where the problematic buttons were located (see Figure 8).

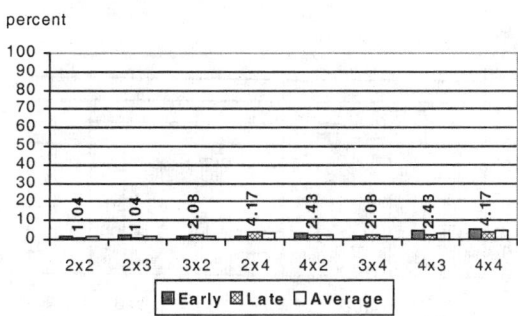

Figure 7. Error rates for each condition of the button task. Numbers shown are for the subjects who saw each later.

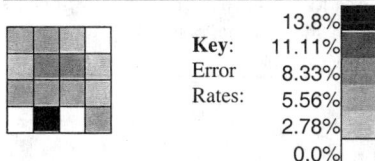

Figure 8. Percent of the taps in each button that were in error in the 4x4 layout.

Homing Speed Task

Figure 9 shows the times for moving each hand in the various conditions of the homing speed task. When moving only one hand at a time (top 4 rows), the subjects took 728 msec to move to the mouse and 701 msec to move back to the keyboard from the mouse. The times to move to the PDA were 744 msec to the PDA and 639 back.

When required to move both hands, the subjects took only slightly longer, requiring about 15% more time to acquire both the PDA and the mouse (838msec), and about 13% more time to acquire the keyboard (791 msec).

1H Keyboard->Mouse	728	
1H Keyboard->PDA	744	
1H Mouse->Keyboard	701	
1H PDA->Keyboard	639	
Keyboard -> Mouse&PDA	838	15.1%
Mouse&PDA -> Keyboard	791	12.8%

Figure 9. Times in milliseconds to move hands. "1H" means when only one hand is moving. The third column shows the percent slowdown of moving both hands compared to the corresponding one-handed mouse time.

Scrolling Task

As in the study we reproduced [15], the time for the first trial with each input device was for practice, so Figure 10 shows the times for the second and third trials.

A repeated measure variance analysis shows that subjects' completion time was significantly affected by input method ($F_{4,44} = 13.3$, $p < .001$). Trial 3 was significantly faster than Trial 2 ($F_{1,11}=17.2$, $p<.001$), showing a learning effect. However, this improvement did not alter the relative performance pattern of the input method ($F_{4,44}=.5$, $p=.73$).

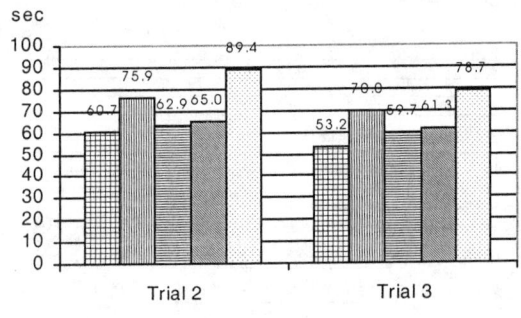

Figure 10. Times in seconds to scroll through 10 pages in trials 2 and 3 of the first version of the web page scrolling task using different input devices.

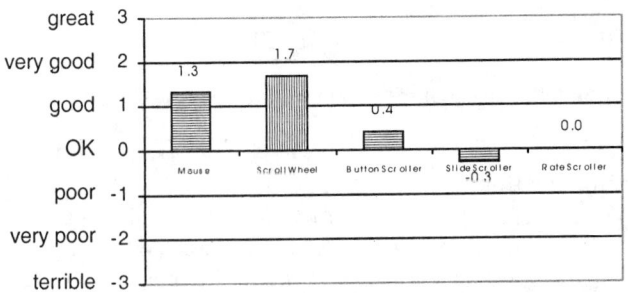

Figure 11. Ratings of the various input methods by the subjects in the first version of the scrolling experiment. We used the same scale as [15].

Taking the Mouse condition as the reference and averaging over both trials, the scroll wheel, the slide scroller, and the rate scroller conditions were 28, 11, and 48 percent slower. The Tukey Test at .05 significant level indicates that the difference between mouse and scroll wheel conditions, between the mouse and button scroller, and between the mouse and slide scroller conditions were not significant, while the difference between mouse and rate scroller conditions was significant.

Figure 11 shows the subjects' ratings of the various scrollers using a rating scale from Zhai *et. al.* [15]. Contrary to the results of that previous study, the Tukey Test at .05 significant level indicates that the difference between ratings of mouse and scroll wheel was not significant. Subjects gave the mouse a significantly higher rating than the slide scroller, while the difference between ratings of mouse and button scroller and the difference between ratings of mouse and rate scroller were not significant. The differences of ratings among the three Pebbles scrollers were not significant.

Revised Scrolling Task

We were not happy with the performance of the scrollers on the PDA, and the subjects provided useful feedback on ways to improve them. Therefore, we performed iterative design on the software, and tried the scrolling task again with 12 new subjects. Figure 12 shows that we were able to improve the performance of the new versions of the button scroller, but the rate scroller may be worse. The new absolute scroller was quite fast. The subjects' ratings of the new scrollers are shown in Figure 13 and parallel the performance.

A repeated measure variance analysis showed that subjects completion time was significantly affected by input method ($F_{5,55} = 29.3$, $p < .001$). Taking the Mouse condition as the reference and averaging over both trials, the button scroller was 8 percent faster but the Tukey Test at .05 significant level indicates that such difference is not significant. The scroll wheel, the absolute scroller, the slide scroller, and the rate scroller conditions were 31, 7, 12, and 64 percent slower than the standard mouse condition. The Tukey Test at .05 significant level indicates that the difference between mouse and scroll wheel conditions and the difference between mouse and rate scroller conditions were significant, while the differences between mouse and absolute scroller conditions and between mouse and slide scroller conditions were not significant.

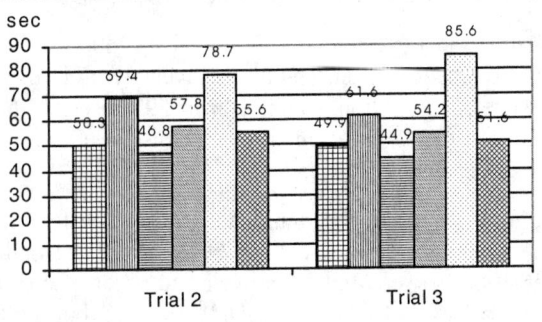

Figure 12. Times in seconds to scroll through 10 pages in trials 2 and 3 of the second version of the web page scrolling task.

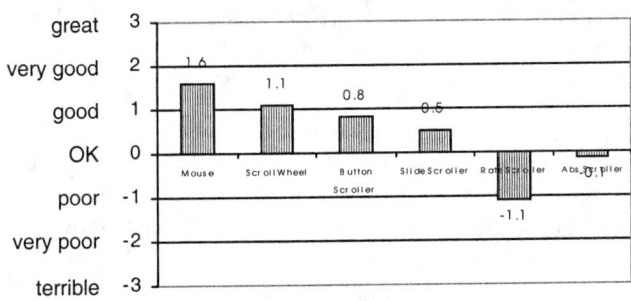

Figure 13. Ratings of the various input methods by the subjects in the second version of the scrolling experiment.

DISCUSSION

Button Size

The subjects were able to hit buttons quite accurately with their left hand, especially for small numbers of buttons. The predicted decrease in performance with decreased button size was observed. There seems to be a threshold of about 12 buttons before there is any affect due to the size.

We believe that we achieved expert performance (the learning curve flattened out) by the end of the experiment, so we tried using the times in models of expert human performance. One candidate is Fitts's law, but we do not know exactly where the subjects' fingers were when they started to move to tap. Assuming a movement of about 2 inches and a target size of 1 inch (in the 2x2 case), Fitts's law as formulated in [6, p. 55] predicts a movement time of about 150msec, compared to our measurement of 593msec. In our task, however, there is also perception and thinking time. For the smaller buttons (½ inch in the 16x16 case), Fitts's law predicts an increase in time of about 100msec, but we saw an increase of about 275msec. We observed that subjects looked back and forth from the monitor to the PDA, at an increasing rate depending on the number of buttons to choose from. Therefore, we believe the performance cannot be modeled simply as a Fitts's law task, but we were unable to find an appropriate alternative model.

Our results showing that users can tap up to 12 buttons accurately and quickly with the left hand is relevant since there are a number of applications where having several buttons on the PDA would be useful. We have created a tool called "Shortcutter" which allows allows the user to design screens of keyboard shortcuts, macros, and strings on the PDA, and use these to control any PC application. The buttons can be big enough to hit with a finger, or tiny so that many will fit on a screen. The Shortcutter[3] can provide customizable interfaces on the PDA even for applications that do not have a customization facility on the PC. Some examples uses include scrolling with buttons (Figure 3a and Figure 4a), and controlling a compiler, playing music on the PC, reading mail, etc.

Homing Times

Our one-handed homing time to move from the mouse to the keyboard (701 msec – see Figure 9) is longer than the time to move from the PDA to the keyboard (639 msec). This may be because the physical distance to the mouse from the home position on the keyboard is longer (14 inches compared to 7 inches) due to the number-pad and arrow keys sections of the PC keyboard. In the other direction, the increased time to acquire the PDA may be due to the unfamiliarity of homing to this kind of device.

In the classic study of text selection devices [6, p. 237], the homing time to move from the space bar to the mouse was measured as 0.36 seconds. This was measured from videotapes of subjects moving. An average homing time of 0.4 seconds was incorporated into the Keystroke Level Model [5]. However, we measured one-handed homing times of around 0.7 seconds, which is substantially longer. Our time was measured from the time of the mouse click to the time that the first keystroke was recognized. Our typing test shows that the average time per keystroke was 0.3 seconds, so this might be subtracted from our measured time to get the predicted 0.4 seconds.

---

[3] Shortcutter and the other Pebbles applications are available from http://www.cs.cmu.edu/~pebbles.

An important observation is that, as predicted, subjects moved both hands simultaneously, and this did not penalize the movement time much. The sum of the one-handed times to move from mouse and PDA to the keyboard is 1340msec (701+639). This is much larger than the time to move from both mouse and PDA to the keyboard in the two handed case which is 791 msec (1340msec is 69% larger). A similar relationship holds for the movement from the keyboard to the PDA and mouse (728+7443=1473 > 838; 76% larger).

Overall, it takes only about 15% longer to acquire both the mouse and the PDA than just to acquire the mouse, and it takes only about 13% longer to get back to the keyboard from both devices than from just the mouse.

We were not able to find any prior studies of the time to acquire two devices at the same time. Most studies of two-handed use of input devices (including our button-size and scrolling tasks) allow the subjects to stay homed on the devices. We found that moving both hands slowed down each hand a little, but there was substantial parallel movement. Realistic tasks are likely to include a mix of keyboard and other input device use, so homing issues may be important.

Scrolling

Our measured times for scrolling the web pages with the mouse (about 60 seconds) is a little faster than the time reported in [15], and in the revised web task, the time for scrolling with the button scroller is 45.9 sec (average of trial 2 and trial 3) which is faster than the time reported in [15] for scrolling with the in-keyboard isometric joystick (around 50 sec). This shows that using the PDA can match or beat the speed of other non-dominant hand devices.

An interesting comparison is between their joystick, our rate scroller (Figure 3c) and the scroll wheel used in its most popular manner as a rate-controlled scroller. All provide the same rate-controlled style of scrolling, but they have significantly different performances and ratings by users. Our attempt to improve the rate scroller obviously did not help, showing that further work is needed to make this scrolling method effective. We observed that the fast speed was much too fast, but the medium speed was too slow. The popularity of the scroll wheel and the success of the pointing stick give us reason to keep trying. Furthermore, IBM did significant experimentation and adjustments before the pointing stick had acceptable performance [13]. Therefore, an important conclusion from the scrolling experiment is that the specific design and pragmatics of the input methods has a very important influence on the performance.

Another interesting result is that our subjects quite liked the scroll wheel (average rating of 1.7 ≈ very good), whereas in the earlier study it was rated much worse (-1 ≈ poor) [15]. This may be due to the increased experience people have with a scroll wheel (many of our subjects have a scroll wheel on their own mouse), and because most of our subjects used it in its rate-controlled joystick mode, whereas most of the earlier study's subjects used the rolling mode.

An interesting observation about this Web scrolling task in general is that it primarily tests scrolling while searching for information, so the scrolling must go slow enough so the subjects can see the content go by. This is why the methods that provided the best control over the speed are preferred. The low rating of the rate scroller on the PDA is because the fastest speed was much too fast to see the text go by, and the medium and slow speeds were rated as too slow. However, other scrolling tasks, such as those tested by [4], require the user to go to a known place in the document, and then a method that can move long distances very quickly may be desirable.

In the future, other kinds of scrollers can be created, that might combine the various modes. For example, it might interpret a tap as a button for line scroll, a "flick" as a page flip, and a drag as a slider. An advantage of using the PDA is that all these are possible, and visible feedback and prompting can be provided.

CONCLUSIONS

Many studies have shown the effectiveness of two-handed input to computers in certain tasks. One hindrance to two-handed applications has been that there may be only a few tasks in which using both hands is beneficial, and the benefits are relatively minor. Another problem is that although it is interesting to study custom devices for use by the non-dominant hand, in order for there to be wide-scale use, it is better to provide mechanisms that users can easily get and configure. Since increasing numbers of people have PDAs that are easy to connect to PCs, it makes sense to see if PDAs can be used effectively in the non-dominant hand. The research presented here shows that PDAs can be used as buttons and scrollers, and that the time to move to two devices is only slightly longer than for one. Our study of one application shows that at least for the scrolling task, a PDA can match or beat other 1-handed and 2-handed techniques. Because there is no incremental cost for the PDA since users already own it, and since the PDA is connected to the PC anyway, even small efficiencies many be sufficient to motivate its use as a device for the non-dominant hand. Our studies and many others have emphasized the importance of the pragmatics and the exact behavior of controls. Because the PDA can be programmed with a variety of controls with various properties, further research is required to determine the most effective ways for a PDA to be used to control the PC in both the dominant and non-dominant hand.

ACKNOWLEDGMENTS

For help with this paper, we would like to thank Rob Miller, Bernita Myers and Shumin Zhai.

The research reported here is supported by grants from DARPA, Microsoft, IBM and 3Com. This research was performed in part in connection with Contract number DAAD17-99-C-0061 with the U.S. Army Research Laboratory. The views and conclusions contained in this document are those of the authors and should not be interpreted as presenting the official policies or position, either expressed or implied, of the U.S. Army Research Laboratory or the U.S. Government unless so designated by other authorized documents. Citation of manufacturer's or trade names does not constitute an official endorsement or approval of the use thereof.

REFERENCES

1. Balakrishnan, R. and Kurtenbach, G. "Exploring Bimanual Camera Control and Object Manipulation in 3D Graphics Interfaces," in *Proceedings SIGCHI'99: Human Factors in Computing Systems*. 1999. Pittsburgh, PA: pp. 56-63.

2. Balakrishnan, R. and Patel, P. "The PadMouse: Facilitating Selection and Spatial Positioning for the Non-Dominant Hand," in *Proceedings SIGCHI'98: Human Factors in Computing Systems*. 1998. Los Angeles, CA: pp. 9-16.

3. Buxton, W., "Lexical and Pragmatic Considerations of Input Structures." *Computer Graphics*, 1983. **17**(1): pp. 31-37.

4. Buxton, W. and Myers, B. "A Study in Two-Handed Input," in *Proceedings SIGCHI'86: Human Factors in Computing Systems*. 1986. Boston, MA: pp. 321-326.

5. Card, S.K., Moran, T.P., and Newell, A., "The Keystroke-Level Model for User Performance Time with Interactive Systems." *Communications of the ACM*, 1980. **23**(7): pp. 396-410. July.

6. Card, S.K., Moran, T.P., and Newell, A., *The Psychology of Human-Computer Interaction*. 1983, Hillsdale, NJ: Lawrence Erlbaum Associates.

7. Guiard, Y., "Asymmetric Divison of Labor in Human Skilled Bimanual Action: The Kinematic Chain as a Model." *Journal of Motor Behavior*, 1987. **19**(4): pp. 486-517.

8. Hinckley, K., *et al.*, "Two-Handed Virtual Manipulation." *ACM Transactions on Computer Human Interaction*, 1998. **5**(3): pp. 260-302. Sept.

9. Jacob, R.J.K., *et al.*, "Integrality and Separability of Input Devices." *ACM Transactions on Computer-Human Interaction*, 1994. **1**(1): pp. 3-26.

10. Kurtenbach, G., *et al.* "The Design of a GUI Paradigm based on Tablets, Two-hands, and Transparency," in *Proceedings, CHI'97: Human Factors in Computing Systems*. 1997. Atlanta, GA: ACM. pp. 35-42.

11. Myers, B.A., Stiel, H., and Gargiulo, R. "Collaboration Using Multiple PDAs Connected to a PC," in *Proceedings CSCW'98: ACM Conference on Computer-Supported Cooperative Work*. 1998. Seattle, WA: pp. 285-294.

12. Rekimoto, J. "A Multiple Device Approach for Supporting Whiteboard-based Interactions," in *Proceedings SIGCHI'98: Human Factors in Computing Systems*. 1998. Los Angeles, CA: pp. 344-351.

13. Rutledge, J. and Selker, T. "In-Keyboard Analog Pointing Device: A Case for the Pointing Stick," in *Technical Video Program of the CHI'90 Conference*. 1990. SIGGRAPH Video Review, Issue 55, No. 1.

14. TestedOK Software, "Speed Typing Test 1.0. Available from http://hometown.aol.com/tokfiles/typetest.html," 1999.

15. Zhai, S., Smith, B.A., and Selker, T. "Improving Browsing Performance: A Study of Four Input Devices for Scrolling and Pointing," in *Proceedings of Interact97: The Sixth IFIP Conference on Human-Computer Interaction*. 1997. Sydney, Australia: pp. 286-292.

The Effects of Animated Characters on Anxiety, Task Performance, and Evaluations of User Interfaces

Raoul Rickenberg and Byron Reeves
Department of Communication
Stanford University
Stanford, CA 94305 USA
+1 650 725 3033
raoul@leland.stanford.edu
reeves@leland.stanford.edu

ABSTRACT
Animated characters are common in user interfaces, but important questions remain about whether characters work in all situations and for all users. This experiment tested the effects of different character presentations on user anxiety, task performance, and subjective evaluations of two commerce websites. There were three character conditions (no character, a character that ignored the user, and a character that closely monitored work on the website). Users were separated into two groups that had different attitudes about accepting help from others: people with control orientations that were *external* (users thought that other people controlled their success) and those with *internal* orientations (users thought they were in control). Results showed that the effects of monitoring and individual differences in thoughts about control worked as they do in real life. Users felt more anxious when characters monitored their website work and this effect was strongest for users with an external control orientation. Monitoring characters also decreased task performance, but increased trust in website content. Results are discussed in terms of design considerations that maximize the positive influence of animated agents.

Keywords
Animated characters, social agents, social facilitation, locus of control.

INTRODUCTION
The history of ideas about animated characters in human-computer interaction is turning a corner. Initial debates concerned the presence of any character performing any kind of behavior. The questions were whether animated characters—as a general concept in interfaces—were good or bad, useful or useless. These debates rarely yielded an answer more satisfying than—"it depends." As has been the case with the introduction of all new media in the 20th century, the initial debate was framed too aggressively to lead to useful answers. The most accurate summary about the impact of all media—film, radio, television, and the internet—is that some effects occur for some people, in some conditions, and for some types of content. The trick in research is to find out which effects, which people, which situations, and which content.

An elaboration of the conditions for animated characters to succeed is underway. There are several new studies that demonstrate the potential for animated characters to automate social interactions in ways that make computing more pleasing, productive, and easy. Research has focused on, for example, character appearance [19, 15], non-verbal behavior [6, 17], personality [12, 31], emotion [3, 4], and speech characteristics [20]. This research is important given the increasing use of animated characters in products and services ranging from search engines to shopping "bots" to virtual employees in commerce transactions.

This experiment tested two new ideas about animated characters that further elaborate our understanding of when and how they affect human-computer interaction. First, we examined whether animated characters have enough social presence—even if virtual—to make users feel that they are being monitored. Second, we examined whether people who respond strongly to the presence of others (because they hold a general belief that others control their destiny) respond differently to animated characters than people who are less affected by the presence of others.

Our ideas about how people would respond to the presence of animated characters came from two significant literatures in psychology. The literature on social facilitation describes how people respond to the presence of others while they work, and how this response is related to the perception of being monitored. The literature on locus of control addresses how individual differences in people's thoughts about personal control affect their reactions to the presence of others. We will briefly review each of these literatures and describe how we used them to study the effects of animated characters on commerce websites.

Social facilitation
Sometimes it's nice to have company—a real person—when you work. Imagine, however, that you're working on a hard problem. Someone enters the room,

takes a seat across from you, and starts to thumb slowly through a magazine. Would the mere presence of this person make you anxious or affect your ability to complete the task at hand? What if the person walked over and stood behind your shoulder to get a better look at what you're doing?

Research on social influence provides answers to these questions. There is strong evidence that the mere presence of another person increases anxiety and lowers levels of performance on complex tasks [33, 9]. If the other person is in a position to evaluate performance, such "social facilitation" effects are strengthened [7, 23].

One explanation for how social facilitation works is Zajonc's [33] drive theory. Zajonc thought that the presence of other people creates a state of increased arousal or generalized drive. The drive produced by the presence of others is an alertness for the unexpected, a preparation to respond to the actions of others. Zajonc thought that drive could be generated by the presence of anyone who has the potential to be active, regardless of the other persons ability to evaluate, reward, or punish.

Cottrell [7] argued, however, that the mere presence of other people is not sufficient, in and of itself, to increase drive. Rather, it is the anticipation of positive or negative outcomes that are associated with the presence of others that cause heightened arousal. Cottrell [8] tested this idea in an experiment where people either completed a task alone, under the attentive gaze of two spectators, or in the presence of two people who were blindfolded. As predicted, the condition in which there were spectators decreased performance, but there was no effect on performance when people were alone or in the presence of people who couldn't see their work [8].

While there is some debate about whether the mere presence of others is sufficient to increase arousal and diminish performance, the conclusion from this literature is clear in those cases when a social actor communicates an intention to monitor someone's work. When the monitoring is obvious, thoughts and behavior change. There is more anxiety and less accurate performance of complex tasks.

Locus of control

Now imagine two different people being monitored while they work on a complex task. The first person believes that she controls her own destiny and that other people have little to do with whether she fails or succeeds. The second person is convinced, however, that he is at the mercy of forces which he doesn't control and that his success depends on the help of others. Would you expect these people to respond differently when someone watches them work?

The answer from research is that you should expect differences. Being monitored is less worrisome for people who believe that they control their own destiny than for those who think that their destiny is in the hands of others. In other words, it depends upon the person's "locus of control" [27, 25, 9]. People tend toward either an *internal* or an *external* locus of control [27]. Those who have an internal locus of control are inclined to believe that rewards are contingent upon their own behavior, whereas those with an external locus of control tend to believe that their fate is either in the hands of others or a product of chance [27, 16, 18].

Several studies indicate that people with an external locus of control have a greater tendency to be influenced by social stimuli in their environment, and to modify their behavior in accordance with the responses and evaluations of others, than people with an internal locus of control [9, 16, 30]. This effect is particularly strong when people perform tasks that are complex or unfamiliar, such as novel math problems. When the locus of control is internal, an audience is no bother; when the control orientation is external, performance suffers [2, 25].

Social reactions to animated characters

Now, keeping in mind the relationship between monitoring and locus of control discussed above, substitute an animated character for the presence of another person. The character is either reading idly in a corner of your computer screen or monitoring your every move as you work on a complex web task. Would your reactions to the animated character be similar to a real person?

There is good reason, from two sources, to assume that this would be the case. First, several studies have found that monitoring can have social effects when electronic equipment is substituted for the presence of real people. In these studies, there is no explicit social actor present, but the results of the research still hold.

Aiello and Svec [2] reported the first empirical demonstration of social facilitation effects in a context that involved electronic monitoring. They showed that complex task performance was impaired for people who are monitored electronically just as for those who were monitored in person. They also showed that people with an external locus of control were more anxious about their performance than those with an internal locus of control, regardless of whether they were monitored interpersonally or electronically, but less anxious than those with an internal locus of control when no monitoring was involved.

Other experiments about electronic monitoring support and extend the conclusions of Aiello and Svec [2]. Kolb and Aiello [1] reported that electronic monitoring decreased task performance and increased anxiety, but that the effects on anxiety can be lessened if people think that they are members of a cohesive workgroup. Stanton and Barnes-Farrell [29] found the same negative effects of electronic monitoring, but also found that the effects can be averted if people perceive that they have the ability to prevent or delay the monitoring—even if they do not exercise this control.

The fact that electronic monitoring often involves evaluation by another person, even if they are not physically present, may signal caution in applying these studies to animated characters. There is a significant body of research, however, that suggests that the psychology of human-human relationships can be applied directly to virtual social actors and their interactions with users. This is not to suggest that people would confuse animated characters with real people, but simply that interactions with such characters may inherently trigger responses that have been well-rehearsed during a lifetime of social relationships.

There is now a substantial body of evidence that adults regularly respond to technology such as computers in a social manner [26, 21]. People develop affiliations with computer "teammates" in a similar manner to the way in which they develop group affiliations with humans [22]. Likewise, people respond to praise and criticism from computers as they respond to these assessments from humans [24, 10], and people are courteous when critiquing computers even though they "know" that computers have no feelings to be hurt [21]. Such interactions with computers are not deliberate, but instead mindless and automatic. Animated characters elicit similar responses.

People confer human personalities upon the simplest of animated characters [14, 26]. Rather than seizing on the differences between such characters and humans—a process that requires thought or deliberation—people slip into social conventions because important features of interactions with animated characters mimic real life. And the more that animations look and act like humans, the stronger this anthropomorphic tendency [23, 11].

Our hypotheses about responses to animated characters follow directly from the psychological literature we reviewed. We expected that the mere presence of such characters would increase the anxiety and decrease the performance of users working on complex tasks. Furthermore, we expected that these effects would be heightened if the animated characters displayed monitoring behavior. We also expected that these effects would be moderated by the locus of control of users. Those with an external locus of control were expected to react more strongly to the animated characters than users with an internal locus of control. In addition to these expectations, we also assessed subjective responses to the context (websites) in which interactions with the animated characters took place. While no previous work has extended social facilitation effects to evaluations of context, ways in which the likability, ease of use, and trustworthiness of websites are affected by animated characters are important to understand—especially in regard to the commercial contexts used in this experiment.

EXPERIMENTAL METHODS

Subjects. Eighty-four people participated in the experiment (60% male and 40% female). An additional 20 people were used to pretest stimulus materials. All subjects were either undergraduate or graduate students recruited at Stanford University. All were experienced computer users (i.e., they knew how to word-process and manage a UNIX email account).

Experimental Design. The experiment was a between-subjects, full-factorial two-by-three design. The two factors were (1) the subjects' locus of control and (2) the monitoring activity of an animated character.

Locus of control was either Internal or External. Forty-two subjects with an internal locus of control and 42 subjects with an external locus of control were chosen from a pool of 159 potential subjects on the basis of a pretest. The monitoring factor consisted of three levels: No Character, Idle Character, and Monitoring Character. Subjects in the No-Character condition completed a series of computer tasks with no animated character present. Subjects in the Idle-Character condition completed the identical tasks, but with an animated character present in the lower right-hand corner of computer screen. This character never made "eye contact" with subjects and appeared to ignore all activity as the tasks were completed. Subjects in the Monitoring-Character condition also had an animated character present on their computer screen as they completed identical tasks, but in this condition the character appeared to look at both the user and at the webpages that users were working on. The character in the Monitoring condition also periodically took photographs of these webpages and took notes on a pad when users submitted information regarding their tasks.

Locus of Control Pretest. Rotter's [27] Locus of Control Scale was used to determine the internal versus external orientations of potential subjects. This instrument consists of 23 forced-choice items that each present a pair of statements. In each pair, one statement expresses an internal viewpoint and the other an external viewpoint. Respondents completed the scale by indicating which of the two statements they agreed with most. The following pair of statements is typical items on the scale:

(a) *Sometimes I can't understand how teachers arrive at the grades they give.*
(b) *There is a direct connection between how hard I study and the grades I get.*

Scores on this scale can range from 0, indicating that no external statements are endorsed, to 23, indicating that all external statements are endorsed. The mean score on the pretest was 13.27 (SD = 3.8). Only the 42 subjects that scored lowest (internal) and highest (external) on the pretest were selected to participate. A two-tailed t-test indicated that the scores on the Locus of Control scale for these two groups was significantly different ($t(82)$ = 18.47, $p <$ .001).

Stimulus Material. The stimulus material consisted of two primary components: the animated characters that comprised the distinction between the Idle-Character and Monitoring-Character conditions and the web-based tasks that all subjects completed during the experiment.

The animated characters were specially developed for this experiment using Microsoft Agent software. The characters used in both the Idle-Character and the Monitoring-Character conditions were based upon Microsoft's "Genius" animations, so their physical features were identical (see Figure 1 for an illustration). The characters were approximately 1.5" tall (1152 x 870 resolution) and appeared in the lower-left corner of the Microsoft Explorer 4.5 browser that people used to view the web pages.

The only distinctions between the idle and monitoring characters lay in the repertoires of their movements and the degree to which these movements were contingent upon the subjects' behavior on the websites. The idle character periodically stretched, scratched his head, and rubbed his eyes, but generally appeared to be preoccupied by reading a book. None of this character's movements or gestures were contingent on the subject's behavior. This was meant to strengthen the perception that it was not paying attention to the subject.

The monitoring character appeared to be watching the subject and would glance at webpages on which the subject was working when the cursor was moved over information pertinent to the tasks. As mentioned above, this character also appeared to take photographs and notes as the user entered information.

Figure 1: Examples of the Monitoring Character (Top Row) and Idle Character (Bottom Row).

Two different websites were created for the experiment. The goal was to provide two prototypical web transactions that required users to solve problems. One of the websites was based upon the Charles Schwab website and the other was based upon the Dell Computer website. Both of these sites were served locally in order to ensure consistent download times as well as track user performance. Subjects viewed the websites above a frame that contained instructions for each task as well as a text field used to submit solutions to the tasks.

The tasks assigned to users included finding specific information, comparing products, and completing forms used to personalize products. Because task difficulty has been identified as a variable that moderates social facilitation [7, 8], it was important to ensure that the tasks were difficult. It was also desirable to balance the level of task difficulty across the two websites in order to facilitate comparisons. This was accomplished by pretesting ten different tasks (with a separate sample of people) and selecting six that were of comparable difficulty (three for each website).

Twenty subjects participated in the task-difficulty pretest. People were given a paper questionnaire that described five tasks on the Dell Computer website and five tasks on the Charles Schwab website, and that contained items to rate the difficulty of these tasks. After finishing each task, subjects completed five 10-point Likert scales that assessed: complexity, confusion, ease, effort, and thought required for the previous task. A Difficulty index was created from these five items ($alpha = .95$) and used to rate the ten tasks. Three tasks on each website were retained for the primary experiment on the basis of this task-difficulty index. A two-tailed t-test showed that the three tasks retained for the Dell Computer website and the three tasks retained for the Charles Schwab website were not significantly different in terms of difficulty ($t(19) = .59$, p = N.S.).

Apparatus. The computers used in the primary experiment were identical Hewlett Packard 440 mhz Kayak XWs with 21" color monitors (1152 x 870 resolution). These computers also had identical keyboards and mice, but were located in different experimental labs. The two labs were similar in terms of size and furnishings and use of the two labs was balanced across conditions.

Procedure. After arriving at a prescheduled time, people were brought to one of the labs in which the experiments were run and given a questionnaire. An experimenter then read an introductory script that was identical for all subjects before leaving the room.

People read specific instructions for each task in a frame that appeared at the bottom of their web-browser. When they found a solution to a task, they keyed their response into a text field in the frame and selected a *Submit* button. This brought up the instructions for the next task.

Half of the subjects worked on the Charles Schwab website and half worked on the Dell Computer website. Before they answered questions about either site, they were asked to view the other site. This was done to give each user experience with their character condition so that they would be introduced to the level of monitoring that would occur when the experiment began.

Anxiety Measure. Anxiety was assessed using a modified version of Speilberger, Gorsuch, & Lushene's [28] State Anxiety Scale. The items were answered on four-point Likert scales on a paper questionnaire. The following items exemplify those that appear on the questionnaire: *I felt calm; I felt secure; I felt strained;* etc.

Performance Measure. Performance was measured by adding the number or tasks completed correctly. The computers used in the experiment recorded people's answers for each task. Performance was calculated on the basis of logfiles compiled by these computers. Tasks included comparing the performance of various mutual funds, configuring computer hardware, etc.

Website Evaluations. Subjective evaluations of the two websites were measured with a sixteen-item questionnaire. Subjects were asked to note their degree of agreement with each of the statements on ten-point Likert scales. Factor analysis was used to build three indices from the total set of items. The first, *Likability*, accounted for 40.4 percent of the variance and included items that assessed levels of enjoyment, fun, and boredom as well as willingness to recommend, expected future use, and likelihood of making purchases. The second factor, *Ease of Use*, accounted for 13.7 percent of the variance and included items that assessed levels of confusion and frustration as well as perceptions concerning the ease of making mistakes, the level of user control, and the quality of the design. The final factor, *Trustworthiness*, accounted for 10.7 percent of the variance and included items concerning objectivity and the degree to which content was opinioned.

Manipulation check. The manipulation check was an index of three items answered by all people who were in one of the conditions that involved characters. People were asked whether the character seemed to be *watching* them, whether the character seemed to *record* their answers, and whether the character seemed to be *judging* them. These items were used to form a Monitoring Index that had a Cronbach's *alpha* of .70.

The manipulation was successful. A planned one-tailed *t*-test on the Monitoring Index showed that subjects in the Monitoring-Character condition reported a higher level of monitoring than subjects in the Idle-Character condition ($t(52) = 4.49$, $p < .001$).

RESULTS

Full factorial ANOVAs were performed on all measures. A summary of these ANOVAs appears in Table 1. The planned tests of all hypotheses are discussed in detail below, as are results pertaining to the relationship of animated characters and locus of control to the evaluation of the websites.

Table 1: Summary of ANOVAs for Dependent Variables

	Independent Variables		
	Character	Locus of Control	Interaction
Anxiety	10.85***	17.09***	3.12*
Accuracy	6.83**	.86	.50
Likability	.16	5.21*	.62
Ease of Use	.76	.05	.68
Trustworthy	3.81*	4.96*	2.42†

*$p < .05$ **$p < .01$ ***$p < .001$ †$p = .10$

Note: All F values for Locus of Control have degrees of freedom $F(1, 83)$, all others have $F(2, 83)$.

Anxiety

As can be seen in Table 1, levels of anxiety differed across the character manipulation ($F(2, 83) = 10.85$, $p < .001$). The mere presence of a character generated more anxiety than no character, but the most anxiety was caused by the monitoring character. Levels of anxiety also varied across the locus of control manipulation ($F(1, 83) = 17.09$, $p < .001$). The most important part of these results, however, is the interaction of the character and locus of control conditions ($F(1, 82) = 3.12$, $p < .05$). The level of anxiety was highest for external subjects who were monitored and lowest for internal subjects who did not see a character.

Figure 2 shows the results for anxiety. One-tailed, *a priori* contrasts showed that people were more anxious when an idle character was present than when no character was present ($t(72) = 1.4$, $p = .08$). Also, people were more anxious if an animated character appeared to monitor them than if no character was present ($t(72) = 4.6$, $p < .001$). And people were more anxious if an animated character monitored them than if an idle character was present ($t(72) = 4.6$, $p < .001$). Users with an external locus of control were also more anxious when monitored by an animated character than were users with an internal locus of control ($t(72) = 4.6$, $p < .001$).

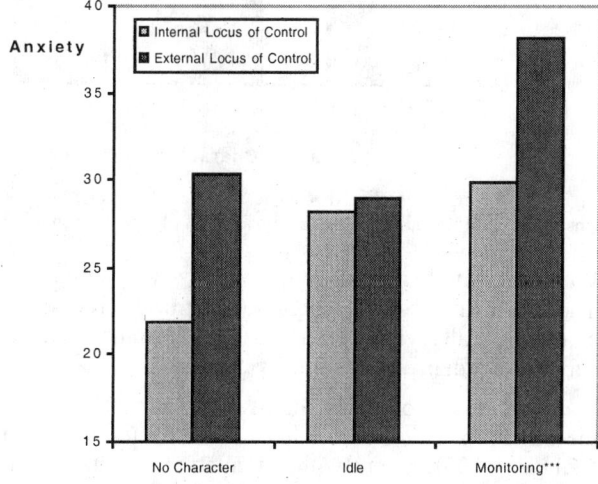

Figure 2: Level of Anxiety in the No Character, Idle Character, Monitoring Character, and Internal as well as External Locus of Control Conditions

Performance

The character manipulation had a significant main effect on the accuracy with which subjects performed tasks ($F(2, 83) = 6.83$, $p < .01$), but as can be seen in Table 1, there were no differences for locus of control nor an interaction effect.

The means for each condition are shown in Figure 3. A *priori* contrasts showed that users completed fewer tasks accurately when they were monitored by an animated character than when no character was present ($t(72) = 4.6$, $p < .01$) or when an "idle" character was present ($t(72) = 3.7$, $p < .001$).

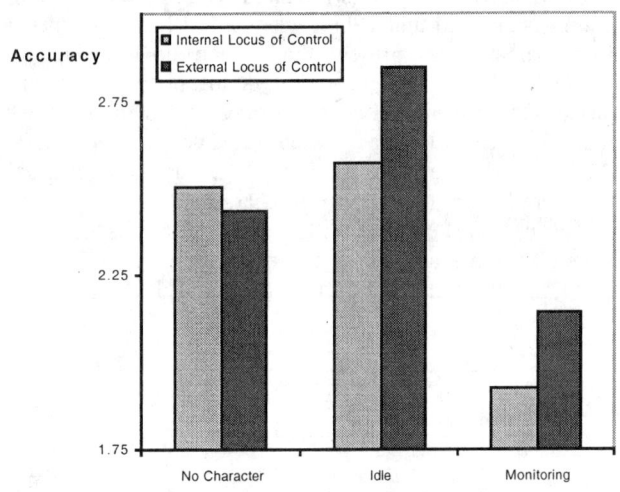

Figure 3: Accuracy in the No Character, Idle Character, Monitoring Character, and Internal as well as External Locus of Control Conditions

Website Evaluations

There were three separate indices (constructed from a factor analysis of all evaluation items) that summarized user's subjective judgments about the websites.

Likability. Locus of control had a significant main effect on the degree to which people liked the websites ($F(1, 83) = 5.21$, $p < .05$). Users with an internal orientation (i.e., those that thought they controlled their own success) liked the websites more than those with an external orientation, especially when there was no character present (Figure 4).

The character manipulation had no significant main effect on liking, and it did not interact with locus of control.

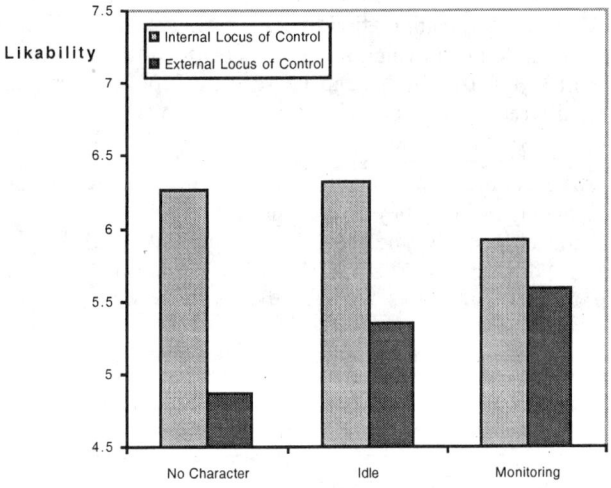

Figure 4: Likability of Website in the No Character, Idle Character, Monitoring Character, and Internal as well as External Locus of Control Conditions

Ease of Use. The evaluations for ease of use showed little change across the experimental conditions (Figure 5). No statistically significant differences were found, either for character or locus of control.

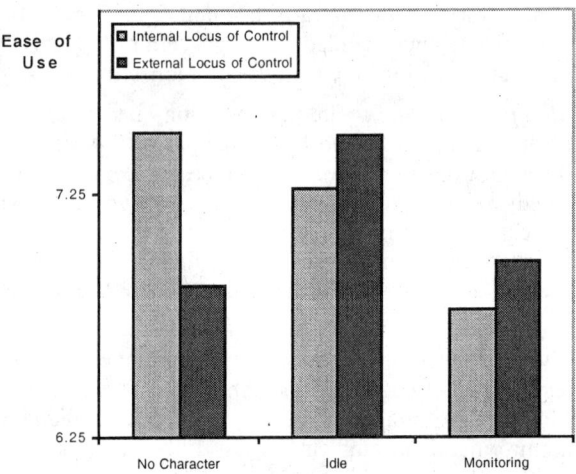

Figure 5: Ease of Use of Website in the No Character, Idle Character, Monitoring Character, and Internal as well as External Locus of Control Conditions

Trustworthiness. The character manipulation had a significant main effect on judgments of trustworthiness ($F(2, 83) = 3.81$, $p < .05$) such that people in the monitoring condition trusted the websites the most, and people who saw no character trusted the website the least (Figure 6). *Post hoc* contrasts showed that the mean level of trust in the Monitoring condition was significantly higher than that in the No-Character condition ($t(83) = 1.2$, $p < .05$).

Locus of Control also affected trustworthiness ($F(1, 83) = 4.96$, $p < .05$) such that people with an internal orientation trusted the websites more than externally oriented people.

The major result, however, is the interaction of character presence and locus of control ($F(2, 83) = 2.42$, $p < .1$), which suggests that internals trusted the websites more than externals, but only in the No-Character and Monitoring-Character conditions.

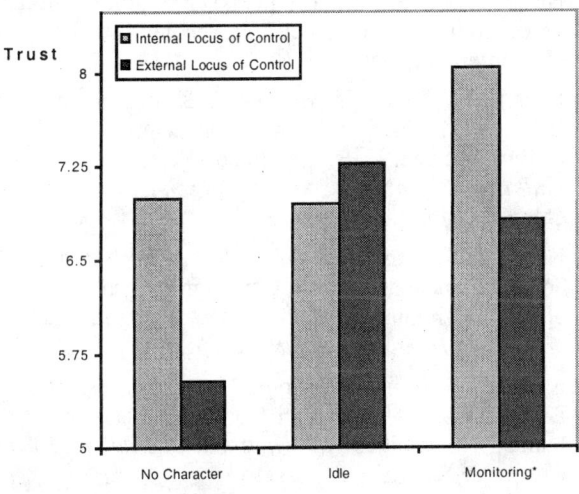

Figure 6: Trustworthiness of Website in No Character, Idle Character, Monitoring Character, and Internal as well as External Locus of Control Conditions

DISCUSSION

The perception of being monitored by an animated character has the same effects on Anxiety and Performance as being monitored by a human, either electronically or in person. When a character watches, users are more likely to feel anxious about their work and to perform less well. This anxiety is most pronounced among users who think that other people control their success.

At the most general level, these results suggests that decisions concerning the use of animated characters should address the details of execution and social presentation. It is not sufficient–for celebration or condemnation–to focus on whether or not an animated character is present. Rather, the ultimate evaluation is similar to those for real people–it depends on what the character does, what it says, and how it presents itself. The effects of animated characters are not unilaterally good or bad; they can be either or both. Using an animated character turns up the volume on social presence, which means that it can accentuate the effects of everything presented.

The possible relationship between anxiety and positive outcomes (in this study increased trustworthiness) is an interesting case in point. Anxiety should not be considered only as a negative response. Anxiety is arousal, the engine of many things psychological. Arousal can determine where we focus our attention as well as what we remember [13]. Highly arousing things can be good or bad (e.g. sexual arousal vs. arousal from witnessing gory surgery), so it should be considered independently of valance.

The conclusion is that *some* arousal in an interface may be useful, a finding consistent with the best preparation methods for exams in school. You don't want to be bored or aroused to the point of distraction–just aroused enough that you pay attention and remember. Characters in interfaces may help designers reach this middle ground. If their presence is executed well, they can increase interest (and a little anxiety) in ways that enhance desirable social responses. In the specific case of trustworthiness, it is easy to imagine, for example, that a financial advisor looking over your shoulder in real life might increase your level of anxiety about money and, thereby, raise your evaluation of the value of his or her advise.

The other significant effect in this study came from the assessment of locus of control. This concept is a new addition to interface research that may help separate those who like and don't like animated characters. There is significant disagreement about the value of animated characters in interfaces, especially in commentary and reviews. Some of these differences in opinion might be explained by internal versus external orientations. Interface designers may want to avoid using animated characters when they know that they are designing for people with an internal locus of control (i.e., when users are confident that they can complete work on their own). The addition of characters may make interactions more robust, however, when users perceive that they lack control over their success. The important thing to remember is that such decisions concerning the use of animated characters should be based on users' (relatively stable) traits regarding perceptions of control rather than on fleeting reactions to a task at hand.

These and other insights point to the value of conceptualizing interaction with computers in terms of interpersonal interaction. One way for designers to develop characters is to reflect on the nature of relationships between people. This study shows that some of the underlying dynamics of such relationships (perceptions that others are paying attention and feelings regarding control) also affect reactions to animated characters. The important conclusion is that these dynamics can be useful when applied to interactions with computers. Having a social actor look over your shoulder–animated or real–is cause for notice.

REFERENCES

1. Aiello, J.R. & Kolb, K.J. (1995). Electronic performance monitoring: A risk factor for workplace stress. In S. Sauter & L. Murphy (Eds.), *Organizational Risk Factors for Job Stress.* Washington, DC: American Psychological Association.

2. Aiello, J.R. & Svec, C.M. (1993). Computer monitoring of work performance: Extending the social facilitation framework to electronic presence. *Journal of Applied Psychology, 23(7).*

3. Ball, G. & Breese, J. (1998). Emotion and personality in a conversational character. *Proceedings of the 1998 Workshop on Embodied Conversational Characters.*

4. Becheiraz, P. & Thalmann, D. (1998). A behavioral animation system for autonomous actors personified by emotions. *Proceedings of the 1998 Workshop on Embodied Conversational Characters.*

5. Bond, C.F. & Titus, L.J. (1983). Social facilitation; A meta-analysis of 241 studies. *Psychological Bulletin, 94(2).*

6. Cassell, J. & Thórisson, K.R. (1999). The power of a nod and a glance: Envelope *vs.* emotional feedback in animated conversational agents. *Applied Artificial Intelligence, 13.*

7. Cottrell, N.B. (1972). Social Facilitation. In C.G. McClintock (Ed.), *Experimental Social Psychology.* New York: Holt, Rinehart, & Winston.

8. Cottrell, N.B., Wack, D.L., Sekerak, G.J., & Rittle, R.H. (1968). Social facilitation of dominant responses by the presence of an audience and the mere presence of others. *Journal of Personality and Social Psychology, 9.*

9. Crown, D.P. & Liverant, S. (1963). Conformity under varying conditions of personal commitment. *Journal of Abnormal and Social Psychology, 66.*

10. Fogg, B.J. & Nass, C.I. (1997). Silicon sycophants: Effects of computers that flatter. *International Journal of Human-Computer Studies, 46.*

11. Isbister, C. & Layton, T. (1995). In J. Nielsen (Ed.), *Advances in Human-Computer Interaction, Volume 5.* Norwood, NJ: Ablex Publishing Corporation.

12. Isbister, C. & Nass, C.I. (1998). Personality in conversational characters: Building better digital interaction partners using knowledge about human personality preferences and perceptions. *Proceedings of the 1998 Workshop on Embodied Conversational Characters.*

13. Lang, P.J. (1995). The emotion probe: Studies of motivation and attention. *American Psychologist, 50(5).*

14. Laurel, B. (Ed.). (1990). *The Art of Human-Computer Interface Design.* Reading, MA: Addison-Wesley.

15. Lee, E-J. & Nass, C.I. (1998). Does the ethnicity of a computer agent matter? An experimental comparison of human-computer interaction and computer-mediated communication. *Proceedings of the 1998 Workshop on Embodied Conversational Characters.*

16. Lefcourt, H.M. (1966). Internal versus external control of reinforcement: A review. *Psychological Bulletin, 65(4).*

17. Lester, J., Towns, S., Callaway, C., & FitzGerald, P. (1998). Deictic and emotive communication in animated pedagogical agents. *Proceedings of the 1998 Workshop on Embodied Conversational Characters.*

18. Martin, S.A. & Knight, J.M. (1986). Social facilitation effects resulting from locus of control using humans and computer experimenters. *Computers in Human Behavior, 1.*

19. Massaro, D.W. (1998). *Perceiving Talking Faces: From Speech Perception to a Behavioral Principle.* Cambridge, MA: MIT Press.

20. Nass, C. & Gong, L. (in press). Maximized modality or constrained consistency? Proceedings of the AVSP 99 Conference, Santa Cruz, CA.

21. Nass, C.I., Moon, Y., Morkes, J., Kim, E-Y., & Fogg, B.J. (1997). Computers are social actors: A review of current research. In B. Friedman (Ed.), *Human Values and the Design of Computer Technology.* Stanford, CA: CSLI Publications.

22. Nass, C., Fogg, B.J., & Moon, Y. (1996). Can computers be teammates? Affiliation and social identity effects in human-computer interaction. *International Journal of Human-Computer Studies, 45.*

23. Nass, C., Moon, Y., Fogg, B.J., Reeves, B.J., & Dryer, D.C. (1995). Can computer personalities be human personalities? *International Journal of Human-Computer Studies, 43.*

24. Nass, C., Steuer, J., & Tauber, E.R. (1994). Computers are social actors. In *CHI '94 Conference Proceedings.*

25. Pines, H.A. (1973). An attributional analysis of locus of control orientation and source of informational dependence. *Journal of Personality and Social Psychology, 26(2).*

26. Reeves, B. & Nass, C. (1996). *The Media Equation: How People Treat Computers, Television, and New Media Like Real People and Places.* Stanford, CA: CSLI Publications.

27. Rotter, J.B. (1966). Generalized expectancies for internal versus external locus of control of reinforcement. *Psychological Monographs: General and Applied 80(1).*

28. Spielberger, C.D., Gorsuch, R.L., & Lushene, R.E. (1970). *Manual for the State-Trait Anxiety Inventory.* Consulting Psychologists Press, Palo Alto CA.

29. Stanton, J.M. & Barnes-Farrell, J.L. (1996). Effects of electronic performance monitoring on personal control, task satisfaction, and task performance. *Journal of Applied Psychology 81(6).*

30. Strickland, B.R. (1965). The prediction of social action from a dimension of internal-external control. *Journal of Social Psychology 66.*

31. Taylor, I.C., McInnes, F.R., Love, S., Foster, J.C., & Mervyn, J. (1998). Providing animated characters with designated personality profiles. *Proceedings of the 1998 Workshop on Embodied Conversational Characters.*

32. Weiss, R.F. & Miller, F.G. (1971). The drive theory of social facilitation. *Psychological Review, 78.*

33. Zajonc, R.B. (1965). Social Facilitation. *Science 149.*

Helper Agent:
Designing an Assistant for Human-Human Interaction in a Virtual Meeting Space

Katherine Isbister[1], Hideyuki Nakanishi[2], Toru Ishida[3], and Cliff Nass[4]
Kyoto University - NTT - Stanford University
Cross-Cultural Digital Environments Project

ABSTRACT
This paper introduces a new application area for agents in the computer interface: the support of human-human interaction. We discuss an interface agent prototype that is designed to support human-human communication in virtual environments. The prototype interacts with users strategically during conversation, spending most of its time listening. The prototype mimics a party host, trying to find a safe common topic for guests whose conversation has lagged. We performed an experimental evaluation of the prototype's ability to assist in cross-cultural conversations. We designed the prototype to introduce safe or unsafe topics to conversation pairs, through a series of questions and suggestions. The agent made positive contributions to participants' experience of the conversation, influenced their perception of each other and of each others' national group, and even seemed to effect their style of behavior. We discuss the implications of our research for the design of social agents to support human-human interaction.

Keywords
Social interface agents, human-human interaction, virtual meeting place, cross-cultural communication.

INTRODUCTION
Communication contexts are becoming an ever more prominent part of the computer interface. People spend a great deal of their computer time communicating with one another. They use a range of tools, from email, to chat, to elaborate 3-D meeting spaces (e.g. Mitsubishi Research Lab's Diamond Park [14], Electric Community's the Palace, and NTT's InterSpace [13]). People are not just talking to those they already know—virtual meeting places make casual meetings between strangers from across town, or even across the world, easy. It is increasingly common for people to meet for the first time through a computer interface. This is a wonderful opportunity to build new social networks based more on common interests than on location, but it is also a tremendous challenge. Virtual meeting places usually provide very little socially meaningful context to use as a basis for finding common ground with each another. Because it easy to arrive at a virtual meeting place from many entry points, it is often hard for visitors to assume much about one anothers' cultural backgrounds, group memberships, and other aspects of social identity. Psychologists have demonstrated that people need this sort of common context in order to build new relationships [2]. Some commercial chat rooms make use of human moderators to help fulfill this need. However, human moderators are a scarce resource.

We believe this is an appropriate new application domain for social interface agents. Social interface agents could provide ongoing, in-context help in forming social connections and building common ground between visitors to virtual environments. Our project is a first step toward exploring this new application space for interface agents.

Social Interface Agents—Related Work
HCI researchers have already discussed and demonstrated some benefits of interface agents in one-on-one task settings, such as taking an educational tutorial [8], going on a tour [6], or looking at real estate [1]. Although the CHI community has a range of opinions about when and where agents should be deployed in the interface, most would agree that there are some beneficial applications. Lester notes that the presence of an agent can lead to "a strong positive effect on students' perception of their learning experience"[p. 364, 8]. Cassell discusses the value of an

---

[1] NTT Open Lab, 2-4 Hikaridai, Seika-cho, Soraku-gun, Kyoto 619-0237 JAPAN, +81 774 93-1946, katherine@warui.com

[2] Department of Social Informatics, Kyoto University, Kyoto 606-8501, JAPAN, +81 75 753-5396, nuka@kuis.kyoto-u.ac.jp

[3] Department of Social Informatics, Kyoto University, Kyoto 606-8501, JAPAN, +81 75 753-4821, ishida@kuis.kyoto-u.ac.jp

[4] Communication Department, Stanford University, Stanford, CA 94305, USA +1 650 723-5499, nass@leland.stanford.edu

embodied conversation partner with the proper human verbal and nonverbal communication skills [1]. However, these findings concern task-support agents.

There are projects which have used text-tracking to create agent-based social support. Julia [3] engages in entertaining one-on-one conversation; the Extempo bartender agent converses with chat visitors, and is designed to enhance the social atmosphere [7]; and there are bots for websites that answer questions and direct visitors in a friendly way (e.g. http://www.artificial-life.com). However, these agents are designed to engage in one-on-one social interactions, rather than facilitating human-human interaction.

Social Interface Agents to Support Human-Human Interaction—Design Concept for a Supporting Role

The agents described above were all designed to be communication partners with human users, and to be present and active at all times during the interaction. We do not believe that an interface agent should be an equal partner in the conversation, or equally active, when supporting human-human communication. This led us to make some important design decisions in creating our agent prototype.

We designed Helper Agent to be a supporting character rather than a central figure in the conversation. Helper's job is to pick up on contextual cues from the conversation, provide help, and then fade back into the background, allowing the central activities of the conversation environment to move forward. As we mentioned in the abstract, this behavior mimics the activities of a host at a party.

Target Application—Cross-cultural Conversations

For our first prototype, we focused on an extreme case of low social context in a virtual meeting space: strangers from different national cultures, meeting for the first time. Even when people can use a common language with reasonable fluency, they do not necessarily have a common context for their conversation. Different cultures have different notions of how to begin and develop conversations. What is a safe topic in one culture, may be very awkward in another culture. For example, in some cultures it is appropriate to ask about family members right away; whereas in other cultures this is private [5, 2].

We developed an agent prototype that could provide safe-topic suggestions, if the conversation was faltering. We focused on conversations between Japanese and Americans. These two national groups are known to have very different interaction styles and cultural norms [5], and so we felt this was a good test case.

In the rest of this paper, we will provide a more detailed overview of the design and features of the prototype, and discuss results of a cross-cultural test of the agent's usefulness. We will conclude with lessons learned for design of this new class of social interface agents, that we hope will be of use to the CHI community.

DESIGN OVERVIEW

The prototype Helper Agent was designed for a minimalist test of the effectiveness of our concept for a human-human communication assistant. The prototype tracks audio from a two-person conversation, looking for longer silences that will trigger its conversation aid. The agent basically acts in the same way a busy human party host does, looking for clues that the guests' conversations are going badly. (Pauses are a powerful cue for what is happening in a conversation [2].) Helper then directs a series of text-based, yes/no questions to both conversation partners in turn, and uses their answers to guide its suggestion for a new topic to talk about. Then the agent retreats until it is needed again.

Communication Environment

Our prototype works within an existing 3-D virtual meeting space called FreeWalk (see Figure 1), which was developed by us (Nakanishi and Ishida) [10]. Using FreeWalk allowed us to track and use audio silences, and supported the metaphor of the agent coming and going from the conversation, rather than becoming a conversation partner.

Figure 1. FreeWalk: Virtual Meeting Place Environment

Users are represented as three-dimensional pyramid objects, with their video image mapped onto one face of the pyramid (see Figure 1). In the lower right corner of the screen there is an overhead view of all avatars; however the user does not see his/her own avatar on the main screen (see Figure 2). The user does see a small video window of themselves in the lower left corner, to help them adjust their camera. Each user's voice is transmitted to others around them in the virtual space. The volume of other peoples' voices is proportional to how close they are to you in the space (farther away is fainter). Users can move around in the space, and rotate the orientation of the 'face' of their own pyramid, using the mouse or the arrow keys.

Helper Agent Features

Embodiment of the Agent in the Virtual Space

Helper Agent is presented on-screen the same way users are (see Figure 2). This allowed us to take advantage of nonverbal cues in designing the agent's behavior, such as turning to face users as it poses a question to them, and approaching and departing the conversation physically.

The agent is an animated dog, done in a style somewhere between typical Japanese and American cartoon dogs. We chose a dog because we wanted users to think of the agent as subservient, friendly, and reasonably socially intelligent. We chose stylized animation instead of more realistic, because we did not want the agent to be interpreted as a specific individual, but as more iconic and minor [9].

Nonverbal Communication Abilities

The dog has a set of animations of the proper nonverbal conversational moves for asking questions, reacting to affirmative or negative responses, and making suggestions. We crafted these animations as a supplement to the agent's speech [1], and focused on making them friendly and submissive in style [12].

The dog orients its face toward the user that it is addressing, and displays the proper animation for each phase: approach, first question, reaction, follow-up question, and finally topic suggestion. After concluding a suggestion cycle, the agent physically departs from the conversation zone, and meanders at a distance from the interaction, until it detects another awkward silence. This makes it clear to the conversation pair that the agent need not be included in their discussion [4].

Topic Knowledge

We gathered topics using an internet survey, that university students from Japan and the United States filled out. We used the collected pool of topics to select common safe and unsafe topics for people from both countries. From these topics, we crafted a set of questions that the agent could ask during interaction, drawing users into conversation. Safe topics included: movies, music, the weather, sports, and what you've been up to. Unsafe topics included: money, politics, and religion. A sample safe question: "Is the weather nice where you are right now?" A sample unsafe question: "So, do you think it is alright for a country to fish for and eat whales?"

Topic-Suggestion Mechanisms

Silence-detection: The agent decides there is silence when the sum of the voice volumes of both participants is below a fixed threshold value. When the agent detects a silence that lasts for more than a certain period of time, it decides the participants arc in an awkward pause.

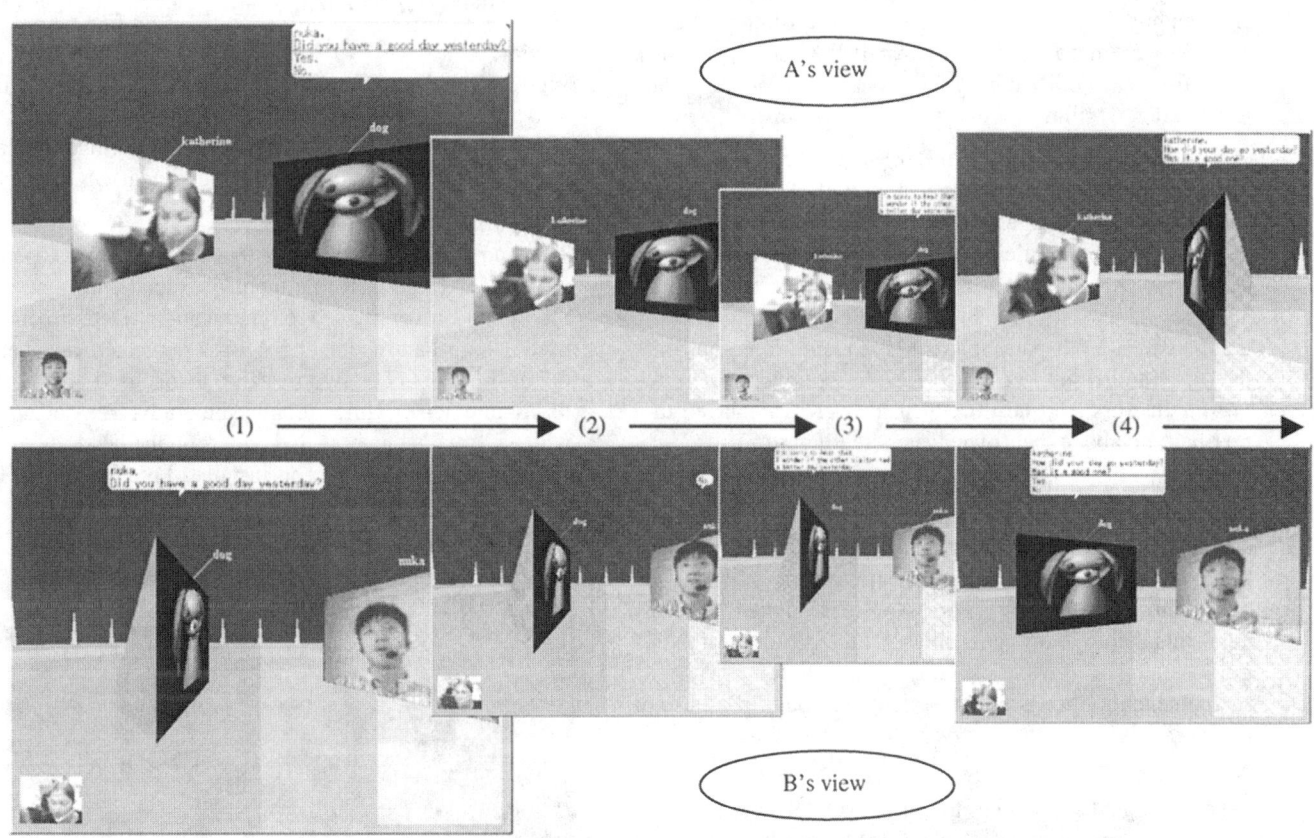

Figure 2. Conversation from both participant's point-of-view: (1) person A is asked the first question (2) and responds, (3) then the agent comments. (4) Next person B is asked a question. Note that the agent faces the person it is addressing.

Positioning: The agent decides how to position itself, based on the location and orientation of each participant. The agent turns toward the participant that it's currently addressing. If the participants move while the agent is talking, the agent adjusts its location and orientation. The agent tries to pick a place where it can be seen well by both people, but also tries to avoid blocking the view between them. If it's hard to find an optimal position, the agent will stand so that it can at least be seen by the participant to whom it is addressing the question.

State-transitions: The agent has three states—idling, approaching, and talking. When idling, the agent strolls at the corner of the virtual space, further away than the normal conversation zone [4]. When the agent detects an awkward pause in the participants' conversation, it begins an approach. Upon reaching the participants, the agent goes into the talking state. However, if the participants start talking again before the agent reaches them, it stops the approach and goes back to idling. (This behavior is strikingly similar to the actions of a hesitant subordinate trying to approach a superior, who is engaged in a conversation with another dominant person.)

The agent will also remain in idling state if the participants are standing far apart from each other (out of conversation range), or are not facing each other. If the participants turn away from each other during the agent's approach, or while it is talking, it will return to idling state, as well.

Conversation Model and Interface

The user interface for communicating with the agent is very easy to learn. The agent does not use voice—it presents questions to the participants in a text-balloon above its head. (We thought text was far less intrusive than audio.) The user indicates 'yes' or 'no' using the mouse to click on their answer. Both participants see all questions, but only the user addressed sees the Yes/No options. When the person answers the question, their answer is displayed in a text-balloon above their own avatar (see Figure 2).

Each topic has a tree structure, with nodes that are: first question for a participant, possible answers by participants, agent's reply to each answer, and flags indicating whether the agent will address its next question to the other person or to the same person. Topics were designed to draw participants into a dialogue, so each turn is tailored for this purpose. The cycle always concludes with a recommendation for how the participants could make use of the particular topic area, given their own answers to the agent.

When the agent approaches to start a cycle, it selects a topic from its repertoire of safe (or unsafe) topics randomly, out of those that have not yet been used. Then it randomly chooses one of the two participants as the target for the first question. Let's call this person A. When A answers, the agent replies to A's answer (see Figure 2). Based on what A answered, the agent then chooses a follow-up question. This question might be directed at A or at B. If it is directed at B, the agent turns to B to pose the question. When B answers, the agent replies to B. Finally, the agent makes a general comment that is meant to guide the participants into using this topic. This general comment is selected based upon the previous answers from the participants, so that it makes sense given their replies. After making this comment, the agent departs. If at any time a user does not respond to the agent's question, the agent will wait for an interval, and then go back into idling mode, without trying to continue its question cycle

EVALUATION OF THE AGENT
Goals

We wanted to test the benefits of our prototype in a controlled setting. Our initial expectations were:

1. The safe-topic agent would create a more satisfying experience, than if there were no agent. Participants would feel they were more similar, would be happier with the interaction and partner, and would form more positive impressions of one another's nationality.

2. The unsafe topic agent would make people uncomfortable, but might lead to a more meaningful and interesting conversation than the safe topic agent.

Design

We designed a 3-condition experiment using pairs of students who were located in the United States and in Japan. Pairs either interacted one-on-one, or had the help of the safe-topic or unsafe-topic Helper Agent.

Procedure

The study was a collaboration between the NTT Open Lab, Kyoto University's Department of Social Informatics, and Stanford University's Communication Department. We used a high-bandwidth (1.5 Mbps) dedicated line between the universities. The two research teams used chat software to communicate while running the study. We set up a PC with a small camera and microphone/headset at each location (see Figure 3), and installed FreeWalk and Helper Agent at both sites.

Figure 3. Set-up for the experiment (Stanford side).

We modified our prototype so that those in the agent conditions would all be exposed to the same number of topics. We divided the conversation session into segments, and forced the agent to display a topic within each time segment. Thus, in the safe-agent condition, the agent introduced all 5 safe topics in random order. In the unsafe-agent condition, the agent introduced all 5 unsafe topics in random order.

Each research team recruited students for the study. The Stanford students were all part of an undergraduate class, which required study participation for credit. The Japanese students were undergraduates from Kyoto University and other nearby universities, who were paid for their participation. Because the study would be held in English, we screened Japanese students and selected those who scored at a reasonably high level on English proficiency tests. Both sets of students were screened for a high level of familiarity with one another's culture, and those with high experience were not asked to participate. In total we had 90 participating students. Due to some problems with equipment, we ended up with data from 45 Japanese students, and 43 American students, for our analysis. Students were assigned randomly to same-gender pairs. Each pair was randomly assigned to one of the three conditions.

Students were told that they would be testing out a new communication environment with a student from the other country. They were asked to talk about anything they liked, just "get to know each other a little bit". They were trained in how to use the system, then left alone to talk for 20 minutes. We made video recordings of all sessions, capturing what was on the screen on the Japan side onto videotape.

After their 20-minute conversation, participants filled out a web-based survey in their native language. (Questionnaire items were translated and then reverse-translated for accuracy.) The questionnaire included questions about the interaction, their conversation partner, the agent (in agent conditions), as well about the participant's own performance. We also asked them to make assessments of themselves, their partner, and the typical person of both participants' cultures on some commonly used stereotypic adjectives.

Results

Safe Agent versus No Agent
We got significant differences when comparing answers in the no agent and safe agent conditions. There was empirical confirmation that the concept of using a social interface agent to support human-human interaction has merit.

American Reaction

The safe agent had positive effects for American participants as we expected (see Figure 4; all items on an 8-point scale, 8 highest):

- **opinion of their own behavior higher**—they rated themselves as more confident, less domineering, and less restrained in the safe agent condition.

- **opinion of partner higher**—they rated their partner as significantly more trustworthy in the safe agent condition.

- **opinion of the typical Japanese person higher**—the safe agent condition had a positive effect on impression of typical Japanese people. Those in the safe agent condition rated the typical Japanese person as more creative and more friendly. However, no-agent condition participants rated the typical Japanese person as more emotionally expressive[5].

variable	safe agent mean	no agent mean	t-value (df=24)
confident	6.46	5.54	-2.33**
domineering	4.00	4.92	2.03*
restrained	3.61	5.00	2.52**
partner trustworthy	6.54	5.91	-2.46**
Japanese creative	5.38	4.54	-2.06*
Japanese friendly	5.92	5.23	-2.08**
Japanese emotionally expressive	3.15	4.23	2.75***

Figure 4. Summary of t-test comparisons of American students' ratings, safe agent versus no agent (*p=.05,**p<.05, ***p=.01).

Japanese reaction

The Japanese participants had a different response to the safe agent's presence—it did not improve their experience. However, it did seem to make them think their partner was more like themselves, as expected (see Figure 5).

- **opinion of the experience lower**—Japanese in the safe agent condition rated the experience as less safe and more uncomfortable. They were less interested in continuing such a conversation, and were less satisfied afterward.

- **opinion of their own behavior lower**—they also rated themselves in a more negative light after the safe agent condition—safe agent condition participants rated themselves as more evasive and more quiet than no agent condition participants did.

- **opinion of their partner mixed**—their ratings of their American partners were mixed. The safe agent condition participants found their partners more talkative and more effusive, and less engaging[6]. Yet, they rated their partners as less typically American and more similar to themselves.

- **opinion of the typical American person lower**—the safe agent condition seemed to exacerbate negative views of Americans for Japanese participants. In the safe agent

---

[5] Americans typically stereotype Japanese people as less creative, less friendly, and less emotionally expressive.

[6] Japanese tend to stereotype Americans as talkative and emotionally effusive.

condition, they rated the typical American as more competitive, more domineering, more selfish, and more effusive than those in the no agent condition[7].

variable	safe agent mean	no agent mean	t-value (df=26)
unsafe	3.29	2.24	-2.05*
uncomfortable	5.14	2.71	-3.9*****
desire to continue	4.86	7.07	3.55****
satisfying	4.79	6.14	2.32**
self evasive	5.86	4.71	-2.09**
self quieter	4.68	3.36	-2.08**
partner talkative	5.00	4.07	-2.06**
partner effusive	2.21	1.50	-2.06**
partner engaging	6.00	6.78	2.47**
partner typically American	5.22	6.14	2.26**
partner similar to self	5.28	4.21	-2.20**
Americans competitive	6.57	5.00	-2.40**
Americans domineering	6.07	3.36	-4.44******
Americans selfish	6.14	4.93	-2.26**
Americans effusive	6.79	5.86	-2.21**

Figure 5. Summary of t-test comparisons of Japanese students' ratings, safe agent versus no agent (*p=.05, **p<.05, ***p=.01, ****p<.01, *****p=.001, ******p<.001).

We cannot be sure why the two groups had such different reactions. One reason may be that the agent's questions were implemented in English. It's possible that Japanese subjects felt it was a two-against-one situation. This might explain why they disliked the interaction, even though it seemed to make them rate their partner as more similar to themselves (our main goal!). We would need to test the system again, using a bilingual agent that address all questions to users with both languages displayed, to be sure. In any case, the positive American reaction was a strong support of our research concept.

Safe Agent versus Unsafe Agent

We got significant and interesting differences in participants' reactions to the safe and unsafe agents.

Awkward isn't necessarily bad

As we had expected, the unsafe agent made things more awkward, but also more interesting.

---

[7] All stereotypical American traits, from the Japanese point of view. It might surprise American readers that 'effusive' is bad; in Japan one is expected to regulate one's emotional expressions, or risk appearing childlike and uncultured [5].

We counted awkward pauses, by observing the videotapes, and found a higher number of awkward pauses in the unsafe versus safe condition (Means = 4.34 and 3.09, t(56)=-3.06, p < .01).

Despite the higher level of awkwardness in these conversations, both Japanese and American participants found the conversation that included the unsafe topic agent more interesting. Americans rated the unsafe agent interaction more interesting; Japanese rated the unsafe agent experience more desirable to continue (see Figures 6 and 7). Japanese participants found the unsafe agent experience more comfortable as well.

variable	safe agent mean	unsafe agent mean	t-value (df=26)
desire to continue	4.86	6.21	-2.00^
uncomfortable	5.14	3.57	2.41**
partner similar to self	5.29	3.64	2.58**
partner considerate	7.31	5.93	3.02****
partner domineering	1.14	2.07	-2.43**
partner friendly	7.29	6.14	2.31**
partner talkative	5.00	3.79	2.14**
self evasive	5.86	4.64	2.03*
self restrained	5.43	3.79	2.37**
self self-abasing	5.54	4.07	2.33**
self team-oriented	4.00	2.64	2.34**
Americans domineering	6.07	5.00	2.26**
agent nice	3.43	5.29	-2.25**
agent competent	4.29	5.57	-2.04*
agent typically Japanese	4.50	3.43	2.71**
agent talkative	5.61	4.36	2.66***
agent nationalistic	1.43	3.29	-2.87***

Figure 6. Summary of t-test comparisons of Japanese students' ratings, safe agent versus unsafe agent conditions (*p=.05, **p<.05, ***p=.01, ****p<.01, ^p=.056).

American partner seemed better in the safe topic condition

Japanese participants rated their partner as less similar to themselves, less considerate, more domineering, less friendly, and less talkative in the unsafe condition. These rankings suggest that the safe agent led to more positive impressions of the partner, for Japanese participants.

Unsafe topics made Japanese act more American

Japanese rated themselves as less evasive, less restrained, less self-abasing, and less team-oriented in the unsafe condition[8]. It seems they thought they acted more American than those in the safe agent condition. Americans rated their partner as more similar to themselves in the unsafe condition. This seems to corroborate the Japanese self-ratings.

variable	safe agent mean	unsafe agent mean	t-value (df=24)
interesting	5.85	6.77	-2.18**
partner similar to self	3.31	4.77	-2.55**
Japanese emotionally expressive	3.15	4.15	-2.16**
Japanese outgoing	4.08	4.77	-2.04*
Japanese talkative	3.77	4.85	-2.30**
Japanese evasive	3.85	4.85	-2.39**
Japanese quiet	6.00	4.38	2.82***
agent blunt	4.69	7.36	-3.84*****
agent domineering	3.38	5.25	-2.31**
agent restrained	3.15	1.92	2.52**
agent friendly	5.46	4.08	2.03*
agent typically American	6.62	4.92	2.40**

Figure 7. Summary of t-test comparisons of American students' ratings, safe agent versus unsafe agent (*p=.05, **p<.05, ***p=.01, *****p=.001).

Safe/unsafe topic choice affected stereotyping in contradictory ways

Japanese participants in the unsafe agent condition thought the typical American was less domineering. This conflicts with their ranking of their own partner's level of domineering-ness.

American participants rated the typical Japanese in conflicting ways: after the unsafe condition, they thought the typical Japanese person was more emotionally expressive, more outgoing, and more talkative; but also more evasive and quieter.

Safe/unsafe agents 'read' differently for Japanese and Americans

The two groups differed in their impressions of the safe and unsafe agents. The Americans formed the intended impression: they rated the unsafe agent's topics as less appropriate, thought it acted more blunt, more domineering, less restrained, and less friendly. They also said it was less typically American, distancing it from their own in-group's behavior. The Japanese thought that the unsafe agent was nicer and more competent than the safe agent. They rated the unsafe agent as less typically Japanese, and as less talkative. They found the unsafe agent more nationalistic, probably because it brought up more political topics than the safe agent.

LESSONS FOR THE DEVELOPMENT OF AGENTS TO SUPPORT HUMAN-HUMAN COMMUNICATION

A social interface agent can help support human-human communication

Our evaluation demonstrated that a human-human communication assistant can have positive effects on perception of the experience, one's own qualities, one's partner, and even one's partner's cultural group.

Provocative help can be good

Our evaluation also suggested that a communication assistant can be helpful both when it offers safe topics to talk about, and when it steers the conversation in less safe directions. In fact, the Japanese participants seemed to prefer the unsafe topic agent, and both groups found it more interesting than the safe topic agent.

For overall conversational support purposes, both kinds of help may be desirable. We suspect that an agent with a model for offering both kinds of topics, depending upon the conversation flow, would be the most desirable.

User-adaptation would make the agent more effective

The two cultural groups had very different impressions of the same agent behaviors, and reacted in different ways. For example, behavior that was perceived as blunt and unfriendly by Americans was seen as nice and competent by Japanese. An effective agent for different types of people will probably need to adapt its behaviors to user subgroups, or perhaps to individuals' own interaction styles and preferences. We believe we created a more American identity for our agent by delivering its topic help in English. In future iterations, we plan to create an agent whose presentation is adapted to different user styles and preferences.

Agent behavior may shift user behavior

Both the Japanese and American participants noted that Japanese seemed to act more American in the unsafe agent condition. This result indicates that it may be possible to mold user behavior with the choices one makes about how the agent will behave, creating a different conversational environment by bringing different traits to the fore. This could have very interesting implications for those interested in setting a specific group conversational tone or style in a virtual meeting space.

Design Notes for creating a supporting character

We believe our prototype's success was partially due to design choices that made it a graceful supporting player in the interaction. We summarize key features here, as suggestions for CHI community members who may be interested in creating this kind of interface agent:

---

[8] These are all stereotypically American traits, for Japanese.

- **Unobtrusive observation of users, and easily visible and controllable approaches.** The users could see when the agent was approaching, and could shoo it away simply by talking before it arrived.
- **Focused interactions with a limited duration.** The agent had a clear purpose in its approach, and the users could quickly grasp its interaction pattern, and knew it would leave after making a topic suggestion. They could see the agent in-between interactions, but did not need to include it in their conversation. This kept the focus on the task at hand: human-human interaction.
- **Easy to ignore.** If users ignored the agent, it simply gave up and went away! It did not hang about forever waiting for them to answer.

Next Steps

Though silence-sensing produced strong results, we would like to incorporate content recognition, to make Helper Agent a more powerful assistant. Also, our first prototype supported only two people. With more participants, we can experiment with additional features, such as recognizing active conversation groups and leading a newcomer to an ongoing conversation. We are also interested in supporting groups with the same base language, but different sub-cultural memberships.

It would of course be interesting to run follow-up studies using pairs from other cultures, and to continue to refine and deepen our understanding of how cultural differences should affect an interface agent's behavior. Our work is really only a beginning in exploring this terrain.

CONCLUSION

We built a social agent prototype, that was designed to facilitate human-human interaction. A cross-cultural evaluation of the prototype demonstrated its effectiveness, and raised interesting considerations for further development of this class of interface agent.

We feel the support of human-human interaction in virtual meeting places is an exciting and useful new domain for interface agents. Given the proliferation of online spaces, and the interest in community formation that far exceeds the industry's ability to staff communities with human hosts, this kind of agent may become a familiar part of the virtual landscape. We hope that CHI community members who are called upon to think about and design these kinds of interface agents will find our prototype design and evaluation results useful for their work.

ACKNOWLEDGMENTS

Thanks to GEMNet for providing the broadband line, and to Eva Jettmar for her assistance in the evaluation.

REFERENCES

1. Cassell, J., Bickmore, T., Billinghurst, M., Campbell, L., Chang, K., Vilhjalmsson, H., and Yan, H., Embodiment in Conversational Interfaces: Rea, *International Conference on Human Factors in Computing Systems (CHI-99)*, pp.520-527, 1999.
2. Clark, H.H., *Using Language,* Cambridge University Press, 1996.
3. Foner, L., Entertaining Agents: A Sociological Case Study, in *Proceedings of the First International Conference on Autonomous Agents,* Marina del Rey, CA, 1997.
4. Hall, E.T., *The Hidden Dimension,* Anchor Books/Doubleday, 1982 (1966).
5. Hall, E.T., and Hall, M.R., *Hidden Differences: Doing Business with the Japanese,* Anchor Books, 1990 reprint.
6. Isbister, K., and Doyle, P. Touring Machines: Guide Agents for Sharing Stories about Digital Places, in *Proceedings of the Workshop on Narrative and Artificial Intelligence,* AAAI Fall Symposium Series, 1999.
7. Isbister, K., and Hayes-Roth, B. Social Implications of Using Synthetic Characters, in *Animated Interface Agents: Making Them Intelligent* (a workshop in IJCAI-97, Nagoya, JAPAN, August 1997), 19-20.
8. Lester, J.C., Barlow, S.T., Converse, S.A., Stone, B.A., Kahler, S.E., and Bhogal, R.S. The Persona Effect: Affective Impact of Animated Pedagogical Agents. *Proceedings of CHI '97* (Atlanta GA, March 1997), ACM Press, 359-366.
9. McCloud, S., *Understanding Comics: The Invisible Art,* Harper Perennial, 1993.
10. Nakanishi, H., Yoshida, C., Nishimura, T., and Ishida, T., FreeWalk: A 3D Virtual Space for Casual Meetings, *IEEE MultiMedia, 6*(2), pp.20-28, 1999.
11. Parise, S., Kiesler, S., Sproull, L., and Waters, K. My Partner is a Real Dog: Cooperation with Social Agents, in *Proceedings of Computer Supported Cooperative Work '96* (Cambridge, MA), ACM Press, 399-408.
12. Reeves, B., and Nass, C., *The Media Equation: How People Treat Computers, Television, and New Media Like Real People and Places*, Cambridge University Press, 1996.
13. Sugawara, S., Suzuki, G., Nagashima, Y., Matsuura, M., Tanigawa, H., and Moriuchi, M., InterSpace: Networked Virtual World for Visual Communication, *IEICE (The Institute of Electronics, Information and Communication Engineers) Transactions on Information and Systems,* E77-D(12), pp.1344-1349, 1994.
14. Waters, R.C., and Barrus, J.W., The Rise of Shared Virtual Environments, *IEEE Spectrum, 34*(3), pp.20-25, 1997.

Agents to Assist in Finding Help

Adriana Vivacqua and Henry Lieberman
Media Laboratory
Massachusetts Institute of Technology
Cambridge, MA 02139 USA
+1 617 253 0315
lieber@media.mit.edu
avivacqua@quark.com

ABSTRACT

When a novice needs help, often the best solution is to find a human expert who is capable of answering the novice's questions. But often, novices have difficulty characterizing their own questions and expertise and finding appropriate experts. Previous attempts to assist expertise location have provided matchmaking services, but leave the task of classifying knowledge and queries to be performed manually by the participants. We introduce *Expert Finder*, an agent that automatically classifies both novice and expert knowledge by autonomously analyzing documents created in the course of routine work. Expert Finder works in the domain of Java programming, where it relates a user's Java class usage to an independent domain model. User models are automatically generated that allow accurate matching of query to expert without either the novice or expert filling out skill questionnaires. Testing showed that automatically generated profiles matched well with experts' own evaluation of their skills, and we achieved a high rate of matching novice questions with appropriate experts.

Keywords

Expertise location, agents, matchmaking, Java, help systems.

INTRODUCTION

Meet Jen: Jen has been in the computer business for a while, doing systems analysis and consulting. She has wide experience in Cobol, mainframes and database programming, but little experience in Java, which her company has now decided to use.

Meet David: David is a hacker. He started programming at the age of 15, and has been playing with Java for a while now. He has worked with user interfaces, computer graphics and client-server systems at one time or another. He now works as a systems programmer for a large software company, which does most of their work in Java.

Jen's new project is a client-server system for a bank: clients of the bank will download software and perform transactions through their computers. The system uses database manipulation and a graphical user interface.

Given that Jen is a novice Java programmer, she has a hard time learning all the existing packages and classes. She breezes through the database part, though, building all the server-side SQL routines without much trouble. Her first problem is the database connection to the program...

The hard way

Jen doesn't know what objects are available to connect her server side routines and database with the front end. She asks around the office, but nobody is familiar enough with the Java language to navigate JDBC objects and connections. She manages to access the database, defines the functionality that should be included in the front end, and now needs to know how it should be done.

She turns to the JDK documentation but is unable to find much information on this new library. She tries to build some of the structures, but finds that testing the objects is a tedious and slow process. She pokes around on the Internet and, lurking in some of the user groups, finds out that there are some books on JDBC which might help her. The book gives her some very basic notions, but not nearly enough to help her build her application. She needs more details on how to call the server-side stored procedures she created.

She wades around the newsgroups, reads their FAQs, and posts a question. Disappointingly, she gets no answers. She finds that most of the newsgroups are tight communities where people tend to get off topic or carried away. She subscribes to a few mailing lists, but traffic is too high. People seem to be more interested in discussing their own problems than addressing the problems of a new user like her.

She finally decides to get in touch with a friend's daughter, Sarah, who studies Computer Science at the local university. Sarah has never programmed in Java, but knows several more advanced students who have. Sarah's boyfriend, David, is experienced in Java. Jen reluctantly sends him an email, to which David replies with a brief explanation and pointers to some websites about JDBC.

Enter the Expert Finder

Let's see how the same scenario goes with our Expert Finder system. Instead of asking around the office, Jen goes to her Expert Finder agent and enters a few keywords.

Expert Finder periodically reads through her Java source files, so it knows how much she knows about certain Java concepts and classes. In fact, it reads through all of the programs she wrote while studying with the "Learn Java in 21 Days" [5] book. Expert Finder verifies what constructs she has used, how often and how extensively, and compares those values to the usage levels for the rest of the participating community to establish her levels of expertise. Jen can see and edit her profile on the profile-editing window, and decides to publish all of it. Table 1 shows Jen's usage for each construct and calculated profile.

Jen types in the keywords "sql", "stored" and "procedure". From the domain model, the agent knows that sql is related to database manipulation – java.sql is a library of objects for database manipulation. From the model, the agent knows which classes are included in this library.

Area	Usage	Expertise Level
java.io	10	Novice
java.util	15	Novice
System	20	Novice
elementAt	5	Novice
println	20	Novice

Table 1: Jen's areas and levels of expertise

The agent communicates with other users' Expert Finder agents calculating their "suitability" by verifying which libraries and classes they know how to use. It picks out David (Table 2), because he has used the "java.sql" library and its objects.

Area	Usage	Expertise Level
java.io	46	Intermediate
java.util	45	Intermediate
Connection	11	Advanced
InputStream	5	Intermediate
CallableStatement	10	Intermediate

Table 2: David's areas and levels of expertise. Note that the levels of expertise are obtained through a comparison with others in the community.

His expertise is higher, but not too distant from Jen's. Jen takes a look at David's published profile, checks his "halo factor" (an indicator of how helpful he is to the community), and sends him a message:

Dear David,

I'm a novice Java programmer and have some problems regarding database connections and manipulation. I have created a series of stored procedures and now need to access them from my program. Is there a way to do that?

Thanks,

Jen

David verifies, based on Jen's "halo factor", that Jen is a new user and decides to answer her question:

Hi Jen,

To call stored procedures you should use a Callable Statement, which can be created with the *prepareCall* method of the Connection class.

Here's a little snippet which might help you:

```
CallableStatement cstmt =
    con.prepareCall("{call MyProc(?, ?)}");
cstmt.registerOutParameter(1,
java.sql.Types.TINYINT);
cstmt.registerOutParameter(2,
java.sql.Types.DECIMAL, 3);
cstmt.executeQuery();
byte x = cstmt.getByte(1);
java.math.BigDecimal n =
    cstmt.getBigDecimal(2, 3);
```

Also, take a look at:
http://java.sun.com/products/jdk/1.2/docs/guide/jdbc/getstart/callablestatement.doc.html

David

With Expert Finder, Jen obtained David's help much faster than she would have otherwise.

Approach

In our approach, each user has his or hers own Expert Finder agent, which builds that user's profile from Java source code files. When necessary, the user can query the agent, which then communicates with other like agents, looking for a user with the appropriate expertise to assist with this problem.

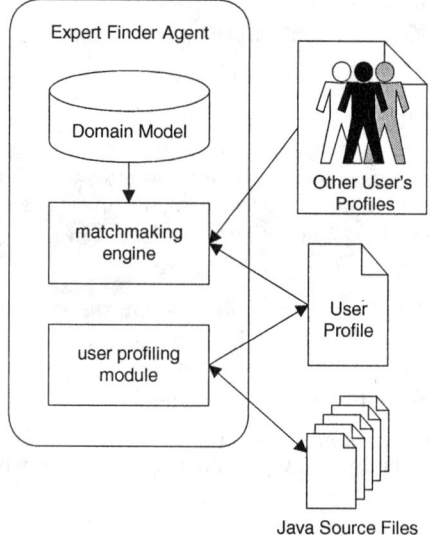

Figure 1: An agent's Internals: Each agent has (1) a profiling module, which builds the user's profile from his/her Java files; (2) a matchmaking engine, which consults and compares other user's profiles and (3) a domain similarity model, used for matchmaking purposes.

Figure 1 shows one agent's internal structure. It is important to note that there are no specialized agents for experts and novices. It often happens that a person might be an expert in one area and a novice in another.

Domain Similarity Model

Our system uses a *similarity model* (rather than an equivalence model) for the Java domain, because an expert whose knowledge lies in a more general or more specific category or related topic to the novice's requirements might still be a good candidate to provide help. In a sophisticated domain like Java programming, there are many overlapping relationships between the knowledge elements. Rather than burden users with the task of manually browsing subject category hierarchies, and judging relevance, we move that task onto the agent.

Even if the agent is not perfectly accurate in its similarity assessment, the agent's model constrains the search space enormously and results in more relevant recommendations than the ones produced without a model. We also provide browsers and editors for the domain model, and for user profiles, allowing any deficiencies in our prior knowledge to be corrected manually.

The Java Programming Domain

Constructs in Java are hierarchically structured into classes and subclasses and organized in packages according to purpose or usage. We built our domain model from the online documentation pages. Many classes also provide an extra hint: the "See also:" entry, which lists related classes, methods or packages. We assigned arbitrary values to each of the relationships between classes. The first step in the process was establishing which items would be taken into account for purposes of determining similarity.

- **Sub/Superclass relationships**: a subclass is fairly similar to its superclass (inheriting methods and properties), but a superclass is less similar to its subclass, since the latter may contain resources not available in the former. For example, the class *Container* is a subclass of class *Component*: it inherits 131 methods and 5 fields. However, *Container* also defines 52 of its own methods. Code: *SUB* or *SUP*.

- **Package coincidence:** Packages group classes by what they are used for. Package *java.awt* contains classes used for graphic interface construction, such as buttons, list boxes, drop-down menus, etc. A person who knows how to use these classes is someone who knows how to build graphical interfaces. Code *PAK*.

- **"See also" entry:** this is a hint which links to other classes that might work similarly or share a purpose. Class *MenuBar*, for instance, is a subclass of class *MenuComponent*, and is related to classes Frame, Menu and MenuItem through the "See Also" relationship. Code: *SEE*.

Thus, one class' similarity to another is determined by

$$\{SUB, SUP\} + PAK + SEE,$$

where the values for each of the variables may vary according to the type of query (free-form keyword based or selected from list.) These values are parameterized: the model holds the different relations, not the numbers.

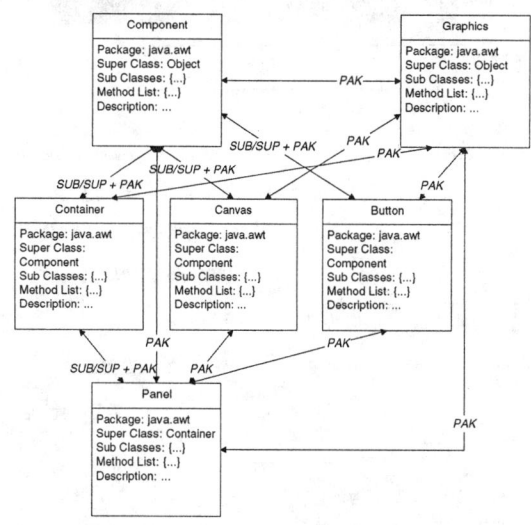

Figure 2: Similarity model for the Java domain (partially shown.)

Moving to another domain would require building a new domain model, from which the agent would derive its knowledge. The rest of the system would remain the same.

Building Profiles

Automatic profiling is important, given that, in general, people dislike filling long forms about their skills. An automated method also reduces the possibility of inaccuracy due to people's opinions of themselves. Another advantage is that automated profiles are dynamic, whereas people rarely update interest or skill questionnaires. However, we acknowledge the fact that the agent might be wrong in its assessment and allow users the option of altering their profiles.

A profile contains a list of the user's areas of expertise, the levels of expertise for each area (novice - beginner - intermediate - advanced - expert) and a flag noting whether or not this information is to be disclosed. Hidden information will still be used in calculations of expertise for a given query. Users might change their profiles at will.

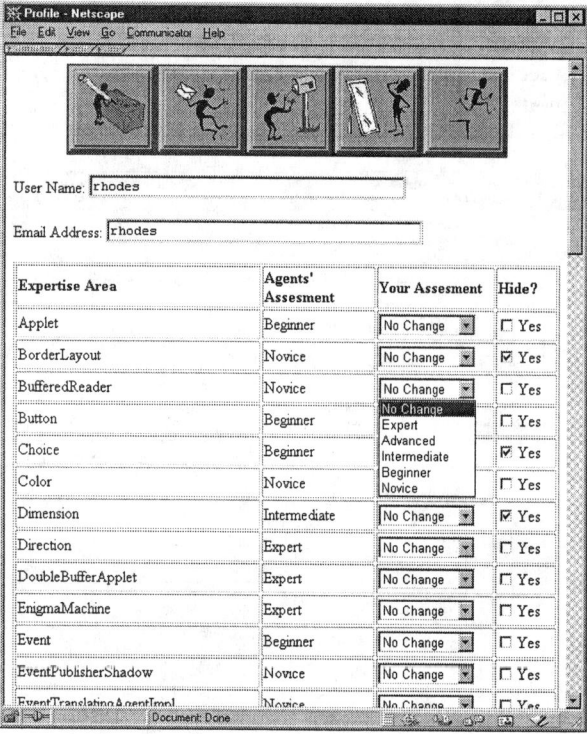

Figure 3: Profile editing window: a user can inspect and edit his or her profile as fit, to compensate for errors in the agent's assessment or hide areas of expertise.

Assessing a user's areas and levels of expertise is done through analysis of his or her Java source files and parsing them, analyzing:

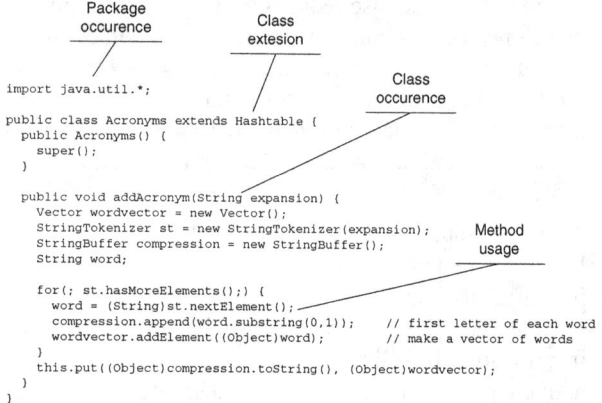

Figure 4: Example code and items analyzed in it.

- **Libraries**: which libraries are being used? How often? Libraries are declared once, usually at the beginning of a file.

- **Classes**: which classes are used? How often? Classes are declared, instantiated and used throughout the file. Classes can also be subclassed, which indicates a deeper knowledge of the class. Implicit in the act of subclassing is the recognition that there is a need for a specialized version of the class and knowledge of how the class works and how it should be changed in each specific case.

- **Methods**: knowing which methods are being used helps us further determine how much he or she knows about a class: Are only a few methods used over and over again? How extensively is the class used?

We verify how often each of these is used and compare these numbers to overall usage (usage by other users in the system). This is similar to Salton's TFiDF algorithm (term frequency inverse document frequency) [9], in that the more a person uses a class that's not generally used, the more relevant it is to his profile. The profile is a list of classes and expertise level for each. Expertise level is initially determined by taking the number of times the user uses each class and dividing by the overall class usage.

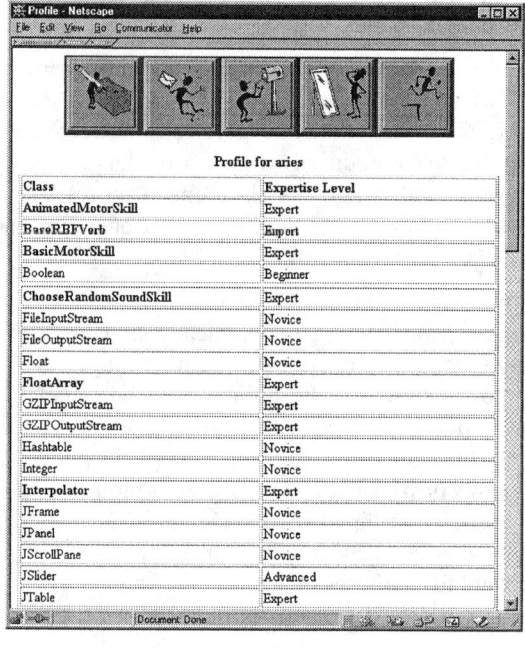

Figure 5: Viewing other users' profiles: the items in bold represent classes that have been subclassed. "Hidden" classes are not shown.

Matching Needs and Profiles

Given a query, related topics are taken from the model and added to the query, thus expanding it. It is then compared to other users' profiles. A query can be formulated as:

- **Keyword entry:** the user enters a set of keywords associated with his or her needs in a text box. The class descriptions are then used to locate appropriate classes from the keywords.

- **Selection of classes** from a list of those existing in the domain model: the user chooses from a list of classes. These are then used to find the experts by doing a cosine similarity vector match on the class list and profiles.

- **A combination of both:** the user chooses some items from the list and enters some keywords.

A screenshot of the query screen can be seen in Figure 6. The rationale for weighing the importance of the type of knowledge is as follows. If a user selects items from the list, it is reasonable to assume that he or she needs help with using these classes specifically. Therefore, sub/superclass relations, denoting structural similarity, are more valuable in finding an expert with the desired knowledge. Entering a few keywords means that the user knows what he or she wants to do, but is uncertain of how to do it. In these cases, functional similarity (packages) is more important. If the user uses a combination of both, both relations can be used, although functional similarity takes precedence over structural: the user almost certainly knows what he or she wants to do, even though he or she may not be doing it correctly (this reflects on picking the wrong items in the list.)

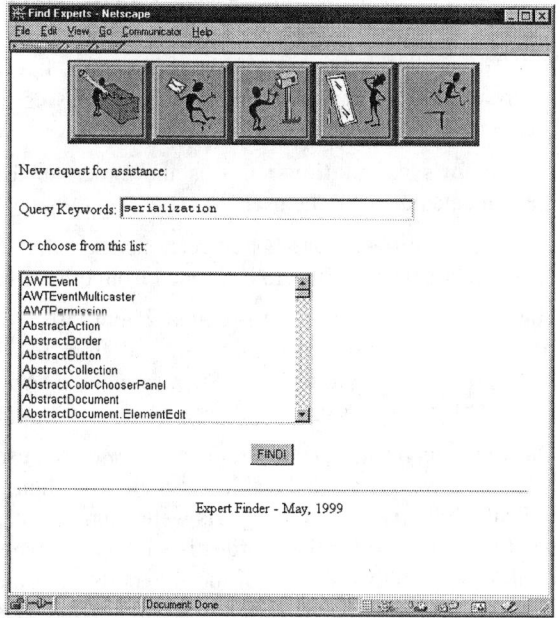

Figure 6: Query screen – a user may choose an item from the list or enter keywords.

A match is made by first finding similar topics in the domain model. The agent then goes on to contact other agents, computing a vector match between its user's needs and other users' expertise. The agent returns a list of potential helpers. We believe that the best person to help is not always the topmost expert, but someone who knows a bit more than the questioner does. First, because the topmost expert is most likely to be unavailable or uninterested in novice questions. But, more importantly, experts and novices have different mental models, as noted by [3] so we are more likely to bring together two people who have similar mental models. Thus, we compute "fitness values" for all of the users, including the questioner. We then take the *n* with closest (but higher) fitness values. The user can inspect each of the experts' profiles before selecting whom he or she would like to contact from that list and send them messages.

Figure 7 shows the screen where users can view a response to their query, listing the experts available.

Incentives

We have built into the system an incentive mechanism to assess the social capital in the community. We keep track of how helpful each person generally is (the *halo factor*). The halo factor of a person is the percentage of questions answered from those received ($[Qa/Qr]*100$). It is displayed every time a person sends or answers a question, and we assume this will motivate both questioner and responder. When a person is new to the system or has never received any questions ($Qr = 0$), the person is billed as being new to the system. We don't want to inhibit a user from asking questions (and asking how many questions one has asked could be interpreted as how much work one is giving others.) As the system keeps track of questions sent and received, we can more evenly distribute questions when there are multiple experts available. We have been considering the possibility of adding a mechanism to account for an expert's refusal to answer a question: a refusal for lack of knowledge would be reflected on the expertise level, a plain refusal with no explanation would impact the halo factor.

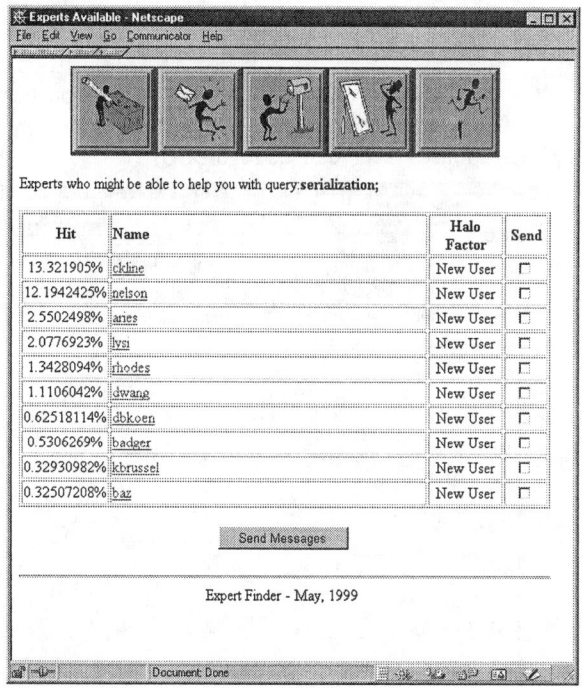

Figure 7: Expert list screen – experts are ranked by appropriateness to a given query.

Interface Overview

A button bar (Figure 8) on the top of each page gives each user the options: making a new query, viewing responses, viewing questions, editing the profile and logging out.

Figure 8: Expert Finder button bar. Left to right: Query, View Responses, View Questions, Edit Profile, Logout.

A user can edit his or her profile on the profile-editing screen (see Figure 3). The queries are submitted to the system via the query interface (see Figure 6). The results of the query are then shown in the result screen (see Figure 7).

Clicking on one of the expert's names, the user may inspect this person's profile in detail, verifying which classes he or she knows how to use. Still on the result screen, the user can select experts and click "Send Message" to go to the message composition screen.

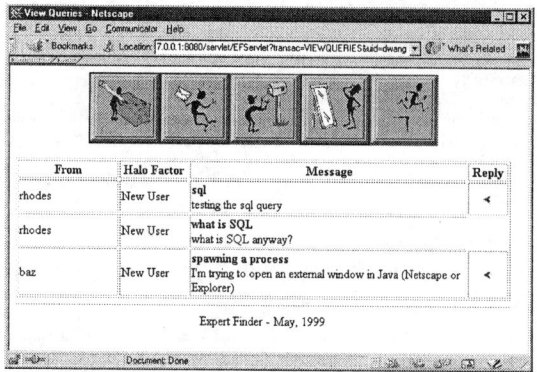

Figure 9: View of the questions received: the expert can click on the blue arrow on the right to start composing a reply.

An expert can view questions sent to him or her, and compose a reply (see Figure 9). The questioner can then view responses received (see Figure 10).

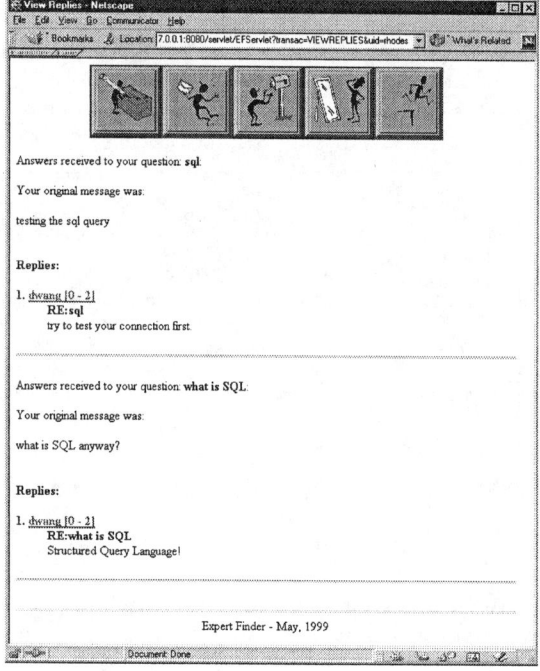

Figure 10: Viewing answers received to one's questions.

Evaluation
As an informal evaluation for this work, we built a prototype system, generated profiles for 10 users, and ran 20 queries through the system. We independently determined whether the experts suggested by the system would be able to answer those questions through a questionnaire. Questions were taken from the Experts-Exchange forum, thus constituting real problems people have. They ranged from very specific ("How do I add an item to a JList") to the more generic ("What are static entries and what are they good for?").

Possible answers from the experts were "I can answer", "I couldn't answer this" and "Not flat out, but I would know where to look". We also showed the users their profiles, so they could verify how well it represented their expertise. We allowed them to edit their profiles and then compared what the agent had said to what the users claimed.

Profiling
To test how well the profiling module worked, we generated profiles for 10 users and had them edit them. We then took the original and edited profiles and checked to see how many items were altered and by how much. Users' profiles are kept in files divided into:

- **Totals:** total number of times the user has used a certain class, library or method, and the classes the user extends in his or her code.
- **Agent's calculations:** this is the expertise level the agent calculated for the user.
- **User values:** User's corrections to the agent's calculations, and values to be hidden from other users.

On average, it seems users edited about 50% of their profiles. The number of changes ranged from 9% for the least altered profile to 63% for the most altered. About one third of all changes were decreases.

On commonly used classes such as *Hashtable,* users felt they were very knowledgeable even though their profile indicated otherwise. Many experts were using this class and what we calculate for the profiles is what percentage of the total usage belongs to each of those experts. If someone is responsible for 55% of the total usage for the Hashtable class, the system will place him or her in the intermediate level. This indicates a lack of variety in the sampling, for all users were reasonably proficient with the Java language.

The decreases for the most part happened when there was only one user who used a given construct, and was therefore deemed the expert. If nobody else is using this class, the user is responsible for 100% of its usage in the community.

31% of changes were 1 step changes (for instance, novice to beginner), 33% 2 step changes, 26% 3 step changes and 10% 4 step changes. These numbers seem to indicate that the agent's calculations weren't so far off the mark.

Matchmaking
Overall, the system performed well, always placing at least one expert who had said he or she could have answered the questions (either right away or looking it up) in the first three recommendations. We now go into more detail about what happened.

Number of success cases (recommending experts who would be able to provide an answer) was around 85%. Breaking these down, 35% were "immediate success" cases (the first expert recommended said he'd be able to answer it right away) and 50% were "delayed success" (the expert answered that he'd be able to answer by looking it up.)

The system performed better for people with at least a little knowledge. Since the system recommends people at a level of expertise close to that of the questioner, if the questioner had little or no expertise, the system did not always recommend people well suited to answer.

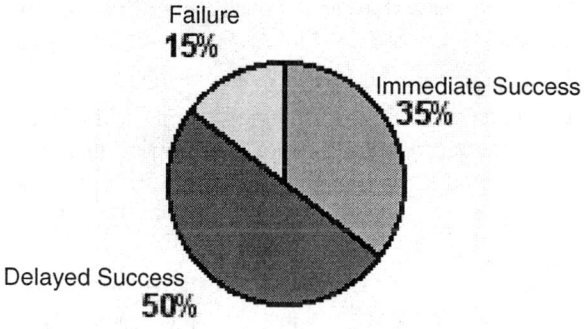

Figure 11: Distribution of Success/Failure cases.

For queries that were more specific, the system performed well. Taking the top 3 experts found (as opposed to the recommended ones) for specific queries, we have 52% said they could answer the question, 19% said they could look it up and 29% said they could not. Analyzing the failure cases, we found that these were either cases in which there was no indication that a user had this knowledge in his profile, and the answer was based solely on the related knowledge model.

In the first situation, no expert said he'd be able to answer the question, although some said they'd know where to look. A quick check of the profiles revealed that none of the experts had these classes in their profiles, either. Therefore, the system had to use the related knowledge model to search for experts. The same happened in the second situation, although this time, despite the fact that a user said he knew how to use a given class, there was no indication in his code to support that statement, and therefore the system couldn't place him very high.

In general, in the cases where related knowledge was needed, Expert Finder produced acceptable results, although not necessarily the optimal choices (once again, ranking experts incorrectly.) This probably means that the model needs to be adjusted to produce better output for the similarity relations, which would result in better matches.

More abstract queries yielded worse results. Once again, taking the top 3 experts found, we have that 45% had claimed they'd be able to answer the questions right away, 30% said they'd have to look them up and 25% said they'd be unable to answer the questions. Despite the apparently good results, we consider these not to be as good as the previous ones. In most cases, Expert Finder placed experts incorrectly, ranking users who had said they couldn't answer higher than others who said they would be able to answer them. This probably happened due to the method used to retrieve keywords (searching through the specification descriptions), since most of these queries were made using keyword entry.

Future Work
Profile Building
More accurate profile building is a major area for future work. Accuracy can be improved by enlisting more sources of information and taking into account other factors such as history of help interactions. We could perform more complex code analysis, which might reveal more about one's programming style, abilities and efficiency or use such techniques as collaborative filtering to rate expertise.

One other consideration on this topic is the issue of time, or what we call "decaying expertise": after a while, people forget how to do things, if they don't keep working on it. As Seifert [10] notes, expertise comes with experience, and memory plays an important part.

Making Expert Finder more proactive
The most immediate next step for Expert Finder would be making it more proactive. A context-aware agent built directly into the development environment could try to figure out the user needs help by watching error messages as he or she writes the program. It could also be done by detecting when the user goes to the help system.

The agent could also help compose the messages by inserting pieces of the questioner's code or the error messages he or she has been getting. It could also help the expert deal with the problem by providing manual pages and other documentation about the classes in question and samples of the expert's own code where the same classes were used to help the expert remember how he or she dealt with this problem before.

Related Work
Information Marketplaces
Experts-Exchange
Experts-Exchange [4] uses a predetermined expertise directory, under which questions and answers are posted. It uses a credit system to provide incentive. Experts-Exchange doesn't automatically generate a user profile and there aren't any recommendations made to the questioner: he or she simply posts a question in a bulletin board-like system and waits for an answer (from a human expert.)

Referral Systems
ReferralWeb
In ReferralWeb [7] a person may look for a chain between him/herself and another individual; specify a topic and radius to be searched ("What colleagues of colleagues of mine know Japanese?"); or take advantage of a known expert in the field to center the search ("List dessert recipes by people close to Martha Stewart"). The system uses the co-occurrence of names in close proximity in public documents as evidence of a relationship. Documents used to obtain this information were links on home pages; co-authorship on papers; etc.

ReferralWeb lacks a domain model or automatic profile construction, but Expert Finder might also benefit from ReferralWeb's social network techniques, since people prefer to ask questions of others who have pre-existing social relationships with them.

Yenta
Yenta [5] is a matchmaking agent that derives users' interest profiles from their email and newsgroup messages. Yenta aims to introduce people who share general interests rather than matching for a specific question or topic. It has doesn't use a domain model. Yenta is notable for its fully decentralized structure, form which Expert Finder could also benefit.

Information Repositories
Answer Garden
Answer Garden, [1] is a system designed to help in environments such as a help desks. It provides a branching network of diagnostic questions through which experts can navigate to match the novice's question. A similar question already in the network may yield the answer, or a new Q&A pair can be saved for future reference. The network can also be edited.

Answer Garden and similar systems look for the contents of the answer rather than the expert, which is harder in some cases, and forgoes the ancillary advantages of locating an expert who might serve as a resource in the future.

Another "Expert Finder"
A MITRE project also called Expert Finder [8] derives expertise estimation from number of mentions in Web-available newsletters, resumés, employee databases and other information. It is a centralized system, which doesn't allow for inclusion of new experts easily and doesn't provide incentive mechanisms as we do. Recent versions are incorporating more proactive elements, bringing it closer in spirit to Expert Finder.

Task-Based Recommendations
PHelpS
The Peer Help System, or PHelpS [2] tracks users who are doing step-by-step tasks, and if a novice runs into difficulty, it matches them with another user who has successfully completed the same or similar sequence of steps. Unlike our system, it's highly task-oriented, which allows it to follow a user's work patterns and check to see when he or she gets stuck. The inspectable user profiles is something we've adopted, but the initial requirement that users fill out (and later maintain) their profiles might prove to be a problem.

Conclusion
We have presented Expert Finder, a user-interface agent that assists a novice user in finding experts to answer a question by matchmaking between profiles automatically constructed by scanning Java programs written by both the novice and the expert. Tests show that the agent does reasonably well compared to human judgment, and Expert Finder obviates the need for skill questionnaires that are daunting to user and hard to maintain over time.

REFERENCES
1. [Ackerman & Malone, 90] – Ackerman, M & Malone, T – Answer Garden: A Tool for Growing Organizational Memory – proceedings of the ACM Conference on Office Information Systems, Cambridge, MA, April 1990

2. [Collins, 97] – Collins, J.A, et al. - Inspectable User Models for Just in Time Workplace Training – User Modelling: Proceedings of the 6th Int. Conference, Springer Wien, NY, 1997

3. [Ericsson & Charness, 97] – Ericsson, K & Charness, N. – Cognitive and Developmental Factors in Expert Performance, in Expertise in Context, Feltovich, Ford & Hoffman (eds.), MIT Press, 1997

4. [Experts, 97] – Experts Exchange – Experts Exchange FAQ - http://www.experts-exchange.com/info/faq.htm

5. [Foner, 97] – Foner, L. – Yenta: A Multi-Agent, referral-based Matchmaking System – The First International Conference on Autonomous Agents, Marina Del Rey, CA, 1997

6. [Lemay & Perkins, 97] – Lemay, L. & Perkins, C. – Teach Yourself Java in 21 Days, second edition – Sams, 1997

7. [Kautz, Selman & Shah, 97] – Kautz, H., Selman, B. & Shah, M. - ReferralWeb: Combining Social Networks and Collaborative Filtering – Communications of the ACM vol 40, no. 3, March 1997

8. [Mattox, 98] – Mattox, D., Maybury, M. & Morey, D. - Enterprise Expert and Knowledge Discovery – MITRE Corporation - 1998

9. [Salton, 88] - Salton, G. - Automatic Text Processing: The Transformation, Analysis and Retrieval of Information by Computer. - Addison-Wesley, Reading, MA, 1988.

10. [Seifert, et. al] – Seifert, C.; Patalano, A; Hammond, K. & Converse, T. – Experience and Expertise: The role of Memory in Planning for Opportunities. In Expertise in Context, Feltovich, Ford & Hoffman (eds.), MIT Press, 1997

Lurker demographics: Counting the silent

Blair Nonnecke
Dept. of Information Systems
1000 Hilltop Circle
Baltimore, MD 21250 USA
+1 410 455 3795
nonnecke@umbc.edu

Jenny Preece
Dept. of Information Systems
1000 Hilltop Circle
Baltimore, MD 21250 USA
+1 410 455 6238
preece@umbc.edu

ABSTRACT

As online groups grow in number and type, understanding lurking is becoming increasingly important. Recent reports indicate that lurkers make up over 90% of online groups, yet little is known about them.

This paper presents a demographic study of lurking in email-based discussion lists (DLs) with an emphasis on health and software-support DLs. Four primary questions are examined. One, how prevalent is lurking, and do health and software-support DLs differ? Two, how do lurking levels vary as the definition is broadened from zero posts in 12 weeks to 3 or fewer posts in 12 weeks? Three, is there a relationship between lurking and the size of the DL, and four, is there a relationship between lurking and traffic level?

When lurking is defined as no posts, the mean lurking level for all DLs is lower than the reported 90%. Health-support DLs have on average significantly fewer lurkers (46%) than software-support DLs (82%). Lurking varies widely ranging from 0 to 99%. The relationships between lurking, group size and traffic are also examined.

Keywords
Lurker, lurking, discussion list, demographic, newsgroup, BBS, email, health-support, traffic, membership

INTRODUCTION
DLs, newsgroups, and Web-based bulletin board systems (BBSs) have experienced rapid growth as the number of Internet users climbs. As of July 1999, there are more than 131,000 DLs using Listserv's® server software. The 69,000,000 members of these DLs receive in excess of 29,000,000 messages per day [6]. Whittaker et al [19] also cite large numbers for Usenet newsgroups. The growth and prevalence of online groups, coupled with the relative ease of gathering persistent and traceable messages, has made online groups a fertile ground for research. The following are a few of the areas so far studied: the development of friendship [12], the perception and quality of community [15], factors affecting interaction within newsgroups [19], and the development of empathy in health-support groups [13, 14]. Each of these studies was based on examining individuals participating in public spaces, i.e., those who post. None examined their chosen area from a lurking perspective, even though lurkers are reported to make up over 90% of several online groups [2, 7].

Given that lurkers are both unstudied and apparently in the majority, knowing more about them will have benefits in many areas. Their sheer number suggests they are important to study. From a usability perspective, improvements in tools and group design will fall out of a better understanding of lurkers and their activities. For lurkers and their communities, self-knowledge of lurking will demystify lurkers' roles, value, and activities. This has already been shown to be the case when a participant in an initial study responded to a draft article on lurking [11]:

> Maybe it's a sign of my own mild discomfort around being a lurker, but I found it reassuring to recognize myself and my behaviour within the continuum you describe, and to see lurking treated seriously, with both acceptance and respect. As a lurker, I'm used to observing from the sidelines and participating vicariously, and it's strangely gratifying to read an article that speaks directly to that experience. It's almost like suddenly feeling part of an (until-now) invisible community of lurkers.

In their pioneering work, Kollock and Smith [3] describe lurkers as free-riders, i.e., noncontributing, resource-taking members. Knowing more about lurkers and their lurking will show whether this is a fitting description.

As group and community development becomes an important component of commerce on the Internet, understanding lurkers will become an essential part of doing business. Every lurker is a potential customer. For example, Amazon.com has been very successful in creating an online retail environment in which lurkers can make purchasing decisions based on how others have purchased in the past and on reviews supplied by other customers.

Amazon.com has leveraged the information gained from those willing to post reviews into purchasing-support tools for the lurker and poster alike.

Definition

The online Jargon Dictionary [1] defines lurker as:

> One of the 'silent majority' in a electronic forum; one who posts occasionally or not at all but is known to read the group's postings regularly. This term is not pejorative and indeed is casually used reflexively: "Oh, I'm just lurking." When a lurker speaks up for the first time, this is called 'delurking'.

This definition suggests that lurking is the normal behaviour of the majority of the population and that lurking can be defined in terms of the level of participation, either as no posting at all or as some minimal level of posting. However, defining lurking is problematic. Should someone who never posts in public spaces, but regularly side-posts to individual group members, be deemed a lurker? If a person posts once and then never again, are they lurking? Is someone lurking when they go on holidays? Is someone lurking when for a period of time they do not post? While these are important considerations, this study takes the simple approach of defining lurking as either no posts or some minimal number of posts over a period of time.

Research questions

The work reported here is the second in a series of studies on lurking [10]. In the first study [11], Internet users were chosen for their membership in online groups, and not for their posting frequency. Given that lurking has been reported as a common means of participation [2, 7], it was assumed that lurkers and their behaviors would be readily encountered within the general Internet population. In the first study, it was found that each participant lurked in at least one online group, and several lurked in all of their online groups. This finding, among others, reinforced the need to better understand lurking. A demographic survey of online discussion groups would provide a different perspective from the first study by emphasizing quantitative measures.

DLs, rather than BBSs or newsgroups, were chosen as the basis of this study for a number of reasons. For the results to have their greatest value, the chosen communication technology needed to be widely used. L-Soft's usage figures show very high levels of use, and of the online discussion groups mentioned by participants in the first study, 25 of the 41 were DLs accessed through email. Just as importantly, DL servers track membership through their subscription mechanism. In turn, DL membership information can be accessed by querying a DL's server. The level of lurking can be measured by tracking posted messages and identifying posters. In contrast, membership levels are unavailable for most BBSs and newsgroups.

This study is an extension of work on online health-support communities [13, 14]. As well, in a study of who pays for content and interactive media, McMillan [8] provides several reasons for studying health-related groups:

> ...health and health related subjects have in the past played a central role in the early financial support in many media; health related sites are the fastest growing topic areas in CMC; health-related sites are heavily used; and this area contains one of the fastest growing categories of consumer advertising.

For these reasons, health-support DLs are the focus of this investigation. For comparison purposes, software-support DLs are also included in this study.

The remainder of this paper examines four main questions:

Q1. How prevalent is lurking, and do health and software-support DLs differ?

Q2. If lurking is defined as no posting, what happens to lurking levels when the definition is broadened to include minimal levels of posting, e.g., 1 post/month?

Q3. Is there a relationship between lurking and the number of members in the DL?

Q4. Is there a relationship between lurking and the traffic level of the DL?

METHOD

The primary aim of this work is to understand how much lurking occurs in DLs, with specific emphasis on health and software-support groups.

Selection of DLs

To select DLs for the study, L-Soft's CataList catalog and DL search facility [4] were used to locate suitable DLs. A search on the word "support" resulted in a listing of 300 DLs and a description of each. From this listing, subscriptions were taken out on all public DLs relating to health or software-support. To increase sample size, additional subscriptions were taken out on a random selection of health (22) and software (10) support DLs. Although the additional DLs provide support for their members, neither their title nor their one-line catalog description contained the term "support". (Note: Analysis comparing these additional support DLs shows their lurking levels are not significantly different from those found through searching on "support", and as a result, they are included in this study.)

In addition to DLs related to health and software, a random set of DLs on other topics were selected for their large size (CataList displays a description of all DLs with membership greater than 1000 [5]). Eleven randomly selected DLs between 1000 and 2000 members were included as a basis for examining whether large DLs have a greater proportion of lurkers than smaller ones (see Q3. above).

DL set	1. Lurking (% of membership)			2. No. of members			3. Traffic (posts/day)		
	Mean	SD	SE	Mean	SD	SE	Mean	SD	SE
All N=109*	55.5	29.6	2.8	551	678.3	65.0	16.2	18.4	1.8
Health N=77*	45.5	28.7	3.3	398.4	439.9	50.1	18.4	18.4	2.1
Software N=21*	82.0	13.9	3.0	662.4	1091.2	238.1	3.1	4.7	1.0

* No. of DLs in set, SD=Standard Deviation, SE=Standard Error of Mean

Table 1: Lurking, no. of members and traffic for the DL sets

Data collection

Messages were collected from the selected DLs over a three-month period at a rate of slightly less than 2000 messages/day. Eudora Pro was used to collect and filter email into separate mailboxes for each list, and to monitor the process on a regular basis. Using CataList, the membership size of each list was determined at the beginning and end of the collection period. Lurking levels reported in this study are based on the lower of the two membership levels recorded for the 12-week period.

At the end of 12 weeks, the DLs were examined to ensure that each DL had sent at least one post a month for the 12 weeks. Of the 135 original subscriptions, 109 DLs are included in the study. DLs were dropped from the study if they stopped sending messages for any reason, e.g., change of server, failure on the part of the researchers to reply to subscription notices, or a non-active DL. Messages from the remaining DLs were then run through a Perl script producing records containing the following fields: list name, date, time, size of message, subject heading, and sender. 147,946 messages were transcribed into records and imported into an SQL database. This provided an effective and flexible means for querying and analyzing the data. The data collected represents over 60,000 members and 19,000 posters.

RESULTS
Lurking levels

Q1. How prevalent is lurking, and do health and software-support DLs differ?

Using information from the SQL database, mean lurking levels were calculated for the set of all DLs, and for each of the health and software DL sets (see Table 1, column 1). Lurking was defined as no posts within the 12-week collection period. The mean lurking level for all DLs is less than the 90% figures reported by Katz and Mason [2, 7]. It should be noted that while the mean was less than 90%, 12% of the DLs had lurking levels higher than 90%.

The differences in mean lurking levels between the health and software-support DLs is significant. Software-support groups had almost double the number of lurkers. Figure 1 shows the distribution of the lurking levels for each DL type using a box and whisker display. (Note: See Sternstein [16, p. 37] for further information on this visual representation.) Each horizontal line represents a boundary for 25% of the DLs in the sample. The thicker line is also the median for each type. Each of the central boxes contains 50% of the DLs.

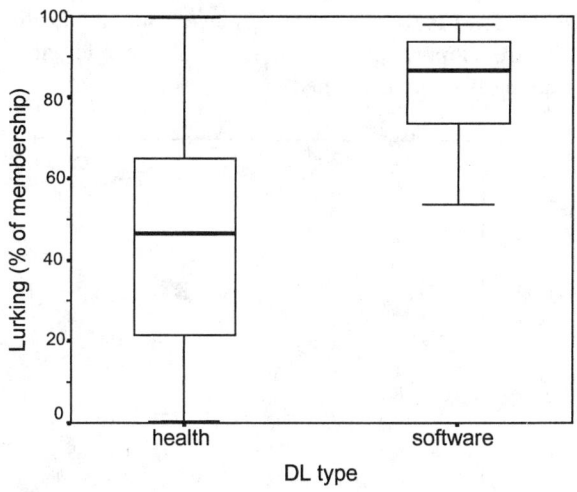

Figure 1: Distribution of lurking levels by quartile for each DL set

Software-support DLs show less variation and none have a lurking level of less than 50%. By contrast, the lurking level of health-support DLs range from zero to 99%. Health-support is a broad umbrella under which to investigate group behaviour. As such, lurking levels may vary according to a number of other factors, e.g., list topic, illness vs. injury, or chronic vs. short term disorders. This difference in variation between the two DL types may be the result of the greater number of health-support DLs in the study, which represents a broader cross-section of their type.

Apart from the low mean number of lurkers in the health-support DLs, what appears most striking about these results is the large variation in lurking levels, and that on average the lurking level for all DLs is lower than the reported 90% figure [2, 7].

Broadening the definition of lurking

Q2. If lurking is defined as no posting, what happens to lurking levels when the definition is broadened to include minimal levels of posting, e.g., 1 post/month?

In Table 1 lurking was defined as no posts during the 12-week collection period. If lurking is examined on a sliding scale where the allowable posting level can grow, a somewhat different picture emerges. In Figure 2, lurking levels were calculated for a range of cumulative posts, from no posts to 3 or fewer posts for the 12-week period (i.e., 1 or fewer posts per month). As the definition broadens to include more posts in the 12-week period (towards the 3 level), lurking levels move higher. At the level of 3 or fewer posts per 12-week period, the mean lurking level for the health DLs is still lower than 90%, while the software DLs' mean has moved above this level. Both the health and software-support DLs behave in a similar manner, and their relative offset is maintained.

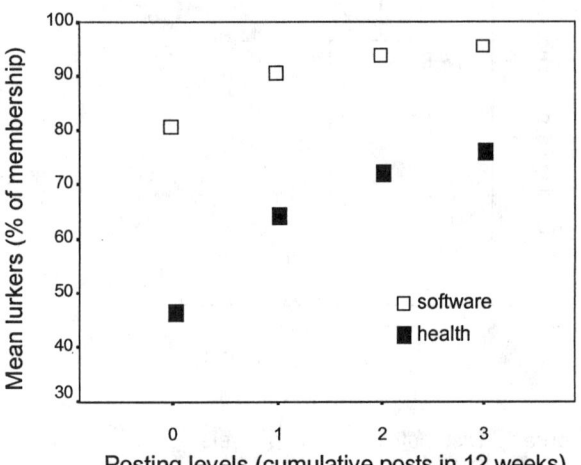

Figure 2: Variation of lurking levels for a range of cumulative posts.

A posting rate of 3 posts in 12 weeks is still an infrequent level of posting. It could be argued that most of what is being done by members at this level is not posting. Presence or visibility of members within a list may be a better indicator of lurking, i.e., is a member known to the other group members in a way that makes them somehow recognizable and thus not lurkers. Defining lurking as a function of the visibility of the poster suggests that other factors would influence this visibility, e.g., the number of members, the number of posters, the activity of the list, and the value and/or notoriety of each participant. It is possible that someone who flames on an irregular basis may be seen as less of a lurker than someone who contributes in a regular but less visible manner. The polar opposite of lurking may be stardom.

Further work is needed in understanding lurking. For example, lurking may not be a continuous state and could be punctuated by periods of public posting based on topic or need. Using the current data set, there is no reason why analyses of this type cannot be carried out in the future. The raw data could also be used from a contextual or ethnographic perspective, one in which content and dialogue analyses could be carried out. Examples of these kinds of analyses can be found in Preece and Ghozati [14], and Worth and Patrick [20].

Lurking and the number of members

Q3. Is there a relationship between lurking and the number of members in the DL?

In large DLs lurking may be easier. As the number of members increases, the need for any given member to participate may decline. In addition, high posting levels could create chaos and lurking in large DLs may be a practical means of reducing the number of posts and maintaining order. If either of these is the case, then large DLs should have a greater proportion of lurkers than smaller ones. As can be seen in Table 1 (column 2) health-support DLs have on average fewer members than the software-support DLs. If increasing membership size has the effect of generating more lurkers, then the difference in mean membership levels could explain why health-support DLs have lower levels of lurking.

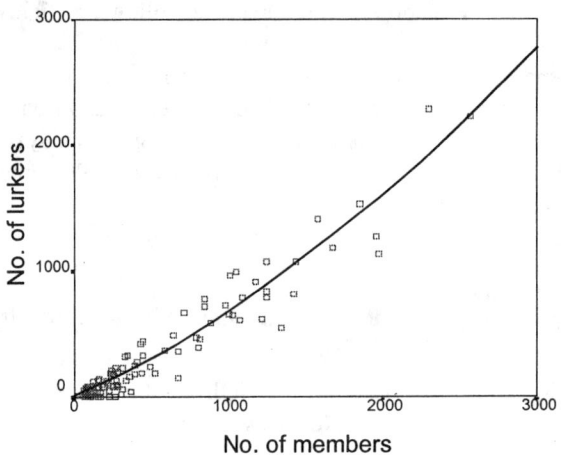

Figure 3: No. of lurkers vs. no. of members for each DL

On examining all 109 DLs in the sample, the anticipated greater incidence of lurking in larger DLs is not strongly shown. Figure 3 shows a strong positive non-linear relationship between the number of lurkers and the size of the DL. A linear regression also fits this data equally well. If this result is taken at face value, membership size does not explain the differences in lurking between the health and software DLs.

The relatively few DLs with over 500 members skews the relationship in favour of the larger DLs. Of the 98 health and software DLs, 74 of them have fewer than 500 members. Figure 4 is a scatter plot of these smaller DLs. The regression line in Figure 4 is a strong positive relationship with a slope less than that in Figure 3. This suggests that for DLs with fewer than 500 members, there are on average fewer lurkers than in the larger DLs. It should also be noted that that the software-support DLs in Figure 4 are distributed in a straight line. This suggests that even when software-support DLs are of equivalent membership size, they will on average have higher lurker levels.

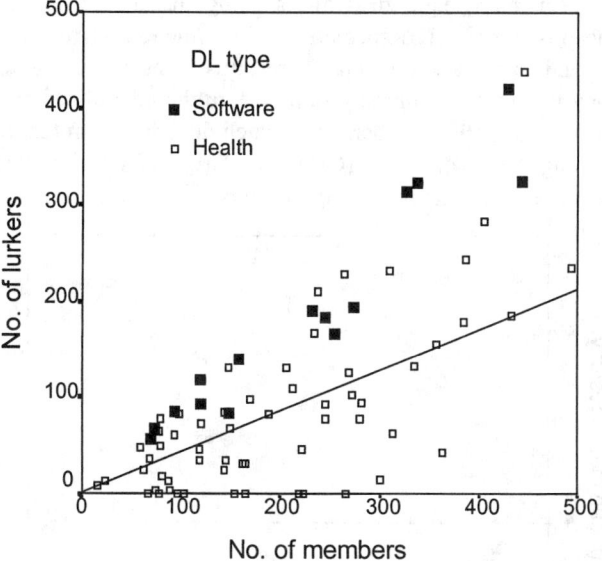

Figure 4: No. of lurkers vs. no. of members for each DL with less than 500 members

DL members receive no direct information about the number of members in a DL. The cues that do exist are indirect, e.g., a query to the server for information, the number of different members posting, the variety of topics covered, and the traffic on the DL. It is possible that a DL of several thousand members could behave like and be indistinguishable from one with only 100 members. More work is required to understand how the size of DLs is perceived by members, and how members respond to this in their various forms of participation.

Lurking and DL traffic levels

Q4. Is there a relationship between lurking and the traffic level of the DL?

From the perspective of personal email management, once message rates get above a comfortable level, participating in a DL may take more effort, i.e., there are more messages to read, skim, reply to, etc. Based on participant input from the first study [9], traffic levels were divided into four categories requiring varying levels of management effort (see Table 2).

Management effort	Traffic level	
	messages/week	messages/day
None	< 1	< 0.14
Low	1-3	0.14-0.5
Medium	4-42	0.5-6.0
High	> 42	>6.0

Table 2: Traffic levels for a DL and the corresponding management efforts

The categorization was done prior to examining the distribution of posting rates from the current study. Several participants in the first study indicated that lists with less than 4 or 5 posts per day were easy to handle. In the current study, more than 50% of the DLs fall in the High category. It should also be noted that what is manageable will vary widely between individuals, and will depend on many factors, including type of email software, experience, demands on time, and interest.

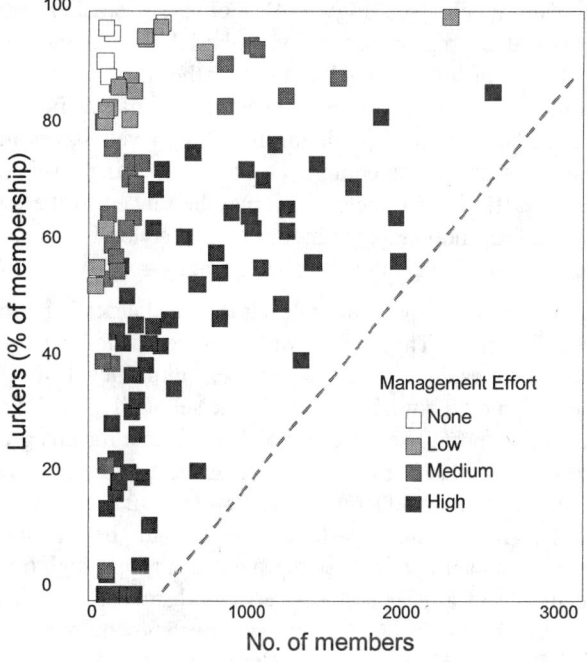

Figure 5: Lurker levels as related to management effort and number of members for each DL

Lurking levels for all DLs were negatively correlated with traffic (Pearson's correlation coefficient of -.426 is significant at the .01 level). Figure 5 shows that for a given DL size, lists with highest traffic levels generally have the lowest lurking levels. Banding by traffic level is visible, starting with the lowest traffic level (None) in the top left hand corner and progressing towards the lower lurking levels and larger DL size. This partially explains the lower levels of lurking in health-support DLs as these had the higher traffic levels (see Table 1, column 3).

Conspicuously absent are DLs in the area below the broken line, which appears to be a kind of interactive no-man's land. Why this should be the case is not known at the present time, but it could be related to the difficulty of making sense out of large DLs with high traffic volumes and large membership levels. At some point, the DL may become unusable and self-adjust through membership attrition and/or a decrease in public posting. It may be that lurking increases under conditions where having a public voice is difficult. In our initial study [11], several participants indicated they knew other people would post opinions similar to their own in active lists, and thus felt no need to post. When traffic is high, there is a sense that adding messages to the list only increases the traffic without improving the quality. For them, lurking was a way of reducing the noise on the list, a civic duty so to speak. It would be interesting to examine DLs that fall near or below the broken line, and determine whether they transform in any way, e.g., split, have high membership turnover, etc.

Below the 500-member level, health-support DLs appear evenly distributed with respect to number lurkers and thus lurking levels (see Figure 4). For these smaller, more personal-sized groups, the size of the DL may be less of an indicator of lurking level and some other factors may be at work. For DLs with fewer than 500 members, traffic levels appear to be a good predictor of lurking levels (see Figure 5). What drives the combination of low lurking levels and high traffic is still unclear, but may be related to the topic of the DL, motivation of members, and style of interaction (e.g., empathy vs. information exchange).

The DLs with high traffic levels are an interesting group (see Figure 6). The 11 DLs with average traffic levels over 40 messages/day had a low average lurking level of 44%. Four of the DLs were from the Large set of DLs and 7 were health-support. The median membership size for this group was high, at 1220. However, three of these high traffic DLs had fewer than 500 members. For the DLs in this high traffic range, lurking levels appear to be randomly distributed across membership size. As a result, high traffic levels don't appear to be a very good indicator of group size. It is possible that group size becomes immaterial to public participation when it isn't readily knowable.

DISCUSSION

Much of the discussion related to the four original questions can be found in the previous section. Therefore, this next section focuses on three important issues: lurkers as free-riders, traffic levels, and lurking elsewhere, i.e., how lurking in DLs may differ from either BBSs or newsgroups.

Lurkers as free-riders

In the Introduction it was mentioned that Kollock and Smith [3] describe lurkers as free-riders. Describing lurkers as free-riders classifies them for their lack of public participation and their use of resources without giving back to the group. Even when lurking is narrowly defined, e.g., less than one post/month, the vast majority of DL members are lurkers. This being the case, how do online groups survive in the face of almost universal free-riding?

One explanation is that lurking is not free-riding, but a form of participation that is both acceptable and beneficial to online groups. Public posting is but one way in which an online group can benefit from its members. Members of a group are part of a large social milieu, and value derived from belonging to a group may have far-reaching consequences, e.g., virus alerts being distributed beyond the posters of a DL specializing in combating viruses. A second explanation is that a resource-constrained model may not apply to online groups where the centralized cost of servicing 100 members isn't much different from that of serving 1000, or even 10,000. In large DLs the danger could be in not having enough lurkers.

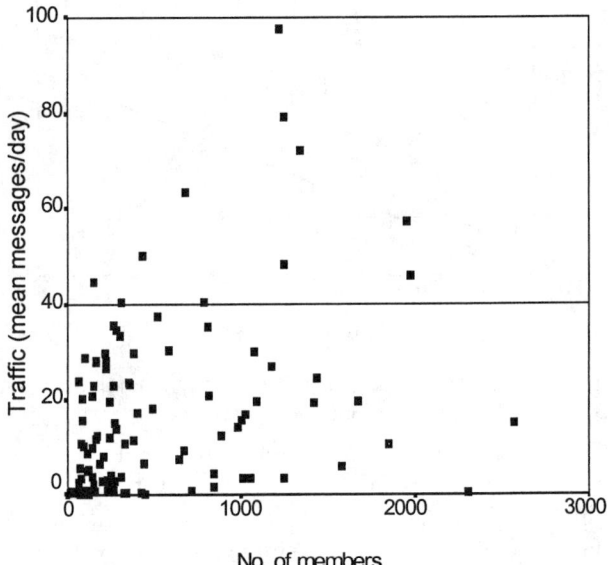

Figure 6: Traffic levels vs. no. of members

Traffic levels

In our first study [11], participants described the effort required to manage DL traffic. If there were few messages, then the DL was effectively out of mind and required little or no effort. If there were many messages, then the DL became burdensome. Several participants cited newsgroups as being less useful because of the large volume of messages. They also mentioned the quality of the messages as being very important, e.g., content, knowledge base of participants, and courtesy. Several participants left newsgroups because of what they perceived as low content quality.

Using figures supplied by participants, DLs with traffic levels of over 6 messages/day were categorized as requiring higher effort to manage. The mean traffic level for the software-support DLs was 3.1 posts/day (see Table 1, column 3). These values fit nicely with our expectations of manageable traffic. However, the mean traffic level of the health-support DLs was 18.4 messages/day, and one DL exceeded an average of 97 posts/day. These higher-than-expected numbers suggest that these DLs are somehow different than the DLs participants described as being ideal in the first study. Why the discrepancy? It is possible that these DLs supply such high-quality content that their members are willing to put in the higher effort to deal with them. It is also possible that high traffic DLs act like many little DLs, each identifiable by a set of subjects and/or authors. The observed high traffic levels suggest that what is an acceptable and perhaps necessary traffic level in one DL may be unacceptable in another. It also suggests that motivation, in addition to quality of messages, is an important facet of acceptable traffic levels.

Understanding what constitutes acceptable traffic rates is an important issues in designing online communities. E-commerce is already running into this problem. For example, when sending promotional materials through distribution lists, it is important to understand how much email can be sent before customers perceive it as a nuisance. Understanding how DL members cope with and make use of high volumes of messages is important for the designers of email-client software. Lastly, messages from DLs are not received in a vacuum; they compete with messages from a variety of other sources, including personal and professional correspondence, and email from other DLs -based email.

Lurking elsewhere

This study focused on lurking in DLs as it would not have been possible to measure lurking levels using posting data from either newsgroups or BBSs. However, it is important to understand the limitations of focusing on DLs by examining some of the differences between DLs and both newsgroups and BBSs.

Perhaps the most important difference is that DL messages are received as email. DL email competes with other types of email for the attention of the subscriber. While it is true that most email clients are capable of filtering and depositing email in separate mail boxes, this has not been shown to be the practice of most email users [9, 18].

In contrast to DLs accessed through email clients, Web-based BBSs and newsgroups are accessed through specially built user interfaces. This separates group communication from other non-group communication. Furthermore, the act of retrieving messages from either a newsgroup or a BBS is conscious and deliberate. Email clients often perform the task of retrieving e-mail automatically, e.g., once every 10 minutes. Email clients can also be used to get or check for email on demand. What is not known is whether an active vs. a passive process of obtaining messages has any impact on participation, e.g., reading, browsing, or posting.

There are two other major differences between DLs and the other tools. Firstly, email-based DLs poorly show conversational threading, and secondly, messages can be received as a digest (a single large email containing a set of messages for the purpose of reducing the volume of email). In both cases, the onus is on the receiver to reconstruct conversational threads. If the continuity of subject headings is to be maintained in the DL, replying to a message received in digest form requires the reply message's subject header to be manually constructed. The lack of visible threading and awkwardness of replying is being addressed by recent advances in digest-reader software [17], but it is not yet a common feature in email clients. In high traffic DLs, the lack of threading and digest format may make it harder to follow conversations. This in turn may make it more difficult to publicly join in the conversation.

In our first study [11], several participants described subscribing to a DL as a form of commitment with associated responsibilities to the other members. They also felt posting to a DL increased their commitment to the group and the presence created through posting should be maintained. Most DLs reinforce this by sending out a welcome message outlining what is expected of members in terms of participation and behaviour. By contrast, there is no subscription process for most BBSs and newsgroups. As a result, participation in DLs may differ from either BBSs or newsgroups, due to a different sense of responsibility to the group.

The effects of different types of email tools and skills have been ignored in this study. However, this could be an important difference between health and software-support DLs and their participants. Software skill and acumen may vary for participants in these DL types. For example, members of software-related DLs may have better computer skills and a greater knowledge of the Internet than those of other DL types.

Personal characteristics that may impact lurking include motivation and comfort in communicating online. To investigate these other approaches are called for. These include member surveys and the examination of DLs from a content and dialogue perspective.

CONCLUSION AND FUTURE WORK

As this study shows, lurkers are everywhere, and that is OK. A case can be made for lurking being normal and public posting being abnormal. After all, if everyone were posting, who would be reading. It is unfortunate that the term lurker, with all of its negative connotations, has gained acceptance. Fortunately, lurking can now be understood as the many activities related to membership in online groups.

Rather than being free-riders, lurkers should be called participants (publicly silent though they may often be).

As a quantitative follow-up of our interview-based study [11], this work proved a capable tool for understanding lurking. There is some irony in studying lurking with a method normally reserved for examining public participation. This work was successful in discovering a number of relationships between lurking levels, DL type, membership levels and traffic. Whether they are causal or not, is left to future work.

The data from this study will continue to be used for follow-up work. Specifically, it will be used to determine whether lurking is related in any way to the diversity of topics within a DL (i.e., breadth vs. depth of the DL), to the distribution of contributions by members of the DL (i.e., the role of stars in a DL), to the response members receive when they delurk, and to the length of messages.

Another area worth pursuing, but perhaps outside of this data set, is the investigation of high-traffic DLs and their members. For cxample, how do members cope with high traffic levels?

ACKNOWLEDGMENTS

Any project of this type requires a wide range of talents. We are indebted to Thawatchai Piyawat and Dick Seabrook for their consummate skills in constructing the Perl scripts and SQL database. Drafts of this paper were considerably improved through the thoughtful comments and editing of Carolyn Davidson and Heather McDonald.

REFERENCES

1. Jargon Dictionary. Lurker Definition. Available at http:// www.netmeg.net/jargon.

2. Katz, J. Luring the Lurkers. Available at http:// slashdot.org/features/98/12/28/1745253.shtml.

3. Kollock, P., and M. Smith. Managing the virtual commons: cooperation and conflict in computer communities. In *Proc. Computer-Mediated Communication: Linguistic, Social, and Cross-Cultural Perspectives*, edited by S. Herring. Amsterdam: John Benjamins, 1996, 109-128.

4. L-Soft International. List search. Available at http:// www.lsoft.com/lists/list_q.html.

5. L-Soft International. List with 1,000 subscribers or more. Available at http://www.lsoft.com/scripts/wl.exe?XS=1000.

6. L-Soft International. LISTSERV -- The mailing list management classic. Available at http://www.lsoft.com/listserv.stm.

7. Mason, B. Issues in virtual ethnography. In *Proc. Ethnographic Studies in Real and Virtual Environments: Inhabited Information Spaces and Connected Communities* (Edinburgh, 1999), 61-69.

8. McMillan, S. J. Who pays for content? Funding in interactive media. *Journal of Computer-Mediated Communication* 4, 1, (1998), Available at http://www.ascusc.org/jcmc/vol4/issue1/mcmillan.html.

9. Nonnecke, B. *Lurking in email-based discussion lists*. Ph.D. thesis: SCISM. London: South Bank University, 2000.

10. Nonnecke, B., and J. Preece. Persistence and lurkers: A pilot study. In *Proc. HICSS-33* (Maui, Hawaii, 2000), IEEE Computer Society.

11. Nonnecke, B., and J. Preece. Shedding light on lurkers in online communities. In *Proc. Ethnographic Studies in Real and Virtual Environments: Inhabited Information Spaces and Connected Communities* (Edinburgh, 1999),123-128.

12. Parks, R. M., and K. Floyd. Making friends in cyberspace. *Journal of Computer-Mediated Communication* 1, 4, (1996), Available at http://www.ascusc.org/jcmc/vol1/issue4/parks.html.

13. Preece, J. Empathic Communities: Balancing emotional and factual communication. *Interacting with Computers: The interdisciplinary Journal of Human-Computer Interaction* 5, 2, (1998), 32-43.

14. Preece, J., and K. Ghozati. Empathy online: A review of 100 online communities. In *Proc. Association for Information Systems, 1998 Americas Conference* (Baltimore, MD, 1998).

15. Roberts, T. L. Are newsgroups virtual communities? In *Proc. CHI'98 Conference on Human Factors in Computing Systems* (Los Angeles, CA, 1998), ACM Press, 360-367.

16. Sternstein, M. *Statistics: Barron's Educational Series*, Inc., Hauppauge, NY, 1996.

17. TECHWR-L. About the Digest Reader. Available at http://www.raycomm.com/techwhirl/digestreader.html.

18. Whittaker, S., and C. Sidner. Email overload: Exploring personal information management of email. In *Proc. CHI'96* (Vancouver, BC, 1996), ACM Press, 276-283.

19. Whittaker, S., L. Terveen, W. Hill, and L. Cherny. The dynamics of mass interaction. In *Proc. CSCW'98* (Seattle, WA, 1998), ACM Press, 257-264.

20. Worth, E. R., and T. B. Patrick. Do electronic mail discussion lists act as virtual colleagues. In *Proc. AMIA Annual Fall Symposium* (1997), 325-329.

Talking In Circles: Designing A Spatially-Grounded Audioconferencing Environment

Roy Rodenstein, Judith S. Donath
Sociable Media Group
MIT Media Lab
20 Ames Street
Cambridge, MA 02142
{royrod, judith}@media.mit.edu
http://www.media.mit.edu/~royrod/projects/TIC

ABSTRACT

This paper presents *Talking in Circles*, a multimodal audioconferencing environment whose novel design emphasizes spatial grounding with the aim of supporting naturalistic group interaction behaviors. Participants communicate primarily by speech and are represented as colored circles in a two-dimensional space. Behaviors such as subgroup conversations and social navigation are supported through circle mobility as mediated by the environment and the crowd and distance-based attenuation of the audio. The circles serve as platforms for the display of identity, presence and activity: graphics are synchronized to participants' speech to aid in speech-source identification and participants can sketch in their circle, allowing a pictorial and gestural channel to complement the audio. We note user experiences through informal studies as well as design challenges we have faced in the creation of a rich environment for computer-mediated communication.

Keywords: Computer-mediated communication, audio, speech, drawing, representation, media space, interaction design, multimodal interfaces, multicast, social navigation, gesture

INTRODUCTION

Communication is one of the primary applications of computing. Electronic mail has become ubiquitous in certain sectors, with synchronous computer-mediated-communication surging in the last decade through networking improvements and critical mass of the online population. Chat, whether purely textual or accompanied by graphics, has ridden the growth of the World-Wide Web to become a popular medium for social interaction. Traditional chat environments, however, are limited by the physical and expressive bounds of typing as input for synchronous communication. Disparities in participants' typing ability and expertise with particular writing conventions of the online medium, as well as the lack of traditional emotional cues such as tone of voice, are among these limitations.

Audio-based communication ameliorates some of these issues, allowing voice interaction to leverage users' experience with spoken conversation. Audio spaces have demonstrated clear potential for fostering rich social interactions [1]. However, though speech is a very natural form of communication and allows a great range of expression, audio-only spaces place multiple conversations into a single audio stream, serializing speaker interactions, and establish user presence only during their transitory speech [26]. We address these issues through the use of simple graphics that display presence, enable parallel conversations, and reveal the evolving interactions within the social space.

With the aim of supporting rich, naturalistic social interaction we have developed *Talking in Circles,* a computer-mediated social environment in which speech is the primary communication channel and graphics convey important expressive and proxemic information. Our environment is a two-dimensional space within which participants are represented as circles of various colors, as shown in Figure 1. Participants can draw on their circle, and the system graphically shows when each is speaking. In addition, users hear those they move close to more clearly than others farther away. This property has supported naturalistic behaviors such as users approaching those they are interested in conversing with, forming conversational subgroups of several users located near each other, and mingling by moving around within the space, listening for topics of interest and moving to join particular groups.

Participants' circles thus serve as indicators of presence and as cues about membership and activity in various conversations, one way our system addresses the problem of observing the state of the media space. The circles are also used as platforms for graphical display. The system visually renders each participant's speech activity on their circle, making it easy to identify individual speakers and differentiate among several overlapping speeches. It also allows users to draw on their own circles, complementing the audio channel with pictorial communication. The screen-shot in Figure 1 shows these graphical features at one instant in time in a *Talking in Circles* chat session.

In the following section we review work which relates to our design for *Talking in Circles*. We then discuss the process of our system's interaction design and analyze it in the context of relevant studies. Next we briefly describe the implementation, noting our responses to technical challenges we encountered. We conclude by noting our ongoing work in this area as well as future directions.

RELATED WORK

Many systems have been used to research computer mediation of audible communication between people. In general these are geared toward computer-supported collaborative work or focus on particular modalities. *Thunderwire* is one audio-only media space studied over several months [10]. Although the system did not foster much work-related communication, it was very successful as a sociable medium based on such criteria as informality and spontaneity of interactions. Interesting usage norms evolved as a result of the lack of visual feedback to deal with finding out who was listening on the system and to indicate lack of desire to converse at particular times. Our work leverages the pliable, sociable quality of the audio medium and uses a graphical interface to resolve participants' lack of knowledge about the space's membership and member interests.

Sun's *Kansas* system uses videoconferencing within small windows and a distance threshold for audio [13]. Geared toward distance learning, it employs a screen-sharing approach to complement videoconferencing with various applications. *FreeWalk* also used video, mapping it onto a flat surface and placing participants in a three-dimensional environment [19]. *FreeWalk* succeeded in fostering certain social behaviors, such as following a participant from afar. Audio was faded based on distance, though not based on the direction participants faced, leading to difficulty in speaker identification in some cases. The excellent infrastructure provided by the more general *MASSIVE* project had similar problems in some of the interfaces developed for it, due to limited representational feedback for participants' speech [8]. *Talking in Circles*' use of animation synchronized with users' speech resolves these sound-source-identification problems.

In terms of point-of-view, both *MASSIVE* and *FreeWalk* focused on close-up views of figurative user representations, differing from our work's emphasis on highly abstract representation and visual overview of the space. Though *MASSIVE* offered flexible viewpoints there were sometimes resultant mismatches in behavior, such as users with disparate views walking right "through" others' bodies. We chose as a tradeoff a single cohesive viewpoint, and address the issue of personal physical boundaries in the section on circle motion.

In the realm of graphical text-based chat, systems such as *The Palace* [21] and *Comic Chat* [16] use typed communication enhanced with changeable displays of avatars for participants. *Babble* shows participants as colored dots which converge toward the center of a circle as they actively take part in a specific chat, selected from a list of topics [7]. Previous work in our group includes *ChatCircles*, a text-based two-dimensional chat space which explored abstract representation of participants as circles as an alternative to the broadly caricatured feel of traditional avatars [32]. *ChatCircles* displays participants' typed messages on their circle, whose expansion thus signals activity. The text is visible only to participants within a threshold distance from the sender.

Our work on *Talking in Circles* makes several contributions to research in computer-mediated communication, including a novel approach to resolving collisions between participants' graphical representations, graphical aids to speech-source indentification, and prominent access to lightweight multimodal communication. More broadly, our work is characterized by a willingness to explore non-traditional representations, exploit rich yet low-overhead interface affordances, and focus specifically on *social*, as opposed to task-oriented, communication. We now discuss this work in detail.

SOCIAL INTERACTION DESIGN

A first issue we dealt with in creating an audio-based communication space was helping users map the voices they hear to the circles representing the respective participants. That is, we wanted to move the user experience from disembodied voices to a more cohesive perception of fellow participants. We thus use a bright inner circle displayed inside their darker circle to represent the instantaneous energy of the speech from a particular participant. Figure 1 shows that Sandy and Al are speaking, with their inner circles' size showing how loud their speech was at that instant. Thus, natural pauses in conversation, which leave only particular circles showing speech activity, make this cognitive matching problem much simpler. Distance-based spatialization, discussed below, also helps, as speech from circles farther from the user's circle sounds fainter. Finally, identity cues such as learning a participant's voice or their circle's labeled name also resolve matchings.

We conducted an informal test of the graphical feedback provided by the dynamically-changing bright inner circle. Six subjects were shown two circles with non-identifying names, equidistant from their own circle and at equal audio volume. The two test circles each played a different RealAudio news stream, and we asked participants about their experience in trying to match the two speakers they heard to the correct circles.

Although this scenario is challenging, with constantly overlapping speech and no individuating cues for the circles, all subjects successfully matched each stream to its corresponding circle within a few seconds. Though simultaneous speech was at first confusing, the subjects mentioned that the occurrence of short natural pauses in the speech soon made the matching apparent, as only one voice was heard and only one circle was bright. A similar situation happened when one speaker said something loud causing one bright inner circle to grow visibly larger than the other. In general, the subjects said the speech-synchronized updating of the bright inner circle's helped them differentiate and identify the speakers by highlighting the matching rhythms of the speech.

We also experimented with graphical display of recency of participant activity, allowing the bright inner circle to fade out slowly over time when a speaker stopped speaking. This provided a slightly enhanced short-term history but it interfered with the real-time feel of the inner circle's rhythm. In addition, we noted that recency of activity is not necessarily equivalent to availability. Although we think availability display is useful, accurate automated detection of participant availability is infeasible while manual control of availability status is also unreliable. As Ackerman found local disruptions can cause frequent changes in listening or speaking availability of users without their remembering to turn off their microphone, even when it resulted in unwanted eavesdropping [1]. Similarly, lack of speech by a user for several minutes does not guarantee that they are not still listening. The system therefore does not currently attempt to display availability status.

Spatialization

While the system supports various user capabilities, some are supported directly while others arise out of a combination of modalities. The most salient behavior is that of circle motion as an indication of interest and membership in a conversation. As Milgram notes, rings are a naturally-emerging configuration of people engaged around a common activity [18]. In *Talking in Circles*, as in face-to-face situations, standing close to someone permits one to hear them clearly while also reducing distraction from other sources farther away, due to distance-based audio attenuation.

This natural tendency toward physical alignment, besides being a functional conversational feature and serving to a limited extent the role of gaze, has additional benefits. It allows other participants to view the formation of groups or crowds around a particular discussion, letting them gauge trends in participants' interest and advising them of conversations that are potentially interesting. Crowd motion does not necessarily require explicit attention; as in real life large gatherings stand out, and can continue to draw people as users notice the traffic and wonder what the fuss is about.

An additional important benefit of this crowd motion is simply the vitality with which it imbues the space. Whyte remarks on the fact that the biggest draw for people is other people, and notes the popularity of people-watching as a form of triangulation— simply stated, a stimulus source which can be observed by multiple members of the population, potentially giving rise to conversation between strangers. The grounding in a 2D space may also bring in features such as traveling conversations, where conversants move across a space to find a comfortable spot along the perimeter [33]. As in real life, it is possible, with some effort, to be near a particular speaker but attend to another, or to stand between two groups and attend to both conversations.

Selective attention, enabled by the physical grounding and audio attenuation, also provides some of the benefits related to the 'cocktail-party effect' [5]. Though audio from those one is closest to is heard most clearly, nearby conversation can be heard more softly within a certain distance threshold. This helps a user concentrate primarily on the conversation group they have joined while preserving peripheral awareness with the possibility of 'overhearing,' such that the mention of a name or keyword of interest can still be noticed. Thus, social mingling is fostered, as participants can move between subgroup conversations as their interest changes or move to an unoccupied physical space and start their own conversation.

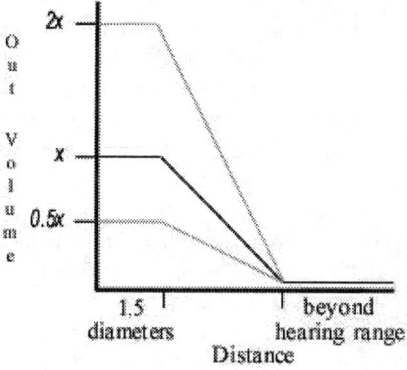

Figure 3: Output volume as a function of distance from a speaker, for input volume x (black line), $0.5x$ and $2x$. The hearing range is 5 diameters.

In order to allow clear audio for participants in a conversation, no audio fading is done within a distance of 1.5 diameters from the center of each speaker's circle. This allows participants located next to or very close to each other to hear the full volume of speakers' speech, while we still perform fading for circles in conversations farther away. Figure 3 plots the shape of the audio-fading function to show how output volume varies by distance from a speaker. The function remains the same but is parameterized by the instantaneous input volume, as shown by the upper and lower lines in the figure. This modification to the spatial rules of our environment preserves the positive qualities of audio fading but helps members of a conversation hear each other clearly; our focus on spatial grounding is always rooted in fostering a sociable space. Though detailed user control over fading parameters could be beneficial, such as in the case of a very widespread conversation group, customizing the physical rules of the space can lead to inconsistent user experiences [29] as well as unnecessary GUI clutter [28].

The distance threshold for audio to be heard, currently 5 diameters, serves multiple functions. Naturally, it aids performance optimization by obviating the need for audio playback for clients beyond the threshold. The major benefit, however, is letting the user know that they cannot hear someone, as activity by those beyond the hearing range is rendered as a hollow circle. For example, screen shots of a *Talking in Circles* chat from the screen of participant Al, the blue circle, shows he has moved from a conversation with Andy and Helen in Figure 1 to one with Josh and Yef in Figure 2. The hollow orange inner circle shows that Al is now beyond the hearing range of Sandy. Since the hollow circle still indicates speech, however, a participant can still note a flurry of activity even if they cannot hear it, and can move closer to see what the discussion is about if they so desire. The audio threshold is symmetric, such that if user X is too far to hear user Y, Y is also too far to hear X. This feature lets a user easily find a spot where they cannot be heard by a certain group, by noting when their inner circles appear hollow. Thus, as in a real cocktail party, one can move to the side to have a semi-private conversation, although this privacy relies only on social rules and is not enforced or guaranteed by the system. These interaction possibilities address some shortcomings of video-mediated conversation, such as the lack of a "negotiated mutual distance" and of a sense of others' perception of one's voice [27].

One group of visitors to our laboratory who tried the system suggested that, in their corporate setting, they would be interested in private breakout rooms for a couple of participants each, as well as a larger full-group meeting room. Although *Talking in Circles* can easily be adapted to support such a mode, our focus on a purely social space makes relying on existing social behaviors more interesting to us than technical enforcement of boundaries. A related concern is that of rudeness or other undesirable behavior by participants. Once again, the system's rich interaction design can support emergent social mores that help sort things out; just as people can move closer to conversations or people they are interested in, they can move away from conversations which become uninteresting or people who show hostility.

Beyond these pragmatic features of distance-based audio attenuation, other potential sociable applications exist. For example, with a low audio distance threshold the popular children's game of "telephone" is playable, in which a large circle is formed by all attendees and a short phrase or story is whispered from person to person around the circle, becoming increasingly distorted, until it gets back to the originator and the starting and ending phrases are revealed to everyone. Though not always desirable, yelling across the room to get someone's attention or to say something to everyone present is also possible.

Circle Motion
Overlap of participants' representations is a problem some graphical chat systems, such as The Palace, have faced. We have observed that avatars partially obscuring each other, whether intentionally or accidentally, can arouse strong responses from those caught underneath. Visibility, as well as bodily integrity of one's representation, are important social factors in graphical chats. *Talking in Circles* deals with both of these factors by not permitting overlap, simultaneously preserving visibility of participants' presence and of their drawing space.

A circle's motion is stopped by the system before it enters another circle's space. In this situation, a participant can drag the pointer inside the circle blocking their path and their own circle follows around at the outer edge of the circle that is in the way, which provides the feel of highly responsive orbiting. Thus, at close quarters, participants still preserve their personal space and can move around in a manner which provides a certain physical interaction with other participants, an attempt at enhancing the feeling of being in a crowd the system provides. Swift motion across large areas is still immediate, as obstructing the participant with all circles along the way to the new location would make for a cumbersome interface. As always, our aim is to leverage spatial grounding with a primary focus on social interaction design, hence our differing policy for motion at close quarters versus over greater distances.

Drawing and Gesture
One major benefit of audioconferencing, of course, is that it frees the hands from being tied to a keyboard. This freedom can be employed to run *Talking in Circles* on a keyboardless LCD-display tablet, with the added benefit of using a more pliant pen for input instead of a mouse, or on a wearable computer. Unlike traditional chat systems we need not display large amounts of text, which takes up a lot of screen real estate, resulting in great freedom in maximizing the potential usage of the space and the graphical area marked off by each circle.

Since the circles' interior space is used only momentarily during speech, this space can be used for drawing. Though the space on one's circle is limited, it is large enough for diagrams, bits of handwriting, and so on. Drawing strokes appear in bright white, visible even over the graphical feedback during speech, and fade away in 30 seconds. Although this makes long-lasting sketching more difficult, a tradeoff worth noting, our design intent is akin to letting people at a cocktail party use a napkin to share sketches on, and obviates potential distraction from cumbersome drawing controls. The relatively fast refresh rate keeps the drawing space available, which is important for drawing to be useful for gesturing.

Drawing faces is a natural tendency, and it is particularly inviting given the circular shape of the user's representation. The circle's space is enough to permit much more than simple emoticons, and even drawing-unskilled users immediately took to writing short phrases and drawing faces. Combined with moving one's circle, drawing a face can be targeted at a particular user both by drawing the face as facing in that user's direction and by moving toward that user with the face showing on one's circle. Coordinating motion with drawing has been popular with users, such as drawing a face with the tongue sticking out and moving quickly up and down next to the intended viewer, enhancing the facial expression with bodily motion.

Shared drawing is also useful for showing explanatory diagrams [30], which Isaacs and Tang note as a user-requested capability in their study [11], for certain kinds of pointing, and potentially for other meta-conversational behaviors such as back channels. These uses are important in creating a social space since studies of telephone conversations have found reduced spontaneity and increased social distance compared to face-to-face discussions [24]. Employing drawing, confusion can be indicated not just by explicit voicing but by a question mark or other self-styled expression on one's circle.

As the system is used in various environments we are very interested in studying the development of novel drawing conventions and gestures for conveying various data. We have already observed novice users effect floor control, for example, by displaying an exclamation mark in their circle upon hearing something surprising, or simply by shaking their circle a bit to indicate they have something to say. Again, although voice by itself is useful in these tasks to some extent, both audio-only and videoconferencing studies have found complex tasks such as floor control to be less effective than can be done in face-to-face communication. Thus, the complementary combination of voice, circle motion and drawing is aimed at overcoming some of these limitations.

In order to make the pictorial modality more accessible, we also provide a set of clickable icons that display drawings in the user's circle, similar to the availability of preset graphics in The Palace. As shown in figure 2, the system currently includes a question mark and exclamation point, as well as expressions indicating happiness, humor, surprise and sadness. The drawings available on the icon bar are standard graphics files editable in any graphics editor, and the drawings the system includes can be removed or modified or new ones added simply by putting them in the *Talking in Circles* directory. The ready access to showing these iconic drawings on one's circle and the ability to customize this set of drawings makes the pictorial channel more available than requiring the user to draw everything from scratch each time. The icon-bar drawings can be clicked on in sequence and are updated immediately, which allows for higher-level expressive sequences such as pictorially sticking one's tongue out while making a humorous remark, followed by displaying the winking face and then the smiling face.

Lastly, drawing can of course be used strictly for doodling, whether out of boredom or to accompany music one is listening to, and for other purely aesthetic ends. Individuals' use of their drawing space —whether they draw constantly or rarely, make abstract doodles, draw faces or words— may provide others a sense of the person's identity. Besides being an expressive channel, these behaviors serve as another form of triangulation, giving participants a spectacle to watch and gather around..

SYSTEM IMPLEMENTATION

The requirements for *Talking in Circles* focused on full-duplex audioconferencing between a substantial number of simultaneous users, where we defined substantial as between a dozen and 20. Even experimentally, we were interested in low-latency audio, as lag is known to be detrimental to the use of speech for social interaction, for example leading to greater formality [20]. However, we were also interested in creating a system that could have as wide a user base as possible, important given our focus on social applications as well as to facilitate wider, extended study of the system's use. This meant that we could not use proprietary broadband networks or high-speed LAN's, as previous systems have typically done.

We initially looked at designing for the internet as a whole, but this proved intractable, as even highly compressed protocols such as RealAudio occasionally suffer from unpredictable network delays and must pause to rebuffer [23]. Next we implemented the system using the Java Sound API [12], but measured end-to-end lag at two to three seconds for machines on the same high-speed LAN subnet. Finally, we settled on adapting RAT, the Robust Audio Tool, an open-source audioconferencing tool from University College, London [31]. RAT uses the MBONE, the internet's multicast backbone, for network transport [25]. Thus we avoided inefficient strictly client-server and peer-to-peer architectures. Each client sends its audio only once and it is then multicast to the other clients.

We modified RAT to support participant state (x/y location, circle color, instantaneous audio energy, and so on) and measured end-to-end lag at approximately 0.3 seconds, considerable but not detrimental unless participants can also hear each other directly [14]. In addition, the bottleneck in our current implementation is screen redraw, as discussed below, and accordingly we have noticed no substantial performance degradation when varying the number of users from one to eleven.

Although the audio code, including compression/decompression and MBONE transport, is written in C, we maintained the user interaction portion in Java using the Java Native Interface in Sun's Java 2 platform. For example, computations including instantaneous audio energy, background noise suppression and logarithmic normalization to map the energy value onto the circle's area are performed in C, and the bright inner circle (see figures 1 and 2) is then updated several times per second in the Java component. Lag was a problem with Java's mouse-motion reporting during freehand drawing, which we adequately resolved using Bresenham's line-interpolation algorithm. Audio bandwidth use is also moderate at 5KB/s per client.

To summarize the data used by our system, each participant multicasts the following:
- circle x/y coordinates
- freehand drawing
- icon selection
- instantaneous audio energy
- speech/audio

The interface is rendered from these features, and all participants' displays share:
- circles' location
- circles' drawing/icon display
- participants' audio

Finally, the local user's relative location produces a subjective rendering of:
- speech volume (audio)
- speech rendering (graphics)

CONCLUSIONS AND FUTURE DIRECTIONS

Approximately thirty people have used the system for periods up to half an hour, and we are working to deploy it for broader testing by a group of steady users. Reactions have been extremely encouraging, with users reporting enjoyment of the cocktail-party-like environment, the drawing capability, and the graphical feedback for speech. The system's fluid representation of participants' speaking and moving about to converse with different groups was cited as the most satisfying aspect, though a shortcoming users noted is slower update rates when multiple people draw or speak while moving around a lot.

In designing *Talking in Circles*, we have strived for a rich medium for communication along dimensions including interactivity, that is, responsiveness, and expressiveness, or "multiplicity of cues, language variety, and the ability to infuse personal feelings and emotions into the communication" [3]. While speech has been found to have reduced cognitive load compared to text generation [15] and to be the key medium for collaborative problem-solving [4], as Chalfonte found text still has certain advantages. For structured data, such as URL's, text is clearly superior to speech due to its permanence and precision. Further, text-based CMC supports threading more so than face-to-face communication [17], which can add fluidity and variety to conversation. Though the ability to draw and do limited handwriting in *Talking in Circles* can help in some cases, textual communication can nevertheless add to the set of useful channels at users' disposal. We are also interested in exploring the usefulness of non-explicit communication through affective channels [22]. Unobtrusive sensing of temperature or skin conductivity, displayed graphically, might add a valuable human element to individuals' representation.

Another view of the system is that of Benford et al's schema for shared spaces. According to their criteria *Talking in Circles* is of medium transportation and artificiality, and of extremely high spatiality, supporting on-going activity, peripheral awareness, navigation and chance encounters, usability through natural metaphors, and a shared frame of reference [2]. We are interested in extending the system's spatiality even further, such as by providing greater persistence and meaning to the space, that is, increasing its sense of place [9]. This might be done by modifying the background of different chat sessions or by permitting wear on the space, such as permanent user drawings or subtly showing which areas of the space get the most use. We are also interested in extending our framework for participants' navigation along architectural notions. The social significance of central and peripheral areas in plazas or the flow of people at street corners [33] indicates potential for interface techniques to make our space more legible and navigable.

A related area for future work is in preserving history in an audio chat. *ChatCircles* has used spatially-useful history mechanisms based on conversation groups at various points during a chat [32], but parsing and browsing of sound remains a major challenge. Braided audio is one interesting recent approach [26].

A suggestion we have heard often is making the audio fully 2D-spatialized rather than the current 1D attenuation. This could certainly add to the system's spatiality, but we first intend to observe how straightforward users' mapping of spatialized audio is onto the flat 2D surface *Talking in Circles* presents. A mechanism for explicitly conveying gaze direction, such as a more abstract analogue to Donath's work with pictures of chat participants facing in different directions [6], might also be of great benefit.

REFERENCES

1. Ackerman, M., Hindus, D., Mainwaring, S., Starr, B. Hanging on the 'wire: A field study of an audio-only media space. In *ACM Transactions on Computer-Human Interaction*, March 1997, 4(1), pp. 39-66.

2. Benford, S., Brown, C., Reynard, G., Greenhalgh, C. Shared Spaces: Transportation, Artificiality and Spatiality. In *Proceedings of CSCW '96*, ACM, 1996, pp. 77-86.

3. Chalfonte, B., Fish, R. S., Kraut, R. E. Expressive Richness: A Comparison of Speech and Text As Media for Revision. In *Proceedings of CHI '91*, ACM, 1991, pp. 21-26.

4. Chapanis, A. Interactive Human Communication. *Scientific American*. 232, 1975, pp. 36-42.

5. Cherry, E. C. Some experiments on the recognition of speech with one and with two ears. *Journal of the Acoustical Society of America*, 25, 1953, pp. 975-979.

6. Donath, J. The Illustrated Conversation. *Multimedia Tools and Applications*, 1, March 1995.

7. Erickson, T. et al. Socially Translucent Systems: Social Proxies, Persistent Conversation, and the Design of "Babble." In *Proceedings of CHI '99*, ACM, 1999, pp. 72-79.

8. Greenhalgh, C., Benford, S. MASSIVE: a collaborative virual environment for teleconferencing. *ACM Transactions on Computer-Human Interaction*. 2(3), 1995.

9. Harrison, S., Dourish, P. Re-Place-ing Space: The Roles of Place and Space in Collaborative Systems. In *Proceedings of CSCW '96*, ACM, 1996, pp. 67-76.

10. Hindus, D., Ackerman, M. S., Mainwaring, S., Starr, B. Thunderwire: A Field Study of an Audio-Only Media Space. In *Proceedings of CSCW '96*, ACM, 1996, pp. 238-247.

11. Isaacs, E. A., Tang, J. C. What Video Can and Can't Do for Collaboration: a Case Study. In *Proceedings of Multimedia '93*, ACM, 1993, pp. 199-206.

12. Java Sound API. Sun Microsystems, Inc. http://java.sun.com/products/java-media/sound

13. Kansas. Sun Microsystems, Inc. http://research.sun.com/research/ics/kansas.html

14. Krauss, R. M., Bricker, P.D. Effects of transmission delay and access delay on the efficiency of verbal communication. *Journal of the Acoustic Society of America*. 41, 1967, pp. 286-292.

15. Kroll, B.M. Cognitive egocentrism and the problem of audience awareness in written discourse. *Research in the Teaching of English*. 12, 1978, pp. 269-281.

16. Kurlander, D., Skelly, T., Salesin, D. Comic Chat. In *Proceedings of SIGGRAPH*, 1996, pp. 225-236.

17. McDaniel, S. E., Olson, G. M., Magee, J. C. Identifying and Analyzing Multiple Threads in Computer-Mediated and Face-to-Face Conversations. In *Proceedings of CSCW '96*, ACM, 1996, pp. 39-47.

18. Milgram, S.. *The Individual in a Social World*. Reading, MA: Addison-Wesley, 1977.

19. Nakanishi, H., Yoshida, C., Nishimura, T., Ishida, T. FreeWalk: Supporting Casual Meetings in a Network. In *Proceedings of CSCW '96*, ACM, 1996, pp. 308-314.

20. O'Conaill, B., Whittaker, S., Wilbur, S. Conversations Over Video Conferences: An Evaluation fo the Spoken Aspects of Video-Mediated Communication. *Human-Computer Interaction*. 8, 1993, pp. 389-428.

21. The Palace, Inc. The Palace. http://www.thepalace.com

22. Picard, R. *Affective Computing*. Cambridge, MA: MIT Press, 1997.

23. Progressive Networks, Inc., http://www.real.com

24. Rutter, D.R. The role of cuelessness in social interaction: An examination of teaching by telephone. In Roger, D. and P. Bull (ed.), *Conversation: An Interdisciplinary Perspective*. Multilingual Matters, 1989.

25. Savetz, K., Randall, N., Lepage, Y. *MBONE: Multicasting Tomorrow's Internet*. IDG, 1996.

26. Schmandt, C. Audio Hallway: a Virtual Acoustic Environment for Browsing. In *Proceedings of UIST '98*, San Francisco, CA: ACM, 1998, pp. 163-170.

27. Sellen, A. J. Speech Patterns In Video-Mediated Conversations. In *Proceedings of CHI '92*, ACM, 1992, pp. 49-59.

28. Singer, A., Hindus, D., Stifelman, L., White, S. Tangible Progress: Less Is More In Somewire Audio Spaces. In *Proceedings of CHI '99*, ACM, 1999, pp. 104-111.

29. Smith, G. Cooperative Virtual Environments: lessons from 2D multi user interfaces. In *Proceedings of CSCW '96*, ACM, 1996, pp. 390-398.

30. Tang, J.C. Findings from observational studies of collaborative work. *International Journal of Man-Machine Studies*. 34(2), 1991, pp. 143-160.

31. University College London. Robust Audio Tool. http://www-mice.cs.ucl.ac.uk/multimedia/software/rat

32. Viegas, F. B., Donath, J. S. *ChatCircles*. In *Proceedings of CHI '99*, ACM, 1999, pp. 9-16.

33. Whyte, W. H. *City: Rediscovering the Center*. NY: Doubleday, 1988

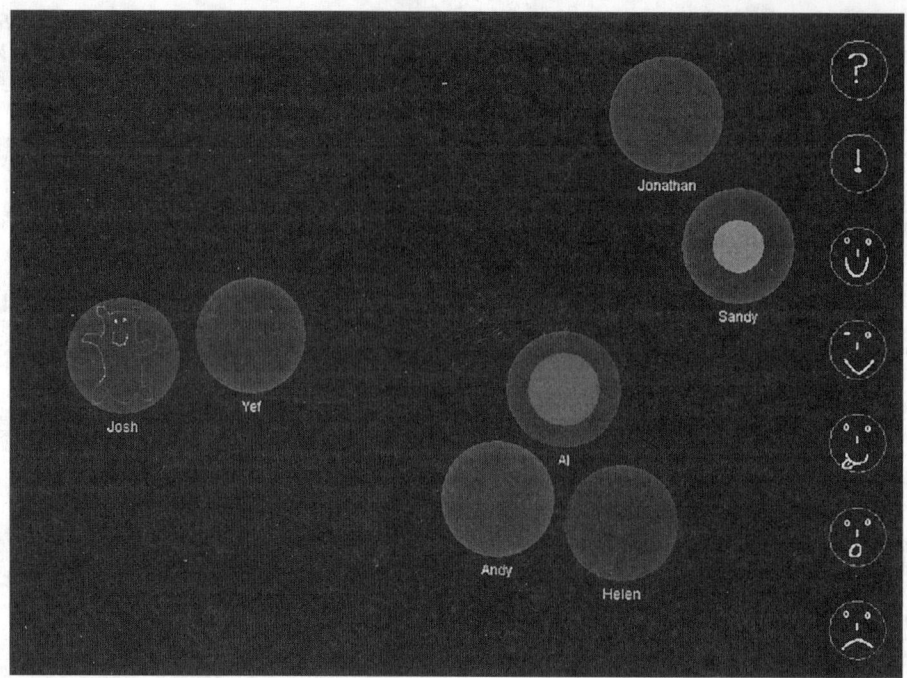

Figure 1: Three conversations in a *Talking in Circles* chat session. Al is talking with Helen and Andy, and Sandy converses with Jonathan. Meanwhile, Josh draws for Yef.

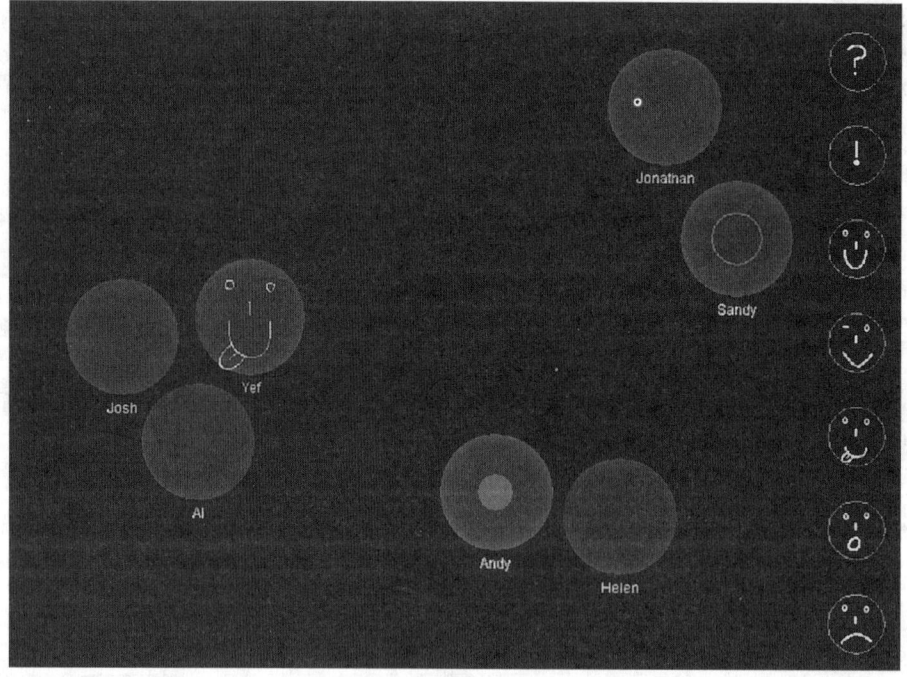

Figure 2: Screenshot from the point of view of Al, the blue circle, after he has wandered over to speak with Josh and Yef. Both Sandy and Andy are speaking, but Andy remains within Al's hearing range while Sandy is now outside it.

Jotmail: a voicemail interface that enables you to see what was said

Steve Whittaker, Richard Davis*, Julia Hirschberg, Urs Muller
ATT Labs-Research
180 Park Avenue
Florham Park, NJ 07932, USA
+1 (973) 360 8339
stevew/julia/urs@research.att.com

*Virtual Ink Corporation
56 Roland St. Suite 306
Boston, MA 02129, USA
+1 (617) 654 0126
richard.davis@virtual-ink.com

ABSTRACT
Voicemail is a pervasive, but under-researched tool for workplace communication. Despite potential advantages of voicemail over email, current phone-based voicemail UIs are highly problematic for users. We present a novel, Web-based, voicemail interface, Jotmail. The design was based on data from several studies of voicemail tasks and user strategies. The GUI has two main elements: (a) *personal annotations* that serve as a visual analogue to underlying speech; (b) automatically derived message header information. We evaluated Jotmail in an 8-week field trial, where people used it as their only means for accessing voicemail. Jotmail was successful in supporting most key voicemail tasks, although users' electronic annotation and archiving behaviors were different from our initial predictions. Our results argue for the utility of a combination of annotation based indexing and automatically derived information, as a general technique for accessing speech archives.

Keywords
Voicemail, annotation, speech access, note-taking, asynchronous communication, "speech as data", empirical evaluation.

INTRODUCTION
Voicemail is a pervasive but under-researched workplace communication technology, with an estimated 68 million users worldwide. Many organizations rely heavily on voicemail for conducting everyday work, and voicemail is often preferred to email [10]. The advantages of voicemail over email are: speech is expressive, easy to produce and critical in many workplace tasks [3,4]. Voicemail is also ubiquitous - any phone acts as an access device. It is also a common feature of most new cellular phones.

In the past, the phone was the only universal access device. As a result, voicemail interfaces were either touchtone or speech-based. However, the Web and PDAs will soon make *graphical* UI methods more widely available for accessing voicemail. Graphical access may have significant advantages: *visual indices* have been used successfully as a general technique to access other types of speech archives

Permission to make digital or hard copies of all or part of this work for personal or classroom use is granted without fee provided that copies are not made or distributed for profit or commercial advantage and that copies bear this notice and the full citation on the first page. To copy otherwise, to republish, to post on servers or to redistribute to lists, requires prior specific permission and/or a fee.
CHI '2000 The Hague, Amsterdam
Copyright ACM 2000 1-58113-216-6/00/04...$5.00

[1,2,5,6,7,9,11,12,15,17]. Visual representation of speech structure allows random access to an inherently serial medium. The aim of this paper is to explore how these new visual indexing techniques can be applied to voicemail access, in particular to address documented problems with current touchtone UIs [13]. We also wanted to evaluate our system with real users: much prior research on speech access has focused on new techniques and not on their evaluation.

The structure of the paper is the following. We present an extended analysis of a previous study of voicemail usage [13], identifying four key user problems: *message scanning, information extraction, status tracking* and *archiving*. A central user strategy for voicemail processing relies on *message indexing by note-taking*. We implement a novel Web-based voicemail GUI that supports annotation for indexing. The UI allows users to take temporally indexed notes associated with individual messages. These notes serve as a *visual analogue* to the underlying speech in the message, allowing straightforward access to message contents, message scanning and status tracking. We also provide people with automatically derived header information for each message. We evaluated Jotmail in an 8-week field trial, where people used it as their only means for accessing voicemail. Jotmail was highly successful in supporting most key voicemail tasks, although users' electronic annotation and archiving behaviors differed from our predictions. Our results argue for the utility of a combination of annotation based indexing and automatically derived header information, as a general method for accessing speech archives.

VOICEMAIL TASKS AND PROCESSING STRATEGIES
We collected qualitative and quantitative data to identify users' key tasks and strategies for processing voicemail, for a typical voicemail system, Audix™, including: (a) server logs from 782 active users; (b) surveys from 133 high volume users (receiving more than 10 messages/day); (c) interviews with 15 high volume users.

The server data showed intensive voicemail use: people accessed the system a mean of 7.1 times each working day, receiving 8.7 messages, and storing 4.8 mins. of messages overnight. Voicemail messages also contained significant amounts of information: about half those surveyed reported average message lengths of between 30-60 secs. and about half reported lengths of 1-2 mins. Our interviews also indicate that voicemail messages contain complex information, not simple *"call me back"* requests: *"[a voicemail message] is really like a whole memo, or a huge email message worth of information."* Furthermore, voicemail often substitutes for a series of face-to-face meetings: *"entire transactions or entire tasks are accomplished by exchanging [voicemail] messages. That is, you will never talk to the person in real time."* Finally people stressed that a key value of voicemail is ubiquity: *"the most important feature for voicemail as opposed to e-mail is that it is easily accessible from any telephone. People tend to respond quicker to voicemail than they do e-mail, because you can access voicemail from anywhere."*

Users report four main tasks when processing voicemail: *scanning* the mailbox to identify important messages; *extracting information* from individual messages; *tracking the status* of current messages; and *managing* their archive of stored messages.

Scanning

Scanning is used for *prioritizing* incoming new messages, and for *locating* valuable saved messages. *Prioritization* is critical for users who have to identify urgent incoming messages, while accessing the mailbox under time constraints (e.g. during a meeting break). These users have to rapidly determine which new messages require their immediate attention. *Message location* occurs when users search for saved messages containing valuable information.

Users' current scanning strategy is to sample all messages in sequence to determine location and status. For prioritization, only 24% of people we surveyed use voicemail message headers to identify urgent messages, reporting they are too slow. Instead they listen to the first few seconds of each message, to the speaker's intonation, to determine whether a message requires immediate action. *"Typically if I am at a meeting, I am on a short break and I find I have 10 or 12 messages ... I only sort of skim them, but listening to the first couple of seconds, as to who it is and what the issue is, to see whether it has to be dealt with immediately. If not, I will just save the message and go on, so I can pick up any priority ones."*

In *locating* stored messages, most users do not retain a detailed model of their archive and 76% of those surveyed report that *"listening to each message in sequence"* is their standard procedure for finding archived messages. However, the linear nature of mailbox search makes location onerous when more than a few messages are stored: *"if I've got 20 messages stored ... and I want that last message, it's a real pain to get to that last message. And ... most of the time I don't even know what message I want to get to".*

Information extraction

When a relevant message is identified, users have to *extract* critical information from it. This is often a laborious process involving repeatedly listening to the same message for verbatim facts such as caller's name and phone number.

Multiple listens are also necessary with vague or highly detailed messages. *"Often you get messages that are somewhat vague... you need to listen to them several times to understand exactly what the agenda is".* Of those surveyed 46% report that they relisten to messages "about half the time".

To reduce repetitive processing, 72% of our survey users report "almost always" taking *written notes*. Users employ two different note-taking strategies. The first strategy is *full transcription*: here users attempt to produce a written transcript of the target message, so as to reduce the need for future access. *"Those notes are like a memory, there are sort of a paraphrase of the message ... sort of a synopsis of what I felt the conversation was about"* The second strategy is to take notes as *indices*. According to our users, voicemail messages have a predictable structure, and the object of this strategy is to abstract the key points of the message (such as caller name, caller number, reason for calling, important dates/times and action items). In most cases, users keep the original voice message as a backup for these incomplete and sometimes sketchy notes. *"I will write a word or two ... something that will jog my mind about the message ... to get the details, I will go back and listen to the voicemail."* These notes are either kept on scraps of paper (75% of users) or in a dedicated note-pad (25% of users), and people refer to them when searching their voicemail archive to locate particular messages.

Status tracking

Workplace tasks are often delegated through voicemail, and a common user problem is *tracking message status*. Status tracking is a prevalent problem for users accessing voicemail under time pressure. They often defer processing a significant number of incoming messages. When accessing voicemail later, they are often unclear about which messages they have dealt with. *"I access my messages in the car, so I don't take any notes and I can't remember which messages I responded to. That makes people very cross".*

There are two main techniques for status tracking. In the first, people use notes taken during *information extraction* as reminders. These notes, taken on scraps of paper, are left around the user's work area to remind them about what needs to be done. *"I sit here with those little stickies and I go, I got to call this person, this person, this person".* One problem with this note-taking strategy is that people sometimes lose these notes, especially when voicemail was originally accessed in a remote location such as a meeting room. Losing notes is less of a problem when people use a dedicated logbook for recording message details.

With the second status tracking strategy, users take no notes but leave undischarged messages in their voicemail mailbox. Reminding takes place when users next scan their archive. *"I will call back to the voicemail, even if the light isn't on, knowing that I have got some messages that I need to respond to".* In the course of scanning they are reminded of outstanding undischarged messages. The weakness of this second strategy is that there is no visible reminding cue, so that if people do not access the voicemail archive they are unaware of the presence of unresolved items.

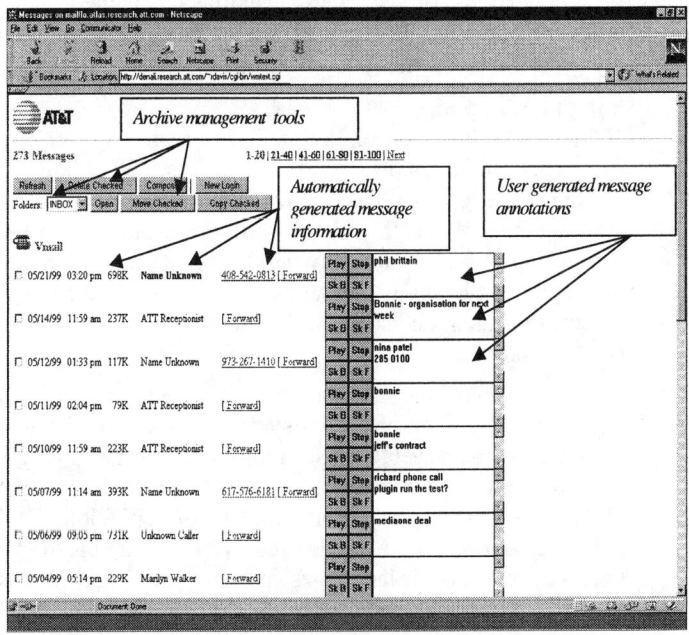

Figure 1 – Jotmail User Interface

Archive management

People also have to *manage their archives*. Given their access strategies, most users' archives consist of a backlog of undischarged messages as well as a store of saved valuable messages. They therefore engage in periodic "clean-ups": accessing each message in sequence to determine whether it should be preserved. *"I do try to schedule at least a half an hour a day where it's not meeting oriented, so I can clean up my messages"*. By removing superfluous messages, users also make it easier both to scan for existing valuable messages, and monitor reminder messages. Those who do not engage in "clean-ups" report being surprised by the extent to which they are accumulating irrelevant messages. *"I will go back and start listening to everything [in the archive] and by and large what I find out is that 90% of the time I delete most of the messages"*.

JOTMAIL USER INTERFACE

We devised a novel Web-based GUI, Jotmail, with the goal of supporting *scanning, information extraction, status tracking* and *archive management* tasks, in response to these findings, along with those obtained from controlled laboratory studies of voicemail access [14]. The design was based around the strategies we had observed being used for processing voicemail, paying particular attention to the critical role of note-taking. A key benefit of voicemail reported by our users is ubiquity. By developing a Web-based UI, we provided access to voicemail in any location where there is a computer with an Internet browser. The UI is shown in Figure 1. The center right of the screen shows text boxes for user generated message annotations, with play control buttons located to the left of them. On the left of the interface is more general header information about each message that has been derived automatically. The upper left of the screen shows archive management tools for creating and managing voicemail folder structures. The design was finalized after several iterations with trial users.

Information extraction using annotations

A key strategy for addressing *information extraction* was the use of personal notes. A central, novel, feature of the UI is therefore support for user annotations. Users can record personal notes, (e.g. "phil brittain", "Bonnie - organisation for next week") in the scrollable text box associated with a given message. One use of annotations is for message summaries. In all our studies users reported the need for repeated replayings of the message to extract critical information. We therefore wanted to provide ways to rapidly identify and replay only the most relevant parts of the message, without having to listen to the entire message. User notes are therefore also *time-indexed* [9,12,15]. The motivation for this came from the observation that handwritten notes serve as an index into the underlying structure of the original voicemail message: *"my notes trigger things - they are ... meant to just give me place holders while I am browsing. Then I have to go back and listen to stuff"*.

Time-based indexing works as follows (see Fig. 2): users take notes as the message is played, and each note is co-indexed with the speech currently being played. If users later click on a given word in their notes, they automatically access the speech that was being played when the note was taken. In this way, notes provide reasonably precise access into the underlying speech, allowing users to focus on areas of specific relevance[1]. To further help information extraction, we provide general play controls for navigating within the message without recourse to notes. These allow people to *play* and *stop* a given message as well as *skip forward* and *skip back* 2 secs. within a given message (Sk B and Sk F buttons).

Scanning using message overview information

Our user studies also revealed the requirement for *scanning* to prioritize and locate important messages. In addition to user generated annotations, an important set of cues to aid scanning is message header information [11]. The UI therefore displays the following information about each message: *date, time, size* in Kbytes, *caller-ID number* (when it is available) and *caller name* (for internal calls

---

[1] There is a delay between the time at which users hear the relevant information and when they enter their related note. In later versions of the UI, we therefore introduced an indexing off-set, so that notes are indexed to material being played two seconds previously. This off-set was determined after iterations with several early users. In later versions we plan to make this interval user configurable. Figure 2 does not show the off-set for ease of exposition.

only, derived by looking up the name for the caller number in the corporate directory). In Fig. 1, the first message in the mailbox was from 408 542 0813, on 5/21/99 at 3.20pm. The message length was 698K, but because the message was from outside the local PBX the system was unable to infer the caller name. By depicting this general information we enable users to visually scan and randomly access messages. They no longer have to access messages in sequence to identify specific messages.

As can be seen from the example of the first message, users manually supplement automatically generated overview information with their own notes. For the first message, the user has added the caller name ("phil brittain"), because the system was unable to infer this, and the caller-ID number was unfamiliar. In this way, annotations were used to support *scanning* as well as *information extraction.*

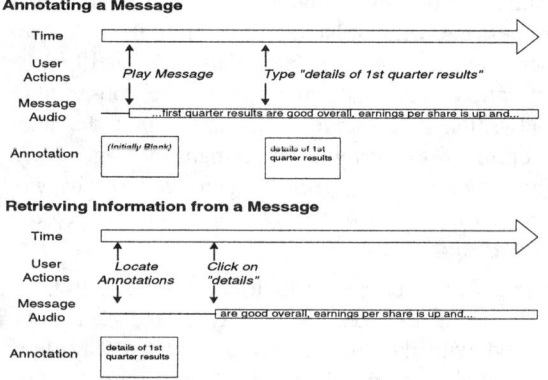

Figure. 2: Time-based indexing

Status tracking using annotations and overview information

Users also reported problems in trying to remember what outstanding actions were required for a given message. The user interface was designed to support status tracking in two ways – again by analogy with people's paper based strategies of leaving themselves visual reminders. Annotations could be used to *explicitly* record the actions necessary for each message. So for example *"Richard phone call plugin run the test"* states the action that was requested in the message, namely to run the relevant test. More *implicitly,* we hoped that the mere fact of having a visual representation of each message visible in the mailbox would serve to remind people of the necessary action whenever they access Jotmail. For example seeing a message from Marilyn Walker (last message in the inbox) might remind me of the action that message requires. A final cue to message status is that unaccessed messages are depicted in **bold** (the first message in Fig.1). Once accessed, their status changes.

Archive management

Users also reported problems in remembering the contents of their archive, and in preventing the build up of irrelevant messages. The Jotmail interface provided them with a set of tools for organizing, managing and deleting voicemail data. Labeled buttons allowed them to create new folders, as well as move, delete and copy information to those folders. More implicit support for archive management is provided by the visibility of messages, enabling the archive to be quickly scanned to identify important messages and filter out superfluous ones.

Implementation

Jotmail is built on top of Webmail, a research system that supports email and voicemail access. Webmail is implemented as a CGI script that connects to a standard mail server. When the script is run, it produces HTML pages with interfaces for viewing, browsing, and archiving messages. Voicemail messages are retrieved from the voicemail system and stored as email messages with special headers and data.

The requirement for broad access influenced our choice of platform. The annotation system in Jotmail was implemented as a web browser plug-in. Our HTML plugin will work on most browsers, but at the same time using a plugin restricts the complexity of possible UI implementations when compared with what could be implemented in other languages such as Java. Webmail was modified to store annotation files alongside voicemail messages, and to display small annotation plug-in windows next to each message. The plug-in application downloads the annotation file and the audio file for each message and allows the user to play the message and type text in the annotation text box. If the message is playing when the user is typing, then the current time index in the message is stored with each word. By holding "Control" and clicking on a word, the user can play the message from that time index. When the plug-in closes, modified annotations are sent back to Webmail so they will be displayed the next time the page is loaded. This gives Jotmail users persistent annotations for information extraction and status tracking.

EVALUATION DESIGN

A major goal of our evaluation was to investigate the experimental system being used by people for their everyday work. We designed the trial to collect the following data to investigate how effectively the system supported the tasks we had identified. We were also concerned with how well the system supported annotation behaviors, given the centrality of this strategy.

Preinstallation survey: Before installing the new system we administered a user questionnaire addressing use of the standard touchtone (TT) voicemail system. We asked people about the number of voicemail messages they currently received and sent, and how they processed these. We also gathered data about *scanning, information extraction, status tracking* and *archiving* tasks, as well as note-taking strategies. We surveyed people about the success of TT features (header information and message operations) in supporting these core tasks.

HYPOTHESIS	MEASURE	TOUCHTONE SYSTEM (MEAN)	JOTMAIL (MEAN)	STATISTICAL DIFFERENCE & HYPOTHESIS CONFIRMED?
O1	Overall ease of processing voicemail (5 = "very easy", 1 = "very hard")	2.6	4.6	$t_{(6)}=4.10, p<0.01$ confirmed
O2	Ease processing voicemail compared with email (5 = "much easier", 1 = "much harder", 3 = "about the same")	1.3	2.7	$t_{(6)}=4.80, p<0.005$ confirmed
S1	Scanning (5 = "very easy", 1 = "very hard")	1.7	4.7	$t_{(6)}=6.22, p<0.0001$ confirmed
S2	Ease of locating a specific message (5 = "very easy", 1 = "very hard")	1.8	5.0	$t_{(6)}=11.50, p<0.0001$ confirmed
S3	Preferred method for locating a message	All users listened to the first few secs.	All users employed visual scanning of the mailbox	Cochran's $Q_{(1)}=7.00, p<0.01$, confirmed
I1	Information extraction (5 = "very easy", 1 = "very hard")	2.1	4.0	$t_{(6)}=5.29, p<0.002$ confirmed
I2	Replay frequency (1 = "very frequently", 5 = "never")	3.3	3.3	$t_{(6)}=0$, ns no effect
I3	Note-taking frequency (1 = "very frequently", 5 = "never")	2.0	3.0	$t_{(6)}=3.57, p<0.02$ disconfirmed
ST1	Ease tracking message status (5 = "easy", 1 = "hard")	3.0	3.9	$t_{(6)}=2.34, p=0.05$ confirmed
ST2	Frequency of losing notes (6 = "never", 1 = "frequently")	3.4	5.3	$t_{(6)}=5.46, p<0.002$ confirmed
A1	Archival behavior (5 = "never archive", 1 = "usually archive")	3.4	2.3	$t_{(6)}=1.80$, ns. no effect

Table 1: Comparison of Jotmail and touchtone user interfaces

Jotmail logs: We logged usage data for 8 weeks. We collected data about: number and duration of Jotmail sessions, messages stored and accessed, operations on messages (stop, play, skip) as well as information about what notes people took and when they used these to replay messages. This data was used to identify the main types and functions of user annotations.

Post-installation survey: After 8 weeks, we took the system away and administered an extended version of the original questionnaire containing additional questions about the basic features of Jotmail, note-taking, and the use of notes for playback.

Interviews: We also carried out two semi-structured interviews with users, one while the system was installed and one after it was taken away. These probed the same issues as the surveys, but we tried to elicit fuller descriptions of the way that the system was being used as well as its main benefits and drawbacks. We also tried to find examples of novel or unexpected uses of the system. We supplemented these interviews with some observations of people using the system.

RESEARCH HYPOTHESES AND RESULTS

We installed the system and collected logs for 9 users for a total of 184 sessions over 935 hours. Our users were researchers and secretarial support staff at AT&T. Of these, 7 completed all questionnaires and interviews. Our logs show frequent system use. People used the system for an average of 20.4 sessions, of mean length 5.1 hours. Jotmail users tended to keep the application constantly running, unlike TT where short sessions were the rule. In each session they would typically access 4.1 messages, and the mean distribution of play operations per session was 2.1 plays, 1.8 skips and 0.3 annotation-based plays. Overall users replayed 36% of messages. On 30% of occasions they listened to the same message two or more times in sequence. Users would typically save 11.0 messages at the end of each session.

Our hypotheses and results were as follows (see Table 1):

Overall preferences

O1: We predicted that people should rate it easier overall to process their messages with Jotmail, given the support it provides for scanning, information extraction, status tracking and archiving.

O2: For the same reasons, people should rate Jotmail processing as closer to email than TT access.

Both these hypotheses were confirmed (see Table 1). User comments also bear out the overall superiority of Jotmail. *"The whole process of dealing with voicemail is that much easier. In the past [with TT] when I came into the office I used to put off dealing with voicemail and always look at my email first, but now I often look at my voicemail first."* Contrast this with comments about TT voicemail access: *"I hate managing voicemail with so little information. What I dislike most is that you cannot "see" messages, whether there are new messages, if so how many, or saved messages.* Users were all highly displeased when we took the system away after 8 weeks.

One repeated observation was about the greater efficiency of Jotmail. People complained that TT required them to execute too many operations to access their voicemail. *"There's too many key presses to get into TT - with Jotmail there's no keys - I like its easy accessibility compared with the time consuming process of voicemail retrieval through the phone."* Another user commented about TT: *"it took me 27 seconds and 16 key presses to access a 5 second message".*

Scanning

S1: Scanning should be rated as easier with Jotmail than TT access, given the visual representation of Jotmail messages, automatically generated information and annotations.

S2: Users should find it easier to locate specific messages with Jotmail than TT access, because of the increased ability to scan.

S3: Users should switch from scanning by sequentially sampling the first few seconds of each message, to using Jotmail to look through headers and notes.

All 3 hypotheses were confirmed. The user logs also showed evidence of using the visual representation for random access: users accessed 36% of messages in an order different from that in which they were received and stored.

By providing the ability to scan messages, Jotmail allows users to prioritize their access. One user said of TT: *"With the old system I never knew which messages were there [in the mailbox] in what order so I couldn't selectively pick the ones I wanted to deal with. I used to put off listening to voicemail until I had the time to go through it all"*. Scanning also allowed rapid access to old messages: *"I can see what's in my mailbox at a glance... the fact that messages are visible means that I can find old ones easily"*.

Information extraction

I1: Users should rate information extraction as easier with Jotmail than TT access, because of the presence of annotations, and the ability to do time-based playback.

I2: Users should replay messages less often with Jotmail - both because they have more information automatically available about each message, and their own notes provide reminders about message contents.

I3: People should take more notes with Jotmail because of the utility of notes as indices.

Table 1 shows that the I1 was confirmed. However there was no difference in the number of replays with the two systems (I2). It may be that greater control over access provided by Jotmail means that users are happy to replay messages. Contrary to our predictions, people took *fewer* notes with Jotmail. The decreased amount of note-taking with Jotmail (I3) may occur because it *automatically* logs relevant information, obviating the need for some manual notes: *"You can take fewer notes with Jotmail because the name of the caller time and date for inside callers is shown on the screen."*

Notes were still useful in Jotmail, however. Their utility was demonstrated by the fact that every user reports taking notes on the system. There was also no reported difference in the types of notes taken with the two systems – a few key words such as name, phone number and action. Nor is there a difference in the reason for taking notes: the majority (86%) of users in both cases said that they take notes as a reminder either about message contents, or about the action a message requires. The fact that Jotmail notes are similar to their paper analogues offers good support for the naturalness of this aspect of the user interface.

Status tracking

ST1: Users should find it easier to track messages with Jotmail, because of the ability to scan outstanding messages at a glance.

ST2: People should be less likely to lose notes with Jotmail than TT voicemail because Jotmail notes are stored at a single on-line location.

Both hypotheses were confirmed, and multiple user comments indicated that status tracking was a critical perceived benefit of Jotmail, especially in the face of constant interruptions: *"I am always being side-tracked and interrupted in my job. Jotmail is like a tickler file. It provides a constant reminder of the things that I have to do"*.

Some users exploited the note-taking features of Jotmail to explicitly add multiple successive comments to the original message *"Fax sent"*, to track progress after each action taken in responding to the call. They also commented on the benefits of having their notes on-line in close association with the original message. *"I used to save post-its as a record of what I was doing, but this way (taking on-line notes), I don't have loose pieces of paper that can get lost"*.

Archiving

A1: Given the increased ease in managing the archive, we expected that users would archive more messages with Jotmail versus TT.

This hypothesis was not confirmed. On the one hand, it was clear that Jotmail made it easier to store and access valued messages. On the other hand, it seemed that the visibility of the archive meant that users were better able to clean up and hence prevent the inadvertent build up of superfluous forgotten messages: *"When I access voicemail over the phone I don't usually access old messages whereas Jotmail provides reminders that I have old messages"*.

We also examined voicemail filing techniques. Somewhat to our surprise, although some users saved many messages, no-one categorized messages. They kept all their messages in the inbox, rather than creating task-specific folders. One user explained this as follows: *"I use the system to track things I have to do. If I haven't done them, I want to keep them in the inbox to remind me that they need attention. If they're done then I delete them. It's just making extra work to file them and have to remember where they are."* The emphasis therefore seems to be on status tracking rather than the construction of a complex archive. Better information extraction in Jotmail may also reduce archiving. One user pointed out the difficulty of information extraction with TT. She didn't keep as many messages with Jotmail because information extraction was more straightforward, and she ended up with paper-based summaries of messages.

Annotations analysis

We also investigated a number of questions concerning annotations, given the centrality of users note-taking strategies in our initial studies:

Types of Jotmail annotations: These tended to be relatively brief (mean of 6.3 words), falling into 6 main categories: caller name, message topic, caller number, time, date and location, with frequencies per message being respectively: caller name (0.75), message topic (0.53), caller number (0.27), time (0.09), date (0.03) and location (0.03). The average message has two of these annotation types and the most frequent combinations of annotation types are: name and topic, name and caller, number and topic. Finally adding further annotations to a previously annotated message occurred relatively frequently, with 44% of annotations being additions to a previously annotated message. According to users, many of these re-annotations were being used to track the status of previously annotated messages.

Functions of Jotmail annotations: Overall, 29% of messages were annotated, with each user annotating 21.8 messages. Annotations were usually associated with messages that were important to the user; annotated messages were played more often than unannotated ones (respective means: 2.79, 0.92, $t_{(458)}=5.08$, $p<0.0001$). Annotations were not widely used to control playback: analysis of play operations indicated that time-indexed playback accounted for only 7% of play operations compared with 51% "play from start of message" operations and 43% skip based plays. 67% of users exploited the time-based indexing feature, but the remainder never did. Non-users argued that they received mainly short messages, reducing the need for controlled access to message contents. With short messages there is little cost to replaying an entire message to extract a single piece of information. This is supported by the fact that people tended to annotate longer rather than shorter messages (respective means: 422.7 and 356.1 Kbytes, $t_{(458)}=2.25$, $p<0.025$). However, it turned out that messages accessed by time-based playback were no longer than messages accessed using "play from start" only ($t_{(200)}=0.28$, ns)

Reasons for annotating specific messages People were more likely to annotate messages from unfamiliar callers (defined as those from outside their immediate workplace), $\chi_{(1)}=6.04$, $p<0.025$. They also made annotations more frequently with messages that had less automatically generated information $\chi_{(1)}=5.61$, $p<0.025$.

Unanticipated uses of Jotmail

Jotmail also led to more call screening. One user pointed out that the ease of accessing messages with Jotmail meant that he fundamentally altered his handling of incoming calls. With Jotmail he was more likely to screen calls by letting them go through to voicemail. *"With Jotmail I let live calls go through to voicemail because I knew I could easily get them later. With [TT] I don't do that because its so time-consuming to go and get them back"*. Another unanticipated use of Jotmail was for playback to a live audience. Two users reported replaying Jotmail messages to others (either face-to-face or over the phone). Again this was facilitated by the greater ease of message access: *"The only confirmation I had of S's promotion was a voicemail message from R., so I replayed that to him. I can't imagine being able to find that message using [TT]"*.

CONCLUSIONS

We built a novel Web-based UI, to voicemail centered on the notion of note-taking, that also provided automatically generated message information and archiving tools. The design was based on requirements data from interviews, experiments, surveys and user logs identifying key user tasks and strategies for voicemail processing. Data from an 8-week field trial showed that Jotmail was much preferred to a TT UI. As predicted, Jotmail improved scanning, information extraction and status tracking tasks.

User archiving and note-taking behaviors were not as expected however. Archive size did not increase with Jotmail, although this may follow from the superior scanning capabilities of Jotmail, preventing the unintentional accumulation of irrelevant messages that often occurred with TT. Removal of superfluous messages may therefore have counterbalanced increases in intentionally archived messages. Archiving behavior may also be influenced by users' prior experience with a previous TT system, which deleted messages after 14 days (a common feature of many such systems). This may have led users to view voicemail data as inherently ephemeral, despite user assertions that some messages had long-term value. A longer field trial might produce more instances of archiving activity once users habituate to the idea of message permanence. Users also failed to exploit the archiving tools provided, preferring to leave all messages in the inbox. Their comments suggested that systematic filing is onerous. Filing may also compromise the ability to track message status [16]: once filed, messages are no longer visible in the inbox, leading them to be forgotten. Given that voicemail volumes are lower than email, it may be possible to keep all current important messages visible in the inbox. Lower message volume in voicemail may reduce the pressure to file that has been reported in email studies [8,16].

Annotation behavior was also not completely as predicted. Users were very positive about their ability to annotate messages and use time-indexed playback. Annotations were used for reminding, status updates and their on-line location meant that they were not mislaid like paper notes.

Screen-based notes were similar to paper ones. As predicted, they were also associated with longer messages that were accessed more often. Despite this, users reported taking fewer notes with Jotmail. Reduced note-taking may have occurred because we automatically generated message header information, and indeed fewer notes were taken for messages with more such information. Time-based indexing was also used infrequently. There are several possible explanations: (a) messages were short enough to replay without undue cost, reducing the need for precise control during information extraction; (b) users found it hard to anticipate what notes would be useful for future retrieval; (c) sparse notes may be sufficient to *remind* users of the contents of the message, without the need for reaccessing the underlying speech. Other work is consistent with the reminding explanation, arguing there are important trade-offs between the *efficiency* of relying solely on hand-written notes as (imperfect) *reminders* versus the *accuracy* of accessing the verbatim speech record itself [9,15].

There are also important system extensions we are currently investigating. These include using automatic speech recognition to produce *transcripts* of voicemail messages. Although the transcripts are errorful, they nevertheless provide a browsable text for each message, allowing users to *read* rather than listen to voicemail. Like user annotations, they also serve as a visual analogue to each voicemail message. We are also exploring techniques for automatically extracting significant information such as names, dates, times and telephone numbers from these transcripts. These new automatic techniques should provide further support for *information extraction* and *scanning tasks*.

Finally, there are both practical and theoretical implications to our results. First our tool successfully addresses a significant problem for many users - namely *efficient* voicemail retrieval at any location where there is Web access. It seems to address many of the problems that users currently experience with TT voicemail. Our data also contribute to a growing body of research on general methods for speech access. We present data showing that, consistent with the claims of prior work, providing a visual analogue as an index into underlying speech structure is important for supporting browsing and retrieval [1,2,5,6,7,9,11,12,15,17]. As with other approaches [6,7], our results suggest that for personal data such as voicemail, a combination of automatically generated data and personal annotations provides a general technique for accessing complex information in speech.

REFERENCES

1. Abowd, G., Atkeson, C., Feinstein, A., Hmelo, C., Kooper, R., Long, S., Sawhney, N., Tani, M. Teaching and learning as multimedia authoring in the classroom 2000 project, *Multimedia96*, 187.

2. Arons, B. Interactively skimming speech. Unpublished PhD thesis, MIT Media Lab, 1994.

3. Chalfonte, B., Fish, R., and Kraut, R. Expressive richness. In *CHI91*, 21-26, 1991.

4. Chapanis, A., Ochsman, R., Parrish, R. and Weeks, G. Studies in interactive communication: I. The effects of four communication modes on the behavior of teams during cooperative problem-solving. *Human Factors*, 14, 487-509, 1972.

5. Degen, L., Mander, R., and Salomon, G. Working with audio. In *CHI92*, 413-418, 1992.

6. Hindus, D., Schmandt, C., and Horner, C. Capturing, structuring and representing ubiquitous audio. *ACM Transactions on Information Systems*, 11, 1993.

7. Kazman, R., Al-Halimi, R., Hunt, W., and Mantei, M. Four paradigms for indexing video conferences. In *IEEE Multimedia*, 3(1), 63-73, 1996.

8. Mackay, W. More than just a communication system: diversity in the use of electronic mail. In *CSCW86*, 344-353, 1986.

9. Moran, T., Palen, L., Harrison, S., Chiu, P., Kimber, D., Minneman, S., van Melle, W., and Zellweger, P. "I'll get that off the audio": salvaging in a multimedia meeting. In *CHI97*, 202-209, 1997.

10. Rice R. and Shook, D. Voice messaging coordination and communication. In C. Egido, J. Galegher and R. Kraut, eds., *Intellectual Teamwork*, Lawrence Erlbaum, NJ, 1990.

11. Schmandt, C. and Arons, B. Phoneslave: a graphical telecommunications system, *SID International Symposium*, 25, 146-149, 1984.

12. Stifelman, L. Augmenting real-world objects: a paper-based audio notebook. In *CHI96*, 199-200, 1996.

13. Whittaker, S., Hirschberg, J. and Nakatani, C. All talk and all action. In *CHI98*, 249-250, 1998.

14. Whittaker, S., Hirschberg, J. and Nakatani, C. Play it again: a study of the factors underlying speech browsing behaviour. In *CHI98*, 247-248, 1998.

15. Whittaker, S., Hyland, P, and Wiley, M. Filochat: handwritten notes provide access to recorded conversations. In *CHI94*, 271-277, 1994.

16. Whittaker, S. and Sidner, C. Email overload: exploring personal information management of email. In *CHI96*, 276-283, 1996.

17. Wilcox, L., Schilit, W., and Sawhney. Dynomite: a dynamically organized ink and audio notebook. In *CHI97*, 186-193, 1997.

Instructional Interventions in Computer-Based Tutoring: Differential Impact on Learning Time and Accuracy

Albert Corbett
Human Computer Interaction Institute
Carnegie Mellon University
Pittsburgh, PA 15213 USA
+1 412 268 8808
corbett@cmu.edu

Holly Trask
American Management Systems
12601 Fair Lakes Circle
Fairfax, VA 22030 USA
+1 703 227 4885
holly_trask@amsinc.com

ABSTRACT
We can reliably build "second generation" intelligent computer tutors that are approximately half as effective as human tutors. This paper evaluates two interface enhancements designed to improve the effectiveness of one successful second generation tutor, the ACT Programming Tutor. One enhancement employs animated feedback to make key data structure relationships salient. The second enhancement employs subgoal scaffolding to support students in developing simple programming plans. Both interventions were successful, but had very different impacts on student effort required to achieve mastery in the tutor environment and on subsequent posttest accuracy. These results represent a step forward in closing the gap between computer tutors and human tutors.

Keywords
Intelligent Tutoring Systems, Instructional Interface Design, Animation, Plan Scaffolding, Student Modeling

INTRODUCTION
Computer-based learning environments first appeared three decades ago that afford one advantage of human tutors: individualized interactive learning support. Fifteen years ago "second generation" computer tutors began to appear that incorporate artificial intelligence technology and we can reliably build intelligent tutors that are about half as effective as human tutors. How will we develop "third generation" tutors that approach the effectiveness of human tutors? A significant effort to bridge the gap focuses on natural language dialogs between student and tutor [8, 12]. In contrast, this paper evaluates two enhancements embedded directly in the problem solving interface. These enhancements are designed to increase the educational efficiency of *cognitive mastery learning* in the ACT Programming Tutor (APT).

APT is a computer-based problem solving environment in which students learn to write short programs. It is a *cognitive tutor* that employs a detailed cognitive model of the programming knowledge students are acquiring to support students in learning. Cognitive tutors [2] are designed to (a) provide students an authentic problem solving interface, (b) provide assistance if needed on each problem solving action, (c) monitor the students growing knowledge in the course of problem solving and (d) provide sufficient learning opportunities for the student to reach mastery.

As described in the first two main sections of the paper, APT is a highly successful learning environment. However, there is evidence that providing more and more problems in the standard problem solving interface ultimately yields diminishing educational gains and some students fall short of genuine mastery. This paper reports two interface modifications that were developed in response to this evidence. As described in the third and fourth main sections these interventions are introduced early in the Lisp curriculum when students confront the first two substantial learning challenges. The first intervention employs animated feedback to help clarify basic operator functionality. The second intervention employs subgoal scaffolding to support students in planning very simple algorithms. The final sections report an empirical evaluation of these enhancements. The two interface interventions proved to be successful, but had very different impacts on learning.

THE ACT PROGRAMMING TUTOR
APT is a *cognitive tutor* which helps students learn to write short programs in Lisp, Prolog or Pascal. Each of these three programming language is constructed around a language-specific cognitive model of the knowledge the student is acquiring. The cognitive model enables the tutor to trace the student's solution path through a complex problem solving space in a process we call *model tracing*. The tutor provides feedback on each problem solving action and, if requested, provides advice on steps that achieve problem solving goals.

Figure 1 displays the APT Lisp Module midway through an exercise. The student has previously read text presented in the window at the lower right and is completing a sequence of corresponding exercises. The problem description appears in the upper left window and the student's solution appears in the code window immediately

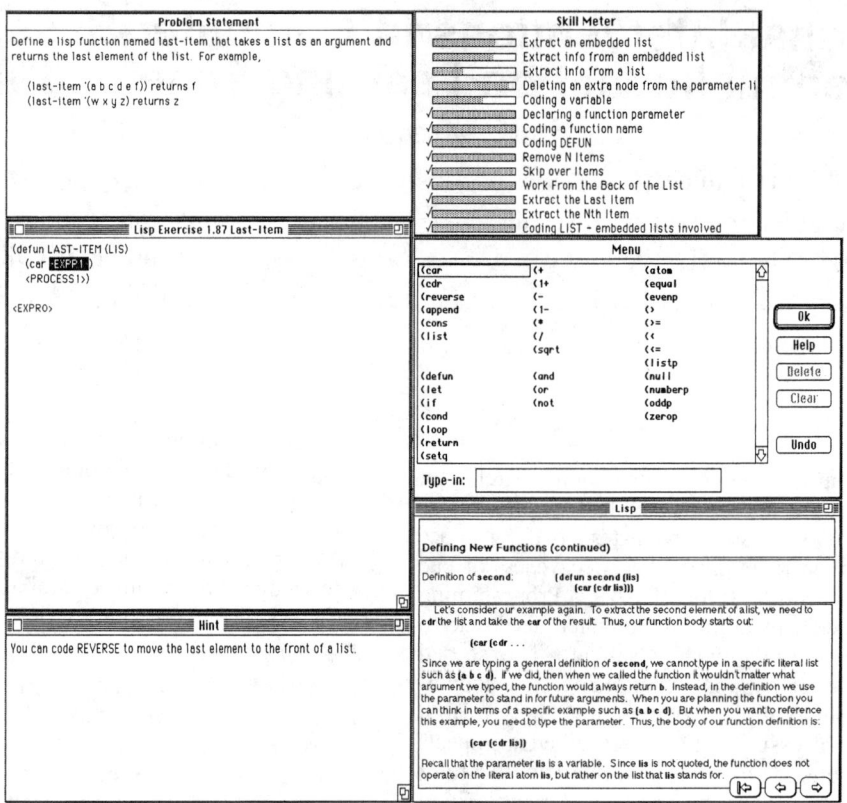

Figure 1. The APT Lisp Tutor interface.

below. The student selects operator templates and types constants and identifiers in the user action window in the middle right. In this figure the student has encoded the operator *defun* used to define a new operator, has entered the operator name, declared input variables and begun coding the body of the definition. The three angle-bracket symbols in the figure, <EXPR1>, <PROCESS1> and <EXPR0> are placeholders which the student will either replace with additional Lisp code or delete. Communications from the tutor appear in the Hint window in the lower left. In this figure the student has asked for a hint on how to proceed.

Cognitive Tutor Effectiveness

We have been developing cognitive tutors for mathematics and programming at Carnegie Mellon for more than a decade. APT has been used to teach self-paced programming courses here since 1984. More recently we have developed cognitive tutor-based algebra and geometry courses that are in use in more than 100 schools around the country. We have completed several summative evaluations comparing the effectiveness of cognitive tutors and traditional problem solving activities. In research which compared college students working with APT to those completing the same programming problems in a conventional programming environment, APT was shown to speed learning by as much as a factor of three and yield small but reliable increases in test performance [2]. Evaluations of the cognitive mathematics tutors have compared classroom use of the cognitive tutors to conventional classroom problem solving activities (seatwork and blackboard work). Time on task is constant in these studies and students in the cognitive tutor condition score about one standard deviation higher (or about a letter grade higher) on achievement tests than students in the conventional classroom condition [2, 9]. A one standard deviation effect size is two or three times larger than the average obtained by conventional computer-based instruction [5], and about half as large as the effect size that can be achieved by individual human tutors [4].

KNOWLEDGE TRACING: COGNITIVE MASTERY

Following the ACT-R theory of skill knowledge [3], the ACT Programming Tutor assumes that goal oriented problem-solving knowledge can be represented as a set of independent production rules that associate problem states and goals with problem solving actions and consequent state changes. For example, Table 1 displays two productions that are acquired early in the Lisp curriculum. The tutor curriculum is structured around production sets. In each curriculum section students read text that introduces a small set of production rules. Then the tutor provides problems that exercise those rules. As the student works, the tutor monitors the student's growing knowledge in a process we *call knowledge tracing*. The tutor employs some simple learning and performance assumptions and a Bayesian computational procedure to estimate the probability that the student has learned each of the productions. As each opportunity arises to apply a rule, the tutor updates its estimate of the probability that the student has learned the rule, contingent on the student's action. (See [6] for more details). The Skill Meter in the upper right corner of Figure 1 depicts the tutor's model of the student's knowledge state. Each entry in the Skill meter represents a production rule in the underlying cognitive model and the shading represents the probability that the student knows the rule. A check mark indicates that the student has mastered the rule.

IF the goal is to return the elements of a list in reverse order,
THEN code the function *reverse*, and set a goal to code the list as an argument.

IF the goal is to insert an expression at the beginning of a list,
THEN code the function *cons*, set a goal to code the expression as the first argument, and set a goal to code the list as the second argument.

Table 1: Two early productions in the Lisp curriculum.

This knowledge tracing process is employed to implement cognitive mastery learning in APT. In each curriculum section the student completes a small fixed set of required problems that introduce all the productions being traced. The tutor then continues providing problems in the section until the student reaches a criterion probability for each rule in the set. This mastery criterion in the tutor is a probability of 0.95.

Cognitive Mastery Learning Successes

Cognitive mastery has proven effective in APT, by several criteria: (a) Average posttest scores are reliably higher for students who work to cognitive mastery than for students who complete a fixed set of required tutor problems [1]; (b) Similarly, twice as many cognitive mastery students reach "A" level performance (at least 90% correct) on posttests in the early Lisp curriculum [6]; (c) Individualized remediation reduces total time invested by a group of students in reaching mastery. (The number of remedial problems varies substantially across students and, in the absence of such individualization, mastery could only be ensured if all students complete the greatest number of problems any student needs); and (d) The knowledge tracing model that guides mastery learning in the tutor reliably predicts individual differences in posttest performance across students [6].

Cognitive Mastery Learning Shortcomings

Nevertheless, there are noteworthy shortcomings in cognitive mastery learning: (a) A substantial proportion of students in the cognitive mastery condition still fall short of "A" level performance in the early Lisp curriculum [6]; (b) Students are investing substantial amounts of time and effort to reach mastery. In a typical study, students in the cognitive mastery condition completed an average about 75% more problems than students who completed the fixed set of required tutor problems [6]; (c) There is an inverse correlation between the number of problems students required to reach mastery and students' test performance. The students who struggle the most and complete the most problems do not do as well on the test as students who perform well in the tutor [6]; and (d) The tutor's knowledge tracing model slightly but systematically overestimates average test performance. Evidence suggests that this occurs because some students are learning sufficient, but suboptimal productions [7].

These shortcomings represent converging evidence that providing more and more remedial problems of the same type yields diminishing educational returns. In this study we explore two forms of augmented support in the problem solving interface designed to speed learning and increase asymptotic test performance in early sections of the Lisp curriculum.

AUGMENTED SUPPORT FOR EARLY CHALLENGES IN THE LISP CURRICULUM

This study focuses on the first four sections in the APT Lisp curriculum. These sections introduce (a) two data types, *atoms* (symbols) and *lists* (hierarchical groupings of symbols), (b) three "extractor" functions, *car*, *cdr* and *reverse*, that take a single list as an argument and return a component or transformation of the argument, (c) three "combiner" functions, *append*, *cons*, and *list*, that take multiple arguments and return new lists, and (d) the syntax and functionality of simple algorithms involving embedded function calls. Table 2 displays sample exercises from each of these sections.

Section 1: Write a lisp function call that takes the list
(c d e) and returns (d e)
Solution: (cdr '(c d e))

Section 2: Write a lisp function call that takes the arguments (a b) c (d) and returns
((a b) c (d))
Solution: (list '(a b) 'c '(d))

Section 3: Write a function call that takes the list
(a b c d) and returns the last element, d.
Solution: (car (reverse '(a b c d)))

Section 4: Write a function call that takes the lists
(a b c) and (d e f) and returns (a f e d).
Solution: (cons (car '(a b c)) (reverse '(d e f)))

Table 2. A sample problem in each of four tutor sections.

Students have little difficulty with the three extractor functions, *car*, *cdr* and *reverse* and with extractor algorithms in section 3. However, many students struggle with the differences among the three combiner functions, *append*, *cons* and *list* in section 2 and with the combiner/extractor algorithm problems in section 4. The two instructional interventions described below focus on these two curriculum sections.

Supporting Data Structure Parsing: Animated Feedback

In curriculum section 2 students are learning to distinguish among three combiner functions and the subtlety of this distinction is captured in the text excerpt displayed in Table 3. There are multiple reasons that this discrimination is difficult. First, the terminology is difficult, with semantically related and confusable terms for novices – atoms, elements, arguments, expressions. Second, students have trouble grasping the hierarchical structure of lists. They may not understand that they need to analyze the structural relationship among the input arguments and results and, finally, students have trouble visually parsing the parentheses when they do try to analyze these structural relationships.

Study the following examples to see how **list**, **append** and **cons** are distinguished from each other.

(list '(a b) '(c (d e) f)) returns ((a b) (c (d e) f))

(append '(a b) '(c (d e) f)) returns (a b c (d e) f)

(cons '(a b) '(c (d e) f)) returns ((a b) c (d e) f)

The function **list** can take one or more arguments, and makes a new list by "wrapping parentheses" around its arguments. The function **append** takes one or more lists as arguments, makes a new list by "removing the parentheses" from around each of its arguments and merging all the elements into one long list. The function **cons** always takes two arguments and inserts the first argument at the beginning of the second.

Table 3. An excerpt from the Lisp text describing the operators *append*, *cons* and *list*.

The college students in our samples do eventually learn this discrimination reasonably well. Students' performance is quite good by conventional standards, but falls short of the mastery ideal that all students may become "A" students. There is evidence that students' posttest performance falls short of true mastery because some students encode rules that are correlated with, but different from optimal rules [7]. For example, students may encode a heuristic that if the arguments in a problem are all lists, then *append* is the appropriate function.

Can we help all students become "A" students? The effectiveness of natural language instruction is limited here by terminology confusion and the perceptual challenge of parsing parentheses. Instead, to highlight the key structural relationships among input arguments and results, we developed an augmented interface for combiner function calls that provides animated graphical feedback, as displayed in Figures 2 and 3.

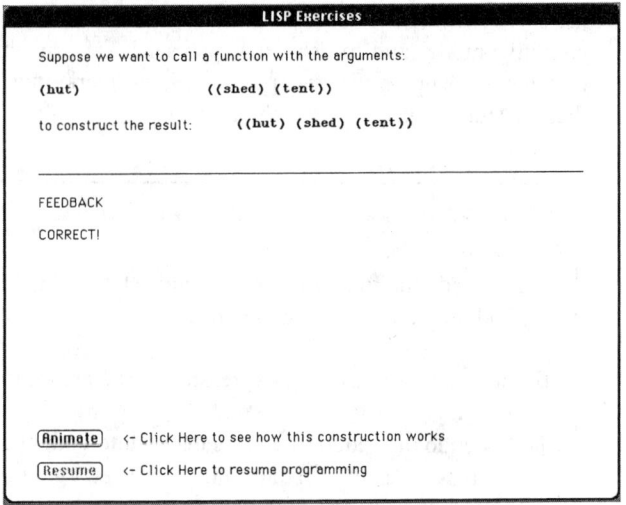

Figure 2 Animated feedback window for combiner function calls in curriculum section 2.

After the student has selected a combiner function in a tutor problem the feedback window displayed in Figure 2 appears. This window appears whether or not the student selects the correct function, since there is a 1/3 chance of selecting the correct one by guessing. When the student clicks the animate button at the bottom of the window, the structural relation between the arguments and result is animated. Arguments and parentheses move on the screen to display this structural relationship. Figure 3 depicts the animation for the function *cons*, in which the initial parenthesis of the second argument slides over to the left and the first argument slides to the right to become the first element of the second argument. In the case of *list*, the input arguments slide together and an encompassing set of parentheses descends. In the case of *append*, the outer parentheses of each input argument descend off the screen, the arguments slide together on the screen, and two new parentheses encompass the result.

Plan Support: Subgoal Scaffolding

Consider the section 4 problem displayed in Table 4. The tutor's structure editor supports top-down programming so under the ACT-R model, students need to satisfy three planning goals before they begin coding. Under cognitive mastery learning, the students should have mastered the final five coding productions at the bottom of the table in the first three curriculum sections. The unique components of this section are the productions needed to satisfy the three planning goals. Again, students master this task reasonably well, but often require many remedial problems to do so.

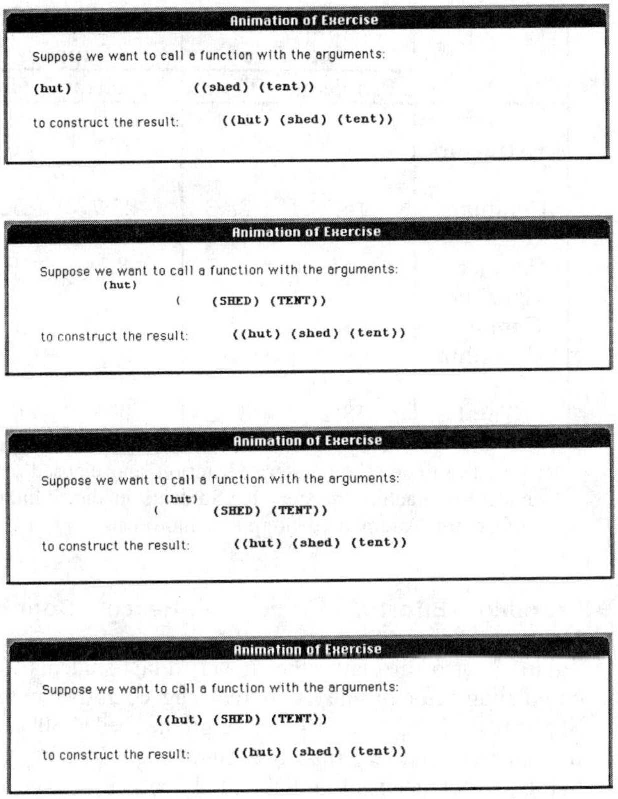

Figure 3. The initial state, two intermediate states and the final state in animating the structural relationship between input arguments and the output list for the function *cons*.

To support more efficient learning in this curriculum section, we developed a plan reification interface as displayed in Figure 4. Panel 4a displays the standard coding interface at the beginning of a programming problem. Students simply select the <code> symbol and begin generating Lisp code. Panel 4b displays the plan reification interface which requires the student to post the two expressions that must be extracted from the given arguments before entering any code. In this example, students must type b for <subgoal1> and (f g h) for <subgoal2> before entering the code (list (car (cdr '(a b c))) (cdr '(e f g h))).

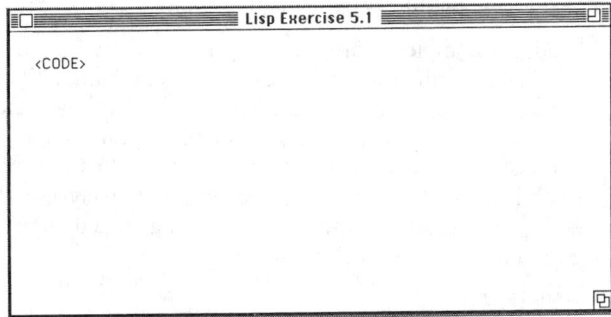

Figure 4a. The standard coding interface.

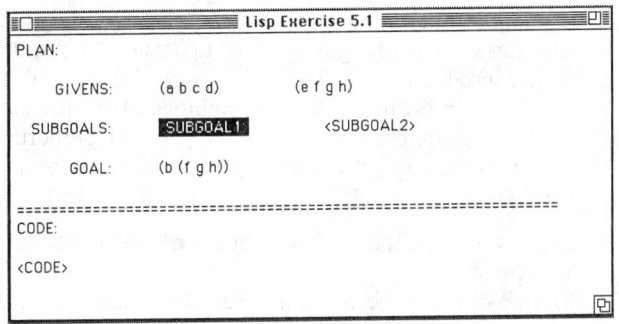

Figure 4b. The plan scaffolding interface

DESIGN OF THE STUDY

In this evaluation of animated parenthesis feedback and subgoal scaffolding interface enhancements, students worked through the first four sections in the ACT Programming Tutor Lisp curriculum and completed three programming tests.

Participants

Thirty nine college students were recruited to participate in the study for pay. These students had an average Mathematics SAT score of 648 and had completed an average of 1.3 programming courses previously, although none had prior experience with Lisp. Both these variables were controlled in assigning students to two groups. Eighteen students completed the study with the standard coding interface. Twenty-one completed the study with augmented feedback and subgoal scaffolding as described above.

Procedure

Students in this study completed the first four sections in the APT Lisp curriculum. In each curriculum section, students read text describing Lisp, completed one or two sets of quiz questions on the text, completed a small fixed set of required programming exercises that covers the rules being introduced, then completed remedial exercises as needed to bring all production rules in the section to a

---

Exercise: Write a function call that takes the lists (a b c) and (d e f) and returns the list (a f e d).

Solution: (cons (car '(a b c)) (reverse '(d e f)))

Recognize that a must be extracted from (a b c).
Recognize that the list (d e f) must be reversed.
Recognize that a and (f e d) must be combined in a list.
Code a call to *cons*.
Code a call to *car*.
Code the first given as the argument to *car*.
Code a call to *reverse*.
Code the second given as the argument to *reverse*.

Table 4. Planning and coding goals for a section 4 problem

mastery criterion (knowledge probability > 0.95). Students in the standard interface condition completed all exercises in the standard APT coding interface. Students in the Augmented Support group received animated parenthesis feedback in the second curriculum section on basic combiner functions and subgoal scaffolding in the fourth section on combiner/extractor algorithms.

Students completed programming tests following the first, third and fourth sections. These tests were cumulative and contained six, twelve and eighteen programming exercises respectively. The test exercises were similar to the tutor exercises and the test interface was identical to the standard tutor interface, except that students could freely edit their code. No tutorial assistance and no augmented support of any kind was available in testing.

RESULTS

Two measures were employed to evaluate the students' learning effort in reaching cognitive mastery: number of problems needed to reach mastery and elapsed time in reaching mastery. The first measure is theoretically relevant, since learning is assumed to occur at opportunities to fire productions. The second measure is of more practical significance. We also employed two measures of learning outcomes, mean test accuracy and proportion of students reaching "A" level performance (at least 90% correct) on the posttests.

Learning Effort: Number of Problems to Mastery

Table 5 displays the mean number of problems required to reach mastery in each curriculum section for students in the two conditions. All students completed 21 required problems in the study. Students in the standard condition needed an additional 37.2 remedial problems to reach mastery, while students in the augmented feedback condition needed 22.8 remedial problems. While this difference is large, the main effect of interface condition is not significant, $F(1,37) = 1.47$. The main effect of curriculum section is significant, $F(3,111) = 18.55$, $p < .01$, confirming that sections 2 and 4 are more challenging than sections 1 and 3. More importantly, the interaction of interface condition and curriculum section is significant, $F(3,111) = 3.09$, $p < .05$.

Students in both groups worked with the standard coding interface in sections 1 and 3 and essentially completed the same number of remedial exercises in those sections. Note that while animated parenthesis feedback was provided to the Augmented Support group in curriculum section 2, this augmented feedback had no impact on the number of problems needed to reach mastery learning. Both groups completed 9 required problems and an average of about 15 remedial problems in section 2. But, subgoal scaffolding in section 4 had a large impact on problems needed to reach mastery. Students in the Standard Coding condition averaged 15.44 remedial problems, while students in the Augmented Support condition averaged only 2.05 problems. This difference is marginally significant, $t(37) = 1.83$, $p < 0.08$.

	Standard Interface		Augmented Support	
	Problems	Time	Problems	Time
Extractors	6.3	9.0	6.3	6.2
Combiners	24.7	28.3	23.7	32.7
Extractor Algorithms	8.8	10.7	8.7	9.0
Combiner Algorithms	18.4	35.2	5.1	15.2
Total	58.2	83.2	43.8	63.1

Table 5. Average Number of APT Tutor Problems Required to Reach the Mastery Criterion and elapsed time (minutes) in reaching mastery for Students in the Standard Interface and Augmented Support conditions.

Learning Effort: Time to Reach Cognitive Mastery

Table 5 also displays the mean time students spent completing tutor problems in reaching cognitive mastery. Students in the Standard Interface group needed 30% more time to reach mastery than the Augmented Support group, but this difference is not reliable, $F(1,37) = 1.2$. The main effect of curriculum section is again significant, $F(3,111) = 13.46$, $p < .01$, and more importantly, the interaction of interface condition and curriculum section is significant, $F(3,111) = 2.9$, $p < .05$).

Note that the two groups spent virtually the same amount of time reaching mastery in the first three curriculum sections (48.0 minutes vs. 47.9 minutes). The entire 20 minute difference in elapsed time occurs in the fourth, combiner algorithm section in which the Augmented Support condition benefited from subgoal scaffolding. However, the difference between the means in curriculum section 4 is not reliable, $t(37) = 1.6$, ns.

While mean learning time did not vary reliably across groups in section 4, the variance in learning times in that section is much larger for the Standard Interface group (3089 vs 160) and this difference is reliable $F(17,20) = 19.3$, $p < .01$. An inspection of the learning time distributions in this section makes the impact of subgoal scaffolding more apparent. In both conditions about 80% of students completed section 4 in 30 minutes or less. The remaining 20% of subgoal scaffolding students finished in less than an hour, while the remaining 20% of standard coding students required between 2 and 3.5 hours. So, the main impact of subgoal scaffolding on learning time is to greatly reduce the time needed for the slowest students to reach cognitive mastery.

To further explore this evidence that subgoal scaffolding primarily helped the slowest students, we divided each group of students in half based on Math SAT scores and reanalyzed the learning effort data for the section 4 combiner algorithms just for the students with lower Math SAT scores. The average Math SAT score for this subset of students was 571. As displayed in Table 6, students with lower SAT scores in the standard interface condition required an average of 33.6 problems and 60.7 minutes to reach mastery while students working in the subgoal scaffolding condition required 6 problems and 19.8 minutes. The difference in number of problems is reliable, t(18) = 2.11, p < .05 and the difference in elapsed time is marginally reliable t(18) = 1.87, p < .08.

	Standard Interface		Augmented Support	
	Lower SATs		Lower SATs	
	Problems	Time	Problems	Time
Combiner Algorithms	33.6	60.7	6.0	19.8

Table 6. Average Number of APT Tutor Problems Required to Reach the Mastery Criterion and elapsed time (minutes) in reaching mastery for Students in the Standard Interface and Augmented Support conditions.

Test Performance

Posttest performance of the two groups is displayed in Table 7. Two performance measures are displayed: (1) mean percent correct and (2) the probability that students reach "A" level performance (90% correct).

	Standard Interface		Augmented Support	
	%C	P> 0.9	%C	P > 0.9
Test 1				
Total	94%	0.72	98%	0.90
Test 2				
Total	86%	0.50	96%	0.90
Extractors	92%	0.83	97%	0.81
Combiners	80%	0.39	95%	0.86
Test 3				
Total	83%	0.40	88%	0.65
Extractors	95%	0.72	95%	0.76
Combiners	81%	0.39	96%	0.76
Combiner Algorithms	71%	0.28	75%	0.24

Table 7. Mean percent correct and probability of reaching "A" level performance on three posttests for the Standard Interface and Augmented Support Groups.

Students in both groups perform very well on the basic extractor problems in Test 1 characteristic of curriculum section 1. Test 2 was administered after curriculum section 3 and contained problems characteristic of the first three sections. Students in the Standard Interface condition perform very well on this test. They average 86% correct and half the students reach "A" level performance. However, students in the Augmented Support condition perform even better. They score 96% correct and 90% of students reach "A" level performance. The difference in average percent correct is reliable, F(1,37) = 8.7, p < .01, as is the difference in proportion of students reaching "A" level (z = 2.8, p < .01). Two subscores are reported for Test 2. The first includes extractor and extractor algorithms questions from sections 1 and 3. As can be seen, there is little difference between the two groups on these questions. However, students in the augmented feedback condition perform substantially better on the combiner problems in testing. The main effect of problem type is reliable F(1,37) = 5.5, p < .05, so the combiner problems are reliably more difficult, and the interaction of interface condition and problem type is marginally reliable, F(1,37) = 2.7, p = 0.10. In pairwise comparisons, the difference between the two interface groups on the extractor problems is non-significant, t(37) = 1.6, while the difference between the two groups on the combiner problems is significant t(37) = 2.7, p < .01. Similarly, the difference in proportion of students reaching "A" level performance in the combiner section is reliable, (0.86 vs. 0.39), z = 3.0, p < .01.

Test 3 was administered following the fourth curriculum section and contained problems characteristic of all four sections. Overall, students again performed quite well on this test. The Augmented Support students performed slightly better than the Standard Interface students, and this overall difference is marginally reliable, F(1,37) = 3.8, p < .06. The proportion of students reaching "A" level performance overall did not vary significantly, z = 1.7. Three test subscales are reported in Table 7, extractor problems including extractor algorithms, basic combiner problems and combiner/extractor algorithm problems. The main effect of problem type is significant, F(2,74) = 326.1, p < .01, and the interaction of interface condition and problem type is reliable, F(2,74) = 5.5, p < .01. Again, there is very little difference between the two groups on the extractor and extractor algorithm problems. The Augmented Support students continue to perform better on the section 2 combiner problems than the Standard Interface students (96% vs 81%). In a pairwise test, this difference is reliable, t(47) = 3.1, p < .01, as is the difference in proportion of students reaching "A" level (76% vs 39%), z = 2.4, p < .05. Finally, performance on the section 4 combiner/extractor algorithm problems is virtually identical for the two groups. Students in the Augmented Support condition completed an average of just 5 tutor problems in section 4, yet reached the same level of test performance on those problems as the Standard Interface students who averaged 18 tutor problems in section 4.

DISCUSSION

The two interface enhancements evaluated in this study led to very different gains in educational efficiency. Animated parenthesis feedback was introduced in curriculum section 2 to make relevant data structures salient in discriminating among three combiner functions. This intervention had no impact on the effort required to satisfy the tutor's cognitive mastery criterion, but this augmented feedback resulted in a substantial gain in test performance, both in mean accuracy and proportion of students reaching "A" level performance. In contrast, the section 4 subgoal scaffolding designed to help students organize productions they mastered in previous lessons, led to a large decrease in number tutor problems required to reach cognitive mastery, but did not lead to increased posttest accuracy. Overall, subgoal scaffolding also sharply reduced the maximum elapsed time that students needed to reach cognitive mastery, but this effect was marginally reliable only for students with lower Math SAT scores.

Although the educational impact of animation in technology enhanced learning has been mixed at best [10, 11], animated parenthesis feedback had a decisive positive impact on test performance in this study. It did not make learning "easier" as measured by time on task or number of tutor problems, but fostered programming knowledge that transferred more successfully to the test environment. This successful transfer implies that students are acquiring a deeper, more optimal encoding of relevant aspects of list structure. Indeed, we believe that animation was successful because it addressed a crucial topic that students find hard to grasp and that does not lend itself well to natural language discussion.

Subgoal scaffolding, in contrast, decreased the average number of problems required to reach mastery in the tutor, but did not reliably reduce average learning time, nor enhance subsequent test performance. This suggests that scaffolding did make it easier for students to organize previously mastered operator knowledge into more complex plans, but did not lead to a substantially deeper knowledge of the operators nor of the plans. Reducing the learning effort needed to reach equivalent performance levels is an important accomplishment, though, and subgoal scaffolding did dramatically reduce section 4 learning time for about 20% of students, from a maximum of 2 to 3.5 hours to a maximum of 1 hour.

This study raises some interesting questions to be pursued. For example, the animated parenthesis feedback was provided *retrospectively*, after the student selected a combiner function, while students posted subgoals *prospectively*, before selecting the subsequent combiner function. Perhaps subgoal scaffolding could lead to deeper understanding if subgoal posting followed rather than preceded the initial combiner selection. Another intriguing question is why the Augmented Support students' superior understanding of basic combiner functionality demonstrated in section 2 problems did not transfer to higher test accuracy in section 4 combiner/extractor algorithm questions.

Nevertheless, these results are very encouraging. Relatively simple interface enhancements can have a substantial impact on learning rate and asymptotic test performance. These results serve as a reminder that as we continue to develop computer-based learning environments, that we should not limit ourselves to studying the strategies that make human tutors effective, but need to assess domain specific challenges and tailor instructional interventions to meet those challenges.

ACKNOWLEDGMENTS

This research was supported by the Office of Naval Research grant number N00014-95-1-0847 and by NSF grant number 9720359 to CIRCLE: Center for Interdisciplinary Research on Constructive Learning Environments. We thank Dana Heath and Susan Klein for help in data collection and Megan McLaughlin for help in manuscript preparation.

REFERENCES

1. Anderson, J.R., Conrad, F. and Corbett, A.T. (1989). Skill acquisition and the LISP Tutor. *Cognitive Science*, *13*, 467-505.

2. Anderson, J.R., Corbett, A.T., Koedinger, K.R., and Pelletier, R. (1995). Cognitive tutors: Lessons learned. *The Journal of the Learning Sciences*, *4*, 167-207.

3. Anderson, J.R., and Lebiere, C. (1998). *The atomic components of thought*. Mahwah, NJ: Erlbaum.

4. Bloom, B. S. (1984). The 2 sigma problem: The search for methods of group instruction as effective as one-to-one tutoring. *Educational Researcher*, *13*, 4-16.

5. Cohen, P. A., Kulik, J. A., & Kulik, C. C. (1982). Educational outcomes of tutoring: A meta-analysis of findings. American *Educational Research Journal*, *19*, 237-248.

6. Corbett, A.T. and Anderson, J.R. (1995). Knowledge tracing: Modeling the acquisition of procedural knowledge. *User modeling and user-adapted interaction*, *4*, 253-278.

7. Corbett, A.T. and Bhatnagar, A. (1997). Student modeling in the ACT Programming Tutor: Adjusting a procedural learning model with declarative knowledge. *Proceedings of the Sixth International Conference on User Modeling*. New York: Springer-Verlag Wein.

8. Graesser, A.C., Person, N.K., & Magliano, J.P. (1995). Collaborative dialogue patterns in naturalistic one-on-one tutoring. *Applied Cognitive Psychology*, 9, 359-387.

9. Koedinger, K.R., Anderson, J.R., Hadley, W.H. & Mark, M.A. (1995). Intelligent tutoring goes to school in the big city. *Proceedings of the 7th World Conference on Artificial Intelligence in Education*.

10. Pane, J.F., Corbett, A.T. and John, B.E. (1996). Assessing dynamics in computer-based instruction. *Proceedings of ACM CHI'96 Conference on Human Factors in Computing Systems*, 197-204.

11. Rieber, L.P., Boyce, M.J., and Assad, C. (1990). The effects of computer animation on adult learning and retrieval tasks. *Journal of Computer-Based Instruction*, 17, 46-52.

12. VanLehn, K., Siler, S., Murray, C. & Bagget, W. (1998). What makes a tutorial event effective? In: M. A. Gernsbacher & S. Derry (Eds.) *Proceedings of the Twenth-first Annual Conference of the Cognitive Science Society*, Hillsdale, NJ: Erlbaum. pp. 1084-1089.

Keystroke Level Analysis of Email Message Organization

Olle Bälter

Interaction and Presentation Laboratory
Nada, Royal Institute of Technology
SE-100 44 Stockholm, Sweden
+46 8 790 9157
balter@nada.kth.se

ABSTRACT

Organization of email messages takes an increasing amount of time for many email users. Research has demonstrated that users develop very different strategies to handle this organization. In this paper, the relationship between the different organization strategies and the time necessary to use a certain strategy is illustrated by a mathematical model based on keystroke-level analysis. The model estimates time usage for archiving and retrieving email messages for individual users. Besides explaining why users develop different strategies to organize email messages, the model can also be used to advise users individually when to start using folders, clean messages, learn the search functionality, and using filters to store messages. Similar models could assist evaluation of different interface designs where the number of items increase with time.

Keywords
Email, model, user, organisation of messages.

INTRODUCTION

Previous research about how email users organize their messages has categorized users into e.g. Prioritizers, Archivers [6], No filers, Spring cleaners, Frequent filers [9], and Folderless cleaners [1]. That research is based on interviews, surveys, and logging of data, describe the current situation for users in these categories. However, it does little to help us understand how and why users develop different organization strategies over time.

Here, another approach is pursued by creating a mathematical model of the time usage for archiving and retrieving email messages. The model can be used for time comparisons of different organization strategies for an individual based on the person's motoric skills. The purpose of the model is to answer questions such as "Would it be time efficient for me to increase the number of folders?" and "If I spend 30 minutes to clean up my folders and delete messages, would I gain time in the long run?".

The model is based on keystroke-level analysis and is thereby limited to the context-independent aspects of email message organization.

The paper begins with a short description of keystroke-level analysis and then the model of email storage and retrieval is presented. The model is illustrated with a description of some fictitious users to explain the development of organization habits. Finally, an analysis of the model is presented by comparing the model's predictions with observed long time user behaviour.

METHODS

Keystroke-level analysis can be used to estimate how much time it will take for a user to accomplish a given task with a given interface. The method has been tested empirically with good results [3].

The execution time of a task can be estimated with the sum of the time for six operators: K (key stroking), P (pointing), H (homing), D (drawing), M (mental preparing), and R (a system response operator), see equation 1.

Eq. 1 Time to execute a task $= T_K + T_P + T_H + T_D + T_M + T_R$

The total time for keystrokes (T_K) can be estimated as the time to perform one keystroke (t_K) multiplied by the number of keystrokes (n_K): $T_K = t_K \cdot n_K$.

The time to move the mouse to point at a target on the screen can be estimated with Fitts's law (Fitts & Posner 1967):

Eq. 2 $T_P = A + B \cdot lg_2 (D/S + C)$

where the value of the constants A, B, and C can be determined experimentally; D is the distance to the target and S is the surface area of the target.

The homing and drawing operators are not used in the model below. The homing time, T_H, is the time to move the user's hand between one physical device and another. The drawing time, T_D, is the time to draw a set of straight line segments.

T_M represents the time the user mentally prepares to execute the physical operators described above. The last operator T_R represents the time for the system to respond to a user action. It is used sparsely in the model below as the systems modeled are considered so fast that this time is negligible for most operations.

Experiments have been made to estimate values of the operators described above [4]. Examples of values are displayed in table 1.

Table 1. Selected operators of keystroke-level analysis [4]

Operator	Description	Time (seconds)
K	Keystroke	0.08 - 1.2
P	Pointing	$0.8 + 0.1 \lg_2 (D/S + 0.5)$
M	Mentally preparing	1.35
Average K for non-secretary typist: 0.28 s		

Restrictions on the model

The model presented below is a simplification of the real world. The model handles only storage and retrieval of email messages in existing folders and is built on the following assumptions:

The user does not make mistakes.

The user is an average non-secretary typist.

Messages are moved to folders with drag-and-drop.

The representations on the screen of folders and messages uses the same font size and/or icons of the same size.

The folder structure is flat (folders do not contain folders or subfolders are visible in the same way as top level folders).

The distribution of messages in folders is even (i.e. all folders contain approximately the same number of messages).

A model of email storage and retrieval

Previous research makes it clear that the number of incoming messages, folders, and the interface affect the strategies used for storing email messages [1,2,9]. A large number of folders increases the time that must be used to store a message, while it may reduce the time to retrieve the same message. If we initially limit this analysis to messages that are archived in folders, the total time spent each day on storing these messages can be expressed according to equation 3.

Eq. 3 Total time spent storing messages =
# of stored new messages ·
average time to store one message

The time spent each day on retrieving messages can be expressed according to equation 4.

Eq. 4 Total time spent retrieving messages =
# of retrieved messages a day ·
average time to retrieve one message

The total time spent on archiving and retrieving messages is of course the sum of equation 3 and equation 4 above. The time to switch context from previous tasks or to following tasks is not a part of the model.

There are only two ways for a user to decrease the time spent on archiving and retrieving messages:

Increased skills in handling the archiving and retrieving facilities of the mail tool.

Use of a more efficient strategy for archiving.

Of these two the choice of strategy is probably dominant, with the possible exception of beginners. A person that can rely on others to remind him or her of information in email messages may choose the strategy to simply delete all messages after reading, and thereby reduce the time spent on archiving and retrieving to practically zero. This strategy is possible for only an exclusive group of users and will not be discussed further. Another extreme strategy is to store all messages in the inbox and thereby reduce the time spent to store messages to zero, but this strategy may have the disadvantage of demanding more time when messages should be retrieved and the stored messages are many. This strategy will be further analyzed below.

In order to construct a model, we have to limit the world it should describe. In this case the model is limited to graphical interfaces where drag-and-drop can be used to move messages between folders. Other limitations are described in their context below.

Storing a message

To store a message in a folder, it is necessary to know the name, position, or the look of the folder, unless all messages are stored in the inbox. First we have to define *known* and *unknown* folder. A known folder is defined as a "folder that the user knows the position and/or the name of". An unknown folder is defined as a "folder that the user has to search for". The user may be aware that the folder exists, and will recognize it in a search, but cannot remember the exact name or position by heart.

Those that do not use folders do not spend any time at all storing a message, but for users with at least one folder the time to find a folder can be estimated as follows. If we assume that the folder structure is flat (i.e. not hierarchical), the time spent on finding a folder can be approximated in two different ways depending on whether the folder is known or not.

For an unknown folder the user has to scan folder names one at a time until the searched folder is found. The approximation will therefore be a linear function of the number of folders. On average, half of these folders must be scanned.

Eq. 5 Time to find an unknown folder =
M + Search constant · # of folders / 2

where the search constant is the inverted number of items (here: folders, further below: messages) on screen that can be processed per second to identify the searched folder.

For a known folder the time to find it can be approximated with Fitts's law.

Eq. 6 Time to find a known folder =
M + 0.8 + Search constant · lg₂ (# of folders + 0.5)

The time for finding a visible folder will be a combination of equation 5 and equation 6 if we assume that the number of known and unknown folders are evenly distributed. We combine the last two equations in equation 7:

Eq. 7 Time to find a visible folder =
(Eq 6 · Known folder % +
Eq 5 · (100− Known folder %)) / 100

The *Known folder %* is the percentage of folder searches that concern known folders.

If the user should search for a folder manually and the number of folders exceeds the number of visible folders on the screen, additional time must be spent to scroll the list of folders. This time is approximated with a constant under the assumption that once the action of moving the cursor to the scroll bar is made, the time for extra scrolling is negligible compared to the time it takes for the user to read the folder names. These two events can also occur simultaneously. We name this time *Scroll constant*. The probability for this, *screenful folder list probability*, can be estimated with:

IF (# of folders) < (# of visible folders) THEN
    screenful folder list probability := 0
ELSE screenful folder list probability :=
    1 − (# of visible folders) / (# of folders).

This time must be added to equation 7:

Eq. 8 Time to find a folder =
Eq 7 + Screenful folder list probability · Scroll constant

Finally, the message must be moved to the folder and this time is approximated with a mental preparation followed by a selection (point and click) and a move (point). In total:

Eq. 9 Time spent to store a message =
Time to find a folder (Eq. 8) + M+P+K+P

In some email tools it is possible to first select the message and then drag it to the folders and by moving the message up or down in the folder list, the list starts scrolling. The term added in the last equation is still a good approximation, as the message still has to be selected (PK) and moved (the second P). It might take additional time to scroll through the folder list compared to "jumping" to a known location of a folder by clicking on the slider. For mathematical simplicity, this type of storing is not handled in the model.

Searching for a message

A message can be searched for manually or by a search function, and search functions can be of many different kinds. Here only two versions will be modeled: search functions that require the user to define the folder to search in, and those that do not require this.

To manually find a message, the folder must be found first (we assume that the message is located in another folder than the currently selected folder). Again this is simplified by not using folders other than the inbox, and the time spent to find a folder can be approximated with equation 8 followed by a selection of the folder (PK). Thereafter the folder content must be scanned for the actual message. We assume here that once the folder is selected, the list of messages in that folder becomes visible and that the searched message can be identified without opening the message (e.g. by reading the information in the list of messages), or that the time to open messages during this search is negligible. In many cases the location in the folder of the message will be known and if we assume that the number of searches in a folder is proportional to the number of messages in a folder and that all folders contain approximately the same number of messages, we can approximate the search time with equation 10, based on Fitt's law.

Eq. 10 Time to find a known message in a folder =
M + 0.8 +
Search constant · lg₂ (# of messages/# of folders+ 0.5)

This simplification gives an upper limit for the time to find a message. Unevenly distributed numbers of messages in folders will give shorter search times (on average), but to estimate that time the distribution of messages on all folders must be known. When searching for messages the inbox should be included in the set of folders.

If the location of the message is completely unknown, on average half of the messages in the folder must be scanned and the time can be approximated with:

Eq. 11 Time to find an unknown message in a folder =
M + Search constant · (# of messages/# of folders) / 2

Once again these two equations can be combined if we can approximate the probability of the two different search methods:

Eq. 12 Time to find a message in a folder =
(Eq. 10 · Known message % +
Eq. 11 · (100− Known message %)) / 100

Where the *Known message %* is defined in a similar way as the *Known folder %*. As in the case of invisible folders, there is a penalty if the number of messages in the folder exceeds the number of visible messages and therefore in analogy with equation 8, we rewrite equation 12 to:

Eq. 13 Time to find a message in a folder =
(Eq. 12 +
Screenful message list probability · Scroll constant

where the screenful message list probability is defined as:

IF (average # of messages in a folder) < (# of visible messages) THEN screenful message list probability := 0
ELSE screenful message list probability :=
1 − (# of visible messages) / (average # of messages in a folder).

Folder-dependent search tools
A tool search for a message where the folder must be specified can be divided into:

1) Find the folder.

2) Formulate a search condition.

3) Wait for the system to search messages

4) Manual search in the resulting selection.

Again, step 1 is described by equation 8 followed by a selection of the folder (P+K). Step 2 varies between tools, individuals and the current search task, but is in general not affected by the number of messages or folders and can therefore be approximated with mental preparation followed by a tool and user-dependent *Query constant*.

To model the time for system response without introducing system dependencies in the model is of course impossible. Therefore we make this model as simple as possible and assume that the system response time is proportional to the number of messages in the folder to search. We introduce a system-dependent constant, *System time*, that represents the time for the system to handle one message in a search.

The fourth step depends on how well the search query is formulated in relation to the variety of communication topics and people, resulting in a large or small number of remaining messages. The number of remaining messages is therefore approximated with an user-dependent constant expressed as a percentage of the number of messages that remain on average after a search, the *Remains constant*. The manual search among these remaining messages is linear, and can be approximated with equation 14.

Eq. 14 Time for manual search in selection =
$M + Search\ constant \cdot Remains\ constant \cdot$
$(\#\ of\ messages\ /\ \#\ of\ folders)\ /\ 2$

Folder-independent search tools
For folder-independent search tools the method to find a message can be simplified to three steps:

1) Formulate a search condition

2) Wait for the system to perform the search

3) Manual search in the resulting selection

In the first step, the user-dependent *Query constant* appears again. The second step is the same as the third in a folder-dependent search, but with the number of messages equal to the total number of stored messages. The third step is the same as equation 14 with the number of folders = 1.

Storing messages in the inbox
For users that store all messages in the inbox, storage time is eliminated since the user do not have to search for a folder or move messages to a folder. However, the time to search for especially unknown messages will increase if the total number of messages is large. Many users have a strategy between the two extremes of storing all messages in inbox and storing no messages there. To include this in the model we assume that the probability of searching for a message in the inbox is proportional of the percentage of messages in the inbox. We also assume that messages are stored in a similar way: proportional to the current distribution of messages between the inbox and other folders.

Time spent on managing email
In total, the time spent on archiving and retrieving messages can be described by the eight constants in table 2 and the five variables in table 3. The division into constants and variables may seem arbitrary (and is not important for the reasoning), but the variables are changing over time, while the constants are approximately constant for a person that masters the mail tool.

Table 2. Message archiving and retrieval constants.

Name	Description	Typical values
Search constant	Seconds to process one item on screen (1 / # of items processed per second)	0.1-1 (1-10 items/s)
Scroll constant	Time to scroll a window	0.5-5s
Query constant	Time to formulate a search query	1-60s
Remains constant	Percentage of messages remaining to search manually among after using a search tool	1-20%
Screenful constant	Number of visible items (folders or messages) on screen	10-60
Known folder %	Percentage of folders used for archiving with known location or name	50-100%
Known message %	Percentage of searched messages with known location	50-95%
System time	Time for system to handle one message in a search	0.1-10ms
Typical values are estimated from informal user trials.		

Table 3. Message archiving and retrieval variables.

Name	Description	Typical values
# of incoming	Number of incoming messages stored each day	0-75
# of folders	Number of categories/folders used for archiving messages	0-200
# of messages	Total number of stored messages	0-100.000
# of searched	Number of searched messages each day	0-20
# in inbox	Number of messages stored in inbox	0-all
The typical values are extreme values from [2, 9].		

The number of incoming messages is beyond most receivers' control, but the number of stored messages can be affected by a choice to store all, most, or only some of the incoming messages. The number of folders is under the control of the user, but the number of stored messages is a consequence of number of incoming messages and time. Note that only the number of incoming messages that should be stored is taken into account. Incoming messages that are deleted are not a part of this model.

Cleaning habits are not a part of the model, but the model could be used to estimate possible time saving of deleting messages. The number of searched messages depends on the current work task. Some tasks require the user to search for information stored in the email messages. It also depends on the number of stored messages. A large number of stored messages may be searched more often for information compared to a few messages.

RESULTS

From the constants in table 2 and variables in table 3 it is possible to estimate the total time spent to archive and retrieve messages, according to Eq. 3 and Eq. 4. For simplicity, the constants are in the following approximated with the values displayed in, table 4.

Table 4. Message archiving and retrieval constants.

Name	Chosen value
Search constant [4], one saccade/item	1/5 s (5 items/s)
Scroll constant [4]	2.6 s
Query constant	5 s
Remains constant	5 %
Screenful constant	30 items
Known folder %	75 %
Known message %	80 %
System time	10 ms

In order to illustrate the influence of the five variables some fictional user data are entered into the equations and the result is displayed in figure 1-7..

The evolution of a fictional email user

A new user of email can be characterized by a small number of incoming messages, few stored messages, and few searched messages a day. In figure 1 the time to store and retrieve a message is displayed as a function of the number of folders. Zero folders are equal to storing all messages in the inbox.

In figure 1, the time to find a folder is estimated with equation 8, store a message equation 9, manual search with equation 13 + P+K, folder-dependent tool search with equation 8 + P+K + Query constant + equation 14 and folder-independent tool search with Query constant + equation 14 with the number of folders = 1.

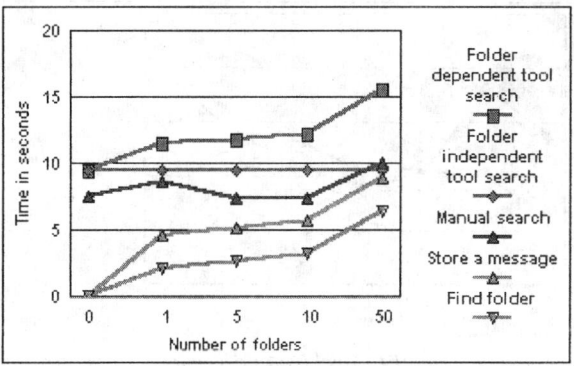

Figure 1. Time in seconds to handle one message for a new user that has 50 stored messages. Inbox contains no messages (with the exception of zero folders that is equal to storing all messages in the inbox).

From figure 1 it is clear that the time spent to find a folder increases slightly with the number of folders. This increase applies also to the time to store a message, and to use a folder-dependent tool to search a message. The time for folderless tool search is of course independent of the number of folders. The time for a manual search is somewhat reduced by using a few folders, but increases again if the number of folders becomes too large.

When fictional numbers for a new email user are entered in equation 3 and equation 4 we get the total time spent on archiving and retrieving messages during one day. This is illustrated in figure 2.

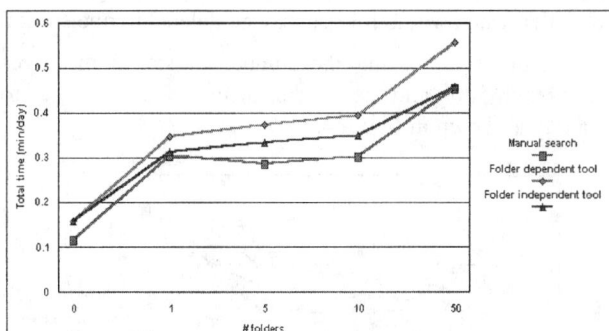

Figure 2. Time in minutes spent to archive and retrieve messages per day. 2 messages to store and 1 searched message a day, 50 stored messages. Inbox contains no messages (with the exception of zero folders that is equal to storing all messages in the inbox).

The graph in figure 2 indicates that the most efficient strategy is not to use folders, and to search manually regardless of the number of folders. Also, the total time is so small that any of the mentioned strategies is acceptable. This situation is typical of a beginner. Many beginners clean their inbox regularly which keeps the number of stored messages low and eliminates the need for folders.

If the number of stored messages is increased to 1000, the number of incoming to 10, and the number of searched messages to 4 we get the results presented in figure 3.

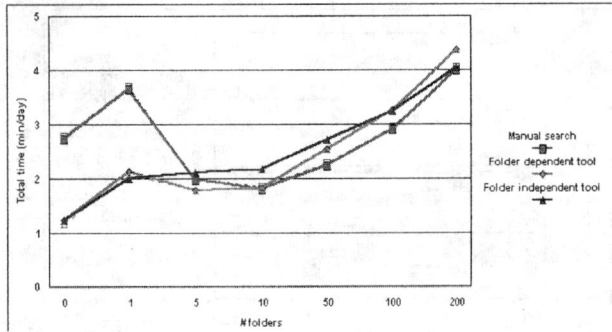

Figure 3. Time spent to archive and retrieve messages per day. 10 messages to store and 4 searched message a day, 1000 stored messages. Inbox contains no messages (with the exception of zero folders that is equal to storing all messages in the inbox).

In figure 3 the advantage of using a few folders in combination with manual search is apparent. However, it would be more efficient to not use folders and use a search tool to retrieve messages. This illustrates the situation when many users start using folders, although the total time spent is still small for all described strategies. There are also other reasons than time saving to use folders. The concept of using folders is well known from the real world. Few offices are equipped with search tools. Also, folders have other advantages that are not visible in the model presented here. A folder provides the user with a context, search for messages belonging to a certain topic is simplified if a folder is used for the actual topic, and cleaning may be simplified when whole folders may be deleted in one stroke.

The result of increasing the number of stored messages increases to 5000 and the number of incoming messages to 40 a day is shown in figure 4.

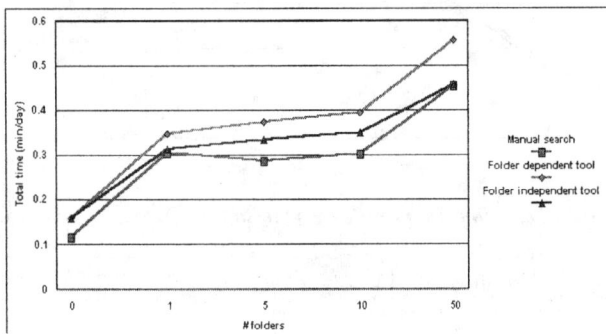

Figure 4. Time spent to archive and retrieve messages per day. 40 messages to store and 4 searched messages a day, 5000 stored messages. Inbox contains no messages (zero folders is equal to storing all messages in the inbox)..

In figure 4, the efficiency of using folders is clearly visible for the case of manual search. We can also note that a folder-dependent search tool is always more efficient than a folder-independent search tool (with the exception of no folders). This situation is typical for a "Frequent filer" [9]. A Frequent filer uses folders and clean often.

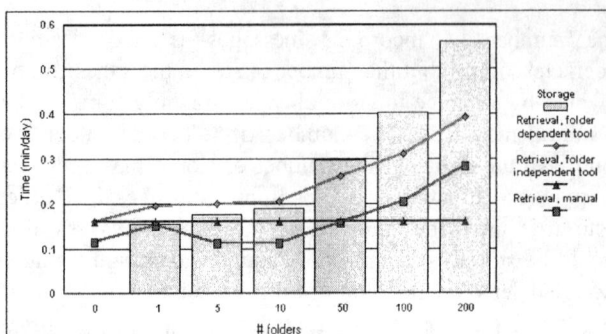

Figure 5. Time spent to archive and retrieve messages per day. 40 messages to store and 4 searched messages a day, 5000 stored messages. Inbox contains no messages (zero folders is equal to storing all messages in the inbox). Storage and retrieval time separated.

In figure 5, the times for archiving and retrieving messages are separated. Here we can clearly see that the overwhelming time for a user with many folders is the storage time. Thus, when a Frequent filer is prevented from handling email for a longer period of time, for example during two weeks of vacation, the number of messages to store has accumulated to approximately 500. According to the model this will take more than half an hour for a user with ten folders and more than an hour for a user with 50 folders. If the user does not have that time to handle email archiving, the user is more or less forced to leave the new messages in the inbox, and at the same time become a "Spring cleaner" [9]. A Spring cleaner uses folders, but cleans irregularly among the messages. If the 500 new messages are stored in the inbox, we get the situation illustrated in figure 6.

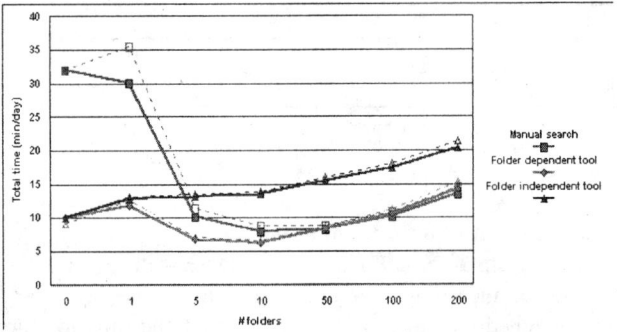

Figure 6. Time spent to archive and retrieve messages per day. 40 messages to store and 4 searched messages a day, 5500 stored messages. Inbox contains 500 messages. Dotted lines represent the time usage if the inbox should contain no messages (zero folders is equal to storing all messages in the inbox).

In figure 6 we can see that storing messages in the inbox is time efficient for all number of folders. This may continue for some time, but when the number of messages in the inbox becomes too large, as illustrated in figure 7, this strategy becomes inefficient. This also illustrates the situation for a "No filer" [9]. A No filer has given up the usage of folders and cleans irregularly, which is an excellent strategy if the user can take advantage of the search functionality.

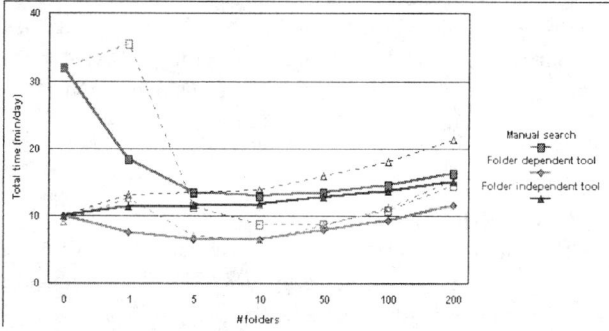

Figure 7. Time spent to archive and retrieve messages per day. 40 messages to store and 4 searched messages a day, 5500 stored messages. Inbox contains 3000 messages. Dotted lines represent the time usage if the inbox should contain the same number of messages as the other folders.

DISCUSSION

The presented model describes some of the context-independent properties of email message storage and retrieval. The model has been illustrated with fictional user data to describe the influence of essential factors such as the number of incoming messages to store, number of folders, the total number of messages, and the number of searched messages each day. According to the model, the best long term strategy is to use folders sparsely in combination with the search functionality.

In what ways do the different variables and constants defined in this chapter affect the total times used for handling email messages? In the following table these variables are analyzed. The analysis is made by changing one variable at a time while the other values are held constant. For the variables in table 5 these values are indicated as *fix values*. The values of the constants are described in table 4.

A similar analysis of the constants gives expected results (e.g. a low value of the Query constant makes tool search more efficient) with exception of the Search constant: A high value makes manual search more inefficient for less than five folders. A low value makes manual search more efficient than tool searches when you have more than one folder besides the inbox.

Besides creating an understanding for the evolution of email users and the strategies used for archiving and retrieving email messages the model could also be used to advise users individually when to start using folders, cleaning messages, learn the search functionality, and using filters to store messages. The model could be used in email systems to give beginners a simple interface to start with and then evolve the interface in pace with the user. As suggested by previous findings that argues for interfaces that change with the users' development [7,8].

Table 5. Analysis of the variables in the model

Name	Fix value	Influence on total time
# of incoming	4	A high number of incoming messages reduces the relative differences between the various search methods (more time is used for storage).
# of folders	10	The time used for handling email messages is minimized by using no folders and search tools, or for manual search between 5 and 25 folders.
# of messages	1500	A low number of stored messages makes manual search more efficient than the tool based searches, especially for the "no folder" strategy. A high number makes few or no (less than five) folders inefficient for manual search.
# of searched	2	A high number of searched messages makes manual search efficient, especially for a moderate number of folders (4-50). Few searched messages equalize the different search strategies.
# in inbox	500	A low number makes many folders somewhat less efficient. A high number makes manual search inefficient.

One weakness of the described model is that the model has been tested with only five real users. However, the model is based on keystroke level analysis that has been used with good results [3]. The informal trials confirmed the model's prediction of these users' time usage, but the model underestimated the time usage for searches among many messages from the same sender with the same subject. These messages appear identical during the search and this increases the time it takes to identify the searched message.

Also, the model gives explanations for the long term behavior of email users, and their development of strategies (conscious or not) from Beginner, over Frequent Filer and Spring Cleaner to No Filer as described in [2,9].

There are limitations of the model: usage of folders is not only a matter of time efficiency of storage and retrieval. Folders have benefits other than reducing the search space. They provide users with the context of other related messages, and may be used to group messages that are difficult to search for with a tool, but still must be read together. Also, in many email tools folders may be used in a hierarchy, and the model does not consider that at all. Nor does the model handle mistakes by the user such as searching in the wrong folder for a message. The folder-independent search tools are not affected by this and would be more efficient compared to the other two methods if this was taken into account.

The many user-dependent variables and constants in the model make it difficult to state general conclusions that would hold for *all* users. However, for users with normal values for the variables and constants that are a part of the model, the following properties hold for the total time for storage and retrieval:

- no folders is an efficient strategy as long as the number of stored messages is less than a few hundred,
- zero to three folders are less efficient than four to approximately twenty folders for manual search,
- using many folders (approximately 30 or more) is not an efficient strategy according to the model, regardless of the values of the variables and constants,
- for many users, the time differences between the various strategies are insignificant as long as the number of stored messages is less than a few thousand,
- the gain in time a reduced number of stored messages gives is less than a few minutes a day for many users (i.e. it is inefficient to do cleanups),
- a folder-dependent search tool is more efficient than a folder-independent one.

The implication that the most efficient strategy for many users would be to use no folders raises demands on search tools that must be as easy to access and use as folders. Customizable search conditions accessible from a menu could reduce the time to search for messages dramatically. Also, for those who want to use folders, agents that automatically suggest folders for archiving could reduce storage times drastically. Storage time is the major time consumer for users with more than a thousand stored messages, according to the model.

While keystroke-level analysis can be used for estimating time usage for single interface actions, this model extends keystroke level analysis by adding the consequences of a particular user behavior over time. Other models based on the same technique could be used to compare time usage for different design solutions in the long run. Natural examples are other data storage such as file systems and databases. This would provide designers with a more holistic view of storage and retrieval and the consequences of design choices over a long period of time.

ACKNOWLEDGMENTS

Many thanks go to Kerstin Severinson-Eklundh and Viggo Kann at the Royal Institute of Technology in Stockholm, Sweden, Candy Sidner and Bob Stachel at Lotus Research in Cambridge, MA, USA, Paul Dickman at the Karolinska Insitute in Stockholm, Sweden, and Katarina Augustsson at Harvard Medical School in Boston, USA, for helpful comments on previous versions of this paper.

REFERENCES

1. Bälter O. (1997): Strategies for organising email messages. In *Proceedings of HCI'97*, Springer, London, United Kingdom, pp 21-38.
2. Bälter O. (1998): Electronic mail in a working context. Doctoral thesis, Nada, Royal Institute of Technology, Stockholm, Sweden.
3. Card S., Moran T. & Newell A. (1980): The keystroke-level model for user performance with interactive systems. *Communications of the ACM*, vol. 23, pp 396-410.
4. Card S., Moran T. & Newell A. (1983): *The Psychology of Human Computer Interaction*. Lawrence Erlbaum Associates, New Jersey, USA.
5. Fitts P. M. & Posner M. I. (1967): *Human Performance*. Wadsworth Publishing.
6. Pliskin N. (1989): Interacting with electronic mail can be a dream or a nightmare: a user's point of view. *Interacting with computers*, vol. 1, no 3, pp 259-272.
7. Trumbly J., Arnett K. & Johnson P. (1994): Productivity gains via an adaptive user interface: an empirical analysis. *International Journal of Human-Computer Studies*, vol. 40, pp 63-81.
8. Trumbly J., Arnett K. & Martin M. (1993): Performance effect of matching computer interface characteristics and user skill level. *International Journal of Man-Machine Studies*, vol. 38, pp 713-724.
9. Whittaker S. & Sidner C. (1996): Email overload: exploring personal information management of email. In *Proceedings of CHI'96*, pp 276-283.

Using Naming Time to Evaluate Quality Predictors for Model Simplification

Benjamin Watson
Dept. Computing Science
615 General Services Bldg.
University of Alberta
Edmonton, Alberta
Canada T6G 2H1
+1 780 492 9918
watsonb@cs.ualberta.ca

Alinda Friedman
Dept. Psychology
University of Alberta
Edmonton, Alberta
Canada T6G 2E9
+1 780 492 2909
alinda@ualberta.ca

Aaron McGaffey
Dept. Psychology
University of Alberta
Edmonton, Alberta
Canada T6G 2E9
+1 780 492 2909
mcgaffey@gpu.srv.ualberta.ca

ABSTRACT
Model simplification researchers require quality heuristics to guide simplification, and quality predictors to allow comparison of different simplification algorithms. However, there has been little evaluation of these heuristics or predictors. We present an evaluation of quality predictors. Our standard of comparison is naming time, a well established measure of recognition from cognitive psychology. Thirty participants named models of familiar objects at three levels of simplification. Results confirm that naming time is sensitive to model simplification. Correlations indicate that view-dependent image quality predictors are most effective for drastic simplifications, while view-independent three-dimensional predictors are better for more moderate simplifications.

Keywords
Model simplification, simplification metrics, image quality, naming time, human vision.

INTRODUCTION
As the number of methods available for constructing or capturing three dimensional (3D) polygonal models proliferates [8,32], so do the models themselves. Often these models are quite large (containing many polygons), allowing high fidelity representation of real world objects. Unfortunately, large 3D models can be quite difficult to display interactively. This has given rise to a significant body of research [14] that attempts to simplify the models; that is, to reduce the number of polygons in the model, while preserving as much as possible the model's fidelity, or quality.

For some applications, purely geometric definitions of quality suffice. But for many other applications, the simplified model must "look like" the original - and by touching upon perceptual issues, the notion of quality becomes much more complex. Ultimately, perceptual quality can only be determined in controlled studies with human observers. But the demands of model simplification do not often allow such involved experimentation. As a simplification algorithm runs, a quality heuristic must predict which of many alternative basic simplifications will have the least impact on visual appearance. During application development, similar quality predictors can be used to predict the perceptual quality of complete simplifications, which represent the repeated application of a heuristic.

Which predictors and heuristics are the best indicators of perceptual quality? Model simplification researchers have understandably avoided this difficult question, typically presenting an array of images depicting the simplifications produced using their algorithms. In the related fields of image compression and rendering, researchers have begun working on a two dimensional (2D) version of this problem. Image compression can be viewed as a 2D analogue of model simplification, with effort directed toward preserving quality while removing pixels, rather than polygons. The result of this research has been several quality predictors [7,23], some of which are also used as quality heuristics [3,26]. Since models have no perceptual qualities until they are turned into images, these results are of particular interest to model simplification researchers.

In this paper, we present an evaluation of various quality predictors for model simplification. As a standard for comparison of various predictions, we use mean naming times for a set of 30 models of common objects. We believe we are the first to use this standard, which is already well established in the field of cognitive psychology.

In the following section, we survey the typical sorts of predictors and heuristics used in model simplification and image compression. We then review the naming time measure, and provide a brief history of its use in cognitive psychology. We continue with a description of the experiments that obtained naming times, and a comparison of these times to several quality predictors. We conclude with a summary and some possible avenues for future research.

QUALITY PREDICTORS AND HEURISTICS

There are three related fields of research that are concerned with visual quality: model simplification, image compression, and image rendering. Below we briefly review the efforts and concerns of each of these fields.

Researchers creating model simplification algorithms often endeavor to preserve the appearance of the models they simplify. To guide the simplifications they make, they employ quality heuristics that predict which of many basic, local simplifications will affect appearance least. To enable comparison and evaluation of simplification results, they may use quality predictors that measure the global similarity in appearance between a simplified model and the unsimplified original. These quality predictors may in turn find use as complex heuristics.

Quality heuristics for model simplification fall primarily into two classes. View-independent heuristics judge quality without reference to the eventual viewing conditions. Most of these heuristics use geometric measurements of distance [5,28] or curvature [15,31]. View-dependent heuristics use a two-stage process to gain more control over appearance. The first view-independent stage produces a nested hierarchy of simplifications. Given knowledge of the current view, the second stage selects an appropriate simplification from the hierarchy, using 2D distance heuristics [21,22] and knowledge of human perception [27,34].

Model simplification researchers have expended little effort on quality prediction, contenting themselves instead with informal examination of a series of views of the simplifications they produce. One notable exception is the view-independent Metro tool from Cignoni et al. [4], which overlays the original and simplified models in 3D space, and uses global geometric measures of the difference between the surfaces in distance and volume. These measures do not reference a particular view or viewpoint. We are not aware of any formal efforts at perceptual evaluation of either quality heuristics or predictors for model simplification.

With some recent exceptions, quality heuristics for image compression [2] have been fairly simple, and often ignore issues of spatial locality and contrast on the image in favor of various measures of distance in color space.

For quality prediction, the field of image compression has long used mean squared error (MSE):

$$\sum\sum ((P_o - P_c)^2 / P_o^2)$$

where P_o is the original pixel value and P_c the compressed pixel value, summed over the image. However, in recent years the shortcomings of MSE have become clear [13], and several predictors based on models of the early stages of the human visual system have been constructed [7,23].

Some initial perceptual evaluation of the Daly predictor was performed by Martens and Myszkowski in [24], who found a high ($R^2 = .83$) correspondence between the predictor and subjective ratings of stimuli masked with texture.

The field of image rendering has long felt the need for a quality heuristic (loosely speaking, rather than knowing which pixel to remove, rendering must know which one to add), and has adopted and extended these quality predictors and used them as quality heuristics [3,26]. Since 3D models have no perceptual qualities until they are viewed as images, image quality predictors might also be used as predictors and heuristics for model simplification.

The goal of our research is to provide a rigorous, perceptually-based evaluation of existing quality predictors and heuristics, especially as applied to the field of model simplification. Our work to date has focussed primarily on evaluation of quality predictors, but it nevertheless has some implications for the design and use of heuristics.

NAMING TIME AS A MEASURE OF QUALITY

As a standard of comparison for our evaluation, we chose naming time, a measure of object identification with a long history of use in cognitive psychology research. Since most interactions with objects begin with their identification, we believed it would be a good perceptual measure of the quality of a simplified model.

Research in cognitive psychology has already shown that the time it takes to name an object can index a number of factors that affect object identification. For example, some linguistic factors that affect naming time are the frequency of an object's name in print, the proportion of individuals who call it by a particular name (percent name agreement), the number of different names it is given, and the age at which the name was acquired (see [33] for review). Among the non-linguistic factors that affect naming times are whether or not it is displayed in its canonical [25] or upright [18,19] orientation, how much prior practice an individual has had naming the same stimulus [19,20], and the degree to which something is visually or structurally similar to other things [1,9,10,16,17].

In a particularly interesting example of the effects of visual similarity, researchers have consistently found that individuals can name manmade artifacts (e.g. furniture, vehicles, musical instruments) faster than they can name natural objects (e.g. animals, fruits, vegetables) [11,17]. Humphreys et al. [16,17] hypothesized that this was due to the similarity of structures existing in nature when compared to the diversity of structures among manmade artifacts. In their explanatory model, information about an object is presumed to be accessed and retrieved in three sequential but overlapping cognitive stages. Initially, visual input undergoes perceptual analyses whose output is used to access one or more representations in a structural description system. Activation of structural descriptions then cascades "forward" to the semantic system, where

Figure 1: One of the experimental stimuli, a standard (0%) bunny model seen in the canonical view.

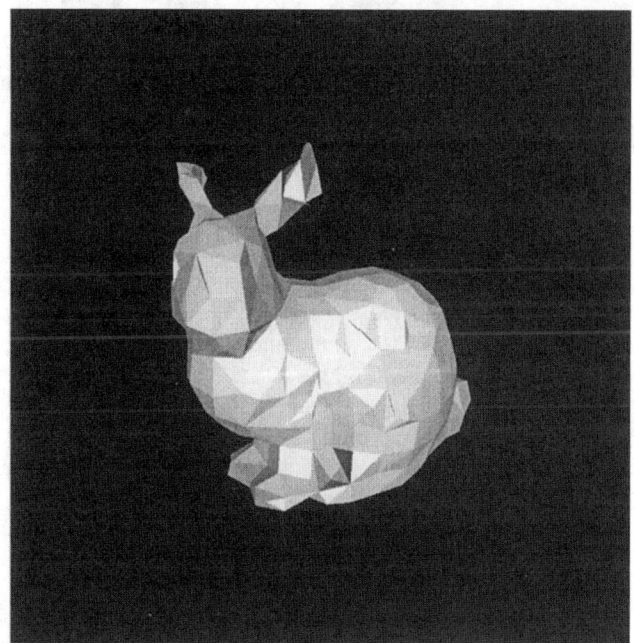

Figure 2: Another experimental stimuli, the same bunny model, this time 80% simplified.

semantic, categorical, and associative characteristics become available. This activation eventually cascades to the phonological name system, where a name or names are activated. An object with many perceptual neighbors (e.g. 4-legged mammals) may activate not only its own structural description, but also structural descriptions of its perceptual neighbors (at least partially). These competing activations then propagate through the semantic to the naming level, leading to naming time delays for members of structurally similar categories.

This structural similarity effect and others like it [9,10,17] led us to expect that naming times would be sensitive to model quality. Since simplification generally results in a stimulus that is less distinctive and more similar to other stimuli, naming times should increase as quality decreases.

For a number of reasons, we believe that naming times provide a more psychologically valid measure of the effects of model or image quality than the subjective ratings sometimes used in image compression research [6,24]. First, it is not known what dimensions people use to construct their ratings, whereas one cannot name a displayed object correctly without having identified it. Second, whereas cognitive variables that are relevant to object recognition affect naming times, variables that are irrelevant to object recognition may affect subjective ratings. For example, ratings may be influenced by identity-irrelevant factors like instructions, task demands, and idiosyncratic strategies; it is unlikely that naming times are affected by these factors.

OBTAINING NAMING TIMES FOR 3D MODELS

Our experiment had two goals. First, we wanted to confirm that naming time was an effective and a valid measure of model quality. Second, having obtained this confirmation, we wanted to use the differences in naming times for models at various levels of simplification to evaluate several quality predictors.

Method

The experiment used a 2 x 3 x 3 x 3 design. There were two types of models: manmade artifacts and animals. The models were presented at three levels of simplification. All participants saw each model at all three levels of simplification, so there were three repetitions. The between-participants independent variable was a counterbalancing factor that controlled for this repeated exposure to the same model.

Stimuli

Stimuli were created from 30 3D polygonal models in the public domain. None of the models contained color, texture, material or vertex normal information. 15 of these models represented manmade artifacts, 15 were representations of animals. Each of these models was then assigned a viewing position, simplified, and displayed, with the resulting digital images being saved to a file for later experimental display.

Model coordinate systems and poses can vary widely, and therefore some viewing parameters had to be defined interactively by the authors. However, certain viewing parameters were constant. Views were always directed towards the mean of a model's vertices, and the virtual field of view was always 40 degrees. The virtual eye point was always at a distance of twice the length of the longest dimension of the bounding box (the smallest axis aligned box that contained a model). The authors then interactively rotated each model about its mean vertex so that it was upright and facing left. Models were next rotated 21 degrees positively in the XZ plane (where X and Y span the image plane, and Z is scene depth), and 18 degrees positively in the YZ plane. The resulting viewpoint was to the right of and slightly above the model. This "canonical" view revealed a reasonable level of detail across the models (see also [25]). Figure 1 shows an example view.

Models were simplified in two stages. First, we ensured that all models had the same number of polygons by using the Qslim algorithm [12] to simplify each model to 3700 polygons (±50). (3700 was the number of polygons in the smallest model in the original set). We will refer to the models resulting from this stage of simplification as the "standards" (0%). Second, the models were further simplified using a vertex clustering [28] algorithm, with each standard model reduced by two levels: 50% and 80% of the original polygons were removed, respectively. Figure 2 shows a standard model with 80% of its polygons removed.

Images were created from the standard and simplified models, creating three exemplars of each object, at 0%, 50%, and 80% simplification, for a total of 90 stimuli. The models were displayed using OpenGL on a Silicon Graphics Crimson RE workstation running the IRIX operating system. Models were illuminated with a single white (RGB = (1,1,1)) light located at the eye point. All models were assigned the same white color as the light source, and flat shaded. The resulting images were saved to a file and regularized for size, with each image scaled down to 591 pixels in the longest dimension, while maintaining aspect ratio.

During experimentation, the images were displayed in grayscale on a black background. All images were centered on the screen.

Design
As described above, we used 30 models (of two types: 15 animals and 15 artifacts) and 90 stimuli (3 levels of simplification for each model). All stimuli were displayed once per session, for a total of 90 trials. Sessions were organized into three blocks of 30 trials, with each model presented once during each block. Blocks were organized into three groups, with each group containing stimuli at the same level of simplification (that is, each block contained 10 stimuli at 0% simplification, 10 at 50%, and 10 at 80%). Models were assigned to groups randomly, and thus model type was not balanced within these groups.

The 30 participants were divided into three participant-groups of 10 individuals. Within each block, the order of presentation of the stimulus-groups was balanced across the three participant-groups. Thus in block 1, depending on which participant-group he was in, a given participant saw a model (e.g. a dolphin) at either 0, 50, or 80 percent simplification, then at the alternate levels in the subsequent two blocks. So, each participant saw all 30 models three times, once at each of the three levels of simplification.

Table 1: Results of experimental statistical analysis.

Variable	Across	ANOVA
simplification	participants	$F(2,58) = 39.81, p<.001$
simplification	items	$F(2,54) = 12.84, p<.001$
model type	participants	$F(1,29) = 21.95, p<.001$
model type	items	$F(1,27) = 2.86, p<.10$
repetition	participants	$F(2,58) = 67.72, p<.001$
repetition	items	$F(2,54) = 41.10, p<.001$
simp x rep	participants	$F(4,116) = 10.42, p<.001$
simp x rep	items	$F(4,108) = 5.90, p<.001$

There were six practice trials. Stimuli for the practice trials were created in a similar manner to the experimental stimuli, however, their original and degraded face counts were not standardized: all the practice models fell below the 3700 original face threshold. The practice stimuli were degraded to the same relative percentages as the experimental stimuli. Specifically, two had 0%, two had 50% and two had 80% of their original faces removed. The practice stimuli were presented in a random order for each participant.

Apparatus and Procedure
The participants were seated approximately 0.7 m from the display, with the stimuli subtending a visual angle of 15 degrees. They performed the task by viewing the pictured stimuli on a computer screen and speaking the names of the modeled objects into a hand-held microphone. The pictures were displayed on a 17-inch Microscan CRT driven by a 166MHz Pentium PC. The experiment was executed under the control of the Microcomputer Experimental Laboratory (MEL) software [29]. Accuracy of the responses was recorded offline by the experimenter.

Participants were told that on each trial of the experiment, they would see a picture of an object, and their task was to name that object as quickly and accurately as they could. They were also told that some pictures would be simplified representations. The experimenter controlled the pace of the trials by pressing the space bar at the start of each trial. After the practice trials, participants were asked if they had any questions regarding the task. The participants then performed the 90 experimental trials.

A trial consisted of the following events: the experimenter pressed a the space-bar, a fixation cross appeared for 750ms, the picture appeared on the screen, the participant named the picture, the picture disappeared as soon as a name was said, the experimenter scored the response and pressed the space-bar again to begin the next trial. Naming times were recorded from stimulus onset to the participant's response.

Participants
Thirty paid undergraduate volunteers from University of Alberta pool participated in the experiment.

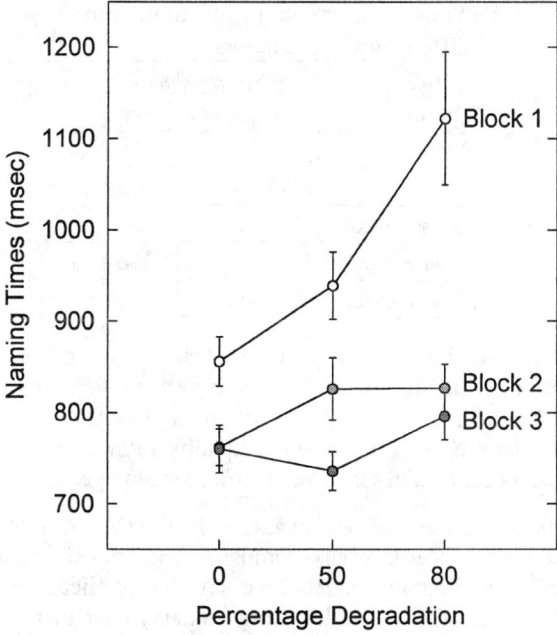

Figure 3: Naming times as a function of model simplification (degradation) and viewing repetition (block).

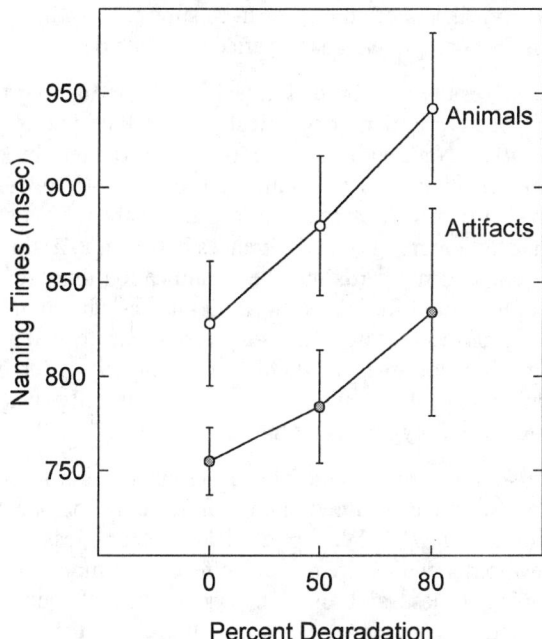

Figure 4: Naming times as a function of model simplification (degradation) and model type.

Results

Two kinds of trials were excluded from analysis. First, we excluded naming times measured during spoiled trials (e.g., trials in which participants failed to trigger the microphone with their first vocalization). Spoiled trials made up 5.1% of all trials. We also excluded naming times from trials in which a participant's response was an error (e.g., calling a picture of a table a "statue"). Only 0.2% of all trials were errors. Near misses inside a semantic category were permissible as correct names (e.g., calling a "hippo" a "rhino"). This is common practice in psychological research [19,20]. However, preliminary analyses showed that the image of the computer monitor was called a "monitor" or "computer monitor" by 47% of the participants, and a "television" or "TV" by 57% of the participants, giving it a percent name agreement score of 57%. Because percent name agreement has been shown to be a strong predictor of naming times [33], and because the computer monitor was the only model with a low percent agreement, we excluded it from further analyses. Only 2.8% of the remaining trials were categorical near misses.

We performed two types of analyses. For the analysis with items as the random factor, we averaged naming times across participants for each combination of simplification and repetition (block); each item thus had nine scores contributing to a mixed design analysis of variance (ANOVA). The within-item factors were repetition and simplification, and the between-item factor was object type. For the analyses with participants as the random factor, we averaged naming times across items for each combination of repetition and simplification for the first ANOVA, and for each combination of object type and simplification for the second.

The results of these analyses are shown in Table 1. The effects of simplification, model type, and repetition were reliable, as was the interaction between percentage simplification and block. No other interactions approached reliability.

Discussion

Figure 3 shows the effects of simplification and repetition on naming time, averaged across participants. Clearly, naming times are sensitive to simplification and model quality. Naming times decrease with repeated viewings of the model. However, simplification has little effect on naming times if the model has previously been seen (explaining the interaction). This finding replicates earlier research in which a manipulation that increases the time it takes to access the structural description of an object (such as rotating the object in the picture plane; [9,10,18,19]) influences naming time on the first viewing, but has a lesser effect on later viewings.

Figure 4 shows the effects of model type and simplification on naming times, averaged across participants. Animals took longer to name than artifacts, replicating results of the psychological research we reviewed earlier. Interestingly, there was no interaction between model type and simplification. This is evidence that the animal-artifact effect is unlikely to be occurring at the perceptual level, but rather, occurs later in the cognitive system. That is, if simplification and model type both affected early perceptual processes, they would be expected to interact within the same "module" (cf. [30]). It is unlikely that other measures of quality, such as subjective ratings, would resolve perceptual/semantic effects in such detail.

These results are compelling evidence that naming time is an effective and psychologically valid measure of model quality. Naming times are strongly affected by model simplification. We also replicated some well-known results from previous research using polygonal models as stimuli. The slight naming time differences between 50% simplified models and standards may be an indication that the task of simplification increases in difficulty as the number of polygons decreases. This may also indicate that naming times are not an ideal standard of comparison for limited application of a quality heuristic, especially when the number of polygons is large.

We should sound a note of caution about these inferences. First, we have obtained naming times only for one almost optimal view. We expect that other, less optimal, viewpoints will increase the effect of simplification on naming times, and thus, the generality of our results. Second, our naming times were obtained with a set of 30 models, which is a moderately small number compared to other naming time studies (although it is within an appropriate range). Finally, models are often used in complex interactive applications, which contain motion, additional color and texture information, and make simultaneous use of several models, not just one in isolation. All of these additional factors are not reflected in our results.

Subject to these concerns, our naming time experiment produced some interesting additional implications for designers of interactive applications. First, designers may wish to take advantage of the simplification-repetition interaction by allowing model quality to degrade when an object has been visible for some time. Also, the animal-artifact effect may indicate that user performance can be improved by simplifying models of natural objects less than models of manmade objects (while keeping in mind that simplification on both types of models has similar effects). Both of these suggestions are early indications that will have to be tested in practice.

EVALUATING MODEL QUALITY PREDICTORS

Having determined that naming times are an effective measure of model quality, we turn to an evaluation of predictors of model quality. We compare two view dependent image quality predictors from image compression and rendering research to a single 3D, view independent predictor.

The image quality predictors we evaluate are MSE [13] and an implementation of the predictor from Bolin and Meyer [3] (BMP), which models the early, perceptual stage of the human visual system. Both compare a standard image to an image that approximates it. We say these two predictors are "view dependent" because the input images must be created from a certain viewpoint. While MSE returns a single difference (quality) measure, BMP returns a

Table 2: Correlations of naming time differences to predictors.

Predictor	50% Simp		80% Simp	
	Corr	Sig?	Corr	Sig?
BMP	0.040	no	0.379	yes
MSE	0.145	no	0.315	marg
Metro max	0.207	no	0.317	marg
Metro mse	0.245	no	0.196	no
Metro mean	0.129	no	0.182	no

difference image. The value at each location of the image predicts the ability of human observers to perceive the difference between the two input images. For our evaluations we need a single quality rating. We use the mean of the values in BMP's difference image.

The 3D predictor we evaluate is the Metro tool from Cignoni et al [4]. Metro compares a 3D model standard to an approximating model. We say this predictor is "view independent" because by using models rather than images, Metro does not reference a certain viewpoint. Metro samples the surface of each model at multiple points, and determines the shortest distance from the surface of one model to the next. Distances are signed depending on which of the two surfaces is outermost. It then returns the mean, mean squared and the maximum of these distances, normalized by the length of the diagonal of the standard model's bounding box.

We used MSE, BMP and Metro to predict the quality difference between the standard (0% simplified) stimuli and each of the corresponding 50% and 80% simplified stimuli, for a total of 60 quality predictions. We compared each of these predictions to differences in mean naming times for corresponding models at the first viewing. For example, we compared the predictions for the quality difference between the standard dolphin and the 80% simplified dolphin to the difference between the mean naming times at first viewing for these two models. We used only the first viewing differences because these were the most affected by simplification.

Table 2 shows the correlations between the predictors and the naming time differences. None of the predictors accounted for much of the difference in naming times between the standard and 50% simplified models. However, several predictors did a reasonable job with the difference between the standard and 80% times, including BMP, MSE and the Metro maximum distance. A step-wise regression with all the predictors as independent variables confirms what is seen in the table; BMP was the best predictor of the difference in naming times, $F(1, 27) = 4.54$, $p<.05$, and none of the other predictors added significantly to the amount of variance accounted for. Each of the MSE and Metro maximum distance predictors accounted for a marginally significant amount of the variance in names times ($p<.10$) if used as the first term in the regression.

Which predictor was most effective? Predictor effectiveness varied with the size of the quality difference being predicted. Below, we present our two recommendations, based on the best available evidence. We follow these recommendations with some discussion of other possible interpretations of our results.

BMP is the best predictor of larger quality differences. BMP predicted the quality of the 80% simplifications reliably, but failed miserably at predicting the quality of the 50% simplifications. This is quite a contrast and it merits further research. It may indicate instead that the naming time measure cannot resolve smaller quality differences. But we think this is unlikely, because the naming time measure is quite sensitive to small featural differences between instances of the same "basic-level" object (e.g., two different cars; Friedman & Rabiau, 1999). However, other predictors did show better correlations to naming times. Perceptually speaking, we feel comfortable calling the standard-50% difference "small". However, from a systems viewpoint, 50% of all polygons is not a small number, a reliable predictor for such quality differences would be very useful.

For smaller quality differences, Metro's view independent predictors were most reliable. Although no predictors reliably accounted for standard-50% variance, it is intriguing to note that Metro's view independent geometric measures had the highest correlations for these smaller quality differences. This may be an indication that geometric measures are good quality heuristics, but this requires confirmation in further experimentation.

Interestingly, for the larger standard-80% quality differences, Metro's maximum distance predictor was fairly effective, while its mean predictors were not. It may be that maximums are more sensitive than means to the changes in large model features that occur during drastic simplifications. However, this possibility also needs experimental confirmation.

CONCLUSIONS AND FUTURE WORK

Model simplification researchers need quality heuristics to guide simplification, and quality predictors to aid in selection of appropriate heuristics and simplification algorithms. We have presented an evaluation of several quality predictors for model simplification using a new perceptual experimental standard, naming times. Naming times have long been used in cognitive psychology research as a measure of recognition. We found that naming times are sensitive to simplification, and we were able to duplicate several well known results from cognitive psychology using polygonal models as stimuli.

We then compared naming times to several model quality predictors. Predictors differed in effectiveness depending on the size of the quality difference they were being asked to predict. Image based predictors such as that from Bolin and Meyer [3] were quite effective for large differences and much less effective for small differences. View independent geometric predictors like Metro were most effective for smaller quality differences, but were even then unreliable.

We plan to increase the reliability of these results by increasing the number of models to be named, and by using additional, non-optimal viewpoints. We will also evaluate the cumulative effect of different quality heuristics embedded in the same simplification algorithm. Ultimately, we also hope to consider colored and textured models in our evaluations.

ACKNOWLEDGEMENTS

Oscar Meruvia wrote indispensible 3D viewing software. Roman Kotovych provided help with simplifications and quality predictors. We thank Greg Turk for his models and geometry filters. Peter Lindstrom participated with a thought provoking correspondence, and by assisting in finding relevant code. Oleg Veryuvka and Lisa Streit provided useful implementations of simple image quality metrics. Peter Johnstone also provided assistance. Stanford University was the source of the bunny model. This research was supported by NSERC grants to the first two authors, and by NSERC Undergraduate Fellowships to the third and Mr. Kotovych.

REFERENCES

1. Bartram, D.J. (1976). Levels of coding in picture-picture comparison tasks. *Memory and Cognition*, 4, 593-602.

2. Baxes, G. (1994). *Digital Image Processing.* John Wiley & Sons, New York.

3. Bolin, M. & Meyer, G. (1998). A perceptually based adaptive sampling algorithm. Proc. of SIGGRAPH 98. In *Computer Graphics* Proceedings, Annual Conference Series, 1998, ACM SIGGRAPH, 299-309.

4. Cignoni, P., Rocchini, C. & Scopigno, R. (1998). Metro: measuring error on simplified surfaces. *Computer Graphics Forum, 17*, 2, 167-174. Available at: http://vcg.iei.pi.cnr.it/metro.html.

5. Cohen, J., Varshney, A., Manocha, D., Turk, G., Weber, H., Agarwal, P., Brooks, F. & Wright, W. (1996). Simplification envelopes. Proc. of SIGGRAPH 96. In *Computer Graphics* Proceedings, Annual Conference Series, ACM SIGGRAPH, 119-128.

6. Cosman, P., Gray, R. & Olshen, R. (1993). Evaluating quality of compressed medical images: SNR, subjective rating and diagnostic accuracy. *Proc. IEEE, 82*, 6, 919-932.

7. Daly, S. (1993). The visible differences predictor: an algorithm for the assessment of image fidelity. In Watson, A. B. (ed.). *Digital Images and Human Vision.* MIT Press, Cambridge, MA, 179-206.

8. DeRose, T., Kass, M. & Truong, T. (1998). Subdivision surfaces in character animation. Proc. of SIGGRAPH 98. In *Computer Graphics* Proceedings, Annual Conference Series, ACM SIGGRAPH, 85-94.

9. Friedman, A., & Rabiau, M. (1999). Lions and tigers and bears: The role of structural similarity and visual detail in naming disoriented objects. Manuscript submitted for publication.

10. Friedman, A., & Vuong, Q. (1999). Cats, cows, cameras, and cars: The role of structural similarity in naming and categorizing upright and disoriented pictures. Manuscript submitted for publication.

11. Gaffen, D., & Heywood, C.A. (1993). A spurious category-specific visual agnosia for living things in normal human and nonhuman primates. *Journal of Cognitive Neuroscience, 5*, 118-128.

12. Garland, M. & Heckbert, P. (1997). Surface simplification using quadric error metrics. Proc. SIGGRAPH 97. In *Computer Graphics* Proceedings, Annual Conference Series, ACM SIGGRAPH, 209-216. Available at: http://www.cs.cmu.edu/~garland/quadrics/qslim.html

13. Girod, B. (1993). What's wrong with mean-squared error? In Watson, A. B. (ed.). *Digital Images and Human Vision.* MIT Press, Cambridge, MA. 207-220.

14. Heckbert, P. & Garland, M. (1997) Survey of polygonal surface simplification algorithms. Technical report, CS Dept., Carnegie Mellon, U. Available at: http://www.cs.cmu.edu/~garland/papers/simp.pdf.

15. Hinker, P. & Hansen, C. (1993). Geometric optimization. *Proc. IEEE Visualization '93*, 189-195.

16. Humphreys, G. W., Lamote, C., & Lloyd-Jones, T. J. (1995). An interactive activation approach to object processing: Effects of structural similarity, name frequency, and task in normality and pathology. *Memory, 3*, 535-586.

17. Humphreys, G. W., Riddoch, M. J., & Quinlin, P. T. (1988). Cascade processes in picture identification. *Cognitive Neuropsychology, 5*, 67-103.

18. Jolicoeur, P. (1985). The time to name disoriented natural objects. *Memory & Cognition, 13*, 289-303.

19. Jolicoeur, P. (1988). Mental rotation and the identification of disoriented objects. *Canadian Journal of Psychology, 42*, 461-478.

20. Jolicoeur, P., & Milliken, B. (1989). Identification of disoriented objects: Effects of context of prior presentation. *Journal of Experimental Psychology: Learning, Memory, and Cognition, 15*, 200-210.

21. Leubke, D. & Erikson, C. (1997). View dependent simplification of arbitrary polygonal environments. Proc. of SIGGRAPH 97. In *Computer Graphics* Proceedings, Annual Conference Series, ACM SIGGRAPH, 199-208.

22. Lindstrom, P., Koller, D., Ribarsky, W., Hodges, L., Faust, N., & Turner, G. (1996). Proc. of SIGGRAPH 96. In *Computer Graphics* Proceedings, Annual Conference Series, ACM SIGGRAPH, 109-118.

23. Lubin, J. (1993). A visual discrimination model for imaging system design and evaluation. In Peli, E. (ed.). *Vision Models for Target Detection and Recognition*, World Scientific, New Jersey, 245-283.

24. Martens, W. & Myszkowski, K. (1993). Psychophysical validation of the visible differences predictor for global illumination applications. *IEEE Visualzation '93*, Late Breaking Topics, 49-52. Also available at: http://wwwsv1.u-aizu.ac.jp/labs/csel/vdp/.

25. Palmer, S., Rosch, E., & Chase, P. (1981). Canonical perspective and the perception of objects. In J. Long & A. Baddelay (Eds.), *Attention & Performance IX*, Hillsdale, NJ : Erlbaum, 135-151

26. Ramasubramanian, M., Pattanaik, S. & Greenberg, D. (1999). A perceptually based physical error metric for realistic image synthesis. Proc. of SIGGRAPH 99. In *Computer Graphics* Proceedings, Annual Conference Series, ACM SIGGRAPH, 73-82.

27. Reddy, M. (1998). Specification and evaluation of level of detail selection criteria. *Virtual Reality: Research, Development and Application, 3*, 2, 132-143.

28. Rossignac, J. & Borrel, P. (1993). Multi resolution 3D approximations for rendering complex scenes. In Falcidieno, B. & Kunii, T. (eds.), *Geometric Modeling in Computer Graphics.* Springer Verlag, 455-465.

29. Schneider, W. (1988). Micro Experimental Laboratory: an integrated system for IBM-PC compatibles. *Behavior Research Methods, Instrumentation, and Computers, 20*, 206-217.

30. Sternberg, S. (1969). The discovery of processing stages: Extensions of Donder's method. *Acta Psychologica, 30*, 276-315.

31. Turk, G. (1992). Re-tiling polygonal surfaces. Proc. of SIGGRAPH 92. In *Computer Graphics, 26,* 2 (July), ACM SIGGRAPH, 55-64.

32. Turk, G. & Levoy, M. (1994). Zippered polygon meshes from range images. Proc. of SIGGRAPH 94. In *Computer Graphics* Proceedings, Annual Conference Series, ACM SIGGRAPH, 311-318.

33. Vitkovitch, M., & Tyrell, L. (1995). Sources of name disagreement in object naming. *Quarterly Journal of Experimental Psychology, 48A*, 822-848.

34. Watson, B., Walker, N., Hodges, L. & Worden, A. (1997). Managing level of detail through peripheral degradation: effects on search performance with a head-mounted display. *ACM Trans. Computer-Human Interaction, 4*, 4, 323-346.

Interactive Textbook and Interactive Venn Diagram: Natural and Intuitive Interfaces on Augmented Desk System

Hideki Koike[†]   Yoichi Sato[‡]   Yoshinori Kobayashi[†]
Hiroaki Tobita[†]   Motoki Kobayashi[†]

[†]Graduate School of Information Systems
University of Electro-Communications
1-5-1, Chofugaoka, Chofu
Tokyo 182-8585, Japan
+81-424-43-5651
koike@acm.org

[‡]Institute of Industrial Engineering
University of Tokyo
7-22-1, Roppongi, Minato-ku
Tokyo 106-0032, Japan
+81-3-3401-1433
ysato@cvl.iis.u-tokyo.ac.jp

ABSTRACT

This paper describes two interface prototypes which we have developed on our augmented desk interface system, EnhancedDesk. The first application is Interactive Textbook, which is aimed at providing an effective learning environment. When a student opens a page which describes experiments or simulations, Interactive Textbook automatically retrieves digital contents from its database and projects them onto the desk. Interactive Textbook also allows the student hands-on ability to interact with the digital contents. The second application is the Interactive Venn Diagram, which is aimed at supporting effective information retrieval. Instead of keywords, the system uses real objects such as books or CDs as keys for retrieval. The system projects a circle around each book; data corresponding the book are then retrieved and projected inside the circle. By moving two or more circles so that the circles intersect each other, the user can compose a Venn diagram interactively on the desk. We also describe the new technologies introduced in EnhancedDesk which enable us to implement these applications.

KEYWORDS: augmented reality, computer vision, finger/hand recognition, information retrieval, Venn diagram, education, computer supported learning,

INTRODUCTION

One of the important goals in Computer Human Interactions is to develop more natural and more intuitive interfaces. Graphical user interface (GUI), which is a current standard interface on personal computers (PCs), is well-matured and provides an efficient interaction for users who have already had experience with PCs. However, it is not true that GUI always provides natural and intuitive interfaces. Consider, for example, a mouse, which is a standard pointing device in GUI. Even though the mouse enables rapid and exact pointing, moving the mouse on a desk in order to move a cursor on computer screen does not come naturally to humans. Very often, users experience confusion when they attempt to use a mouse for the first time. To select an item from an item list on display or to point rough position on display, it is more natural and more intuitive for them to use their index finger. This statement is supported by the fact that there are many touch panel interfaces (e.g., ATMs) which are used by ordinary people.

There are many information systems which can provide more natural and more intuitive interface when we remove the restrictions imposed by the GUI. For example, Bolt demonstrated effectiveness of a multi modal interaction with gesture and speech in SDMS [3]. It is mainly the high cost of hardware along with immature sensing technology that have prevented us from developing such systems. Now the cost of hardware has been reduced and many sensing technologies have been developed. It is time to discuss and develop alternative interfaces. Some researchers have proposed new interface frameworks. For example, Turk [18] proposed the Perceptual User Interface (PUI) which uses several sensing technologies to interact with computers. Fitzmaurice et al. [4] and Ishii et al. [6, 7, 19, 20] proposed the Graspable/Tangible Interface which uses real world objects to manipulate digital information; that group developed many prototype systems.

We have developed an augmented desk interface by using computer vision as a key technology. EnhancedDesk [9] is an infrastructure to develop applications supporting work in the office. EnhancedDesk is influenced by Wellner's DigitalDesk [21]. The basic hardware configuration such as the use of a desk, a CCD camera, and a video projector is similar to that of DigitalDesk. However, some novel technologies are introduced in EnhancedDesk to enable more advanced interaction.

This paper describes two interface prototypes developed on EnhancedDesk. One is the Interactive Textbook, which is intended to provide and effective learning support environment. The other is the Interactive Venn Diagram which is aimed at supporting effective information retrieval. The next section describes the Interactive Textbook. Section 3 describes the Interactive Venn Diagram. The implementation of EnhancedDesk is discussed in detail in Section 4. In Section 5, we discuss advantages and limitations of our systems. Section 6 concludes the paper.

COMPUTER SUPPORTED LEARNING ENVIRONMENT
Issues in Computer Supported Learning Environment
In any course of study, textbooks are generally the tools that are used. It is, however, difficult to learn correct pronunciation by relying only on a textbook. It is also difficult to understand dynamic behavior in scientific experiments merely by reading text and by looking at static figures in the textbook.

Video or multimedia teaching materials make up for these textbook shortcomings. For example, foreign language teaching materials provide aids to correct pronunciation, and those for science demonstrate experiments by the use of video clips and/or computer graphics.

The problem with such materials is that they require students to perform additional tasks which are not directly related to the learning tasks. For example, students might be required to execute an application program in a CD-ROM whenever they read certain pages. To accomplish the execution, they have to insert the CD-ROM into the CD-ROM drive, search the application program, and then execute the program. The students' main purpose is to understand the experiment rather than understand how to use a computer. Such additional tasks might well cause them to lose their concentration.

On another perspective, unnatural interaction might disturb effective learning. Consider, for example, an interactive application which enables users to see a weight-spring experiment in Physics. Even though the students can manipulate a weight by using a mouse, it is unnatural for them to manipulate the weight on the screen indirectly.

Design Approach
As one solution to the issues described in the previous section, we propose a computer vision supported learning environment.

- Automatic retrieval and execution
  In the study of Physics, the main task of students is to understand the experiments described in the textbook, not to demonstrate their ability to use a computer. Therefore, it is more convenient for the students if the system recognizes what the student is currently studying and then shows corresponding digital contents automatically.
- Dynamic manipulation
  Whether or not the student understands a particular subject and remembers it longer depends on how real their experience is. As we described previously, direct manipulation by hand or finger will give more real experience than indirect manipulation by mouse. Although mouse is useful in practical office work, hand/finger manipulation is much more effective in educational environment. We developed novel techniques to recognize two-dimensional matrix code and users' hand/finger as is described in later section. These techniques were introduced in EnhancedDesk.

Interactive Textbook
In Interactive Textbook, a matrix code [12] is attached to the page which has digital contents as shown in Figure 1. Each matrix code corresponds to each application program. When a student opens a page containing a matrix code, the system recognizes the matrix code and projects the digital content onto the desk. The system not only identifies the unique ID of the matrix code but also recognizes its size and orientation. By using such size and orientation information, the system decides where to project the digital contents. For example, if the book is inclined on the desk, the digital contents are projected to the right position, i.e., the correct slant.

Figure 1 shows a student reading a textbook of Physics. When the student reaches the page describing the spring-weight experiment, computer graphics simulation of the spring-weight experiment is automatically projected onto the right side of the textbook. The student can manipulate the weight by his or her hand and observe the dynamic behavior of the spring and the weight. By exchanging the weight, the student can compare the different dynamics of the spring. When the student opens a page describing the pendulum experiment, CG simulation is projected onto the desk and the student can drag the pendulum to see its dynamic behavior.

INFORMATION RETRIEVAL
The most popular technique in current information retrieval is keyword searching. In keyword searching, people use one or more keywords and formulate queries by combining the keywords with Boolean operators such as AND/OR/NOT. Historically, such keyword searching has been used by a small number of people such as database operators who are trained to use database systems, computer-related people who are knowledgeable about database systems, or people who use online retrieval systems at libraries. However, the widespread World Wide Web (WWW) has enabled the general public, who not necessarily familiar with database systems, to become adept at using keyword searching via WWW search engines.

Issues in Keyword Searching
- Selection of appropriate keywords
  The lone trigger in keyword searching is the keyword; therefore, inappropriate keywords make it impossible to retrieve relevant information. Also, because they produce a huge number of hits, keywords that are too generic make it difficult to retrieve relevant information. The key to effective retrieval is how to select

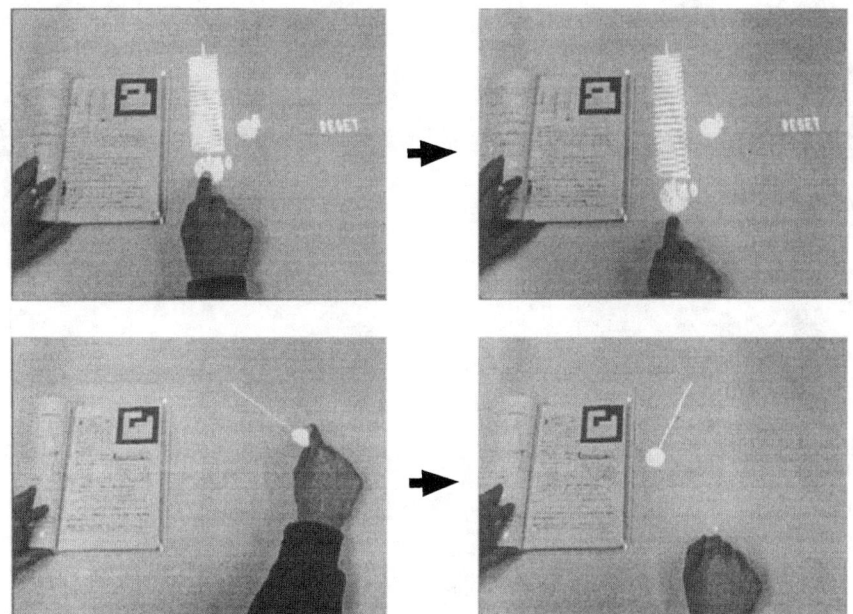

Figure 1: Interactive Textbook. When a student reads a page in a physics textbook that describes a spring-weight experiment, computer graphics simulation is automatically projected onto the desk. The student can drag the weight by using his or her finger. Then, the student opens another page describing a pendulum experiment. The system projected another CG simulation.

appropriate keywords.

Such keyword selection requires skill and knowledge for target domain. For example, when we use a keyword *information retrieval* to find this paper in a database, a huge number of papers corresponding to information retrieval would be retrieved. People who know that the phrase *information retrieval* is too generic would avoid using it. If, however, they use the keywords *augmented reality*, they could find this paper more efficiently. Such keyword selection is, however, a little difficult for novice users.

- Formulating complex queries

When too much information is retrieved, we could refine the query retrieval by combining two or more keywords with Boolean operators such as AND/OR/NOT. However, such query formulation is a little difficult for the general public. For example, when we retrieve papers which have keywords *information retrieval* and *augmented reality* and papers which have keywords *information retrieval* and *visualization*, we would form the query such as:

(*information retrieval* AND (*augmented reality* OR *visualization*)). —(A)

The difficulty of formulating such a query increases as the number of keywords increases.

- Observing different conditions

In most information retrieval systems, each query produces one result. When we refine the query, the previous result is cleared and a new result is displayed on the screen. If the refinement is so strict that the result does not contain relevant information, the user

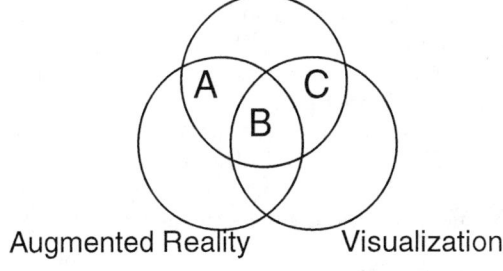

Figure 2: Venn diagram.

has to search again to return to the previous result. To see a result with a slightly different condition, the user has to formulate a new query and search again. Moreover, it is difficult to compare both results.

For example, when the user who performed the previous query (A) would perform another query with slightly different condition such as:

(*information retrieval* AND *augmented reality* AND *visualization*). —(B)

Although people might notice that (B) is a subset of (A), it is relatively difficult to recognize data which are not in (B) but do appear in (A).

Design Approach

As one solution to this issue, we propose a visual interface for an information retrieval system.

- Visualization

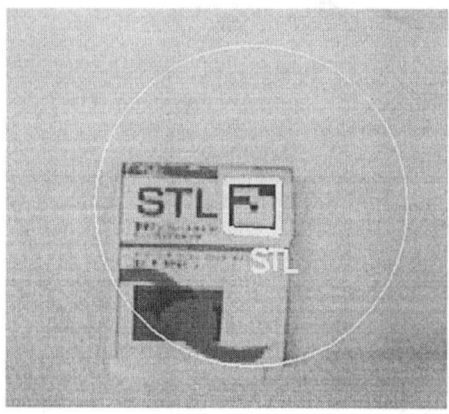

Figure 3: A book with two-dimensional matrix code. When a user put the book on the desk, a circle is projected around the book.

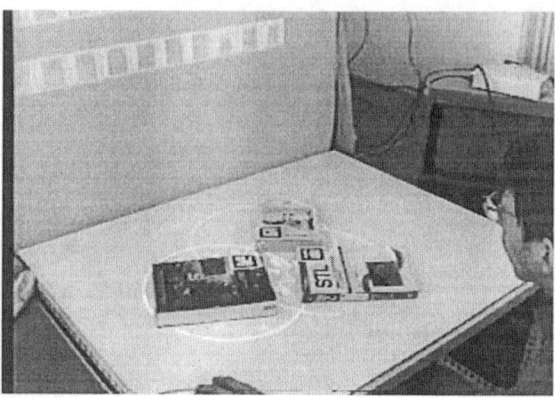

Figure 4: Interactive Venn Diagram in use. When a user selects one of the regions in the Venn diagram, the images of the books in that region are displayed on from screen of EnhancedDesk.

To represent a concept of mathematical set intuitively, the Venn diagram such as shown is Figure 2 is very popular. A set represented by the query (A) is shown as region A and B and C, and a set represented by the query (B) is shown as region B.

In the Venn diagram, each set is visualized intuitively. Moreover, even people who are not familiar with database query can understand each subset visually.

- Augmented reality
  Suppose, for example, a user wants to buy unknown music CDs. It seems difficult to express his or her tastes by the use of appropriate keywords. The user's collection of CDs, however, would be a representative example of those tastes.
  Our idea is to introduce this concept to information retrieval. That is, instead of giving keywords to the information retrieval system, the user would show real objects such as CDs or books to the system.

Interactive Venn Diagram

In the Interactive Venn Diagram, a two-dimensional matrix code is attached to each book cover. Each matrix code corresponds to multiple keywords. For example, a matrix code on the book "C Programming Language" is associated with two keywords, *C Language* and *structured programming*. A matrix code on the book "OpenGL Programming Guide" is associated with two keywords *OpenGL* and *computer graphics*.

When a user puts a book on EnhancedDesk, the system recognizes the book's matrix code and projects a circle around the book as shown in Figure 3. Then the system searches the database using keywords associated with the matrix code. Finally, the retrieved data are projected onto the desk as an icon and they move inside the circle.

In the same way, when the user puts another book on the desk, another circle and retrieved results are projected onto the desk. If the user moves these books so that the two circles intersect one another, data corresponding to both books are displayed within the intersection of the two circles.

Figure 5 shows that three books have been put on the desk. Those three circles are projected onto the desk and finally the Venn diagram, as shown in the figure, is completed (Figure 5(B)).

When the user points to one of the subregions, the details of the data in the region are displayed on the front screen of EnhancedDesk 4. If the user selects one of the books, a textured image of the book is projected onto the screen (Figure 5(C)). Then a circle is projected around this textured image and data associated with this book moves inside the circle. The user can perform further retrieval using three real books and one virtual book.

ENHANCEDDESK: IMPLEMENTATION DETAIL

In this work, we propose a new method for tracking a user's palm center and fingertips by using an infrared camera and template matching based on normalized correlation.

Unlike regular CCD cameras which detect lights in visible wavelength, an infrared camera can detect lights emitted from a surface with a certain range of temperature. Thus, by setting the temperature range to approximate the human body tempature, image regions corresponding to human skin appear particularly bright in input images from an infrared camera.

The use of an infrared camera is especially advantageous for our application, EnhancedDesk, in which a user can manipulate both physical objects and electrically projected objects on a desk. In this situation, the previously proposed methods would fail to find human skin regions. Because a LCD projector projects various kinds of objects such as text or figures even onto human skin, the observed color of the human skin can be altered completely, and the background of the input image changes

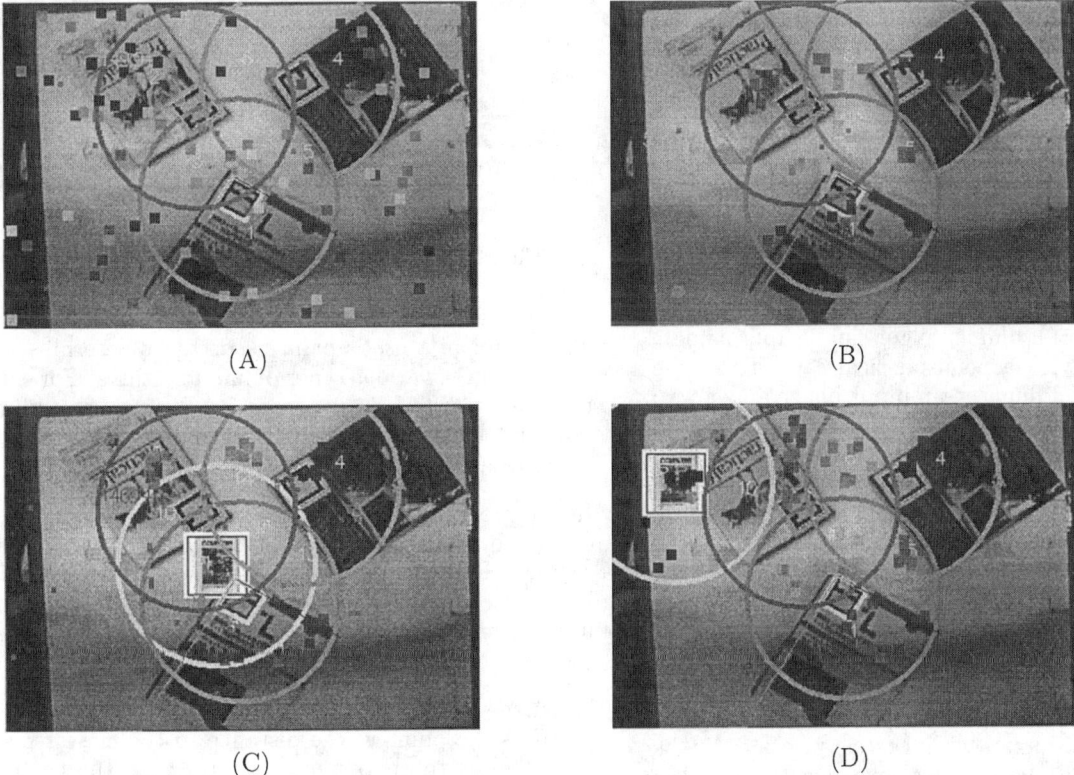

Figure 5: Information retrieval on Interactive Venn Diagram. (A) When a user puts three books onto the desk, retrieved data are first displayed randomly. (B) The data "flies" into each circle. (C) When the user selects one book from the front screen, the book is projected onto the desk with the circle. (D) The user can perform further retrieval using both real and virtual books.

dynamically. As a result, the previously proposed methods which are typically based on color segmentation or background subtraction do not work well.

Extraction of Left and Right Arms

First an infrared camera is installed with a surface mirror so that a user's hands on a desk can be observed by the camera.

The video output from the infrared camera is digitized as a gray-scale image with 256 × 220 pixel resolution by a frame grabber on a PC. Because, the infrared camera is adjusted to measure a range of temperatures near human body temperature, e.g., typically between 30° and 34° beforehand, values of image pixels corresponding to human skin are higher than other image pixels. Therefore, image regions corresponding to human skin can be easily identified by binarizing the input image with a threshold value. In our experiments, we found that a fixed threshold value for image binarization works well for finding human skin regions regardless of room temperatures. Figure 6(a) and (b) show one example of an input image from the infrared camera, and a region of human skin extracted by binarization of the input image.

If some other objects on a desk have temperatures similar to that of human skin, e.g., a warm cup or a note PC, image regions corresponding to those objects as well as to human skin are found by image binarization. To remove those regions other than human skin, we first remove small regions, and then select the two regions with the largest size. If only one region is found, we consider that only one arm is observed on the desk.

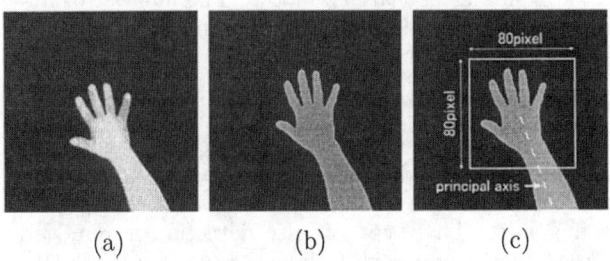

Figure 6: Extraction of hand region

Finding Fingertips

Once regions of a user's arms are found in an input image, fingertips are searched for in those regions. Compared to extraction of user's arms, this search process is computationally more expensive. Therefore, a search window is defined in our method, and fingertips are searched for only within the window instead of being searched for over the entire region of a user's arm.

A search window is determined based on the orientation of each arm which is given as the principal axis of inertia of the extracted arm region. The orientation of the principal axis can be computed from the image moments up to the second order as described in [5]. Then a search window of a fixed size, i.e., 80 × 80 pixels in our current implementation, is set so that it includes a hand part of the arm region based on the orientation of the arm. (Figure 6(c)) We found that a fixed size for the search window works reliably because the distance from the infrared camera to a user's hand on a desk changes little.

Once a search window is determined for each hand region, fingertips are searched for within that window. The overall shape of a human finger can be approximated by a cylinder with a hemispherical cap. Thus, the projected shape of a finger in an input image appears to be a rectangle with a semi-circle at its tip.

Based on this observation, fingertips are searched for by template matching with a circular template as shown in Figure 7 (a). In our proposed method, normalized correlation with a template of a fixed-size circle is used for the template matching. Ideally, the size of the template should differ for different fingers and different users. In our experiments, however, we found that the fixed size of template works reliably for various users. For instance, a square of 15 × 15 pixels with a circle whose radius is 7 pixels is used as a template for normalized correlation in our current implementation.

While a semi-circle is a reasonably good approximation of the projected shape of a fingertip, we have to consider false detection from the template matching. For this reason, we first find a sufficiently large number of candidates. In our current implementation of the system, 20 candidates with the highest matching scores are selected inside each search window. The number of initially selected candidates has to be sufficiently large to include all true fingertips.

After the fingertip candidates are selected, false candidates are removed by means of two types of false detection. One is multiple matching around the true location of a fingertip. This type of false detection is removed by suppressing neighbor candidates around a candidate with the highest matching score.

The other type of false detection is a matching happening in the middle of fingers as illustrated in Figure 7(b). This type of false detections is removed by examining surrounding pixels around the center of a matched template. If multiple pixels in a diagonal direction are inside the hand region, then it is considered not to exist at a fingertip, and therefore the candidate is discarded.

By removing these two types of false matchings, we can successfully find correct fingertips as shown in Figure 7(c).

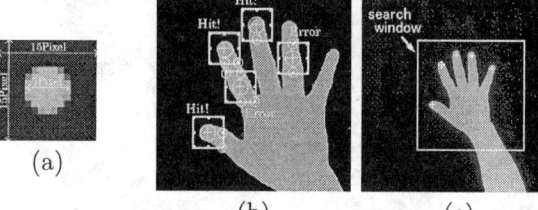

Figure 7: Template matching for fingertips

Finding Centers of Palms

The center of a user's palm needs to be determined for recognition of various types of hand gestures. For example, the location of the center is necessary to estimate how extended each finger is, and therefore it is essential for recognizing basic gestures such as click and drag.

In the previously proposed methods, the center of a user's hand is often given as the center of mass of a hand region. However, the center of mass moves significantly when opening and closing a hand or by including a user's arm in the hand region. Therefore, we cannot estimate the center of a user's hand.

In our proposed method, the center of a user's hand is given as the point whose distance to the closest region boundary is the maximum. In this way, the center of the hand becomes insensitive to various changes such as opening and closing of the hand. Such a location for the hand's center is computed by morphological erosion operation of an extracted hand region. First, a rough shape of the user's palm is obtained by cutting out the hand region at the estimated wrist as shown in Fig.8 (a). The location of the wrist is assumed to be at the pre-determined distance, e.g., 60 pixels in our case, from the top of the search window and perpendicular to the principal direction of the hand region.

Then, a morphological erosion operator is applied to the obtained shape of the user's palm until the area of the region becomes small enough. As a result, a small region at the center of the palm is obtained. Finally, the center of the hand region is given as the center of mass of the resulting region as shown in Figure 8(c).

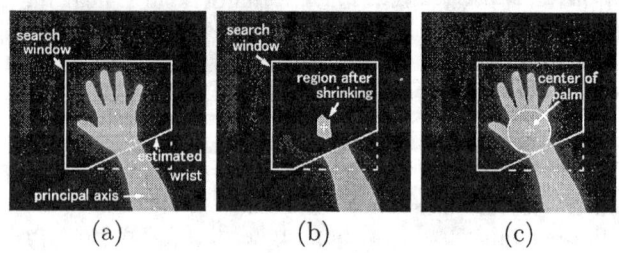

Figure 8: Center of a user's palm

EnhancedDesk System

The proposed method for tracking hands and fingertips in infrared images was successfully used for our EnhancedDesk system. The system is equipped with a LCD projector, an infrared camera, and a pan-tilt camera. The LCD projector is used for projecting various kinds of digital information such as computer graphics objects, texts, or a WWW browser on a desk.

For alignment between an image projected onto a desk by the LCD projector and an input image from the infrared camera, we determine a projective transformation between those two images through initial calibration of the system. The use of projective transformation is enough for calibration of our system since imaging/projection targets can be approximated as to be planer due to the nature of our application. In addition, a similar calibration is also carried out for the pan-tilt camera so that the camera can be controlled to look toward a desired position on the desk.

The pan-tilt camera is controlled to follow a user's fingertip whenever the user points at a particular location on the desk with one finger. This is necessary to obtain enough image resolution to recognize real objects near a user's pointing finger. Currently-available video cameras simply do not provide enough image resolution when the entire table is observed. In our current implementation of the interface system, a two dimensional matrix code [12] is used for identifying objects on the desk. More sophisticated computer vision methods would be necessary for recognizing real objects without any markers.

DISCUSSION AND FUTURE WORK

Interactive Textbook

Some people might think that everything in textbooks (e.g., texts, figures, etc.) could be included in a CD-ROM with multimedia contents so that students could learn everything just by using a computer. However, the students cannot use the computer everywhere, for example, in crowded trains. They still need textbooks. As we discussed in [9], people use paper or digital media depending on their situations. Interactive Textbook provides a way to integrate both media smoothly on the desk.

In this paper, we showed an application to physics. Interactive Textbook can be applied other subjects such as foreign languages, mathematics, geology, history, and so on. We are planning to develop more practical teaching materials. Then the system should be evaluated by students.

Interactive Venn Diagram

One of advantages of the Interactive Venn Diagram is that it can perform retrieval just by showing books even if a user cannot find appropriate keywords. However, it is inconvenient when the user knows appropriate keywords or when there are not appropriate books around the user. Therefore, keyword searching should be integrated into the system. For example, when users type in a keyword, the system would project the keyword and a circle on the desk. Then, the users could manipulate the keyword as they manipulate virtual book.

User Testing

Formal user studies have not been done yet. However, the system was demonstrated in some places and we received many useful comments. Visitors of our laboratory used Interactive Textbook and they commented that they preferred the automatic execution of applications. Most of them enjoyed interaction with CG simulation by their own hand.

Core technologies of EnhancedDesk were also applied to some media art exhibitions. At Haishi (Mirage City) exhibition at NTT Inter Communication Center on July 1997, the technologies were extended from the desktop to the floor [8]. A CCD camera and a video projector were mounted on the ceiling. The camera captured visitors of the exhibition and the projector projected a ripple around each visitor. (Although this visual effect is similar to Ishii's PingPongPlus [7], our system used only computer vision to identify visitors' position.) On February 1998, EnhancedDesk was exhibited at Bauhaus Museum in Berlin [17]. EnhancedDesk was applied to the guide/navigation system of the museum. Snapshots of exhibits were projected onto the desk. When someone touched one of snapshots, the detail of the exhibit was shown on the desk. Although no instructions for the system were given, the museum visitors soon learned how to use and enjoy the tactile system.

RELATED WORK

InfoCrystal [16] is a visual tool for information retrieval. InfoCrystal proposed to use visualization to formulate complex queries and showed its effectiveness. Instead of using the original Venn diagram, InfoCrystal used unique visualization to improve the visibility of intersections.

The two-dimensional matrix code used in our system was developed by Rekimoto [12]. It identifies 2^{16} bits of information. Also, it can be used to determine the size and orientation.

The pioneering work of an augmented desk interface was done in DigitalDesk [21]. DigitalDesk proposed a basic hardware setup which has been used in later research. DigitalDesk also experimented with basic finger recognition. Kruger [10] and MacKay [11] also experimented augmented desk systems.

InteractiveDesk [2] used a one-dimensional bar code to link from a real paper folder to email or web pages which were related to the documents in the folder. Arai also developed PaperLink [1] which links paper to electronic content. Robinson et al. also used a 1D bar-code to link from a printed web page to the original web page [15]. In these works, interaction with digital information was done by using the traditional mouse and keyboard. Although the use of bar-code is similar to ours, our two-dimensional matrix code can offer size and orientation information of the paper.

MetaDesk [6] used real objects (Phicons) to manipulate

digital information such as electronic maps. Rekimoto's Augmented Surfaces [14] enable users to smoothly interchange digital information among PCs, table, wall, and so on. However, users' finger and hand recognition was not explored in both systems.

HoloWall [13] used IR lights and a video camera with IR filter. However, it detects not only human skins but also objects near the surface. On the other hand, our technologies enable to detect only human skins.

CONCLUSIONS

This paper described two interface prototypes on our augmented desk system named EnhancedDesk. Interactive Textbook automatically retrieves multi media teaching materials and projects them onto the desk surface when students open the corresponding page. The system also allows the students to interact by using their hands or fingers. It enables the students to concentrate on learning process. The Interactive Venn Diagram recognizes real books on the desk and retrieves a database without requiring users to supply any keywords. The system projects a circle around each book and the retrieved results are projected inside the circle. By manipulating two or more circles so that they intersect each other, users can perform AND-search and OR-search simultaneously without formulating complex queries. We also presented some new technologies which were introduced to EnhancedDesk. Those technologies will be useful for many researchers who are now working on the similar augmented desk interfaces.

ACKNOWLEDGMENTS

The authors would like to thank Dr. Jun Rekimoto of Sony Computer Science Laboratory who provided us his two-dimensional matrix code.

REFERENCES

1. T. Arai, D. Aust, and S. Hudson. Paperlink: A technique for hyperlinking from real paper to electronic content. In *Proceedings of the ACM Conference on Human Factors in Computing Systems (CHI'97)*, 1997.

2. T. Arai, K. Machii, and S. Kuzunuki. Retrieving electronic documents with real-world objects on interacivedesk. In *Proceedings of the ACM Symposium on User Interface Software and Technology (UIST'95)*, pages 37–38, 1995.

3. R. A. Bolt. *The Human Interface*. Lifetime Learning Publications, Belmont, Calif., 1984.

4. G.W. Fitzmaurice, H. Ishii, and W Buxton. Bricks: Laying the foundations for graspable user interfaces. In *Proceedings of the ACM Conference on Human Factors in Computing System (CHI'95)*, pages 442–449, 1995.

5. W. T. Freeman, D. B. Anderson, P. A. Beardsley, C. N. Dodge, M. Roth, C. D. Weissman, and W. S. Yerazunis. Computer vision for interactive computer graphics. *IEEE Computer Graphics and Applications*, pages 42–53, May-June 1998.

6. H. Ishii and B. Ullmer. Tangible bits: Towards seamless interface between people, bits and atoms. In *Proceedings of the ACM Conference on Human Factors in Computing System (CHI'97)*, pages 234–241, 1997.

7. H. Ishii, C. Wisneski, J. Orbanes, B. Chun, and J Paradiso. Pingpongplus: design of an athletic-tangible interface for computer-supported cooperative play. In *Proceedings of the ACM Conference on Human Factors in Computing System (CHI'99)*, pages 394–401, 1999.

8. M. Kobayashi and H. Koike. http://www.ntticc.or.jp/special/utopia/index_e.html.

9. M. Kobayashi and H. Koike. Enhanceddesk: Integrating paper documents and digital documents. In *Proceedings of 1998 Asia Pacific Computer Human Interaction (APCHI'98)*. IEEE CS, 1998.

10. M. Kruger. *Artificial Reality*. Addison-Wesley, 2nd edition, 1991.

11. W. MacKay. Augmenting reality: Adding computational dimensions to paper. *CACM*, 36(7):96–97, 1993.

12. J. Rekimoto. Matrix: a realtime object identification and registration method for augmented reality. In *Proc. 1998 Asia Pacific Computer Human Interaction (APCHI'98)*, 1998.

13. J. Rekimoto, M. Oka, N. Matsushita, and H. Koike. Holowall: interactive digital surfaces. In *Proceedings of the Conference on SIGGRAPH 98: conference abstracts and applications*, page 108, 1998.

14. J. Rekimoto and M. Saito. Augmented surfaces: a spatially continuous work space for hybrid computing environments. In *Proceedings of the ACM Conference on Human Factors in Computing System (CHI'99)*, pages 378–385, 1999.

15. P. Robinson. Animated paper documents. In *Proceedings of HCI'97(21B)*, pages 655–658, 1997.

16. A. Spoerri. Visual tools for information retrieval. In *Proceedings of 1993 IEEE/CS Symposium on Visual Languages (VL'93)*, pages 160–168. IEEE CS Press, 1993.

17. T. Takada and H. Koike. http://www.vogue.is.uec.ac.jp/~zetaka/Public/kenkyu/table/berlintable.html.

18. M. Turk. Moving from GUIs to PUIs. In *Proc. of Fourth Symposium on Intelligent Information Media*, 1998.

19. J. Underkoffler and H. Ishii. Illuminating light: An optical design tool with a luminous-tangible interface. In *Proceedings of the ACM Conference on Human Factors in Computing System (CHI'98)*, pages 542–549, 1998.

20. J. Underkoffler and H. Ishii. Urp: A luminous-tangible workbench for urban planning and design. In *Proceedings of the ACM Conference on Human Factors in Computing System (CHI'99)*, pages 386–393, 1999.

21. P. Wellner. Interacting with the paper on the DigitalDesk. *CACM*, 36(7):87–96, 1993.

curlybot: Designing a New Class of Computational Toys

Phil Frei, Victor Su, Bakhtiar Mikhak*, and Hiroshi Ishii
Tangible Media Group
*Epistemology and Learning Group
MIT Media Laboratory
20 Ames Street
Cambridge, MA 02139
+1 617.253.9401
{frei, vsu, mikhak, ishii}@media.mit.edu

ABSTRACT
We introduce an educational toy, called *curlybot*, as the basis for a new class of toys aimed at children in their early stages of development – ages four and up. *curlybot* is an autonomous two-wheeled vehicle with embedded electronics that can record how it has been moved on any flat surface and then play back that motion accurately and repeatedly. Children can use *curlybot* to develop intuitions for advanced mathematical and computational concepts, like differential geometry, through play away from a traditional computer.

In our preliminary studies, we found that children learn to use *curlybot* quickly. They readily establish an affective and body syntonic connection with *curlybot*, because of its ability to remember all of the intricacies of their original gesture; every pause, acceleration, and even the shaking in their hand is recorded. Programming by example in this context makes the educational ideas implicit in the design of *curlybot* accessible to young children.

Keywords
Education, learning, children, tangible interface, toy

INTRODUCTION
The role of physical objects in the development of young children has been studied extensively in the past. In particular, it has been shown that a careful choice of materials can enhance children's learning. A particularly notable example of such materials is Friedrich Froebel's collection of twenty physical objects (so called "gifts"), each designed with the purpose of making a particular concept accessible to and manipulable by children [5]. The presence of objects inspired by Froebel in almost all kindergartens today is a reflection of their recognized value in the development of young children.

Permission to make digital or hard copies of all or part of this work for personal or classroom use is granted without fee provided that copies are not made or distributed for profit or commercial advantage and that copies bear this notice and the full citation on the first page. To copy otherwise, to republish, to post on servers or to redistribute to lists, requires prior specific permission and/or a fee.
CHI '2000 The Hague, Amsterdam
Copyright ACM 2000 1-58113-216-6/00/04...$5.00

Figure 1: Three palm-sized *curlybots* (each with a large record/playback button and a small indicator light).

Most recently, Mitchel Resnick and his Lifelong Kindergarten Group at the MIT Media Laboratory have introduced a collection of digital manipulatives that builds on Froebel's work, taking full advantage of computational ideas and resources not available until recently [12,14]. Much like Froebel's gifts, these tools attempt to make new domains of knowledge accessible to children.

In this paper, we contribute to this initiative a new class of computational toys that is aimed at children as young as four. *curlybot*, the first instantiation and the basis of this class of toys, is a two-wheeled toy that can record and play back physical motion replicating every intricacy of the original motion. It is a smooth, easily graspable curved object with a button and an LED for indicating whether the device is in record (red) or playback (green) mode. To record a gesture, a child presses the button and moves *curlybot* through a desired path. A child presses the button a second time, to stop recording and begin playback of the recorded gesture. The playback mode repeats the gesture indefinitely until the button is pressed again. Because of the simplicity of the interface, children quickly learn to create intricate gestures with *curlybot*, which they can refine through an iterative process.

This version of *curlybot* also has a mode called "boomerang," in which a *curlybot* will move backwards through its path and then forward again. Holding the button when the toy is turned on activates this mode. *curlybot* also has a pen attachment to create drawing of gestures (see Figure 2).

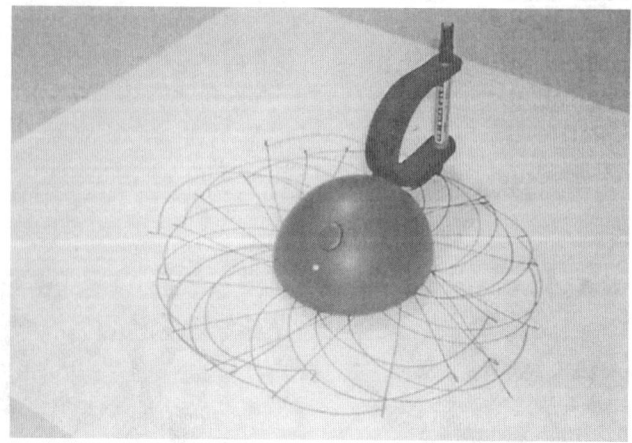

Figure 2: *curlybot* with a pen attachment.

We believe that *curlybot* can play a significant role in mathematics education research. Therefore, a part of this paper will be devoted to the discussion of the educational issues that *curlybot* is designed to address. We will propose a set of possible play scenarios for *curlybot* to highlight different educational ideas, which extend *curlybot* into a class of new toys that support multiple forms of play. Finally, we will review the user testing and present the design and implementation of the current system.

MOTIVATION

Many of the computational environments designed for children have been thus far limited to activities on the computer screen. One notable example that has enjoyed great recognition in and out of the classroom is graphical Logo. The main computational object in Logo is a turtle, whose heading and trajectory can be controlled by simple programs written by children. Graphical Logo was inspired by a small robot (about one cubic foot in size) built at the MIT Artificial Intelligence Laboratory by Seymour Papert and collaborators. This robot, called the Floor Turtle, was quite heavy and tethered to a mainframe computer. By typing commands at a terminal, children controlled the turtle and its pen to draw geometric patterns on large sheets of paper on the floor. As Graphical Logo was developed and soon became widely used, the Foor Turtle was put on hold.

In the 1980's, Fred Martin, Seymour Papert and Mitchel Resnick resurrected this work at the MIT Media Lab by building computation and programmability into the familiar LEGO bricks. Children could build these Programmable Bricks into their robots and program them to bring their creations to life. The most recent member of the Programmable Brick family is the Cricket, which encapsulates the core functionalities of the previous generation into a much smaller package and makes the system expandable through a unique bus structure. The Programmable Brick inspired the LEGO Mindstorms Robotic Invention System [10].

Robots built with the Programmable Bricks and Crickets are currently programmed in text-based or graphical programming languages that are dialects of Logo. Research has shown that children as young as ten years can successfully use Programmable Bricks and traditional construction material to build and program their own robots to exhibit the behavior they are looking for. Extending these types of activities to younger children is an active area of research [13].

The design of *curlybot* is also inspired by the natural and expressive quality of Golan Levin's gesture-based animation environment system called Curly [8], which builds on Scott Snibbe's Motion Phone system [19]. These systems capture the gestures of the computer mouse on the screen and replay it graphically.

	Programming/Input	
	Digital	Physical
Execution/Output Digital	Turtle Graphics	Motion Phone Curly
Execution/Output Physical	Floor Turtle LEGO/Logo Programmable Bricks Crickets LEGO Mindstorms BigTrak	curlybot

Table 1: Input vs. Output Interaction Space.

Table 1 summarizes the differences in the ways in which children interact with the various design and expression media we have discussed so far. This parameterization of possible modes of programming and interacting with computational media highlights *curlybot's* significance, including the coincidence of the input and the output space.

INTERACTION & DISCUSSION

We envision children having several basic interactions with the *curlybot* family of toys. The current design of *curlybot* can be extended in multiple directions. Some of the activities in this section rely on augmentations to the current system that we will discuss in conjunction with the various activities.

Repetition – How do you keep the toy repeating a gesture while not falling off a table? A child, in this case, would learn to create repetitive patterns that as a rule would end up at the origin or circle around a central point. Through this direct manipulation, a child can learn many lessons by just playing and experimenting with movement, spatialization and repetition. Another example is a child trying to create a star with three gestures. This activity introduces a child to the idea of building complex shapes by combining simpler elements. A child is also exploring computational and mathematical ideas, like loops and vectors. To create the star, you have to be concerned with elements of a vector, such as point of origin, direction, and magnitude. When *curlybot* loops the recorded vector, it is critical to start and finish with correct orientation, not just position. A pen can be attached to *curlybot* to leave a trail of its path and make the visualization of more complex pattern easier (see Figure 3).

 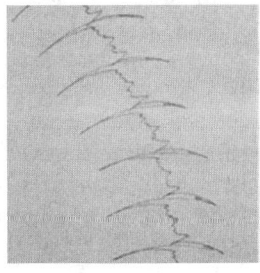

Figure 3: Examples of pen drawings.

Making it possible to record and play back whether the pen is up or down would allow for a broader range of designs that include discontinuous lines, like dotted or dashed lines.

Pen Position - The use of pens introduces additional mathematical concepts, since the pen can be placed in different locations relative to the wheels of the vehicle. For example, a *curlybot* asked to move forward and turn 90 degrees, will create a square, if the pen is placed exactly between the two wheels (see first pattern in Figure 4). However, a different pattern will emerge, if the pen is placed farther from the center. This should be contrasted with the graphical turtle, which is assumed to be a point-like object with its pen located at its center. *curlybot* allows for more surprising patterns to emerge which encourages a child to think about the distinction between point-like and extended objects. A child might not mathematically understand the concept, but will have at least developed intuitions for relative position and motion of points.

If one adds an additional degree of freedom and had the pen move independently in a circle around *curlybot*, one can create more complex patterns that begin to mimic orbital patterns. If we then moved *curlybot* in a circle and had the pen move at a higher frequency around *curlybot*, one creates the orbital pattern of the moon relative to the sun.

Conditional Behavior – Additional sensors could be added to *curlybot*, like bump and light sensors, in order to program conditional behavior. For instance, one could teach *curlybot* to move forward and it would then drive straight until hitting a wall with one of its bump sensors. At that point, the toy would stop moving. The LED on the device would turn yellow, prompting the user to record a sequence in response to hitting the wall. One could then record, going backwards a little and turning, which would now be *curlybot's* standard response to hitting an obstacle with that particular sensor. This type of conditional programming would allow *curlybot* to respond to its environment instead of simply playing back a recorded gesture allowing *curlybot* to act as an autonomous creature with complex behavior.

This type of behavior is the same as that of creatures made with Programmable Bricks. However, *curlybots* are programmed by example rather than using traditional programming. Nonetheless, a child can still learn about "if" and "while" statements.

Recording Primitives – If one records several sequences and stores them on a computer under different names, like circle, box, and line, they could later be used as procedures in a programming language such as Logo. Separate gestures could then be combined together in a computer program and sent back out to *curlybot*. This added functionality leverages the simplicity of physical programming and gestural output with the added flexibility of a computer program. This is a concrete example of procedural abstraction.

Gesture and Narrative - Since *curlybot* captures not only the trajectory of movement but also velocity and acceleration, it can be used to express gesture. For example, a child could record a nervous shaking, and *curlybot* would do just that. This gestural expression is also useful in playing out a story with the toy. It is common for children to act out stories with toys, but in this case the toy could boomerang back through all the obstacles and start replaying the interaction physically - pausing and accelerating in all the right places. One could even add audio recording and playback to the device and synchronize what the children say with their movements [16]. The child could learn aspects of storytelling and gesture by watching his or her own actions from the point of view of an observer.

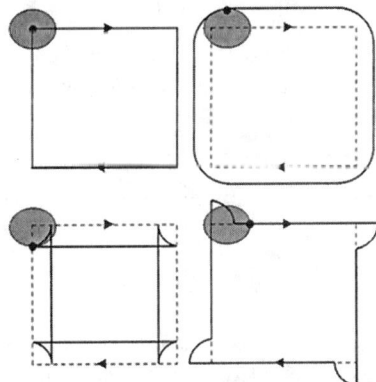

Figure 4: Four different pen positions and their resultant trails on the same *curlybot* pattern.

Synchronization – Two or more *curlybots* could be synchronized to create a medium for haptic communication, like the inTouch [4] or HandJive [6], or a display of remote physical activity, like Ambient Displays [7].

Music – One could map different gestures created with *curlybot* to musical sounds and have them loop like *curlybot's* physical motion in a rhythmic pattern. Alternatively, one could affect a musical piece that is played.

Trading – The exchange of digital information is a very rich area of research. *curlybot* supports the exchange of its information because its memory is physically removable and can be used to save a session or exchange it with someone else's *curlybot*. The memory could also be exchanged with a personal computer, where the recorded path could be displayed, or potentially altered and resaved to the memory. The file could be sent to distant friends to be played on their *curlybots*. Some interesting issues arise in the representation of the information, since we are not only recording a trajectory, but also velocity and acceleration. There could be some interesting challenges in the visualization and editing of this information.

The exchange could also happen without a physical exchange of memory, but rather through infrared (IR) or radio frequency (RF). One could have one *curlybot* teach another *curlybot* an interesting gesture and these could be passed on and saved It would be interesting to see how a particular pattern spreads and to examine which gestures people felt compelled to pass on to others. One could also introduce evolutionary ideas, involving the progressive alteration of patterns over time via an exchange with different patterns [2,3].

Different Personalities - One can change the variables in the control algorithm to create different "personalities" in the *curlybots*. For example, a *curlybot* could be designed to be better at fast motions, while another could be designed to be better at slow motions. Creating these distinct *curlybots* gives them personality outside the recorded gesture, making them individual characters which children will be drawn to in different ways.

Editing - There could be other forms of input, like electric field sensing, that could be used to change the motion as it is playing it back

EDUCATIONAL IMPLICATIONS

In this section, we will discuss the complementary educational opportunities afforded by the *curlybot* family of toys in light of its potential to:

- serve as objects-to-think-with [11]
- make new domains of knowledge accessible or old domains of knowledge approachable in new ways
- support multiple learning styles [15]

curlybot as an object-to-think-with

curlybot's physical form, size and weight makes it a natural extension of the hand which a child can program by example. A child can map ideas from his or her mind directly into a clear physical instantiation of the ideas. The process and validity of the execution is transparent because the motion involved in the act of programming is bodily syntonic. The immediate feedback from the observed behavior of the robot allows children to examine and reflect on their initial mental models with respect to the outcomes they observe and gives them a chance to debug and extend their thinking.

In *Mindstorms*, Papert eloquently describes the significance of programming as a tool for thinking about one's own thinking [11]. The very process of externalizing models and concepts in ones mind into the physical world allows for the critical evaluation of the validity of the models by oneself and others against easily understandable physical behavior. In turn, the external instantiation of an idea can be internalized again to modify the initial models. *curlybot* takes advantage of the rich educational opportunities afforded by creating and supporting such internalization/externalization feedback loops.

curlybot and new domains of knowledge

curlybot makes the core ideas in Logo accessible to even younger children. *curlybot* can provide a tangible way of exploring many important ideas that have been studied extensively within the Logo community. For example, moving forward a little and turning a little will result in a circle, if one repeats it over and over again. This will result in a more even circle than if the child tried to create the circle out of a single gesture. This is a concrete instantiation of the idea of differential calculus as well as local representation of a circle.

In addition to differential calculus or local and intrinsic representation of curves, *curlybot* could be used as a tool to gain intuitions for turtle geometry [1], Aristotelian and Newtonian physics [12], and the law of large numbers and probability [19], to name a few. Many of these topics are ordinarily considered too advanced for children, but interacting with carefully designed objects can make this material accessible to them.

curlybot and multiple styles of play and learning

An equally important component of any powerful learning experience is the affective quality of the relationship between the learner and the material. Studies have shown that children's learning and play patterns can be divided into two overlapping categories, namely patterners and dramatists [17]. In the design of *curlybot*, we were conscious of supporting both forms of play. Whether a child is a planner or a dramatist, he or she will connect to the same mathematical ideas but in ways that are more natural for them (see section on user interaction for more examples).

curlybot can engage children who are more artistic and expressive. The entry point into mathematics for these children is through their artistic involvement with a tool and a medium. In this case, the critical feature of *curlybot* is that it lives up to children's expressive expectations. This allows children to make a strong affective, as well as intellectual, connection to *curlybot*.

Papert makes a strong argument for the importance of this last idea in his book *Mindstorms*. He begins by explaining how gears taught him "advanced" mathematical ideas as a child, but then claims that giving sets of gears to children will not necessarily result in the same learning experience for most of them. The success is in part due to the child's personal attachment to the gears - Papert "fell in love" with his gears. He could project himself into the gears and "be the gear," which is what "gives the gear the power to carry powerful mathematics into the mind." If a child is not completely engrossed in playing with a toy, they will not learn very much from it [11].

USER INTERACTION

We began by allowing several hundred adults to play with *curlybot* and found that many of them discovered new gestures and patterns that we had not anticipated. This, of course, was a promising result, since our hope was to design an open-ended toy that would continue to be interesting over time.

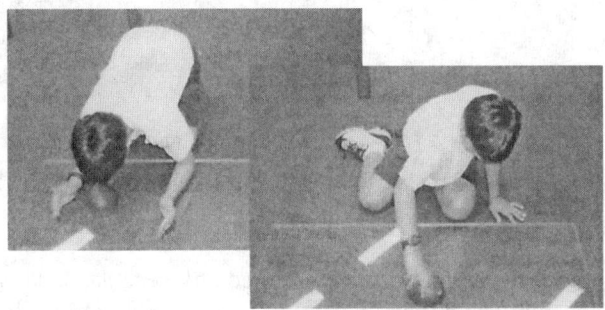

Figure 5: Child playing with *curlybot* on Plexiglas.

In particular, it was interesting to see people take advantage of the fact that *curlybot* records every pause one makes. In one case, someone had *curlybot* do nothing for a long time and then shake around. This resulted in an interesting behavior during playback: *curlybot* would appear inactive or off, but then surprise the audience by suddenly starting to shake. Another user recorded a pause, a shake forward and back, a pause, and then a shake from side to side. When playing back, he asked *curlybot* if it liked him, and it moved forward and back. He then asked if it liked his friend, and it shook from side to side. By having others play with *curlybot*, we discovered a real satisfaction in learning a new behavior or pattern.

First Study with Children

The first study was conducted at the Science Museum in Boston, Massachusetts. Though this was not a completely random cross section of children, it easily provided us with a large group (81) of children for initial tests. The Science Museum is a good environment for making observations, since the children are prepared to play with things and generally do not notice when someone is observing them.

In the entrance to the Discovery Center of the Museum, we set up a 3'x 4' piece of Plexiglas to clearly demarcate the play space. This was also done to observe if the children would learn how to keep *curlybot* in that space. The play was forced to remain on the Plexiglas, since we used a version of *curlybot* that was not designed to run on the surrounding carpet.

Very little instruction was given to the children, in order to learn how effective the interface was from the start. We then were interested in monitoring what children did with *curlybot*. Did children figure out how to keep *curlybot* from running outside of the demarcated area? Were they more interested in geometric designs, gestures, or narratives? How long did it take them to figure it out? Can we generalize the responses of different age groups? Is there an age where children cannot interact with the device at all? It should be noted that our results are based on qualitative observations, and subjective categorization. These results are nonetheless interesting, because they provide us with a rough guide for further study.

Most of the children knew what to do with *curlybot* by observing how others had used it. If they did not, we would ask one of the other children to explain it to them. Through this, we were able to observe if they had learned something beyond the basic functionality of how to record and play. Out of the twenty-two children who were asked to explain how to use the toy to someone else, only three of them described how to keep *curlybot* on the platform in addition to explaining the basic functionality.

About a quarter of the children (21 out of 81), explicitly created geometric shapes. Four children did what we considered to be explicitly gestural recordings, while the rest did narrative recordings. It was difficult to draw lines between the different interactions, since there was some overlap between the categories. One ten-year-old girl, for example, recorded a beautiful geometric piece after observing four boys of her age record strictly geometric shapes. However, unlike the boys, her geometric shapes had accelerations and pauses, which created a more gestural pattern. This made us categorize her actions as gestural rather than geometric, even though she was also very successful at keeping *curlybot* on the platform through a geometric pattern. It is interesting to note that the boys were impressed and tried to create some more gestural patterns after her performance. This also shows that a child can be affected by another child's interaction with *curlybot*. Our results are heavily affected by this fact, since we were not working in a controlled environment where children were isolated from one another while playing with *curlybot*.

We hoped to see trends in play between the different age groups, however the only conclusive results we found were that children under the age of four generally could not meaningfully interact with *curlybot*. We also thought that older children might not learn much from the interaction, but that did not seem to be the case. Older children spent just as much time as younger ones trying to figure out how to design a pattern that would stay on the platform.

It was interesting to observe that the children had a tendency to make large and fast gestures with *curlybot*. This caused two problems. One, because there was a constrained play area, large motions, that did not end exactly where they began, made *curlybot* fall off the platform. Two, this version of *curlybot* was not designed to reproduce fast motions as accurately as slow ones and, as a result, *curlybot* did not repeat geometric shapes perfectly. Overall, the children were not concerned with these problems and continued to play with *curlybot* anyway. For future tests, though, we will redesign the control algorithm.

It usually was not possible to have children perform specific tasks given the informal environment of the study. However, there was one seven-year-old girl that played with *curlybot* for an extended period of time and accepted our challenge to create a few geometric shapes out of their most basic elements. We found that she needed us to provide an example before being able to create the shapes on her own. We showed her how to create a square and let her try it on her own. When we asked her to create a circle, she started by designing it with very large arcs. She needed additional help to understand that a circle could be created from a very small segment. Later on, the same girl came back, and asked if she could try a shape she had been thinking about. We were pleased to see that she continued to process her new knowledge about shapes even outside the play area. *curlybot* appears to have become an object-to-think-with for her.

Though this user test was not conclusive, it confirmed that *curlybot* is fun for children and that our questions were indeed relevant in view of the children's interactions with the toy. In particular, we would like to present children with specific design challenges, like creating a complex pattern out of simple elements and study how they would perform and their thinking process.

IMPLEMENTATION
Current Implementation
The *curlybot*'s two wheels have independent drive and sensing capabilities that are controlled by a microprocessor. Mechanically, the toy consists of two 10 Watt Maxon motors with Hewlett-Packard Optical encoders. They are mounted on the bottom of *curlybot* in such a way that, after gearing the torque up 4:1, the shafts of both wheels are co-linear. This allows it to not only move forward and back, but also rotate freely about its center. This is also the most compact design that allows the device to easily fit in the user's hand. The physical configuration also simplifies what needs to be recorded. If both motors are moving forward, the device is moving forward. If they are moving in opposite directions, then the device is turning.

The 10 Watt motors are very efficient and power is not lost in heat dissipation. The use of these large motors gives us additional mass, which is useful in creating sufficient friction for the drive wheels. In this way, the user can feel resistance when they push against the direction of the wheels. Also, the additional weight creates a good inertia for play.

A 20MHz Microchip microprocessor with built-in pulse width modulation controls the motors. The encoders available to us had 500 counts/revolution. Because of the gearing, the resolution of the wheel is 2000 counts/revolution. If *curlybot* is moving quickly, the encoder interrupts the microprocessor continuously, which does not allow other processes to be run. To overcome this, we divide the encoder information by four using a counter, so that the resolution of the wheel is only 500 counts/revolution.

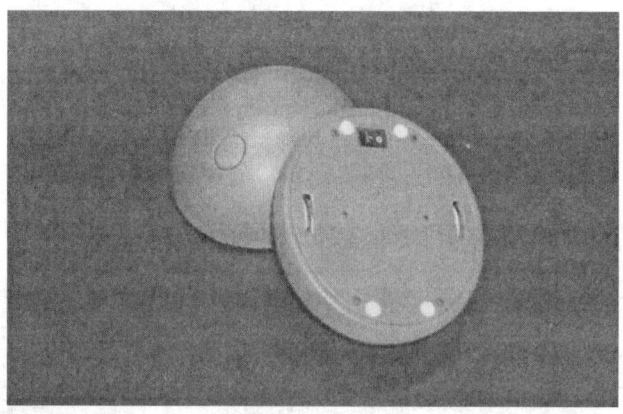

Figure 6: Top and Bottom of *curlybot*.

The encoder information is stored on a separate 32 kilobyte memory chip (256 kilobits) at a rate of 100Hz. At this rate, we can record the encoder information of both motors for about two and a half minutes. The device currently runs on six AAA batteries – four for the motors and two for the circuit board.

Originally, we used two 9 Volt batteries in parallel for the whole system, but there were two problems. One, the capacity of 9 Volt batteries is much less than that of AAA, so a *curlybot* would not run continuously for more than two hours. Second, when the motors draw a lot of current, the voltage for the circuit board drops below 5 Volts and the circuit resets. We also originally used a one-megabit serial eeprom memory chip, since we were not sure with what frequency we wanted to record. When we finally decided that 100Hz would be enough, this memory chip gave us about ten minutes of recording time, which is much more than what we needed. We then switched to our current eight-pin 256 kilobit eeprom memory chip that can be easily removed from the board and replaced with any other

8 pin eeprom. It also has a fraction of the leads, since one reads and writes to it serially.

The motor is run on pulse width modulation with feedback only from the encoder. The performance of the playback could be improved by monitoring the current feedback from the motor.

To record, the user presses a button that lights up a red indicator LED. When the user is done recording a sequence, the button is pressed again and the indicator LED turns green. At this point, the processor runs a PID control function that calculates the force that the motors need to exert to reach the recorded position. The processor compares its current position (from the encoder) to the desired position (from the memory) and then applies the necessary force to move from one to the other. When the button is pressed again, the indicator LED turns off and *curlybot* is in neutral mode. Here it is free to roll around and nothing is recorded or played back. The sequence can be started again by pressing the button one more time.

We can also switch *curlybot* into boomerang mode, by pressing the button while turning the device on. In this mode, the toy boomerangs back through its recorded path to its starting position, where it then begins to repeat the motion again.

Another Implementation

In order to test some of our other interface ideas, we decided to design another version of *curlybot*. First, we added a two-button interface with separate record and playback buttons. This allows users to re-record a motion without playing it back or, likewise, to stop playing a motion and then start again without re-recording. We have also explored using a double-click on the single button interface to click over the record or playback mode. This provides the additional functionality of the two-button interface without making it more confusing for novice users.

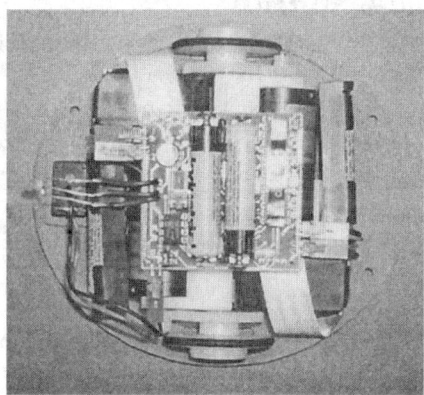

Figure 7: Inside *curlybot* (top view).

We have also reduced the size of *curlybot* to something smaller than a computer mouse. This version uses 1 Watt Maxon motors that are about the size of a AAA battery, including a 12 count/revolution encoder and 4:1 gearhead. Though the resolution of the encoder is lower, we still managed to maintain about the same resolution on the wheel circumference. To keep the toy small, we used two AAA batteries to run both the circuit board and the motors, even though we knew we could run into problems with high current draws. The main problem with this prototype was that, because it was lighter and smaller, the wheels' traction was not enough when a user pushed against the direction the wheels turn.

RELATED WORK

The Epistemology and Learning Group at the Media Lab, as mentioned in the Motivation section, has done very closely related work for many years, spearheaded by Seymour Papert, Mitchel Resnick and Fred Martin. This work includes Logo, LEGO/Logo, Programmable Bricks, Crickets, and LEGO Mindstorms Robotic Invention System. The ideas for trading information between *curlybots*, mentioned earlier, are based on the research of Rick Borovoy, such as his MemeTags and Dance Craze Buggies [2,3].

The work of Kimiko Roykai and Justine Cassell called StoryMat is about creating a space that encourages children to tell stories with a plush toy and later have them replayed. The replay is not in physical form, but occurs with a moving projection of the toy on the StoryMat accompanied by the recorded audio [16].

Microsoft's ActiMates Barney, like *curlybot*, attracts the child's attention by being a character that exists in the child's physical rather than virtual space. One of the major differences, though, is that Barney is a story-based toy. This means that the child's interactions with Barney are limited by a preprogrammed or uploaded set of stories. *curlybot* on the other hand, invites the child to discover by playing. Instead of being told a story or being given a specific task, the child learns through teaching *curlybot* and exploring the results. Because this interaction is more complex, Barney is still easier to use for very young children [20].

In manufacturing, to save time programming robotic arms in assembly lines, the robot is physically given end points for its trajectory and is then allowed to calculate the optimal path. If there are obstacles for the robot arm to avoid, extra points are added to create the desired trajectory. Like *curlybot*, this is an example of physical programming.

Similarly, in robotic artificial intelligence, researchers have for some time used techniques of programming a robot by recording the motion it should perform. For example, with the help of a human hand, one can quickly program the many degrees of freedom in three dimensions of a robotic arm picking up a cup.

FUTURE WORK & CONCLUSION

The first and most important next step is to perform controlled user studies with children in order to determine what types of things children learn from their interaction with *curlybot*. This means that we will need to give children a longer time to play with *curlybot*, so that they have time to explore its full range of possibilities. We will make systematic observations of this interaction. We will also present children with specific design challenges, in order to determine what they are learning and how they are thinking about accomplishing the goals of the challenge. We will follow the activities with interviews. A longer term study would be needed to reveal if and to what extent interacting with *curlybot* prepares children for working in text-based or graphical programming environments, such as Logo. These types of studies are much more challenging, since it is difficult to isolate the contributions from a specific source to a child's future abilities.

Currently, we are focusing on the implementation of the augmentations mentioned in the Interaction & Discussion section. These would provide different computational and mathematical concepts for children to explore, which could confirm *curlybot* as a toy capable of supporting multiple learning and play styles. Furthermore, these new implementations may lead us to discover new directions for this research.

In conclusion, our preliminary results show that *curlybot* succeeds in engaging children ages four and above to play around with advanced mathematical and computational concepts (previously learned at a later age and often with the aid of a traditional computer) in a much more fluid and expressive fashion. *curlybot* is an introductory tool, much like Logo, that can help build a child's basic mathematical intuition by engaging them in genuine mathematical activities. Once this basic understanding has been established, children will have an easier time moving into computer programming and formal mathematics as they get older. The example interactions presented in the paper, position *curlybot* as the basis for an entire class of computationally enhanced educational toys, which we are actively designing.

ACKNOWLEDGMENTS

I would like to thank Brad Niven of Interval Reseach Corporation, Ali Malazek, Megan Galbraith, Rujira Hongladaromp and all the other members of the Tangible Media Group at the Media Lab who have contributed their ideas and time to this project. This project has been supported by the MIT Media Lab's Things That Think consortium and Interval Research Corporation.

REFERENCES

1. Abelson H., and diSessa A. (1981). *Turtle Geometry*. MIT Press.
2. Borovoy R., and Martin F. (1999). Tradable Bits. <http://el.www.media.mit.edu/people/borovoy/cars/>.
3. Borovoy R., Martin F., Vemuri S., Resnick M., Silverman B., and Hancock C. (1998). Meme Tags and Community Mirrors: Moving from Conferences to Collaboration. *Proceedings CSCW '98*, ACM Press, 159-168.
4. Brave, S., and Dahley, D (1997). inTouch: a medium for haptic interpersonal communication. *Extended Abstracts of CHI'97*, ACM Press, 363-364.
5. Brosterman, N. (1997). *Inventing Kindergarten*. New York, Harry N. Adams Inc.
6. Fogg, B.J., Cutler, L., Arnold, P., and Eisback, C. (1998) HandJive: A device for interpersonal haptic entertainment. *Proceedings of CHI'98*, ACM Press, 57-64.
7. Ishii, H. and Ullmer, B. (1997). Tangible Bits: Towards Seamless Interfaces between People, Bits and Atoms. *Proceedings of CHI'97*, ACM Press, 234-241.
8. Levin, Golan. Curly. <http://acg.media.mit.edu/people/golan/curly/>.
9. Kafai Y., and Resnick, M., eds. (1996). *Constructionism in Practice:* Designing, Thinking, and Learning in a Digital World. Mahwah, NJ, Lawrence Erlbaum.
10. Martin F., Mikhak B., Resnick M., Silverman B., and Berg R. (1999). To Mindstorms and Beyond: Evolution of a Construction Kit for Magical Machines. *Robots for Kids* edited by Alison Druin and James Hendler, Morgan Kaufmann Publishers, Inc.
11. Papert, Seymour (1980). *Mindstorms*: Children Computers and Powerful Ideas. BasicBooks.
12. Resnick, M. (1998). Technologies for Lifelong Kindergarten. *Educational Technology Research and Development*, vol. 46, no. 4.
13. Resnick, M., Eisenberg, M., Berg, R., and Martin, F. (1999). Learning with Digital Manipulatives: A New Generation of Froebel Gifts for Exploring "Advanced" Mathematical and Scientific Concepts. Research proposal, May 1999.
14. Resnik A., Martin F., Berg R., Borovoy R., Colella V., Kramer K., Silverman B (1998). Digital Manipulatives: New Toys to Think With. Paper Session, *Proceedings of CHI'98*, ACM Press, 281-287.
15. Turkle, S., and Papert, S. (1990). Epistemological Pluralism. *Signs 16*, 1, 128-157.
16. Ryokai K., and Cassell J. (1999). StoryMat: A Play Space for Collaborative Storytelling. *Extended Abstracts of CHI'99*, ACM, 201.
17. Shotwell, J., Wolf, D., and Gardner, H. (1979). Exploring Early Symbolization. In B. Sutton-Smith (ed.), *Play and Learning*.
18. Silverman B., and Tempel M. (1991). Fuzzy Logo. Logo Foundation Memo. <http://el.www.media.mit.edu/groups/logo-foundation/Publications/Fuzzy-Logo.html>.
19. Snibbe, Scott (1995). Motion Phone. Interactive Communities, SIGGRAPH '95.
20. Strommen, Eric (1999). When the Interface is a Talking Dinosaur: Learning Across Media with ActiMates Barney. *Proceedings of CHI'99*, ACM Press, 288-295.

HandSCAPE: A Vectorizing Tape Measure for On-Site Measuring Applications

Jay Lee, Victor Su, Sandia Ren, and Hiroshi Ishii
MIT Media Laboratory, Tangible Media Group
20 Ames Street, E15 {-444, -433, -485}
Cambridge, MA 02139 USA
{jaylee, vsu, sren, ishii}@media.mit.edu

ABSTRACT

We introduce HandSCAPE, an orientation-aware digital tape measure, as an input device for digitizing field measurements, and visualizing the volume of the resulting vectors with computer graphics. Using embedded orientation-sensing hardware, HandSCAPE captures relevant vectors on each linear measurements and transmits this data wirelessly to a remote computer in real-time. To guide us in design, we have closely studied the intended users, their tasks, and the physical workplaces to extract the needs from real worlds. In this paper, we first describe the potential utility of HandSCAPE for three on site application areas: archeological surveys, interior design, and storage space allocation. We then describe the overall system which includes orientation sensing, vector calculation, and primitive modeling. With exploratory usage results, we conclude our paper for interface design issues and future developments.

Keywords

input device, field measurement tool, on-site applications, orientation-aware, physical interaction, tangible interface

INTRODUCTION

The act of measuring is a human task that dates back thousands of years, and evolved from a need to describe physical structures for the purpose of construction or surveying. The act of measuring is often associated with body gestures, such as the way we stretch our arms when describing the length of an object (e.g. "I caught a fish this big"). This body aspect of measurement also gave rise to the first primitive units of measurement, such as the *cubit*, which was defined the distance between the elbow and the tip of one's fingers.

Measurement Tools

Over the years, measurement tools have evolved from using our arms and feet, to more quantitative tools such as rulers and tape measures. Throughout these evolutions of technology tools were augmented with new capabilities for surveys and constructions for greater accuracy and ease of use. More recently, the advent of electronics and computing has given rise to a great number of new tools and systems for digitizing and measuring 3D objects.

Figure 1. The HandSCAPE tool (12 x 12 x 4.5 cm)

Examples of these are commercially available 3D digitizing and modeling products include *Polhemus 3Ball*^{TM}[11], *Monkey2*^{TM} [10], *Digibot3*^{TM} [4], etc. Despite the accuracy and versatile usage of the device, many of these new tools are not practical or appropriate (and often too expensive) for common and portable measurement tasks; we have employed in such as use warehouses, construction sites, shipping yards, sporting events, and archeological field surveys.

More relevant measuring and testing input device is *SHAPE TAPE*^{TM} [12] which tracks the shape of tape while the tape is twisting and bending. It has 6DOF position and orientation of the two. Although the devices and systems provide accuracy and speed to digitize physical 3D, these are designed for a desktop application.

In terms of computer modeling of 3D environment, there are few programs that perform rapid modeling of primitive 3D scene [16] and allow sketching of freeform objects [8], against the complicated CAD like 3D modeling applications. Their emphasis is on ease of low-level correction and simplicity of interface for sketching. These are well developed in combining the ideal sketched by hands and computer-based modeling programs to improve the efficiency of sketching approximate models.

The Basic HandSCAPE Concept

The motivating concept for HandSCAPE was to create a simple handheld tool which would allow workers in the field to digitize their measurements and gain the productivity and efficiency from modern-day computer technology. We considered several possible approaches, such as computer vision, ultrasonics, or lasers, but the requirements of portability and robustness led us to investigate what could be done with the existing tool commonly known as the tape measure. It is portable, simple to use, and low-cost. For this reason, the traditional measuring tape continues to be the tool of choice for common everyday measuring tasks. We set out to augment a traditional tape measure with digital functionality, so that we could employ peoples' existing skills and familiarity in physical environments with this classic tool. The key innovation of HandSCAPE is that it seamlessly combines a 3D input device with a rapid 3D modeling visualization of non-desktop measuring application.

Orientation Awareness

Since a handheld tape measure simply measures a linear distance, and it is not apparent how such a device could be utilized to capture the necessary spatial information. However, if orientation sensors are added, angular information can be measured as well. Then, knowing the distance and direction enables such a device to measure *vectors*. By measuring a series of vectors, the spatial dimensions of a physical object can be recorded and reconstructed in digital domain automatically. Similarly, by choosing a fixed reference point for the entire space, then the relative position and orientation of each of the objects can be recorded by simply measuring the distance from one object to the next. This approach formed the basis of the HandSCAPE concept and proceeded the technology.

REAL WORLD NEEDS

In this paper, we present HandSCAPE as an example of a wireless technology to enhance the efficiency of on-site measuring tasks. We have closely studied the intended users, their tasks, and the physical workplaces to extract the needs from real world measuring tasks on-site. In the following sections, we first describe the potential utility of our device for each application. We then describe the overall system, which includes vector calculation, relative positioning, and generating 3D models. Finally we conclude our paper with reporting results from an exploratory usage and discussing intuitive observations. Furthermore we consider the future plan in further development of HandSCAPE.

Our research concentrates on field applications where getting information from the physical world becomes more complicated. The following three application areas will comparatively examine the aim of this research as a solution to particular measuring problems. Each application demonstrates issues in both hardware and software that arise in using HandSCAPE.

1. Optimal Packing in Storage and Transportation

Efficient use of space for packing, storing, and transporting goods is vital in the industrial sector [5]. As demand on space and distribution operation increases, investing in material handling and storage equipment to improve efficiency, profit margins, and reduce distribution and storage cost becomes crucial. High costs result from a lack of coordination between transport and storage systems, and between a storage system and its enclosure. The task of space optimization in terms of volume in bulk storage packing is usually a multi-step process involving measurements and calculations done by hand. HandSCAPE has transformed this process into a one step interaction with clear visualization. With HandSCAPE the measuring has also been directly coupled with space optimization, thus allowing the users to perform the physical task efficiently.

Figure 2. Trucking at a loading dock

Scenario: Trucking with Packing Simulation

Imagine two truck drivers who need to transport hundreds of differently sized boxes in a truck container (Fig. 2). Assuming i) it is not necessary to unload any of these boxes until arriving at the destination, ii) the weight of box is determined by the volume, iii) repeated access to a box is limited due to the truck driver's tight pick-up schedule. How, then, can they pack as many boxes as possible as fast as possible?

Given a known volume of available storage, the user employs HandSCAPE to measure the dimension of each box. Whenever a box is measured, the host computer determines the best-fit position of the current box to minimize the use of the space. After measuring all the boxes (Fig. 3), the optimal packing configuration is' visualized on the screen (Fig. 4) and the actual physical packing can be performed according to this result.

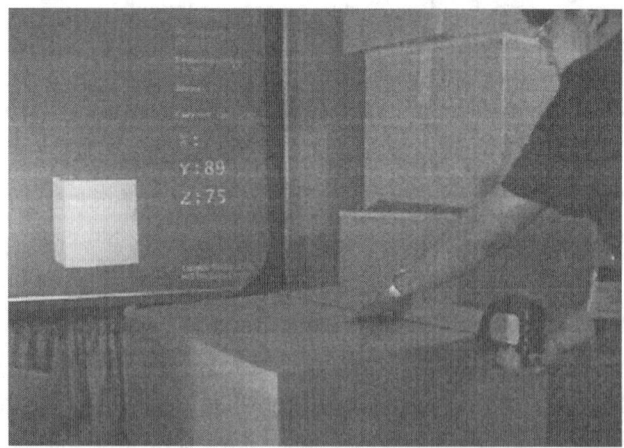

Figure 3. Using HandSCAPE for on-site packing optimization configuration displayed on screen.

Software: Optimal Packing Algorithm

The goal of our packing application is to determine the most efficient way of packing all the user's boxes. We surveyed approximation algorithms for some well-known and very natural combinatorial optimization problems, such as minimum set covering, minimum vertex covering, maximum set packing, and maximum independent set problems. The algorithm we derived for the application is an offline algorithm that achieves this goal by performing two steps. First, it sorts the boxes by volume, from largest to smallest. It then uses a greedy algorithm and packs the boxes in sorted order, using dynamic programming to determine the best position for each box. The result is an optimal solution for packing all the boxes.

- *Volume Sorting*

The first step in the algorithm is to sort the boxes by volume from largest to smallest. Since the boxes are then packed in sorted order, this causes large boxes to be packed first. This conserves space because larger boxes tend to have a larger top surface area than smaller boxes so packing them first allows more stacking to occur because the smaller boxes can be packed on top of them. Additionally, we assume that the larger boxes are heavier than the smaller boxes, so packing the larger boxes first will prevent a heavier box from being packed on top of a lighter one. The second step in the algorithm is where the packing is actually performed. A greedy algorithm is used, so each box is packed in the currently optimal place. Since each box could potentially be in many places, we use dynamic programming to determine.

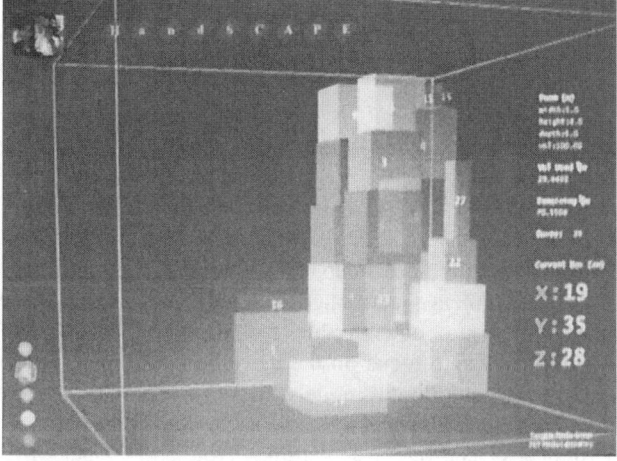

Figure 4. Optimal packing configuration simulation

- *Determining the Best-Fit Position*

The given space is first broken up into *height* number of floors, where *height* is the height of the space. The algorithm then tries to place the box on each floor by starting from the back right corner and moving outwards, checking at each position whether or not there is space to place the whole box. Once a space is found, the placement is recorded and the algorithm attempts to place the box on the next floor. It does this for each floor and at the end, all possible placements for the box are compared, and the one that minimizes the amount of space used is the placement that used for the box. This is done for each box, in sorted order. The result is that each box is placed in an optimal position and the user can then efficiently pack his or her boxes according to the consequent configuration.

2. Collecting Measurements On-Site Excavations

The domain of archaeological field excavation promotes the comprehensive and interdisciplinary study of the human past. The goal is to preserve the great quantity of irreplaceable information associated with archeological excavations. For archeologists, there is the added responsibility of taking primary field data — the innumerable photographs, maps, drawings, and notebooks that make up the archeological excavation record [3]. Therefore, accurate recordings of the field datum of the specimens are crucial at the on-site excavations.

Figure 5. Three units in the process of excavation (above). Measuring the level and angle in a dig (right) [3]. Photo Courtesy of 1999 © Cotsen Institute of Archaeology at UCLA.

By taking all data on the site, a more dynamic and complex consideration of archaeological data results, and much of the primary data is returned to the archeological community for further study. The goal of field measuring also becomes increasingly important in a large deposit because having to take the same repetitive set of measurements on hundreds of specimens each day quickly becomes so tedious that even the most conscientious workers may become careless. In order to reduce innumerous measurement errors, positioning measurements in the field needs a good measuring system to facilitate the accurate determination of the position of any object.

Existing Field Measurements

To achieve the least error, archaeologists traditionally used a primary point for the vertical coordinate as a surface to the ground and lay out a grid system on which lines are no more than a meter apart. The excavators are locating an artifact with respect to two walls of the unit using the two tape measures laying in the pit. They are also using a third tape measure and a level string to measure depth. The level string is being stretched diagonally across the pit from the "candy cane" - a red and white metal pin near the bottom corner of the dig is set at a certain depth and does not move during the course of the excavation. (Fig. 5, above: Morse Point, Santa Cruz Island, Ca.). There is also a technique to measure the angle and position of the placement with a rigid level. It also consists of a carriage for the level bubble, and a pivot bracket to adjust the angle of the bar (Fig. 5, right). Despite various efforts and other on-site measuring techniques, it is difficult to reduce chances of error. Besides accuracy, the principal factors in choosing a measuring technique should be speed and possibility of error [3]. These objectives are way for a new measuring technology.

New Approach
Regarding the accuracy and speed of measuring excavation measurement in a field deposit, the awareness of orientation and position of artifacts is a key in getting surface data informatively. As an orientation-aware measuring device, HandSCAPE provides a way for the user to get an accurate relative location of the excavated fossil as well as a visualization of the artifacts found as the archaeologists dig further and further down. This information is determined using the vectors produced by HandSCAPE measurements.

- *Positioning accurate location of artifacts*

HandSCAPE first begins by drawing out a rectangular area that is to be excavated. Once an artifact is found, HandSCAPE then sketches a rectangle that encircles the artifact. The corners of the original rectangular area can then be used as an anchor point in positioning the new artifact. This is simply done by using HandSCAPE to measure the distance between one corner of the outer rectangle and one corner of the outlining rectangle.

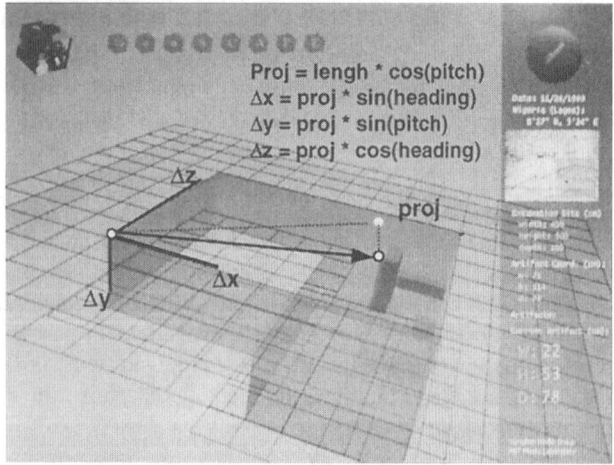

Figure 6. Accurate positioning of the location of an artifact in an archaeological excavation, on-screen visualization.

Taking the linear *length*, along with the *pitch* and *heading* calculations (discussed in implementation section) the user can determine the relative position of the artifact in a dig. Once the relative position of the artifact has been determined, two simple measurements in the x and z direction creates the small rectangular area outlining the artifact. One last measurement of the height of the object completes the representative cube of the artifact.

The remote computer displays the cube on computer screen (Fig. 6) in which the artifact is located within an outer three-dimensional box representing the area that has been excavated. As the excavator digs deeper, the outer box will become larger and any additional artifacts that are found will also be represented as a small box within the space. The overall result, therefore, is a 3D visualization of the area excavated containing accurate representations in terms of size and position of any artifacts that were found in that area. In addition to storing measurement data on a computer, HandSCAPE is also able to incorporate geographical and historical reference data while archaeologists perform the excavations. It enhances and verifies innovative archeological research with dynamic and interactive on-site data interpretation.

3. Modeling Architectural Interior Surfaces
Traditionally people have represented architectural interior surfaces on paper by measuring and drawing approximate sketches of the measurement. When the user needs to create a 3D model of the space for purposes of construction and furniture allocation, the usual approach involves typing all the measurements into a computer-modeling program to visualize the space. With this methodology, these two tasks (measuring and modeling) are always performed separately.

 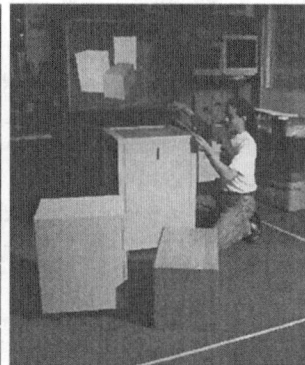

Figure 7. Measuring physical objects and generating the corresponding 3D models

To overcome the need to abstract physical space into units of measurement and subsequently translate those measurements to units usable by graphics, HandSCAPE allows the user to focus on the task of measurement alone when generating digital models. Moreover, it is very complicated to model relations between the objects and spaces without measuring orientation. Note the fact that orientation measure is even more complicated if modeling requires accurate visualization of a physical space that contains several objects.

Scenario: Primitive Modeling of Interior Spaces
An interior designer steps into a room containing several pieces of furniture that he or she wishes to model. The primary interaction involves taking measurements of these objects and the distances between them. Once the user measures an object, its representative three-dimensional model with corresponding vectors immediately appears on the host computer in real-time. The coordinate defines the

endpoint of the vector that originates from the base of the tape measure and passes through this point in the three dimensional space. Now the user measures the vector, (x, y, z) (r, θ, ϕ) between the first and second objects. The user can measure multiple physical objects in the space. It is also possible to make a procedure such that the user can capture the measurement vectors in any arbitrary order as natural as we used to perform.

IMPLEMENTATIONS

HandSCAPE is a single unit consisting of a measuring tape along with custom sensing electronics located on a printed circuit board. It communicates through a RF signal to perform graphics rendering with *TGS Open Inventor*[TM] and using customized Visual C++ programs.

Figure 8. System Components

To increase the mobility of the system, we especially implemented the whole system on a 400Mhz laptop (Fig. 8). The digital model generated by the measuring input is displayed on the computer screen and made available for manipulation by a keyboard and a mouse.

• *Data Processing*

Technically, the digital tape measure tracks the length of the measuring tape by means of a linear optical encoder. The handheld electronics module also includes a two-axis micro-machined accelerometer made by Analog Devices. The accelerometer acts as a tilt sensor that indicates the displacement of the HandSCAPE device from the horizontal plane. To measure the final degree of freedom, rotation about the vertical axis (heading), we have used a three-axis magnetometer responsive to the Earth's magnetic field in its three sensing axes. The Microchip PIC[TM] controller, PIC16C715 compiles the sensor data described above and transmits it through a RF interface to a host computer.

• *Wireless Communication*

To increase the mobility of HandSCAPE for on-site applications, wireless Radio Frequency communication is employed. The RF unit is composed of two bi-directional parts, an on-board RF transmitter/receiver and an external RF receiver/transmitter base unit which communicates with the host computer via a RS-232 serial interface. A cyclic redundancy check has been incorporated to ensure accurate data transmission between HandSCAPE and the base unit. The RF unit utilizes the ultra-compact, low-cost LC-series transmitter and receiver, communicating at 315, 418, 433MHz from Linx Technologies. The range of this communication is 30 feet to 50 feet, based on the magnetic interference on the site. Using RF technology also allows

HandSCAPE to handle multiple users with just one base unit since the data from each handheld unit can be tagged and processed accordingly by the host computer.

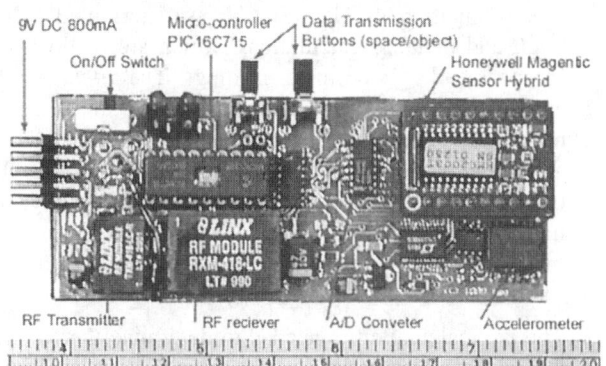

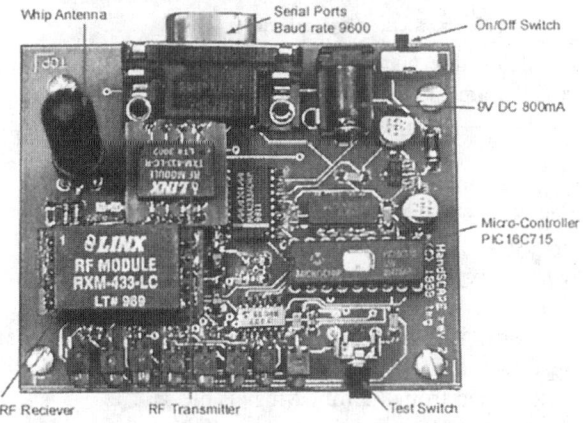

Figure 9. Handheld Board (9 x 4 x 1.5 cm, top) and Base Unit (6 x 8 x 2 cm, bottom)

Vector Calculations

Although the vector is a quantitative description of the physical dimension, the task of estimation frequently does not require this numerical abstraction. In actual measuring tasks therefore, we measure linear distances with a series of vectors to recognize the volume of objects and spaces. Thus, orientation awareness is a significant contribution to the demand of measuring tasks. This embedded function enhances the capability of the hand-held tool effectively.

• *Coordinate System*

Azimuth information can be extracted from its readings much in the same way as a digital compass. We chose to represent 3D orientation in spherical coordinates, as it was the most natural choice given the sensor data. Note that r, the radius, is obtained from the length of the tape, θ is produced by

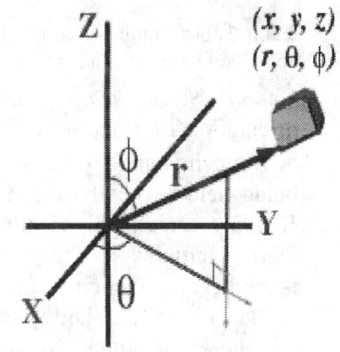

Figure 9. Coordinate System

the magnetometer data, and ϕ is derived from the accelerometer data.

• Pitch and Heading

The orientation of HandSCAPE is determined from the *pitch* and *heading* calculations, which are made from the compass and accelerometer readings. The *pitch* calculation is how much HandSCAPE is being tilted from the horizontal. The *heading* calculation corresponds to what angle, with respect to magnetic north, HandSCAPE is being held. The compass and accelerometer readings are all normalized to 512. The *pitch* calculation is the inverse *sin* of the normalized x reading from the accelerometer:

```
pitch = sin-1(ax_norm)
```

The *roll* calculation is needed to determine the heading. It is simply the inverse sin of the normalized y reading from the accelerometer:

```
roll = sin-1(ay_norm)
```

To determine *heading*, the normalized horizontal compass x and y are first calculated, using the following

```
cx_norm_horz = cx_norm * cos(pitch) +
     cy_norm * sin(roll) * sin(pitch);
cz_norm * cos(roll) * sin(pitch);
cy_norm_horz = cy_norm * cos(roll) +
     cz_norm * sin(roll);
```

where cx_norm is the normalized x reading from the compass, cy_norm is the normalized y reading. and cz_norm is the normalized z reading. In all cases, the readings are normalized to 512. The *heading* is then determined from the following [Table 1].

cx_norm_horz	cy_norm_horz	Heading
0	< 0	$\pi/2$
0	> 0	$3\pi/2$
< 0	Anything	$\pi - \tan^{-1}$(cy_norm_horz/ cx_norm_horz)
	< 0	$-\tan^{-1}$(cy_norm_horz/ cx_norm_horz)
> 0	≥ 0	$2\pi - \tan^{-1}$(cy_norm_horz/ cx_norm_horz)

Table 1. Determination of heading from normalized horizontal x and y compass calculations

Once *pitch* and *heading* have been calculated, the current orientation of HandSCAPE can be determined. *Pitch* is how much HandSCAPE is tilted. If it is at $\pi/2$ or $-\pi/2$, it is being held vertically and a *y* measurement is being read. *Heading* corresponds to how much HandSCAPE is turned. During initialization, the room heading is calculated. This is the heading that corresponds to a perfect *x* measurement. If the difference between the room heading and the calculated *heading* is greater than $\pi/4$, then a *z* measurement is being made. Otherwise, the reading is of a *x* measurement. Thus, with the *pitch* and *heading* calculations, HandSCAPE can determine its orientation and therefore determine which dimension of the box is being measured.

Relative Positioning 3D Objects

• Generating 3D models

The room modeling application involves measuring the contents of a room as boxes in their correct orientations and relative positions. The orientation of a box can easily be determined from the heading of the *x* measurement. The angle a box is turned is the difference between the heading of the room and the heading of the *x* measurement.

```
rotateAngle = heading - room_heading;
```

The placement of a box depends on its relative positions to the box that was measured before it. Each distance measurement is taken from the back right corner of the previous box to the front left corner of the next box. We can again use the *pitch* and *heading* calculations (discussed in Implementation section) to determine the relative position of the new box.

```
x_translation= length x cos(heading -
     room_heading);
y_translation= lengh x sin(pitch);
z_translation= lengh x sin(heading
     room_heading);
```

x_translation and z_translation are the distances between the two corners in the x and z direction respectively. y_translation is how much higher (or lower) the corner of the new box is from the corner of the old. From these three calculations, we can determine the position of the new box with respect to the box measured before it.

Modeling a room begins by first determining the room heading, for later orientation calculations. The first box is then measured, then the distance between the first and second, then the second, etc. By going through the room and measuring all the boxes and the distances between them, we can rapidly model the room. Later on, the user can move to the computer and manipulate the objects in digital space. For example, he/she can rotate the model in order to gain a greater understanding of the spatial relationships through views from different angle. The user can also simulate object allocation and space configuration.

• Object mode vs. Space mode

HandSCAPE generates a series of vectors, which can be used to produce a digital representation of the physical object. The vectors generate the correctly oriented models in accurate relative positions to each other. A frame of reference is established for each new object in relation to objects that have already been measured. As long as each set of measurements originates from a certain reference point in both physical and virtual space, and this relationship is consistently observed, the objects will be modeled correctly relative to each other. Each vector is tagged as an *object* or *space* measurement by the micro-controller, selectable by a button on the device. In *object* mode, the vectors are taken to be *the dimensions* of a parallelepiped, which represents the object being measured. In *space* mode, the vector is taken to indicate *the spatial relationship* between objects, and serves as the reference point for the next object to be measured. Whenever a new measurement is transmitted to the computer, HandSCAPE emits an audible tone through the onboard buzzer and light signals so that the user recognizes the data transmission.

EXPLORATORY USAGE RESULTS

We exhibited HandSCAPE in the Emerging Technology Pavilion at SIGGRAPH'99, demonstrating the optimal packing application (Fig.3). We let hundreds of users from diverse research backgrounds use HandSCAPE. It was very easy to inform them about how the device works, because they were familiar with using measuring tapes in everyday life, and the size and shape of HandSCAPE is that of a standard tape measure. Secondly, most users were surprised at the tight coupling of visual cues with transmission of data. The interaction was very natural and users were convinced of the efficiency in speed and accuracy of what they were measuring.

Feedback

By running the optimal space configuration software, users could clearly see that the current box they measured was packed in the best fit position just as it would be in normal real world optimal packing. We proved that HandSCAPE was working in both hardware and software for the application. Although we only presented the packing optimization application, users responded well to adopting the device to various other uses. They even mentioned themselves that HandSCAPE is a very practical device that could be applied immediately to certain real world needs. Some users asked us about the capability of HandSCAPE to measure not only straight lines, but also curves.

Comparative Experiment

We conducted a simple experiment to study two factors of interest, *speed* and *accuracy*. Each subject performed the experiment using two different tape measures --one being HandSCAPE and the other being a normal measuring tape. We asked 15 non-professional users to perform a 3D modeling task, (i.e. measuring a box and modeling it on the computer screen). Using the normal tape measure, users followed the standard approach of entering the measuring value into the modeling program to visualize the box. With HandSCAPE, on the other hand, users simply needed to measure the three dimensions of the box and its visualization immediately appeared on the screen. Results of this experiment are shown in Table 2. From the data we observed 77% of users were more accurate when they used HandSCAPE. Measuring with HandSCAPE was also an average of 2.1 times faster than measuring with a normal tape.

LIMITING FACTORS and FUTURE WORK

Despite the potential improvements, we have also observed errors caused by pushing the button two or three times repeatedly. These errors usually resulted in graphical errors. We also found that the accuracy of orientation is greatly affected by nearby magnetic fields. Definitely the archaeology and interior design applications need more accurate and precise orientation measurements. For these limiting factors, we have plans to add new features such as add and delete buttons to correct the transmission. To increase the accuracy of the orientation measurements, we plan to explore another technology to improve the current sensing hardware.

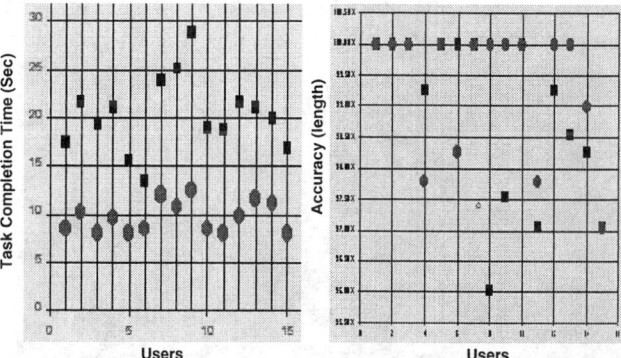

Table 2. User data showing subject accuracy and completion time without HandSCAPE (■) and with HandSCAPE (●).

We have also learned from each application how generally useful HandSCAPE is not only in hardware but also in software. For instance, archaeologists mentioned that we needed to give more control over the visualization on computer screen of the "pointclouds" – i.e. 3D positioning visualization of artifacts inside of the dig. Therefore we plans to adjust the software interface. Another challenge is to explore modeling curves and manipulating surfaces by integrating a high dimensional input device [1]. We will look carefully into the possibility since the aim of this research is for existing real world applications and user population. We will look more closely at the applications and reflect the needs in our future direction.

DISCUSSION

At the beginning of this paper we proposed a new interface design for addressing real world needs. By examining the comparative efficiency issues on each on-site application, we were able to verify our work in certain intuitive observation:

- *Value of traditional interaction techniques:* Without the need to learn and adjust to a new interface, making use of existing physical skills preserves the human senses and interactions that are commonly employed in everyday life [6], [9], [14]. In this case, feedback from physical tasks can have an immediate influence on the user's activities. Note that HandSCAPE knows what the user is measuring in real-time. This real-time feedback reinforces the means of understanding physical space in the digital domain, not just through measurements but also perceptual, tactile, and kinesthetic feedback in physical environments. In addition two-handed interactions in measuring use the natural way of using motion to recognize 3D spaces and object.

- *Application-specific:* As opposed to designing a general-purpose interface, there seems to be a clear advantage in designing interfaces that are very application-specific, [2],

[7], [13], [15]. Bridging the gap between existing measuring techniques and new features of digital modeling functions provide a comprehensive interaction and enhances the consistency of the measuring workload incorporated with the modeling tasks on-site. This particular interaction technique examines designing interfaces to solve the real world problems in both measuring and modeling that has always been separated.

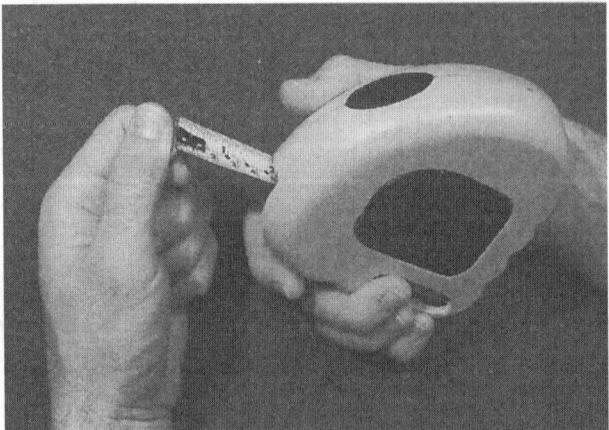

Figure 10. The HandSCAPE in two hands

CONCLUSIONS
We have presented HandSCAPE, a computer augmented measuring tape for digitizing filed measurements. By using orientation-sensing technology, the device can provide the field workers with access to the efficiency of complex measuring tasks such as loading, packing, locating, and configuration in the manner of traditional measuring techniques. Although HandSCAPE an input device is not designed to the most known 3D modeling and animation applications, we believe that modeling primitive 3D objects cooperated with measuring for the specific applications enhances the efficiency of real word measuring tasks.

In the beginning, HandSCAPE attempted to improve the primary concerns of complex measuring tasks along the digital functionality. However the result from combining measuring and modeling as a single seamless step without any extra effort beyond measuring presents exciting avenues for improving comparative efficiency of traditional on-site measuring tasks. Through this project, we realize that augmenting digital functionality on top of familiar tools increases the efficiency of real world tasks that develop with human senses and skills.

ACKNOWLEDGEMENT
We thank Rujira Hongladaromp, Blair Dunn, James Hsiao for designing prototypes. We also thank Rich Fletcher, Paul Yarin and Joe Paradiso who provide an initiative technical advice on the result. We would like to thank Joseph Branc at Steelcase Inc., who encourages us in design with valuable real world comments and Dr. Louise Krasniewicz in the Cotsen Institute of Archaeology at UCLA, for the discussion on the archaeological application. Finally we thank the members of the Tangible Media Group and our colleagues of the Things That Think consortium at the MIT Media Lab for their support of this research.

REFERENCES
[1] Balakrishnan, R., Fitzmaurice, G.W. and Kurtenbach, G., Singh, K. (1999) Exploring Interactive Curve and Surface Manipulation Using a Bend and Twist Sensitive Input Strip, In Proceedings of ACM Symposium on Interactive 3D Graphics (I3DG'99), ACM Press, pp. 111-118

[2] Balakrishnan, R., Fitzmaurice, G.W. and Kurtenbach, G., Buxton W. (1999) Digital Tape Drawing, In Proceedings of ACM Symposium on User Interface Software and Technology, UIST'99, pp. 161-169,

[3] Brien D. Dillon (eds.) (1993) Practical Archaeology: Filed and Laboratory Techniques and Archaeological Logistics, Institute of Archaeology, University of California Press: Los Angeles, pp. 33-38

[4] Digibot3™ *Product Information*, Digibotics, Inc. (http://www.digibot.com)

[5] Falconer, P., drury, J. (1975) Building and Planning for Industrial Storage and Distribution, The Architectural Press: London, Introduction

[6] Fitzmaurice, G.W., Ishii, H., and Buxton, W. (1995). Bricks: Laying the foundation for Graspable User Interfaces. In Proceedings of CHI'95, pp. 422-449.

[7] Hinckley, K., Pausch, R., Proffitt, D., and Kassell, N. (1998) Two-Handed Virtual Manipulation, ACM Transactions on Computer–Human Interaction, 260-362

[8] Igarashi, T., Matsuoka, S., and Tanaka, H. (1999) "Teddy; A Sketching Interface for 3D Freeform Desing, In Proceeding of SIGGRAPH'99, pp. 409-416

[9] Ishii, H. and Ullmer, B. (1997) Tangible Bits: Toward seamless Interfaces between People, Bits, and Atoms, in *proceeding of CHI '97*, ACM Press, pp. 234-241

[10] Monkey2™. Product Information, Digital Image Design Incorporated, New York. (http://www.didi.com)

[11] Polhemus 3Ball™, *Product Information*, Polhemus Inc. , Chchester, VT (http://www.polhemus.com)

[12] SHAPE TAPE ™, *Product Information,* Measurand Inc., Fredericton, New Brunswick, Canada. (http://www.measurand.com)

[13] Underkoffler J. and Ishii, H. (1999) Urp: A Luminous-Tangible Workbench for Urban Planning and Design, In *Proceedings of the CHI '99*, ACM Press, pp. 386-393

[14] Weiser, M. "The Computer for the 21 Century." In *Scientific American*, 265 (3), pp. 94-104

[15] Yarin, P. and Ishii, H. (1999) TouchCounters: Designing Interactive Electronic Labels for Physical Containers, in *Proceeding of the CHI '99*, ACM Press, pp. 362-369

[16] Zeleznik R.C., Herndon K.P., and Hughs J.F (1996) SKETCH: An Interface for Sketching 3D Scenes", In *Proceedings of SIGGRAPH '96,* pp. 163-170

Bringing Order to the Web:
Automatically Categorizing Search Results

Hao Chen
School of Information Management & Systems
University of California
Berkeley, CA 94720 USA
hchen@sims.berkeley.edu

Susan Dumais
Microsoft Research
One Microsoft Way
Redmond, WA 99802 USA
sdumais@microsoft.com

ABSTRACT
We developed a user interface that organizes Web search results into hierarchical categories. Text classification algorithms were used to automatically classify arbitrary search results into an existing category structure on-the-fly. A user study compared our new category interface with the typical ranked list interface of search results. The study showed that the category interface is superior both in objective and subjective measures. Subjects liked the category interface much better than the list interface, and they were 50% faster at finding information that was organized into categories. Organizing search results allows users to focus on items in categories of interest rather than having to browse through all the results sequentially.

Keywords
User Interface, World Wide Web, Search, User Study, Text Categorization, Classification, Support Vector Machine

INTRODUCTION
With the exponential growth of the Internet, it has become more and more difficult to find information. *Web search* services such as AltaVista, InfoSeek, and MSNWebSearch were introduced to help people find information on the web. Most of these systems return a ranked list of web pages in response to a user's search request. Web pages on different topics or different aspects of the same topic are mixed together in the returned list. The user has to sift through a long list to locate pages of interest. Since the 19th century, librarians have used classification systems like Dewey and Library of Congress classification to organize vast amounts of information. More recently, *Web directories* such as Yahoo! and LookSmart have been used to classify Web pages. The manual nature of the directory compiling process makes it impossible to have as broad coverage as the search engines, or to apply the same structure to intranet or local files without additional manual effort.

To combine the advantage of structured topic information in directories and broad coverage in search engines, we built a system that takes the web pages returned by a search engine and classifies them into a known hierarchical structure such as LookSmart's Web directory [24]. The system consists of two main components: 1) a text classifier that categorizes web pages on-the-fly, and 2) a user interface that presents the web pages within the category structure and allows the user to manipulate the structured view (Figure 1).

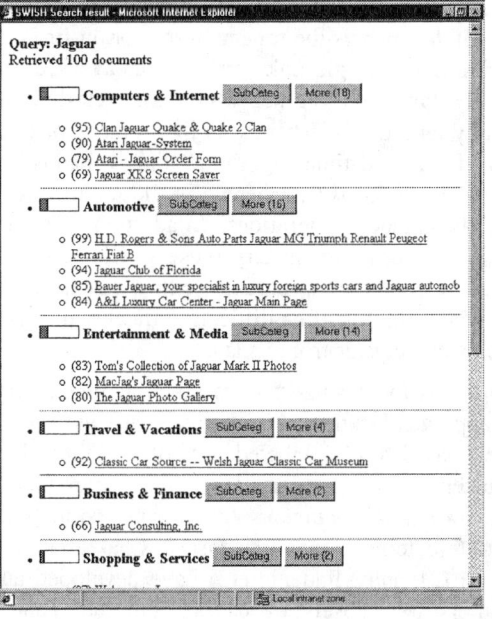

Figure 1: Presenting web pages within category structure

RELATED WORK
Generating structure
Three general techniques have been used to organize documents into topical contexts. The first one uses *structural information* (meta data) associated with each document. The DynaCat system by Pratt [15] used meta data from the UMLS medical thesaurus to organize search results. Two prototypes developed by Allen [1] used meta data from the Dewey Decimal System for organizing results. In the SuperBook project [10], paragraphs of texts were organized into an author-created hierarchical table of contents. Marchionini et al. [12] also used table of content

views for structuring information from searches in the Library of Congress digital library. Others have used the link structure of Web pages to automatically generate structured views of Web sites. Maarek et al.'s WebCutter system [11] displayed a site map tailored to the user's search query. Wittenburg and Sigman's AMIT system [18] showed search results in the context of an automatically derived Web site structure. Chen et al.'s Cha-Cha system [4] also organized search results into automatically derived site structures using the shortest path from the root to the retrieved page. Manually-created systems are quite useful but require a lot of initial effort to create and are difficult to maintain. Automatically-derived structures often result in heterogeneous criteria for category membership and can be difficult to understand.

A second way to organize documents is by *clustering*. Documents are organized into groups based their overall similarity to one another. Zamir et al. [19, 20] grouped Web search results using suffix tree clustering. Hearst et al. [7, 8] used the scatter/gather technique to organize and browse documents. One problem with organizing search results in this way is the time required for on-line clustering algorithms. Single-link and group-average methods typically take $O(n^2)$ time, while complete-link methods typically take $O(n^3)$, where n is the number of documents returned. Linear-time algorithms like k-means are more efficient being $O(nkT)$, where k is the number of clusters and T the number of iterations. In addition, it is difficult to describe the resulting clusters to users. Clusters are usually labeled by common phrases extracted from member documents, but it is often difficult to quickly understand the contents of a cluster from its label.

A third way to organize documents is by *classification*. In this approach, statistical techniques are used to learn a model based on a labeled set of training documents (documents with category labels). The model is then applied to new documents (documents without category labels) to determine their categories. Chakrabarti et al. [2], Chekuri [3], and Mladenic [13] have developed automatic classifiers for subsets of pages from the Yahoo! Web directory. Only a small number of high level categories were used in their published results. And, the focus of these papers was on the underlying text classification algorithms and not on user interfaces that exploit the results. Recently, Inktomi [22] announced that it had developed techniques for automatic classification of web pages. However, its technical details were not disclosed and we are not aware of any search services employing this technology.

Using structure to support search
A number of web search services use category information to organize the search results. Yahoo! [27], Snap [26] and LookSmart [24] show the category label associated with each retrieved page. Results are still shown as a ranked list with grouping occurring only at the lowest level of the hierarchy (for Yahoo! and Snap). There is, for example, no way to know that 70% of the matches fell into a single top-level category. In addition, these systems require pre-tagged content. Before any new content can be used, it must be categorized by hand. Northern Light [25] provides Custom Folders in which the retrieved documents are organized hierarchically. The folders are organized according to several dimensions -- source (sites, domains), type (personal page, product review), language, and subject. Individual categories can be explored one at a time. But, again no global information is provided about the distribution of search results across categories.

The most common interface for manipulating hierarchical category structures is a hierarchical tree control, but other techniques have been explored as well. Johnson et al. [9] used a treemap that partitioned the display into rectangular bounding boxes representing the tree structure. Characteristics of the categories and their relationships were indicated by their sizes, shapes, colors, and relative positions. Shneiderman et al. [17] have recently developed a two-dimensional category display that uses categorical and hierarchical axes, called hieraxes, for showing large results sets in the context of categories. Hearst et al. [5] used three-dimensional graphics to display categories together with their documents. Multiple categories could be displayed simultaneously along with their hierarchical context. In all of these systems, documents must have pre-assigned category tags.

Few studies have evaluated the effectiveness of different interfaces for structuring information. Landauer et al. [10] compared two search interfaces for accessing chemistry information -- SuperBook which used a hierarchical table of contents, and PixLook which used a traditional ranked list. Browsing accuracy was higher for SuperBook than PixLook. Search accuracy and search times were the same for the two interfaces. However, different text pre-processing and search algorithms were used in the two systems so it is difficult to compare precisely. More recently, Pratt et al. [16] compared DynaCat, a tool that automatically categorized results using knowledge of query types and a model of domain terminology, with a ranked list and clustering. Subjects liked DynaCat's category organization of search results. Subjects found somewhat more new answers using DynaCat, but the results were not reliable statistically, presumably because there were only 15 subjects and 3 queries in the experiment.

In this paper we describe a new system showing how automatic text classification techniques can be used to organize search results. A statistical text classification model is trained offline on a representative sample of Web pages with known category labels. At query time, new search results are quickly classified on-the-fly into the learned category structure. This approach has the benefit of using known and consistent category labels, while easily incorporating new items into the structure. The user interface compactly displays web pages in a hierarchical category structure. Heuristics are used to order categories and select results within categories for display. Users can further expand categories on demand. Tooltip-like overlays are used to convey additional information about individual web pages or categories on demand. We compared our category interface with a traditional list interface under exactly the same search conditions. We now describe each of these components in more detail.

TEXT CLASSIFICATION

Text classification involves a training phase and a testing phase. During the training phase, web pages with known category labels are used to train a classifier. During the testing or operational phase, the learned classifier is used to categorize or tag new web pages.

Data Set

For training purposes, we used a collection of web pages from LookSmart's Web directory [24]. LookSmart's directory is created and maintained by 180 professional Web editors. For our experiments, we used the directory as it existed in May 1999. At that time there were 13 top-level categories, 150 second-level categories, and over 17,000 categories in total. On average each web page was classified into 1.2 categories.

Pre-processing

A text pre-processing module extracted plain text from each web page. In addition, the title, description, keyword, and image tag fields were also extracted if they existed. A vector was created for each page indicating which terms appeared in that page.

The results returned by search engines contain a short summary of information about each result. Although it is possible to download the entire contents of each web page, it is too time consuming to be applicable in a networked environment. Therefore, in our prototype, the initial training and subsequent classification are performed using only summaries of each web page. The training summaries were created using the title, the keyword tag, and either the description tag if it existed or the first 40 words otherwise. When classifying search results we use the summary provided in the search results.

Classification

A Support Vector Machine (SVM) algorithm was used as the classifier, because it has been shown in previous work to be both very fast and effective for text classification problems [5][14]. Roughly speaking, a linear SVM is a hyperplane that separates a set of positive examples (i.e., pages in a category) from a set of negative examples (i.e., pages not in the category). The SVM algorithm maximizes the margin between the two classes; other popular learning algorithms minimize different objective functions like the sum of squared errors. Web pages were pre-processed as described above. For each category we used the 1000 terms that were most predictive of the category as features. Vectors for positive and negative examples were input into the SVM learning algorithm. The resulting SVM model for each category is a vector of 1000 terms and associated weights that define the hyperplane for that category.

We used 13,352 pre-classified web pages to train the model for the 13 top-level categories, and between 1,985 and 10,431 examples for each of these categories to train the appropriate second-level category models. The total time to learn all 13 top-level categories and 150 second-level categories was only a few hours. Once the categories are learned, the results from any user query can be classified. At query time, each page summary returned by the search engine is compared to the 13 top-level category models. A page is placed into one or more categories, if it exceeds a pre-determined threshold for category membership. Pages are classified into second-level categories only on demand using the same procedure.

We explored a number of parameter settings and text representations and used the optimal ones for classification in our experiment. Our fully automatic methods for assigning category labels agreed with the human-assigned labels almost 70% of the time. Most of the disagreements were because additional labels were assigned (in addition to the correct one), or no labels were assigned. This is good accuracy given that we were working with only short summaries and very heterogeneous web content. Although classification accuracy is not perfect, we believe it can still be useful for organizing Web search results.

USER INTERFACE

The search interface accepted query keywords, passed them to a search engine selected by the user, and parsed the returned pages. Each page was classified into one or more categories using the learned SVM classifier. The search results were organized into hierarchical categories as shown in Figure 1. Under each category, web pages belonging to that category were listed. The category could be expanded (or collapsed) on demand by the user. To save screen space, only the title of each page was shown (the summary can be viewed by hover text, to be discussed later). Clicking on the title hyperlink brought up the full content of the web page in another browser window, so that the category structure and the full-text of pages were simultaneously visible.

Information Overlays

There is a constant conflict between the large amount of information we want to present and the limited screen real estate. We presented the most important information (titles of web pages and category labels) as text in the interface, and showed other information using small icons or transient visual overlays. The techniques we used included:

- A partially filled green bar in front of each category label showed the percentage of documents falling into the category. This provided users with an overview of the distribution of matches across categories.

- We presented additional category information (parent and child category labels) as hover text when the mouse hovered over a category title. This allowed users to see the subcategories for category as well as the higher-level context for each page.

- The summaries of the web pages returned by search engines provide users with additional information about the page helping them decide which pages to explore in greater depth. In order to present category context along with the search results, we displayed only titles by default and showed summaries as hover text when the mouse hovered over the titles of web pages.

Distilled Information Display

Even with the help of information overlays, there is still more information than a single screen can accommodate. We developed heuristics to selectively present a small portion of the most useful information on the first screen. The first screen is so important that it usually determines whether the user will continue working on this search or abandon it all together. We wanted to enable the user to either find the information there or identify a path for further exploration. In order to do this effectively we must decide: how many categories to present, how many pages to present in each category, how to rank pages within a category, and how to rank categories.

We presented only top-level categories on the first screen. There were several reasons for this. First, the small number of top level categories helped the user identify domains of interest quickly. Second, it saved a lot of screen space. Third, classification accuracy was usually higher in top level categories. Fourth, it was computationally faster to match only the top-level categories. Fifth, subcategories did not help much when there were only a few pages in the category. The user can expand any category into subcategories by clicking a button.

In each category, we showed only a subset of pages in that category. We decided to show a fixed number of pages (20) across all categories, and divided them in proportion to the number of pages in that category. So, if one category contained 50% of results, we would show 10 pages from that category in the initial view. The user can see all pages in a category by clicking a button.

Three parameters affected how pages are ordered within a category: its original ranking order in the results, its match score (if returned by the search engine), and the probability that it belongs to the category according to the classifier. For the experiment, we used only the rank order in the original search results to determine the order of items within each category. Thus if all the search result fall into one category the category organization returns the same items in the same order as the ranked list.

The categories can be ordered either in a static alphabetical order, or dynamically according to some importance score. The advantage of dynamic ranking is to present the most likely category first. The disadvantage is that it prevents the user from establishing a mental model of the relative position of each category in the browser window. For our experiment, importance was determined by the number of pages in the category. The category with the most items in it was shown first, and so on.

USER STUDY

A user study was conducted to compare the category-based interface (referred to as "Category Interface" henceforth) with the conventional search interface where pages are arranged in a ranked list (referred to as "List Interface" henceforth). The two interfaces are shown in Figure 2.

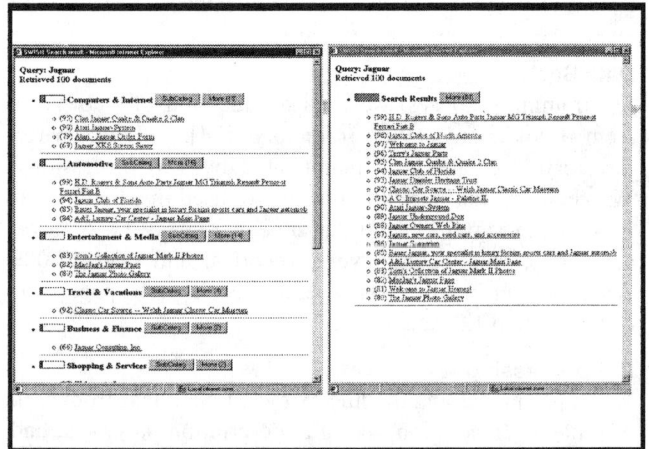

Figure 2: Category vs. List Interface

The top 100 search results for the query *"jaguar"* are used in this example. Twenty items are shown initially in both interfaces. In the List interface the 20 items can be seen without scrolling; in the Category interface scrolling is always required in spite of our attempt to conserve screen space. In both interfaces, summaries are shown on hover. Both interfaces contain a ShowMore button which is used to show the remaining items in the category; in the case of the List interface the remaining 80 items are shown. In addition, in the Category interface a SubCategory button is used to sub-categorize the pages within that category. The same control program is used in both cases, so timing is the same in both interfaces.

Methods

Subjects

Eighteen subjects of intermediate web ability participated in the experiment. Subjects were adult residents of the Seattle area recruited by the Microsoft usability lab, and represent a range of ages, backgrounds, jobs and education level.

Procedure

The experiment was divided into two sessions with a voluntary break between. Subjects used the Category interface in one session and the List interface in the other. The user read a short tutorial before each session began. During each session, the user performed 15 web search tasks, for a total of 30 search tasks. At the end of the experiment, the user completed an online questionnaire giving his/her subjective rating of the two interfaces. The total time for the experiment was about 2 hours.

During the experiment, the subject worked with three windows (Figure 3). The control window on the top shows the task and the query keywords. In this example, the *task* is to find out about *"renting a Jaguar car"* and the *query* we automatically issued is *"jaguar"*. The search results were displayed in the left bottom window. In the Category interface, the results were automatically organized into different categories, and in the List interface, the top 20 items were shown on the initial screen.

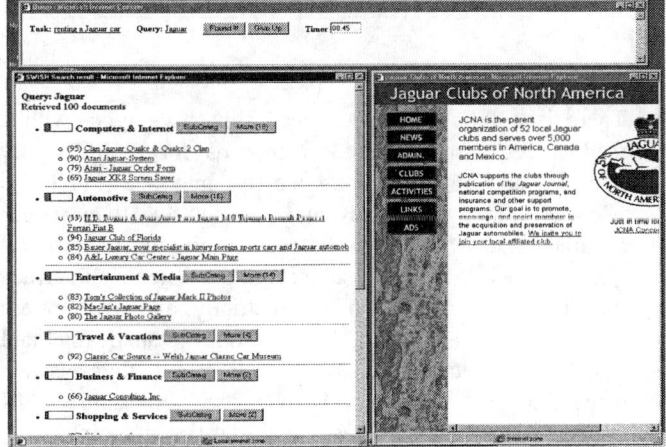

Figure 3: Screen of the User Study

When the subject clicked on a hyperlink, the page opened in the right window. When the subject found an answer, s/he clicked on the "Found It!" button in the control window. If no answer could be found, s/he clicked on the "Give Up" button. There was a timer in the control window that reminded the subject after five minutes had passed. If a reminder occurred, the subject could continue searching or move on to the next task. User events such as hovering over a hyperlink to read the summary, clicking on a hyperlink to read the page, expanding or collapsing the list were logged.

Search Tasks

The 30 search tasks were selected from a broad range of topics, including sports, movies, travel, news, computers, literature, automotive, local interest, etc. Ten of the queries were popular queries from users of MSNWebSearch. In order to facilitate evaluation we selected tasks that had reasonably unambiguous answers in the top 100 returned pages (a kind of known-item search). The tasks varied in difficulty – 17 had answers in the top 20 items returned (on the first page in the List interface), and 13 had answers between ranks 21 and 100. The tasks also varied in how much manipulation was required in the Category interface – 10 required subjects to use ShowMore or SubCategory expansion, and 10 required some scrolling because the correct category was not near the top.

To ensure that results from different subjects were comparable, we fixed the keywords for each query in the experiment. We also cached the search results before the experiments so that each subject got the same results for the same query. The MSNWebSearch engine [22] was used to generate the search results.

Each subject performed the same 30 search tasks. For 15 tasks they used the Category interface and for 15 they used the List interface. The order in which queries were presented and whether the Category or List interface was used first was counterbalanced across subjects. Nine lists of tasks were used -- each list contained all the tasks in a different order and was assigned to a pair of subjects, one in the Category-first condition and one in the List-first condition. This yoking of presentation orders reduces error variance which is desirable given the relatively small number of subjects and tasks we used.

Results

The main independent variable is the Category interface vs. the List interface. The order of presentation (List first or Category first) is a between subject variable. We analyzed both subjective questionnaire measures and objective measures (search time, accuracy, and interactions with the interface such as hovering, and displaying Web pages).

Subjective questionnaire measures

After the experiment, subjects completed a brief online questionnaire. The questionnaire covered prior experience with Web searching, ratings of the two interfaces (on a 7-point scale), and open-ended questions about the best and worst aspects of each interface. Seventeen of the eighteen subjects used the Web at least every week, and eleven of the eighteen subjects searched for information on the Web at least every week. The most popular Web search service among our subjects was Yahoo!.

Subjects reported that the Category interface was "easy to use" (6.4 vs. 3.9, $t(17) = 6.41$; $p<<0.001$), they "liked using it" (6.7 vs. 4.3, $t(17) = 6.01$; $p<<0.001$), they were "confident that I could find the information if it was there" (6.3 vs. 4.4, $t(17) = 4.91$, $p<<0.001$), that it was "easy to get a good sense of the range of alternatives" (6.4 vs. 4.2, $t(17) = 6.22$; $p<<0.001$), and that they "prefer this to my usual search engine" (6.4 vs. 4.3, $t(17) = 4.13$; $p<<0.001$). On all of our overall measures subjects much preferred the Category interface.

For the two questions that asked about the usefulness of interface features (hover text and ShowMore), there were no reliable differences between interfaces, suggesting that subjects did not simply have an overall positive bias in responding to questions about the Category interface. Subjects thought the display of page summaries in hover text was useful in both interfaces (6.5 Category vs. 6.4 List, t(17) = 0.36; p<0.72), and that the ShowMore option was useful (6.5 Category vs. 6.1 List, t(17) = 1.94; p<0.07).

Accuracy/GiveUp.
When creating search tasks, we had a target correct answer in mind. However, other pages might be relevant as well, so we examined all pages that subjects said were relevant to see if they in fact answered the search task. We looked at performance with strict and liberal scoring of accuracy. For strict scoring only pages that were deemed by the experimenters to be relevant (after including additional pages found by subjects that we had missed) were counted as relevant. Using the strict criterion, there were slightly more wrong answers in the List interface (1.72 out of 30) than in the Category interface (1.06 out of 30), but this difference is not reliable statistically using a paired t-test (t(17) = -1.59; p<.13). The lack of difference between interfaces is not surprising, since it reflects a difference in criterion about what the correct answer is rather than task difficulty per se. For liberal scoring, any answer that subjects said was relevant was deemed relevant, so by definition, there were no wrong answers in either interface. We used the liberal scoring in subsequent analyses.

Subjects were allowed to give up if they could not find an answer. They could do this at any time during a trial. After 5 minutes had elapsed for a task, subjects were notified and encouraged to move onto the next task. Some subjects continued searching, but most gave up at this time. There are significantly more tasks on which subjects gave up in the List interface than in the Category interface (t(17) = -2.41; p<.027), although the absolute number of failures is small in both interfaces (0.77 in List and 0.33 in Category).

Search Time
We used the median search time across queries for statistical tests, because reaction time distributions are often skewed and statistical tests can be influenced by outliers. (We also find exactly the same results using mean reaction times, so outliers were not a problem in this experiment.) A 2x2 mixed design was used to measure differences in search time. The between subjects factor is whether subjects saw the List or Category interface first, and the within subjects factor is List or Category interface. Median search times are shown in Figure 4.

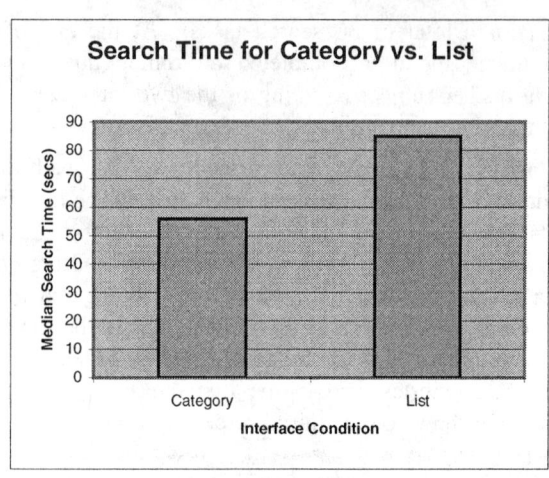

Figure 4: Search time by interface type

There is a reliable main effect of interface type, with a median response time of 56 seconds for the Category interface and 85 seconds for the List interface (F(1,16) = 12.94; p=.002). The advantage is not due to a speed-accuracy tradeoff or to a tendency to give up on difficult queries, since if anything subjects in the Category interface were more accurate (when scored strictly) and gave up less often. This is a large effect both statistically and practically. It takes subjects 50% longer to find answers using the List interface. On average it took subjects 14 minutes to complete 15 tasks with the Category interface, and 21 minutes with the List interface. There is no effect of the order in which interfaces were shown, list first or category first (F(1,16) = 0.26; p=0.62). And, there is no interaction between order and interface (F(1,16) = 1.23; p=0.28), which shows that results are not biased by order of presentation.

There are large individual differences in search time. The fastest subject finished the 30 search tasks in a median of 37 seconds, and the slowest in 142 seconds. But, the advantage of the Category interface is consistent across subjects.

There are also large differences across tasks or queries. The easiest task was completed in a median of 22.5 seconds, and the most difficult task required 166 seconds to complete. We divided the queries into those whose answers were on the first screen of the List interface (i.e., in the Top20 returned by the search engine) and those whose answers were not in the Top20. The search times are shown in Figure 5. Not surprisingly, there is a reliable main effect of whether the answer is in the Top20 or not -- median time for Top20 (57 seconds) and NotTop20 (98 seconds), F(1,56) = 16.5; p<<.001.

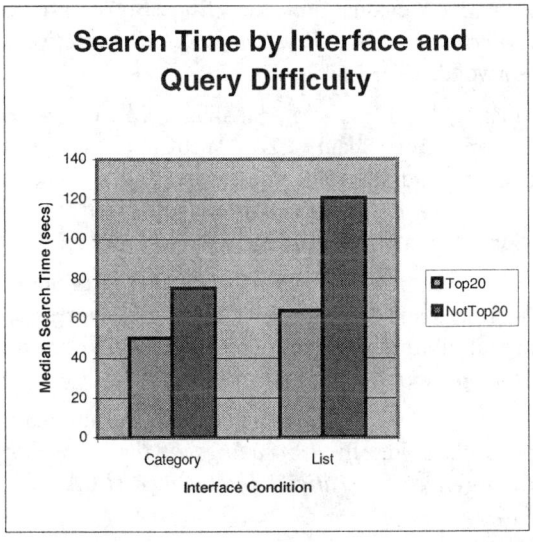

Figure 5: Search time by interface type and query difficulty

There is no interaction between query difficulty and interface (F(1,56)=2.52; p=.12). The Category interface is beneficial for both easy and hard queries. Although there is a hint that the category interface is more helpful for difficult queries, the interaction is not reliable. The Category interface is still beneficial even when the matching web page is in the first page of results. In our List interface items which were in the Top20 did not require any scrolling, whereas several of the Category interfaces for these items did. The advantage appears to be due to the way in which the category interface breaks the list of returned items down into easily scanable semantic chunks.

Interaction Style – Hovering, Page Views, ShowMore, SubCategory

We measured the number of hovering and page viewing actions subjects performed in the course of finding the answers. Subjects in the List interface hovered on more items than those in the Category interface (4.60 vs. 2.99; t(17) = -5.54; p<<0.001). The number of pages that subjects actually viewed in the right window is somewhat larger in the List interface (1.41 List vs. 1.23 Category; t(17) = -2.08; p<0.053). Although the difference is not large, it suggests that the category structure can help disambiguate the summary in the hover text. It is interesting to note that the average number of page views is close to 1, suggesting that users could narrow down their search by reading just the titles and summaries. Subjects read the full pages mostly to confirm what they found in the summary. This significantly reduces search time because the short summaries can be read faster than a full page of text, and there is no network latency for accessing summaries (summaries were stored locally, but retrieving the full-text of the pages required net access).

We also measured the expansion operations that subjects used in searching for information. In the List interface, subjects could expand this list of results by ShowMore. In the Category interface, subjects could ShowMore within each category, or they could break down categories into SubCategories. Overall, subjects in the Category interface used more expansion operations (0.78 ShowMore + SubCategories in Category vs. 0.48 Show More in List; t(17) = 3.54; p<0.003). So, subjects performed more expansion operations in the Category interface, but the selective nature of the operations (i.e., they applied to only a single category) meant that they were nonetheless more efficient overall in finding things.

CONCLUSION

We developed and evaluated a user interface that organizes search results into a hierarchical category structure. Support Vector Machine classifiers were built offline using manually classified web pages. This model was then used to classify new web pages returned from search engines on-the-fly. This approach has the advantage of leveraging known and consistent category information to assist the user in quickly focusing in on task-relevant information. The interface allows users to browse and manipulate categories, and to view documents in the context of the category structure. Only a small portion of the most important and representative information is displayed in the initial screen, and hover text and overlay techniques are used to convey more detailed information on demand. A user study compared the category interface with traditional list interface using the same set of tasks, search engine, and search results. The results convincingly demonstrate that the category interface is superior to the list interface in both subjective and objective measures.

There are many directions for further research. One issue to explore is how the results generalize to other domains and task scenarios. The categories used in our experiment were designed to cover the full range of Web content. Nonetheless, not all user queries will match the category structure to the same extent. Results for some queries may fall entirely within one category (e.g., results for the query *"used parts for Jaguar XJ6L"*, would likely fall entirely within the Automobile category). In such cases, the Category interface (given our current display heuristics) is exactly the same as the List interface, so we are no worse off. Results for other queries may not match any of the categories very well. In our current interface we have a "NotCategorized" group at the bottom. In our experiment 5-40% of the results for each query were NotCategorized, but few of the answers were in the NotCategorized group. We hope to deploy our system more widely to look at this issue by getting a large sample of typical user queries. This would also allow us to explore a wider range of user tasks in addition to the known-item scenario we used.

There are also many interesting issues concerning how best to present concise views of search results in their category contexts. We chose to order categories by the number of matches and within each category to order the pages by search rank. Our text classification algorithms can easily handle thousands of categories, and we may have to move beyond our simple display heuristics for such cases.

ACKNOWLEDGMENTS

We are grateful to John Platt for help with the Support Vector Machine code, to Kirsten Risden for help in setting up the user study, and to reviewers for helpful suggestions.

REFERENCES

1. Allen, R. B., Two digital library interfaces that exploit hierarchical structure. In *Proceedings of DAGS95: Electronic Publishing and the Information Superhighway* (1995).

2. Chakrabarti, S., Dom, B., Agrawal, R., and Raghavan, P. Scalable feature selection, classification and signature generation for organizing large text databases into hierarchical topic taxonomies. *The VLDB Journal 7*, (1998), 163-178.

3. Chekuri, C., Goldwasser, M., Raghavan, P. and Upfal, E. Web search using automated classification. In *Sixth International World Wide Web Conference*, Santa Clara, California, Apr. 1997, Poster POS725.

4. Chen, M., Hearst, M., Hong, J., and Lin, J. Cha-Cha: a system for organizing intranet search results. In *Proceedings of the 2nd USENIX Symposium on Internet Technologies and SYSTEMS (USITS)* (Boulder CO, October 1999) (to appear).

5. Dumais, S. T., Platt, J., Heckerman, D. and Sahami, M. Inductive learning algorithms and representations for text categorization. In *Proceedings of ACM-CIKM98*, Nov. 1998.

6. Hearst, M., and Karadi, C. Searching and browsing text collections with large category hierarchies. In *Proceedings of the ACM SIGCHI Conference on Human Factors in Computing Systems (CHI), Conference Companion* (Atlanta GA, March 1997).

7. Hearst, M., and Pedersen, P. Reexamining the cluster hypothesis: scatter/gather on retrieval results. In *Proceedings of 19th Annual International ACM/SIGIR Conference* (Zurich 1996).

8. Hearst, M., Pedersen, J., and Karger, D. Scatter/gather as a tool for the analysis of retrieval results. *Working Notes of the AAAI Fall Symposium on AI Applications in Knowledge Navigation* (Cambridge MA, November 1995).

9. Johnson, B., and Shneiderman, B. Treemaps: a space-filling approach to the visualization of hierarchical information structures. In *Sparks of Innovation in Human-Computer Interaction*. Ablex Publishing Corporation, Norwood NJ, 1993

10. Landauer, T., Egan, D., Remde, J., Lesk, M., Lochbaum, C., and Ketchum, D. Enhancing the usability of text through computer delivery and formative evaluation: the SuperBook project. In *Hypertext – A Psychological Perspective*. Ellis Horwood, 1993.

11. Maarek, Y., Jacovi, M., Shtalhaim, M., Ur, S., Zernik, D., and Ben Shaul, I.Z. WebCutter: a system for dynamic and tailorable site mapping. In *Proceedings of the 6th International World Wide Web Conference* (Santa-Clara CA, April 1997).

12. Marchionini, G., Plaisant, C., and Komlodi, A. Interfaces and tools for the Library of Congress national digital library program. *Information Processing and Management, 34*, 535-555, 1998.

13. Mladenic, D. Turning Yahoo into an automatic web page classifier. In *Proceedings of the 13th European Conference on Artificial Intelligence (ECAI'98)* 473-474.

14. Platt, J. Fast training of support vector machines using sequential minimal optimization. In *Advances in Kernel Methods –Support Vector Learning*. B. Schölkopf, C. Burges, and A. Smola, eds., MIT Press, (1999).

15. Pratt, W. Dynamic organization of search results using the umls. In *American Medical Informatics Association Fall Symposium*, 1997.

16. Pratt, W., Hearst, M. and Fagan, L. A knowledge-based approach to organizing retrieved documents. In *Proceedings of AAAI-99*.

17. Shneiderman, B., Feldman, D. and Rose, A. Visualizing digital library search results with categorical and hierarchical axes. CS-TR-3993, UMIACS-TR-99-12. ftp://ftp.cs.umd.edu/pub/hcil/Reports-Abstracts-Bibliography/99-03html/99-03.html

18. Wittenburg, K. and Sigman, E. Integration of browsing, searching and filtering in an applet for information access. In *Proceedings of ACM CHI97: Human Factors in Computing Systems,* (Atlanta GA, March 1997).

19. Zamir, O., and Etzioni, O. Grouper: A dynamic clustering interface to web search results. In *Proceedings of WWW8* (Toronto, Canada, May 1999).

20. Zamir, O., and Etzioni, O. Web document clustering: a feasibility demonstration. In *Proceedings of the 19th International ACM SIGIR Conference on Research and Development in Information Retrieval (SIGIR '98)*, 46-54.

21. http://cha-cha.berkeley.edu/
22. http://search.msn.com/
23. http://www.inktomi.com/new/press/directory.html
24. http://www.looksmart.com/
25. http://www.northernlight.com/
26. http://www.snap.com/
27. http://www.yahoo.com/

Enhancing a Digital Book with a Reading Recommender

Allison Woodruff, Rich Gossweiler, James Pitkow, Ed H. Chi, Stuart K. Card
Xerox Palo Alto Research Center
3333 Coyote Hill Road
Palo Alto, CA 94304 USA
+1 650 812 4429
{woodruff, rcg, pitkow, echi, card}@parc.xerox.com

ABSTRACT
Digital books can significantly enhance the reading experience, providing many functions not available in printed books. In this paper we study a particular augmentation of digital books that provides readers with customized recommendations. We systematically explore the application of spreading activation over text and citation data to generate useful recommendations. Our findings reveal that for the tasks performed in our corpus, spreading activation over text is more useful than citation data. Further, fusing text and citation data via spreading activation results in the most useful recommendations. The fused spreading activation techniques outperform traditional text-based retrieval methods. Finally, we introduce a preliminary user interface for the display of recommendations from these algorithms.

Keywords
Spreading activation, bibliometrics, recommendations, information visualization, 3D book, degree of interest

INTRODUCTION
The digital nature of online books enables various enhancements to the reading experience that are not afforded by printed books. For example, the reader of a digital book can easily locate all occurrences of a given keyword. A primary advantage of the digital book as compared to its printed cousin is that it can be customized to suit the interests of a particular reader.

In this paper, we restrict ourselves to one enhancement of digital books—providing readers with personalized recommendations. Consider the case of an edited collection of academic papers. The reader does not always wish to read these papers in the order in which they appear in the book. Further, advances in information technology make it possible to construct an information environment in which the reader has immediate access to all of the literature referenced in a digital book. However, the related literature is often extensive, and the reader is ill equipped to choose which documents to read next, as the knowledge required to make sensible decisions is precisely the content of the unfamiliar corpus. Readers can be aided by algorithms that attempt to predict users' changing *degree of interest* [1] in information space and by user interfaces that use these predictions to direct user attention in visualizations [2] of *information scent* [3].

To address this problem, we study the application of text- and citation-based spreading activation algorithms to the reading recommendation problem. Spreading activation is a mathematical technique for determining the relatedness of items based on their degree of association [4] and has certain properties we felt were well suited to this problem. We make several contributions:

- We present a model for recommendation that uses documents rather than terms as inputs, i.e., the inputs are the set of documents the reader has read instead of user-specified keywords. This approach reduces the reader's burden and allows us to take advantage of the extensive information available about the document.
- We present novel results in the application of spreading activation to text. The state-of-the-art in spreading activation has advanced significantly since the last (negative) results in this area [5] and we have found that the issue is worth revisiting.
- We introduce novel algorithms for the fusion of text and citation data.
- We provide a systematic evaluation of different recommendation strategies and the effects of using various data.

Our findings reveal that for the tasks performed in our corpus, spreading activation over text is more useful than over citation data. However, the fusion of text and citation data through spreading activation proves to be the most effective technique. Moreover, the new fused spreading activation text-citation technique outperforms traditional text-based retrieval methods.

We explore the utility of a reading recommender by constructing a corpus of the complete text of nearly all the documents contained in or cited by a printed book, *Readings in Information Visualization: Using Vision to Think* by Card, Mackinlay, and Shneiderman [2], hereafter referred to as RIV.

Permission to make digital or hard copies of all or part of this work for personal or classroom use is granted without fee provided that copies are not made or distributed for profit or commercial advantage and that copies bear this notice and the full citation on the first page. To copy otherwise, to republish, to post on servers or to redistribute to lists, requires prior specific permission and/or a fee.
CHI '2000 The Hague, Amsterdam
Copyright ACM 2000 1-58113-216-6/00/04...$5.00

In the remainder of this section, we discuss related work. In the next section, we discuss our approach. In subsequent sections, we describe our algorithms, discuss the evaluation, and introduce a preliminary interface for presenting recommendations to the reader.

Related Work

Tracing the history of related work is a complicated task given that our methods combine three rather disparate fields. Given this and the space limitations, we focus only on attempts that fuse a) citation methods, b) text and citation methods, and c) spreading activation with either text or citations. We further restrict ourselves to examining only related research that uses citation data, leaving out similar work currently being done with the WWW, since citations are in many ways quite different from hypertext links, e.g., a primary function of links is navigation.

In what may be the first attempt to fuse citation and text representations, Salton [6] in 1963 demonstrated that citation and term data could be integrated effectively into a vector space model. Eight years later, using the SMART retrieval system, Salton [7] showed that using citations plus index terms to represent documents and queries resulted in better results than using index terms alone.

A year later in 1972, Robert Amsler [8] proposed what may be the first attempt to fuse bibliographic coupling and cocitation measures (defined in the following sections) to determine subject similarity between document pairs. Interestingly, Amsler's work predates the use of cocitation as a standalone measure of topic and document similarity. The work did not integrate textual data into the measure of document similarity. This pioneering work later inspired Bichteler and Eaton 1980 [9] who showed that re-ranking query results by the combined use of bibliographic coupling and cocitation techniques improved precision over using just bibliographic strengths. They noticed considerable variations in performance between queries. As we shall see, our work confirms this variation and provides insight into why it occurs.

Cohen and Kjeldsen 1987 [10] used a constrained spreading activation network over a knowledge base of topics to show enhanced precision and recall over keyword text methods. The knowledge base and activation networks were constructed manually. There have been several attempts to refine term and query expansion methods by various flavors of spreading activation techniques. In what appears to have placed a cap on that line of research, Salton and Buckley 1988 [5] showed that vector models performed better than several spreading activation methods using term by document matrices. While the study evaluated various normalization schemes for the spreading activation models, the spreading activation models were quite simple. As we shall see, our work is noteworthy in that it challenges this study's findings by showing the added benefits of using spreading activation over text methods.

The methods proposed in this paper differ significantly from the prior work. First, the spreading activation model used in this paper is more sophisticated and general than earlier models. These earlier models were limited by theoretical and computational constraints. Second, most prior work focused on retrieving a set of relevant documents given a particular set of terms as the query. The methods tested in this paper deal with finding relevant documents given a set of one or more documents. As a result, the prior methods tend to use term by document representations, whereas the methods investigated in this paper use document by document matrices. Third, while a few efforts have attempted to fuse various citation data, none have attempted to combine citations, bibliographic coupling strengths, and cocitation strengths, let alone all of these with text. Finally, we are unaware of any prior work that provides a systematic investigation of the fusion of citation and text data using spreading activation.

APPROACH

The purpose of our system is to make personalized recommendations about what documents to read next. We make these recommendations using only the citation and text data from the corpus and a set of input documents of interest. We assume scenarios in which readers take different paths through the set of collected articles and have individualized interests within the field of information visualization. Traditionally, readers confronted with these choices simply make "best-guess" decisions about what to read next. These decisions are often inefficient, wasting the reader's time and energy. Alternatively, a small number seek customized recommendations from an expert on the subject, requesting the recommendations on what to read next in person, by phone, or by email. While rich in interaction and nuance, this form of interaction does not scale gracefully to handle hundreds or thousands of readers with a limited number of experts. The critical question we seek to explore is how best to provide readers with relevant individualized recommendations without having to constantly pester the experts.

In the remainder of this section, we describe the corpus and alternative approaches. This section is followed by a discussion of the algorithms used to simulate expert recommendations.

Corpus

Our study examines recommendations in the context of RIV. The printed book contains 43 articles and 9 chapter commentaries totaling 686 pages. By including all documents to which any of these articles or commentaries refer, the total expanded RIV document count rises to 719. This expanded document, including its citation data, will be abbreviated by RIV* to avoid confusion.

We acquired and extracted text from 653 of the 719 documents, scanning and applying optical character recognition (OCR) software to the large fraction of the documents that were only available in printed form. Textual proxies (usually tables of contents, bibliographies and indices) represent some books and dissertations. The intent is for the proxies to contain most of the appropriate key words and terms. The text portion of RIV* contains some 5 million words (12,000 pages). The RIV* citation data consists of 1151 references from documents in the printed book to the 719 RIV* documents. Citations in documents not in the printed book are not considered.

Alternative Approaches

Several alternative recommendation approaches are worth considering. First, there are manual approaches. For example, one could imagine asking experts to define a fixed set of reading paths through the book and use these manually constructed paths in a digital book. This approach has several limitations. Foremost, it is limited by the predefinition of reading paths, which may or may not correspond to the actual paths of users through the book. An alternative approach would be to provide recommendations for every possible path through the book. Unfortunately, the number of paths scales as the factorial of the number of articles in the book. For the RIV book of 43 articles, the total number of unordered paths is $6.04*10^{52}$, a completely unreasonable number of recommendations to generate manually.

Second, one could use social filtering, where the usage patterns of other readers through the book are harvested to make recommendations. In today's digital and networked world, one could imagine the book's usage information collected via the Internet, recommendations generated on demand and shipped to the reader requesting a recommendation. While such a system would be dynamic and evolve over time, some form of bootstrapping would be required to seed the recommendations. That is, the usage data required to form the recommendations can not be known until some readers actually use the system to generate recommendations. Such social filtering systems also raise complicated issues such as data privacy, incentives, and biasing/spamming. Given the resource intensive nature of manual solutions and the bootstrapping and complexity issues of social filtering, we sought to systematically explore methods that would rely only upon the information contained in the RIV* book.

RECOMMENDATION ALGORITHMS

To leverage these data, we introduce a set of new methods similar to our prior work with hyperlinks and the WWW of combining citation and text data with spreading activation. We provide background material in citation analysis and spreading activation algorithms. We then describe the algorithms used in our study, including several novel algorithms.

Citation Analysis

We now define the key concepts of citation analysis, cocitation analysis and bibliographic coupling. An intuitive treatment of each will be presented, as well as formal definitions in terms of linear algebra.

Dating back to the use of the 1873 Shepard's Citations in the legal community, citation indexing has been used to harness the decisions made by authors to include references to relevant previously recorded information. Within the scientific community, these references tend to identify prior research whose methods, equipment, results, etc. influenced the current work. By capturing the semantic judgement of authors and the works of others, citation indexes create a powerful tool that serves three main application areas [11]: 1) information search and retrieval, 2) qualitative and quantitative evaluations of scientists, publications, and scientific institutions, and 3) modeling of the historical development of science.

A citation index can be represented as a directed graph (*citation network*) as well as the corresponding incidence matrix for the graph (*citation matrix*). In the former case, a directed edge between node D_i and D_j indicates that D_i references D_j and that D_j is referenced by D_i. In the latter case, the value of the cell for row D_i and column D_j denotes the number of times document D_i refers to document D_j. This number of times a document is cited is called the *citation frequency*. In this manner, the citation matrix C illustrates the "cites" relationships and the transpose of the citation matrix C^T illustrates the "is-cited-by" relationships.

The cocitation matrix and bibliographic coupling matrix (as well as a number of other interesting properties) can be readily computed from the citation matrix. If we have m source documents that contain references to n other documents with the corresponding citation matrix $C=(c_{ij})$, then

- the number of references of a given document D_i is the sum of the row vector for D_i or $(CC^T)_{ii}$;
- the number of references that documents D_i and D_j share in common (called the *bibliographic coupling strength* [12]) is given by the equation:

$$\sum_{k=1}^{n} C_{ik}C_{jk} = (CC^T)_{ij} ;$$

- the number of citations received by document D_i is the sum of the column vector for D_i or $(C^TC)_{ii}$;

- the number of citations which documents D_i and D_j share in common (called the *cocitation strength* [13, 14]) is given by the equation:

$$\sum_{k=1}^{m} C_{ki} C_{kj} = (C^T C)_{ij} \; ;$$

The intuition behind the value of cocitation and bibliographic coupling is as follows. Once written, the references a document D_i makes to other papers are fixed, yet additional papers can be written that reference D_i as well as cite the references in D_i. At any given point in time, one can inspect the bibliographic coupling strengths for a set of documents to gain insight into what awareness authors had of each others' work. Cocitation identifies pairs of documents that are referenced together. Frequently citing documents together implies the shared semantic judgement of authors that the pair of documents $D_i D_j$ are related—even though the two documents may not contain a reference to each other. Cocitation strengths vary over time and can provide a glimpse into the papers that influence a particular field at any given time.

Spreading Activation

Spreading activation refers to a class of algorithms that propagate numerical values among a set of connected items. For any source of interest, activation can be spread though an association network. The highest values of the resulting activation vector represent the items most closely associated with the item of interest. Additionally, multiple sources of activation can be used to compute the interest function over several items at once. As we shall see in the next section, this feature enables the degree-of-interest function to individualize recommendations.

The particular version of spreading activation we use is the leaky capacitor model [4], which has been studied parametrically by [15]. Specifically, an activation network can be represented as a matrix R, where each element $R_{i,j}$ contains the strength of association between nodes i and j, and the diagonal contains zeros. The amount of activation that flows between nodes is determined by the activation strengths, which for our purposes correspond to bibliographic coupling and cocitation strengths. Source activation is represented by a vector C, where C_i represents the activation pumped in by node i. The dynamics of activation can be modeled over discrete steps $t = 1, 2, ...N$, with activation at step t represented by a vector $A(t)$, with element $A(t, i)$ representing the activation at node i at step t. The evolution of the flow of activation is determined by:

$$A(t) = C + MA(t-1)$$
$$M = (1-\gamma)I + \alpha R$$

where M is a matrix that determines the flow and decay of activation among nodes, with γ determining the relaxation of node activation back to zero when it receives no additional activation input, and α denoting the amount of activation spread from a node to its neighbors. I is the identity matrix.

The Algorithms

We introduce seven algorithms (Table 1). One (Text) is a non-spreading activation algorithm used as a baseline for comparison. The remaining algorithms are spreading activation algorithms using different association matrices.

For the comparative baseline for the various spreading activation algorithms, we used a standard text-based vector space model developed internally called Pipes. Pipes uses the standard term frequency inverse document frequency method (TFIDF) for normalization. A document by document matrix was constructed using the cosine dot product between document pairs. The resulting document similarity vectors were used to make recommendations. We refer to this non-spreading activation algorithm as the Text algorithm, and view it as the simple but traditional method with which to gauge the effectiveness of the spreading activation algorithms.

Table 1. Algorithms and association matrices.

Algorithm	Association Matrix (R)
Cite	C = Citation
BibCoup	BC = Bibliographic Coupling
Cocite	CC = Cocitation
Fused Citation	FC = C+BC+(3*CC)
Text	none input; outputs Document x Document
SAText	SAT = Document x Document
SATextFC	SATFC = T+(3*FC)

To leverage the citations and text data in RIV*, we created six spreading activation algorithms. The data used in each association matrix R is described in Table 1. Four methods use individual association matrices, i.e., Cite, BibCoup, Cocite, and SAText use the matrices listed in the table. Note that the document by document matrix input to SAText is the one generated by Text. The remaining two methods, Fused Citation and SAText + Fused Citation (SATextFC), use a weighted combination of the other matrices to produce their final association matrices. The weightings appear in Table 1. The weights were selected manually[1] to provide normalization across matrices. For example, with the Fused Citation method, the average cocitation strength needed to be increased by a factor of three to contribute equally with the other methods. The weighting of the matrices is supported theoretically by the additive properties of the underlying spreading activation algorithm.

---

[1] The Fisheyes scenario described below required different weightings due to abnormalities in the citation graph. We leave it to future work to develop methods to automatically correct the weights.

Introduction	*I'm a computer graphics researcher. I'm generally interested in information visualization, but I don't know much about it. I bought the book Readings in Information Visualization: Using Vision to Think, and I've read the first chapter, as listed below. What should I read next?* Card, S. K., Mackinlay, J. D., & Shneiderman, B. (1999). Information Visualization (Chapter 1). In S. K. Card, J. D. Mackinlay, & B. Shneiderman (Eds.), Readings in Information Visualization: Using Vision to Think. San Francisco, CA: Morgan Kaufman.
Fisheyes	*I am a VLSI chip designer. I'm writing a tool that uses fisheyes to show circuit layout. I want to read everything I can about fisheyes. I have already read Furnas' paper listed below. What should I read next?* Furnas, G. W. (1981). The FISHEYE view: a new look at structured files. Murray Hill, NJ: Bell Laboratories.
Networks	*I work at a networking company. I'm developing a tool for visualizing large networks, and I want to read about appropriate techniques. I have read the papers in Readings in Information Visualization: Using Vision to Think, Section 2.5, listed below. What should I read next?* Becker, R. A., Eick, S. G., & Wilks, A. R. (1995). Visualizing Network Data. IEEE Transactions on Visualization and Computer Graphics, 1(1 March), 16–28. Eick, S. G., & Wills, G. J. (1993, October 25-29). Navigating Large Networks with Hierarchies. Proceedings of IEEE Visualization'93 Conference, San Jose, CA, 204-210. Fairchild, K. M., Poltrock, S. E., & Furnas, G. W. (1988). SemNet: Three-dimensional representations of Large Knowledge Bases. In R. Guindon (Ed.), Cognitive Science and Its Applications for Human-Computer Interaction (pp. 201-233). Hillsdale, New Jersey: Lawrence Erlbaum Associates.
Techniques	*I'm a researcher in the area of user interfaces. Once in a while I use an information visualization technique in my work, so I'd like to learn more about information visualization. I've read the papers listed below. What should I read next?* Bederson, B. B., & Hollan, J. D. (1994). Pad++: A Zooming Graphical Interface for Exploring Alternate Interface Physics. Proceedings of UIST'94, ACM Symposium on User Interface Software and Technology, Marina del Rey, California, 17-26. Fishkin, K., & Stone, M. C. (1995, May 7-11 1995). Enhanced Dynamic Queries via Movable Filters. Proceedings of CHI'95, ACM Conference on Human Factors in Computing Systems, Denver, CO, 415-420. Furnas, G. W. (1981). The FISHEYE view: a new look at structured files. Murray Hill, NJ: Bell Laboratories. Lamping, J., & Rao, R. (1996). The Hyperbolic Browser: A Focus+Context Technique for Visualizing Large Hierarchies. Journal of Visual Languages and Computing, 7(1), 33-55. Milash, B., Plaisant, C., & Rose, A. (1996). LifeLines: Visualizing Personal Histories (Video). Conference Companion of CHI'96, ACM Conference on Human Factors in Computing Systems, Vancouver, Canada, 392-393. Plaisant, C., Rose, A., Milash, B., Widoff, S., & Shneiderman, B. (1996). LifeLines: Visualizing personal histories. Proceedings of CHI'96, ACM Conference on Human Factors in Computing Systems, Vancouver, Canada, 221–227. Rao, R., & Card, S. K. (1994). The Table Lens: Merging graphical and symbolic representations in an interactive focus + context visualization for tabular information. Proceedings of CHI'94, ACM Conference on Human Factors in Computing Systems, Boston, MA, 318–322 and 481-482. Shneiderman, B. (1994). Dynamic queries for visual information seeking. IEEE Software, 11(6), 70-77. Spoerri, A. (1993). InfoCrystal: A visual tool for information retrieval. Proceedings of IEEE Visualization'93 Conference, San Jose, CA, 150-157.

Table 2. Scenarios used in evaluation.

EVALUATION

Experimental Design

To evaluate the effectiveness of the algorithms, we constructed four reading scenarios (Table 2), generated recommendations, and had experts rate the relevance of the recommendations. Each scenario represents a hypothetical task, and the documents listed below each scenario represent relevant documents a reader might have read up to the point of requesting a recommendation. We deliberately choose to vary both the number of documents read as well as whether the documents were from the RIV book or referenced documents to gain insight into the behavior of the algorithms under various conditions.

The recommendations for Text were generated by the following method: the row vector corresponding to the read document was sorted and the top documents selected. In cases in which more than one document was read, sorted vectors for each read document were merged into a single list from which the top ranked documents were selected.

For the spreading activation algorithms, the documents read by the hypothetical reader were used in the source activation vector to pump activation. Since the average values in the association matrices for the citation matrices were much lower than in the document by document matrix, we used an alpha of 1 for the citation methods and an alpha of 0.01 for the SAText methods. Activation was spread for ten iterations and in all cases quickly converged. The top ten values from the final activation vector were selected as the final recommendations. As a further baseline for the comparisons, we included a random recommendation generator that generated random documents.

In order to determine the ability of each algorithm, we had three information experts rate the relevance of the recommendations. The evaluators included two authors of the RIV book and an expert in the area of information visualization. Each scenario was presented to the evaluators with a random permutation of the recommended documents. Evaluators were asked to rate each document's "usefulness" to the reader described in the scenario on a scale of 1 (least useful) to 5 (most useful). The evaluators took approximately half an hour to complete the task. The correlation between the rankings of the evaluators ranged from 0.50 to 0.59.

Analysis of Precision

To assess the quality of each recommendation, we computed the geometric mean of the experts' ratings for each document within the context of a given scenario. The mean was then compared against a threshold value to assess whether the recommendation was *useful* (geometric mean 4 or higher) or *somewhat useful* (geometric mean 3 or higher), e.g., if the geometric mean of the experts' ratings of a document was 3.9, it was a *somewhat useful* recommendation.

From the mean, we computed various precision metrics in a similar manner to [16]. Recommendations are presented in ranked lists from 1 to 10, with 1 being the highest recommended. For a given algorithm, the precision at some ranking *r* is the total number of useful recommendations within the first *r* answers divided by *r* (we also calculate the precision for somewhat useful recommendations). For example, suppose an algorithm had a *not useful* first recommendation and a *useful* second recommendation. In this case, its precision at 1 would be 0 and its precision at 2 would be 1/2. Intuitively, precision tells us the percentage of *useful* versus *not useful* rankings for the first *r* recommendations made by the algorithm.

We next define an aggregate metric, the average precision of an algorithm. The average precision of an algorithm is the sum of its precision at all ranks divided by the total number of ranks. In our case, since each algorithm made 10 recommendations, the average precision is the sum of the precision at ranks 1 to 10 divided by 10. Intuitively, this metric assigns higher values to algorithms that get useful recommendations early in the rankings. For example, suppose algorithm A had 5 *useful* rankings followed by 5 *not useful* rankings and algorithm B had 5 *not useful* rankings followed by 5 *useful* rankings. In this case, algorithm A would have an average precision of 0.86 while algorithm B would have an average precision of 0.18.

The first seven data columns in Table 3 contain the average precision for each of the algorithms across all scenarios. The random algorithm had an average precision of zero in almost every case, so we omit it from the table. The final column compares the relative performance of SATextFC and Text (computed by dividing the average precision of SATextFC by the average precision of Text and subtracting 1).

Discussion of Average Precision
We begin with several general observations. First, we see that the average precision values vary by scenario as was noted by Salton and Buckley 1988 [5]. In particular, the algorithms performed quite differently on the Networks scenario, as we will discuss further below. Second, the average precision is much higher for the *somewhat useful* than for the *useful* conditions, as is to be expected. The relative orderings of the algorithms remain fairly stable across these conditions, with the exception of Cocite, which drops dramatically in the more stringent *useful* condition. Third, there is considerable variability in the precision of each of the individual citation methods. With the less stringent criteria, the Cocite method performs the best, but in the more stringent condition, bibliographic coupling performs the best. As one would expect, the Fused Citation method provides a smoother precision across scenarios. Finally, we see that the algorithms that use text compare favorably to those that use citation structure.

More specifically, we observe that our new algorithms SATextFC and SAText both compare favorably to Text and all citation algorithms, particularly in the demanding *useful* condition, in which they yield a 47% improvement over Text overall. The SATextFC fusion method appears somewhat superior to SAText in almost all cases except the Network scenario; this decreases its performance so that on average these two algorithms have very similar behavior. If the network scenario is not considered, SATextFC yields on average a 71% improvement over Text in the *useful* condition and a 22% improvement over SAText.

Discussion of Average Precision by Scenario
To gain more insights into variation by task, we intentionally chose scenarios with different characteristics. Below, we make observations about the behavior of these algorithms for these particular scenarios.

The Introduction scenario is a broad, vague query. The Introduction contains a large amount of highly relevant text and a large number of citations. The citation algorithms (C, BC, CC) did generate a number of unique recommendations that were *useful* or *somewhat useful*. The text similarity algorithms did very well in this scenario, presumably because the input document was a strong indicator of the reader's interests. SATextFC did slightly better than either strategy.

Table 3. Average precision for all scenarios.

	Cite	BibCoup	Cocite	FC	Text	SAText	SATextFC	SATextFC vs Text
Somewhat Useful								
Introduction	0.52	0.46	—	0.24	0.86	0.86	0.93	9%
Fisheyes	0.49	0.80	0.50	0.64	0.87	0.84	0.97	11%
Networks	0.14	0.05	0.49	0.20	0.38	0.62	0.34	-11%
Techniques	0.70	0.58	0.91	0.84	0.72	0.73	0.95	32%
Average	0.46	0.47	0.63	0.48	0.71	0.76	0.80	13%
Useful								
Introduction	0.15	0.17	0.00	0.19	0.37	0.48	0.56	52%
Fisheyes	0.49	0.73	0.37	0.47	0.77	0.79	0.95	22%
Networks	0.00	0.00	0.00	0.01	0.36	0.59	0.26	-29%
Techniques	0.19	0.02	0.05	0.14	0.00	0.33	0.44	—
Average	0.21	0.23	0.14	0.20	0.38	0.55	0.55	47%

The Fisheyes scenario is the most focussed query. The input document contains specific terminology and is cited by many other documents in the collection. Given the strong text and citation cues, it is not surprising that citation and text algorithms both performed well. Most interesting, the fused SATextFC did better than both the citation and text measures independently, successfully integrating, which resulted in near perfect precision. For example, SATextFC ranked as its second choice a document that had received lower rankings from the text and citation measures (Cocite 5; Text 4; SAText 4).

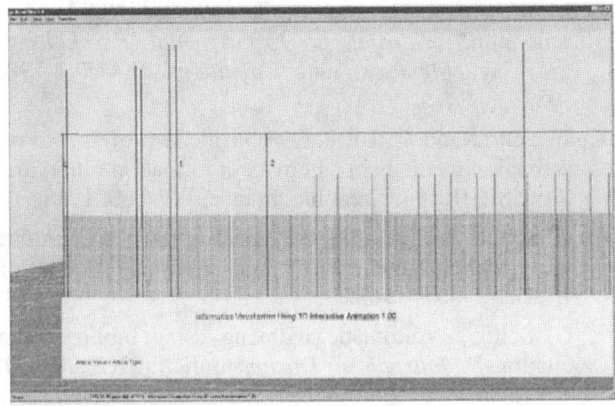

Figure 1. Close-up of a section of the Book Ruler.

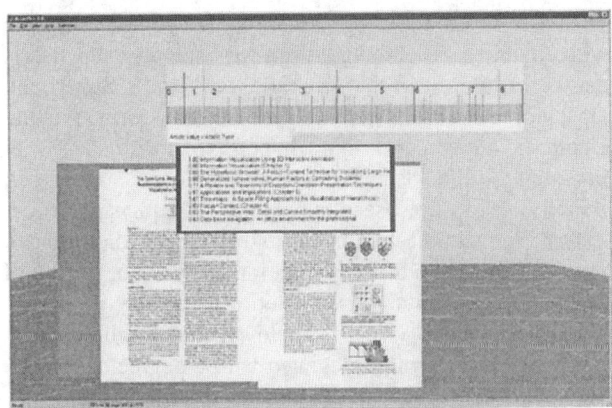

Figure 2. Book, Book Ruler, and recommendation list.

The Network query is a less focussed text query. This was the most difficult scenario for all the algorithms. The citation algorithms performed particularly poorly on this scenario, which we believe results from the lack of cohesive citations in this sub-discipline. In this instance, the seminal papers do not cite the same authoritative references (low BibCoup), but are cocited by a fair number of later papers (high Cocite). We were surprised and excited to find this behavior of the citation methods. The document cosine algorithm was outperformed by the SAText algorithm by 62% for *somewhat useful* recommendations and 65% for *useful* recommendations. This is the only case in which SATextFC performed worse than the other text algorithms. This occurred because the citation recommendations were so poor that they lowered the quality of SATextFC recommendations.

The Techniques query is very broad query, using nine input documents. In the *somewhat useful* conditions, both the citation algorithms and the text algorithms did well, with SATextFC having the best performance, closely followed by Cocite. The citation methods performed their best as a result of the well-defined input citation structure. In the *useful* condition, the citation algorithms and Text have very poor performance, indicating that the overall quality of the recommendations was moderately useful. The SAText and SATextFC were able to produce higher quality recommendations than the other algorithms. We credit the improved performance to the manner in which the spreading activation algorithm reinforces useful documents during each iteration.

Comparison of Algorithms

To assess the statistical significance of our findings, we performed the Wilcoxon Sums of Ranks Tests for each pair of algorithms using a significance level of 0.01. For each algorithm, the input was an ordered list of the expert ratings for the algorithms' recommendations (recall that values for expert ratings were computed by taking the geometric mean of all experts). From this analysis, we can conclude that the difference between SATextFC and Text was statistically significant. However, the differences between SATextFC by SAText and SAText by Text were not statistically significant. All text methods were significantly better than the citation and fused citation methods.

USER INTERFACE

Having determined that the recommendation algorithms, especially SATextFC, produce reasonable recommendations, we now discuss a simple user interface prototype that uses the recommendations to augment the user's reading experience. Our prototype has a recommendation engine and two major graphical components: the 3D Book [17] and the Book Ruler. The 3D Book presents a graphical representation of RIV*. The reader can interact with the book, e.g., to turn pages. The Book Ruler uses graduated lines to give the reader an overview of the contents of the 3D Book. Each major graduation on the Book Ruler represents a chapter commentary, the next finer divisions represent articles in the corresponding chapter, and the finest graduations represent references for each article (these graduated lines appear below the horizontal line in Figure 1). Selecting a graduation turns the pages of the book to that article. Much like the Data Slider [18], the Book Ruler allows the reader to brush over the graduations. Brushing highlights the graduation and highlights any other locations where the same document appears, allowing the reader to see the pattern of references to a particular document.

By interacting with the Book Ruler or by issuing queries to the system, the reader can select a set of documents. A simple user interface gesture feeds the current selection set into the recommendation engine to indicate, "Given I've read this, what should I read next?" In response, the system performs a recommendation analysis and displays the results as taller, colored bars above the horizontal line in the Book Ruler. The reader can see where recommendations appear in the book and make quick comparisons of relative value. Results also appear in a pop-up text box (Figure 2).

3D book metaphors have been used in user interfaces for some time (for a review, see [17]). The 3D book metaphor helps orient users and maintains correspondence between physical and digital versions of documents. What we believe is new is the integration of a 3D book on this scale with citation links and algorithms for computing the user's degree of interest in the material. BellCore's Superbook [19], for example, was designed to transform existing electronic documents into hypertext documents with indexing and a fisheye table of contents. However, Superbook did not use a simulation of a physical book, did not include the text of all its references, and did not use our algorithms based on citation analysis for degree of interest, although it did have a notion of degree of interest. Our earlier WebBook [17] could not handle the data on this scale and was not integrated with a degree of interest algorithm.

CONCLUSION

The reading experience of digital books can be enhanced in a number of ways. In this paper, we have focused on automatically generating recommendations of further reading material based on various user scenarios. In our evaluation, we compared our algorithms to standard information retrieval algorithms. We demonstrated that spreading activation over text and fusing text with citation data techniques are very effective, according to expert evaluation. While our algorithms make successful recommendations using only text and citation data, we believe we could enhance these algorithms by adding usage data, and spreading activation is a natural mechanism for fusing these three disparate data types.

The corpus we considered is in the research literature genre, where a particular paper or book often serves as an initial exploration into a group or groups of documents. We believe the results support the exploration of the new techniques in other linked information environments such as the digital libraries or the Web.

ACKNOWLEDGMENTS

We thank the many individuals who were invaluable in assembling and processing the corpus: Eytan Adar, Lisa Alfke, Michelle Baldonado, Richard Burton, Amy Hurst, Dan Larner, Sally Peters, Ken Pier, and Aaron Solle. We are also grateful to Jock Mackinlay for generating the bibliography and to the experts for participating in our evaluation. Portions of this work were supported by ONR contract No. N00014-96-C-0097.

REFERENCES

1. G. W. Furnas, "The FISHEYE view: a new look at structured files," Bell Laboratories Technical Report, reproduced in *Readings in Information Visualization: Using Vision to Think*, S. K. Card, J. D. Mackinlay, and B. Shneiderman, Eds. San Francisco: Morgan Kaufmann Publishers, Inc., 1981, 312-330.

2. S. K. Card, J. D. Mackinlay, and B. Shneiderman, *Readings in Information Visualization: Using Vision to Think*. San Francisco, California: Morgan-Kaufmann, 1999.

3. P. Pirolli and S. K. Card, "Information foraging," *Psychological Review* 106(4):643-675, 1999.

4. J. R. Anderson and P. L. Pirolli, "Spread of activation," *Journal of Experimental Psychology: Learning, Memory, and Cognition* 10(4):791-798, 1984.

5. G. Salton and C. Buckley, "On the use of spreading activation methods in automatic information retrieval," Proc. SIGIR '88, Grenoble, France, 147-160, 1988.

6. G. Salton, "Associative document retrieval techniques using bibliographic information," *Journal of the ACM* 10(4):440-457, 1963.

7. G. Salton, "Automatic indexing using bibliographic citations," *Journal of Documentation* 27(2):98-110, 1971.

8. R. Amsler, "Applications of citation-based automatic classification," Linguistics Research Center, Univ. Texas at Austin, Technical Report 72-14, Dec. 1972.

9. J. Bichteler and E. A. Eaton, "The combined use of bibliographic coupling and cocitation for document retrieval," *Journal of the American Society for Information Science* 31(4):278-282, 1980.

10. P. R. Cohen and R. Kjeldsen, "Information retrieval by constrained spreading activation in semantic networks," *Information Processing and Management* 23(4):255-268, 1987.

11. P. Zunde, "Structural models of complex information sources," *Information Storage and Retrieval*, 7(1):1-18, 1971.

12. M. M. Kessler, "Bibliographic coupling between scientific papers," *American Documentation* 14:10-25, 1963.

13. I. V. Marshakova, "System of document connectionism based on references," *Nauchno-Teknicheskaya Informatsiya*, Series 2:2-6, 1973.

14. H. Small, "Co-citation in the scientific literature: a new measure of the relationship between two documents," *Journal of the American Society for Information Science* 24(4):265-269, 1973.

15. B. A. Huberman and T. Hogg, "Phase transitions in artificial intelligence systems," *Artificial Intelligence* 33(2):155-171, 1987.

16. J. Dean and M. R. Henzinger, "Finding related pages in the World-Wide Web," *Computer Networks & ISDN Systems* 31(11-16):1467-1479, 1999.

17. S. K. Card, G. G. Robertson, and W. York, "The WebBook and the Web Forager: an information workspace for the World-Wide Web," Proc. SIGCHI '96, 111-117, Vancouver, BC, Canada, 1996.

18. S. G. Eick, "Data visualization sliders," Proc. ACM UIST '94, 119-120, Marina del Rey, CA, 1994.

19. J. R. Remde, L. M. Gomez, and T. K. Landauer, "Superbook: an automatic tool for information exploration," Proc. ACM Hypertext '87 Conference, 187-188, Chapel Hill, NC 1987.

The Scent of a Site: A System for Analyzing and Predicting Information Scent, Usage, and Usability of a Web Site

Ed H. Chi, Peter Pirolli, James Pitkow

Xerox Palo Alto Research Center
3333 Coyote Hill Road, Palo Alto, CA 94304
{echi,pirolli,pitkow}@parc.xerox.com

ABSTRACT

Designers and researchers of users' interactions with the World Wide Web need tools that permit the rapid exploration of hypotheses about complex interactions of user goals, user behaviors, and Web site designs. We present an architecture and system for the analysis and prediction of user behavior and Web site usability. The system integrates research on human information foraging theory, a reference model of information visualization and Web data-mining techniques. The system also incorporates new methods of Web site visualization (Dome Tree, Usage Based Layouts), a new predictive modeling technique for Web site use (Web User Flow by Information Scent, WUFIS), and new Web usability metrics.

Keywords

Information foraging, information scent, World Wide Web, usability, information visualization, data mining, longest repeated subsequences, Dome Tree, Usage-Based Layout.

INTRODUCTION

The World Wide Web is a complex information ecology consisting of several hundred million Web pages and over a hundred million users. Each day these users generate over a billion clicks through the myriad of accessible Web sites. Naturally, Web site designers and content providers seek to understand the information needs and activities of their users and to understand the impact of their designs. Given the magnitude of user interaction data, there exists a need for more efficient and automated methods to (a) analyze the goals and behaviors of Web site visitors, and (b) analyze and predict Web site usability. Simpler, effective, and efficient toolkits need to be developed to explore and refine predictive models, user analysis techniques, and Web site usability metrics.

Here we present an architecture and system for exploratory data analysis and predictive modeling of Web site use. The architecture and system integrates research on human information foraging theory [6], a reference model of information visualization [3], and Web data-mining techniques [9]. The system also incorporates new methods of Web site visualization, a new predictive modeling technique for Web site use, and new Web usability metrics. The system is currently being developed for researchers interested in modeling users within a site and investigating Web site usability; however, the ultimate goal is to evolve the system so that it can be effectively employed by practicing Web site designers and content providers.

WEB SITE ANALYSIS AND PREDICTION

Most Web sites record visitor interaction data in some form. Since the inception of the Web, a variety of tools have been developed to extract information from usage data. Although the degree of reliability varies widely based upon the different heuristics used, metrics like the number of unique users, number of page visits, reading times, session lengths, and user paths are commonly computed. While some tools have evolved into products[1], most Web log file analysis consists of simple descriptive statistics, providing little insight into the users and use of Web sites.

A new emerging approach is to employ software agents as surrogate users to traverse a Web site and derive various usability metrics. WebCriteria SiteProfile uses a browsing agent to navigate a Web site using a modified GOMS model and record download times and other information. The data are integrated into metrics that assess: (a) the load times associated with pages on the site, and (b) the accessibility of content (ease of finding content). The accessibility metric is based upon the hyperlink structure of the site and the amount of content. An analysis of the actual content is not performed.

Current approaches to Web site analysis are aimed at the Webmasters who are interested in exploring questions about the *current* design of a Web site and the *current* set of users. However, Webmasters must also be interested in *predicting* the usability of *alternative designs* of their Web sites. They also seek to answer these same questions for new kinds of (hypothetical) users who have slightly different interests than the current users.

Our work aims to develop predictive models capable of simulating hypothetical users and alternative Web Site design. Using these models, we also seek to develop means for the automatic calculation of usability metrics. Our

---

[1] For instance, Accrue Insight (http://www.accrue.com), Astra SiteManager (http://www.merc-int.com), and WebCriteria SiteProfile (http://www.webcriteria.com).

research on new analysis models, predictive models, and usability metrics contribute to the development of tools for the practicing Web site designer interested in exploring "what-if" Web site designs.

Our system was developed to answer questions beyond those answered by basic descriptive statistics. Specifically, we sought to answer questions concerning the entire Web site, specific pages, and the users:

- *Overall site.* What is the overall current traffic flow? What are the actual and predicted surfing traffic routes (e.g., branching patterns, pass-through points)? How does the site measure on ease of access (finding information) and cost?
- *Given page.* Where do the visitors come from (i.e., what routes do they follow)? Where do they actually go? What other pages are related?
- *Users.* What are the interests of the visitors (real or simulated) to this page? Where do we think they should go given their interests? Do actual usage data match these predictions, and why? What is the cost (e.g., in terms of download time) of surfing for these visitors?

INFORMATION FORAGING AT WEB SITES

Information foraging theory [6] has been developed as way of explaining human information-seeking and sense-making behavior. Here we use the theoretical notion of *information scent* developed in this theory [5,6] as the basis for several analysis techniques, metrics, and predictive modeling. We also employ a data mining technique involving the identification of *longest repeated subsequences* (LRS, [9]) to extract the surfing patterns of users foraging for information on the Web. This fusion of methods provides a novel way of capturing user information goals, the affordances of Web sites, and user behavior.

Information Goals and Information Scent

On the Web, users typically forage for information by navigating from page to page along Web links. The content of pages associated with these links is usually presented to the user by some snippets of text or graphic. Foragers use these *proximal* cues (snippets; graphics) to assess the *distal* content (page at the other end of the link).[2] Information scent is the imperfect, subjective, perception of the value, cost, or access path of information sources obtained from proximal cues, such as Web links, or icons representing the content sources.

In the current system, we have developed a variety of approximations for analyzing and predicting information scent. These techniques are based on psychological models [6], which are closely related to standard information retrieval techniques, and Web data mining techniques based on the analysis of Content, Usage, and hyperlink Topology (CUT, [3,10]). For more details, see [1].

---

[2] Furnas referred to such intermediate information as "residue" [4].

Reverse Scent Flow to Identify Information Need

A well-traveled path may indicate a group of users who have very similar information goals and are guided by the scent of the environment. Therefore, given a path, we would like to know the information goal expressed by that path. We have developed a new algorithm called Inferring User Need by Information Scent (IUNIS) that uses the Scent Flow model in reverse to determine users' information goals [1]. Such goals can be described by a sorted list of weighted keywords, which can be skimmed by an analyst to estimate and understand the goals of users traversing a particular path.

Mining Web Site Foraging Patterns

Pitkow and Pirolli [9] systematically investigated the utility of a Web-mining technique that extracts significant surfing paths by the identification of longest repeating subsequences (LRS). They found that the LRS technique serves to reduce the complexity of the surfing path model required to represent a set of raw surfing data, while maintaining an accurate profile of future usage patterns. In essence, the LRS technique extracts surfing paths that are likely to re-occur and ignores noise in the usage data. We use the LRS data mining technique to identify significant surfing paths in real and simulated data.

Overview of the Analysis Approach

Our assumption is that, for the purposes of many analyses, users have some information goal, and their surfing patterns through the site are guided by information scent. Given this framing assumption we have developed techniques for answering a variety of Web site usability questions. First, for a particular pattern of surfing, we seek to infer the associated information goal. Second, given an information goal, some pages as starting points, and the information scent associated with all the links emanating from all the pages, we attempt to predict the expected surfing patterns, and thereby simulate Web site usage. Finally, we develop metrics concerning the overall goodness of the information scent that leads users to goal content (cf. [11]). Using these methods, we analyze the quality of Web links in providing good proximal scent that leads users to the distal content that they seek.

ARCHITECTURE

The architecture of the system is based on the Information Visualization Reference Model [3]. Figure 1 shows the architecture of the system using the Data State Model and the associated operators. The figure summarizes the data states and the operators defined by the system components, which we describe in detail below.

In this figure, circles represent the data states, while edges represent operators. There are four major data state stages: Value, Analytical Abstraction, Visualization Abstraction, and View. There are three major operator types: Data, Visualization, and Visual Mapping Transformations. The right side of the figure depicts these stages and types.

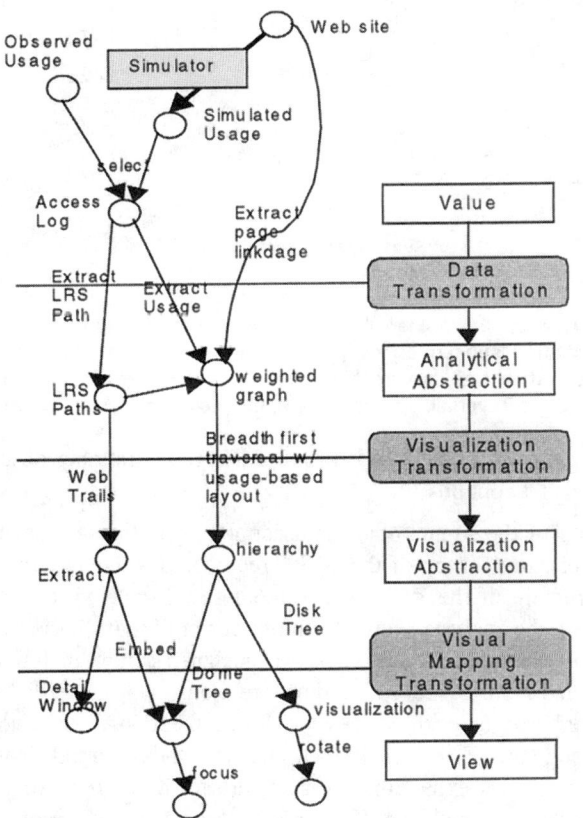

Figure 1: Data State Model for Web Scent Visualization

At the conceptual level, Figure 1 shows an important feature of the architecture: the actual observed usage data can be seamlessly replaced by simulated usage data, without disturbing other parts of the system. By pushing the observed or the simulated surfing data through the system, we obtain visualizations of actual or simulated usage. By providing this capability, users of the system can quickly test hypothetical cases against actual usage in a real-time, iterative manner, thus supporting detailed investigation into a site's usability.

SYSTEM FOR WEB SCENT VISUALIZATION

Using the reference model, we constructed a system for visualizing and analyzing a site's information scent, user trails, and usability. In the next sections, we describe the system components, followed by a series of cases illustrating the utility of the system.

Web Site and Observed Data

To develop and test the system, we used data collected at www.xerox.com on May 18th, 1998. Although slightly dated, the data set has been explored for a variety of other purposes [8,9] and was chosen to enable cross study comparisons and validation. The snapshot consists of roughly 15,000 pages and its associated Content, Usage, Topology (CUT) data. Content and topology data were extracted from the actual Web site using the techniques outlined in [7]. Usage data were extracted from the Extended Common Log Format access logs using the Timeout-Cookie method [8] to identify individual paths of contiguous surfing of Web pages by individual users.

Simulated Data

For the simulated data we developed a new technique called *Web User Flow by Information Scent* (WUFIS) [1]. Conceptually, the simulation models an arbitrary number of agents traversing the links and content of a Web site. The agents have information goals that are represented by strings of content words such as "Xerox scanning products." For each simulated agent, at each page visit, the model assesses the information scent associated with linked pages. The information scent is computed by comparing the agents' information goals against the pages' contents. This computation is a variation of the computational cognitive model for information scent developed in [6]. The information scent used by the simulation may be the distal scent of the actual linked content, or the proximal scent of the linked pages as represented by a text snippet or icon. For the cases examined in this paper, we used simulations based on the distal information scent, but, as we shall illustrate, this turns out to be fruitful way of identifying problems with the way pages are presenting proximal information scent.

Usability Metrics

We are developing metrics to assess the quality of scent at a Web site in leading users to information they are seeking, and the cost of finding such information. One of these metrics involves (a) the specification of a user information goal (e.g., "Xerox products"), (b) the specification of one or more hypothetical starting pages (e.g., the Xerox home page), and (c) one or more target pages (e.g., a Xerox product index). Using the WUFIS simulator, agents traverse the Web site making navigation decisions based on the information scent associated with links on each page. The navigation decisions are stochastic, such that more agents traverse higher-scent links, but some agents traverse lower-scent links [1]. The simulation assumes that the agents either stop at the target page when found, or failing to find the page they surf up to some arbitrary amount of effort. We then assess the proportion of simulated agents that find the target page.

Network Representations of CUT

CUT graphs and various derivatives are readily extractable from most Web sites and the corresponding usage logs. In this representation, nodes in the graph correspond to Web pages and weighted directed edges correspond to the strength of association between any pair of nodes. For the analyses in this paper, we extracted the following graphs:

- *content similarity* graph [7], represents the similarity between Web pages as determined by the textual content of the pages. The edge values provide an approximate measure of the *topical relevance* of one page to another.
- *usage* graph [7], represents the proportion of surfers that traverse the hyperlinks between pages. The edge

values reflect how users "voted with their clicks" in finding relevant information.
- *co-citation* graph [10], reflects the frequency that two nodes were linked to by the same page. The edge values provide an indication of the *authoritative relevance* of pages to one another.

Spreading Activation Assessments of Scent

We use a spreading activation algorithm [7] on the various graphs to compute relevance or scent over a Web site. Conceptually, spreading activation pumps a metric called activation through one or more of the graphs. Activation flows from a set of source nodes through the edges in the graph. The amount of activation flow among nodes is modulated by the edge strengths. In this model, source nodes correspond to Web pages for which we want to identify related pages. After a few iterations, (subject to the selection of the appropriate spreading activation parameters), the activation levels settle into a stable state. The final activation vector over nodes defines the degree of relevance for a set of pages to the source pages.

Surfer Patterns Identified by LRS

A longest repeating subsequence (LRS) is a sequence of items where (1) subsequence means a set of consecutive items, (2) repeated means the item occurs more than some threshold T, where T typically equals one, and (3) longest means that although a subsequence may be part of another repeated subsequence, there is at least once occurrence of this subsequence where this is the longest repeating.

To help illustrate, suppose we have the case where a site contains the pages A, B, C, D, where A contains a hyperlink to B and B contains hyperlinks to both C and D. As shown in Figure 2, if users repeatedly navigate from A to B, but only one user clicks through to C and only one user clicks through to D (as in Case 1), the LRS is AB. If however more than one user clicks through from B to D (as in Case 2), then both AB and ABD are LRS. In this event, AB is a LRS since on at least one other occasion, AB was not followed by ABD. In Case 3, both ABC and ABD are LRS since both occur more than once and are the longest subsequences. Note that AB is not a LRS since it is never the longest repeating subsequence as in Case 4 for the LRS ABD.

Dome Tree Visualization

Chi et al. [2], developed a visualization called the Disk Tree to map large Web sites. See the right side of Figure 3 for an example. At the center of the Disk Tree is the root node, and successive levels of the tree are mapped to new rings expanding from the center. The amount of space

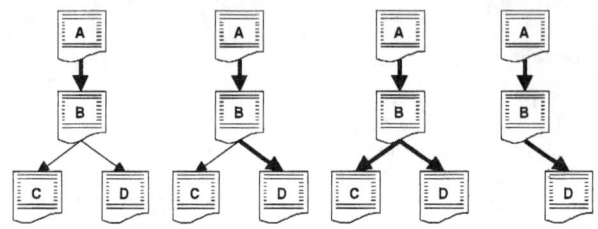

Longest Repeating Subsequences (LRS)
Case 1: AB   Case 3: ABC, ABD
Case 2: AB, ABD   Case 4: ABD

Figure 2. Examples illustrating the formation of longest repeating subsequences (LRS). Thick-lined arrows indicate more than one traversal whereas thin-lined arrows indicate only one traversal. For each case, the resulting LRS are listed.

given to each sub-tree is proportional to the number of leaf nodes it contains.

One of the limitations of this approach is that overlaying user paths on top of the Disk Tree occludes the underlying structure of the Web site, removing important visual data from the analyst's view. With our current focus on the flow of users through web sites, we designed a new technique called Dome Tree. In a Dome Tree, only 3/4 of the disk is used and at each successive level, the disk is extruded along the Z dimension. The rationale behind using extrusion is expanding the structure to 3D so that we can embed user paths in 3D rather than on the surface of the Disk Tree. By using only 3/4 of the disk, we can peer into the Dome through the opening like a door, without being occluded by the object itself. While this provided a useful layout, we sought to further minimize the impact of path crossings inherent in visualizing Web trails.

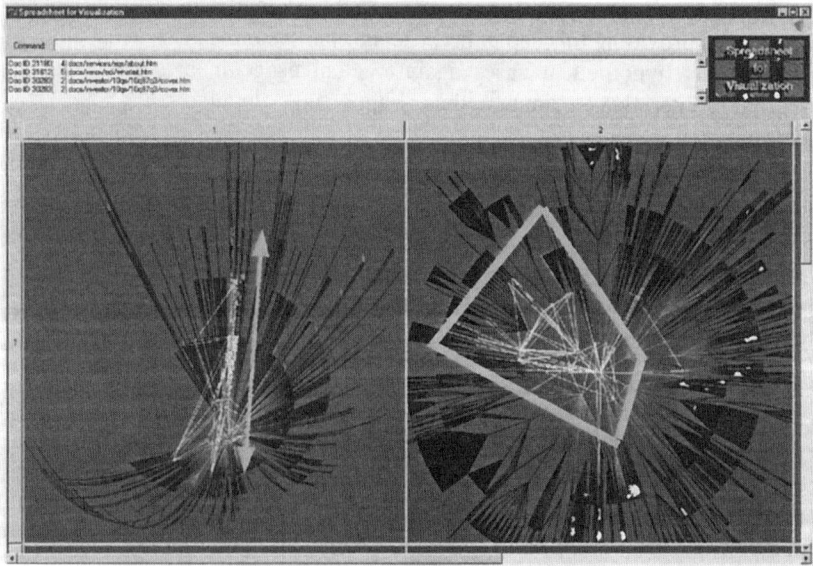

Figure 3: Dome Tree with Usage-Based Layout (left) shows that links (shown in yellow) are laid along significant paths (shown by orange arrow), eliminating crossings. In comparison, the traditional Disk Tree approach (right) has many crossing yellow links (shown in enclosed orange box). White arrows point to the current document being examined (investor.html).

Usage-Based Layout

To provide a visualization of Web paths with less path crossings, we developed new layout methods called Usage-Based Layout (UBL). UBL algorithms determine hierarchical relationships by various popularity metrics derived from user's paths and usage data. These methods represent a departure from traditional graph layout methods that rely exclusively upon the traversal of structural relationships.

Applied to the Web, UBL can also identify user paths between two pages even though no explicit hyperlink exists between the two pages. We call this *link induction*. Link induction finds usage paths that arise from the use of history buttons, search result pages, other dynamic pages, and so forth, which cannot be obtained by crawling the site.

To determine the hierarchical relationships between documents, we conduct a priority-based traversal based on usage data. Starting from the root node, its children are determined by looking at the existing hyperlink structure as well as the inducted links. Instead of using a simple queue as in a breadth-first traversal algorithm, we use a priority queue, where the top-most used page is chosen as the next node to expand. The expanded children list is then sorted in increasing usage order, and then inserted at the end of the queue along with their usage data. Then we proceed to the next highest-used child of the root node, which is at the top of the queue.

Figure 3 displays LRS user trails using the Dome Tree with UBL as compared with the Disk Tree. The green structure is the map of the Web site, and the yellow and blue lines represent user trails. This example demonstrates that we are able reduce trail crossings by using UBL.

By using a mouse-brushing technique, we highlight each node and show its URL and frequency of usage as the mouse moves over the documents on the Dome Tree (left of Figure 3). An orange ball highlights the current document of interest. The user is then allowed to pick a particular document to bring up additional details on it.

Web Trails

One of the details shown is the extracted Web Trails that are made by the users. All paths that lead into this document are called *History Trails*, which are shown in blue. All paths that spread out from this document are *Future Trails*, which are shown in yellow.

A dialog box also pops up, containing trail information related to this document (See Figure 4 for an example). The dialog shows the history and future portion of each path, along with its length and frequency. A scrollbar on the right enables the user to graze over this list. The bottom of the dialog box shows the documents that are on these paths, with their frequency of access, size, and URL.

Clicking on a path or a portion of a path narrows the list of documents to just the documents on that particular path. In this way, we enable analysts to drill down to specific paths of interest. Selecting a path also highlights it in the Dome Tree visualization in red.

Clicking on the Reverse Scent button in the dialog box dynamically computes a set of keywords using IUNIS that describes the information needs expressed by that path. The list is shown to the user in sorted order, with the most diagnostic words at the top.

We also compute and show an estimated download time of a user traversing this path using a modem. The estimation is derived from the total bytes of the files on the path. Analysts can therefore quickly judge the cost of traversing this path, and make appropriate judgements on the path's usability.

Scent Visualization

The user can choose to show several kinds of scent related to the selected document, including spreading activation based on Content Similarity, Co-citation, and Usage graphs, and WUFIS-computed scent flow. The system dynamically computes these scent assessments for each document and shows the result using red bar lines on the Dome Tree. The taller the red bar, the higher the scent.

By visually comparing the documents that lie on user trails and the computed scent, we can see whether users are finding the information that they need. This gives us a direct visual evidence of goodness of the design of the Web site. If the paths and the scents match, then users are navigating the Web site with success. If the paths and the scents mismatch, then it is possible that users are not finding the information because the Web site design gives inappropriate scents.

In practice, we have found Spreading Activation based on Content Similarity and Scent Flow computed by WUFIS to be very useful. Therefore, we have included this information as a column in the bottom portion of the dialog box. A mark of "C" means Content Spreading Activation predicted its relevance, and a mark of "S" means Scent Flow predicted its relevance.

Overview

This section described our system for the analysis and visualization of information scent, user surfing, and Web usability. The interactions between the different pieces of the components enable analysts to mine both the actual and predicted usage data of a large Web site. Looking at the architecture depicted in Figure 1, one important data flow through the components is

Log+Web site → LRS+Graph → Hierarchy → Dome Tree. This path uses the Usage-based Layout to compute a Dome Tree, which visualizes the whole site, with room to accommodate the Web Trails. Another data flow is

Log → LRS Paths → Web Trails → Embed on Dome Tree, which computes the appropriate trails that are to be embedded on the Dome Tree.

In the next section, we will show the tool in action, and present a number of case studies.

CASE STUDIES

Earlier in this paper, we presented questions that might be posed about surfers and Web sites. In this section, we illustrate the system by various analysis scenarios of the Xerox Web site[3]. Specifically, we will attempt to answer the following questions:

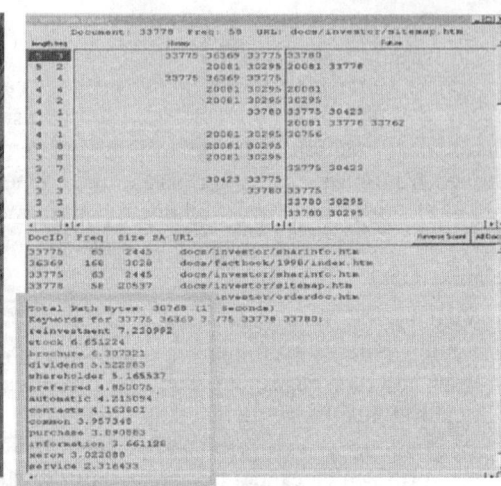

Figure 4: Multi-way Branching Point (investor/sitemap.htm) shown enclosed by orange lines, and Web Path detail dialog box (orange box shows the inferred user information need keywords, which are reinvestment, stock, brochure, dividend, and shareholder).

1. What pages act as *multi-way branching* points for user traversals? Do users branch on these pages? What pages behave as *pass-through* points?
2. For a page, what are the *well-traveled paths*? Do users find the desired information on these paths?
3. For well-traveled paths, what is the *users' information goal*? How can this information goal be extracted?
4. What are the *predicted useful information destinations*, given a specific information need? Does actual usage conform to these predictions? Why, or why not?

Page Types

Some pages act as indices, serving as *way points* in navigation patterns. Other pages act as *conduits* in a set of serially organized pages. Given these and other page types, the question arises, "How are users actually surfing these pages?" One may posit the design principle that effective way points should be kept around for good navigational scent. Once identified, ineffective way points can be redesigned, integrated with other content, or removed.

Figure 4 reveals a multi-way branching point where a few history paths lead into the branching point and result in a few well-traveled future paths. Upon drill-down, we discover that the branching page leads to several important destination pages, including the shareholder information page, the 1998 Xerox Fact Book, and a financial document-ordering page. While the page is relatively under-utilized (~60 accesses/day), our analysis shows it to be a very effective local sitemap. Within a few clicks, users are able to access the desired content.

Figure 5 shows an example of a pass-through point where UBL has laid out the pages in path-priority order. In traversing this path, some users leave the serial organization of the pages to find a related page (yellow path going to the red Content Spreading Activation page,

Figure 5: Pass-through Point in a series of pages (marked by orange arrows and current page pointed by white arrow is annualreport/1997/market.htm)

bottom right). Users then backtrack to continue surfing the serial links. From this inspection, we conclude that while it is a fairly well designed pass-through point, the page could potentially be improved to incorporate the related content directly. The tradeoff may be between coherence of the pages and navigational effort.

Well-Traveled Paths

Currently, most Web site visualizations focus on the identification of high usage areas. Our system identifies well-traveled paths by using a combination of two methods. First, the LRS computation reduces the number and complexity of user paths into manageable chunks. Second, embedding the paths onto the Dome Tree facilitates the visual extraction of well-traveled paths. We do not consider these methods perfect, rather they permit investigations that are otherwise difficult to attempt.

The left-hand image of Figure 6 illustrates the well-traveled paths related to a specific Web page (the TextBridge Pro 98 product page). As evidenced by the myriad of yellow future paths, related information is laid out across many different areas of the Web site, suggesting a possible redesign to bring more cohesion into the site. One

---

[3] Since the system is built for displays exceeding the resolution of paper, we have placed a copy of the figures for inspection online at: http://www.parc.xerox.com/uir/pubs/chi2000-scent.

interesting well-worn path is the serial pattern on the left (long arching yellow and blue path) that corresponds to the product tutorial pages. The right-hand image of Figure 6 shows the well-traveled paths extending from the Pagis product page. The zigzagging paths near the page indicate surfing between popular sibling pages. Many users travel the software demo tour and this is made explicit by the large blue path radiating upwards.

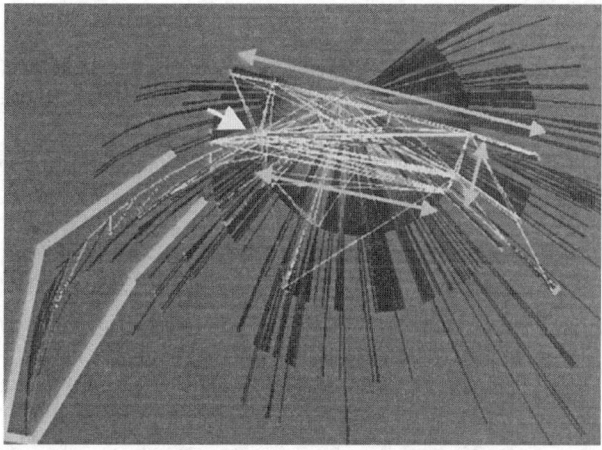

Figure 6: Well-traveled paths related to scansoft/tbpro98win/index.htm (left) and scansoft/pagis/index.htm (right), where major traffic routes are marked by orange lines.

In both of these examples, the red bar marks throughout the Dome Tree indicate the related pages to TextBridge and Pagis as computed by the Scent model. The correspondence of predicted related content to actual user paths suggests that the related content is not only reachable, but also well traveled by users. Visually, the yellow user paths that connect the red bars extending from the related pages reveal this correlation.

Identifying Information Need

Since well-traveled paths indicate items that compete well against other items for users' attention, it is important to find out, given a well-traveled path, what information need has the user expressed in that path. The bottom of Figure 4 shows the information need of a well-traveled path as computed by the reverse scent algorithm. The example is taken from a path related to investor/sitemap.htm. The top keywords computed by the reverse scent algorithm are reinvestment, stock, brochure, dividend, and shareholder. These keywords represent the goal of the users that traverse the path from the Shareinfo to the Orderdoc Web pages.

Figure 7 (corresponding to Figure 5) shows a more specific information need for the highly traversed path that starts at the employment recruitment page and winds through the 1997 Annual Report. In this case, some of the top keywords are reexamine, employment, socially, and morals, suggesting that potential Xerox employee are investigating the attitudes and culture of Xerox as expressed in the Annual Report. Another possible interpretation is that researchers are examining the correspondence between Xerox's employment policy and its social/moral position. In Figure 8, a large number of paths relate to how to upgrade previous versions of TextBridge96. A representative path shows top keywords as TextBridge, upgrade, OCR, Pro, bundled, software, windows, and resellers.

These examples suggest that we are able to automatically identify the information goals of users by first discovering

Figure 7: from annualreport/1997/market.htm (Figure 5)

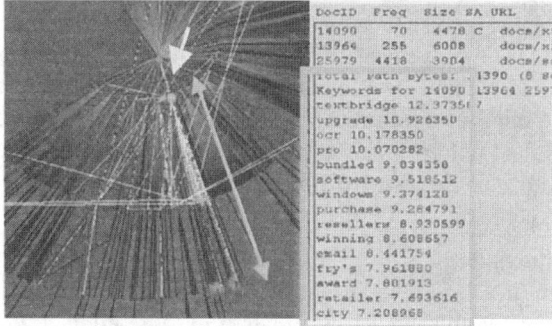

Figure 8: from xis/tbpro96win/index.htm

the well-traveled paths and then computing the informative keywords using the Scent Flow model. These examples help demonstrate that the Scent Model is not only good at predicting future user surfing behavior given a starting page, but also good at determining the information needs of a set of users given their paths through a site.

Predicted Destinations Based on Scent

One analysis centers on the differences between the WUFIS Scent Flow Model and actual user behaviors. We seek to answer the question, "Where in the Web site does the Scent Flow model differ from observed data, and why?"

For example, 100 hypothetical users interested in information related to Pagis product were simulated to flow through the Web site from two different entry points. Figure 9 shows the result of these two simulations, where actual user paths are encoded with yellow lines and the frequency of visit by the simulated users is encoded by the height of red bars. In the left-hand image, we placed users at scansoft/pagis/index.html, and watched the users surf to various points in the Web site, including pages relating to a tour of the software, release notes, and software registration pages. The correspondence of the yellow trails to the red pages reveals a match between the flow of real and simulated users. The right-hand image of Figure 9 displays the result of simulating users from products.html. It is immediately clear from the picture that many pages containing information relating to "Pagis" are found by the simulation, but real users are not finding pages.

 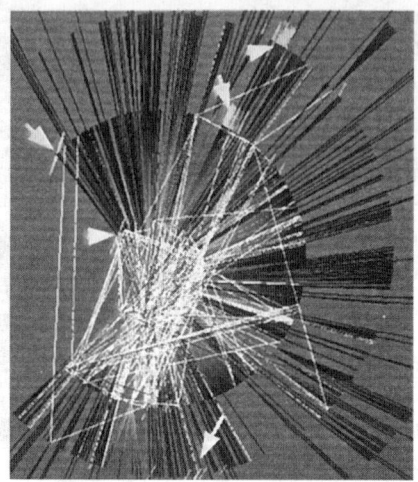

Figure 9: Given an information need related to "Pagis", Scent Flow simulation results in good match in scansoft/pagis/index.html (left, good match points pointed by orange arrows), but poor match from products.html (right, bad match points pointed to by purple arrows).

Upon careful examination we discovered that while the "Pagis" scent is contained near products.html, the scent is buried in layers of graphics and texts. The example shows that products.html does not adequately provide access to information relating to "Pagis".

There remain many limitations to the current system that remain for future work. Although we have ameliorated some of the visual clutter problems associated with visualizing Web sites and user paths, there is clearly much room for improvement. Techniques such as animation might aid in showing and comparing Web Trails. Another way to improve the current visualization of Web Trails is to fade colors out as we move into history or future portion of the path. To do this, we would have to first compute the aggregate path flow down each section over all paths.

CONCLUSION

Within the last few years, we have seen an explosive growth in Web usability as a field. Given its infancy, it is not surprising that there are so few tools to assist Web analysts. We presented a Scent Flow model for predicting and analyzing Web site usability. The analysis and visualization system presented in this paper is aimed at improving the design of Web sites, and at improving our understanding how users forage for information in the vast ecology of the Web.

Acknowledgement

This research was supported in part by an Office of Naval Research grant No. N00014-96-C-0097 to Peter Pirolli and Stuart Card.

REFERENCES

1. Chi, E. H., Pirolli, P., Pitkow, J. (1999) Using Information Scent to Model User Information Needs and Actions on the Web. (submitted).
2. Chi, E.H., Pitkow, J., Mackinlay, J., Pirolli, P., Gossweiler, R., and Card, S. (1998). Visualizing the Evolution of Web Ecologies. *Proceedings of the Human Factors in Computing Systems, CHI '98*. (pp. 400-407). Los Angles, CA.
3. Chi, E.H. and Riedl, J.T. (1998). An operator interaction framework for visualization systems. *Proceedings of the IEEE Information Visualization Symposium.* (pp. 63-70).
4. Furnas, G.W. (1997). Effective view navigation. *Proceedings of the Human Factors in Computing Systems, CHI '97* (pp. 367-374), Atlanta, GA.
5. Pirolli, P. (1997). Computational models of information scent-following in a very large browsable text collection. *Proceedings of the Conference on Human Factors in Computing Systems, CHI '97* (pp. 3-10), Atlanta, GA.
6. Pirolli, P. and Card, S.K. (in press). Information foraging. *Psychological Review.*
7. Pirolli, P., Pitkow, J., and Rao, R. (1996). Silk from a sow's ear: Extracting usable structures from the web. *Proceedings of the Conference on Human Factors in Computing Systems, CHI '96* Vancouver, Canada.
8. Pirolli, P. and Pitkow, J.E. (1999). Distributions of surfers' paths through the World Wide Web: Empirical characterization. *World Wide Web, 1*, 1-17.
9. Pitkow, J. and Piroll, P. (1999, in press). Mining longest repeated subsequences to predict World Wide Web surfing. *Proceedings of the USENIX Conference on Internet.*
10. Pitkow, J. and Pirolli, P. (1997). Life, death, and lawfulness on the electronic frontier. *Proceedings of the Conference on Human Factors in Computing Systems, CHI '97* (pp. 383-390).
11. Spool, J.M., Scanlon, T., Snyder, C., and Schroeder, W. (1998). Measuring Website usability. *Proceedings of the Conference on Human Factors in Computing Systems, CHI '98* (pp. 390), Los Angeles, CA.

Browsing Digital Video

Francis C. Li[2], Anoop Gupta[1], Elizabeth Sanocki[1], Li-wei He[1], Yong Rui[1]

[1]Collaboration and Multimedia
Microsoft Research
Redmond, WA 98052
{anoop, a-elisan, lhe, yongrui}@microsoft.com

[2]Group for User Interface Research, EECS Dept.
University of California, Berkeley
Berkeley, CA 94720-1776
fli@cs.berkeley.edu

ABSTRACT
Video in digital format played on programmable devices presents opportunities for significantly enhancing the user's viewing experience. For example, time compression and pause removal can shorten the viewing time for a video, textual and visual indices can allow personalized navigation through the content, and random-access digital storage allows instantaneous seeks into the content. To understand user behavior when such capabilities are available, we built a software video browsing application that combines many such features. We present results from a user study where users browsed video in six different categories: classroom lectures, conference presentations, entertainment shows, news, sports, and travel. Our results show that the most frequently used features were time compression, pause removal, and navigation using shot boundaries. Also, the behavior was different depending on the content type, and we present a classification. Finally, the users found the browser to be very useful. Two main reasons were: i) the ability to save time and ii) the feeling of control over what content they watched.

Keywords
Digital video, video browsing, video indexing, time compression, pause removal, next-generation video playback interfaces.

INTRODUCTION
One of the primary mediums for content creation and distribution is video. However, the way we watch video has not changed significantly since the invention of the analog video-cassette recorder (VCR) in the 1970s and 80s. The VCR makes it possible to watch video on-demand with the additional ability to pause, fast-forward, and rewind.

Today, Internet video streaming and set-top devices like ReplayTV [19] and TiVo [22] are technologies that are defining a new platform for interactive video playback. Unlike traditional VCRs, ReplayTV and TiVo store video in digital form (MPEG-2) on large hard disks. With digital video stored on hard disks and/or as Internet-based streaming media, instant random access into the content is possible. Seeking to a random location was possible with VCRs but had a large delay associated with it due to the use of tape storage. The instant random access facilitates features such as instant replay of just-observed action and rich indices into the content such as the chapter lists found on digital versatile disc (DVD) videos [7]. In addition, as computing costs continue to drop, processing techniques can be utilized to automatically generate indices or increase the playback speed while maintaining intelligibility. Such features potentially allow a viewer to save significant amounts of time watching a video and more effectively filter the content during playback.

Given this emerging new platform for interactive video playback, we explore the following questions in this paper:

- What potentially high-value features can we provide for browsing digital video?
- Will users derive significant benefits from their use and availability?
- How will the benefits vary with the task and type of content being watched?

We designed and implemented a prototype software video browsing application that provides a wide array of features enabled by digital video technologies. In addition to traditional VCR controls, the prototype provides rich indices for navigation (e.g., table of contents and video shot boundaries), speeded-up playback features (e.g., time compression and pause removal), the ability to make personal annotations that are anchored to the video timeline, and other advanced browsing controls.

Some of these features have been studied previously, but primarily in isolation and only for a narrow set of video content types. We evaluated the combined use of these features across a wide range of video content types: classroom lectures, conference presentations, sports, television dramas, news, and travel. This paper quantifies the use of the various features for the different content types and also documents viewers' subjective experiences. We also present an informal classification of video content types that helps predict the usefulness and applicability of the different browsing features and their impact on the viewing experience.

Permission to make digital or hard copies of all or part of this work for personal or classroom use is granted without fee provided that copies are not made or distributed for profit or commercial advantage and that copies bear this notice and the full citation on the first page. To copy otherwise, to republish, to post on servers or to redistribute to lists, requires prior specific permission and/or a fee.
CHI '2000 The Hague, Amsterdam
Copyright ACM 2000 1-58113-216-6/00/04...$5.00

RELATED WORK

Previous research in browsing digital media has often focused on either audio or video, but not both. The SpeechSkimmer [3,4] provided an interface for selecting time compressed and pause removed audio playback and for jumping back and forward between pre-defined segments of the recording. The Audio Notebook [21] used time-stamps of pen strokes to index audio and allowed time compressed playback.

For browsing video, the Hierarchical Video Magnifier [15] displayed frames near the current video position to provide context. Arman et al [2] improved the frame selection methods by detecting shot boundaries, which were found useful in editing systems [13]. The Classroom 2000 project at Georgia Tech [5] investigated richly indexed videos of lectures, including indexing based on strokes drawn on a black-board. However, none of these systems explore the wide range of browsing techniques and/or video content types explored here.

Christel et al [6] and He et al [8] have discussed techniques for shortening the viewing time of a video based on audio and/or video analysis. Such techniques, used in systems like CueVideo [17], condense the content into a shortened video summary intended to be watched in its entirety. The user does not control what is deleted to create the shortened summary and cannot browse the resulting video, the focus of this study.

The Informedia [9] project at CMU has performed substantial research in indexing and searching video in the context of information retrieval and digital library systems. Companies like Virage [22] and MediaSite [12] are currently providing these services for finding video on the Internet. Others have used domain knowledge to improve these services for specific video content types like news [11]. Such work focuses on query-based searching of collections of video content rather than on browsing an individual video that is the focus of this study.

The computer software industry has quickly embraced the Internet as a platform for digital video. However, the main focus of industry development has been the creation and distribution of content, not viewing or browsing. As a result, the leading software playback applications such as the Real Networks RealPlayer [18], Apple QuickTime Player [1], and Microsoft Windows Media Player [14] offer relatively few controls for browsing. In addition to the controls found on a VCR, these applications add a seek bar allowing random access via a "thumb" and a table of contents index.

The consumer electronics industry has begun to incorporate more advanced browsing features in the next generation of hardware video playback devices. DVD Video players support random access using a table of contents index. ReplayTV and TiVo set-top boxes offer an index to recorded shows. In addition, they provide the ability to jump forward by 30 or 60 seconds, possibly allowing skipping of commercials, and back 8–10 seconds for "instant replays." However, these devices do not currently provide features like time compression or shot boundary frames. The user interface design is also quite different as input must be performed using a remote control. Finally, no public data is available on how these devices are actually being used.

PROTOTYPE FEATURES AND FUNCTIONALTIY

Our study used two video browsers: "Basic" and "Enhanced." The enhanced browser was developed using a modified version of the Microsoft Windows Media Player. The basic browser leveraged the same software, but displayed only a subset of the functionality.

Basic browser controls: The basic controls provide the features typically found on current software video playback applications. They include *Play*, *Pause*, *Fast-forward*, *Seek*, *Skip-to-beginning* of video, and *Skip-to-end* of video. No audio was played during fast-forward as is common with current media players, and seek was accomplished by dragging the seek thumb on the timeline in the interface. Due to limitations of the Windows Media Player, a traditional rewind feature could not be provided.

Enhanced browser controls: Figure 1 shows the user interface for the enhanced browser. The following additional controls were provided:

- Speed-up controls: *Time compression (TC), Pause removal (PR)*
- Textual indices: *Table of contents (TOC), Notes*
- Visual indices: *Shot boundary (SB) frames, Timeline markers*
- Jump controls: *Jump-back, Jump-next*

The speed-up controls allow the user to shorten the viewing time of a video. *Time compression* (TC) uses signal processing techniques to increase the playback speed while preserving the pitch of the audio. *Pause removal* (PR) detects pauses and silence in continuous speech and removes both the audio and video segments associated with them.

The textual indices are lists of text entries that describe locations in the video. The user can seek to the location in the video by selecting the associated entry. The *table of contents* (TOC) index is a pre-generated list of entries that cannot be modified. In contrast, the *notes* index is generated from end-user annotations. When the user creates a note, the comment entered by the user is anchored to the current position of the video. We expected that users might use the notes feature to bookmark significant parts of the video for later reference and to record their thoughts regarding the content of the video.

The visual indices are the shot boundary frames and the timeline markers. The numbered *shot boundary frames* (SB) allow the user to visually identify and then seek to a

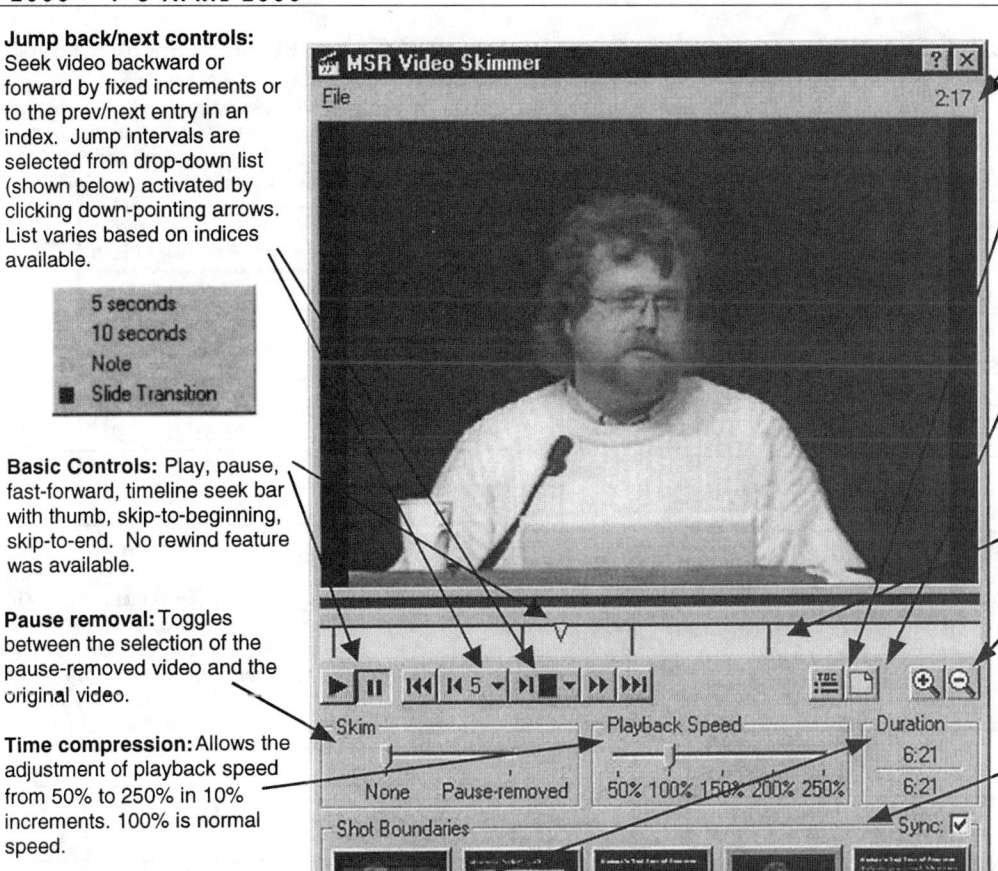

Jump back/next controls: Seek video backward or forward by fixed increments or to the prev/next entry in an index. Jump intervals are selected from drop-down list (shown below) activated by clicking down-pointing arrows. List varies based on indices available.

Basic Controls: Play, pause, fast-forward, timeline seek bar with thumb, skip-to-beginning, skip-to-end. No rewind feature was available.

Pause removal: Toggles between the selection of the pause-removed video and the original video.

Time compression: Allows the adjustment of playback speed from 50% to 250% in 10% increments. 100% is normal speed.

Duration: Displays the length of the video taking into account the combined setting of Pause-removal and Time compression controls.

Elapsed time indicator

Table of contents: Opens separate dialog with textual listings of significant points in the video. Contains "seek" feature allowing user to seek to points in the video. Index entries are also indicated on the Timeline seek bar.

Personal notes button: Opens separate dialog with user-generated personal notes index. Contains "seek" feature allowing user to seek to the points in video. Notes index entries also indicated on Timeline seek bar.

Timeline Markers: Indicate placement of entries for TOC, shot boundaries, and personal notes.

Timeline zoom: Zoom in and zoom out.

Shot boundary frames: Index of video. Shot is an unbroken sequence of frames recorded from a single camera. Shot boundaries are generated from a detection algorithm that identifies such transitions between shots and records their location into an index. Current shot is highlighted as video plays (when sync box is checked). User can seek to selected part of video by clicking on shot.

Figure 1. Enhanced Browser User Interface

particular shot by clicking on it. As the video plays, the frame corresponding to the currently playing shot is highlighted. The *timeline markers* show the location of the TOC and notes entries with color coded bars. They can be used to judge the location of entries relative to the current position of the video (shown by the thumb).

The *jump-back* and *jump-next* controls seek the video backward or forward, respectively, by a fixed interval or to entries in an index. Users can jump by 5 seconds, 10 seconds, TOC entry, note, or shot boundary. It was hypothesized, for example, that a user might jump back 5 or 10 seconds to repeat parts of the video, whereas the jump next TOC entry control might be used to preview the first few minutes of each consecutive entry in the TOC. Also, it is very difficult to do these operations using the seek thumb. For example, a one-hour video (3600 seconds) spread across roughly 400 pixels (width of our browser) means that moving the thumb one pixel seeks 9 seconds.

Our goal for the prototype was to expose video browsing functionality with a user interface adequate for evaluation. Both the basic and enhanced browsers were instrumented to record the usage of each feature during the study.

USER STUDY DESIGN

The user study was designed to evaluate both feature usage and overall experience with the enhanced browser. Participants were presented with a scenario and browsing task related to one of six video content types: classroom lectures, conference presentations, sports, television dramas, news, and travel.

Each participant completed his or her video browsing task three times. The task was first completed using the basic browser. Then, after a short practice tutorial, the enhanced browser was used for the last two tasks. To encourage browsing behavior, the participants were limited to 30 minutes to browse a 45–60 minute video.

In addition to pre- and post-study surveys, the participants completed a survey after each task. They were asked to describe their browsing strategy and rate their interest in the content of the video, the quality of their experience, and the usefulness of the available features.

The participants were recruited from a pool of non-employees that expressed interest in usability studies at Microsoft. They were screened for two years of computer experience and matching interests with one of the scenarios.

Table 1. **Average Ratings of Feature Usefulness.** The highest rated Enhanced browser feature for each scenario is highlighted. FF = Fast forward, SB = Shot boundaries, TC = Time compression, PR = Pause removal, Jmp = Jump-back & -next, TOC = Table of contents, Bas = Basic browser, Enh = Enhanced browser. Scale: 1 = strongly disagree, 4 = neutral, 7 = strongly agree.

	Seek		FF		SB	TC	PR	Jmp	TOC	Note
	Bas	Enh	Bas	Enh	Enhanced Browser					
Classroom	4.8	5.6	4.4	4.1	5.0	5.4	5.1	4.8	6.8	3.5
Conference	5.6	4.1	3.6	3.3	4.9	6.9	6.5	5.1	N/A	3.8
Sports	5.2	4.7	5.6	5.9	6.1	5.7	4.3	5.6	5.3	4.5
Shows	5.0	3.6	4.4	4.3	5.1	6.0	4.3	2.8	N/A	2.5
News	5.8	4.9	5.4	4.3	6.4	6.7	6.6	5.6	6.6	4.6
Travel	5.2	5.7	5.4	4.2	6.3	6.6	6.0	6.3	N/A	6.4
Overall	5.3	4.8	4.8	4.4	5.6	6.2	5.5	5.0	6.2	4.1

Table 2. **Average Number of Times Feature Used per Participant per Video.** The most frequently used Enhanced browser feature for each scenario is highlighted. SB Sk = Shot boundary seek, Jmp Bck/Nxt = Jump Back/Next, TOC Sk = Table of contents seek, Note Sk = Note Seek

	Seek		FF		SB Sk	Jmp Bck	Jmp Nxt	TOC Sk	Note Add	Note Sk
	Bas	Enh	Bas	Enh	Enhanced Browser					
Classroom	21.6	0.0	10.8	0.0	1.5	4.5	2.0	12.5	0.0	0.0
Conference	15.7	0.5	4.2	0.0	2.0	0.5	7.0	N/A	3.0	1.0
Sports	20.0	7.0	12.8	4.5	26.5	0.0	4.0	1.5	2.0	0.5
Shows	14.8	3.0	9.8	1.0	4.5	0.0	11.0	N/A	0.0	0.0
News	34.0	0.5	10.2	0.0	9.5	2.0	10.5	3.5	1.0	0.0
Travel	51.8	3.0	11.0	0.0	55.0	14.5	4.5	N/A	9.5	5.0
Overall	26.3	2.3	9.8	0.9	16.5	3.6	6.5	5.8	2.6	1.1

Five participants per scenario completed the study for a total of 30 participants. Each participant received a Microsoft software product for his or her involvement.

SCENARIOS AND RESULTS

In this section, we describe the browsing scenarios in detail and discuss the corresponding results of the study, but first we present the data that we will reference.

Table 1 presents the average rating of feature usefulness over the participants in each scenario and overall, calculated from surveys completed after each task. Table 2 presents the average number of times features were used by a participant while watching a video. Table 3 shows the average effective playback speed attained using time compression and the combination of time compression and pause removal. Table 4 shows the average percentage of a video watched and decomposes that value into the percentage of video watched only once, exactly twice, and three or more times. Finally, Table 5 shows, on average, what percentage of the task time was spent with a video in different playback modes.

Table 3. **Average Effective Playback Speed Attained with the Enhanced Browser.** Gain indicates percentage increase over time compressed with no pause removal. The first column is calculated by taking the total length of video watched divided by the total actual viewing time. The effects of pause removal are added by including the length of the deleted pauses into the total length of video.

	Time Comp.	Time Comp. and Pause Removed (Gain)
Classroom	123.4%	137.1% (11.1%)
Conference	122.0%	147.1% (20.6%)
Sports	116.8%	137.1% (17.4%)
Shows	132.6%	146.1% (10.2%)
News	117.8%	138.5% (17.5%)
Travel	132.0%	138.9% (5.2%)
Overall	124.1%	140.8% (13.5%)

Table 4. **Average Percentage of Video Watched.** This table shows the average percentage of a video watched (%W) and decomposes that value into the percentage of video watched only once (%W1), exactly twice (%W2), and three or more times (%W+). The highlighted entries show that nearly 20% more of a video was watched with the Enhanced browser than with the Basic browser.

	Basic				Enhanced			
	%W	%W1	%W2	%W+	%W	%W1	%W2	%W+
Classroom	33.0	32.5	0.5	0.0	48.2	41.1	6.3	0.8
Conference	64.4	62.6	1.6	0.2	86.1	75.2	10.0	0.9
Sports	21.8	20.7	1.1	0.0	41.3	34.1	5.6	1.6
Shows	40.5	40.3	0.2	0.0	53.8	53.0	0.8	0.0
News	35.0	33.5	1.5	0.0	51.9	46.7	4.8	0.4
Travel	26.5	20.1	5.7	0.7	57.2	30.9	11.6	14.7
Overall	36.9	35.0	1.7	0.2	56.4	46.8	6.5	3.1

Table 5. **Average Percentage of Task Time Spent in Playback Modes.** Using the Enhanced browser participants spent considerably less time watching video at normal playback speeds than with the Basic browser. Playing = Normal playback speed, FF = Fast forward, TC = Time compressed, PR = Pause removed.

	Paused		Playing		FF		TC	PR	TC&PR
	Bas	Enh	Bas	Enh	Bas	Enh	Enhanced		
Classroom	15.7	12.5	74.4	33.4	9.9	0.4	24.7	5.6	23.4
Conference	14.4	16.1	83.4	9.8	2.2	0.0	13.4	2.6	58.1
Sports	9.1	4.5	53.2	35.8	37.7	10.7	21.1	6.5	21.4
Shows	15.7	4.8	72.0	21.5	12.3	0.9	33.0	0.2	39.6
News	10.8	10.0	73.9	10.8	15.3	0.0	22.5	12.3	44.4
Travel	14.3	21.5	67.2	21.6	18.5	0.0	23.9	1.1	31.9
Overall	13.3	11.6	70.7	22.2	16.0	2.0	23.1	4.7	36.5

Classroom Lecture
Many educational institutions videotape courses for archival and rebroadcast. Stanford University, for example, offers hundreds of courses each year, both live and on-demand, via television broadcast, videotape, and Internet delivery [20]. In the classroom lecture scenario, the participants were asked to imagine they were taking a C programming class. A quiz was going to be administered in ½ hour but they did not attend the previous one-hour lecture. The task was to watch an archived video of the lecture and summarize the main points of the content.

The time constraint ensured that the participants would not be able to watch the entire video. However, the participants were selected based upon previous programming experience in a language other than C. Since many programming concepts are similar across different languages, it was presumed that the participants could effectively skim the video based upon previous knowledge.

Using the basic browser, though, the participants had a difficult time skimming the video. The participants fast-forwarded through topics and skipped topics using the seek thumb. However, with no indication of the position of topic changes, the participants made random guesses to seek. Table 2 shows they used the seek thumb an average of 21.6 times in the half hour, or roughly once every 1.5 minutes.

Using the enhanced browser, the participants in this scenario used the TOC to seek the video with greater frequency than any other scenario (avg. use 12.5 times, Table 2). The TOC was generated from slides used in the lecture. In addition, they also made considerable use of TC and PR. This increased the fraction of content they watched once or more from 33.0% to 48.2% (Table 4), corresponding to an effective playback speed of 137.1% (Table 3). The TOC, TC, and PR were the highest rated controls (6.8, 5.4, and 5.1 respectively, Table 1).[1]

The basic browser interface is not unlike that of the VCRs that many Stanford students use in campus libraries to watch videotaped courses. Providing a TOC index as well as TC and PR could make video a more useful and efficient tool for reviewing courses.

Conference Presentation
Conferences and seminars are valuable means for keeping up with the latest trends. Electronically accessible on-demand presentations provide the flexibility of anytime, anywhere viewing. We believe browsing capabilities can potentially be of great value when time is limited and as the number of presentations increases.

The participants were asked to pretend they had ½ hour before attending a meeting with co-workers to discuss a conference they had attended. The participants did not attend the same presentations as their co-workers, but would still like to take part in the discussion. The task was to review a video of a missed presentation and summarize the main points in preparation for the meeting.

The videos were selected from the ACM 97 presentations of "The Next 50 Years of Computing" and ranged in length between 40–50 minutes. Participants were recruited based upon background interests in the future of computing and education. Unlike the classroom lecture scenario, the contents of videos were not technical or highly structured, so a TOC was not generated for the enhanced browser.

Using the basic browser, the participants used the seek thumb and fast forward to skim the video much like in the classroom scenario.

Using the enhanced browser, the highest rated controls were TC and PR (6.9 and 6.5, Table 1). On average, an effective playback speed of 147.1% was attained by the participants (Table 3) and, as compared to the basic browser, they covered 86.1% of the content instead of 64.4% (Table 4). Shot boundary frames were used twice on average, usually to skip lengthy introductions as the transition between the host and the speaker could be seen in the frames.

Although the average rating was neutral (3.8, Table 1), personal notes were used by several participants. Two of the five participants used notes to mark interesting locations in the video. One of them included the shot boundary frame number in the title of her notes, providing a visual indicator for their location. Both participants used their notes to review the main points of the video for their summary. A third participant used the notes feature to bookmark the start and end of video segments he skipped to review them later if time allowed. This behavior suggests the need for a quick bookmark feature that does not require typing a title for a note and/or a logging feature that can automatically marks the portions of a video skipped.

Overall, as in the classroom scenario, TC and PR made it possible to watch substantially more of the presentation in the limited time available. In the absence of a TOC, shot boundaries and notes were utilized to effectively mark and jump to locations in the video.

Sports
Sports programming is one of the most popular forms of video entertainment. In the sports scenario, we wanted to see how participants would react to the added ability to browse and skim events. Each participant reported that they watched sports or sports news shows regularly.

The task was to find highlights in a baseball game to discuss with friends at the health club in ½ hour. A single baseball game was divided into three one-hour segments and presented in order to the participants. Since baseball can have long periods of little or no scoring activity, it was expected that there would be ample opportunity to skim the video. As an aid, a TOC was generated for the enhanced

---
[1] Although "Seek" is rated high in Table 1, notice that it is used zero times in the enhanced browser (Table 2). The high rating is due to the fact that the participants thought of TOC also as a seek mechanism.

browser indexing the top and bottom of each inning in the video (about 6 entries per video).

Using the basic browser, most of the participants started out by using fast-forward to skip commercials and dead time between plays. The participants spent an average of 37.7% of their time watching the game in fast-forward (Table 5), more than any other scenario. Play highlights can be identified visually, so the lack of audio during fast-forward was not a deterrent. Fast-forward, however, was not enough to skim the video in ½ hour. As a result, the participants also frequently used the seek thumb (average 15.7 times, Table 2).

Using the enhanced browser, the participants most frequently used the shot-boundary frames to seek the video (average 26.5 times, Table 2) and rated it highest in surveys (average 6.1, Table 1). Using the five frames at the bottom of the browser, the participants could determine the outcome of the current play. By scrolling the frames ahead, the participants could preview and seek to successive plays. In contrast, the TOC inning index was only used once or twice, mainly to skip the ads at the end of an inning.

TC, PR, and fast-forward were also very popular in the enhanced browser. Unlike other scenarios, fast-forward was still useful as it allowed greater speed-up than time compression, and key information was in the video channel anyway. TC and PR combined resulted in an effective playback speed of 137.1% (Table 3).

In this scenario, we saw the development of more sophisticated strategies over time. For example, when watching the second video using the enhanced browser, two participants chose to watch the home team at bat while *completely* skipping the visitors. Another two participants used the notes feature to bookmark interesting plays for later reference. Both strategies demonstrate the user being in control over the game, unlike watching a set of highlights from a news show.

When asked if the availability of the enhanced browser would affect how they watched television, the average response increased from 4.2 (neutral) using the basic browser, to 6.0 (agree) after the second use of the enhanced browser (scale of 1–7, 7 being strongly agree). Similarly, when asked about the quality of their experience, ratings increased from 4.8 to 6.0 (scale of 1–7, 7 being best).

The participants in this scenario found the enhanced browser to be a useful tool and enjoyed being able to browse the game. Features that support skimming visually, such as shot boundaries, were more useful here than in previous scenarios. TC and PR continued to be of high value too.

Shows

As in the sports scenario, we wanted to see how participants would react to the ability to browse a one-hour television show. Each participant regularly watched at least one weekly television show. They were asked to pretend that the final episode of their favorite show was airing in ½ hour, but they still needed to watch the previous episode that they had recorded.

The task was to review the major events in the show before watching the final episode. Each participant watched a full episode of "E.R.," "Ally McBeal," and "Babylon 5" (including commercials). We knew that the browsing features would be used to skip commercials. However, how each participant might choose to browse the content of the shows might depend heavily upon personal preference.

Using the basic browser, it was not possible for the participants to watch the entire show in ½ hour even if they skipped commercials. The seek thumb was used 14 times on average, or roughly one seek every 2 minutes (Table 2). The participants reported that they seeked randomly.

Using the enhanced browser, TC was the highest rated feature (6.0, Table 1). It was used to increase the amount of the show watched from an average of 40.5% in the basic condition to 53.8% over the enhanced conditions (Table 4). The second highest rated feature was shot boundaries (5.1, Table 1). By scrolling the shot boundary frames, the participants could instantly and accurately skip commercials. The average use of 5 shot boundary seeks corresponds roughly to the number of commercials in a one-hour show (Table 2). Otherwise, the participants did not exhibit any particular browsing behavior.

When asked to rate satisfaction of coverage of the show, the participants reported an increase from 3.4 using the basic browser to 5.4 after the second use of the enhanced browser (scale of 1 to 7, 7 being best coverage). However, unlike the sports condition, the participants did not agree that the availability of a video browser would affect the way they watched television, reporting an average 3.6 for the basic browser and 4.3 for enhanced browser (scale of 1 to 7, 7 being strongly agree). The participants all reported that they would not regularly watch a show under such time constraints. One participant complained that watching a show with TC and PR was "mentally fatiguing."

News

The participants in the news scenario were asked to pretend that they were forced by family members to spend less time watching the news. The task was to watch a one-hour news show in the ½ hour before dinner and summarize the program for discussion at the table. Each participants reported that they watched at least ½ hour of news daily.

The participants were presented three consecutive airings of "The News Hour with Jim Lehrer" which consists of a general news summary followed by five in-depth reports. Since the content is highly structured into discrete story segments, we expected that the participants would want to choose the stories they were interested in watching. A TOC was generated for the enhanced browser to index the beginning of the news summary and each story.

Using the basic browser, the seek thumb was used heavily (34.0 times, Table 2). The participants had to make many guesses to find the beginnings of stories in the video.

Using the enhanced browser, participants were able to use TC and PR to watch more of the video (35.0% watched with basic vs. 51.9% with enhanced, Table 4). In addition, the TOC made it possible for participants to "select which one [story to watch] or in which order I watched them". Like the classroom scenario, TC, PR, and the TOC were the highest rated features (6.7, 6.6, 6.6, respectively, Table 1).

Unlike the classroom scenario, though, shot boundary frames proved to be a useful preview feature for the participants as they watched the video (average rating of 6.4 versus 5.0 for classroom, Table 1). Participants would scroll the frames to get an overview of the contents of a story, using the jump-next button or clicking on a frame to skip ahead.

Ultimately, all the participants felt that they could better cover the news program using the enhanced browser, with an average satisfaction of coverage rating of 6.6 versus an average rating of 4.8 in the basic (scale of 1–7, 7 being best coverage). When asked if a video browser would affect the way they watched television, the participants were more enthusiastic than those in the sports scenario, rating an average of 6.9 (scale of 1–7, 7 being strongly agree).

Overall, as in the sports scenario, the participants enjoyed the additional control the browser afforded them. News is a very rich video content type, and browsers can take advantage of both textual and visual indices for searching as well as TC and PR for saving time.

Travel

Travel videos are often used to preview destination getaways. The participants in this scenario were asked to form a five minute summary of a travel video by identifying the begin and end points of interesting clips. The summary would be used as a potential travel itinerary to convince their families where they wanted to go on their vacation.

Each participant reported an interest in travel as well as having planned or taken a vacation in recent years. The travel videos contained tourist points of interest and used narrator voice-overs to describe the scenes.

Using the basic browser, the seek thumb was used nearly twice as often as in the next most used scenario (average 51.8 times vs. 34.0 for news, Table 2). The greater accuracy needed for finding the begin and end points of clips required many adjustments using the seek thumb.

In the enhanced condition, the participants relied on the shot boundary frames to navigate the videos, using them to identify interesting looking destinations. They used the shot boundary frames to seek an average of 55.0 times versus an average of 16.5 over all the scenarios (Table 2) and rated it the third most useful feature (6.3, Table 1).

The notes were invaluable for marking the start and end points of clips, receiving its highest rating across the scenarios (6.4 versus 4.1 overall, Table 1). An average of 9.5 notes were added by each participant versus 2.6 overall (Table 2). They often positioned their notes by hitting the jump-back button after noticing an interesting landmark. Jump-back was also used the most (14.5 times, Table 2) and rated the highest in this scenario (6.3, Table 1).

Ultimately, the participants rated TC the highest in this scenario (6.6, Table 1). TC and PR made it possible to watch almost twice as much of the video, increasing from 26.5% using the basic browser to 57.2% using the enhanced (Table 4). When asked to rate the quality of their itinerary summaries, the participants reported an increase from 4.4 using the basic browser to 5.8 by the second use of the enhanced browser (scale of 1–7, 7 being best).

As for other scenarios, TC and PR were used very effectively. However, the rich visual content of the videos also allowed for effective shot boundary based navigation. In addition, with the added requirement of precise positioning, the participants found notes and the jump buttons very useful.

CONCLUDING REMARKS

Overall, the participants exhibited substantially different viewing behavior when they used the enhanced video browser. For example, the traditional seek and fast-forward controls were almost never used. Instead, features like the table of contents and shot boundaries were used to more accurately jump to locations in the video. In addition, the participants spent a substantial amount of time watching the video with time compression and pause removal, increasing the amount of video they watched by 20%.

In the sports and news scenarios, the participants enthusiastically agreed that having enhanced browsing features would affect the way they watched television. In these scenarios, we observed content-specific browsing behavior using the enhanced features such as watching only the home team of a sports game or choosing the order of stories to watch in a news show.

Based on such patterns of feature use and experience across the scenarios, we can also informally classify our six video content types into different categories: informational audio-centric, informational video-centric, and narrative-entertainment.

Informational audio-centric videos like classroom lectures and conference presentations contain most of their content in the audio channel and usually have little visual activity. As such, visual browsing features like shot boundary frames provide minimal cues. For structured content, a TOC provides a valuable index, although users can take advantage of notes and shot boundaries to form their own ad-hoc index when it is unavailable.

With *informational video-centric* content like travel and sports videos, the rich video content makes shot boundary frames an effective navigation tool. When combined with notes and the jump-back button, it was possible to accurately position the video. News can fall equally into both the informational audio-centric and informational video-centric categories, and can take advantage of a combination of the different indices for effective browsing.

When watching *narrative-entertainment* like television dramas, the viewing experience was affected when the participants were forced to use browsing features like TC and PR. One participant succinctly stated the general sentiment: "I saved time but I would seldom want to watch a show in a fast version."

However, when watching news and sports, the participants reported the opposite response. A sports participant remarked that "anything to remove excess time from viewing is positive." A news participant went further to say that "saving time isn't the best part—being in control is". The features provided the ability to "move to what interested me most and then return to the other segments as time permitted."

In the travel scenario, the participants believed that the enhanced features could be useful for editing. When asked about the technology, one participant responded: "It's exciting. I think editing home movies would be fun." Another remarked, "I would buy this software in a minute if it would allow me to edit video."

In general, we are greatly encouraged by the participants' positive reaction to the enhanced browser. However, this study is only the first step in evaluating the potential of these advanced browsing and skimming features. We need to look at the results from a larger number of subjects, ideally in more natural task environments where we can measure usage over a longer period of time. Having performed this broad study across the six different content types, we can now design for specific video browsing tasks with a better understanding of how these features can be applied and evaluated.

ACKNOWLEDGMENTS
We would like to thank Asta Glazer and Cindy Solomon for helping with the initial design of the browser prototype.

REFERENCES
1. Apple QuickTime Player, Apple Corporation Inc., http://www.apple.com/quicktime/
2. Arman, F., Depommier, R., Hsu, A., and Chiu, M.-Y. "Content-based browsing of video sequences." In *Proceedings of the second ACM international conference on Multimedia '94*, 1994, Page 97
3. Arons, B. "SpeechSkimmer: A System for Interactively Skimming Recorded Speech." *ACM Transactions on Computer Human Interaction, 4,* 1, 1997, 3-38.
4. Arons, B. "Techniques, Perception, and Applications of Time-Compressed Speech." In *Proceedings of 1992 Conference, American Voice I/O Society,* Sep. 1992, pp. 169-177.
5. Brotherton, J. A., Bhalodia, J. R., and Abowd, G. D. "Automated Capture, Integration, and Visualization of Multiple Media Streams." In the *Proceedings of IEEE Multimedia '98,* July, 1998.
6. Chistel, M. G., Smith, M., Taylor, C. R., and Winkler, D. B., "Evolving video skims into useful multimedia abstractions". In *Proceedings of CHI '98* (Los Angeles, CA, 1998), ACM Press, 171-178.
7. DVD Video Group, http://www.dvdvideogroup.com/
8. He, L., Sanocki, E., Gupta, A., and Grudin, J., "Auto-summarization of audio-video presentations". In *Proceedings of the Conference on ACM Multimedia 99,* 1999, Pages 489-498.
9. Informedia, http://www.informedia.cs.cmu.edu/
10. Komlodi, A. and Marchionini, G. "Key frame preview techniques for video browsing." In *Proceedings of the third ACM Conference on Digital libraries,* 1998, Pages 118 – 125.
11. Low, C. Y., Tian, Q., and Zhang, H. "An automatic news video parsing, indexing and browsing system." In *Proceedings ACM Multimedia '96,* 1996, Page 425.
12. MediaSite, http://www.mediasite.com/
13. Meng, J. and Chang, S. "CVEPS - a compressed video editing and parsing system." In *Proceedings ACM Multimedia '96,* 1996, Page 43.
14. Microsoft Windows Media, Microsoft Corporation Inc., http://www.microsoft.com/windows/windowsmedia/
15. Mills, M., Cohen, J., and Wong, Y. Y., A magnifier tool for video data, in *Proceedings of CHI '92,* 1992, ACM Press, 93-98.
16. Omoigui, N., He, L., Gupta, A., Grudin, J., and Sanocki, E., Time-Compression: Systems Concerns, Usage, and Benefits, in *Proceedings of CHI '99* (Pittsburgh, PA, 1999), ACM Press, 136-143.
17. Ponceleon, D., Strinivasan, S., Amir, A., Dragutin, P., and Diklic, D. "Key to effective video retrieval: effective cataloging and browsing." In *Proceedings of the 6th ACM international conference on Multimedia,* 1998, Pages 99 – 107.
18. Real Networks RealPlayer, http://www.real.com/
19. Replay Networks ReplayTV, http://www.replaytv.com/
20. Stanford Online, http://stanford-online.stanford.edu/
21. Stifelman, L. "The Audio Notebook: Paper and Pen Interaction with Structured Speech" *Ph.D. dissertation, MIT Media Laboratory,* 1997.
22. TiVo Inc., http://www.tivo.com/
23. Virage, http://www.virage.com/

Comparing Presentation Summaries:
Slides vs. Reading vs. Listening

Liwei He, Elizabeth Sanocki, Anoop Gupta, Jonathan Grudin
Microsoft Research
One Microsoft Way, Redmond, WA 98052
+1 (425) 703-6259
{lhe,a-elisan,anoop,jgrudin}@microsoft.com

ABSTRACT
As more audio and video technical presentations go online, it becomes imperative to give users effective summarization and skimming tools so that they can find the presentation they want and browse through it quickly. In a previous study, we reported three automated methods for generating audio-video summaries and a user evaluation of those methods. An open question remained about how well various text/image only techniques will compare to the audio-video summarizations. This study attempts to fill that gap.

This paper reports a user study that compares four possible ways of allowing a user to skim a presentation: 1) PowerPoint slides used by the speaker during the presentation, 2) the text transcript created by professional transcribers from the presentation, 3) the transcript with important points highlighted by the speaker, and 4) a audio-video summary created by the speaker. Results show that although some text-only conditions can match the audio-video summary, users have a marginal preference for audio-video (ANOVA f=3.067, p=0.087). Furthermore, different styles of slide-authoring (e.g., detailed vs. big-points only) can have a big impact on their effectiveness as summaries, raising a dilemma for some speakers in authoring for on-demand previewing versus that for live audiences.

Keywords
Video abstraction, video summarization, digital video library, video browsing, video skim, multimedia.

INTRODUCTION
Digital multimedia content is becoming pervasive both on corporate intranets and on the Internet. Many corporations are making audio and video of internal seminars available online for both live and on-demand viewing, and many academic institutions are making lecture videos and seminars available online. For example, research seminars from Stanford, Xerox PARC, University of Washington

Permission to make digital or hard copies of all or part of this work for personal or classroom use is granted without fee provided that copies are not made or distributed for profit or commercial advantage and that copies bear this notice and the full citation on the first page. To copy otherwise, to republish, to post on servers or to redistribute to lists, requires prior specific permission and/or a fee.
CHI '2000 The Hague, Amsterdam
Copyright ACM 2000 1-58113-216-6/00/04...$5.00

and other sites can be watched at the MURL Seminar Site (http://murl.microsoft.com). Microsoft's corporate intranet has hundreds of presentations available. Close to 10,000 employees have watched one or more presentation [7]. These numbers are likely to grow dramatically in the near future. With thousands of hours of such content available on-demand, it becomes imperative to give users necessary summarization and skimming tools so that they can find the content they want and browse through it quickly.

One solution technique that can help in browsing is *time compression* [3,12]. It allows the *complete* audio-video to be watched in a shorter amount of time by speeding up the playback with no pitch distortion. The achievable speed-up is about 1.5-2.5 fold depending on speaker [3], beyond which the speech starts to become incomprehensible. Higher speed-ups are possible [5], but at cost of increased software complexity and listeners' concentration and stress level.

Getting a much higher time saving factor (2.5+) requires creating an audio-video summary of the presentation. A summary by definition implies that portions of the content are thrown away. For example, we may select only the first 30 seconds of audio-video after each slide transition in a presentation, or have a human identify key portions of the talk and include only those segments, or base it on the access patterns of users who have watched the talk before us.

In an earlier paper [8], we studied three automatic methods for creating audio-video summaries for presentations with slides. These were compared to author-generated summaries. While users preferred author-generated summaries, as may be expected, they showed good comprehension with automated summaries and were overall quite positive about automated methods. The study reported in this paper extends our earlier work by experimenting with non-video summarization abstractions to address the following questions:

- Since all of the audio-video summaries included slides, how much of the performance/comprehension increment was due to slides alone? In fact, this is the most common way in which presentation are archived on the web today—people simply post their slides. What is gained by skimming just the slides?

- How will people perform with the *full text transcripts* of the presentation, in contrast to the audio-video summaries? Two factors motivate this. First, speech-to-text technology is getting good enough that this may become feasible in the not-so-distant future. Second, people are great at skimming text to discover relevance and key points. Perhaps given a fixed time to browse the presentation, they can gain more from skimming a full text transcript than spending the same time on an audio-video summary.

- If we highlight the parts of the transcript that a speaker included in a video summary, would performance be comparable to or better than the performance with the video summary? The highlighted transcript and the video summary would each provide the information that a speaker thinks is important. Would users prefer reading the highlighted text summary or watching the audio-video summary?

Motivated by these questions, we compare four conditions: slides-only, full text-transcript with no highlights, full text-transcript with highlights, and audio-video summary. We also compare the results of this study and our earlier one [8]. We find that although comprehension is no different for full text transcript with highlights condition and audio-video summary condition, users have a subjective marginal preference for audio-video summary (ANOVA f=3.067, p=0.087). Furthermore, different styles of slide-authoring (e.g., detailed vs. big-points only) can have a big impact on their effectiveness as summaries, raising a dilemma for some speakers in authoring for on-demand previewing versus that for live audiences.

The paper is organized as follows: The next section describes the previous work on automatic summarization that this study extends. Next, the experimental design of the current study is presented, followed by the results section. Finally, we discuss related work and draw conclusions.

AUTOMATIC AUDIO-VIDEO SUMMARIZATION

We briefly summarize our earlier study on automated audio-video summarization methods [8]. The combination of the current study and this older study enable us to build a more complete picture of the overall tradeoffs.

Our study used a combination of information sources in talks to determine segments to be included in the summary. These were: 1) analysis of speech signal, for example, analysis of pitch, pauses, loudness over time; 2) slide-transition points, i.e., where the speaker switched slides; and 3) information about other users access patterns (we used details logs indicating segments that were watched or skipped by previous viewers).

We experimented with three algorithms based on these sources of information: 1) *slide-transition* points based (S); 2) *pitch activity* based (P), using a modified version of algorithm introduced by Arons [3]; 3) a combination of slide transitions, pitch activity, and previous user access patterns (SPU). In addition, we obtained a human-generated video summary (A) by asking the author-instructor for the talk to highlight segments of transcript[1]. Below we describe each of these algorithms in slightly more detail.

The slide-transition-based algorithm (S) uses the heuristics that slide transitions indicate change of topic, and that relative time spent by speaker on a slide indicates its relative importance. Given a target summarization ratio of N%, the algorithm selects the first N% of audio-video associated with each slide for the summary.

The pitch-based algorithm (P) is based on research indicating that pitch changes when people emphasize points. In presentations, the introduction of a new topic often corresponds with an increased relative pitch, and pitch is more discriminating than loudness, etc. The pitch-based summary uses a slight variation of the algorithm proposed by Arons [3] to identify emphasis in speech. The use of this algorithm allows us to compare the new algorithms with this seminal earlier work.

The third algorithm (SPU) combines the information sources used by the two algorithms above and adds information about previous users viewing patterns, in particular, the number of distinct users that had watched each second of the talk. We used the two heuristics: 1) If there is a surge in the viewer-count of a slide relative to the previous slide, it likely indicates a major topic transition. 2) If the viewer-count within a slide falls quickly over time, the slide is likely not interesting. These heuristics are used to determine a priority for the slide, and each slide in the talk gets a fraction of the total time based on its priority. Given a time quota for a slide, we use the "pitch-activity" based algorithm to pick the segments included in the summary.

For our study, four presentations were obtained from an internal training web site. Each author was given the text transcript of the talk with slide transition points marked. They marked summary segments with a highlighting pen. These sections were then assembled into a video summary by aligning the highlighted sentences with the corresponding video. A study of 24 subjects was then conducted to compare the summaries created by the authors to the three automatically generated summaries.

---

[1] In one case, the author was unavailable and designated another expert to highlight the summary.

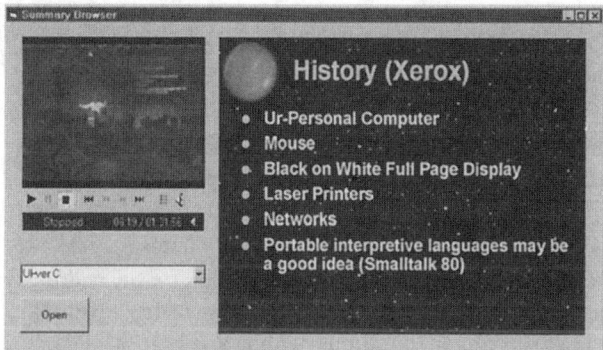

Figure 1: The interface for the experimental software.

Figure 1 shows the display seen by subjects watching the summaries. All video summaries are shown with the associated slides. As the video window, shown in the upper-left, plays the segments in the summary, the slides in the right pane change in synchrony.

We used two measures for our evaluations: performance improvement on quizzes before and after watching the video summary, and ratings on an opinion survey.

The outcome for the first measure was that author-generated summaries resulted in significantly greater improvement than computer-generated summaries (ANOVA f=16.088, p=0.000). The automated methods also resulted in substantial quiz score improvement, but the amount of improvement was statistically indistinguishable from each other (ANOVA f=0.324, p=0.724).

One hypothesis for lack of significant difference between the automated methods was that most of the useful information may be coming from the slides. Although the audio-video segments selected for summary were quite different for the different methods, the slides shown were substantially the same (as slide transitions are infrequent). However, participants estimated that slides carried only 46% of the information and audio-video carried 54%. So the hypothesis is not quite justified; one goal of the current study is to help resolve this issue.

Survey responses also indicated a preference for author-generated summaries (see details later in Table 4). While along one dimension—the summary provides a good synopsis of the talk—the author-generated and automated summaries did comparably (ANOVA f=0.521, p=0.472), the author-generated summaries were preferred along other dimensions (all p's were less than 0.005 using ANOVA).

Overall, the computer-generated summaries were well received by participants, many of whom expressed surprise upon being told afterwards that a computer generated them.

NEW SUMMARIZATION ABSTRACTIONS

In this study, we extend the previous work by examining three non-video summarizations or abstractions: slides only (SO), text transcripts with slides (T), and transcripts highlighted by the authors with slides (TH). Author-generated video summaries with slides (A), which are the same used in the earlier study, are included so that a comparison with the results of previous study is possible.

Slides Only (SO)

Technical presentations are usually accompanied with slides that set the context for the audience, indicating what was just said and what will be addressed next. Speakers also use the slides as cues for themselves. Normally, majority of the time preparing for a talk goes into preparing the slides: deciding how many slides, which ideas go onto which slides, and so forth. Because so much energy is put into the slides, it seems natural to use them as a summary. Furthermore, slides is what people frequently post on the web, slides is what they send around in email, so it is useful to understand how well people comprehend just using slides.

Text Transcript with Slides (T)

People are great at skimming text to discover relevance and key points. Perhaps given a fixed time to browse the presentation, they can gain more from skimming a full text transcript than spending the same time on an audio-video summary. Text transcripts are also interesting because commercial dictation software, such as ViaVoice from IBM and NaturallySpeaking from Dragon Systems, can produce text transcript automatically. The error rates are high without training, but close to 5% with proper training and recording condition. Speech-to-text will continue to improve and may become feasible for lecture transcription in the not-so-distant future.

For this condition we assumed the ideal case, and had all of the presentations fully transcribed by human. We then manually segment the text into one paragraph per slide. The title of the slide is also inserted in front of each paragraph. The process can be made fully automatic if we later use speech-to-text software, which gives the timing information of the text output, and have the slide transition times. The slides were also made available to the subjects in this condition.

Transcript with Key Points Highlighted and Slides (TH)

The benefit of providing the full text transcript is that every word that was said during the presentation is captured. The disadvantage is that spoken language is informal, it contains filler words and repetitions, and it may often be grammatically incorrect. It can be longer and harder to read than a paper or a book that is written specifically for reading and has the formatting and structuring elements to assist reading and skimming.

Viewers could benefit from having key points in the transcript highlighted. We had the option of having key points marked by human experts or use the ones generated by our automated algorithms. We chose to use the ideal case, the transcript highlighted by an author or expert. The reasons were two fold. First, this choice allowed us to compare the effectiveness of exactly the same summary when delivered with and without audio-video. Second, we believed this choice increases the longevity of results of this paper, as the quality of automated summaries will keep evolving (making future comparisons difficult), while those generated by humans should be quite stable.

Each author was given the text transcript of the talk with slide transition points marked. They marked summary segments with a highlighting pen. The same sections were also assembled into the video summary (A) by aligning the highlighted sentences with the corresponding video. The highlighted parts are presented to the subjects as bold and underlined text on screen. Again, slides were also made available to the subjects in this condition.

Talks Used in the Study

We reused the four presentations and quizzes from the previous study to permit comparison of the results. The four talks were on user-interface design (UI), Dynamic HTML (DH), Internet Explorer 5.0 (IE), and Microsoft Transaction Server (MT), respectively.

Table 1: Information associated with each presentation.

	UI	DH	IE	MT
Duration (mm:ss)	71:59	40:32	47:01	71:03
# of slides in talk	17	18	27	52
# of slides / min	0.2	0.4	0.6	0.7
# of words in transcript	15229	8081	6760	11578
# of pages in transcript	15	10	8	15
Highlighted words	19%	24%	25%	20%
Duaration of AV summary (mm:ss)	13:44	9:59	11:37	14:20

Table 1 shows some general information associated with each talk. It is interesting to note the wide disparity in number of slides associated with each talk. For example, although UI and MT are both around 70 minutes long, one has 17 slides and the other 52. The raw transcription texts are quite voluminous -- between 8 to 15 pages. They are hard to read even with the paragraph breaks on the slide transition points. In the summaries, the fraction of words highlighted by the speaker in the summaries is about 19 to 25%. Obviously, the end results may be different if much less or much higher summarization factors were chosen. A factor of 4 to 5 summarization seemed an interesting middle ground to us.

EXPERIMENTAL DESIGN

The same measures were used for outcome as for the first study: quizzes for objective learning and surveys to gauge subjective reactions.

Each presentation author had written 9 to 15 quiz questions that required participants to draw inferences from the content of the summary or to relay factual information contained in the summary. We selected 8 from each to construct a 32-question multiple-choice test. The questions from different talks were jumbled up to reduce memorization of questions.

The 24 participants were employees and contingent staff members of Microsoft working in technical job positions. All lacked expertise in these four topic areas. Participants were given a gratuity upon completing the tasks.

Participants first completed a background survey and took the quiz to document their initial knowledge level. We randomly ordered questions within and across talks, so that people would have less ability to be guided by questions while watching the talk summaries.

Each participant watched or read four summaries, one for each talk and one with each summarization technique. Talk order and summarization technique were counterbalanced to control for order effects.

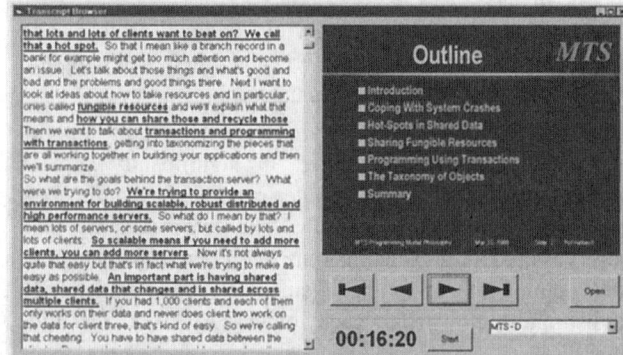

Figure 2: Interface for the conditions SO, T, and TH. The participant can use the vertical scroll bar to navigate the text transcript or use the four control buttons (shown below the slides) to navigate the slides. However, the current slide and the displayed text transcript are not linked. This allows the participant to view the slide in one part and review the transcript in another area. The countdown timer below the slide-navigation controls serves as a reminder of how much time left to review the current summary.

The display for video summary condition (A) was the same as for our previous study (see Figure 1). Figure 2 shows the interface for the other three conditions. In the slide-only condition, the left transcript pane is blank. While watching or reading a summary, a participant was given the same time as the duration of the audio-video summary of corresponding talk (see Table 1). They were free to navigate within the slides and transcript. Once finished, however, participants were instructed not to review portions of the summary. Participants were provided pen and paper to take notes. After each summary, participants filled out the subjective survey and retook the quiz.

RESULTS

Evaluating summarization algorithms is a fundamentally difficult task, as the critical attributes are highly complex and difficult to quantify computationally. We use a combination of performance on a quiz and ratings on an opinion survey for our evaluation.

Quiz Results

We expected the author-generated summaries (TH and A) to produce the highest quiz scores, as the quizzes were created by the authors. However, we wanted to know:

1. Are there significant differences between the author-generated video summary and the text transcript with the same portion highlighted? What is the value of audio-video?

2. How much worse are SO and T compared to A and TH? This focuses on understanding the value of effort put in by authors in identifying highlights.

3. Are there performance differences across the talks? Is there something in the structure and/or organization of talks that affects performance?

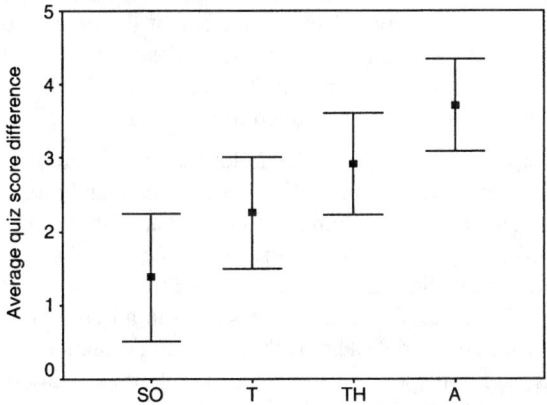

Figure 3: Average quiz score improvement by condition with 95% confidence intervals. The apparent linearity of the quiz score improvement is coincidental.

Figure 3 shows the average difference between pre-summary and post-summary quiz scores as a function of the conditions. Quiz scores were improved most by the audio-video summaries (A). To a lesser extent, quiz scores were improved by the summaries that combined highlighted transcripts and slides (TH). The smallest improvements were obtained from the slides alone (SO) and transcript with slides (T) versions.

To more specifically answer the first question raised above, we compared A and TH conditions. Analysis of data shows that there is only a marginal preference for audio-video (ANOVA f=3.067, p=0.087). We also had subjective comments from users about added value from hearing the speaker's voice and intonation. We present these comments in "User Comments" subsection below.

To answer the second question, we analyzed the data with quiz scores from conditions A and TH as one group, and those from T and SO as another group. Data show that A and TH are significantly better than SO and T as a group (ANOVA f=16.829, p=0.000). Thus, the author effort in identifying highlights does add significant value to the viewers.

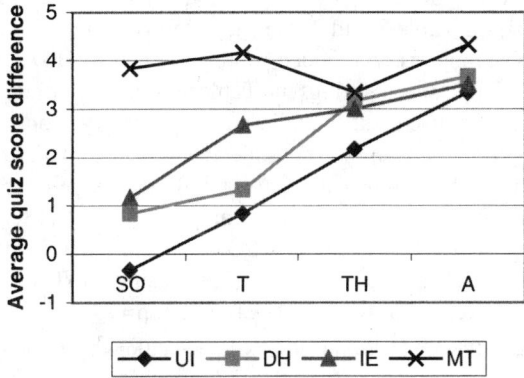

Figure 4: Variation in quiz score for each talk as a function of summary method.

To answer the third question, we show quiz score improvement as a function of summary method for each of the talks in Figure 4. Two things stand out. First, the variation in score improvement across summary methods is the least for the MT talk and the most for the UI talk. Second, the variation in score improvement across talks is the least for summary method A and the most for summary method SO.

As we looked deeper, we found the amount of variability in the quiz score improvements across methods seemed to correlate with the extent of information present in the slides. For example, one measure of information density in slides is the number of slides per minute.[2] By this metric (see Table 1) the talks are ordered as MT highest (0.7 slides/min), IE (0.6 slides/min), DH (0.4 slides/min) and UI (0.2 slides/min). This corresponds to the observation that the variance as a function of summarization method is the least for MT talk and the most for UI talk.

Our intuition regarding this is that slide content is a key source of information for the viewers. If there is lots of detail in the slides (e.g., MT has 52 slides) then the summary method matters much less than if there is little information in the slides (e.g., UI has 17 slides). Conversely, a good summary can compensate for lack of information in the slides by suitably providing audio-video segments where key points are made.

Survey Results

Participants completed a short survey after watching each summary. The surveys were administered prior to repeating the quiz so that quiz performance would not affect their opinions on the surveys. The goal of the surveys was to get subjective reaction of users to the summary methods.

---

[2] Of course, this does not take into account the amount of information within each slide.

User Ratings

The pattern of responses was similar to that of the quiz scores (see Table 2 and Table 3). Average ratings for the video summaries (A) tended to be the highest. However, none of the seven ratings in Table 2 were significantly different between the audio-video summary and the highlighted transcript (TH) using ANOVA at p=0.05. For A and TH as a group, all of the ratings were significantly greater than those for the slides-only (SO) and transcript (T) using ANOVA at p=0.01. SO was significantly worse than the other three on all ratings, using ANOVA at p=0.05, except Enjoy (ANOVA f=2.131, p=0.148).

Table 2: Post-quiz survey results by conditions[3]. Responses were from 1 ("strongly disagree") to 7 ("strongly agree").

By condition	Synop.	Effic.	Enjoy	Key points (%)	Skip talk	Concise	Coher.
A	4.96	5.04	4.78	68.91	4.41	5.13	4.13
TH	4.70	4.61	3.83	64.13	4.52	4.52	4.35
T	3.58	3.25	3.29	61.67	3.83	3.50	4.17
SO	3.13	3.38	3.33	41.25	1.96	2.92	2.83

Also following the quiz score trend is the fact that the average ratings for the MT talk were consistently higher than the others (see Table 3), independent of the summary method. Again this is likely due to the fact that the slides were sufficiently detailed so that they could "stand alone" and be interpreted without the speaker's audio-video/text-transcript.

Table 3: Post-quiz survey results by talks. Responses were from 1 ("strongly disagree") to 7 ("strongly agree").

By talk	Synop.	Effic.	Enjoy	Key points (%)	Skip talk	Concise	Coher.
UI	3.79	3.96	3.92	52.50	3.04	3.71	3.88
DH	4.17	3.96	3.74	60.22	3.78	3.91	3.73
IE	3.83	3.96	3.30	51.09	3.36	3.83	3.96
MT	4.50	4.33	4.21	71.25	4.42	4.54	4.61

User Comments

MT talk aside, most of the participants found that the slides only condition (SO) lacked sufficient information. They also felt scanning the full text in condition T tedious.

---

[3] Complete wording: 1) Synopsis: "I feel that the condition gave an excellent synopsis of the talk." 2) Efficient: "I feel that the condition is an efficient way to summarize talks." 3) Enjoyed: "I enjoyed reading through (or watching) the condition to get my information." 4) Key points: "My confidence that I was presented with the key points of the condition is:" 5) Skip talk: "I feel that I could skip the full-length video-taped talk because I read (or watch) the condition." 6) Concise: "I feel that the condition captured the essence of the video-taped talk in a concise manner." 7) Coherent: "I feel that the condition was coherent–it provided reasonable context, transitions, and sentence flow so that the points of the talk were understandable."

Thirteen of the 24 participants rated the audio-video summary (A) as their favorite summary abstraction, while eleven chose the highlighted transcript with slides (TH).

Participants liking the audio-video summary did so mainly because it allowed them to listen passively, it was self-contained, and multi-modal. One participant said, "It felt like you were at the presentation. You could hear the speaker's emphasis and inflections upon what was important. It was much easier to listen and read slides versus reading transcripts and reading slides." Another commented, "It kept my interest high. It is more enjoyable listening and seeing the presenter."

Participants liking the highlighted transcript with slides condition most did so because it gave them more control over the pace and allowed them to read what they considered important. One participant liking the highlighted transcript most commented, "I felt this was a more efficient way to get a summary of the presentation. ... I could re-read the portions I was interested in or unclear about." Another said, "I like having the option of being able to get more detailed info when I need it."

Comparison with the Automatic Summary Study

There are several similarities between this study and our previous study on automatic summary algorithms: 1) The talks and quiz questions were the same; 2) The author-generated audio-video summary (condition A) was present in both studies; 3) Slides were shown in all conditions in both studies; and 4) The method used to evaluate outcome was essentially the same. These similarities allow us to compare the results from these two studies.

Figure 5 shows the average quiz score difference by conditions for each talk from the automatic summary study. Compared with Figure 4, there is no clear correlation between the variability among the talks and conditions. It may be because the differences between the computer-generated video summaries are not as big as the differences between conditions S, T, and TH. Also, audio-video is present in all conditions in the old study, compensating for level-of-detail differences in the slides.

In Table 4, we list the post-quiz ratings that are in common between the two studies. The top half of the table shows the ratings for the previous study, while the bottom half shows the ratings for this study.

Condition A was included in both studies, though we see its ratings are consistently lower in the present study. One hypothesis is that the ratings are relative to the quality of other conditions explored in the same study. The closeness of TH to A in current study might have resulted in A getting these slightly lower scores.

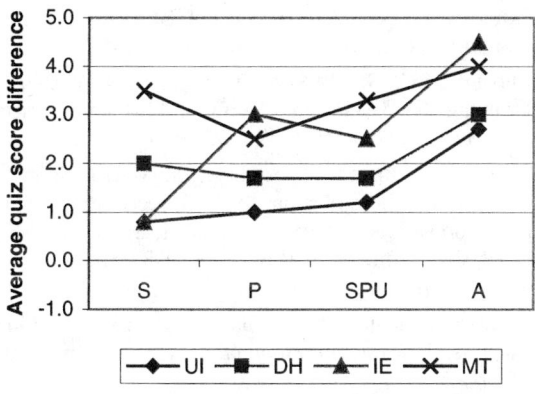

Figure 5: Average quiz score difference by conditions from the automatic summary study.

It is interesting to compare the S and SO conditions across the two studies. The slide-transition-based summary in the previous study (Condition S) assembled a summary by allocating time to each slide in proportion to the amount of time that the speaker spent on it in the full-length talk. Thus condition S differed from the slide-only condition (SO) in the present study by showing audio-video in addition to all the slides of the talk. From Table 4 we see that the ratings for condition S are consistently higher than condition SO[4], suggesting that providing an audio-video summary can add a lot of value to the slides, even when the summary is created with a simple summarization technique.

Table 4: Responses to quality of summary for various methods for the automatic summary study (top half) and the current study (bottom half).

		Synopsis	Key points (%)	Skip talk	Concise	Coherent
Old Study	SPU	4.92	64.17	3.54	4.63	3.58
	P	4.83	62.50	3.04	4.13	3.46
	S	4.33	56.25	3.21	4.08	3.57
	A	5.00	76.25	4.96	5.63	5.33
Current Study	A	4.96	68.91	4.41	5.13	4.13
	TH	4.70	64.13	4.52	4.52	4.35
	T	3.58	61.67	3.83	3.50	4.17
	SO	3.13	41.25	1.96	2.92	2.83

One surprising result in the previous study was that participants rated the computer-generated summaries more positively as they progressed through the study. The summary shown to the participants last in each session was consistently rated as being clearer (ANOVA p=0.048), less choppy (ANOVA p=0.001), and of higher quality (ANONA p=0.013) than were the first three summaries in the same session independent of condition. The study was designed so that each of the four summary methods was presented equally often in each position in the sequence. We found no such effect in the current study. However, summary presentation styles varied more in the current study, possibly reducing the chance for the participants to habituate to disadvantages of each abstraction.

RELATED WORK

There has been considerable research on indexing and searching the rapidly expanding sources of digital videos [1,2,5,10,11,13,15,18,19]. All these systems use automatic techniques based on the visual aspect of the media, primarily employing image-recognition and image-processing techniques (e.g., shot boundaries). Some of them [11,15] use textual information from speech-to-text software or closed captions. While these systems focus on the technical aspect, our study focuses on the human side: comparing the effectiveness of different summary abstractions for audio-video presentations.

Christel et al. [4] report a subjective evaluation of video summaries created from image analysis, keyword speech recognition, and combinations, for general-purpose videos. Based on analysis, summaries (or what they call skims) are constructed by concatenating 3-5 second video segments. They tested the quality of skims using image recognition and text-phrase recognition tasks. Performance and subjective satisfaction of all skimming approaches contrasted unfavorably with viewing the full video. This paper, in contrast, focuses on presentations allowing domain specific knowledge to be used (e.g., 3-5 second speech segments used by Christel et al. are too short for comprehension).

The interfaces we used in the user study were simple (see Figure 1). One can imagine more sophisticated interfaces that takes the advantage of the digital media. Barry Arons' SpeechSkimmer [3] allows audio to be played at normal speed, or speeded-up with no pitch distortion, with pauses removed, or restricted to phrases emphasized by the speaker. Lisa Stifelman introduced Audio Notebook [16,17], a prototype note-pad combining pen-and-paper and audio recording. Audio Notebook allowed the user to use notes made on paper (ink marks) as an index into the recorded audio presentation. In contrast to SpeechSkimmer and Audio Notebook that focused on audio alone, Li et al. [9] explored interfaces addressing both audio *and* video, providing enhanced features such as time-compression, pause removal, navigation using shot boundaries and table-of-content. Integrating these enhanced browsing features with the summary methods studied here could substantially enrich the end user's experience.

CONCLUDING REMARKS

As storage cost drops, network bandwidth increases, and inexpensive video cameras becomes available, more audio

---

[4] This is true even if we adjust the ratings of S using S'=S*A/A* because of the difference between A and A* in the two studies.

and video technical presentations will go online. Given this expected explosion, it becomes imperative to give users effective summarization and skimming tools so that they can find the presentation they want and browse through it quickly.

This paper reports a study that extends our previous work by comparing three non-video summarization abstractions with an audio-video summary created by the speaker. The three non-video summary techniques are: 1) PowerPoint slides in the presentation, 2) a text transcript created from the presentation, and 3) the transcript with important points highlighted by the speaker.

We show that slides-only (S) and plain transcript (T) summary methods are significantly worse than author generated transcripts with highlights (TH) and audio-video (A) summary methods. Author participation clearly adds value. We also show that while comprehension given transcripts with highlights method (TH) can match the audio-video summary (A), there is a marginal preference for audio-video (ANOVA $f=3.067$, $p=0.087$). Furthermore, we observe that diffcrent styles of slide-authoring (e.g., detailed vs. major points only) can have a large impact on their effectiveness as summaries. The results conflict with the common advice that slides should contain only the major points to retain attention of live audience. This raises a dilemma for speakers who are authoring for both on-demand and live audiences. On solution might be to create two versions of slides. The succinct version can be used in the live presentation, while the more detailed version is placed online.

The two-versions of slides solution, of course, requires cooperation from the authors. As the technology for creating computer-generated summaries improves, the amount of author work in the creation of summaries should be reduced. At the same time, as more people browse audio-video online, authors may often be more willing to contribute to improving their experience. An interesting future direction is technology-assisted tools that allow authors to very quickly indicate important segments (e.g., speech-to-text transcript marked by author in 5 minutes using a tool).

ACKNOWLEDGMENT

Thanks to the Microsoft Usability Labs for use of their lab facilities. Steve Capps, Pat Helland, Dave Massy, and Briand Sanderson gave their valuable time to create the summaries and quiz questions for their presentations. Gayna Williams and JJ Cadiz reviewed the paper and gave us valuable suggestions for improvement.

REFERENCES

1. Aoki, H., Shimotsuji, S. & Hori, O. A Shot Classification Method of Selecting Effective Key-frames for Video Browsing. In *Proceedings of Multimedia'96*, pp 1-10. ACM.
2. Arman, F., Depommier, R., Hsu, A. & Chiu M.Y. Content-based Browsing of Video Sequences, In *Proceedings of Multimedia'94*, pp 97-103. ACM.
3. Arons, B. SpeechSkimmer: A System for Interactively Skimming Recorded Speech. *ACM Transactions on Computer Human Interaction, 4*, 1, 1997, 3-38.
4. Christel, M.G., Smith, M.A., Taylor, C.R. & Winkler, D.B. Evolving Video Skims into Useful Multimedia Abstractions. In *Proceedings of CHI, April 1998*, pp. 171-178.
5. Covell, M., Withgott, M., & Slaney, M. Mach1: Nonuniform Time-Scale Modification of Speech. Proc. IEEE International Conference on Acoustics, Speech, and Signal Processing, Seattle WA, May 12-15 1998.
6. Foote, J., Boreczky, J., Girgensohn, A. & Wilcox, L. An Intelligent Media Brower using Automatic Multimodal Analysis. In *Proceedings of Multimedia'98*, pp. 375-380. ACM.
7. He, L., Gupta, A., White, S.A. & Grudin, J., 1999. Design lessons from deployment of on-demand video. *CHI'99 Extended Abstracts*, 276-277. ACM.
8. He, L., Sanocki, E., Gupta, A. & Grudin, J., 1999. Auto-summarization of audio-video presentations. In *Proc. Multimedia'99*. ACM.
9. Li, F.C., Gupta, A., Sanocki, E., He, L. & Rui, Y., 1999. Browsing Digital Video. In *Proc. CHI 2000*. ACM.
10. Lienhart, R., Pfeiffer, S., Fischer S. & Effelsberg, W. Video Abstracting, *ACM Communications*, December 1997.
11. Merlino, A., Morey, D. & Maybury, M. Broadcast News Navigation Using Story Segmentation. In Proceedings of the 6th ACM international conference on Multimedia, 1997.
12. Omoigui, N., He, L., Gupta, A., Grudin, J. & Sanocki, E. Time-compression: System Concerns, Usage, and Benefits. Proceedings of *ACM Conference on Computer Human Interaction, 1999*.
13. Ponceleon, D., Srinivasan, S., Amir, A., Petkovic, D. & Diklic, D. Key to Effective Video Retrieval: Effective Cataloging and Browsing. In Proceedings of the 6th ACM international conference on Multimedia, September 1998.
14. Stanford Online: Masters in Electrical Engineering, 1998. http://scpd.stanford.edu/cee/telecom/onlinedegree.html
15. Smith M. and Kanade T. Video skimming and characterization through the combination of image and language understanding techniques. Proceedings of IEEE Computer Vision and Pattern Recognition, 775-781. 1997.
16. Stifelman, L. The Audio Notebook: Paper and Pen Interaction with Structured Speech *Ph.D. dissertation, MIT Media Laboratory*, 1997.
17. Stifelman, L.J., Arons, B., Schmandt, C. & Hulteen, E.A. VoiceNotes: A Speech Interface for a Hand-Held Voice Notetaker. *Proc. INTERCHI'93 (Amsterdam, 1993)*, ACM.
18. Tonomura, Y. & Abe, S., Content Oriented Visual Interface Using Video Icons for Visual Database Systems. In *Journal of Visual Languages and Computing*, vol. 1, 1990. pp 183-198.
19. Zhang, H.J., Low, C.Y., Smoliar, S.W. and Wu, J.H. Video parsing, retrieval and browsing: an integrated and content-based solution. In *Proceedings of ACM Multimedia, September 1995*, pp. 15-24.

An Interactive Comic Book Presentation for Exploring Video

John Boreczky, Andreas Girgensohn, Gene Golovchinsky, and Shingo Uchihashi

FX Palo Alto Laboratory, Inc.
3400 Hillview Avenue, Bldg. 4
Palo Alto, CA 94304, USA
{johnb, andreasg, gene, shingo}@pal.xerox.com

ABSTRACT

This paper presents a method for generating compact pictorial summarizations of video. We developed a novel approach for selecting still images from a video suitable for summarizing the video and for providing entry points into it. Images are laid out in a compact, visually pleasing display reminiscent of a comic book or Japanese manga. Users can explore the video by interacting with the presented summary. Links from each keyframe start video playback and/or present additional detail. Captions can be added to presentation frames to include commentary or descriptions such as the minutes of a recorded meeting. We conducted a study to compare variants of our summarization technique. The study participants judged the manga summary to be significantly better than the other two conditions with respect to their suitability for summaries and navigation, and their visual appeal.

KEYWORDS: Video summarization, video browsing, keyframe extraction.

INTRODUCTION

As video is used more and more as the official record of meetings, teleconferences, and other events, the ability to locate relevant passages or even entire meetings becomes important. To this end, we want to give users visual summaries and help them locate specific video passages quickly. Such a system is useful in settings that require a quick overview of video to identify potentially useful or relevant segments. Examples include recordings of meetings and presentations, home movies, and domain-specific video such as recordings used in surgery or in insurance. These techniques are also effective when applied to commercials and to films. These seemingly different forms of video are related because they consist of multiple shots, shot at different times, perhaps by different cameras or by a hand-held camera, but the segments are often not clearly separable from the user's perspective (and thus are not readily accessible through an index or a table of contents).

We developed a novel approach for selecting still images from a video suitable for summarizing the video and providing entry points into it. Images are laid out in a compact, visually pleasing display reminiscent of a comic book or

Japanese *manga*. For further exploration of the video, we created an interactive version of the pictorial summary. Playback can be started at a desired time by clicking on a particular frame; additional detail is available via hyperlinks. Textual information can be added to the presentation in the form of captions. Meeting minutes, for example, may be integrated with the video recording of the meeting.

The video summarization component has been used as part of a larger video database system [6] for over a year. The video database contains a large collection of video-taped meetings and presentations, as well as videos from other sources, and is used regularly by the employees of our company.

In another paper [12], we described the technical details for creating a manga summary of a video. In the meantime, we have added several new user interface features. This paper focusses on the user interface and its use for finding information in the video. We describe the user interface features and discuss how they help users find the information they are looking for.

To determine the effectiveness of our image selection and layout techniques, we conducted an experiment to test the different components of our system. To test the suitability for navigation within a video, we asked the study participants to perform several tasks of finding specific information in several videos. In addition, we also presented the participants with pair-wise comparisons of the different combinations of our techniques and asked them to rate them according to their suitability for summaries and navigation, their visual appeal, and to judge their overall preferences.

In the next section, we discuss the elements of our approach for presenting a compact summary of a video. After that, we present the user interface of the video summary and the interaction style for exploring the video. Next, we describe the setup and the results of the experiment. We conclude with a discussion of directions for future work.

CREATING VIDEO SUMMARIES

Typically, a video is summarized by segmenting it into shots demarcated by camera changes. The entire video can then be represented as a collection of keyframes, one for each shot. Although this can reduce the amount of information a user must go through to find the desired segment, it may still be too much data. Furthermore, the relative similarity of keyframes, coupled with a large and regular display, may make it harder to spot the desired one. In contrast, the system presented here abstracts video by selectively discarding or de-emphasizing redundant information (such as repeated or alternating shots). The goal of this approach is to present a concise summary of a video, a summary that can be used to get an overview of the video and one that also serves as a navigation tool.

One possible way of identifying the desired segment is to fast-forward through a video, producing a temporal rather than a spatial summary. Unfortunately, this approach does not work with streaming video, whereas the techniques described here are applicable to all types of video.[1]

Given an existing segmentation, we calculate shot importance for each segment. By thresholding the importance score, less important shots are pruned, leaving a concise and visually varied summary.

Elements in our summaries vary not only in their content, but also in the size of the keyframe. In our system, keyframes are displayed in different sizes according to their importance: less important keyframes are displayed in a smaller size. The algorithm used in our system packs different-sized keyframes into a compact representation, maintaining their time order. The result is a compact and visually pleasing summary reminiscent of a comic book or Japanese *manga*.

In this section, we provide a brief description of the steps for creating video summaries. More of the technical details are described in [12].

Video Segmentation by Clustering

Many techniques exist to automatically segment video into its component shots, typically by finding large frame differences that correspond to cuts, or shot boundaries. Once detected, shots can be clustered by similarity such that related shots (e.g. similar camera angles or subjects) are considered to be one shot or cluster. For example, a film dialog where the camera repeatedly alternates between two actors would typically consist of two clusters, one for each actor. Many systems break videos into shots and use a constant number of keyframes for each shot. For example, Ferman *et al.* [4] cluster the frames within each shot. The frame closest to the center of the largest cluster is selected as the keyframe for that shot.

Other systems use more keyframes to represent shots that have more interesting visual content. Zhang *et al.* [15] segment the video into shots and select the first frame after the completed shot transition as a keyframe. The rest of the shot is examined, and frames that are sufficiently different from the last keyframe are marked as keyframes as well. Zhuang *et al.* [16] use a clustering approach to determine keyframes for a shot. They still extract at least one keyframe per shot. Cherfaoui *et al.* [1] segment the video into shots and determine if there is camera motion or zooming in each shot. Shots with camera motion are represented with three frames. Zooms and fixed shots are represented as single frames with graphical annotations that describe object motion or the zoom parameters.

---

1. A separate channel that delivers pre-computed keyframes may be used to construct a summary of the video.

These existing systems either provide only limited control over the number of keyframes or do not perform an adequate job of finding truly representative frames. Instead of segmenting the video into shots, we cluster all the frames of the video. Frames are clustered using the complete link method of the hierarchical agglomerative clustering technique [8]. This method uses the maximum of the pair-wise distances between frames to determine the intercluster similarity, and produces small, tightly bound clusters. For pictorial summaries described in this paper, we use smoothed three-dimensional color histograms in the YUV color space for comparing video frames. Once clusters have been selected, each frame is labeled with its corresponding cluster. Uninterrupted frame sequences belonging to the same cluster are considered to be segments.

Determining Keyframes and Their Sizes

To produce a good summary, we must still discard or de-emphasize many segments. To select appropriate keyframes for a compact pictorial summary, we use the importance measure described in [12]. This calculates an importance score for each segment based on its rarity and duration. Longer shots are preferred because they are likely to be important in the video. This preference also avoids the inclusion of video artifacts such as synchronization problems after camera switches. At the same time, shots that are repeated over and over again (for example wide-angle shots of the room) do not add much to the summary even if they are long. Therefore, repeated shots receive lower scores. The clustering approach discussed in the previous section can easily identify repeated shots.

Segments with an importance score higher than a threshold are selected to generate a pictorial summary. For each segment chosen, the frame nearest the center of the segment is extracted as a representative keyframe.[2] Frames are sized according to the importance measure of their originating segments, so that higher importance segments are represented with larger keyframes. This draws the attention to the more important portions of the video. The keyframe packing algorithm described in the next section creates a compact and attractive presentation that summarizes the whole video.

Keyframe Packing

Once frames have been selected, they need to be arranged in a logical order. A number of layouts have been proposed. Shahraray *et al.* [9] at AT&T Research use keyframes to represent an HTML presentation of video. One keyframe was selected for each shot; uniformly sized keyframes are laid out in a column along closed-caption text. Taniguchi *et al.* [11] summarize video using a 2-D packing of "panoramas" which are large images formed by compositing video pans. In this work, keyframes are extracted from every shot and used for a 2-D representation of the video content. Because frame sizes are not adjusted for better packing,

---

2. Although frames early in the shot may be preferable semantically in some cases, camera motion and noise due to camera switching may make these frames unreliable.

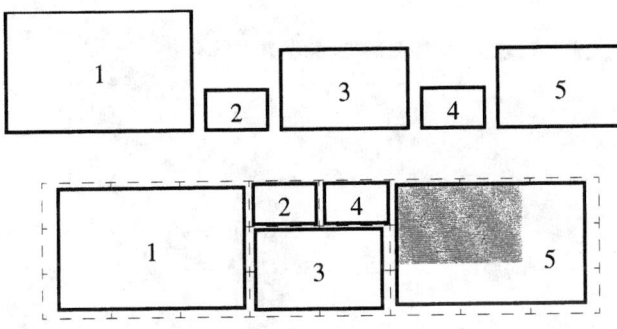

Figure 1: Packing keyframes into a row block

much white space appears in the result, making the summary inefficient.

Yeung et al. [14] use clustering techniques to create pictorial summaries of videos. They select one keyframe for each video shot and cluster those frames on the basis of visual similarity and temporal distance to groups of similar shots. They create pictorial summaries of video using a "dominance score" for each shot. Though they work towards a similar goal, their implementation and results are substantially different. The sizes and the positions of the still frames are determined only by the dominance scores, and are not time-ordered.

In our system, keyframes are laid out in two dimensions to form a pictorial abstract of the video similar to the panels in a comic book. Thresholding the importance score allows the desired number of frames to be displayed. To facilitate layout, frames can be displayed in smaller or larger sizes, depending on their importance score, and are resized to best fit the available space.

Given a set of frames, we have to find a sequence of frame sizes that both fills space efficiently and provides a good representation of the original video. We use a near-optimal "row block" packing procedure that we described in more detail in [12]. An example of this packing method is depicted in Figure 1. The rectangles at the top are the original frame sequence, sized by importance. The bottom picture illustrates the frames packed into a row block. Note that the size of Frame 5 has been increased from the original size (indicated as a gray rectangle) for a better packing with minimal white space.

Figure 2 shows the algorithm applied to the video of a staff meeting. A larger gap between rows graphically emphasizes that the temporal order is row by row. Within a row, frames go from left to right with occasional excursions up and down. Lines are provided between frames to unambiguously indicate their order.

The smallest image size and row width in the manga display default to useful values (64x48 pixels, 8 columns). These can be changed on-the-fly if users prefer a different-size presentation, perhaps to better fill their web browser window. The packing algorithm is very efficient so that a new layout can be produced on request.

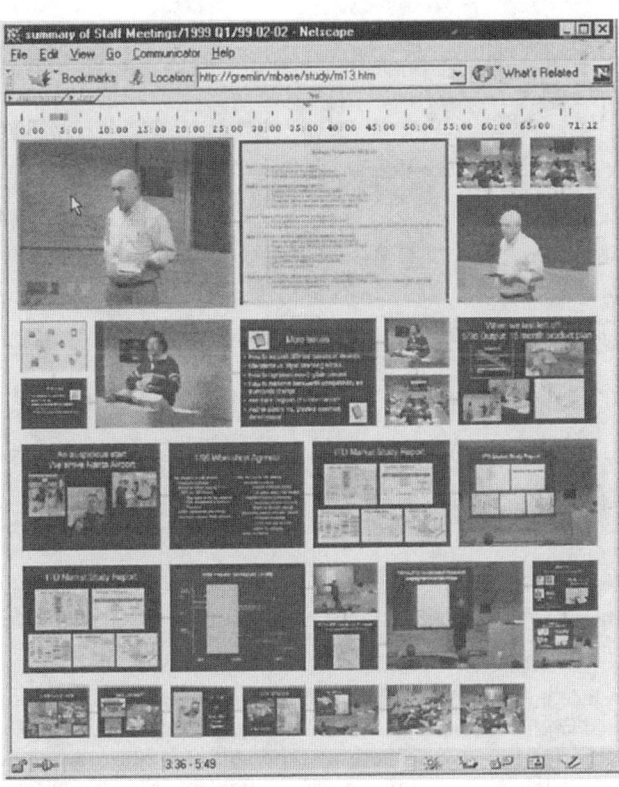

Figure 2: Pictorial summary of a video

EXPLORING VIDEOS

A video summary may help a user identify the desired video simply by showing relevant images. Often, that is not enough, however. Even after finding an appropriate video, it may be difficult to locate desired passages within it simply by inspection. As video is used more often as a permanent, often official, record of events, it becomes more important to be able to locate relevant passages. As video collections grow, the time spent looking for material becomes a scarce commodity.

We believe that interactive browsing of video summaries can shorten the time required to find the desired segment. Guided by this principle, we have implemented a web-based interactive version of the pictorial summary. Users can play the video corresponding to a keyframe by clicking on it. They can request more information in the form of a "tool tips" pop-up (see Figure 3). Finally, they can request a more detailed view that shows additional shots in the selected segment and select one of those shots for playback (Figure 5). These features are described in more detail in the following sections.

Interactive Video Summary

Users may browse the video based either on the keyframes or on the video's timeline. The two views are always synchronized: the timeline shows the duration of the segment that corresponds to the frame under the cursor (see Figure 3). Similarly, when the cursor is over the timeline, the corresponding keyframe is highlighted. This display allows users to explore the temporal properties of a video. At a glance, they can see both the visual representation of an important segment and its corresponding time interval, and can inter-

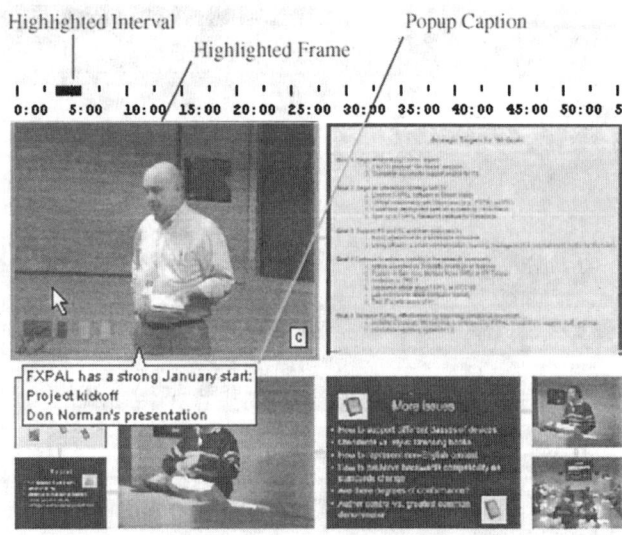

Figure 3: Highlighted frames and embedded captions

Figure 4: Playing the video

act with the system in manner that best suits their particular needs.

Once an interesting segment has been identified, clicking on its keyframe starts video playback from the beginning of that segment (see Figure 4). The ability to start the video playback at the beginning of a segment is important for exploring a video in-depth. It is not likely that the informal video material such as captured meetings would be segmented by hand, as is common with more formal material such as feature films. Our automatic clustering approach, combined with importance scores, yields good enough segment boundaries to aid the exploration of a video. This interface makes it easy to check promising passages of a video. If a passage turns out to be uninteresting after all, other segments can be easily reviewed just by clicking on their keyframes.

Several other systems use keyframes to support browsing of videos. The system described in Christel *et al.* [2] produces video highlights by selecting short segments of video. For each shot they select a keyframe that emphasizes moving objects, faces, and text. Shot keyframes are selected in rank order and a short segment of video surrounding the selected keyframe is added to the video highlight until the desired length of video is reached. Other tools have been built for browsing video content [13, 15]. These do not attempt to summarize video but rather present video content "as is." Therefore, keyframes are typically extracted from every shot and not selected to reduce redundancy. In most approaches, the frame presentation is basically linear, although some approaches occasionally use other structures [13].

Video Captions

Many videos in our video database depict meetings. If meeting minutes exist, they can be included in the manga display as captions. Such captions are expected to increase the value of video summaries. Ding *et al.* [3], for example, report that summaries of videos consisting of keyframes and coordinated captions were preferred by participants, and led to better predictions of video content. Huang *et al.* have created summaries of news broadcasts, as reported in [7]. Story boundaries were pre-determined based on audio and visual characteristics. For each news story, a keyframe was extracted from a portion of video where keywords were detected the most. Their method nicely integrated information available for news materials, but relies heavily on the structured nature of broadcast news and would not apply to general videos.

Initially, we chose to place captions on top of the corresponding images in the summary. This caused problems because small images did not provide enough room for captions. Also, if there were a large number of captions (e.g., extracted from the close caption track of a video), many of the images were partially obscured. To address these issues, we decided to pop up captions as the mouse moves over an image (see Figure 3). Small icons (the letter "C") indicate which images have attached captions.

Textual annotations can enhance the quality of the video summary [3]. The pictorial layout captioned with text from minutes is a better summary than either of the individual parts. If the meeting minutes are taken on a computer and time-stamped, they can be automatically time-aligned with the video. Otherwise, the minutes may be aligned by hand. Such video captions provide additional information about the scenes represented by the keyframes. For the study described in the next section, we did not provide any captions.

Exploring Additional Details

While the initial display provides a good summary of the video, it is helpful to be able to explore additional details without having to play back all likely segments. Furht *et al.* [5, p. 311] describe an approach for showing additional detail while exploring a video. Equal numbers of shots are grouped and represented by a keyframe. Expanding a keyframe recursively reveals keyframes for each shot in the subgroup until each group consists only of a single shot.

Figure 5: Drilling down

Their approach treats all shots equally and does not make any judgements about shot importance.

Our approach for exploring a video takes the structure of the video into account. We first look at the neighboring segments of the segment being explored. There might be several neighbors that have importance scores too low to be represented by keyframes. Showing keyframes for such segments can provide additional context for the segment being explored. If a direct neighbor of the segment is already represented by a keyframe, a different algorithm (described below) can be used to show additional structure within the segment.

The overlaid detailed view of a segment consists of some frames selected from shots in the selected segment. We show at least one image from the segment being explored. If, after showing the neighbors, there is space for additional images, we select subclusters of the cluster the segment belongs to by descending into the hierarchical cluster tree. As before, cluster membership is used for segmentation. We increase the number of clusters until we have as many subsegments as required. As before, we choose one keyframe from each subsegment.

Once identified, the new, detailed images are overlaid on a faded-out manga display. Lines are drawn from its representative frames to the four newly shown frames to show the connection to the segment being explored. The new frames have the same relation to the timeline as the original ones, and video playback can be started from them as well.

Figure 5 shows the results of exploring the segment shown in the top-left of Figure 3. The top two frames represent the two segments that precede the expanded segment. They provide context that may help the user understand the selected segment. The remaining two frames show additional detail about the expanded segment.

The interaction technique for exploring video segments in greater detail was used extensively in the experiment described in the next section. It is an effective means for exploring a video without having to wait for playback.

EXPERIMENT

We conducted an experiment to test the different components of our system. Participants performed several tasks of finding specific information in several videos. In the second part of the study, we asked the participants for their judgements regarding the suitability of combinations of our techniques for summaries and navigation, and regarding their visual appeal, and also asked them to report their overall preferences. We expected that our summarization techniques would provide an advantage in terms of time to complete the task, but we wanted to assess the contributions of individual components. In addition, the study was intended to provide general insights for how users use the summaries, and to identify potential usability problems.

Participants

Twenty-four participants (17 male, 7 female) participated in the study. Participants included researchers, support staff, and administrative staff at our company; they had varying degrees of expertise with using digital video. The videos used in the study were recordings of staff meetings so that the participants were exactly our target users who might have missed a staff meeting or who need to go back to find some information. For their efforts and time, the participants were rewarded with a free lunch.

Materials

Visual summaries of three videos of staff meetings were created. Three different styles of summaries were created for each video. The different styles were designed to explore two manga features: image selection by importance score and variable-size image layout. All styles were displayed in the same screen area. The first style (control) used neither feature: the display consisted of fixed-size images sampled at regular time intervals. The second style (selected) used manga image selection, but retained fixed-size images. The third style (manga) used both features.

The image size of the video player was 352x240 pixels. Images were presented in three different sizes (64x48, 136x102, and 208x156 pixels, respectively). The first two styles only presented images in the intermediate size. They both placed 20 images in a 4x5 grid. For the third style, the largest number of images was used that still fit in the same area as used for the other two summaries. That led to summaries for the three videos using 20, 23, and 30 images, respectively.

All summaries used the same user interface with the same size popup images (208x156 pixels). For the control condition, popup images were determined by sampling in smaller regular intervals. The selected and manga conditions used the normal manga algorithm for determining popup images.

Procedure

The experiment used a within-subject design in which each participant saw all videos and all summary styles (just not the same combinations of the two). Each participant saw

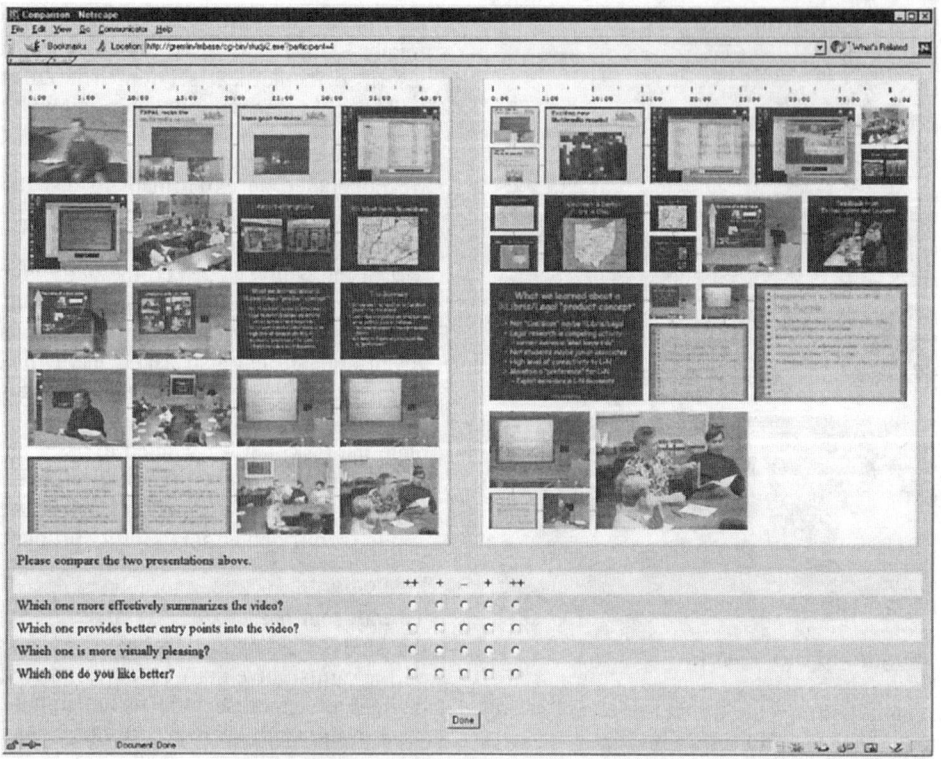

Figure 6: Comparison of visual summaries

each video and each summary style once. We went through all permutations across the participants for the order of the summary styles, the assignment of videos to summary styles, and the order of videos so that the different conditions were balanced for multiples of 12 participants. Each participant was given a training session with a manga summary of a fourth video.

For each video, participants had to read three questions, find the relevant video segments, and then write the answers on a sheet of paper. If a question could not be answered within three minutes, the participants were shown the answer and asked to move on. All mouse actions were recorded and time-stamped. An experimenter noted the strategy used to discover the information required for each task. The completion time for each video was measured.

For the second part of the experiment, participants were asked to report preferences for the different summary styles. They were presented with pair-wise comparisons of different summaries for the same video. For all three videos, all three summaries were compared to each other for a total of nine pair-wise comparisons. Each summary was presented the same number of times on the left and the right side. The order of videos and summaries within each video was randomized across the participants. For each pair of presentations, participants answered the following questions:

1. Which one more effectively summarizes the video?
2. Which one provides better entry points into the video?
3. Which one is more visually pleasing?
4. Which one do you like better?

For each question, a preference could be indicated by selecting "++" or "+" for the summary on the left or on the right. A subject could also select "–" (no difference) if there was no preference (see Figure 6).

Results

The average task completion time of 273 seconds across all summaries is much shorter than the average length of the three videos (3184 seconds) so that the tasks were completed in a small fraction of the time for watching the whole videos. This corresponds to watching the whole video at 12-times speed without ever stopping. Of the 216 total questions, participants were unable to answer eight, of which five were in the control condition. We do not have task completion times for people just using a video player because we assumed that those times would be much longer and we did not want to impose too much on the study participants.

Analysis of variance found no statistically-significant differences in time to complete the tasks among the three conditions ($p > 0.10$). We also calculated the fraction of total time spent watching the video, and the frequency of selecting video to watch. No differences between conditions were found for any of these measures. We may attribute this lack of difference in part due to a bug in the algorithm that made the start of a desired segment of one of the videos unavailable when a detailed view for that segment was displayed. The start of the segment was required to answer the question. As a result, participants spent quite some time watching the video, although it was already past the required point.

In addition to the task completion time data described above, we collected subjects' comparisons of the interfaces (control, selected, manga) for each video. Subjects answered four questions, and their pair-wise judgments on a five point scale were binned into three categories: prefers interface shown on the left, neutral, prefers interface on the right. (Interface condition was counterbalanced across presentations.) We then counted responses in each bin, and classified responses as neutral either if the neutral category was selected more than either interface, or if opinion was split between the two interfaces. As shown in Table 1, the selected condition was indistinguishable from the control: the χ^2 test yielded $p > 0.10$ for Questions 1 and 2, and the neutral response dominated in Questions 3 and 4. Subjects preferred the manga condition to the others for all questions ($\chi^2(2) \geq 9.1, p \leq 0.01$), the manga condition being preferred from 1.9 to 5.8 times more frequently than either control or selected. Based on these results, we can conclude that subjects believe the manga condition to be a more effective summary, to provide better entry points, and to be more visually pleasing. Overall, they preferred it to the other two categories by a count of 97 to 24, with 23 neutral judgments.

Comparison	Q	$\chi^2(2)$	– %	0 %	+ %
Control vs. Selected	1	3.1	23	38	39
	2	3.3	24	40	36
	3	42.6	7	68	25
	4	10.3	17	47	36
Control vs. Manga	1	50.3	19	8	72
	2	25.8	24	15	61
	3	50.3	19	8	72
	4	49.1	13	15	72
Selected vs. Manga	1	18.6	25	18	57
	2	9.1	26	24	50
	3	40.1	26	7	67
	4	27.8	21	17	62

Table 1: Results of χ^2 test[1]

Discussion

The lack of significant differences among the conditions in time to complete tasks may be due in part to the tasks we assigned. We expect that more naturalistic use (when subjects have their own search needs and are better able to understand when the desired passage has been found) will reveal differences in the interface. Furthermore, addressing some of the usability issues discovered during the experiment should improve performance of the manga interface.

In addition to collecting performance and opinion data in the experiment, we observed subjects in their use of the interface, looking for usability defects. Some subjects also volunteered their impressions about the interfaces. Finally, a preliminary analysis of the log also indicated some usability

1. Although we show percentages, raw frequencies were used to calculate the χ^2 values.

Figure 7: Mockup showing details in a focus+context view

issues. The problems fell into two categories: input and output.

On the input side, many participants were confused by the right/left mouse button (drill down vs. play video). Although some people only made the mistake initially, two participants had the problem throughout their sessions. A related suggestion was to expand the view when the mouse hovered over an image, rather than requiring a click. There was also an inconsistency in the way the timeline responded to mouse events: both left and right clicks caused playback, which was not consistent with the keyframe interface.

Output problems were related to the design of the detailed view. The detailed (overlaid) view did not contain the selected image, but did contain other, seemingly unrelated, keyframes (see Figure 5). The main problem with our design was that it was difficult to distinguish context frames from the actual expansion. One way to solve the problem is to make the context keyframes smaller, as shown in Figure 7. Finally, a number of comments centered around image size: some thought that the small (64x48 pixel) images were too small, and others wanted slide images to be large enough to be legible.

CONCLUSIONS

In this paper we described an approach for summarizing videos and for navigating the summaries. We presented a method for generating compact pictorial summarizations of video based on scores that determined the most important shots in a video and assigned sizes to extracted keyframes. Different-sized keyframes were packed into a compact, comic-book-like visual summary. Links from each keyframe allowed video playback and the exploration of additional detail.

We tested our approach in an experiment in which we had participants perform tasks using different summary styles with the manga interface. We also asked the participants to compare the different summary styles to each other with respect to different criteria. The study participants judged the manga summary to be significantly better than the other two conditions with respect to their suitability for summaries and navigation, and their visual appeal.

A system like the one described here can be useful in situations where large collections of video must be accessed. Examples include an application that presents a library of videos such as our staff meetings, or a search engine application that returns a large list of videos that must be examined. The interactive summary described here is part of a larger system for browsing video collections [6] used for the past two years in our laboratory.

Future work will improve summaries by including automatically generated information, such as face recognition data, and automatic captioning from different sources. We also plan to use other sources such as speaker identification and notification of slide changes to improve the segmentation of the videos.

ACKNOWLEDGMENTS

Thanks to John Doherty for producing the meeting videos in our corpus.

REFERENCES

1. Cherfaoui, M. and Bertin, C. (1994). Two-Stage Strategy for Indexing and Presenting Video. In *Proc. Storage and Retrieval for Still Image and Video Databases II, SPIE 2185,* San Jose, CA, pp. 174-184.

2. Christel, M.G., Smith, M.A., Taylor, C.R., and Winkler, D.B. (1998). Evolving Video Skims into Useful Multimedia Abstractions. In *Proc. ACM CHI 98,* Los Angeles, CA, pp. 171-178.

3. Ding, W., Marcionini, G., and Soergel, D. (1999). Multimodal Surrogates for Video Browsing. In *Proceedings of Digital Libraries 99,* ACM, pp. 85-93.

4. Ferman, A.M. and Tekalp, A.M. (1997). Multiscale Content Extraction and Representation for Video Indexing. In *Proc. Multimedia Storage and Archiving Systems II, SPIE 3229, (Dallas, TX),* pp. 23-31.

5. Furht, B., Smoliar, S.W., and Zhang, H.J. (1995). *Video and Image Processing in Multimedia Systems.* Boston: Kluwer Academic Publishers.

6. Girgensohn, A., Boreczky, J., Wilcox, L., and Foote, J. (1999). Facilitating Video Access by Visualizing Automatic Analysis. In *Proc. INTERACT'99,* pp. 205-213.

7. Huang, Q., Liu, Z. and Rosenberg, A. (1999). Automated Semantic Structure Reconstruction and Representation Generation for Broadcast News. In *Proc. IS&T/SPIE Conference on Storage and Retrieval for Image and Video Databases VII,* Vol. 3656, pp. 50-62.

8. Rasmussen, E. (1992). Clustering Algorithms. In W. B. Frakes & R. Baeza-Yates (Eds.), *Information Retrieval: Data Structures and Algorithms,* Prentice Hall, pp. 419-442.

9. Shahraray, B. and Gibbon, D.C. (1995). Automated Authoring of Hypermedia Documents of Video Programs. In *Proc. ACM Multimedia 95,* pp. 401-409.

10. Snedecor, G.W. and Cochran, W.G. (1989). *Statistical Methods.* Ames: Iowa State University Press.

11. Taniguchi, Y., Akutsu, A., and Tonomura, Y. (1997). PanoramaExcerpts: Extracting and Packing Panoramas for Video Browsing. In *Proc. ACM Multimedia 97,* pp. 427-436.

12. Uchihashi, S., Foote, J., Girgensohn, A., and Boreczky, J. (1999). Video Manga: Generating Semantically Meaningful Video Summaries. In *Proc. ACM Multimedia 99,* pp. 383-392.

13. Yeo, B.-L. and Yeung, M. (1998). Classification, Simplification and Dynamic Visualization of Scene Transition Graphs for Video Browsing. In *Proc. IS&T/SPIE Electronic Imaging '98: Storage and Retrieval for Image and Video Databases VI,* pp. 60-70.

14. Yeung, M. and Yeo, B.-L. (1997). Video Visualization for Compact Presentation and Fast Browsing of Pictorial Content. *IEEE Trans. Circuits and Sys. for Video Technology,* 7(5), pp. 771-785.

15. Zhang, H.J., Low, C.Y., Smoliar, S.W., and Wu, J.H. (1995). Video Parsing, Retrieval and Browsing: An Integrated and Content-Based Solution. In *Proc. ACM Multimedia 95,* pp. 15-24.

16. Zhuang, Y., Rui, Y., Huang, T.S., and Mehrotra, S. (1998). Adaptive Key Frame Extraction Using Unsupervised Clustering. In *Proc. ICIP '98,* Vol. I, pp. 866-870.

Face to InterFace:
Facial Affect in (Hu)Man and Machine

Diane J. Schiano, Sheryl M. Ehrlich, Krisnawan Rahardja & Kyle Sheridan
Interval Research Corporation
1801 Page Mill Road, Palo Alto, CA 94304
+ 1 650 842 6099
{schiano, ehrlich, rahardja, sheridan}@interval.com

ABSTRACT

Facial expression of emotion (or "facial affect") is rapidly becoming an area of intense interest in the computer science and interaction design communities. Ironically, this interest comes at a time when the classic findings on perception of human facial affect are being challenged in the psychological research literature, largely on methodological grounds. This paper presents two studies on perception of facial affect. Experiment 1 provides new data on the recognition of human facial expressions, using experimental methods and analyses designed to systematically address the criticisms and help resolve this controversy. Experiment 2 is a user study on affect in a prototype robot face; the results are compared to the human data of Experiment 1. Together they provide a demonstration of how basic and more applied research can mutually contribute to this rapidly developing field.

Keywords
Affective computing, facial affect, facial expression of emotion, affect, emotion, face, nonverbal communication.

INTRODUCTION

Emotion (or "affect") is central to human experience, and facial expressions are our primary means of communicating emotion. Face-to-face communication is inherently natural and social for human-human interactions, and substantial evidence suggests this may also be true for human-computer interactions. That is, people appear to regard computers as social agents with whom "face-to-interface" interaction may be most easy and efficacious [10, 12]. In addition, human (or human-like) faces have been found to provide natural and compelling computer interfaces [e.g., 7, 9, 18]. These findings, together with advances in display and recognition technologies, have produced a surge of interest in facial affect by researchers in human-computer interaction (HCI) and artificial intelligence (AI) alike. Indeed, the burgeoning new field of "affective computing" focuses on computational modeling of human perception and display of emotion, and on the design of affect-based computer interfaces [9, 13].

The question of how to best characterize perception of facial expressions has clearly become an important concern for many researchers in affective computing. Ironically, this growing applied interest is coming at a time when the established wisdom on human facial affect is being strongly challenged in the basic research literature. In particular, recent methodological criticisms have thrown suspicion on a large body of long-accepted data.

The classic psychological research on facial expression of emotion was performed by psychologist Paul Ekman and colleagues, beginning in the 1960s [see 5 for a review]. A substantial body of evidence has been gathered in over three decades, identifying a small number of so-called "basic" emotions: anger, disgust, fear, happiness, sadness and surprise (contempt was tentatively added only recently). In Ekman's theory, the basic emotions are considered to be the building blocks of more complex feeling states [5]. Ekman's data showed that each of these emotions was recognized cross-culturally with substantial consensus among study participants. Ekman and Friesen [4] developed the "facial action coding system" (FACS), a method for quantifying facial movement in terms of component muscle actions. The FACS is a highly complex coding system which requires extensive training to use appropriately. Recently automated [1], the FACS remains the single most comprehensive and commonly accepted method for measuring emotion from the visual observation of faces.

In the past few years, psychologist James Russell and colleagues [14, 15] have strongly challenged the classic data, largely on methodological grounds. Russell argues that emotion in general (and facial expression of emotion in particular) can be best characterized in terms of a multi-dimensional affect space, rather than discrete emotion categories (such as Ekman's "fear" or "happiness"). More specifically, Russell claims that two dimensions--"pleasure" (or "valence") and "arousal"--are sufficient to characterize facial affect space [13, 15]. He calls for new research on perception of facial affect using improved methods and multi-dimensional analyses [15]. The results of such a research program could have profound implications for the psychological understanding of facial affect and direct implications for the field of affective computing. For example, a two-dimensional (2D) characterization of emotion related to Russell's model is assumed in many studies on affective computing, especially those employing physiological measures of arousal [see 11]. On the other hand, many AI models of facial affect recognition rely upon the FACS and Ekman's classic data as the means by which to assess performance [see 9].

This paper presents two studies on perception of facial affect. Experiment 1 provides new data on the recognition of human facial expressions, using experimental methods designed to systematically address Russell's criticisms and help resolve this controversy. Experiment 2 is a user study on affect in a prototype robot face; the results are compared to the human data of Experiment 1. Taken together, they provide a demonstration of how basic and more applied research can mutually contribute to this rapidly developing field.

EXPERIMENT 1: HUMAN FACIAL AFFECT

Russell attacks the methods used in the classic studies on facial affect on various grounds. He points out that a great deal of the data from a large number of studies was generated using a single corpus of (fairly unnatural) stimuli. Russell also indicates certain experimental design flaws in the previous research (e.g., failure to properly randomize stimuli, biasing procedures for practice trials, small numbers of trials) and makes an argument against the common reliance on "within-subject" designs. Russell's primary criticism, however, concerns response format. The vast majority of previous studies employed the forced-choice response format, in which participants were presented with a list of emotion labels and were asked to pick the one that best matches the expression on the stimulus image. Russell's critique of the forced-choice format is that: 1) it tends to be implemented in an all-or-none fashion, which is insensitive to perceived differences in intensity of emotion, and 2) if participants were free to pick multiple responses or to provide their own emotion terms, the results might well look very different.

Our laboratory has recently begun to focus its efforts on generating a credible new body of data on the perception of human facial affect, using improved experimental methods and a variety of stimulus and viewing conditions. An additional goal is to empirically assess the Ekman-Russell controversy: Will the methodological improvements suggested by Russell make the data on perception of human facial affect differ fundamentally from Ekman's classic findings or not? In two previous studies [see 3], we asked independent groups of participants to judge images of the six basic emotions, using "multiple-choice" (i.e., more than one emotion label could be chosen from the response list) and "open-ended" (i.e., participants freely provided the emotion label) response formats, as suggested by Russell. In Experiment 1 of this paper, we complete the response-format assessment by performing a final study using the forced-choice format. All three studies were identical except for the response format manipulation, and they incorporated various methodological improvements over previous research. Several experimental design flaws were eliminated and appropriate techniques for stimulus randomization and presentation were used. We constructed a new (more naturalistic) stimulus set rather than re-using the standard corpus. A rating scale was provided with each forced-choice response so that degrees of perceived emotional intensity could be indicated. Finally, the response format studies comprise a "between-subjects" manipulation, which permits direct comparison of results to the classic data, and additional independent analyses.

Method

Participants

Eighteen Stanford University students between the ages of 18 and 35 participated in this study. All were naïve to the purposes of the study.

Stimuli

Four drama students (2 female, 2 male) produced the facial expression stimuli. To promote naturalness, the actors were briefly familiarized with the facial expressions of interest, and then simply instructed to imagine a time when each emotion was felt strongly [see 3, 5]. Each actor provided a total of 14 different front-view exemplars of the six basic emotions (anger, disgust, fear, happiness, sadness, surprise). Figure 1 provides an exemplar of each emotion (in alphabetical order), and also shows each of the actors' faces. Each participant viewed all the stimuli created by two (randomly chosen) actors. The high-resolution digital images were shown on a 13" Panasonic color TV monitor connected to a PowerMac computer.

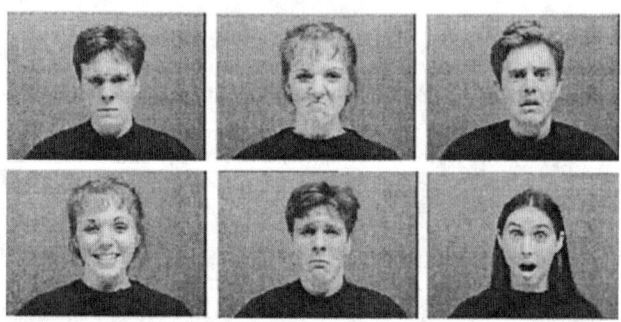

Figure 1: Sample stimulus images for each emotion in Expt. 1: Anger, disgust, fear, happiness, sadness and surprise.

Procedure

Participants viewed stimulus images depicting emotional expressions and responded with an emotion label for each image, using the forced-choice (plus intensity ratings) response format. The images were viewed in random order, at a distance of about 30 inches. In each trial, the participant was shown the stimulus image on the TV monitor and simultaneously viewed a response screen on the computer monitor. The response format was implemented in the following manner: An alphabetized list of labels for the basic emotions was displayed on the computer monitor. Participants chose the one label that best corresponded to the depicted emotion in each image, and then rated the degree to which that emotion was present in the image on a scale of 0 (not at all) to 6 (extremely high). The rating scale appeared as radio buttons adjacent to the selected emotion label. Ten initial (non-feedback) practice trials used randomly selected images from actors not viewed during the test trials. The experimental protocol was implemented in HyperCard. Participants proceeded at their own pace; the entire procedure lasted under one hour.

Results and Discussion

Figure 2 presents the "correct recognition" scores (i.e., the proportion of trials on which the participant responded with the expression portrayed by the actor) for each expression in Experiment 1 (forced-choice response format). These results are shown in context with comparable data from the alternative response format studies (in which the highest rated response was used). Correct recognition scores are the standard form of data presentation in the classic studies. As the Figure shows, correct recognition in Experiment 1 was highest for happiness (mean = 99%) and lowest for fear (mean = 82%), with the remaining emotions falling in between. Performance was uniformly high. These findings closely replicate Ekman's classic results, both in terms of relative pattern and absolute levels of performance. The similarity of results is especially impressive considering the differences not only in methods but also in the stimuli used in this research. The classic stimuli were created in a painstaking fashion by highly trained actors moving specific muscle groups; ours were made in a much more natural way.

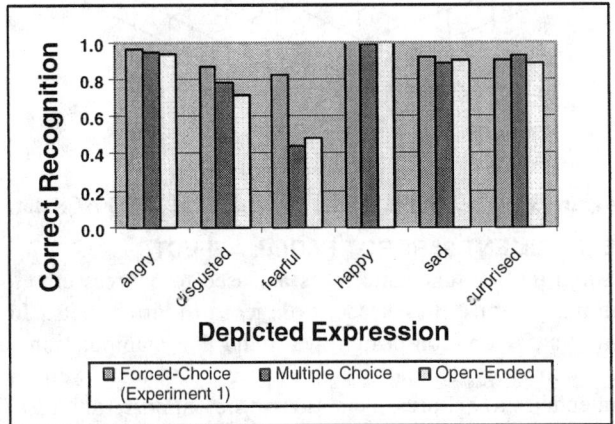

Figure 2: Recognition scores for each emotional expression by response format. Forced-Choice format was used in Expt. 1.

The rating scale data serve as a manipulation check in this study, ensuring that the recognized emotion was in fact seen as present in the stimulus image to at least a moderate degree. If Russell's critique of the forced-choice method is correct and the recognition scores are inflated due to constrained response options, extremely low ratings might be expected for at least some expressions. However, mean ratings for the emotional expressions was 3.93 (out of 6) overall; the mean ratings did not fall below the moderate level for any of the expressions. This suggests that, overall, the participants did see the depicted emotion in the images to at least a moderate degree.

A brief comparison across the datasets for the forced-choice (Experiment 1), multiple-choice and open-ended response format studies is illuminating. (In the multiple-choice study, participants could respond with--and rate--more than one emotion label on the response list, or could press "other" and type in a preferred term. In the open-ended study, participants were given an open text window and simply asked to type in their own responses [see 3]. Contrary to Russell's predictions, the results for the three response formats show a strikingly similar pattern. Correct recognition was generally quite high, highest for happiness and lowest for fear. The one exception to this rule is found in the scores for fear using the multiple-choice and open-ended formats. When the alternative response formats were used, fear was often "misrecognized" as surprise (mean = 25%) or sadness (mean = 17%). This confusion pattern is consistent with similarities in the FACS codes for these emotions. Why fear alone should show such a performance decrement with response format is not quite clear. Fear may be the least compelling emotion under posed conditions, and because it is one of the more ambiguous expressions in terms of FACS code overlap, when observers are encouraged to give multiple responses, they may tend to do so more for fear. Still, even the "misrecognition" of fear was highly systematic, not simply showing greater variability. Taken together, the results of these studies generally support Ekman's classic findings. (The results of an extensive series of additional analyses of the data from these studies are available upon request.)

To perform multi-dimensional scaling (MDS) analyses on the data from Experiment 1, a confusion matrix was generated from the number of times each emotion was mistakenly recognized as any other emotion. This was used to create the similarity space for the analysis. In addition, we asked a trained FACS coder to independently create a FACS confusability index for the basic emotions, based on the degree of overlap of FACS codes between all pairs of emotional expressions (number of overlapping codes over total number of codes for the two emotions, x 100%). The specific expressions were those of the "Directed Facial Action" (DFA) task [8]. While this confusability index is a fairly crude measure (e.g., it does not permit differential weighting of codes), such quantification permits the specification of a FACS prediction pattern that can serve as a benchmark for comparison with our data.

Russell's model of affect space is a "circumplex" about two axes, identified as pleasure and arousal. A schematization of this model (using Russell's terms) is shown in Figure 3.

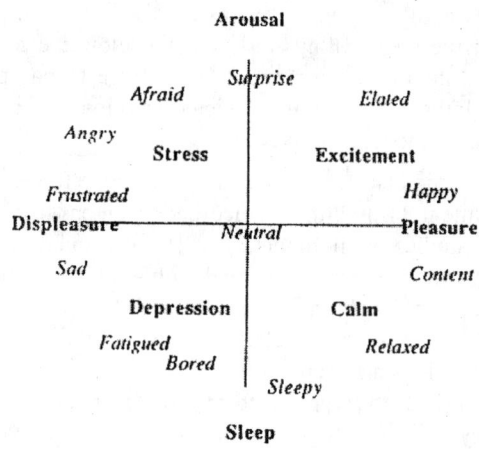

Figure 3: Russell's circumplex model of affect space.

The results of the MDS analysis showed that a two-dimensional (2D) solution accounted for 85% of the variance of the Experiment 1 dataset (and 76% of the FACS dataset). Figure 4 shows the 2D solution for the Experiment 1 dataset (the FACS pattern is highly similar). At first glance, this 2D space looks similar to Russell's model. The datapoints show a roughly circular arrangement. Since rigid rotations of the dataset are permissible in MDS, rotating our results for optimal fit with Russell's schematization does find a fair degree of overlap.

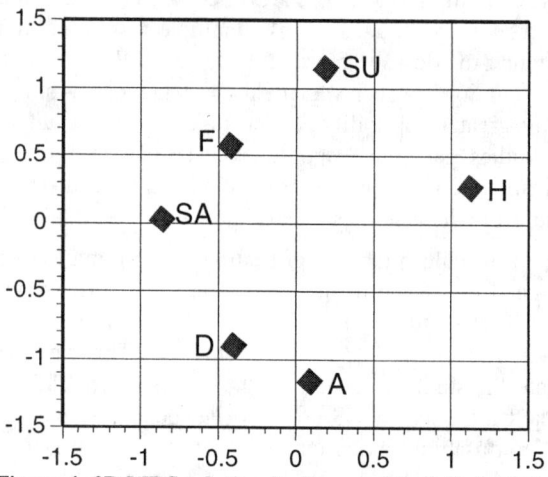

Figure 4: 2D MDS solution for human (Expt. 1) data

However, the ordering of emotions on the circle does not quite match; in particular, the relative ordering of anger and sadness--and the question of where to put disgust--is problematic. The relative ordering of the datapoints is the primary result of the 2D MDS solution, and determines the interpretation of the orthogonal dimensions. Our space could perhaps be interpreted as showing an axis corresponding to pleasure, but identifying an arousal dimension is less plausible.

A 3D MDS solution accounted for 96% of the variance of the Experiment 1 data and 90% of the FACS dataset, a substantial improvement over the 2D approach. Figure 5 presents the 3D solutions for both datasets. A strikingly similar pattern is shown, which Russell's 2D model cannot match. Similar MDS analyses on the multiple-choice and open-ended format data show generally similar results. The fact that the FACS codes are based on physical facial features and movements suggests that the facial affect space's dimensions may correspond more to physical or image parameters than to feeling states (or at least to salient physical parameters associated with feeling states). However, the FACS codes are extremely complex, and at this point it is difficult to speculate on exactly what the dimensions may refer to. Further research is needed. One approach would be to explore facial affect under simplified conditions, which may shed some light on the most salient cues for emotional expression. This is the approach taken in the study described below.

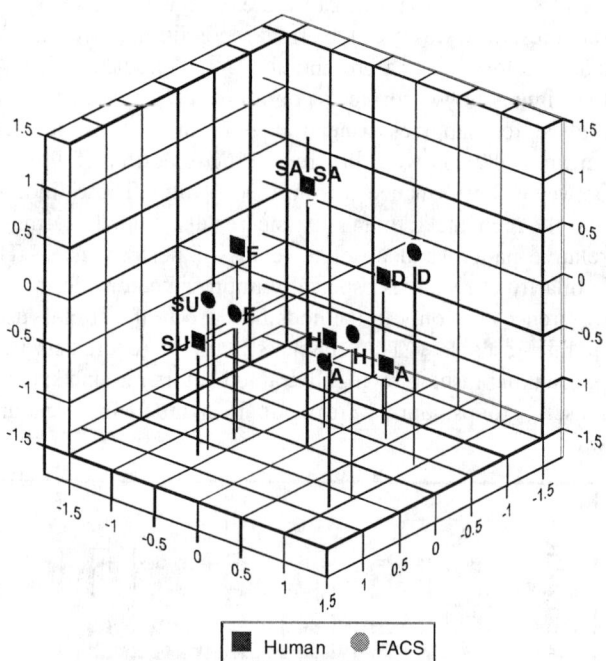

Figure 5: 3D MDS solution for human (Expt 1) & FACS data

EXPERIMENT 2: ROBOT FACIAL AFFECT

Simplified or schematic faces are used to express emotion in many forms of interface, from icons to virtual pets. In a pilot study on icon design, we found that manipulations of an extremely simple facial feature set--which were not intended to express emotion--were almost universally interpreted in emotional terms. Recent work elsewhere shows that cartoon-like facial icons are sufficient to serve as an affective interface for interactions with avatars and autonomous computer agents [e.g., 7]. The present study was performed in the context of user-testing an early design prototype of a mechanical robot face. While made of metal, the face was intended to be cartoon-like in nature. Created by an independent group of researchers at Interval Research [Tow & Scheeff, personal communication; see

17], it was designed largely through intuition inspired by cartoon animation principles [6] and some reading of Darwin [2] on emotions. This initial prototype was constructed largely as a "proof of concept" to demonstrate that a very simple robot face could effectively express affect. A similar face would be incorporated into a later, more complete prototype, capable of whole-body movement and expression.

The initial prototype consisted of a box-like face containing eyes with moveable lids, tilting eyebrows, and an upper and lower lip which could be independently raised or lowered from the center. Figure 6 shows the face displaying various emotional expressions. Compared to the highly detailed human faces of Figure 1, the robot facial features are extremely sparse and their motion is highly constrained. The face has no skin, so the telltale folds, lines and wrinkles specifying many FACS codes are simply not available. And the motion of the features (especially the eyebrows and lips) is only schematically related to human facial muscle movements. Experiment 2 was conducted in collaboration with the robot design team. They primarily wanted some assurance that users felt satisfied with the robot's ability to express a range of basic emotions. Another goal was to obtain feature settings for various emotional displays, to be stored as templates so that (at least in principle) the prototype could be set up to quickly display appropriate emotional responses as needed. We were especially interested in comparing the robot "affect space" with our human data, after verifying that the displays were indeed correctly recognized by an independent group of observers.

METHOD

Participants
Eighteen participants between the ages of 18 and 35 were involved in this study. Nine Interval Research Corporation employees participated in the first condition of the study, and 9 Stanford University students in the second condition. All were naïve to the purposes of the study.

Materials
The robot face consisted of a 12 cm x 14 cm mechanical metal (primarily aluminum) face with independently moveable eyelids, eyebrows, upper and lower lips (see Figure 6). The eyelids were small metal sheets that could move up or down. The eyebrows were metal bars, placed with a pivot point towards each side of the face, to allow rotations between horizontal and vertical positions. Each lip consisted of a spring fixed at both ends and with a tie in the center that could be pulled up or down (stretching the spring on both sides). Each feature was controlled by a computerized motor with 255 possible positions. For all features except the eyelids, the "neutral" position was in the center of the range of motion. The neutral position for the eyelids was fully open (or up).

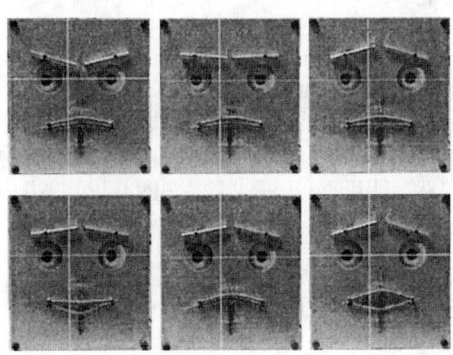

Figure 6: "Average" robot face display for each emotion in Expt 2: Anger, disgust, fear, happiness, sadness, surprise.

Procedure
Condition 1: Feature Setting Adjustments. In each of two blocks of trials, instructions on the computer monitor asked the participant to set the features of the robot face to express each of the 6 basic emotions (angry, disgusted, fearful, happy, sad, or surprised) at each of 3 degrees of intensity (slightly, moderately, or very), twice, in random order. Participants adjusted feature settings by pressing keys on a computer keyboard; "up" and "down" keys were labeled on the keyboard for each of the 4 features. Upon completion of each expression, participants rated their overall satisfaction with the expression on a scale of 1 (least) to 5 (most). Each trial began with the features in the neutral position, except for the eyelids, which were closed. Each testing session began with 10 randomly chosen (non-feedback) practice trials. Participants proceeded at their own pace, and the entire procedure took less than one hour. The robot face was attached to a Toshiba laptop PC, which implemented the experimental protocol.

Condition 2: Recognition Validation. Participants viewed the robot facial expressions obtained from Condition 1. The expressions were given by the mean feature settings for each emotion (angry, disgusted, fearful, happy, sad, or surprised) at each of the three degrees of intensity (slightly, moderately, or very), plus the "average" setting. The average setting for each emotion is depicted in Figure 6. In all, 4 exemplars of each emotion were shown, 3 times, in random order. For each trial, participants used the forced-choice (plus ratings) response method of Experiment 1, choosing the one term from an alphabetized list of basic emotion labels that best described the expression of the robot face. They then rated the degree to which the emotion was present in the robot face on a scale of 0 (not at all) to 6 (extremely high). In this study, the robot facial expressions were controlled by a Toshiba laptop PC while the rest of the experimental protocol was implemented in HyperCard on a PowerMac computer. Participants proceeded at their own pace; the entire procedure took about 30 minutes.

RESULTS AND DISCUSSION

The results for Condition 1 are given in terms of numerical setting values for each feature, which are difficult to summarize succinctly except pictorially. Figure 6 illustrates the "average" display for each emotion, derived from the mean feature settings for each participant for each degree of emotion. The figure does suggest that the interface was capable of expressing various emotions, although perhaps not all equally well. While participants' satisfaction ratings were fairly high overall (3.7 out of 5, for all degrees of intensity), satisfaction with disgust, in particular, was fairly low (2.9). The FACS codes characterize disgust by the drawing up of the nasal-labial muscles, producing striking patterns of wrinkles and folds around the mouth and nose. However, the robot face has neither nose nor skin. That disgust was found to be particularly difficult to express was not especially surprising.

The results for Condition 2 are shown in Figure 7, in terms of mean correct recognition scores for each of the emotions (averaged over degree of intensity). To aid comparison, the human data from Experiment 1 are also provided in the figure. The scores for the robot are generally somewhat lower than those for human faces (especially for disgust), but this is not very surprising. First, the schematic nature of the robot face (as described above) should have made it more difficult to express emotion than human faces. Secondly, these scores were averaged over stimuli intended to depict emotion at varying intensity levels, while the human actors presumably at least intended to create stimuli that showed each emotion to a high degree of intensity. Third, due to time constraints, our sample size was small and so the dataset is fairly variable. That the human and robot results are nonetheless so close is noteworthy.

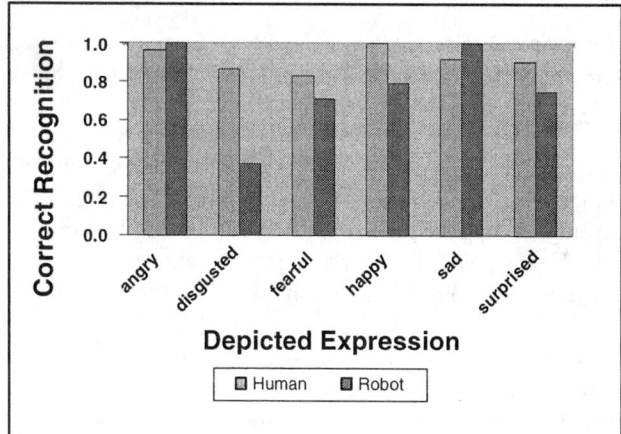

Figure 7: Recognition scores for each emotional expression for human (Expt. 1) and robot (Expt. 2) stimuli.

As in Experiment 1, intensity ratings served as a manipulation check in this experiment. The mean ratings over all emotions were moderately high (mean = 4.34 out of 6). Further analyses (available upon request), found that the ratings did generally vary with the intensity of the depicted emotion. And, as expected, recognition of the emotions tended to increase with rated intensity.

MDS analyses were performed on the robot feature setting data. The mean direction and amount of movement of each of the 4 facial features for each emotion (taken from the average of all expressions) were used to generate 4D vectors; the distance between each vector gave the (dis-) similarity matrix for the MDS analysis of the robot dataset. Ninety-seven percent of the variance of the robot data was accounted for by a 2D MDS solution. Figure 8 presents the 2D solutions for the robot and human (Experiment 1) datasets. The close similarity of the patterns is immediately obvious, with ordering of emotions identical for the two datasets.

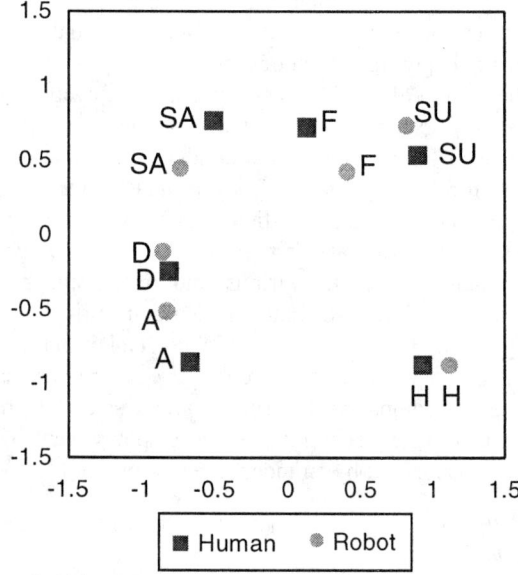

Figure 8: 2D MDS solution for human (Expt 1) & robot data

Ninety-nine percent of the variance of the robot dataset was accounted for by a 3D MDS solution. Figure 9 shows the 3D solutions for the human and robot data. The similarity of the patterns is remarkable when considering the disparity of the stimuli and the fact that the robot data were plotted directly from the feature setting parameters. As in the case of the human data, the robot results map easily onto the FACS index pattern (see Figure 5) but not so easily onto Russell's model (see Figure 3). Moreover, the similarity of findings across the human, FACS and robot datasets further supports the notion that the dimensions of facial affect space might correspond most closely to physical or image parameters; indeed, to very simple ones [see also 19]. Our initial speculation is that the primary axis may correspond to concavity/convexity of the lips, the second to the upward/downward tilt of the eyebrows. The third dimension is less clear, but may be related to the set of the mouth, perhaps its degree of openness (note that many of our disgust stimuli had open mouths). Further research is clearly needed, but these results do suggest implications for many applications in which the complexity of the face is constrained or compressed. We are currently looking at human facial affect under a variety of compressed image conditions, to see if a similar affect space is found.

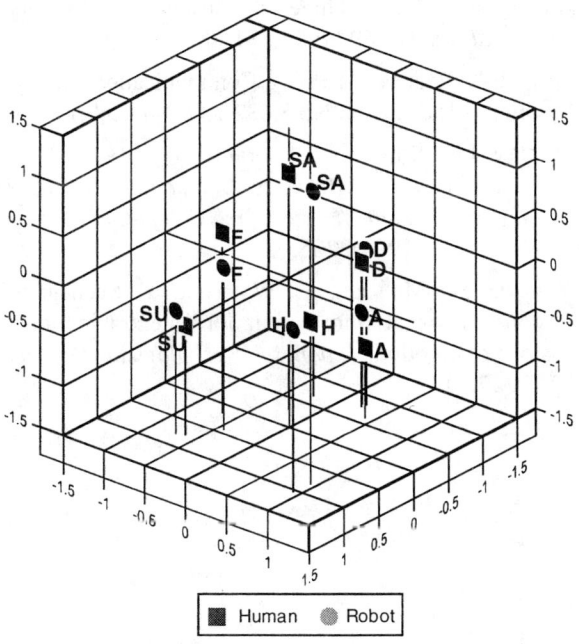

Figure 9: 3D MDS solution for human (Expt 1) & robot data

GENERAL DISCUSSION

Experiment 1 provides new baseline data on human facial affect recognition, using improved experimental methods and somewhat more naturalistic stimuli than those of the classic studies. The pattern of results for the forced-choice response format closely replicate Ekman's classic findings, and (except for fear), this was generally true for the alternative response formats as well. Thus, on the whole, Russell's criticisms are not borne out by the data. Our MDS analyses suggest that 3 dimensions are substantially better than 2 in specifying facial affect space. However, even the 2D solution does not match Russell's model. Indeed, our data match the FACS-based solutions much more closely. We find this intriguing, suggesting that the dimensions of facial affect may be based more on physical or image parameters than on feeling states (such as pleasure and arousal) per se.

Experiment 2 was performed in the context of a user test; its primary aim was to inform the designers of the affective robot face. We succeeded both in demonstrating that the face was sufficient to communicate various emotional expressions, and in providing feature setting templates for specific emotions of varying intensities. The revised prototype of the affective robot incorporates a face very similar to the one we tested. The pattern of results for this experiment was strikingly similar to our human data, despite extreme schematization of the robot face. The similarity of the MDS solutions for robot, human and FACS-based data underscore the notion that physical or image-based parameters--perhaps very simple ones--could be used to interpret the dimensions of facial affect space. Some speculations on what those parameters may be were provided above, and we note that our research on this topic is continuing. In addition to exploring human facial affect under various compression conditions (using both still and moving images), we are also collaborating with another laboratory in training a neutral-net AI model on our stimuli, to see what features it picks up [see 9]. Interestingly, 3-D models of affect have been suggested before, largely based on feeling states [e.g., 16; see 15], but no consensus in axis interpretation was achieved in that earlier research. This is an exciting time for new research on facial affect in both humans and machines. We hope this paper helps demonstrate the need for both basic and applied contributions to this rapidly developing field.

ACKNOWLEDGEMENTS

We would like to thank Diane Beck, John Pinto, Mark Scheeff, and Rob Tow for their contributions to this project.

REFERENCES

1. Bartlett, M.S., Hager, J.C., Ekman, P. & Sejnowski, T.J. *Psychophysiology, 36* (1999), 253-263.

2. Darwin, C. *The Expression of Emotion in Man and Animals.* Univ. of Chicago Press: Chicago, IL, 1965 (originally 1872).

3. Ehrlich, S., Schiano, D.J., Sheridan, K., Beck, D. & Pinto, J. Facing the issues: Methods matter *Psychonomic Society* (Dallas, TX, May 1998).

4. Ekman, P. and Friesen,W.V. *The Facial Action Coding System.* Consulting Psychologists Press: Palo Alto, CA, 1978.

5. Ekman, P., Friesen, W.V. & Ellsworth, P. *Emotion in the Human Face.* Pergamon, New York, 1972.

6. Hamm, J. *Cartooning the Head and Figure.* Perigree, New York, 1982.

7. Kurlander, D., Skelly, T. & Salesin, D. "Comic chat," *Computer Graphics Proceedings*, (1996), ACM Press, 225-236.

8. Levenson, R. W., Ekman, P., Friesen, W. V. Voluntary facial action generates emotion-specific autonomic nervous system activity. *Psychophysiology, 27* (1990), 363-384.

9. Lisetti, C.L. & Schiano, D.J. Automatic facial expression interpretation: Where Human-Computer Interaction, Artificial Intelligence and Cognitive Science intersect. *Pragmatics and Cognition* (Special Issue on Facial Information Processing, a Multidisciplinary Perspective). 1999, in press.

10. Nass, C.I., Steuer, J.S. & Tauber, E. "Computers are social actors", in *Proceedings of CHI'94* (Boston MA, April 1994), 72-78.

11. Picard, R. *Affective Computing.* MIT Press: Cambridge, MA, 1997.

12. Reeves, B. & Nass, C.I. *The Media Equation: How People Treat Computers, Television, and New Media Like Real People and Places.* Cambridge University Press: New York, 1996.

13. Russell, J.A. A circumplex model of affect. *Journal of Personality and Social Psychology*, 39(6) (1980), 1161-1178.

14. Russell, J.A. Is there universal recognition of emotion from facial expression? *Psychological Bulletin, 95* (1994), 102-141.

15. Russell, J.A and Fernandez-Dols, J.M. *The Psychology of Facial Expression.* Cambridge University Press: New York, 1997.

16. Schlosberg, H. Three dimensions of emotion. *Psychological Review*, 69 (1954), 81-88.

17. Tow, R. "Affect-based Robot Communication Methods and Systems" - U.S. Patent No 5,832,189, November, 1998.

18. Walker, J.H., Sproull, L., Subramani, R. (1994). Using a human face in an interface, in *Companion of CHI'94 Conference on Human Factors in Computing Systems* (Boston MA, April 1994), 205.

19. Yamada, H., Matsuda, T., Watari, C. & Suenaga, T. Dimensions of visual information for categorizing facial expressions of emotion. *Japanese Psychological Research*, 35(1993), 172-181.

Hedonic *and* Ergonomic Quality Aspects Determine a Software's Appeal

Marc Hassenzahl, Axel Platz, Michael Burmester and Katrin Lehner
Corporate Technology - User Interface Design, Siemens AG
81730 Munich, Germany
+49 (0) 89 636-49653
marc.hassenzahl@mchp.siemens.de

> "we continue to see [...] the prospect of a decade of research analysis of usability possibly failing to provide the leverage it could on designing systems people will really want to use by ignoring what could be a very potent determination of subjective judgements of usability - fun" (p. 23) [6].

ABSTRACT

The present study examines the role of subjectively perceived *ergonomic quality* (e.g. simplicity, controllability) and *hedonic quality* (e.g. novelty, originality) of a software system in forming a judgement of appeal. A hypothesised research model is presented. The two main research question are: (1) Are ergonomic and hedonic quality subjectively different quality aspects that can be independently perceived by the users? and (2) Is the judgement of appeal formed by combining and weighting ergonomic and hedonic quality and which weights are assigned?

The results suggest that both quality aspects can be independently perceived by users. Moreover, they almost equally contributed to the appeal of the tested software prototypes. A simple averaging model implies that both quality aspects will compensate each other.

Limitations and practical implication of the results are discussed.

Keywords

perceived software quality, emotional usability, hedonic components, joy of use

INTRODUCTION

In 1988, Carroll and Thomas [6] admonished us not to confuse the concepts *easy to use* and *fun to use* when talking about software quality. They argued that ease of use implies simplicity, which in turn is partly incompatible with fun. By making something as simple as possible, there is a good chance to make it boring as well. On the other hand, fun requires a subtle balance of not being too simple and not being too challenging (see [7]). As Carroll and Thomas put it: "we do not necessarily want to merely make things as simple as possible. We ought to make them fun" (p. 22).

They call for the scientific study of fun and specialised design methods for promoting fun.

Since then *fun of use* did not receive that much attention in the field of Human-Computer Interaction. It plays a minor role in Technology Acceptance literature. For example, Davis et al. [8] showed that perceived fun (defined as "the extent to which using a software system is enjoyable in its own right") can accelerate usage intention if the software system is already perceived as useful ("the extent to which a person believes that using a particular software system would enhance his or her job performance"). Fun had no effect on usage intentions if a software system was not regarded as useful.

Igbaria, Schiffman and Wieckowski [12] studied the impact of perceived usefulness and perceived fun on both system usage and user satisfaction in a work context. Their results showed an almost equal effect of perceived fun and perceived usefulness on system usage. Perceived fun had even a stronger effect on user satisfaction than perceived usefulness. User satisfaction in turn had a clear effect on system usage. It can be concluded that enhancing perceived fun will primarily lead to increased time spent with a software system. This in turn may lead to a better understanding and/or a more productive use of the system. Furthermore, enhancing perceived fun should be a valuable road to directly increase user satisfaction.

Both studies succeed in demonstrating the positive impact of perceived fun - a factor independent of the efficiency and effectiveness of a software system - on usage intentions and user satisfaction. What these studies do not provide is an idea which design features will increase perceived fun.

One approach to find design principles which have the power to promote fun/enjoyment of a software system is to analyse what makes computer games fun [6;19;20]. Malone [19] identifies the following three broad design categories: "Challenge", "Fantasy" and "Curiosity". Each consists of several principles and recommendations for designing an appealing computer game. Among those, principles both consistent and contradictory to the notion of usability can be found. For example, the recommendation "provide a fantasy" is generally consistent with the idea of using a "metaphor" to increase familiarity and thereby the usability of a system.

However, besides the cognitive aspects (increased familiarity) Malone stresses the emotional aspects of a metaphor used, i.e. its power to satisfy the user's emotional needs. Mainly contradictory to usability is the principle to foster curiosity by designing a system that is novel and surprising (but not completely incomprehensible). In order to work, novelty and surprise must impair at least the external consistency of the software, a core principle of usability. In usability design, this might conflict with the task-related efficiency and effectiveness of the software system. This raises the question, whether it is appropriate to call for enjoyment and fun when it comes to work-related software systems (e.g. [11]).

The strong focus on task-related efficiency and effectiveness arises from the implicit notion that the computer is a tool - or even stronger - *has to be* a tool when applied at the workplace. This somewhat conservative perspective where "things must be taken seriously" is predominant in business and might be at least partly wrong considering the role of perceived fun/enjoyment in the context of technology acceptance at the workplace. The somewhat narrow focus of usability on task-related efficiency and effectiveness is eventually criticised in studies concerning consumer products [1;14;18].

What is needed is an expanded concept of usability which adopts enjoyment and satisfaction of the user as the major design goal[1]. Obviously, in a work context designing for efficiency and effectiveness (i.e. traditional usability) will be still a valuable way to reach the goal. Nevertheless, something should be added which makes the software system interesting, novel, surprising etc. Being both usable and interesting, a software system might be regarded as appealing and as a consequence the user may enjoy using it. Such an expanded perspective on usability would take us a step further toward designing *user experiences* [16] instead of merely making a software usable.

An expanded concept of usability was suggested by Logan [17]. He has developed a two-component usability concept that consists of behavioural and emotional usability. Behavioural usability refers to more or less traditional usability. Emotional usability refers "to the degree to which a product is desirable or serves a need(s) beyond the traditional functional objective" (p. 61). Although important, the concept of emotional usability still leaves many questions unanswered. For example, Logan provides no model or data about whether and how behavioural and emotional usability influence each other.

Another line of research to broaden the traditional usability concept mainly considers the impact of visual design on the expected (apparent) usability of a system *before* actually using it [5;15;22]. By demonstrating the impact of visual design on expected (apparent) usability, these studies succeed in establishing visual design as a key design factor. However, this is rather done by regressing visual design on the traditional usability concept than expanding usability itself.

This paper attempts to provide and test a model for an expanded concept of usability that incorporates key factors for designing appealing, enjoyable software interfaces and systems.

APPEALING SOFTWARE SYSTEMS: A HYPOTHESIZED MODEL

Figure 1 shows the elements of the hypothesised model for appealing software systems, i.e. quality aspects of the software system, evaluation of the system and consequences of using it.

Fig. 1: Research Model

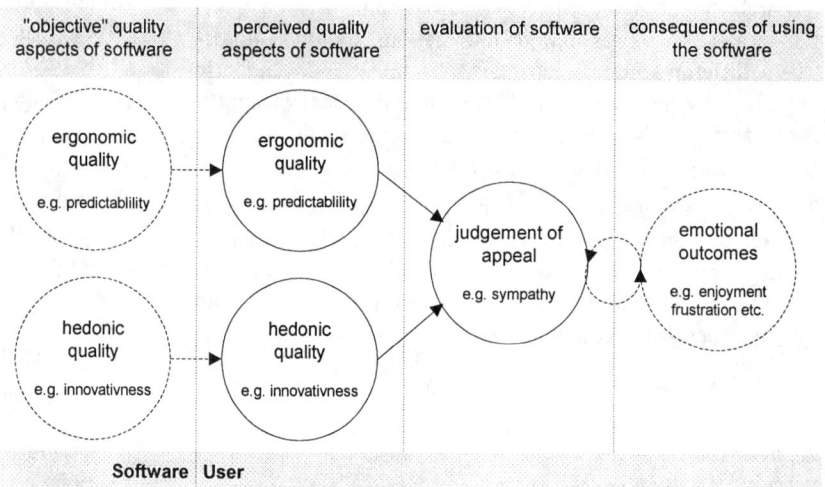

A software system can be described on different quality dimensions such as its predictability, controllability etc. The model distinguishes between two groups of quality dimensions (i.e. aspects): *ergonomic* and *hedonic* quality.

Ergonomic quality (EQ) comprises quality dimensions that are related to traditional usability, i.e. efficiency and effectiveness (e.g. [13]). EQ focuses on task-related functions or design issues.

Hedonic quality (HQ) comprises quality dimensions with no obvious relation to the task the user wants to accomplish with the system, such as originality, innovativeness, beauty etc. Although not task-related, the users may regard HQ as an important quality aspect for its own sake.

---

[1] We are aware that user satisfaction is a part of the usability concept provided by ISO 9241-11. However, it seems as if satisfaction is conceived as a consequence of user experienced effectiveness and efficiency rather than a design goal in itself (see [10] for an example). This implies that assuring efficiency and effectiveness alone guarantees user satisfaction.

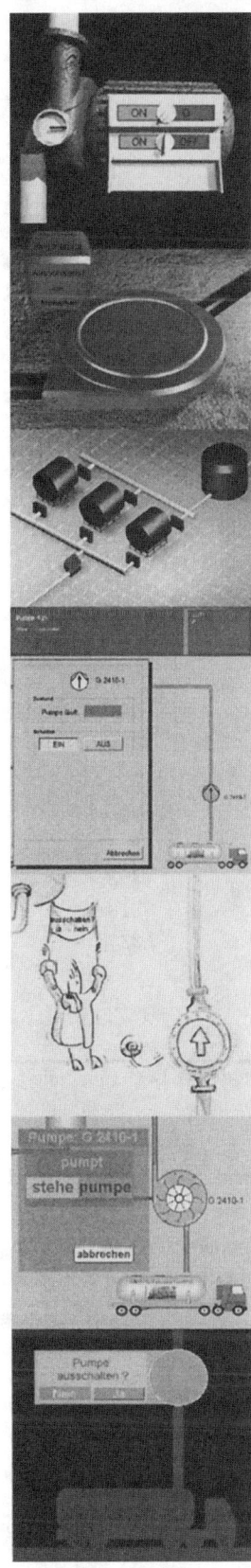

Fig. 2: Prototypes

Whether a software system is viewed as appealing by the users or not depends heavily on the users' perception of the quality aspects. For example, "objective" usability (i.e. inherent or intended by the software designer) is not always also perceived by the users [15]. Clearly, a software system designed to be as simple as possible fails if this simplicity can not be perceived or experienced by the users. For this reason, we focus on the subjective perceptions of EQ and HQ provided by the users.

The user's *judgement of appeal* (APPEAL) differs from the mere quality aspects of an additional evaluation. The judgement may be formed by weighting and combining different perceptions of the system's quality aspects (EQ and HQ, respectively).

Emotional outcomes are conceptualised as a consequence of using a software system, rather than as a design goal in itself. If the software system is appealing, the user may experience enjoyment or even fun (if not, she may experience frustration or anger). For this reason, the emotional outcome is not explicitly included in the present study. However, it remains important and should be included in further studies.

The present paper will see to the following research questions.

Q1: Are ergonomic (EQ) and hedonic quality (HQ) subjectively different quality aspects that can be independently perceived by the users?

Q2: Is the judgement of appeal (APPEAL) formed by combining and weighting EQ and HQ? Which weights are assigned to EQ and HQ?

Q3: The research concerning apparent usability [5;15;22] emphasises user expectations concerning the quality aspects of a software system *before* actually using it. Based on this general idea, we additionally investigated EQ, HQ and APPEAL under both conditions - before and after using a software system. Furthermore, the stability of EQ, HQ and APPEAL will be examined.

METHOD
Participants
Twenty individuals (6 women, 14 men) participated in the study. They were recruited among the employees of Siemens Corporate Technology in Munich. The sample's mean age was 32.5 years (Min 25, Max 57). Computer expertise varied from moderate to high.

Prototypes
Seven software prototypes were designed and implemented (see Figure 2). The participant's task was to switch off a pump in an assumed industry plant. It was a very simple, but realistic task from a domain unknown to the participants. Thereby, unwanted comparison to existing software systems should be reduced. Each prototype allowed the user to accomplish the same task but strongly varied in design and interaction style. The prototypes were designed and implemented by students of visual, industrial and ergonomic design. Multiple design dimensions were varied (e.g. colours, design style). Six out of the seven prototypes had animated parts. The predominant design principle was heterogeneity, i.e. maximising variance. Hence, flaws in the visual and ergonomic design were not corrected.

Semantic differential
A semantic differential (see [9] p. 73) was constructed to measure EQ, HQ and APPEAL. It consists of 23 seven-point scale items with bipolar verbal anchors (see Table 1).

Each scale item was selected beforehand to represent a facet of the quality aspect to be measured (EQ, HQ) and the judgement of appeal (APPEAL) according to the definitions given above. Informal expert reviews were conducted to assure measurement quality of the differential's initial version. The scales were presented in random order.

Pre and Post Usage
All measurements were obtained twice: before and after using the software system. EQ, HQ and APPEAL values obtained *before* actual use are named *expected* (i.e. expected by the user); EQ, HQ and APPEAL values obtained *after* using the system are called *experienced*.

Procedure
The study was carried out in the usability laboratory of the User Interface Design department of Siemens Corporate Technology. Each participant was separately led into the lab by the experimenter. After a short instruction, the experimenter showed the first prototype to the participant for approximately one minute. The prototypes were presented in random order. The participant was then asked to make an initial assessment with the semantic differential before actually using the prototype (*expected* values).

At a signal of the experimenter, the participant had to select the running pump on the screen with the mouse. Depending on the prototype, a question appeared asking whether the participant really wanted to switch off the pump. After a confirmation and a safety check the pump could be turned off. The participant then had to wait until the pump stopped. The end of the task was reached when the participant indicated that s/he was convinced that the pump came to a halt. The whole interaction per prototype took about two minutes.

After finishing the task, the participant was requested to revise his/her initial assessment (*experienced* values). This procedure was repeated for all seven prototypes.

Questionnaires concerning demographics, computer expertise and computer anxiety were handed out. S/He was then shortly debriefed and led out of the laboratory. A session took about an hour.

Tab. 1: Bipolar verbal scale anchors per factor

Scale Item	Anchors	
EQ 1	comprehensible	incomprehensible
EQ 2	supporting	obstructing
EQ 3	simple	complex
EQ 4	predictable	unpredictable
EQ 5	clear	confusing
EQ 6	trustworthy	shady
EQ 7	controllable	uncontrollable
EQ 8	familiar	strange
HQ 1	interesting	boring
HQ 2	costly	cheap
HQ 3	exciting	dull
HQ 4	exclusive	standard
HQ 5	impressive	nondescript
HQ 6	original	ordinary
HQ 7	innovative	conservative
APPEAL 1	pleasant	unpleasant
APPEAL 2	good	bad
APPEAL 3	aesthetic	unaesthetic
APPEAL 4	inviting	rejecting
APPEAL 5	attractive	unattractive
APPEAL 6	sympathetic	unsympathetic
APPEAL 7	motivating	discouraging
APPEAL 8	desirable	undesirable

Note: verbal anchors of the differential are originally in German

RESULTS
Factorial validity and scale reliability of the semantic differential (Q1)

The first question to be answered is whether ergonomic (EQ) and hedonic quality (HQ) are independently perceived by the participants (Q1).

A factor analysis (Principal Components, Varimax rotation) of the *experienced* EQ and HQ items of the semantic differential extracted two factors with an Eigenvalue higher than 1 (see Table 2). Together both factors explain approx. 68% percent of the total variance. This analysis confirms the initial version of the differential. It shows high consistency with the theoretically assumed factors EQ and HQ. Furthermore, the successfully attained simple structure (e.g. [21]) by applying Varimax rotation shows that EQ and HQ are perceived as independent quality concepts by the participants. For the remainder of the paper, a mean ergonomic and hedonic quality value is computed from the respective single scale item values.

Tab. 2: Factorial validity of experienced EQ and HQ

Scale Item	Principal Components with Varimax	
	Factor 1	Factor 2
EQ 1	.783	
EQ 2	.816	
EQ 3	.715	-.233
EQ 4	.880	
EQ 5	.893	
EQ 6	.824	
EQ 7	.885	
EQ 8	.679	
HQ 1		.805
HQ 2		.731
HQ 3		.722
HQ 4		.854
HQ 5		.892
HQ 6	-.226	.818
HQ 7	-.218	.831
Eigenvalue	5.44	4.76
% Variance explained	36.27	31.72

Note: N=140 (20 participants x 7 prototypes), EQ: ergonomic quality, HQ: hedonic quality, factor loadings < .20 are omitted

An analysis of the *expected* EQ and HQ items revealed similar results.

A factor analysis (Principal Components) of the *experienced* APPEAL items of the semantic differential extracted only one factor with an Eigenvalue higher than 1 (see Table 3).

This confirms the initial selection of items for the APPEAL scale, its univariate character and internal consistency (see scale reliability). For the remainder of the paper a mean judgement of appeal is computed from the values of the single scale items.

An analysis of the *expected* APPEAL items revealed similar results.

Tab. 3: Factorial validity of experienced APPEAL

Scale Item	Principal Components
	Factor 1
APPEAL 1	.868
APPEAL 2	.794
APPEAL 3	.683
APPEAL 4	.878
APPEAL 5	.865
APPEAL 6	.903
APPEAL 7	.800
APPEAL 8	.775
Eigenvalue	5.43
% Variance explained	68.20

Note: N=140 (20 participants x 7 prototypes), APPEAL: judgement of appeal

Table 4 summarises the characteristics of the computed scales.

Tab. 4: Scale characteristics

Scale	Cronbach's Alpha	Mean	Stand. Dev.	Min	Max
expected (pre usage)					
EQ	.93	0.87	1.37	-2.88	3.00
HQ	.93	-0.19	1.47	-3.00	2.75
APPEAL	.95	0.38	1.40	-3.00	2.86
experienced (post usage)					
EQ	.93	0.46	1.50	-2.75	3.00
HQ	.92	0.02	1.40	-2.88	3.00
APPEAL	.93	0.31	1.37	-3.00	3.00

Note: N=140 (20 participants x 7 prototypes), EQ: ergonomic quality, HQ: hedonic quality, APPEAL: judgement of appeal

Predicting the participant's judgement of appeal (Q2)

The judgement of appeal is conceptualised as being formed on the basis of the individual's perceptions of EQ and HQ. To check this assumption two regression analysis were performed in order to predict expected APPEAL from expected EQ and HQ and experienced APPEAL from experienced EQ and HQ (see Table 5).

In both analyses, the variables succeed in predicting APPEAL: EQ and HQ almost equally contribute to the judgement of appeal. No interaction of EQ and HQ was found. This pattern implies an averaging model, with the final judgement of appeal depending on both, the perception of ergonomic *and* hedonic quality.

Tab. 5: Regression analysis of EQ and HQ on APPEAL (expected and experienced)

Criterion	Adjusted R^2	Predictors	Beta	Std. Error	Sig.
expected (pre usage)					
APPEAL	.74***	EQ***	.51	.05	.000
		HQ***	.62	.04	.000
APPEAL	.74***	EQ***	.40	.14	.003
		HQ***	.53	.10	.000
		EQ x HQ	.16	.03	n.s.
experienced (post usage)					
APPEAL	.67***	EQ***	.64	.05	.000
		HQ***	.58	.05	.000
APPEAL	.67***	EQ***	.64	.12	.000
		HQ***	.72	.14	.000
		EQ x HQ	-.09	.03	n.s.

Note: N=140 (20 participants x 7 prototypes), EQ: ergonomic quality, HQ: hedonic quality, APPEAL: judgement of appeal, ***p<.000

Expected vs. Experienced EQ, HQ and APPEAL (Q3)

Figure 3 shows expected and experienced EQ, HQ and APPEAL (i.e. pre and post usage, see also Table 4 for scale characteristics).

A repeated measurements analysis of variance with the within-subject factors "time of measurement" (pre usage, post usage) and "type of measurement" (EQ, HQ, APPEAL) revealed a highly significant main effect of "type" (F=11.30, d.f.=2, p=.000) and a highly significant "type" and "time" interaction (F=19.92, d.f.=2, p=.000). No main effect was found for "time" (F=1.71, d.f.=1, n.s.).

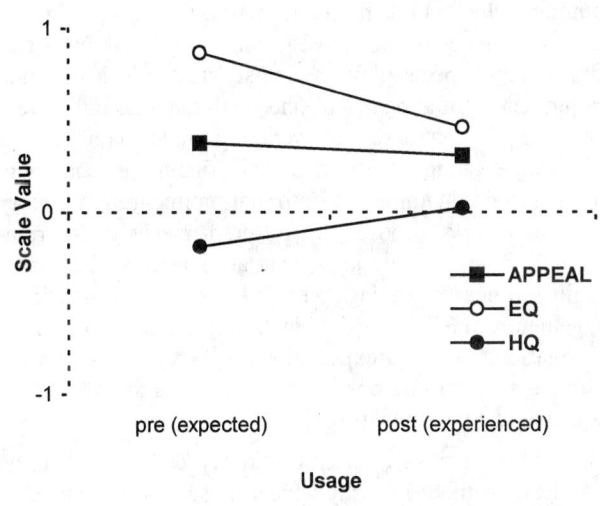

Fig. 2: Expected and experienced EQ, HQ and APPEAL

Single paired sample t-tests for each measure showed a highly significant increase of HQ (diff=.21 scale units, t=3.66, p=.000) and a significant decrease of EQ (diff.=-.41 scale units, t=-3.94, p=.000) after using the prototypes. APPEAL remained stable (diff.=-0.07 scale units, t=-0.87, n.s.).

Values of expected and experienced EQ, HQ and APPEAL were all significantly correlated (expected EQ - experienced EQ: r=.64, N=139, p=.000; expected HQ - experienced HQ: r=.89, N=139, p=.000; expected APPEAL - experienced APPEAL: r=.78, N=139, p=.000).

DISCUSSION

The results of the factor analysis show that the two assumed quality aspects EQ and HQ can be perceived consistently and independently by users. This is true for both expected and experienced values. In other words, users are able to distinguish task-related aspects from non task-related aspects.

Carroll and Thomas' [6] argument that ease of use implies simplicity, which in turn is partly incompatible with fun, is apparent in the data (see Table 2). The items EQ 3 "simple - complex", HQ 6 "original - ordinary" and HQ 7 "innovative - conservative" are negatively correlated with the opposing aspect (factor). This points to the fact that from a software design perspective it might be impossible to have both aspects maximised. In other words, making it innovative (increase HQ) may result in increased perceived complexity (decreased EQ) and making it simple (increase EQ) may lead to a boring software system (decreased HQ). These findings are consistent with the idea that curiosity can by stimulated by providing an optimal level of informational complexity (e.g. [3]). If the software system is either too simple or too complex, the user will feel bored or overloaded, respectively.

The judgement of appeal (APPEAL) demonstrates to be a highly consistent construct. Even such different dimensions as motivation and aesthetics have the evaluational aspect in common, which is taken into account by the users. The regression analysis of EQ and HQ on APPEAL showed that both quality aspects play an almost equal role in forming the judgement of appeal. Together with the missing interaction, this points at a simple averaging model with EQ and HQ compensating each other. This finding is consistent with the averaging model of Information Integration Theory [2]. This theory proposes that different pieces of information in a multi-attribute judgement are integrated by an averaging process. Taking APPEAL as a multi-attribute judgement and EQ and HQ as internally generated pieces of information (based on expectations or experiences with the software system), the observed averaging is simply a consequence of a cognitive process.

From these results we may cautiously conclude that the hypothesised model for the appeal of software is valuable for guiding future research.

No substantial differences can be found when comparing the way the judgement of appeal is formed before and after using the software. It strikes the eye that the expected APPEAL seems to be more based on HQ, whereas the experienced APPEAL seems to be based on both HQ and EQ. This may be due to the fact that the HQ quality aspect is much easier to perceive without using the software than EQ.

A comparison of expected and experienced EQ and HQ shows that in general HQ increases whereas EQ decreases. The suggestion of this result is two-fold: First, it shows that HQ and EQ are based on more than simply the appearance of the software system. Both can be influenced by the experiences individuals have with the system. Second, an increase in one quality aspect may always lead to a decrease in the other. This may again be due to the partly incompatibility of HQ and EQ already discussed above.

The fact that APPEAL remains stable might again be due to the averaging process where the decrease in EQ is compensated for by the increase in HQ.

Looking at the correlations of expected with experienced values we find relatively high correlations. This may be a consequence of the relatively short interaction time in our study (see the following section for a discussion).

Limitations

In the following some possible limitations of the present study will be discussed:

Stimulus dependency: The observed effects may depend on the stimulus material provided (the prototypes), hence generalisation may be limited. To reduce this problem beforehand, the prototypes were designed in a way to elicit a wide variety of positive and negative reactions. The scale characteristics (see Table 4) lend support to the notion that this strategy succeeded: minimum and maximum values show that the whole scale was used, the overall means tend to be around the theoretical mean of the scale and the standard deviations for each scale are similar. Nevertheless, there remains the possibility that the presented results can not be generalised to other stimuli. Future studies should address this problem.

Validation of the semantic differential: Internal consistency and factorial validity are important indicators of the reliability of the semantic differential. However, whether the scales really measure the hypothesised quality aspects and the user's evaluation (i.e. the validity of the scales) remains unanswered. Future studies should attempt to validate the scales. For EQ this can be easily achieved by correlating available usability questionnaires. For HQ and APPEAL other ways have to be found.

Lack of judgmental context: The procedure employed in the present study gives no explicit information about the context for the judgement of appeal. The question on hand is whether individuals are able to form a judgement about a software system without thinking of a context, i.e. where to use it. If a context-free judgement is not possible, the participants might simply have induced their own contexts. For example, a group of participants may have thought of how it would be like to use a software system akin to presented prototypes in their daily work. These participants may base

APPEAL more strongly on EQ than on HQ. On the other hand, a second group of participants may have thought that there is a low probability of using a software system akin to prototypes in the future or in their daily work. These participants may emphasise the importance of HQ. The overall result of the two possible strategies would be similar to the observed averaging model with both HQ and EQ being of equal importance.

Future studies should seek to control the context by providing an explicit situation where to apply the software system (e.g. at work). This would change the context-free judgement of appeal to an evaluation that takes the software system's intended "context of use" [4] into account.

Short interaction time: The direction of the expected vs. experienced effect for HQ and EQ (increase of HQ vs. decrease of EQ) seems to be counter-intuitive. We expected an increase in EQ stemming from a sense of reduced complexity and increased familiarity induced by getting to know the system and a decrease in HQ stemming from reduced novelty. The actual results may point at a limitation of the study: the short interaction time. The interaction time of about two minutes might have been too short to change much in the participants' perception and evaluation. The same explanation holds for the relatively strong pre and post usage correlation of HQ, EQ and APPEAL. Further studies should provide more complex systems and should make a longer interaction time (maybe with several measurements over time) possible. This will help to understand how perceptions and evaluations of users change over time.

Practical implications for the design of appealing software systems

Since hedonic quality is perceived by the users and plays a substantial role in forming their judgement of appeal, it should be explicitly taken into account when designing a software system. Without considering hedonic quality a potential source for increased software quality is neglected.

Due to the partial incompatibility of hedonic and ergonomic quality, software designers should try to find a subtle balance of both quality aspects rather than to independently maximise them. Especially interface designers must identify ways to introduce novelty and surprise with their interfaces (and the behaviour of the software system) without sacrificing to much ergonomic quality (e.g. familiarity). From this perspective the impact of hedonic quality on the appeal of a software system may be the rationale for introducing new interface elements (or even completely new metaphors) and to justify the risk of impaired ergonomic quality.

The averaging model implies that a lack of hedonic quality can be compensated by increased ergonomic quality and vice versa. This means also that good usability can be cancelled out by a lack of hedonic quality. Therefore, the primary strategy should be to have both aspects covered. If this is not possible, the designer may concentrate on maximising only one quality aspect. Which one may depend on the software systems purpose and context of use.

CONCLUSION

Despite the possible limitations of the presented study the results look promising and should stimulate further research. The quantitative research approach presented herein should be complemented by a qualitative approach that seeks to identify specific design factors that are able to stimulate the perception of hedonic quality. The major goal should be to provide a better understanding of what makes a software system appealing to people. Such an understanding may guide software design and thus may have positive effects on the acceptance of software, its creative use and the well-being of the users among us who have to spend a large proportion of their professional lives in front of computer screens.

ACKNOWLEDGMENTS

We like to thank the MediaPlant project group, especially Stefan Hofmann, Alard Weisscher, Jochen Klein and Tobias Komischke for designing and implementing the prototypes used in the present study.

REFERENCES

1. Adams, E. and Sanders, E. An evaluation of the fun factor for the Microsoft EasyBall Mouse, in *Proc. of the 39th Human Factors and Ergonomics Society Annual Meeting* (1995), 311-315.

2. Anderson, N. H. *Foundations of information integration theory*. Academic Press, New York, NY, 1981.

3. Berlyne, D. E. Curiosity and exploration. *Science 153* (1968), 25-33.

4. Bevan, N. and Macleod, M. Usability measurement in context. *Behaviour & Information Technology 13 1&2* (1994), 132-145.

5. Burmester, M., Platz, A., Rudolph, U., and Wild, B. Aesthetic design - just an add on? in *Proc. of the HCI '99* (1999), 671-675.

6. Carroll, J. M. and Thomas, J. C. Fun. *SIGCHI Bulletin 19 3* (1988), 21-24.

7. Csikszentmihalyi, M. *Beyond Boredom and Anxiety*. Jossey-Bass, San Francisco, 1975.

8. Davis, F. D., Bagozzi, R. P., and Warshaw, P. R. Extrinsic and Intrinsic Motivation to Use Computers in the Workplace. *Journal of Applied Psychology 22 14* (1992), 1111-1132.

9. Fishbein, M. and Ajzen, I. *Belief, Attitude, Intention and Behavior*. Addison-Wesley, Reading, MA, 1975.

10. Harrison, A. W. and Rainer, R. K. A general measure of user computing satisfaction. *Computers in Human Behavior 12 1* (1996), 79-92.

11. Hollnagel, E. Keep cool: The value of affective computer interfaces in a rational world. In *Proc. of HCI International '99* (1999), 676-680.

12. Igbaria, M., Schiffman, S. J., and Wieckowski, T. J. The respective roles of perceived usefulness and perceived fun in the acceptance of microcomputer technology.

Behaviour & Information Technology 13 6 (1994), 349-361.

13. ISO. ISO 9241: Ergonomic requirements for office work with visual display terminals. Part 11: Guidance on usability (1996), International Organization for Standardization.

14. Kim, J. and Moon, J. Y. Designing towards emotional usability in customer interfaces - trustworthiness of cyber-banking system interfaces. *Interacting with Computers 10* (1998), 1-29.

15. Kurosu, M. and Kashimura, K.: Apparent usability vs. inherent usability. In *CHI '95 Conference Companion*. (1995), 292-293.

16. Laurel B. *Computers as Theatre*. Addison-Wesley, Reading, MA, 1993.

17. Logan, R. J. Behavioral and emotional usability: Thomson Consumer Electronics, in M. Wiklund (ed.) *Usability in Practice*. Academic Press, Cambridge, MA, 1994.

18. Logan, R. J., Augaitis, S., and Renk, T. Design of simplified television remote controls: a case for behavioral and emotional usability, in *Proc. of the 38th Human Factors and Ergonomics Society Annual Meeting* (1994), 365-369.

19. Malone, T. W. Toward a theory of intrinsically motivating instruction. *Cognitive Science 4* (1981), 333-369.

20. Malone, T. W. Heuristics for designing enjoyable user interfaces: Lessons from computer games, in J. C. Thomas and M. L. Schneider (eds.) *Human Factors in Computer Systems*. Ablex, Norwood, NJ, 1984.

21. Thurstone, L. L. Multiple factor analysis: A development and expansion of vectors of the mind. University of Chicago Press, Chicago, 1947.

22. Tractinsky, N. Aesthetics and Apparent Usability: Empirically Assessing Cultural and Methodological Issues, in *Proc. of the CHI '97* (1997), 115-122.

ALTERNATIVES
Exploring Information Appliances through Conceptual Design Proposals

Bill Gaver and Heather Martin
Royal College of Art, London
Kensington Gore, SW7 2DEU
w.gaver / h.martin@rca.ac.uk

ABSTRACT
As a way of mapping a design space for a project on information appliances, we produced a workbook describing about twenty conceptual design proposals. On the one hand, they serve as suggestions that digital devices might embody values apart from those traditionally associated with functionality and usefulness. On the other, they are examples of research through design, balancing concreteness with openness to spur the imagination, and using multiplicity to allow the emergence of a new design space. Here we describe them both in terms of content and process, discussing first the values they address and then how they were crafted to encourage a broad discussion with our partners that could inform future stages of design.

Keywords: design research, information appliances, home, conceptual design

INTRODUCTION
As digital technologies migrate into our everyday lives, we expect their forms, functions, and values to expand beyond those embodied by the desktop PC. Recently we have joined in the formation of the Information Appliance Studio, a virtual organisation headed by the newly created Appliance Studio Ltd. which spans Hewlett-Packard, IDEO Product Development, and the Computer Related Design department at the Royal College of Art, to explore and shape new possibilities for everyday technologies.

Information appliances, as described for instance by Norman [7], are devices that perform a single function (or closely related cluster of functions) with simplicity and elegance. Networking, often assumed to be wireless, allows new synergies to form among them, recreating the complex possibilities of traditional computing while offering new affordances for interaction. Because specialised appliances can take a wide variety of forms, their benefits can be fluidly integrated in peoples' everyday lives, without requiring that they withdraw to a desktop computer.

The notion of information appliances is inspiring but fuzzy, defined by a combination of abstract vision, technological infrastructure, and only occasional exemplars. For instance, the Hewlett Packard CapShare, a handheld scanner that allows images to be captured by passing the device over a page and wirelessly transmitted to a PC or printer, is a good example of an information appliance. The Palm Pilot, on the other hand, is usually held to perform too many functions to qualify. Other possibilities that have been suggested as "good" information appliances include digital cameras, cookbooks, or gardening appliances. So far, however, the field has been defined largely at a conceptual level, with the space of devices that might populate it little defined either analytically or by example.

In order to better understand the range of information appliances that might evolve, we developed a large number of conceptual design proposals which we presented in a workbook produced for our partners. These speculations were intended to open a conversation with the group about the values that might characterise everyday technologies—values seldom reflected in existing products.

The goal of this paper is to describe the Alternatives workbook both in terms of the proposals it made for future information appliances and as a method for pursuing design. First, almost half the paper is devoted to the presentation of reduced versions of the workbook pages, slightly modified to retain legibility, as a way of simulating their impact directly. Second, we discuss the ideas in terms of the cultural values they suggest for technologies meant to be integrated in everyday life. Finally, we describe the proposals as an example of research through design, describing how they were designed to balance concreteness, openness and multiplicity to allow the emergence of a design space that could be developed with our partners.

ALTERNATIVE VALUES
Suggestions for how digital technologies might be employed in everyday settings tend to represent a narrow range of cultural possibilities, reinforcing a simple dichotomy between work and play. Many devices import values from the workplace into the home, emphasising the requirements of "domestic work" by allowing chores to be done more efficiently or productively. Others emphasise the desirability of taking "time off," allowing people to play unproductive games or access new forms of broadcast media. Other values seem rarely to be addressed at all.

One exception to this generalisation is in telecommunications, which has long been appropriated for domestic use. More recently, products have explored the potential for supporting emotional communication, often without explicit messages. For instance, the Lovegety [4] signals romantic availability to nearby users via lights and sounds when their simple profiles match. While this may seem a crude reflection of the subtleties of everyday courting behaviour, more sophisticated forms of emotional communication are starting to be explored by research groups [e.g., 1, 9, 11]. Still in early forms of development, their explorations of sensual aesthetics and implicit expression, coupled with the value they place on emotional connections, suggest new roles for technologies in our personal lives.

Permission to make digital or hard copies of all or part of this work for personal or classroom use is granted without fee provided that copies are not made or distributed for profit or commercial advantage and that copies bear this notice and the full citation on the first page. To copy otherwise, to republish, to post on servers or to redistribute to lists, requires prior specific permission and/or a fee.
CHI '2000 The Hague, Amsterdam
Copyright ACM 2000 1-58113-216-6/00/04...$5.00

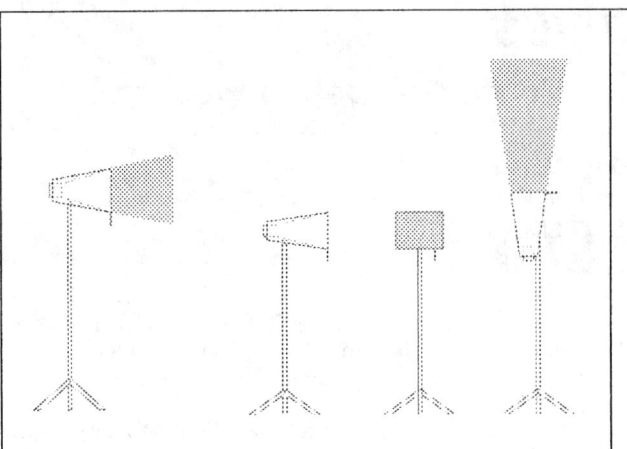

Data Lamp
The Data Lamp is an alternative display device which allows imagery to become an ambient addition to the home. Images are backprojected onto a translucent screen at the front of the lamp for localised viewing. The screen, however, can be made transparent: made of LCD film, its opacity varies as different voltages are applied to it. In this way, images can be contained by the device, or released into the household.

Although the Lamp can be used as a generic display device, it is designed to be used to show slowly moving, ambient images. With forms and colours drifting over the screen, the speed and tone of the images can be altered to create varying moods, and released to paint a wall, the ceiling, or perhaps some hidden nook.

1. Data Lamp: The front LCD screen can be made opaque or transparent, allowing images to be contained or escape.

Dawn Chorus
It is pleasant to be awakened by the sound of local songbirds, but how much more enjoyable it would be if they knew our favourite music.

This could be made possible by an artificially intelligent birdfeeder. Joining a microphone, speaker, pitch-tracker, and software, it would use behaviourist principles to teach the birds new songs, first playing an example of the tune, then progressively rewarding them as they learned to sing in response, to sing small phrases of the song, and finally to match it in tune and tempo.

With more sophisticated bird-recognition software (perhaps based on weight, voice-recognition, or image-recognition), individuals could be taught to take different harmonic roles. The process could take months, but in the end a polyphonic dawn chorus might be achieved.

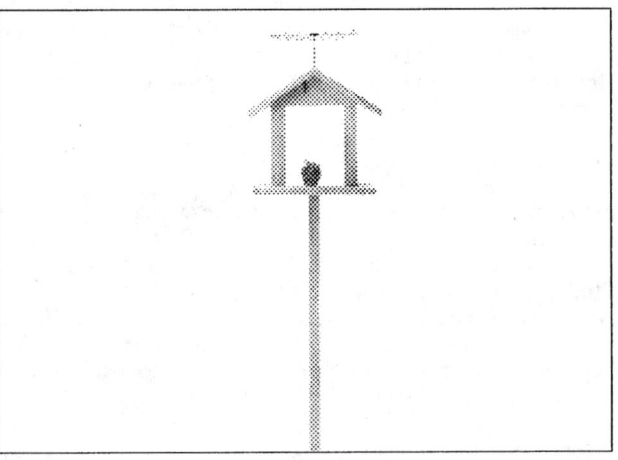

2. Dawn Chorus: An intellegent birdfeeder uses behaviourist training to teach local songbirds the owner's favorite songs.

(De)tour Guide
Exploring a strange city, is an enjoyable experience. But maps force a certain way of seeing the city, from the top, as an objective and officially sanctioned network of junctions and road names. The necessity of orientation can interfere with the pleasant feeling of immersion in the city, leading, at worst, to the problem of mistaking the map for the terrain.

The (De)Tour Guide would use audio and tactile prompts to help users navigate the city. The device could use GPS and vector sensors to determine the users' location and orientation, or alternatively prompt the user to capture images of street signs for comparison against an internal database. Based on this information, the Guide would offer an audio commentary about the immediate surroundings, as well as suggestions about routes to follow.

The Guide would permit a variety of functions, from leading users to a designated location to encouraging them to become totally lost in unfamiliar districts. Different tours might be available, including idiosyncratic routes allowing users to explore good skateboarding sites, places where ufos had been sighted, or the routes and preferences of a local eccentric.

3. The (De)Tour Guide uses a global positioning device to give audio and tactile directions—or misdirections—in the city.

While emotional interfaces have been increasingly recognised as a meaningful alternative to more traditional perspectives on functionality, there are many other examples of noninstrumental values that technologies might support. The speculative design examples presented in the Alternatives workbook are intended to greatly expand the ways that people find meaning in and through technology.

In presenting these proposals we are deliberately noncommittal about the exact technologies that might be used in their implementation. As Fiona Raby has noted [personal communication], in the technology industry a prototype "works" when the technology has been implemented, even if aesthetic and cultural issues are neglected. In design the opposite is true: A prototype "works" when it successfully captures the experience of using a given device, even if implementation issues are not fully resolved. At the same time, we see little value in "science fiction" concepts which rely on technological effects that can or do not exist. Instead, the proposals are intended to be *technologically plausible*, in the sense that it seems likely that they can be realised even if the exact means are unknown or unspecified. In practice, achieving plausibility depends on designers' knowledge and judgement, while an evaluation of the results may depend on discussions with technical experts.

While maintaining technical plausibility, however, we do propose several systems that seem socially implausible. Proposals such as the Dawn Chorus (figure 2) and Democratic Advertising (Figure 4) may be seen as examples of what Tony Dunne calls *value fictions* [3]. Unlike science fiction, in which implausible technologies are invented to support recognisable cultural activities, value fictions propose practical technologies for implausible social goals. They can be valuable as criticisms of culture and technology—in the case of Democratic Advertising, pointing out the overwhelming degree to which public spaces are controlled by commercial and governmental interests; while the Dawn Chorus might be seen as a comment on our desire to tame nature.

In the following, we discuss some of the values our designs are meant to encourage. Analysed post hoc, this may be a somewhat incomplete list: As we discuss later, the best embodiment of the values we have been exploring are the proposals themselves.

Impressionistic Displays
One of the values the proposals speak to is people's desire for attention to and variety in the aesthetics of devices they use in their daily lives. Many of the proposed designs seek to move away from the precise symbolic displays often associated with computers and provide impressionistic, ambient information more normally provided by analogue devices. For instance, the Datalamp (Figure 1) allows images to spill out of the device into its surrounds—perhaps onto a screen, but equally possibly onto a corner of the room, the ceiling, or a piece of furniture. The soft aesthetics of displays such as this seem well-suited for domestic environments. Perhaps less demanding than displays providing more precise information, they also permit a degree of ambiguity that might encourage imagination and speculation.

In encouraging the use of these sorts of aesthetics, we join researchers such as Weiser [12], Ishii [see 1], and Philips Corporate Design [9] in embracing a softer approach to technologies designed for home environments. For such an aesthetics to be effective, however, the mapping between information and display must be functionally appropriate. It is no use using an imprecise display to convey information that people might want or need to inspect closely. Equally, the mapping must be appropriate emotionally. Using display techniques which evoke a calm and reflective experience may seem slightly ridiculous when linked, for instance, with urgent or detailed technical data. While an artist-designer approach may introduce a wide range of new aesthetics to technologies, the underlying skill lies in crafting these aesthetics with respect to the functions and cultural roles they are meant to support.

A crafted aesthetics is often the most obvious feature of the design approach to technology, but for many designers this is only a surface feature of a process primarily concerned with the conceptual design of devices meant to fit everyday life. This involves shaping the functionality and cultural roles of technology in conjunction with their physical form [see also 13]. Shaping the appearance of devices was not enough; they also had to be designed to provide functions meaningful in those parts of life that are not dominated by productive work or unproductive entertainment. This was the primary focus of the work reported here.

Diversions
A conceptual analog to the value of impressionistic displays is the value people find in being diverted from their normal patterns of perceiving and behaving in the world. Rather than pursuing clear and precise goals, we often find enjoyment and meaning in experiencing the world in novel or surprising ways.

The (De)Tour Guide proposal (figure 3) expresses this value clearly. The device would use location and direction information to play audio and tactile cues to travellers in the city. In moving away from the kind of omniscient overview provided by maps, the (De)Tour Guide would allow districts and landmarks to be discovered only upon approach, as if by chance. Beyond this, however, people might sometimes use the device to get lost on purpose, or to follow the idiosyncratic paths of unusual strangers. In providing for these sorts of possibilities, the device gives technological support for Situationist detournés [8], wanders through the city's emotional and cultural topology.

Other concepts, too, explore the notion of diversions in everyday life. For example, several of our proposals concerned psychological exploration, using alterations of normal perception as a way of evoking insight. The Gestalt Camera (figure 10), for instance, would manipulate images of the users' surroundings to produce ambiguous stimuli upon which viewers might project meaning. In producing new, accidental patterns, it is hoped, the device would stimulate novel interpretations of the viewer's relationship with the surrounding social and physical environment.

Using technology to surprise or distract runs contrary to the usual efforts to increase efficiency and productivity through predictability and control. Yet there are values to be gained from diversion such as the chance to discover new places, people, or ways of looking at the world. Such benefits are recognised implicitly within the CSCW field, where support for "peripheral awareness" of colleagues, without explicit purpose or function, is advocated as leading to opportunities for serendipitous communication and a general increase in the coherence of work groups [e.g., Kraut,

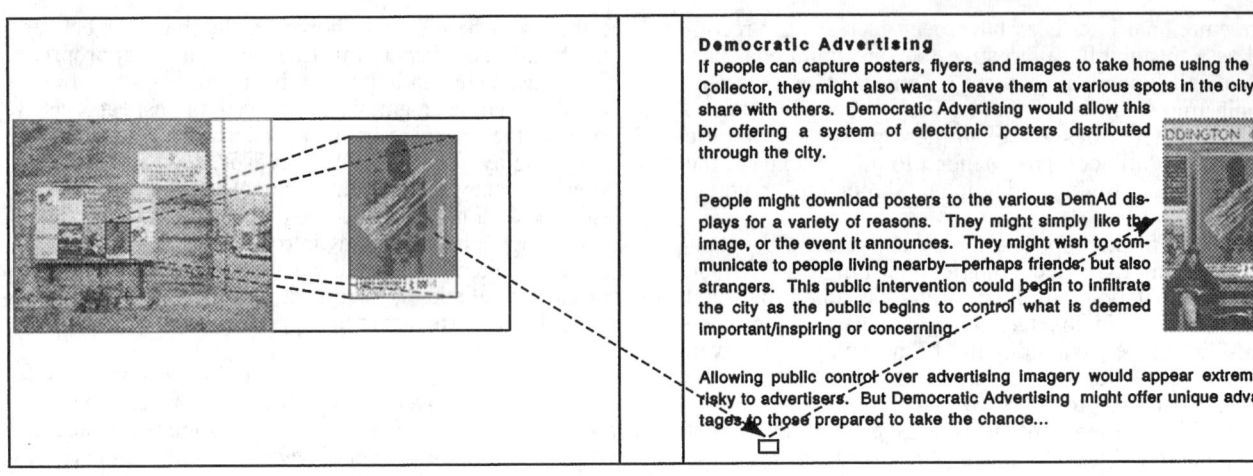

Democratic Advertising

If people can capture posters, flyers, and images to take home using the Ad Collector, they might also want to leave them at various spots in the city to share with others. Democratic Advertising would allow this by offering a system of electronic posters distributed through the city.

People might download posters to the various DemAd displays for a variety of reasons. They might simply like the image, or the event it announces. They might wish to communicate to people living nearby—perhaps friends, but also strangers. This public intervention could begin to infiltrate the city as the public begins to control what is deemed important/inspiring or concerning.

Allowing public control over advertising imagery would appear extremely risky to advertisers. But Democratic Advertising might offer unique advantages to those prepared to take the chance...

4. Democratic Advertising allows people to propagate advertising imagery to new sites in the city.

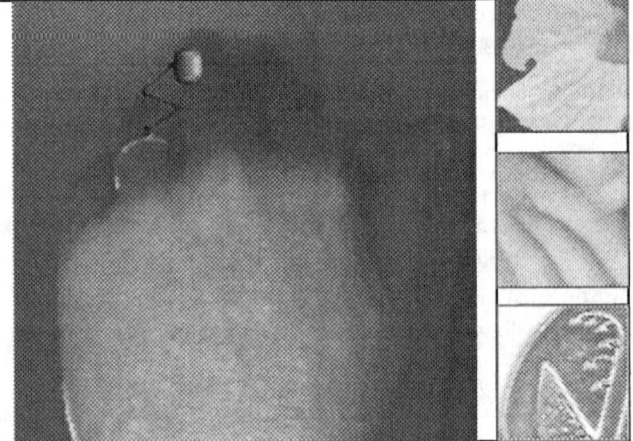

Intimate View

The objective nature of text, words, even video images is not only unsatisfactory for supporting distant love relationships, but even seems to interfere with the deeper, more subtle forms of interaction that create intimacy.

The Intimate View seeks to build a visual connection between lovers, but constrained so that shared perceptions, rather than visual facts, become the object. This is achieved by transmitting images from a tiny macro camera, or personal scanner, worn on the body.

By sharing only tightly focused portion of the local environment--the veins of a leaf, a drop of water, the corner of one's smile--the system would encourage partners to join together in a moment of highly focused mutual perception. Used playfully, aesthetically, or erotically, the device would permit rich new forms of loving communication to exist even over great distances.

5. The Intimate View allows lovers to send each other extremely magnified images, sharing moments of mutual focus.

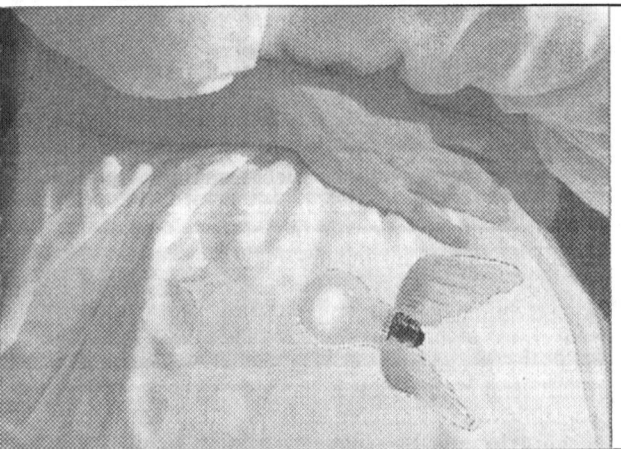

Dream Communicator

Years of our lives are spent in dreaming. Dreams have long been considered important sources of emotional and spiritual information, but currently clinical psychoanalysis is the primary inheritor of their insights. There are very few products that allow a wider audience to tap the possibilities of our dreams.

The Dream Communicator would build on our tendency to incorporate external stimuli into our dreams, to allow lovers to contact each other's dream-selves. Distant lovers might be alerted when their partners enters REM sleep, for instance, and be allowed to stimulate their dreams with sounds or speech. The intimacy and trust implied by such a device would only be found in the closest of relationships.

6. The Dream Communicator allows lovers to influence one another's dreams.

6]. Supporting such accidental encounters—with other places or perceptions, as well as people—can thus be seen as a sort of "serious play" fitting between the usual categories of work and time off.

Influence
People seek influence over their environments, and digital technology have traditionally extended possibilities to extend knowledge and control. Two of our proposals emphasise this value as well, but in new domains and surprising ways.

First, the Dawn Chorus (figure 2) offers the possibility that people might control the songs sung by local birds. Although this may seem far-fetched, with the use of a pitch detector and simple machine intelligence it seems plausible that an automatic behavioural paradigm could be arranged to shape birds' behaviour as their songs increasingly approximate a target tune. Individual bird recognition, moreover, could allow the different birds to be trained to take different harmonic roles in an overall composition.

The Dawn Chorus speaks to the value people find in appreciating—and domesticating—nature. Just as the burgeoning garden industry has allowed people to extend control from their living rooms to their gardens [10], so the Dawn Chorus would extend this control to the very wildlife that shares domestic neighbourhoods.

Of course, the control offered by the Dawn Chorus is not unproblematic. Apart from the problem of neighbours having incompatible musical tastes, interfering with bird's natural songs might adversely affect their mating behaviour. Perhaps it is fortunate that not all songbirds are susceptible to training. Nonetheless, the proposal often seems to delight people, at least conceptually, because of the surprising influence it offers.

In another sphere, Democratic Advertising (figure 6) would allow people to copy advertisements between various sites using portable capture devices. By spreading advertisements that they found beautiful, useful, or socially important, people could influence elements of the public sphere which are normally beyond their reach. Advertisers, in turn, would find advantages if their advertisements spread for no extra cost, and direct feedback if they failed.

Intimacy
People also find meaning in nonverbal, inexplicit forms of communication that few technologies support well. Whereas emotional communication systems described earlier in this paper have often sought to make the qualities of the medium itself mimic or reflect those of intimacy—using softness, tactility, and so on—the proposals here use constraints on the media to suggest or encourage intimate forms of communication.

The Intimate View, for instance, proposes that sharing extremely constrained images of magnified details might allow lovers to create moments of mutual focus. Intimate View is deliberately open about the kinds of images that might be shared, so that partners might share erotic explorations as easily as mundane details of their surroundings.

The Dream Communicator encourages an even more intimate form of communication. Building on the idea that people incorporate external stimuli into their dreams, and the situation in which people in different time zones are awake when their partners are asleep, the device monitors the sleeper and signals the user when their partner enters REM sleep. At this point, the traveller may seek to influence his or her partner's dreams, transmitting sounds, or perhaps tactile stimuli, lights, or scents.

Proposals like the Dream Communicator may raise questions about the degree to which people will accept technological mediation or support of their intimacy. However, it may not be the technology per se that is at issue, but the degree of intimacy it implies: allowing one's dreams to be influenced by somebody else would seem to require an extraordinary amount of trust, and desire for togetherness.

Insight
Many people desire to understand and change their experience of their selves and the world around them. When reflected by phenomena such as the sales of self-help books or new age materials, such values are clearly widespread, yet can be easy to dismiss. Nonetheless, designs might reflect them in ways that can be meaningful without being solemn, externalising psychological mechanisms into digital technologies either to escape or encourage them.

For instance, the Worry Stone (figure 7) is basically an electronic to-do list that uses its processing power to endlessly and visibly rehearse entries. By taking on the fretting of the user, the intention is that it should free time for less neurotic activities. This proposal may be seen as a comment on peoples' behaviour, but also as an appropriate use of digital technologies to perform tasks that are onerous or difficult for humans.

The Gestalt Camera would allow people to capture and manipulate images of the environment around them. This might be done merely for entertainment. But the forms of manipulations possible would be guided by the projective tests of psychoanalysis such as Rorschach inkblots. The interpretations people make of these ambiguous stimuli can give insights into their preoccupations or desires; using the Gestalt Camera, then, might similarly provide a pleasurable foundation for people to reflect on their attitudes towards their current situation. Related to this, the Daydreamer would present pictures to users and prompt them to write short interpretive phrases. Using both the image and the phrase, the system would search for a new picture in an internal database or over the internet. A chain of images and words would emerge, providing support for and a record of an extended revery.

These proposals, and others like them, suggest that technology might work with the tendency for people to structure their worlds meaningfully. On the one hand, they might lead users to reflect on the ways they interpret the world. But even used unreflectively, such devices might help relieve stress (e.g., the Worry Stone) or provide new forms of meaningful play (e.g., the Gestalt Camera).

Mystery
People can find value and solace in contemplating the

Worry Stone

In stressful times, many people maintain internal lists of the tasks they face. As the list grows, they spend more time fretting about the things they have to do than they spend on getting them done. The problem is compounded by the tendency of worries to surface into consciousness without reason, at inconvenient times. The distraction this causes is often counterproductive and even neurotic.

The Worry Stone would be a device that allows people to externalise their cares so that they need not dwell on them. Speaking into the device would offload a task to a small database via a speech convertor. The user could scroll through the list, delete obsolete entries, or print out a 'to do' list.

But the primary function of the Worry Stone would be to take over the user's anxious fretting. To show that the device had truly internalised the use's concerns, its processing power would be dedicated to displaying the list of worries in random order, as quickly as possible. The user could let go of the problem for the moment, knowing that it was safely held in the Worry Stone's memory.

7. The Worry Stone internalises the user's concerns and frets over them without pause.

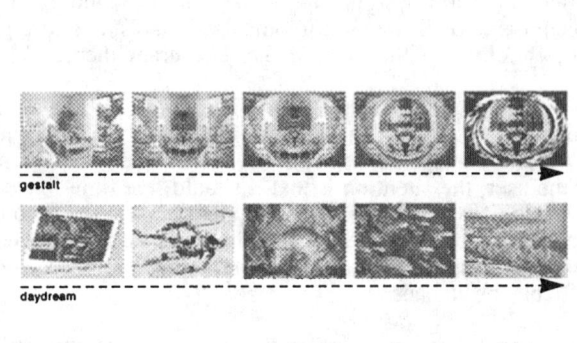

Gestalt Camera / Daydreamer

Imagination could be more rewarding than TV. Rather than providing us with scripted narratives crafted for the widest possible audiences, devices might support us in interactively developing our own particular stories.

The Gestalt Camera, for example, would manipulate captured images to produce abstract, suggestive results. Like the projective tests used by psychologists, these could spur interpretations and ideas beyond the original scene.

The Daydreamer would allow people to create a narrative series of images. As new images appeared, users could enter their associations or reactions. Their words, as well as the original image itself, would form the basis for a new image search. The result would be a semi-structured, impressionistic daydream.

These devices could provide enough support to allow entertaining imaginative fantasies. Beyond a pastime, however, they also set the stage for reflection upon ones preoccupations, concerns, hopes and desires. Over time, they might lead to a useful self-awareness, not using the methods of psychotherapy, but those of engagement and delight.

8. The Gestalt Camera manipulates images forming projective tests to help people understand their attitudes.

Prayer Device

Spirituality, the need to connect to forces beyond the familiar and mundane, should not be discounted in modern times. Despite the apparent rationality and even cynicism of our age, large numbers of people persist in a search for transcendental meaning through psychics, exorcists, and a belief in extraterrestrial lifeforms.

It is surprising, from this perspective, that technology has not been employed to support spiritual quests. The Prayer Device would be a first attempt to rectify this omission. Most likely deployed in public spaces, it would serve as a kind of telephone booth to heaven. People could speak privately into the mouthpiece, and their prayers, wishes, or confessions would be transmitted via a highly focused transmission to the skies.

The recipient of these words might never be known. Individual users might hope, however, that they would be picked up by God, or by benevolent aliens. The use of a potent technology to reinforce their thoughts might strengthen their faith that somebody, somewhere, might hear them.

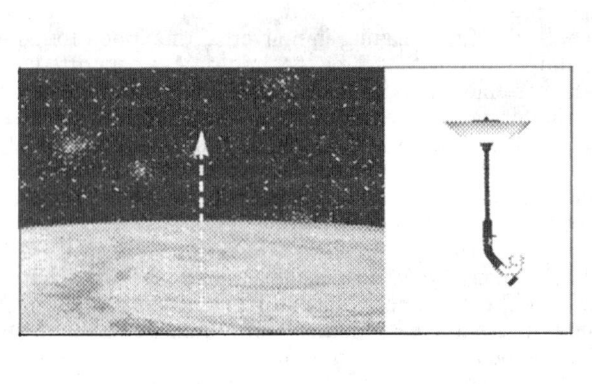

9. The Prayer Device allows people to transmit their voices to the skies.

unknown. Just as technologies tend to overlook peoples' interest in the psyche, so do they neglect their enthusiasm for speculations about spirituality, extraterrestrial life, and parapsychology. Without committing ourselves to belief or disbelief, we developed several proposals to explore these issues.

For instance, the Prayer Device would simply transmit a speakers voice to a tightly focused beam of energy—perhaps radio waves—to be transmitted into the heavens. Perhaps installed in public spaces like a new kind of telephone booth, the device would allow people to send their thoughts or supplications to whomever or whatever they thought might listen. No provision is made for any kind of feedback, or for devices allowing others to listen in. However, it might be imagined that some proportion of transmissions would be heard accidently via radios, TV's, or even telephones. This would allow people to overhear and perhaps gain appreciation for their fellow humans' deepest concerns.

Other proposals explored paranormal phenomena, suggesting, for instance, the creation of a Psy Exerciser that would allow people to work through game-like exercises to increase their paranormal powers. A culmination of this line of thought is the Telegotchi, which is similar to the popular children's toy but without any controls.

These proposals do not depend on a belief in numinous phenomena to be meaningful. Instead, like many of the proposals aimed at exploring psychological phenomena, they depend on peoples' tendency to project meaning onto ambiguous stimuli such as chance events. Although the devices may suggest spiritual or paranormal interpretations, they require merely an open mind and a willingness to experiment.

A BOOK OF SHORT STORIES

The proposals presented in the Alternatives workbook were developed over a period of two to three months working part-time. They were informed by research into information appliances in particular, and new electronic products in general. But they were also inspired by other eclectic influences—stories in the popular press, past experiences, and other projects from design and the conceptual arts. Most of all, they were the products of a kind of daydreaming, in which we imagined the devices that might be found in our own and other peoples' everyday lives. This process—left implicit and undervalued by the sciences—is a fundamental part of our design practice.

We designed the workbook to encourage a similar process of imagination in our partners. Presented as fictional products, they encourage people to imagine living with them, raising many of the sorts of reactions that might be encountered if they actually existed. This encourages both a more integrative and a more emotional approach than might be encouraged by formal analyses, and one marked by imaginative engagement, so that the proposals act as tools for brainstorming new ideas. In this way, the proposals acted as probes into the values and beliefs of our design partners, eliciting a conversation about the directions we might take in pursuing information appliances.

In order to spark conversations without overly constraining them, the proposals were presented with attention to several factors. Individually, the proposals required a balance between *concreteness* and *openness:* they needed to be specific enough to evoke intuitive reactions, yet indefinite enough to encourage imaginative extensions. In addition, the group of proposals was *numerous* enough to avoid undue focus on any single one of them, allowing a design space to be an emergent outcome of the process. In the following sections, we discuss these issues in more detail.

Concreteness: (Fictional) Artifact as Theory Nexus

Synthesising ideas in the form of design proposals is an efficient and effective way to promote discussion of a wide range of design issues. As Carroll and Kellogg [2] have pointed out, the design of an artifact involves commitment to stances on many potentially articulable theoretical issues, and these commitments may be—often are—implicit. This is no less true for imaginary artifacts than for realised ones. Beyond serving as suggestions for development, then, design proposals can also be seen as complex hypothetical statements for debate.

While Carroll and Kellogg's analysis focused on cognitive aspects of artifacts' usability, their insight is more generally applicable to questions of technology, aesthetics, and (most importantly here) the psychological, social, and cultural effects that systems might make. Some of these issues are difficult to articulate and seldom discussed in the HCI community. This makes design proposals particularly valuable: insofar as they address such issues at all, even if implicitly, they provide a ground for discussing them, whether explicitly or through intuitive reactions ("that looks too nostalgic").

In addition, artifacts—imaginary or real—take on a reality apart from their creators. On the one hand, they can serve as a representative of a theoretical stance without implying the commitment of the designer. On the other hand, they do not enforce a single theoretical framework, but allow multiple perspectives. This again makes proposals a valuable basis for communication among disparate partners.

The intuitive approach we took to developing and presenting our ideas does not preclude a more analytic perspective. The dimensions and decisions implicit in the proposals can be identified via a post-hoc analysis, as we have done to some degree in the first part of this paper. As a tool for promoting discussion, however, it seems more useful to leave such aspects unarticulated. This may help avoid premature and distracting discussions of abstractions, and allow different participants to develop their own views about the issues involved. In addition, it may be more efficient to work directly on the synthesis of issues in artifacts rather than via a mediating layer of abstraction and articulation.

Openness: Proposals as Prototypes

Presenting ideas as narrative proposals allows their concreteness to be balanced with openness, because many details of their implementation, aesthetics, or functionality do not need to be resolved. This allows them to remain open to imaginary extensions, developments, and modifications in a way that would be difficult to achieve with more finished examples.

Balancing concreteness with openness depended crucially on the presentation of our ideas in the Alternatives workbook. Each of the proposals relied on images and text to suggest how the idea might be developed. Both were meant to be suggestive but clearly uncommitted as to details of form, function, or technological implementation. We sought to develop a narrative feel to the proposals, similar to commercial advertisements or science fiction stories which describe enough to imagine a device without necessarily specifying either its form or underlying mechanisms.

A variety of imagery was used to convey the basic concepts. Many used collages of disparate elements to suggest facets of the designs' forms, aesthetics, or potential technologies. Using juxtapositions of found elements allowed us to evoke new relationships of aesthetics and function. For instance, the collage for the Dawn Chorus (figure 2), an intelligent feeder that teaches local birds to sing favourite songs, combines images of a birdhouse to indicate the product genre, a strawberry suggesting the desirability of the birds' rewards, and an antennae pointing to the electronic nature of the product. None of these elements are to be taken literally, of course. Nonetheless, the combination of elements is effective in suggesting both the kind of artifact being proposed as well as its playful and somewhat surreal nature. We also used simple diagrams to illustrate some of our ideas. as well as found imagery conveying the context of the proposal. This range of collages, diagrams, and found imagery allowed us to suggest aspects of our design concepts without prematurely committing ourselves to details.

The writing was styled to support the impressionistic feel of the images. Most of the texts briefly set a context for the ideas and then described the concept with pointers towards possible technologies and forms. But the voice used in doing this is meant to seem slightly estranged. In this we were inspired, in part, by Kabakov's [5] "Palace of Projects," a book describing a series of installations in which many are purported to be written by fictional authors—teachers, chauffeurs, pensioners, and the like—to emphasise the ordinary if seldom articulated values they address. Without going so far as to attribute the concepts to fictional characters, we did try to write the descriptions as if borrowing the voices of people with value systems and technological knowledge slightly less cynical than our own.

The text and images were mutually dependant in communicating the proposals. The text set the scene and described the functions and roles of the proposed artifacts, while the images suggested their aesthetic and cultural feel in a more intuitive way, opening a space for imagination to fill in details or extend the concepts beyond what was written. The surreality of collages, writing style, and value fictions all contributed to the openness of individual proposals. Moreover, the range of proposals, from practical to poetic, contributed to their openness as a group of ideas.

THE ROLE OF SPECULATIVE DESIGN: PLACEHOLDERS AS LANDMARKS

While we believe that even the most speculative of these proposals has merit, their overriding function was to serve as landmarks opening a space of design possibilities for future information appliances. As such, the concepts are *placeholders*, occupying points in the design space without necessarily being the best devices to populate it.

The sheer number of proposals encouraged them to be treated as a group defining a broad territory, rather than as a number of proposals to be evaluated separately. While individual ideas might be taken forward, the existence of many alternatives discouraged too much weight being placed on any single idea. Though some might stand out more than others, all contributed to a rough sense of a larger domain.

The design proposals are primarily intended to explore a number of positions on issues concerning future technologies. This might have been done by addressing the space more analytically, perhaps trying to identify its important dimensions before creating examples within the space. Instead, we followed our intuitions and interests in developing the proposals, and allowed the space to emerge from the territories they covered. While the space thus defined is necessarily biased by our desires and interests, this approach has strengths in allowing the discovery of new areas and dimensions as unarticulated interests guide the introduction of new ideas.

Proposals such as these might also form the basis for new kinds of user studies. Aimed at a middle ground between research into peoples' general lifestyles, and studies more focused on their reactions to product concepts, concept proposals could introduce speculative new ideas to potential users in such a way as to evoke general insights into their attitudes as well as more specific reactions. Perhaps most promising, by inherently acknowledging seemingly unusual values, the proposals might encourage people to admit to pleasures and desires that the high technology industry often seems to dismiss as unworthy or nonexistent.

ACKNOWLEDGEMENTS

We are grateful to our partners from Hewlett Packard, IDEO Product Design, and Appliance Design Inc. for their collaboration on this project, and in particular Colin Burns, Ray Crispin, David Frolich, Bill Sharpe, and Phil Stenton. In addition, we thank Tony Dunne, Fiona Raby, and Anne Schlottmann for valuable conversations about these ideas.

REFERENCES

1. Brave, S., and Dahley, A. inTouch: A medium for haptic interpersonal communication. *Extended proceedings of CHI'97*. ACM Press, New York, 1997.

2. Carroll, J. and Kellogg, W. Artifact as theory-nexus: Hermeneutics meets theory-based design. *Proceedings of CHI'89*. ACM Press, New York, 1989.

3. Dunne, A. *Hertzian tales: Electronic products, aesthetic experience and critical design.* RCACRD Research Publications, London, 1999.

4. http://www.lovegety.cc

5. Kabakov, I. A palace of projects. Roundhouse Gallery, London, 1997.

6. Kraut, R. and Egido, C. Patterns of contact and communication in scientific research collaboration. *Proceedings of CSCW'88*. ACM Press, New York, 1988.

7. Norman, D. A. *The invisible computer.* MIT Press, Cambridge MA, 1998.

8. Plant, S. *The most radical gesture: The Situationist International in a postmodern age.* Routledge, London, 1992.

9. Philips Corporate Design. *Vision of the Future.* V+K Publishing, The Bossum, 1996.

10. Rogers, R. *Technological landscapes.* RCACRD Research Publications, London, 1999.

11. Strong, R., and Gaver, W. Feather, scent, and shaker: Supporting simple intimacy. *Proceedings of CSCW'96.* ACM Press, New York, 1996.

12. Weiser, M. The world is not a desktop. *Interactions*, January, 1994, pp 7-8.

13. Winograd, T. (ed.) *Bringing design to software.* ACM Press, New York, 1996.

An Observational Study of How Objects Support Engineering Design Thinking and Communication:
Implications for the design of tangible media

Margot Brereton and Ben McGarry
Department of Computer Science and Electrical Engineering
University of Queensland
Brisbane, QLD 4072 Australia
+61-7-3365-4194
margot@csee.uq.edu.au, mcgarry@csee.uq.edu.au

ABSTRACT
There has been an increasing interest in objects within the HCI field particularly with a view to designing tangible interfaces. However, little is known about how people make sense of objects and how objects support thinking. This paper presents a study of groups of engineers using physical objects to prototype designs, and articulates the roles that physical objects play in supporting their design thinking and communications. The study finds that design thinking is heavily dependent upon physical objects, that designers are active and opportunistic in seeking out physical props and that the interpretation and use of an object depends heavily on the activity. The paper discusses the trade-offs that designers make between speed and accuracy of models, and specificity and generality in choice of representations. Implications for design of tangible interfaces are discussed.

Keywords
Tangible media, augmented reality, interaction design, design thinking, user models, cognitive models.

INTRODUCTION
There has been an increasing interest in objects within the HCI field. This is manifested both in the increasing popularity of virtual objects, and the increasing interest in augmented reality and tangible media. Although the utility of physical objects is often asserted, we know relatively little about how they help us to think and to solve problems. In particular, we do not understand much about how physical and virtual objects differ in their abilities to support our activities. In this paper we study groups of engineers using physical objects to prototype a design, and show what roles physical objects play in supporting their design thinking and communications. In the discussion, we analyse the difference between how objects, CAD tools and sketching support thinking by drawing comparisons with the literature. The extent to which object supported thinking in engineering design transfers to everyday thinking with objects is considered. We draw conclusions about how to design tangible interfaces that support thinking and are easy to use.

BACKGROUND LITERATURE
Our research is based on the premise that if we can understand how humans interpret and use objects in activity, we will be better positioned to design devices and tangible interfaces to support human activity. The literature on interpretation and use of objects stems from a number of fields. Studies of human memory such as those by Rosch [17] have examined how humans categorise objects in order to reduce the cognitive burden of discriminating between the large number of objects in the world. In studies of designers at work, Harrison and Minneman [9] identified that objects are an integral part of design communications, altering the dynamics in multi designer settings and forming part of the pool of representations that are drawn upon by designers. Brereton [2][3] and Miller[13] have reported on how engineering students learn by prototyping with hardware. Writings on distributed cognition by Hutchins [10], and Chaiklin and Lave [6] report that cognitive achievements derive not only from the internal thought processes of people but also from the material systems and information technologies with which they work. Literature in HCI reports on the design of tangible interfaces. Examples are Weiser and Jeremijenko's Livewire [23], Brave and Ishii's Psybench and Intouch systems [1] and Bishop's graspable telephone answering machine in which messages are represented as marbles, reported in [5].

Norman's "Psychology of Everyday Things" [14][15] took the approach of observing how well we manage to use a variety of devices in a world of tens of thousands of different objects, many of which we would only encounter once. Norman's answer was that the appearance and feel of the device and the context in which it appeared should provide the critical clues required for its proper operation. Norman laid out a framework for designing intelligible devices that consists of devising a conceptual model and implementing it physically with appropriate physical mappings, physical constraints and physical affordances.

DATA COLLECTION

The research in this paper is drawn from a multi-year study of engineering students and professional designers engaging in design project work [2,3]. The broader study investigated how engineers learn through designing. The research highlighted that students and professional engineers are heavily reliant on the physical world and strongly influenced by things that they can see and feel. When their knowledge of theory suggests that a literal interpretation of what they are seeing and feeling is erroneous, they are inclined to disregard the theory rather than to disbelieve what they see. The study demonstrated that engineers learn by paying attention to discrepancies between the physical world and the conceptual model. Through continually challenging abstract representations against material representations they advance their conceptual model of the design in progress, their repertoire of familiar physical objects and behaviours, and their understanding of technical fundamentals. The comparison of representations reveals gaps in understanding which inspire further design and analytic activity.

This paper draws upon a single design exercise, which is very limited in scope, in order to illustrate the role that physical objects play in supporting thinking and to develop a basis for a discussion of the roles that physical objects can play in augmented reality systems.

The design problem given was to develop a conceptual design for an internal mechanism for a kitchen weighing scale shown below. This problem is constrained in scope as the external appearance of the scale is already determined. (The essence of the problem posed is to devise some kind of mechanism that converts a vertical linear motion of the weight pushing onto a spring, into a horizontal rotational pointer motion that is proportional to the weight on the scale pan.) So this design brief does not embrace the full design project lifecycle of problem identification and framing, but rather it focuses upon a subset of design activity in which when one seeks to identify and embody possible working concepts to meet the constraints as currently understood. Such constraints are quite common (if temporary and tentative) once one gets to designing sub-systems. (In practice, if elegant embodiments of concepts cannot be found, the designer generally examine ways of redefining the problem in order to change the constraints.)

The data in this paper is drawn from nine 40-minute sessions in which groups of three undergraduate students were observed and videotaped developing a conceptual design for a kitchen weighing scale. Students were asked to develop ideas and present them on a sketch pad. Students were not given any prototyping materials for conceptual design, (since studying use of objects was not our original intent), however students were found to opportunistically seek out all sorts of miscellaneous objects to support their thinking. In a barren design environment consisting of a classroom full of chairs, tables, sketch pad and pens, students sought out inspiration from gesturing with pens, pulling and twisting a rubber band that was spotted lying on the floor and dissecting a ballpoint pen dug out from a student's back pack. They made numerous references to prior experiences with objects. These references are an illustrative subset of those observed in longer (weeks and months) design projects undertaken by both students and professionals in which prototypes are produced. Such projects included motorised self-navigating all-terrain vehicles, a passive restraint system for a truck etc.

Having noticed the extent to which objects were used to support thinking, particularly in conceptual design, we examined the videotape records, to look at how the different objects were drawn into the activity and used to support design thinking and communication. In all cases, videotaping was implemented with a single unmanned camera supported by a desk microphone.

Problem Brief:  Design an internal mechanism for the kitchen scale concept shown left. The mechanism should transfer the weight from the scale pan in the vertical plane to the rotation of a pointer in the horizontal plane.

Pedagogical aims: to give experience in developing and embodying concepts; to generate curiosity about how everyday objects such as kitchen scales work inside; to give practical knowledge of the various mechanisms employed in scales by taking several scales apart.

DATA ANALYSIS

The technique of Video Interaction Analysis was used to identify typical ways in which objects are used to support cognition, [11]. An interdisciplinary team (of engineers, a linguistics expert, a sociologist, an anthropologist and a computer scientist) viewed segments of tapes selected by the primary investigator and identified routine practices, routine problems and resources for their solution. Only those practices confirmed by the raw data that occurred repeatedly in different parts of the tape were considered admissible in the analysis. This exploratory research technique [18] was adopted because it supports formulation of our understanding of natural activity. Such an approach does not attempt to define a controlled setting that affords comparison and statistical proof, because such experimental design and analysis tends to overlook some of the interesting, unforeseen natural activity. The examples presented in the paper are representative of activity in that they have been observed in many different groups and in many different segments of videotaped footage.

FINDINGS

The fundamental finding of our inquiry is that design thinking is heavily dependent upon experiences with physical objects and materials, as evidenced by the frequent references in design conversations.

Hardware[1] is a compelling medium because:
1. It is tangible -- can be seen and touched -- and is thus appreciated by at least two of our senses, often more.
2. It gives physical presence to conceptual models
3. Its behaviour reveals errors in conceptual models
4. It behaves in unpredicted ways which provokes the user to explore it
5. It behaves in different ways in different environments and different contexts of use.
6. Interaction with hardware, and integration of hardware components reveals properties and limits of the hardware and hardware components.
7. It is integral to communications, affecting the course of inquiry, idea generation, discovery and the dynamics of group interaction. Physical objects are used to command attention, to demonstrate and to persuade.

The probability that a designer will draw upon a particular object to support their thinking depends on:
1. The context (in particular the design problem at hand)
2. The physical availability of an object
3. The likelihood that the designer identifies an attribute or affordance of the object that facilitates thinking in the context. This thinking could be tacit or explicit.
4. Availability of the object in memory - ability to recall an episode of using the object that has some link to the current context
5. The likelihood of the designer making a connection between an object and another object already being used to support thinking. Designers often make connections between objects tacitly or explicitly by identifying common affordances, common physical attributes, common geometric attributes, common sub-components etc.

The roles played by objects in supporting design thinking are summarized in Table 1.

Hardware Starting Points

Hardware and prior experiences with hardware are often the starting points from which students develop design proposals. Students look for possibilities in existing hardware to meet design requirements. The videotape revealed that in order to support conceptual design, students draw upon memory of experiences with hardware and are opportunistic in seeking out any kind of miscellaneous hardware to think with, such as rubber bands, pencils and a ballpoint pen which they dissected to examine its internal workings.

The Roles of Physical Objects and Prototyping Materials in Supporting Design Thinking and Communication
Hardware as a Starting Point   Hardware is tangible. It exists. It serves as a starting point it easily noticed, remembered, seen and touched. It offers a basis for comparison.
Hardware as Chameleon   Hardware is always in a context of use. What the hardware reveals depends upon the context of use. A variety of informal experiments in different contexts reveals different facts.
Hardware as Thinking Prop   Hardware objects have all sorts of properties that afford different actions. Hardware that was easily accessible and had a useful property was adopted as a gestural aid to support thinking.
Hardware as an Episodic Memory Trigger   Episodes of experiences with physical objects serve as memory devices.
Hardware as Embodiment of Abstract Concepts (Functional and Theoretical)   Observing and testing hardware reveals fundamental concepts, physical embodiments of abstract concepts; and unanticipated design issues in hardware behaviour
Hardware as Adversary   Challenging theoretical model predictions against hardware behaviour reveals discrepancies and provides clues to modelling errors. This reveals theoretical assumptions, and causal relations
Hardware as Prompt   Device behaviour prompts student questions and suggests experiments. Through repetitive interaction with hardware students bring order, distilling out key operational parameters and their relationships.
Hardware as a Medium for Integration   Integrating components in their functional context reveals practical limits of use, characteristics of operation, methods of connection, causal relations, and physical quantities. This empirical knowledge extends the student's hardware repertoire.
Hardware as a Communication Medium   Hardware is integral to learning communications, affecting the course of inquiry, idea generation, discovery and the dynamics of group interaction. Hardware is used to command attention, to demonstrate and to persuade.

Table 1: The roles of physical objects and prototyping materials in supporting design thinking and communication.

---

[1] We use the term 'hardware' to refer collectively to physical objects and physical prototyping materials.

Hardware as Chameleon

Hardware relies upon its context of use for its functional meaning[2]. In this design context a pen was used as
- Something to prototype a linkage with (see below)
- Something to take apart (see below)
- Something to write with
- Something to point with.

In other situations one could imagine a pen used as
- Something to prod with
- Something to prototype an axle with etc
- Something you use to press a recessed button in order to restart a computer etc.

The selection of a device for design prototyping relies upon the context of use and the device having an attribute relevant to the context of use. Once a prototype design has been built, what the prototype reveals depends upon its context of use. Different experiments in different contexts reveals different facts.

The way that the device may be classified generally (out of the context of use) has little or no bearing on how it is used to support design activity. This has important implications for figuring out good mappings when designing objects for use, a point that will be considered in the Discussion section.

Hardware as a Thinking Prop

What hardware was adopted to support thinking depended upon (a) availability and (b) whether it had any attributes that afforded exploring the design space. Hardware objects have all sorts of properties that afford different actions as the example of pen usage above indicates. Students adopted hardware tools that were easily accessible and also had affordances or convenient properties that supported thinking and communicating. Pens were long and slender like linkage links and afforded gesturing the workings of a linkage mechanism in space. Rubber bands were stretchy like springs. One group explored a rubber band to see if it could provide the kind of motion they were looking for in the scale -- a vertical linear motion of the load translated into a horizontal rotational motion of a scale pointer. They relaxed a stretched, twisted rubber band to see if it would tend to unravel when tension was released.

Several groups adopted pens and used the pens as links in a linkage. They held two pens tightly together at one point (the pivot in the diagram of Figure 1) and tried to figure out if moving one pen in one direction would cause the other pen to move in the desired direction. Pens were adopted because they were long, skinny and rigid and afforded thinking about linkage mechanism motions. Figure 1 below shows a design developed by a group who gestured with pens in their hands to develop a linkage mechanism. The pens formed the two levers of the linkage mechanism.

The properties of appropriated objects were not necessarily optimal or entirely suitable for supporting design thinking. It is probably not possible to specify in advance of being in the design situation what kind of props would be desirable. The hardware was simply conveniently available and had some attribute that meant students found it helpful to gesture and think with. This behaviour lends support to the idea that an accurate model is less important than a quick model that helps the designer to explore the space.

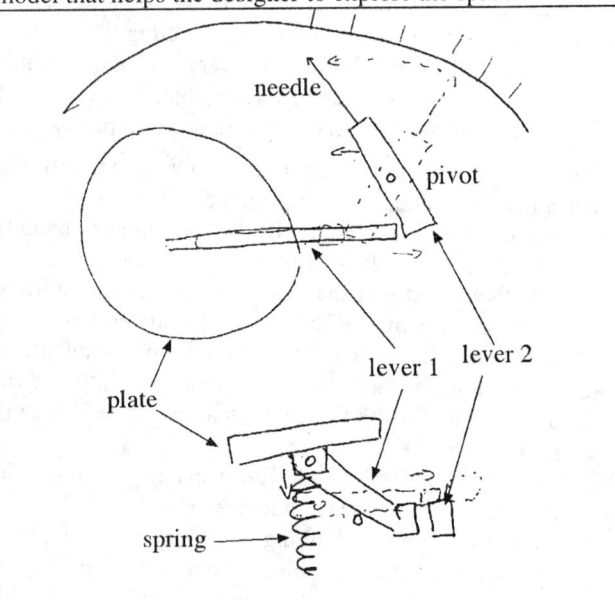

Above: Student concept sketch of a kitchen scale designed by gesturing with pens.

Below: Transcript of student describing his group's design to the class. While describing the design, the student gestures with pens and points at the sketch. Typed labels and arrows are added for clarity by author

Daniel:	*"I was working on the kitchen scale design.* *There's a spring there and there's a plate pushing down on top Ok. And that's pushing down this lever which, as that is pushed down on [the top] end, this [lower end] is moved out.* *And that movement in that direction is pushing the base of another lever here which is pivoted over here. So as that moves out that moves the needle around. So it's a really simple design."*

Figure 1. Hardware as thinking prop: students appropriate convenient hardware to support their thinking.

---

[2] We could say that hardware function is indexical. Expressions that rely on their situation for significance are commonly called indexical, after the "indexes" of Charles Pierce [15], [from p58 Suchman, Plans and Situated Actions, 1987]. Heritage (1984) p158 offers as an example the indexical expression "that's a nice one," pointing out that the significance of the descriptor "nice" has different meanings if it refers to a photograph or to a head of lettuce.

mechanisms worked. In doing so most students did not make any explicit reference to the abstract function or actual mechanisms employed in these devices. Rather, they referred to the way they experienced the devices, referring to the kinds of movements they made, how they were operated and how they felt. For example, in the case of a ballpoint pen, one student stated, "you know how with a ballpoint pen, when you push the button down it turns around." Hardware behaviour was often referred to using the linguistic expression "like a". The scale mechanism could be "like a ballpoint pen" or "like a wind up toy" (See Transcript 1). These observations provide evidence that one way in which we think about designing devices is through analogy to other experiences with devices. That is if we try to design a catch mechanism, one way to go about it is to seek inspiration from all sorts of things that we open and close; our umbrellas, CD holders, doors, egg cartons, brief cases, computer lap tops, and VCRs. Novice designers do not appear to store a library of different kinds of abstract catch mechanisms where they remember the particular geometric configurations of each catch; whether or not experts do is an open question. Rather, novices recall experiences of products that need catches to keep them open or shut. And they recall the catch in its particular context of use, remembering the feel of opening it. Transcript 1 illustrates how novices designers refer to prior experiences with hardware.

Vivian:	If this is em you know like one of those farmer toys where you pull the string and it rolls back, maybe it's something like that where it's maybe a spring loaded coil or a spring loaded em disk with a thing attached to it
Vivian:	Did you ever watch a music box unravel? Like you know these kinds of springs like this, so if you squish it causes some kind of rotation
Juan:	Mmmm right
Vivian:	And if you have rotation in one orientation you can usually translate it into another
Juan:	Yeah you're right I guess watched have those too
Vivian:	Yeah exactly right... right

Transcript 1: Students Draw Heavily on Existing Experience with Devices as they Design Scale Mechanisms.

Figure 2 shows a design developed by one student who searched his bag to find a ballpoint pen or Biro ® when he noticed that the desired motion of the scale was similar to his recollection of the motion of a retractable Biro ®. In many retractable ballpoint pens or Biros ® when you push the button down to eject the point, the point twists around as it ejects and retracts. The student proceeded to dissect the Biro ® in order to learn how it worked. The students' design builds heavily upon the ballpoint pen deployment-retraction mechanism.

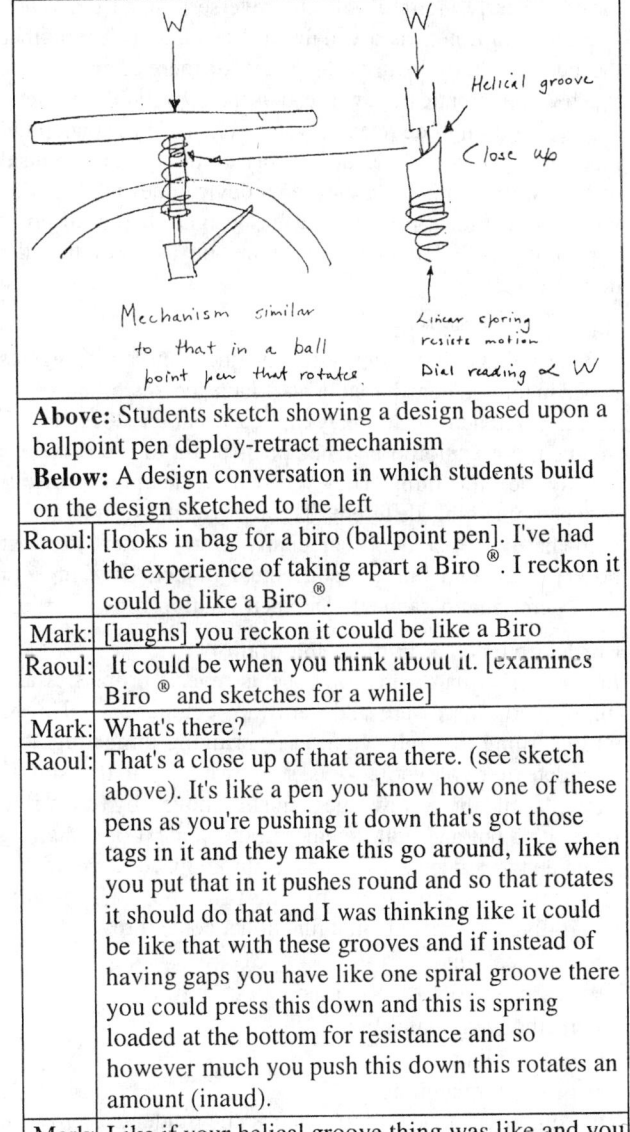

Above: Students sketch showing a design based upon a ballpoint pen deploy-retract mechanism
Below: A design conversation in which students build on the design sketched to the left

Raoul:	[looks in bag for a biro (ballpoint pen)]. I've had the experience of taking apart a Biro ®. I reckon it could be like a Biro ®.
Mark:	[laughs] you reckon it could be like a Biro
Raoul:	It could be when you think about it. [examines Biro ® and sketches for a while]
Mark:	What's there?
Raoul:	That's a close up of that area there. (see sketch above). It's like a pen you know how one of these pens as you're pushing it down that's got those tags in it and they make this go around, like when you put that in it pushes round and so that rotates it should do that and I was thinking like it could be like that with these grooves and if instead of having gaps you have like one spiral groove there you could press this down and this is spring loaded at the bottom for resistance and so however much you push this down this rotates an amount (inaud).
Mark:	Like if your helical groove thing was like and you had your tongue thing sitting out here on the groove like a screw thread that screws it around

Figure 2.. Hardware as Starting Point and Episodic Memory Trigger: designs build on experience with existing hardware devices.

Hardware as Embodiment of Abstract Concepts and Conceptual Models

Hardware gives physical tangible presence to conceptual models. This helps students to remember theories. Many students did not know what a "pin-joint" that they had seen in numerous abstract diagrams might look like in real life. When they dissected bathroom scales they found many examples of "pin-joints" that looked like knife-edges resting in grooves.

Hardware as Adversary

Hardware behaviour reveals errors in conceptual models. One group proposed a screw-type mechanism for their scale, based on the idea that as a nut turns it moves linearly along a bolt. On implementing this model in hardware the team discovered that even though when you turn a nut it moves linearly along a bolt, the reverse is not true. When you push on a nut (as a weight pushes on a scale), neither the nut nor the bolt turns, they just sit there. There is too much friction unless the thread is at a particularly steep angle. Building devices reveals errors in conceptual models, particularly errors in assumptions and causal relations. In a similar way, when a device does not work as you expect it to, it reveals that the designer had a different conceptual model than the one that you used, or that the device is broken.

Hardware as Prompt
Device behaviour prompts student questions and suggests experiments. When the students touch the scale pan of an existing kitchen scale, they notice what is does and it invites more exploration. They press again to see if it does exactly the same thing. They vary where they press to see if it does something different. Through repetitive interaction with hardware students gradually bring order to their observations and build a conceptual model, distilling out key operational parameters and their relationships.

Hardware as a Medium for Integration
Integration of hardware components reveals properties and limits of the hardware and hardware components. This is why learning through synthesis -- bringing things together to create a new whole -- is so powerful. In the simple example of the screw mechanism above, by trying to integrate a nut and bolt together in the context of designing a scale, the students extended their knowledge about the characteristics and limits of nuts and bolts. Typically, integrating components in a functional context reveals:
- practical limits of use,
- characteristics of operation,
- methods of connection,
- causal relations,
- physical quantities.

This empirical knowledge extends the student's hardware repertoire -- the students knowledge of devices and their experiences from which to draw episodic knowledge.

Hardware as a Communication Medium
Hardware is integral to learning communications. Hardware is used to command attention, to demonstrate and to persuade. Whoever holds the hardware tends to command attention, particularly if the hardware is a mouse or remote control that provides a means of control. Thus objects affect the dynamics of group interaction. Further, hardware objects affect the course of inquiry, idea generation and discovery. The groups that accessed the rubber band and the pens to support their thinking would have had different design processes, different ideas generated and made different discoveries had the available hardware been different.

DISCUSSION
In this discussion we consider these findings and draw comparisons with the literature on how objects and other representations such as sketches and CAD primitives support thinking.

Our primary finding of significance to HCI and tangible media is that the interpretation and use of an object depends heavily on the context in which it placed. The problem context drives what attributes of an object people notice and in which ways they try to use an object.

This finding, derived from exploring how objects are used, differs quite sharply from the findings of memory studies, which explore how people categorize objects outside of any particular context of use. Rosch's studies of memory [17] showed that people categorize objects based upon similarity or dissimilarity of attributes. She found that categories of objects become definitively structured because they are coded in cognition in terms of prototypes of the most characteristic members of the category. The most cognitively economical code for a category is a concrete image of an average category member. Although general categorization provides for cognitive economy in recognition or description of objects presented out of context, this does not apply once we consider objects in a context of use. Further work is needed to determine how people understand objects that they encounter and the extent to which understanding is governed by context, by particular attributes, or by belonging to a common class of objects.

Our study confirms Norman's observation that the appearance and feel of the device and the context in which it appeared should provide the critical clues required for its proper operation. Norman argued for appropriate cognitive models, physical mappings, physical constraints and physical affordances . The question this raises, is how to identify appropriate physical mapping or affordances Should the guide to what is appropriate be the context of use of the object or the class of the object as it would generally be categorised. Our work suggests that in some contexts of use (designing being the case in point), the context of use is the dominant factor and should be the guide for what is considered an appropriate conceptual model and mapping.

The effect of contexts of use on tangible interface design needs to be explored through designing and testing tangible interfaces. This paper identifies the kinds of trade-offs between context of use and object category that need to be considered in order to design useable systems.

A second primary finding of this paper is the large extent to which designers appropriate objects to help them think. This section of the discussion considers why this is so. It draws upon other studies of CAD tools and sketching in design to explain this phenomena and to identify what it is about objects that supports design thinking and communication.

It is helpful to begin with Goel's [7] comparison of sketching and CAD. Goel's empirical studies of designers found that sketches facilitate design idea generation and concept development while CAD does not. Goel presents evidence that sketches support design thinking because sketches are a dense and ambiguous symbol system. A line made using an informal representation method such as sketching could represent an edge, a piece of rope, or a rod. Designers working around a sketch are constantly asking for clarification of the sketcher's intent and also suggesting new designs based upon misinterpretations of the sketch. The ambiguity in the sketched representation is appropriate for designers when developing concepts because it represents the level of the designer's thinking when they draw the sketch The sketcher draws a shape paying attention to one aspect or attribute of the shape, but because they are in the process of defining the design they have not fully determined what they are drawing. They can pay attention to only one, or at most a few attributes at a time. Schon's [19][20][21] observations of designers sketching reveals a process of negotiation with the sketch, in which the designer draws, then interprets their own sketch then continues drawing out the idea.

Although CAD primitives (lines, circles, cuboids etc.) have potential for misinterpretation in the same way as do sketch elements, the designer using CAD is forced to pay explicit attention to which geometric primitives s/he will use to represent an idea. The representation forces attention to the internal logical consistency of drawing rather than just what it looks like, for example to draw a circle one has to pick a centre point and then a radius. Further, CAD tools focus attention to precision so that the drawing looks neat. As a result of this cognitive overhead, designers tend to work out what they will draw before they begin in CAD. Goel's experimental data showed that designers who worked with CAD were forced to make commitments early in the process and that CAD drafting facilitated fewer instances of concept generation and development than sketching.

CAD systems are designed to produce drawings to engineering and architectural standards that minimise ambiguity in representation, because by allowing only one possible interpretation, parts can be built from drawings without making errors. Perhaps as a consequence, CAD systems have focused on accuracy and integrity of representation of the final design, rather than supporting the fluid process of idea development.

The fluid process of sketching and the ambiguity of the sketched representation have analogues in physical prototyping. Because physical objects can be interpreted in multiple ways depending upon their context of use, they too are ambiguous and facilitate context-dependent interpretation as do sketch elements. The physical prototype helps the engineer by bringing to the fore specific attributes of an object. The particular attributes that are emphasised depend on the context of use. Objects and physical prototyping materials give rapid visual and tactile feedback and afford gesturing in 3D space, physical testing and so on. Our research has indicated that quick rough prototypes are often preferable to time consuming accurate prototypes because they allow the designer to explore the space quickly. Professional engineers often use Lego or Meccano to prototype ideas, yet while these models are fairly quick to construct, designs are always constrained by the limiting ways in which joints between parts can be made. This is in some ways analogous to the attention to detail one has to pay when producing a CAD drawing. This suggests that there is some in-between ground to be explored for tangible prototyping interfaces. For example a system could take advantage of gestures and affordances of objects in physical space, combining them with a variety of joining methods implemented in virtual form so that such procedures facilitate rapid explorations and give rapid feedback, but do not force an exact representation.

CONCLUSION

This paper has articulated with examples, the roles that hardware objects play in supporting engineering designers thinking and communication. The fundamental findings of our inquiry are that:

1. Design thinking is heavily dependent upon references to physical objects and gesturing with physical objects. Designers are active and opportunistic in seeking out physical props to help them think through design problems and communicate design ideas.
2. The interpretation and use of an object depends heavily on the context in which it placed.
3. Quick rough prototypes that model key attributes of designs are often preferable to time-consuming accurate prototypes
4. Tangible interfaces need to make a trade off between exploiting the ambiguity and varied affordances of specific physical objects and exploiting the power of general representations

ACKNOWLEDGEMENTS

We sincerely thank Ben Matthews, Tom Erickson and the CHI reviewers for their insightful comments and discussion. The authors are responsible for shortcomings.

REFERENCES

1. Brave, S., Ishii, H. and Dahley, A. Tangible Interfaces for Remote Collaboration and Communication, *Proceedings of CSCW*, 1998.
2. Brereton M.F. (1999), " Distributed Cognition in Design: Negotiating Between Abstract and Material Representations" in *Proceedings of the 4th Design Thinking Research Symposium*, held MIT, Boston USA, April 23-25th 1999.
3. Brereton, M.F. (1998) *"The Role of Hardware in Learning Engineering Fundamentals: An Empirical Study of Engineering Design and Dissection Activity,"* PhD Dissertation, Stanford University.
4. Brown, J.S., Collins, A., and Duguid, P.: Situated Cognition and the Culture of Learning, *Educational Researcher*, Vol 18 (1), pp 32-42 (1989).
5. Crampton Smith, G., The Hand that Rocks the Cradle, *I.D.* May/June 1995, pp60-65

6. Chaiklin, S. and Lave J.: *Understanding practice : perspectives* on *activity and context.* Cambridge University Press, (1993).

7. Goel, V. Cognitive role of ill-structured representations in preliminary design, in *Proceedings of Visual and Spatial Reasoning in Design* (MIT, Cambridge MA, June 1999), University of Sydney, 131-145.

8. Goodwin, C. Seeing in Depth, *in Social Studies of Science,* Vol 25, 237-74.

9. Harrison, S. and S. Minneman: *A Bike in Hand: a study of 3D* objects *in design.* in Cross N., Christiaans, H. and K. Dorst (Eds): *Analysing Design Activity*, Wiley, (1996)

10. Hutchins, E. *Cognition in the Wild.* MIT Press, (1995).

11. Jordan, B. and Henderson, A.(1995). Interaction Analysis: Foundations and Practice. *The Journal of the Learning Sciences*. Vol 4 No.1.

12. MacNeil, D., (1992) *Hand and Mind,* University of Chicago Press.

13. Miller, C.M.: *So Can You Build One? Learning Through* Designing*: connecting theory with hardware in engineering education*, PhD Thesis, MIT (1995).

14. Norman, D. (1988) , *The Psychology of Everyday Things,* New York:Basic Books.

15. Norman, D., (1999) "Affordance, Conventions and Design" *Interactions* Vol.6 No.3 May 1999.

16. Peirce, Charles Sanders (1931) *Collected Papers,* Cambridge, MA: Harvard University Press (excerpted in Buchler, Justus, ed., Philosophical Writings of Peirce, New York: Dover, 1955)

17. Rosch, E., and Lloyd, B. (eds.). *Cognition and Categorization*. Lawrence Erlbaum Associates, Hillsdale NJ, 1978.

18. Sanderson, P.M. and C. Fisher (1994) "Exploratory Sequential Data Analysis: Foundations*," Human-Computer Interaction*, Volume 9, pp251-317.

19. Schon, D.A. "Designing as a reflective conversation with the materials of a design situation*," Research in Engineering Design*, Vol 3 pp131-147 New York: Springer Verlag, (1994)..

20. Schön, Donald A., (1983).*The reflective practitioner: how professionals think in action*, New York: Basic Books.

21. Schon, D.A.(1990). "The Design Process" in (ed.) V.A. Howard *Varieties of Thinking*, Routledge.

22. Suchman, L.: *Plans and Situated Actions, The Problem of Human* Machine *Communication*, Cambridge University Press (1987).

23. Weiser, M. (1993) "Hot Topics: Ubiquitous Computing" *IEEE Computer*, October.

See also:
http://www.ubiq.com/hypertext/weiser/UbiHome.html

Tagged Handles:
Merging Discrete and Continuous Manual Control

Karon E. MacLean
Interval Research Corp.
1801 C Page Mill Rd.
Palo Alto, CA 94304 USA
maclean@interval.com

Scott S. Snibbe
Interval Research Corp.
1801 C Page Mill Rd.
Palo Alto, CA 94304 USA
snibbe@interval.com

Golan Levin
MIT Media Lab E15-447
20 Ames Street
Cambridge, MA 02139 USA
golan@mit.media.edu

ABSTRACT

Discrete and continuous modes of manual control are fundamentally different: buttons select or change state, while handles persistently modulate an analog parameter. User interfaces for many electronically aided tasks afford only one of these modes when both are needed. We describe an integration of two kinds of physical interfaces (tagged objects and force feedback) that enables seamless execution of such multimodal tasks while applying the benefits of physicality; and demonstrate application scenarios with conceptual and engineering prototypes. Our emphasis is on sharing insights gained in a design case study, including expert user reactions.

Keywords

Discrete, continuous, haptic, force feedback, tagged object, tangible, tool, token, container, design process.

INTRODUCTION

Tagged objects and haptic force feedback are two means of bringing tangibility to user interfaces. Complementary in their control affordances, one facilitates discrete selection and the other enables continuous manipulation. Both allow a user to employ his or her hands, and to manipulate media in ways that can be more intuitive and convenient than a keyboard and screen. However, the most natural functions of these tangible mediators are different, indeed orthogonal. Tagged objects (physical icons marked with electronic ID or memory) are relevant as tangible references to virtual information, representing data or operations. Haptic force-feedback interfaces (actuated robotic devices through which a user feels computer-generated environments) are used to handle, navigate and sculpt virtual terrains.

Many computer-aided tasks have components of both discrete and continuous manual control. Here we describe a new interaction concept that unites the two, placing both discrete and continuous manual control into a single consistent model. We will develop the percept of using physical selectors to change force-feedback behavior, setting context both electronically and physically: tagged objects have specific shape and action, while haptic objects have a general shape and many actions. This approach connects specific shapes to specific actions while maintaining generality, and we believe that the result's elegance can ease the introduction of continuous control into digital interfaces.

We present several versions of the idea and scenarios for its use, in two primary vehicles:

I. Tagged objects are handles interchangeably plugged into a force-feedback interface, switching the display's dynamic behavior while simultaneously changing its appearance and grip;

II. Distinctive discrete selectors are permanently integrated into the force-feedback display and can be manually activated to set a particular function mode and dynamic behavior.

We then describe an iterative conceptual and engineering prototyping process pursuing one branch of this concept in a constrained application space, and reflect on our insights from developing these prototypes and sharing them with several expert users.

BACKGROUND

Discrete and Continuous Control

User interfaces relate to discrete or continuous information and control. While these terms really compose a continuum rather than disjoint spaces, they are a helpful way of looking at the world in the sense of manual control.

Buttons, switches and tagged objects are discrete controllers: they trigger something to happen automatically and beyond the user's immediate sphere of influence. Flipping a light switch causes the light to come on. The information age brings new kinds of discrete controls with complex responses: opening the door of a "smart" apartment might cause lights to come on, music to play and the oven to warm up.

A handle is a continuous controller: you grab it and maintain direct, active authority. Since the handle couples you to the environment, you can quickly adjust your own motion to its physical signals, forming a tightly closed control loop. One does this when penning a curve, drawing the aftertone out of a piano key or steering through a virtual game-world. Force feedback interfaces share these attributes, and earn their keep when constant, two-way engagement with the environment is needed.

Tagged Objects

From wedding rings to ATM cards, physical tokens have always played key roles in our social and information processes. Holmquist offers a useful classification that distinguishes containers (generic data collections), tokens (data representations whose content is physically demonstrated), and tools (objects that set a function or state for operating on data selected in some other way) [6]. Tagged objects live in the margin between electronic and physical worlds, used to turn things on, select, combine, physically move data, change its context or trigger common tasks [1, 3, 18]. They appeal to those who like to touch as part of doing, and enjoy variety in shape and texture and heft. Arbitrary objects can be tagged; they easily broadcast their function. They rely on cheap, mature technology such as bar codes and radio and magnetic sensing.

But tokens and buttons command a repertoire of behavior more limited than that of many other real objects: they are inherently discrete. While computers are binary creations, the natural world is as often composed of and perceived as an infinity of continuums. People who have grown up within it are accustomed to moving and deforming and creating new possibilities from malleable media. In this sense, tagged objects do not ease the language barrier between humans and electronic devices.

Some prior research has special bearing here. Ullmer's slideshow browser [17] illustrates an intriguing use of a media-container as a handle. However, individual slides merge into a stream as their number increases, and overwhelm the browsing capabilities of a passive slider. Gorbet's Triangles [4] have discrete display capability (flashing LEDs), and a magnetic tug between Triangles confirms connection. Yarin has placed multiple tags in a single object to indicate a series of discrete states [19].

Haptic Force-Feedback Interfaces

Force-feedback interfaces are the children of robot technology and biomedical inspiration, and have evolved from the powerful, expensive systems of the 80's [5, 12] to today's commercial desktop and gaming devices [7, 13]. Active force feedback can unload other senses, reduce ergonomic strain, support expressive abstract input, and enable continuous manual control. Interfaces assume many configurations – knob, mouse, joystick, tactile array – and use many mechanisms and actuators. They simultaneously provide input and output, informing the user of a system's state while transmitting her intent.

Current devices have problems with both cost and usability. The technological challenge of generating significant, controllable forces in a small, low-power package maintains a tough tradeoff between expense and quality. Therefore psychophysical illusions that heighten perception of force magnitude, fidelity or excursion can leverage cheaper devices (e.g. [15, 16]).

The point of a force-feedback display is to execute arbitrary functions and feels with the same interface. Unfortunately this means the user might not know what to expect, and the handle might not be the right one. Users may be disoriented and even fearful of "invisible" environments without sufficient context, particularly upon changes in environment or control mode. Additionally, variety in tactile shape and heft of the tool cannot be exploited to overcome limits on environment fidelity imposed by expense and actuator technology. Psychophysical studies as early the '50s [2] substantiate the role of distinctive shape and location of console handles. More recent work (e.g. Rock [11] and Srinivasan [15]) indicates that visual stimuli overpower tactile in most cases, and thus a tool's appearance is central to its function. Nevertheless, economics support the pervasive generic and multi-purpose handle.

TAGGED HANDLES CONCEPT

We feel that physical interfaces are most valuable when customized for a given application at every level –the grasped object, the perceived forces, and in simultaneity and integration with other sensory displays. When both continuous and discrete (modal) control ability are desirable, and where having the right handle matters, use of tagged objects or force feedback alone fails to encompass the whole control task at hand, resulting in

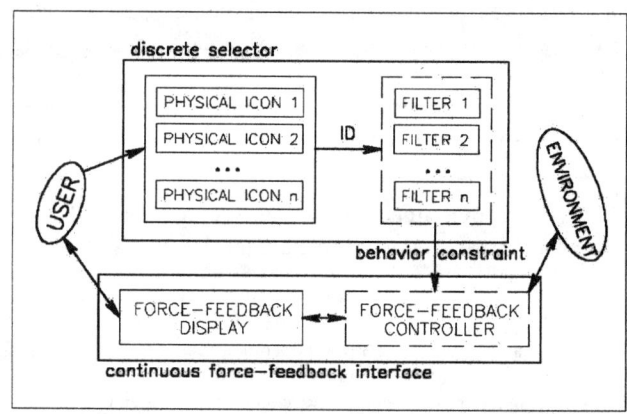

Figure 1: Tagged Handles concept components.

user frustration and disorientation. Here we combine bidirectional, continuous manual communication with changeable, appropriate handles and physical context.

Figure 1 demonstrates this schematically with four components: user and environment mediated by the Discrete Selector (DS) and the Continuous Force-Feedback Interface (CFFI). A user interacts with the environment by choosing a physical icon, and grasping a force feedback display – usually *through* the physical icon. The computer recognizes the physical handle, and applies its associated behavior filter to the force-feedback controller. The DS thus consists of the set of physical icons and their corresponding filters; the CFFI is the force-feedback display and controller. Of these, the physical icons and force display are tangible, and the filters and the force controller are generally implemented in code. The environment may be a virtual model, a spreadsheet, streaming media such as video or audio, a remote environment or a multitude of other contexts.

By "filter", we mean the system's interpretation of the discrete selector at a given time. It might determine the data the system will operate on, the device that a controller targets, the function executed on an environment, or a user's system setup preferences.

Example Scenarios

The following scenarios are representative contexts where the handle matters and continuous bidirectional control is beneficial; many other applications exist.

A: Drawing Implements

To benefit fully from force feedback while using a drawing program (Figure 2), the user selects a drawing implement from a tagged collection of real implements (e.g. paintbrushes, pencils and chalks) and plugs it into the force display. The system recognizes the implement and supplies forces that convey the sense of that tool, while translating the user's motion into an element in a digital drawing. The tool also sets the line's properties and the function used to interpret the controller's motion, creating a fuzzy chalk line or a calligraphic swathe.

Multiple discrete selectors can be used in combination, and a user can generate new interpretations of the objects. Another set of tagged objects (not shown) can indicate the drawing surface – smooth vellum or pebbly handmade paper. Haptic properties of the virtual paper may be automatically extracted from the real sample (e.g [8, 10]) and linked to the sample with a barcode sticker. The chosen surface will be reflected in the haptic interaction between it and the drawing implement, and in the line drawn in the application.

B: Slide Show

A slide presentation is an example of a set of discrete elements that a user must access and manipulate in a many contexts (whiteboard, scanner, printer, projector). As demonstrated by Ullmer [17], a container-type tagged object is a good way to represent and transfer the slide show; but it is cumbersome to browse a large set. Force feedback renders it as a continuous, feature-rich stream.

Figure 3 shows a user scrolling a slideshow by plugging the token into a force-feedback slider and moving it back and forth, feeling detents that mark slide or section edges and annotations. It is easy to step through slides and stop at an intended destination. In this context, a linear, limited range-of-motion slider (shown) has the benefit of clearly demarcating the beginning and end of the slide show – they correspond to the physical endpoints of the input device, and orient the user spatially.

C: Browsing Digital Media

Figure 4 illustrates a means of media navigation and editing. In this scenario, the DS could represent either a set of media-containers (e.g., video clips or shows) or tools (functions to be applied to the media stream, e.g.

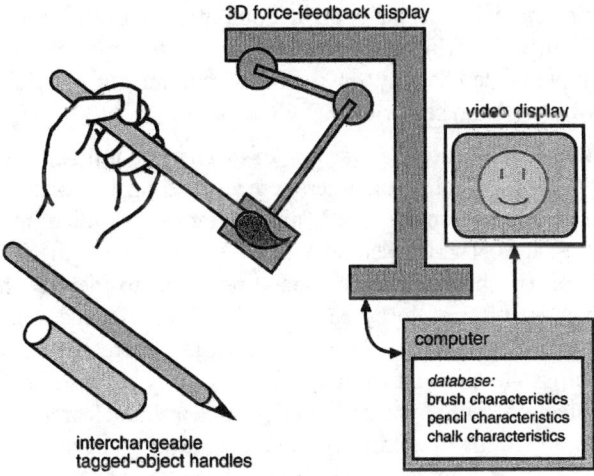

Figure 2: Drawing program. Processor recognizes tagged implement and provides appropriate force-feedback.

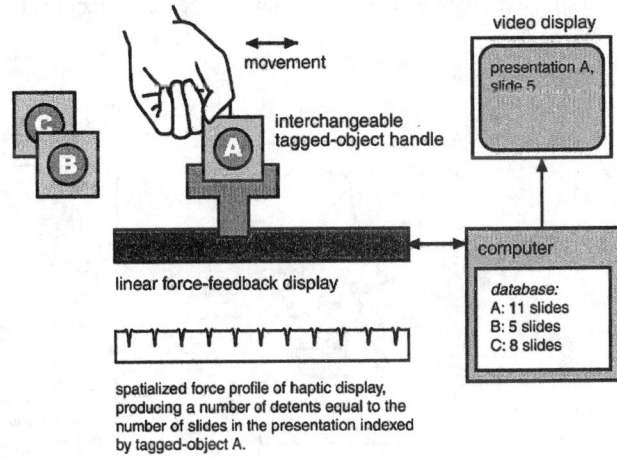

Figure 3: Slideshow browser

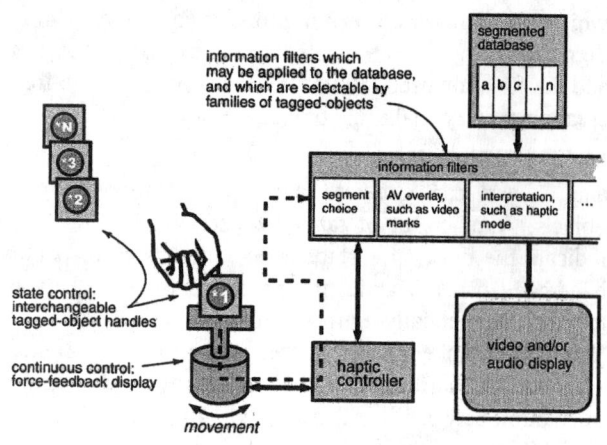

Figure 4: Browsing digital media.

view, edit, zoom, or feel a set of annotations).

One interesting tool is the *content filter*, which causes the force feedback to emphasize a particular kind of content. This could be a violence detector for child-safe movie watching, or a person detector that uses signal processing to detect appearances of Grandma in a home video. When the stream is lengthy, innovative interaction techniques may help to navigate it – e.g. the "haptic clutch" metaphor mentioned in [9].

D: Browsing Cartographic Data

Figure 5 demonstrates another variation of a token-type handle, where the tagged object is linked to a parameter in a database (e.g. cartographic) and used with a planar haptic display to explore that parameter.

E: Deep-Parameter Marking in Graphics Applications

Graphics professionals editing images and animations tend to repeatedly modify a small set of parameters, e.g. image brightness or joint position. These parameters are buried in modal dialog boxes and once accessed, difficult to set precisely with a mouse. Here, a user can temporarily associate a deep parameter and a single knob controller using a tagged icon to easily access the parameter and receive specific haptic feedback.

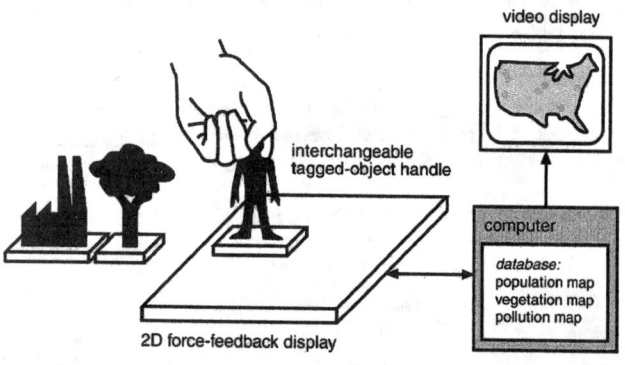

Figure 5: Tagged tokens access cartographic parameters.

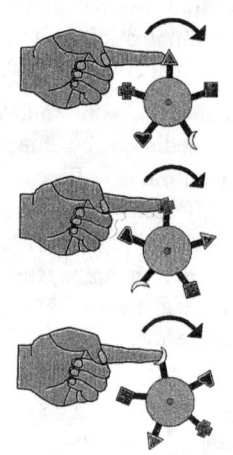

Figure 6: Integral Handles being used for media browsing.

F: Force-Feedback Home Remote with Integral Selectors

The last example demonstrates a different take, wherein the Discrete Selectors remain attached to the force-feedback controller at all times and selection is made by grasping or activating one of them (Figure 6). In the case of our Home Remote prototypes described later, selectors are mapped to control functions such as video selection, play mode and volume control.

PROTOTYPES

We constructed many conceptual and engineering prototypes to explore two branches of the Tagged Handles concept. The first is the Exchangeable Handles configuration (Examples A-D above), and the second is a series of iterations on the Integral Handles variation described in Example E.

Prototype I: Exchangeable Tagged Handles

Our first prototype was an engineering mockup with no context. Exchangeable knobs (distinguished by abstract shapes) with electronic tags are plugged singly onto a mechanical receptacle on a computer-controlled motor (Figure 7). Object recognition and motor control were implemented locally on a PIC, communicating serially with an audio server on a QNX Pentium.

We implemented a variety of scenarios utilizing the handles as audio containers, corresponding for example to a collection of MP3 music tracks or voice mail messages. The user could select a track by plugging in a knob (in this first try, we made no effort to identify the content referred to by each handle). Force feedback was supplied according to a set of rules that were consistent across the different audio types: the user could browse the selection with functions such as a continuous range of scrub speeds in both directions, defaulting to steady play when the handle was released. Mode changes were accomplished using haptic dynamic models previously developed and demonstrated by the lab [14].

Figure 9: Functional Integral Handles prototype.

Figure 10: Conceptual side-selection prototypes. The force-feedback wheels mode is chosen by pressing the wheel at a specific position (left and right prototypes) or a position on the wheels base (center). Since activation location is grounded relative to hand position, the selector is easy to find.

built would permit only one selector to be depressed at a time, reminiscent of an old car radio. However, we flunked this configuration based on the form mockup – it lost the "handleness" of the basic idea. In Figure 8c, the selectors have moved to the top surface and are rotated with the non-supporting hand like an old rotary phone dial, the feel of which delighted us. Concurrent experiments with the basic shape resulted in a hemi-egg form (Figure 8c) that was pleasing to hold while retaining a definite "pointing" direction.

The informal user reactions detailed below were collected while we built a functional prototype of Figure 8c, shown in Figure 9, used only by team members. It employs distinctive fabric swatches to abstractly "label" the selectors (a more semantic mapping is obviously needed eventually), and was programmed to browse digital video and cable TV in realtime. We soon discerned that another iteration was in order: it was hard to find the selectors when they moved relative to the base. This led to the concept series shown in Figure 10, which utilizes muscle memory to find the selectors in locations fixed relative to "ground". At this point the project was terminated for unrelated reasons, and thus a final working prototype was never built and tested.

EARLY-CONCEPT USER REACTIONS

We informally shared our conceptual and engineering prototypes (Figure 7-Figure 8 and others) with several usability-savvy potential users for observation and brainstorms. We feared that these abstract designs would confuse subjects of real usability studies, but we wanted a larger perspective as we produced a version that *would* be suitable for wider exposure. The following comments are distilled from sessions with four colleagues unfamiliar with the project, chosen for diversity in background, aesthetic leaning and attitude towards pervasive technology. Despite obvious usability issues, we used our entire prototype collection to stimulate discussion, and emerged with more ideas and questions.

General Issues and Observations

A few consistent and unsurprising themes surfaced. Perhaps most significantly, our participants unanimously approved of bringing more continuous control to digital interfaces; one observed that those with limited dexterity – e.g. the elderly – might find handles more manageable than buttons. They reinforced our intuition that users will react strongly to a satisfying dynamic feel: multiple subjects preferred nice-feeling prototypes, explaining that their enjoyment came from a combination of the motion and the heft of the object in their hand. One claimed "Make one that feels like this and I'll buy it".

They agreed that removable handles should express what they do and how. The usual concerns with tagged objects came up – losing, organizing and locating them; remembering what they represent; indicating changeable contents. Participants observed that integrating the selectors into the force-feedback handle solved many of the concerns with tagged objects, but introduced other issues of selector usability and findability.

Exchangeable Handles

Reactions to Media-Container Scenarios

The removable media-container scenarios suffered the brunt of general tagged-object criticisms. However, participants perceived the potential of simplification in using physical containers; e.g. they might replace a graphic list display in portable or embedded devices. One questioned the benefit of exchanging a handle if its shape did not markedly change; another predicted confusion in operating a single handle in different modes, but proposed clever means of expressing mode.

There was general agreement on the desirable physical attributes of media-container handles: each must indicate both its contents and how to use them. They should exhibit handleness, stackability and findability (because there are many of them) – a tall order. However, all data containers of a given *type* should share a similar shape, and thus will only need be designed once.

Reactions to Tool-Use Scenarios

Participants responded positively to the idea of choosing (thereby displaying) mode by inserting a physically distinctive handle. Several were taken with the idea of an exchangeable handle as a filter for content selected some other way, as long as the content demanded continuous manipulation; e.g. using a small handle for precise operations and a larger one for powerful movements.

Reactions to some specific tool-use scenarios (e.g. for drawing and surgical applications) were evenly mixed. Half applauded; the rest wondered if the additional functionality of a switched handle justified the effort.

Integral Handles

Interaction

All participants found a set of attributes among the remote control prototypes that pleased them enough to "buy". The whimsical spokes were admired but rejected as awkward because grasp changes with wheel position; but none thought them dangerous. Recessed selectors distinguished by texture (velvet, leather, velcro) were easier to manipulate, but too abstract.

Participants confirmed the need for spatial grounding of the rotating selectors noted earlier, and approved of sketches of the prototypes shown in Figure 10.

Task Allocation, Apparency and Flow

Participants arrived at similar optimums of simplicity and direct access: employ selectable handles to determine an operation, and pointing or a small set of nearby buttons to choose the active device from a set. Operational rules can be consistent if the set shares attributes, such as the need for volume and rate control.

One was willing to learn abstract associations; the rest felt the form should clearly indicate its function, through label or shape. No one felt this would be hard to do.

Random access will sometimes be desirable: "If you know what you want, you should be able to go straight there." That is, it must be easy to modify and configure what is accessible in that continuous range. E.g., if one "function" is a collection of preset channels, it must be easy to add and subtract channels from that set. Otherwise, the ease of traversing the set will be countered by the annoyance of the set being too large.

DISCUSSION

Our insights are a mix of personal intuition and experience, outsider reactions, and awareness of application contexts. Despite prototype usability problems, the process converged towards an interesting solution. Of greatest importance are the notion's value and most useful contexts. The configuration and design parameters we have come to consider significant include:

- Selectors exchanged, or integrated and chosen by holding or touching.
- Force feedback received directly through the selector, or through a separate physical connection. The former may be best for tool use, the latter for containers.
- The user must be able to locate and understand the selector via shape/texture and location.

What Price Physicality and Continuity?

As usual, users want it all. Touch, continuous control, low complexity and intuitive appearance are valued, but cannot replace random access. They prefer the austere elegance of a single lovely knob; but it should have all the capability of a fifty-button universal remote. Functions should be apparent, as long as none are lost.

Some of our scenarios traded too much pushbutton convenience for physical handle affordance. A good compromise for the Home Remote (Example F) may be an ergonomic Integral Handle for function selection and continuous browsing, combined with a few buttons on the base to choose a device.

In other cases a physical content selector may truly simplify an interface by eliminating the need for an on-screen interface. This will be most true when there is no inherent visual content in the media (e.g. an audio application) or when a visual interface would intrude on a principally visual content (e.g. video browsing).

Exchangeable Handles

The Exchangeable Handles concept may be better suited to choosing functions than media targets, because of the difficulty and redundancy of physically distinguishing media-container tagged objects while shaping them as good handles. The container notion shares many drawbacks of simple tagged objects, while the related continuous-control requirements do not seem urgent.

For tool-type Exchangeable applications, having the right handle is valuable; but switching must be no harder than mouse-clicking a screen icon. Handles as media filters that highlight or obscure garnered enthusiasm, perhaps because of philosophic consistency with viewing *through* a visual filter. In such situations, it must be desirable to control and handle the media as well as observe it.

Integral Handles

The Integral Handles concepts make selecting handles easier (you don't have to dig them out of the couch or plug them in) but loses them locally. We think we can make Integral Handles easy to find *and* easy to turn. Textural distinction will not suffice for blind-use handle detection because it requires serial exploration; but shape and texture confounded with static positional cues should.

SUMMARY AND DIRECTIONS

We have described a new interaction percept: the integration of discrete and continuous control capability into a single seamless interface. We believe that designs of this sort are will alleviate stress from forcing inherently continuous tasks into the discrete affordance of prevalent button interfaces, while bringing aesthetic and functional benefits of physicality. Exploration of several application areas helped to trigger the ideas as well as stimulate their development.

We have presented design case studies of two versions of the concept. The process's emphasis on quick iteration in building, trying and discussing, and on relation to promising applications has resulted in satisfying prototype variety and evolution. Informal expert-user responses to these prototypes have validated some of our starting premises – perceived value of physicality, desire for more continuous control and the importance of function apparency. They have pruned others: e.g., the incremental benefit arising from custom handles will not always outweigh inconveniences of organizing and swapping them. Based on this work, the underlying premise appears sufficiently strong enough to continue development in multiple directions.

ACKNOWLEDGMENTS

This research was conducted at Interval Research Corporation. We are grateful to the Haptics and Tangible Interfaces teams for their support, and in particular to Eric Dishman, Terry Winograd, Bill Verplank, Cy De Groat, John Anany, Brad Niven and Mark McCabe.

REFERENCES

[1] J. Cohen, M. Withgott, and P. Piernot, "Logjam: a tangible multi-person interface for video logging," in Proc. of the Conference on Human Factors in Computing Systems (CHI '99), Pittsburgh, PA, 1999.

[2] P. M. Fitts, "Engineering psychology and equipment design," in *Handbook of experimental psychology*, S. S. Stevens, Ed. New York: John Wiley & Sons, 1951, pp. 1287-1339.

[3] G. Fitzmaurice, H. Ishii, and W. Buxton, "Bricks: laying the foundations for graspable user interfaces," in Proc. of Conference on Human Factors in Computing Systems (CHI '95), Denver, CO, pp. 442-449, 1995.

[4] M. G. Gorbet, M. Orth, and H. Ishii, "Triangles: Tangible Interface for Manipulation and Exploration of Digital Information Topography," in Proc. of the Conference on Human Factors in Computing Systems (CHI '98), Los Angeles, CA, pp. 49-56, 1998.

[5] J. M. Hollerbach and S. C. Jacobsen, "Haptic interfaces for teleoperation and virtual environments," in Proc. of First Workshop on Simulation and Interaction in Virtual Environments, Iowa City, 1995.

[6] L. Holmquist, J. Redström, and P. Ljungstrand, "Token-Based Access to Digital Information," in Proc. of First International Symposium on Handheld and Ubiquitous Computing (HUC) '99, Karlsruhe, Germany, 1999.

[7] Immersion Corporation, *The Immersion Feel-It Mouse & I-Force Game Controllers*. San Jose, CA, 1999.

[8] K. E. MacLean, "The Haptic Camera: A Technique for Characterizing and Playing Back Haptic Environments," in Proc. of the 5th Ann. Symp. on Haptic Interfaces for Virtual Environments and Teleoperator Systems, ASME/IMECE, Atlanta, GA, DSC-Vol. 58, 1996.

[9] K. E. MacLean and S. S. Snibbe, "An Architecture for Haptic Control of Media," in Proc. of the 8th Ann. Symp. on Haptic Interfaces for Virtual Environment and Teleoperator Systems, ASME / IMECE, Nashville, TN, DSC-5B-3, 1999.

[10] C. Richard, M. Cutkosky, and K. MacLean, "Friction Identification for Haptic Display," in Proc. of the 8th Ann. Symp. on Haptic Interfaces for Virtual Environment and Teleoperator Systems, ASME/IMECE, Nashville, TN, 1999.

[11] I. Rock and C. S. Harris, "Vision and Touch," *Scientific American*, vol. 216 (5), pp. 96-104, 1967.

[12] M. J. Rosen and B. D. Adelstein, "Design of a two degree-of-freedom manipulandum for tremor research," in Proc. of IEEE Frontiers of Engineering and Computing in Health Care, Los Angeles, CA, pp. 47-51, 1984.

[13] Sensable Technologies, *PHANTOM Desktop 3D Touch System*. Cambridge, MA, 1998.

[14] S. S. Snibbe, R. Shaw, K. E. MacLean, J. B. Roderick, K. Johnson, O. Bayley, and W. L. Verplank, "Haptic Metaphors for Digital Media," Interval Research Corp., Palo Alto, TR IRC #1999-071 (in preparation), 1999.

[15] M. A. Srinivasan, G. L. Beauregard, and D. Brock, "The Impact of Visual Information on the Haptic Perception of Stiffness in Virtual Environments," in Proc. of the 5th Ann. Symp. on Haptic Interfaces for Virtual Environment and Teleoperator Systems, ASME/IMECE, Atlanta, GA, DSC:58, 1996.

[16] H. Z. Tan, M. A. Srinivasan, B. Eberman, and B. Cheng, "Human Factors for the Design of Force-Reflecting Haptic Interfaces," in Proc. of the 3rd Ann. Symp. on Haptic Interfaces for Virtual Environment and Teleoperator Systems, ASME/IMECE, Chicago, IL, DSC:55-1, 1994.

[17] B. Ullmer, H. Ishii, and D. Glass, "mediaBlocks: Physical Containers, Transports, and Controls for Online Media," in Proc. of Siggraph '98, 1998.

[18] R. Want, K. P. Fishkin, A. Gujar, and B. Harrison, "Bridging Physical and Virtual Worlds with Electronic Tags," in Proc. of The Conference on Human Factors in Computing Systems (CHI '99), Pittsburgh, PA, pp. 370-377, 1999.

[19] P. M. Yarin, *Towards the Distributed Visualization of Usage History*, M.S. Thesis, MIT, 1999.

//kzcHZ
Traversable Interfaces Between Real and Virtual Worlds

Boriana Koleva, Holger Schnädelbach, Steve Benford and Chris Greenhalgh
School of Computer Science & Information Technology
The University of Nottingham
Nottingham NG8 1BB, UK
+44 115 951 4203
{bnk, hms sdb, cmg}@cs.nott.ac.uk

ABSTRACT

Traversable interfaces establish the illusion that virtual and physical worlds are joined together and that users can physically cross from one to the other. Our design for a traversable interface combines work on tele-embodiment, mixed reality boundaries and virtual environments. It also exploits non-solid projection surfaces, of which we describe four examples. Our design accommodates the perspectives of users who traverse the interface and also observers who are present in the connected physical and virtual worlds, an important consideration for performance and entertainment applications. A demonstrator supports encounters between members of our laboratory and remote visitors.

Keywords

Mixed Reality, Virtual Environments, Augmented Reality, Tele-presence, Tele-embodiment

INTRODUCTION

Various technologies have been developed to allow people to experience remote environments. These might be virtual environments that are experienced through virtual reality technologies or physical environments that are experienced through tele-embodiment and tele-presence technologies. A thread running through this research is the idea of using immersive technologies to establish the illusion of entering the remote environment, resulting in a sense of presence.

A major weakness in this illusion is that users clearly do not leave their physical environment behind them when they enter a remote environment. They remain firmly and visibly present within their local physical space. This is a problem for two reasons. First, their own illusion of remote presence may be destroyed by distractions from the local physical space. Examples can be found in previous experiments with virtual reality. In studies of presence in single user virtual environments, users reported that 'breaks in presence' were caused by background noise and interference from hardware such as cables [11]. Bowers *et al.* note how conduct in a collaborative virtual environment was disrupted by events in the physical environments of the participants [4]. Second,

observers of the interaction can clearly see that participating users have not gone anywhere. This is a particular problem if the interaction is being staged at least in part for the benefit of these observers, for example as part of an entertainment or performance application. It might also be a problem if these observers may themselves become participants at a later date. For example, if they are waiting their turn in an entertainment application or in a shared working environment due to the limited availability of equipment.

Our response to these problems is the concept of traversable interfaces. These enhance the illusion of immersion by making it appear that participants leave their local physical environment in order to enter into a new remote environment. They aim to do this in a way that makes sense to the participants who are entering the remote environment, to observers who are already in the remote environment, and to observers who remain behind in the local physical environment. Our discussion will focus on traversal between physical and virtual environments. However, a traversable interface could also be used in a tele-presence application to link a local physical environment to a remote physical environment.

Further motivation for traversable interfaces is provided by recent work on mixed reality. Paul Milgram has classified mixed reality technologies according to a 'virtuality continuum' [9]. At one extreme of this continuum we find purely physical environments and at the other purely virtual environments. In between, we find augmented reality where physical environments are enhanced with digital information, and augmented virtuality, where virtual environments are enhanced with physical information.

Traversable interfaces provide a mechanism for people to dynamically relocate themselves along this continuum. At one moment they may be primarily located in augmented reality, with a view into an adjoining virtual environment. They may then traverse the interface and find themselves primarily located within an augmented virtuality, with a view back into a physical environment. Traversal allows people to move back and forward between primarily real and primarily virtual environments, repositioning themselves along the virtuality continuum, according to their interest and whether they want the physical or virtual to be their primary focus.

TRAVERSABLE INTERFACES

We begin with a general design for a traversable interface. Figure 1 summarises the illusion that we wish to create. On the left we see a physical environment that is connected to the virtual environment on the right. Our design needs to consider the perspectives of the four classes of participant, A, B, C and D. A is an observer in the physical environment. B is an observer in the virtual environment. C is crossing from physical to virtual, and D is crossing from virtual to physical.

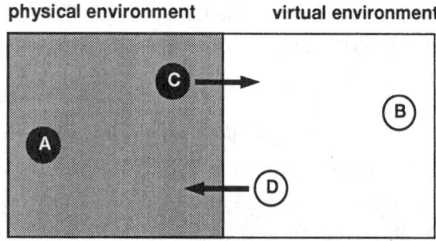

Figure 1: the illusion that we wish to create

An important point is that the illusion should potentially work for all of these classes of participants, although some applications may give priority to one class over another. For example, a performance might require that the audience believe the illusion, while the performers could be aware of the mechanisms involved. This observation challenges traditional approaches to interface design that have focussed on the experience of the direct participant, but have tended to neglect the experience of observers. We suggest that this is an important consideration for any application where an interface is deployed in a shared or "public" environment, including office environments as well as performance and entertainment applications.

Two other general points should be noted. First, objects as well as participants might traverse the interface. Second, partial traversal might be possible, for example pushing a limb through the interface. However, in this paper we restrict our consideration to complete traversal by humans.

Our general design for a traversable interface integrates a number of techniques:

- mixed reality boundaries [2] for creating windows between physical and virtual environments.
- tele-embodiment for allowing remote virtual participants to enter a physical environment [8,10].
- immersive interfaces for accessing virtual environments, including head-mounted displays (HMD) and projected displays ranging from single screens up to multi-surface CAVEs [5].
- non-solid projection surfaces to allow participants to seemingly pass through a projected image, moving from a public to a more private physical space.

The following sections describe how these are integrated into an overall design, beginning with the idea of mixed reality boundaries.

Mixed reality boundaries

Mixed reality boundaries represent a specific approach to mixed reality that involves creating transparent windows between physical and virtual environments so that occupants of each can communicate with the other [2]. In contrast to other approaches that focus on superimposing the two environments on top of one another (e.g., augmented reality typically overlays a virtual environment on top of a physical environment), the spaces on either side of the boundary are adjacent, but remain distinct. A feature of this approach is that multiple boundaries might be used to join together many different physical and virtual environments into a larger mixed reality structure.

Figure 2 shows how a simple mixed reality boundary can be created. On the left is a physical environment and on the right a virtual environment. An image of the virtual environment is projected into the physical environment and an image of the physical environment captured from a video camera is displayed as a live video texture within the virtual environment. The physical and virtual cameras and projections are aligned so that the images appear to be the reverse sides of a common boundary.

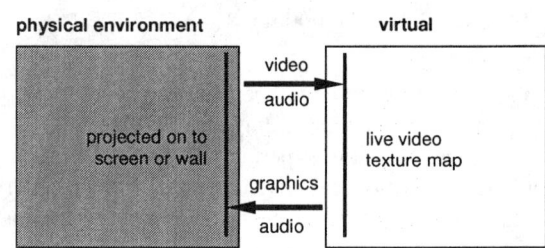

Figure 2: a simple mixed reality boundary (from [2])

A variety of mixed reality boundaries might be created with different properties in terms of their 'permeability', the extent to which they allow information and objects to pass across them; 'situation', their spatial relationship to the connected spaces; 'dynamics', their temporal properties; and 'symmetry' [7]. Permeability properties are particularly interesting here because they include the sub-property of 'solidity', the extent to which a boundary allows objects and participants to pass through it. This can be broken down into two issues, how to allow participants and objects to enter the remote environment and how to create the illusion that they have left their current environment when doing so.

Entering the remote environment

Entering a remote physical environment can be achieved by taking control of a remote physical proxy such as a robot. The field of tele-robotics is well established, particularly in areas such as working in hazardous environments such as outer space and the deep ocean. Of more direct relevance here is recent work on tele-embodiment in collaborative settings, where participants take control of a physical proxy or surrogate [8]. In one recent example, participants control a tele-embodiment called a Personal Roving Presence (PRoP) that is armed with a video camera, microphones and speakers, and steer it round a remote environment in order to meet and converse with others [10]. Designs for early PRoPs include 'space browsers', helium filled blimps that act as airborne tele-robots and ground based platforms called 'surface cruisers'. By placing a PRoP on the physical side of a mixed reality boundary and integrating the controls for this PRoP and the video and audio from it within the virtual environment, participants on the virtual side could enter the physical.

An alternative approach towards introducing remote virtual participants into a physical environment would be to use shared augmented reality technology such as [3]. See-through HMDs could display avatars superimposed onto the physical scene. In fact, this could be combined with the use of PRoPs. The position of the PRoP could be tracked and the image of the avatar superimposed upon it.

Techniques that allow a user in a physical environment to enter a remote virtual environment are well known and include a range of immersive displays including HMDs and different tracking and interaction mechanisms for interacting with a projected image of a virtual environment.

Leaving the current environment

The illusion of traversal requires that a user is seen to leave their current local environment when they enter the remote one. We propose that this may be achieved by using non-solid projection surfaces so that the user can appear to directly step into and through the image of the remote environment.

This is straightforward in the virtual environment. The image of the remote physical environment is displayed as a video texture attached to a graphical object. This can be non-solid, enabling avatars to pass through it.

It is more difficult in the physical environment. Later on, we shall describe four different approaches that we have implemented involving projection onto non-solid materials such as water, the use of fabric curtains as well as mechanical devices such as sliding doors and movable screens. For the remainder of this section we shall assume the existence of such technologies.

It should be noted that in all cases, what actually happens is that the user passes from a public space through the image, into a more private space beyond. From the physical environment they move to a physical antechamber beyond the screen where they find the immersive technology required to enter the virtual environment. From the virtual environment, their avatar moves to a virtual antechamber beyond the screen where they may find the controls to access a PRoP. The physical antechamber may take on a variety of forms. In a performance, the public space will be the key focus of activity, with the antechamber being 'the wings' or behind the scenes. Conversely, the antechamber might be the main focus of the activity, for example it might be a CAVE installation [5], with the traversable interface providing an entry point to and from the outside world.

An integrated design

Figure 3 shows how the above techniques for entering and leaving physical and virtual environments can be integrated into a general traversable mixed reality boundary.

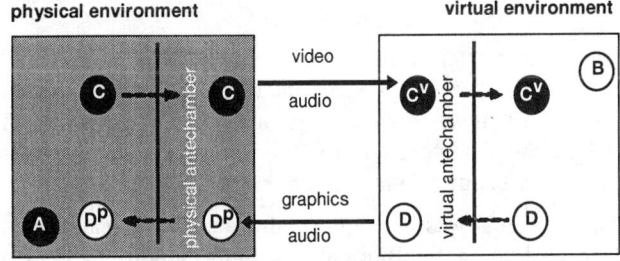

Figure 3: creating a traversable boundary

On the left is a physical environment containing a non-solid projection surface onto which is projected a view of the remote virtual environment. Behind this is an antechamber containing the immersive technology required to become embodied within the virtual environment. On the right is a virtual environment containing the video view into the physical environment. Behind this is a virtual antechamber that contains controls for a remote PRoP and that also contains a second video texture showing the view from this PRoP as it moves around the physical environment.

We can now consider how the four participants A, B, C and D from figure 1 will experience this design. Participant A is the observer in the physical environment. They will see participant B through the mixed reality boundary. They will see participant C step through the physical projection screen, apparently into the virtual world. At the same time, they will see C's virtual avatar, C^V emerge into the virtual world. They will see participant D's avatar approach the projection screen and then disappear from view. D's PRoP, D^P, will then emerge through the physical screen.

Participant B is the observer in the virtual environment. They will see A through the boundary and will see C approach them in the video view, disappear and then replaced by C's avatar, C^V, appearing through the video texture. They will see D's avatar approach the video texture, pass into it and then see D's physical proxy, D^P, appear in the video image.

Participant C traverses from the physical to the virtual. They will step through the physical projection screen, entering the physical antechamber. There they will find the technology required to independently access the virtual environment. This might be a headmounted display, desktop computer, CAVE, specialised vehicle (for example, a 'pod' in a simulation ride) or further projected display. Their avatar will initially appear in the virtual antechamber and they will then steer it through the video texture into the public virtual environment.

Participant D traverses from the virtual to the physical. They will steer their avatar through the video image of the remote physical environment, entering the virtual antechamber. Here they will find the virtual controls for the remote PRoP, D^P, as well as a further video texture showing the view from its onboard camera. They will then be able to steer the PRoP from the physical antechamber, through the physical projection surface into the public physical environment.

Design considerations

This design for a traversable interface is a general one. A particular realisation will have to make a number of specific design choices in order to meet the two goals of traversable interfaces as outlined in the introduction.

The first goal was to minimize distractions for participants who wish to become present in a remote environment. This is achieved by locating the VR equipment required to access this environment in a private antechamber. This can be designed to provide an optimal operating environment for this equipment, for example, being painted and lit to support video tracking, being free of other equipment that might interfere with electromagnetic tracking, and generally being free of clutter on which the user might snag themselves.

The second goal was to create the illusion of physically leaving the current environment in order to enter a new remote environment. Successfully meeting this goal will require considering the following design issues.

The physical and virtual antechambers can be decorated to support the transition to the new environment. For example, in a theme park ride, the physical antechamber might be modeled to match the virtual world. If the user thinks that they were going to pass into a virtual cave, then this antechamber should look like that cave. The physical and graphical design of PRoPs and avatars can also support the illusion of traversal. In a theme-park ride, the PRoP might be a sophisticated animatronic figure (such figures are already used in theme-parks). Likewise, the positions of physical bodies, PRoPs and avatars at the key transition points will be important. With careful design, it may be possible to make them appear to directly replace one another, to be overlaid on one another, or to time the sequence of appearances and disappearances to reinforce the illusion of traversal.

Traditional theatrical techniques may be used to enhance or alter the illusion of traversal, including changes in lighting, the use of smoke and sound effects. Another key effect is the use of shadows. Several of the non-solid projection surfaces that we introduce below can be configured to show the physical user beyond the screen as shadow. In some cases it will be important to avoid shadows so as to maximise the illusion of traversal. In others, the silhouette of a participant's body seen against the image of the virtual environment may be used for its artistic effect (see figure 8) or as one way of overlaying participants' physical and virtual bodies as noted above.

DESIGNS FOR NON-SOLID PROJECTION SURFACES

The use of non-solid projection surfaces is an essential part of our design. It has also been the most challenging part to realise. This section describes four attempts to construct such surfaces: fabric curtains, water curtains, a sliding door, and a flip-up screen. Figure 4 summarises the four designs and shows examples of each.

Fabric curtain

Curtains are familiar devices for partitioning physical space. Curtains can provide privacy and can be readily traversed, introduced and removed. They have been extensively used in theatre to hide and reveal actors and objects and to give the illusion of transitions between scenes. There are a wide variety of familiar designs of curtains; they can be pulled back, raised, vertically slit and be formed into blinds.

Curtains can be made of materials that can hold a projected image and so represent a natural choice for creating non-solid projection surfaces. Our initial design as shown in figure 4 (a) is based around a number of vertical segments of projection screen fabric, weighted at the bottom to hold their shape. A user can easily push through these and the curtain settles down to its regular shape within a few seconds. We back project the image onto the curtain by bouncing it off of a mirror on the ceiling. This creates an area in the antechamber where a participant may stand or sit without casting a shadow onto the screen. Conversely, they may be deliberately positioned so as to create a shadow for artistic effect as noted above. Figure 4 (b) shows a participant emerging through the curtain.

Water curtain

We have also experimented with a second curtain – a curtain of water. In 1998 we began collaborating with the performing arts company Blast Theory who were already experimenting with projecting images and video into a vertical curtain of water. Projection into water has also been explored in other contexts. For example, Disney-MGM studios projected film clips into fountains and a water screen as part of a dream sequence in their "Fantasmic" show in their October 1998 program.

The overall design of the water curtain is shown in figure 4 (c). The curtain is produced by several fine spray nozzles (originally designed for spraying pesticide) attached to a metal pipe that is suspended roughly two meters above a trough on the ground. Water is pumped through the pipe, descends as a fine spray about half a meter thick and is collected from the trough and recycled. Figure 4 (d) shows this physical infrastructure. The water curtain holds a back-projected image surprisingly well, although early experimentation showed that the projector needs to point straight at the curtain, making shadows unavoidable as participants pass through it.

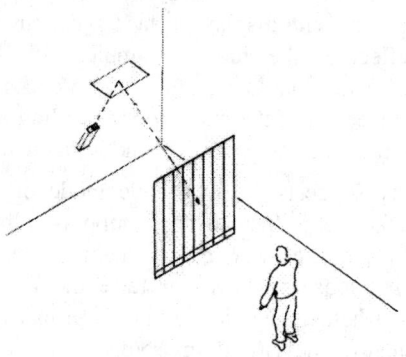
(a) Fabric curtain design

(b) Emerging through the fabric curtain

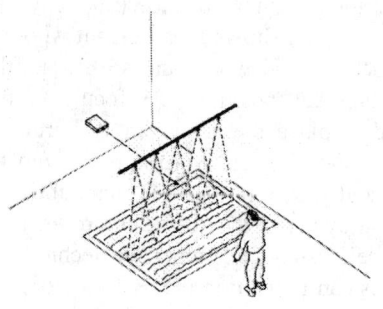
(c) Rain curtain design

(d) Rain curtain infrastructure   (e) Rain curtain in use

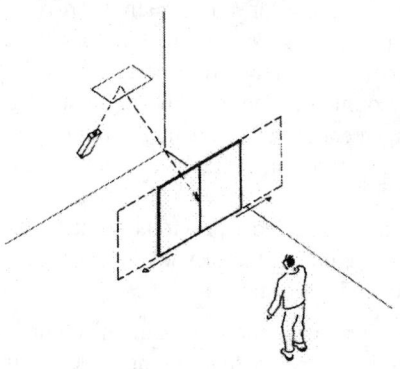
(f) Sliding door design

(g) Opening the sliding door

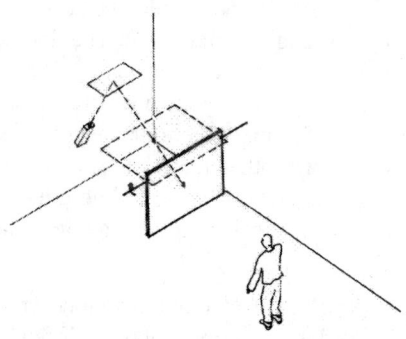
(h) Flip-up screen design

(i) Raised as ambient display   (j) Raising the screen

Figure 4: four designs for non-solid projection surfaces

Being completely fluid, a person or object can pass through the water curtain much more seamlessly than they can with a fabric curtain (so long as they are prepared to get wet!). It is also transparent when viewed from behind, allowing for easy observation of its users (e.g., by performers who can then time their emergence through the curtain to match the user's actions). Like a fabric curtain, the water curtain can be readily introduced and removed by switching the pump on and off. Holes can be dynamically punched through it by using solid objects to interrupt the flow of the water. Finally, it has a powerful aesthetic, in terms of the continually shifting quality of the visual image, the sound of the water and its physical feel.

In January 1999 we staged a public demonstration of using a water curtain as an interface to a virtual environment. Participants undertook a journey through a virtual world, during which they were interrupted by a performer emerging through the curtain – an event that had a significant theatrical impact. Figure 4 (e) shows the performer emerging through the water curtain. We are currently developing a full-scale public performance that will involve the use of six rain curtains to allow an audience to experience a shared virtual environment.

Sliding door

Unlike a curtain, a door is a solid projection surface that is traversed by physically moving a large section of it. As with conventional doors, there are many potential designs including hinged, sliding and rotating. Our first design has been a sliding door made from perspex as shown in figure 4 (f). Figure 4 (g) shows a participant opening the door in order to step through it.

The sliding door has several interesting properties. Being solid, it can more easily be locked than a curtain, allowing participants to minimise possible interruptions. Its solidity also favours applications where it is part of a more permanent architectural framework. Early tests suggest that our sliding door can simultaneously hold two different images, one on each side, provided that the images that have similar contrasts (otherwise there is too much visible interference between the two). This potentially saves space, as it only requires one projection surface to display both public observers' and immersed participants' views of the virtual environment. The properties of solidity and holding multiple images could usefully be combined in using a sliding door as the entrance to a CAVE. One surface of the CAVE could be slid open to allow participants to enter. Visitors remaining on the outside could see a specially tailored (e.g., without head-tracking) public view of the activity in the CAVE on the outside of the door.

Flip-up screen

Our final example is a flip-up screen as shown in figure 4 (h). This is a screen that can be moved from a vertical to a horizontal position at ceiling height, allowing people to pass underneath it. Figure 4 (i) shows a participant raising the screen. The flip-up screen is essentially a specialised form of door. However, it has the additional property of being able to act as an ambient display surface when in the raised position, reflecting the idea of ambient display media proposed in [6]. This is possible because the projected image is bounced off of the mirror on the ceiling and hits the screen when it is in both its vertical and horizontal positions.

This property suggests an alternative mode of use to the previous examples. Instead of stepping through the projected image, the user may remain in one physical location, but choose to lower or raise the flip-up screen according to whether their interaction is primarily focussed in the physical or the virtual environment. To focus on the physical environment, the user raises the screen, opening up their physical space to the public space beyond and displaying a peripheral image of the virtual environment on the ceiling. Figure 4 (j) shows a participant who is focussed on a task in the physical world and so has set the flip-up screen to its ambient position. To focus on the virtual environment, they place the screen in its vertical position, shielding their local physical environment from the public space beyond, and providing users in this public space with an image of their avatar in the virtual environment instead of their physical self using the immersive technology. In this way participants can reposition themselves along Milgram's virtuality continuum as noted in the introduction.

An extension to this approach would be to use the physical raising and lowering of the screen to drive a switch to automatically configure a user's local environment according to whether they were currently in the physical or virtual environment. The switch might configure lighting and tracking technologies and might minimise distractions, for example by routing the user's phone to their voice mailbox when they were immersed in the virtual environment. This reflects previous work on using physical doors to manage electronic privacy in an office environment, using a so-called "doormouse" [12].

In summary, we have realised four different kinds of non-solid projection surface that might be used in traversable interfaces. These can be broadly grouped into the two categories of curtains (fabric and water) and doors (sliding and flip-up). The curtains potentially offer the most seamless illusion of traversal and could be especially suited to performance, art and entertainment. The doors provide a less fluid illusion of traversal, but may offer some practical advantages for use in everyday environments such as offices and the home. Of course, there are many other possibilities. Perhaps we can use other materials such as smoke to create highly fluid projection surfaces, and no doubt there are other possible mechanical designs based on doors and curtains.

DEMONSTRATION

We have developed a demonstration of a traversable interface in order to show that our design is technically implementable. It should be noted that we do not claim that this is yet a real or effective application, although our future plans involve developing and evaluating such an application.

Our demonstration has been constructed in our laboratory. Its aim is to provide a social space where lab members can meet with visitors who "drop in" over the Internet. A mixed reality boundary allows lab members and visitors to see and talk to one another. Both can also traverse this boundary. A single visitor at a time can take charge of a simple PRoP and use it to explore an area of the laboratory. A single lab member at a time can step into the virtual world to become part of a virtual meeting. Figure 5 shows the collaborative virtual environment that we are using in our demonstrator. This has been realised using the MASSIVE-2 system [1]. The image shows the video texture that forms half of the mixed reality boundary with the physical environment.

Figure 5: the virtual environment with video texture

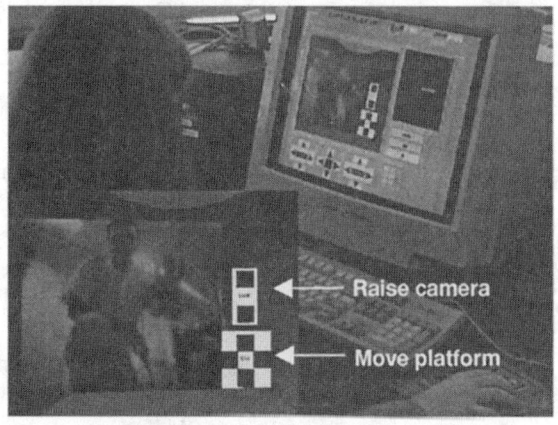

Figure 6: controlling the PRoP from MASSIVE-2.

Avatars can step through this boundary to enter a small virtual antechamber where they find the interface to control our remote PRoP. This consists of a second video texture that shows the view from the PRoP's on board camera as well as six buttons, four to move the PRoP forwards and backwards and to rotate it left and right, and two to tilt the camera up and down vertically. Figure 6 shows the view over a remote user's shoulder when they have just entered this antechamber. Inset is a close up of the virtual controls for the PRoP.

The PRoP itself is a small wireless robot that has been constructed using a LEGO Mindstorm kit (see figure 7). This platform can be moved around the floor, includes a raisable arm for the camera and can be controlled over an infrared link. A small wireless video camera and microphone have been mounted on the platform along with a pen-torch to illuminate nearby objects. The wireless connections currently have a limited range and there are as yet no on-board speakers (so the PRoP can see and hear, but not talk). The PRoP is also rather small, standing at approximately one foot tall. However, it does provide an inexpensive workable solution for initial demonstrations and application development.

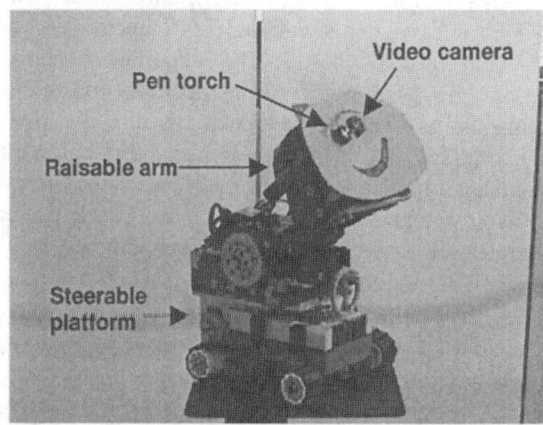

Figure 7: the PRoP

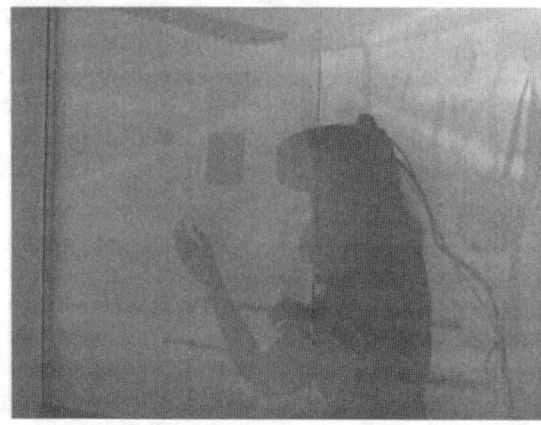

Figure 8: immersed in the virtual environment

The physical side of the boundary can utilise the fabric curtain, sliding door or flip-up screen designs. The images in figures 4 (b), 4 (g) and 4 (j) all show examples of the view looking into our virtual environment, as if from out of the video texture. In each case a video camera is mounted on the top of the frame of the boundary to provide the video view shown in figure 5. This positioning is less than ideal as the two sides of the boundary are not strictly spatially aligned and a solution that allows a small camera to be located in the centre of the screen is ideal. Mounting the camera in the centre of the flip-up screen would also allow it to provide a peripheral view from above the user's workspace when in the raised (ambient) position. Having traversed the physical screen, the user enters the physical antechamber and finds equipment to access the virtual world. Figure 8 shows an example where the user has stepped through the screen and has donned a head-mounted display. In this case, they have been deliberately positioned so that we see their shadow.

SUMMARY

This paper has developed the idea of traversable interfaces that give the illusion that participants in a local physical environment can completely cross into a remote virtual (or indeed physical) environment and vice versa. The key innovation in the paper is the extension of the now familiar illusion of entering a remote environment to include appearing to leave one's current environment. We have argued that this is particularly important when the interaction may be observed by people in the two environments as well as experienced directly by the participants. This will be the case in many performance and entertainment applications, but will also be relevant whenever virtual environments and tele-presence technologies are deployed in shared environments, be they public, working or domestic.

We have presented a general design for a traversable interface between a physical and a virtual environment that combines three key components. The first is the use of Physical Roving Proxies (PRoPs) to allow a virtual participant to enter a physical environment. The second is the use of VR technologies to allow a physical participant to enter a virtual environment. The third is the use of non-solid projection surfaces to allow a participant to seemingly step into a projected image of a remote environment. We have presented four early designs for non-solid projection surfaces, a fabric curtain, a water curtain, a sliding door and a flip-up screen. Finally, we have described a demonstrator that shows one possible realisation of our design.

Among the most obvious applications of traversable interfaces are entertainment applications where it may be important to establish a strong illusion of entering a virtual environment. VR-based theme park rides that wish to smooth the transition between watching the ride while waiting for a turn and entering the ride as a participant are a particularly strong candidate, especially as such rides already use animatronic figures and participants occasionally get wet! We also anticipate that our design might be incorporated into more general immersive interfaces. For example, a traversable interface based on our sliding door design might form one side of a CAVE facility, allowing passage to and from the CAVE and providing an external public view of the activity inside.

Our future plans involve developing and evaluating real applications of traversable interfaces. Evaluation will employ ethnographic techniques of the kind that have been previously used to study social interaction in collaborative virtual environments (e.g., [4]).

We would like to finish by reinforcing two points that have more general relevance to human-computer interaction.

First, is the idea that shared and public interfaces need to be designed with third party observers in mind as well as direct participants. Second, is the observation that virtual reality and telepresence technologies have always been concerned with creating an illusion – the illusion of entering a new and remote environment. This paper has explored how more traditional theatrical effects, such as moving screens and curtains, and changes in lighting might enhance this illusion, an approach that might be applied to the design of a wide range of human-computer interfaces.

ACKNOWLEDGMENTS

We thank the ESPRIT IV I^3 programme for supporting this work through the eRENA project and the EPSRC for their support through PhD studentship awards. We are indebted to Blast Theory for their collaboration on the rain curtain.

REFERENCES

1. Benford, S. D., Greenhalgh, C. M. and Lloyd, D, Crowded Collaborative Virtual Environments, *Proc. CHI'97*, pp. 59-66, 1997

2. Benford, S., Greenhalgh, C., Reynard, G., Brown, C., and Koleva, B., Understanding and Constructing Shared Spaces with Mixed Reality Boundaries, ACM ToCHI, 5 (3), 185-223, ACM Press.

3. Billinghurst, M., Weghorst, S., Furness III, T., Shared Space: An Augmented Reality Interface for Computer Supported Collaborative Work, Proc. CVE'96, Nottingham University, UK, Sept 1996.

4. Bowers, J., Pycock, J. and O'Brien, J., Talk and Embodiment in Collaborative Virtual Environments, Proc. CHI'96, 1996.

5. Cruz-Neira, C., Sandin, D. J., DeFant, T. A., Kenyon, R. V. and Hart, J. C., The Cave - Audio Visual Experience Virtual Environment, CACM, 35 (6), pp 65-72, 1992.

6. Ishii, H. and Ullmer, B, Tangible bits: Towards seemless interfaces between People, Bits and Atoms, Proc. CHI'97, pp 234-241, ACM Press, 1997.

7. Koleva, B., Benford, S. and Greenhalgh, C, The Properties of Mixed Reality Boundaries, Proc. ECSCW'99, Copenhagen, September 1999, Kluwer.

8. Kuzuoka, H., Ishimoda, G. and Nishimura, T., Can the GestureCam be a surrogate?, Proc. ECSCW'95, Stockholm, Sweden, September 1995, Kluwer.

9. Milgram, P. and Kishino, F., A Taxonomy of Mixed Reality Visual Displays, IEICE Transactions on Information Systems, Vol E77-D (12), Dec. 1994.

10. Paulos, E. and Canny, J., PRop: Personal Roving Presence, Proc. CHI'98, pp 296-303, 1998.

11. Usoh, M., Arthur, K., Whitton, M., Bastos, R., Steed, A., Slater, M. and Brooks, F., Walking > Walking-in-Place > Flying, in Virtual Environments, *Proc. SIGGRAPH'99*, pp359-364.

12. Buxton, B., Living in Augmented Reality: Ubiquitous Media and Reactive Environments, *Proc. Imagina'95*. Monte Carlo, February 1995.

Tradeoffs in Displaying Peripheral Information

Paul P. Maglio   Christopher S. Campbell
IBM Almaden Research Center
650 Harry Rd.
San Jose, CA 95120 USA
408-927-2857
{pmaglio, ccampbel}@almaden.ibm.com

ABSTRACT
Peripheral information is information that is not central to a person's current task, but provides the person the opportunity to learn more, to do a better job, or to keep track of less important tasks. Though peripheral information displays are ubiquitous, they have been rarely studied. For computer users, a common peripheral display is a scrolling text display that provides announcements, sports scores, stock prices, or other news. In this paper, we investigate how to design peripheral displays so that they provide the most information while having the least impact on the user's performance on the main task. We report a series of experiments on scrolling displays aimed at examining tradeoffs between distraction of scrolling motion and memorability of information displayed. Overall, we found that continuously scrolling displays are more distracting than displays that start and stop, but information in both is remembered equally well. These results are summarized in a set of design recommendations.

Keywords
Peripheral information, dual-task tradeoffs, user interface design.

INTRODUCTION
With the widespread use of advertising banners on web pages designed to distract users and capture their attention, it is becoming increasingly important to understand the nature of interruptions and distraction in computer interfaces. Advertising is particularly insidious, as users have little control over what is displayed. In general, though, because computer users routinely leave open many applications while they work on one thing at a time, only a small amount of the information available on the computer screen is central to the user's current task [3]. But when mail arrives, when print jobs are finished, or when an application abnormally terminates, users often like to be informed. To take other examples, tips on how to use the current application displayed in a pop up window or news headlines displayed in a scrolling text display might be helpful, but are not strictly necessary. In configuring their systems to provide this sort of helpful information, users routinely balance their desire to be informed with their tolerance for being interrupted.

Some interfaces designed to assist users in automobile driving [1] or web navigation [14,15,17] display recommendations through highlighting, annotation, sound, or speech. In these cases, the normal mechanism for driving or web browsing is still available, but the interfaces attempt to make the task easier and richer by providing additional information. Annotation in particular can be an effective means for conveying extra information about links on web pages without being distracting [2].

In designing a variety of user interfaces, the challenge is to create information displays that maximize information delivery while at the same time minimize intrusiveness or distraction. We call nonessential information *peripheral information* because it is not central to the current task, but might be helpful to it or informative in other ways. There are two ways in which information might be considered peripheral, in content and in display. Although peripheral information is ubiquitous, neither peripheral content nor peripheral displays have been systematically explored (but see [20]). In this paper, we are concerned mainly with issues of display.

In terms of content, the key to peripheral information is that it is not critical to task performance. Unlike what is generally studied in the literature on monitoring and supervisory control (e.g., [18]), inattention to peripheral information does not result in catastrophe, such as a nuclear meltdown or a plane crash. However, by providing peripheral information, a system offers a user the opportunity to learn more, to do a better job, or to keep track of less important tasks.

The term *ambient information* has been used to refer to subtle environmental cues when designed into systems to peripherally convey information such as network traffic (e.g., [9]). Specific environmental changes, such as the frequency of background noise or amount of background lighting, are associated with specific changes in system status. Of course, the design problem here is to make the mapping from system state to environmental state as obvious as possible and without interfering with the primary task [19].

Our goal in this paper is to consider what it takes to inform peripherally. Our approach is to view peripheral informing as imposing extra-task demands on a user's cognitive resources. In a computer context, this means a user attends to a primary task, such as reading an online document or editing text, but occasionally another item, such as a news story, becomes important and is briefly attended to. Specifically, the present study was performed to assess the mental workload tradeoffs and effectiveness of three single-line text displays (which we generally refer to as *tickers*) in conveying information under dual-task demands. Participants in our studies performed a primary text-editing task while at the same time monitoring headlines displayed on a ticker. We measured how distracting the tickers were on editing performance, and we measured how memorable the headlines were on a post-experiment test. In the end, we found that constantly scrolling tickers distract more than those that start and stop, but that some motion actually provides effective feedback, helping users efficiently schedule glances at the peripheral display. These results have implications for the design of a variety of user interfaces, including web pages, notification services, and help systems.

The paper is organized as follows. We first discuss scrolling ticker displays and some background theory. We next detail the three experiments run to test the tickers under dual-task conditions. We finally summarize the results of the experiments and draw conclusions or the design of systems that incorporate peripheral information.

TICKER DISPLAYS

The displays tested included a variety of tickers in three broad families: (a) continuous scrolling text (CS), (b) discrete scrolling text (DS), and (c) serial presentation (SP). In the CS case, text scrolls at a constant rate either from right to left (horizontal) or from top to bottom (vertical). In the DS case, text scrolls quickly to the center of display (either horizontally or vertically) where it stops for some period before scrolling off the display. In the SP case, the text does not scroll at all; rather, it is displayed in a constant position in the center of the display, each update replacing the last text with new text.

Despite the large number of possible tickers, only two classes have been systematically explored: horizontal CS and SP. Studies directly comparing horizontal CS and SP found no difference in comprehension for a display reading task [12]. Other studies comparing CS to static displays found that text is read more slowly [22] and is less comprehensible [8] when scrolling horizontally than when displayed on a static page. Further work showed that this effect does not depend on the number of words on the screen or on window size [5]. Studies comparing SP and static pages of text show that comprehension performance is about the same [10], and that reading latency is about the same as well [21]. Overall, these studies fail to show strong differences in comprehension among CS, SP, and static displays, and therefore cannot be used to make strong recommendations about display design.

A practical evaluation of scrolling displays as peripheral information displays requires an analysis of the mental workload involved in performing a primary task (such as text editing), and in shifting to a secondary task (reading the ticker). Since the tasks are performed concurrently, some type of scheduling strategy must be devised to determine the length of time to perform each task and the frequency of switching between tasks. Scheduling strategy or timesharing skills are one important determinant of performance in multi-task situations [23]. Evidence that scheduling strategy can influence multi-task performance independent of resource limitations comes from research showing improved dual-task performance after strategy training [13]. Additional evidence comes from studies showing improved performance after extended experience managing two tasks concurrently but with no improvements in performance on each task alone [4,7]. Task priority (relative payoff or relevance, see [10]) and bandwidth (frequency or pace of display) also influence scheduling strategy [24].

The present study concerns how visual display parameters can influence performance in a dual-task setting. Importantly, our participants were not told how to prioritize the two tasks or at what speed the display updates, either of which would help them create a strategy for scheduling their glances at the scrolling display. Many scheduling strategies are possible. One would be to watch the first few updates of the ticker display to establish a time span in which to edit then glance down at the ticker. Another strategy would be to edit until movement is detected in the periphery and then to look down to read the new headline, effectively using motion of the text displayed in the ticker as feedback to cue looking at the ticker.

Feedback the tickers provide, including motion and flashing, can help cue glances at the scrolling display. So we can order the tickers by the quality of feedback they provide users. Because the update of DS occurs gradually as the headline scrolls down to the center of the display, we believe that DS provides a better cue for shifting gaze to the ticker than SP, in which there is no motion. Because SP updates the display instantly and CS updates the display constantly, we believe that SP provides better feedback than CS. Thus, DS gives the best feedback, followed by SP, and then CS.

Along these lines, we hypothesized that (a) tickers with more motion would be more distracting, leading to fewer edits; and (b) tickers with less feedback would provide fewer cues for scheduling, leading either to fewer edits or to lower comprehension and retention of the ticker-displayed information.

The basic plan of our experiments is as follows. First, we tested the effects of three common tickers (horizontal continuous scrolling, vertical discrete scrolling, and serial presentation) on editing and memory performance to see whether the predicted ordering would obtain. Second, we switched the direction of motion for the scrolling displays to determine whether direction was relevant. Third, we tried alternative methods for providing update feedback–auditory and visual cues other than scrolling motion–to see whether feedback or motion dominates performance.

EXPERIMENT 1

The first experiment examined distraction versus memorability of three tickers. Distraction was defined as the change in performance for text editing alone versus text editing while concurrently reading the ticker display. The memorability of each display was defined as the number of displayed headlines recognized on a post-experiment multiple-choice test.

Method

Participants first performed the text-editing task without the headline-reading task. Participants then performed both tasks. Performance on the editing task alone served as the baseline to compare to editing performance in the dual-task condition. After simultaneously performing the text-editing and headline-reading tasks, memorability was measured by a headline recognition test. More precisely, one between-subjects factor, ticker type, was manipulated across three levels: horizontal continuous scrolling (HCS), serial presentation (SP), and vertical discrete scrolling (VDS). Two dependent measures were collected for each participant: number of corrections made to the document and number of correct responses on the recognition test.

Participants

Twenty-nine participants were recruited from IBM Almaden Research Center and compensated for their time.

Materials

The text editor and the ticker appeared as two separate displays on a computer screen (see Figure 1). The text edited by the participants was taken from a chapter of a scientific dissertation [16] and broken into two parts, which were counterbalanced in the two phases (i.e., with ticker/without ticker) of the experiment. As shown in Table 1, both parts were of similar reading difficulty, as given by the Flesch ease of reading index and the Flesch-Kincad grade level [6].

Errors were introduced into the texts by hand according to three rules: (a) between 0 and 2 errors were put in each sentence; (b) errors were evenly spaced throughout the entire document; (c) errors included only subject-verb agreement, word order, and inconsistent verb tense. These error types were chosen (rather than spelling and other typographical errors) so that the editing task would be sufficiently demanding and thus produce a dual task

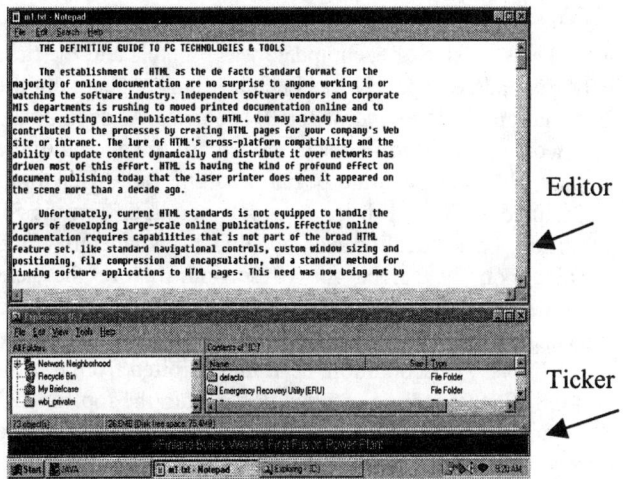

Figure 1: Screen shot showing editing window and ticker display for the dual-task conditions.

performance trade-off. Including typographical errors would make possible a text-skimming strategy. Skimming likely requires less effort than reading and is not the task we intended to test.

Table 1: Reading difficulty of edited documents.

	Experiment 1		Experiments 2 and 3			
Document	1	2	1	2	3	4
Total words	1760	2027	2467	1580	2365	2060
Words/sentence	5.9	5.5	6.0	5.2	6.4	6.4
Reading ease⁺ (Flesch)	38.3	39.1	63.6	47.1	62.0	44.0
Grade level (Flesch-Kincaid)	9.3	9.1	5.8	7.9	6.2	8.7

⁺Reading ease ranges from 0 to 100, with 0 being most difficult.

The tickers displayed thirty single-line headlines, averaging seven words apiece (see Table 2). The headlines were constructed as concise and self-contained summaries of news stories. The topics of the headlines were fictional but plausible. Each of the thirty headlines was displayed once in random order, and then this sequence repeated so that overall, each headline was displayed twice.

Table 2: Example headlines.

President Names Hoffman Secretary of Defense
French F-10 Fighter Jet Down Over Turkey
Finland Builds World's First Fusion Power Plant

The text for all ticker displays appeared in 17-point font with a cyan foreground and a black background. Each of these displays was 796 pixels wide by 24 pixels high. In the HCS condition, text continuously scrolled from right-to-left. The step size of each update was 5 pixels and the time between updates was 132 ms. Each headline appeared in the information display window for 10 seconds. As soon as the last word of a headline had moved 10 pixels onto the display, the next headline began scrolling out. Headlines maintained a 10-pixel distance as they scrolled. For the SP condition, each headline was updated instantaneously and remained on the screen for 10 seconds before the next headline was presented. Each headline completely replaced the last one so that no two were on the screen at the same time. In the VDS condition, headlines scrolled from top-to-bottom. Each headline scrolled down from the top for 333 ms and remained in the horizontal and vertical center of the screen for 9.666 seconds. When the next headline scrolled down from the top, the current headline scrolled from the center to the bottom and off the screen over a period of 333 ms.

Procedure

In the first phase, participants edited one of the two documents, making as many corrections as possible within ten minutes. They were told about the types of errors they could expect to find. After ten minutes, the second phase began, in which participants edited the other document and read the ticker display at the same time. The order in which the documents were presented was balanced across participants. Participants were told to make as many corrections as possible while reading the headlines. The importance of performing both tasks to the best of their ability was stressed, as well as the fact that they would be tested on their memory of the headlines. At the end of the ten-minute time limit, the experimenter administered the multiple-choice test. No time limit was enforced for completing this test.

Results

Editing performance was calculated as the percentage decrease in number of correct edits from the no ticker condition to the ticker condition for each participant. Based on this measure, three outliers—exceeding two standard deviations from the mean—were removed from consideration, leaving 10 participants in each of the HCS and SP groups, and 6 participants in the VDS group. The alpha level for this and all experiments was set at 0.05.

A paired samples t-test showed that number of corrections decreased significantly from the single- to the dual-task conditions, $t(25) = 4.38, p < .001$. To test for differences in the effects of the three information displays, a one-way analysis of variance was calculated using percentage decrease in number of corrections as the dependent measure. A main effect for information display was marginally significant, $F(2,23) = 3.20, p = 0.059$. Figure 2 shows the mean percent decrease in the number of corrections made from the single-task to the dual-task condition. Planned comparisons using independent samples

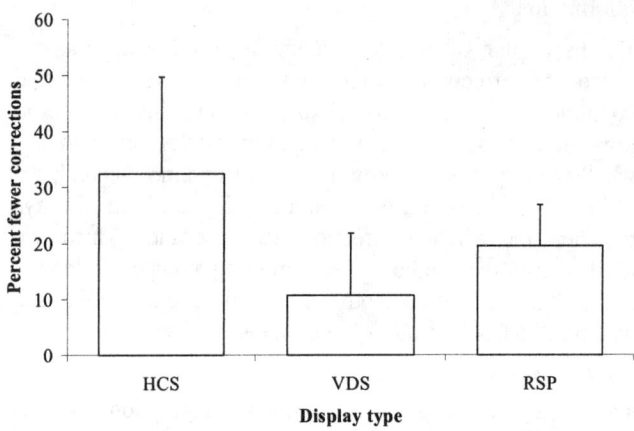

Figure 2: Mean percentage decrease in corrections between single- and dual-task conditions in Experiment 1.

t-tests (assuming unequal variance) indicated: (a) continuous scrolling (HCS) led to worse editing performance than discrete scrolling (VDS), $t(13) = 2.46, p = 0.014$; (b) continuous scrolling (HCS) was marginally worse than serial presentation (SP), $t(12) = 1.54, p = 0.075$; and (c) serial presentation (SP) was marginally worse than discrete scrolling (VDS), $t(10) = 1.64, p = 0.066$.

To determine the memorability of the tickers, a one-way analysis of variance was calculated with percent correct on the multiple-choice test as the dependent variable. No difference was found among the displays, $F(2,23) = 0.58$, NS. The failure to find significant differences was not the result of floor or ceiling effects, as scores spanned a normal range from 30% to 100% correct, with means of 70% for HCS, 67% for VDS, and 76% for SP.

Discussion

Though the differences in the number of corrections among information displays appear small during a 10-minute period, over the course of a full workday, productivity could be greatly influenced by the type of display chosen. For instance, using a continuous display rather than a discrete scrolling display, five fewer corrections were made every 10 minutes. If this rate of distraction is constant during four hours of editing, this would result in 100 fewer corrections for the continuous ticker, which could amount to several documents.

As mentioned, ordering the tickers by the quality of update feedback they provide, discrete scrolling gives the best feedback, followed by serial presentation, and then continuous scrolling. The ordering shown in Figure 2 is consistent with these predictions. More movement in the display (continuous scrolling) produces greater distraction and less feedback. At the same time, less movement (discrete scrolling) results in less distraction and more update feedback. These relationships suggest that distraction is a function of several factors, including amount of display movement and amount of feedback.

EXPERIMENT 2

The results of Experiment 1 suggest that the discrete scrolling ticker provides the best balance of motion and update feedback, leading to least impact on concurrent task performance. However, because the discrete display scrolled vertically and the continuous display scrolled horizontally, we cannot rule out scroll *direction* as another factor affecting performance. The purpose of the second experiment was to rule out direction by swapping movement direction and movement amount. Specifically, in this experiment, the discrete display scrolled horizontally (HDS) and the continuous display scrolled vertically (VCS). If scroll direction influences editing performance, then the results should be the opposite from those obtained in Experiment 1: more corrections in the continuous scrolling condition than in the discrete scrolling condition. However, if the amount of movement dominates editing performance, then the results should be the same those obtained in Experiment 1: more corrections in the discrete scrolling condition than in the continuous scrolling condition.

In addition, to add more update feedback to the serial presentation display, we modified the SP ticker to gradually fade headlines in and out. This effectively eliminates both motion and update feedback. If motion were more detrimental to performance than the lack of feedback, we would expect performance to be better for the fading display than the other tickers.

Method
As in Experiment 1, participants first performed the editing task without the headline-reading task. In the dual task condition, participants performed both tasks. Performance on the editing task alone served as the baseline to compare to editing performance in the dual-task conditions. After simultaneously performing the text-editing and headline-reading tasks, headline memorability was measured by a headline recall test. Unlike in the first experiment, participants in the second experiment saw all three tickers, as well as the no ticker condition. Thus, one within-subjects factor, ticker type, was manipulated across three levels: vertical continuous scrolling (VCS), fading serial presentation (FSP), and horizontal discrete scrolling (HDS).

Participants
Twenty-three participants were recruited from a temporary employment agency and were compensated for their participation.

Design and Dual-task Construction
Each participant was exposed to all three tickers as well as the no ticker condition. For this design, three different sets of headlines were constructed along with three recall tests. Additionally, four different documents were created for editing (one in each condition). With the new design came three new factors, display order, document, and headline, which were randomized across participants. Distraction was measured as the number of corrections made within the ten-minute time period, and memorability was measured by the percent correct answers on a headline recall test.

Materials
Because differences had not been observed in memorability for Experiment 1, we attempted to increase the sensitivity of the test by changing it to a short answer fill-in-the-blank test (cued recall test). The documents used for editing in Experiment 1 were of moderately high difficulty. This may have contributed to somewhat higher variation in editing performance than expected. For Experiment 2, four new documents were chosen from popular press articles concerning current software tools or Internet applications. These documents were far easier to read (see Table 1). The method of introducing errors to the documents was the same as that used in the first experiment.

The FSP updated headlines centered in the display window. Headlines faded in by increasing the brightness of the text according to an exponential function, and then faded out by decreasing brightness according to the same function. The location and size of the text editor and headline display windows were the same as Experiment 1.

Procedure
Participants were presented with all four conditions of the experiment—no display, HDS, VCS, and FSP—in random order. The order of the four documents and the three sets of headlines were both randomized before the experiment began. In each condition, participants edited a document for ten minutes. After all but the "no ticker" condition, the appropriate recall test was administered. Participants were instructed to do their best on the recall test and to guess if they did not know the answer.

Results
Participants were required to surpass a minimum criterion of editing competency to be included in the analysis. Even reading at a slow pace, a participant should be able to read about four sentences per minute. Because each sentence contained between zero and two errors and the errors were evenly spaced throughout each document, participants would be expected to find at least one error every four sentences. Thus, our minimum performance criterion was 1 error per minute (and 10 errors in ten minutes) in the no display condition. Five of the twenty-three participants failed to meet this minimum criterion and were not considered further.

A one-way repeated measures ANOVA was calculated across the four within-subjects conditions, showing a significant effect on number of corrections, $F(3,51) = 20.659$, $p < 0.0001$. The average number of corrections dropped from 24.67 in the baseline condition to 15.00 in the ticker conditions. The percentage decrease in the number of corrections from the baseline to the dual-task conditions is shown in Figure 3. Paired samples *t*-tests showed significantly more corrections in the HDS condition than the VCS condition, $t(17) = 267$, $p = 0.016$, but neither of the other comparisons (HDS vs. FSP, and VCS vs. FSP) were significant.

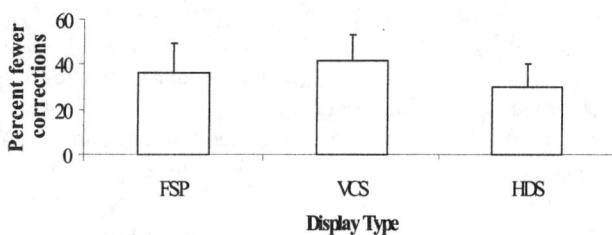

Figure 3: Percentage decrease in corrections between single- and dual-task conditions in Experiment 2.

Scores on the recall test measured memorability of the headlines. Once again, no difference was found among the displays, $F(2,51) = 0.6$, NS. Mean scores were 58% for FSP, 56% VCS, and 49% for HDS.

Discussion
The results here support our original interpretation of the results of Experiment 1: Display movement and update feedback—rather than display direction—affect editing performance. As discussed, displays with less movement ought to be less distracting, enabling participants to remain primarily focused on the editing task. At the same time, tickers with less movement ought to provide more update feedback, thus helping participants efficiently schedule their glances at the ticker.

Although Experiment 2 ruled out motion direction, distraction and feedback were not clearly distinguished. The FSP condition in this experiment was intended to provide somewhat less feedback than the original SP condition of the first experiment. Yet, because the headlines on the FSP display does not provide update feedback, it is possible that FSP is more distracting than SP. Because of their different designs and different details, it would be inappropriate to compare conditions across these experiments directly. However, the ordering of the conditions by editing performance obtained in Experiment 1 (Figure 2) can be informally compared to the ordering obtained in Experiment 2 (Figure 3), revealing the same relative ordering. Although SP clearly differed from the discrete and continuous tickers in Experiment 1, FSP did not differ significantly from either the discrete or continuous conditions in Experiment 2. Thus, the fading condition served to muddle rather than to clarify the relationship between distraction and feedback.

EXPERIMENT 3
The third experiment tested whether distraction or feedback dominates performance by introducing conditions that provide feedback independent of motion. In particular, conditions in which simple visual and auditory highlighting provided update feedback were created. Whereas continuous scrolling displays move without providing explicit update feedback, and discrete scrolling displays move and provide explicit update feedback at the same time, visual and auditory highlighting provide explicit update feedback to a continuous scrolling display independent of display motion. In the visual highlighting condition, the background of the ticker display flashed briefly when a new headline was fully visible. In the auditory highlighting condition, a simple beep alerted participants when a new headline was fully visible. Because these methods for highlighting or providing update feedback do not involve motion, it should be possible to determine whether feedback or distraction dominates performance.

Method
As in the first experiment, a between-subjects design was used. In this case, participants first edited text without reading headlines. In the dual-task condition, participants saw one of four possible tickers: discrete scrolling (DS), continuous scrolling (CS), continuous scrolling with visual feedback (VF), and continuous scrolling with auditory feedback (AF). As in the two previous experiments, the effect of the headline-reading task was measured as the decrease in performance in the dual-task compared to the single-task conditions. As in Experiment 2, memorability of the headlines was measured by a recall test administered after the dual-task condition.

Participants
Forty undergraduates at the University of California, Santa Cruz participated for psychology course credit.

Materials and Procedure
The materials and procedure were the same as in the previous experiments with the following exceptions. Only articles 1 and 2 from Experiment 2 and the two sets of headlines from Experiment 1 were used. The procedure was the same as that of Experiment 1.

In all conditions, headlines scrolled from right to left. In the CS condition, text scrolled continuously, as in the HCS condition of Experiment 1. In the DS condition, headlines scrolled discretely, as in the HDS condition of Experiment 2. The VF condition was similar to the CS condition in that the text scrolled continuously left to right, but differed in that the background black turned yellow for 500 ms when a headline was fully displayed. Visually, the AF condition was identical to the CS condition, but when a headline was fully displayed, a low beep sounded.

Results
Four outliers—whose editing performance exceeded two standard deviations from the mean—were removed from further analysis, leaving nine participants in each of the four groups.

To test for differences among the four information displays, a one-way analysis of variance was calculated with percent decrease in number of edits as the dependent measure. A main effect was found for editing performance, $F(3,32) =$

10.30, $p < 0.0001$. Figure 4 shows the mean percent decrease in number of corrections made between single-task and dual-task conditions. Planned comparisons using independent samples t-tests indicated: (a) discrete scrolling (DS) differed from continuous scrolling (CS), $t(16) = 2.44$, $p = 0.027$; (b) discrete scrolling (DS) differed from visual feedback (VF), $t(16) = 4.01$, $p = 0.001$; (c) continuous scrolling (CS) did not differ from visual feedback (VF), $t(16) = 0.30$, NS; (d) auditory feedback (AF) differed from continuous scrolling (CS), $t(16) = 2.37$, $p = 0.03$; and (e) visual feedback (VF) differed from auditory feedback (AF), $t(16) = 2.86$, $p = 0.01$. Overall, DS decreased performance least (26%), VF (42%) and CS (43%) had a greater impact, and AF (55%) led to the largest decrease in performance.

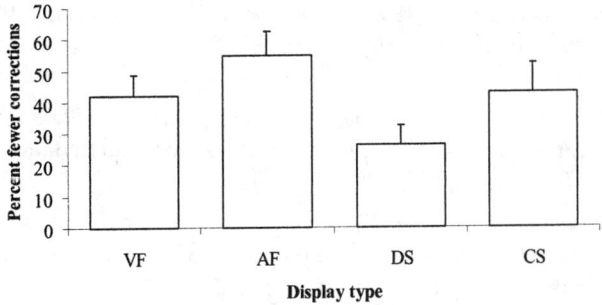

Figure 4 Percentage decrease in corrections between single- and dual-task conditions in Experiment 3.

As in the first two experiments, no difference was found among the displays in memorability, $F(3, 32) = 1.16$, NS. Mean scores were 43% for DS, 36% for CS, 42% for AF, 53% for VF.

Discussion
The effect of display type shows that display motion rather than update feedback dominates performance. Because the number of corrections decreased less (relative to baseline) for the discrete scrolling display than for continuous scrolling displays in which update feedback and motion were separate, we can conclude that continuous motion is more distracting than discrete motion. Moreover, because the auditory and visual feedback conditions led to worse editing performance than the discrete scrolling condition, we can conclude that update feedback is less important than display motion or distraction. Finally, because the headlines presented by all displays were equally memorable (as determined by the headline recall test), we can tentatively conclude that effective information delivery does not depend on distraction or feedback.

GENERAL DISCUSSION
This investigation was aimed at determining the properties of peripheral information displays that minimize distraction, maximize memorability, and cue effective multi-task scheduling strategies. Experiment 1 showed that a vertical, discrete scrolling headline display was as memorable as a horizontal, continuous scrolling display, but that the discrete display had less negative impact on a concurrent text-editing task. Experiment 2 showed that a horizontal, discrete scrolling display was as memorable as a vertical, continuous scrolling display, but that the discrete display had less impact on concurrent text editing. Experiment 3 showed that a discrete display was as memorable as a continuous display and less intrusive than a continuous display, regardless of external visual or auditory signals that new information is displayed. Taken together, these results demonstrate that motion of a scrolling display dominates performance on concurrent tasks. That is, constant motion of continuous scrolling displays distracts users from effectively performing other jobs at the same time. The start and stop motion of discrete scrolling displays distracts users far less under dual-task demands. What is more, the start and stop motion of the discrete display provides effective update feedback, enabling users to schedule their glances at the display.

Because continuous motion distracts computer users regardless of update feedback, it is likely that motion in the periphery captures users' attention too easily, leading to constant and costly task switching. Because discrete motion is not continual and the start-and-stop motion naturally provides update feedback, it does not constantly draw attention from other concurrent tasks and at the same time facilitates scheduling glances at the display. It seems as though discrete displays effectively balance motion and feedback in displaying peripheral information.

Although the serial presentation (SP) of Experiment 1 and the fading serial presentation (FSP) of Experiment 2 provide update feedback similar to that of discrete scrolling displays without adding motion, they led to worse editing performance than the discrete scrolling displays. One possible reason is that updates of the SP display were too brief to be noticed. Alternatively, the discrete updates of SP may have been too disruptive, creating higher task-switching costs. Updates of the FSP display, by contrast, might have been too subtle, fading in so gradually that they did not provide appropriate update feedback. In any event, our data cannot decide among these alternatives.

CONCLUSION
As user interfaces attempt to present more and more information, computer screens often become cluttered and distracting, effectively eliminating a user's ability to accomplish a single task. As stated, the key to peripheral informing is to maximize the information presented while minimizing the presentation's impact on ongoing activities. In this paper, we have considered the peripheral display of information by investigating mental workload tradeoffs among several scrolling ticker displays. Our finding that distraction dominates performance in this context marks only the first step toward a full understanding peripheral information displays. Additional studies are needed to understand, for instance, the nature of interruptions and the details of how users actually shift attention to peripheral

information. Nevertheless, based on our results, we can suggest some guidelines for designing peripheral displays.

1. Motion or animation should be kept at a minimum.
2. Motion or animation (when used) should not be continuous.
3. Discrete animation seems ideal for update feedback.
4. Visual feedback is better than auditory (for visual information).
5. Scrolling direction does not seem to matter.

Overall, we recommend the discrete scrolling display as the best way to convey memorable yet peripheral information.

ACKNOWLEDGMENTS

Thanks to Teenie Matlock, Michael Wenger, and Shumin Zhai for helpful comments on a draft of this paper.

REFERENCES

1. Adorni, G., & Poggi, A. (1996). Route guidance as a just-in-time multiagent task. *Applied Artificial Intelligence, 10*, 95-120.
2. Campbell, C. S. & Maglio, P. P. (1999). Facilitating navigation in information spaces: Road signs on the World Wide Web. *International Journal of Human-Computer Studies, 50*, 309--327.
3. Cypher, A. (1986). The structure of users' activities. In D. A. Norman & S. W. Draper (Eds.), *User centered system design* (pp. 243-263). Hillsdale, NJ: LEA.
4. Damos, D. L. & Wickens, C. D. (1980). The identification and transfer of timesharing skills. *Acta Psychologica, 46*, 15-39.
5. Duchnicky, R. L. & Kolers, P. A. (1983). Readability of text scrolled on visual display terminals as a function of window size. *Human Factors, 25*, 683-692.
6. Flesch, R. (1948). A new readability yardstick. *Journal of Applied Psychology, 32*, 221-233.
7. Gopher, D. & North, R. A. (1977). Manipulating the conditions of training in time-sharing performance. *Human Factors, 19*, 583-594.
8. Granaas, M. M., McKay, T. D., Laham, R. D., Hurt, L. D. & Juola, J. F. (1984). Reading moving text on a CRT screen. *Human Factors, 26*, 97-104.
9. Ishii, H. & Ullmer, B. (1997). Tangible bits: Towards seamless interfaces between people, bits, and atoms. In *Proceedings of CHI '97* (pp. 234-241), New York: ACM Press.
10. Juola, J. F., Ward, N. J., & McNamara, T. (1982). Visual search and reading of rapid serial presentations of letter strings, words, and text. *Journal of Experimental Psychology: General, 111*, 208-227.
11. Kanarick, A. F. & Petersen, R. C. (1969). Effects of value on the monitoring of multi-channel displays. *Human Factors, 11*, 313-320.
12. Kang, T. J. & Muter, P. (1989). Reading dynamically displayed text. *Behaviour & Information Technology, 8*, 33-42.
13. Kramer, A. F., Larish, J. F., & Strayer, D. L. (1995). Training for attentional control in dual-task settings: A comparison of young and old adults. *Journal of Experimental Psychology: Applied, 1*, 50-76
14. Lee, E. S., Okada, R., & Jeon, I. G. (1997). Agent-based support for personalized information with Web search engines. In *Proceedings of HCI International 97* (pp. 783-786), Elsevier.
15. Lieberman, H. (1997). Autonomous interface agents. In *Proceedings of CHI 97* (pp. 67-74), New York: ACM Press.
16. Maglio, P. P. (1995). *The computational basis of interactive skill*. Unpublished doctoral dissertation, University of California, San Diego.
17. Maglio, P. P. & Barrett, R. (1997). How to build modeling agents to support web searchers. In *Proceedings of User Modeling 97* (pp. 5-16), New York: Springer Wein.
18. Moray, N. (1986). Monitoring behavior and supervisory control. In K. R. Boff, L. Kaufman, & J.P. Thomas (Eds.), *Handbook of Perception and Human Performance: v. II*, New York: Wiley.
19. Norman, D. A. (1986). Cognitive engineering. In D. A. Norman & S. W. Draper (Eds.), *User centered system design*, Hillsdale NJ: LEA.
20. Owen, D. (1986). Answers first, then questions. In D. A. Norman & S. W. Draper (Eds.), *User centered system design* (pp. 361-375). Hillsdale, NJ: LEA.
21. Potter, M. C., Kroll, J. F. & Harris, C. (1980). Comprehension and memory in rapid sequential reading. In R. Nickerson (Ed.), *Attention and Performance VIII*. Hillsdale, NJ: Erlbaum.
22. Sekey, A. & Tietz, J. (1982). Text display by 'saccadic scrolling'. *Visible Language, 16*, 62-76.
23. Wickens, C. D. (1992). *Engineering psychology and human performance* (2nd Edition). New York, NY: Harper Collins.
24. Wickens, C. D. & Seidler, K. S. (1997). Information access in a dual-task context: Testing a model of optimal strategy selection. *Journal of Experimental Psychology: Applied, 3*, 196-215.

The Impact of Fluid Documents on Reading and Browsing: An Observational Study

Polle T. Zellweger, Susan Harkness Regli[*], Jock D. Mackinlay, Bay-Wei Chang

Xerox PARC, 3333 Coyote Hill Road, Palo Alto, CA 94304
{zellweger, mackinlay, bchang}@parc.xerox.com, shr1@cmu.edu

ABSTRACT

Fluid Documents incorporate additional information into a page by adjusting typography using interactive animation. One application is to support hypertext browsing by providing glosses for link anchors. This paper describes an observational study of the impact of Fluid Documents on reading and browsing. The study involved six conditions that differ along several dimensions, including the degree of typographic adjustment and the distance glosses are placed from anchors. Six subjects read and answered questions about two hypertext corpora while being monitored by an eyetracker. The eyetracking data revealed no substantial differences in eye behavior between conditions. Gloss placement was significant: subjects required less time to use nearby glosses. Finally, the reaction to the conditions was highly varied, with several conditions receiving both a best and worst rating on the subjective questionnaires. These results suggest implications for the design of dynamic reading environments.

Keywords

Fluid user interfaces, fluid documents, focus+context, hypertext navigation, on-line reading, eye tracking, studies of dynamic user interfaces

INTRODUCTION

Typographic conventions such as footnotes and sidebars are often used to keep the main body of a document clear and succinct while still allowing the reader to access additional details. In an electronic document, hypertext can be used to provide more details than can fit typographically on a page. However, locating details elsewhere makes them more difficult to compare with the source document. Furthermore, hypertext requires users to navigate while reading, which is known to be cognitively difficult [5].

The Fluid Documents project has been exploring how electronic documents can provide more details on a page by fluidly adjusting typography on demand [4, 8, 20, 21]. We have developed a range of techniques that vary by how radically they adjust document typography. One focus of these techniques is to support hypertext browsing by providing additional information (termed a *gloss*) at a link anchor, in order to help readers in choosing among links and understanding the structure of a hypertext [20].

While placing glosses close to their anchors seems most beneficial, we were concerned that the necessary radical adjustments of typography might be disruptive. In addition, we wanted to find out how different techniques affect how glosses are used, and how readers react to the availability of glosses and the techniques used to display them.

To shed light on these questions, we carried out an observational study exploring the impact of Fluid Documents on reading and hypertext browsing. We compared three fluid techniques with two established techniques for incorporating details into a document, as well as a conventional hypertext condition that displayed no glosses. Subject comments conveyed a wide range of reactions to the fluid techniques: some subjects loved it for reading and browsing; some hated it. However, a visualization of eyetracking data collected during the experiment showed no substantial differences in eye behavior between conditions, indicating that the fluid techniques did not create visual disruptions. Furthermore, we compared the average length of gloss events. In conditions with glosses near the source anchor, subjects kept glosses open on average for significantly shorter intervals than in conditions with distant glosses. Finally, we observed some subjects actively using the novel capabilities of Fluid Documents to freeze and thaw glosses while browsing.

RELATED WORK

Providing information about links that can hint at the content of the destination has been proposed in hypertext [11] and in the general field of information navigation [7]. The Hyperties system [12] used gloss-like information that appeared at the bottom of the page to help readers choose which links to follow. More recently, World Wide Web browsers have begun to provide facilities that can be used to display glosses: Microsoft Internet Explorer displays popup ToolTips to show a link's "title" attribute [15]. In addition, researchers have modified WWW browsers to show information about links and their destinations [10, 19].

In addition to providing information about link destinations, supplementary material can eliminate the need to follow a

---

[*] Current address: Baker Hall 259, Carnegie Mellon University, 5000 Forbes Ave, Pittsburgh, PA 15213

link altogether and give readers the opportunity to read additional information in its related context. Classic and recent research in reading comprehension processes has upheld the result that the context in which a passage is encountered can have a significant effect on a reader's comprehension (e.g., [3, 9]). A potential hypertext strategy, then, would be to provide a way for additional information to be brought into view in proximity to the primary material so that the elaboration can be read in context.

Fluid Documents bring more detail into view while preserving the surrounding information. This shares characteristics with fisheye views [6], zooming user interfaces [2], and other focus+context interfaces [16].

FLUID LINKS

We have implemented a hypertext browser, the Fluid Links browser, to experiment with techniques for fluidly displaying glosses in textual documents. The Fluid Links browser displays documents in pages of text, surrounded by margins on all four sides. In this study, each hypertext document fit fully on the page, so no scrolling was required.

Underlined text indicates the presence of a gloss; when a reader places the mouse over that text, the gloss smoothly and quickly expands nearby while the surrounding information alters its typography to create the needed visual space. The technique uses perceptually based animation to provide a lightweight and natural feeling to readers as a gloss appears and disappears.

Fluid techniques

The Fluid Links browser can be configured to display glosses in several ways. We briefly describe the techniques here; more detail can be found in [4, 20, 21].

The *fluid interline* technique displays the gloss directly below the anchor (see Figure 1a on separate color plate). When the mouse moves over the anchor, the gloss grows from an invisible, tiny size to its full size, while intensifying into its final color. At the same time, the primary text moves apart to make space below the anchor into which the gloss can expand, using the top and bottom margin space as well as compressing the interline spacing throughout the page.

The *fluid margin* technique displays the gloss in the margin (see Figure 1b). When the mouse moves over the anchor, a line expands from the anchor to the margin, and the gloss grows into the margin.

The *fluid overlay* technique grows the gloss directly below the anchor, as in the interline technique (see Figure 1c). However, the primary text does not move. Instead, any overlapped lines of text fade to a light gray color, allowing the overlaying gloss to be readable.

The fluid techniques display gloss text in red so as to more clearly set them off from the primary text. Furthermore, readers can freeze glosses in place to leave them open on the page while other glosses are examined, and then remove or "thaw" the glosses when the information is no longer useful.

Conventional techniques

The Fluid Document techniques introduce many dynamic elements: text moves around on the page, grows and shrinks, changes in color, and can overlap. While these animated effects are often seen in television commercials, they have not been common in "serious" reading interfaces.

To compare this activity with established techniques for displaying additional text, we augmented the Fluid Links browser to display glosses in two other ways. The first of these is familiar from paper documents, and was employed in the Hyperties system: displaying the gloss material in a *footnote* area at the bottom of the page (see Figure 1e).

The second technique shows the gloss material in a *popup* window that appears at the cursor, as in Microsoft Internet Explorer's implementation of link titles (see Figure 1d).

Because footnotes and popups were intended to mirror existing implementations, they do not support freezing or animated growing and shrinking—they simply appear and disappear immediately as the cursor dwells on the anchor and leaves the anchor. Furthermore, their text is shown in black.

The final version of the browser does not show glosses at all, mimicking a conventional hypertext or WWW browser.

Dimensions of variation

The various gloss techniques that the Fluid Links browser can exhibit allowed us to examine three major dimensions of variation.

Placement of gloss. Fluid Documents attempt to place glosses close to the anchor to minimize the reader's eye movement and allow easy comparison to the primary text. Fluid interline, fluid overlay, and (non-fluid) popups all display the glosses directly below the anchor. To avoid disrupting the primary text, the fluid margin technique places glosses in the side margins, still relatively close to the anchor. However, a typical margin is narrow and requires a significantly different aspect ratio for the gloss (tall and narrow vs. short and wide). Finally, footnote placement in the bottom margin has the advantages of a predictable location, normal aspect ratio, and possible familiarity from footnotes on printed pages.

Modifications to the primary material. Placing glosses near the anchor usually requires some kind of modification to the primary material to make the gloss material readable. Fluid interline moves the primary text apart, and fluid overlay fades part of the primary text and covers it with the gloss. The popup technique completely obscures part of the primary text, but this disruption is likely to be familiar to most users of modern computer interfaces. The fluid margin technique interferes minimally, by drawing a line in between the text lines; otherwise it simply uses existing white space to display the gloss. Finally, the footnote technique does not interfere with the primary text at all, with glosses appearing in a dedicated region of the page.

Transitions. We intend for the animation in the three fluid techniques to guide the reader's eye to a gloss as it is displayed, and then back again to the primary text afterwards. We believe this smooth movement mitigates the concomitant changes in the primary text's typography. The transitions for the non-fluid techniques were not animated.

These dimensions allowed us to explore three questions:

1. Do large fluid-style changes in the page (text moving apart or fading in color, lines shooting to the margin) impact eye behavior?
2. Do different techniques affect how readers use glosses?
3. Given the technique variations (in the use of animation, in ways they alter the primary text, and in where the gloss is displayed), do readers have preferences?

To answer these questions, we created an experimental situation in which a large amount of usage data could be collected about a set of readers, using a within-subjects design in which each subject saw and interacted with each of the different conditions.

METHODS

Task
Subjects were asked to browse through hypertext documents in order to answer a series of 18 questions.

Subjects
Eight study subjects were recruited from a pool of research interns at Xerox PARC, ranging in age from 17 to 40; data from two subjects had to be discarded due to equipment problems (see Table 1). Subjects regularly read documents and browsed the web on computer screens. No subject had any prior knowledge of the Fluid Documents interface.

gender	5 male, 1 female
education	2 recent high school grad, 1 undergrad, 3 grad student
field	3 computer science, 1 chemistry, 1 architecture & product design, 1 undecided

Table 1. Selected characteristics of the final six subjects.

Materials
Two sets of linked documents and questions were constructed from materials used to conduct previous experiments: the Spoofer Corpus described how to identify someone sending forged email [14], and the Wine Corpus gave information about wine [13]. Each corpus consisted of 30-40 pages of interlinked material. No table of contents was provided. To make substantially the same corpus available to subjects even in the No Gloss condition, glosses were constructed in one of two forms. The text of *complete glosses* was repeated on the destination page for all conditions; these glosses functioned as annotations to the primary text and were identified with a distinctive end marker (||). *Leading glosses*, available only in gloss conditions, gave a short summary of the information that would be found on the destination page, along with a double arrow (>>) end marker.

To encourage subjects to read and engage fully with the material, answers to questions were located sometimes in the primary text and sometimes in the text of a complete gloss.

Conditions
Subjects interacted with the documents in six conditions: Fluid Interline, Fluid Margin, Fluid Overlay, Popup, Footnote, and No Gloss. These conditions employ the corresponding gloss techniques described earlier. The order of conditions among the six subjects was balanced such that subjects saw the conditions in different ordinal positions and no two adjacent condition pairs were repeated.

In gloss conditions, dwelling the mouse cursor over an anchor for 200 msec triggered the display of the gloss. (We chose a relatively short dwell as a starting point to investigate easy gloss access in a reading environment.) Once triggered, the animation to a readable gloss in fluid conditions took 300 msec (including all text movement, lines to margin, etc.).

Measures
Application logs. Logs of application activity recorded all user and application events to the nearest millisecond.

Eye logs. We used the ISCAN RK-426PC pupil/corneal reflection tracking system to track the eye movement of subjects as they read and searched for information on screen. Eye logs written by the tracking system captured the point of regard of the eye on the display at 60 Hz. Similar systems have been used in previous user interface studies [1].

Think-aloud protocol. Subjects were asked to speak aloud as they navigated the documents; specifically, each was asked to talk about "what you are looking for as you read a particular document, what links you are planning to explore and why, and what information you are trying to find if you explore a particular link." In a conventional think-aloud protocol for reading research, subjects would have been asked to speak continuously and read everything out loud; we varied the method by allowing silent reading, so as to minimize the disruption of reading patterns for the eyetracking data.

Video of screen, eye movement, and think-aloud protocol. The video signal to the subject's display was captured and composed with a video signal from the eyetracker that showed the real-time point of regard of the eye by a small white annular marker. The audio of the think-aloud protocol was included in this recording as well.

Subjective response. Subjects completed a questionnaire collecting subjective responses about effect of gloss placement, effect of gloss movement, usefulness of the glosses, and ratings for each condition. They could interact with each interface as they filled out the questionnaire. In addition, we conducted a semi-structured interview with each subject after the questionnaire was complete.

Procedure

The eyetracking equipment was calibrated at the start of the experiment session and re-calibrated two or three times throughout the experiment to adjust for expected drift.

The subject began by using the experimental browser to read a page of text that contained no link anchors, for baseline measurement of reading. The following steps were then repeated for each of the six conditions: First, the subject used the browser to find the answer to one question about the Spoofer Corpus. This pre-instruction phase of each condition was designed to capture the subject's untrained reaction to a condition's display changes. Next, subjects were instructed in the operation of the condition, including (if permitted) how to freeze and thaw glosses, and that multiple glosses could be frozen in view at once. Subjects were encouraged to try these actions during the instruction period. Finally, the subject used the browser to find the answers to two questions about the Wine Corpus. If the subject took longer than five minutes to answer any question, hints were provided at increasing frequency until the subject found the location of the answer. This was intended to ensure that no subject fell significantly behind others in their experience with either corpus.

After all conditions had been encountered, the subject answered a subjective questionnaire, and engaged in a semi-structured interview about subjective responses to the variations among interface behaviors. The entire experimental session lasted 2-3 hours.

ANALYSIS

Gloss event table

We were most interested in the instances that subjects caused a gloss to be displayed, which we call *gloss events* (see Figure 2). A gloss event begins the moment when the gloss display is triggered, after the mouse has dwelled sufficiently on the anchor. It extends through the time the gross grows to full size (in the fluid conditions) and until either (a) the gloss begins to shrink back down or (b) a new page is visited.

Application logs were processed to construct a table of gloss events. 889 gloss events were recorded for the six subjects in all conditions.

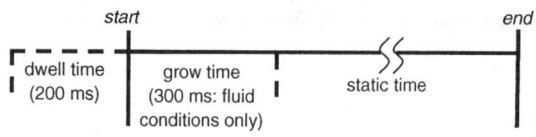

Figure 2. Gloss event.

Log simulation and visualization

Application and eye logs were temporally and spatially synchronized using a custom simulator matched with the videotape record. We also instrumented the simulator to generate a visualization, which we call a *gaze region*, to show the impact of a gloss event on the point of regard of the eye. In particular, we were concerned that fluid animations that affect large portions of the screen (e.g., moving all of the text apart in the fluid interline condition) might draw the subject's eyes to screen locations distant from the gloss or its related primary text. Even a momentary glance could indicate that the animation was distracting the subject from the reading task. Because the gaze region visualization fills in the entire area that the eye has observed, even brief glances to distant parts of the screen are quite obvious.

The gaze region visualization uses a black line to trace the path of the point of regard during a gloss event. (Because we were interested in possibly rapid behavior of the eye, point-of-regard data was used rather than an analysis of eye fixation.) The temporal aspect of the path is described by dividing the first eight seconds of the gloss event into one-second intervals. Each interval is summarized by a convex hull (i.e., a rubber-banded line enclosing all points during that interval) that is overlapped and shaded from red to blue to show the temporal behavior of the eye in a static visualization (see Figure 3 on color plate). Most gloss events last less than eight seconds (see Table 2); later eye movement is shown only by the black line. Gaze region visualizations were generated for all 889 gloss events.

Pruned gloss event table

Initial examination of the raw gloss events revealed that there were many very short events (80 under 300 msec, 141 under 500 msec, 253 under 1 sec). Analysis of the raw application logs and the videotapes yielded the explanation that these short events were a result of the lightweight mouseover UI, in which the system interprets mouse dwell as an implicit request to view a gloss. One way such an "inadvertent" gloss event can occur is if the subject clicks on an anchor to follow the link, but happens to dwell on the anchor long enough to begin the gloss event. Subjects who track their reading location with the mouse can also inadvertently trigger short gloss events that they immediately cancel by moving the mouse away.

Another kind of inadvertent gloss event can occur if the subject is attending to one region of the screen and their mouse happens to dwell on an anchor in a different region of the screen. In this case, it is possible that the subject will not notice the gloss event.

Because we wanted to focus on how subjects used glosses they intentionally opened and (to some extent) read, we pruned the gloss event table of inadvertent gloss events. We carried this out in two steps. First, to eliminate "unnoticed" inadvertent gloss events, two experimenters independently coded all 889 gaze region visualizations to determine whether the eye went into the gloss at all. 156 gloss events in which the eye did not enter the gloss were pruned. (121 of these were less than 1 second long; the maximum of the remaining 35 was a Footnote gloss event lasting 5.318 seconds). Two such gloss events were retained because those glosses had been intentionally frozen and were thus not inadvertent. Second, to eliminate "short" inadvertent gloss events, we calculated a figure for each gloss event based on subject reading speed, gloss word count, gloss event static time, and an assumption that subjects need only skim a gloss to derive value from it. A cutoff time of 1 second could conservatively account for these figures, so all gloss events whose length was shorter than 1 second were pruned, removing 133 additional gloss events.

One final adjustment was made to the gloss event data. Glosses in animated (i.e., fluid) conditions were not available for attention to anything other than their existence (i.e., no reading or skimming activity) for the first 300 msec after the gloss event, because they were moving and growing. We thus subtracted 300 msec from those raw recorded durations to yield a *static time*, which could then be compared on an equal footing with the unaltered durations in the non-animated conditions. Pruning adjusted gloss events under 1 second in length removed 26 more events, yielding a final count of 574 gloss events.

Transcription

The audio recording of the think-aloud protocols and the subjective discussions for three out of six subjects were transcribed in detail with the goal of understanding self-reported variations in preference and usage. The resulting transcripts were mined for comments that clustered around common issues, such as distraction, disturbing text (broken into obscuring and moving text), reactions to animations, use of screen space and spatial arranging, locality of gloss to anchor text, and the amount of visible information on a page. Twenty-eight clusters arose out of the data; some quotations were relevant to more than one cluster. Related statements from different subjects were then analyzed for areas of strong agreement or disagreement. On the basis of these relations we were able to describe several dimensions along which user preference varied.

OBSERVATIONS

Impact on eye behavior

After comparing the 889 gloss events with videotape, simulation, and visualization, we did not discern any substantial differences in eye behavior between conditions except those due to gloss location and shape. Furthermore, pre-instruction gloss events, which were the first ones seen by a subject, looked just like post-instruction events. Figure 3b (on color plate) is an example of an interline gloss where the eye moves to the gloss during the grow even though the rest of the text on the page is also moving.

Gloss usage

Analysis of gloss event length

To test the hypothesis that the condition would have a significant effect on the gloss event length, repeated measures ANOVA tests were conducted for both the full gloss event table and the pruned gloss event table (analyses were performed on a log_{10} transformation of the data to satisfy the assumption of a normal distribution). In order to account for the unequal number of measures in each condition, the ANOVA was completed using 30 mean calculations as the dependent variable of interest:[1] the mean was computed for each subject in each condition, as an estimate of how long a subject keeps a gloss open in a given condition. To determine whether a condition had a significant effect on the number of gloss events, a similar repeated measures ANOVA was conducted for the count of gloss events. This analysis revealed no main effects for count (total sample: $F(4,20)=1.435$, $p=.2590$; pruned sample: $F(4,20)=1.510$, $p=.2371$).

Main effects for condition were found in the full gloss event table, $F(4,20)=3.469$, $p=.0262$. The observed effect was even stronger in the pruned gloss event table, $F(4,20)=11.374$, $p=.0001$. The Fluid Margin and Footnote conditions share the quality that they are more distant from the anchor text than Fluid Overlay, Popup, and Fluid Interline, which are local to the anchor text. A contrast analysis [17] of the pruned gloss event table revealed a significant difference between the distant Fluid Margin and Footnote conditions versus the local Fluid Overlay, Popup Gloss, and Fluid Interline conditions, $F(1,20)=42.015$, $p=.0001$.

	Total Sample	Pruned Sample
Fluid Margin	.426 (.107)	.565 (.126)
Footnote	.214 (.115)	.496 (.072)
Fluid Overlay	.297 (.218)	.324 (.136)
Popup	.195 (.184)	.246 (.158)
Fluid Interline	.251 (.074)	.239 (.075)

Table 2. Means (S.D.s) for length of gloss opens (log_{10} seconds) for total and pruned event tables (N=30).

These results suggest that subjects were able to read local glosses more quickly than distant ones. However, it should be noted that some subjects reported that text was harder to read in the Fluid Margin condition due to its narrow formatting (see Figure 3c). Moreover, subjects may be motivated to dismiss local glosses more quickly to remove them from their region of focus, while distant glosses may require less urgency since they are not in the way.

Attention to glosses

Analysis of the gaze region visualizations revealed that when a gloss opened on a page, the reader did not always look at the content of the gloss; footnote glosses in particular tended not to be noticed. Footnote gloss events were both most numerous and longest among the events that were pruned. Figure 3a illustrates an instance in which the eye never moved anywhere near the footnote gloss text during the gloss event, thus indicating that the gloss was apparently not noticed at all.

Subjective data supported the observation that footnotes tended not to be seen: responses to the questionnaire item "I

---

[1] We would like to acknowledge Dr. Joel Greenhouse for valuable assistance in determining how to analyze the gloss event tables.

found the placement of glosses (bottom of the page, side margin, between lines, etc.) in this condition to be hard to find or easy to find" indicated that footnote glosses were hardest to find.

In subjective discussion transcripts, subjects also reported that footnote glosses required extensive eye movement: "the one irritant is it requires a fairly drastic eye movement"; "you really have to jump your eyes down. I don't care so much about that, but there's something nice about having the information appear near the link"; and "I tend to be very focused at where I am, and this thing happening at the peripheral vision was not very fun."

Value of glosses

Figure 4 (on color plate) shows the aggregated time that each subject spent answering all three questions in each question set in a given condition, with the total time subdivided into time when different numbers of glosses were open. No subject had more than five glosses open at one time. To facilitate comparison of conditions across question sets, conditions are shown in the same order in each group, while the order of subjects varies.

Glosses were strongly but not universally valued during hypertext browsing. Across all subjects, questions, and gloss conditions, subjects had at least one gloss open for 26.6% of the time; per-subject gloss usage ranged from a minimum of 22.4% to a maximum of 32.1%. Since all of the information to answer the questions could be accessed via hypertext jumps rather than by viewing glosses, this number suggests that subjects found them useful. Responses to the subjective questionnaires support this observation: subjects reported that having information in the glosses was helpful, clear, and helped them follow useful links. One subject stated that "after using the others, it (the No Gloss condition)'s like surfing blind."

User preference

User preference varied widely along several dimensions. The only two areas of strong agreement supported the original dilemma which Fluid Documents were designed to address; namely, that it is desirable to keep the gloss local to the anchor text and that the gloss should not disturb the primary text. For some readers, however, "disturbing" the text in a negative way meant moving the primary text in any way; for others it meant occluding the primary text so they couldn't read it. This reflected one of the most extreme observed variations: between readers who had a strong need for the primary text to remain completely static (and thus occluding the text would be preferable to moving it) and readers who genuinely appreciated the ability to vary the spatial arrangement of the primary text while seeing as much information as possible (thus moving the text would be preferable to occluding it).

Reactions to the animation of text in the Fluid interfaces varied from extreme dislike to extreme appreciation. In the words of one subject, the practice of animating the text was "bad, because again I can't read things while it's animated so it's a waste of time and a distraction" and "irritating"; for another, the animation was "aesthetic," "useful," and (discussing the interline condition) "Really really cool. … it's very intuitive for me. The whole thing of the text parting and new thing coming on is a great metaphor for deeper information." Preference for animation was related overall to the desire for a static versus a variable page layout, the time it takes for the animation to occur, the control it exerts over their focus, and the aesthetic of the text shifting to accommodate more visible information.

Preference for speed of the gloss appearance was not uniform, even in terms of what one reader might have desired. For one reader popup was too fast: "I tend to move my mouse around a lot. And even though this has a slight, you know, pause before it brings it on, I find the popup message far too demanding of attention." For another the fast speed of the popup appearance was good, but might be affected by task: "when I'm doing something like this [the experimental task], it's nice for it to be fast because I'm using it all the time but if I'm actually reading and mousing I might want something . . . that's a little slower, so that if I just go over a thing it doesn't activate."

Preference for degree of control varied: in general readers didn't want the interface to be too automated and demanding of attention, but some appreciated automated assistance like the guidance the animation can provide to the eye as it moves to find the gloss text. Figure 3c presents a good example of the amount of control the animated interface can exert on the eye movement of a reader. When the gloss appeared, the brightest red gaze region shows that the eye was immediately drawn to the gloss text. This effect may be perceived as either positive or negative, depending on whether the reader likes the interface to exert control over the focus of the eye.

Preference for speed and control seemed to be related partially to the way a reader uses the mouse while reading (e.g., following the reading location, doodling elsewhere, etc.), which differed noticeably among our subjects. Some of these behaviors can interact poorly with the implicit interpretation of mouse dwell as a command, especially when the primary text begins to shift inadvertently as a result.

Freezing and thawing behavior

The ability to freeze glosses in view is a novel functionality offered by the fluid interfaces. To give subjects a potential motivation for exercising this new behavior during the short experiment, the materials were designed to require subjects to compare information in two different glosses to answer four of the questions. All subjects were guaranteed to see at least one of these questions in a fluid condition, where freezing the glosses to view their contents simultaneously could be accomplished. Two of the six subjects nonetheless never froze glosses, preferring to view the individual glosses in succession, often repeatedly, to compare their contents. One other subject froze glosses only during active comparison (when the condition permitted) to answer a focused question.

One subject froze glosses even before any comparison questions were asked, but only opened more than one gloss at a time during active comparison. This subject began to open glosses in a frozen state, which could be done by right-clicking on the link anchor. This usage style removes much of the implicit grow/shrink behavior of the interfaces, which is based on mouse position and movement. It has the dual advantages of opening the gloss more quickly, without any dwell time, and tacking the gloss down so that it will not vanish prematurely even if the mouse is moved. It thus allows for more dynamic, less deliberate, mouse movement.

Two subjects adopted a freezing strategy that involved multiple open glosses without a comparison prompt. One subject developed a breadth-first approach to reading and hypertext browsing, choosing to simultaneously open and freeze many or all of the glosses on a page shortly after arrival, and then to examine those glosses and possibly their destinations as appropriate as a later step. Two other subjects offered variants of the breadth-first approach during the subjective discussion.

CONCLUSION

This study suggests that care must be taken when designing dynamic documents for reading and browsing. Subject preferences were complex and intense. Subjects were even sensitive to subtle distinctions such as whether a gloss occludes or disturbs a document (popup or overlay). We were gratified to see in the eyetracker data that dynamic adjustments to document typography did not cause the eye's point of regard to shift wildly across the page. We were also gratified to see that gloss placement, a central issue for the Fluid Documents project, had significant impact. However, we caution the reader not to interpret this result as biased for one group over the other. When glosses must be processed quickly, typographic adjustments should be used to put them close to their source anchor. Otherwise, placing glosses outside the primary text can reduce negative reactions to typographic adjustments. Interaction details about how glosses are invoked, frozen, and removed are also important in the design of dynamic documents. Lightweight interaction, in particular, may lead to inadvertent invocations of glosses.

Clearly, the increasing use of computer-based documents is changing how we read. Our subjects ignored footnotes, even though they are the conventional way to put details in paper documents. Perhaps, as one subject suggested, Web browsers are training us to ignore text popping up at the bottom of the screen. We fully expect that changes in reading styles will continue. For example, frozen glosses were used by our subjects to compare glosses and to search the hypertext in a breadth-first order. We have high expectations that future authors and readers will want to use dynamic documents as long as issues such as the ones raised by this observational study are addressed.

ACKNOWLEDGMENTS

Cliff McKnight and Andrew Dillon graciously supplied materials from previous hypertext studies [13, 18] for use in constructing the study materials. We'd like to thank David Fleet and his team for last minute video magic to recover data. The ISCAN eyetracker was made available through the Office of Naval Research contract N00014-96-C-0097 with Peter Pirolli and Stuart Card.

REFERENCES

1. A. Aaltonen, A. Hyrskykari, K. Raiha. 101 spots, or how do users read menus? *Proc CHI'98*, 132-139.
2. B. Bederson, J. Hollan. Pad++: A zooming graphical interface for exploring alternate interface physics. Proc. *UIST'94*, 17-26.
3. J. Bransford, M. Johnson. Considerations of some problems of comprehension. In W.G. Chase (ed.), *Visual information processing*. New York: Academic Press, 1973, 383-438.
4. B. Chang, J. Mackinlay, P. Zellweger, T. Igarashi. A negotiation architecture for fluid documents. *Proc. UIST'98*, 123-132.
5. J. Conklin. Hypertext: an introduction and survey. *IEEE Computer*, Sept 1987, 17-41.
6. G. Furnas. Generalized fisheye views. *Proc. CHI'86*, 16-23.
7. G. Furnas. Effective view navigation. *Proc. CHI'97*, 367-374.
8. T. Igarashi, J. Mackinlay, B. Chang, P. Zellweger. Fluid visualization of spreadsheet structures. *Proc. Visual Languages'98*.
9. W. Kintsch. *Comprehension: A paradigm for cognition*. Cambridge: Cambridge University Press, 1998.
10. T. Kopetzky, M. Mühlhäuser. Visual preview for link traversal on the World Wide Web. *Proc WWW8*, 1999.
11. G. Landow. Relationally encoded links and the rhetoric of hypertext. *Proc. Hypertext'87*, 331-338.
12. G. Marchionini, B. Shneiderman. Finding facts vs. browsing knowledge in hypertext systems. *IEEE Computer*, Jan 1988, 70-80.
13. C. McKnight, A. Dillon, J. Richardson. A comparison of linear and hypertext formats in information retrieval. In R. McAlesse, C. Green (eds.), *Hypertext: State of the art*. Oxford: Intellect Books, Ltd., 1990, 10-19.
14. C. Neuwirth, J. Morris, S. Regli, R. Chandhok, G. Wenger. Envisioning communication: Task-tailorable representations of communication in asynchronous work. *Proc. CSCW'98*, 265-274.
15. J. Nielsen. Jakob Nielsen's Alertbox for January 11, 1998. http://www.useit.com/alertbox/980111.html
16. G. Robertson, S. Card, J. Mackinlay. Information visualization using 3D interactive animation. *CACM*, 36 (4), 1993, 57-71.
17. R. Rosenthal, R.L. Rosnow. *Contrast Analysis: Focused comparisons in the analysis of variance*. Cambridge: Cambridge University Press, 1985.
18. J. Rouet, J. Levonen, A. Dillon, R. Spiro, eds. *Hypertext and Cognition*. Lawrence Erlbaum Associates, 1996.
19. D. Stanyer, R. Procter. Improving Web usability with the link lens. *Proc. WWW8*, 1999.
20. P. Zellweger, B. Chang, J. Mackinlay. Fluid links for informed and incremental link transitions. *Proc. Hypertext'98*, 50-57.
21. P. Zellweger, B. Chang, J. Mackinlay. Fluid links for informed and incremental hypertext browsing. *CHI'99 Extended Abstracts (Video Program, 6:38 minutes)*, 7-8.

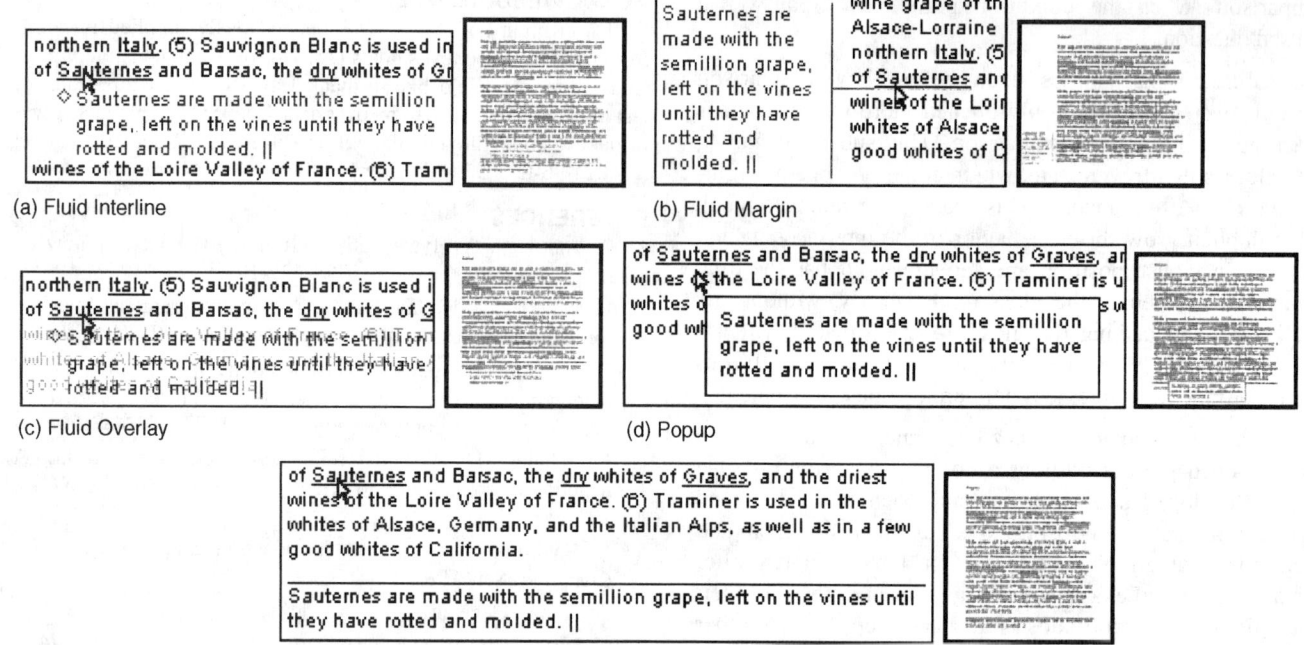

Figure 1. Fluid and conventional glosses, along with thumbnails of the pages of text from which they are excerpted. Font sizes have been greatly reduced for this figure; in the study, the primary text size was chosen to have a visual size of 12 points.

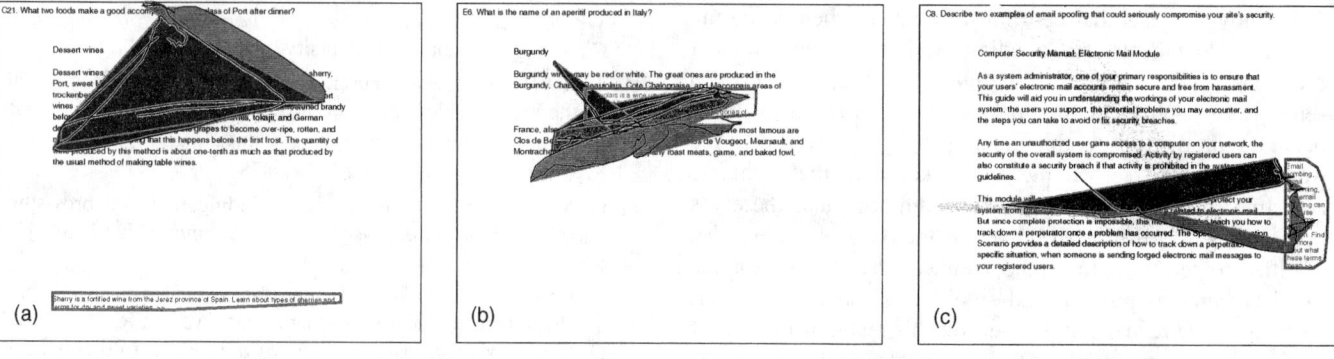

Figure 3. Gaze region visualizations of gloss events. The bold green line outlines the gloss, while the thin black line traces the eye's point of regard within the gaze region. The gaze region during the first second of each event is shown in bright red and may occlude later regions; gaze regions for successive seconds are shown in darker colors. The pale colors behind the text show gaze regions for three seconds before the event. (a) shows a Footnote event that the subject does not see. (b) shows a Fluid Interline event that the subject sees and reads. (c) shows a Fluid Margin event that the subject reads with seeming difficulty (note twists in black line).

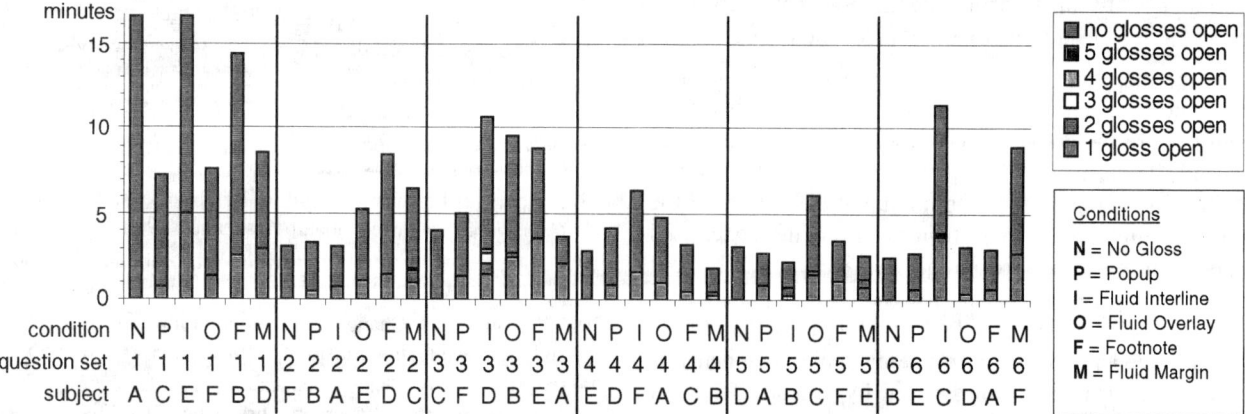

Figure 4. Gloss usage (pruned static time) per subject in each question set.

Effects of Contextual Navigation Aids on Browsing Diverse Web Systems

Joonah Park

Jinwoo Kim

Human Computer Interaction Lab
Department of Cognitive Science, Yonsei University
Seoul, 120-749, Korea
+82 2 361 2528

juna@ccs.yonsei.ac.kr

jinwoo@base.yonsei.ac.kr

ABSTRACT

In spite of the radical enhancement of web technologies, many users still continue to experience severe difficulties in navigating web systems. One way to reduce the navigation difficulties is to provide context information that explains the current situation of users in the web systems. In this study, we empirically examined the effects of two types of context information, namely, structural and temporal context. In the experiment, we evaluated the effectiveness of the contextual navigation aids in two different types of web systems: an electronic commerce system and a content dissemination system. In our experiment, subjects performed several browsing tasks and answered a set of post-questionnaires. The results of the experiment reveal that the two types of contextual navigation aids significantly improved the performance of browsing tasks regardless of different web systems. Moreover, context information changed the users' navigation patterns, and increased their subjective ease of navigation. This study concludes with implications for understanding the users' browsing patterns and for developing effective navigation systems.

Keywords
Context information, navigation, web systems, structure, browsing, hypertext.

INTRODUCTION

The emergence of Internet technologies has created new opportunities for changing our everyday life such as gathering information and purchasing products. However, as various kinds of information have been dispersed across numerous web sites, users continue to experience severe difficulties in navigating through the web sites. For example, users cannot identify where they are, return to previously visited locations, find the information that they believe to exist somewhere, and finally they cannot remember the key points they have learned [16]. These difficulties have been summarized as "lost in hyperspace" phenomenon, which have been classified as either disorientation or cognitive overhead [1].

Disorientation is defined as "the tendency to lose one's sense of location and direction in nonlinear document" [1]. There are four kinds of disorientation that are particularly relevant to web systems: not knowing where to go next, not knowing how one arrived at a particular node, not knowing where the information is, and finally not knowing how to get there [2]. On the other hand, cognitive overhead is defined as "the additional effort and concentration necessary to maintain several tasks or trails at one time" [1]. Users have to perform many tasks simultaneously such as remembering the tasks and sequences, searching the target items, browsing the general topics and related items, surfing the items of interest, comparing between items, moving from one item to others, and so on. Performing all these tasks simultaneously causes users to experience cognitive overhead, which leads them to get lost in hyperspace [8].

A plausible reason for these problems is the lack of context information in web systems. Context information is defined as the explanation of users' current situation in the web environment. The context information is important for effective navigation because each navigation process takes place in a particular information environment and is inextricably tied to the specificity of the environment [7]. If users do not have appropriate context information they could experience disorientation, because context information provides the temporal and structural cues of locations. At the same time, users without context information tend to experience cognitive overload, because context information also provides valuable cues for users' actions and task flows. Users frequently experience cognitive overload and disorientation in web systems because the non-linearity of hypertext systems hinders users maintaining the context information. Hypertext systems have the ability to produce complex, richly interconnected, and cross-referenced bodies of various information [17]. However, at the same time, hypertext can also produce complicated and disorganized tangles of haphazardly connected web sites, which do not provide users with appropriate context information [17].

Although several previous studies have been conducted to design effective web systems [4] [5] [7], there has been little research on the relation between web systems and the context information in spite of the importance of context in hypertext systems. In this study, we propose a simple method to provide two types of contextual navigation aids by changing link properties of web systems. We evaluated their effectiveness empirically in two different web systems. The objective of this study is to investigate empirically the impact of context information on various browsing tasks, and thereby to provide a basis for the design of an effective navigation aid.

CONTEXT

Context refers to the information surrounding stimuli being recognized, categorized, or searched for. Therefore, the context can provide feedback that tells the user where he or she is in the process, what the past choices and outcomes were, and possibly how much further it is to the terminal node [11].

As explained in the introduction section, context information is more important on the web than other applications, because each navigation process on the web takes place in a particular information environment of temporal-spatial context. In particular, context information is extremely important in the navigation process where the user experiences disorientation and cognitive overload.

First, disorientation can occur when users fail to compute the temporal-spatial contextual coordinates of the current information. Therefore, disoriented users need context information to reestablish a sense of location. Context information that provides the structure of the document, size of the document, and the way they respond to given functions may help users reestablish their sense of location [3]. Especially when information is scattered across wide areas of web sites, context information helps users to orient their locations more effectively [14].

Second, cognitive overhead can occur when users fail to remember their actions and task flows. Therefore, users under cognitive overhead need context information to identify the location and to understand current task flows at any time and place in hyperspace. Context information is especially important for web systems, because delays inherent in web navigation add to the user's cognitive load since they have to keep contextual information in their minds while waiting [14].

In particular, two types of context are needed: spatial context and temporal context. First, spatial context is related to the question "Where can I go from here?" [17]. Several alternative ways can be used to answer this question. In this study, we provided the structural context to give a preview extended from the current position to all other positions that can be reached within two clicks. The structural context is expected to facilitate forward navigation, that is, making predictions of what will come next [12] [13]. By showing the structural context information, we expected users to reestablish orientation and identify the location of the target item more easily. Second, temporal context is related with the question "How did I get here?" [17]. In order to answer this question, we provided the temporal context that contains all the distinctive locations that users have visited until the current time. Therefore, the temporal context can facilitate backward navigation by facilitating the search for previously encountered information. By showing the temporal context, we expected users to remember less about their past trails and perceive the task flow at the current time and location more easily.

CONTEXTUAL NAVIGATION AIDS

Various contextual navigation aids have been developed in order to solve the severe problems in browsing web systems [17]. The navigation aids can be classified into two categories: one with the global context and the other with the local context. The first category of navigation aids includes overviews of the node structure of the entire web system. A comprehensive overview can help users understand the entire system structure. However, as the size of web systems is increasing exponentially, providing the entire structure of web systems in a single page becomes extremely difficult, if not impossible. Therefore, we need the second category of navigation aids that provide only the local context for web systems. In this study, we provided the local context by developing two types of add-on links in the navigation design. The navigation design is to develop a link structure through which users can move from one page to another with minimum effort. Links are generally categorized as either basic or add-on links. Basic links are a set of minimum connections that enable users to visit any node in a web system. Basic links are mostly pre-determined by its node structure. On the other hand, add-on links are additionally provided to improve the navigation behavior. These add-on links are particularly useful when the basic links do not facilitate effective navigation by themselves [4]. However, too many add-on links will not only confuse the user's logical understanding of information [10], but also lead the users to unexpected destinations in the site [2]. Moreover, if add-on links are not properly presented within the site, users may rather experience disorientation and cognitive overload [9].

In this study, we provided two types of contextual navigation aids (structural and temporal) by using two sets of add-on links on top of the basic links. Two basic links are provided in all conditions: Up-To-Parent links (UTP) and Down-To-Child links (DTC). These links allow subjects to move one level up or down along the hierarchy from the current position (CP). The basic and add-on links provided in this study are shown in Figure 1.

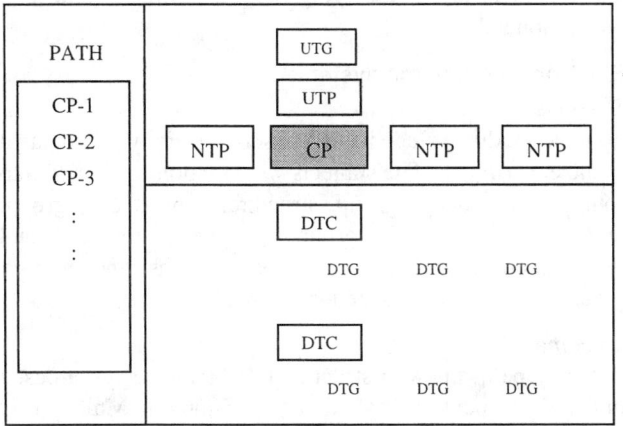

Figure 1. Display layout of context information

To provide the structural context, we developed three kinds of add-on links: Next-To-Peer links (NTP), UP-To-Grandparent links (UTG), Down-To-Grandchild links (DTG). We expected that all the links for the structural context would have two functions: to jump between distant locations and to preview the local context information. First, NTP (Next-To-Peer links) support horizontal navigation among peer levels in the hierarchical structure. When a subject is viewing one page, he or she can move directly to an adjacent page within the same level using NTP links. UTG (Up-To-Grandparent links) help subjects to locate at a certain level move two levels directly upward, while DTG (Down-To-Grandchild links) help them to move two levels directly downward. Therefore, these links enable users to jump to the distant locations directly without stopping at the intermediate level. Second, the set of all these links (NTP, UTG and DTG) provides the structural context by previewing all the nodes that are two levels upward, two levels downward, and horizontally at the same level with the current location. Consequently, the combination of the three additional links and two basic links can provide the information about the local structure around the current location.

At the same time, in order to provide the temporal context, we implemented the historical information mechanism, which we call the PATH links. The PATH links used in this study is a type of recency-based history saving the URL only in its latest position [15]. Recency is a simple yet effective temporal navigation aid, especially when duplicates are saved only in their latest position [6]. Similar to the links for the structural context, PATH links providing for the temporal context would have two functions: to jump to the distant nodes visited before and to review the local context information. First, the PATH links enable users to move to distant nodes that they have visited in the past without stopping at the intermediate locations. Second, the PATH links maintain the local context because

URLs that have been visited recently will stay at the top of the list [6].

Subjects with both context information actually saw links on the content pages as shown in Figure 1. In control group, subjects saw only two basic links (UTC, DTC). In temporal-context-only group, subjects saw the PATH links in addition to the two basic links. Finally, in structural-context-only group, subjects saw five links in total, UTG, DTG, and NTP links in addition to the two basic links. The actual user interface for the two different web systems is presented in Figure 2.

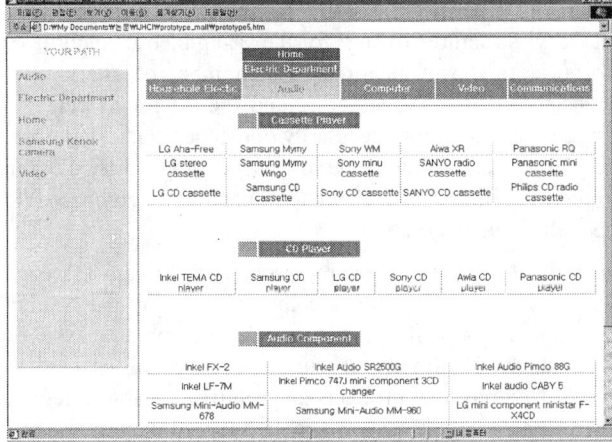

The e-commerce system

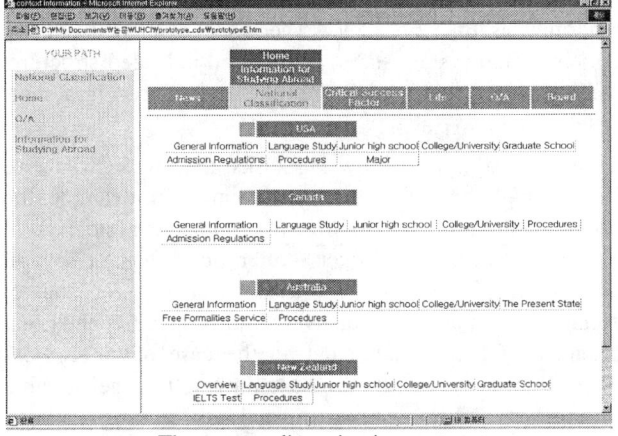

The content dissemination system

Figure 2. Actual user interface with context information for two different web systems

OVERVIEW OF STUDY

The aim of this study is to empirically examine the effects of two types of context information, structural and temporal context. We evaluated the effectiveness of contextual navigation aids in browsing diverse web systems.

In order to increase the external validity, we designed our experiment in the following three steps.

First, we tried to select two domains that are radically different from each other along a dimension that are supposed to influence the effectiveness of contextual navigation aids. We decided that the degree of structuredness is an important dimension because different degrees of structuredness can affect the navigation patterns [5]. Two extreme cases were selected along the degree of structuredness. One extreme is a well-defined structure and the other extreme is an ill-defined structure. The well-defined structure has a simple balanced category structure that has the same number of levels across all branches. The advantage of well-structured hierarchies is that users can expect the category of alternatives at each level and predict the local category of the system in any node. On the other hand, an ill-defined structure has a complex unbalanced category structure owing to the asymmetric nature of categories and subcategories. The number of alternatives varies throughout the structure and the length of paths to terminal nodes is not necessarily constant [11]. Therefore, the ill-defined structure gives users more difficulty in predicting subcategories and terminal nodes compared to the well-defined structure.

Second, we chose a good representative domain for each type of the two structures: e-commerce system and content dissemination system. In general, the e-commerce systems have a well-defined structure because of its well-categorized physical products and relatively stable category sizes. On the other hand, the content dissemination systems usually have an ill-defined structure because relevant information for sub-categories is expanding almost randomly as time goes by. Therefore, even if a content dissemination system starts with a good balanced structure, the entire structure tends to become unbalanced in a short time.

Third, we designed browsing tasks that were what we believe might be typical for each domain. For example, the user's goal of browsing in an e-commerce system is to purchase specific products after comparing several alternatives. Therefore, the tasks designed for the e-commerce domain are relatively more specific and goal oriented. On the other hand, in the case of the content dissemination system, the user's main goal is to get suitable information after gathering and comparing related information as much as possible. Therefore, the tasks designed for the content dissemination system are relatively more general and interest oriented.

EXPERIMENT

A set of two experiments was conducted to test the effects of context information at each domain. First, the effects of the context information on browsing were examined in the e-commerce system. Second, the effects of the context information were examined in the content dissemination system. We hypothesized that context information would be useful in both the e-commerce system and the content dissemination system.

In each experiment, four different versions of web systems were constructed. All were identical in terms of the content and graphical layout, differing only in terms of contextual navigation aids.

Electronic commerce system

Subjects
40 undergraduate students at Yonsei University participated in the experiment. The subjects were randomly divided into four groups: control group, structural context only group, temporal context only group, and both structural and temporal context group. Ten subjects were assigned to each group. Subjects were tested individually.

Task materials
A prototype of the well-structured e-commerce system used in the experiment was called "Cyber Shopping Mall." This prototype was a hypertext document with a well-balanced hierarchical structure similar to common Internet shopping malls. With the owner's consent, this prototype was constructed based on a cyber shopping mall that had actually been conducting business for several months. This mall consists of seven departments, 18 stores, 66 corners, and 674 products in total.

Experimental procedure
The experimental sessions were divided into three sections. First, subjects were given instructions about the general nature of the experiment and were required to fill out a pre-questionnaire. The pre-questionnaire measured the ability of subjects to navigate the WWW. The questionnaire asked the subjects if they had any prior experience using computers as well as the WWW. After the pre-questionnaire, thirteen browsing tasks were provided in random order. We set thirteen tasks similar to various actual situations in the e-commerce system. Subjects were asked to perform the given browsing tasks. The navigation behaviors of each subject were recorded in system log files. After performing each task, subjects were asked to answer a post-questionnaire. The post-questionnaire consisted of three questions to measure the ease of navigation, the ease of locating the items, and the ease of comparison with multiple items. Under each item a seven point scale was presented, ranging from strongly disagree to strongly agree.

The analysis was performed for each of the three experimental sections. First, the results from the pre-questionnaire revealed that the four experimental groups were not different from each other in terms of their ability to navigate the WWW. Second, three kinds of data were analyzed using the system log files: usage of the context information, number of nodes visited and number of nodes repeatedly visited. The number of times subjects used the contextual aids was first analyzed for the manipulation check. Next, we measured the total number of nodes visited and number of repeatedly visited nodes, since disorientation and cognitive overhead can make the subjects miss out sections of the site and therefor to open specific nodes repeatedly. This leads to an increase in navigational pages, which in turn increases the total number of nodes visited. Finally, the post questionnaires were analyzed to investigate the subjective ratings for navigation convenience.

Content dissemination system

Subjects

64 undergraduate students at Yonsei University participated in the experiment. The subjects were randomly divided into four groups: control group, structural context only group, temporal context only group, and both structural and temporal context group. Sixteen subjects were assigned to the each group. Subjects were tested individually.

Task materials

A prototype of the unstructured content dissemination system used in the experiment was called "Studying Abroad Site." We chose the specific domain of studying abroad among various domains, because subjects participating in the experiment were undergraduate students who are familiar with studying abroad. This prototype was an unstructured hypertext document with an unbalanced hierarchical structure similar to common content dissemination systems. In order to increase the external validity of the study, this prototype was constructed based on actual studying abroad sites. This site consists of nine levels of depth with 733 related nodes in total.

Experimental procedure

All the experimental procedure except tasks was identical to that of structured e-commerce system. Subjects were asked to perform six browsing tasks. We set the six tasks to be similar to typical questions that people may have for the studying abroad sites.

All the analysis procedure was identical to that of the structured e-commerce system. Similar to the electronic commerce system case, the results from the pre-questionnaire revealed that the four experimental groups were similar in terms of their ability to navigate the WWW.

RESULTS

Electronic commerce system

Use of the context aids

Table 1 presents the mean use rate of the context aids, which is the frequency of each link usage in the log data. The mean rate of using structural context links (UTG, DTG, and NTP links) is 79%, and the mean rate of using temporal context links (PATH links) is 15%. The results indicate that most subjects were aware of the given contextual aids and used them throughout the experimental sessions.

Table 1. Mean use rate of the context aids in e-commerce system (%)

Link type	Control	TC	SC	Both
UTP	40	21	10	4
DTC	60	62	8	10
UTG	X	X	15	9
DTG	X	X	36	36
NTP	X	X	31	30
PATH	X	17	X	12

[1] TC = Temporal context only

[2] SC = Structure context only

Total number of nodes visited

Figure 3 presents the total number of nodes visited during browsing tasks for all four groups. A between subjects ANOVA revealed main effects of structural context ($F(1,36)=60.2$, $p<.001$) and temporal context ($F(1,36)=11.27$, $p<.005$). Subjects with the structural context visited significantly fewer nodes than those without the structural context. Also, when subjects were browsing with temporal context they visited fewer nodes than non-temporal context group. There was a significant interaction between the structural context and temporal context ($F(1,36)=5.65$, $p<.05$). When structural context was provided, there was no difference between subjects with the temporal context and those without the temporal context. However, when the structural context was not provided, there was a considerable difference between subjects with the temporal context and without the temporal context. In other words, the temporal context was found to influence the navigation patterns only when the structural context was not provided.

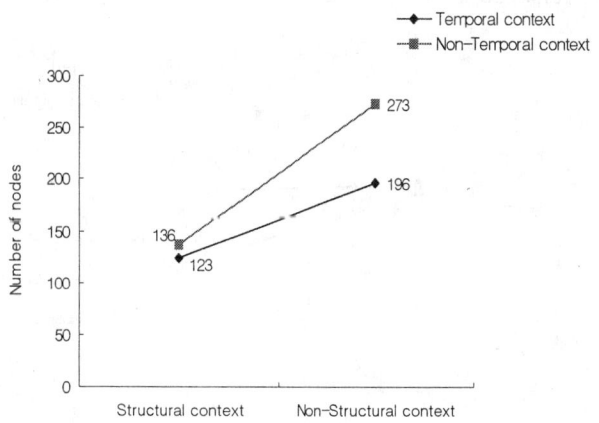

Figure 3. Number of nodes visited in the e-commerce system

Number of nodes repeatedly visited

Figure 4 presents the number of nodes repeatedly visited for all four groups. A between subjects ANOVA revealed main effects of structural context ($F(1,36)=86.54$, $p<.001$) and temporal context ($F(1,36)=21.24$, $p<.001$). The main effect for the context information revealed that subjects visited fewer repeated nodes with the context information than those without the context information. These results indicate that the structural and temporal context significantly reduced the number of repeatedly visited pages (70 pages by the structural context, 35 pages by the temporal context). There was also a significant interaction effect between the structural context and temporal context ($F(1,36)=13.73$, $p<.005$). When the structural context was provided, there was no difference between subjects with the temporal context and without the temporal context. However, if the structural context was not provided, there was a considerable difference between subjects with the temporal context and without the temporal context. Therefore, the temporal context was found to play its roles only when the structural context was not provided.

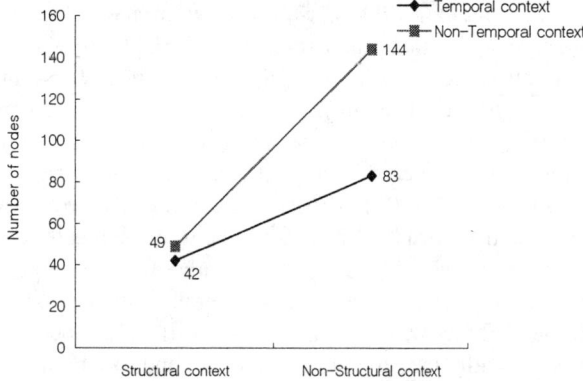

Figure 4. Number of nodes repeatedly visited in the e-commerce system

Post-questionnaire result: subjective ratings for navigation convenience

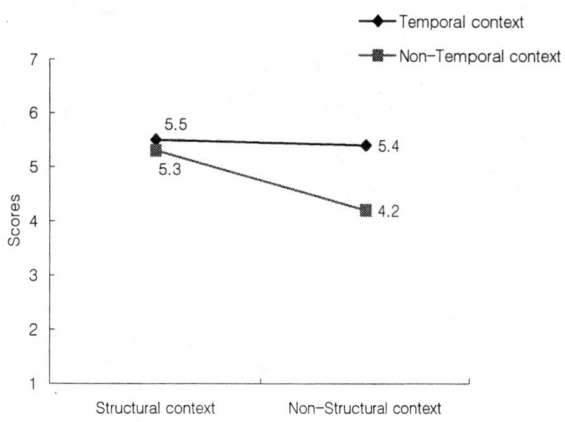

Figure 5. Subjective ratings for navigation convenience of the e-commerce system

Figure 5 presents the mean scores of three questions per condition for the navigation convenience. The higher the scores resulted in more convenient perceived navigation. An ANOVA showed that there was a significant main effect of the structural context ($F(1,36)=6.28$, $p<.05$) and the temporal context ($F(1,36)=7.75$, $p<.05$). The results indicate that subjects perceived the e-commerce system with the structural context as being more convenient for navigation than those without the structural context. Subjects with the temporal context also reported feeling more convenient for navigation than those without the temporal context. There was no significant interaction effect between the structural context and temporal context.

Content dissemination system

Use of the context aids

Table 2 presents the mean use rate of the context aids. The mean rate of using structural context links (UTG, DTG, and NTP links) is 63%, and the mean rate of using temporal context links (PATH links) is 18%. Similar to the e-commerce case, the results indicate that most subjects knew the given contextual aids and used them frequently during the experimental sessions.

Table 2. Mean use rate of the context aids in content dissemination system (%)

Link type	Control	TC	SC	Both
UTP	29	10	18	6
DTC	71	70	17	18
UTG	X	X	16	8
DTG	X	X	21	28
NTP	X	X	28	24
PATH	X	20	X	16

[1] TC = Temporal context only

[2] SC = Structure context only

Total number of nodes visited

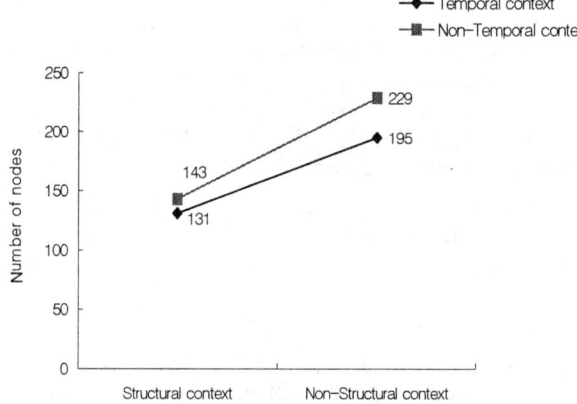

Figure 6. Number of nodes visited in the content dissemination system

Figure 6 presents the total number of nodes visited during browsing tasks for all four groups. A between subjects ANOVA revealed main effects of structural context

($F(1,60)=52.9$, $p<.001$), and temporal context ($F(1,60)=5.07$, $p<.05$). Subjects with the structural context condition visited significantly fewer nodes than those without the structural context. Also, when subjects were browsing with the temporal context they visited fewer nodes than the non-temporal context group. There was no significant interaction effect between the structural context and temporal context.

Number of nodes repeatedly visited

Figure 7 presents the number of nodes repeatedly visited for all four conditions. A between subjects ANOVA revealed main effects of structural context ($F(1,60)=28.56$, $p<.001$), and temporal context ($F(1,60)=8.15$, $p<.01$). The main effect for the context information revealed that subjects visited fewer repeated nodes with the context information than those without the context information. These results indicate that the structural and temporal context significantly decreased the number of repeatedly visited pages (27 pages by the structural context, 15 pages by the temporal context). There was no significant interaction effect between the structural context and temporal context.

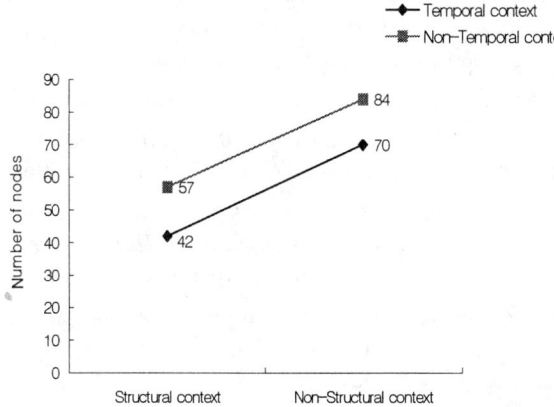

Figure 7. Number of nodes repeatedly visited in the content dissemination system

Post-questionnaire result: subjective ratings for navigation convenience

Figure 8 presents the mean scores of three questions per condition for the navigation convenience. An ANOVA revealed significant main effects of structural context ($F(1,60)=9.76$, $p<.005$), and temporal context ($F(1,60)=10.34$, $p<.005$). These results indicate that subjects perceived the content dissemination system with the structural context as being more convenient for navigation than those without the structural context. Subjects with the temporal context also reported feeling more convenient for navigation than those without the temporal context. There was no significant interaction effect between the structural context and temporal context.

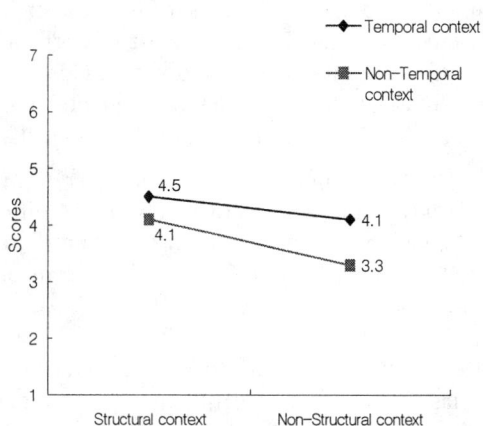

Figure 8. Subjective ratings for navigation convenience of the content dissemination system

CONCLUSION AND DISCUSSION

In summary, subjects perceived the web system with the context information as being more convenient for navigation than those without the context information. Subjects with the context information visited fewer nodes repeatedly, and therefore visited fewer pages in total. These results indicate that subjects benefit from context information by increasing the efficiency of browsing. Moreover, these benefits were observed across two different types of web systems. These results indicate that the contextual navigation aids play a powerful role regardless of different levels of structuredness and different task characteristics. Therefore, even though different web systems might have different navigation patterns, the context information turned out to be useful equally. The structural context could help subjects to reduce navigation problems by previewing information, while the temporal context may contribute to reducing the navigation problems by reviewing information.

Even though the two systems were almost similar in terms of the effects of contextual navigation aids upon navigation behaviors, a couple of differences were observed between the two systems. For example, a strong interaction effect between the structural and temporal context was observed in the e-commerce system but not in the content dissemination system. In other words, the temporal context affected the navigation patterns of the subjects in the content dissemination system even though the structural context was provided simultaneously. On the contrary, in the e-commerce system, the temporal context could play its role only when the structural context was not provided. This difference can be explained in terms of the difference of tasks in the two domains. In terms of the e-commerce system, most tasks are goal-oriented and most relevant information is usually located within two clicks. Therefore, most of the relevant information could be provided by the structural context alone in the e-commerce system. On the other hand, most tasks are interest-oriented in the content dissemination system, and relevant information was dispersed across wide areas further than two clicks away. Therefore, the structural context could not provide most of the needed information alone, and the temporal context could still play its role even with the structural context.

There are several limitations in this study. First, even though this study conducted experiments with two different systems, it could not compare the two systems directly because the two domains are different both in terms of the level of structuredness and the specificity of tasks. We selected more specific goal-oriented tasks for the well-structured e-commerce system, and more general interest-oriented tasks for the ill-structured information dissemination system. Therefore, we cannot decide whether the node structuredness or the task specificity influenced the differences between the two groups. Our selection was made to maximize the external validity of our study so that study results can be used more widely in various web systems. However, in order to investigate the effects of structuredness and task specificity upon the contextual information, further studies should conduct more controlled experiments using homogeneous domains and tasks.

Second, we tested only one type of structure, the hierarchical structure. However, other types of structures may result in different effects of context information. Future studies should include other types of structures such as linear, grid and network structures.

Third, task type in this study is browsing general or specific topics. However, other types of tasks may have different effects of context information. Future study should be extended to other types of tasks, such as searching.

In conclusion, the impact of contextual navigation aids is substantial in enhancing the navigation patterns and consequently decreasing the disorientation and cognitive overhead problems. This effect is more meaningful because it can be observed across two radically different domains and our method of providing the contextual aids is by simply changing add-on links. A wide use of the contextual navigation aids suggested in this study is expected to solve the recurrent navigation problems and increase the overall usability of most web systems.

ACKNOWLEDGEMENT

This research was supported by the Korean Institute of Information Technology Assessment grant #99-10.

REFERENCES

1. Conklin, J. Hypertext: an introduction and survey, *IEEE Computer, 20-7*, (1987), 17-41.
2. Edward, D. M., & Hardman, L. Lost in hyperspace: Cognitive mapping and navigation in a hypertext environment. In R. McAleese (Ed.), *Hypertext: Theory and pratice* (pp. 105-125). Oxford, England: Intellect Books. 1989.
3. Fillion, F. M. and Boyle, C.D.B. Important issues in hypertext documentation usability. *Proceeding of the 9th ACM Annual International Conference on Systems Documentation, SIGDOC'91* (U.S.A., 1991).
4. Furnas, G. Effectiveness view navigation. *Conference Proceedings of Human Factors in Computing System.* Atlanta, Georgia, (April 1997), 367-374.
5. Glenn, B. T., & Chignell, M. H. Hypermedia: Design for browsing. In H. Rex Hartson and D. Hix (Eds.), *Advances in Human-Computer Interaction, Vol. 3.* (p. 143-183). Norwood, NJ: Ablex Publishing Corp, 1992.
6. Greenberg, S. *The Computer User as Toolsmith: The Use, Reuse, and Organization of Computer-based Tools*. Cambridge, MA: Cambridge University Press. 1993.
7. Jul, S. & Furnas, G. W. Navigation in electronic worlds. *Nav 97 Report.* 1997.
8. Kim, H. & Hirtle, S. Spatial metaphors and disorientation in hypertext browsing. *Behavior and Information Technology, 14(4)*, (1995). 239-250.
9. Lynch, P. J. & Horton, S. Imprudent liking weaves a tangled web. *Computer, 30, 7,* (July 1997), 115-117.
10. Morris, M. and Hinrichs, R. *Web Page Design.* Prentice-Hall, Englewood Cliffs, New Jersey, 1996.
11. Norman, K. L. *The Psychology of Menu Selection: Designing Cognitive Control at the Human/Computer Interface.* Ablex Publishing Corporation, 1991.
12. Perrig, W. & Kintsch, W. Propositional and situational representation of text. *Journal of Memory and Language*, 24, (1982). 503-518.
13. Pohl, R. F. Acceptability of story continuation. In A. Flammer and W. Kintsch (Eds.), *Discourse Processing.* Amsterdam: North-Holland, 1982.
14. Shubin, H. and Meehan, M. M. Navigation in Web Applications. *Interactions.* (November and December), 1997.
15. Tauscher, L., & Greenberg, S. How people revisit web pages: empirical findings and implications for the design of history systems. *International Jounal of Human – Computer Studies, 47,* (1997). 97-137.
16. Theng, Y. L., Thimbleby, H.& Jones, M. 'Lost in hyperspace': Psychological problem or bad design?, *APCHI'96* (Singapore, 1996), 387-396.
17. Utting, K., & Yankelovich, N. Context and orientation in hypermedia networks. *ACM Transactions on Information Systems, 7(1)*, (1989). 58-84.

INTERACTING WITH EYE MOVEMENTS IN VIRTUAL ENVIRONMENTS

Vildan Tanriverdi and Robert J.K. Jacob

Department of Electrical Engineering and Computer Science

Tufts University

Medford, MA 02155, USA

{vildan,jacob}@eecs.tufts.edu

ABSTRACT

Eye movement-based interaction offers the potential of easy, natural, and fast ways of interacting in virtual environments. However, there is little empirical evidence about the advantages or disadvantages of this approach. We developed a new interaction technique for eye movement interaction in a virtual environment and compared it to more conventional 3-D pointing. We conducted an experiment to compare performance of the two interaction types and to assess their impacts on spatial memory of subjects and to explore subjects' satisfaction with the two types of interactions. We found that the eye movement-based interaction was faster than pointing, especially for distant objects. However, subjects' ability to recall spatial information was weaker in the eye condition than the pointing one. Subjects reported equal satisfaction with both types of interactions, despite the technology limitations of current eye tracking equipment.

Keywords

Eye movements, eye tracking, Polhemus tracker, virtual reality, virtual environments, interaction techniques.

INTRODUCTION

Virtual environments can now display realistic, immersive graphical worlds. However, interacting with such a world can still be difficult. They usually lack haptic feedback to guide the hand and to support it in space. We thus require new interaction techniques to provide users with easy and natural ways of interacting in virtual environments. This is particularly important for dealing with displayed objects beyond the reach of the user's arm or the range of a short walk. Some recent studies have focused on developing interaction techniques using arm, hand, or head movements in virtual environments [3, 6, 12, 13, 15], but the field is still in its infancy. We believe eye movement-based interaction can provide easy, natural, and fast ways of interacting in virtual environments. Work on eye movement-based interaction has thus far focused on desktop display user interfaces [1, 2, 7, 8, 17, 18, 19], while eye movement-based interaction in virtual reality has hardly been explored. In this study, we develop a VR interaction technique using eye movements and compare its performance to more conventional pointing.

EYE MOVEMENT-BASED INTERACTION IN VIRTUAL REALITY

Eye movement-based interaction is an example of the emerging non-command based interaction style [14]. In this type of interaction, the computer observes and interprets user actions instead of waiting for explicit commands. Interactions become more natural and easier to use. One system that suggests such advantages is a screen-based system developed by Starker and Bolt [18]. It monitors eye movements of the user, interprets which objects attract the user's interest, and responds with narration about the selected objects. It minimizes the physical effort required to interact with the system and increases interactivity. High interactivity is even more important in VR applications where users often deal with more dynamic and complex environments.

Our overall approach in designing eye movement-based interaction techniques is, where possible, to obtain information from a user's natural eye movements while viewing the display, rather than requiring the user to make specific trained eye movements to actuate the system. This approach fits particularly well with virtual reality interaction, because the essence of a VR interface is that it exploits the user's pre-existing abilities and expectations. Navigating through a conventional computer system requires a set of learned, unnatural commands, such as keywords to be typed in, or function keys to be pressed. Navigating through a virtual environment exploits the user's existing "navigational commands," such as positioning his or her head and eyes, turning his or her body, or walking toward something of interest. By exploiting skills that the user already possesses, VR interfaces hold out the promise of reducing the cognitive burden of interacting with a computer by making such interactions more like interacting with the rest of the world. An approach to eye movement interaction that relies upon natural eye movements as a source of user input extends this philosophy. Here, too, the goal is to exploit more of the user's pre-existing abilities to perform interactions with the computer.

Another reason that eye tracking may be a particularly good match for VR is found in Sibert and Jacob's [17] study on direct manipulation style user interfaces. They found eye movement-based interaction to be faster than interaction with the mouse, especially in distant regions. The eye movement-based object selection task was not well modeled by Fitts' Law, or, equivalently, that the Fitts' Law model for the eye would have a very small slope. That is, the time required to move the eye is only slightly related to the distance to be moved. This suggests eye gaze interaction will be particularly beneficial when users need to interact with distant objects, and this is often the case in a virtual environment.

Finally, combining the eye tracker hardware with the head-mounted display allows using the more robust head-mounted eye tracker without the inconvenience usually associated with that type of eye tracker. For many applications, the head-mounted camera assembly, while not heavy, is much more awkward to use than the remote configuration. However, in a virtual environment display, if the user is already wearing a head-mounted display device, the head-mounted eye tracker adds very little extra weight or complexity. The eye tracker camera obtains its view of the eye through a beam splitter, without obscuring any part of the user's field of view.

In this study, our first goal was to test the hypothesis that eye movement-based interactions would perform better in virtual environments than other natural interaction types. In order to test it, we need to compare against a more conventional interaction technique as a yardstick. We used hand movement for comparison, to resemble pointing or grabbing interaction that would commonly be found in virtual environments today. In addition, we investigated whether there would be performance differences between "close" and "distant" virtual environments, i.e., where objects are respectively within and beyond the reach of the user. Since pointing-based interaction requires the user to use hand and arm movements, the user has to move forward in order to reach and select the objects in the distant virtual environment. Eye movement-based interaction, however, allows the user to interact with objects naturally using only eye movements in both close and distant virtual environments. Therefore, we expected eye movement-based interactions to be faster especially in distant virtual environments.

Despite its potential advantages, eye movement-based interactions might have a drawback in terms of the users' ability to retain spatial information in virtual environments. Search tasks can be cumbersome for users in virtual environments, especially in large virtual worlds. They may fail to remember the places they previously visited, and may have to visit them again. Therefore, we looked for the effect of eye movement-based interactions on spatial memory, i.e., the ability of the user to recall where objects reside in space. For this, we compared users' ability to recall spatial information in the eye movement vs. pointing conditions. As argued above, eye movement-based interaction decreases the effort required for interaction, whereas pointing based interaction engages the user more in the interaction. This reduced level of engagement might also reduce the user's ability to retain spatial information about the objects during eye movement-based interactions. Hence, we expected spatial memory to be weaker in eye movement than in pointing based interactions in virtual environments.

A NEW INTERACTION TECHNIQUE

Using natural eye movements in virtual environments requires development of appropriate interaction techniques. In this study, we developed an interaction technique that combines features of eye movements and non-command based interactions in a virtual environment. Our objective is to enable users to interact with eye movements, without explicit commands where possible. However, we should also avoid the Midas Touch problem, i.e., unwanted activation of commands every time user looks at something [9]. Our approach here was for the computer to respond to the user's glances about the virtual environment with continuous, gradual changes. Imagine a histogram that represents the accumulation of eye fixations on each possible target object in the VR environment. As the user keeps looking at an object, histogram value of the object increases steadily, while histogram values of all other objects slowly decrease. At any moment we thus have a profile of the user's "recent interest" in the various displayed objects.

In our design, we respond to those histogram values by allowing the user to select and examine the objects of interest. When the user shows interest in a 3D object by looking at it, our program responds by enlarging the object, fading its surface color out to expose its internals, and hence selecting it. When the user looks away from the object, the program gradually zooms the object out, restores its initial color, and hence deselects it. The program uses the histogram values to calculate factors for zooming and fading continuously.

As Figures 1, 2, and 3 suggest, there is too much information in our virtual environment scene to show it all at once, i.e., with all the objects zoomed in. It is necessary for the user to select objects and expand them individually to avoid display clutter. This is intended to simulate a realistic complex environment, where it is typically impossible to fit all the information on a single display; user interaction is thus required to get all of the information. We developed an alternate version of our design for use with the hand, using the Polhemus sensor. In this version, the user indicates interest in a displayed object by placing a virtual pointer on it. When the user moves the virtual pointer into an object, our program responds by zooming in the object, fading out its color, and hence selecting the object. When the user moves the pointer away from the object, the program deselects it.

METHOD

Apparatus

We conducted this study in the Human Computer Interaction Laboratory in the Tufts University Electrical Engineering and Computer Science Department. We used a Silicon Graphics Indigo2 High Impact workstation, Virtual i-Glasses head mounted display, ISCAN eye tracker, and Polhemus 3Space FASTRAK magnetic tracker. One of the Polhemus receivers was on the head mounted display to provide VR camera positioning, and one was on a cardboard ring around the subject's finger for pointing in the virtual environment.

The ISCAN eye tracker system consists of the eye tracker, eye and scene monitors, and ISCAN Headhunter Line of Sight Computation and Plane Intersection Software (version 1.0), and its own separate computer. The software monitors eye movements, performs calibrations, and processes eye images. It runs on a separate Pentium 100MHz personal computer. The eye tracker hardware is built into the head mounted display. It has a tiny eye camera, infrared (IR) light source, and a dichroic mirror. The IR light source creates the corneal reflection on the eye. The eye camera captures the eye image reflected in the mirror and sends it to a frame grabber in the PC. The PC software calculates the visual line of gaze using the relationship between the center of the pupil and corneal reflection point. Unlike more conventional usage of head-mounted eye trackers, in VR we only require the eye position relative to the head, not the world, since the VR display moves with the head. The PC sends the stream of processed eye data to the Silicon Graphics computer, where the main VR system is running via a serial port.

Stimuli

Our virtual environment displays a virtual room that contains fifteen geometrical objects (spheres and polygons) in five different colors (blue, pink, green, yellow and orange) (Figure 1). Each object contains four cylinders that are textured with a letter. In one version ("close"), objects are within the reach of the subject's arm; in the other ("distant"), subjects need to move 5-15 inches in order to reach the objects. Our interaction technique allows subjects to see the cylinders and letters inside the objects by selecting the objects. Initially, the cylinders and letters are not visible (Figure 2). When the subject selects an object (by either Polhemus or eye tracker), the object starts growing and its color fades out, making the internal cylinders and letters visible (Figure 3). When the subject deselects the object, it starts shrinking, its color fades back, and it returns to its initial state, with the cylinders again invisible. The program allows subjects to select only one object at a time. We set the time constants for fading in and out based on informal trials with the eye and Polhemus versions to optimize each version separately; the result was that eye movement response is set about half as fast as the pointing-based version, mainly to avoid the Midas Touch problem and to balance the time spent in locating the Polhemus.

Virtual Reality Software

We implemented our VR software using the new PMIW user interface management system for non-WIMP user interfaces being developed in our lab [10, 11]. Our system is particularly intended for user interfaces with continuous inputs and outputs, and is thus well suited to the histogram interaction technique, with its gradual, continuous fading. The VR programs run on the Silicon Graphics workstation, using SGI Performer to display the graphics and enable users to interact in the virtual environment. There are four versions of the program, for eye movement vs. pointing based interactions and for close vs. distant virtual environments.

Experimental design

We used a within-subjects experimental design for devices. All subjects used both eye movement and pointing interactions. The order of the two was counterbalanced to eliminate differences in fatigue and learning. We also investigated how performance varies between the "close" and "distant" conditions with a between-subjects design, by dividing the subjects into two groups; one interacted in close virtual environments and the other, in distant.

Participants

Thirty-one subjects volunteered to participate in the experiment. They were undergraduate and graduate students in our department. Their average age was 22. Twenty-eight subjects had no prior experience with VR while three subjects had some insignificant experience: they had used a VR system only once. We eliminated seven subjects from the sample before beginning their sessions, because we could not successfully calibrate the eye tracker to their pupils. The remaining 24 subjects (twenty male, four female) successfully completed the experiment.

Experimental setting

Figure 4 is an illustration of the devices used in the experiment. The subject used the VR programs by standing next to the Polhemus transmitter, which was placed on an aluminum camera tripod, 8 inches below the subject's head. We removed all furniture and equipment in the physical area used by the subjects in order to prevent distraction and facilitate movements especially during their interaction with distant virtual environments. We used a table lamp for lighting.

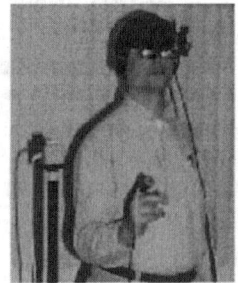

Figure 4. Illustration of the devices

Procedure

One day before the experiment, the subject attended a training session to familiarize him or her with the head mounted display, eye tracker, Polhemus, and the interaction techniques. First, the experimenter introduced the devices and explained their functionality. Then, she calibrated the eye tracker for the subject's pupil using the ISCAN software. This calibration took approximately 2-3 minutes. Next, the subject used the eye tracker, the Polhemus, and the programs to practice the interaction technique. He or she learned how to select and deselect objects and see the letters in the virtual environments using eye movement and pointing based interactions. On average, the whole training session took about 20 minutes[1].

On the next day, the subject was asked to do two search tasks, one with the eye tracker, the other with the Polhemus. First, the experimenter read a script explaining that the search task is to find the two geometrical objects[2] that contain a particular target letter on their internal cylinders, and asked the subject to inform the experimenter as soon as he or she found the objects[3]. Then she adjusted the physical position of the subject in order to ensure that all subjects start the experiment in a standard position. Next, she initialized the program. At this point, the virtual environment was not visible to the subject yet. Second, she pronounced the target letter and pressed a mouse button to make the virtual environment visible to the subject and record the start time, and the subject started searching for the target letter. The subject notified the experimenter each time he or she found an object containing the target letter. The experimenter pressed the mouse button to record completion times for the first and second objects. When the subject found the second object, the program and the search were terminated. In the analysis, we used the elapsed time to find both of the objects as the performance measure.

The next task was the memory task. The goal was to record how well the subject could recall which two objects contained the target letter. The experimenter started a program that displayed to the user the virtual environment he or she had just interacted with. This display enabled the subject to see all of the objects that were present in the virtual environment, but not to access the internal cylinders or letters. The experimenter asked the subject which two objects had contained the target letter in the search task, and recorded his or her responses. In the analysis, we used the number of correctly recalled objects as the measure of spatial memory.

Finally, the subject filled out a survey containing questions about satisfaction with the eye tracker and Polhemus[4]. The experimenter also encouraged the subjects to give verbal feedback about their experience with the two technologies, and recorded the responses.

RESULTS

Our first hypothesis was that in virtual environments subjects would perform better with eye movement-based interactions than with hand based interactions. As a sub-hypothesis, we also expected an interaction effect, that the performance difference between eye and Polhemus would be greater in the distant virtual environment compared to the close one. Our independent variables were interaction type (eye movement, pointing) and distance (close, distant). Our dependent variable was performance (time to complete task). Table 1 provides the descriptive statistics for our measurements. We tested the first hypothesis by comparing the means of the pooled performance scores (performance scores in close and distant virtual environments) using one-way analysis of variance (ANOVA). We found that performance with eye movement-based interactions was indeed significantly faster than with pointing ($F[1,46]=10.82$, $p = 0.002$). The order in which the subjects used eye movement and pointing based interactions did not indicate an important effect on the performance: 21 out of the 24 subjects performed faster with eye movement-based interaction. Next, we separated the performance scores in close and distant virtual environments into two subgroups,

---

[1] At the end of the experiment, we asked subjects to respond to the following statement on a 7-point Likert scale (1: strongly disagree..7: strongly agree) in order to check if the training achieved its goal: "The training session familiarized me with using eye tracker/polhemus." The responses show that they were sufficiently familiarized with eye tracker (Mean = 6.08; Standard deviation = 1.02) and polhemus (Mean = 6.00; Standard deviation = 1.14).

[2] We used 2 objects to test if there is a learning effect, but we could not find any.

[3] At the end of the experiment, we asked subjects to respond to the following statement on a 7-point Likert scale (1: strongly disagree..7: strongly agree) in order to check if they understood the task: "I understood the task before starting with the experiment." The responses show that they understood the task (Mean = 6.29; Standard deviation = 0.95).

[4] The survey contained questions adapted from Shneiderman and Norman's [15] questionnaire for user interface satisfaction (QUIS), and Doll and Torkzadeh's [5] end-user satisfaction scale.

Table 1. Performance of eye movement and pointing based interactions

	Time to complete the task (seconds)					
	Eye movements			Pointing		
	n	M	SD	N	M	SD
Close virtual environments	12	67.50	23.86	12	96.56	46.76
Distant virtual environments	12	56.89	19.73	12	90.28	36.16
Overall	24	62.19	22.09	24	93.47	41.01

Notes: n, M, and SD represent number of subjects, mean, and standard deviation respectively.

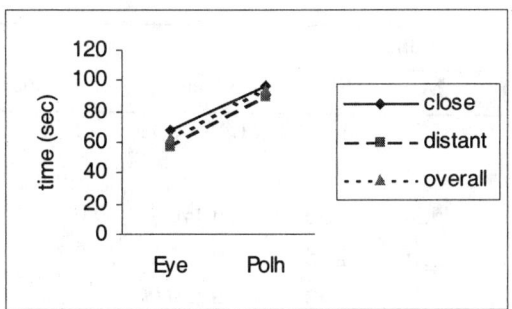

Figure 5. Mean times of performance

and repeated the ANOVA for the two subgroups. We found that in distant virtual environments, performance with eye movement-based interactions was indeed significantly higher than performance with pointing ($F[1,22] = 7.89$, $p = 0.010$). However, performance between the two types of interactions was not significantly different in close virtual environments ($F[1,22] = 3.7$, $p = 0.067$). Figure 5 shows the graphical representation of the means of pooled, distant and close performance for both eye movement and pointing based interactions.

Our second hypothesis was that in virtual environments, spatial memory of subjects would be weaker in eye movement-based interactions than that in pointing based interactions. We measured spatial memory by the number of objects correctly recalled after completion of the memory task as our dependent variable; our independent variable was the type of interaction as before. Table 2 shows the descriptive statistics for our measurements. Comparing the means of the spatial memory scores in the two types of interactions using one-way ANOVA, we found that the number of correctly recalled objects in eye movement-based interactions was indeed significantly lower than that in pointing ($F[1,46] = 7.3$, $p = 0.010$).

Finally, we were interested in exploring subjects' satisfaction with interacting with eye movements and pointing. We asked subjects about ease of getting started, ease of use, accuracy, and fatigue felt with eye tracker and Polhemus, and whether they found these technologies useful in VR systems. We were particularly interested in the users' reaction to eye tracking, because this is an immature technology, and eye trackers do not yet work as steadily or reliably as mice or Polhemus trackers. Table 3 shows the questions and subjects' responses. As the ANOVA results in the last two columns of the table indicate, subjects' satisfaction with eye tracker and Polhemus were not statistically different. We further posed the subjects the following open-ended question: "Do you prefer eye tracker or Polhemus for this task? Please explain why." Eighteen of the 24 subjects specified that overall they preferred the eye tracker to the Polhemus. They said that using eye movements was natural, easy, faster, less tiring. Six subjects preferred the Polhemus because they found that it was accurate and easy to adapt because of its similarity to the mouse.

DISCUSSION AND CONCLUSIONS

By developing an interaction technique that allows the use of natural eye movements in virtual environments, we were able to compare the performance of eye movement and pointing based interactions in close and distant virtual environments. The results show that interaction with eye movements was faster than interaction with pointing. They further indicate that the speed advantage of eye movements was more significant in distant virtual environments. Our findings suggest that eye movement-based interactions could become a viable interaction type in virtual environments provided that proper interaction techniques are developed.

However, our data also point to a price paid for this increased speed when the task requires spatial memory. Our subjects had more difficulty in recalling the locations of the objects they interacted with when they used eye movement-based interactions. They recalled significantly more objects when they used pointing. One possible explanation for this result may be the ease of use of eye movement-based interactions. Our subjects explained that they were "just looking," as in the real world, when they interacted with eye movements. Hence, the cognitive burden of interacting was low. They did not have to spend any extra effort for locating the objects or manipulating them. When they used pointing, however, they had to spend extra physical effort to

Table 2. Spatial memory in eye movement and pointing based interactions

	N	Spatial memory measures	
		M	SD
Eye movements	24	0.79	0.83
Pointing	24	1.5	0.98

Notes: Spatial memory is measured by the number of letters recalled by subjects after completion of the task. n, M, and SD represent number of subjects, mean, and standard deviation respectively.

Table 3. Satisfaction of subjects with eye tracker and Polhemus

Survey Questions	Eye-tracker		Polhemus		F-value	p-value
	M	SD	M	SD		
Getting started with eye tracker/polhemus is easy.	5.33	1.69	5.54	1.59	(1,46)=0.19	0.66
The eye tracker/polhemus is easy to use.	5.83	1.34	5.54	1.56	(1,46)=0.48	0.49
The eye tracker/polhemus is accurate.	5.29	1.33	5.79	1.38	(1,46)=1.63	0.21
Are you satisfied with the accuracy of eye tracker/polhemus?	5.83	1.05	5.83	1.43	(1,46)=0.00	1.00
Did you feel fatigued when searching with eye tracker/polhemus?	3.21	1.74	3.29	1.94	(1,46)=0.02	0.88
I would find the eye tracking/pointing useful in virtual reality systems.	5.75	1.33	5.46	1.47	(1,46)=0.52	0.48

Notes: All questions were posed using a 7-point Likert scale ranging from strongly disagree (1) to strongly agree (7). M and SD represent mean and standard deviation respectively. F and p-values are from ANOVA analyses that compare the means of the answers given for eye-tracker and polhemus.

reach out to the objects and interact with them. Spending this extra effort may have helped them retain the spatial information of objects in their memory. This finding has implications for the choice of interaction technique in a virtual environment: eye is a particularly good choice if later spatial recall is not necessary. It may also be possible to design new eye movement-based interactions that facilitate the user in retaining the spatial information of objects after interacting with them. One approach to address this weakness might be to incorporate more spatial cues to help users recognize and retain spatial information in the virtual environments [4].

Currently, eye tracker technology is not as mature and reliable as the Polhemus-type magnetic tracker. This applies to the technology in general, other available eye trackers we have used in other work have given roughly similar performance. Therefore, we had expected that the subjects would be less satisfied with eye tracker technology than with Polhemus. However, our satisfaction survey and interviews with subjects after the experiment showed that they were equally satisfied with eye tracker and Polhemus. They stated that they liked the idea of using eye movements without having to think of initiating a command. This provides support for our claim that eye movements fit well into a non-command style user interface.

This study has shown that eye movement-based interactions are promising in virtual environments. We believe that by developing proper interaction techniques, eye movement-based interactions can address weaknesses of extant interaction techniques in virtual environments. In subsequent work, we hope to examine still more subtle or "lightweight" interaction techniques using eye movements, to compare "interaction by staring at" with a more subtle "interaction by looking around."

ACKNOWLEDGMENTS

We want to thank Prof. Sal Soraci of the Psychology Department at Tufts for valuable discussions about the spatial memory issue. We thank each of our subjects who volunteered to participate in this experiment.

This work was supported by National Science Foundation Grant IRI-9625573 and Office of Naval Research Grant N00014-95-1-1099. We gratefully acknowledge their support.

REFERENCES

1. Bolt, R.A. Gaze-Orchestrated Dynamic Windows, *Computer Graphics*, vol. 15, no. 3, pp. 109-119, August 1981.

2. Bolt, R.A. Eyes at the Interface, *Proc. ACM Human Factors in Computer Systems Conference*, pp. 360-362, 1982.

3. Bowman, D.A., Hodges, L.F., An Evaluation of Techniques for Grabbing and Manipulating Remote Objects in Immersive Virtual Environments, *Proceedings of the 1997 Symposium on Interactive 3D Graphics*, 1997, pp.35-38

4. Darken, R., Sibert, J., Wayfinding Strategies and Behaviors in Large Virtual Worlds, *Proceedings on Human Factors in Computing Systems*, CHI'96 pp. 142-149

5. Doll, W.J., Torkzadeh, G., The measurement of end-user computing satisfaction, *MIS-Quarterly*, Vol.12, n12, pp.259-274. (1988).

6. Forsberg A., Herndon K., Zeleznik R., Effective Techniques for Selecting Objects in Immersive Virtual Environments, *Proc. ACM UIST'96 Symposium on User Interface Software and Technology (UIST)*, 1996.

7. Glenn, F.A., and others, Eye-voice-controlled Interface, *Proc. 30th Annual Meeting of the Human Factors Society*, pp. 322-326, Santa Monica, Calif., 1986.

8. Jacob, R.J.K., The use of eye movements in human-computer interaction techniques: what you look at is what you get, *ACM Transactions on Information Systems*, 9, 3, pp. 152-169(April 1991).

9. Jacob, R.J.K., Eye Tracking in Advanced Interface Design, in *Advanced Interface Design and Virtual Environments*, ed. W. Barfield and T. Furness, Oxford University Press, Oxford (1994)

10. Jacob, R.J.K., A Visual Language for Non-WIMP User Interfaces, *Proc. IEEE Symposium on Visual Languages* pp. 231-238, IEEE Computer Society Press (1996).

11. Jacob, R.J.K., Deligiannidis, L., Morrison, S., A Software Model and Specification Language for Non-WIMP User Interfaces, *ACM Transactions on Computer-Human Interaction*, Vol. 6(1) pp.1-46 (March 1999).

12. Koller, D., Mine, M., Hudson, S., Head-Tracked Orbital Viewing: An Interaction Technique for Immersive Virtual Environments, *Proc. ACM UIST'96 Symposium on User Interface Software and Technology (UIST)*, 1996.

13. Mine M., Brooks Jr.,F.P., Sequin C., Moving Objects in Space; Exploiting Proprioception in Virtual-Environment Interaction, *Proceedings SIGGRAPH 97*, Los Angeles, CA, pp. 19-26

14. Nielsen, J., Noncommand User Interfaces, *Communications of the ACM* 36, 4 pp. 83-99 (April 1993)

15. Poupyrev I., Billinghurst M., Weghorst S., Ichikawa T., The Go-Go Interaction Technique: Non-linear Mapping for Direct Manipulation in VR, *Proc. ACM Symposium on User Interface Software and Technology (UIST)*, 1996.

16. Shneiderman, B., Norman, K., Questionnaire for User Interface Satisfaction (QUIS), *Designing the User Interface, Strategies for Effective Human-Computer Interaction*, Second Edition, Addison-Wesley press (1992).

17. Sibert L.E., Jacob, R.J.K., Evaluation of Eye Gaze Interaction Techniques, Proc. ACM CHI'2000 Human Factors in Computing Systems Conference, Addison-Wesley/ACM Press (2000). (in press)

18. Starker, I., Bolt, R.A., A Gaze-Responsive Self-Disclosing Display, *Proc. ACM CHI'90 Human Factors in Computing Systems Conference*, pp.3-9, Addison-Wesley / ACM Press (1990).

19. Ware C., Mikaelian, H.T., An Evaluation of an Eye Tracker as a Device for Computer Input, *Proc. ACM CHI+GI'87 Human Factors in Computing Systems Conference*, pp. 183-188, 1987..

Figure 1. The entire virtual room.

Figure 2. A portion of the virtual room. None of the objects is selected in this view.

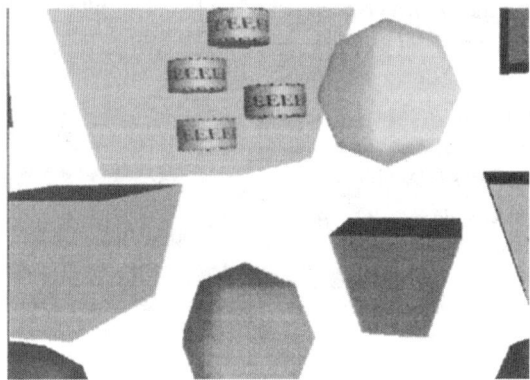

Figure3. The purple object near the top of the display is selected, and its internal details have become visible.

Intelligent Gaze-Added Interfaces

Dario D. Salvucci
Cambridge Basic Research
Four Cambridge Center
Cambridge, MA 02142
+1 617 374 9669
dario@cbr.com

John R. Anderson
Department of Psychology
Carnegie Mellon University
Pittsburgh, PA 15213 USA
+1 412 268 2788
ja+@cmu.edu

ABSTRACT

We discuss a novel type of interface, the intelligent gaze-added interface, and describe the design and evaluation of a sample gaze-added operating-system interface. Gaze-added interfaces, like current gaze-based systems, allow users to execute commands using their eyes. However, while most gaze-based systems replace the functionality of other inputs with that of gaze, gaze-added interfaces simply add gaze functionality that the user can employ if and when desired. Intelligent gaze-added interfaces utilize a probabilistic algorithm and user model to interpret gaze focus and alleviate typical problems with eye-tracking data. We extended a standard WIMP operating-system interface into a new interface, IGO, that incorporates intelligent gaze-added input. In a user study, we found that users quickly adapted to the new interface and utilized gaze effectively both alone and with other inputs.

Keywords

Gaze-added interfaces, gaze-based interfaces, intelligent interfaces, eye movements, user models.

INTRODUCTION

In the quest to facilitate human-computer interaction, a number of researchers have developed *gaze-based* interfaces in which a user controls the computer using his/her eye movements [e.g., 2, 3, 11]. Gaze-based interfaces have proven especially useful for physically-disabled users, for whom gaze control is the only, or easiest, available method of input. Such interfaces cover a wide range of applications, including typing and word processing [e.g., 2, 5, 10] and locomotion and control [e.g., 12]. While gaze-based interfaces have also shown promise for able-bodied users in specific contexts [e.g., 10, 13], they have yet to make a significant impact on the design and development of today's most common user interfaces.

In this paper, we describe the design and evaluation of a special type of gaze-based interface that we call the *gaze-added* interface. Most existing gaze-based interfaces replace the functionality of certain input(s), such as the mouse, with gaze input. In contrast, gaze-added interfaces provide exactly the same functionality as similar standard (non-gaze) interfaces but also add the ability to utilize gaze input when desired. In doing so, gaze-added interfaces give users more flexibility in choosing when and how to employ gaze input. The few existing systems that could be categorized as gaze-added interfaces [e.g., 13] have manifested the large potential benefits for these types of interfaces.

All gaze-based interfaces must deal with a significant problem in both their design and implementation: the difficulty of interpreting user eye movements. The interpretation of gaze input requires an assignment of each gaze to the visual target to which the user is attending during the gaze. This assignment is often complex for at least two reasons: inherent noise in the eye-tracking equipment, and the dissociation between the gaze point and the user's visual attention. To alleviate this problem, we developed an *intelligent* gaze-added interface that incorporates a probabilistic model of user behavior. The model helps to guide the interface to more likely interpretations of observed gazes, thus creating a significantly more responsive and user-friendly interface.

To demonstrate the power of intelligent gaze-added interfaces, we extended a standard WIMP (Window, Icon, Menu, Pointer) operating-system interface to incorporate intelligent gaze-added input. The WIMP operating-system interface was chosen for several reasons. Such interfaces are extremely common in today's most popular operating systems, such as Apple Mac OS and Microsoft Windows. In addition, these interfaces typically do not require the accurate pointing that interfaces for other applications, such as word processing or drawing, normally require; this feature facilitates the use of gaze input given the noise and variability in eye-tracking data. Our novel operating-system interface, IGO (Intelligent Gaze-added Operating system), thus serves as a realistic but manageable example of a gaze-added interface that has important implications for the development of future gaze-based interfaces.

INTERFACE DESIGN AND IMPLEMENTATION

IGO, the intelligent gaze-added operating-system interface, is modeled primarily on the Mac OS system but has a great degree of overlap with Windows and related systems. We now describe IGO in three components: the basic non-gaze interface, the gaze-added interface, and the intelligent gaze interpretation as determined by the probabilistic user model.

Permission to make digital or hard copies of all or part of this work for personal or classroom use is granted without fee provided that copies are not made or distributed for profit or commercial advantage and that copies bear this notice and the full citation on the first page. To copy otherwise, to republish, to post on servers or to redistribute to lists, requires prior specific permission and/or a fee.
CHI '2000 The Hague, Amsterdam
Copyright ACM 2000 1-58113-216-6/00/04...$5.00

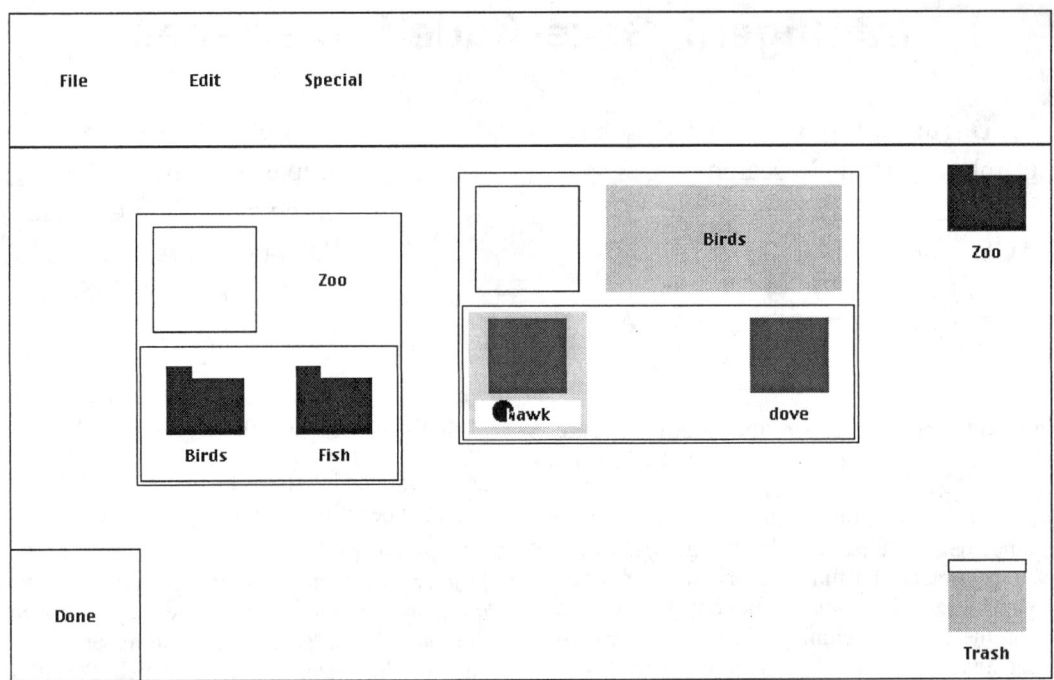

Figure 1: Sample screen from IGO. The inverted circle on the **hawk** icon marks the user's current gaze point, and the light (yellow) highlighting around the icon indicates that this icon is the current gaze focus.

The Basic Non-Gaze Interface

In essence, IGO is extremely similar to common existing interfaces with respect to the look and feel of the interface. A sample interface screen is shown in Figure 1. The screen includes windows, icons for folders (e.g., **Birds**) and files (e.g., **hawk**), and a menu bar at the top of the screen. Windows include a title bar at the top with a close box in the upper-left. Icons are color-coded such that folders appear in red and files in blue. Note that some items (e.g., the close boxes) are larger than in typical operating-system interfaces because current limitations in eye-tracking equipment preclude closely-spaced items; however, the design will generalize nicely as eye trackers improve and allow for more tightly-spaced items.

The basic, non-gaze functionality of IGO is identical to that in typical WIMP operating systems. By controlling an on-screen pointer with a mouse, users can select icons, drag icons to other parts of the screen, open and close folder windows, and select various menu options. Opening folders requires a double-click on a folder icon, and dragging requires a click-down on the icon to be dragged and a click-up at the new location of the icon. Menu selection can be performed in two ways: with a click-down on the menu name and click-up on the menu item, or with two separate clicks on the menu name and item ("sticky menus" in Macintosh terminology). Icons can be renamed by selecting the icon and typing its new name on the keyboard.

The interface as currently implemented contains three menus based on the Mac OS menu set. The File menu includes three items: Create Folder, which creates an untitled folder in the current window; Print, which is currently unused; and Duplicate, which duplicates a selected icon. The Edit menu includes three items: Cut, Copy, and Paste. The Special menu contains a single item, Empty Trash, which clears the contents of the special **Trash** folder. These menu items are clearly a subset of those needed for a full-fledged system but do provide a diverse set of commands for use in the evaluation stage.

The Gaze-Added Interface

IGO augments the basic interface to allow for gaze input in a way fully analogous to mouse input. The system employs an eye tracker (described later) to send gaze information to the interface. As the user looks around the screen, the item to which the user is attending is highlighted with a yellow background; this highlighting is analogous to the on-screen mouse pointer in that it shows the current focus of attention for the gaze modality. The user can then use the Control key to actuate a command for the highlighted item. This *gaze button* is completely analogous to the mouse button and has the exact same functionality — that is, single clicks for mouse correspond to single clicks for gaze, double-clicks correspond to double-clicks, drags to drags, etc.

To better illustrate the gaze-based functionality of IGO, let us briefly consider how a user might perform the task of throwing away a file by dragging it to the **Trash** folder. To open the appropriate folder(s), the user first looks at the folder's icon (thus highlighting it) and double-clicks the gaze button to open its window. Then, to drag a file to the trash, the user looks at the file's icon, clicks and holds the gaze button, looks at the trash icon, and release the gaze button. Finally, the user closes all windows by looking at their close boxes and clicking the gaze button. Thus, the task is identical to that with the mouse except that gaze controls the current focus and the gaze button executes a command for the current focus.

We implemented IGO on the Macintosh platform in Macintosh Common LISP. This system interfaces with an IScan (Cambridge, MA) eye tracker to capture gaze data. This eye tracker includes a head-mounted camera and specialized hardware and produces a sampling rate of 60 Hz and an accuracy of approximately 1° of visual angle.

Intelligent Gaze Interpretation

IGO requires some way of interpreting user gazes — that is, mapping gaze points to the items to which the user is likely attending. Although we could incorporate a naive algorithm that maps gazes to the nearest items, this algorithm often fails because of equipment noise and individual variability [9]. Thus, we employ an intelligent probabilistic algorithm that determines the best interpretation based on two criteria: the location of the gaze point and the current task context. The algorithm is somewhat similar in essence to algorithms previously employed in building gaze-based interfaces [10] and in analyzing gaze data from psychological experiments [9]. This body of research has shown that such probabilistic algorithms can interpret gaze data as accurately as human experts in real time [9], thus making them an excellent choice for IGO.

The interpretation algorithm takes a given gaze location g and target items $i \in I$ and returns the item i_{best} that most likely corresponds to g. More formally, the algorithm finds the item i_{best} that maximizes the probability $\Pr(i|g)$:

$$\begin{aligned} i_{best} &= \arg\max_{i \in I}\left[\Pr(i|g)\right] \\ &= \arg\max_{i \in I}\left[\frac{\Pr(g|i)\cdot \Pr(i)}{\Pr(g)}\right] \\ &= \arg\max_{i \in I}\left[\Pr(g|i)\cdot \Pr(i)\right] \end{aligned}$$

The first step in this process involves calculating the probabilities $\Pr(g|i)$ of producing a gaze at g given the intention to attend each item i. The gaze location g simply denotes the estimated point-of-regard coordinates $<x,y>$ as determined by the eye tracker. Each item i can be described as a rectangle $<cx,cy,sx,sy>$, where cx and cy describe the coordinates of the center of the rectangle and sx and sy describe the size of the rectangle as the distance from cx and cy to the edges of the rectangle. To compute $\Pr(g|i)$, we multiply the probability of each coordinate given a Gaussian distribution around the item's rectangle:

$$\Pr(g|i) = G(x,cx,sx)\cdot G(y,cy,sy)$$

Here the function $G(x,\mu,\sigma)$ denotes the probability of observing the value x in a Gaussian distribution with mean μ and standard deviation σ.

The second step in the process involves calculating prior probabilities $\Pr(i)$ of attending each item i. We compute these probabilities from the current state of the interface by assigning *prior scores* to various items and then normalizing the scores into probabilities. The various items are assigned scores as follows:

- File menu: 1 if the last action was an open or select, 1/5 otherwise. This models the fact that users tend to use this menu immediately after opening or selecting folders or files.
- Edit menu: 1/5. This models the fact that the menu is unused in the current implementation.
- Special menu: 1 if there are items in the Trash, 1/5 otherwise. This models the fact that users only empty the trash if items have already been thrown away.
- Window close box: 1/5 if the last action selected an icon in the window, 1 otherwise. This models the fact that users typically do not close a window immediately after selecting something within it.
- Window: 1/25. This models the fact that windows are very infrequently the focus of attention, with the exception of dragging an icon into a window.
- Other: 1. This models the default case.

Thus, while items have a default prior score of 1, the score is reduced for items that are unlikely in the current situation. The prior scores are then normalized to produce prior probabilities $\Pr(i)$. While this preliminary design estimated these priors informally in pilot studies of the interface, a more rigorous design could determine better priors empirically by observing long-term behavior in the interface.

Given $\Pr(g|i)$ and $\Pr(i)$, we can determine i_{best} — the i that maximizes $\Pr(i|g)$. However, we would like to give the interface the option of not assigning a gaze to any item if this probability is too low. For this purpose, we ensure that the value $log\ \Pr(i_{best}|g)$ is above a minimum threshold; below threshold, the interface considers there to be no current focus. In pilot testing we found that a threshold of –20 works well for our implementation.

To illustrate the behavior of the gaze interpretation algorithm, let us consider the situation where a gaze point falls near in or around a window close box. Figure 2 shows the assignment of gazes at various points for two possible cases; filled circles represent gaze points assigned to the close box, open circles to the icon, and X's to nothing. In Figure 2(a), the close box and the icon have equal prior scores, and thus are given equal weight in the interpretation algorithm. In this case gaze points are assigned to the nearest item above threshold, which even allows points somewhat far from the close box to be assigned to the close box. In Figure 2(b), the close box has only 1/5 the prior score of the icon (assuming that the last action selected an icon in the window). Here gaze points between the close box and the icon are more likely assigned to the icon, and gaze points farther from the close box are no longer assigned to it. Thus, as the prior probability of the close box decreases, the area in which gaze points are assigned to it shrinks and other assignments become more likely.

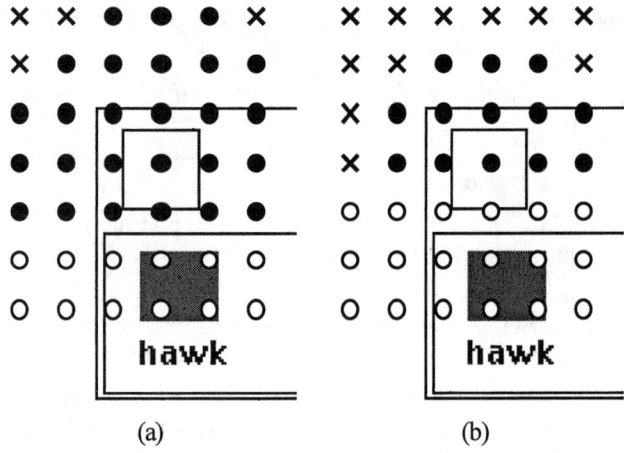

Figure 2: Probabilistic assignment for the close box at various points for (a) the default prior and (b) a smaller prior. Filled circles represent points assigned to the close box, open circles to the icon, and X's to no item.

Discussion

We would like to emphasize three important aspects of IGO. First, we have not removed any functionality from the mouse or other inputs; instead, we have simply added the ability to utilize gaze as a complementary source of input. In contrast, most existing gaze-based interfaces replace the functionality of the mouse and/or keyboard with gaze, typically because the interfaces are intended for physically-disabled people who have difficulty using, or cannot use, other forms of input [e.g., 3, 5]. Some existing systems intended more for able-bodied users add gaze functionality that is always "on" [e.g., 13]. IGO allows for total flexibility in using the gaze functionality as little or as much as desired.

Another important aspect of the interface involves the use of a gaze button to invoke commands. All gaze-based interfaces must deal with the so-called "Midas touch" problem [6]: users focus on many items, some of which are intended for commands and some of which are not. Of the ways of managing this problem (see [6] for a review), the use of a "dwell threshold" is the most common in current interfaces [e.g., 2, 11]. However, gaze buttons, in various forms, are also employed in some interfaces [e.g., 6]. We chose to use a gaze button for two primary reasons: it allows gaze functionality fully analogous to the mouse, and it can be implemented on a keyboard to provide fast and convenient access.

A third important aspect of the interface is its interpretation algorithm, which utilizes a probabilistic model of gaze location and context to assign gazes to a current focus of attention. The vast majority of existing gaze-based interfaces employ a naive method of gaze interpretation, namely mapping gazes to the nearest targets. A few systems employ probabilistic models to choose what commands to present next but not actually to interpret gaze input [e.g., 2]. Only one system incorporates a sophisticated model of user behavior, implemented as a hidden Markov model [10]; however, this system requires a detailed sequential model and a division of continuous input into subsequences for analysis. The probabilistic model presented here represents a balance of the naive and complex probabilistic algorithms, allowing for fast and robust interpretation of continuous input while avoiding the complexities of more fully-specified models.

INTERFACE EVALUATION

Several pilot trials with IGO showed that, with little practice, users could successfully use the system and perform the basic functions with the gaze modality easily and efficiently. To evaluate IGO more quantitatively, we ran a study with several interface tasks and asked users to perform the tasks as quickly and accurately as possible. We had two primary goals in this study. First, we wished to train users on the gaze and mouse modalities separately to determine how their performance improves at various points in the training. Second, we wanted to analyze how users integrate gaze input with mouse input after a period of training in each modality. While more rigorous evaluation would require a long-term study of interface use, our study sheds light on a number of interesting aspects of the system that can guide future development of this and similar systems.

Interface Setup and Tasks

We set up IGO so that users could perform a number of basic tasks. We first specified the file system as a three-level hierarchy of folders and files. The following lists each folder in the system along with its contents; folders are capitalized, files are in lower-case:

- **Zoo**: { **Birds** , **Fish** }
- **Birds**: { hawk , dove }
- **Fish**: { minnow , trout }

Given this file hierarchy, we defined five tasks for users to perform. The tasks, along with sample instructions as given to users and the basic actions required in the tasks, are shown in Table 1. For instance, the Move task involves dragging a file from one folder to another; this task comprises five basic actions: two open actions (for the **Zoo** and **Fish** folders), one drag action (from **Fish** to **Birds**), and two close actions. Similarly, the Create task involves creating and naming a new folder; the Duplicate task, duplicate a file through the File menu; the Rename task, selecting and renaming a file; and the Trash task, dragging a file to the trash and emptying the trash contents through the Special menu. All tasks involve one or two open actions, followed by some combination of the drag, select, type, and menu actions, followed by one or two close actions.

Method

Subjects

Ten users (three women and seven men) successfully participated in the experiment. An additional three users participated but were omitted from data analysis because of extreme noise in their eye-tracking data. All participants had at least three years of experience with either the Mac OS or Windows operating system. None of the participants had any prior experience with a gaze-based interface or with eye-tracking equipment.

Table 1: Interface tasks with instructions and actions.

Task	Sample Instructions	Actions
Move	In **Zoo**/ **Fish**/ , Move **trout** to **Zoo**/ **Birds**/ and close all windows.	Open, open, drag, close, close
Create	In **Zoo**/ , Create folder **Dogs** and close all windows.	Open, menu, type, close
Duplicate	In **Zoo**/ **Fish**/ , Duplicate **trout** and close all windows.	Open, open, menu, close, close
Rename	In **Zoo**/ **Birds**/ , Rename **hawk** to **owl** and close all windows.	Open, open, select, type, close, close
Trash	In **Zoo**/ **Birds**/ , Trash **dove**, empty trash, and close all windows.	Open, open, drag, menu, close, close

Materials

The experiment included two stages: a *training stage* and a *free stage*. The training stage comprised eight blocks of 10 trials each. The blocks alternated between *gaze blocks* using gaze input alone and *mouse blocks* using mouse input alone; the starting block type was counterbalanced across users. The free stage comprised two *free blocks* of 10 trials each where users could employ both inputs freely as desired. All blocks included two instances of each of the five tasks in a randomized order.

Procedure

Users were first introduced to the workings of the eye-tracking equipment and the gaze-based interface. After being calibrated on the eye tracker, the users completed five mouse trials followed by five gaze trials with help from the experimenter to become acquainted to the interface and experimental tasks. Finally, they completed the 10 blocks of experimental trials. Comments were gathered from users after the experiment to note their impressions of the ease of the system and any specific strategies they may have utilized.

Each trial comprised two parts: first, the user would read the on-screen instructions and click the gaze or mouse button when the instructions were understood; and second, the user would perform the task and click in a special **Done** region in the lower-left of the screen. Although the instructions remained on the screen in the second part, users were encouraged to use them as little as possible to provide a more accurate estimate of performance.

Results

Training Stage Results

The training stage allowed users to improve their skills with each modality separately. We examine user behavior in this stage beginning with how accurately users performed the given tasks. We utilize two criteria for determining whether a user's task behavior was correct: whether the user's actions include all the necessary actions for the task (as shown in Table 1), and whether there were two or less actions in addition to the necessary actions. Figure 3 shows the percentage of tasks classified as correct for the four blocks in the training stage. (We analyze the free stage in the next section.) Users consistently make more errors in the gaze trials than in the mouse trials and the number of errors remains fairly constant throughout the four blocks. A repeated-measures ANOVA with within-user factors of modality and block confirms these observations, showing a significant effect of modality, $F(1,9)=22.07$, $p<.01$, but no effect of block or their interaction, $p>.5$.

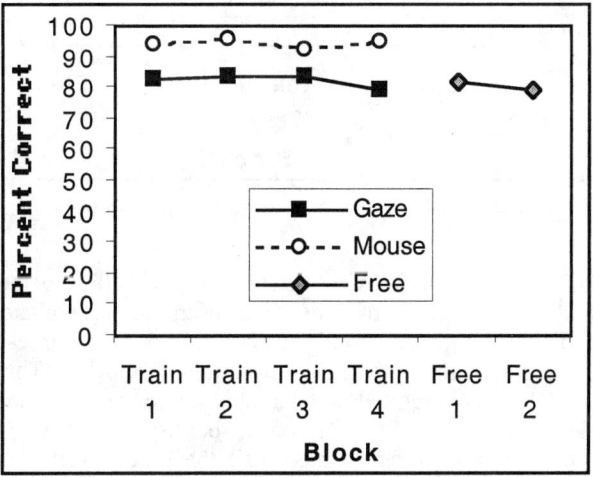

Figure 3: Percent correct across experiment blocks.

User errors with the gaze modality in both the gaze and free trials could largely be attributed to a specific type of error that we call a "leave-before-click" error. This error occurred when the user looked at an item, tried to click the gaze button, but looked away before the button was actually pressed. Ironically, users were *less* prone to commit this error as total novices because they fixated items more deliberately; however, as their confidence in the interface grew, they performed actions faster and became more prone to the error. Further practice helped users to understand the temporal interaction of gazing and clicking and better cope with the problem.

Considering only correct trials, Figure 4 shows the average time needed for subjects to complete a single task in each of the training blocks. For both modalities, users exhibit a nice learning curve, with rapid improvement in the first blocks and more gradual improvement in later blocks. Although the gaze blocks show consistently longer times than the mouse blocks, a repeated-measures ANOVA shows that this difference is not significant, $p>.1$. The effect of block is very significant, $F(1,9)=22.16$, $p<.001$, confirming the learning trend. Although we might expect that, given users' familiarity with the mouse, the gaze learning curve would be steeper than the mouse curve, the modality-block interaction is not significant, $p>.2$ — in other words, users improved at approximately the same rate with both modalities. Thus, much of the learning in the interface seems to have arisen from familiarity with the screen and tasks rather than use of the gaze or mouse input, suggesting that users had little trouble becoming adept with the gaze modality.

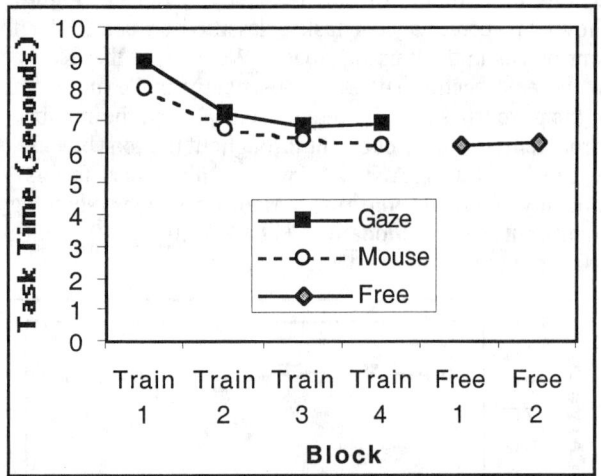

Figure 4: Task times across experiment blocks, in seconds.

Taking this analysis a step further, it is interesting to look at the performance of individual users in the different modalities. Figure 5 shows the average task times for each user and modality in the final two training blocks. Three users (3, 4, 7) required an additional one or more seconds with the gaze modality. However, four users (1, 8, 9, 10) actually exhibited faster times with the gaze modality — even with little practice in this modality and years of practice with the mouse modality. Again we have evidence that users quickly and easily learned to employ the gaze modality in IGO.

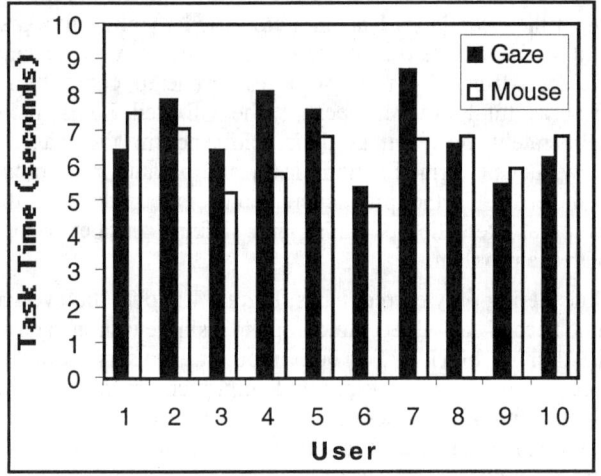

Figure 5: Task times for individual users in the final two training blocks.

We should note that these results comparing gaze and mouse performance in IGO do not necessarily reflect the raw physical performance of the eyes and hands. Rather, the results reflect a complex interaction between raw performance and a number of other factors, including the interface's ability to interpret focus (e.g., using intelligent gaze interpretation), the user's ability to coordinate focus with the gaze/mouse button, and user's ability to adapt to all these factors. In the general discussion, we mention how we can utilize our understanding of raw physical performance to improve the system through detailed cognitive modeling and analysis.

One final important aspect of user's interaction with IGO involves the effect of intelligent interpretation: did the intelligent algorithm have a significant impact on how gazes were interpreted? To answer this question, we re-ran user protocols under two other versions of IGO with simpler methods of gaze interpretation: *basic* interpretation, where gazes directly over a target are assigned to the target and gazes not over a target are assigned to nothing (i.e., the same algorithm used for the mouse); and *no-context* interpretation, where the prior scores are eliminated from the probabilistic intelligent algorithm. We then analyzed correctness (as defined earlier) of user protocols when interpreted by these methods versus the proposed intelligent method. While 83% of the protocols were correct with intelligent interpretation, 65% were correct with no-context interpretation, and only 17% were correct with basic interpretation. Thus, the intelligent algorithm greatly assisted in interpreting user gaze and thus facilitated interaction with the system.

Free Stage Results

The free stage allowed users to employ either modality whenever and however they wished. Overall, users clearly liked the gaze modality, employing it in 67% of all task actions. Figures 2 and 3 include the percent correct and task times, respectively, for the two free blocks. Users' task times in the free blocks were slightly faster than the final two blocks in for mouse and gaze alone. Thus, users were successfully able to integrate the two modalities in terms of overall speed. Users' percent correct in the free blocks was nearest that in the gaze training blocks. This result is due in part to users' extensive use of the gaze modality. Also, users experienced some difficulty in dealing with two foci (i.e., gaze and mouse), causing them to use the wrong button for the intended focus (e.g., using the mouse button when only gaze focus was over an item).

We can also look at how individual users employed the two modalities in the free stage. Figure 6 shows each user's gaze use defined as the percentage of task actions using the gaze modality. A number of users (e.g., 1, 6, 10) employ gaze in a vast majority of actions. Only two users (5, 7) employ gaze less than half the time, and one of these users (7) avoids gaze completely. Users thus exhibited a fair amount of variability in the amount of their gaze use. However, their gaze use was not completely random: Gaze use was closely correlated to the difference between task times for gaze and mouse, $R=.70$; thus, users who performed better with gaze relative to mouse in the training blocks generally preferred gaze in the free blocks. While some of users' gaze use could be attributed to experimenting with the two modalities, it is clear that users appreciated the gaze modality and made good use of it to achieve fast performance.

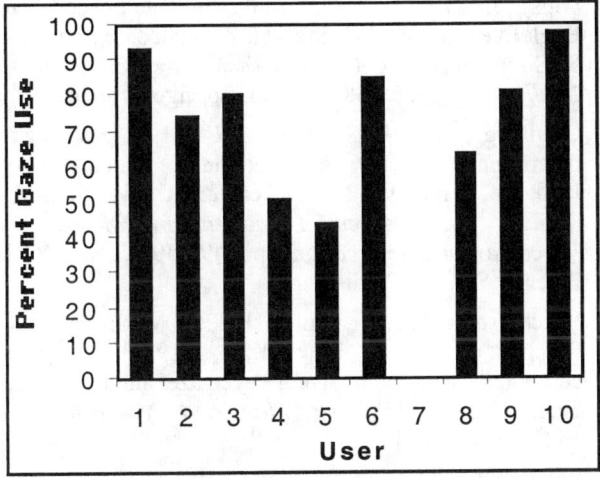

Figure 6: Percent gaze use for individual users in the free stage.

Because the task actions varied in complexity, we can analyze how this complexity affected gaze use. Figure 7 shows average gaze use for each of the five action types. The action types are shown left-to-right in terms of increasing complexity: Select and Close require a single click, Open requires a double-click, Menu requires a drag or two clicks, and Move requires a drag. As the figure shows, users exhibited a clear tendency to favor gaze use for simpler actions over complex actions. Users employed gaze approximately 75% of the time for the simplest actions, Select and Close. They employed gaze approximately 40% of the time for the most complex action, Move, and 50-60% of the time for actions in between, Open and Menu.

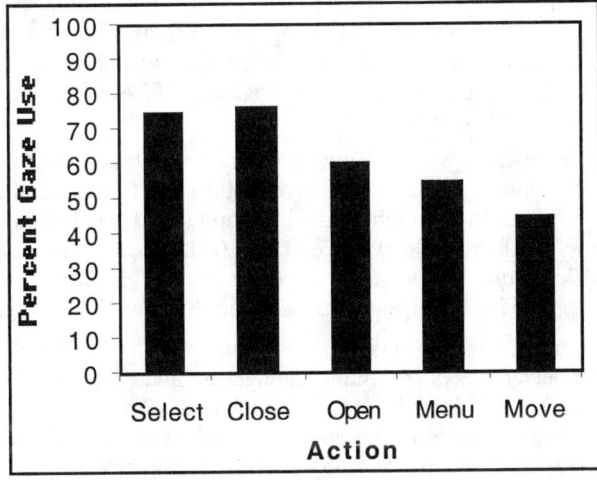

Figure 7: Percent gaze use for the five action types in order of decreasing complexity.

User comments after the experiment and our own qualitative impressions of their behavior manifested a number of interesting aspects of the system. Several users reported specific strategies to employ gaze only for simpler actions and to avoid it for complex actions, particularly dragging: "the main difficulty [with gaze] is the drag-and-drop"; "I didn't like dragging things [with gaze]". Users often seemed to utilize a strategy in which they used mouse when the mouse pointer was near the focus of the intended action and gaze otherwise; one user explicitly noted this strategy and termed it "selection based on distance". Users also tended to use gaze after typing on the keyboard to avoid long latencies to shift from keyboard to mouse: "it took a long time to move to the mouse". In addition, they often employed gaze until it "failed" them (i.e., they made an error), then switched to mouse temporarily, then returned to gaze after a short time. Thus, users derived several interesting strategies both implicitly and explicitly to help them cope with the integration of the gaze and mouse modalities.

GENERAL DISCUSSION

Beyond WIMP Interfaces

The overall success of IGO demonstrates the power of intelligent gaze-added interfaces for everyday applications, such as those that utilize WIMP interfaces. However, the potential for intelligent gaze-added interfaces goes well beyond WIMP interfaces. Jacob et al. [7] and others have outlined a variety of "non-WIMP" interfaces that allow for more flexible interface design and use. These researchers cite a number of features important to non-WIMP interfaces, including multimodal interaction, parallel input streams, and continuous-valued input.

Intelligent gaze-added interfaces offer exciting potential for future non-WIMP interfaces. As IGO and other systems [e.g., 13] demonstrate, users can learn to integrate gaze input with other modalities quickly and naturally. Gaze input is inherently stream-like, and thus provides a range of possibilities for uses in parallel with other input/output streams. In addition, gaze input can be incorporated into a "non-command" interface that acts on implicit rather than explicit user commands [6]. However, like speech and handwriting input, gaze input is not only continuous-valued but also noisy. This fact often makes it difficult to infer user intent, emphasizing the need for intelligent interpretation algorithms such as that in IGO or similar algorithms [e.g., 10].

Importance of Intelligent Gaze Interpretation

IGO, like all gaze-based interfaces, is only as user-friendly as the eye-tracking equipment allows. When we were able to calibrate users accurately on the eye tracker, they reported that the gaze modality felt smooth and seamless. However, when our calibration was somewhat problematic, users reported some amount of difficulty and frustration with the system. Other researchers [e.g., 6] have noted similar experiences with other gaze-based interfaces. In IGO, intelligent gaze interpretation clearly helps a great deal in alleviating these problems with the eye tracker. In the near future, we hope to conduct a detailed study that quantitatively measures the effect of intelligent gaze interpretation on user performance and ease of use.

The problem of interpreting gaze goes beyond the accuracy of eye tracking, however. Even with perfect eye tracking, we cannot know exactly what users are attending to based on the estimated gaze point, since users often view items in the parafovea and periphery (roughly speaking, farther than

1° of visual angle outside the line of sight). This dissociation between gaze and attention makes gaze interpretation significantly more difficult with tighter spacing between items and faster user input [9, 10]. Thus, systems must make use of as much predictive information as possible to maximize the likelihood of correct interpretations. The interpretation algorithm instantiated in IGO demonstrates the power of probabilistic algorithms for building intelligent, accurate, and more user-friendly gaze-based systems.

Gaze-Based Interfaces and Cognitive Modeling
The design and implementation of gaze-based interfaces such as IGO incorporates a number of major and minor design decisions, some of which can be difficult to make. For instance, for IGO, we considered incorporating a lag in which gaze focus on an item would last some time after the gaze has left the item. This change might help to alleviate the "leave-before-click" problem but also might decrease the responsiveness of the system as observed by users. The impact of such design options is often unclear, leaving the system developer to choose between user studies or ad hoc implementation decisions.

As an alternative to these choices, we have started considering how to employ cognitive modeling to improve current gaze-based interfaces. Cognitive modeling provides a rigorous way to express user behavior and test design options without the need for a full-scale user study. CPM-GOMS [8] is one framework that allows for fast, convenient modeling of the "critical paths" in an interface. Such a framework may help to identify and eliminate bottlenecks in the processes required by gaze-based interfaces based on knowledge of characteristics of raw eye and hand performance. ACT-R [1] is another framework that allows for detailed modeling of behavior at the level of keystrokes and eye movements. Using ACT-R, system designers can not only evaluate the times needed for various actions but also predict user learning trends and performance improvement with practice. We hope to soon model user behavior in IGO to attempt to determine how modeling can inform the design of this and similar interfaces.

ACKNOWLEDGMENTS
This work was done at Carnegie Mellon University and was supported in part by Office of Naval Research grant N00014-95-10223 awarded to John R. Anderson. We thank Rob Jacob for helpful comments in the early development of this work.

REFERENCES
1. Anderson, J. R., & Lebiere, C. (1998). *The atomic components of thought*. Hillsdale, NJ: Lawrence Erlbaum Associates.
2. Frey, L. A., White, K. P., & Hutchinson, T. E. (1990). Eye-gaze word processing. *IEEE Transactions on Systems, Man, and Cybernetics, 20*, 944-950.
3. Gips, J. (1998). On building intelligence into EagleEyes. In V. O. Mittal, H. A. Yanco, J. Aronis, & R. Simpson (Eds.), *Assistive Technology and Artificial Intelligence* (pp. 50-58). Berlin: Springer-Verlag.
4. Goldberg, J. H., & Schryver, J. C. (1995). Eye-gaze determination of user intent at the computer interface. In J. M. Findlay, R. Walker, & R. W. Kentridge (Eds.), *Eye Movement Research: Mechanisms, Processes, and Applications* (pp. 491-502). New York: Elsevier Science Publishing.
5. Hutchinson, T. E., White, K. P., Martin, W. N., Reichert, K. C., & Frey, L. A. (1989). Human-computer interaction using eye-gaze input. *IEEE Transactions on Systems, Man, and Cybernetics, 19*, 1527-1534.
6. Jacob, R. J. K. (1995). Eye tracking in advanced interface design. In W. Barfield & T. A. Furness (Eds.), *Virtual Environments and Advanced Interface Design* (pp. 258-288). New York: Oxford University Press.
7. Jacob, R. J. K., Deligiannidis, L., & Morrison, S. (1999). A software model and specification language for non-WIMP user interfaces. *ACM Transactions on Computer-Human Interaction, 6*, 1-46.
8. John, B. E. (1990). Extensions of GOMS analyses to expert performance requiring perception of dynamic visual and auditory information. In *Proceedings of CHI 90* (pp. 107-115). New York: ACM Press.
9. Salvucci, D. D. (1999). *Mapping eye movements to cognitive processes*. Doctoral Dissertation, Department of Computer Science, Carnegie Mellon University.
10. Salvucci, D. D. (1999). Inferring intent in eye-movement interfaces: Tracing user actions with process models. In *Human Factors in Computing Systems: CHI 99 Conference Proceedings* (pp. 254-261). New York: ACM Press.
11. Stampe, D. M., & Reingold, E. M. (1995). Selection by looking: A novel computer interface and its application to psychological research. In J. M. Findlay, R. Walker, & R. W. Kentridge (Eds.), *Eye Movement Research: Mechanisms, Processes, and Applications* (pp. 467-478). New York: Elsevier Science Publishing.
12. Yanco, H. A. (1998). Wheelesley: A robotic wheelchair system: Indoor navigation and user interface. In V. O. Mittal, H. A. Yanco, J. Aronis, & R. Simpson (Eds.), *Assistive Technology and Artificial Intelligence* (pp. 256-268). Berlin: Springer-Verlag.
13. Zhai, S., Morimoto, C., & Ihde, S. (1999). Manual and gaze input cascaded (MAGIC) pointing. In *Human Factors in Computing Systems: CHI 99 Conference Proceedings* (pp. 246-253). New York: ACM Press.

Evaluation of Eye Gaze Interaction

Linda E. Sibert
Human-Computer Interaction Lab
NCARAI
Naval Research Laboratory
Washington DC 20375 USA
202 767-0824
sibert@itd.nrl.navy.mil

Robert J.K. Jacob
Department of Electrical Engineering &
Computer Science
Tufts University
Medford MA 02155 USA
617 627-3217
jacob@eecs.tufts.edu

ABSTRACT

Eye gaze interaction can provide a convenient and natural addition to user-computer dialogues. We have previously reported on our interaction techniques using eye gaze [10]. While our techniques seemed useful in demonstration, we now investigate their strengths and weaknesses in a controlled setting. In this paper, we present two experiments that compare an interaction technique we developed for object selection based on a where a person is looking with the most commonly used selection method using a mouse. We find that our eye gaze interaction technique is faster than selection with a mouse. The results show that our algorithm, which makes use of knowledge about how the eyes behave, preserves the natural quickness of the eye. Eye gaze interaction is a reasonable addition to computer interaction and is convenient in situations where it is important to use the hands for other tasks. It is particularly beneficial for the larger screen workspaces and virtual environments of the future, and it will become increasingly practical as eye tracker technology matures.

Keywords

Eye movements, eye tracking, user interfaces, interaction techniques

INTRODUCTION

We describe two experiments that compare our eye gaze object selection technique with conventional selection using a mouse. We have previously found that people perform well with eye gaze interaction in demonstrations. The next step is to show that our technique can stand up to more rigorous use and that people are comfortable selecting objects using eye gaze over a more extended period of time. We compare the performance of eye gaze interaction with that of a widely used, general-purpose device: the mouse. Eye gaze interaction requires special hardware and software. The question is whether it is worth the extra effort. If it performs adequately, we can also gain some hard-to-quantify side benefits of an additional, passive or lightweight input channel. For example, we have found that when eye gaze interaction is working well, the system can feel as though it is anticipating the user's commands, almost as if it were reading the user's mind. It requires no manual input, which frees the hands for other tasks. A reasonable definition of performing well is if eye gaze interaction does not slow down interaction and can "break even" with the mouse in a straightforward experimental comparison, despite the immaturity of today's eye tracker technology. If the eye gaze interaction technique is faster, we consider it a bonus, but not the primary motivation for using eye tracking in most settings.

Our experiments measured time to perform simple, representative direct manipulation computer tasks. The first required the subject to select a highlighted circle from a grid of circles. The second had the subject select the letter named over an audio speaker from a grid of letters. Our results show a distinct, measurable speed advantage for eye gaze interaction over the mouse in the same experimental setting, consistently in both experiments.

The details of the experiment give insight into how our eye gaze interaction technique works and why it is effective. It is not surprising that the technique is somewhat faster than the mouse. Our research tells us the eye can move faster than the hand. The test of our approach is how our entire interaction technique and algorithm preserves this speed advantage of the eye in an actual object selection task. We studied the physiology of the eye and used that information to extract useful information about the user's higher-level intentions from noisy, jittery eye movement data. Even though our algorithm is based on an understanding of how eyes move, it was unclear that our eye gaze interaction technique would preserve the quickness of the eye because the eye tracking hardware introduces additional latencies. Performance of any interaction technique is the product of both its software and hardware. The experiments show that we have been successful.

RELATED WORK

People continuously explore their environment by moving their eyes. They look around quickly and with little conscious effort. With tasks that are well-structured and speeded, research has shown that people look at what they are working on [17]; the eyes do not wander randomly. Both normal and abnormal eye movements have been recorded and studied to understand processes like reading [16] and diagnosing medical conditions (for example, a link between vestibular dysfunction and schizophrenia shows up in smooth pursuit). People naturally gaze at the world in conjunction with other activities such as manipulating objects; eye movements require little conscious effort; and eye gaze contains information about the current task and the well-being of the individual. These facts suggest eye gaze is a good candidate computer input method.

A number of researchers have recognized the utility of using eye gaze for interacting with a graphical interface. Some have also made use of a person's natural ways of looking at the world as we do. In particular, Bolt suggests that the computer should capture and understand a person's natural modes of expression [5]. His *World of Windows* presents a wall of windows selectable by eye gaze [4, 6]. The object is to create a comfortable way for decision-makers to deal with large quantities of information. A screen containing many windows covers one wall of an office. The observer sits comfortably in a chair and examines the display. The system organizes the display by using eye gaze as an indication of the user's attention. Windows that receive little attention disappear; those that receive more grow in size and loudness. Gaze as an indication of attention is also used in the self-disclosing system that tells the story of *The Little Prince* [23]. A picture of a revolving world containing several features such as staircases is shown while the story is told. The order of the narration is determined by which features of the image capture the listener's attention as indicated by where he or she looks.

Eye gaze combined with other modes can help disambiguate user input and enrich output. Questions of how to combine eye data with other input and output are important issues and require appropriate software strategies [24]. Combining eye with speech using the OASIS system allows an operator's verbal commands to be directed to the appropriate receiver, simplifying complex system control [8]. Ware and Mikaelian [25] conducted two studies, one that investgated three types of selection methods, the other that looked at target size. Their results showed that eye selection can be fast provided the target size is not too small. Zhai, Morimoto, and Ihde [27] have recently developed an innovative approach that combines eye movements with manual pointing.

In general, systems that use eye gaze are attractive because people naturally look at the object of interest. They are used to performing other tasks while looking, so combining eye gaze interaction with other input techniques requires little additional effort.

DEMONSTRATION SYSTEM AND SOFTWARE ARCHITECTURE

Incorporating eye gaze into an interactive computer system requires technology to measure eye position, a finely tuned computer architecture that recognizes meaningful eye gazes in real time, and appropriate interaction techniques that are convenient to use. In previous research, we developed a basic testbed system configured with a commercial eye tracker to investigate interfaces operated by eye gaze [9, 10, 11, 12, 13, 14]. We designed a number of interaction techniques and tested them through informal trial and error evaluation. We learned that people prefer techniques that use natural not deliberate eye movements. Observers found our demonstration eye gaze interface fast, easy, and intuitive. In fact, when our system is working well, people even suggest that it is responding to their intentions rather than to their explicit commands. In the current work, we extended our testbed and evaluated our eye gaze selection technique through a formal experiment.

Previous work in our lab has demonstrated the usefulness of using natural eye movements for computer input. We have developed interaction techniques for object selection, data base retrieval, moving an object, eye-controlled scrolling, menu selection, and listener window selection. We use context to determine which gazes are meaningful within a task. We have built the demonstration system on top of our real-time architecture that processes eye events. The interface consists of a geographic display showing the location of several ships and a text area to the left (see Figure 1) for performing four basic tasks: selecting a ship, reading information about it, adding overlays, and repositioning objects.

The software structure underlying our demonstration system and adapted for the experiments is a real-time architecture that incorporates knowledge about how the eyes move. The algorithm processes a stream of eye position data (a datum every 1/60 of a sec.) and recognizes meaningful events. There are many categories of eye movements that can be tapped. We use events related to a saccadic eye movement and fixations, the general mechanism used to search and explore the visual scene. Other types of eye movements are more specialized and might prove useful for other applications, but we have not made use of them here. For example, smooth pursuit motion partially stabilizes a slow moving target or background on the fovea and optokinetic nystagmus (i.e., train nystagmus) has a characteristic sawtooth pattern of eye motion in response to a moving visual field containing repeated patterns [26]. These movements would not be expected to occur with a static display.

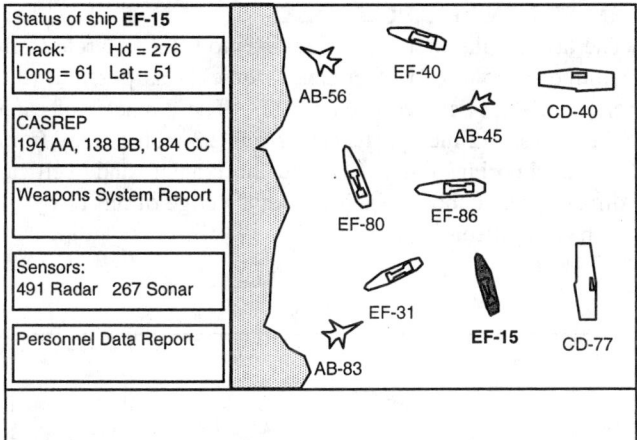

Figure 1. Display from eye tracker demonstration system. Whenever a user looks at a ship in the right window, the ship (highlighted) is selected and information about it is displayed in the left window.

The eyes are rarely still because, in order to see clearly, we must position the image of an object of interest on our fovea, the high-acuity region of the retina that covers approximately one degree of visual arc (an area slightly less than the width of the thumb held at the end of the extended arm). For normal viewing, eyes dart from one fixation to another in a saccade. Saccades are the rapid ballistic movements of the eye from one point of interest to another, whose trajectory cannot be altered once begun. During a saccadic eye movement, vision is suppressed. Saccades take between 30 and 120 msec. and cover a range between 1 to 40 degrees of visual angle (average 15 to 20 degrees). The latency period of the eye before it moves to the next object of interest is at least 100 to 200 msec., and after a saccade, the eyes will fixate (view) an object between 200 to 600 msec. Even when a person thinks they are looking steadily at an object, the eyes make small, jittery motions, generally less than one degree in size. One type is high frequency tremor. Another is drift or the slow random motion of the eye away from a fixation that is corrected with a microsaccade. Microsaccades may improve visibility since an image that is stationary on the retina soon fades [3]. Likewise, it is difficult to maintain eye position without a visual stimulus or to direct a fixation at a position in empty space.

At the lowest level, our algorithm tries to identify fixation events in the data stream and records the start and approximate location in the event queue. Our algorithm is based on that used for analyzing previously recorded files of raw eye movement data [7, 18] and on the known properties of fixations and saccades. A new requirement is that the algorithm must keep up with events in real time. The fixation recognition algorithm declares the start of a fixation after the eye position remains within approximately 0.5 degrees for 100 msec. (the spatial and temporal thresholds are set to take into account jitter and stationarity of the eye). Further eye positions within approximately one degree are assumed to represent continuations of the same fixation. To terminate a fixation requires 50 msec. of data lying outside one degree of the current fixation. Blinks and artifacts of up to 200 msec. may occur during a fixation without terminating it. The application does not need to respond during a blink because the user cannot see visual changes because vision is suppressed.

Tokens for eye events - for start, continuation (every 50 msec. in case the dialogue is waiting to respond to a fixation of a certain duration), end of a fixation, raw eye position (not used currently), failure to locate eye position for 200 msec., resumption of tracking after failure, and entering monitored regions (a strategy typically used for mouse interaction) - are multiplexed into the same event queue stream as those generated by other input devices. These tokens carry information about the screen object being fixated. Eye position is associated with currently displayed objects and their screen extents using a nearest neighbor approach. The algorithm will select the object that is reasonably close to the fixation and reasonably far from all other objects. It does not choose when the position is halfway between two objects. This technique not only improves performance of the eye tracker (which has difficulty tracking at the edges of the screen, see discussion of the range of the eye tracker in the *Apparatus* section) but also mirrors the accuracy of the fovea. A fixation does not tell us precisely where the user is looking because the fovea (the sharp area of focus) covers approximately one degree of visual arc. The image of an object falling on any part of the fovea can be seen clearly. Choosing the nearest neighbor to a fixation recognizes that the resolution of eye gaze is approximately one degree.

The interaction is handled by a User Interface Management System that consists of an executive and a collection of simple individual dialogues with retained state, which behave like coroutines. Each object displayed on the screen is implemented as an interaction object and has a helper interaction object associated with it that translates fixations into the higher unit of gazes. This approach is more than an efficiency. It reflects that the eye does not remain still but changes the point of fixation around the area of interest. (Further details on the software are found in [13].)

STUDY OF EYE GAZE VERSUS MOUSE SELECTION

In developing our demonstration system, we have been struck by how fast and effortless selecting with the eye can be. We developed the interaction techniques and software system after much studying, tinkering, and informal testing. The next step was to study the eye gaze technique under more rigorous conditions. For eye gaze interaction to be useful, it must hold up under more demanding use than a demonstration and operate with reasonable responsiveness. We conducted two experiments that compared the time to select with our eye gaze selection technique and with a mouse. Our research hypothesis states that selecting with eye gaze selection is faster than selecting with a mouse.

Our hypothesis hardly seems surprising. After all, we designed our algorithm from a understanding of how eyes move. Physiological evidence suggests that saccades should be faster than arm movements. Saccades are ballistic in nature and have nearly linear biomechanical characteristics [1, 2, 20]. The mass of the eye is primarily from fluids and the eyeball can be moved easily in any direction, in general. In contrast, arm and hand movements require moving the combined mass of joints, muscles, tendons, and bones. Movement is restricted by the structure of the arm. A limb is maneuvered by a series of controlled movements carried out under visually guided feedback [21]. Furthermore, we must move our eyes to the target before we move the mouse.

However, we were not comparing the behavior of the eye with that of the arm in these experiments. We were comparing two complete interaction techniques with their associated hardware, algorithms, and time delays. For our research hypothesis to be true, our algorithm, built from an understanding of eye movements, plus the eye tracker which adds its own delay, must not cancel out the inherent speed advantage of the eye.

METHOD

To test whether eye gaze selection is faster than selecting with a mouse, we performed two experiments that compared the two techniques. Each experiment tried to simulate a real user selecting a real object based on his or her interest, stimulated by the task being performed. In both experiments, the subject selected one circle from a grid of circles shown on the screen. The first was a quick selection task, which measured "raw" selection speed. The circle to be selected was highlighted. The second experiment added a cognitive load. Each circle contained a letter, and the spoken name of the letter to be selected was played over an audio speaker. The two experiments differed only in their task. The underlying software, equipment, dependent measures, protocol, and subjects were the same.

Interaction Techniques

Our eye gaze selection technique is based on dwell time. We compared that with the standard mouse button-click selection technique found in direct manipulation interfaces. We chose eye dwell time rather than a manual button press as the most effective selection method for the eye based on previous work [9]. A user gazes at an object for a sufficiently long time to indicate attention and the object responds, in this case by highlighting. A quick glance has no effect because it implies that the user is surveying the scene rather than attending to the object. Requiring a long gaze is awkward and unnatural so we set our dwell time to 150 msec., based on previous informal testing, to respond quickly with only a few false positive detections. The mouse was a standard Sun mouse without acceleration.

EXPERIMENT 1: CIRCLE TASK

The task for the first experiment was to select a circle from a three by four grid of circles as quickly as possible (the arrangement is shown in Figure 2). The diameter of each circle was 1.12 inches. Its center was 2.3 inches away from its neighboring circles in the horizontal and vertical directions and about 3 inches from the edge of the 11 by 14 inch CRT screen.

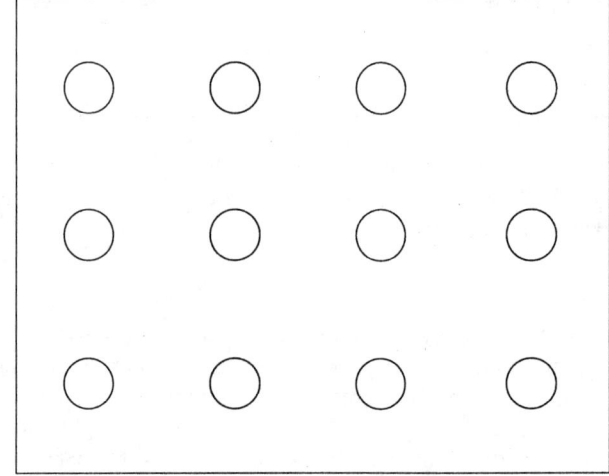

Figure 2. Screen from the circle experiment. The letter experiment has the same arrangement with the letters inscribed alphabetically in the circles, left to right, top to bottom.

Targets were presented in sequences of 11 trials. The first trial was used for homing to a known start position and was not scored. The target sets were randomly generated and scripted. One restriction was imposed that no target was repeated twice in a row. The same target scripts were presented to each subject. A target highlighted at the start of a trial; when it was selected, it de-highlighted and the next target in the sequence highlighted immediately. In this way, the end position of the eye or mouse for one trial became the start position for the next. No circle other than the target was selectable (although information about wrong tries was recorded in the data file). We presented the trials serially rather than as discrete trials to capture the essence of a real user selecting a real object based on his or her own interest. The goal was to test our interaction technique in as natural a setting as possible within a laboratory experiment.

Apparatus

The subject sat in a straight-backed stationary chair in front of a table (29.5 inches tall) that held a Sun 20-inch color monitor. The eye to screen distance was approximately three feet. The mouse rested on a 15-inch square table (28.5 inches tall) that the subject could position. The eye tracker hardware and experimenter were located to the subject's left, which dictated that only individuals that use the mouse right-handed could be subjects (otherwise we would have had to rearrange the equipment and recalibrate). The operator stood in front of the eye tracker console to adjust the eye image when needed and control the order of the experiment. The subject wore a thin, lightweight velcro band around the forehead with a Polhemus 3SPACE Tracker sensor attached above the left eye, which allowed a little larger range of head motion with the eye tracker.

The eye tracker was an Applied Science Laboratories (Bedford, MA) Model 3250R corneal reflection eye tracker that shines an on-axis beam of infrared light to illuminate the pupil and produce a glint on the cornea. These two features - the pupil and corneal reflection - are used to determine the x and y coordinates of the user's visual line of gaze every 1/60 second. Temporal resolution is limited to the video frame rate so that some dynamics of a saccade are lost. The measurable field of view is 20 degrees of visual angle to either side of the optics, about 25 degrees above and about 10-degrees below. Tracking two features allows some head movement because it is possible to distinguish head movements (corneal reflection and center of pupil move together) from eye movements (the two features move in opposition to one another). We extended the allowable range that a subject could move from one square inch to 36 square inches by adding mirror tracking (a servo-controlled mirror allows +/-6 inches of lateral and vertical head motion). Mirror tracking allows automatic or joystick controlled head tracking. We enabled magnetic head tracking (using head movement data from the Polhemus mounted over the subject's left eye) for autofocusing.

The position of gaze was transmitted to a stand-alone Sun SPARCserver 670 MP through a serial port. The Sun performed additional filtering, fixation, and gaze recognition, and some further calibration, as well as running the experiments. The mouse was a standard Sun optical mouse. Current eye tracking technology is relatively immature, and we did have some equipment problems, including the expected problem of the eye tracker not working with all subjects. Our eye tracker has difficulties with hard contact lenses, dry eyes, glasses that turn dark in bright light, and certain corneas that produce only a dim glint when a light is shown from below. Eye trackers are improving, and we expect newer models will someday solve many of these problems.

Our laboratory's standard procedure for collecting data is to write every timestamped event to disk as rapidly as possible for later analysis, rather than to perform any data reduction on the fly [15]. Trials on which the mouse was used for selection tracked the eye as well, for future analysis. We stored mouse motion, mouse button events, eye fixation (start, continuation, end), eye lost and found, eye gaze (start, continuation, end), start of experiment, eye and mouse wrong choices, eye and mouse correct choices, and timeout (when the subject could not complete a trial and the experiment moved on). All time was in milliseconds, either from the eye tracker clock (at 1/60 sec. resolution) or the Sun system clock (at 10 msec. resolution). We isolated the Sun from our network to eliminate outside influences on the system timing.

Subjects

Twenty-six technical personnel from the Information Technology Division of the Naval Research Laboratory volunteered to participate in the experiment without compensation. We tested them to find 16 for whom the eye tracker worked well. All had normal or corrected vision and used the mouse right-handed in their daily work (required because the eye tracker and experimenter occupied the space to the left). All participants were male, but this was not by design. The four women volunteers fell into the group whom the eye tracker failed to track, though women have successfully used our system in the past. The major problems were hard contact lenses and weak corneal reflections that did not work well with our system.

Procedure

Each subject first completed an eye tracker calibration program. The subject looked, in turn, at a grid of nine points numbered in order, left to right, top to bottom. This calibration was checked against a program on the Sun and further adjustments to the calibration were made, if needed, by recording the subject's eye position as they looked at 12 offset points, one at each target location. These two steps were repeated until the subject was able to select all the letters on the test grid without difficulty. The subject then practiced the task, first with the mouse and then the eye gaze selection technique. The idea was to teach the underlying task with the more familiar device. The subject completed six sets of 11 trials (each including the initial homing trial) with each interaction device. Practice was followed by a 1.5 minute break in which the subject was encouraged to look around; the eye was always tracked and the subject needed to move away from the infrared light of the eye tracker (the light dries the eye, but less than going to the beach) as well as to rest from concentrating on the task. In summary, the targets were presented in blocks of 66 (six sequences of 11), mouse followed by eye. All subjects followed the same order of mouse block, eye block, 1.5 minute rest, mouse block, eye block. Because of difficulties with our set-up, we chose to run only one order. We felt this to be an acceptable, although not perfect solution, because the two techniques use different muscle groups, suggesting that the physical technique for manipulating the input should not transfer. Because of blocking in the design, we were able to test for learning and fatigue. Each experiment lasted approximately one hour.

Results

The results show that it was significantly faster to select a series of circle targets with eye gaze selection than with a mouse. The mean time for selection is shown in Figure 3.

Device	Experiments			
	Circle		Letter	
	Mean (ms.)	Std. Dev.	Mean (ms.)	Std. dev.
Eye gaze	503.7	50.56	1103.0	115.93
Mouse	931.9	97.64	1441.0	114.57

Figure 3. Time per trial in msec.

Performance with eye gaze averaged 428 msec. faster. These observations were evaluated with a repeated-measures mixed model analysis of variance. Device effect was highly significant at $F(1,15) = 293.334$, $p < 0.0001$. The eye gaze and mouse selection techniques were presented in two blocks. While there was no significant learning or fatigue, the mouse did show a more typical learning pattern (performance on the second block averages 43 msec. faster) while eye gaze selection remained about the same (about 4 msec. slower).

Only performance on correct trials was included in the analysis. We also observed that excessively long or short trials were generally caused by momentary equipment problems (primarily with the eye tracker; 11% of eye trials and 3% of mouse) and were therefore not good indications of performance. We removed these outliers using the common interquartile range criterion (any observation that is 1.5 times the interquartile range either above the third quartile or below the first was eliminated).

An issue is whether the stopping criteria, dwell time for the eye and click for the mouse, can be fairly compared. Does one take much more time than the other? When we first researched the question, we thought we would have to set our dwell time higher than 150 msec. because Olson and Olson [19] reported that it takes 230 msec. to click a mouse. When we tested a click (mouse down - mouse up), we found it took less time in our setting. We confirmed our decision that using 150 msec. dwell time is reasonable by analyzing the time it actually took subjects to click the mouse in our circle experiment using the time-stamped data records we had collected. It took an average of 116 msec. Only four subjects averaged more than 150 msec., the highest being 165 msec. The fastest time was 83 msec. Olson and Olson's figure probably includes more than just the end condition we needed. We concluded that the 150 msec. dwell time compared with an average 116 msec. click for the mouse would, if anything, penalize performance in the eye condition rather than the mouse.

EXPERIMENT 2: LETTER TASK

The task for the second experiment was to select a letter from a grid of letters. Each letter was enclosed in a circle, and the circles were the same size and arrangement as in Experiment 1. The letters fit just inside the circles; each character was approximately 0.6 inches high, in a large Times font. The subject was told which letter to select by means of a prerecorded speech segment played through an audio speaker positioned to their right. When a letter was selected, it highlighted. If the choice was correct, the next letter was presented. If incorrect, a "bong" tone was presented after 1250 msec. so that a subject who misheard the audio letter name could realize his or her mistake. We set the length of the delay through a series of informal tests. The delay we chose is fairly long, but we found if the signal came more quickly in the eye condition, it was annoying. (One pilot subject reported feeling like a human pin ball machine at a shorter duration.)

The apparatus used was the same as in the circle experiment with the addition of an audio speaker placed two feet to the right of the subject. The names of the letters were recorded on an EMU Emulator III Sampler and played via a MIDI command from the Sun. Playing the digitized audio, therefore, put no load on the main computer and did not affect the timing of the experiment. The internal software was the same and the same data were written to disk. The timing of the experiment was the same for the eye gaze selection condition and the mouse condition.

The subjects were the same 16 technical personnel. All completed the letter experiment within a few days after the circle experiment. The protocol for the letter experiment was identical to the first experiment: calibration, practice, alternating mouse and eye gaze blocks, all interspersed with breaks. The difference between the two experiments was the cognitive load added by having the subject first hear and understand a letter, and then find it. The purpose of the task was to approximate a real-world one of thinking of something and then acting on it.

Results

The results show that it was significantly faster to hear a letter and select it by eye gaze selection than with the mouse. The mean time for selection is shown in Figure 3. Performance with eye gaze averaged 338 msec. faster. These observations were evaluated with a repeated-measures analysis of variance. Device effect was highly significant at $F(1,15) = 292.016$, $p < 0.0001$. The eye gaze and mouse selection techniques also were presented in two blocks. Again, there was no significant interaction. The mouse showed typical learning (performance in the second block averaged 17 msec. faster). Eye gaze selection showed some slowing (by 17 msec.). Again, only performance on correct trials was included in the analysis and outliers were removed as before (5% of eye trials and 3% of mouse).

DISCUSSION

Our experiments show that our eye gaze selection technique is faster than selecting with a mouse on two basic tasks. Despite some difficulties with the immature eye tracking technology, eye selection held up well. Our subjects were comfortable selecting with their eyes. There was some slight slowing of performance with eye gaze that might indicate fatigue, but there is not enough evidence to draw a conclusion.

We do not claim that the speed advantage we obtained is sufficient reason to use this technology. What the speed advantage shows is that our eye gaze interaction technique and the hardware we used works well. Our algorithm maintains the speed advantage of the eye. Our previous experience suggests benefits for eye gaze interaction in naturalness and ease. It is a good additional input channel, and we have now shown that its claimed benefits can be obtained without incurring any performance penalty.

In making our comparisons, we explicitly excluded the higher cost of the eye tracking equipment, which we view as a temporary obstacle as these costs continue to fall. We are concerned with the potential of eye movement-based interaction in general, rather than the performance of a particular eye tracker. For our results to be useful in practical settings, we postulate a better and cheaper eye tracker becoming available, but we simulate such with the hardware available today. Except for the most severely time-critical applications, we would not suggest deploying a duplicate of our laboratory configuration yet.

Because both experiments used the same design and subjects, we can say something about how the two different tasks responded to our techniques. The increment in time from the circle experiment to the letter experiment was similar for each device: 599 msec. for the eye and 509 msec. for the mouse. We suggest that this increment might account for a comprehension and search subtask in the letter experiment, which was not required in the circle one. That subtask is likely to be similar regardless of whether mouse or eye gaze is used. The speed advantage for eye gaze in the selection phase is about the same across tasks.

As a byproduct of this experiment, we also analyzed our data from the circle experiment with respect to Fitts' law, to investigate whether this eye movement-based interaction technique follows the model, as manual interaction techniques typically do. Previous research by Ware and Mikaelian [25] suggests that it does. Their eye interaction techniques produce slopes almost like those for a mouse. However, they include either a long dwell time (400 msec.) or a button press in their times. In contrast, Abrams, Meyer, and Kornblum [1], studying pure eye movement, show only a little increase in time of saccadic eye movements with movement distance and a noticeable increase in velocity.

While we did not have a range of target sizes in this experiment, we did have a range of distances, from adjacent targets to opposite ends of the screen. The distance for each trial is known because the starting point for a trial is the target from the preceding trial (provide the user hit the correct target on the preceding trial; trials preceded by errors were thus deleted from this analysis).

The results of this analysis are given elsewhere [22], but our overall finding is that our eye gaze results are more similar to those of Abrams, et. al. (The mouse results are similar to other Fitts' studies.) The circle task was a fairly pure movement one, and our technique does not involve a long dwell time. The result shows that our algorithm preserves the speed advantage of the eye. What our Fitts analysis points out is that, within the range we have tested, the further you need to move, the greater the advantage of the eye because its cost is nearly constant.

CONCLUSIONS

Eye gaze interaction is a useful source of additional input and should be considered when designing advanced interfaces in the future. Moving the eyes is natural, requires little conscious effort, and frees the hands for other tasks. People easily gaze at the world while performing other tasks so eye gaze combined with other input techniques requires little additional effort. An important side benefit is that eye position implicitly indicates the area of the user's attention.

We argue for using natural eye movements and demonstrate interaction techniques based on an understanding of the physiology of the eye. Our algorithm extracts useful information about the user's higher-level intentions from noisy, jittery eye movement data. Our approach is successful because it preserves the advantages of the natural quickness of the eye.

We presented two experiments that demonstrate that using a person's natural eye gaze as a source of computer input is feasible. The circle experiment attempted to measure *raw* performance, while the letter experiment simulated a real task in which the user first decides which object to select and then finds it. Our experimental results show that selecting with our eye gaze technique is indeed faster than selecting with a mouse. The speed difference with the eye is most evident in the circle experiment. Selecting a sequence of targets was so quick and effortless that one subject reported that it almost felt like watching a moving target, rather than actively selecting it.

ACKNOWLEDGMENTS

We thank Jim Templeman and Debby Hix for all kinds of help with this research; Astrid Schmidt-Nielsen for help and advice on experimental design and data analysis; and our colleagues at the Naval Research Laboratory who took time from their work to serve as experimental subjects. We also thank the CHI reviewers for their thoughtful comments. Portions of this work were sponsored by the Office of Naval Research (at NRL) and ONR Grant N00014-95-1-1099, NSF Grant IRI-9625573, and NRL Grant N00014-95-1-G014 (at Tufts University).

REFERENCES

1. Abrams, R. A., Meyer, D. E., and Kornblum, S. Speed and accuracy of saccadic eye movements: Characteristics of impulse variability in the oculomotor system. *Journal of Experimental Psychology: Human Perception and Performance 15*, 3 (1989), 529-543.

2. Bartz, A. E. Eye-movement latency, duration, and response time as a function of angular displacement. *Journal of Experimental Psychology 64*, 3 (1962), 318-324.

3. Boff, K. R. and Lincoln, J. E., eds. *Engineering Data Compendium: Human Perception and Performance*. AAMRL, Wright-Patterson AFB, OH, 1988.

4. Bolt, R. A. Gaze-orchestrated dynamic windows. *Computer Graphics 15*, 3 (Aug. 1981), 109-119.

5. Bolt, R. A. Eyes at the interface. *Proc. ACM CHI'82* (1982), 360-362.

6. Bolt, R. A. *Eyemovements in Human/Computer Dialogue*. AHIG Research Report 92-1, MIT Media Laboratory, 1992.

7. Flagg, B. N. *Children and Television: Effects of Stimulus Repetition on Eye Activity*. Thesis, Doctor of Education degree, Graduate School of Education, Harvard University, 1977.

8. Glenn III, F. A., Iavecchia, H. P., Ross, L. V., Stokes, J. M., Weiland, W. J., Weiss, D., and Zakland, A. L. Eye-voice-controlled interface, *Proc. of the Human Factors Society* (1986), 322-326.

9. Jacob, R. J. K. *The use of eye movements in human-computer interaction techniques: What you look at is what you get*. ACM Transactions on Information Systems 9, 3 (April 1991), 152-169.

10. Jacob, R. J. K. What you look at is what you get: Eye movement-based interaction techniques. *Proc. ACM CHI'90* (1992), 11-18.

11. Jacob, R. J. K. Eye-gaze computer interfaces: What you look at is what you get. *IEEE Computer 26*, 7 (July 1993), 65-67.

12. Jacob, R. J. K. What you look at is what you get: Using eye movements as computer input. *Proc. Virtual Reality Systems'93* (1993), 164-166.

13. Jacob, R. J. K. Eye movement-based human-computer interaction techniques: Toward non-command interfaces. In *Advances in Human-Computer Interaction*, Vol. 4, ed. H. R. Hartson and D. Hix, Ablex Publishing Co. (1993), Norwood, N.J., 151-190.

14. Jacob, R. J. K. Eye tracking in advanced interface design. In *Advanced Interface Design and Virtual Environments*, ed. W. Barfield and T. Furness, Oxford University Press, Oxford, 1994.

15. Jacob, R. J. K., Sibert, L. E., Mcfarlane, D. C., and Mullen, M. P. 1994. Integrality and Separability of Input Devices. *ACM Transactions on Computer-Human Interaction 1*, 1 (1994), 3-26.

16. Just, M. A. and Carpenter, P. A. A theory of reading: from eye fixations to comprehension. *Psychological Review 87*, 4 (1980), 329-354.

17. Just, M. A. and Carpenter, P. A. Eye fixations and cognitive processes. *Cognitive Psychology 8* (1976), 441-480.

18. Lambert, R.H., Monty, R.A., and Hall, R.J. High-speed Data Processing and Unobtrusive Monitoring of Eye Movements. *Behavior Research Methods and Instrumentation 6*,6 (1974), 525-530.

19. Olson, J. R., and Olson, G. M. The growth of cognitive modeling in human-computer interaction since goms. *Human-Computer Interaction 5* (1990), 221-265.

20. Prablanc, C. and Pelisson, D. Gaze saccade orienting and hand pointing are locked to their goal by quick internal loops. In *Attention and Performance XIII: Motor Representation and Control*, M. Jeannerof (ed.). Lawrence Erlbaum Associates (1990), Hillsdale, N.J., 653-676.

21. Sheridan, M. R. A reappraisal of fitts' law. *Journal of Motor Behavior 11*, 3 (1979), 179-188.

22. Sibert, L. E., Templeman, J. N., and Jacob, R. J. K. *Evaluation and Analysis of Eye Gaze Interaction*. NRL Report, Naval Research Laboratory, Washington, D.C., 2000, in press.

23. Starker, I., and Bolt, R. A. A gaze-responsive self-disclosing display. *Proc. ACM CHI'90* (1990), 3-9.

24. Thorisson, K. R., and Koons, D. B. *Unconstrained Eye Tracking in Multi-Modal Natural Dialogue*. AHIG Research Report 92-4, MIT Media Laboratory, 1992.

25. Ware, C. and Mikaelian, H. H. An evaluation of an eye tracker as a device for computer input. *Proc. ACM CHI'87* (1987), pp. 183-188.

26. Young, L. R. and Sheena, C. Methods and designs: survey of eye movement recording methods. *Behavior Research Methods & Instrumentation 7*, 5 (1975), 397-429

27. Zhai, S., Morimoto, C., and Ihde, S. Manual and Gaze Input Cascaded (MAGIC) Pointing. *Proc. ACM CHI'99* (1999), pp. 246-253.

Enriching buyers' experiences: the SmartClient approach

Pearl Pu
Ergonomics of Intelligent Systems
ISR/DMT
Swiss Institute of Technology Lausanne
CH-1015 Ecublens EPFL, Switzerland
+41 21 693 6081
pearl.pu@epfl.ch

Boi Faltings
Artificial Intelligence Laboratory
Computer Science Department
Swiss Insitute of Technology Lausanne
CH-1015 Ecublens EPFL, Switzerland
+41 21 693 2738
boi.faltings@epfl.ch

ABSTRACT

In electronic commerce, a satisfying buyer experience is a key competitive element. We show new techniques for better adapting interaction with an electronic catalog system to actual buying behavior. Our model replaces the sequential separation of needs identification and product brokering with a conversation in which both processes occur simultaneously. This conversation supports the buyer in formulating his or her needs, and in deciding which criteria to apply in selecting a product to buy. We have experimented with this approach in the area of travel planning and developed a system called SmartClient Travel which supports this process. It includes tools for need identification, visualization of alternatives, and choosing the most suitable one. We describe the system and its implementation, and report on user studies showing its advantages for electronic catalogs.

Keywords

eCommerce, on-line travel planning systems, visual overview, client-server architecture, constraint solver

INTRODUCTION

A common assumption in electronic commerce is that buying starts from clearly identified needs that the buyer is able to articulate. According to [14], there are 6 stages of consumer buying behavior: need identification, product brokering, merchant brokering, negotiation and purchase and delivery. Activities migrate from one stage to another and some stages are iterative processes. However, most e-commerce user interaction separates the first two stages: first the buyer states her criteria, then an initial set of products is shown, followed by possibilities for comparison shopping, negotiation and placing an order.

We believe that in most cases, needs define themselves as a result of the products being offered. For example, originally we might have decided that a 300MHz processor was what we needed for our new PC. When we find out that we can get a 366Mhz and a CD-Rom drive at almost the same price, these might suddenly become part of our needs as well. Conversely, when we find out that the 19" screen we had asked for is very ugly, we might not want it anymore. Needs can end up completely redefined, and not just for irrational reasons: the design of the famous Sydney Opera house was selected in spite of violating *all* the criteria stated in the design competition.

Another important aspect in buyer decision making is that everyone wants to be convinced of getting a *good deal*. The buyer has to be convinced that among all products that can be obtained, what she is buying is an outstanding choice with respect to some criteria. One obvious criterion is price, but often it is also another feature or combination of features. Such a conviction cannot be achieved without comparing an item with its alternatives.

As a consequence, we believe that needs identification, product brokering and comparison should in fact be an iterative *conversation* where criteria and proposals are being exchanged. The buyer takes part in this dialogue by formulating needs and optimization criteria. The catalog should propose solutions and compare them to alternatives, and thus elicit refinement of the buyer's specifications.

We have explored such interfaces in the example of airline flight catalogs. In most current e-commerce sites, the traveler needs to enter dates and times and is then directly led to a small choice of flights. In our approach, the buyer initially specifies only a set of possible destinations and ranges of dates. The catalog then proposes a large set of possible products, using three different displays:

- an overview display which allows comparing the entire range of possibilities according to selected criteria,
- specific example products to elicit further constraints on their attributes, and
- a visualization comparing small sets of alternatives in their attributes.

In all these three displays, buyers identify their needs and evaluate them constantly until they reach an optimal solution. Flexible interaction sequences are supported through the use of constraint satisfaction as a basic selection mechanism. This allows us to model a buyer's criteria accurately and explicitly. Constraints can be posted and retracted in any order, resulting in a flexible conversation that rapidly leads the buyer to refine his needs. Additionally, the constraint satisfaction paradigm provides support for visualizing and comparing the entire space of possibilities, thus ending up with a solution that appears to be the best deal and that he is thus ready to buy. Finally, it is the basis for our SmartClient architecture that implements this rich user experience with lightweight applets.

BUYER EXPERIENCE IN CURRENT TRAVEL E-COMMERCE

Jurca surveyed 10 commercial on-line flight reservation systems [10]. Almost all of them impose a fixed decision-making sequence on the user. As an example, consider Travelocity [19], a popular site for air travel. It requires a user to first fill in a form with the following information:

- Itinerary
- Dates and times of travel
- Class of travel, number of seats required
- Preferences of airline companies

Then it returns a list of possibilities for the first leg of the trip, of which we have to choose one, then possibilities for the second leg, etc.. When all flights have been selected, the systems shows the total price and offers to reserve. Alternatively, we can have a display of 9 different complete itineraries, or an entirely different model where we start by selecting a fare and then choose dates and flights that fit it. In all cases, the buyer has to fit a particular sequence of decision-making in which needs specification and browsing the product offering are two distinct stages.

We think that this fixed sequence makes the site cumbersome to use. If we discover that there are no good connections for the return date we have chosen, we have to restart the entire process from the beginning. If we find that business class is terribly expensive, we have to plan the trip all over again to see what the price in economy would be. As a consequence, it is not uncommon to spend more than 1 hour planning a trip using this site! The process would be much more efficient if it were possible to specify and compare criteria while browsing the available flights.

Furthermore, for longer trips there are many criteria that are important to the traveler but for which the site offers no optimization possibilities: total flying time, departure time, transfer airports, total ground time at transfer airports, etc.

RELATED WORK ON PRODUCT BROKERING

Electronic catalogs are examples of decision-making problems: the task is to make a decision among a set of alternatives. Decisions are made as a result of constraints, preferences, or optimization criteria. In most systems, constraints, preferences and criteria are not modeled explicitly, but remain implicit in selections the user makes.

In most existing systems for e-commerce, buyers have to commit to their needs early on in the buying process, and information about products can only be displayed in response to this. As a result, when buyers' needs change, it results in frequent information exchanges between buyer and seller, with the associated delays and server load. This problem is not uncommon in other practices of e-commerce. The dilemma encountered by system designers is how to best support decision making with sufficient information while guaranteeing a reasonable speed to download it. So far, few solutions have been proposed to implement a rich user experience in an efficient way. Most of the existing electronic catalogs fall into four types: using hierarchies, filtering, preferences, or configuration.

In hierarchically organized catalogs (e.g., PC-Zone [15]), buyers first answer a fixed sequence of questions corresponding to how databases are organized. This questionnaire form is then sent to the seller's catalog server. Product information is retrieved and sent to the buyer. Two problems can be identified: 1) this model is inadequate for natural interaction between buyers and product servers since buyers do not identify their needs in any particular order; 2) only information on a few products is given to the buyers, leaving them wondering what other alternatives are available.

Filtering catalogs (e.g., Personalogic[16], Automated Travel Assistant[12]) allow users to explicitly formulate constraints on what acceptable products are. These act as a filter: the catalog only shows products that satisfy these constraints. Rather than distinguishing acceptable from unacceptable products, constraints could also return a number that reflects the degree to which they are satisfied. It then models users' criteria for optimization. Many catalogs allow optimizing such criteria or combinations of them.

Soft constraint techniques, that is expressing users' criteria as a scale of preferences using weights [11], are more flexible for navigating in large product catalogs [17,18]. There is no sequential order to define criteria. At the same time, soft constraints allow sub-optimal solutions to remain in the navigation space, thus making a large enough portion of the catalog available. Our work is based on the similar observation that larger product space encourages serendipitous buying opportunity. The difference is that we use partial constraint satisfaction techniques to handle soft constraints instead of weights. In domains such as travel, it is hard to attach importance to a criterion before hand without knowing what are the influences of other criteria in the same problem.

Finally for complex products, the approach is very different. Instead of representing every data item in the catalog, configuration techniques [7] are used to propose products to customers according to his current preferences and constraints. Some of the configurable electronic catalogues are currently employed by e-commerce sites at Dell [6] and Cisco[5], and several e-commerce solutions are offered by Calico Commerce[3] and by ILOG [8]. Most of the configuration catalogs, however, do not provide enough interaction techniques for browsing alternatives, nor supporting tradeoff analysis.

As we can see from the above examples, providing a rich user experience by integrating needs identification, product brokering, and product comparison poses challenges for the system architecture and interaction design.

SMARTCLIENT ARCHITECTURE

We have patented a technique that formulates travel planning as a constraint satisfaction problem (CSP [13,20]). It allows transferring product information between product server and buyers through a skinny data connection. At the buyer's side, information is assembled into product configurations according to their constraints and preferences. In travel planning, when a buyer contacts the flight server, he only defines a range of destinations and possible dates. From this information, the system constructs a CSP model which gets shipped to the customer computer. In addition, a constraint solver and a graphical user interface are also included. These generate solutions according to the customer's stated needs and preferences. An advantage of our technique is that the code is very lightweight and can be efficiently packaged into small Java applets. Thus, the download of a typical travel example requires about 500 Kbytes, corresponding to the size of 8 average web pages of Travelocity. Interaction with the visualizations is then instantaneous, providing a rich and satisfying user experience.

We call this architecture SmartClient because it involves clients that are both thin (smart) and intelligent (smart) at the same time. The possible space of products offered at the client sites can go up to thousands.

INTERACTION DESIGN FOR SMARTCLIENT

Buyers enter the interaction with often very vague ideas of what their actual needs are. Thus, in air travel, they usually know where and roughly when they want to travel, but they do not think of many other secondary criteria, such as departure times, airports and airlines they like to avoid, etc. Only the destination airports and ranges of departure and arrival dates are required.

Defining initial needs

We use the world map as a metaphor for defining origination and destination airports.

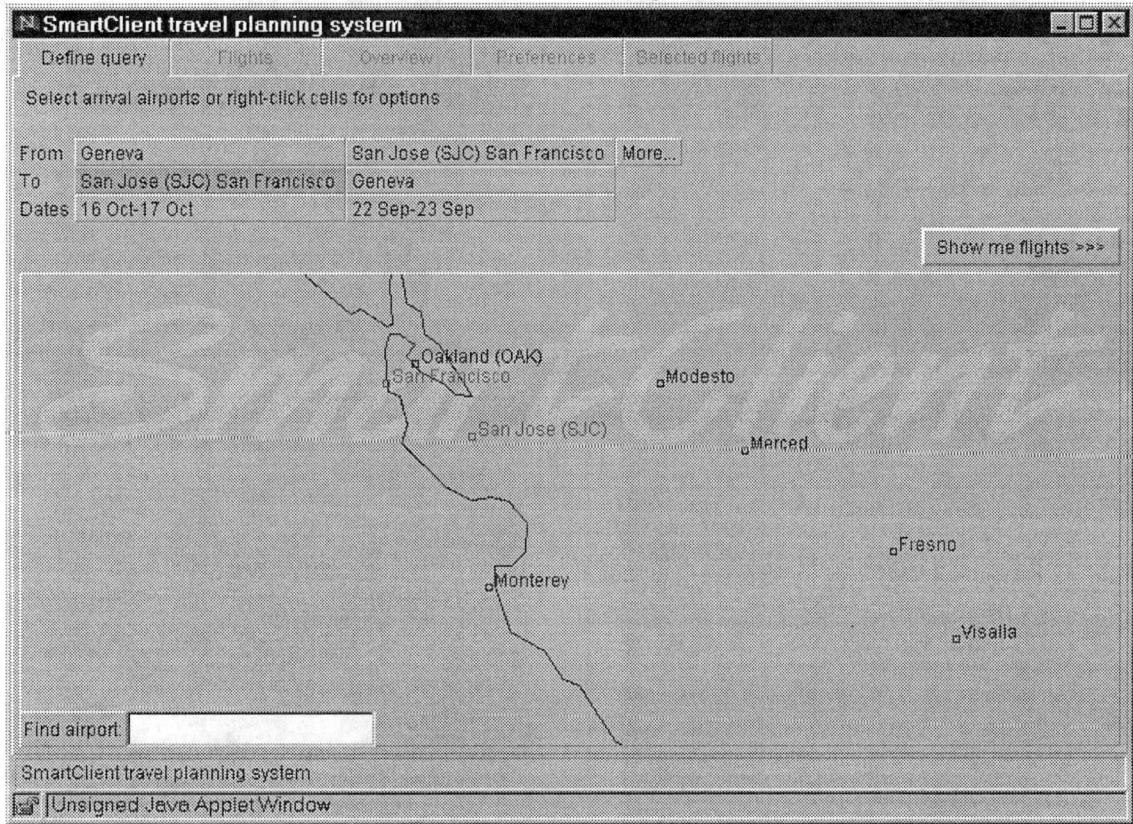

Figure 1: Query display with world map.

As a user zooms in, detailed information appears, such as each country's contours and the names of available airports. Clicking any name will enter the corresponding airport into the itinerary definition panel located on the upper-left corner. An example shown in Figure 1 shows the itinerary data of the following example:

A professor living in Geneva, Switzerland, who wants to spend a week in Silicon Valley to meet with his colleagues. The best airport for his trip is in San Jose.

In SmartClient travel, one can also type the names of the initial airports. This leads to displays of the regional map that shows the selected city among its neighbors (see Figure 1). When selecting the destination, it thus becomes obvious that flying into San Francisco could also be a good alternative, so this city gets selected as well.

The "show me flights" button generates a solution space (not solutions themselves) which is then shipped to the customer's side, whose constraints and preferences are used as guidelines to define an initial focus on the solution space.

Getting an overview of the available products

Often, the range of choices that a buyer might consider is bewildering. When there are many competing and possibly conflicting evaluation criteria, there can be a huge number of relevant choices, each optimal for some of the criteria and suboptimal for others. For example, the cheapest flight may require three plane changes. A non-stop flight, on the other hand, is expensive. In trip planning, the complexity and richness of such decision problems are further compounded by the conditional nature of users' criteria. That is for certain flights, they'd prefer the cheapest, while for others, they'd prefer the non-stop feature.

Some catalog systems attempt to solve this situation by requiring a numerical weighting of the criteria and using the weighted sum of the different criteria to rank the solutions. When criteria depend on context, it is easy to find situations where no weighting can accurately model the correct preference structure. For example, for a flight leaving from Geneva, our professor might like a departure as early as 8 o'clock, while with a departure from Zurich this should be 11 o'clock to account for the train ride there – but this interaction cannot be modeled by feature weights alone. For this reason, it is not very realistic to expect buyers to quantify tradeoffs in this way. More importantly, it leaves the buyer with the uneasy feeling of choosing a product without knowing why.

We believe that a better approach is to help the buyer find the criterion in which one choice clearly stands out as the best one. For example, if all flights leave between 8 and 9 am, this is not a useful criterion for comparing them. On the other hand, if the price varies between 300 and 1500 Francs, this could be a much more important attribute to look at. A buyer who chooses a flight because it costs only half of similar alternatives feels that he is getting a good deal, whereas if he should take a flight because it leaves at 8:45 instead of 8:30 he might not be very sure of his choice. This is an important element in convincing the buyer to actually go ahead with the purchase.

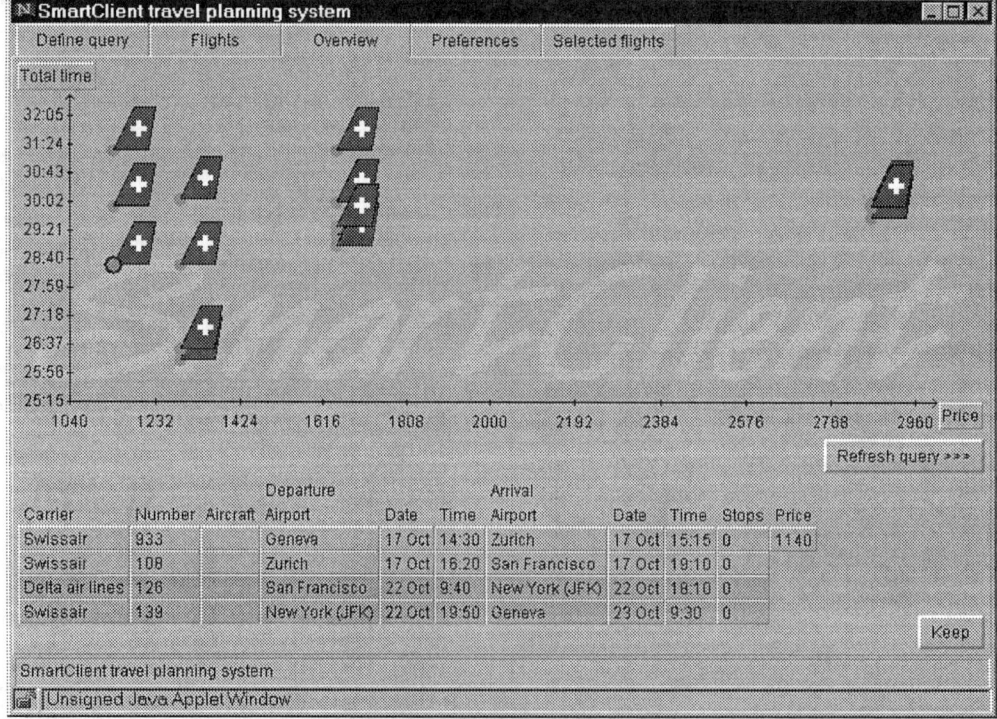

Figure 2: Tradeoff between price and total flying time in Overview.

Such multi-criteria analysis can be performed in the overview display (Figure 2), showing a scatter-plot of a sample set of solutions according to fare (horizontal axis) and total travel time (vertical axis). This technique is similar to the starfield display used in data base query systems described in [1]. However, only a focused set of solutions are displayed. As users change their criteria and preferences, this overview shifts its viewing area to other solutions. Therefore, it implements a type of semantic fish-eye, as opposed to a normal fish-eye view [2]. This scatterplot is useful to see that:

- There is quite a variation in fares, so fare should be a criterion we check for.
- Paying a higher fare does not seem to allow us much shorter flying times.

It is possible to inspect each of the possibilities in the display below, and use this to make an initial choice which we then further inspect in other displays. Here, we select the shortest flight but with the lowest possible fare. This flight is then shown in detail below, and provides a good starting point for further selection. Solutions can be selected using the "Keep" button and are then stored in the "Selected flights" folder.

At any moment during solution space navigation and browsing, users can go to the overview area to further compare trips. Overviews can be provided for any combination of price, total travel time, number of intermediate stops and solution quality regarding to users' criteria. A trip ranks low on solution quality if it violates many of the criteria. Additional information such as the main carrier's flags are denoted by the graphical forms of each node.

Eliciting further needs and constraints

A typical buyer has many constraints that are not stated up front. He becomes aware of these only when solutions are proposed that violate them. In our example, the most cost-effective solution in fact has several problems:

- It leaves too early to allow finishing up the last breakfast meeting on Oct. 22nd.
- It transits in New York JFK airport, which the customer would like to avoid.

The solution display, shown in Figure 3, allows posting

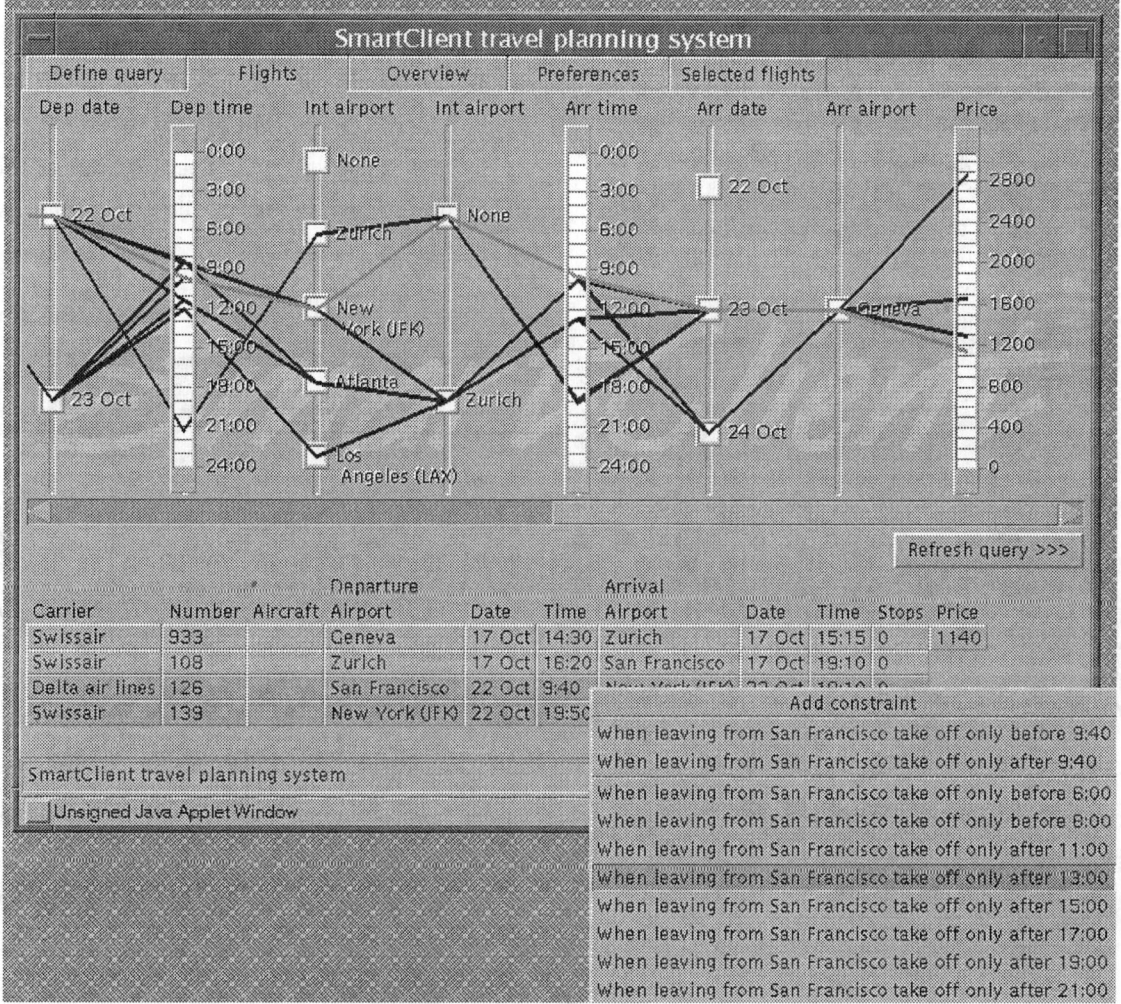

Figure 3: Posting constraints

constraints on any item in the display. In the textual display, they can be posted by clicking on the respective cell and thus activating a menu, as shown in the figure. Similarly it is possible to post constraints on any of the following attributes:

- Price
- Airlines
- Aircraft types
- Departure and arrival dates
- Departure and arrival times
- Intermediate airports
- Direct or non-direct flights

When constraints are posted in this way, they are automatically restricted to the context in which they were posted. For example, if I post a constraint on a departure time, it will by default be applied to flights for that particular leg and leaving from that particular airport only. Applicability can be further restricted by selecting cells as a context, for example only when leaving from San Francisco airport because of the longer driving time.

Constraints can also be posted using sliders in the graphical tracer display above the textual display, which is discussed in more detail later.

Posting constraints in this manner eliminates one major difficulty with conversational interfaces: it makes it impossible for the user to input constraints that cannot be understood. Since the display does not show attributes or values that do not exist, it is not possible to post constraints on them. Thus, we cannot post a constraint on the type of food served on the airline, nor on the size of the seats (unfortunately), since these attributes are not available in airline reservation systems.

At any time, the user can request the system to compute a new set of solutions that satisfy all constraints posted so far, and will usually obtain immediate response. In our example, this shifts the most advantageous trip to one where the return is now from San Jose and through Chicago.

Comparing alternatives in detail

Our users studies found that once a buyer is familiar with such navigation techniques in product space, he'll want faster interaction. When he has narrowed down the choices by posting constraints, he is ready to select particular solutions. We applied the parallel coordinate display method [4,9] to the travel domain. A tracer display (Figure 4) shows each solution as a trace through the set of flight attributes comprising a trip itinerary. For each attribute, there is one vertical bar with its possible values. A solution is a trace that links the values of the different attributes. An individual solution is selected whenever the mouse is moved over it and is then displayed in detail at the bottom of the display.

The tracer display quickly makes apparent the differences among a set of possibilities. Here, we can see in particular that given our constraints, there is a choice of returning via a variety of airports: Chicago, Los Angeles, New York and Zurich, and a variety of times (morning to afternoon).

The tracer display also allows posting additional constraints by using sliders on each attribute. This gives users who prefer to work with more graphical abstractions another way of declaring needs and criteria to the system.

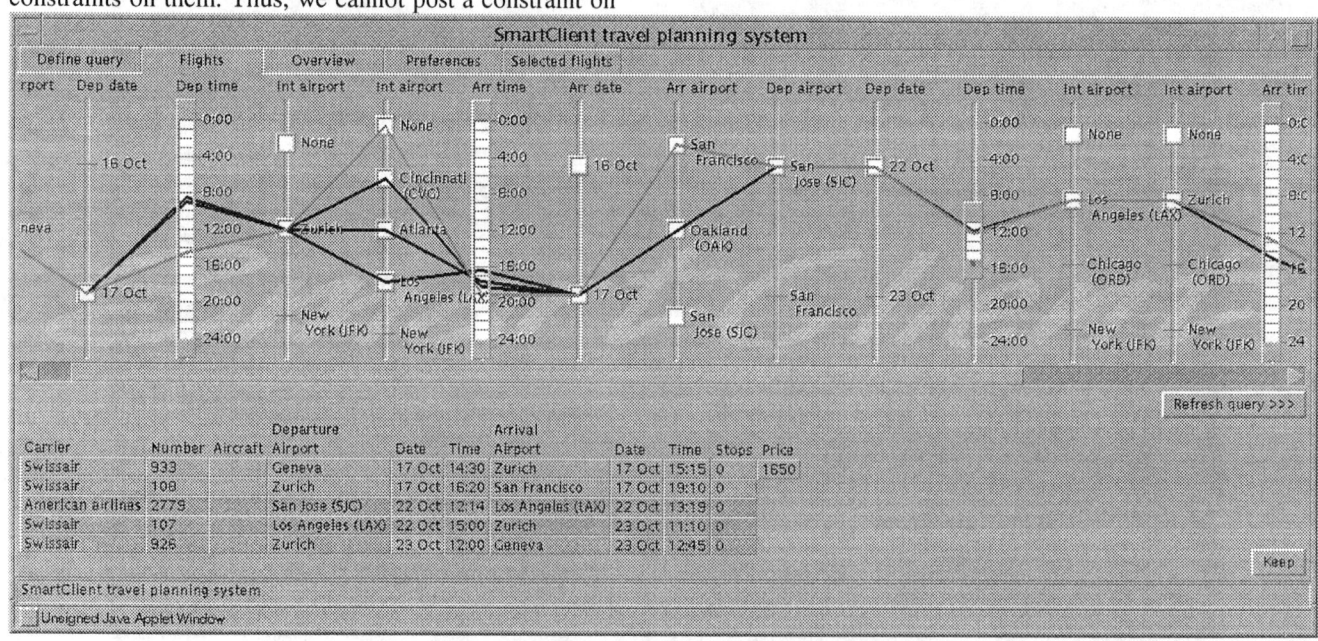

Figure 4: Tracer display with final solution.

Sliders on attribute bars can further provide rapid specification of ranges of data for dates and time. Clicking on the other hand allows easy interaction for choices, such as whether someone wants to stop in Zurich or not. Even though many users are first shocked to see what usually looks like stock market graphs, they appreciate the power of visualizing many options in a compact display area.

Figure 4 displays 17 possibilities as curves mapped onto the multi-value graphs for our example trip. The canvas area for the detailed flight information is sensitive to the current mouse position on the curves. Thus, each solution is compared with the rest and viewed in detail at the same time.

In our example, the user may be particularly sensitive to transfer airports, and inspect solutions based on that criterion. This will lead to further narrowing of the space by:

- Disallowing transfer in Chicago
- Constraining departure time to be after 11 am, from San Jose only

In the final display, shown in Figure 4, the customer now only sees solutions that satisfy all the posted constraints, and can manually compare them to find the truly best one.

As an alternative way, this interaction design is particularly useful when travel is less constrained, allowing buyers to quickly decide that they can only expect small differences in cost, but potentially large gains in travel time. This can help buyers define criteria to find the "good deal".

All of the above can be done without further contacts with the server. Using a common travel site such as Travelocity, the same trip requires more than 10 client-server contacts, at the end of which the customer is still not sure how good the solution he is getting really is.

USER STUDIES

We have evaluated SmartClient with 43 users, all of them students of our university. Being in the 20-23 years old range, this youth group is one of the most targeted groups for airline eCommerce in Europe. The students range from computer science, electrical engineering, industrial engineering, to civil engineering majors. They formed into a team of two students and each team was asked to use SmartClient Travel to complete three specified trips and one trip of their own choice. They were to compare the experience of SmartClient Travel to a commercially available system. One person was to work on the problem using the computer, while the other recorded the time needed for each trip planning, the usability of the software, and the usability of the constraint editing features. Based on the findings, we have the following conclusions:

- None of them had trouble discovering the criteria editing feature offered by SmartClient Travel.
- All of them agreed that SmartClient Travel allows them to examine a much larger space of solutions than other tools.
- Most of them complained that the speed in getting the initial data from SmartClient Travel is a problem (this was due to the low quality connection to the airline information system that was available for the study).
- When trips are simple and needs are known up front, they noted that SmartClient Travel and others are more or less equally powerful.

CONCLUSION

We have shown an approach to electronic catalogs where user criteria and preferences are explicitly modeled as constraint satisfaction. This simple and general formalism is the basis for SmartClients, lightweight applets that allow browsing a space of solutions in an intelligent way. This offers important practical advantages for electronic catalogs:

- Criteria can be given and modified in any order, rather than following a predefined dialogue model. Product selection can become a flexible conversation where customers discover their criteria through inspection on the available choices..
- Using overview displays, users can get a quick idea of the importance of different criteria, and understand tradeoffs between them.
- Different solutions can be compared in a single framework using the tracer display. This lets users make a final their choice that they are confident about and ready to buy.

The SmartClient approach has been implemented as a prototype system. We are currently working on a system to be put in practical use. We are also exploring application of the same technique for electronic commerce of insurance products.

PATENTS

Two patent applications have been submitted for the SmartClient system architecture and the intelligent interface technology.

ACKNOWLEDGEMENT

We thank the Swiss National Science foundation for funding the initial phase of SmartClient. Iconomic Systems SA (http://www.iconomic.com) developed the second generation software architecture and user interface, and is currently commercializing the software.

We thank the undergraduate students in our university for participating in our user studies. We also thank Marc Torrens, Adriana Jurca, and Sebastian Gerlach for the implementation of SmartClient at various stages.

REFERENCES

1. Christopher Ahlberg and Ben Shneiderman, Visual Information Seeking: Tight Coupling of Dynamic

Query Filters with Starfield Displays, Human Factors in Computing Systems. Conference Proceedings CHI'94, pp. 313-317, 1994.

2. Marc H. Brown and James R. Meehan and Manojit Sarkar Browsing Graphs Using a Fisheye View, in *Proceedings of ACM INTERCHI'93 Conference on Human Factors in Computing Systems*, Formal Video Programme: Visualisation, p. 516, 1993.

3. Calico Commerce, http://www.calicotech.com

4. Card, S. Eick, S., and Nahum G., Information Visualization Tutorial, in *CHI'99 tutorial notes*, 1999.

5. Cisco Connection Online, //www.cisco.com

6. Dell Inc., http://www.dell.com

7. Faltings, B. and Freuder, E., Configuration, in *IEEE Intelligent Systems and their applications*, as guest editors' introduction, July/August 1998.

8. ILOG S.A., http://www.ilog.com

9. Insclburg, A. Dimsdale, B., "Parallel Coordinates: A Tool for Visualizing Multi- Dimensional Geometry," in *Proceedings of the First IEEE Conference on Visualization*, 1990.

10. Jurca, A.., Survey of Online Travel Planning Systems, Technical Report (No. 99-01), ISR/DMT, Swiss Federal Institute of Technology Lausanne, 1999.

11. Keeney, R.L., Faiffa H., Decision Making with Multiple Objectives: Preferences and Value Tradeoofs. Cambridge University Press, Cambridge, UK, 1993.

12. G. Linden and S. Hanks and N. Lesh, Interactive Assessment of User Preference Models: The Automated Travel Assistant, in Proceedings of the 6th International Conference on User Modeling (UM-97), CISM, Vol. 383, pp. 67-78, Springer, June 02-05 1997.

13. Mackworth, A., Constraint Satisfaction, *Encyclopedia of Artificial Intelligence*, pp. 205-211, John Wiley and Sons, 1987.

14. Pattie Maes and Robert H. Guttman and Alexandros G. Moukas, Agents that buy and sell, Communications of the ACM, 42(3), pp. 81-91, March 1999.

15. PC-Zone, http://www.pc-zone.com

16. Personalogic/AOL, http://www.personalogic.com

17. Stolze, M. Soft Navigation in Product Catalogs, in *Proceedings of the Second European Conference on Research and Advanced Technology for Digital Libraries*. C.Nikolaou and C. Stephanidis. Heraklion, GR, Springer, Berlin: 385-396, 1998

18. Stolze, M. Comparative Study of Analytical Product Selection Support Mechanisms, in *Proceedings INTERACT 99*, August 1999,

19. Travelocity, http://www.travelocity.com

20. Edward Tsang, Foundations of Constraint Satisfaction, In *Academic Press*, 1993.

Quality is in the Eye of the Beholder: Meeting Users' Requirements for Internet Quality of Service

Anna Bouch
Computer Science Department
University College London
Gower Street, London WC1E 6BT UK
+44 171 504 4436
A.Bouch@cs.ucl.ac.uk

Allan Kuchinsky, Nina Bhatti
Hewlett Packard Labs
1501 Page Mill Road
Palo Alto, CA 94304 USA
+1 650 857 7423
+1 650 857 5610
{kuchinsk,nina@hpl.hp.com}

ABSTRACT

Growing usage and diversity of applications on the Internet makes Quality of Service (QoS) increasingly critical [15]. To date, the majority of research on QoS is systems oriented, focusing on traffic analysis, scheduling, and routing. Relatively minor attention has been paid to user-level QoS issues. It is not yet known how objective system quality relates to users' subjective perceptions of quality. This paper presents the results of quantitative experiments that establish a mapping between objective and perceived QoS in the context of Internet commerce. We also conducted focus groups to determine how contextual factors influence users' perceptions of QoS. We show that, while users' perceptions of World Wide Web QoS are influenced by a number of contextual factors, it is possible to correlate objective measures of QoS with subjective judgements made by users, and therefore influence system design. We argue that only by integrating users' requirements for QoS into system design can the utility of the future Internet be maximized.

Keywords
Internet, Quality of Service, User perception.

INTRODUCTION

The population of users of the World Wide Web is expected to grow from 100 million in 1998 to 320 million by 2002 [24]. While a vision of the future Internet offers the potential to break traditional barriers in communications and commerce, the current service quality to users is often unacceptable [7] [19]. With increasing usage of Internet services, the topic of providing adequate Quality of Service (QoS) for the Internet has become a focus of research. Traditional QoS metrics such as response time and delay no longer suffice to fully describe quality of service as perceived by users. The success of any scheme that attempts to deliver desirable levels of QoS for the future Internet must be based, not only on the progress of technology, but on users' requirements [10].

An increase in demand for access to network bandwidth has led to suggestions that the Internet should implement classes of service according to the QoS needs of applications ([11],[12],[25]). Although resource allocation schemes differ in details, the premise from which they are constructed is fundamentally the same. This premise is to allocate service resources according to the assumed objective QoS requirements of applications. Network service providers are indeed now able to offer a wide range of facilities and services designed to address variability in the QoS requirements of applications. An example is the deployment of adaptive content delivery, where content is altered to take into account application QoS requirements and varying network conditions[2][3]. However, it is currently not known to what extent changes in objective QoS metrics are perceived by, and impact the behavior of users.

Previous research shows that users may judge a relatively fast service to be unacceptable unless it is also predictable, visually appealing and reliable [5]. Additionally, many QoS parameters have been found to interact in users' judgements of quality. For example, Ramsay et. al. [23] found that Web pages that were retrieved faster were judged to be significantly more interesting than their slower counterparts. Although it is often recognized that a measurement of user satisfaction must be included in the assessment of the efficiency of the network as a whole [16], research is needed to identify how the cost of resource allocation for the service provider relates to the value of that resource to users.

We present results from our study into how users define and perceive Internet QoS. We chose to study the influence of different levels of latency. Latency is defined as the delay between a request for a Web page and receiving that page. Previous research has established the salience of such parameters to users [5][8].

RESEARCH QUESTIONS

When designing technologies for electronic services, it is important to understand the interplay between different stakeholders and the extent to which providing benefits to one stakeholder may result in increased costs to another. The questions that we asked in this study examine the requirements of users while considering the need to provide system-level developers with information from which to make design decisions that meet users needs.

We asked the following main questions:

To what extent is there a mapping between objective and subjective QoS?

The definition of QoS from the perspective of users and systems-level network designers diverges. From the perspective of the Internet Service Provider (ISP), QoS approaches that focus on optimizing objective QoS may inform resource allocation mechanisms at a systems level. However, it is not clear to what degree such improvements result in perceived improvements from a user perspective. This problem is compounded by the observation that any definition of objective thresholds that might be established is subject to context-dependent behavior. Subjective QoS thresholds are therefore not fixed throughout a user's interaction. For example, users tolerance for quality is influenced by the particular task in which they are engaged at a particular time.

In addition to research into the partitioning of network resources are claims that levying a charge for premium services is the only way to reflect the value of quality to users and provide economic incentives to service providers [21]. One of the approaches for deploying different levels of QoS is the use of differentiated service mechanisms [25]. A strength of this approach is that it allows traffic flows generated from applications to be aggregated. However, in order to allocate resources to aggregate flows in the appropriate manner, it is first necessary to identify common requirements for QoS that can be associated with those flows. It is not clear that users are willing to pay a premium for any perceived improvements to quality of service, nor have the thresholds above which users perceive the QoS they receive to be qualitatively superior been established.

This question was posed to investigate the level of objective QoS that was considered acceptable to users and to find out if general subjective thresholds could be identified below which an objective level of quality would not be tolerated, in the context of Internet commerce.

What contextual factors influence users' tolerance of Internet QoS?

Our perspective challenges the assumption that there is a strict correlation between objective levels of quality received by users and their perceptions of that quality. For example, demands for a certain type and level of network performance have been shown to vary widely depending on the amount of browser feedback provided [4]. A consistent finding is that QoS received by users should concur with their expectations but that these expectations change according to the pattern of quality received [5], [8]. We set out to investigate the factors that influence users' perception of QoS by asking if users' tolerance for QoS:

- Depends upon the type of task in which they are engaged?
- Changes as the time they spend interacting with the Web site increases?
- Changes if new Web pages are brought up incrementally, or all at once?

What underlying conceptual models influence users' judgements of QoS?

Previous research has established users' models of network operations and how those models influence the levels of quality users are willing to tolerate [5]. Several factors intervene in users' judgements of network quality and play a large part in forming their expectations of future QoS. In addition to investigating contextual factors, we wanted to understand the reasons behind the influence of such factors on users' judgements of QoS. This question, therefore, was designed to investigate how users' conceptual models related to their perceptions of QoS.

METHOD

Research approach

Our research approach combined the gathering of quantitative and qualitative data. This approach was adopted to determine if and where thresholds exist below which users will not tolerate levels of QoS. Objective metrics such as these are the most direct way of informing resource allocation mechanisms. We conducted experimental work to provide information on tolerance thresholds. Additionally, participants were asked to speak aloud about the QoS they received as they performed the task. Through the capture of these verbal protocols, we hoped to record participants' dynamic opinions of QoS. Focus group studies were conducted to gain qualitative data in order to address how contextual factors that influence the definition of thresholds relate to users' conceptual models.

The next section describes the methods we used to answer these research questions.

Participants

There were 30 male participants, aged between 18 and 68, in the study. Previous research indicated that no control was required for age [6]. It was essential that a appropriately homogenous group of users was selected. This was essential because users with different amounts of knowledge and experience of Web QoS have different expectations of QoS [5]. The following criteria were applied when selecting participants. All participants must:

- Use the Internet for at least 2 hours per week

- Have made at least 2 purchases on the Internet in the last year.
- Have at least an intermediate level of self-assessed skill with using computers.

Male participants were selected for the study as it has been shown that there are gender differences in visual perception and learning [17]. Males were identified as the most frequent users of Internet commerce services [22].

Task

The task involved purchasing a home computer system using the HP Shopping Village Web site [14]. This is a frequently used site ranked first for retail revenue generated by e-commerce [13]. This grounded the task in a real-world context. During the task participants were asked to purchase each component of the computer system separately. Participants accessed 22 Web pages during the task. To study whether users' requirements for QoS were similar for similar sub-tasks, a set pattern of actions was repeated through the task. For each component purchased participants were required to:

1. View a class of similar products.
2. Select a specific product from a class of products.
3. Add the chosen product to their shopping cart.
4. View the contents of their shopping cart.

Having a set pattern was important in determining if users' tolerance changed over time. If the tasks had been widely variable then any change in tolerance could be ascribed to the variation in what participants were asked to do, and not to a genuine accumulation of frustration.

The task was designed so that all participants followed the same path through the Web site.

Experimental conditions

During the experiments, users were presented with Web pages that had predetermined delays ranging from 2 to 73 seconds. The choice of this range of speeds was guided by speeds that users had perceived to be qualitatively different in previous research [8][23]. Each user took part in all 3 conditions. The same task was used for each condition.

Investigating Latency: Non-incremental loading

This part of the study investigated whether the latency between requesting a page and receiving it influenced user perceptions of the delay of page delivery. Varying the latency in this condition had the effect that the page where the link had been clicked remained displayed in the browser until the next page had been loaded. This next page was then brought up in its entirety. Predetermined delays were injected into the page loading process. There were two patterns of variation applied to each of two conditions where latency was investigated.

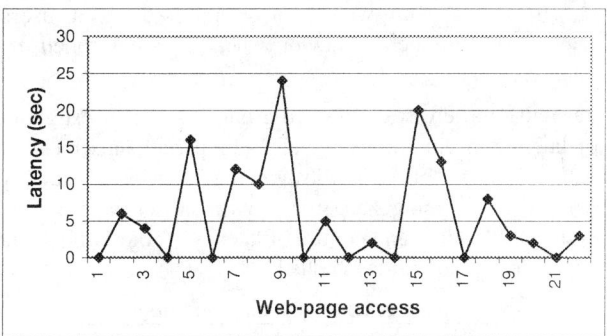

Figure 1: Latency pattern A: Random quality

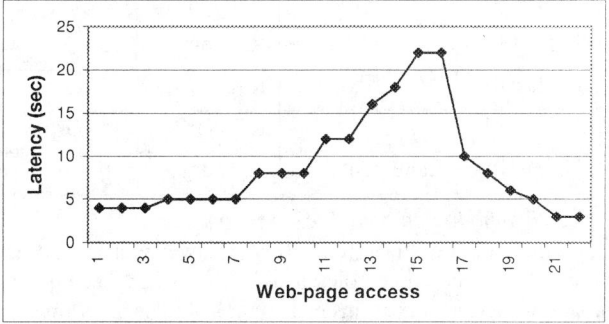

Figure 2: Latency pattern B: Regular quality

Pattern 1 (Figure 1) mimicked a random pattern of delay. In pattern 2 (Figure 2) the delay generated on the Web pages formed a more regular, relatively smooth pattern.

1. Classification of latency (condition 1)

During condition 1, participants were asked to perform the task and rate the latency received for each Web page access. An interface (Figure 3) was developed to register ratings. Participants were directed to click one of the buttons in this interface for each Web page accessed. Participants were told that the black button should be used to indicate that the quality was totally unacceptable.

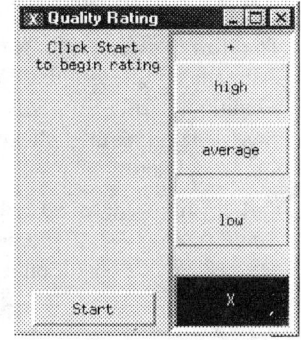

Figure 3: UI

2. Control of latency (condition 2)

During condition 2, participants were again asked to assess the latency of each Web page. However, in this condition they were told that if they found the delay of the Web page unacceptable, they could click a button labeled 'Increase Quality'. The effect of this button was to immediately bring up the requested page. Previous research suggested that this would be a valid measure of users' requirements for speed [4]. This experimental set-up contrasted users' opinions about tolerance of QoS, captured in their direct classifications, with what can be inferred about users' tolerance from their behavior when they controlled the quality.

Participants were split into two groups for the investigation of latency in conditions 1 and 2. 15 participants received pattern 1 for both classification and control of latency, while the remaining 15 participants received pattern 2 for both classification and control of latency. Table 1 shows the experimental conditions applied.

Table 1: Experimental conditions

Condition	Page loading	Pattern	Participants
1: Classify	All at once	Random	15 (Grp 1)
2: Control	All at once	Random	15 (Grp 1)
1: Classify	All at once	Regular	15 (Grp 2)
2: Control	All at once	Regular	15 (Grp 2)
3: Classify	Incremental	Random	30

Investigating Incremental Loading (condition 3)

This part of the study (condition 3) investigated whether users would be more tolerant of delay when Web pages loaded incrementally instead of all at once. Previous research suggests that providing continuous feedback reassures users that the system is working and gives them something to look at while waiting [18]. However, Nielson [20] points out that standard browser feedback, provided in the form of progress bars, fails to communicate the amount of the page that has been completed. Loading Web pages incrementally can address this shortcoming, while providing users with visually useful feedback.

The flow of information between Web server and client was manipulated to cause the Web pages to load in parts. In this condition, participants would receive the banner of the next page as soon as they clicked a link. This was followed by text, and, later, graphics.

Participants were asked to evaluate the time it took for the whole Web page to complete. All participants received pattern 1 in this condition.

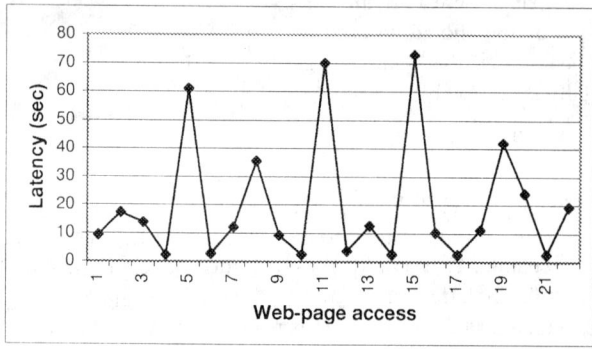

Figure 4: Latency pattern (incremental loading)

Figure 4 shows the mean delay taken by each web-page to complete in condition 3. These measurements were taken using client-based software that captures latency received by users with 100% accuracy [9].

FINDINGS

There were 3 key findings to the research:

- A mapping between objective QoS and users' subjective perceptions of that QoS can be identified and quantified.

- This mapping is influenced by a number of contextual factors including the type of task in which the user is engaged, the method of page loading, and cumulative time of interaction.

- Users' conceptual models underlie the influence of contextual factors on subjective perceptions of QoS.

To what extent is there a mapping between objective and subjective QoS?

Verbal protocols indicated that participants used the 'Low' button when they found the QoS was unacceptable; very few participants used the black button. We aggregated these responses when conducting a set of Chi-squared tests for statistical significance.

Classification of Latency (condition 1)

A significant number of participants assigned certain levels of latency to particular levels of service. Table 2 shows this classification. Figure 5 shows the number of users that classified particular levels of delay in certain categories. For example, it shows that 14 participants classified an 8 second delay as 'High' while 4 participants classified it as 'Low'. Not every participant registered a rating for every level of delay.

Table 2: Classification of latency

Rating	Range of latency Condition 1	Range of latency Condition 3
High	0 – 5 sec	0 – 39 sec
Average	> 5 sec	> 39 sec
Low	> 11 sec	> 56 sec

Previous Web usage studies have found that delays greater than 10 seconds encourage users to believe that an error has occurred in the processing of their request [20]. Our finding that the threshold where QoS is judged as 'Low' is around 11 seconds is therefore consistent with previous work.

Control of Latency (condition 2)

We observed, in condition 2, that there was a large standard deviation among participants in terms of their tolerance of latency. Although the average tolerance was 8.57 seconds in this condition, the standard deviation was 5.85 seconds.

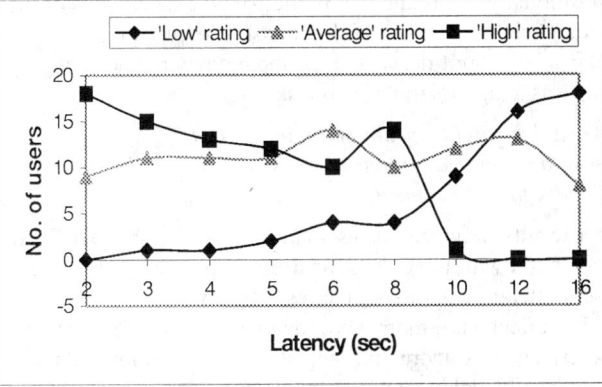

Figure 5: Ratings of latency (condition 1)

It is not possible for us to conclude from this condition that users will tolerate a specific amount of latency before finding that QoS unacceptable. Multiple regression analysis revealed that the number of hours participants used the Web significantly influenced their tolerance for latency in this condition. Higher levels of Web usage were associated with less tolerance for delay during interaction ($p<0.01$).

The large standard deviation observed when participants were asked to control latency may have been due to differences amongst participants in terms of their risk-taking behavior. Participants differed in terms of whether they took advantage of the fact that there was no penalty for pushing the button to increase quality. This difference is also suggested by the fact that there was no correlation between participants' tolerance when classifying latency and their tolerance when controlling latency.

Classification of Latency (condition 3)
The upper bound of latency assigned by participants to each classification in condition 3 are almost 6 times higher in each case compared to the classifications made in condition 1 (see Table 2). This indicates that users are more tolerant of latency when Web pages load incrementally than when there is a delay followed by the display of the page in its entirety.

What contextual factors influence users' tolerance of Internet QoS?

The data we gained from verbal protocols and focus groups indicates that participants were strongly influenced by their expectations of delay when responding to the QoS they received in the experiment. We found that there was almost unanimous agreement amongst participants concerning the factors that help form these expectations.

The amount of time users allocate to the task
Our findings suggest that users anticipate the time it will take them to perform on-line tasks. This anticipation helps form their expectations of the time it should take them to complete the task. Our results suggest that when the process of completing a task is disrupted by unanticipated delays a conflict arises between users' expectations and the QoS received. That QoS is likely to be rejected:

- *'So I'll be sitting there for half an hour so I'm set for that...so a lot of it depends on the time I anticipated I had when I set out'*.
- *'If I'm going to buy something that I need to do research on, mentally I'll allocate more time'*.

Understanding systems-level operations
Since our participants were selected on the basis of their experience in Internet shopping, they were likely to possess an understanding of the manner in which networks store and route data. This understanding was shown to influence the QoS participants expect from the network. Specifically, users are likely to tolerate less delay in receiving a Web page if they believe that it should be easily accessible from memory:

- *'(It should be fast because) I've already been there, it should be cached'*

Users are more likely to be tolerant of delay at certain times of the day, or for 'busy' Web sites:

- *'Understanding that there's a lot of people coming together on the process makes us more tolerant'*.

The company whose products are advertised
Participants in our study believed that companies that are more commercially successful should possess the financial means to supply at least adequate levels of QoS, 100% of the time. This expectation means that users are less likely to accept delays incurred while interacting with a site that promotes the products of such companies:

- *'Because the companies are so huge they should pour money into their web-sites, should have fast sites. If I try to get on those sites and they're slow then I'm not as patient'*.
- *'This is the way the consumer sees the company...it should look good, it should be fast'*.

Duration of interaction
In all conditions we found that users' tolerance for delay decreased as the length of time they spent interacting with the system increased. In all cases this finding is statistically significant ($p<0.01$). The effect is more powerfully significant for condition 3 ($p<0.001$). Figure 6 shows the maximum delay tolerated by a participant in condition 2. Maximum tolerance is here indicated by the point at which the participant clicked the 'Increase Quality' button so that the Web page would appear.

Task variation
For users who possess an understanding of the way that networks store and route data, the type of real-world task in which they are engaged is likely to be involved in forming expectations of QoS and therefore have an influence on the amount of delay tolerated:

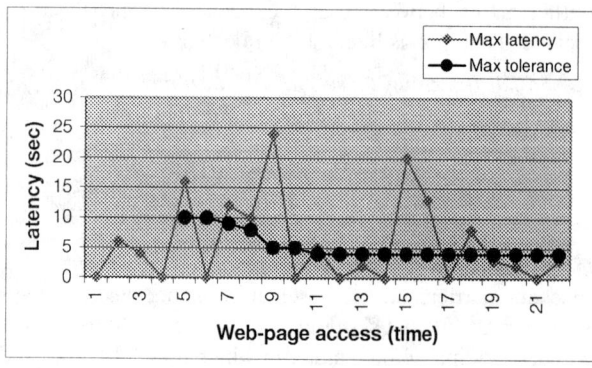

Figure 6: Latency tolerated over time (participant 3)

- *'Like when you're comparing I expect that to take a little longer because it's going to have to go out and get information'.*

Qualitative data suggests that participants expect different tasks to take longer than others. From this information, we were able to classify tasks according to participants' expectations of the latency each task should incur. High tolerance tasks were:

- Comparing several items.
- Viewing the shopping cart.

By comparison, low tolerance tasks were:

- Returning to a previously accessed page.
- Viewing a class of products.
- Adding to the shopping cart.

During the experiments we found that users tolerated different levels of latency depending on what they were doing. A closer look at Figure 5, for example, shows that more participants classified an 8 second delay, corresponding to comparing different printers, as 'High' quality, compared to an objectively lower 6 second delay, corresponding to viewing a class of monitors. Statistical tests show that users will accept more delay when they are comparing products or viewing the contents of their shopping cart than when they are viewing a class of products or adding to the shopping cart, ($p<0.01$).

What underlying conceptual models influence users' judgements of QoS?

Qualitative data showed that participants possess a conceptual model of the way that networks store and access information. This conception influenced their tolerance for delay. For example, our results showed that tolerance for delay associated with specific tasks was dependent on *a)* if the task required accessing a database, e.g. from which to compare products, or *b)* if the task involved a calculation to be made, e.g. in calculating the total spent from the items placed in the shopping cart. Although no pages were cached in the experiment, it was participants' conception of this technology that made them relatively intolerant of delay when re-visiting previous pages.

Additionally, we found that participants' expectations of the commercial setting in which the task was placed influenced their tolerance of delay. For example, viewing the shopping cart was a high tolerance sub-task:

'when I brought up my shopping cart I figured it would have to compile a bit longer so I was more willing to wait a little bit for it to come up'.

Our results show that, as users interact with a Web site their frustration with the delays incurred accumulates; the longer users interact with a site, the less latency they will tolerate. This effect is more pronounced when the latency experienced is more unacceptable. In condition 3, users tolerance of delay declined more rapidly than in condition 1, where the latency received was lower.

Qualitative data suggests that participants in this study felt that any cumulative slowness on Web pages suggested, not only that the products being sold were of inferior quality, but that the security of their purchase was compromised:

'If it's slow I won't give my credit card number'.

'I'd say, you haven't got your resources figured out, you're a poorly managed outfit, I don't trust you any longer'.

Once users perceive that their security has been compromised, no purchase will be made and the main purpose of any commercial Web site becomes critically damaged. It is therefore crucial for systems designers to understand the effect of cumulative frustration, especially as it is typically in the later stages of interaction that users are likely to commit to a purchase.

Qualitative data may explain the reasons behind increased tolerance for delay when Web pages load incrementally. These reasons center around the utility of the feedback provided by incremental loading. Participants were less frustrated when feedback in the browser showed them that the network was processing information:

'As long as you see things coming up it's not nearly as bad as just sitting there waiting and again you don't know whether you're stuck'.

Some participants in our study used the standard browser feedback to indicate activity in the network:

'I think if I had a way to know that it loading I wouldn't mind the old page intact until it was ready to switch'.

Typically, these participants did not prefer incremental loading. This finding suggests that users require feedback to be assured that the network is continuing to process their request. Either browser feedback or incremental loading can provide this feedback.

Our results show that users' conceptual models of the way in which networks operate can significantly influence their tolerance of QoS in predictable ways. Consequently, an understanding of users' conceptual models, and, perhaps more importantly, the behavior which is driven by them, is a crucial step in accommodating user demand.

IMPLICATIONS FOR SYSTEMS DESIGN

The perspective of this work is not only to understand user behavior relating to QoS but to interpret those findings into solutions for real-world problems.

Servers implement mechanisms that dynamically control the processing and delivery of information in response to users' requests. These functions are performed by using scheduling algorithms that allow requests to be processed in a specific order, for example, by Earliest Deadline First scheduling. By prioritizing requests in this way, the server can process the highest number of requests whose delivery still has utility to the user. Our research has defined objective thresholds below which delivering information has less utility. Service providers should design with these kinds of thresholds in mind.

Knowing the latency users will tolerate also allows servers to provide the appropriate feedback to the browser. Our results have shown that providing information concerning the processing of a request can significantly increase users' tolerance of poor QoS. For example, if a server receives a computationally intensive request it can calculate if it will be possible to process the request within the specified deadline. The server is then able to calculate the trade-offs concerning the number of requests that can be delivered to users within the required time-frame. Informing the user if their request will be delayed above the established threshold for tolerance implicates that the QoS of the task as a whole is seen as better:

'I think it's great...saying we are unusually busy, there may be some delays, you might want to visit later. You've told me now. If I decide to go ahead, that's my choice'.

Our results have shown that there are contextual factors that influence users' tolerance for latency These factors can be used in prioritizing requests in the server. Users will accept more delay when performing tasks that they believe to be computationally intense. Technology exists that can tag packets of information that are generated from a certain task [2]. The server could then queue and service tagged packets accordingly. Tasks perceived as computationally intense, for example, could be tagged according to a certain profile and placed near the back of the server's queue. This technology can also be used to implement differentiated service schemes. We have been able to show that a significant number of participants classify certain levels of QoS into different classes of service. This information can be used by the designers of priority service schemes in installing processing deadlines for each class of service.

A central finding in our study was that users' tolerance for latency decreases over the duration of interaction with a Web site. This phenomenon can be considered when performing server scheduling optimizations.

IMPLICATIONS FOR FURTHER HCI RESEARCH

Traffic smoothing techniques exist that can shape the pattern of latency received by users [3]. Participants who were given latency pattern 2 (Figure 2) made classifications that were lower than participants given pattern 1 ($p < 0.05$). This effect was found for both short, e.g. 3 seconds, and long, e.g. 16 second delays. Our comparisons were limited in that we investigated this affect by comparing randomly generated QoS with a single arbitrarily selected pattern. With the goal of gauging the effect of a predictable service, further work is needed to establish this trend for other patterns of latency where the magnitude of delay is more precisely controlled.

All participants in our study were given the conditions in the same order. Further work is needed to establish the effect of the order of exposure on acceptance of delay. Exposure to delay in former conditions could introduce memory effects and effect participants' subsequent expectations of delay:

'You get a bit spoiled I guess once you're used to the quickness, then you want it all the time'.

Our results have shown that users are more tolerant of delay when the Web pages load incrementally. However, the range of objective delays given in this condition was wider than those given for the condition used as a comparison, that where the page was displayed all-at-once. Although our results possess a high level of statistical significance one has to consider that the classifications given by participants in our study were relative. Further work is therefore needed to establish the effect of incremental loading between conditions with an identical range of latency.

Our study was specific to a Web shopping task. Further studies of users' perceptions of QoS should investigate the validity of our findings in different domains, such as an entertainment Web site.

For real-time audio and video applications it has also been shown that users' conceptual models affect subjective perceptions of QoS [5]. For example, users judge the acceptability of real-time audio quality based on visual interface feedback, as opposed to making direct evaluations of objective levels of quality [4]. Reliance on continuous feedback, and preferences for stable – even if lower – levels of QoS in real-time applications results from a need for predictable levels of service. However, further work is needed to investigate the influence of the other contextual factors reported in this paper on users' perceptions of real-time QoS.

The combination of results from different domains, and for different applications, would make it possible to create generalized conceptual models for predicting how tolerance changes according to a number of contextual factors. Our research represents an important first step in identifying that such relationships exist, and therefore indicates the need for technology to meet user requirements.

CONCLUSIONS

This study was designed to investigate users' requirements for Internet QoS. We specified a set of objective thresholds that reflect users' subjective assessments of quality. We showed that:

- The task in which users are engaged, the length of time they have been interacting with a site, and the method of page loading affects the acceptability of QoS.
- Tolerance of delay is influenced by users' conceptual models of how the system works.
- Poor Web site performance leads to poor company image and often compromises users' conceptions of the security of the site.

There are several stakeholders in the design of Internet services: server designers, network providers, advertisers, companies whose products are sold on-line, and consumers themselves. A failure to understand users' on-line QoS requirements may effect users' conception of a company's stature and commercial viability which, in turn, affects the business interests of service providers and advertisers. The future Internet will have more users and support a greater diversity of Internet applications. It has the potential to change the way in which consumers interact with companies. Our research has shown that it is possible to integrate users' requirements into systems design. Only through such integration will it be possible to achieve the customer satisfaction that leads to the success of any commercial system.

REFERENCES

1. Abdelzaher, T., and Bhatti, N. Adaptive content delivery for web server QoS. *Proceedings of IWQoS'99* (London, May 1999).
2. Bernet, Y. A Framework for End-to-End QoS Combining RSVP/Intserv and Differentiated Services. *IETF* (March 1998).
3. Bhatti, N., and Friedrich, R. Web Server Support for Tiered Services. *IEEE Network* (September/October 1999).
4. Bouch, A., and Sasse, M.A. Network QoS: What do users need. *Proceedings of IDC'99* (Madrid, September 1999).
5. Bouch, A., and Sasse, M.A. It ain't what you charge it's the way that you do it: A user perspective of network QoS and pricing. *Proceedings of IM'99* (Boston MA, May 1999).
6. Boyer, D.L., Pollack, J.G., and Eggemeier, T.F. Effects of Aging on Subjective Workload and Performance: Determinants of Age Differences in Cognitive Performance. *Proceedings of Human Factors Society 36th Annual Meeting 1*, (1992).
7. Cullinane, P. Ready, set, crash. *Telephony 3*, (Nov 1998).
8. Dunlop, M.D., and Johnson, C. Subjectivity and notions of time and value in interactive information retrieval. *Interacting with Computers 10*, 1, (1998).
9. ETEWatch for Web Browsers. http://www.candle.com/etewatch.
10. Fox, R. News Track. *Communications of the ACM* (May 1999), 9-10.
11. Fishburn, P.C., and Odlyzko, O.M. Dynamic behavior of differential pricing and Quality of Service options for the Internet. *Proceedings of ICE-98*, ACM Press.
12. Gupta, A., Stahl, D.O., and Whinston, A.B. Pricing of services on the Internet. http://cism.bus.utexas.edu/alok/pricing.html.
13. Hogan, M. The first ever report on the top 100 e-commerce businesses and the secrets of their success. *PC Computing Magazine*, June 8, 1999.
14. HP Shopping Village. http://www.shopping.hp.com.
15. Keller, J.J. Ex-MFs managers plan to build global network based on Internet. *Wall Street Journal*. January 20, 1998.
16. Mackie-Mason, J.K., and Varian, H. Economic FAQs about the Internet. In McKnight, L.W., and Bailey, J.P. (eds.). *Internet Economics*. MIT Press, 1997.
17. Morgan, K., Morris, R.L., Macleod, H., and Gibbs, S. Gender Differences and Cognitive Style in Human-Computer Interaction. *Proceedings of EWHCI'92* (Moscow, Russia, August1992).
18. Myers, B. A. The Importance of Percent-Done Progress Indicators for Computer-Human Interfaces. *Proceedings of CHI' 85* (San Francisco CA, April 1985).
19. Network Reliability Steering Committee. *Annual Report 1998*. http://www.nric.org.
20. Nielson, J. *Usability Engineering*. AP Professional Press, Boston MA, 1994.
21. Odlyzko, O.M. The Internet and other networks: Utilization rates and their implications. http://www.research.att.com/~amo.
22. Perry, C. Travelers on the Internet: A survey of Internet users. *Online 19*, 2, (1995).
23. Ramsay, J., Barbesi, A., and Preece, J. Psychological Investigation of Long Retrieval Times on the World Wide Web. *Interacting with Computers 10*, 1, (1998).
24. The Internet, Technology 1999, Analysis and Forecast, *IEEE Spectrum* (January 1999).
25. Wang, Z. USD: Scalable bandwidth for differentiated services. http://www.ietf.org/drafts-wang-00.txt.

What makes Internet Users visit Cyber Stores again? Key Design Factors for Customer Loyalty

Jungwon Lee, Jinwoo Kim
Human Computer Interaction Lab,
Yonsei University, Seoul, Korea
{jinwoo, jwon}@base.yonsei.ac.kr

Jae Yun Moon
Information Systems Department,
New York University
jmoon@stern.nyu.edu

ABSTRACT

Retaining customer loyalty is crucial in electronic commerce because the value of an Internet store is largely determined by the number of its loyal customers. This paper proposes a multi-phased model of customer loyalty for Internet shopping, which fully takes the characteristics of the Internet and cyber shopping into consideration. In order to validate the model, we conducted a web-based survey of the customers of various Internet stores, and the data was processed using structural equation analysis. The results indicate that several factors can effectively increase customer loyalty towards an Internet store and that the relative importance of the identified factors varies according to the level of involvement with the product purchased through the store. We suggest several managerial implications in developing Internet stores for higher customer loyalty based on these results.

Keywords

Customer loyalty, customer interface, transaction cost, trust, involvement, Internet shopping, electronic commerce.

INTRODUCTION

The proliferation of the Internet and WWW has influenced various aspects of our daily lives, not the least of which is Electronic Commerce (EC). Electronic commerce has been defined most commonly as the transactions between two or more parties through an electronic medium [14]. One of the critical requirements for the success of electronic commerce is the appropriate customer interface that is defined as the user interface of e-commerce systems [16]. The main difference between ordinary user interfaces and customer interfaces is the shift of the perspective from the user of a computer system to the customer of commercial transactions. In real world commerce, the emotions and impressions evoked by the sales agent or commercial institution as a whole influence the overall satisfaction of the customer, which is usually implemented by the customer interface in cyber world commerce.

In order for the Internet stores to be successful, the customer interface should be able to not only entice customers into visiting their stores for the first time, but also make the customers visit their stores again in the future [22]. Ensuring that customers visit their stores repeatedly is especially important for Internet stores because the stores' value is determined mostly by the number of loyal customers in the context of the web. If none of the customers are willing to visit the site again, its business value becomes zero regardless of its technical or managerial assets.

Therefore, securing loyal customers has become one of the most important goals for many Internet stores [14], and most Internet stores invest a considerable amounts of money and effort in order to maintain customer loyalty by providing various functions and interfaces [23, 27]. However, few studies have investigated the effective strategies for increasing customer loyalty in the design of customer interface in the Internet. Although the concept of customer loyalty has been a subject of much research in the marketing literature, the unique nature of electronic commerce in the WWW environment warrants a different framework than has been conventional for customer loyalty in the physical stores. Moreover, selecting the most effective strategy becomes more difficult because a myriad of new technologies and system techniques are proposed to increase customer loyalty in Internet shopping. Some focus on the online community approach [23] or the more conventional relationship management approach using customer relationship software [27], while others argue that effective interface design must be an integral part of the Internet store [16,18]. However, it is not well understood when a particular technology will be more effective, and how the technologies are related to retaining customer loyalty. Furthermore, will it be the same for all types of products?

In order to answer the above questions, we need to fully understand the characteristics of the Internet. The Internet has dramatically reduced the transaction costs for consumers by making it easier to search for different alternatives [3]. However, the ease with which new information can get disseminated over the WWW also creates problems in the form of information overload. Since consumers are limited in their ability to process information, they will actively process such new information only when it is relevant and/or important to them. The consumer's perceived relevance or importance of a product, service or situation is defined as the consumer's level of involvement with the object [8, 10]. Prior studies have found that the consumer's level of involvement with the product has a significant impact on the amount of evaluation and search that he is willing to perform [1, 5]. This implies that customer loyalty will be influenced depending on the customer's extent of perceived involvement with the particular product. In this paper, we try to model the different factors that affect customer loyalty for products that are either high or low in the level of involvement perceived by the customer in Internet stores.

The main goal of this paper is to identify important factors that affect customer loyalty in Internet shopping, and to determine how the impacts of the identified factors change with respect to the type of product serviced by Internet stores. In order to achieve the goal, we first propose a general multi-phased model of Internet customer loyalty [2, 9]. The model is validated through a web-based survey administered to the customers of various Internet stores. The survey results are analyzed further to compare how the level of involvement with the different products affects the importance of the factors in determining customer loyalty. Finally, based on the study results, we suggest several implications in developing Internet stores that will increase customer loyalty.

A MULTI-PHASED MODEL OF CUSTOMER LOYALTY

The proposed model of customer loyalty for Internet shopping consists of nine constructs. The constructs and their proposed relations are presented in Figure 1.

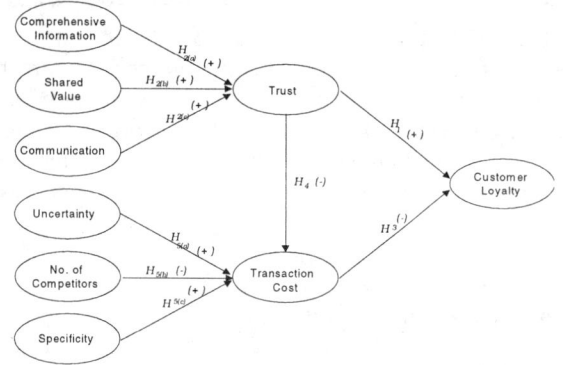

Figure 1. Multi-Phased Model of Internet Customer Loyalty

Customer loyalty, the final dependent variable in the proposed multi-phased model, is defined as the customers' intention to revisit the Internet stores again based on their prior experience and expectation of the future. This definition is based on prior studies, where customer loyalty has been regarded as an important source of sustainable competitive advantage [7]. Among the various factors that have been suggested as the important factors for retaining customer loyalty, this study selected "trust" and "transaction cost" as the two most important variables because they are closely related to Internet shopping.

Trust, defined as the willingness to rely on an exchanging partner in whom one has confidence [20], has been touted as one of the most critical factors for increasing customer loyalty [25]. Trust has been regarded as one of the most important variables in the relations between human and computer [9] as well as between human and human [24]. The importance of trust has been especially emphasized in the context of electronic commerce [13]. For example, lack of trust has been found to be the most important barrier in buying anything from the Internet stores, and most people were afraid of the possibility that somebody might have exploited their trust in Internet shopping [6]. Thus, we hypothesize the following:

H_1: *Trust has a positive impact on customer loyalty, i.e., his intention to revisit the store.*

Three variables are hypothesized as important precursors of trust: comprehensive information, shared value, and communication. They were selected based on prior studies of Internet shopping (e.g., [14]) and trust (e.g., [21]).

Comprehensive Information is defined as the extent to which a customer has enough information to make a purchase decision. With comprehensive information, the customer can predict the consequences of the purchase decision and use the information provided to predict the behavior of the Internet store. For example, a clear statement of the fulfillment procedures of an Internet store has been found to be an important factor for developing trust. Therefore, we hypothesize that:

H_{2a}: *Comprehensive information increases customers' trust*

Shared value is defined as the extent to which customers have beliefs in common about the type of behaviors, goals, and policies that are important and appropriate [26]. Shared value has become a variable of great interest especially in the Internet because the development of online community has become one of the most effective ways to increase the level of credibility of an Internet store[2]. Therefore, Shared values is hypothesized to be related to the trust since customers' attitudes such as trust result from having the same values with other people or groups.

H_{2b}: *The perception of shared values with the Internet store and its customers will have a positive impact on trust.*

Finally, *communication* can be defined as the formal as well as the informal sharing of meaningful and timely information between buyers and sellers. Communication fosters trust by resolving disputes and aligning expectations. Communication is especially important in

Internet shopping because the customer interface in the Internet can provide a variety of new communication channels. A customer's perception of the quality and frequency of communication by the Internet store will have a positive impact on trust. Therefore, we hypothesize that:

H_{2c}: *The perception of high quality communication will increase trust.*

Transaction costs include negotiating, monitoring, and enforcing costs that are accrued during the two-way communications between sellers and buyers [28]. Transaction cost has been regarded as one of the main theoretical constructs to explain various kinds of relationships such as transactions for durable goods as well as long term relations within a single organization. We believe that transaction costs are especially important in the context of Internet shopping, since opportunism and bounded rationality, the main source of transaction costs, can occur more easily in the cyber space compared to the real world [6]. Therefore, higher transaction costs will decrease the intention to revisit the Internet stores again.

H_3: *Transaction costs have a negative impact on customer loyalty.*

Furthermore, transaction costs occur because of the lack of trust between the sellers and buyers, as well as their bounded rationality and opportunism. If the sellers and buyers cannot trust each other, their costs for negotiating, monitoring and enforcing the terms and conditions of the transaction will increase naturally. Conversely, high levels of trust between the parties of the exchange will reduce the tendency for opportunistic behavior, thereby eliminating the need to incur such transaction costs related to monitoring the fulfillment of the respective transaction contract. Therefore, we hypothesize that:

H_4: *Trust reduces transaction costs.*

The model also hypothesizes three variables to be related to transaction costs: uncertainty, number of competitors, and Internet store specificity. First, *uncertainty* is defined as the potential of mis-interpreting overall product quality in Internet stores [12]. The level of uncertainty is high, when a product becomes more complex and difficult to evaluate based solely on its search qualities, such as price [5]. Services such as Internet shopping can only be evaluated based on experience qualities that can only be known after the purchase and credence qualities that refer to such intangible attributes as reputation [5]. In other words, the customer will only be able to measure the final outcome when the product is delivered. The performance ambiguity of service exchanges in Internet shopping stem largely from the characteristics of the customer interfaces. In general, the higher the performance ambiguity or uncertainty is, the greater the difficulty of negotiating, monitoring, and enforcing exchanges between the organization and the customer, and the more complex the governance mechanism required will be. Thus, increased uncertainty is expected to raise the transaction costs because the lack of accurate evaluation regarding product quality may motivate the Internet store to behave opportunistically.

$H_{5(a)}$: *Uncertainty is positively related to transaction costs.*

Second, the *number of competitors* can be defined as the number of Internet stores that provide the same service and products [28]. Transaction costs will increase when there are only a few alternative Internet stores, where they may behave more opportunistically by using their bargaining power to gain the maximum profits.

$H_{5(b)}$: *Number of stores is negatively related to transaction costs.*

Finally, the Internet *store specificity* is related to the lock-in customers and can be defined as the inability of customers to transfer the skills, knowledge and value created at the Internet store to other similar Internet stores [1]. If Internet stores is high in specificity, it may behave more opportunistically, because it knows that their customers have already invested a significant amount of efforts that would be less valuable if their customers switch to different Internet stores. Therefore, we hypothesize the following:

$H_{5(c)}$: *Internet store specificity is positively related to transaction costs.*

The model proposed in Figure 1 does not take into account the fact that consumers process information selectively based on their level of involvement with the purchase decision, largely determined by the extent to which they perceive that the product they purchase is relevant or important to them [11,30]. Visitors to a web site can be classified into two broad categories, low-involvement surfers and high-involvement searchers [26]. Prior research in marketing has shown that level of product involvement is an important factor determining consumer behavior [11,26,30]. For example, low involvement with products was found to lead to a lack of active information seeking about brands, little comparison among product attributes and no special preference for a particular brand [30]. Involvement has also been shown to play a role in consumers' commitment to a brand [11]. Products of low involvement were more likely to have positive relationships with brand commitment than were products of high involvement. In simpler terms, users do not care much and will take whatever they can get for low involvement products whereas they will be highly selective in purchasing high involvement products. In spite of the differences in consumer behavior and preferences depending on the different involvement levels, little research has been done to address how involvement would affect consumers' perception of transaction costs and trust, and its subsequent effect on customer loyalty. Therefore, we propose that:

H_6: *The level of product involvement will moderate the hypothesized relationships between the transaction cost, trust, its respective antecedents and customer loyalty.*

SURVEY AND ANALYSIS PROCEDURE

A web-based survey was conducted to test the proposed model of customer loyalty. The survey questionnaires were posted at several Internet stores in operation at the time of study, and participants were solicited by providing monetary compensation. A total of 289 people were included in the final data set after deleting those who responded more than once or those who submitted incomplete answers. Among the final respondents, 233 were males (80%) and 56 were females (20%). Most of them were in their twenties (196, 68%), followed by those in their thirties (77, 26%), forties (12, 4%), and younger than twenty (4, 2%). Almost half of the total respondents (140, 48%) reported that they had prior experience in shopping on the Internet at least once, while the other half had none (149, 52%).

The initial survey questions were adopted from prior studies in customer involvement (e.g., [17]), trust (e.g., [20]), and transaction costs (e.g., [1]). In order to test the reliability of the initial survey questions, the Cronbach alpha coefficients for each set of the questions were calculated, and the results are presented in Table 1.

The involvement level was manipulated by asking people to have a product in their mind while answering the survey questions. The first three questions at the beginning of the questionnaire asked subjects about the level of involvement they felt about the product they were thinking about. A cluster analysis was performed using the participants' responses to the first three questions to classify the data set along the dimension of product involvement. The results indicate that the respondents can be classified into two groups. The classification through cluster analysis ensured that the members of each cluster were as different as possible. One hundred sixty three (56%) among the 289 respondents were classified as the high involvement group (HIGH), while the rest (126, 44%) were classified as the low involvement group (LOW).

The survey questions that did not contribute to the convergence and consistency of theoretical constructs were deleted from further analysis. The final survey questionnaire consists of 2 questions for comprehensive information, 3 for shared value, 3 for communication, 2 for uncertainty, 2 for number of competitors, 2 for specificity, 3 for trust, 9 for transaction costs, and 2 for customer loyalty.

Variables	Initial questions	Final questions	Cronbach alpha
Comprehensiveness	2	2	.7467
Shared Value	3	3	.5837
Communication	3	3	.6108
Uncertainty	2	2	.8039
Number of Competitors	2	2	.6088
Specificity	2	2	.7350
Trust	3	3	.6187
Transaction costs	9	8	.6140
Customer Loyalty	3	2	.7288
Involvement	3	3	.7195

Table 1. Reliability of Final Scales[1]

SURVEY RESULTS

The LISREL analysis was conducted three times using the final data set. First, the total set of responses was used to construct the general model of customer loyalty to Internet store. Second, according to the results of the cluster analysis on the involvement questions, a LISREL model was computed for each of the two involvement groups. Finally, the corresponding path coefficients between the two involvement groups were compared using the standard t test.

A General LISREL Model of Internet Customer Loyalty

Using structural equation modeling analysis, the hypothesized sequence of relationships of the models was tested as a set [19]. Figure 2 presents the LISREL results for the general model based on the total responses.

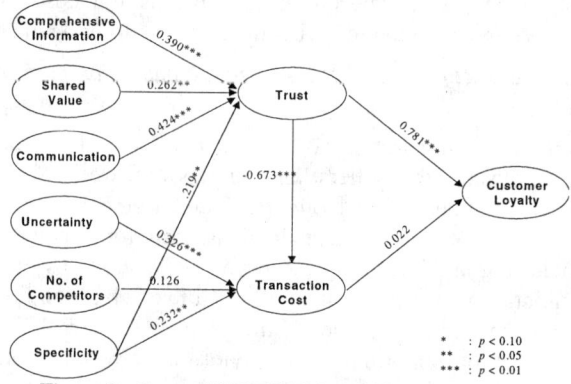

Figure 2. General LISREL Model (Structural Model)

The coefficients for the nine paths in Figure 2 represent the proposed relations between the latent constructs. The fit of the model was assessed using several indicators, including the chi-square to degrees of freedom ratio (*Chi-square*=194.56, df=166, p=0.064), adjusted goodness of fit test (AGFI=0.917), and root mean square residuals (RMR=0.045) [4]. These results clearly indicate that the model explains the overall relations among the constructs faithfully. Therefore, it was deemed sufficient to proceed and test the set of paths hypothesized by the model using maximum likelihood estimation.

All variables hypothesized to be related to trust were found to have significant relations with trust; comprehensive information (.390, $p<0.01$), shared value (.262, $p<0.05$), and communication (.424, $p<0.01$). The final model depicted in Figure 2 indicates that the customer interface specificity also had a significant positive relationship with the trust that customers placed on the Internet store (.219, $p<0.05$). On hindsight, this is in accordance with what

---

[1] More detailed descriptions about the final verbal scales are presented in [15].

might be expected. The more time and effort that customers have invested in a particular customer interface, the higher their trust in the system would be, in order to reduce any cognitive dissonance that would result from not trusting the system that they had invested so much in. Therefore, more comprehensive information about the Internet store, greater shared value, and more diverse communication tools as well as the customer interface specificity are positively related to the higher level of trust on the Internet store ($H_{2(a)}$, $H_{2(b)}$, $H_{2(c)}$).

Among the three variables that were expected to be related to transaction costs, only two were found to be significantly related to transaction costs – uncertainty (.326, $p<0.01$; $H_{5(a)}$) and specificity (.232, $p<0.05$; $H_{5(c)}$). In addition, trust was found to be negatively related to the transaction costs (-0.673, $p<0.01$; H_4). The number of competitors was not significantly related to transaction costs ($H_{5(b)}$). Therefore, higher difficulty to measure the quality of products and higher specificity of the customer interface were found to cause higher transaction costs, whereas higher credibility decreases transaction costs. Among the two variables that were hypothesized to be related to customer loyalty, only trust was found to have a strong impact on customer loyalty (.781, $p<0.01$; H_1). Therefore, more comprehensive information, higher shared value, and diverse communication tools lead to a higher level of trust, which in turn induce a higher level of customer loyalty.

Two Sub LISREL Models for the Low and High Involvement Groups

Two LISREL models were computed, one for the group of respondents who were thinking about low involvement products while they were responding to the questionnaire (the LOW group), and the other for the respondents who were considering high involvement products (the HIGH group). The two models are presented in Figure 3 and Figure 4.

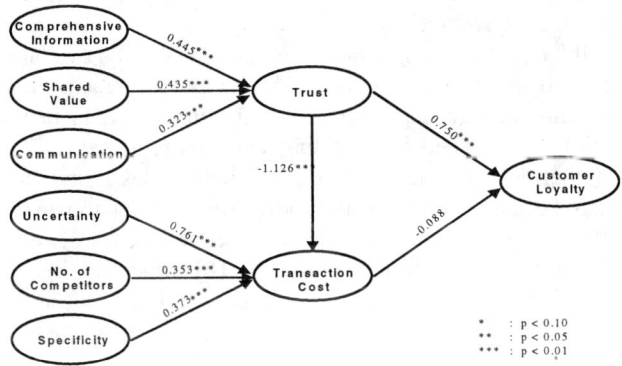

Figure 3. LISREL Structural Model (HIGH Involvement)

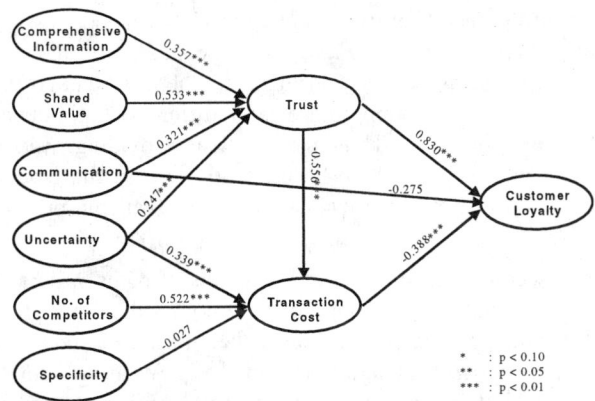

Figure 4. LISREL Structural Model (LOW Involvement)

Similar to the general LISREL model, the fits of the two sub models were assessed using the chi-square to degrees of freedom ratio (LOW group, *Chi-square*=170.88, $df = 164$, $p = 0.340$; HIGH group, *Chi-square* = 205.28, $df=174$, $p = 0.053$), adjusted goodness of fit test (LOW group = 0.847; HIGH group = 0.861), and root mean square residuals (LOW group = 0.064; HIGH group = 0.066). Again, these results clearly indicate that the two sub models faithfully explain the relations among the constructs. However, in the LISREL structural model for the LOW group, it must be noted that there are two more paths that are not present for the structural model of the HIGH group. In the low involvement group, uncertainty had a positive impact on trust, while communication had a negative impact on customer loyalty. The two relationships are somewhat counterintuitive, with the former being more so in that it has such high statistical significance. Perhaps, customers experience more trust towards Internet stores when the product is more difficult to evaluate. In other words, having no other means of justifying the purchase, customers will rely on the credence qualities of the Internet stores, and trust them when the product uncertainty is high. This is especially relevant for low involvement products, where the costs of an erroneous decision are not as high. Therefore, a cost-benefit analysis of the relative value of meticulous evaluation of alternatives might yield the conclusion that it is better to simply trust the Internet stores than to expend the enormous amount of effort required to evaluate the alternatives, especially when *uncertainty* is high.

The different paths for the sub models for the LOW and HIGH involvement groups respectively, indicate that there are some underlying differences between the constructs depending on the level of involvement. Therefore, we proceed to compare the paths coefficients of the two models to investigate how the LOW and HIGH groups were different in terms of the relative importance of the latent constructs toward customer loyalty. Table 2 summarizes the results of the *t* test to compare corresponding path coefficients between the LOW and HIGH groups.

The differences between the two models can be classified into three categories. First, only one path, the one from *communication* to *trust*, belongs to the first category in which the relative importance is not different at all between the two groups. This means that providing various communication channels in order to increase the level of trust works for the low and high involvement products alike.

The second category includes two sets of paths, where the path is significant for one group but not for the other group. For example, the path between the *transaction costs* and *customer loyalty* is significant for the LOW group, but not for the HIGH group. In other words, higher transaction costs decrease customer loyalty when a customer is looking for low involvement products, whereas transaction costs do not affect customer loyalty for high involvement products. This is in agreement with prior research in the effect of involvement [17]. When customers consider a product to be of greater relevance to them, the presence of transaction costs does not deter them from their intended purchase, whereas for products of low involvement the presence of transaction costs will easily sway the customer to switch to different stores.

The other path that belongs to the second category is the one from *specificity* to *transaction costs*. This path is significant for the HIGH group, but not for the LOW group. In other words, higher specificity increases transaction costs when a customer is looking for high involvement products, whereas specificity does not affect transaction costs for low involvement products. The customers did not take specificity into consideration in calculating transaction costs for low involvement products because they might not invest enough period of time to build specificity in buying low involvement products. Therefore, since they did not spend enough time to adjust themselves to the specificity of individual malls, specificity might not affect the transaction costs for the low involvement products.

No. of significant paths	Path's coefficient comparison			Path			Path's Coefficient		T value	
							Low	High		
No Path	Low	<	High	Communication	→	Trust	0.3210	0.3230	-0.1077	
Either Path	Low	>	High	Transaction Cost	→	Customer Loyalty	-0.3880	-0.0880	-25.7188	***
	Low	<	High	Specificity	→	Transaction Cost	-0.0270	0.3730	-52.6749	***
Both Paths	Low	>	High	Shared Values	→	Trust	0.5330	0.4350	8.6932	***
				No. of Competitors	→	Transaction Cost	0.5220	0.3530	12.1278	***
				Trust	→	Customer Loyalty	0.8300	0.7500	2.5107	**
				Trust	→	Transaction Cost	-0.5560	-1.1260	24.6813	***
	Low	<	High	Comprehensive Information	→	Trust	0.3570	0.4450	-14.4113	***
				Uncertainty	→	Transaction Cost	0.3390	0.760	-38.7805	***

* : $p<0.1$, ** : $p<0.05$, *** : $p<0.01$

Table 2. Summary of Comparison Between the Low and High Involvement Groups

The third category includes six sets of paths, where both paths from the two groups are significant altogether, but the degree of significance is different from each other. Two sets of the paths are more substantial for the HIGH group compared to the LOW group; the path from *comprehensive information* to *trust*, and the path from *uncertainty* to *transaction cost*. This means that in order to increase customer loyalty, providing comprehensive information and decreasing uncertainty are more effective for the high involvement group than for the low involvement group.

The remaining four sets of paths were more substantial for the LOW than for the HIGH group; the path from *shared value* to *trust*, from the *number of competitors* to *transaction cost*, from *trust* to *customer loyalty*, and from *trust* to *transaction cost*. This means that increasing shared value is a more effective way to increase trust level, which in turn is found to be more effective method to increase customer loyalty for the low involvement products. Special attention should be paid in interpreting the results about the number of competitors. In our hypothetical model, the path from *number of competitors* to *transaction cost* was expected to be negative, whereas the path coefficients from the two sub models turned out to be significantly positive. In other words, we expect that more competitors would decrease transaction costs, whereas empirical results indicate the reverse. This result might be explained based on the relatively large number of shopping malls available in the Internet. The number of stores that are reachable at one time is significantly more in the Internet than in the real world. Therefore, adding even more stores in the Internet does not decrease but rather increases transaction costs because browsing and comparing among too many stores might cause customers to experience cognitive overload and get lost in cyberspace [16], which might also explain why the path was not found to be significant in the general model in Figure 2.

CONCLUSION AND DISCUSSION

The results from the general LISREL model indicate that most of the hypotheses in the general model sustain except for two paths: the path from *number of competitors* to *transaction costs* and the one from the *transaction costs* to *customer loyalty*. Therefore, comprehensive information, shared value, and diverse communication affect the level of trust, which in turn influence customer loyalty. If customers have more comprehensive information about the products in the store, if they share more values with other customers in the store, if they have more diverse means of communication, and if they adjust themselves more highly to the specificity of individual shopping malls, they would feel a higher level of trust in the Internet store. Finally, if the customer feels a higher level of trust in the store, he/she would intend to revisit the site more repeatedly. The reason why the path from *transaction cost* to *customer loyalty* and the one from *number of competitors* to *transaction costs* were not significant in the general LISREL model becomes clearer when the results of the two sub LISREL models are considered.

One of the main differences between the model for the high involvement products and the model for the low involvement products is the path from *transaction costs* to *customer loyalty*. For the low involvement products, the transaction cost is found to have a strong impact upon customer loyalty, whereas it cannot affect customer loyalty for the high involvement products. Therefore, low transaction cost can increase customer loyalty only for the low involvement products, but not for the high involvement products, which might make the path from the transaction cost to customer loyalty insignificant in the general LISREL model. Customers may not regard transaction costs seriously for high involvement products because the cost itself might not be an important factor compared to other important factors such as personal tastes and preferences. On the other hand, they may be concerned very much with the cost factors for the low involvement products, since the cost may be the single most important factor for low involvement products.

Another important difference between the hypothetical model and the two sub models is the path from *number of competitors* to *transaction costs*. This path was hypothesized to be negatively related, but it was found to be positively related in the two sub models. As was mentioned in the result section this finding might result from the relatively large number of Internet stores available at one time. Therefore, more number of competitors might decrease bargaining and negotiation risks but at the same time increase search and browsing costs, which might result in insignificant path coefficient in the general model.

The differences between the low and high involvement groups provide several interesting implications in the development of the Internet stores. Internet stores that mostly deal with high involvement products should focus on the trust factors in order to increase customer loyalty because transaction costs do not have a significant effect on customer loyalty. Among the three factors related to trust, the stores may focus on providing more comprehensive information to the customer because the comprehensive information has the greatest impact upon the customer loyalty, as well as it is significantly more important for the high involvement products compared to the low involvement products. Therefore, Internet stores selling high involvement products should invest more in providing comprehensive information by several methods such as increasing the number of data fields for each product and providing powerful search engines for exhaustive browsing. The Internet stores that mostly sell low involvement products should balance between the trust and transaction costs because both of them are important for customer loyalty. Among the six factors, they should pay more attention to increasing shared value among other customers, because shared value has the biggest impact upon trust and it is also significantly more important for the low involvement products compared to the high involvement products. Therefore, the Internet stores selling low involvement products should pay more attention to increase shared value by providing convenient bulletin boards and encouraging open feedback from customers.

This study has several limitations, which must be taken into consideration in interpreting the results. First, the model does not include many of the more obvious factors that will influence customer loyalty, such as the product prices in the Internet store, or the quality of the product purchased through the Internet store. These were omitted from the model since they were seen as *necessary* but not *sufficient* conditions for retaining customer loyalty. Low price and high quality are without a doubt important, and confirming these beliefs would have had little managerial implications. Therefore, in our model, only trust and transaction cost are proposed to have a direct impact on customer loyalty. However, several other factors such as product availability, delivery time, and personalization should be added to the basic model in a future study in order to enlarge the scope of the theoretical model even though they may not be as important as trust and transaction costs. Second, the survey in this study was administered to customers of Korean Internet stores. It is not clear how generalizable the results would be to the larger Internet population. Even though the survey results from the general LISREL model indicate that our survey participants are not different from the Internet stores and customers in other countries, a future study should include international stores and customers in order to identify cultural effects on the findings. Third, the study was based on a snap-shot survey analysis, and hence fails to take into account the dynamic nature of trust formation and the nonlinear trends in transaction costs. For example, we need to have more longitudinal data in order to fully explain the path from the number of customers to transaction costs. An additional limitation of the survey approach lies in the difficulty in establishing causal relations between the constructs. Fourth, the comparison of the LOW involvement and HIGH involvement groups was based on comparisons of two different structural models in which the paths were not constrained to be equal. This might have an adverse effect on the reliability of the results. Fifth, the level of involvement was implemented in this study by asking subjects to think about certain products that they had purchased most recently or that they wanted to buy most from the Internet stores. However, actually buying a product may be different from remembering or reminding the product. Therefore, a future study should employ a more direct way of manipulating the level of product involvement. For example, we can directly compare the customers who bought high involvement products and those who bought low involvement ones by tracking customer navigation within the Internet stores. Finally, this study focuses only on the level of product involvement, as a factor that influences the overall impacts of trust and transaction costs upon customer loyalty. However, in order to build a more comprehensive model of the Internet customer loyalty, a future study needs to consider other important factors such as customer types in addition to product types.

ACKNOWLEDGEMENTS

This study was funded by the Korean Ministry of Information and Communication. The authors thank Dr. Minsoo Kim at the National Computation Agency and Mr. Ki-Joo Lee of the Ministry of Information and Communication for supporting the study, and members of the Human Computer Interaction Lab at Yonsei University for their comments on an early draft of this paper.

REFERENCES

1. Anderson, E. The salesperson as outside agent or employee: A transaction cost analysis. *Marketing Science*, 4, 3, (1985), 243-245.

2. Armstrong, A., and Hagel, J. The real value of on-line communities. *Harvard Business Review*, (May-June, 1996), 134-141.

3. Bakos, Y. Reducing buyer search costs: Implications for electronic marketplaces. *Management Science*, 43, 12, (1997), .

4. Bentler, P. M., and Bonnett, D. G. Significance tests and goodness of fit in the analysis of covariance structures. *Psychological Bulletin*, 88, (1980), 588-606.

5. Bowen, D. E., and Jones, G. R. Transaction cost analysis of service organization-customer exchange. *Academy of Management Review*, 11, 2, (1986), 428-441.

6. Cheskin Research and Studio Archetype/Sapient. *eCommerce Trust Study*. Cheskin Research, 1999.

7. Czepiel, J., and Gilmore, R. Exploring the concept of loyalty in service. In Czepiel, J. A., Congram, C. A., and Shanahan, J. (eds.) *The Service Challenge: Integrating for Competitive Advantage*. Chicago, IL: American Marketing Association, (1987), pp. 91-94.

8. Day, E.;Stafford, M. R., and Camacho, A. Research note: Opportunities for involvement research - A scale development. *Journal of Advertising*, 24, 3, (1995), 69-.

9. Fogg, B., and Tseng, H. The elements of computer credibility. *Communications of the ACM*, (1999), .

10. Gilles, L., and Kapferer, J.-N. Measuring consumer involvement profiles. *Journal of Marketing Research*, 22, (February, 1985), 41-53.

11. Gordon, M. E.;McKeage, K., and Fox, M. A. Relationship marketing effectiveness: The role of involvement. *Psychology & Marketing*, 15, 5, (1998), 443-459.

12. Ha, H., and Hoch, S. Ambiguity, procession strategy, and advertising evidence interactions. *Journal of Consumer Research*, 16, (December, 1989), 334-360.

13. Handy, C. Trust and the virtual organization. *Harvard Business Review*, (May-June, 1995), 40.

14. Kalakota, R., and Whinston, A. *Electronic Commerce: A Manager's Guide*: Addison Wesley Longman Inc., 1997.

15. Kim, J.;Lee, J., and Moon, J. Y. *A multi-phased model of customer loyalty for Internet shopping*. HCI Lab, Yonsei University, Seoul, Korea, Technical Report #99-3, 1999.

16. Kim, J., and Moon, J. Y. Designing towards emotional usability in customer interface: Trustworthiness of cyber-banking system interfaces. *Interacting with Computers*, 10, (1998), 1-29.

17. Laurent, G., and Kapferer, J.-N. Measuring consumer involvement profiles. *Journal of Marketing Research*, 22, 1, (1985), 41-53.

18. Lohse, G. L., and Spiller, P. Electronic shopping. *Communications of the ACM*, 41, 7, (1998), 81-88.

19. Mittal, B. Testing consumer behavior theories: LISREL is not a panacea. *Advances in Consumer Research*, 20, (1993), 647-653.

20. Moorman, C.;Deshpande, R., and Zaltman, G. Factors affecting trust in market research relationships. *Journal of Marketing*, 57, (January, 1993), 81-101.

21. Morgan, R. M., and Shelby, D. H. The communication-trust theory of relationship marketing. *Journal of Marketing*, 58, (July, 1994), 20-38.

22. Nielsen, J. Loyalty on the Web. *Alertbox*, useit.com, (August 1, 1997) [URL] http://www.useit.com/alertbox/9708a.html.

23. Pelton, C. Back to basics: Customer focus on the E-commerce frontier. *Information Week*, (December 21, 1998).

24. Rempel, J. K.;Holmes, J. G., and Zanna, M. P. Trust in close relationships. *Journal of Personality and Social Psychology*, 49, 1, (1985), 95-112.

25. Rotter, J. B. Interpersonal trust, trustworthiness, and gullibility. *American Psychologist*, 35, (1980), 1-7.

26. Singh, S. N., and Dalal, N. P. Web home pages as advertisements. *Communications of the ACM*, 42, 8, (1999), 91-98.

27. Stein, T. Service on the net., *Information Week*, (December 21, 1998).

28. Williamson, O. E. *Markets and Hierarchies: Analysis and Anti-trust Implicatoins*. New York: The Free Press, 1975.

29. Zaichkowsky, J. L. Measuring the involvement construct. *Journal of Consumer Research*, 12, (December, 1985), 341-352.

Speak Out and Annoy Someone: Experiences with Intelligent Kiosks

Andrew D. Christian and Brian L. Avery
Cambridge Research Laboratory, Compaq Computer Corporation
One Kendall Square, Building 700, Cambridge, MA 02139
+1 (617) 551-7600   {Andrew.Christian, Brian.Avery}@compaq.com

ABSTRACT

An intelligent kiosk is a public information kiosk that senses the presence of humans and communicates in a natural way. To examine issues of human-kiosk interaction, we have built and deployed two versions of intelligent kiosks. The first kiosk design combines machine vision to locate and track people in the vicinity with an animated talking head that focuses on clients and talks to them. The second kiosk design uses infrared and sonar sensors to sense clients and multiple interacting agents to communicate with the client.

The foremost lessons learned from public trials include (1) people are attracted to an animated face that watches them, (2) small mobile agents interact better with kiosk content than a single fixed face, (3) speaker-independent speech recognition is only useful in targeted applications, and (4) the quality of the content on the kiosk strongly influences the client's evaluation of the quality of the technology.

Keywords

Public kiosk, talking avatar, speech recognition, machine vision, user interface design, information display.

INTRODUCTION

Have you ever walked up to a touchscreen-based public kiosk and started punching away only to discover that it was broken? Have you ever walked away from a kiosk wondering if the next person to use it would see your confidential information? Do you feel that public kiosks are cold, impersonal creations?

Our research goal is to build computerized public kiosks that act in a friendly and inviting way. We believe this can best be accomplished by making kiosks that can recognize the presence of a client and interact intelligently. Our methodology is to build kiosks, deploy them in public spaces, and gather feedback from real-world users. We have found that long-term public deployments of kiosks are critical to test interaction technologies; what works well in the lab often fails in public due to unforeseen difficulties.

Permission to make digital or hard copies of all or part of this work for personal or classroom use is granted without fee provided that copies are not made or distributed for profit or commercial advantage and that copies bear this notice and the full citation on the first page. To copy otherwise, to republish, to post on servers or to redistribute to lists, requires prior specific permission and/or a fee.
CHI '2000 The Hague, Amsterdam
Copyright ACM 2000 1-58113-216-6/00/04...$5.00

Figure 1: Photo of the vision kiosk

The first kiosk uses machine vision and an animated talking head to verbally communicate and convey client awareness. The second kiosk senses the client's presence with active infrared and sonar sensors, and communicates via multiple interacting on-screen agents and speech recognition.

The intent of this paper is to describe our experiences and lessons learned from deploying two different public kiosk designs. This paper does not include numerical results from data analyses of surveys and logged kiosk data.

VISION KIOSK

Our first kiosk design combines machine vision with an animated talking head. This kiosk, the "Vision Kiosk" was publicly deployed starting in August 1997. The vision kiosk design has been described in [1]. This section of the paper gives a brief overview of the technology in the vision kiosk and then discusses the results of the public deployments.

System Overview

The design of the vision kiosk is centered around the use of an animated talking head as the focus of user interaction. We'll refer to this talking head as the "avatar". The avatar turns and watches approaching clients. When the client gets close enough and appears interested in the kiosk, the avatar speaks a greeting. As the client navigates the information on the kiosk, the avatar provides useful information, witty repartee, and assistance. When the session is ended by the client's departure, the avatar bids the client farewell.

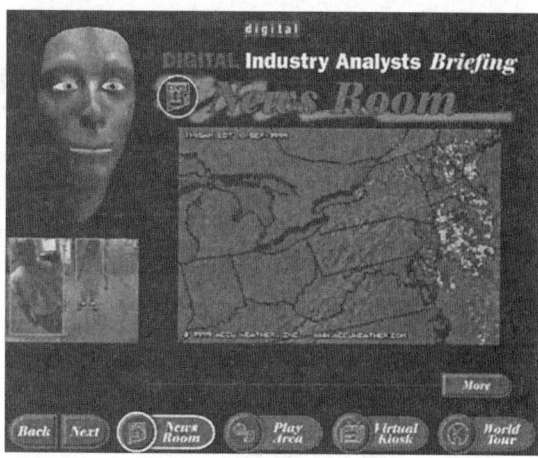

Figure 2: Image of the kiosk screen

Figure 1 is a photograph of the vision kiosk assembly. The kiosk has a 21" touchscreen monitor, a sound system, and a single video camera at the top. The kiosk is self-contained; two AlphaStation 500 computers inside handle the image processing, store the kiosk content, and record interactions.

Figure 2 is a screen capture from the kiosk interacting with two people. This figure shows the basic elements of the kiosk. The avatar appears in the upper-left corner of the screen. Below the avatar is a window that displays what the kiosk sees and provides feedback as to where the kiosk believes people are standing. Note that the avatar is turned to face the person standing closest to the kiosk. The remainder of the kiosk screen contains a web browser and a navigation bar. The client uses large finger-friendly buttons in the navigation bar and the web browser to travel through the content stored on the kiosk web site. Although the kiosk connects to the Internet, we normally do not allow clients to leave the local, kiosk-friendly, web site. Instead the kiosk extracts items of interest from the Internet such as weather, news, and stock quotes, and formats them for local display.

Machine vision

To create an avatar that behaves realistically, the kiosk must quickly locate and accurately track people in the vicinity. To smoothly move the avatar, the tracking rate must be at least 10 Hz with an accuracy of 5–10 degrees.

A solution is a single-camera machine vision tracker. A camera with a wide-angle lens sits on top of the kiosk pointed approximately 30 degrees down from the horizontal. This camera orientation was selected so that the feet of people standing near the kiosk would be visible. The location of a person's feet in the visual field of the kiosk and the knowledge that the floor is flat is used to estimate the person's location in space.

The vision system works as follows: A known image of the background scenery is stored and periodically updated. Possible targets are identified as large regions in the current image that differ in color and luminosity from the stored background image. Targets are screened based on size and shape; only people-sized targets are tracked. Targets are tracked through time by comparing position, estimated velocity, and color. On an AlphaStation 500/400, the vision system can track 6 targets at 15 Hz.

Avatar

The kiosk avatar performs two functions. First it provides the kiosk client with a sense of "awareness" by using tracking information from the vision system. The client can tell that the kiosk is aware of her presence because the avatar "looks" at her; using the tracking data and a dynamic model of head motion, the avatar rotates to appear to be facing the client and the avatar's eyes focus on the client's face. Second, the avatar provides verbal information and assistance to the client. This verbal information complements the kiosk's visual information.

The avatar is an enhanced version of DECface [5–6]. DECface is a three-dimensional texture-mapped polygonal model of a head with an articulated mouth. By varying the texture map and the geometry, any number of different faces may be displayed on the avatar. DECface is designed to lip-synch with synthesized [7] or recorded [4] speech, creating the illusion that the avatar is talking. DECface has movable eyes, eyelashes, and facial muscles that allow it to display a wide variety of expressions and emotions.

Behavior System

The kiosk behavior system controls what information is displayed and how the kiosk responds to client inputs. At the top level, the behavior system operates in either "attraction" or "interaction" mode. When no client is present, the kiosk runs in attraction mode. In this mode, the web browser displays a slide show of interesting images and the avatar turns and watches people as they walk past. If a person appears to be interested in the kiosk, the kiosk switches to interaction mode. In interaction mode, the avatar greets the new client, and the kiosk displays an introductory page. The client may now browse the site, viewing new pages accompanied by informative comments from the avatar. Finally, when the client walks away, the avatar bids the client farewell and the kiosk returns to the attraction mode.

The specific behaviors of the avatar and web browser are scripted by the kiosk content developer. The scripts control what pages are displayed, what the avatar says, and how the kiosk responds to button pushes. For example, a script may state that the kiosk should display a certain web page and avatar, pause for 10 seconds, have the avatar say something, switch to a different avatar, pause again for two minutes, and then jump to a different script.

System Deployment

Over a six month period, three complete vision kiosks were constructed and publicly deployed at four separate venues. The kiosk content was completely revised for each venue to provide a new look and feel, content appropriate to the location, and new services. New avatar personas were created for each location.

Our longest deployment was at the Cybersmith Café in Harvard Square, Cambridge, Massachusetts. A vision kiosk sat in the coffee shop for 4 ½ months showing timely and relevant content including current events, store contests, and birthday parties that week. Over four thousand people interacted with the kiosk at Cybersmith during that period, where a "kiosk client" is a person who pauses in front of the kiosk for at least 10 seconds and pushes at least one button.

All kiosk states, actions, and sensor inputs are recorded in textual and graphical log files. Processed textual log information can be used to evaluate what types of content are the most popular, how people navigate (or fail to navigate) the site, how long people use the kiosk, and how many people pass in front of the kiosk. A second way of checking kiosk interactions is through the visual log. The kiosk stores snapshots at 5 second intervals from the vision system. These snapshots are time and date stamped and compiled into movies.

In addition to logged information, we conducted informal interviews of kiosk clients and set up on-screen surveys. On-screen surveys proved to be of little use; subjects rarely had the interest or patience to answer more than one or two questions. Informal interviews with subjects after kiosk use proved to be the most valuable source of feedback.

Lessons Learned

We learned a number of lessons from our public kiosk deployments. These can be divided into thoughts about the vision system, avatar, overall behavior, and content.

Vision System

The first half of any human-computer interaction is the sensing of the human. Machine vision was used by this kiosk to locate and track potential clients. The advantage of machine vision is that when it works, it is fast, accurate, and informative. In all installations, the kiosk accurately tracked from one to five people moving in front of cluttered and visually noisy backgrounds. However, the kiosk vision system has three major obstacles to commercial viability: crowds, location customization, and robustness.

1. Crowds

The vision system was designed to sense small numbers of people standing in the vicinity of the kiosk while ignoring background clutter. When large numbers of people cluster closely about the kiosk, the vision system reacts to the crowd by turning the avatar to face the center of the crowd and occasionally glancing around.

This assumption of "crowds = people standing in front" fails in what we call the "caribou herd" problem. The caribou herd occurs in installations where the space in front of the kiosk is packed with a crowd of people all moving in the same direction. Because the kiosk cannot distinguish individuals within the crowd and hence track their motion, it assumes that the crowd is stationary. This mistaken assumption annoys people because (a) the avatar stares straight ahead and does not track individuals walking by, and (b) the avatar will attempt to hold a conversation with what it considers to be a stationary crowd. This problem could be solved by more sophisticated machine vision algorithms that estimated crowd motion, or by adding a velocity sensor such as a Doppler radar.

2. Location customization

A kiosk must be correctly configured and adapted for its visual environment. The visual environment changes as people and other objects move around in the background and as the lighting changes during the course of the day. Ideally the kiosk would automatically learn how the visual background could vary over the course a few days, and then use that information to optimize the vision system. Unfortunately, this approach works only if the vision system can initially disambiguate between valid clients and the visual environment.

One challenge occurs when the background on the floor near the kiosk contains dynamic elements. For example, our kiosks have encountered long, dark shadows cast by the setting sun, the bright illumination of a carpeted floor by car headlights shining through colored glass windows, sunlight reflecting brightly off of a tiled lobby floor, and moving images deliberately projected onto the floor in a tradeshow; in this case, a slowly revolving projection of the company logo in red with a white background. If these floor-based dynamic elements line up with objects moving in the distance, they may appear to be people standing in front of the kiosk.

The prevalence of floor-based dynamic elements has prevented us from implementing a completely automatic adaptation system. Instead, the kiosk installer must initially hand tune the vision system to disambiguate valid targets from the background clutter. Once tuned, the kiosk self-adapts to normal environmental changes such as the movement of furniture or the daily cycle of light levels.

The location customization problem could be solved by adding complementary forms of sensing that could disambiguate people from images cast upon the floor; for example, stereo machine vision or infrared range detectors.

3. Robustness

The vision system should gracefully handle sudden and strange changes to its visual environment. For example, one of our kiosks was placed at a trade show in Florida for a few days facing a lobby with tall glass windows. Day after sunny day the vision system correctly compensated for bright sunlight and dark shadows creeping across the floor. One day, towering thunderstorms descended upon the hotel. The hotel was plunged into darkness interrupted by bolts of lightning. Through the storm, the kiosk could be seen alternately talking to lighting strikes and ignoring people who were attempting to retrieve the latest weather reports.

In short, the sudden, dramatic, and frequent light level changes threw off the vision adaptation algorithms which had handled the more gradual effects of intermittent clouds and sunshine. The people trying to use the kiosk to retrieve the weather forecasts were out of luck. This problem could be solved by adding other types of sensors to back up the vision system; for example, active infrared range detectors.

Avatar

The kiosk avatar provides a focal point of attention for the kiosk. It attracts people to the kiosk area, prompts them to use the kiosk, and entertains them with humorous remarks. It is by far the most remarked upon feature of the kiosk and has been a major source of amusement for the clients.

People's attention is attracted when the avatar turns and watches them walk by. We have asked many people why they stopped to use the kiosk; one of the more common answers is "it looked at me, so I went over to see what it was." Curiously, it sometimes doesn't occur to these same people that the avatar had actually turned to watch them. When it was pointed out to them that the avatar would turn as they moved from side to side in front of the kiosk, most of them were surprised and excited by the discovery.

We learned a few lessons that helped improve client-avatar interactions. First, synthesized speech was less intelligible than recorded speech, particularly in noisy public spaces. People also found the recorded speech to be more engaging. The synthesized speech should be reserved strictly for material that can't be recorded such as news stories or weather forecasts. Second, each individual avatar should speak with a single unique voice. Customers were disconcerted when an avatar which had been speaking in one voice suddenly switched to a completely different voice, particularly if one voice was recorded speech and the second was synthesized.

An unfortunate implication of the second lesson is the difficulty of handling one of the most popular activities on the kiosk: poking the face. Some kiosk clients experimentally poke the avatar to see how it responds. We programmed the avatar to respond to pokes with a randomly selected humorous comment, complaint, threat, or physical reaction (e.g., poking an eye may cause it to close). Once the kiosk client has discovered that poking the avatar provokes a reaction, he/she will often poke the avatar over and over again. The difficulty in avatar poking is that as each avatar should speak with only one voice, an avatar that uses recorded speech must be fleshed out with dozens or hundreds of sayings. This adds greatly to the time required to build a recorded-voice avatar.

The avatar has several features that could be improved. First, our implementation allows only one avatar on the screen at a time, and the position of the avatar is fixed. This limits the functionality of the avatar as a disembodied commentator; it can't interact with other avatars or easily point to screen content. Second, the emotive mode of the avatar hasn't proven useful except for entertainment. A smiling avatar is a happy avatar, but it doesn't add much to the user experience.

A third problem with the avatar is actually a sensing problem. A common kiosk action in response to a button press is to display a new web page and have the avatar say something. Kiosk clients who are reading information on the web page do not like to be interrupted by the talking avatar; similarly, kiosk clients who are watching and listening to the avatar are not reading the visual information. Ideally the kiosk's vision system would identify where the client's attention was focussed and adjust the avatar's behavior accordingly. The current vision system cannot provide this information. As a result, sections of the kiosk that present a great deal of both textual and verbal information are difficult to follow by clients.

To summarize our experiences: the avatar does well at attracting and entertaining people, but does poorly at conveying or interacting with content.

Overall behavior

After four public installations and thousands of hours of usage, people's overall rating of the vision kiosk has been positive. We were initially concerned that the technical elements of machine vision and a talking avatar would scare off prospective clients; this turned out to not be the case.

When described verbally, the talking disembodied avatar staring at you from a kiosk screen sounds disturbing. In practice we found that people did not react to the avatar negatively; sometimes they were tentative, but that reaction turned quickly into curiosity. We were also concerned that a video camera on the kiosk would frighten off customers because they would think that they were being remotely monitored (which, in fact, they frequently were). However, this did not disturb anyone we talked with. In fact, the combination of the video camera and the displayed video camera image on the kiosk screen intrigued people. We have video footage of clients dancing in front of the kiosk and watching how the vision system tracked them.

Perhaps the ultimate proof of the approachability of the vision kiosk comes from its tradeshow installations. Our proudest kiosk moments have occurred when we found members of the hotel cleaning staff playing with the kiosk late at night and laughing.

The Most Important Lesson: Content is King!

The quality and quantity of the content on a kiosk and the ease of creating interesting content are critical to the perceived value of the kiosk itself. The novelty of the vision kiosk technology attracts people and gets them to use the kiosk for the first minute or so. After that, the kiosk must be entertaining and informative, or the client will depart. Moreover, the client's subjective evaluation of the quality of the technology is strongly influenced by the quality of the content. This coupling causes no end of trouble; the

Figure 3: Photograph of the agent kiosk

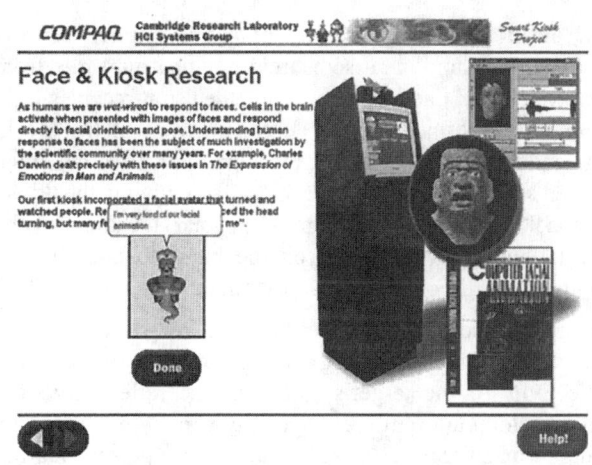

Figure 4: Image of the agent kiosk screen

average subject finds it difficult to separate the utility of the technology from the quality of the presentation.

Directions for the Future

At the end of the vision kiosk project we compiled a list of things we'd like to change, improve, or add to the kiosk. This list includes common technology requests. Apart from lower system cost, the most important items listed are:

(1) More flexible on-screen characters. The fixed screen format of the vision kiosk forces the content developer to always use a single talking avatar, which is frequently inappropriate for the displayed content.

(2) A holistic approach to client sensing, incorporating other types of sensors that would complement the machine vision system and improve its accuracy.

(3) An improved behavior model that simplifies the development of complex kiosk interactions.

(4) Speech recognition. Speech is frequently requested by kiosk clients; after all, if the kiosk is so interactive, why can't it understand what you say?

AGENT KIOSK

To address issues raised by the vision kiosk, we decided to develop a completely different kiosk architecture. The vision kiosk was centered around the use of the talking avatar as the focus of user interaction, and led to a machine vision system to provide fast real-time tracking of clients. The primary disadvantage of this approach is that there is always one disembodied avatar fixed on the screen as a constant observer of all that happens in the kiosk.

Our second kiosk architecture uses multiple interacting on-screen agents and active sensing systems; for the purposes of this paper we will refer to it as the "agent kiosk".

System Overview

Figure 3 is a photograph of the agent kiosk. The kiosk consists of a 15" flat panel touchscreen, a Pentium computer, and an array of active sensors. This particular kiosk was designed to have a minimum footprint, suitable for installation in commercial locations.

The agent kiosk has three main components: the display, the sensing system, and the central authority.

Display

The kiosk content consists of full-screen web pages stored on a local web server. Kiosk agents float above the web pages, interacting with the page content and with each other. Figure 4 is a typical screen capture of the kiosk. The kiosk behavior is controlled by JavaScript programs contained within the web pages.

Microsoft Agent [2] technology is used for the agents themselves. The Microsoft Agent technology allows multiple on-screen graphical agents to move about the screen, express emotions, point to objects, and speak using speech synthesis. The visual appearance of a Microsoft Agent can be thought of as being created by stitching together a series of short, pre-recorded movie clips. This approach allows the agent creator full control over the appearance and actions of the agent, subject to size limitations. Agents come with canned "idle" routines; that is, sequences of animations to play when they are not performing any scripted gestures. This gives them the appearance of being alive.

The flexible visual appearance of the Microsoft Agent and its ability to move arbitrarily around on the screen have great advantages over the talking avatar of the vision kiosk. However, this flexibility comes at a price; as pre-recorded bits of movie footage, Microsoft Agents cannot synthesize arbitrary facial expressions or precisely turn and face a kiosk client. These restrictions severely limit the agent kiosk's ability to project an awareness of the kiosk client.

Sensing System

Given the restricted visual feedback possible from the on-screen agents, we decided that client sensing for the agent kiosk could be adequately accomplished using inexpensive active sensors, rather than a machine vision system. The two types of active sensors we considered were infrared and sonar. In the end, it proved necessary to use both types.

Active infrared sensors work by flashing an infrared LED and measuring the strength of the reflected infrared light.

The agent kiosk uses a series of active infrared sensors arranged in a horizontal semicircle pointing outwards from the kiosk front. Each receiver is shielded so it receives light from a well-defined angular region in front of the kiosk. The angular regions overlap, providing complete coverage of the space in front of the kiosk and giving the kiosk feedback as to the approximate angular location of the client, thus allowing agents on the kiosk to point in the approximate direction of the kiosk client.

Infrared sensors cannot measure the range to a target accurately without knowing the size of the target and the reflectivity of the target's clothing. The reflectivity of a target's clothing to infrared light may vary by more than an order of magnitude. To an infrared sensor, a person wearing a reflective raincoat 6 feet away may be indistinguishable from a person wearing a black T-shirt 2 feet away.

We chose to solve this difficulty by using sonar sensors aligned with the infrared sensors. The sonar sensors accurately measure the range to a client to within a few inches. The sonar sensors alone cannot drive the kiosk because they are too slow. Practical limits on the sonar circuitry limit the update rate to 5 Hz, which is not fast enough to register the presence of a target, track the target's position, and promote the target to kiosk client (people complain that the kiosk seems "sluggish" when the sensor rate goes too low).

The combination of infrared and sonar sensors can be used to accurately and quickly track the target's position. The infrared sensors continuously monitor the angular position of the target; the sonar sensor data is used to calibrate the infrared sensor's measurement of target distance.

Central Authority

An individual web page displayed on the kiosk is necessarily transient in nature; it goes away as soon as the kiosk client moves to a new page. Hence kiosk state information and programming logic are not stored in the web page. Instead, a central authority program stores program state and manages all kiosk behavior. Individual web pages which wish to make use of higher kiosk behaviors open a connection to the central authority.

The central authority provides the following services: (1) direct agent control, agent scripting and "poking" behaviors, (2) sensor data, including behaviors to be taken when targets enter or exit, (3) background music and related audio services, (4) persistent data storage, and (5) speech recognition services.

The agent interfaces provide a good example of how the central authority handles the complex problem of appropriate client-kiosk interaction. Consider the "help" screen shown in figure 5. The purpose of this screen is to demonstrate how to use a touchscreen. Surprisingly, a small but significant percentage of kiosk clients are unfamiliar with touchscreens. On the vision kiosk, we found that the avatar had to verbally prompt clients to "go ahead, push a

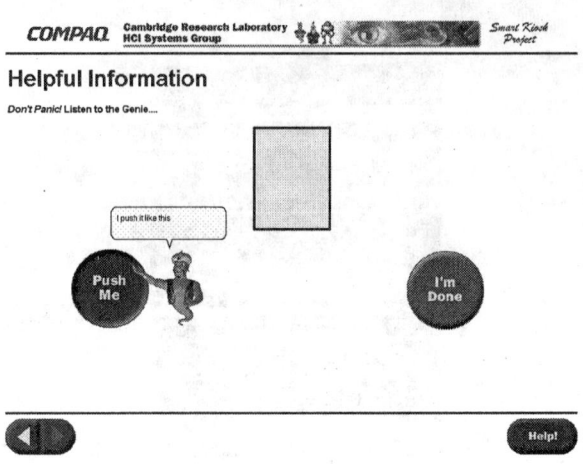

Figure 5: Help screen

button". Unfortunately, not all clients understood that the avatar was referring to a button on the screen. The agent kiosk goes one step further. When a new client fails to press a button, the kiosk displays the help screen and the "Genie" agent visually demonstrates how buttons work.

The basic script for the help screen is as follows: the genie appears on the screen and flies to the center. With gestures and speech, the genie explains the use of a touchscreen and demonstrates how to push a button. A large purple button appears; the genie flies down and demonstrates how to push the button. Finally, a green "exit" button appears and the genie thanks the client for his/her patience. If the kiosk client is well behaved, this little "help" script will be carried out without interruption. But, as is more likely, if the client begins pushing buttons randomly or poking the genie in the nose, then the kiosk must respond appropriately and yet still work through the planned script.

Scripting behavior is handled by the central authority. A script is registered with the authority as a sequence of actions and callbacks to be performed by one or more agents. Valid actions include agent expressions, agent movement on the screen, agent speech, deliberate pauses, and callbacks to the web page (which allow the web page to synchronize dynamic HTML content with agent behavior). Actions may be carried out in parallel and synchronized where appropriate; for example, one agent may gesture and speak while another moves across the screen. Actions are annotated with notes on how to handle an interruption. For example, if an agent is speaking and is interrupted by a client's button press, once the button push has been handled the agent may resume speaking. The speech is not resumed from the point of interruption, but from the beginning of the interrupted sentence.

All scripts registered with the central authority run at a designated priority level. Higher priority scripts interrupt lower priority scripts. This combination of scripts and priority levels simplifies the control of a complicated web page like the "help" page. The basic help script runs as a low-priority script that lasts for about a minute and contains

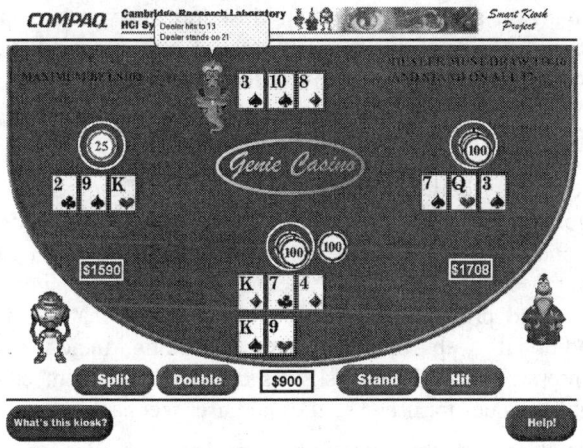

Figure 6: Blackjack

a complicated series of movements, expressions, and speech by the Genie combined with callbacks to dynamic HTML elements of the web page. A kiosk client pressing buttons inappropriately causes the web page to dispatch higher priority scripts that ask the kiosk client to "stop it" in progressively stronger language. These scripts briefly interrupt the lower priority help script. They may be in turn interrupted by still higher priority scripts that are dispatched by the central authority if the client pokes an agent.

Finally, long term kiosk behavior on the help page can be adjusted by storing help page usage in the central authority's persistent store for this kiosk client. If the client should return to the help page, the kiosk remembers that this particular client has been here before and runs an abbreviated version of the help script.

Integrating Speech Recognition
As noted previously, content is the key to a good kiosk. Our challenge with the agent kiosk was to create interesting content that would use the full potential of the machine; in particular, speech recognition and interacting agents.

Speaker independent speech recognition is only reliable with a small vocabulary and clear audio. To get the kiosk client to use that small vocabulary, you must either pick an application area where the vocabulary is well known or prompt the client. Verbal prompting slows down the pace of interaction. Visual prompting can be frustrating because if a client can see the word or phrase displayed on the screen, the client frequently finds that just pushing a button works better than repeating the phrase.

Our solution is a kiosk implementation of the casino game Blackjack (for the rules of blackjack, see [3]). Figure 6 is a screen capture of the kiosk playing blackjack. The genie acts as the dealer; the kiosk client plays the middle hand. Other agents come and go during the course of the game; each agent has a unique personality and style of play. Blackjack works well as an example of multi-agent interaction and speech recognition. Having several players in the game adds to its entertainment value by providing friendly competition. Speech is a natural interface for blackjack; people find it more comfortable than pushing buttons and the required vocabulary is small.

Speech-Enabled Game Design
As a result of testing early versions of blackjack, we incorporated or improved a number of features in the game such as vocabulary prompting and silly move checking.

Initially we chose a relatively small vocabulary for the game, based on our own experience playing blackjack. This vocabulary was deliberately mirrored by buttons, thus allowing players to either press the button or say the command. When we invited other people to play the game, we discovered that each person had their own Blackjack vocabulary and that the buttons were completely ignored. For example, the "stand" command was also given in 9 other common ways: "stick", "stay", "stop", "no more", "no more cards", "finished", "done", "I'm done", and "I'll stand". As a result, the speech recognition failed.

We addressed this problem using two standard techniques: first, a series of test subjects provided a list of common utterances for the various Blackjack commands; the most common were incorporated into the speech recognition engine. This expanded the vocabulary by roughly 300%. Second, the agents playing the game prompt the kiosk player to use the valid vocabulary by example. Agents always use valid utterances when speaking to the dealer. These utterances are randomly selected from the game vocabulary. This indirect prompting works remarkably well; players quickly adopt the agent's vocabulary.

Even with improvements to the vocabulary, the speech recognition engine can still fail. Failure can take one of three forms: not recognizing a valid utterance, incorrectly recognizing a valid utterance, and recognizing an invalid utterance. Missing a valid utterance only annoys players, but incorrectly interpreting "stand" for "hit me" angers players. To avoid angering our players, we incorporated a "silly move" checker. If the kiosk believes that the player commanded it do something clearly foolish (such as standing on a 5), the dealer asks the player to confirm the command ("do you really want to stand on a 5?").

Error Rates with Speech Recognition
The current version of the agent kiosk uses the Microsoft Speech Recognition engine, version 3. We ran trials to measure the accuracy rate of the speech recognition. It varies widely between speakers, from 37% to 96% correct. The most common error is a run of misses; the kiosk accurately recognizes a long string of commands and then repeatedly fails to recognize a single command. Fortunately, most players recover from this situation by pushing a button on the screen (in the speech trials, we forbid the use of buttons). In practice this means that the effective accuracy rate is 85% to 97%, which is sufficient for a research kiosk but too low for a commercial kiosk.

We have been assured by the speech recognition research group at Microsoft that the newer versions of the speech recognition engine are more reliable, however we haven't had the opportunity to test this. We are also looking into other speech recognition engines and modifications to our microphone and speaker set up. Speech recognition improves dramatically if the kiosk client wears a headset. Unfortunately, headsets are inappropriate for public kiosk installations, so we restrict ourselves to stem microphones.

Lessons Learned

The agent kiosk has been to one tradeshow and has run continuously in our laboratory for over a year. This section summarizes the most important lessons we've learned:

Agents

The multiple interacting agents are excellent for "close-up" work. A carefully designed interactive web page (such as blackjack) holds the kiosk client's interest. The agent characters are particularly evocative when they are designed to have individualized behaviors. For example, after a few rounds, players at the blackjack table begin to notice that Robby the Robot suggests plays based on the odds whereas Merlin the Magician seems to have prescient knowledge. It's true; Robby counts cards, whereas Merlin, being a magician, knows the top card on the deck. In the long run, Merlin's magic makes a considerable difference in winning. Blackjack players get excited about this ability and often report back to the researcher that "that Merlin guy cheats".

A disadvantage of the multiple interacting agents is that they are programmed using JavaScript on web pages. This is a tedious and error-prone process; for example, the entire blackjack game takes over 3600 lines of code, much of which is concerned with agent behavior. Additionally, JavaScript programming requires more technical skill making it difficult to find good content developers.

In the future, we hope that better tools and programming languages will simplify the content creation. SMIL is a step in this direction. [http://www.w3.org/AudioVideo/]

Sensors

The active infrared and sonar sensors on the agent kiosk reliably detect and follow clients. However, the sensor update rate and positional accuracy is limited; in the future, we believe truly interactive kiosks will need to fuse fast machine vision with robust infrared and sonar sensors.

Speech Recognition

Speech recognition has not yet proven to be useful in a kiosk, except in limited circumstances. The blackjack game is an entertaining demonstration, but it still suffers from inaccurate recognition. A more general application with a larger vocabulary would be more error prone.

Overall Impressions

When asked to compare the agent kiosk to the vision kiosk, the common response is that the agent kiosk is fun and interesting, but it isn't as compelling as the vision kiosk. Having the avatar on the vision kiosk turn and watch people walking by does an excellent job of attracting people's interest and getting them to use the kiosk. The agent kiosk simply doesn't have the same draw.

CONCLUSIONS

We have built and deployed two different public information kiosk architectures. The "vision" kiosk uses machine vision to locate prospective clients, demonstrates awareness of prospects by tracking them with an animated face, and provides information using a combination of a traditional web browser and a talking, emotionally expressive face. This kiosk was deployed in a number of public venues for usage studies and user feedback.

The "agent" kiosk was built in response to requests for multiple on-screen interacting agents and speech recognition. This kiosk locates and tracks clients using inexpensive active infrared and sonar sensors. We found that a multi-player version of the casino game blackjack worked as a compelling speech-enabled application.

In summary, the avatar on the vision kiosk attracted attention and projected a great sense of "awareness" of the client, but was not useful in the interaction mode (except for entertainment). The multiple agents on the agent kiosk did not do a good job of attracting attention or projecting a sense of awareness, but were more useful than the avatar in the interaction mode. Future kiosk work will concentrate on combining the best aspects of both kiosks.

ACKNOWLEDGMENTS

We would also like to thank the research and administrative staff at CRL for their help on the project and tolerance in listening to the kiosks blather in the corridors and their willingness to play blackjack for hours on end.

REFERENCES

1. Christian, A. and Avery, B., Digital Smart Kiosk Project. ACM Proceedings of the SIGCHI Conference, Los Angeles, California, April, 1998. 155–162.

2. *Developing for Microsoft Agent*, Microsoft Press, 1997.

3. Frey, R. L. and Hoyle, E., *According to Hoyle: Official Rules of More Than 200 Popular Games of Skill and Chance*, Ballantine Books, 1996.

4. Goldenthal, W., Waters, K., Van Thong, J.M., and Glickman, O. Driving Synthetic Mouth Gestures: Phonetic Recognition for FaceMe!, EuroSpeech, 1997.

5. Lee, Y. Terzopoulos, D., and Waters, K. Realistic modeling for facial animation. *Computer Graphics* (SIGGRAPH'95), 1995, 55–62.

6. Parke, F., and Waters, K. *Computer Facial Animation*. A. K. Peters, Ltd., 1996.

7. Waters, K. and Levergood, T. An automatic lip-synchronization algorithm for synthetic faces. ACM Proceedings of the Multimedia Conference, San Francisco, California, September, 1994. 149–156.

The Effect of Task Conditions on the Comprehensibility of Synthetic Speech

Jennifer Lai, David Wood
IBM Corporation/T.J. Watson Research Center
30 Saw Mill River Road
Hawthorne, New York 10532
lai, dawood@watson.ibm.com

Michael Considine
Rice University
6300 S. Main Street
Houston, Texas 77005
consid24@rice.edu

ABSTRACT
A study was conducted with 78 subjects to evaluate the comprehensibility of synthetic speech for various tasks ranging from short, simple e-mail messages to longer news articles on mostly obscure topics. Comprehension accuracy for each subject was measured for synthetic speech and for recorded human speech. Half the subjects were allowed to take notes while listening, the other half were not. Findings show that there was no significant difference in comprehension of synthetic speech among the five different text-to-speech engines used. Those subjects that did not take notes performed significantly worse for all synthetic voice tasks when compared to recorded speech tasks. Performance for synthetic speech in the non note-taking condition degraded as the task got longer and more complex. When taking notes, subjects also did significantly worse within the synthetic voice condition averaged across all six tasks. However, average performance scores for the last three tasks in this condition show comparable results for human and synthetic speech, reflective of a training effect.

KEYWORDS
Text-to-Speech, Synthetic Speech, User Study, Comprehension

INTRODUCTION
For many years now, people have been predicting that the use of computer generated speech would become widespread; showing up in applications ranging from talking appliances and vending machines, to fully conversational data entry and retrieval systems. For the most part however it seems that the forecasted wide-ranging use of synthetic speech has not materialized, and that we encounter it primarily in situations where the use of recorded human speech is not practical, as in access to large and frequently changing databases of information such as stock quotes or mutual fund prices. This may be because people report that synthetic speech sounds unnatural and is unpleasant to listen to [1, 7, 8]. More recently though, synthetic speech has been applied to the retrieval of e-mail messages and the reading of news articles in products such as Portico by General Magic [10] and Webley by Webley Systems Incorporated [11].

Synthetic speech, also known as text-to-speech (TTS), is speech produced by a computer. It is often referred to as "rule-based" speech because the computer uses a series of rules to convert text to the sounds that are generated. The text goes through several stages of transformation prior to the actual synthesis itself. The type of synthethis varies from one TTS engine to another and can be formant-based, articulation-based, or concatenative [3]. Such systems are capable of producing an unlimited number of messages without storage constraints, and are absolutely necessary when reading dynamically created text such as e-mail messages and late breaking news stories.

As part of a group of researchers in Pervasive Computing Solutions, facing the prospect of incorporating synthetic speech into several of our prototypes, my colleagues and I have a keen interest in measuring how well synthesized speech can be understood. One of the areas in which we are focusing our research involves the use of a virtual assistant, which would allow users to call from any telephone and stay in touch with their office by interacting with an "assistant" which conveys information about messages and select news articles. Although we believe that digitally recorded human speech would be subjectively preferred over synthetic speech in such an application, the dynamic nature of the messages and file size considerations make this solution impractical if not impossible.

Our goal was not to produce a rank-ordering of the five synthetic speech engines measured, but to understand if there were optimal conditions for the comprehension of synthesized speech, and to what degree comprehension would degrade when these conditions varied. The conditions we looked at were the nature of the task and the effect of note-taking while listening. The tasks consisted of

Permission to make digital or hard copies of all or part of this work for personal or classroom use is granted without fee provided that copies are not made or distributed for profit or commercial advantage and that copies bear this notice and the full citation on the first page. To copy otherwise, to republish, to post on servers or to redistribute to lists, requires prior specific permission and/or a fee.
CHI '2000 The Hague, Amsterdam
Copyright ACM 2000 1-58113-216-6/00/04...$5.00

short informal e-mail messages, longer messages, and CNN news stories. We used comprehension levels of the passages when read by a professional voice talent as a reference condition for each subject.

PRIOR WORK

Most of the prior evaluations of text-to-speech systems have concentrated on either segmental intelligibility (at the word or sentence level) or listener preference (subjective evaluation of the acceptability of synthetic speech). The two tests most often used for word level intelligibility are the Modified Rhyme Test (MRT) and the Diagnostic Rhyme Test (DRT) [1]. Both are closed response tests where the user hears a word and is asked to select the word from a list of possible choices. Sentence level intelligibility has been measured by using either the Harvard Psychoacoustic sentences or the Haskins Syntactic sentences [8]. The Harvard sentences are a fixed set of semantically and syntactically normal sentences such as *"Add salt before you fry the egg"*. The Haskins sentences are also a fixed set of 100 sentences, but in this case they are semantically unpredictable as in *"The old farm cost the blood"*. Although findings indicate synthetic speech is less intelligible than natural speech and that it takes longer to process [5, 6], a wide variance in the performance of synthesizers has been reported.

Even though segmental intelligibility is a critical component in the overall acceptability of a speech synthesizer, Ralston et. al [7] point out the significant difference between the task of recognizing words and sentences, and the task of understanding the intended meaning of the message or spoken passage. When listening to longer passages of synthetic speech, issues of context, task, prosody, fluency (timing) and intonation come into play. These issues may have either a negative impact on comprehension or a positive one, based on the task and synthesizer.

Very few studies have assessed the comprehension of synthetically produced passages of meaningful continuous speech. One study often referred to in the literature, by Pisoni and Hunnicutt in 1980 [4], measured comprehension based on 15 narrative passages of text obtained from a standardized adult reading test. Comprehension performance was measured based on the ability to correctly answer multiple choice questions when listening to synthetic speech, natural speech and reading the text. They found that subjects did significantly better if reading the passage rather than hearing it with either natural or synthetic speech. This effect may have been skewed by the fact that the passages were designed to measure reading comprehension, not auditory comprehension. The study also showed that subjects did significantly better on the second half of the test when listening to synthetic speech, a gain that was not seen in the other media. Overall comprehension levels for synthetic speech were comparable to natural speech for the MITalk synthetic speech system (the only TTS system measured in this study).

We were motivated to conduct a study on the comprehensibility of synthetic speech for several reasons. The first is that most of the findings reported in the literature evaluate the perception of synthetic speech based on TTS engines from 1979 to 1986. Given the progress in technology over the past 13 years we believed it would be valuable to reassess comprehension levels. The second reason is that we agree with Ralston et. al [7] who indicate after a comprehensive review of the research over the past 15 years, that evidence in the matter of the comprehension of synthetic speech is not totally compelling. While a body of findings indicate reliable differences in levels of comprehensibility between synthetic and natural speech, results vary considerably across different studies. Our third motivation was a desire to understand if certain tasks were better suited than others for listening to with synthetic speech on the phone. We structured our study based on findings in Pisoni et al. [5] that there are five major factors that influence listener's performance in laboratory situations 1) quality of speech signal 2) size and complexity of message set 3) short-term memory capacity of listener 4) complexity of the listening task or other concurrent tasks 5) the listener's previous experience with the system.

SYSTEMS

Five commercially available text-to-speech systems were used in the study. While it was tempting to use more "cutting edge" prototype systems available either from Universities or Research organizations, the decision was made to draw the line at commercial products. It was not possible given constraints on time and resources, to include every commercial TTS engine available in the market today. Ultimately we decided to use: DECtalk for Windows 95 v 4.4, AcuVoice AV1700 text reader, IBM Via Voice Outloud (from VV '98), L&H TTS engine version 6.03, and Lucent Release 2. The first two products were purchased, the IBM product was acquired internally, and the latter two were sent to us as evaluator products, free of charge.

METHOD

Experimental Design

Each subject was randomly assigned to one of the five text-to-speech products. Half the subjects heard synthesized speech first and half heard recorded human speech first. The order of passage presentation was counterbalanced using a Latin Square design. Dependent measures included: (1) response accuracy, and (2) time on task (the time from when a subject turned over the questionnaire to view the questions, until completion of responses). Digitized natural speech was used a reference condition for every subject.

Subjects

All 78 subjects were IBM employees and native English speakers recruited from within IBM Research or the neighboring building which houses the Internet division. Subjects were screened to ensure that they did not report any prior history of hearing problems and that they were not working in the field of synthetic speech or otherwise overly familiarized with text-to-speech. Half of the subjects were female and half were male. All subjects volunteered for the study and received a $20 gift certificate for their participation in the 45 minute session. Consent was obtained prior to videotaping the session.

Procedure

To familiarize subjects with the task at hand, each subject was given a practice task prior to experiment participation. The training passage was recorded using a human voice, so that no expertise could be acquired with synthetic speech prior to the actual test itself. A human voice other than the one used in the experiment was used for the training passage so that there too, there would be no adaptation or familiarization. After hearing the training passage and answering the practice questions the subjects were given the correct answers to self-check their work. They were asked if they had any questions about the procedure and if the volume was appropriate for them.

The subject was then instructed to begin the test. Listening to the test passages required following the same process the subject had used for the training passage, namely dialing a four digit telephone number (which was posted next to the phone), listening to the text, hanging up the phone, and turning over the sheet of questions for the text. Thus, the subject had no knowledge of the questions prior to hearing the text, and was not allowed to hear any portion of the passage again. The subject was given as much time as needed to answer as many questions as possible. The amount of time spent answering the questions was recorded for each passage. Once the subject was done with a passage, the process started over again until all six test tasks were completed.

The test consisted of two sets of data. Each set was made up of a short (S), a medium (M) and a long (L) passage. The passages in each set were similar but different. Each subject heard one set "read" by a synthetic engine, and a set with a recorded human voice. The order of voice presentation was alternated such that subject N heard set 1 with human voice and set 2 with synthetic voice and subject N+1 heard set 1 with synthetic and set 2 with human. As mentioned earlier, the order of passage presentation within each set was counterbalanced using a Latin Square. The first 36 subjects always heard set 1 first (regardless of the voice type) and were allowed to take notes. The next 36 subjects always heard data set 2 first and were not allowed to take notes.

The comprehension questions consisted of multiple choice questions of both general comprehension, and specific recognition.

Sample general comprehension question:

2) What is the main topic of this story?
    a.) Indian Foreign Minister Singh's request to endorse the inviolability of the cease fire line;
    b.) Pakistan's attempts to disguise their support of Islamic militants;
    c.) The breaching of the 1972 cease-fire line;
    d.) The tenuous peace between Pakistan and India

Sample specific recognition question:

1) What is the room number for the meeting?
    a. H1-D20
    b. H1-D30
    c. H3-D20
    d. H2-B30

One quick word on our choice of an offline/successive measure rather than a simultaneous measure such as word monitoring. A successive measure is one that is made after presentation of the material, while a successive test measures comprehension as it is taking place [7]. We opted for multiple choice questions which are one of the oldest and most commonly used recognition tests for comprehension. While it does not allow for measuring of processing load during comprehension, it is a good test for determining what has been extracted and retained from the text passage. This choice corresponds to our interest in understanding how well pertinent information presented in an application can be remembered.

Task

Six passages of text were selected for the study. Our goal was twofold in the selection of passages. The first was to match the text in the test to the type of text we expect will be used in the virtual assistant (VA) application. These are primarily e-mail messages of varying length and news stories. The second goal was to select from standardized tests created to measure listening comprehension, not reading comprehension. To this end, we had three sources for the three types of tasks. The short passages were approximately 100 words long and consisted of informal and fictitious e-mail messages. The medium passages (about 300 words) were derived from the Test of English as a Foreign Language (TOEFL) and the long passages (about 500 words) from CNN news stories.

The TOEFL [9] is an English proficiency standardized exam, which has sections on listening comprehension, written expression and reading comprehension. The listening comprehension section is divided into three parts: the first consisting of short conversations, the second of longer conversations and the third of short talks. We selected our passages from the short talks. We took talks from sample exams, that could conceivably be received as messages or obtained as information from the Web, such as weather reports, or biographical information on guest speakers.

The CNN stories that we selected were chosen as ones which we believed would be mostly unfamiliar material for many of our subjects. These were stories which only "news hounds" or people with a particular interest in the subject matter would be closely familiar with. The questions were designed so that even a person generally familiar with the topic, would be unable to answer them correctly without having heard the particular article. For example:

1. Which is NOT mentioned in this article as a reaction to the verdict?

    a.) security was reinforced at airports, prisons and embassies;
    b.) certain activists were placed under surveillance;
    c.) violent protest broke out in several European cities;
    d.) Turks sang the national anthem;

As a check for this, we had 10 subjects, independent of those used in the comprehension study answer all six sets of questions without ever hearing the text. In spite of all these precautions however, we realized that some subjects could have more familiarity with the news articles than others. We felt that this would also be true of users listening to newstories in the virtual assistant, and since a goal of ours was to recreate conditions of application usage, we decided to include real news articles instead of fictitious ones.

We expected that listeners would be able to apply a top-down real-world knowledge to deciphering in context, words and names which might have otherwise been unintelligible. For example the word "mudanya" would be difficult to perceive in a word level intelligibility test, but is easy to comprehend as part of the sentence "Families of soldiers killed in fighting against the PKK gathered at Mudanya, the port closest to the prison island where the trial is being held."

Listening Environment
The experiment was carried out in what could be considered optimum listening circumstances, especially when compared to a potential real-world setting such as being dialed in over a cell phone while driving one's car. The subjects were seated alone in the observation area of a usability lab. There were no interruptions or distractions.

Since the conditions of usage of the VA are potentially quite varied (one could call from a phone anywhere) it was difficult to duplicate the environment for testing purposes.

However, certain factors that are constant for the VA were reproduced. The subjects were required to call in using a telephone, and listened to text that was of a similar nature to that of the application. The level of experience, motivation and demographic characteristics of the subjects fell into the same range as the user set we believe would adopt usage of this type of application.

For each engine (except AcuVoice) we selected the "adult male" voice and the "adult female" voice, using them with their default settings for pitch and breathiness. With AcuVoice we only had male voices to select from so no female voice was used. We modified, where necessary, the words per minute (wpm) rate so that all engines "read" the text at 175 wpm.

The passages of text recorded by a professional voice talent were timed and were shown to have a rate of 200 wpm. The voice talent we hired is male. The human voice was recorded at a sampling rate of 44,100 and a 16 bit resolution. For the synthetic speech passages, the text was "read" each time rather than being played as a stored .wav file.

Evaluation Measures
Recognition memory was measured at the end of each set by presenting the subject with a list of 20 sentences. Ten of the sentences were actually spoken during the set, the other 10 sentences had semantically the same meaning but were presented with a change in the wording or grammar.
For example:
a. The rival nuclear powers have fought three wars since 1947, two of them over Kashmir.
b. Since 1947 three wars have erupted between these rival nuclear powers, two of which have been fought over Kashmir.

In addition to measuring comprehension performance and time on task, subjective measures were obtained by calculating the Mean Opinion Score [1], the use of an exit questionnaire and exit interview. Here again our goal was not a rank-ordering of the five engines, but to find out if there was a large variance between the subject's perception of his or her performance and the actual level of performance itself. We were also looking for any gender preference issues that might surface.

At the end of the study, subjects were asked to complete a Mean Opinion Score (MOS) survey and exit survey. They were instructed that the MOS was only pertinent to the passages they had heard in synthetic speech. The MOS survey is comprised of seven evaluative 5-point scales measuring:
1. Global Impression; 2. Listening Effort; 3. Comprehension Problems; 4. Voice Pleasantness; 5.Speech Sound Articulation; 6. Pronunciation; 7. Speaking Rate

The first four can be viewed as measuring the overall quality of the speech while the last three scales relate to specific aspects of the output and require more analytical listening.

FINDINGS

There was no significant difference for comprehension performance of synthetic speech among the five different TTS engines used (F(4,70) = .106, p = .980).

Nor was there an effect found for time on task. The amount of time required to answer questions in the synthetic set was nearly equal to the time required for the human set (see table 1). The difference in time between the note conditions suggests that subjects looked at their notes to a minor degree while answering questions. We speculate that the lack of difference in timings is due primarily to the off-line nature of the multiple choice questions.

	Notes	No Notes
Hum. Av. Time	199 seconds	174 seconds
Syn. Av. Time	202 seconds	175 seconds

Table 1: Average time on task for the set as a function of voice type and note-taking.

The Training Effect

An important factor with TTS is the rate at which humans adapt their ear to the sound of synthetic speech (or human speech when spoken by a foreigner) and the effect that this adaptation has on levels of comprehension. The amount of training necessary to see a change in the perception of synthetic speech ranges in various studies from a few minutes of exposure [8] to four hours of training [2]. Also when subjects improve performance during the course of a study, it is hard to differentiate what portion is due to a familiarization with the process, and what is due to training on synthetic speech.

Subjects in this study showed an ability to quickly learn both cognitive and perceptual strategies to improve performance. As seen in Table 2, subjects within the note taking condition, who heard the human voice first achieved a comprehension performance of 82% for set 1 and 73% for set 2. Those that heard synthetic first scored 59% for set 1 and 67% for set 2.

	Set 1	Set 2
Hum.	82%	67%
Syn.	59%	73%

Table 2: results by set for the note taking condition (N=36) Results are shown as a percentage of correct answers.

Even though the first group was hearing set 2 in synthetic speech, they scored higher on this set than the group that heard it with natural speech. A post-study analysis showed that there were no significant demographic differences among the four groups of subjects in our study.

This slightly counterintuitive finding could be explained by the fact that subjects try harder when listening to synthetic speech and that the task training which had occurred by the second set had a stronger effect in the synthetic condition. We believe that many subjects trained very rapidly, perhaps within the first sentence or so of synthetic speech. One subject commented "I found I'd miss the first sentence while I trained on the voice, so perhaps some introductory preamble would help." Another said that "by the end of the session I found myself less distracted by the idiosyncrasies of the synthesized voice and better able to focus on the content."

The Effects of Age, Education and Prior TTS Exposure

Even though subjects were screened to eliminate any people who had significant prior exposure or interaction with synthetic speech, we still had 13 (5 for the notes condition and 8 for the non note-taking condition) who arrived for the study and indicated on their data sheet that they used it with "some regularity". Each subject was asked to report his or her level of experience with TTS using the following scale:
1. know what it is but have not heard it
2. have heard it once or twice on the radio or TV
3. have cause to listen to it with some regularity
4. hear it at least once a week
5. work with it

No subjects were included that reported prior exposure at levels 4 or 5.

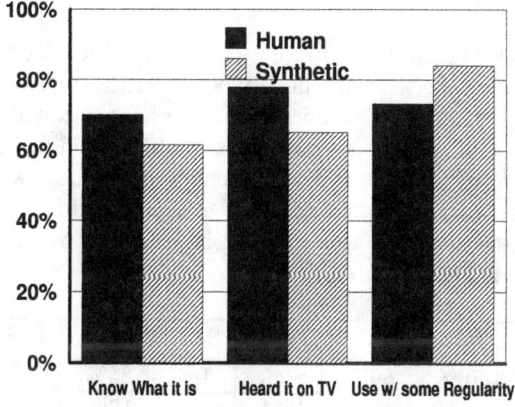

Figure 1: Performance as a function of voice type and experience with TTS for the notes condition.

As Figure 1 shows, there was an interaction between voice type and TTS experience within the notes condition (F(2, 35) = 5.240, p = .01). Subjects who had cause to listen to TTS "with some regularity" did better for the synthetic voice than for human, as well as doing better than those subjects who had less exposure. The advantage these subjects had was lost in the non note-taking condition, as shown in Figure 2.

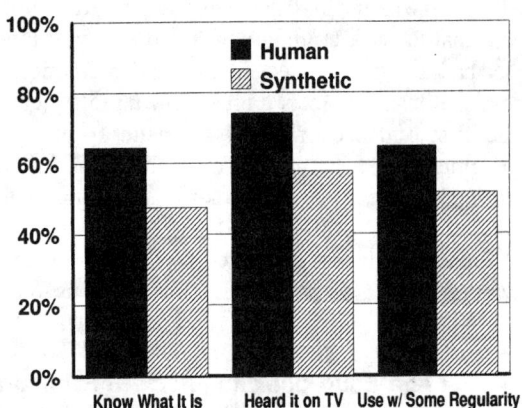

Figure 2: Performance as a function of voice type and experience with TTS for the non note-taking condition.

As expected, subjects who had completed a higher level of education performed better overall ($F(3,71) = 8.200$, $p < .001$). However there were no significant interactions with other variables such as note-taking and voice type. Across both note-taking conditions, there was no effect observed for age of subjects ($F(3,71) = 1.991$, $p = .13$).

The Effect of Note Taking

We observed for the most part, that subjects who were told they could take notes, took a large amount of notes. When answering the questions however, only rarely did we see the subjects referring back to their notes.

While we expected to see significant differences in performance between the three types of tasks, we did not observe this effect within the note taking condition. Certainly the ability to take notes while listening is not reflective of task conditions if listening to messages while driving a car, however we wanted to see what differences would surface that were totally independent of any short-term memory issues.

	S2	M2	L2	S1	M1	L1
Hum.	70%	69%	51%	70%	85%	71%
Syn.	65%	48%	38%	61%	69%	42%

Table 3: performance by task for the non note-taking condition (N=36)

The notation in Table 3 corresponds to the task (S,M,L) and the data set. Thus S2 is the short passage in set 2.
As seen in Table 3, subjects performed worse for all passages with synthetic speech in the non note-taking condition. Performance deteriorated as the task became longer and less familiar.

Comparisons in Figure 3 of comprehension performance for the note-taking and non note-taking the condition show that (perhaps not surprisingly) subjects did better overall when allowed to takes notes ($F(1,73) = 8.026$, $p = .006$). Subjects performed better on the medium ($t(73) = 3.146$, $p=.002$) and long ($t(73) = 2.789$, $p=.007$) synthetic passages in the note-taking condition. What is more interesting is that there was also a significant interaction between note-taking and voice type ($F(1,73) = 4.175$, $p<.05$). This indicates that the effect of note-taking is significantly stronger when listening to synthetic speech rather than human.

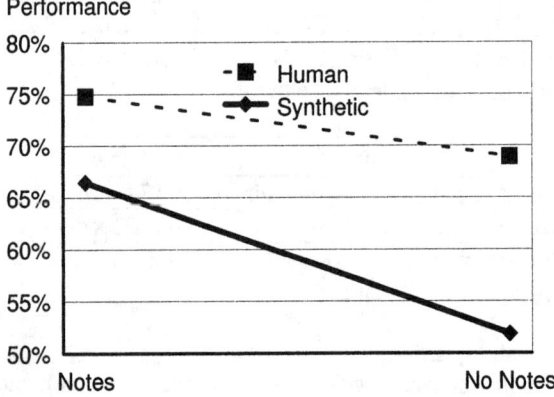

Figure 3: performance by voice type for the notes and non note-taking condition

Results were similar for the recognition memory task. As seen in Table 4, note-taking had no effect on the subjects' ability to recognize phrases with human speech. However, with synthetic speech, recognition memory was improved with notes ($t(63)=2.329$, $p=.023$).

	Notes	No Notes
Hum. Rec. Mem.	70%	70%
Syn. Rec. Mem.	66%	58%

Table 4: Recognition Memory performance as a function of voice type and note-taking

Gender Issues

MOS scores for the four engines that had both male and female voices, showed that female subjects showed a slight preference for the male voice over the female voice (preference seen in 3 out of 4 engines). There was no clear preference for male subjects.

Although there was no significant overall subjective preference for engines with male voices over engines with female one's ($t(73)=.687$, $p=.494$), subjects performed better in the synthetic voice condition when listening to a TTS engine with a male voice. Table 5 shows that subjects in both engine gender groups performed equally well on the comprehension of human speech, but that the subjects who listened to a male synthetic voice performed significantly better on the synthetic section than those who listened to a female voice ($F(1,73) = 9.624$, $p = .003$).

	Human	Synthetic
Male TTS Gp.	72%	65%
Female TTS Gp.	72%	51%

Table 5: comprehension performance by voice type for subjects hearing male and female synthetic voices

Subjective Findings

Regarding subjects' perception of performance, only 16 reported that listening to the TTS was much harder than the human voice, yet scored higher (or equal) in synthetic comprehension than in human. For the majority of subjects (N=49), perception of the increased difficulty of synthetic was correlated to lower performance. Two subjects thought that synthetic speech was actually easier (even though they scored lower with synthetic). The balance (N=11) felt the difficulty was about the same for both voices. A representative comment was "I'm a technophile, but the synthetic speech was too difficult to comprehend. It required much more concentration to listen to synthetic speech than to human speech". Or "I found the synthesized speech rather painful to listen to requiring a great deal of concentration (much like listening to someone with a very heavy accent)."

Even though the human speech was read at a faster word per minute rate than the synthetic (200 wpm vs. 175 wpm), subjects often commented that the speaking rate for TTS was too fast (but did not comment equally on the speed of the human speech). MOS scores for the speaking rate question showed that 36 subjects felt the speaking rate for synthetic was adequate, while 33 felt that it was either a little fast (N=22) or much too fast (N=11). Only one subject thought the rate was too slow (and we believe this subject was thinking more of the process of accessing messages via synthetic speech) and 8 felt the rate was a little slow.

An Evaluation of the Data Sets

As mentioned earlier, we had 10 subjects, independent of those taking the comprehension test, answer all six sets of questions without ever hearing the text. Those subjects answered on average one third of the questions correctly. This result is slightly higher than the 25% that we would have expected given the probability of guessing correctly among the possible 4 choices for each question. When we looked at the results, it was somewhat baffling to us the questions that a majority of these 10 subjects guessed correctly such as "Which three items were you asked to pick up from the store?" a) milk, butter and eggs b) milk, orange juice and butter c) eggs, milk and orange juice d) butter, bread and eggs. The correct answer is c) which does not appear to us any more likely than any of the other choices.

Our goal was to create two sets of data, each consisting of a short, medium and long task, that would be equal in terms of complexity. In reviewing the results by task, it appears that M1 may have been easier than M2. Certainly for the human voice condition, performance scores were the highest for M1 in both the note-taking and non note-taking conditions. (See Tables 6 and 3).

	S1	M1	L1	S2	M2	L2
Hum.	65%	94%	79%	70%	72%	65%
Syn.	52%	77%	48%	75%	77%	70%

Table 6: performance by task for the note-taking condition (N=36)

FUTURE WORK

Something we did not measure in this particular study was the degree to which complex cognitive operations such as decision making or perceptual processing are affected while listening to synthetic speech. If as we suspect, the perception of TTS imposes a greater demand on cognitive resources than perception of human speech, performance on a secondary task would most likely be affected. Given one of the possible scenarios for the application we have in mind, the secondary task could be driving, which could have serious ramifications. Alternatively, a more likely scenario is that the perception of the text-to-speech would decrease due to the concentration effort given to driving. In our future work we want to conduct an experiment that uses similar passages of text, but has the subjects listening to them while in a driving simulator.

Additionally, we would like to run a study that examines the effect of intermingling synthetic speech with human speech. There have been indications in the past that intermingling the voice types decreases comprehension significantly. Because of the need to use synthesized speech for dynamically created text such as e-mail, and the desire to minimize the use of synthetic speech, many systems use a combination of recorded human prompts with synthetically generated speech. It is not clear that we do our users a favor by juxtaposing human speech with synthetic in such systems. We speculate that if we were to run the same test again, the only difference being the intermingling of a synthetic passage with a recorded human passage, the differences in comprehension accuracy between the voice types would be greater.

Another issue that we did not deal with in this experiment, but that needs to be addressed is the pre-processing step that is required for optimal comprehension of synthesized e-mail messages. For our study this was not a problem since we did not include in the messages any headers, footers, imbedded objects or multiple levels of forwarded messages or replies with history. Markup languages exist for synthetic speech and can be used in a pre-processing step with varying degrees of effectiveness. We would like measure how effectively these conditions can be handled and what the resulting effect on comprehension is.

Lastly we would like to measure, given conditions similar to the study we just ran, how long it would take for naive subjects to achieve comparable comprehension levels between synthetic speech and human, and if this ever happens when listening without notes.

CONCLUSION

While performance for the second set was comparable between synthetic speech and human speech within the note taking condition, it is unlikely that users will have a pad and pen at hand when using pervasive (i.e. ubiquitous) computing devices. One subject commented "I doubt I would recall facts without taking notes and it is doubtful you'd have paper and pencil available all the time (so general comprehension is more germane to the use of this technology)". Given the inability to take notes, we can expect that a majority of users will experience some comprehension problems when listening to messages generated with synthetic speech. Also, given that there was no significant difference in performance across all five engines, these problems will exist regardless of the chosen engine. Since we saw that comprehension difficulties were exarcerbated when the text was longer and more unfamiliar to the subject, we can conclude that a remote access system for messages would have the greatest chance of success if messages were brief and their delivery conducted concisely.

Several people indicated that based on this experience they would not be inclined to use a text-to-speech system for listening to news articles. A representative comment was "Listening to my e-mail would be fine, but not for longer items. I didn't enjoy listening to synthesized news". One of the common culprits mentioned was the difficulty subjects had with hearing and understanding names. Several subjects mentioned that numbers were easily recognizable within the text.

With regard to the potential usefulness of such a system, many subjects commented that such a system would be worth the additional mental effort of listening to synthetic speech for access to critical information such as urgent messages, or for the convenience of rapid access (no computer connections) to e-mail messages. Like any other system, the willingness to submit to the rigors of the system (whatever they be) depends on the value proposition for the end user. This aspect is highlighted in the following comment "I found understanding somewhat difficult, so use of such a system would be highly dependent on what other features/conveniences were included in the system."

ACKNOWLEDGMENTS

Thanks to R. Middlebrooks for invaluable assistance in running subjects as well as in the transcription and analysis of data. Thanks also to J. Karat for his guidance and feedback on the design of the study.

REFERENCES

1. Francis, A.L., Nusbaum, H.C. (1999). Evaluating the quality of Synthetic Speech. In Gardner-Bonneau, D. (Ed.), *Human Factors and Voice Interactive Systems* (pp. 63 - 97)

2. Greenspan, S. L., Nusbaum, H.C., and Pisoni D.B. (1988). Perception of synthetic speech produced by rule: Intelligibility of eight text-to-speech systems. *Behavioral Research Methods, Instruments, and Computers*, 18 100-107

3. Klatt, D. H. (1987). Review of text-to-speech conversion for English. *The Journal of the Acoustical Society of America*. September 1987 (pp. 737 - 793)

4. Pisoni, D.B. and Hunnicutt, S. (1980). Perceptual Evaluation of MITalk: The MIT unrestricted text-to-speech system. IEEE *International Conference on Acoustics, Speech and Signal Processing* (pp. 572-575) New York

5. Pisoni D.B., Nusbaum H.C. & Greene B.G. (1985). Perception of synthetic speech generated by rule. *Proceedings of the IEEE 73: 1665-1676*

6. Ralston, J.V., Pisoni, D.B., Lively, S.E., Greene, B.G. & Mullennix, J.W. (1991). Comprehension of Synthetic Speech Produced by Rule: Word Monitoring and Sentence-by-Sentence Listening Times. *Human Factors* 1991, 33(4) (pp. 471-491)

7. Ralston, J.V., Pisoni, D.B., and Mullennix, J.W. (1995). Perception and comprehension of speech. In Syrdal, A.K., Bennett, R.W., Greenspan, S.L. (Eds.), *Applied Speech Technology* (pp. 233-288) Boca Raton: CRC Press

8. Van Bezooijen, R. & van Heuven, V. (1998) Assessment of Synthesis System. In Gibbon, D., Moore, R. and Winski, R. (Eds.), *Volume III: Spoken Language System Assessment* (pp. 167[481] - 249 [563]) Mouton de Gruyer

9. www.toefl.org

10. www.genmagic.com/portico/portico_home.shtml

11. www.webley.com

Does Computer-Generated Speech Manifest Personality? An Experimental Test of Similarity-Attraction

Clifford Nass
Department of Communication
Stanford University
Stanford, CA 94305-2050
+1 650-723-5499
nass@stanford.edu

Kwan Min Lee
Department of Communication
Stanford University
Stanford, CA 94305-2050
+1 650-497-7357
kmlee@stanford.edu

ABSTRACT
This study examines whether people would interpret and respond to paralinguistic personality cues in computer-generated speech in the same way as they do human speech. Participants used a book-buying website and heard five book reviews in a 2 (synthesized voice personality: extrovert vs. introvert) by 2 (participant personality: extrovert vs. introvert) balanced, between-subjects experiment. Participants accurately recognized personality cues in TTS and showed strong similarity-attraction effects. Although the content was the same for all participants, when the personality of the computer voice matched their own personality: 1) participants regarded the computer voice as more attractive, credible, and informative; 2) the book review was evaluated more positively; 3) the reviewer was more attractive and credible; and 4) participants were more likely to buy the book. Match of user voice characteristics with TTS had no effect, confirming the social nature of the interaction. We discuss implications for HCI theory and design.

Keywords
TTS (Text-to-Speech), CASA (Computers are social actors), Speech User Interfaces, personality, similarity-attraction effect.

INTRODUCTION
In recent years, there has been increasing interest in and use of Text-to-Speech (TTS) technologies (see [13] for a review). This is due to several factors. First, there has been growing demand for Speech User Interfaces (SUIs). SUIs are needed as an alternative/complement to Graphical User Interfaces (GUI), because GUIs have several limitations

(see [19] for a review). For example, GUIs are not appropriate for hands-occupied or eyes-occupied environments such as driving [18]. GUIs also cannot be effectively implemented on technologies such as telephones or PDAs, because of screen-size limitations [5].

Second, TTS systems are useful for presenting text (via synthesized speech) and voice on telecommunication devices, such as cellular phones [6].

Third, for some people, it is hard or even impossible to use GUI-based systems. For example, about 11 million people have some form of visual impairment and about 1.5 million are totally blind in the U.S. alone [5]. TTS technologies provide a new opportunity to use the power of the computer and the Internet for the disabled. In addition, TTS technologies also make it easier for the illiterate and pre-literate (e.g., children) to use computers.

Finally, TTS systems are increasingly used to develop more human-like interfaces. The basic assumption is that by incorporating anthropomorphic indicators such as speech, users will feel more comfortable with computers and perceive them as more intelligent [10, 11, 14, 17].

Assessment Criteria for TTS
In general, TTS systems are evaluated according to two dimensions: 1) intelligibility and 2) naturalness of the resulting speech [1, 6]. There is relatively little difference in intelligibility across systems [1]; indeed, word intelligibility scores for the best TTS systems are close to 97%, suggesting that the intelligibility of the best TTS systems approaches that of real human speech [6].

Naturalness scores for the best TTS systems are in the fair-to-good range, indicating that even the best TTS systems still do not match the quality and prosody of natural human speech [6]. Consequently, the speech generated by current TTS systems would never be confused with real human speech, even though it is clearly understandable [10, 13].

Given the un-naturalness of synthesized speech, it would seem absurd for users to respond to speech characteristics that suggest essentially human attributes. That is, although it would not be surprising for users to be influenced by the comprehensibility of speech (e.g., difficult-to-understand speech leading to greater frustration) or the naturalness of speech (e.g., more natural speech leading to greater feelings of comfort and social presence [10]), the obviously synthetic nature of TTS should make the non-verbal aspects that influence *social* assessment irrelevant to users' attitudes and behaviors. For example, odd hesitations in human speech are interpreted as attempts at deception [7]. These inappropriate pauses are a ubiquitous aspect of all synthesized speech engines, but it is unlikely that users would respond to all TTS engines as highly deceptive.

Even the literature that most dramatically demonstrates that individuals consistently apply social rules to computers (although users know that it is absurd to do so) — the Computers are Social Actors (CASA) paradigm [11,17] — cannot be directly applied to assignment of human attributes for systems employing text-to-speech.

In traditional CASA experiments, the user is presented with a computer that does nothing to remind the user that it is *not* human or, more specifically, that it is an ersatz human. For example, the vast majority of studies in the CASA paradigm involve plain text presenting commonplace language and omitting any behavior uniquely associated with computers (e.g., error messages, "crashing,"). Thus, the content produced is as consistent with computer-mediated communication as it is with human-computer interaction. Even the use of the word "I" is eschewed in the experiments to ensure that the user is not reminded that there is an incongruity in a machine using the term for human identity.

In contrast to the failure to mark non-humanness in the previous CASA studies, TTS is a "liminal" case [21] which sits at the boundary of human and machine, although it is clearly on the machine side of the divide. A demonstration that users employ social cues even with highly-present reminders that the voice is not coming from a person would represent a critical test of the CASA framework, as well as having critical implications for HCI design.

PERSONALITY MARKERS IN SPEECH

Among humans, paralinguistic aspects of speech convey a wide range of trait information about the speaker, such as gender and age. In the present study, we attempt to determine whether the vocal characteristics of speech can convey "personality." That is, given identical content, will the particular settings in a TTS engine lead individuals to identify and respond to the voice as if it had a personality?

We chose personality for a number of reasons. First, personality is more subtle and complex than the more obvious characteristics of gender and age. Second, there is a large literature demonstrating that individuals will assess computers according to their personality [4,11,12,17], although this literature has focused primarily on the message's linguistic content [see 4 for an exception]. Third, the term "personality" comes from the Latin *personare*, to *sound* through, referring to the mouth opening in a mask worn by an actor [2]; thus, it is natural to focus on the "sound" characteristics of personality. Fourth, the field of personality psychology provides a rich set of predictions concerning how individuals will respond to various personality types. Thus, we are not limited to merely determining whether individuals can *identify* the personality of a TTS voice; instead, we can also determine whether these personality markers will influence users' attitudes and behaviors. Fifth, personality is an important and easily-measured individual difference among users, so the parallels to computers, as well as the interaction between user personality and computer personality, become particularly interesting. Finally, and most importantly, there is a significant literature that outlines the specific linkages between the characteristics of the vocal channel and the assignments of personality that result.

Numerous dimensions of personality have been identified [3, 12, 17, 20, 23]. The present study focuses on the extroversion/introversion dimension, both because it is the dimension most strongly marked by paralinguistic cues and because it has proven to be important in the HCI literature [4, 12,17].

Four readily-manipulable aspects of TTS engines are associated with characteristics that individuals use to distinguish introversion from extroversion in human-human interaction:

Speech rate: Extroverts speak more rapidly than introverts [15, 20, 24].

Volume: Extroverts speak more loudly than introverts [15].

Pitch: Extroverts speak with higher pitch than introverts [15]

Pitch range: Extroverts speak with more pitch variation than do introverts [3].

To ensure a successful manipulation, consistency among the features, and consistency with humans, we manipulated all four attributes simultaneously.

THEORETICAL EFFECTS OF TTS PERSONALITY

To determine whether paralinguistic cues in TTS lead users to respond to the voice as if it manifests a "personality," we used three different approaches. First, we determined whether users could simply identify whether the voice was "introverted" or "extroverted." This assessment not only involved direct questions about the voice, but also whether the personality of the voice affected assessments of the *content* that the voice presented. This is a much stronger test than the direct test, in that it requires users to think of the voice as the *source* of the content.

A yet more stringent test is to determine whether the users draw on social rules to respond to the voice. In this study, we examined the principle of similarity-attraction.

According to this principle, people like others who possess a personality that is similar to their own [12, 17]. Similarity-attraction has proven to be a very robust finding when a computer's personality is manifest through verbal/textual cues [e.g., 4, 12, 17]. However, text-to-speech is so obviously *not similar* to a human voice that to be able to find second-order similarity effects would be remarkably strong evidence for the assignment of personality to TTS.

SIMILARITY-ATTRACTION EXPERIMENT

This study was executed in the context of a book-buying website site (an example can be found at www.stanford.edu/class/comm169/Kwan/1-1.fft). Each web page had an identical visual interface based on Amazon's book description. Each page included the titles, the authors (in text) and the pictures of five books. Instead of having the book description in text, there was a link to an audio (.wav) files; clicking on the link would play the review

Hypotheses

Based on the idea that paralinguistic cues will be used to assess personality even in TTS, we can draw the following conclusions.

H1: Users will recognize vocal cues indicating personality even when they hear computer-synthesized speech. TTS that has a speech rate, volume, pitch, and pitch rate associated with extroversion or introversion in humans will be perceived to be extroverted or introverted, respectively.

Although the correlation between people's voices and their personality is not very strong [see below], individuals nonetheless use voice characteristics to assess personality, especially when other cues are absent. Despite the fact that all of the individuals in our experiment understood that the book descriptions were written by actual people, the tendency toward proximate source orientation [see 17, chap. 16 for a review] may lead users to assign the characteristics of the voice to the *writer* of the review, especially when the text is not strongly revealing of the author's personality. That is, absent other cues, individuals decide on the attributes of the distant source based on the most-readily available cues [17, see Ch. 16]. This leads to the following hypothesis:

H2: Users will infer the personality of a review writer based on the personality of the TTS voice they heard.

The principle of similarity-attraction is not only powerful; it is generalized from mere attraction to a variety of other positive attributes. Thus, we can derive the following hypotheses:

H3: Users will be more attracted to a TTS voice that exhibits similar personality to their own than a TTS voice that exhibits a dissimilar personality.

H4: Users will regard a TTS voice that exhibits similar personality to their own as more credible than a TTS voice that exhibits a dissimilar personality.

If we combine the literature on similarity-attraction and proximate source orientation, then the match between TTS personality and user personality should affect the reactions to the reviewer and the review itself, rather than merely affecting direct assessments of the voice. This leads to the following hypotheses:

H5: Users will evaluate a book review more positively if the review is narrated by a TTS voice that matches their own personality.

H6: Users will like a review writer more if the writer's review is narrated by a TTS voice that matches their own personality.

H7: Users will regard a review writer as more credible if the writer's review is narrated by a TTS voice that matches their own personality.

Finally, from a web design perspective, one would like to influence behavior as well as attitude. Given the above, we can predict:

H8: Users will show more buying intention for a book whose review is narrated by a TTS voice that matches their own personality.

Method

Participants

Several weeks prior to the study, a web-based personality survey was administered to students who registered for a large introductory communication course. Both Myers-Briggs (see [9] for review) and Wiggins [23] personality tests were administered to maximize the likelihood of correctly assessing the personality of students. From a total of approximately 150 undergraduate students, a total of 72 participants—36 extrovert and 36 introvert students—who had English as a first language were invited to participate in the study. Participants were randomly assigned to condition, with gender approximately balanced across conditions. All participants signed informed consent forms, were debriefed at the end of the experiment session, and received class credit for their participation.

Procedure

The experiment was a 2 (computer voice personality: extrovert vs. introvert) by 2 (participant personality: extrovert vs. introvert) balanced, between-subjects design, with five book descriptions as a repeated factor. Upon arrival to the lab, each participant was assigned to a computer equipped with a headphone and an Internet Explorer 4.0 browser. Participants were instructed to use the headphones during the whole experiment and not to adjust the volume level of either the headphone or the computer.

The experimenter opened either the extroverted TTS or introverted TTS web page containing the first book description. As noted earlier, each book description page consisted of a picture of the book, a title and author, and a wave file description of the book. The descriptions were edited version of the actual descriptions on the Amazon.com site. The books were chosen to have the following characteristics: 1) the books and their authors had not sold very well (so that our participants would not be familiar with them; this was confirmed in debriefing); 2) the descriptions had a single main character (so that we could unambiguously ask questions about the characters in the book), and 3) all books were fiction (so that affective criteria would be more important).

Below the icon for the audio file, there was an eight-question questionnaire. Below the questionnaire was a button that allowed the user to progress to the next book description page (which employed the same voice). Except for the personality of the voice, the visual layout, textual information, and book description content were identical across conditions.

Participants read the instruction on the web page and heard five book descriptions. After hearing each book review, they provided answers for questions regarding the book reviewed, main character(s) in that book, and the review itself. After hearing all five book descriptions, participants were presented with a web-based questionnaire. Participants were then told that all five book descriptions had been written by the same person (although they were actually written by different people) and were asked questions about the reviewer. They were also asked questions about the voice they had heard. After participants filled out the web-based questionnaire, we made an audio-recording of them giving their name and describing their experience in the experiment; this allowed us to code each participants' voice as extroverted or introverted (see below). Finally, all participants were debriefed and thanked.

Manipulation

The CSLU Toolkit was used to produce and manipulate the parameters for the two synthesized voices. The four voice parameters discussed above were simultaneously manipulated to instantiate the personality of the voice. The extrovert voice had a speech rate of 216 words per minute, the maximum volume level possible with the Toolkit; a fundamental frequency of 140Hz; and a pitch range of 40Hz. The introvert voice had a speech rate of 184 words per minute; the volume level set at 15% of the maximum; a fundamental frequency of 84 Hz; and a pitch range of 16 Hz. Pre-tests ensured that the voices manifested the appropriate personality and did not differ in intelligence (e.g., too slow speech was classified as both introverted and very unintelligent).

Measures

All dependent measures were based on items from the web-based, textual questionnaires. Participants used radio buttons to indicate their responses. Each question had an independent, ten-point Likert scale.

Questions concerning book quality, the main character, and review quality were asked for each of the different book descriptions. General questions regarding the reviewer and voice were asked once at the end of the complete hearing session. One set of these questions asked, "How well do each of these adjectives describe the reviewer" (identical questions were asked about the voice itself), followed by a list of adjectives. The response scales were anchored by "Describes Very Poorly" (1) and "Describes Very Well" (10). Other sets of questions were based on standard scales.

A number of indices were created, based on theory and factor analysis. All indices were highly reliable.

Introvertedness was an index composed of seven Wiggins [23] introvert adjectives items: Bashful, Introverted, Inward, Shy, Undemonstrative, Unrevealing, and Unsparkling. It was used for assessments of both the voice (Cronbach's alpha=.80) and the reviewer (alpha = .83).

Extrovertedness was an index composed of seven Wiggins [23] extrovert adjective items: Cheerful, Enthusiastic, Extroverted, Jovial, Outgoing, Perky, and Vivacious. It also was used for assessments of both the voice (alpha = .95) and the reviewer (alpha = .94).

Voice attractiveness was an index composed of the items, "How much did you enjoy hearing the computer voice?," "How likely would you be to have the voice read you other descriptions?," and the following adjectives: enjoyable, likable, and satisfying (alpha = .89).

Voice credibility was an index composed of three adjectives: credible, reliable, and trustworthy (alpha = .89).

Quality of the review was an index composed of three items: "What was the quality of the review that you just heard?," "How much did you like the review?," and "How trustworthy was the review?" (alpha ranged from .75 to .91, with a mean of .86).

Liking of the reviewer was an index composed of three adjectives: enjoyable, likable, and satisfying (alpha = .92).

Credibility of the reviewer was measured by a standardized trust scale [22] (alpha = .88).

Users' buying intention was measured by a single item: "How likely would you be to buy this book?"

Results

For the measures that were asked for each book, we used a full-factorial repeated measure ANCOVA with book as the repeated factor and computer voice personality and subject personality as the between-subjects factors. For the items that were only asked once, we used a full-factorial 2x2 ANCOVA. Gender was used as a covariate for all analyses.[1]

---

[1] We also conducted a full-factorial repeated measure ANOVA and a full factorial 2x2 ANOVA for all hypotheses. The results were substantively identical in all cases, indicating that the

Consistent with Hypothesis 1, users applied vocal stereotypes to computer-synthesized voices. Specifically, the extrovert computer voice was perceived as being more extroverted ($M = 4.3$) than the introvert computer voice ($M = 2.5$), $F(1, 67) = 18.01$, $p < .001$, $\eta^2 = .21$, and the introvert computer voice was perceived as more introverted ($M=6.1$) than the extrovert computer voice ($M = 5.2$), $F(1, 67) = 7.56$, $p < .01$, $\eta^2 = .10$. There was also a cross-over interaction for extroversion, $F(1,67) = 6.60$, $p < .05$, $\eta^2 = .10$.

Consistent with Hypothesis 2, the personality of the voice influenced perceptions of the personality of the reviewer, even though the content of the review was held constant. Specifically, the reviewer was perceived as being more extroverted when the descriptions were narrated by the extrovert computer voice ($M = 5.7$) than by the introvert computer voice ($M = 4.3$), $F(1, 67) = 8.7$, $p < .01$, $\eta^2 = .16$. Conversely, the reviewer was perceived as being more introverted when the descriptions were narrated by the introvert computer voice ($M = 6.2$) than the extrovert computer voice ($M = 5.1$), $F(1, 67) = 7.7$, $p < .01$, $\eta^2 = .10$.

Consistent with the literature on similarity-attraction and its application to human-computer interaction (Hypothesis 3), there was a significant cross-over interaction between computer voice personality and subject personality for voice attractiveness, such that introverts preferred the introvert voice and extroverts preferred the extrovert voice, $F(1, 67) = 14.6$, $p < .001$, $\eta^2 = .18$ (see Figure 1).

Figure 1. Voice attractiveness

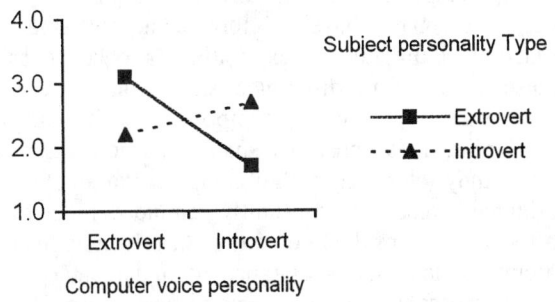

Similarly, extroverts found the extroverted voice more credible than the introverted voice, while introverts felt the opposite (Hypothesis 4), as indicated by the significant computer voice personality by subject personality interaction, $F(1, 67) = 7.86$, $p < .01$, $\eta^2 = .11$ (see Figure 2).

Figure 2. Voice credibility

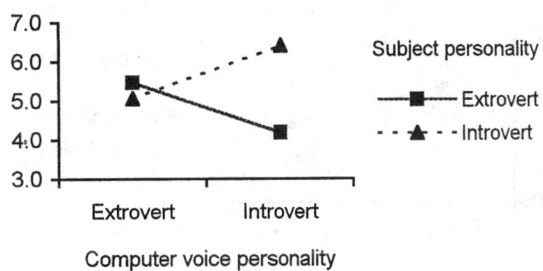

Similarity-attraction extended beyond the voice. There was a significant crossover interaction, with introverts preferring the review read by the introverted voice, and extroverts, the extroverted voice, $F(1, 67)=3.62$, $p<.06$, $\eta^2 = .05$ (Hypothesis 5; see Figure 3). Introverted subjects evaluated the book descriptions more positively in general than did extrovert subjects, $F(1, 67) = 6.31$, $p < .05$, $\eta^2 = .09$.

Figure 3. Quality of the review

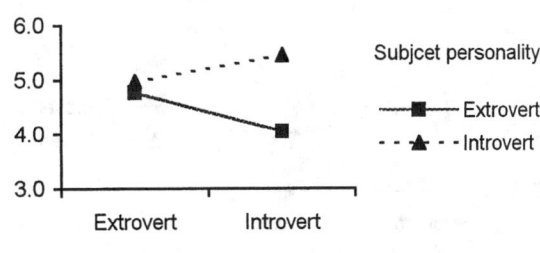

Consistent with proximate source orientation, there was a significant crossover interaction for reviewer attractiveness: Introverts found the reviewer represented by the introverted voice to be more attractive, while extroverts preferred the reviewer when presented with an extrovert voice, $F(1, 67) = 8.35$, $p < .01$, $\eta^2 = .11$ (Hypothesis 6; see Figure 4). Introverts liked the reviewers more in general, $F(1, 67) = 4.87$, $p < .05$, $\eta^2 = .07$.

Figure 4. Liking of the reviewer

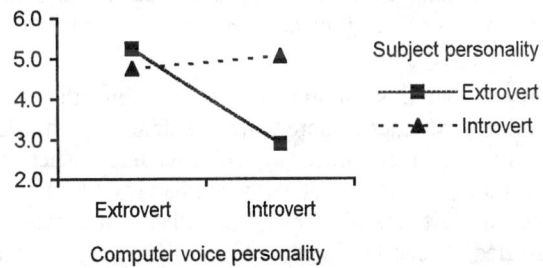

There was a significant cross-over interaction in the expected direction with respect to trust, $F(1, 67) = 10.88$, $p < .01$, $\eta^2 = .14$. (Hypothesis 7 and Figure 5).

---

gender of subject was not an influential factor in participants' responses to TTS voices.

Figure 5. Credibility of the reviewer

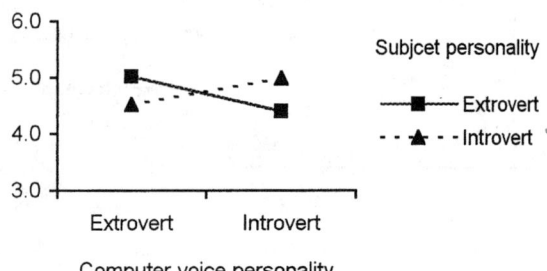

Finally, consistent with Hypothesis 8, there was a significant computer voice personality X subject personality crossover interaction in the predicted direction, $F(1, 67) = 5.45$, $p < .05$, $\eta^2 = .08$ (see Figure 6), such that introverts were more willing to buy the book when it was presented with an introverted voice, while extroverts were more willing to buy the book from an extroverted voice.

Figure 6. Buying intention

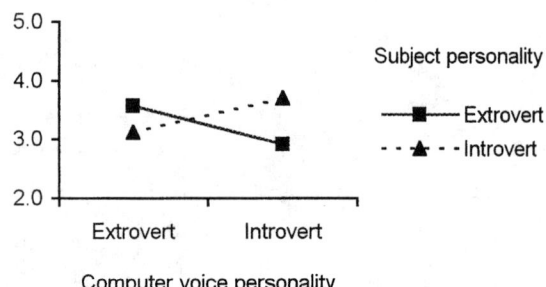

Addressing an alternative explanation

One compelling alternative explanation to the above results is that while similarity-attraction does occur, the similarity is actually between the TTS voice parameters and the *user's* speech parameters. That is, rather than a personality-based explanation for these phenomena, this alternative explanation implies a pure cognitive approach. In addition to the theoretical implications of this explanation, there is also a practical one: Should one measure the user's personality via questionnaires and adapt to personality, or should one instead measure voice characteristics of the user and set the TTS parameters to match?

To ensure that this question is worth asking, that is, to determine whether there are differences between personality as determined by voice characteristics and actual (questionnaire) personality, we had two blind coders rate each participant's voice as either extroverted or introverted. Consistent with the literature [15,16], there was extremely high agreement between the two coders (inter-coder reliability = .90; disagreements were resolved by discussion). Approximately 41 of the participants were labeled as having an extroverted voice, and 30 were labeled as having an introverted voice.[2] However, the correlation between the questionnaire personality and the coded voice personality was a remarkably low (though significant) $r = .24$, $p < .05$. Thus, there are clear and possibly consequential differences between user personality and user voice.

To determine which is the basis for similarity-attraction, we repeated all of the statistical analyses using user voice personality rather than questionnaire personality. (Because there were approximately equal numbers of extroverted and introverted participants, this was a viable strategy). In contrast to the extremely strong results for questionnaire personality, there were *no* similarity-attraction effects, as indicated by *no* interactions for user voice personality by TTS voice personality on any of the dependent measures.[3]

As further evidence of the fundamental aspects of actual (questionnaire) personality, we created a new variable which indicated the (mis)match between the personality of the participants' voice and the personality of the computer voice. We used this variable as a covariate in all of the original ANCOVAs. The covariate had *no* effect on the results; that is, all of the original similarity-attraction effects were obtained. This provides additional evidence that it is the personality, not the cognitive, aspects of similarity-attraction that are relevant to users.

DISCUSSION

One can critique previous experimental tests of the Computers are Social Actors paradigm by arguing that they are too liberal. In previous studies, there were no strong cues to remind individuals that they were working with a computer (other than the monitor). Thus, participants may have simply conceptualized the interaction, implicitly or explicitly, as chat or email, because there was no difference between how one would interact with a person *via* computer and the computer itself. Unlike these previous studies, participants in the present study interacted with a computer via an obviously non-human voice that constantly reminded them of the non-social nature of the interaction. Surprisingly, then, the participants nonetheless assigned a fundamental human property (personality) to the voice and were strongly influenced by that human characteristic. Thus, this study demonstrates that the CASA paradigm is very robust.

The result that even a TTS voice can have a personality has critical implications for the design of TTS systems. Developers of TTS systems have been focused on those aspects of speech that increase intelligibility and naturalness. Now, however, intelligibility is not a serious

---

[2] One subject refused to record his/her voice.

[3] Of course, it could be argued that the relevant similarity is between the voice that people hear inside their head and the TTS voice, but this information is not readily accessible and hence beyond the scope of experimental research.

problem, and there are a large number of very difficult problems to solve (natural language understanding, emotional understanding, better voice models) before naturalness is achieved. The present study demonstrates that at present, the maximum leverage can be achieved by focusing on the social and personality aspects of speech. Virtually all TTS systems provide socially-relevant parameters that are readily manipulable and remarkably powerful.

The user assignment of personality (and other social attributes) to TTS voices suggests that TTS presents a "casting" problem. For example, when a voice is combined with a character, the personality of the two must be consistent: Pictorial representations of extroverted characters (e.g., open postures; expressive faces; see [4]) should be combined with an extroverted voice, for example.

The proximate source orientation results suggest that companies or individuals should be very careful in choosing their representing voice. For example, if a company or the product of the company is oriented toward extroversion (e.g., a match-making service), the company should employ an extrovert voice, even though the voice would seem to have no relationship with the product. The good news is that it is surprisingly easy to crate a perception of personality in computer voice. With the simple manipulation of accessible vocal dimensions, a designer can easily generate a clear and strong personality. The result that users are attracted to voices that are similar to themselves (similarity-attraction) means that customization of a computer voice according to the users' personality will increase the attractiveness, credibility and informativeness of computer speech output. Previous research [12, 17] has argued that one should match the textual content of an interface to the personality of the user. However, because of the limitations of natural-language production software, this has required designers to build two or more versions of the site, a generally impractical task. Because TTS operates in real-time, a simple change of the speech parameters immediately changes the personality of the entire site. The similarity-attraction principle is so powerful that interface designers can increase the positive evaluation of a product and even the company that makes the product by simply matching the personality of the voice and user.

The fact that questionnaires rather than voice characteristics are the critical determinant of user personality with respect to similarity-attraction is good news for interface designers, as the former information is often easier to obtain than the latter. Many systems are not equipped with microphones or have microphones with insufficient resolution to determine voice parameters.

A key limitation of the current study is that we used personality-neutral content to control the influence of linguistic cues on users' psychological responses to computer voice. However, text frequently manifests a clear and consistent personality. This is a common problem in email readers, branded products, various categories of tasks, etc.. Future studies should examine the tradeoff between similarity-attraction and users' desire for consistency when the voice, the content, and the user all have personality characteristics.

In sum, the results from the present study provide very strong evidence that despite its failure to seem human, individuals nonetheless respond as if text-to-speech has a personality. This fact is both an opportunity and a problem for interface designers.

Reference

1. Beutnagel, M., Conkie, A., Schroeter, J., Stylianou, Y. & Syrdal, A. *The AT&T Next-Gen TTS system.* Paper presented at joint meeting of ASA, EAA, and DAGA, Berlin, Germany, March , 1999. [Online] Available at http://www.research.att.com/projects/tts/papers/1999_ASA_Berlin/nextgen.pdf

2. Giles, H. & Powesland, P. F. *Speech style and social evaluation.* Academic Press, London. 1975.

3. Hall, J. A. *Nonverbal sex differences.* Johns Hopkins University Press, Baltimore, MD, 1980.

4. Isbister, K. & Nass, C. Consistency of personality in interactive characters: verbal cues, non-verbal cues, and user characteristics. *International Journal of Human-Computer Interaction*, in press.

5. James, F. *Representing structured information in audio interfaces: A framework for selecting audio marking techniques to represent document structures.* Ph. D. Dissertation, Stanford University, 1998.

6. Kamm, C., Walker, M. and Rabiner, L. *The role of speech processing in human-computer intelligent communication.* Paper presented at NSF Workshop on human-centered systems: Information, interactivity, and intelligence, February, 1997. [Online] Available at http://www.ifp.uiuc.edu/nsfhcs/talks/rabiner.html

7. Massaro, D. M. *Perceiving talking faces: From speech perception to a behavioral principle.* MIT Press, Cambridge, MA, 1997

8. Moon, Y. When the computer is the "salesperson": Computer responses to computer "personalities" in interactive marketing situations. *Working paper No. 99-041*, Harvard Business School. Boston, MA, 1998.

9. Murray, J. B. Review of research on the Myers-Briggs type indicator. *Perceptual and Motor Skills, 70.* 1187-1202, 1990.

10. Nass, C. & Gong, L. *Maximized modality or constrained consistence?* Paper presented at audio-visual speech processing conference, August, 1999.

11. Nass, C. & Moon, Y. Machines and mindlessness: Social responses to computers. *Journal of Social Issues*, in press.

12. Nass, C., Moon, Y., Fogg, B. J., Reeves, B., & Dryer, D. C. Can computer personalities be human personalities? *International Journal of Human-Computer Studies, 43*, 223-239, 1995.

13. Olive, J. P. The talking computer: Text to speech synthesis. in D. Stork (ed.) *Hal's legacy: 2001's computer as dream and reality.* 101-130. The MIT Press, Cambridge, MA, 1997.

14. Oren, T., Salomon, G., Kreitman, K., & Don, A. Guides: Characterizing the interface. In B. Laurel (Ed.) *The art of human-computer interface design.* Addison-Wesley, Reading, MA, 1990.

15. Pittam, J. *Voice in social interaction: An interdisciplinary approach.* Sage, Thousand Oaks, CA, 1994.

16. Ramsay, R. W. Personality and speech. *Journal of Personality and Social Psychology, 4*, 116-118. 1966.

17. Reeves, B. & Nass, C. *The media equation: How people treat computers, television, and new media like real people and places.* Cambridge University Press, New York, NY, 1996.

18. Sawhney, N. and Schmandt, C. *Design of spatialized audio in nomadic environments.* Paper presented at the international conference on auditory display (ICAD'97), November, 1997. [Online] Available at http://www.media.mit.edu/~nitin/NomadicRadio/ICAD97/ICAD97_paper/ICAD97.html

19. Shneiderman, B. *Designing the user Interface: Strategies for effective HCI (3rd ed.).* Addison-Wesley, Reading, MA, 1997.

20. Smith, B. L., Brown, B.L., Strong, W.J., & Rencher, A.C. Effects of speech rate on personality perceptions. *Language and Speech, 18*, 145-152, 1975.

21. Turkle, S., *The second self: Computers and the human spirit.* Simon and Schuster, New York, NY, 1984.

22. Wheeless, L. R. & Grotz, J. The measurement of trust and its relationship to self-disclosure. *Human Communication Research, 3(3), 250-257*, 1977.

23. Wiggins, J. S. A psychological taxonomy of trait-descriptive terms: The interpersonal domain. *Journal of Personality and Social Psychology*, 37(3), 395-412, 1979.

24. Woodall, W. G., & Burgoon, J.K. Talking fast and changing attitudes: A critique and clarification. *Journal of Nonverbal Behavior, 8*, 126-142, 1983

The research was supported in part by NSF CARE and Challenge grants awarded to the Oregon Graduate Institute (R. Cole, PI) and a grant from the Center for the Study of Language and Information (CSLI) at Stanford University. The views expressed in this article do not necessarily represent the views of the National Science Foundation or CSLI.

A Toolkit for Strategic Usability: Results from Workshops, Panels, and Surveys

Stephanie Rosenbaum, President
Tec-Ed, Inc.
P.O. Box 1905
Ann Arbor, MI 48106
1-734-995-1010 voice
1-734-995-1025 fax
1-650-493-1010 California
stephanie@teced.com

Janice Anne Rohn, Director
User Experience
Siebel Systems, Inc.
1855 South Grant Street
San Mateo, CA 94402-2667
1-650-295-5852 voice
1-650-295-5114 fax
jrohn@siebel.com

Judee Humburg, Principal
JL Humburg Associates
480 Lytton Avenue, Suite 7
Palo Alto, CA 94301
1-650-462-0798 voice
1-650-462-0797 fax
judee@pacbell.net

ABSTRACT

This paper describes the organizational approaches and usability methodologies considered by HCI professionals to increase the strategic impact of usability research within companies. We collected the data from 134 HCI professionals at three conferences: CHI 98, CHI 99, and the Usability Professionals' Association 1999 conference. The results are the first steps towards a toolkit for the usability community that can help HCI practitioners learn from the experiences of others in similar situations.

Keywords

Usability, strategic usability, corporate planning, methodology, HCI professionals, organizational change

INTRODUCTION

Usability organizations often say they would like to be more effective and influential in how corporations develop products. Since usability organizations have grown within many companies over the last 5-10 years, we now have the opportunity to examine how influential various approaches have been at the more strategic level.

For this research, we defined "strategic usability" as embedding usability engineering in the organizational processes, culture, and product roadmaps. In strategic usability, usability data contributes to corporate-wide decision-making, such as product priorities and make vs. buy decisions.

Since we couldn't realistically conduct a controlled study in dozens of organizations, instead we asked usability experts their perceptions of the strategic effectiveness of a variety of approaches. These approaches included organizational approaches (such as leveraging existing initiatives) and methodological activities (such as usability testing in a lab).

In addition, we theorized that organizational demographics such as the size of the company, the size of the usability organization, or the type of company might affect the perceived effectiveness of the various approaches.

We hoped to produce a toolkit for the usability community, to enable usability practitioners to learn from the experiences of others in similar situations. This toolkit would consist of the organizational and methodological approaches, the organizational demographics, the usage rating (how frequently the approach is used), the effectiveness rating, and whether any of the correlations were statistically significant.

This work is a continuation of the research from the CHI 98 workshop the authors organized on "Unpacking Strategic Usability: Corporate Strategy and Usability Research. [3]" Other CHI workshops and panels [1, 4, 5], along with presentations at the Usability Professionals' Association (UPA) [2], contributed to the research, starting with a CHI 96 workshop organized by the authors and Judith Ramey. This paper discusses the sequence of data gathering and proposed theories, and the state of the toolkit to date.

CHI 98 WORKSHOP LED TO PILOT SURVEY

While preparing for the CHI 98 workshop, we observed that the organizational backgrounds of the participants would be closely tied to their workshop contributions. Therefore, we asked the participants to fill out a pre-workshop profile, with 20 detailed questions about their companies and their HCI groups. We distributed the answers to all participants before the workshop, along with the position papers, and this information informed everyone's participation.

During the workshop, the 13 participants and 3 organizers identified and described 17 organizational approaches and 10 usability methods that they believed contribute to strategic usability. However, after the workshop was over, the organizers recognized an opportunity to gather additional data and perform more analysis in two areas:

- How did the workshop participants' opinions of the organizational approaches and usability methods compare with opinions from other members of the HCI community?

- Was there a relationship between the organizational backgrounds of HCI practitioners and their opinions of what activities contribute to strategic usability?

To begin answering these questions, the authors edited the pre-workshop profile questionnaire in two ways. We simplified the questions somewhat (although they were still quite elaborate), and we asked for effectiveness ratings of the 27 activities identified during the workshop. We then asked the workshop participants to update their responses, and we solicited more respondents from a usability listserv. The 23 responses received from these two groups—workshop participants and listserv respondents—became our pilot sample; they included 9 respondents from large organizations (over 1,000 employees), 8 respondents from smaller firms, and 6 HCI/usability consultants.

The pilot survey questions are listed in Appendix A at the end of this volume. The complete survey listed the activities in Question 19 and included explanations of Questions 3, 6, 10, and 20; this text is available on the first author's website (www.teced.com) under Courses and Papers.

Results of Pilot Survey

Survey respondents were asked to rate only those organizational approaches and usability methods that they had actually used, to avoid obtaining ratings based on hearsay. Table 1 shows how the pilot survey participants rated the organizational approaches and usability methods as contributing to strategic usability. Note that the number of respondents reporting use of each of the organizational approaches and usability methods varies considerably.

We then looked at the collected data describing all 23 organizations to see if there were any correlations between that information and the toolkit ratings. For the pilot data, we looked at the answers to Questions 1, 2, 5, 6, 9, 10, 12, 14, 15, and 17, which we had captured in spreadsheets. When considering organizations' size (Question 1), we categorized respondents as being from large companies (more than 1,000 employees), small companies (<1,000 employees), and consultancies (all with <100 employees).

Table 1: Ratings by Pilot Survey Participants

Organizational Approaches (O) or Usability Methodology (M)	Mean Score[1]	# Reporting Use
Design Café[2] (M)	1.43	7
Partnering with Marketing (O)	1.64	14
Field Studies (M)	1.65	20
Usage Scenarios (M)	1.69	18
High-Level/Founder Support (O)	1.73	15
Usability Testing Without a Lab (M)	1.76	19
Task Analysis (M)	1.78	18
High-Profile Projects (O)	1.79	14
Lab Usability Testing (M)	1.83	18
Usability Advocates/Champions (O)	1.92	12
Fit into Current Engineering (O)	1.95	20
Heuristic Evaluation (M)	1.95	20
Organizational Usability Planning (O)	2.00	7
Leveraging Related Initiatives (O)	2.00	11
Contextual Inquiry (M)	2.18	17
Educate/Train: Development &	2.23	13
UI Group Reports to UI (O)	2.25	8
Usability Open Houses (O)	2.29	7
Focus Groups (M)	2.36	18
Coach/Supt Grass Roots Efforts (O)	2.38	8
Corp. Mandates/Usability Objectives (O)	2.42	12
Internal Task Forces (O)	2.43	7
Organizational Audits (O)	2.50	4
User Interface Committee (O)	2.89	9
Surveys (M)	2.92	19
Communities of Practice (O)	3.09	11
Design Review Boards (O)	3.33	6

[1] Low numbers are high ratings; the lowest mean scores were rated as being most effective at contributing to strategic usability
[2] Interpreted by some respondents as a cross-functional team approach, such as participatory design

There were no statistically significant correlations between any of the answers to the above questions and the respondents' toolkit ratings. However, when we looked at the organizations' size categories, there were suggestions that the effectiveness of some approaches might vary by company size. In particular, smaller settings (taking consultancies and smaller firms together) might find the following approaches better: high-level/founder support, task analysis, and contextual inquiry. Consultants might find the following better: usability testing without a lab, lab usability testing, and communities of practice (alliances with academia and industry).

Activities with means less than 2.00 might be considered promising based on consistently good experiences, and those with means over 2.50 might be risky based on consistently bad experiences. In addition, a few activities received highly variable scores: fit into current engineering processes, contextual inquiry, UI group reports to UI, focus groups, and corporate mandates.

After the pilot survey, we removed "design cafe" from the usability methods; it was suggested by one CHI 98 workshop participant and was unfamiliar to the others. We believe some pilot test respondents interpreted it as a cross-functional team approach, such as participatory design.

STREAMLINED SURVEY FOR LARGER POPULATION

Based on the pilot study data, the authors decided to collect data from a larger sample of HCI practitioners. We refined and scaled down the original questionnaire to a one-page, 10-question version that we administered at CHI 99 and at the Usability Professionals' Association 1999 conference.

Design and Administration of the CHI/UPA Survey

We began the new questionnaire with the definition of strategic usability given in the Introduction. We wanted to emphasize for the wider audience (who had not been workshop participants) that we weren't asking what approaches and methods were effective when used for product-design decisions, but rather which ones had an impact on corporate decision-making. We then asked the 10 questions listed in Appendix B at the end of this volume; the complete text of the CHI/UPA survey can be reviewed on the first author's website (www.teced.com) under Courses and Papers.

For the questions that were retained from the pilot survey, we made a few changes. We removed "design cafe" from the usability methods. Since the pilot list did not include participatory design, we added it. We also added "UI staff members co-located with engineering" to the organizational approaches. Finally, we listed three kinds of "usability testing" (in a lab, using portable equipment, and outside a lab) rather than the two in the pilot test.

Therefore, when considering the number of respondents citing any of these activities, note that some of them appeared only in the pilot survey and others only in the CHI and UPA surveys. In addition to these changes, we clarified the distinction between consultants and corporate practitioners by changing the "HCI/Usability" entry under company categories to read "HCI/Usability Consulting."

We administered the survey at two conference sessions about strategic usability: the CHI 99 Panel on "What Makes Strategic Usability Fail? Lessons Learned from the Field [4]" and a UPA session on "What Makes Strategic Usability Succeed or Fail? Lessons from the Field [2]." Note that the CHI surveys were administered at the end of the session, during which we gave a brief overview of the pilot data. The UPA surveys were distributed and collected at the beginning of the session, preceded only by an oral definition of strategic usability. We collected a total of 111 surveys, 31 from CHI 99 and 80 from UPA 99.

The pilot survey data and the UPA survey data included most respondents' affiliations; the CHI survey data did not. Reviewing the pilot and UPA surveys, only one company (Hewlett-Packard) was represented by more than three respondents; the large majority of respondents were the only ones from their organization.

ANALYSIS OF QUESTIONNAIRE DATA

After the CHI and UPA surveys, we tabulated the data from all three groups (134 respondents). Tables 2 and 3 show the results for Questions 1 and 2.

Table 4 shows the results for Question 4, in which we asked respondents to select a category describing their company; note that the total of responses is 147 rather than 134 or fewer, because some respondents selected more than one category. We also asked (in Question 3) for a free-form description of what their company does, but the answers were so ambiguous, disparate, and incomplete that a meaningful compilation wasn't possible. Instead, we relied on Question 4 for our insights about this topic (see Discussion of Results, next).

Table 5 summarizes the answers to the question about funding (Question 6 in the CHI/UPA survey; Question 17 in the pilot survey). Usability groups are typically funded by either an annual budget (either their own or part of another department) or on a bill-back by project basis. Sixty percent of the survey respondents who answered the question are funded by an annual budget. Another 15 respondents were funded by another budget (either R&D, Marketing, IT, or "salary paid"). The second most common approach to funding is bill-back by project, with 21 respondents. Some groups had a blend of funding, combining an annual budget or government funding with project funding. Some survey respondents did not know their funding source.

Table 2: Sizes of Survey Participants' Companies

Company Size	# Responding
1	3
2-5	5
6-10	2
11-25	4
26-50	5
51-100	5
101-250	5
251-500	9
501-1000	11
1001-5000	22
5001-10,000	14
over 10,000	47

Table 3: Ages of Participants' Companies

Years in business	# Responding
< 1	5
1 - 2	5
3 - 5	11
6 - 10	11
11 - 15	17
16 - 20	8
21 - 30	25
31 - 40	6
41 - 50	1
over 50	41

Table 4: Company Categories

Category	# Responding
Aerospace	2
Automotive	1
Computer	46
Education/Training	5
Financial Services	14
Government	9
Health/Medical Services	7
HCI/Usability Consulting	10
Internet/E-Commerce	11
Manufacturing	5
Telecommunications	17
Publishing	2
Retail/Wholesale	1
Other	17

Table 5: How HCI/Usability Groups are Funded

Type of Funding	# Responding
Annual Budget	73
By Project (Bill-Back)	21
Part of R&D Budget	10
Annual Budget and By Project	3
Salary Paid	3
R&D Budget and By Project	2
Government Funding and By Project	2
Part of Marketing Budget	1
Government Funding	1
Part of IT Budget	1

Table 6 summarizes the respondents' reports of the top two obstacles they perceive to creating greater strategic impact for usability engineering/HCI within their organizations (Question 7 in the CHI/UPA survey; Question 18 in the pilot survey). See the discussion of these responses under Discussion of Results, next.

Table 6: Obstacles to Strategic Usability

Categories, Descriptions, and Frequency Cited
Resource Constraints: 28.6%
• "perceive usability as taking more time in schedule"
• "time to market is tight" or "too fast turnaround between revs"
• "schedule limitations"
• "lack of budget — no money to hire usability, need money to act"
• "too much to do & too few employees to handle projects"
Resistance to User-Centered Design/Usability: 26.0%
• "resistance among engineers and/or management to usability"
• "no see the value of usability/HCI"
• "lack of management interest/respect/support"
• "organizational inertia – we've always done things this way"
• "engineers believe they already know and understand HCI/usability—they have (HCI/usability) skills"
Lack of Understanding/Knowledge about What Usability Is: 17.3%
• "need education"
• "seen as only testing activity"
• "role of HCI not specifically known"
Better Ways to Communicate Impact of Work and Results: 13.3%
• "need cost-benefit analysis – unable to prove link to what happens in the market/with the user from our recommendations"
• "visibility of impact of results"
• "need to differentiate (usability) from systems development – what's our value-add"
• "credibility of our impact"
Lack of Trained Usability/HCI Engineers: 6.1%
• "can't find people with the technical expertise'"
• "lack of experience in the field/corporate practice of usability/HCI"
Lack of Early Involvement: 5.1%
• "need more partnerships with marketing earlier in the cycle"
• "strategic usability overlaps with marketing's role – we need to coordinate with them more"
• "we're brought in too late to have real impact"
• "impact in limited due to mostly usability testing input later (in cycle)"
No Economic Need — Customers Not Asking for Usability: 3.6%
• "no customers asking for greater usability – products are successful in the marketplace without it"
• "no negative market consequences identified for not including usability in our process/consequences for not including usability don't exist"

When asked to describe the top two obstacles to achieving strategic impact in their respective organizations, respondents most often cited resource constraints and resistance to changing the status quo of "we've always done things this way" (without usability). The resource limitations most frequently listed related to the perception that "usability takes too much time" in an already tight schedule and to lack of budget to hire trained specialists, allocate facilities, or purchase equipment. The responses comprising the "resistance to usability" category tended to be more energetically negative, describing organizational climates that included "disinterest" and "lack of any management support" to expressed resistance in the form of "engineers who feel they have HCI skills and don't need any usability" and "no one seems to see the value."

In compiling these responses, we minimized inferences about the causes of the obstacles mentioned in the open-ended answers. Rather, we preserved differences in phrasing that reflect how these professionals perceived the issues in their organizations.

Table 7 shows the results for Question 10 and includes only CHI and UPA data; the question wasn't asked in the pilot survey. The ratings and use of organizational approaches and usability methodologies (Questions 8 and 9 in the CHI/UPA survey; 19A and 19B in the pilot survey) are listed in Table 8.

Table 7: How Successful is Strategic Usability?

Rating	# Responding
Very successful	0
Quite successful	14
Somewhat successful	55
Neutral	10
Somewhat unsuccessful	9
Quite unsuccessful	14
Very unsuccessful	3

Table 8: Ratings and Use of Organizational Approaches and Usability Methodologies

Organizational Approaches (O) or Usability Methodology (M)	Mean Score[1]	# Rptg Use	% Rptg Use
Lab Usability Testing (M)	1.61	87	65%
Usability Testing Without a Lab or Outside of Lab Facility[2] (M)	1.69	73	55%
High-Profile Projects (O)	1.71	75	56%
Usability Test. w. Portable lab. Equipment[3] (M)	1.76	29	26%
UI Staff Members Co-located with Engineering[3] (O)	1.77	49	45%
Field Studies or Field Studies other than CI (M)	1.77	56	42%
High-Level/Founder Support (O)	1.79	61	46%
Usage Scenarios (M)	1.83	60	45%
Task Analysis (M)	1.83	82	62%
Participatory Design[3] (M)	1.86	44	40%
Usability Advocates/Champions (O)	1.99	72	54%
Contextual Inquiry (M)	2.02	63	47%
Leveraging Related Initiatives (O)	2.03	41	31%
Fit into Current Engineering Processes (O)	2.07	84	63%
Partnering/Collaborating with Marketing on Projects (O)	2.15	47	35%
Heuristic Evaluation (M)	2.16	93	70%
Organizational Usability Planning (O)	2.17	35	26%
Coach/Support Grass Roots Efforts (O)	2.25	48	36%
Educate/Train Other Functional Groups (e.g., Mktg., Development, Doc.) (O)	2.45	63	47%
Focus Groups (M)	2.45	55	41%
Surveys (M)	2.52	69	52%
Usability Open Houses (O)	2.57	30	23%
Internal Task Forces (O)	2.57	28	21%
Corporate Mandates / Usability Objectives (O)	2.6	48	36%
UI Group Reports to UI not Development (O)	2.67	42	32%
Organizational Audits (O)	2.69	16	12%
User Interface Committees (O)	2.77	30	23%
Communities of Practice - Alliance w. Academia / Industry (O)	2.84	31	23%
Design Review Boards (O)	2.93	27	20%

[1] Low numbers are high ratings; the lowest mean scores were rated as being most effective at contributing to strategic usability
[2] First wording is pilot test
[3] CHI and UPA only

Based on the possible relationships suggested by the pilot data, we looked for any correlations between organization size and respondents' ratings of the organizational approaches and usability methods. First, we categorized the firms into the same three groups as we did for the pilot data. However, for usability consultancies, all had less than 250 employees. Large firms continued to be those with more than 1,000 employees. The number of respondents in each category is shown in Table 9.

Table 9: Categorizations of Respondents by Organization Size

	HCI/Usability Consultants	Smaller Firms	Large Firms	Total
Pilot data	6	8	9	23
CHI 99	0	9	22	31
UPA 99	6	21	53	80
Total	12	38	84	134

Looking at the total of 134 respondents, there are no statistically significant correlations between any of the organizational approaches or usability methods and an organization's size (or whether consulting or corporate). We did observe, not surprisingly, that the number of people reporting use of an approach or method goes down as its perceived lack of contribution to strategic usability goes up. This information is shown in Figure 1.

Figure 1: Relationship of Effectiveness Ratings and Number of People Reporting Use

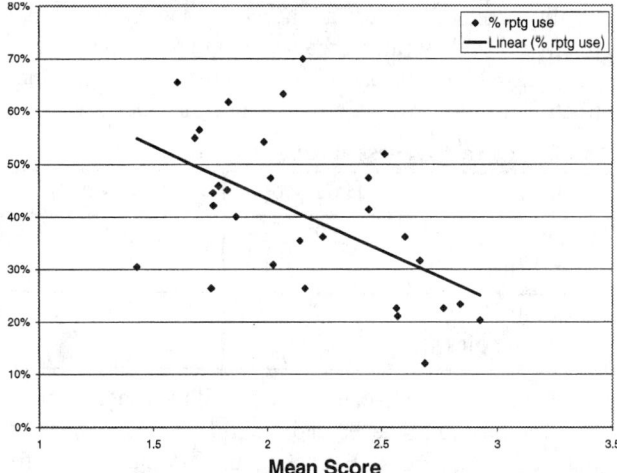

DISCUSSION OF RESULTS

Since the large survey sample did not show correlations between the organization size and organizational approaches or usability methods, we looked for other possible relationships. We considered whether the size of the usability group correlated with the effectiveness ratings.

In the pilot questionnaire, we asked for the size of the respondents' groups and also for the number of HCI/ usability people in the company, and we offered categories (Question 10, Appendix A). In the CHI and UPA questionnaires, we simply asked how many HCI/usability people were in the company (Question 5, Appendix B). We looked for correlations between how many HCI/usability people were in the company and what toolkit items were rated more effective, and we found no statistically significant correlations.

We suggested another hypothesis, that the HCI people who are a low percentage of their company's size might rate the usability methods as being more effective than they rate organizational approaches. If their small numbers meant they couldn't change their organizations much, they might believe that seeing usability methods used well would have more strategic impact. Overall we might expect both organizational approaches and usability methods to be rated less effective by the smaller HCI populations than by the larger ones, with organizational approaches rated less effective than usability methods.

A reliable ratio of HCI professionals to company size could not be constructed from the data. So to explore this hypothesis, we looked only at the 84 people from large companies. We divided these 84 people into two groups, those with fewer than 20 HCI people in the company and those with 20 or more. Then, if we combine all the ratings of each type, we find the averages in Table 10.

The sample sizes in all cells are in hundreds, and the averages, according to the central limit theorem, closely follow a Gaussian, or "normal" distribution. Using the observed means and variances and the properties of the normal distribution, three of these relationships are statistically significant: both large and small HCI groups rated usability methodologies as a whole more strategically effective than they rated organizational approaches. And as we expected, organizational approaches were rated more effective overall by large groups than by smaller ones.

Table 10: Effectiveness Ratings by HCI Group Size

Toolkit Type	<20 HCI people	>=20 HCI people
Organizational approaches	2.28	2.11
Usability methodologies	1.98	1.92

Looking at the effectiveness ratings and the usage of all the organizational approaches and usability methodologies, we compiled two groups. Excluding the "design café," five items had fairly high ratings and fairly low usage, as shown in Table 11; these might be investigated for more extensive use. Another five items had fairly low ratings and fairly high usage, as shown in Table 12; these might be investigated for less extensive use. See also the Summary and Conclusions, next.

Table 11: Investigate for More Extensive Use

Approaches and Methodologies	Mean Score	# Rptg Use	% Rptg Use
Usability Test. w. Portable lab. Equip.[2]	1.76	29	26%
UI Staff Members Co-located with Engineering[2]	1.77	49	45%
Field Studies or Field Studies other than CI[1]	1.77	56	42%
High-Level/Founder Support	1.79	61	46%
Usage Scenarios	1.83	60	45%
Participatory Design[2]	1.86	44	40%

[1] First wording is pilot test
[2] CHI and UPA only

Table 12: Investigate for Less Extensive Use

Approaches and Methodologies	Mean Score	# Rptg Use	% Rptg Use
Educate/Train Other Functional Groups (e.g. Mktg., Development, Doc.)	2.45	63	47%
Focus Groups	2.45	55	41%
Surveys	2.52	69	52%
Corporate Mandates / Usability Objectives	2.60	48	36%
UI Group Reports to UI not Development	2.67	42	32%

The authors identified several other questions that we hoped to answer with the survey data:

- Do usability consultancies rank some or all usability methods as more effective than do in-house usability professionals?

- Do smaller companies have a better focus on their customer populations, and thus find contextual inquiries and task analysis more effective?

- Is there a connection between certain company categories and how successful respondents from these companies rate organizational approaches and usability methods?

The twelve respondents from consultancies gave 97 scores to the 12 usability methods in the survey, with a mean score of 1.74. In contrast, 122 respondents from in-house staffs gave 621 scores to the same usability methods, with a mean score of 1.98. For 10 of the 12 usability methods, the consultancy score was better (lower) than that from in-house professionals. This is statistically significant at the 5% level. Consultancies do rate usability methods more useful than in-house staffs.

Investigating the second question, we found that 38 respondents from small firms gave 38 scores to contextual inquiry and task analysis, with a mean score of 1.95. In contrast, 84 respondents from larger firms gave 85 scores to the same usability methods, with a mean score of 1.98. The difference is not statistically significant, nor are these scores significantly different from the average score given to the other usability methods.

With the ever-increasing blurring of the lines between business and consumer products, hardware and software, and products and technologies, the categorization of companies has become more complex. Given this complexity, it was difficult to theorize on which types of companies might have significant differences in their effectiveness ratings.

To address the third question, we hypothesized that respondents in certain categories of companies—those often considered more innovative in their processes and products—would rate the organizational approaches and usability methods as more effective than respondents from other categories. In particular, we hypothesized that the Internet/E-commerce, Computer, and Consulting categories (considered as a group and called Group 1) would have higher effectiveness ratings than would the Financial Services, health/medical, and Government categories (Group 2).

The mean score for usability methods was 2.01 for Group 1 and 2.00 for Group 2. These do not differ in a statistically significant way. Both are actually slightly greater (i.e., worse ratings) than the average for all other company categories, but this is also not statistically significant. With respect to organizational approaches, the mean scores were 2.30 for Group 1 and 2.18 for Group 2, but these are still well within the limits of variation which should be expected to arise from strictly random effects (the estimated standard error of the difference of these means is 0.15). There is no statistically significant difference in the ratings of these items either.

SUMMARY AND CONCLUSIONS

Our goal in the initial pilot and subsequent CHI 99 and UPA 99 surveys has been to evolve a toolkit of organizational approaches and usability methodologies that contribute to making usability activities and data have strategic impact in corporate decision-making. We began with a loosely constructed hypothesis that the demographics of an organization might affect what approaches and methodologies would work best to create strategic impact.

Organization size did not affect what organizational approaches and usability methods were rated most effective in achieving strategic usability. Also, the results of a chi-squared test showed no statistically significant differences in the rates at which obstacles were cited by respondents from large and small companies. The size of organizations appears to have no impact on people's perception of factors that are inhibiting their ability to contribute at the strategic level in their respective environments.

The CHI 99 panel participants offered some insights and advice related to the specific obstacles cited by survey respondents: resistance, lack of understanding of HCI/usability, and lack of ability to communicate cost-benefit/impact of usability results. The CHI 99 panelists tended to agree with the advice offered by Don Norman in his CHI 99 session with the second author that we should "learn to speak the business language" of our internal functional area partners in marketing and management.

The CHI 99 panelists also said HCI practitioners should develop a business case for usability and learn enough of the technical constraints of any recommendation to communicate with engineers in their own language about the best ways to implement suggested product changes. Using creative and innovative ways to distribute findings throughout our organizations and making them accessible on-demand from colleagues' desktops via intranet sites were still other recommendations from the CHI 99 panel.

Based on the authors' experiences in a variety of large and small companies, the same methodologies and organizational approaches were equally effective. What mattered most was that the usability professionals worked to involve the cross-functional teams directly in the research effort, through firsthand observation followed by participation in some form of summary activity. In addition, consistent and visible management support at the highest levels of the organizations gave usability greater credibility and perceived importance to overall product and company success in the marketplace.

Choosing high profile projects and having high level, or company founder support were the organizational approaches that were ranked as most effective across all three survey groups. These activities allow greater visibility across functional and organizational boundaries and lend credibility in the form of expressed management support. The high effectiveness rating of High-Profile Projects is another argument that a more effective strategy may be to select projects carefully and staff them with sufficient HCI resources, rather than spreading limited resources too thin and providing only "Band-Aid" improvements to a larger number of projects.

Of note also is that although 47% of the respondents had utilized the organizational approach of Educate/Train Other Functional Groups (e.g., Marketing, Development, Documentation), this approach was one of the lower rated in its contribution to strategic usability. This runs counter to the belief of many HCI practitioners that building usability literacy within organizations is very helpful. On the other hand, when we look at the obstacles respondents cited to achieving strategic usability, several of them—particularly lack of understanding of what HCI is, and lack of ways to communicate the value of results—imply a need to educate internal groups about the benefits of usability.

Usability testing—whether inside a lab facility, using portable equipment, or outside of a lab facility—was rated highest as an effective usability methodology to create greater strategic impact. One reason for this high rating might be that the activity of product, or prototype, testing affords more team members the chance to observe firsthand how users can and cannot interact with their designs. Even if members miss the sessions, videotapes can provide the immediate experience of product usage.

Laboratory usability testing was also a widely used methodology. Results from usability tests tend to be immediately implementable and focused on specific changes to improve ease of use or effectiveness of the product. In comparison, field studies often yield robust descriptive data that requires greater interpretation and is more subjective. Applying the results, even if well categorized and tabulated, can be difficult because they often must be applied to future releases.

Surveys are widely used (52%), despite their lower effectiveness rating. This indicates that surveys provide some benefit (namely, a larger sample size) that HCI practitioners want, while not providing data in the most effective form. Thus, there is an opportunity for improvement in the methods used with larger sample sizes.

It's also interesting to note that the most commonly used method is heuristic evaluation, even though its effectiveness to strategic usability is ranked far below usability testing, field studies, usage scenarios, task analysis, and participatory design. This is probably because it is relatively quick and easy to perform a heuristic evaluation, and HCI practitioners are often under pressure to provide feedback to a product that will soon be released.

There appears to be an interesting contradiction in the survey data between the usability methodologies that respondents felt were not as effective and the obstacles they cited as inhibiting their ability to have strategic impact in their organization's decision-making. While education and training on usability were frequently mentioned, these methods were cited as less effective. Yet several of the most often cited obstacles seem to call for more education of our internal clients and partners. Is the content of our current educational and training efforts at the core of this seeming contradiction? Or are our customers not demanding greater usability, and thus our organizations are not being driven to greater action?

The CHI 97 and CHI 99 panelists (including the authors) agreed on the importance of building partnerships early in the product planning and design process with our internal colleagues in marketing, engineering, and corporate management. They recommended that we apply usability methods to our internal clients and partners and learn more about their goals, priorities, customer contacts, and customer data. The panelists believed that these activities and partnerships can help the strategic penetration of usability within organizations.

Assessing the effectiveness of strategies and tactics in the real world is not simple. The authors could not perform a controlled study, and companies do not track all the metrics necessary to compare the various approaches more objectively. Given these constraints, we judged that our best approach was to gather the opinions of practitioners. Since we do see some trends and commonality, we believe that this is a valid approach. However, it should be emphasized that all the data in this paper is based on the perceptions of the respondents, not on any direct knowledge the authors have of the respondents' activities.

ACKNOWLEDGMENTS & RESEARCH CONTRIBUTORS

Data compilation, text production, and graphics for this paper were performed by Jacie Wedberg, Josh Mills, William Cotter, Beth Almay, and Melani Cantlin. Robert Farrell generously contributed the statistical analysis.

The authors would also like to thank the following workshop participants and pilot testers for their contributions to this research: Mark Anderson, Sarah Bloomer, Charlie Breuninger, Annette Bryce, Donald Chartier, Scott Clifford, Herb Colston, Betsy Comstock, Mary Czerwinski, Terri Ducay, Ken Dye, Chris Farnes, Alicia Flanders, Shannon Halgren, Avi Harel, Tom Holzman, Ed Israelski, Masako Itoh, Janice James, Caroline Jarrett, Irena Jaska, Michael King, O-Seong Kweon, Lori Landesman, JoAnn Lotfi, Jon Meads, Jeff McCartney, Ian McClelland, Tim McCollum, Debbie Mrazek, Michael Muller, Jakob Nielsen, Hanna Parrish, Amanda Prail, Judith Ramey, Mary Beth Raven, Ginny Redish, Mary Beth Rettger, Carol Righi, Dave Rinehart, Steve Sato, Karen Shor, Tammy te Winkel, John Thomas, Karel Vredenburg, Dennis Wixon, Cathleen Wharton, Susan Wolfe, and Ron Zeno.

REFERENCES

1. Humburg, J., Rosenbaum, S., and Ramey J. "Corporate Strategy and Usability Research: A New Partnership," in *CHI 96 Conference Companion* (Vancouver, BC, April 1996), ACM Press, 428.

2. Rohn, J., Rosenbaum, S., and Humburg, J. "What Makes Strategic Usability Succeed or Fail? Lessons from the Field," in *Proceedings of the Usability Professionals' Association 8th Annual Conference* (Scottsdale, AZ, June-July 1999), Usability Professionals' Association, Inc. 225-248.

3. Rosenbaum, S., Humburg, J., and Rohn, J. "Unpacking Strategic Usability: Corporate Strategy and Usability Research." in *CHI 98 Summary*, (Los Angeles, CA, April 1998), ACM Press, 205-206.

4. Rosenbaum, S., Rohn, J., Humburg J., Bloomer, S., Dye, K., Nielsen, J., Rinehart, D., and Wixon, D. "What Makes Strategic Usability Fail? Lessons Learned from the Field," in *CHI 99 Extended Abstracts*, (Pittsburgh, PA, May 1999), ACM Press, 93-94.

5. Rosenbaum, S., Rohn, J., Thomas, J., Humburg, J., Bloomer S., and Czerwinski, M. "Corporate Strategy and Usability Research: A New Partnership," in *CHI 97 Extended Abstracts* (Atlanta, GA, March 1997), ACM Press, 115-116.

Measuring Usability: Are Effectiveness, Efficiency, and Satisfaction Really Correlated?

Erik Frøkjær
Dept. of Computing
University of Copenhagen
Copenhagen Ø, Denmark
+45 3532 1456
erikf@diku.dk

Morten Hertzum
Centre for Human-Machine Interaction
Risø National Laboratory
Roskilde, Denmark
+45 4677 5145
morten.hertzum@risoe.dk

Kasper Hornbæk
Dept. of Computing
University of Copenhagen
Copenhagen Ø, Denmark
+45 3532 1452
kash@diku.dk

ABSTRACT

Usability comprises the aspects effectiveness, efficiency, and satisfaction. The correlations between these aspects are not well understood for complex tasks. We present data from an experiment where 87 subjects solved 20 information retrieval tasks concerning programming problems. The correlation between efficiency, as indicated by task completion time, and effectiveness, as indicated by quality of solution, was negligible. Generally, the correlations among the usability aspects depend in a complex way on the application domain, the user's experience, and the use context. Going through three years of CHI Proceedings, we find that 11 out of 19 experimental studies involving complex tasks account for only one or two aspects of usability. When these studies make claims concerning overall usability, they rely on risky assumptions about correlations between usability aspects. Unless domain specific studies suggest otherwise, effectiveness, efficiency, and satisfaction should be considered independent aspect of usability and all be included in usability testing.

Keywords

Usability measures, effectiveness, efficiency, satisfaction, information retrieval, usability testing, user studies

INTRODUCTION

Although the importance of usability is gaining widespread recognition, considerable confusion exists over the actual meaning of the term. Sometimes usability is defined quite narrowly and distinguished from, for example, utility [11], on other occasions usability is defined as a broad concept synonymous to quality in use [2]. We adopt ISO's broad definition of usability [7] as consisting of three distinct aspects:

- *Effectiveness*, which is the accuracy and completeness with which users achieve certain goals. Indicators of effectiveness include quality of solution and error rates. In this study, we use quality of solution as the primary indicator of effectiveness, i.e. a measure of the outcome of the user's interaction with the system.

- *Efficiency*, which is the relation between (1) the accuracy and completeness with which users achieve certain goals and (2) the resources expended in achieving them. Indicators of efficiency include task completion time and learning time. In this study, we use task completion time as the primary indicator of efficiency.

- *Satisfaction*, which is the users' comfort with and positive attitudes towards the use of the system. Users' satisfaction can be measured by attitude rating scales such as SUMI [8]. In this study, we use preference as the primary indicator of satisfaction.

While it is tempting to assume simple, general relations between effectiveness, efficiency, and satisfaction, any relations between them seem to depend on a range of issues such as application domain, use context, user experience, and task complexity. For routine tasks good performance depends on the efficient, well-trained execution of a sequence of actions which is known to yield stable, high-quality results [3]. For such tasks high-quality results are routinely achieved, and task completion time may therefore be used as an indicator of overall usability. For non-routine, i.e. complex tasks, there is no preconceived route to high-quality results, and good performance is primarily dependent on conceiving a viable way of solving the task [9, 14]. The efficient execution of the sequence of actions is of secondary importance. Consequently, efficient execution of the actions may or may not lead to high-quality results, and diligence is not even guaranteed to lead to task completion. This suggests that, at least for complex tasks, efficiency measures are useless as indicators of usability unless effectiveness is controlled.

Nielsen & Levy [12] analyzed the relation between efficiency and user preference in 113 cases extracted from 57 HCI studies. Their general finding was that preference predicts efficiency quite well. However, in 25% of the cases the users did not prefer the system they were more efficient in using. The ambition of finding a simple, general relationship between efficiency and satisfaction is therefore questionable [see also 1]. Studies of, for example, specific application domains may yield more precise and informative models. With respect to the relationship between satisfaction and effectiveness, Nielsen & Levy [12] note that their very comprehensive literature survey did not encounter a single study that compared indicators of these two aspects of usability.

In this paper we investigate the connection between efficiency, indicated by task completion time, and effectiveness, indicated by quality of solution. This is done by reanalyzing data from the TeSS-experiment [6] where 87 subjects solved a number of information retrieval tasks, using four different modes of the TeSS system and programming manuals in hard copy. In analyzing the data we look for correlations between efficiency and effectiveness across retrieval modes, tasks, and individual subjects.

The purpose of this paper is to emphasize the importance of accounting for all three aspects of usability in studies that assess system usability, for example to compare the usability of different designs. Effectiveness is often difficult to measure in a robust way. This may be the reason why several studies involving complex tasks refrain from accounting for effectiveness and settle for measures of the efficiency of the interaction process [for example, 5, 13]. These studies rest on the assumption that an efficient interaction process indicates that the user also performed well in terms of crucial effectiveness indicators such as solution quality. The TeSS-experiment illustrates that this assumption is not warranted—unless it can be supported by an argument that effectiveness is controlled.

The first two sections present the method and results from the TeSS-experiment, establishing the argument that efficiency and effectiveness are weakly—if at all—correlated. Next, we discuss the general relationship between the three aspects of usability, exemplifying the impact of our findings by studies from the CHI Proceedings of the years 1997-99. We then discuss the implications of our findings with regard to the selection of usability measures. In the final section, we outline our main conclusions concerning the weak and context-dependent relation between the usability aspects.

THE TESS-EXPERIMENT

The purpose of the TeSS-experiment was to compare the usage effectiveness of browsing and different forms of querying in information retrieval tasks concerning programming problems. Further, the experiment aimed at establishing a detailed description of the subjects' interaction with the TeSS system.

Experimental Conditions

To solve the tasks the subjects needed information concerning the development of graphical user interfaces in the X Window System. Access to the necessary documentation (approximately 3 Mb of text) was provided through an experimental text retrieval system called TeSS and by means of manuals in hard copy. TeSS can be operated in four different modes, each providing the user with a different set of retrieval facilities. Thus, the experiment involves five retrieval modes:

- BROWSE. In TeSS, browsing can be done by expanding and collapsing entries in the table of contents and by searching the table of contents for specific strings. The text itself is presented in separate windows.

- LOGICAL. A mode of TeSS offering conventional Boolean retrieval where queries are logical expressions built of query terms, ANDs, ORs, NOTs, parentheses, and wildcards.

- VENN. In this mode of TeSS queries are expressed by means of a Venn diagram which replaces Boolean operators with a, supposedly, more immediately understandable graphical image of intersecting sets.

- ALL. The whole of TeSS offering the combination of BROWSE, LOGICAL, and VENN.

- PAPER. In this mode searching is done in hard copies of the programming manuals, i.e. independently of TeSS.

Subjects

The subjects were 87 students in their third year of a bachelor degree in computer science. While the project was a mandatory part of the students' education, participation in the experiment by allowing the data collection to take place was voluntary and anonymous. The subjects were first-time users of TeSS and had no prior knowledge of the programming tools on which the tasks were based.

Tasks

In the TeSS-experiment each subject solved 20 information retrieval tasks. As preparation, the subject completed two practice tasks. The 20 tasks concerned whether and how certain interface properties could be achieved in a graphical user interface. To answer the tasks the subjects had to identify the relevant user interface objects, e.g. widgets, methods, and resources, and outline an implementation. As the subjects were unfamiliar with the X Window System, the tasks involved a substantial element of learning in addition to the need for retrieving specific pieces of information. Some tasks were formulated in the context of the X Window System in general; others took the user interface of TeSS as their point of departure. Two examples of tasks used in the TeSS-experiment are:

Task 5. Radio buttons are used in situations where exactly one option must be chosen from a group of options. Which widget class is used to implement radio buttons?

Task 11. The caption on the button "done" should be changed to "quit". How is that done?

Procedure

The experiment was explained to the subjects at a lecture, after which the subjects had ten days to complete the tasks. The subjects received a manual for TeSS and a two-page walk-up-and-use introduction. The system itself was available on terminals to which students have access 24 hours a day. The manual searching was done in the library where one of the authors was present three hours a day to hand out tasks and receive solutions. Upon entering the library, the subjects received hard copies of the three manuals, a sheet with the proper task, and a log sheet with fields for starting time, finishing time, and solution.

The experiment employed a within-groups design where all subjects solved the tasks in the same sequence and each subject was required to use all retrieval modes. To avoid order effects, the subjects were exposed to the retrieval modes in a systematically varied order. The 20 information retrieval tasks were clustered into five blocks. The first block was solved with one of the five retrieval modes, the second block with one of the remaining four retrieval modes. Thus the permutations of the modes on the two first blocks divided the subjects into 20 groups. The number of subjects did not allow all 5! sequences of the five modes to be included, and the 20 groups were not divided further. Rather, the order of the three remaining modes was kept the same within each group.

Data Collection and Analysis

The data collected in the experiment include a detailed log of the subjects' interaction with TeSS. The interaction log gives a time-stamped account of the commands executed by the subjects. It also includes task demarcation and solutions reached, both obtained from a separate module governing the subjects' access to TeSS. This Task Handling Module makes it possible to let the subjects work unsupervised while at the same time enforcing a strict experimental procedure. The Task Handling Module presents the tasks to the subject one at a time, gives access to the retrieval mode to be used by that subject when solving that particular task, and records his or her solution. For the PAPER retrieval mode, the subjects recorded their starting time, finishing time, and task solution on the log sheets.

The 87 subjects received 20 information retrieval tasks each, giving a potential total of 1740 answers. However, 113 answers were not submitted; 19 were excluded because they included a more than one hour long period with no logged user activity; 17 were excluded due to technical problems with TeSS; 14 were excluded because it was impossible to judge the quality of the answer; and 2 were

Grade	Mnemonic	Description
1	Very low	Failure, a completely wrong answer
2	Low	Inadequate or partially wrong answer
3	Medium	Reasonable but incomplete answer
4	High	Good and adequate answer
5	Very high	Brilliant answer

Table 1—The five-point scale used to grade the tasks

excluded because they were solved poorly in less than two minutes, i.e., without any attempt to reach a solution. Finally, 4 subjects were excluded because they clearly did not take the experiment seriously. Thus, 11% of the answers were not submitted or excluded. The analysis is based on the remaining 1555 answers, the results of 648 hours of work performed by 83 subjects.

In this paper we focus on two aspects of the usability of TeSS:

- Efficiency measured as task completion time, which is extracted from the interaction log or the log sheets.

- Effectiveness measured as the quality of the solution, which was assessed by one of the authors and expressed by a grade on a five-point scale, see Table 1. As an example, a medium and a high quality solution to task 5 (see above) must identify toggle widgets as the relevant widget class. A brilliant answer also explains the use of radio groups to cluster the toggle widgets.

The following analysis is restricted to the 20 information retrieval tasks—the bulk of our data. Data concerning user satisfaction, measured as subjects' preference for one or the other retrieval mode, were collected for three implementation tasks, which followed the information retrieval tasks. The preference data show that the subjects did not prefer the retrieval mode with which they performed best. Rather, they overwhelmingly preferred ALL, the retrieval mode where they did not exclude themselves from any of the search facilities available in BROWSE, BOOLEAN, or VENN [6]. This suggests that user satisfaction is not simply correlated with performance measures such as task completion time and grade. Thus, the TeSS-experiment was another exception to the general finding of Nielsen & Levy [12] that users prefer the objectively best system.

RESULTS OF THE TESS-EXPERIMENT

Table 2 shows the relation between task completion time and grade for the 1555 tasks solved in the TeSS-experiment. A contingency analysis of this table suggests that task completion time and grade are not independent ($\chi^2[16, N=1555]=47.81$, $p<0.001$).

Task completion time for subjects receiving a certain grade varies much, as can be seen from the large standard deviations in Table 2. An analysis of variance shows

Task completion time Grade (no. of observations)	<P_{20}	P_{20}-P_{40}	P_{40}-P_{60}	P_{60}-P_{80}	>P_{80}	Mean time for grade (SD)
5 (N=147)	17	35	33	31	31	24.27 (20.62)
4 (N=566)	170	121	92	96	87	21.71 (38.80)
3 (N=216)	37	48	55	38	38	24.70 (26.18)
2 (N=192)	29	35	46	36	46	26.72 (32.60)
1 (N=434)	58	72	85	110	109	28.94 (27.35)
Median grade (P_{25}-P_{75})	4 (2-4)	4 (2-4)	3 (1-4)	3 (1-4)	3 (1-4)	

Table 2—Distribution of task completion time and grade for all tasks in the TeSS-experiment (N=1555). The column to the left shows the five grades given to the tasks, cf. Table 1. The next columns show the number of tasks in each of five intervals based on the 20, 40, 60, and 80 percentiles of task completion time. The rightmost column shows the mean time in minutes for a certain grade and, in parentheses, the standard deviation. The bottom row shows the median grade for each time interval, indicating the variation in grades by the 25- and 75-percentile.

significant variation in task completion times between different grades (F[4,1550]=3.31, p<0.01). However, we did not find any pairwise differences between grades using Tukey's post hoc test at a five-percent significance level.

The tasks in any of the five intervals of task completion times shown in Table 2 received markedly different grades. Between time intervals there is significant variation in grades (analysis of variance with time interval as the independent and grade as the dependent variables, F[4,1550]=9.10, p<0.001). Pairwise comparisons of the five time intervals using Tukey's post hoc test show that the 20% fastest solved task receive significantly higher grades than the 60% slowest solved tasks. Similarly, solutions to tasks in the P_{20}-P_{40} time interval receive significantly higher grades than solutions in the time intervals P_{60}-P_{80} and >P_{80}.

Spearman's rank order correlation analysis shows that task completion time and grade are significantly correlated in tasks solved in the TeSS-experiment (r_s=-0.156, two-tailed p-level <0.001). Using more time for completing a task is thus correlated with receiving a lower grade. However, the correlation between time and grade is weak; only two percent of the variation in grade can be predicted from task completion time (r_s^2=0.024). According to [4] a correlation of this magnitude is negligible.

Retrieval mode (no. of observations)	Mean time (SD)	Median grade (P_{25}-P_{75})	r_s	p	r_s^2%
Browse (N=310)	22.88 (20.89)	3 (1-4)	-0.150	0.008	2.2
Logical (N=307)	30.15 (34.70)	3 (1-4)	-0.089	0.119	-
Venn (N=305)	25.79 (25.45)	3 (1-4)	-0.107	0.062	-
All (N=314)	30.80 (51.84)	3 (1-4)	-0.128	0.030	1.6
Paper (N=319)	15.66 (11.27)	4 (2-4)	-0.265	0.001	7.0

Table 3—Correlation between time and grade in different retrieval modes. The first column shows the retrieval modes, and the second and third columns the mean time in minutes and the median grade for each mode. Columns four to six show the Spearman correlation coefficient between time and grade r_s, the significance level for the correlation p, and the strength of the correlations at a five-percent significance level r_s^2%.

To control for interplay between the design of the experiment and the weak correlation found, we performed a partial correlation analysis of the TeSS data. In the partial correlation analysis, the influence from different tasks and retrieval modes is removed from the correlation coefficient between time and grade [4]. This analysis also reveals a weak but statistically significant correlation between task completion time and grade (Spearman's partial correlation coefficient r_s[time,grade| configuration,task]=-0.170, p<0.001).

These analyses show that at the general level efficiency and effectiveness are only weakly correlated. In spite of this, time and grade could be correlated at a more detailed level of analysis, hereby undermining the conclusion at the general level. In the following sections we therefore analyze whether time and grade are correlated for specific retrieval modes, tasks, or subjects.

Correlation between Time and Grade for Different Retrieval Modes

The retrieval modes LOGICAL and VENN—the only retrieval modes requiring the subjects to formulate queries—do not show a significant correlation between time and grade (see Table 3). The retrieval modes BROWSE, ALL, and PAPER all show a statistically significant but weak correlation between task completion time and grade (r_s^2% between 1.6 and 7.0). The tasks solved in the retrieval mode PAPER have a numerically larger correlation between time and grade than the other retrieval modes. However, the correlation for PAPER is still weak and not significantly different from the correlations for BROWSE and ALL (Fisher's r-to-z

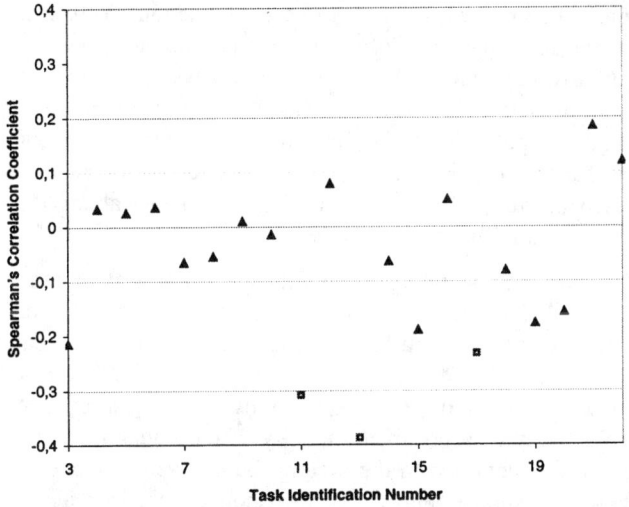
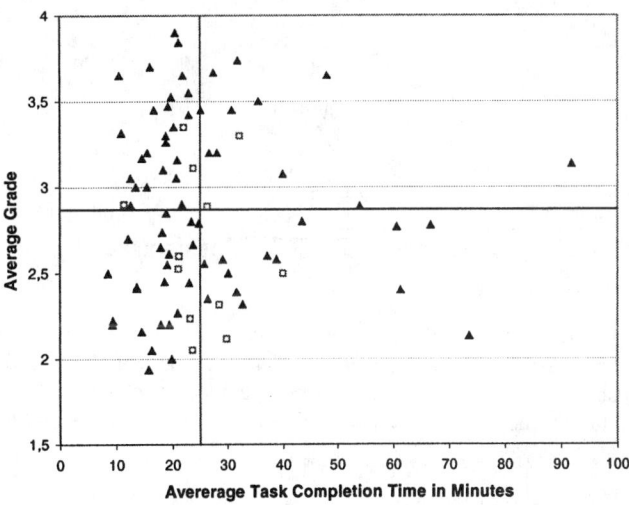

Figure 1—Correlation between time and grade for different tasks. The figure shows Spearman's correlation coefficient (r_s) for each of the 20 information retrieval tasks. Each task has been solved by between 69 and 81 subjects. Time and grade are significantly correlated for tasks 11, 13, and 17. These tasks appear as squares in the figure. The task identification numbers begin at 3, because tasks 1 and 2 are tasks used for training [6].

transformation, ALL vs. PAPER: z=-1.783, p>0.075, BROWSE vs. PAPER, z=-1.504, p>0.133).

Correlation between Time and Grade for Different Tasks

The correlation between task completion time and grade varies somewhat across the tasks (see Figure 1). For 85% of the tasks there is no correlation between time and grade. However, three tasks show a significant correlation between time and grade: task 11 (r_s=-0.308, p<0.007), task 13 (r_s=-0.387, p<0.001), and task 17 (r_s=-0.232, p<0.040). For these tasks between 5% and 15% of the variation in grade can be predicted from time, where more time spent is correlated with lower grade.

Task 11 and task 13 have a higher average grade than the other tasks (task 11: mean grade 3.42, t[1393]=-3.734, p<0.001; task 13: mean grade 3.72, t[1398]=-5.739, p<0.001). Task 13 is also solved faster than the other tasks (mean completion time 13.43 minutes, t[1398]=3.316, p<0.001). The description of these tasks given to the subjects specifies in detail some of the central interface objects of the tasks (see for example the wording of task 11 showed earlier). For task 17 it is only the relation between time and grade that is significant, individually neither time nor grade differs significantly from the other tasks.

Correlation between Time and Grade for Different Subjects

Looking at the average performance of subjects, the tasks solved by 12 of the subjects show a significant correlation between time and grade (see Figure 2). These correlation

Figure 2—Average time and grade for each of the 83 subjects included in the data analysis. The horizontal line indicates the overall mean grade (2.87), the vertical line the overall mean time (25 min.). Subjects with a significant correlation between time and grade appear as squares, other subjects appear as triangles.

coefficients are all negative, suggesting that more time spent is correlated with lower grade (r_s between –0.758 and –0.453). For 86% of the subjects, time does not predict grade at all.

It is difficult to find a common denominator for the subjects where time and grade are correlated. The average time and grade of those subjects vary above and below the mean time and grade for subjects (see Figure 2). However, there is a significant difference between the grade for subjects with a significant correlation between time and grade and those without (Wilcoxon test, z=2.393, p<0.017). Subjects who obtain a correlation between time and grade did not use a specific retrieval mode for certain tasks (Chi-square test of which retrieval mode was first used, χ^2[4, N=12]=3.833, p>0.05).

Summary of Correlations between Usability Measures

Our analysis of the TeSS-experiment shows that efficiency (measured as task completion time) and effectiveness (measured as grade) are either not correlated or correlated so weakly that the correlation is negligible for all practical purposes. For the individual retrieval modes, a weak correlation is found for three of the modes, while two of the modes do not show any significant correlation between task completion time and grade. Task completion time and grade are not correlated for 85% of the tasks. Finally, only 14% of the subjects display a significant correlation between time and grade—for the large majority no correlation is found. These results and the previous results [6] concerning satisfaction and effectiveness (cf. the section Data Collection and Analysis, last paragraph) show that assumptions about correlations between effectiveness,

Figure 3—The usability aspects measured in the 19 studies of complex tasks from CHI '97 to CHI '99. Eight of these CHI-studies include measures of all three usability aspects, seven CHI-studies measure two aspects, and four CHI-studies only one aspect.

efficiency, and satisfaction do not seem to hold in the context of the TeSS-experiment.

CORRELATIONS BETWEEN ASPECTS OF USABILITY

We now extend the discussion of correlations between aspects of usability by including studies of computer support for complex tasks published in the CHI Proceedings for the years 1997-99. A total of 19 studies investigate aspects of usability in sufficient detail to enable an analysis of their choice of usability measures, see Figure 3. Eight (42%) of the 19 studies cover all three usability aspects. The other 11 studies, implicitly or explicitly, rely on assumptions of correlations between the different usability aspects, or seem confident that their choice of only one or two aspects of usability is sufficient to capture overall usability.

The only CHI-study with an analysis of correlations between the three aspects of usability

Of the eight studies including measures of all three usability aspects, only the study by Walker et al. [17] has analyzed the correlations between the aspects. Let us summarize their study, so the reader can see that the correlation analysis pays off.

Walker et al. compare two different designs of a spoken language interface to email: (a) a mixed-initiative dialogue, where the users can flexibly control the dialogue, and (b) a system-initiative dialogue, where the system controls the dialogue. The study measures effectiveness by qualitative measures such as automatic speech recognition rejects, efficiency by number of dialogue turns and task completion time, and user satisfaction by a multiple-choice survey. The results show that even though the mixed-initiative dialogue is more efficient, as measured by task completion time and number of turns, users prefer the system-initiative dialogue.

A correlation analysis with user satisfaction as the dependent variable uncovers how "…users' preferences are not determined by efficiency per se, as has been commonly assumed. One interpretation of our results is that users are more attuned to qualitative aspects of the interaction." [17, p. 587]. The number of automatic speech recognition rejects contributes the most to user satisfaction. Walker et al. suggest that the users' preference for the system-initiative dialogue arises from it being easier to learn and more predictable. This result was contrary to the authors' initial hypothesis and illustrates the importance of measuring efficiency, effectiveness, and satisfaction independently, as opposed to basing conclusions about one of them on measures of the others.

Two CHI-studies without any measure of effectiveness

Two CHI-studies concerning computer support for complex tasks, entitled "Time-compression: systems concerns, usage, and benefits" [13] and "Effects of awareness support on groupware usability" [5], do not include any measure of the quality of the outcome of the users' interaction with the system. Below we comment on these two studies, and show how their conclusions about overall usability are jeopardized by their incomplete choice of usability measures.

In the first study, Omoigui et al. [13] analyze how time-compression can be used to enable quick video browsing. An experimental time-compression system was used for comparing different granularities of the time-compression (discrete vs. continuous) and differences in the latency (long wait-time vs. no wait-time) experienced by users after adjusting the degree of time-compression. Omoigui et al. measure efficiency by savings in task time and the use of time-compression, and they measure satisfaction by, e.g., user feedback and preference indicated by usage of time-compression during video browse sessions. As already mentioned, no effectiveness measures were employed, although effectiveness could have been measured as the accuracy and completeness of the subjects' verbal summary of each video. In the concluding remarks, Omoigui et al. emphasize efficiency as the important aspect of time-compression systems: "Quite surprisingly though, there are no significant differences in the time-savings under the three conditions. Thus the implementers are free to choose the simplest solution..." [13, p. 142]. This conclusion neglects the satisfaction measures, which indicate that real differences might exist between the experimental conditions: "… several subjects commented in post-study debriefing that the long latency and discrete granularity conditions had affected their use of the time compression feature. The subjects felt that they made fewer adjustments and watched at a lower compression rate when long latency and discrete granularity were used." [13, p. 141]. An analysis of the correlations between the efficiency and satisfaction measures might have shed further light on the differences between conditions, as might solid measures of effectiveness.

In the second study, Gutwin and Greenberg [5] analyze whether enhanced support for workspace awareness improves collaboration. In an experiment, they compare users' performance on two real-time groupware systems where workspace miniatures were used to support workspace awareness. The basic miniature shows information only about the local user, the enhanced miniature about others in the workspace as well. Efficiency is measured by task completion time and communication efficiency; satisfaction is measured as preference for one or the other system. The correlations between the measures are not analyzed, and no measure of effectiveness is employed. The overall conclusion of the study is that workspace-awareness information reduces task completion time, and increases communicative efficiency and user satisfaction. The support for this conclusion is weak. For one out of the three task types, task completion time was not reduced. For two task types out of the three, the communicative efficiency was not increased. All 38 participants preferred the awareness-enhanced system, suggesting that the employed measures of usability are incomplete: "The overwhelming preference for the interface with the added awareness information also suggests that there were real differences in the experience of using the system, but that our measures were insensitive to these differences." [5, p. 517]. These differences might have been more explainable if the study had included measures of effectiveness, making possible an analysis of how users' preferences were affected by the quality of the outcome of their activities.

SELECTION OF USABILITY MEASURES

We believe that the weak correlation between effectiveness, efficiency, and satisfaction has three implications regarding the choice of measures in evaluations of system usability.

First, it is in general recommendable to measure efficiency, effectiveness as well as satisfaction. When researchers or developers use a narrower selection of usability measures for evaluating a system they either (a) make some implicit or explicit assumptions about relations between usability measures in the specific context, or (b) run the risk of ignoring important aspects of usability. In our analysis of the CHI-studies we have shown how interpretation of experimental data based on only one or two usability aspects leads to unreliable conclusions about overall usability. Given that the three usability aspects capture different constituents of usability—we have not seen arguments to the contrary for complex tasks—there is no substitute for including all three aspects in usability evaluations.

Second, at the moment no clear-cut advice can be given about which usability measures to use in a particular situation. On the contrary, identifying the usability measures that are critical in the particular situation should be recognized as a central part of any evaluation of system usability. This requires a firm understanding of how tasks, users, and technology interact in constituting the use situations within the particular application domain [10, 16]. The study by Su [15] is an illustrative example of the kind of work needed to distinguish and refine performance measures. Su investigated the correlation between 20 measures of information retrieval performance in an academic setting, and suggests a best single measure (the user's perception of the value of the search result as a whole) and best pairs of measures of information retrieval performance. Such work may lead to the development of reliable, domain-specific collections of critical performance measures. General descriptions of the relation between usability aspects [e.g. 12] will not aid the selection of usability measures, since there is no way of knowing in advance whether efficiency, effectiveness, and satisfaction are actually correlated in a particular situation.

Third, effectiveness measures oriented toward the outcome of the user's interaction with the system are gaining attention in usability evaluation [2], although two of the CHI-studies discussed earlier did not include such measures. The development of valid and reliable outcome measures is a prerequisite for assessing overall system usability and is necessary for working systematically with improving the usability of systems supporting users in solving complex tasks.

CONCLUSION

The relations between efficiency, effectiveness, and satisfaction—the three aspects of usability—are not well understood. We have analyzed data from a study of information retrieval and found only a weak correlation between measures of the three usability aspects. Other studies imply that for complex tasks in other domains, a similarly weak correlation between usability measures is to be expected. In general, we suggest that efficiency, effectiveness, and satisfaction should be considered independent aspects of usability, unless domain specific studies suggest otherwise.

Studies that employ measures of only a subset of the three usability aspects assume either that this subset is sufficient as an indicator of overall usability or that the selected measures are correlated with measures covering the other aspects of usability. As we have exemplified with an analysis of studies from previous CHI Proceedings, such assumptions are often unsupported. Hence, these studies jump to conclusions regarding overall usability while measuring, say, efficiency only. This is a problem for the HCI community, since more than half of the last three years of CHI-studies concerning complex tasks do not measure all aspects of usability.

Usability testing of computer systems for complex tasks should include measures of efficiency, effectiveness, and user satisfaction. In selecting these measures, the application domain and context of use have to be taken into account so as to uncover the measures that are critical in the particular situation. Discovering solid measures of effectiveness seems especially critical.

ACKNOWLEDGEMENTS

Morten Hertzum was supported by a grant from the Danish National Research Foundation. The design and implementation of TeSS as well as the design and execution of the experiment were done in collaboration with Jette Broløs, Marta Lárusdóttir, Kristian Pilgaard, and Flemming Sørensen. We wish to thank the students who participated in the experiment as subjects, and the CHI-reviewers. We are indebted to Per Settergren Sørensen for his support on statistical issues and to Peter Naur for many judicious proposals for clarification.

REFERENCES

1. Bailey, R.W. Performance vs. preference, in *Proceedings of the Human Factors and Ergonomics Society 37th Annual Meeting* (Seattle WA, October 1993), HFES, 282-285.

2. Bevan, N. Measuring usability as quality of use. *Software Quality Journal* 4 (1995), 115-150.

3. Card, S.K., Moran, T.P. and Newell, A. The keystroke-level model for user performance time with interactive systems. *Communications of the ACM* 23, 7 (1980), 396-410.

4. Cohen, J. and Cohen, P. *Applied Multiple Regression/Correlation Analysis for the Behavioral Sciences*. Lawrence Erlbaum Associates, Hillsdale NJ, 1975.

5. Gutwin, C. and Greenberg, S. Effects of awareness support on groupware usability, in *Proceedings of CHI '98* (Los Angeles CA, May 1998), ACM Press, 511-518.

6. Hertzum, M. and Frøkjær, E. Browsing and querying in online documentation: A study of user interfaces and the interaction process. *ACM Transactions on Computer-Human Interaction* 3, 2 (1996), 136-161.

7. ISO 9241-11. Ergonomic requirements for office work with visual display terminals (VDTs) - Part 11: Guidance on usability (1998).

8. Kirakowski, J. and Corbett, M. SUMI: The software usability measurement inventory. *British Journal of Educational Technology* 24, 3 (1993), 210-212.

9. Naur, P. Intuition in software development, in H. Ehring et al. *Formal Methods and Software Development*, Vol. 2, Lecture Notes in Computer Science 186, Springer, Berlin, 1985. Also in Naur, P. *Computing: A Human Activity*. ACM Press/Addison-Wesley, New York, 1992, 449-466.

10. Newman, W. and Taylor, A. Toward a methodology employing critical parameters to deliver performance improvements in interactive systems, in *Proceedings of INTERACT '99* (Edinburgh, August 1999), IOS Press, 605-612.

11. Nielsen, J. *Usability Engineering*, Academic Press, Boston, 1993.

12. Nielsen, J. and Levy, J. Measuring usability: Preference vs. performance, *Communications of the ACM* 37, 4 (1994), 66-75.

13. Omoigui, N., He, L., Gupta, A., Grudin, J. and Sanocki, E. Time-compression: Systems concerns, usage, and benefits, in *Proceedings of CHI '99* (Pittsburg PA, May 1999), ACM Press, 136-143.

14. Rasmussen, J. Skills, rules, and knowledge; signals, signs, and symbols, and other distinctions in human performance models. *IEEE Transactions on Systems, Man, and Cybernetics* SMC-13, 3 (1983), 257-266.

15. Su, L.T. Evaluation measures for interactive information retrieval. *Information Processing & Management* 28, 4 (1992), 503-516.

16. Van Welie, M., van der Veer, G.C. and Eliëns, A. Breaking Down Usability, in *Proceedings of INTERACT '99* (Edinburgh, August 1999), IOS Press, 613-620.

17. Walker, M.A., Fromer, J., Di Fabbrizio, G., Mestel, C. and Hindle, D. What can I say?: Evaluating a spoken language interface to email, in *Proceedings of CHI '98* (Los Angeles CA, May 1998), ACM Press, 582-589.

The Streamlined Cognitive Walkthrough Method, Working Around Social Constraints Encountered in a Software Development Company

Rick Spencer
Microsoft Corporation
One Microsoft Way
Redmond, WA 98052-6399 USA
(425) 936-1138
ricksp@microsoft.com

ABSTRACT

The cognitive walkthrough method described by Wharton et al. may be difficult to apply in a large software development company because of social constraints that exist in such companies. Managers, developers, and other team members are pressured for time, tend to lapse into lengthy design discussions, and are sometimes defensive about their user-interface designs. By enforcing four ground rules, explicitly defusing defensiveness, and streamlining the cognitive walkthrough method and data collection procedures, these social constraints can be overcome, and useful, valid data can be obtained. This paper describes a modified cognitive walkthrough process that accomplishes these goals, and has been applied in a large software development company.

Keywords
Cognitive Walkthrough, Usability Inspection

INTRODUCTION

The cognitive walkthrough (CW) method was designed to evaluate the learnability of software interfaces without the overhead of full-blown empirical usability lab testing. The CW can be applied early in the design process because it can be applied when only the user interface is specified. As a result, the CW method is valuable for evaluating learnability of the integration of features when those features are at various stages of development.

I have used the cognitive walkthrough method to evaluate software at a software company under the constraints of an actual software development cycle. This paper describes the problems I encountered applying the Wharton, et al. cognitive walkthrough method (WCW) [4] within these constraints, and outlines a modified CW process that works better. The effectiveness and validity of the modified method, is examined as well.

The CW method was used several times over the course of one software development cycle to evaluate user-interface specification for the latest release of an established Integrated Development Environment (IDE). An IDE allows programmers to edit code, design user interfaces, compile code, debug code, and perform most other programming tasks within a single computer program. An IDE, therefore, tends to be extremely feature rich, and requires a fairly large team, with distributed design responsibilities, to design and implement [1].

The design team for this product is large – approximately 50-100 people – and many different people have responsibility for specifying, designing, and implementing the UI. For many tasks that users will perform with the IDE, no one specification describes completely how that task will be accomplished. Rather, a set of specifications, each written and owned by a different person, must be understood and evaluated to understand how users will accomplish tasks with the IDE when it ships.

The CW method seemed quite well suited to evaluating the learnability of the IDE, since a single evaluation could tie together the work of several sub teams and the specifications for which they were responsible.

Wharton et al. proscribe performing a task analysis for the UI elements in question, and walking through the task analysis step by step with the team. For each step, the team attempts to tell a plausible story for each of four questions (See Table 1). For steps where plausible answers are generated by the team, the team records those stories, otherwise they record that a plausible story could not be told.

Table 1

<u>4 questions from Wharton et al. (1994)</u>

1. Will the user try to achieve the right effect?

2. Will the user notice that the correct action is available?

3. Will the user associate the correct action with the effect that user is trying to achieve?

4. If the correct action is performed, will the user see that progress is being made toward the solution of the task?

PROBLEMS APPLYING THE CW METHOD

In the first application of the CW method the cognitive walkthrough procedure detailed in Wharton et al [4] was followed. However, the method failed to produce good results and the development team did not perceive it as useful, at least in part because the WCW method did not accommodate some of the social constraints (SC) of a large software development company.

Table 2

Social Constraints That Hampered the Effectiveness of the CW Method

SC 1 – Time pressure

SC 2 – Lengthy design discussions

SC 3 – Design defensiveness

SC 1 – Time Pressure

Generally, many managers and developers feel pressure to make very good use of their time, and any activities perceived to take more time than justified by the results are avoided. When developing software, requirements, constraints, and changes are applied to a specification over time. As a result, numerous design iterations occur late in the development process, when a development team usually feels considerable pressure to actually implement specifications, and may not think they have the time to evaluate them properly.

Even user interface specialists often feel that there is not sufficient time in the software development cycle to perform their jobs well [1].

Voluminous output of WCW sessions

Following the procedure outlined in Wharton, et al., a plausible story is written down at every step where a plausible story is produced. As a result, the team writes down things like, "From previous experience, the user knows that the Print command is available in the File menu" and "From previous experience, the user knows that the Print command will activate the Print dialog."

Recording such obvious observations may not be perceived by the team as a good use of their limited time. Furthermore, what purpose such written comments will serve in the future may not be clear, and the ratio of useful written comments (problems, design ideas, and design gaps) will be diminished, diminishing the perception of usefulness for the CW exercise.

Finally, sorting the potential problems identified from the plausible stories may lengthen the time from the CW to when the problems are reported.

CW sessions seem to go too sowly

When doing a WCW the team answers four questions for each step. This results in the team dwelling on each step, even ones that are obviously designed correctly because they follow standard user-interface convention, or because that part of the user interface has been working for users over many previous versions. Furthermore, similar plausible stories are often repeated multiple times for the same step.

For steps with obvious problems, asking the four questions can seem redundant, especially if the team has had difficulty distinguishing between the first three questions and has brought up the issues pertinent to questions 2 and 3 while attempting to answer question 1.

Others have also found the CW method to be too slow. For instance, the Cognitive Jogthrough method was developed for the sole reason that one development team found that they were unable to cover enough material while using the CW method [3].

SC 2 - Lengthy design discussions

WHEN A design or user-interface problem is identified by a design team, that team will often attempt to resolve the problem "in-line" during the CW session. Time allotted to evaluation is spent designing instead.

SC 3 - Design Defensiveness

As surprising as it may seem, it turns out that not everyone who puts considerable effort into creating user interface specifications enjoys having those specifications publicly evaluated by others. Specification writers may appear personally offended by the suggestion that there specifications should undergo an evaluation process in the first place, because, after all, they may have been working on those specifications for many months, or even a year or more.

Since, in the short term, problems that are identified may result in more work for a team that could already be under considerable time pressure, some team members may try to defend their designs and specifications during the CW, may be argumentative, and may reject seemingly obvious observations as being opinions that lack data to support them.

CONDUCTING A STREAM-LINED COGNITIVE WALKTHROUGH

The three social constraints that limit the effectiveness of the WCW method have been addressed through a combination of approaches. Namely, properly preparing the team to perform a CW, modifications to the CW method itself, and strong leadership during the CW process to keep team members from dwelling on design discussions or defending their designs.

Table 3

Overview of the Cognitive Walkthrough process, adapted directly from Wharton, et al [4]

1. Define inputs to the walkthrough
    a. Identification of users
    b. Sample tasks for evaluation
    c. Action sequences for completing the tasks
    d. Description or implementation of interface
2. Convene the walkthrough
    a. Describe the goals of the walkthrough
    b. Describe what will be done during the CW
    c. Describe what will not be done during the walkthrough
    d. Explicitly defuse defensiveness
    e. Post ground rules in a visible place
    f. Assign roles
    g. Appeal for submission to leadership
3. Walkthrough the action sequences for each task
    a. Tell a credible story for these two questions:
    - Will the user know what to do at this step?
    - If the user does the right thing, will they know that they did the right thing, and are making progress towards their goal?
    b. Maintain control of the CW, enforce the ground rules
4. Record critical information
    a. Possible learnability problems
    b. Design ideas
    c. Design gaps
    d. Problems in the Task Analysis
5. Revise the interface to fix the problems

DETAILED DESCRIPTION OF THE STREAMLINED WALKTHROUGH PROCEDURE

1. Define inputs to walkthrough

Before the CW session, the usability professional is responsible for defining the important user task scenario or scenarios and producing a task analysis of those scenarios by explicating the action sequences necessary for accomplishing the tasks in the scenarios. Wharton et al. should be used as a resource for determining how to decide on the scenarios and how to describe the task sequence.

2. Convene the walkthrough

The first step is to describe the goals of performing the walkthrough. Namely, the walkthrough is an opportunity to evaluate the user interface in terms of learnability. This is the first opportunity to address SC 3 – Design defensiveness, by defusing defensiveness on the part of any team members. It is important that the usability professional points out that learnability is only one aspect of usability, and that the team recognizes that learnability may have been traded off for other aspects of usability. Nonetheless, there is inherent value in knowing when users may encounter problems learning an interface as the issue could be explicitly addressed elsewhere, for example, though marketing or the help system.

A CW session is analytical in nature, and therefore lacks the definitiveness of an empirical usability tests. In light of the CW method's tentative nature, specification owners may resent absolute proclamations that "this is a problem". The usability specialist should, therefore, take care to use softer language, like "this is a potential problem, we need to think about it". Constant reference to the tentative nature of the finding should help defuse defensiveness.

The usability professional then points out for the first time that the CW is an evaluation session, not a design session, and goes on to describe the process of walking through the task sequence and answering the two questions for each step (See table 4). The usability professional then gives an example of an action sequence from software not currently under consideration and that has plausible answers to the two questions, and the team is encouraged to produce those answers. Then the usability professional gives another example, one without plausible answers, and the team is prompted to try to provide answers. For each example, the usability professional should model the format that the data is captured in before proceeding with preparing the team for the CW.

Table 4

2 questions from the streamlined CW

1. Will the user know what to do at this step?
2. If the user does the right thing, will they know that they did the right thing, and are making progress towards their goal?

After describing what the team will do during the walkthrough, describe what the team will *not* do during the walkthrough. This is the first opportunity to directly address SC 2 – Lengthy design discussions, and indirectly address SC 3 – Design defensiveness. In particular, the usability professional explains that if the team finds a step with possible learnability issues, they will note the possible problem and move on to the next step, but they won't redesign the interface. Furthermore, the usability professional should explain that if the team encounters a gap in the design (for example when it is not clear from the specification what action sequence the user is supposed to perform), the team will note the gap and move on, but they won't stop and design the missing actions. Also, if a design idea is suggested, the team may briefly discuss the design idea, note it, and then move on, but the team will not flesh it out. Lastly, if the task analysis appears to be faulty, or only describes one of multiple possible paths towards achieving the goal, then the problem will be noted and the team will move on, but the task analysis will not be revised during the CW session.

After the team understands what will and will not be done during the walkthrough session, the usability engineer explicitly addresses SC 3 – Design defensiveness. Since the CW session is an evaluation session, and not a design session, no changes will be made during the CW session. If changes are going to be made to specifications, they will be made later. Therefore, if anyone feels that a specification needs to be defended, the time to do so is later. More importantly, defending designs or specifications during the CW is not productive and will distract the team from completing the evaluation.

After the CW process is explained and defensiveness defused, the usability professional should post the ground rules for conducting the walkthrough. These ground rules can then be referred to later in order to keep the team on track.

Table 5
Ground rules for conducting a streamlined CW
1. No designing
2. No defending a design
3. No debating cognitive theory
4. The usability specialist is the leader of the session

Ground rule 3 is a further effort to address SC 3 – Design defensiveness. In my experience, team members who feel that their designs are threatened may appeal to esoteric cognitive theories in order to justify their designs or to explain away a possible issue. The usability professional should state that as long as a significant number of team members feel there is a possible issue, it should be noted

Responsibility for collecting data from the walkthrough is distributed across the team participating in the walkthrough. Four kinds of data will be collected, design ideas, design gaps, potential learnability problems, and flaws in the task analysis. If the team participating in the CW is large enough, and the scope of the tasks under scrutiny cover multiple areas of user interface design, then different team members can be assigned the role of collecting potential learnability problems for different areas. Generally, team members can be assigned to collecting data on the areas for which they are responsible. Usually, one person is sufficient for collecting both design ideas and design gaps; however, if there are enough team members, then one person can collect design ideas and one person can collect design gaps.

It is important to explicitly assign a role for collecting flaws in the task analysis. This helps address SC 3 – Design defensiveness, by modeling a willingness to admit to and take responsibility for oversights and mistakes.

Lastly, the usability professional should explain that in order for the CW to proceed efficiently, the team must follow ground rule 4 and submit to the usability professional's leadership. Then an explicit appeal to submit to leadership should be made.

3. Walkthrough the action sequences for each task

The proposed modified CW severely prunes the evaluation procedure for each action sequence. For each action sequence, the usability professional first describes the action sequence and the system state after the correct action is performed. Then the team tries to answer the two questions for each action sequence.

Wharton et al.'s questions 1-3 evaluate whether the user will know what the next appropriate step is, and how to do it. For the streamlined CW, these three questions have been collapsed into one question, "Can you tell a credible story that the user will know what to do?" Question 4 is slightly recast, "If the user does the step correctly, and <describe system response>, is there a credible story to explain that they knew they did the right thing?

During the course of walking through the action sequences, the usability professional must take care to enforce the ground rules, making sure that the CW session does not lapse into a design session, that team members do not stop the process to defend their designs, and that the team does not get wrapped up in debates over cognitive issues.

4. Record critical information

If, for a particular action sequence, a plausible story is told for both questions, then nothing is recorded, and the team moves on to the next task. This helps address SC 1 – Time pressure, by spending minimal time on steps that appear to be properly designed.

Sometimes a plausible story cannot be told because the interface design assumes knowledge that users might not have. In such cases, record the failure and the knowledge that the user will need to formulate the goal, for instance, "Users might not click the ellipsis button to modify the list, because they might not know that they can modify the list."

Often, the team will not try to tell a plausible story, but instead will skip right to pointing out a design flaw. In this case, a bullet should be captured that represents the problem for the user and the design flaw. For example, "Users might not know that they can modify the auto generated code because it looks the same as read only code."

During a CW session, team members will likely bring design ideas that address problems encountered or suggest totally alternative designs. It is important to capture these design ideas because they may prove valuable later, and summarizing a design idea as a bullet is a good way to end design discussions and move on. For example, a design idea bullet may read "DI: Automatically generate the code for the user, instead of having the user issue the command." This also allows team members who raise design ideas to feel that their contribution was valued, while allowing the team to continue with the evaluation.

The usability professional is responsible for stopping a design discussion before it gets out of hand. When exactly to intercede may be affected by many factors. Generally, a good time to intercede and move on is as soon as a design idea is well-formed enough to be expressed in a bullet, but before other team members begin pointing out flaws in the design idea, or elaborating on it.

During the course of the CW session, the team may find that they forgot to design some important functionality. "How does the user do such and such a setting?" Such design gaps should be captured as a bullet, but definitely not designed during the CW session. If that gap is encountered again during the CW, the team must agree to hand wave over it, and assume that the ultimate design will support the functionality.

When a bullet is being captured, it is helpful for the usability professional to rephrase it in the form of a hypothesis. This allows the team to express consensus on the bullet that is being captured, as well as giving the person capturing that bullet a head start on writing it down properly.

If the usability professional made a mistake in the task analysis used in the CW, the problems in the analysis should be noted, and the team should move on. It is important to not try to retool the task analysis during the CW session. It is far better either skip the part that is wrong and cover it during a later session, or if the mistake was substantial, the usability professional should apologize for the misunderstanding, and reconvene the CW at a later time, after the task analysis is done properly.

Attempting to retool a task analysis in a few minutes is not likely to lead to a quality analysis. Major mistakes should be rare if the task analysis is checked for accuracy by the specification owners before the CW session.

IMPACT ON THE EFFECTIVENESS AND VALIDITY OF THE CW METHOD

Preparing the team, clearly laying out ground rules, and defusing defensiveness are steps that can be confidently added to the CW method without too much fear of decreasing the effectiveness or validity of the CW method. But what about the more radical changes, such as collapsing three questions into one question, disallowing design discussions, and capturing less data during the walkthrough? I will discuss these in turn, in order of probable severity in negative impact on the method.

Collapsing three questions into one

This modification to the CW method probably leads to a coarser-grained evaluation of the user interface under scrutiny. Logically, asking only half as many questions about an action sequence will probably lead to fewer problems identified for each step, presumably by both a function of time, less time is spent on each question, and a function of detail, each action sequence is examined in less detail.

Furthermore, when problems are encountered, the cause of the problem may not be revealed as effectively as in the WCW method. For example, the result of the first question from the streamlined CW might read, "Users might not know that the Print command will bring up the print dialog." However, the same datum from the WCW method might read, "Users might not associate the Print command with activating the Print Dialog, because the word 'Print' implies an action." The datum from the WCW seems to imply a cause and solution to the problem, where the streamlined method merely identifies the problem.

In practice, however, I have not found that the WCW does not lead to better data because the team is generally interested only in identifying problems and getting enough information to fix them. Furthermore, many people have trouble understanding the nuances of Wharton, et al.'s four questions [2].

Disallowing Design discussions

Given the best of all possible worlds, letting design discussion play out at the time that potential learnability problems are identified could lead to some very effective design sessions. When design discussions are blocked from the evaluation session, the usability professional is trading off identifying possible solutions to identified problems for coverage.

However, open-ended design sessions take an indeterminate amount of time, and may or may not result in a workable redesign. Worse stills, if the team redesigns as they proceed through the CW, then problems will be resolved in the order that they occur in the task sequence, with no necessary relationship to the severity of the problems. Time taken up by redesign may result in truncation of the CW because the team simply runs out of time. Important steps involving completing tasks may therefore be missed. In other words, redesigning during the CW violates a basic dictum of software design, to profile *before* you optimize.

However, this bias against design discussions during CW sessions is not universally shared. In fact, Rowley and Rhoades specifically created the Cognitive Jogthrough method because they found that a WCW did not allow *enough* time for design discussions [3]!

Capturing Less Data

Using the streamlined method, the plausible stories are not captured. Though this will clearly speed the process, it means that design rationale is not captured. Later in the process, if there are questions about the data from the CW, it may be difficult to remember why a particular step was considered to be acceptable by the team. Furthermore, the design rational will not be available later in the design process.

EFFECTIVENESS OF THE STREAMLINED CW

The streamlined CW method was used to evaluate the learnability of an IDE under development. Specifically, the task of performing a series of programming tasks necessary to create and deploy the basics of a one kind of computer program was evaluated. The programming task was considered to be one of the most important programming tasks for the version of the IDE under development, and it incorporated functionality described in multiple specifications. At the time of the CW, the necessary functionality was months from being sufficiently well implemented for traditional usability lab testing.

Methods

The task analysis took one usability professional with a background in task analysis about 25 hours to complete. Since the specifications for the task were distributed across many documents, and not all of the documents were up to date, most of the 25 hours was spent researching specifications and interviewing program managers responsible for designing various parts of the user interface.

A team of approximately 8 people was convened to conduct the CW, including 3 usability specialists, 1 graphic designer, and 4 project managers responsible for various aspects of the user-interface specification. Only one usability professional leading the session was familiar with CW methods. The walkthrough itself took about 2.5 hours, and was conducted over 2 sessions, separated in time by about a week. Only the results from the first session, which took about 1.5 hours and covered 32 action sequences, are discussed here. About 20 minutes of the first session were used to prepare the team, assign roles, and defuse defensiveness. Since the task analysis spanned many specifications, most team members were not familiar with all of the action sequences. Therefore, many minutes were required to explain the user action and the system response for each of the 32 action sequences covered.

Results

Twenty-four potential problems and 11 design ideas were identified during the first 1.5-hour session. Of the 24 potential problems generated during the CW, 14 suggested that users would lack sufficient knowledge to take the correct action, and 10 suggested that the IDE did not provide good feedback to the user when the correct action was taken.. Six of the 11 design ideas were specific solutions to one or more of the potential issues.

These results are consistent with the hypothesis that the streamlined CW method trades granularity for coverage. The 10 potential usability problems identified that lacked an explicit cause also did not suggest an explicit solution. In other words, for many of the identified problems, the team agreed that they were potential problems, but the cause of the problems were attributed only to a lack of knowledge on the part of users, and not necessarily to a mismatch between the users' knowledge and the user interface. Had the team considered each of the first three questions from the Wharton method, it is possible that a better understanding of the causes of the problems would have surfaced.

Generally, the efforts to defuse defensiveness on the part of the team members were successful, as the team did not spend much time defending design decisions, and an atmosphere of cooperation seemed to prevail. When program managers found themselves defending their design, they tended to remember ground rule 2, drop the discussion, and allow the CW to continue. After the CW session, the team members expressed that it was a useful exercise, and in fact many CW sessions have been conducted since.

EXTERNAL VALIDITY

The difficulty in assessing the external validity of any usability inspection method extends, of course, to the streamlined CW as well. Ideally, the results of the CW session would be evaluated against the results of an empirical usability test, where the usability specialists conducting the test were not aware of the CW results. Naturally, such a luxury is not afforded within the scope of this effort. However, an opportunity to at least roughly assess the streamlined CW method did present itself.

Methods

Usability tests on the IDE have been conducted, and one of the usability tests did include some overlap with the material covered in the CW session. While the usability specialist who conducted the usability test was present at the CW session, he was not the usability specialist who led it. Fortunately, design changes suggested from the results of the CW sessions had not yet been implemented in the area of the IDE user interface that was being tested.

Results

Nine problems were discovered in the usability test of the UI covered in the CW session. Of those 9 problems, the CW identified 6. However, for the same user-interface elements, the usability test covered more functionality than the CW session did. For example, in most of the action sequences in the CW session, the user could accept default values to be successful, but in the user test participants needed to adjust those defaults to be successful.

This comparison suggests that, a team can expect to uncover with a streamlined CW many issues that would also be uncovered in an empirical usability test.

For the portions of the user interface covered by both the CW session and the usability test, 13 potential problems were predicted by the CW sessions. Of those, 7 were directly related to findings in the usability report, 4 were indirectly related to problems in the usability report, and 2 were not related to problems in the usability report.

The result of this comparison suggests that, for those parts of the user interface covered by a streamlined CW session, the team can expect to hit a few false positives, get a sense of which steps may cause some problems for users, and accurately predict many learnability problems.

CONCLUSION

The Wharton, et al cognitive walkthrough method does not take into account several social realities of large software companies. The method can be applied successfully if the usability specialist takes care to properly prepare the team for the walkthrough, avoids design discussions during the walkthrough, explicitly defuses defensiveness among team members, and streamlines the procedure by collapsing the first three questions into one question, and captures data selectively.

Streamlining the walkthrough may trade-off granularity for coverage, but without that trade off, program managers and developers may perceive the walkthrough as being an inefficient use of time. Performing a streamlined CW is a good way to profile a user interface for potential problem areas, identify many steps that may be problematic for users, and accurately predict many usability problems. However the method will probably result in a few false positives.

REFERENCES

1. Grudin, J. Systemic sources of suboptimal interface design in large product development organizations. *Human-Computer Interactions 6*, 2 (1991), 147-196.

2. John, B. E, & Packer, H. Learning and using the cognitive walkthrough method: A case study approach, in *Proceedings of CHI '95* (Denver CO, May 1995), ACM Press, 429-436.

3. Rowley, D. E., and Rhoades, D. G. The cognitive jogthrough: A fast-paced user interface evaluation procedure. *Proceedings of CHI '92* (May 1992), ACM Press, 389-395.

4. Wharton, C. W., Reiman, J., Lewis, C. & Polson, P. (1994) The cognitive walkthrough method: A practitioner's guide. In J. K. Nielson, & R. L. Mack (Eds.) Usability Inspection Methods. Wiley, New York, 1994.

Visual Similarity of Pen Gestures

A. Chris Long, Jr., James A. Landay, Lawrence A. Rowe, and Joseph Michiels

Department of Electrical Engineering and Computer Science
University of California at Berkeley
Berkeley, CA 94720-1776
{allanl,landay,rowe,cujoe}@cs.berkeley.edu
+1-510-643-7106
http://www.cs.berkeley.edu/~{allanl,landay,rowe}

ABSTRACT

Pen-based user interfaces are becoming ever more popular. Gestures (i.e., marks made with a pen to invoke a command) are a valuable aspect of pen-based UIs, but they also have drawbacks. The challenge in designing good gestures is to make them easy for people to learn and remember. With the goal of better gesture design, we performed a pair of experiments to determine why users find gestures similar. From these experiments, we have derived a computational model for predicting perceived gesture similarity that correlates 0.56 with observation. We will incorporate the results of these experiments into a gesture design tool, which will aid the pen-based UI designer in creating gesture sets that are easier to learn and more memorable.

Keywords

Pen-based user interfaces, pen gestures, multi-dimensional scaling, similarity, perception

INTRODUCTION

Pen and paper is a versatile, powerful, and ubiquitous technology [17]. Pen-based user interfaces are becoming more widespread [9] and have great promise in power and versatility [4, 8, 17, 25]. Gestures, or commands issued with a pen, are one desirable feature of pen-based UIs. Because command and operand can be specified in one stroke, they are fast [5]. They also are commonly used and iconic,[1] which makes them easier to remember than textual commands [19]. Gestures are useful on displays ranging from the very small, where screen space is at a premium, to the very large, where controls can be more than arm's reach away [20].

A survey of PDA users [14] showed that users think gestures are powerful, efficient, and convenient. They want more gestures in applications and the ability to define their own gestures. However, the survey also revealed problems with gestures. Specifically, users often find gestures difficult to remember, and they become frustrated when the computer misrecognizes gestures. Users of other systems have also found gestures to be awkward [7].

---

1. By "iconic", we mean that the gesture shape suits or suggests its meaning.

Permission to make digital or hard copies of all or part of this work for personal or classroom use is granted without fee provided that copies are not made or distributed for profit or commercial advantage and that copies bear this notice and the full citation on the first page. To copy otherwise, to republish, to post on servers or to redistribute to lists, requires prior specific permission and/or a fee.
CHI '2000 The Hague, Amsterdam
Copyright ACM 2000 1-58113-216-6/00/04...$5.00

We believe gestures can be difficult to use because they are difficult to design. We are developing a tool to assist pen-based UI designers in creating and evaluating gestures for pen-based UIs [15]. The primary benefit of the tool will be to advise designers about how to improve their gesture set. For example, it will notify the designer of gestures that: 1) are likely to be perceived as similar by users, 2) may be difficult for users to learn and remember, and 3) may be misrecognized by the computer. This advice will enable the designer to improve their gestures early in the design process before investing in expensive user studies.

The current work is an investigation into gesture similarity. The goal of this work is to develop a computable, quantitative model of gesture similarity that can be incorporated into the gesture design tool. Perceived similarity is useful for designers to know because it affects how easily users can learn and remember the gestures. We contend that similar operations with a clear spatial mapping, such as scroll up and scroll down, should be assigned similar gestures. Conversely, gestures for more abstract operations that are similar, such as cut and paste, may be easily confused if they are visually similar.

To determine what features affect perceived gesture similarity, we ran a pair of experiments to measure the similarity of a variety of gestures. The data collected in these experiments enabled us to derive an algorithm for computing how similar novel gestures are. In conjunction with information about the impact of gesture similarity on their ease of learning and recall (from another experiment whose results are being analyzed), this will allow our gesture design tool to provide better advice.

The remainder of the paper is organized as follows. The first section discusses related work. The next section describes the gesture similarity experiments. A discussion of the results of both experiments follows. Finally, we present future work and conclusions.

RELATED WORK

This section discusses relevant prior work. The first subsection gives some background on pen-based user interfaces, which is the context for this work. The second discusses prior work on perceptual similarity of gesture-like shapes. The last section introduces multi-dimensional scaling, a data analysis technique used in our experiments.

Pen-based interfaces

The device that popularized pen input was the Apple Newton MessagePad. It was designed primarily for pen input. It minimized the use of overlapping windows and encouraged the user to focus on one document at a time. Its core applications were a notepad, to-do list, calendar/scheduler, and address book. By default it recognized text as it was entered, but the user could elect to enter ink and recognize it later. It also supported a small number of gestures for editing text and drawings. Our work deals with the same type of gesture used on the Newton: single-stroke and iconic. The Newton's handwriting recognition was widely criticized when it was first introduced, but by the last model the recognition had greatly improved.

More recently, the 3Com PalmPilot has become a popular pen-based platform. Its display is even smaller than the Newton's. Its applications have very few on-screen controls, and the pulldown menu at the top of the screen is normally hidden. Its core applications are the same as the Newton's. The Pilot does not recognize normal English letters, but uses the Graffiti character set [12], which must be entered in a dedicated area of the screen. It does not use gestures of the kind we focus on, but instead uses a keyboard accelerator facility. The user may write a special "command" stroke to indicate that the next stroke is a command. For example, in many applications "command" followed by the "d" stroke invokes the delete operation.

Pens have been used in many different computer applications, on desktop- and on wall-sized displays. These applications have included spreadsheets, word processors, a disk manager, music editors, an equation editor, a GUI design tool, an air-traffic control UI, and note-taking applications [4, 6, 7, 10, 18, 25, 27].

Pens performed well for many of these applications. In their standard office applications, Briggs and associates found that users liked using a pen for navigation and positional control, although not for text entry [4]. This is in spite of a lack of pen-specific interaction techniques.

Wolf and colleagues added pen input and gestures for editing operations to a drawing application, a spreadsheet program, a music editor, and an equation editor [25]. Users reported that gestures were easier to recall than keyboard commands, and edited spreadsheet documents 30% faster with the pen than with keyboard.

Zhao used a combination of gestures and menus to facilitate object creation in a structured drawing program [27]. It allowed users to draw the object creation gesture and select from the object type menu in either order. This technique could allow rapid access to many commands while limiting the number of gestures that the user is required to learn.

Landay developed an interface design tool based on sketching that was well-suited for pen input and gestures [10]. It used iconic gestures of the same type as the ones in our study for creating and editing the UI elements.

Chatty and Lecoanet discussed how pen input and gestures are useful for air-traffic control [6]. Their interface allowed controllers to navigate the airspace and change aircraft attributes such as speed and heading. An evaluation of their system showed that although a few gestures had to be modified because of confusion, they were still useful.

Lopresti and Tomkins advocated treating electronic ink as a first class datatype [16]. They developed a prototype system that supported ink input and searching based on matching feature vectors. This technique was used to search a library of gestures using the gesture as the key.

We concentrate on gestures in the spirit of copy editing [13, 22] rather than marking menus [23], because we believe that traditional marks are more useful in some circumstances. For example, they can specify operands at the same time as the operation, and they can be iconic. For simplicity of analysis and because of restrictions in our gesture recognizer, we use only single-stroke gestures.

Perceptual similarity

Psychologists have investigated similarity of simple geometric shapes, which are less complex than our gestures. Attneave studied how changes in geometric and perceptual properties of different kinds of simple figures influenced their perceived similarity [2]. Participants in one experiment reported how similar they perceived parallelograms of differing sizes and tilts to be. Attneave found that similarity was correlated with the log of the area of the parallelograms and with their tilt. Also, parallelograms that were mirror images of one another were perceived as similar.

Another study by Attneave indirectly measured perceived similarity by measuring how easily names of triangles and squares were remembered. The assumption was that similar shapes will be misremembered. The squares varied in reflectance and area; the triangles varied in tilt and area. The result of these experiments indicated that similarity of form caused more confusion than similarity of area.

Based on these experiments, Attneave concluded the following. In general, the logarithm of quantitative metrics was found to correlate with similarity. Also, if the range of differences in stimuli is small, these differences are linearly related to perceived similarity, and when multiple dimensions of the stimuli change, their dimensions combine nearly linearly to give the change in perceived similarity. When the range of differences is large, the relationship between stimuli value and similarity is not linear, and the stimuli do not combine linearly.

Lazarte and colleagues studied how rectangle height and width affected perceived similarity [11]. They found that reported similarity was related to rectangle width and height and they derived a model to fit the reported similarity data. Also, they found that not only did different people use different similarity metrics, but that the same participant may have used different metrics for different stimuli.

Multi-dimensional scaling

Multi-dimensional scaling (MDS) is a technique for reducing the number of dimensions of a data set so that patterns can be more easily seen by viewing a plot of the data, usually in two or three dimensions. It takes as input one or more sets of pairwise proximity measurements of the stimuli. It outputs coordinates and/or a plot of the stimuli in a predetermined number of dimensions (typically 2–3) such that the pairwise inter-stimuli distances in the new space correlate with the input proximities of the stimuli.

There are several decisions to make in using MDS. One is how to use data from multiple participants. A simple method is to average the pairwise proximities and analyze the resulting proximity matrix as if it came from a single participant. However, there is evidence that this method does not give good results [1], and it also prohibits analyzing the differences among participants. Fortunately, we were able to use a version of MDS, INDSCAL [26], that takes as input a proximity matrix for each participant and takes individual differences into account.

Another decision when using MDS is how many dimensions to use in the analysis. Like other MDS methods, INDSCAL outputs how well the distances it produces correlate with the input proximities. A graph of this correlation vs. dimension ideally has an obvious "knee" in the graph, which indicates the number of dimensions to use.

Also, a rule of thumb for standard MDS is to use no more than a quarter as many dimensions as there are stimuli.[2]

How to measure distance is another issue in doing MDS analysis. The most often used metric is Euclidean distance, which is a special case of the Minkowski distance metric:

$$d_{ij}^p = \sum_a |x_{ia} - x_{ja}|^p$$

where d is the distance, there are r dimensions, and x_{ia} is the coordinate of point i on dimension a [26]. When p is 2, this is Euclidean distance. Another common p value is 1, in which case it is called city-block or Manhattan distance. Infinity is also sometimes used for p, which makes the sum equal to the distance along the dimension that differs most. There are sometimes psychological reasons to prefer city-block or Euclidean [24], but generally researchers use the metric that fits their data best.

The final step in MDS analysis is assigning meaning to the axes. Sometimes the experimenter may know the axes in advance. In this work that was not the case, and so two methods were used to determine the axes: 1) inspecting plots of the stimuli and 2) correlation with measurable quantities. More details of MDS analysis are available in [26].

GESTURE SIMILARITY EXPERIMENTS

To better understand the principles people use to judge gesture similarity, we performed an experiment to measure how people perceived similarity among gestures in a predefined set. We hoped that the experiment would enable us to derive metrics for predicting the human-perceived similarity of gestures.

We ran this experiment twice with different data sets and different subjects in order to confirm our results. The two trials are described below.

Similarity Trial 1

In the first experiment, we attempted to make a gesture set consisting of gestures that varied widely in terms of how people would perceive them. The gesture set is shown in Figure 1. It was designed by one of the authors (Long) based on personal intuition with two criteria in mind: 1) to span a wide range of possible gesture types, and 2) to have differences in orientation (e.g., 14 and 15, 20 and 22).

We considered whether the participants should draw the gestures or not. Drawing them would mimic actual usage more closely, but it would have lengthened the experiment. In order to accommodate more participants and more gestures, we decided not to have participants draw the gestures. Instead, the test program animated the gestures to show participants the dynamic nature of the gestures.

Participants

We recruited twenty-one people from the general student population to participate in the experiment. We only required that they be able to operate the computer and tablet. Each participant was paid $10 (US).

---

2. ALSCAL uses more information than standard MDS, so it is reasonable to think that more dimensions would be valuable. Unfortunately, we were unable to find an analysis of how many.

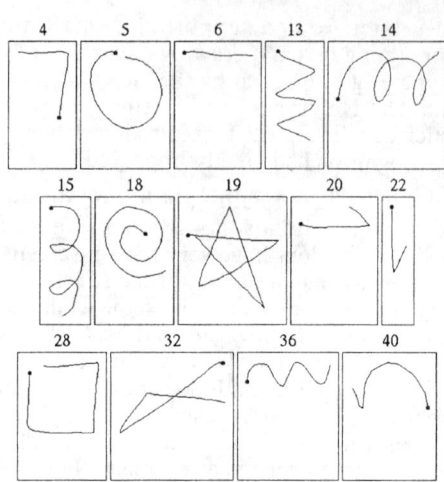

Figure 1 Gesture set for first gesture similarity experiment. A dot indicates where each gesture starts. (Gestures are not numbered 1, 2, 3... because they were chosen from a larger set.)

Equipment

For the main part of the experiment participants used a display tablet (a Mutoh MVT-12P) attached to an IBM PC compatible computer running Windows NT. The experimental application was written in Java.

Procedure

Participants were first shown an overview of the experiment, which outlined the tasks they would perform. Participants were then shown the tablet and the task was explained to them. The program displayed all possible combinations of three gestures, called triads. The order of the triads was randomized independently for each participant, as was the on-screen ordering of gestures within each triad. Figure 2 is a representative screen shot of the triad program. For each triad, participants selected the gesture that seemed most different from the others by tapping on it with a pen. The program recorded the selections of the users and computed the dissimilarity matrix.

The program was run using a practice gesture set of five gestures, shown in Figure 3, so that participants could become familiar with the program and the tablet. Participants were asked to select the gesture in each triad that seemed most different to them. After the practice, they performed the task again using the experimental gesture set of fourteen gestures (Figure 1). All participants saw all

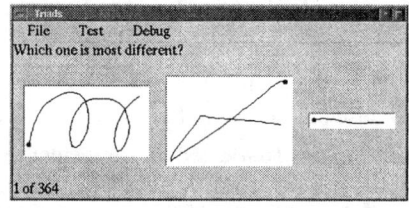

Figure 2 Triad program.

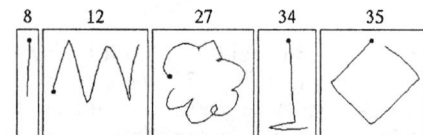

Figure 3 Practice gesture set.

possible triads of gestures exactly once, for a total of $\binom{14}{3} = 364$ triads.

When the experimental task was completed, participants filled out a questionnaire which asked: 1) their impressions of the task and program and 2) demographic information about themselves. This questionnaire was a web form that was filled out on a different computer than the one used for the experimental task [21].

Analysis

The goals of the analysis were: 1) to determine what measurable geometric properties of the gestures influenced their perceived similarity and 2) to produce a model of gesture similarity that, given two gestures, could predict how similar people would perceive those two gestures to be.

The first goal was addressed using plots of gestures generated by MDS. In these plots, the Euclidean inter-gesture distances corresponded to the inter-gesture dissimilarities reported by the participants. By examining these plots, we were able to determine some geometric features that contributed to similarity. To determine the number of dimensions to use, we did MDS in two through six dimensions and examined plots of stress and goodness-of-fit (r^2) versus dimension to find the "knee" in the curve[3]. Similarity data was analyzed with MDS as interval/ratio and as ordinal. The ordinal model gave a better fit, so it was used for all subsequent analysis. We used Euclidean distances since it provided a better fit to our data than the city-block metric did.

The second goal was achieved by running regression analysis to determine which of many measurable geometric features of a gesture correlated with the reported similarity. Regression also produced weights indicating how much each feature contributed to the similarity. To compute the similarity of two gestures, their feature values are computed. The feature values and weights together give the positions of the gestures in feature space. The similarity of the gestures is given by the Euclidean distance between the two gestures in the feature space, where smaller distance means greater similarity.

The features used in our regression analysis came from a few sources. Some features were taken from Rubine's gesture recognizer [22]. Others were inspired by plots from the MDS analysis. The list of features that we thought might predict similarity is given in Table 1. We wanted our model to be computable, so we did not include in the final regression analysis features whose values were only obtainable by subjective judgement.

Results

We were able to derive a model of gesture similarity that correlated 0.74 ($p < 0.003$, 2-tailed t-test) with the reported gesture similarities.

The multi-dimensional scaling indicated that the optimal number of dimensions was five (S-stress = .13, $r^2 = 0.76$). For ease of comprehension, we plotted the gesture positions two dimensions at a time. Examination of the plot of dimensions 1 and 2 (shown in Figure 4) quickly showed that dimension 1 was strongly correlated with how "curvy" the gestures were — for example, g5 and g40 are curvy and g32 and g28 are straight. The curviness metric was derived in an

---

3. Examination of stress vs. dimension and r^2 vs. dimension is a standard MDS technique for determining dimensionality.

1. **Cosine of initial angle**
2. **Sine of initial angle**
3. Size of bounding box
4. **Angle of bounding box**
5. Distance between first and last points
6. **Cosine of angle between first and last points**
7. Sine of angle between first and last points
8. Total length
9. Total angle
10. Total absolute angle
11. Sharpness
12. **Aspect [abs(45° – #4)]**
13. **Curviness**
14. Total angle traversed / total length
15. Density metric 1 [#8 / #5]
16. Density metric 2 [#8 / #3]
17. Non-subjective "openness" [#5 / #3]
18. Area of bounding box
19. Log(area)
20. Total angle / total absolute angle
21. **Log(total length)**
22. **Log(aspect)**

Table 1 Possible predictors for similarity. Features 1-11 are taken from Rubine's recognizer. Bold features were found to be significant and so are used in the model.

attempt to capture our intuitive notion of curviness and to match this dimension from the MDS plot.[4]

It was observed in the MDS plots that short, wide gestures were perceived as being very similar to narrow, tall ones and that both types were perceived as different from square gestures. Angle of bounding box represented the difference between thin and square gestures, but not the similarity of tall vertical and short horizontal ones. We created the aspect feature to represent this relationship.

Table 2 shows which features strongly correlate with each dimension, based on a regression analysis. Although the most important (i.e., lower numbered) dimensions are predicted by relatively few features, the other dimensions require many features.

A separate regression analysis was done for each dimension, using the computed feature values as the independent variables. From these regressions we derived a set of equations to predict the position of an arbitrary gesture in the feature space. Given the predicted positions of two gestures, the degree of similarity humans would

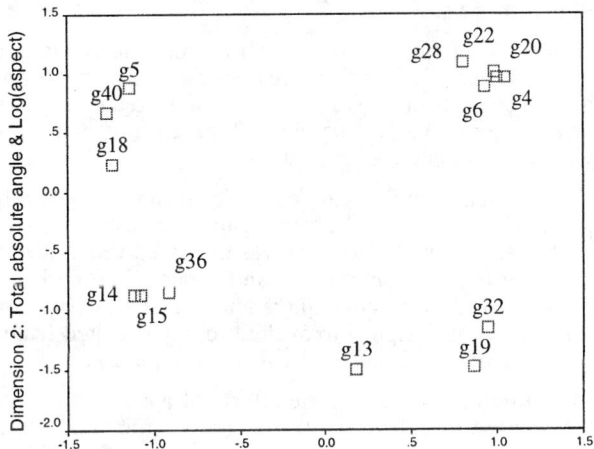

Figure 4 MDS plot of dimensions 1 and 2 for first similarity experiment.

---

4. Curviness of a gesture was computed by adding up all inter-segment angles within the gesture whose absolute value was below a threshold (19°). The threshold was chosen so that the metric would agree with the author's curviness judgements of gestures in trial 1.

Dimension	Correlated features (in order of decreasing importance)
1	Curviness, Angle / distance
2	Total absolute angle, Log(aspect)
3	Density 1, Cosine of initial angle
4	Cosine of angle between first and last points, Cosine of initial angle, Sine of initial angle, Distance between first and last points, Angle of bounding box
5	Aspect, Sharpness, Cosine of initial angle, Total angle

Table 2 Predictor features for similarity trial 1, listed in decreasing order of importance for each dimension.

perceive is approximated by the Euclidean distance between the gestures in the feature space. The derived model predicts the reported gesture similarities with correlation 0.74 ($p < 0.003$, 2-tailed t-test). The MDS model upon which it is based fits the data only slightly better, so this is a good fit.

Another interesting aspect of trial 1 was the differences among participants. As expected, the degree that different features affected similarity judgements varied across participants. This disparity is consistent with the finding in other perception experiments that different people judge similarity using different features [11]. What was surprising was that the participants seemed to be clumped into two distinct groups. We separated the data for the two groups of participants and analyzed them separately. However, the resulting MDS models were not as good as the original, combined model, so we did not pursue the analysis further. It would be interesting to see if more participants reinforced this trend and illuminated a pattern.

Participants took an average of 26 minutes to complete the experimental task. The total time for each participant was approximately 40 minutes.

Similarity Trial 2

The results of the first similarity trial were encouraging, but we wanted to test the predictive power of our model for new people and gestures. We also were interested in exploring how systematically varying different types of features would affect perceived similarity.

To investigate how varying particular features would affect perceived similarity, three new gesture sets of nine gestures each were created. The first was designed to explore the effect of total absolute angle and aspect (Figure 5). The second was designed to explore length and area (Figure 6). The third was designed to explore rotation-related features such as cosine and sine of initial angle (Figure 7).

In addition to examining the effects of particular features, we wanted to determine the relative importance of the features. The most straightforward way to perform this test is to combine all gestures into one big set and have participants look at all possible triads from the combined set. Unfortunately, combining all of these gesture sets into one results in far too many triads, based on the time per triad taken for the first experiment. To allow us to compare the three sets against one another without a prohibitively large gesture set, two gestures from each of the three gesture sets were chosen and added to a fourth set of gestures (see Figure 8). All participants were shown all possible triads from all four gesture sets.

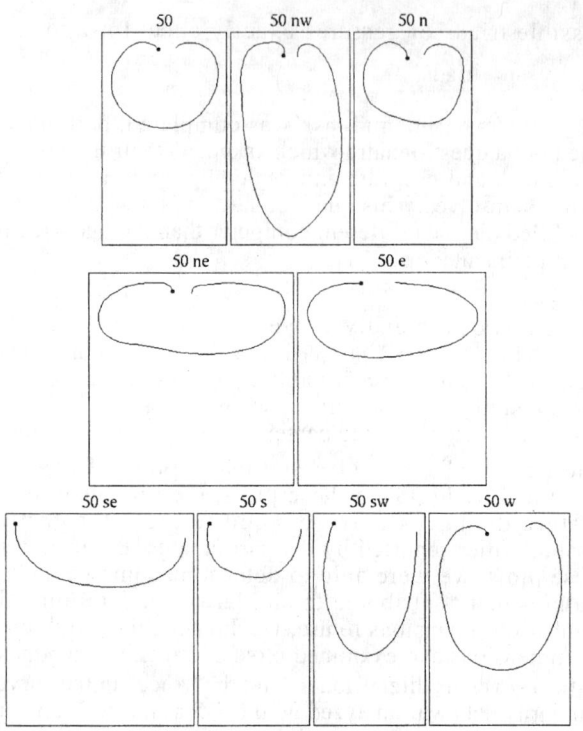

Figure 5 Similarity trial 2, gesture set 1. It was used to explore absolute angle and aspect.

Participants
Twenty new participants were recruited from the general student population. As in trial one, the only requirement was that they be physically able to use the computer and stylus. Each participant was paid $15.

Equipment
The same equipment was used as in the first trial.

Procedure
The procedure was the same as in trial one, except that participants saw triads from four gesture sets rather than one. Each participant saw all possible triads of gestures from each set, for a total of $3\binom{9}{3}+\binom{13}{3} = 538$ triads.

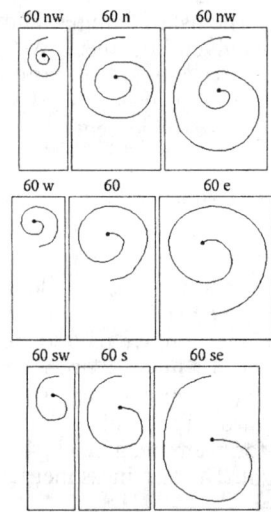

Figure 6 Similarity trial 2, gesture set 2. It was used to explore length and area.

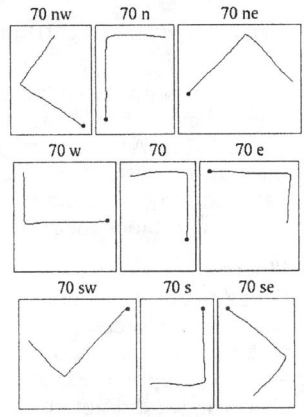

Figure 7 Similarity trial 2, gesture set 3. It was used to explore rotation.

Analysis

This trial was analyzed using the same techniques as the first trial, MDS and regression. First, a combined analysis was done, using the data from all four gesture sets. The goal of the combined analysis was the same as trial one: to determine what features were used for similarity judgements and to derive a model for predicting similarity. Many pairwise dissimilarity measures were missing from the data, because not all possible triads of all gestures were presented to the participants. However, this was not a problem because MDS can accommodate missing data.

In addition to the combined analysis, data from each of the first three sets was analyzed independently. The focus of the independent analyses was to determine how the targeted features affected similarity judgements.

Lastly, the results of trial 1 were used to predict the perceived similarity of the gestures in trial 2. These

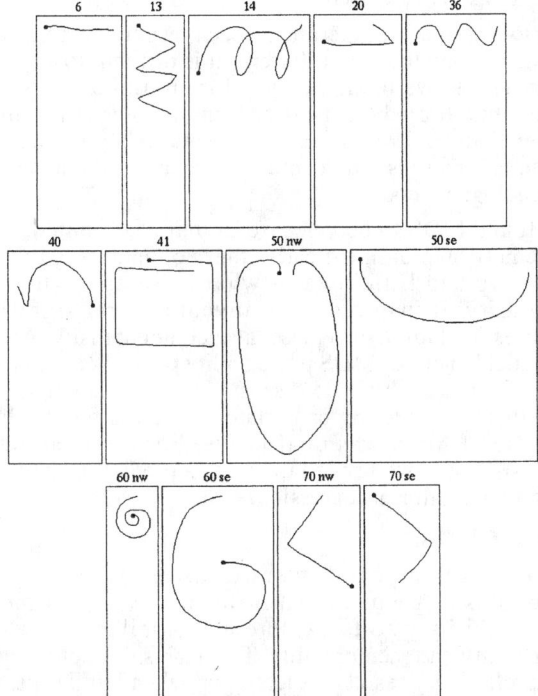

Figure 8 Similarity trial 2, gesture set 4. It includes gestures from Figure 5, Figure 6, and Figure 7.

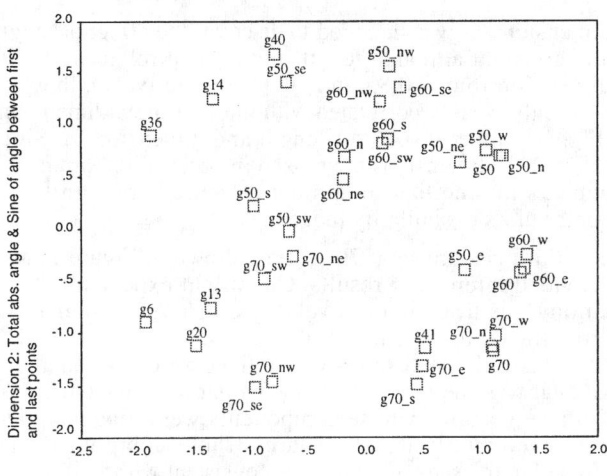

Figure 9 MDS plot of dimensions 1 and 2 for second trial in similarity experiment (combined data).

predictions were compared with the similarities reported by participants in trial 2.

Results

The best number of dimensions for MDS for the trial 2 data was 3 (S-stress = 0.08663, r^2 = 0.89539). Unfortunately, when the data was plotted, the meaning of the dimensions was not as obvious as in trial 1. (A plot of dimensions 1 and 2 is shown in Figure 9).

Table 3 shows which features correlate with each dimension, based on a regression analysis. The derived model predicts the reported gesture similarities with correlation 0.71 ($p < 0.000003$, 2-tailed t-test). Separate analyses of individual gesture sets (shown in Figures 5, 6, and 7) revealed that: 1) bounding box angle is an important feature and 2) alignment or non-alignment with the normal coordinate axes is significant for similarity.

Analysis of the first gesture set ("50" series, shown in Figure 5) gave a three dimensional MDS plot. This gesture set was intended to show the effects of absolute angle and aspect. We found that absolute angle is highly correlated with the first dimension (−0.81) and aspect is highly correlated with the second (−0.77). Unfortunately, the absolute angles of gestures in this set covaried greatly with the values of several other features, so it was not possible to determine whether absolute angle was significant. Strong covariance with other features was not a problem for aspect. However, bounding box angle correlated even more strongly with dimension two (−0.92) than aspect did.

Data from the second gesture set ("60" series, shown in Figure 6) were surprising. Its analysis was done in four

Dimension	Correlated features (in order of decreasing importance)
1	Log(aspect), density 1
2	Total absolute angle, Sine of angle between first & last points
3	Density 2, Non-subjective openness

Table 3 Predictor features for similarity trial 2 (using data from *all* trial 2 gesture sets).

dimensions. It was intended to discover the effect of length and area, but although length and area correlate well with dimension four (–0.83 and –0.92, respectively), they are both only weakly correlated with the first three dimensions (highest value of 0.46 for length and dimension 3). Since INDSCAL dimensions are ranked in order of importance, it appears that neither length nor area are very significant contributors to similarity judgement.

The third gesture set ("70" series, shown in Figure 7) also provided interesting results. One might expect similarity among gestures that are rotations of one another to be proportional to the amount of rotation, but this was not the case. Instead, the gestures whose lines were horizontal and vertical were perceived as more similar to one another than to those gestures whose components were diagonal. The perceived similarity of gestures whose components are aligned in the same directions is consistent with findings on texture in the vision community [3]. This set was analyzed in five dimensions.

As in trial one, there were differences among participants in their similarity judgements. Unlike trial one, the participants did not separate easily into two groups, but consisted instead of one clump with outliers trailing off. We experimented with removing outliers from consideration, but they did not appreciably improve the MDS model.

Similarity analysis

To validate the models produced by the two trials, each model was used to predict the similarities between all pairs of gestures used in the other trial. These predictions were compared with the reported similarities from the other trial. The correlation between the prediction of trial 1 and the data from trial 2 was 0.56 ($p < 0.0005$, 2-tailed t-test). The correlation between the prediction of trial 2 and the data from trial 1 was 0.51 ($p < 0.058$, 2-tailed t-test). Based on these correlations, the model derived from experiment 1 is a slightly better predictor of gesture similarity than the model from experiment 2.

DISCUSSION

This section discusses the results of the experiments described above and the challenges involved in designing the experiments and analyzing their results.

Results

Human perception of similarity is very complicated, even for simple shapes [11]. Shapes like pen gestures can be viewed as similar or dissimilar based on many different perceptual cues. In the face of this difficulty, we are pleased at how well our model predicts similarity. It correlates with its own data 0.74 ($p < 0.003$, 2-tailed t-test) and with novel (i.e., trial two) data 0.56 ($p < 0.0005$, 2-tailed t-test).

We were pleased to find that a small number of features explain the three most salient dimensions. In experiment 1, we saw that dimensions one through three can be predicted based on only two features each. Several possible explanations exist for the larger number of features needed for dimensions four and five. One is that the underlying perceptual model is complex. Another is that the gesture set used in the experiment was not complex enough or did not vary in the right way to illuminate those dimensions.

It was surprising to us that neither length nor area were significant factors in experiment 1, so the "60" series (Figure 6) in trial 2 was designed to investigate the effect of these two features. Trial 2 confirmed the results of trial 1; neither length nor area was a significant feature for similarity.

Our results are consistent with Attneave's [2]. We found that the logarithm of aspect had more influence on similarity than aspect itself. Also, the range of distances among feature values of our gestures was large and did not combine linearly (as shown by the better fit of the Euclidean distance metric over the city-block metric).

Design and Analysis

The primary challenge in designing both the similarity and memorability experiments was creating good stimuli (i.e., gesture sets). For the first similarity experiment, we wanted the stimuli to span the perceptual feature space. However, this was difficult because we did not know the structure of the perceptual feature space in advance. We culled gestures from an informal survey of colleagues and from another experiment [15] in an attempt to create a "well-rounded" gesture set.

For the second similarity trial, we wanted gestures that varied with respect to particular features. Our gesture design tool [15] was modified to display values for these features, but the process was still difficult. In particular, some of the features we wanted to investigate covaried with other features, which made the results difficult to interpret.

We were concerned at the outset that developing a model for similarity would be complicated by differences among the participants. However, in spite of the individual differences the model does have predictive power. Although analyzing different groups of participants separately was not useful for our data, more data might make it feasible to create multiple models, each of which models a subset of users well. In that case, a gesture design tool could use multiple similarity metrics and notify the designer about similar gestures along any metric. The designer may want two gestures to be similar or dissimilar, depending on the semantics of the operations they are used for.

The two similarity experiments each resulted in a model for similarity, and they are different. It is difficult to say which is better, but we think the model from trial one is slightly preferable. It predicts the data from the other trial slightly better than trial two predicts its data. Also, it uses more features, and thus may capture more about the underlying psychological model.

We found MDS to be very useful, but also limited. It was extremely helpful in the early stages of analyzing trial one, when we had little idea of what features might affect similarity. It inspired us to invent several significant features, including curviness, aspect, and density. Another potential benefit of MDS is the ability to analyze differences in participants, which are discussed above. Although it was useful for discovering candidate predictors for similarity, our use of MDS was qualitative. For the quantitative analysis, when we needed to create a predictive model, we used standard linear regression.

FUTURE WORK

We have run an experiment to investigate geometric properties of gestures that influence learning time and memorability, and to explore how similarity relates to learnability and memorability. The analysis is not complete, but preliminary results indicate that when similar gestures are used for similar operations, they are easier to remember.

We plan to incorporate these discoveries about gesture similarity into a tool for gesture set design that we are developing [15]. Gestures are an important part of pen-based UIs, and we believe that designers of pen-based UIs could greatly benefit from a gesture design tool that informed them of gestures that may be difficult for the computer to recognize or for people to learn and remember.

These experiments also suggest avenues for more psychological experiments. For example, it would be interesting to measure memorability and similarity of other gestures to validate and/or refine our current models. Also, in the two experiments described here, participants saw animated gestures but did not draw them. It is possible that different similarity criteria would emerge if participants drew the candidate gestures before judging their similarity.

CONCLUSIONS

Gesture set designers may want their gestures to be similar or dissimilar depending on the semantics of the operations. We have shown that perceptual similarity of gestures is correlated with well-defined computable features such as curviness. With these features, we have derived a computable, quantitative model for perceptual similarity of gestures that correlates 0.56 with reported similarity. Using our model, we can predict how similar people will perceive gestures to be. We expect similarity predictions to be a useful addition to our gesture design tool.

Our model and our experiences of experimental design and analysis should provide an excellent starting point for further investigation into gesture similarity, memorability, and learnability. When integrated with our gesture design tool, our model will allow designers to create gestures that are more memorable and learnable by users.

ACKNOWLEDGMENTS

The authors would like to thank Richard Ivry for his invaluable advice on experimental design and analysis, and the experimental participants for their time.

REFERENCES

[1] Ashby, F. G., Maddox, W. T., and Lee, W. W. On the dangers of averaging across subjects when using multidimensional scaling or the similarity-choice model. *Psychological Science*, May 1994. 5, 3, 144–151.

[2] Attneave, F. Dimensions of similarity. *American Journal of Psychology*, 1950. 63, 516–556.

[3] Beck, J., Prazdny, K., and Rosenfeld, A. A theory of textural segmentation. In *Human and Machine Vision* (Beck, J., Hope, B., and Rosenfeld, A., eds.), vol. 8 of *Notes and Reports in Computer Science and Applied Mathematics*. Academic Press, New York, NY, 1983 1–38. Proceedings of the Conference on Human and Machine Vision, Aug. 1981.

[4] Briggs, R., Dennis, A., Beck, B., and Nunamaker, Jr., J. Whither the pen-based interface? *Journal of Management Information Systems*, 1992-1993. 9, 3, 71–90.

[5] Buxton, W. There's more to interaction than meets the eye: Some issues in manual input. In *User Centered System Design: New Perspectives on Human-Computer Interaction* (Norman, D. A. and Draper, S. W., eds.). Lawrence Erlbaum Associates, Hillsdale, N.J, 1986 319–337.

[6] Chatty, S. and Lecoanet, P. Pen computing for air traffic control. In *Human Factors in Computing Systems (SIGCHI Proceedings)*. ACM, Addison-Wesley, Apr. 1996 87–94.

[7] Forsberg, A., Dieterich, M., and Zeleznik, R. The music notepad. In *Proceedings of the ACM Symposium on User Interface and Software Technology (UIST)*. ACM, ACM Press, New York, NY, Nov. 1998 203–210.

[8] Frankish, C., Hull, R., and Morgan, P. Recognition accuracy and user acceptance of pen interfaces. In *Human Factors in Computing Systems (SIGCHI Proceedings)*. ACM, Addison-Wesley, Apr. 1995 503–510.

[9] International Data Corporation. Smart handheld device market will enjoy robust growth in 1999 and 2000, May 1999. Available at http://www.idc.com/Data/Personal/content/PS052699PR.htm.

[10] Landay, J. and Myers, B. Interactive sketching for the early stages of user interface design. In *Human Factors in Computing Systems (SIGCHI Proceedings)*. ACM, Addison-Wesley, Apr. 1995 43–50.

[11] Lazarte, A. A. and Schonemann, P. H. Saliency metric for subadditive dissimilarity judgments on rectangles. *Perception and Psychophysics*, Feb. 1991. 49, 2, 142–158.

[12] Lee, Y. L. PDA users can express themselves with Graffiti. *InfoWorld*, Oct 3 1994. 16, 40, 30.

[13] Lipscomb, J. A trainable gesture recognizer. *Pattern Recognition*, Sep. 1991. 24, 9, 895–907.

[14] Long, Jr., A. C., Landay, J. A., and Rowe, L. A. PDA and gesture use in practice: Insights for designers of pen-based user interfaces. Tech. Rep. UCB//CSD-97-976, U.C. Berkeley, 1997. Available at http://bmrc.berkeley.edu/papers/1997/142/142.html.

[15] Long, Jr., A. C., Landay, J. A., and Rowe, L. A. Implications for a gesture design tool. In *Human Factors in Computing Systems (SIGCHI Proceedings)*. ACM, ACM Press, May 1999 40–47.

[16] Lopresti, D. and Tomkins, A. Computing in the ink domain. In *International Conference on Human-Computer Interaction* (Anzai, Y., Ogawa, K., and Mori, H., eds.), vol. 1 of *Advances in Human Factors/Ergonomics*. Information Processing Society of Japan and others, Elsevier Science, Jul. 1995 543–548.

[17] Meyer, A. Pen computing. *SIGCHI Bulletin*, Jul. 1995. 27, 3, 46–90.

[18] Moran, T. et al. Implicit structures for pen-based systems within a freeform interaction paradigm. In *Human Factors in Computing Systems (SIGCHI Proceedings)*. ACM, Addison-Wesley, Apr. 1995 487–494.

[19] Morrel-Samuels, P. Clarifying the distinction between lexical and gestural commands. *International Journal of Man-Machine Studies*, 1990. 32, 581–590.

[20] Pier, K. and Landay, J. A. Issues for location-independent interfaces. Tech. Rep. ISTL92-4, Xerox Palo Alto Research Center, Dec 1992.

[21] Reeves, B. and Nass, C. *The media equation: how people treat computers, television, and new media like real people and places*. Center for the Study of Language and Information; Cambridge University Press, Stanford, Calif.: Cambridge [England]; New York, 1996.

[22] Rubine, D. Specifying gestures by example. In *Computer Graphics*. ACM SIGGRAPH, Addison Wesley, Jul. 1991 329–337.

[23] Tapia, M. and Kurtenbach, G. Some design refinements and principles on the appearance and behavior of marking menus. In *Proceedings of the ACM Symposium on User Interface and Software Technology (UIST)*. ACM, Nov. 1995 189–195.

[24] Thomas, H. Spatial models and multidimensional scaling of random shapes. *American Journal of Psychology*, 1968. 81, 4, 551–558.

[25] Wolf, C., Rhyne, J., and Ellozy, H. The paper-like interface. In *Designing and Using Human-Computer Interfaces and Knowledge Based Systems* (Salvendy, G. and Smith, M., eds.), vol. 12B of *Advances in Human Factors/Ergonomics*. Elsevier, Sep. 1989 494–501.

[26] Young, F. W. *Multidimensional Scaling: History, Theory, and Applications*. Lawrence Erlbaum Associates, Hillsdale, NJ, 1987.

[27] Zhao, R., Kaufmann, H.-J., Kern, T., and Müller, W. Pen-based interfaces in engineering environments. In *International Conference on Human-Computer Interaction* (Anzai, Y., Ogawa, K., and Mori, H., eds.), vol. 20B of *Advances in Human Factors/Ergonomics*. Information Processing Society of Japan and others, Elsevier Science, Jul. 1995 531–536.

Providing Integrated Toolkit-Level Support for Ambiguity in Recognition-Based Interfaces

Jennifer Mankoff[1]   **Scott E. Hudson**[2]   **Gregory D. Abowd**[1]

[1] College of Computing & GVU Center
Georgia Institute of Technology
Atlanta, GA 30332-0280, USA
{jmankoff, abowd}@cc.gatech.edu

[2] Human-Computer Interaction Institute
Carnegie Mellon University
Pittsburgh, PA 15213, USA
hudson@cs.cmu.edu

ABSTRACT

Interfaces based on recognition technologies are used extensively in both the commercial and research worlds. But recognizers are still error-prone, and this results in human performance problems, brittle dialogues, and other barriers to acceptance and utility of recognition systems. Interface techniques specialized to recognition systems can help reduce the burden of recognition errors, but building these interfaces depends on knowledge about the ambiguity inherent in recognition. We have extended a user interface toolkit in order to model and to provide structured support for ambiguity at the input event level. This makes it possible to build re-usable interface components for resolving ambiguity and dealing with recognition errors. These interfaces can help to reduce the negative effects of recognition errors. By providing these components at a toolkit level, we make it easier for application writers to provide good support for error handling. Further, with this robust support, we are able to explore new types of interfaces for resolving a more varied range of ambiguity.

Keywords
Recognition-based interfaces, ambiguous input, toolkits, input models, interaction techniques, pen-based interfaces, speech recognition, recognition errors.

INTRODUCTION

Recognition technologies have made great strides in recent years. By providing support for more natural forms of communication, recognition can make computers more accessible. Recognition is particularly useful in settings where a keyboard and mouse are not available, such as very large or very small displays, and mobile and ubiquitous computing.

However, systems that make use of recognition technology still labor under a serious burden. Recognizers are error-prone—they make mistakes. This can confuse the user, cause human performance problems, and result in brittle interaction dialogues which can fail due to seemingly small recognition errors. In addition to simple mis-recognition, the inherent ambiguity of language can lead to errors. Ambiguity arises when there is more than one possible way to interpret the user's input. Even a perfect recognizer cannot always eliminate ambiguity, because human language is inherently ambiguous. For example, ambiguity arises in determining the antecedent for words like 'he,' and 'it.'

When humans communicate with computers using speech, gesture, or handwriting, recognition errors can reduce the effectiveness of an otherwise natural input medium. For example, Suhm found that the effective speed of spoken input to computers was only 30 words per minute (wpm) because of recognition errors, even though humans speak at 120 wpm [25]. Similarly, some of the slow success of the PDA market has been attributed to error-prone recognizers.

Although we cannot eliminate ambiguity we can build interfaces which reduce its negative effects on users. For example, Suhm found that user satisfaction and input speed both increased when he added support for multimodal error handling to a speech dictation system [25]. McGee et al. found that they were able to reduce ambiguity in a multimodal system by combining results from different recognizers [21]. In the pen and speech communities, the pros and cons of n-best lists and repetition [4,26,1,9], as well as secondary input modes (such as soft keyboards) [20,3], have been investigated.

Interfaces which help the user deal with ambiguity depend on having knowledge about the ambiguity resulting from the recognition process. Yet existing user interface toolkits have no way to model ambiguity, much less expose it to the interface components, nor do they provide explicit support for resolving ambiguity. Instead, it is often up to the application developer to gather the information needed by the recognizer, invoke the recognizer, handle the results of recognition, and decide what to do about any ambiguity.

Our primary innovation is to model ambiguity explicitly in the user interface toolkit, thus making ambiguity accessible to interface components. Our toolkit-level solution makes it possible to build re-usable components

---

Permission to make digital or hard copies of all or part of this work for personal or classroom use is granted without fee provided that copies are not made or distributed for profit or commercial advantage and that copies bear this notice and the full citation on the first page. To copy otherwise, to republish, to post on servers or to redistribute to lists, requires prior specific permission and/or a fee.
CHI '2000 The Hague, Amsterdam
Copyright ACM 2000 1-58113-216-6/00/04...$5.00

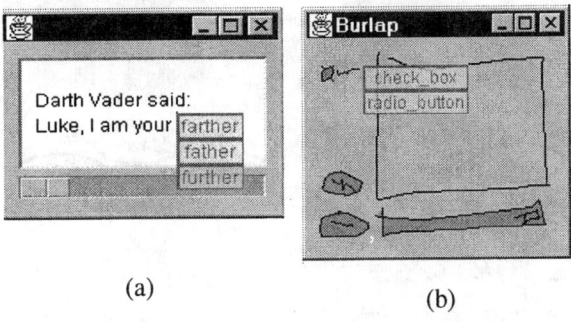

Figure 1: Two examples of a menu of possible interpretations of the user's input. (a) The user is dictating text. (b) The user is drawing a button in Burlap, a user interface sketching tool based on SILK[16].

and strategies for resolving ambiguity. It also allows us to give the user feedback about how their input is being interpreted before we know for sure exactly which interpretation is correct. At the same time, by carefully separating and structuring different tasks within the toolkit, neither application nor interface need necessarily know anything about recognition or ambiguity in order to use recognized input, or to make use of predefined interface components for resolving ambiguity.

Our secondary innovation is to support ambiguity and recognition at the input level, so that objects receive recognition results through the same mechanism as mouse and keyboard events. This allows the application to make use of recognized input without having to deal with it directly or to rewrite any interface components. Our approach differs from other toolkits that support recognized input such as Amulet [17] and various multimodal toolkits [21, 23]. These toolkits provide varying amounts of support for ambiguity, but require the application writer to deal with recognition results directly.

For example, consider a word-prediction application [2]. As the user types each character, the system tries to predict what words may come next. Our toolkit's input system automatically delivers any text generated by the word predictor to the current text focus, just as it would do with keyboard input. The text focus may not even know that a word predictor is involved. If there is more than one possible completion (the text is ambiguous), the toolkit will maintain the relationship between the possible alternatives and any interactive components using that input. The toolkit will automatically pass the ambiguous input to the *mediation* subsystem. This subsystem decides when and how to resolve ambiguity. If necessary, it will involve the user via interfaces that deal with mediation, called *mediators*. For example, it may bring up an *n*-best list, or menu, of alternatives from which the user can select the correct choice (see Figure 1a).

We have extended the user interface toolkit subArctic [8,14] to provide explicit access to ambiguity at an input level, and to provide support for resolving ambiguity through the mediation subsystem. In particular, we have populated the toolkit with some generic mediators that handle a wide variety of types of input, output, and ambiguity.

In order to demonstrate the effectiveness of our ambiguity-aware user interface toolkit, we have built Burlap, a simplified version of the sketch-based user interface builder, SILK (Sketching Interfaces Like Krazy [16]). Burlap, like SILK, allows the user to sketch interactive components (or what we will call *interactors*) such as buttons, scrollbars, and checkboxes on the screen. Figure 1b shows a sample user interface sketched in Burlap. The user can click on the sketched buttons, move the sketched scrollbar, and so on. In this example, the user has drawn a radiobutton, which is easily confused with a checkbox. The system has brought up a default mediator (the same as is shown in Figure 1a), asking the user which interpretation is correct.

The next section describes how ambiguous input is modeled at the toolkit level. We then discuss how ambiguity is handled at the user interface level, using Burlap as a demonstration medium. Burlap illustrates several types of ambiguity, makes use of mediators provided by default in the toolkit, and also demonstrates the benefits of being able to extend these mediators to take advantage of application-specific information. The following section shows how our toolkit-level mechanisms for hiding ambiguity from the application reduce the burden of writing applications that use recognition. We conclude with a description of the variety of recognizers and mediators used in Burlap and some other very simple applications we have built.

MODELING AMBIGUOUS INPUT

Existing user interface toolkits handle input from traditional input sources by breaking it into a stream of autonomous input events. Most toolkits determine a series of eligible objects that may *consume* (use) each event. The toolkit will normally *dispatch* (deliver) an event to each such eligible object in series until the event is consumed. After that, the event is not dispatched to any further objects. Variations on this basic scheme show up in a wide range of user interface toolkits.

In order to extend this model to handle recognized input, we first need to allow the notion of an input source to be more inclusive than in most modern user-interface toolkits. Traditionally only keyboard and mouse (or pen) input is supported. We allow input from any source that can deliver a stream of individual events. In particular, we have made use of input from a speech recognizer (each time text is recognized a new event is generated) and from a context toolkit [24]. The context toolkit handles work of

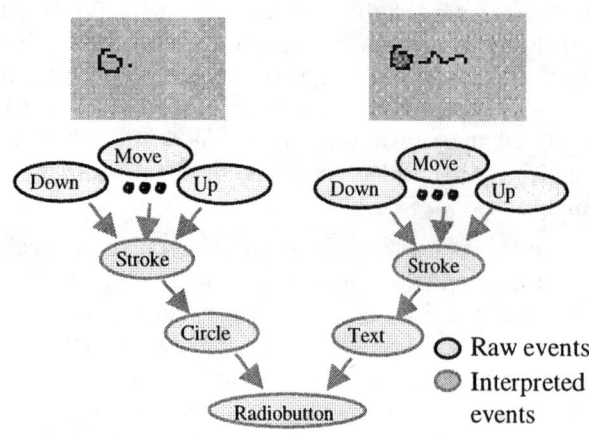

Figure 2: Two mouse strokes and the resulting hierarchical events

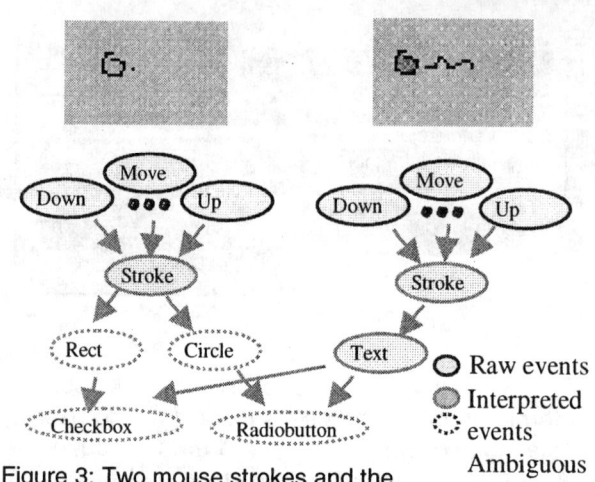

Figure 3: Two mouse strokes and the resulting hierarchical **ambiguous** events

gathering information from sensors and interpreting, or abstracting it to provide information such as the identity of users in a certain location (e.g., near the interface).

The next step in supporting recognized input requires us to track explicitly the interpretation and use of input events. This is done using hierarchical events, which have been used in the past in a number of systems (see for example Green's survey [10] and the work by Myers and Kosbie [22]). Hierarchical events contain information about traditional input events (mouse or pen and keyboard) are used, or in the case of recognition, *interpreted*. They can be represented as a directed acyclic graph. Nodes in the graph represent events. Arrows into a node show what events it was derived from. Arrows out of a node show what events were derived from it. Figure 2 shows what this graph looks like when the user draws a radiobutton in Burlap. This graph allows us to explicitly model the relationship between raw events (such as mouse events), intermediate values (such as strokes), derived values (such as recognition results) and what the application does with those values (such as create a radiobutton).

Hierarchical events are generated by requiring each event consumer to return an object containing information about how it interpreted the consumed event [22]. For example, the circle in Figure 2 was created by a gesture recognizer that consumed the stroke the circle derived from and returned the circle as its interpretation of that stroke. Each derived interpretation becomes a new node in the graph. Nodes are dealt with just like raw input events: they are dispatched to consumers, which may derive new interpretations, and so on.

Hierarchical events can also be used to model ambiguity explicitly. Ambiguity arises whenever two events are in conflict (both cannot be correct). For example, the gesture recognizer from Figure 2 may return two possible interpretations (a rectangle and a circle), only one of which can be correct. Figure 3 shows the hierarchical events from Figure 2 with the addition of recognizer-generated ambiguity. The circle, the rectangle, and any interpretation of either are ambiguous.

Just as in the unambiguous case, an ambiguous hierarchy is generated as events are dispatched and consumed. However, unlike most existing systems, we dispatch an event to each consumer in sequence even after the event has been consumed. If an event is consumed by more than one consumer, or a consumer generates multiple interpretations, ambiguity results.

We dispatch every new event (node in the graph) *even if it is ambiguous*. This is critical, because it allows us to put off the process of resolving ambiguity while giving the user early feedback about how events will be used if they are chosen. Remember that ambiguity arises when events are in conflict. We resolve ambiguity by *accepting* one of the possible conflicting events, and *rejecting* the others. The accepted event is correct if it is the interpretation that the user intended the system to make. It is wrong otherwise. It is the job of the mediation subsystem (described in the next section) to resolve ambiguity by deciding which events to accept or reject.

Since we dispatch events even when they are ambiguous, consumers are expected to treat events as tentative, and to provide feedback about them with the expectation that they may be rejected. In particular, we require that feedback related to events be separated from the application actions resulting from those events. By separating feedback about events from action on those events, we can support more flexible strategies for resolving ambiguity. (As described later, a mechanism is also provided for existing interactors which are not ambiguity-aware to operate without this separation, and hence without change).

When ambiguity arises, it is resolved by the mediation subsystem of our extended toolkit. The mediation subsystem consists of a set of objects, called *mediators*, representing different mediation strategies. These mediator objects, may ignore ambiguity (of types they are not designed to handle), directly resolve ambiguity, or defer resolution of ambiguity until more information is available (or they are forced to make a choice). The ability to defer decisions when appropriate is important to providing robust interaction with ambiguous inputs.

HANDLING AMBIGUITY IN THE INTERFACE

As stated above, the correct interpretation is defined by the user's intentions. It is often appropriate for mediators to resolve ambiguity at the interface level by asking the user which interpretation is correct. For example, the menu of choices shown in Figures 1a and 1b were created by a re-usable mediator that is asking the user to indicate the correct interpretation of their input. This type of mediator is called an *n-best list*, and it is a very common interface technique in both speech and handwriting, as documented in our survey on interfaces for error handling in recognition systems [18]. When the user selects a choice from the list, she is telling the system which of several possible interpretations of her input should be accepted. The system, in turn, informs any object which consumed that input which alternative was accepted.

An *n*-best list is a specific instance of a mediation strategy uncovered by our survey called a *multiple alternative display*. This class of interface techniques gives the user feedback about more than one potential interpretation of her input. In recent work focussing specifically on support for this class of techniques, we identified several dimensions of multiple alternative displays [19] including layout, instantiation time, additional context, interaction, and feedback. For example, in the Pegasus drawing beautification system, user input is recognized as lines. These lines are simply displayed in the location they will appear if selected, rather than converted into a text format and displayed in a menu.

A second common mediation strategy, also discussed in our survey, is *repetition*. A simple type of repetition occurs when automatic mediation results in an error, and the user undoes the error and repeats her input. Alternatively, the user may simply edit some portion of her input. Repetition may take place in the same input mode as the original input, or in a mode with orthogonal or fewer errors. For example, Suhm found that an effective strategy for correcting spoken dictation was to simply edit the resulting text with a pen [25].

One limitation of multiple alternative displays is that they often do not allow the user to specify an interpretation not in the list of alternatives. This can be addressed by combining multiple alternative displays with repetition strategies. Alternatively, when recognition involves graphical objects such as lines or selections, it is enough to simply make the multiple alternative display more interactive. For example, the corner of a selection could be draggable.

Although the examples given above are limited to systems with graphical output, it is also possible to do mediation in other media. For example, we have built a speech I/O *n*-best list style mediator that is equivalent to the mediator in Figure 1b. When the user clicks on an ambiguous button, it asks "Is that a radiobutton?." The user can give one of several answers including "yes" or "it's a checkbox."

Two important design heuristics affect the usability of both types of mediators. The first is to provide sensible defaults. If the top choice is probably right, it should be selected by default so that the user does not have to do extra work. On the other hand, if the recognition process is very ambiguous, such as word prediction, the right default is to do nothing until the user selects an alternative.

The second heuristic is to be lazy. There is no reason to make choices (which may result in errors) until necessary. Later input may provide information that can further disambiguate things. Some input may not need to be disambiguated. Perhaps it is intended solely as a comment or an annotation. For example, in Burlap, there is no need to disambiguate a sketch until the user actually tries to interact with it. Its presence as a drawing on the screen serves a purpose though while ambiguous.

Multiple alternative displays and repetition represent two very general mediation strategies within which there are a huge number of design choices. The decision of when and how to involve the user in mediation can be complex and application-specific. Horvitz's work in decision theory provides a structured mechanism for making decisions about when and how to do mediation [12]. In contrast, our work focuses on supporting a wide range of possible interface strategies.

Modeling Different Types of Ambiguity

The user interface techniques for mediation uncovered in our survey dealt exclusively with one common type of ambiguity, which we call *recognition ambiguity*. However, other types of ambiguity can lead to errors that could be reduced by involving the user. In addition to recognition ambiguity, we often see examples *target ambiguity*, and *segmentation ambiguity*.

Recognition ambiguity results when a recognizer returns more than one possible interpretation of the user's input. For example, Burlap depends upon a recognizer that can turn strokes into interactors. The recognizer is error-prone: If the user draws a rectangle on the screen followed by the symbol for text, the recognizer may mistake the rectangle for a circle (an error), resulting in a radiobutton instead of a checkbox. It is also ambiguous, because it returns multiple guesses in the hope that at least one of them is right and can be selected by mediation. This type of ambiguity is called *recognition ambiguity* (See Figure 1b for an example). In our model of hierarchical ambiguous events, recognition ambiguity is represented by

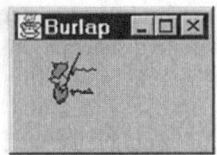

Figure 4—Target ambiguity: Which radiobutton did the user intend to check?

an event with multiple conflicting derivations. Figure 3 shows a case of this.

Target ambiguity arises when the target of the user's input is unclear. For example, the checkmark in Figure 4 crosses two radiobuttons. Which should be selected? A classic example from the world of multimodal computing involves target ambiguity: If the user of a multimodal systems says, *'put that there,'* what does *'that'* and *'there'* refer to? Like recognition ambiguity, target ambiguity results in multiple interpretations being derived from the same source. In the case of recognition ambiguity, one event consumer (a recognizer) generates multiple interpretations. In the case of target ambiguity, the interpretations of an event are generated by multiple potential consumers.

Our third type of ambiguity, segmentation ambiguity, arises when there is more than one possible way to group input events. Did the user really intend the rectangle and squiggle to be grouped as a radiobutton, or was she perhaps intending to draw two separate interactors--a button, and a label? If she draws more than one radiobutton, as in Figure 5, is she intending them to be grouped so that only one can be selected at a time, or to be separate? Should new radiobuttons be added to the original group, or should a new group be created? Similarly, if she writes **'a r o u n d'** does she mean **'a round'** or **'around'**? Most systems provide little or no feedback to users about segmentation ambiguity even though it has a significant effect on the final recognition results.

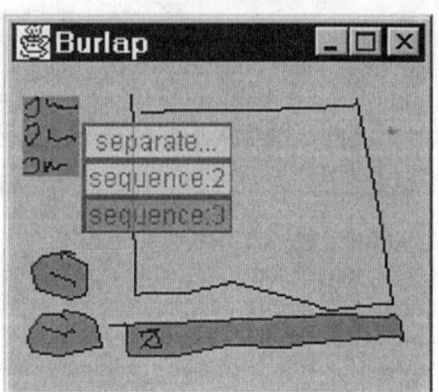

Figure 5--Segmentation ambiguity: The user telling Burlap to group 3 radiobuttons together. When grouped, only one button can be selected at a time.

In the toolkit, segmentation ambiguity corresponds to a situation when two conflicting events are derived from one or more of the same source events. Figure 6 shows the hierarchical events that correspond to the sketch in Figure 5. The user has drawn two radiobuttons on the screen. Just as in Figure 3, these radiobuttons themselves are ambiguous. However, since there are *three* radiobuttons, the interactor recognizer associated with Burlap generates two new interpretations—the first two radiobuttons may be grouped together, or all three may be grouped together. We modified our mediator to add a third possibility—the radiobuttons may all be separate.

Choosing the wrong possibility from a set of ambiguous alternatives causes a recognition error. Many common recognition errors can be traced to one or more of the three types of ambiguity described above. Yet user interfaces for dealing with errors and ambiguity almost exclusively target recognition ambiguity while ignoring segmentation and target ambiguity. For example, of the systems

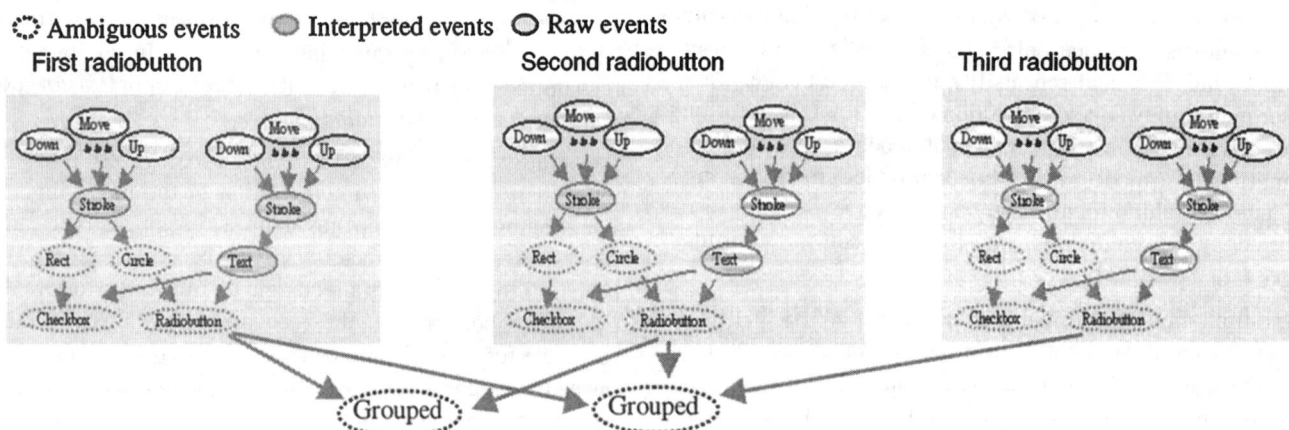

Figure 6: Three radiobuttons resulting in **Segmentation ambiguity**. The grayed out portions of the tree are identical to that in Figure 3.

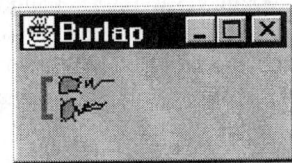

Figure 7: An application—specific mediator asking whether these radiobuttons should be grouped together or separately. If the user draws a slash through the mediator, the buttons will be separated. Any other action will cause them to be grouped (this is the default).

surveyed [18], none dealt directly with segmentation ambiguity. In one case, researchers draw lines on the screen to encourage the user to segment their input appropriately, but they give no dynamic feedback about segmentation [9]. None dealt with target ambiguity.

Extending Mediators

We have extended the concept of mediation. Our re-usable multiple alternative display, shown working with recognition ambiguity resulting from word-prediction and sketches in Figures 1a and 1b, also handles target and segmentation ambiguity (Figure 5).

In addition, we have begun to explore the possibilities for designing better interaction techniques for mediation by making use of application specific information, something that is lacking in many of the most common mediation interfaces found in the literature and in products. For example, most *n*-best lists simply show the alternatives generated by the recognizer, without indicating how different alternatives may affect the application.

Figure 7 demonstrates an application-specific mediator for segmentation ambiguity in Burlap. This mediator handles the same situation as that shown in Figure 5, but it makes use of task-specific knowledge. It knows that radiobuttons drawn near each other are usually intended to be grouped together, so it provides a sensible default: if the user continues to sketch interactors, the mediator will automatically accept the grouping as the correct option. It also knows enough about the task to pick a simple graphical representation of its options rather than trying to pick a word for a graphical relationship without an obvious short verbal description. This also allows it to take up less space and be less obtrusive (since the user probably will pick its top choice anyway). The user need only click on the mediator to reject its suggestion that the radiobuttons be grouped.

Mediation in the Toolkit

Ambiguous events are automatically identified by the toolkit and sent to the mediation subsystem. This system keeps an extensible list of mediators. When an ambiguous event arrives, it passes that event to each mediator in turn until the ambiguity is resolved.

If a mediator is not interested in the event, it simply returns PASS. For example, the *n*-best list mediator in Figure 5 only deals with segmentation ambiguity in radiobuttons—other kinds of ambiguity it simply ignores by returning PASS. If a mediator resolves part of the event graph—by accepting one interpretation and rejecting the rest—it returns RESOLVE. Finally, if a mediator wishes to handle ambiguity, but defer final decision on how it will be resolved, it may return PAUSE. Each mediator is tried in turn until some mediator signifies that it will handle mediation by returning PAUSE or a mediator returns RESOLVE and the complete event graph is no longer ambiguous. The toolkit provides a fallback mediator that will always resolve an ambiguous event (by taking the first choice at each node). This guarantees that every ambiguous event will be mediated in some way.

Not all mediators have user interfaces—sometimes mediation is done automatically. In this case, errors may result (the system may select a different choice than the user would have selected). An example of a mediator without a user interface is the *stroke pauser*. The stroke pauser returns PAUSE for the mouse events associated with partially completed strokes, and caches those events until the mouse button is released. It then allows mediation to continue, and each cached mouse event is passed to the next mediator in the list. Feedback about each stroke or partial stroke still appears on the screen since they are delivered to every interested interactor *before* mediation. However, no actions are taken until the mouse is released (and mediation completed).

Interactive mediators are treated no differently than automatic mediators. The mediators shown in Figure 1b and Figure 5 get events after the *stroke pauser*. When they see an event with ambiguity they know how to handle, they add the menu shown to the interface. They then return PAUSE for that event. Once the user selects an alternative, they accept the corresponding choice; and allow mediation to continue.

The toolkit enforces certain simple guarantees about events that are accepted or rejected. All interpretations of rejected events are also rejected. All sources of accepted events are also accepted. Further, any event that conflicts with an accepted event is rejected.

HIDING AMBIGUITY

We have described an architecture that makes information about ambiguity explicit and accessible. At the same time, we do not wish to make the job of dealing with ambiguity onerous for the application developer. The solution is to be selective about how and when we involve the application. This is possible because we separate the task of mediation from the application and from other parts of the toolkit. Mediation, which happens separately, determines what is accepted or rejected. Note that mediation can be integrated into the interface even though it is handled separately in the implementation.

It is sometimes too costly to handle the separation of feedback and action. Some things, such as third party software, or large existing interactor libraries, may be hard to rewrite. Also, in certain cases, a rejected event may break user expectations (e.g., a button that depresses yet in the end does nothing).

In order to handle these cases, we provide two simple abstractions. An interactor can be *first-dibs*, in which case it sees events first, and if it consumes an event no one else gets it. Thus ambiguity can never arise. Alternatively, a *last-dibs* interactor only gets an event if no one else wants it, and therefore is not ambiguous. Again, ambiguity will not arise. These abstractions can handle most situations where ambiguity-aware and ambiguity-blind interactors are mixed. This allows conventional interactors (e.g., the buttons, sliders, checkboxes, etc. of typical graphical user interfaces) to function in the toolkit as they always have, and to work along side ambiguity-aware interactors when that is appropriate. This property is important in practice, because we do not wish to force either interface developers, or users, to completely abandon their familiar interfaces in order to make use of recognition technology.

CURRENT STATUS AND FUTURE WORK

We have implemented the infrastructure described in this paper as an extension to the subArctic user interface toolkit. With this extended system, we have built two applications—Burlap, which makes use of pen, speech, and identity of its users as input, and a simple word-prediction text entry system.

In order to build these applications, we have connected an off-the-shelf speech recognizer (IBM's ViaVoice™), a context toolkit [24], and two pen recognizers (A 3rd party unistroke gesture recognizer [6] and an interactor-recognizer based directly on the design used for SILK[16] to our toolkit. Note that in order to handle speech and identity, we had to add two new event types to the toolkit's input system (which handles mouse and keyboard input only by default). The speech system simply creates "speech" events whenever ViaVoice™ generates some interpretations. The context system creates an identity event each time a piece of relevant identity context arrives. We also used Pegasus, Igarashi's drawing beautification system [15], with a previous version of the extended toolkit, described in [19].

We have populated the toolkit with default mediators for pen (Figure 1a), speech, and identity. These mediators address segmentation and recognition ambiguity. We also built application-specific mediators for segmentation of pen input (Figure 1b) and for identity.

Our work differs from other GUI toolkits that support recognized input such as Artkit [11], and Amulet [17], which require recognizers to return a single, unambiguous result. We track the relationship between input and recognized input using an extended version of hierarchical events [22] that supports ambiguity. Hudson and Newell also present work that tracks ambiguity [13], but their work was not integrated with a toolkit, and confined to individual input handlers rather than supporting ambiguity throughout the input cycle. Multimodal toolkits that provide support for ambiguity do not integrate this within the input handling mechanism of a GUI toolkit and thus lack the flexibility to provide re-usable support for mediation at an interface level [23,21].

In the future, we plan to extend the mediation infrastructure to support fluid negotiation for space on the screen [5]. We will investigate mediators for target ambiguity. We will provide better support for repetition-based mediation, including undo. Once we can undo events, we can allow interpretations to be created after mediation is complete. For example, if additional, relevant information arrives, an accepted alternative may need to be undone and mediation repeated.

CONCLUSIONS

We have provided principled, toolkit-level support for a model of ambiguity in recognition. The purpose of building toolkit-level support for ambiguity is to allow us to build better interfaces to recognition systems. As stated in the introduction, user interface techniques can help reduce the negative effects of ambiguity on user productivity and input speed, among other things. Our solution addresses a wide range of user interface techniques found in the literature, and allows us to experiment with mediators that take advantage of application-specific knowledge to weave mediation more naturally into the flow of the user's interaction.

By making ambiguity explicit, we were able to identify two types of ambiguity not normally dealt with—segmentation and target ambiguity. In addition, because we separate feedback about what *may* be done with user input from action on that input, we can support a much larger range of mediation strategies including very lazy mediation.

Our work is demonstrated through Burlap, for which we have built a series of mediators for segmentation, recognition, and target ambiguity.

ACKNOWLEDGMENTS

We thank Anind Dey for his help with implementation and as a sounding board for these ideas. This work was supported in part by the National Science Foundation under grants IRI-9703384, EIA-9806822, IRI-9500942 and IIS-9800597.

REFERENCES

1. Ainsworth, W.A. and Pratt, S.R. Feedback strategies for error correction in speech recognition systems. *International Journal of Man-Machine Studies,* **39**(6), pp. 833-842.

2. Alm, N., Arnott, J.L. and Newell, A.F. Prediction and Conversational momentum in an augmentative communication system. *Communications of the ACM,* **35**(6), pp. 833-842.

3. Apple Computer, Inc. The Newton MessagePad

4. Baber, C. and Hone, K.S. Modeling error recovery and repair in automatic speech recognition. *International Journal of Man-Machine Studies,* **39**(3), pp. 495-515.

5. Chang, B., Mackinlay, J.D., Zellweger, P. and Igarashi, T. A negotiation architecture for fluid documents. In *Proceedings of UIST'98,* ACM, November 1998, pp.123–132.

6. Christian, A. Long, J.R. Landay, J.A. and Rowe, L.A. Implications for a gesture design tool. In *Proceeding of CHI'99,* May, 1999, pp. 40-47.

7. DragonDictate product Web page. Available at: http://www.dragonsystems.com/products/dragondictate/index.html

8. Edwards, W. K., Hudson, S., Rodenstein, R., Smith, I. and Rodrigues, T. Systematic output modification in a 2D UI Toolkit. In *Proceedings of UIST'97,* ACM, October 1997, pp. 151-158.

9. Goldberg, D. and Goodisman, A. Stylus user interfaces for manipulating text, *in Proceedings of UIST'91,* ACM, 1991, pp.127-135.

10. Green, M. A survey of three dialogue models. *ACM Transactions on Graphics,* **5**(3), July, 1986, pp. 244-275.

11. Henry, T.R., Hudson, S.E., and Newell, G.L. Integrating gesture and snapping into a user interface toolkit. In *Proceedings of UIST'90.* October 1990. pp.112-122.

12. Horvitz, E. Principles of mixed-initiative user interfaces. In *Proceeding of CHI'99,* May, 1999, pp. 159-166.

13. Hudson, S., Newell, G. Probabilistic state machines: Dialog management for inputs with uncertainty. In *Proceedings of UIST'92,* ACM, November 1992, pp. 199-208.

14. Hudson, S. and Smith, I. Supporting dynamic downloadable appearances in an extensible user interface toolkit. In *Proceedings of UIST'97,* ACM, October 1997, pp. 159-168.

15. Igarashi, T., Matsuoka, S., Kawachiya, S. and Tanaka, H. Interactive beautification: A technique for rapid geometric design. In *Proceedings of UIST'97,* ACM, October 1997, pp. 105-114.

16. Landay, J. A. and Myers, B.A. Interactive sketching for the early stages of user interface design. In *Proceedings of CHI '95,* 1995, pp.43-50.

17. Landay, J.A. and Myers, B.A. Extending an existing user interface toolkit to support gesture recognition. In *Proceedings of INTERCHI'93,* ACM Press, pp. 91-92.

18. Mankoff, J. and Abowd, G.D. Error correction techniques for handwriting, speech, and other ambiguous or error prone systems. Georgia Tech GVU Center Technical Report, GIT-GVU-99-18, 1999.

19. Mankoff, J. Abowd, G.D. and Hudson, S.E. Interacting with multiple alternatives generated by recognition technologies. Georgia Tech GVU Center Technical Repor*t,* GIT-GVU-99-26, 1999.

20. Marx, M. and Schmandt, C. Putting people first: Specifying proper names in speech interfaces. In *Proceedings of UIST'94,* ACM, N.Y. November 1994, pp. 30–37.

21. McGee, D. R., Cohen, P.R., and Oviatt, S. Confirmation in multimodal systems. In *Proceedings of the International Joint Conference of the Association for Computational Linguistics and the International Committee on Computational Linguistics (COLING-ACL '98),* Montreal, Quebec, Canada.

22. Myers, B.A. and Kosbie, D.S. Reusable hierarchical command objects. In *Proceedings of CHI'96,* 1996, pp. 260-267.

23. Nigay, l. and Coutaz, J. A Generic platform addressing the multimodal challenge. In *Proceedings. of CHI'95,* pp.98-105.

24. Salber, D. Dey, A.K. and Abowd, G.D. The context toolkit: Aiding the development of context-enabled applications. In *Proceeding of CHI'99,* May, 1999, pp. 434-441.

25. Suhm, B. Myers, B. and Waibel, A. Model-based and empirical evaluation of multimodal interactive error correction. In *Proceeding of CHI'99,* May, 1999, pp. 584-591.

26. Zajicek, M. and Hewitt, J. An investigation into the use of error recovery dialogues in a user interface management system for speech recognition. In *Proceedings of IFIP Interact'90,* pp. 755-760.

Programming and Enjoying Music with Your Eyes Closed

Steffen Pauws, Don Bouwhuis
IPO, Center for User-System Interaction
Den Dolech 2
5600 MB Eindhoven, the Netherlands
+31 40 2475250
{S.C.Pauws, D.G.Bouwhuis}@tue.nl

Berry Eggen
Philips Research Laboratories Eindhoven
Prof. Holstlaan 4
5656 AA Eindhoven, the Netherlands
+31 40 2745160
berry.eggen@philips.com

ABSTRACT

Design and user evaluation of a multimodal interaction style for music programming is described. User requirements were *instant usability* and *optional use of a visual display*. The interaction style consists of a visual roller metaphor. User control of the rollers proceeds by manipulating a force feedback trackball. Tactual and auditory cues strengthen the roller impression and support use without a visual display. The evaluation investigated task performance and procedural learning when performing music programming tasks with and without a visual display. No procedural instructions were provided. Tasks could be completed successfully with and without a visual display, though programming without a display needed more time to complete. Prior experience with a visual display did not improve performance without a visual display. When working without a display, procedures have to be acquired and remembered explicitly, as more procedures were remembered after working without a visual display. It is demonstrated that multimodality provides new ways to interact with music.

Keywords

multimodal interaction, nonvisual interaction, interface design, user evaluation, interactive music system

INTRODUCTION

Considering the wide assortment of music available, instant access to a large music collection and, particularly, the task of music programming becomes increasingly important. Music listeners are already tempted to organize their music electronically and prefer extended play facilities without the need to handle physical storage media. For instance, music encoded in computer files, jukeboxes and portable players are becoming increasingly popular. However, it takes time to explore a music collection and select the favourite recordings with only sequential access to the music. In addition, current jukebox players intended for home use are inconvenient to operate [1,2]. They generally contain numerous control elements. They often have inadequate visual displays that lack relevant information required for music programming, or that are poorly legible in dimly lit situations or from a large viewing distance. Multimodality may enhance interaction with music.

This paper describes the design, implementation and a user evaluation of a multimodal interaction style for music programming. Music programming is defined here as the serial selection of multiple music recordings from a music collection. The evaluation was focused on assessing whether or not the user requirements were met by the interaction style.

REQUIREMENTS

The most important user requirements for the multimodal interaction style are *instant usability* and *optional use of a visual display*.

Instant usability

Usually, first-time users of a home device attempt to operate the device immediately without the aid of procedural instructions. A user manual is sometimes never consulted or is simply lost. Since users of home devices have no opportunities for training or are not willing to take these opportunities, learnability is considered a fundamental usability criterion for home devices [3]. Therefore, an interaction style for a home device should be as transparent, intuitive or self-explanatory as possible meaning that users are able to perceive at a glance the most effective and efficient ways to use the interaction style without procedural instructions. In other words, rather than learnability, a home device should allow *instant usability*.

Some methods to design for *instant usability* are consistency of operation, minimality of features, the use of a small number of control elements and the use of a conceptual metaphor for interaction. Consistency of operation (or similarity of protocols [3]) means that the same pattern of actions can be used in different situations, allowing users to learn such a pattern only once. Minimality of features means that infrequently used or more complex functions are implemented in such a way that they do not interfere with initial learning. A small number of control elements allows users to learn the meaning of only a small

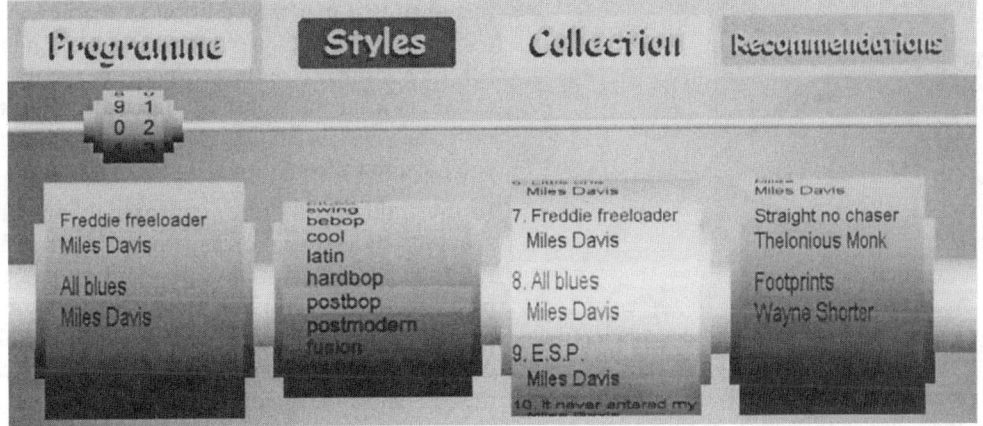

Figure 1: Visual roller metaphor of the interaction style. The music programme contains two recordings. The currently selected music style is 'postbop' and a recording of Miles Davis is playing. Three visible recommendations are linked to this recording.

set of controls or actions, for instance, the meaning of a few buttons. The use of a conceptual metaphor may form a starting point to understand an interaction style. It may aid learning domain knowledge [4], may explain the expression and meaning of actions [5] or may facilitate the learning of procedures [6,7].

Optional use of a visual display

Current CD jukebox players have an inadequate presentation of visual information [1,2]. Some visual displays lack the presentation of relevant information required to operate the jukebox effectively. They are often too small or do not have enough contrast to be legible in dim light or from a large viewing distance. Portable players even lack a convenient visual display. The need for visual inspection of information can also be less desirable, for instance, when relaxing while going through a music collection. It is for these reasons that the use of a visual display should be *optional* in a music programming task, without sacrificing instant usability.

DESIGN

The interaction style comprises a manual input modality and various output modalities: the visual, tactual and auditory modalities. User control proceeds entirely by manipulating a force feedback trackball. The auditory modality consists of three different audio streams: synthetic speech, non-speech audio and music audio.

Visual roller metaphor

The concept of a music programming for the purpose of selecting favourite music from a collection is generally self-evident, irrespective of whether it is a request or carried out at will. However, first-time users of a home device often find difficulties to transfer declarative domain knowledge and a task purpose into procedural knowledge on *how* to achieve the desired result, though they generally like to know *how-to-use-a-device* first. Without procedural instructions, the initial problem that first-time users have to overcome is to discover what exactly constitutes an action, what consequences can be expected from an action, and whether or not these consequences are effective with respect to their purpose [8]. In order to partly solve this initial problem for users, a metaphor is used as a conceptual interaction model.

During an iterative design process, several metaphors starting from a spherical object were considered as an conceptual interaction model. The use of a spherical object was prompted by the use of a trackball as input device (see Figure 3). While adding actions that are essential to navigation and selection, the model was assessed on appropriateness. The main criteria for assessment were consistency of operation, a minimum number of actions required and compliance with the envisaged metaphor. Secondary criteria were implementation feasibility and predicted computational resources required.

As shown in Figure 1, the final interaction model resembles a fruit machine consisting of four rollers on which title and artist of music recordings are projected. The left-hand roller represents a music programme that the user creates. A counter is positioned over this roller and displays the number of recordings added to the programme. The next roller represents the music styles in the collection. Music styles are arranged in a chronological order, that is, in the order in which the music styles (in this case, jazz styles) have emerged in time. Next, the music collection roller displays all available recordings in the collection or just the subset of recordings that belongs to a particular music style. Figure 1 shows that this roller has input focus and is therefore high-lighted. Recordings on this roller are first alphabetically ordered on artist, then grouped by album and placed in the order in which they came out on the album. Finally, albums are chronologically ordered on year of publication. The numbers displayed on this roller are intended to indicate the number of recordings available in a music style. The right-hand roller contains a list of music recommendations that corresponds to the music recording that is at the front of the music collection roller. Music recommendations are intended to ease and speed up the music selection process.

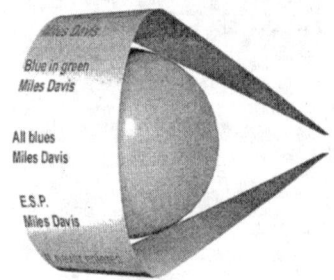

Figure 2: An indefinitely long list of music recordings tapered along a ball.

In order to coincide any rotation of the trackball with a proportional turn of the roller on the screen, the list of items (music recordings and styles) is virtually tapered round a ball, analogous to wrapping a sheet of paper around a cylindric object (see Figure 2). In that configuration, a forward ball rotation turns a roller clock-wise or upward.

Force feedback trackball

The sole control element of the interaction style is the IPO force feedback trackball device [9]. This trackball is a ground-based device (see Figure 3). A user can freely rotate a hard-plastic ball (diameter: 57 mm.), which is placed in a housing. Force feedback is mediated by two motor-driven wheels which touch the ball at its x and y axes. The force can be made dependent on the current context of interaction. Besides ball rotations, a contact switch, which is mounted underneath the ball bearing system, notifies ball presses.

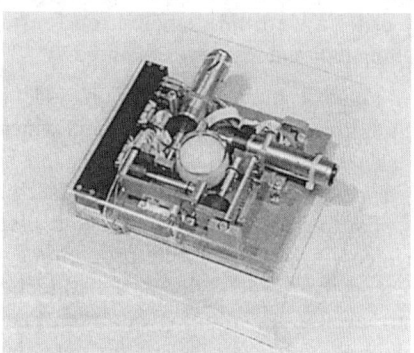

Figure 3: The IPO force feedback trackball device.

In current graphical user interfaces, the hand and eyes are decoupled. For instance, working with a trackball requires continuous visual attention to inspect the state of interaction, rather than attention to the wrist, hand and fingers that control the device. A trackball is also notorious for its inefficient performance in target acquisition tasks [10], which makes moving a mediating cursor back and forth in a complex navigation task an effortful activity.

In order to diminish these problems, actions on the trackball are directly mapped to roller behaviour. There is, for instance, no mediating cursor. Rolling the ball back or forth can be directly felt and directly corresponds to a proportional like-wise rotation of a roller. Thus, user actions have direct meaning which may increase usability. The use of touch feedback is intended to guide users in performing an action, even without using a visual display. It has been demonstrated that the addition of touch feedback to the trackball used makes users perform actions more efficiently [9]. As a result, a user may experience the trackball device rather as a means to feel and act directly on rollers by instantaneous hand-finger movements, than a means to roll and point to rollers.

In the interaction style, contextual force feedback is specified by a spatial arrangement of force fields. These fields generate a force dependent on the current ball position. The force is mediated by a motor-driven ball rotation. For instance, the force field of a 'hole' evokes a directional pulling-force towards the 'hole's centre, when the ball enters the region of the 'hole'. This feels like being captured in a region when rolling the ball, and needs some additional hand force to leave this region.

Figure 4: Spatial arrangement of force fields.

The layout of force fields in the interaction style is shown in Figure 4. Each ball movement starts with the ball at the centre of the layout. A circular force field pushes the ball slightly towards the centre. Grooves, placed in a cross, guide ball movement along a straight line (forward, backward and lateral movements). The ends of the grooves are marked by raised edges; these edges slightly hold back a continuous roll movement. When moving across an edge, the ball will be captured by a small hole which typifies the conclusion of a roll movement. This configuration of force fields mimics the felt sensation of a 'mechanical click'. If the ball is moved across an edge with some more hand force, the ball will miss the first hole and will end in the next hole. A small forward (backward) hand movement brings the next (previous) item to the front. Moving the ball with some more hand force skips the next or previous two items. Lateral ball movements corresponds to hopping from one roller to another. Again, a small movement makes a hop to the next roller, and a movement with some more hand force skips a roller.

It may be clear that a user can, for instance, first choose a particular music style on the music styles roller and continue searching for music recordings within that style by going to the music collection and recommendation rollers. A user can either quickly go to familiar recordings by skipping irrelevant ones or proceed sequentially through all recordings on the roller. Double-pressing the trackball means adding or removing a music recording to or from the music programme.

Auditory feedback

Speech synthesis

Speech output is used to convey information about the current state of interaction, primarily intended to allow the interaction style to be used without a visual display. For instance, after hopping from one roller to another, spoken feedback indicates to which roller was moved to. When entering the music programme roller, it indicates how many music recordings have been added to the programme. When the user rolls through the music styles roller, it tells the user which music style is left and which is entered (e.g., 'from bebop to hardbop'). By pressing the ball once, the user can ask what the currently selected music style is or to what music style a selected recording belongs.

Non-speech sound

Every user action produces a software-generated sound known as an *auditory icon*. Auditory icons are probably best described as caricatures of naturally occurring sounds. A benefit of using auditory icons is that they may have an intuitive appeal to the listener [11]. A special class of auditory icons is used, namely, impact sounds of material being struck. Each roller and user action has its own set of qualitatively different sounds intended to distinguish the rollers and the actions by sound alone. The music programme roller is featured by sounds that are best described as sounds produced when a glass-like object is struck by another solid object. The music styles roller produces sounds as hitting a metal object. The music collection roller produces wood-like sounds, and the recommendations roller sounds like rubber. Valid parameters to generate these material-like sounds were determined by user experiments [12], though it was not intended that users should be able to tell what the roller would be made of by the sound that it makes. When a roller is rotated step-wise (by a small hand force) a single click is produced. Rolling more vigorously produces a spectrally broader click followed by a rattling sound to represent the scroll of the roller. Adding or removing a music recording to or from a music programme produces a 'creaking' or a 'bouncing' sound respectively.

IMPLEMENTATION

As shown in Figure 5, a component-based software architecture based on Microsoft ActiveX technology was used to implement the interaction style. The graphical notation used is adopted from the Unified Modeling Language [13]. Music recommendations are generated by an in-house developed *music recommender system* [14]. The manual input and tactual output of the force feedback trackball is controlled by three components: *LightHole*, *TacServer*, and *TacServer Extension*. Software control of the trackball is distributed over two serially connected PC platforms. Data of the music collection is contained in a database component. Music is played back by an *MPEG Audio Player*. Non-speech audio is generated by a component *Impact Sound*; this component contains an in-house implementation of the Constrained Additive Synthesis technique for generating impact sounds [11]. Synthetic speech is generated by a *Text-to-speech* component. The interaction style requires simultaneous streaming and mixing of audio from different sources and audio output formats with low latency and maximum control. Therefore, an *Audio Device* component for all audio output streaming services encapsulates and extends on Microsoft's DirectSound technology. Visualisation of the rollers is implemented in the *Roller* component. All components starts their own threads which can be interrupted or stopped, when other services are asked from them or the state of the interaction changes. However, threads between components are not synchronized.

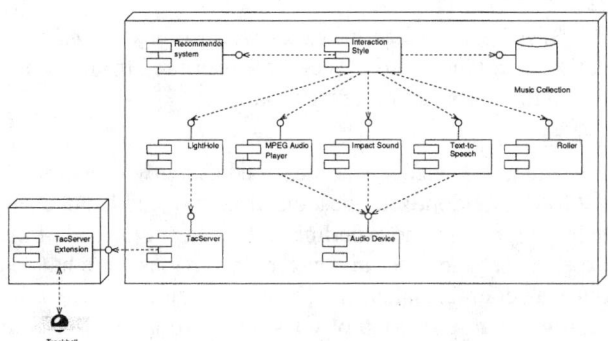

Figure 5: Component-based software architecture. The 'lollipops' represent software interfaces.

USER EVALUATION

A user evaluation was carried out to assess instant usability with and without a visual display of the interaction style. Particularly, task performance and the learning of procedures with and without a visual display were compared. After three minutes of free exploration of the interaction style, participants performed pre-defined music programming tasks. They were not given procedural instructions on *how-to-use* the interaction style. Task performance was measured in real-time. Procedural learning was measured by a post-task questionnaire.

Hypotheses

As tactual and auditory feedback are transient in the interaction style and visual feedback is continuous and persistent, the critical factor between visual and nonvisual interaction, that is, only tactual and auditory interaction, is human memory. Hence, it is hypothesized that:

(i) *Visual* interaction is more efficient, that is, requires less time and fewer actions, than *nonvisual* interaction, while leaving the level of auditory and tactual feedback in both conditions constant.

Nonvisual interaction requires the explicit acquisition and remembering of procedures. Consequently, it is likely that users who interact nonvisually develop a substantial body of procedural knowledge. It is therefore hypothesized that:

(ii) Users who have worked *without* a visual display have a higher score on *procedural knowledge* than users who have worked *with* a visual display, while leaving the level of auditory and tactual feedback in both conditions constant.

Something that is learnt while doing tasks with a visual display (e.g., an imagery of the visual display) is likely to be carried over ('transferred') to the same type of tasks without a visual display, which may facilitate a more efficient performance of the latter tasks. It is therefore hypothesized that:

(iii) Users who worked *with* a visual display in one condition and are subsequently transferred to another condition *without* a visual display perform more efficiently, that is, need to spend less time and fewer actions, than users who start working *without* a visual display.

Measures

Task performance

Two task performance measures were defined: *number of actions* and *time on task*. They were measured from the first action to the last action of the task.

Questionnaire score

Procedural knowledge was measured using two versions of a 20-item randomised questionnaire; one was handed out half-way through the experiment, the other at the end of the experiment. The order in which both versions were handed out was counter-balanced. The two versions contained a relatively large overlap of questions (16) as well as four distinct questions. Questions asked what sequence of actions, that is, procedure, most efficiently and successfully transformed a given initial state of interaction into a given final state of interaction. Half of the questions concerned a single action (single-step interaction). The other half concerned a sequence of two or three actions (multiple-step interaction). Answers were judged correct if the responded action sequence successfully transformed the initial state of interaction into the final one and the action sequence was of minimal length, that is, was most efficient. Answers were otherwise judged incorrect. All correct answers were added up to arrive at a *questionnaire score* (maximum: 20).

Method

Instruction

Participants read a short text about the music programming domain to provide them with the necessary amount of declarative knowledge. No reference was made to the interaction style; no instructions about procedures or the roller metaphor were given. Participants were asked to rephrase the given text in their own words. Any misconception of the music programming domain was corrected by the test supervisor.

At the outset of each music programming task, participants received a written task description. They were asked to rephrase the task instruction to avoid any misconceptions of the task.

Task

The task was to select 10 distinct music recordings, equally drawn from two pre-defined music styles, as quickly as possible while paying no attention to personal preferences or to the order of the selection process. Four music programming tasks were defined and their order of presentation was counterbalanced. The tasks were designed to be equally difficult; a successful and most efficient task completion demanded 23 actions.

Design

Four conditions were applied: two control and two experimental conditions. In one control condition, denoted by VAT (Visual, Auditory, Tactual feedback), participants completed four tasks by using the complete multimodal interaction style. The four consecutive tasks were denoted by *task repetition*. In the other control condition, denoted by AT (Auditory, Tactual feedback), participants only worked with the interaction style without visual display for all four tasks; the monitor was physically removed. In the two experimental conditions, participants worked with both interaction styles, one after the other. In the experimental condition denoted by VAT $\rightarrow$ AT, there was no visual display for the last two tasks. In the other experimental condition, denoted by AT $\rightarrow$ VAT, this was reversed.

Test material and equipment

A music collection comprising 480 one-minute excerpts of jazz music recordings (MPEG-1 Part 2 Layer II 128 Kbps stereo) from 12 different music styles served as test material. The interaction style was implemented on a PC, running under Windows 95. MPEG data was stored on the hard disk. Real-time MPEG decoding was done by software. Music was amplified by a mid-range audio amplifier (Philips Integrated Digital Amplifier DFA888) and played through a pair of high-quality loudspeakers (Philips 9818 multi-linear 4-way). A second PC was used for controlling the force feedback trackball.

Participants were seated in a comfortable chair in a non-reverberant studio. The visual display was a 17-inch colour monitor. They could adjust the audio volume to a preferred level. The trackball was placed on a small table next to the chair, in such a way that the participant could control the trackball in a comfortable way.

Procedure

Twenty-four participants performed two experimental sessions on two separate days. They were randomly assigned to one of the four conditions: four participants were assigned to each of the control conditions, and eight participants were assigned to each of the experimental conditions. At the first session only, a 15-minute familiarisation phase let participants accustom to the force feedback and the required fine motor skills to control the trackball. Subsequently, participants were informed about the music programming domain. In both sessions, they were allowed to freely explore the interaction style (with or without a visual display) for three minutes. Then the tasks were executed.

The participants completed two music programming tasks during each session. Then, participants completed a version of the questionnaire at a desk from which it was impossible to view the test equipment. The questionnaire was completed in a dialogue with the test supervisor.

Participants

The average age of the 24 participants (18 males, 6 females) was 28 (min.: 21, max.: 45). Half of them were recruited by advertisements and got a fixed fee for expenses (16 Dutch guilders = 7.24 euro). The other half were colleague researchers who participated voluntarily. All participants had completed higher vocational education. They were not selected based on their musical preferences or musical education.

Results
Task performance data of one participant in the AT → VAT condition were excluded from the analyses; values for *time on task* and *number of actions* were three times as high as the mean values of other participants. The participant admitted that he had selected music by taking care of his personal preferences which was not in accordance with the task instruction.

Number of actions

The results on *number of actions* are shown in Figure 6. In order to compare the difference between working with and without a visual display in the first two tasks, a new variable called *visual display condition* was created. In this variable, the performance in the VAT and VAT → AT conditions (*task repetition* 1 and 2) was separated from the performance in the AT and AT → VAT conditions.

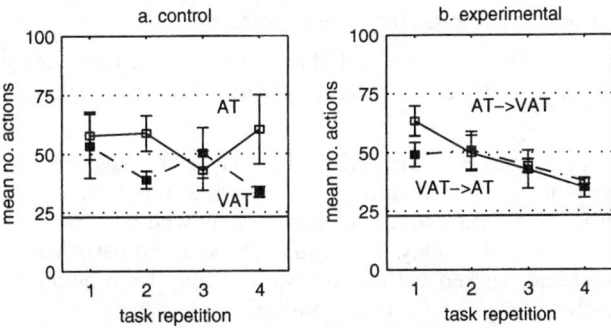

Figure 6: Mean *number of actions* and standard error. The left-hand panel (a) shows means for four tasks in the control conditions. The right-hand panel (b) shows means for four tasks in the experimental conditions. The minimal number of actions to complete the task is 23.

An ANOVA with repeated measures was used with *task repetition* (2) as a within-subject independent variable and *visual display condition* (2) as a between-subject independent variable, and *number of actions* as the dependent variable. No significant effects were found. The mean *number of actions* performed in the first two tasks was 52.7.

In order to compare the differences over all four tasks, a MANOVA with repeated measures was conducted in which *task repetition* (4) was a within-subject independent variable, *condition* (4) was a between-subject independent variable and *number of actions* was the dependent variable. Only a significant main effect for *task repetition* was found ($F(3,57) = 3.79$, $p < 0.05$). A linear trend in the data was significant ($F(1,19) = 7.92$, $p < 0.05$). Participants performed fewer actions for each successive task (mean *number of actions* across successive tasks: 55.7, 49.8, 44.6 and 40.0).

In order to assess the 'transfer' effect for *number of actions*, *transfer* was defined as the comparison between the *number of actions* performed in *task repetition* 3 and 4 in the VAT → AT condition and the *number of actions* performed in *task repetition* 1 and 2 in both the AT and AT → VAT conditions.

An ANOVA with repeated measures in which *transfer* was treated as a between-subject variable, and *task repetition* (2) as a within-subject variable was carried out. *Number of actions* was the dependent variable. A main effect for *task repetition* was found to be just not significant ($F(1,17) = 4.11$, $p = 0.06$). A significant *transfer* effect was found ($F(1,17) = 8.81$, $p < 0.01$). It appeared that almost 17 *fewer* actions were required for participants who had done two previous tasks with a visual display, than for those who started to work without a visual display. This suggests that performance without a visual display improved, when two previous tasks with a visual display were done.

However, performance improvement can also be caused by mere practice. Therefore, *number of actions* performed in *task repetition* 3 and 4 in the VAT → AT condition was compared with *number of actions* performed in *task repetition* 3 and 4 in the AT condition. In an ANOVA analysis, no significant effects were found; improvement resulted from practice.

Time on task

The results on *time on task* are shown in Figure 7. Similar to the *number of actions* analysis, a new variable *visual display condition* was created.

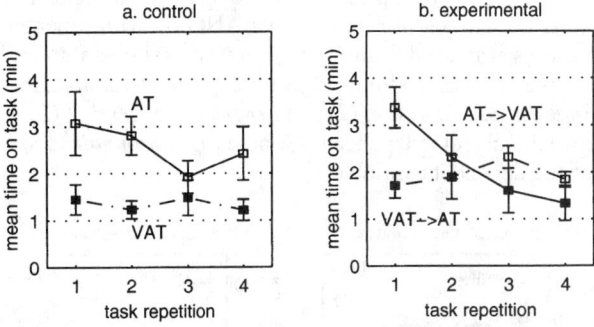

Figure 7: Mean *time on task* (minutes) and standard error. The left-hand panel (a) shows means for four tasks of the two control conditions. The right-hand panel (b) shows means for four tasks of the two experimental conditions.

An ANOVA with repeated measures was used with *task repetition* (2) as a within-subject independent variable and *visual display condition* (2) as a between-subject independent variable, and *time on task* as the dependent variable. A main effect for *task repetition* was found to be nearly significant ($F(1,21) = 4.32$, $p = 0.05$). Fewer time was required to perform the second task. A significant main effect for *visual display condition* was found ($F(1,21) = 9.78$, $p < 0.01$). Participants who worked without a visual display required almost twice as much time than participants who worked with a visual display (mean: 2 min. 53 sec. (without display), 1 min. 36 sec. (with display)).

In order to compare differences over all four tasks, a MANOVA with repeated measures was used with *task repetition* (4) as a within-subject independent variable and *conditions* (4) as a between-subject independent variable. *Time on task* was the dependent variable. A significant main effect for *task repetition* was found ($F(3,57) = 3.64$, $p < $

0.05). A linear trend in the data was significant (F(1,19) = 5.97, p < 0.05). Participants needed less time to compile each successive music programme (mean *time on task* across successive tasks: 2 min. 25 sec., 2 min. 4 sec., 1 min. 53 sec. and 1 min. 41 sec.). A significant *task repetition* by *condition* interaction effect was found (F(9,57) = 3.26, p < 0.005). When means were compared, it was found that *task repetition* 1 in the AT → VAT condition required more time than in the other conditions (F(3,19) = 4.17, p < 0.05).

The 'transfer' effect was investigated in the same way as for the *number of actions* analysis. An ANOVA with repeated measures in which *transfer* was treated as a between-subject variable, and *task repetition* (2) as a within-subject variable was carried out. *Time on task* was the dependent variable. A significant main effect for *task repetition* was found (F(1,17) = 10.12, p < 0.01). Participants needed less time to complete *task repetition* 2 and 4 than for respectively *task repetition* 1 and 3. A main effect for *transfer* was found to be just not significant (F(1,17) = 4.20, p = 0.06), however, participants who had worked with the visual display before needed 48 *fewer* seconds in a non-visual task, than participants who started to work without a visual display. Because no significant 'transfer' effect was found, any performance improvement resulted from practice.

Questionnaire score

Data on *questionnaire score* in the four conditions were re-coded to create two different *visual display condition*s. The results are shown in Figure 8. An ANOVA with repeated measures was used with *kind of question (*single-step and multiple-step) as a within-subject independent variable, and *visual display condition* (2) and *experimental session* (2) as between-subject independent variables. *Questionnaire score* was the dependent variable.

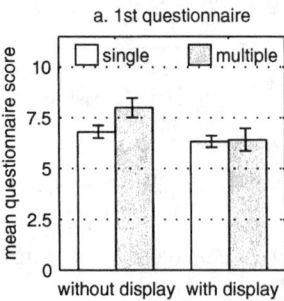
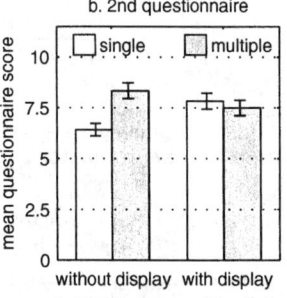

Figure 8: Mean *questionnaire score* and standard error. The left-hand panel (a) shows means divided up in scores for single step and multiple step interactions (maximum: 10 both), across visual display conditions after two tasks. The right-hand panel (b) shows means after four tasks.

A significant main effect for *experimental session* was found (F(1,44) = 5.96, p < 0.05). Participants were better in completing the questionnaire for the second time (mean *questionnaire score* across successive sessions: 13.4, 15.0). A significant main effect for *kind of question* was found (F(1,44) = 17.76, p < 0.001). Participants were better in answering the questions concerning the multiple-step interactions than the questions concerning the single-step interactions (mean *questionnaire score* for *kind of question*: 6.7 (single-step), 7.6 (multiple-step)). An interaction effect for *kind of question* by *visual display condition* was found to be significant (F(1,44) = 23.07, p < 0.001). Participants who had worked without a visual display were better in answering questions concerning multiple-step interactions (mean *questionnaire score* concerning multiple-step interactions for *visual display conditions*: 8.2 (without display), 7 (with display)).

Discussion

Participants were instructed to program music as efficiently as possible without taking notice of personal music preferences. All twenty-four participants were able to perform a music programming task successfully, right from the start, with or without a visual display. They were given three minutes of free exploration with the interaction style before the tasks started. They had received no procedural instruction on how to work with the interaction style.

According to Hypothesis (i), music programming with a visual display should be more efficient than programming without a visual display. The results showed that programming without a visual display needed significantly more time to complete. However, it did not require the execution of significantly more actions. On the basis of these results, Hypothesis (i) cannot be rejected.

The results also showed that participants were steep learners; they needed increasingly less time for successive task.

According to Hypothesis (ii), participants who have worked without a visual display should have a higher score on procedural knowledge than participants who have worked with a visual display. The results showed that participants who had worked without a visual display were better in answering questions about multiple-step interactions. On the basis of these results, Hypothesis (ii) cannot be rejected. Procedures had to be acquired and remembered explicitly, when there was no visual display was available.

According to Hypothesis (iii), participants who worked with a visual display in one condition and are subsequently transferred to another condition without a visual display should perform more efficiently than participants who start working without a visual display. The results showed that performance improvement was caused by practice, primarily, and not by previous experience with a visual display. In other words, participants who started to work without a display performed less efficiently than participants who had earlier experience with the display, only because they had performed fewer tasks. On the basis of these results, Hypothesis (iii) must be rejected.

In the first task, it also appeared that programming without a visual display took more time per action (mean time per action: 3.2 sec. (without display), 1.9 sec. (with display); t = -7.08, p < 0.001) and more lateral trackball movements (mean number of lateral movements: 19.1 (without display), 7.9 (with display); t = -3.11, p < 0.01), whereas other types of actions did not differ across visual display conditions. Lateral movements were actions to hop from one roller to another and hence were suitable for exploring the spatial relations between rollers. It is therefore likely that extra time and extra lateral ball movements are linked to acquisition of spatial and procedural knowledge when there is no visual display available. When there is no visual display and the user is unfamiliar with the interaction style, it is suggested that interaction follows a time-consuming *dead-reckoning process* in which each current state of interaction has to be inspected *explicitly*, using other means than vision, to be able to do further action. If, on the other hand, a visual display is available, which shows all information for interaction, actions can be *planned ahead*, which is more efficient.

CONCLUSION

Instant usability is particularly important for interactive home devices, as these devices are typically used intermittently without training. Therefore, any interaction style for music programming should be self-explanatory. Selecting and listening to music should also be possible without a visual display. Think of, for instance, the use of remote controls or portable devices, a dimly lit situation while having a party or the desire to enjoy music without a need for visual inspection of information.

The presented interaction style combines a manual input modality with various output modalities: the visual, tactual and auditory modalities. It consists of a self-explanatory visual roller metaphor which is controlled by a force feedback trackball. Tactual and auditory cues are used to strengthen the impression of rollers, but also to enable music programming without a visual display.

The user evaluation assessed instant usability of the interaction style with and without a visual display. Users were able to complete a given music programming task successfully with or without a visual display after three minutes of free exploration and without procedural instruction. Over time, they learned to perform the tasks more efficiently; less time and fewer actions were required for each successive task.

It is obvious that purposeful nonvisual interaction grows out of knowledge about the interaction space. For the interaction style, this knowledge includes declarative, procedural and spatial knowledge. Therefore, working without a visual display involves explicit acquisition and remembering of procedures and the spatial relationships of task objects. This acquisition limits efficient *first-time* use of the interaction style, but does not impede a successful completion of the task. Participants who worked with both versions of the interaction style preferred a visual display just for convenience. Only one participant preferred the displayless interface; he commented that 'it contains all attractive features that a novel has and a movie picture lacks; one is free to make an own interpretation and devise an own world.'

In summary, it has been demonstrated that users who work with only a tactual and auditory interface are able to operate a new interaction style successfully, only less efficient in time. Tactual and auditory feedback makes interaction possible in contexts-of-use in which information on a visual display is poorly legible or even absent. Working without a visual display is not a commonly preferred method of operation, due to the need of the explicit acquisition and remembering of procedures. Concluding, the multimodal interaction style for music programming meets its user requirements on *instant usability* and *optional use of a visual display*. If hand-held trackball devices with force feedback mature to fully usable and low-cost input devices consuming only a little power, this interface to music programming may be an desirable feature on, for instance, jukeboxes, portable players, remote controls and car audio equipment.

ACKNOWLEDGMENTS

We would like to thank Henk Korteweg for helping us with creating Figures 2 and 4.

REFERENCES

1. Consumer Reports (1997). Jukebox time? *Consumer Reports, 62*, 2, February 1997, 34-38.

2. Kumin, D., (1994). Ch-ch-changers! *Stereo Review, 59*, 60-66.

3. Eggen, J.H., Haakma, R., and Westerink, J.H.D.M. (1996). Layered Protocols: hands-on experience. *International Journal of Human-Computer Studies, 44*, 45-72.

4. Veer, G.C., van der (1990). *Human computer interaction: learning, individual differences, and design recommendations*. Doctoral Thesis, Vrije Universiteit, Amsterdam, the Netherlands.

5. Hutchins, E. (1989). Metaphors for Interface Design. In: Taylor, M.M., Néel, F., and Bouwhuis, D.G. (Eds.). *The Structure of Multimodal Dialogue*, Amsterdam: North-Holland, 11-28.

6. Kieras, D.E., and Bovair, S. (1984). The role of a mental model in learning to operate a device. *Cognitive Science, 8*, 255-273.

7. Payne, S.J., (1988). Metaphorical instruction and the early learning of an abbreviated-command computer system, *Acta Psychologica, 69*, 207-230.

8. Shrager, J., and Klahr, D. (1986). Instructionless learning about a complex device: the paradigm and observations. *International Journal of Man-Machine Studies, 25*, 153-189.

9. Engel, F.L., Goossens, P.H., and Haakma, R., (1994). Improved efficiency through I- and E-feedback: a trackball with contextual force feedback, *International Journal of Human-Computer Studies, 41*, 949-974.

10. MacKenzie, I.S., Sellen, A., and Buxton, W. (1991). A comparison of input devices in elemental pointing and dragging tasks, *Human Factors in Computing Systems: CHI'91 Conference Proceedings*. New York: ACM.

11. Gaver, W. (1993). What in the world do we hear? An ecological approach to auditory event perception. *Ecological Psychology, 5*, 4, 285-313.

12. Hermes, D.J. (1998). Auditory material perception. *IPO Annual Progress Report, 33*, 95-102.

13. Rumbaugh, J., Jacobson, I., and Booch, G. (1999). *The Unified Modeling Language Reference Manual*. Amsterdam: Addison-Wesley.

14. Pauws, S.C. (2000). *Music and choice: Adaptive systems and multimodal interaction*, Doctoral Thesis, Eindhoven University of Technology, the Netherlands.

Presenting to Local and Remote Audiences: Design and Use of the TELEP System

Gavin Jancke, Jonathan Grudin, Anoop Gupta
Microsoft Research
One Microsoft Way
Redmond, WA 98052-6399
{gavinj; jgrudin; anoop}@microsoft.com

ABSTRACT
The current generation of desktop computers and networks are bringing streaming audio and video into widespread use. A small investment allows presentations or lectures to be multicast, enabling passive viewing from offices or rooms. We surveyed experienced viewers of multicast presentations and designed a lightweight system that creates greater awareness in the presentation room of remote viewers and allows remote viewers to interact with each other and the speaker. We report on the design, use, and modification of the system, and discuss design tradeoffs.

Keywords
Tele-Presentation, Streaming Media

INTRODUCTION
The well-publicized availability of audio and video over the Internet and intranets ushers in new uses for digital technology, ranging from entertainment to distance education. Desktop computers can handle real-time audio and video. Many networks (including the Internet) require upgrading, but the technology is available. If streaming media prove to be of value, they can be delivered.

At Microsoft, as at many large corporations, presentations are now broadcast "live" over the corporate intranet, 5-10 research lectures and an equal number of more general presentations every week. Broadcasts include audio, video and slides. By clicking on web pages that list talks, employees can attend from their offices or even from home.

Clearly, there are potential benefits for remote viewers. They do not have to travel to attend the talk; if the talk is uninteresting they can quit without wasting time or risking offending a speaker or host, and if parts of the talk are uninteresting, they can multitask with other work (e.g., read email). However, there are also potential disadvantages.

First, from a speaker's perspective, remote viewing can result in fewer people attending live in the lecture room. To the extent that speakers are unaware of the remote audience, they may perceive a small live audience as lack of interest in their work. They may be offended or less motivated to deliver a good talk. It is not uncommon to hear a host reassure a speaker facing a small live audience with words to the effect of "Don't be deceived by the small audience in the room, lots of people are watching remotely."

Second, remote-viewers do not experience the ambience and subtlety of the live talk and audience. They cannot see the expressions of other audience members, whisper a question to a colleague, or direct questions to a speaker. Live audience members' questions are often inaudible to remote users unless repeated by the speaker.

Finally, a live audience member, like the speaker, may infer from a small live audience a lack of interest in the topic (which is generally of greater interest to those who traveled to the lecture room). The live experience is also diminished by the inability of remote viewers to ask questions.

Although one obvious way to eliminate these disadvantages is to disallow broadcast of talks, in this paper we explore how technology could enhance the benefits and minimize the disadvantages. We report on TELEP (short for telepresence), a system designed to provide speakers and local audiences with greater awareness of remote viewers, to provide remote viewers with a means to interact with speakers and each other, and to do this in a lightweight manner that requires little of remote viewers and almost no additional work by speakers.

TELEP is a working system currently used for seminars. In this paper we report on its design—the system components, the user interface and interaction paradigm—and the design tradeoffs we faced. We also report on audience behavior before and after the deployment of TELEP.

The paper is organized as follows. Following a section on related work, we present design goals and a TELEP system overview. The next section presents the TELEP interface and design tradeoffs. We then describe experience with broadcast presentations before and after deployment. The final sections highlight lessons learned.

RELATED WORK
For decades, videoconferencing systems (e.g., PictureTel [16]) have linked two or more rooms with bi-directional audio-video channels and split-screen displays or multiple television monitors. Our design focus is different. There may be scores of people attending remotely, each from an

Permission to make digital or hard copies of all or part of this work for personal or classroom use is granted without fee provided that copies are not made or distributed for profit or commercial advantage and that copies bear this notice and the full citation on the first page. To copy otherwise, to republish, to post on servers or to redistribute to lists, requires prior specific permission and/or a fee.
CHI '2000 The Hague, Amsterdam
Copyright ACM 2000 1-58113-216-6/00/04...$5.00

office. An office may or may not have a camera or microphone. The situation is much more asymmetric; consequently, the tradeoffs differ.

Distance education programs face similar challenge. Stanford University's SITN program has offered courses to students at Bay Area companies for over 25 years [18], broadcasting from a classroom via a microwave channel. The students sit at designated conference rooms within their companies to watch lectures. They can ask questions by a telephone call to the classroom.

We can confirm from personal experience teaching at Stanford that lecturer awareness of remote students is minimal. One has no idea how many are attending "live" remotely or how many have a VCR turned on to record for later viewing. Remote students do not have precise control over when to interrupt, so their questions occur as "crackling voices" in the middle of a lecturer's sentences.

TELEP is designed for a different context. Research seminars are usually given by visitors who use the system only once. Classroom instructors will use a system repeatedly, giving instructor and students more time and greater incentive to interact and establish a relationship. Remote students often have an investment in understanding the material that is equal to live students, which is often not the case in the situation we target. TELEP also differs in assuming more technology infrastructure, through which it can provide significantly greater awareness of remote viewers.

Closest to our work are systems targeted for desktop-to-desktop presentations (i.e., all viewers are remote, the speaker is in an office or studio without a local audience). Examples include Forum from Sun [4-6], Flatland from MSR [13], and commercial products such as Centra [14], NetPodium [15], and PlaceWare [17]. They may broadcast a speaker's audio-video and slides, and include additional capabilities for asking and responding to multiple-choice questions. Viewers can raise hands, ask questions via audio-channel or chat, and vote. A list of attendees is available to the speaker and viewers. Restriction to text is common, as the systems are often designed to support very large audiences and do not assume high connection bandwidth.

TELEP also provides awareness and interactivity, but the circumstances and features differ. The systems above were built for speakers with no local audience—they could devote more attention to complex software interfaces. Rich back-channels and awareness were particularly important because the speakers had no live audiences. Experiments have shown that although remote viewers like the systems, speakers are unsettled by the lack of feedback they would get from a local audience; the software interaction channels that have been tried did not fully compensate [5, 13].

In contrast, TELEP focuses on mixed live (local) and remote audiences. Because speakers must devote considerable attention to the live audience, we have kept their interface simple, requiring no keyboard use. Presence of a live audience also affects how the remote audiences can be displayed to the speaker. By assuming higher bandwidth connectivity, we can evaluate graphical representations of remote viewers (image or video) for the first time in this context. The existence of a live audience may reduce the pressure on the technology and increase the chance of success. Consider, by analogy, early radio, which started without studio audiences but introduced them because performers preferred a live audience.

In an extension to their work on Forum, Sun researchers conducted unpublished studies of "Forum Studio" with mixed live and remote audiences (John Tang, Rick Levenson, Ellen Isaacs, personal communications, 1999). Speakers stood before a podium containing a recessed computer monitor and used the Forum software to interact with remote viewers. Preliminary results contrasting local-only, remote-only, and mixed audiences showed that mixed audiences may learn less. Engaging with two audiences can distract speakers. Distant audience members may feel excluded, and a live audience may be distracted by a speaker's efforts to deal with the technology.

TELEP differs in two important ways: it does not require speakers to use technology, and our initial remote viewers have had up to two years experience passively attending lectures. TELEP can only maintain or increase their sense of inclusion, making successful reception easier to achieve.

Finally, research on supporting informal interaction (e.g., Bellcore Cruiser[8], Xerox PARC and NYNEX Portholes [1, 2], Sun Montage [11], University of Toronto CAVECAT [9], NTT Clearboard [7], and University of Calgary prototypes [3]) has addressed quite different scenarios, yet it has influenced our work.

TELEP OVERVIEW

Prior to deploying TELEP, we examined the use of the preexisting passive viewing system through observation and surveys of speakers, live audiences, and remote viewers. These data are discussed later. In this section we cover the design goals and provide an overview of the system

Design Goals and Constraints

- Presentations with a "live" audience in the lecture room and a remote audience attending from desktops.

- The lecture room interface should benefit both the speaker and the live audience.

- Medium-sized (fewer than 100) remote audiences, with access to computer but not necessarily a microphone or camera.

- Support for one-time visiting speakers. They should not have to use a keyboard. Suitable protocols for interaction should arise as naturally as possible.

- Assumption of adequate network bandwidth and computation, so it is feasible to multicast and render low-resolution video of remote viewers.

- Until proven to be reliable and acceptable, TELEP should be decoupled from pre-existing software used to watch the video of speakers and slides. A TELEP failure should not prevent people from viewing talks in the familiar non-interactive fashion.

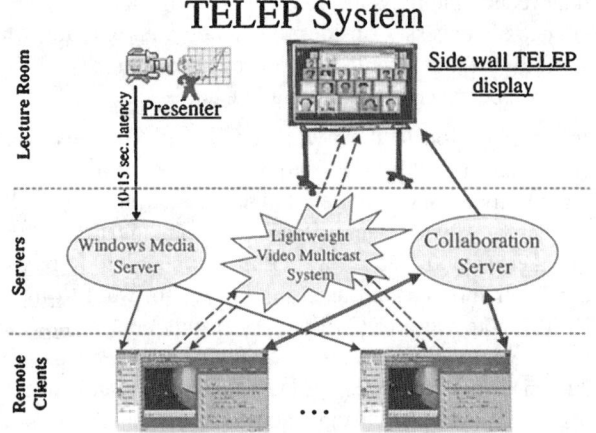

Figure 1: TELEP System Overview

TELEP System Overview

Figure 1 illustrates the interrelation of TELEP components. There are two parallel systems. The first, shown on the left, is the pre-existing system that multicasts presentations for passive viewing. Based on Microsoft Windows Media Server, it broadcasts a speaker's audio, video and slides. The corresponding display seen by remote viewers appears in the right window of Figure 2. The slides switch automatically as the speaker advances them. A key aspect is that the video-encoder, video-server, and client-side buffering, introduce a 10-15 second delay before the audio-video is received by the remote audience. This is not an issue for purely passive remote viewing, but it clearly constrains interaction between the speaker and remote audience members using TELEP.

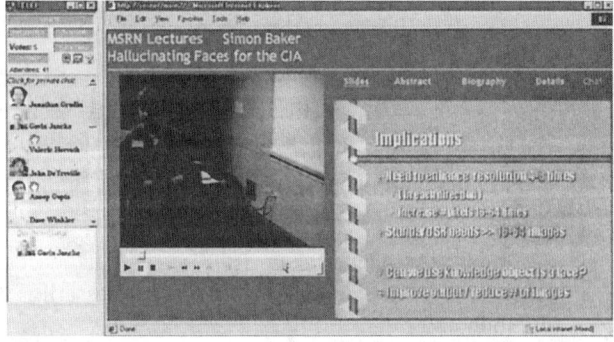

Figure 2: Remote User Layout: TELEP window on left, web page with speaker video and slides on right

The second component is shown on the top-right in Figure 1. It produces the features in the left window of Figure 2 on remote users' desktops, and a lecture-room display as shown in Figure 3. We discuss these interfaces in detail in the next section.

Underlying the TELEP system is a collaboration server, built on MSR's Virtual World Server [12]. It communicates remote viewer actions (e.g., raising hand, voting, chat) to other remote viewers and in some cases to the lecture room display. To give remote attendees with cameras the option of using streaming video for their representation, we added a custom lightweight video multicast system. This distributes video (no audio) of remote viewers to other remote viewers and the lecture room.

The video encoder is designed to consume minimal processor cycles as it extracts and compresses live video frames from the video capture hardware. Multicast IP was chosen as an efficient network transport to distribute the video streams between remote clients and the lecture room client. The collaboration server manages the IP addresses and ports required for multiple concurrent streams.

The video stream decoder reads multicast video frames, decompresses them, and displays them in real-time. It is sufficiently lightweight that thirty or more videos can be played without saturating the processor. The decoder is adaptive: If processor use exceeds a threshold, the frame rate is decreased to avoid overwhelming the system.

DESIGN OF TELEP INTERFACE

In this section we present initial designs of interfaces for the lecture room and for remote viewers, along with considerations that affected the designs.

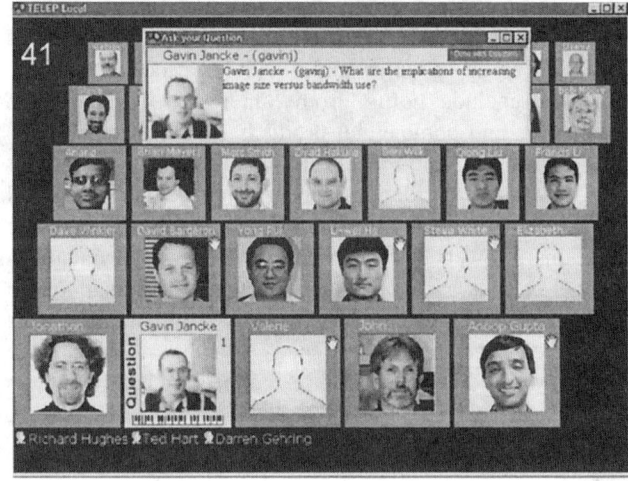

Figure 3: TELEP Lecture Room Display

TELEP Lecture Room Interface

In the lecture room, a dynamic, high-quality image is projected onto a large screen to the speaker's left (Figure 3). This TELEP display, visible to all in the room, is distinct from the normal projection of slides or overheads onto a screen behind the speaker. It constantly displays a representation of the remote audience.

Remote viewers can choose to appear as a live video feed from a desktop camera (for those who have one), a static digital image (for those with images in the system), a generic head and shoulders profile, or their logon alias at the bottom of the display (this is how users of the passive viewing system are represented).

An image is accompanied by a viewer's full name or first name if the name is long. The total number of remote viewers (including passive viewers) appears in the upper left. The images fill from the bottom and diminish in size in subsequent rows, giving a front-to-back impression. They range from 96x96 to 32x32 pixels, fonts from 11 to 8 pt Verdana. With this system, up to 38 images can be displayed; additional viewers can only watch. Overflow mechanisms are considered in the final section.

The black background was chosen to minimize increases in ambient light in the darkened lecture hall. However, a result is that the appearance or disappearance of images is quite noticeable to the live audience.

Remote viewers can control their representation on the lecture-room display in several ways. The border around the first author's image in the bottom row indicates that he has begun typing a question (it is yellow on the actual display). The number on the right indicates its position in the question queue. The animated keyboard beneath the image signals typing. When sent, a question appears in a large box, possibly overlaying other images until closed. Remote viewers can "raise a hand," as five viewers have done, enabling a speaker to verbally poll the entire audience. (The total of remote hands raised is not provided, though it could be if there is demand.) Viewers can change their representation (camera, still image, generic) at any time, or close TELEP and disappear from view. Of course, remote viewers can stop paying attention without visibly exiting by simply muting the sound.

Given our goal of minimizing speaker training, speakers do not directly manipulate the lecture-room interface. They use the standard forward audio-video channel to encourage questions, respond to questions when asked, and they can only verbally manage the question queue if conflicts arise.

TELEP Remote Viewer Interface

As noted above, TELEP currently runs alongside the pre-existing unidirectional presentation application, a web page consisting of two frames: one for the video of the speaker and one for slides. The slide frame can alternatively display other details: talk abstract, speaker biography, and so forth. Audio, video, and slide transitions are synchronized. Figure 2 is a typical arrangement, with these two frames in the center and right, and the TELEP window on the left.

The TELEP window, shown in detail in Figure 4, is divided into three main sections. The upper section has controls and indicators for the interactive features, system configuration and state information. The scrollable central section displays the representations chosen by the other remote attendees currently using the system. The lower section shows viewers who are preparing or waiting to send

questions to the presenter. This question queue is intended to facilitate the development of social protocols to govern turn-taking.

The principal interaction features (asking questions, chatting, raising a hand) are described in the next subsection. The Configure button allows viewers to select or change representation forms. They can select live video if they have a camera. Most employees in the research division have photo images in a departmental database, which TELEP can locate. Many viewers are outside Microsoft Research, so we are developing a way for anyone to provide an image. A viewer sees a preview of their image before it is sent.

The icons to the right of the Configure button launch a TELEP feedback window, a dynamically captured snapshot of the lecture-room display, and TELEP Help respectively. The only information provided uniquely by the snapshot is the arrangement of the remote viewers.

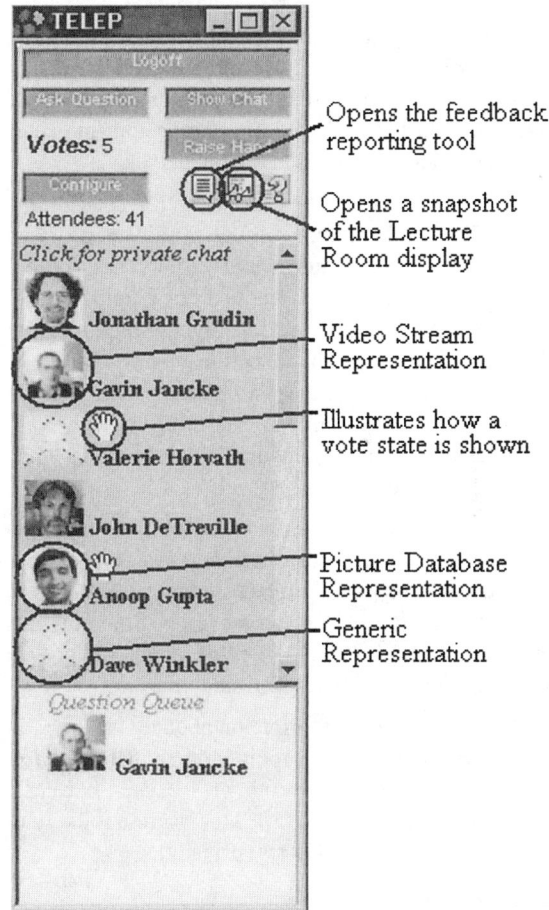

Figure 4: TELEP window for remote viewer

In the middle section, the number of remote viewers visible without scrolling would be greater if images were not displayed. The images could create more of a sense of co-presence. With viewers using photos they could also serve a minor community-building role–many remote viewers are not acquainted but could easily cross paths in the future.

Asking the Presenter a Question

When the Ask Question button is clicked, a window appears on the viewers display (Figure 5), a yellow border and question queue number immediately appears around the image in the lecture room (Figure 3), and an entry appears in the question queue on all remote displays. A prompt at the bottom of the window informs the viewer how to proceed based on their queue position and current state.

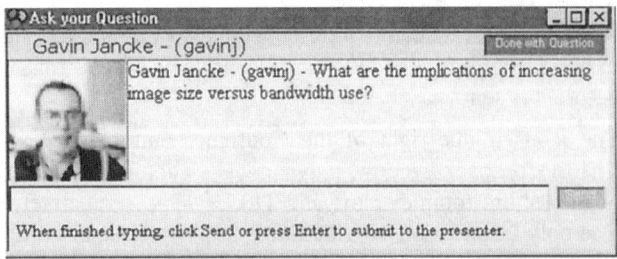

Figure 5: "Ask Question" window (questioner's view).

The remote viewer types text in the edit field at the bottom. If no other question is queued, the Send button is green and the prompt indicates that the question may be sent. Otherwise the Send button is red and the prompt indicates that another questioner is ahead in the queue.

When a question is sent, the text moves to the central area (as in Figure 5). At this point, a similar window appears on all other displays. (The lecture room display has no text entry field.) On remote displays the text entry field appears and the button to its right is labeled Reply, inviting others to respond to the question. The questioner may clarify or follow up the question, or thank the speaker, after hearing the response. Upon sending a question, a viewer is prompted to use the button in the upper right to close when done, to free the queue.

If a remote viewer sends a question when the Send button is red, it appears and the previously visible question is closed. This potentially anti-social queue-jumping feature is provided so that the discussion can move on if the previous questioner forgot to close and free the queue. This is a consequence of the minimal speaker-side interface.

We initially included more information about questioners in the window, drawn from the corporate personnel database. It was thought this might be useful for speakers, but early tests of the system indicated that speakers were not likely to use it, and it annoyed some viewers (privacy).

Remote Viewer Chat Feature

TELEP has a chat facility for use among remote viewers, not shown in the lecture room [cf. 9]. Invoked using the Chat button (Figure 4), a window appears (Figure 6). Clicking on a remote viewer's image opens another chat window for a private message. To reduce window clutter, when a message is typed and sent, the private chat window disappears and the message appears in the public chat window prefaced by "(person A to person B)" to signal that only the two can see it.

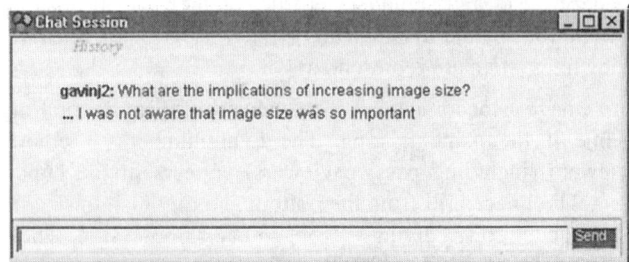

Figure 6: Chat window (remote only)

Hand Raising or Voting

A presenter may request a show of hands. As the local audience responds, remote viewers can click a button, causing hands to appear by their images (Figures 3 and 4). The vote tally is incremented. After thirty seconds, the hands disappear.

TELEP Installation, Invocation, and Maintenance

Ease of discovery and installation were considered to be critically important. For this reason, both email announcements of talks and the web calendar provide live links to the TELEP installation guide (or to TELEP once it is installed). Installation requires one button click, and subsequent modifications are automatically installed when TELEP is launched. This feature has been critical to adoption.

USER EXPERIENCE PRIOR TO TELEP

Within Microsoft Research, over 500 presentations to live audiences were multicast over the preceding two years. A distribution list of 1500 people receives talk announcements, which contain live links for viewing the presentation (and now for TELEP). Thus, many employees are fully familiar with viewing presentations live on their desktop, without interactivity. For them, the obvious comparison with TELEP is not attending in person, but between attending passively and attending with the interactivity TELEP affords. We were able to collect baseline data on how people attending in person (speakers and audience) regarded the remote viewers before and after TELEP was introduced, and how remote viewers assessed their experience before and after introduction of the system.

Initial Survey of Remote Viewing Experience

Prior to the release of TELEP, we prepared a web-based survey and emailed its URL to the presentation announcement distribution list.

This survey was designed to assess overall levels of satisfaction and problems with the pre-existing passive remote viewing system. We do not know how many of the recipients had used the system. We received 182 replies, primarily from people with an active interest in viewing presentations remotely.

The average number of presentations they reported watching remotely was 9.7; the median was 5. On average, they reported staying tuned through 54% of a presentation.

They were asked to indicate their satisfaction with the system on a 0 = Not at all to 6 = Extremely satisfied scale. The average was 3.65, slightly above the midpoint, with eight zeros and twelve 6's.

Respondents were asked how the system could be improved. The most frequent responses were requests for improved audio (in particular, for microphones that could capture live audience questions and comments), improved video, improved slide quality (if speakers do not make slides available in advance, they are simply shown in the video window and are not very readable) and greater system reliability. The most frequently requested software functionality was for remote viewer interaction with the speaker, requested by 18 respondents.

Baseline Survey of Local and Remote Experience

Next, prior to the announcement of TELEP, a paper survey was given to 11 speakers following their presentations to gauge their awareness of remote viewers and cameras, and to guess at the size of the remote audience. The local audiences ranged from 15 to 100, the remote audiences from 8 to 57, on average about 60% the local audience.

101 live audience members from eight of the talks filled out paper surveys that asked the same questions, as well as how much they had attended to the talk, daydreamed, did other work, and so forth. As noted in the introduction, remote viewing could increase multitasking or openness to distraction. We also measured live audience attrition. For four talks, we asked remote viewers to fill out a web survey that addressed the same issues. We received 31 responses.

Speakers
- Speakers were oblivious to the remote audience.

Although informed prior to talks of the ceiling-mounted cameras, nine of eleven speakers rated their awareness of remote viewers as 0 on a 0-to-6 scale, with one 1 and one 2. Ten rated the effect on their behavior at 0, with one 1. All speakers reported never looking at a camera.

- Speakers underestimated the remote audience size.

Might speakers imagine a large remote audience and be disturbed to have TELEP reveal its size? This concern appears to be unfounded: 9 of 11 speakers underestimated the remote audience size; only one greatly exaggerated it. (Actual average was 29, estimates averaged 27.)

Local Audience
- Local audiences are oblivious to remote audience.

Local audience members know that lectures are broadcast, but reported not being aware during a talk: their average rating was 0.5 on a 0 to 6 scale, with four in five rating it 0. They rated the effect on their behavior even lower at 0.2.

- They slightly underestimate remote audience size.

In only one case did an audience overestimate the remote audience size. The consensus was extremely close, but low.

- They report focusing on the talk 82% of the time.

The speaker had the viewers' attention 81.6% of the time, thinking or daydreaming 15.6%, reading or working 1.3%, and other (sleeping, looking at people, etc.) 1.6%.

Remote Audience
- They reported higher attrition than in the room.

For the live audiences measured, 65% to 90% of attendees stayed to the end. Remote viewers reported making it through only on average 37% to 67% of different talks.

- They reported greater awareness of local audience.

The average across presentations was 3.2, with behavior affected rated at 1.4. These are low, but much higher than the local audience awareness. Several specified the benefits of hearing audience questions when they were audible or repeated by the speaker, and frustration when not.

- They overestimated remote audience size and underestimated live audience size.

Remote viewers were the only group to overestimate remote attendance. When averaged, they were close, but overestimated every talk. Their estimates of live audiences were low for all talks except one. They do see occasional camera shots of the audience, but not of the whole room.

- When they were watching, they reported focusing on talks 56% of the time.

The speaker received 55.6% of their attention, thinking or daydreaming 9.8%, reading or other work up to 32% and "other" 2.6%.

For several talks, one author attended and observed interaction. No speaker was seen to poll the audience. Many questions included clarification or follow-up, which TELEP supports but the audio delay makes difficult. Many questions or comments were longer than we would expect people to type. Occasionally a discussion broke out.

VIEWER EXPERIENCES WITH TELEP

The first formal use of TELEP was a presentation to introduce TELEP itself. It was treated as a pilot and to obtain feedback. Some of the design features described above were influenced by this feedback.

TELEP has since been in regular use. It is described briefly to speakers along with the usual A/V preparation, typically a few minutes before the presentation. The authors have not intervened appreciably, other than to observe and collect data. TELEP participation in talks has ranged from 2 to 40.

Survey data addressing awareness issues are discussed below. The interaction to date has consisted of spontaneous questions from remote viewers and a little chat among remote viewers: there has been no polling.

Questions have ranged from zero to three for a talk. To date, remote questions have not coincided or required queuing. The appearance of questions has not generally been noted by speakers, but audience members (not the authors) have pointed them out. The audience has explained the latency, but speakers have to decide how to handle it. The appearance of the "question being typed" indication forces speakers to decide whether to wait or continue—and questions have been longer than we anticipated, longer than our fixed-size window could handle on occasion.

To date, chat has been used more among remote viewers, the camera operator, and the author-observers to discuss TELEP than for content. Placing private chat (appropriately labeled) in the same window as public chat has resulted in replies to private messages almost invariably being made in that window, meaning that they were made public.

Speakers were surveyed immediately following nine talks. For 8 of these, paper surveys were distributed to the live audience; 82 were filled out. 15 remote TELEP viewers responded to a request to fill out a web survey.

During two recent talks, email was sent to 36 people using the passive system only, asking them to select among alternative explanations for why they were not using TELEP. This timely intrusion yielded a 70% response rate, including a few lengthy discussions. (Response rates to other surveys: 100% of speakers, over 50% of local audiences and TELEP users, and about 25% of remote audiences when polled after a talk had completed.)

Speaker Reactions to TELEP
- Speakers generally found TELEP interesting.

They did not seem bothered, although two wrote that some training would be useful, presumably for handling questions and the 15-second latency.

- Speakers became aware of the remote audience.

Awareness rose from 0.3 to 2.2 on the 0 to 6 scale, with no presenter indicating zero. 5 of 9 reported an effect on their behavior, but not much: the average rose from 0.1 from 0.8.

- Speakers equated the remote audience to images.

Speakers estimated the remote audience size to be roughly the maximum number of images at any one time. They overlooked the aliases of passive viewers, even when these had been explained, and did not consider remote viewer turnover. (The total number of remote viewers could be twice the number appearing at any one time.)

- Speakers equated the display with the camera.

Speakers reported looking at a camera 2.6 times (versus 0 pre-TELEP). They actually were looking at the display, which was not near a camera.

Local Audience Reactions to TELEP
- The audience generally found TELEP interesting.

Most comments were positive, but some reported being distracted by changing images, especially video.

- They became more aware of the remote audience.

Their awareness rose from 0.5 to 2.9 on the 0 to 6 scale. About half reported some effect on their behavior, with the average rising to 1.0 from 0.2.

- Their remote audience size estimates reflected the number watching at one time.

Their estimates reflected the total shown on the display when around its peak. Given the relatively high turnover, this is considerably less than the total present overall.

- Their focus on the talk may have dropped slightly.

They reported 77% of their attention on the speaker (down 5%), 14.8% daydreaming or thinking, 4.6% other work (up 4%), and 2.5% "other" (up 1%), with many attributing this last to the TELEP display. This is a possible negative effect, but it is small and may decline as familiarity rises.

Remote Viewer Reactions to TELEP
- Satisfaction reported for TELEP is quite high.

TELEP received 4.4 on the 0-6 scale, up from 3.6 for the passive viewing system. But there were few 6's and numerous suggestions for improvement.

- Their estimates of remote audience size dropped.

They appeared to base the estimate on the number of TELEP viewers, not considering the passive viewers.

- Attention to speakers dropped somewhat.

TELEP users reported attending to the speaker 44% of the time, down 12% from passive viewers. Most of this was a 350% increase in "Other" activity, which several identified as being TELEP experimentation. Future polling will determine whether or not this will drop with experience.

- Some remote viewers prefer anonymity.

Several of those still watching passively mentioned the desire to be invisible, particularly when attending in the background. "More often I'm watching it (a presentation) in the background, and so prefer to remain in the background. There's a certain symmetry to it." "I would use Telep, if my identity were only revealed when I asked a question."

LESSONS LEARNED: REDESIGNING TELEP

TELEP is in routine use, requires little maintenance, and is liked by its self-selected users. Nevertheless, many of the features were not used as expected; these lessons guided the design of the next version.

Use of live video by remote viewers for representation on the lecture-room display was not successful. The live audience found it distracting, and remote viewers with cameras did not want to be seen multitasking, on the phone, and so forth. It appears though, they may be willing to show this view to other remote viewers, and they may like to turn it on when directing a question to the speaker.

Anonymous representations are needed, perhaps as an unlabeled smaller image to the back of the display. All remote viewers should probably be represented to restore the relatively accurate estimates of remote attendance. Arguably, remote questioners should have to be identified.

A camera should be placed near the display, since speakers assume one is there. The arrival of a question should be signaled by a sound. Possibly the projection should be behind the audience rather than to its side, or in both places. Some remote viewers noted that just as local audiences can see them, they would like a camera view of the audience.

The signaling of a question being typed was disruptive: speakers did not know whether to wait or continue. We also found that remote viewers are willing to ask questions, but they do so rarely and almost never queue questions. Design should focus on simplifying the initial-question case.

Displaying public and private chat messages in one window was confusing and lead to inadvertent exposure.

For the lectures we observed, audience polling is extremely rare. More complex methods of presenting alternatives and tallying votes may be useful in other settings.

Reducing the 15-second delay in presentations reaching remote viewers to a few seconds would enable remote viewers with microphones to have an audio channel to speakers. This was a feature of the Sun Forum system. However, it is more complicated than it first seems. Questioners often prefer to catch a speaker's attention before speaking. Remote viewers may rarely use this without a more complex interface.

A New Version of TELEP

We have now released a version of TELEP designed to profit from the experiences described above. It is integrated into the viewing system, so all remote viewers participate.

Representation options have been expanded: Viewers can identify an image or use a camera snapshot as a still image. They can independently choose to show their name or be anonymous. The display can accommodate up to 60 images, moving anonymous images to the rear and moving images forward as vacancies occur to reduce visual disruption.

Questions are not foreshadowed in the lecture room until sent, at which point a sound signals their arrival. Dynamic messages guide viewers when multiple questions are being queued. Public and private chat are distinct windows. A header indicates that chat is not seen in the lecture room.

We are starting to collect data on the use of the new system. With users no longer self-selected and the basic problems addressed, we will observe how use of the system evolves.

CONCLUDING REMARKS

Although TELEP has enabled more interaction, a larger purpose was to raise mutual awareness of local and remote participants, and among remote participants. Indications are that it is succeeding in this. This could have important indirect consequences. Our initial survey found that major dissatisfactions of remote viewers included not having questions asked loudly enough or repeated by the speaker, not having slides delivered early enough to put online, not having legible overheads or whiteboard writing. As speakers and (equally importantly) their local hosts become more aware of the remote viewers, these problems will be more naturally addressed.

With use of TELEP, will more attention to remote viewers be at the expense of the local audience? Will it lead more people to attend remotely, where they are subject to more distractions? Will smaller live audiences demotivate speakers, or will more interaction with remote, often large audiences compensate? These are interesting questions that longer-term studies with TELEP will help answer.

ACKNOWLEDGMENTS

Steve White worked on early version of prortotype, Jeremy Crawford and James Crawford of MSRN helped with deployment, and Harry Chesley and Lili Cheng of the MSR Virtual Worlds Group contributed.

REFERENCES

[1] Dourish, P. and Bly, S. (1992). Portholes: Supporting awareness in a distributed work group. *Proc. CHI'92*, 541-547. ACM.

[2] Girgensohn, A., Lee, A. and Turner, T. (1999). Being in public and reciprocity: Design for Portholes and user preference. *Proc. INTERACT'99*, 459-465.

[3] Gutwin, C., Roseman, M. and Greenberg, S. (1996). A usability study of awareness widgets in a shared workspace groupware system. *Proc. CSCW'96*, 258-267. ACM.

[4] Isaacs, E.A., Morris, T., and Rodriguez, T.K. (1994). A forum for supporting interactive presentations to distributed audiences. *Proc. CSCW'94*, 405-416. ACM.

[5] Isaacs, E.A., Morris, T., Rodriguez, T.K., and Tang, J.C. (1995). A comparison of face-to-face and distributed presentations. *Proc. CHI'95*, 354-361, ACM.

[6] Isaacs, E.A., and Tang, J.C. (1997). Studying video-based collaboration in context: From small workgroups to large organizations. In K.E. Finn, A.J. Sellen & S.B. Wilbur (Eds.), *Video-Mediated Communication*, 173-197. Erlbaum.

[7] Ishii, H., Kobayashi, M., and Arita, K. (1994). Iterative design of seamless collaboration media: From Team-WorkStation to ClearBoard. *Comm. of ACM, 37*, 8, 83-97.

[8] Mantei, M.M., Baecker, R.M. Sellen, A.J., Buxton, W.A.S. and Milligan, T. (1991). Experiences in the use of a media space. *Proc. CHI'91*, 203-208.

[9] Rekimoto, J., Ayatsuka, Y., Uoi, H. & Arai, T. (1998). Adding another communication channel to reality: An experience with a chat-augmented conference. *CHI'98 Summary*, 271-272.

[10] Root, R.W. (1998). Design of a multi-media vehicle for social browsing. *Proc. CSCW'88*, 25-38. ACM.

[11] Tang, J.C and Rua, M. (1994). Montage: Providing teleproximity for distributed groups. *Proc. CHI'94*, 37-43.

[12] Vellon, M., Marple, K. Mitchell, D. and Drucker, S 1998. The Architecture of a Distributed Virtual Worlds System. *Proc. of the 4th Conference on Object-Oriented Technologies and Systems (COOTS)*. 1998.

[13] White, S.A., Gupta, A., Grudin, J., Chesley, H., Kimberly, G. and Sanocki, E. (2000). Evolving use of a system to support education at a distance. To appear in *Proc. HICSS-33*. IEEE.

[14] Centra Symposium Software. http://www.centra.com./

[15] NetPodium. http://www.netpodium.com/

[16] PictureTel Systems. http://www.picturetel.com/

[17] Placeware Conference Center. http://www.placeware.com/

[18] Stanford Instructional Television Network. http://www-sitn.stanford.edu

Coming to the Wrong Decision Quickly: Why Awareness Tools Must be Matched with Appropriate Tasks

Alberto Espinosa, Jonathan Cadiz, Luis Rico-Gutierrez, Robert Kraut, William Scherlis
Carnegie Mellon University
Pittsburgh, PA 15213 USA
+1 412 268 7694
robert.kraut@cmu.edu

Glenn Lautenbacher
School of Information Science
University of Pittsburgh
Pittsburgh PA 15217 USA
glenn@sis.pitt.edu

ABSTRACT
This paper presents an awareness tool designed to help distributed, asynchronous groups solve problems quickly. Using a lab study, it was found that groups that used the awareness tool tended to converge and agree upon a solution more quickly. However, it was also found that individuals who did not use the awareness tool got closer to the correct solution. Implications for the design of awareness tools are discussed, with particular attention paid to the importance of matching the features of an awareness tool with a workgroup's tasks and goals.

Keywords
Task awareness, workgroups, awareness devices, computer-mediated communication, distributed work, asynchronous work

INTRODUCTION
The group is the fundamental unit of work in organizations [15, 25]. However, recent trends in globalization, downsizing, and outsourcing are changing the look of group work. It used to be a safe assumption that most members of a group worked in the same office at the same time; however, it is not unusual today to find members of groups distributed throughout the world, forcing team members to work in different places and often at different times.

One of the key issues for these distributed, asynchronous groups is how to solve problems quickly and efficiently. Problem-solving often involves exploring and researching several options, weighing the advantages and disadvantages of each, and then choosing the best option based on all the information available. However, when team members are not working in the same place at the same time, important awareness information is often lost. Examples of lost information include who has explored which options, who knows what information, and what people's lines of reasoning are about the best possible solution.

Thus, the goal of this research is to design and test a tool to help distributed, asynchronous groups solve problems by providing task awareness. It is hypothesized that by providing task awareness, group members' lines of reasoning about the best possible solution will converge more quickly, thus helping the group to solve the problem faster.

Groups provided with a task awareness tool did exactly as we predicted: they came to decisions more quickly. However, groups without the tool came closer to the correct solution, necessitating a closer look at how awareness information can affect behavior in asynchronous, distributed group work.

In the following sections, we discuss the research on which this project is based, the design of the awareness tool, the method used to test the awareness tool, and our findings. We will also discuss the implications of our findings, especially in regard to designing awareness tools with features that are appropriate for the nature of a group's task.

PREVIOUS RESEARCH
Group awareness has been defined as "an understanding of the activities of others, which provides a context for your own activity" [7]. The value of providing awareness to teams is suggested in the literature, which indicates that members of workgroups will be more successful if they maintain awareness of the state of the team, task, and environment [4, 20]. It is also suggested that simple awareness of one's colleagues is a strong predictor of success in collaborations, thus highlighting the importance of awareness for team performance [12].

The Task: Distributed Problem-Solving
However, the context of a group's work cannot be ignored when discussing the merits of awareness. The nature of the task makes a difference when studying group performance [18], particularly with computer mediated group interaction [19].

Thus, it is important to note that our research focuses on distributed problem-solving under asynchronous conditions. Schlichter, Koch, and Bürger [23] define distributed problem-solving as the cooperative activity of several decentralized and loosely coupled problem-solvers acting in separated environments. A key point for successful collaboration in distributed problem-solving has to do with how well individual work relates to the objectives of the group as a whole.

We believe one way to help distributed problem-solving groups is by replacing some of the awareness information that is lost when team members do not work in the same location. However, choosing what kind of awareness information to provide can be difficult since the literature defines several types and taxonomies of awareness that could be given to groups [1, 5, 9, 10, 13, 14, 26]. Among these descriptions, the most useful model for this research is the framework of task awareness provided by Chen and Gains [5].

The Focus: Peripheral, Chronological Awareness

Chen and Gaines [5] identify two analogous forms of awareness based on a collective intelligence model:

1) Collective awareness: awareness of the group's collective, long-term memory
2) Awareness of teammates

The first form, collective awareness, exists in groups at three levels of detail: deep awareness (a highly detailed level of information), peripheral awareness (a less detailed, but still substantial level of information), and global awareness (a relatively low level of detail). Choosing the correct level of detail is important since providing too little information could be useless, while providing too much information could overload a person and reduce the amount of time and cognitive resources available to work on the primary task [11, 25, 26]. Taking this into consideration, we believe peripheral awareness is the correct level of detail for asynchronous, distributed, problem-solving groups. Peripheral awareness can help to provide some of the contextual information typically unavailable in the absence of physical proximity and face-to-face interaction.

The second form, awareness of teammates, can be of three types: resource awareness (who has what expertise within the group), task-socio awareness (information about social/political dynamics within the group), and chronological awareness. Chronological awareness is the instantaneous awareness that an individual has regarding the activities of others and knowing that something has changed—i.e., what, when and by whom.

We believe chronological awareness has the most potential to help asynchronous, distributed, problem-solving teams. First, knowing the activities of one's teammates can help the team to divide labor quickly and efficiently. If a person knows that a team member has already explored a certain solution, then a duplication of effort can be avoided. Second, a lesser amount of direct communication is necessary if team members know what each other are doing. Third, knowing the activities of one's teammates can help a team to form a shared mental model of the work, which may trigger new lines of reasoning about how to solve a problem [17].

For these reasons, the awareness tool we developed provides members of asynchronous, distributed, problem-solving teams with peripheral, chronological awareness.

THE AWARENESS TOOL

Figure 1 shows the awareness tool we designed. (This tool was adapted from a similar system developed to deliver awareness information to MBA students at Carnegie Mellon University [11]). Participants in our lab study had to decide how to treat a cancer patient by exploring a set of documents, each with information that may or may not be helpful. Examples of documents include x-rays, results of blood tests, and articles describing types of therapies.

In this study, the set of documents was large enough such that one person could not read and evaluate all the documents in the time allowed. Furthermore, participants could not read any document any time they wished. Sometimes documents were "denied" to participants, meaning that the group had not fulfilled the prerequisites to view a document. For example, an x-ray should not be ordered for a woman before checking her medical history to ensure she is not pregnant, or an expensive test should not be run before reviewing the patient's basic history. Prerequisites are applied at the group level and structured such that using common sense and good judgment should result in fewer denials.

The awareness tool provides information about what

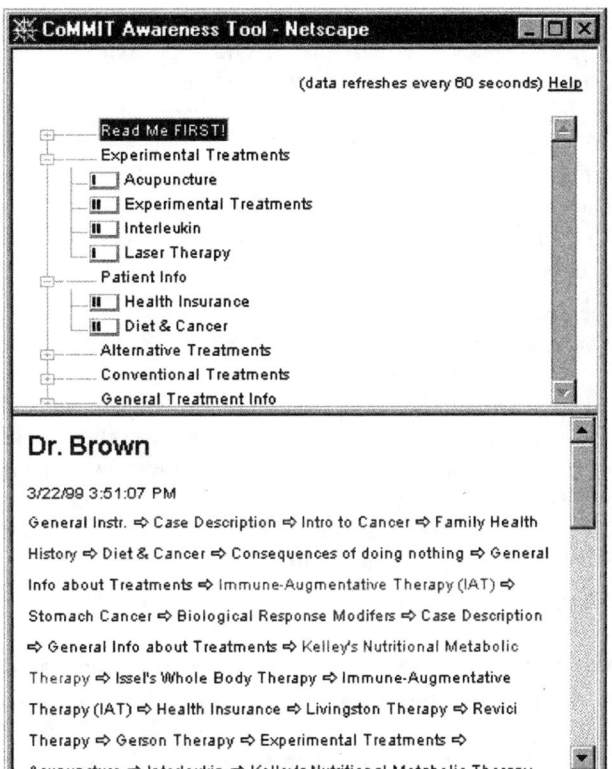

Figure 1: The tool designed to provide peripheral, chronological awareness. Items are color coded so that successful requests are blue while denials are red.

documents have been explored and denied using two displays. The top half of the window shown in Figure 1 contains all the documents in the document universe, structured as a two-level tree. The first level has all the document groups while the second level displays all the document names. An icon with zero to five bars (corresponding to the amount of activity associated with the document) is displayed along with the document name. If the last request for the document was denied, the bars are red. If the last request was successful, the bars are blue. Activity is represented on a five-point scale where each attempt to view a document increases the amount of bars shown. Activity is adjusted for time on a logarithmic scale; thus, recent activity will generate more bars.

The bottom section of Figure 1 contains a chronological listing of all the documents explored by one's team members. As in the top section, successful document requests are blue and denials are red. Such peripheral, chronological awareness information is intended to help individuals infer and follow their team members' lines of reasoning.

The awareness tool was implemented using Microsoft's Visual Basic and Active Server Pages. The data were accessed from a relational database kept by CoMMIT, the system participants used to explore documents and discuss solutions with team members.

The COMMIT System

CoMMIT (Collaborative MultiMedia Instructional Toolkit) is a computer system that facilitates asynchronous, distributed problem-solving. It is a web-based learning system designed to help groups solve diagnostic problems [16]. CoMMIT is used to explore documents with information about medical patients and to discuss possible diagnoses with teammates. As team members move through the documents, they can record notes on the documents and read comments from other team members.

The main window of the CoMMIT system is shown in Figure 2. Documents are structured into groups. To access a document, the user first clicks on one of the groups (labeled area 1 in Figure 2), which causes the documents in that group to display below (area 2). The user then clicks one of the documents in area 2, which causes the contents of the document to appear in area 3. Each document has three portions: the material with information about the problem (area 3), comments from team members (area 4), and a text box where additional comments can be made by the user (area 5).

Note that the group labels in area 1 of Figure 2 correspond to the groups shown in the awareness tool in Figure 1. Similarly, the document names in area 2 of Figure 2 correspond to the document names in the awareness tool.

In addition to the window shown in Figure 2, CoMMIT provides users with an additional "Notepad" window (not shown) that lists all team members' comments on all

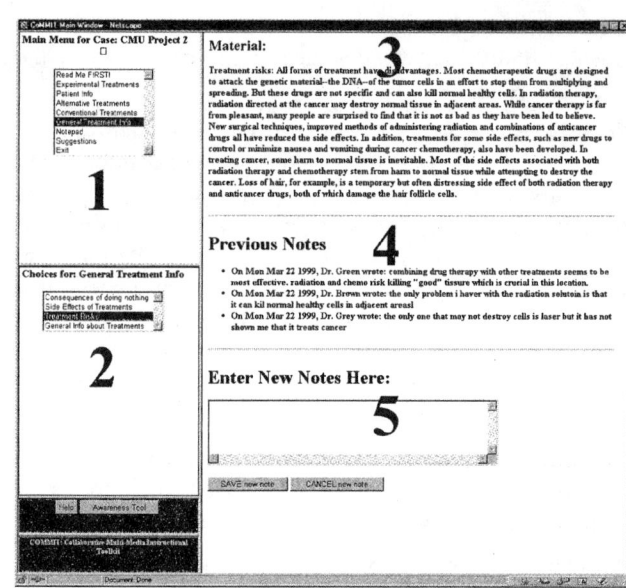

Figure 2: The main window of the CoMMIT system. The "Awareness Tool" button was only visible for participants in the experiment condition.

documents. The notepad window also provides a separate discussion area where team members can talk about strategies, share observations and thoughts, and decide upon the final solution.

Even though CoMMIT provides the team with some degree of awareness about each other's thinking about specific documents and the problem in general, it does not help members to know easily who has explored which documents and in what order documents were explored. For this reason, the CoMMIT system was chosen as the focus for this research.

HYPOTHESES

Technology affects social interaction [2, 3, 6]. As such, we speculate that an information technology that provides peripheral and chronological information of team member's document reading behavior will have an effect on team interaction. More specifically, we speculate that peripheral, chronological awareness of team members' information search patterns will help people to infer and follow lines of reasoning of other teammates, thus helping the team converge toward a joint solution.

H1: Awareness of peer document search patterns helps convergence of lines of reasoning of team members, thus helping them achieve consensus on joint team solutions of diagnostic problems.

We anticipate that this more effective team problem-solving process will translate into higher team performance. Furthermore, we also anticipate that peripheral and chronological awareness of information search patterns will help teams develop more effective problem-solving strategies, such as dividing labor (e.g., deciding who needs to read what), developing shared mental models (e.g., following another team member's information search pattern), and reducing communication time (e.g., knowing who read what reduces the need for asking about it).

H2: Tools that provide peripheral, chronological awareness have a positive effect on asynchronous, distributed team performance in diagnostic, problem-solving tasks.

METHOD

To test these hypotheses, 60 participants were recruited from the Pittsburgh community, including students from the University of Pittsburgh and Carnegie Mellon University, as well as some non-student Pittsburgh residents. Participants were randomly grouped into 20 teams of three people each. Half of the teams were provided with the awareness tool.

Team members worked on the diagnostic problem at the same time but were randomly assigned to a lab at either the University of Pittsburgh or Carnegie Mellon University. Participants were identified to each other by pseudonyms and had no way of knowing who their teammates were or where they were located. Participants were trained on how to use the system and tools and then were given one hour to work on the problem.

Even though team members worked on the problem during the same time period, steps were taken to make the interaction asynchronous. These include not guaranteeing that all team members would start or end at the same time, periodically asking participants to stop working and write down their current thoughts about the problem, and having them fill out a survey halfway through the session. In addition, the interaction was made more asynchronous by inherent latencies within the CoMMIT system, as well as the substantial amount of individual text reading and notepad writing required of participants.

Participants were paid $15 for participating in the experiment. As additional motivation, two $150 prizes (one prize for each experimental condition) were awarded to the teams that came up with the best solution to the problem. To make the task more realistic and discourage haphazard document reading, participants were told that minimizing document requests and document denials would improve their chances of winning the $150 prize.

The Problem: How to Treat a Cancer Patient

Participants were told that all of them were medical doctors with the task of determining how to treat a patient with stomach cancer. The problem was created such that no medical knowledge was required to solve the problem. In fact, the problem is a reformulated version of Dunker's radiation problem [8], which is widely used with cognitive science students. The original problem involves the eradication of a malignant tumor without killing the patient, but our reformulated version extended the problem by introducing issues such as medical insurance, side effects, and family health history.

Even though the solution was contained in the information provided to participants, teams had difficulty finding the correct solution within the allotted time. Furthermore, a reading of the participants' chat text reveals that even when a member of a group found the correct solution and suggested it to the group, the other group members often did not agree. Only one group agreed upon the correct solution after substantial discussion.

RESULTS

Where applicable, most statistical tests were conducted at both the individual and group levels of analysis. Aside from usage data gathered by the CoMMIT system, data were also collected via two surveys. The first survey was given after participants had worked on the problem for approximately 30 minutes (session 1). After completing the first survey, participants were given 30 more minutes to work on the problem (session 2). At the conclusion of session 2, a second survey was given.

H1 Results: Convergence of Lines of Reasoning

Convergence of lines of reasoning was defined for our purposes as the degree of team agreement on the final solution. Except where noted otherwise, one-way ANOVA results are reported. The main variable studied captures solution agreement by team members from the second survey. On that survey, after recording the team's solution, participants were asked to rate the degree to which they agreed with the solution on a five point scale (1=strongly agree, 5=strongly disagree).

We did not find a significant difference between conditions with respect to the mean agreement level. The average in both conditions centered around 1.5 indicating a good deal of agreement in both conditions. However, as illustrated in the box plot in Figure 3, an F-ratio test revealed that the teams in the tool condition had significantly less variance in solution agreement ($p<0.01$, individual level; $p=0.25$, team level). This provides some evidence that teams with the awareness tool have a narrower range of solution agreement ratings.

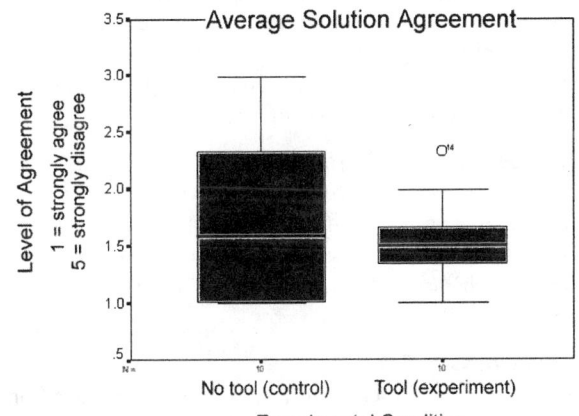

Figure 3: Box plot showing how much individuals agreed with their group's solution to the problem.

Agreement with Solution		Number of People		
		Without Tool	With Tool	Total
Strongly Agree	1	15	18	33
	2	10	6	16
	3	0	5	5
	4	0	0	0
Strongly Disagree	5	3	0	3
Total		28	29	57

Table 1: Degree of individual agreement with group solution.

In other words, teams in the no-tool condition have more extreme levels of disagreement. This is apparent from the chi-square analysis in Table 1. A chi-square test of independence at the individual level provided evidence that there is a significant difference in the pattern of solution agreement between conditions (χ^2=9.258, p=0.026).

As Table 1 illustrates, more individuals in the tool condition highly agreed with their team solution, while more teams in the no-tool condition highly disagreed with the team solution. No individuals in the tool condition either disagreed or highly disagreed with their team solution. Because we observed that some participants in the tool group did not use the awareness tool frequently, we split individuals in the tool condition at the median into high and low tool users based on how much time they had the tool active on their screen. We found that the high tool usage group had significantly higher levels of agreement (p=0.016) than the low tool users and moderately higher levels of agreement than everyone else (p=0.103). However, these effects were not significant at the team level, providing some evidence that the negative effects of low tool usage by some individuals within the team may offset the benefits of high tool usage by others. Thus, high tool usage by all team members is important when seeking a unified team solution.

In sum, the results above provide adequate support for the first hypothesis: teams that used the awareness tool more had a higher level of convergence toward a team solution. Furthermore, high tool usage accentuates this effect and increases the level of team agreement, provided that most team members are high tool users.

H2 Results: Team Performance

To test the second hypothesis, two types of performance measures were established. The first measure dealt with the correctness of a team's final solution; the second measure dealt with how efficiently the team searched for the solution.

Correctness of Solution

How close a team came to finding the correct solution was measured in two ways. First, at the beginning of the second survey, all participants had to type their group's solution. No team in the tool condition and only one team in the no-tool condition figured out the right solution. This was expected to some extent since the problem had been purposely developed to be difficult to resolve in the time provided. Even the one team that found the right solution had a substantial amount of debate before reaching an agreement. Members of two other teams discussed the right solution but did not select it.

Second, also in the second survey, participants had to rate several possible hypotheses on a five point scale indicating the degree to which they believed the hypotheses were correct. (Most of the analysis for the second hypothesis was done using one-way analysis of variance. Similar results were obtained using two-way analyses of variance and ordinary least squares regression models using a number of control factors. Thus one-way ANOVA results are provided here, except where noted otherwise.) Surprisingly, analyzing these data we found that individuals and teams in the no-tool condition were significantly closer to the right solution (p=0.001, individual level; p=0.17, group level).

We suspected these results could be due to the process constraints imposed by the system and the experimental setting. The system denies documents when not requested in a logical order, while the experimental setting rewards minimal denials and minimal document requests. Thus, an alternative explanation for our results may have been that the awareness tool helps teams cope more effectively with process constraints by providing awareness information about these constraints, at the expense of not focusing as much on the information provided to resolve the case. However, we tested the effect of tool use on denials and document requests and did not find a significant effect.

Search Efficiency

The second type of performance measure was intended to evaluate the team's solution search efficiency in terms of document requests, document denials, and division of labor. Chronbach's α reliability scores were computed for each team on the number of times each document had been requested by each team member. This was done to find the similarity of search patterns and used as a proxy for division of labor within the team. The teams in the no-tool condition had significantly higher reliability scores (p=0.012) indicating that teams in the tool condition had less overlap in their requests for documents, suggesting a higher level of division of labor. Further analysis of document requests, number of document denials, and number of entries into comment notepads revealed no significant differences between the two conditions.

We speculated that these results could be due to the fact that some individuals in the tool condition did not actively use the tool. Thus, once again we split the tool group into high and low tool users. One interesting result found was that high tool users made more entries in the notepads (p=0.02) but used fewer words (p=0.03) than low tool users, providing some evidence that high tool users interact more synchronously. Although not highly significant, a moderate interaction effect between high tool usage and session was found (p=0.151), suggesting that high tool users may become more synchronous than low tool users as the deadline for solution submission approaches.

In sum, these results provide some evidence that teams in the tool condition are somewhat more efficient in their solution search by better dividing labor. Teams with high tool usage were also more efficient by interacting more frequently and synchronously, while using fewer words. Other than these moderate results, the data did not adequately support the second hypothesis. Furthermore, teams in the no-tool condition were more likely to select the right solution as a plausible solution.

Additional Results
We conducted further analysis to investigate the effects of the awareness tool on team processes. Several team process questionnaire items on team satisfaction, strategy strength, team spirit, and team communication were reduced using the Principal Components method of factor analysis. We found no significant tool effects on any of these factors.

Further tests were then conducted to evaluate the effect of the tool on the team interaction process using objective data collected from notepad entries. Two-way analysis of variance models were formulated to test differences between conditions and between the two sessions in the average number of notepad entries made by each team and the average number of words used per notepad entry. The respective differences in variance were also tested for these two variables to evaluate the effect of the awareness tool on homogeneity of interaction within the team.

As Figure 4a (top) illustrates, teams in the tool condition made fewer entries on the notepads, but this difference was not significant (p=0.210). Overall, teams made more entries in the second session (p=0.071). However, as Figure 4b (middle) illustrates, when the same test was done between high tool user groups against all other groups, the difference between groups became moderately significant (p=0.088). Finally, as illustrated in Figure 4c (bottom), teams in the tool condition exhibited more evenness in the use of notepads, but this difference was not significant (p=0.129). As the plot illustrates, the difference in evenness in notepad use between conditions becomes more marked in the second session. In fact, teams in the tool condition tended to become more even, while teams in the no-tool condition tend to become less even towards the end. Interestingly, there is no difference in evenness of notepad use between high tool usage teams and low tool usage teams. Also, the team variance in the number of words per notepad entry tends to be larger for teams in the tool condition than for teams in the no-tool condition, but this difference is not significant (p=0.226). It is interesting to note that teams in the tool condition become more even in the number of entries in notepads, but more variable in the number of words per notepad entry, whereas the opposite happens with teams in the no-tool condition.

In order to gain a deeper understanding of how teams communicated and processed information, we carefully reviewed all chat text entered by teammates in the notepads, which was the only communication channel available. This qualitative review of notepad interaction revealed that teams without the awareness tool did a more thorough job of inspecting and discussing substantive issues regarding document contents, while teams with the awareness tool seemed to be more concerned with discussions of process to help them converge towards a solution. This is consistent

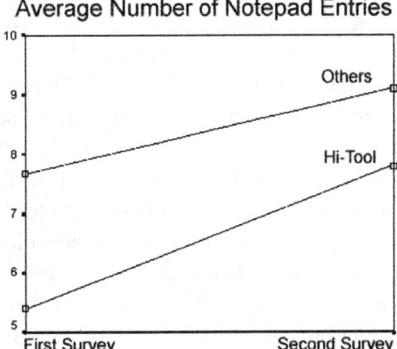

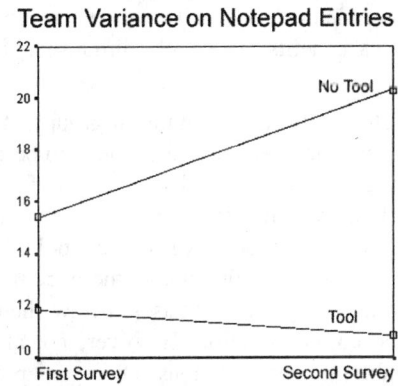

Figure 4: A) Interaction Plot—Notepad Entries Against Tool Condition by Survey (top). B) Interaction Plots—Notepad Entries Against High Tool Usage by Survey (middle). C) Team Variance on Notepad Entries Against Tool by Survey (bottom).

with the fact that teams in the no-tool condition had more overlap in the documents they read, which is perhaps what helped them have more common ground to do a more thorough job of discussing substance. The fact that teams in the tool condition had less overlap in the documents they read means that they were likely to have less common ground for discussions about substance, thus making them focus more on discussions about process.

DISCUSSION

Although the results obtained provide some encouraging evidence about the benefits of awareness tool use, they also make evident how the availability of such tools can be more of a distraction when available but not properly used. While peripheral, chronological awareness seems to help teams converge their lines of reasoning and reach joint team solutions more efficiently, it seems that this comes at the expense of a less thorough review of available information. Similarly, while the awareness tool seems to have contributed to a more efficient division of labor, the resulting reduced overlap in documents read by team members seems to result in a loss of common ground, thus foregoing the benefits of shared mental model formation. This is consistent with the fact that teams in the no-tool condition were less efficient in reaching consensus about the joint solution, but got closer to the right solution than teams in the tool condition.

Similarly, it seems that simply having the awareness tool without using it frequently may make some of these negative effects materialize more often, without capitalizing on some of the benefits of the tool. Furthermore, the negative effects of low tool use may offset some of the benefits of high tool use within a team when only one or two members use the tool actively. Consequently, although it is quite possible that the benefits of high tool usage may be attributed to a self-selection bias, it seems clear that such benefits can only materialize when all members use the tool actively and effectively. This effect is similar to the negative externalities commonly observed in other groupware systems like group schedulers and discussion databases, in which if one or two members do not use a given tool, its benefits are greatly diminished for all [21, 22].

Also, consistent with the literature on groups, it is evident from our results that awareness tools need to be matched to appropriate tasks [18, 19]. The primary focus of our awareness tool was helping teams to solve a problem quickly. This is precisely what the tool did in our experiment, but this benefit came at the expense of people not inspecting enough documents to allow them to come closer to the correct solution. However, not all problems have verifiable, correct solutions. Divergent problems, as discussed by Senge [24], do not have a single correct solution. With divergent problems, coming to a strong consensus about the course of action to take is more important than finding a "correct" solution, since no such solution exists.

Therefore, the features implemented in our awareness tool are adequate for a divergent problem in which there is no apparent right solution, and in which reaching a unified team solution is important. Strategic planning, sports team strategies, surgical teams in the operating room, and economic planning committees are examples of situations in which awareness tools of this type can help. However, in order to provide support for problems in which a correct solution does exist, different types of awareness information would have to be presented to the user. This highlights the all familiar tradeoff between general awareness tools that provide a little help for many types of tasks, and specific awareness tools that significantly help only one type of task. It also highlights the need to find the optimal amount and type of awareness information to make available without creating unnecessary distractions and information overload.

CONCLUSION

Perhaps the most useful lesson learned from this experiment is that although task awareness can be very beneficial to team performance, it may actually be detrimental to the team if the task awareness information provided is not properly matched to the needs of the specific task. Thus, in addition to the obvious conclusion that our tool could have been more helpful had the design matched the task, we believe our tool could also be more helpful in the following conditions:

1) Truly asynchronous conditions: although an asynchronous environment was simulated, we believe our awareness tool would be more effective had the teams interacted asynchronously over a longer period of time. Results of an earlier pilot test where interactions were truly asynchronous suggest that task awareness may yield more benefits under more asynchronous conditions. For example, we noticed the awareness tool was very useful in detecting social loafing, which was not as helpful in the shorter time period of our lab experiment.

2) More dynamic tasks: the literature on situation awareness suggests that general awareness is more critical under dynamic conditions. Our task and tool could be reformulated such that the patient's health deteriorates over time and each piece of information available has a cost to the team. The task used for the experiment was static, thus making situation awareness less important.

ACKNOWLEDGMENTS

This material is based on work supported by the Defense Advanced Research Projects Agency and administered by the Office of Naval Research under agreement number N66001-96-C-8506. The U.S. Government is authorized to reproduce and distribute reprints for government purposes notwithstanding any copyright annotation thereon. The views and conclusions contained herein are those of the authors and should not be interpreted as necessarily representing the official policies or endorsements, either expressed or implied, of the Defense Advanced Research Projects Agency or the U.S. Government.

REFERENCES

1. Adams, M., Tenney, Y., and Pew, R. Situation Awareness and the Cognitive Management of Complex Systems. *Human Factors*, 37(1), 1995.

2. Barley, S. The Alignment of Technology and Structure Through Roles and Networks. *Administrative Science Quarterly*, 1990, 61-103.

3. Barley, S. Technology as an Occasion for Structuring: Evidence from Observations of CT Scanners and the Social Order of Radiology Departments. *Administrative Science Quarterly*, 1986, 78-108.

4. Cannon-Bowers, J., Salas, E., and Converse, S. Shared mental models in expert decision-making teams. In N. J. Castellan, Jr. (Ed.), *Current Issues in Individual and Group Decision Making*. Erlbaum: Hillsdale, NJ, 1993, 221-246.

5. Chen, L., and Gaines, B. A Cyber-Organism Model for Awareness in Collaborative Communities on the Internet. *International Journal of Intelligent Systems*, 12(1), 1997.

6. DeSanctis, G., and Poole, M. Capturing the Complexity in Advanced Technology Use: Adaptive Structuration Theory. *Organization Science*, May 1994.

7. Dourish, P. and Bly, S. Portholes: Supporting Awareness in a Distributed Work Group. *Proceedings of the ACM Conference on Human Factors in Computing Systems* (INTERCHI '92).

8. Dunker, K. On Problem Solving. *Psychological Monographs*, 58 (Whole No. 270), 1945.

9. Endsley, M. Toward a Theory of Situation Awareness in Dynamic Systems. *Human Factors*, 37(1), 1995.

10. Fuchs L., Pankoke-Babatz U., and Prinz W. Supporting Cooperative Awareness with Local Event Mechanisms: The Groupdesk System. *Proceedings of the 4th European Conference on Computer Supported Cooperative Work* (ECSCW '95), 247-262.

11. Fussell, S., Kraut, R., Lerch, F, Scherlis, W., McNally, M., and Cadiz, J. Coordination, Overload and Team Performance: Effects of Team Communication Strategies. *Proceedings of the 1998 Conference on Computer Supported Cooperative Work* (CSCW '98), 275-284.

12. Gaver, W. The Affordance of Media Spaces for Collaboration. *Proceedings of the 1992 ACM Conference on Computer Supported Collaborative Work* (CSCW '92).

13. Greenberg S., Gutwin C., and Cockburn, A. Using Distortion-Oriented Displays to Support Workspace Awareness. Technical report, Dept of Comp. Science, Univ. of Calgary, Canada, Jan. 1996.

14. Gutwin, C., Stark, G., and Greenberg, S. Support for Workspace Awareness in Educational GroupWare. *Proceedings of the 1995 Conference on Computer Supported Cooperative Learning* (CSCL '95).

15. Hackman, R. The Design of Work Teams. In Lorsch, J. (Ed.), *Handbook of Organizational Behavior*, Prentice-Hall, 1987.

16. Lautenbacher, G., Campbell, J., Sorrows, B., and Mahling, D. Supporting Collaborative, Problem-Based Learning Through Information Systems Technology. *IEEE Conference on Frontiers in Education*, 1997.

17. Lautenbacher, G., and Mahling D. Interface Design Issues for Web-Based Collaborative Learning Systems. *Proceedings of the 1997 WebNet Conference*.

18. McGrath, J. *Groups: Interaction and Performance*, Prentice Hall: Englewood Cliffs, N.J., 1984.

19. McGrath, J. and Hollingshead, A. *Groups Interacting With Technology*, Sage Publications: Thousand Oaks, California, 1994.

20. Orasanu, J. and Salas, E. Team Decision Making in Complex Environments. In Klein, G., Orasanu, J., and Calderwood, R. (Eds.), *Decision Making in Action: Models and Methods,* Ablex Publishing Co: Norwood, NJ, 1993.

21. Orlikowski, W. Learning From Notes: Organizational Issues in Groupware Implementation. *Proceedings of the 1992 Conference on Computer-Supported Cooperative Work* (CSCW '92).

22. Orlikowski, W. Improvising Organizational Transformation Over Time: A Situated Change Perspective. *Information Systems Research*, 1996, 63-92.

23. Schlichter, J., Koch, M., and Bürger, M. Workspace Awareness for Distributed Teams. *Proceedings of the Workshop Coordination Technology for Collaborative Applications*, Singapore, 1997.

24. Senge, Peter. *The Fifth Discipline*. Doubleday: New York, 1990.

25. Sproull, L. and Kiesler, S. *Connections: New Ways of Working in the Networked Organization*. MIT Press: Cambridge, Massachusetts, 1991.

26. Wellens, R. Group Situation Awareness and Distributed Decision-Making: From Military to Civilian Applications. In Castellan, J. (Ed.), *Individual and Group Decision Making: Current Issues*, LEA Publishers, 1993.

Gaze Communication using Semantically Consistent Spaces

Michael J Taylor and Simon M Rowe
Canon Research Centre Europe
Guildford GU2 5YJ, UK
+44 1483 448882
{mjt,simonr}@cre.canon.co.uk

ABSTRACT

This paper presents a design for a user interface that supports improved gaze communication in multi-point video conferencing. We set out to use traditional computer displays to mediate the gaze of remote participants in a realistic manner. Previous approaches typically assume immersive displays, and use live video to animate avatars in a shared 3D virtual world. This shared world is then rendered from the viewpoint of the appropriate avatar to yield the required views of the virtual meeting. We show why such views of a shared space do not convey gaze information realistically when using traditional computer displays. We describe a new approach that uses a different arrangement of the avatars for each participant in order to preserve the semantic significance of gaze. We present a design process for arranging these avatars. Finally, we demonstrate the effectiveness of the new interface with experimental results.

Keywords
Gaze, avatar, animation, virtual meeting, videophones.

INTRODUCTION

The telephone is a remarkably successful means of communicating with one other remote person. However, communicating with more than one person using audio alone is much less successful [7]. In this situation, the appropriate use of video can help by supporting some of the rich set of visual cues that aid effective communication in face-to-face meetings [3]. This paper pursues the vision of making the multi-point virtual meeting as convenient as a telephone call [16] and as effective as a face-to-face meeting.

The design of such an interface needs to address many relevant cues, which can be broadly divided into two categories: spatial and non-spatial. Both gaze and gesture, together with stereo sound, are inherently 3D spatial cues. Unfortunately, simplistic videoconferencing interfaces (Figure 1a) destroy the 3D context within which such cues originate. Misinterpretation is inevitable, and users quickly become confused about when to speak and who is speaking [18]. The result is unsatisfactory group communication.

The second category of cues covers the non-spatial aspects of the interface. Examples include audio latency [11,7], audio-visual synchronization and fidelity, conference set-up convenience, and ease of integration with other collaborative tools [20]. These issues are important but beyond the scope of the paper.

a) b)

Figure 1: **Multi-point videophone interfaces.** *a) A typical simple interface. b) The Hydra interface [17]: each remote participant is represented by a surrogate unit consisting of a camera, display, microphone and speaker. This enables gaze communication.*

This paper describes a new interactive system that supports the communication of gaze more effectively than previously possible using traditional computer displays. Previous approaches typically assume immersive displays, and use live video to animate virtual participants (VPs) in a shared 3D virtual world. This shared world is then rendered from the viewpoint of the appropriate VP to yield the required view of the virtual meeting.

The next section summarises previous work on interfaces that support the communication of gaze. We then describe the disadvantages of using shared spaces with traditional computer displays. Under these circumstances, we make the novel observation that such views of a shared space do not convey gaze realistically. We explain why it is necessary to render the scene from the viewpoint of the real participant (RP). Consequently, we then show that it is necessary to animate the VPs differently at each node (RP location) for gaze to be communicated effectively. Next, we describe a

Permission to make digital or hard copies of all or part of this work for personal or classroom use is granted without fee provided that copies are not made or distributed for profit or commercial advantage and that copies bear this notice and the full citation on the first page. To copy otherwise, to republish, to post on servers or to redistribute to lists, requires prior specific permission and/or a fee.
CHI '2000 The Hague, Amsterdam
Copyright ACM 2000 1-58113-216-6/00/04...$5.00

design process for optimally arranging the VPs at each node, in order to preserve the semantic significance of gaze. Finally, we demonstrate the improved effectiveness of gaze communication using the new asymmetric interface with a user study.

THE EFFECT OF GAZE ON GROUP COMMUNICATION

In the context of this paper, we define *gaze* as the person at whom a single participant is looking. The *semantic gaze configuration* of a meeting is a list of who each participant is looking at, *i.e.* a list of each participant's gaze.

Gaze has been known to be an important cue for effective communication by researchers for some time [1]. When combined with spatial audio, it is especially important for multi-party meetings, where it supports a participant's ability catch the eye of other participants and to tell when other participants are looking at them [18]. This is typically observed via conversational phenomena such as length of turn and degree of formality when switching speakers [11]. For example, compared with face-to-face interaction, participants using the interface of Figure 1a are more likely to name the next speaker. This process is known as explicit handover.

Sellen has compared these phenomena across various videoconferencing interfaces, including one called Hydra (Figure 1b) that supports gaze communication [18]. Her measurements showed that in fact there was no significant difference in the frequency of explicit handovers between the system of the type shown in Figure 1a and the gaze-preserving Hyrda. Sellen attributes this to the fact that the users often felt psychologically disconnected from the situation, and therefore they might have compensated for this by behaving in a generally more explicit manner [5]. As expected, the frequency of parallel conversations was much greater with the Hydra interface.

The questionnaire results of her study favored the Hydra system as expected. The most frequently stated reason was that "they could selectively attend to people, and could tell when people were attending to them." Another common comment was "that participants liked the multiple sources of audio...[that] helped them keep track of one thread of the conversation when people talked simultaneously."

The conclusion that we draw from this work is that, despite the failure of the Hydra system to support implicit handover, we believe that the positive questionnaire results and the occurrence of parallel conversations to be significant evidence for the utility of gaze communication. We believe that the fact that the participants were strangers and also inexperienced users could have contributed to the feeling of disconnection, and thus increased the explicit handover frequency.

Vertegaal has performed experiments to isolate the effect of gaze from spatial audio. In [20] he concludes that "conveying gaze direction – especially gaze at the facial region – eases turn-taking, allowing more speaker turns and more effective use of deictic [depends on the circumstances of use] verbal references."

To summarise, we have cited research that indicates that gaze communication helps participants to:

- Establish when to talk more naturally and effectively [20,18],
- Hold parallel conversations and make side comments to other participants [18].

It is on this basis that we claim that the technique we later describe for improved gaze communication will, in turn, lead to more effective video-mediated communication.

EXISTING INTERFACES THAT SUPPORT GAZE

This section describes recent approaches to gaze communication. In order to present this work, it is convenient first to describe the important properties of three aspects of all video conferencing systems: display type, VP representation and the meeting space geometry.

Display Type

In this paper, we make the distinction between *traditional* and *immersive* displays. We use the term *traditional display* to refer to normal desk-top or lap-top computer screens or TVs. Immersive displays come in two forms. The first is the head mounted display (HMD). The second is a very large screen, typically projected onto a wall and referred to as a spatially immersive display [16,4]. The scene containing the VPs can be displayed on either a traditional display or an immersive display.

We believe that traditional displays have many advantages over immersive displays. HMDs have the following disadvantages: they disassociate the user from their familiar environment, make it hard to acquire images of the participant's face, and are rather cumbersome and potentially uncomfortable for spectacle users. Spatially immersive displays are typically expensive room-based installations. In contrast, traditional displays are relatively cheap and ubiquitous at work, on the road and at home. There is therefore considerable potential benefit in designing video communication interfaces for traditional displays.

VP Representation

The simplest representation of the VPs is a window in which raw video taken from the VP's camera is rendered (Figure 1). A more sophisticated representation is a 3D avatar. The avatar can be animated by the real motions of the participant obtained from tracking algorithms running on the video from the participant's camera [13,19]. The avatar can then be rendered realistically from *any view*, enabling for example, a profile view to be generated from a captured frontal view. In this way, a single camera can be used to generate the different views of a head necessary for gaze communication.

This approach also enables the provision of a richer set of 3D cues to the user. Motion parallax – the visual cue that encodes scene depth as image motion with respect to viewpoint change – can be generated as the user moves their head. In addition, if auto-stereoscopic displays are used, static binocular stereo effects can also be provided for greater realism.

Meeting Space Geometry

At each node the virtual meeting is necessarily conducted in a 3D space defined by the positions of the VPs in the display and the relative position of the RP in front of the display. Examples range from spaces like Figure 1a where raw video window VPs are tiled arbitrarily in the display, to interfaces where 3D avatar VPs are embedded in a 3D world which is rendered on the display surface, simulating a view of a real meeting [4,13,19].

The Hydra Interface

Figure 1b shows the first type of interface to address the multi-point communication of gaze [2,17]. It uses surrogate participants to represent the VPs. Each surrogate unit contains a small display, camera, microphone and loudspeaker. This interface is studied in considerable detail by Sellen in [18].

A schematic showing how the surrogates are arranged symmetrically at each node is shown in Figure 2. In this manner, all the views of each participant necessary for gaze communication are obtained from the multiple cameras at each node.

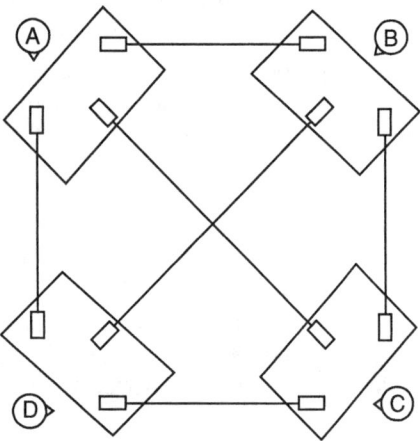

Figure 2: **Arrangement for a 4-way Hyrda conference**. *Using such an arrangement, it is possible for each participant to see at whom all the other participants are looking.*

In order for gaze to be successfully communicated in this type of system, the angular separation between the optical axis of the camera and the line of sight of the RP must be kept small. This is achieved by using small displays in the surrogate units. If larger displays were used, Hydra VPs would not appear to be looking in quite the right directions.

Thus, the Hydra system supports effective gaze communication when small displays are used. The principal disadvantage of the system is cost and scalability, requiring $N{\times}(N-1)$ surrogate units, where N is the number of participants in the virtual meeting.

The MAJIC Interfaces

The MAJIC system [14] follows the same principles as Hydra. It improves upon the Hydra interface by allowing life-size VPs. It positions the cameras behind large half-transparent curved screens, thus enabling the camera axes to be approximately aligned with the gaze of the RPs. MAJIC projects the VPs onto the screen. Like Hydra, it requires $N{\times}(N-1)$ cameras and projectors, and is consequently very expensive.

Desktop MAJIC [15] uses traditional displays. It represents a VP with a set of static images of the particular participant looking in a range of directions within the plane of the image (e.g. points of the clock). Consequently, if more than two VPs are placed in a line, the gaze communication is ambiguous. This means that for meetings with more than two VPs, they cannot be distributed along a line. Instead, they have to be distributed over the width and height of the display surface, which looks very unnatural, as VPs have to be rendered looking up and down, as well as from side-to-side as usual.

Interfaces using 3D Models

Ohya *et al.* [13] propose the use of large immersive, wall-sized displays. They take the first steps towards a system that does not display raw video. Instead they track the motion of the RPs. The extracted motion parameters are used to animate 3D avatars – the VPs. These VPs can be rendered from any viewpoint, and so the cameras no longer need to be positioned close to the display as before.

The VPs are placed in a shared 3D meeting space. They can be rendered from any view as required by the VP's relative position in the virtual meeting. As with MAJIC, the use of an immersive display means that the rendered VPs can be positioned where they would have been if the meeting were real. However, Ohya *et al.* track markers attached to the user's face, which has obvious practical disadvantages.

The TRAIVI project [19] again projects views from a shared space onto a wall-sized immersive display. They improve upon Ohya's approach by using a tracking technique based upon features already in the face, such as eyes and the mouth. They also consider approaches to animating the 3D model of the head with the measured movements of the mouth and expression changes.

PANORAMA [12] displays VPs in a shared space using auto-stereoscopic displays (usually HMDs), providing both parallax and binocular stereo cues. To achieve this they use real-time acquisition of the shape of the participants from a triplet of cameras in the form of a depth map (an image where the intensity at each pixel site represents the distance from the camera to the scene). TELEPORT [4] uses wall-sized displays but does not communicate gaze.

Another style of interaction that is common for internet meeting places [10] is based on the interface typically used for multi-user, networked computer games. Each participant controls their position and orientation (and hence view) within a shared 3D world with controls issued from a keyboard or joystick. This type of interface is perfect for controlling racing cars or spaceships, but less appropriate, we feel, for mediating group conversation in virtual meetings in a natural manner.

The GAZE Groupware System

Vertegaal [20] adopts a more straightforward approach. He uses traditional displays. Live video is not used. Instead, he uses a static 'snapshot' for the VPs, obtained when looking directly into the camera. This, he argues, reduces the bandwidth necessary for communication and also avoids the apparent directional gaze error caused by the angle between camera and display.

These snapshots are rendered on planes, once more in a shared 3D space (Figure 3). The orientation of these planes is driven by the gaze of the particular participant, which is in turn measured by tracking the orientation of the eye, using an infrared camera. More precisely, when the RP looks at a VP for more than a certain time, the system alters the state of a central 3D model such that the corresponding image plane is rotated to face the appropriate direction (Figure 3).

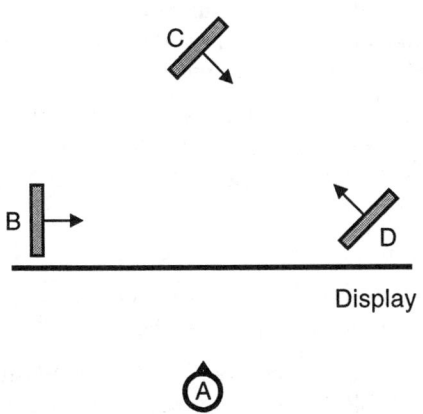

Figure 3: **Using rotating image planes for VPs.**

We believe that it is important to communicate the expressions of VPs. It seems clear therefore, that it is desirable to send real-time video instead of using static snapshots. Of course, this should be done without compromising the quality of the gaze communication.

The use of animated planes instead of 3D head models is problematic for two reasons. First, consider the 4-way conference shown in Figure 3. We define subscripts R and V to indicate real and virtual participants respectively. The RP A (henceforth referred to as A_R) should see a profile view of D_V, who should appear to be looking into the screen at C_V. This profile is correctly provided by Hydra (Figure 1b,2) because it is a multiple camera system. However, because the GAZE system only uses a single camera, it is not capable, for example, of capturing a natural profile view. Further, because it has no 3D model to reanimate from a different viewpoint, it is not able to generate an artificial profile view. Hence, with the symmetric (square) configuration shown in Figure 3, the closest view to the profile view available of A_R from the display-top camera is one at 45° to the plane of the face of A_R, obtainable when A_R is looking at B_V or D_V.

This limits the natural gaze communication capabilities of this type of interface. This is related to what we call the "Mona Lisa" effect. The angle at which the image is positioned in the virtual space has little affect upon the perceived gaze direction. In other words, a frontal image of a person rendered at 45° in a virtual world still looks like a frontal image (albeit slightly smaller), and so the person appears to be looking out of the screen. What dictates the perceived gaze direction is the actual view of the subject that was painted, in our case by a graphics engine, not the position of the viewer relative to the painting. The GAZE system mounts the snapshot images on 3D solid blocks. The orientation of these blocks can be perceived, and hence, the intended gaze direction of the VP can be inferred implicitly.

Finally, we note that in Figure 3, the view that A_R has of B_V or C_V is foreshortened, reducing the resolution of the face, so that expression and mood become harder to assess, and hence communication impaired. This problem is most acute with A_R's view of D_V, which in this case is completely degenerate, the image plane of D_V being viewed end-on, and thus disappearing. Again, these problems can be solved with the use of 3D avatars as VPs, instead of planes.

THE PROBLEM WITH SHARED SPACES

This section shows why the use of shared spaces yields poor gaze communication under practical viewing conditions using traditional displays.

All virtual meeting systems known to us that use animated models, render the VPs in a common shared 3D space. Each node then renders the model from the viewpoint of their VP. Thus, A_R's view of the meeting is invariably obtained by rendering the shared virtual world from the viewpoint of A_V (Figure 4). Since the shared virtual space is naturally symmetric, the VPs are positioned around a regular polygon. For this scene to be rendered from the viewpoint of A_V, it is natural to place the display such that the nearest VPs are at either side of the display, as shown in Figure 4.

This presents no particular problem if immersive displays are used, which handle such wide-angle scenes. However, in the case of a desk-top computer, we note that people usually sit about 20" away from their displays, which are typically 14" wide. Figure 4 shows how a practical asymmetric viewing situation (solid A) differs from the symmetric regular polygonal position used in shared space systems (dashed A). The angle θ_R is the angular gaze range of the practical RP, and θ_N is the gaze range for the symmetric position. Figure 4 shows how θ_N gets larger as N

increases, as $\theta_N = 180(1-2/N)°$. Now in our example, $\theta_R \approx 2\arctan(7/20) = 40°$, which is significantly less than $\theta_3 =$ 60°, $\theta_4 = 90°$ (Figure 4a) and $\theta_5 = 108°$ (Figure 4b).

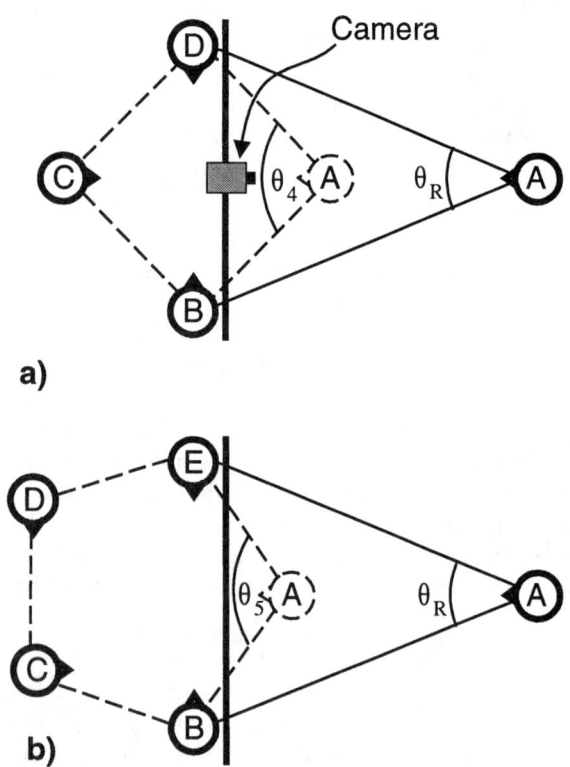

Figure 4: **Symmetric and realistic viewpoints.** *The dotted polygons represent the symmetric arrangement of participants in a shared space. The dashed (solid) circles in front of the displays show the symmetric (realistic) position of the RP respectively. a) A 4-way meeting where $\theta_4=90°$. b) A 5-way meeting, where the symmetric RP position is closer to the display, and $\theta_5=108°$.*

Intuitively, if views are rendered using a shared symmetric model from the symmetric (dashed) viewpoint, we would expect the effectiveness of the gaze communication to be reduced because the RP is in a different place. More precisely, we expect the gaze communication performance to degrade with increasing angular gaze error $\theta_N - \theta_R$, as N increases. The next section examines this effect in a little more detail.

VP-VP Gaze Accuracy Requires Asymmetric Spaces

We first consider the accuracy of the perceived gaze of VPs looking at other VPs when views of a shared symmetric scene are rendered from the symmetric viewpoint. We assume that there is an accurate gaze tracking process in operation at each node [6]. Each RP's gaze is sent to the other nodes over a multicast channel. Thus, each node has a dynamically updated representation of the semantic gaze configuration.

With reference to Figure 4a, we consider how B_V and D_V looking at each other appear to A_R in the asymmetric (practical) viewing position. Figure 5a shows the view that is obtained by rendering the scene from the symmetric (dotted) viewpoint. It shows views of the faces of B_V and D_V at ±45° to the frontal view. The result is a view that shows them looking out of the screen - they do not appear to be looking at each other. These views are not correct for the asymmetric viewer, who expects to see views of the head nearer to the profile view shown in Figure 5b.

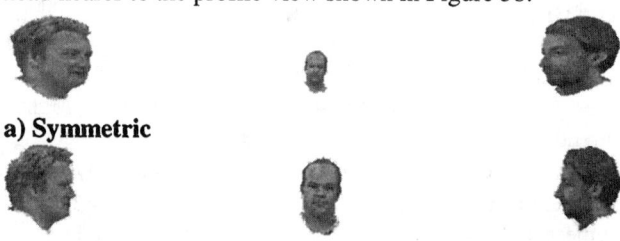

a) **Symmetric**

b) **Asymmetric**

Figure 5: **VP-VP gaze accuracy for a 4-way meeting.** *Left and right VPs are looking at each other. a) A view from the symmetric viewpoint (Figure 4a dashed) shows the VPs appearing to look out of the screen. b) A view from the asymmetric viewpoint gives the correct appearance. Note the favourable relative scale in the asymmetric case – the middle head is larger.*

We also note that the symmetric case generates a greater range in scale of the projections of the VPs, which is again unnatural and makes it difficult to discern both the gaze and the expressions of the smaller VPs.

We conclude that to convey VP-VP gaze effectively we should *not* render from the symmetric viewpoint (dashed positions in Figure 4). Correct VP-VP gaze communication is obtained by:

- Removing the RP's VP from the shared 3D scene.
- Rendering the remaining avatars from the asymmetric viewpoint that corresponds to the position of the RP.

Asymmetric Spaces Require Asymmetric Animation

We have established that we need to render from the asymmetric viewpoint of the RP for VP-VP gaze to appear realistic. We now consider how this affects the rendering of VP's when they look out at the asymmetric RP.

Figure 6a shows the asymmetric situation where D_V is looking out of the screen at A_R. The scene is rendered from A_R's viewpoint. Therefore, in order for A_R to perceive that D_V is looking directly at A_R, D_V needs to be facing ~70° out of the display from B_V.

Figure 6b shows the same situation from C_R's viewpoint. Here, D_V needs to be facing 45° into the display (or equivalently, away from B_V) for C_R to perceive that D_V is looking at A_V. This means that D_V needs to be in a different orientation relative to the other VPs, depending upon whether it is viewed from A_R or C_R.

Figure 7 illustrates this with head models. It shows both top and RP views for the symmetric (a,c) and the asymmetric (b,d) cases respectively. We note that the views in Figure 7c,d are not too different. The point is that the VPs have to be orientated differently to achieve the same apparent gaze.

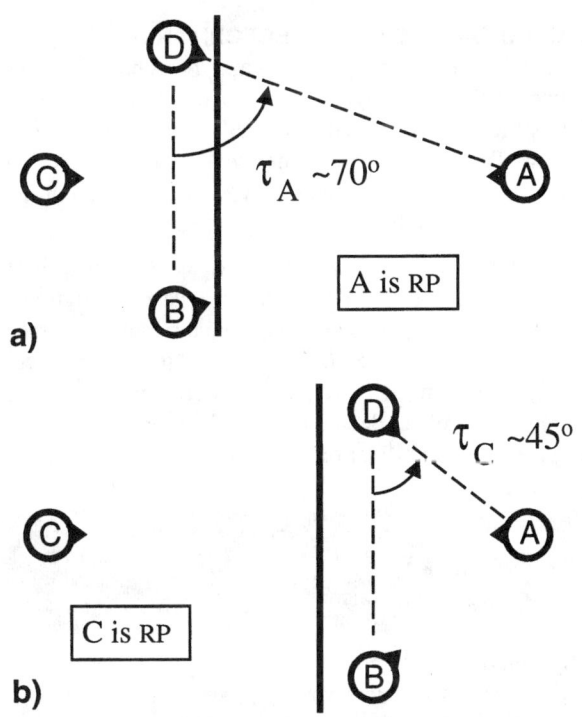

Figure 6: **Asymmetric animation.** *Given that we are compelled to render the scene from the viewpoint of the asymmetric RP for VP-VP gaze to appear realistic, the orientations of the VPs in the scene need to be different to make VP-RP gaze work effectively. The figure shows how the system has to render D looking at A. a) VP-RP: For A_R, D_V is rendered at ~70° to the display. b) VP-VP: For C_R, D_V is rendered at ~45° to the display.*

In general then, VPs need to be in different relative orientations depending on the RP for which they are being rendered, in order for VP-RP gaze to be communicated effectively. They need to be animated asymmetrically.

Shared Spaces do not Support Gaze Communication

We have shown that rendering views of a shared 3D model cannot yield effective gaze communication. More precisely, we have shown two corollaries of asymmetric viewing conditions caused by using traditional computer displays:

1. The VPs must be rendered from the RP viewpoint for VP-VP gaze to operate effectively.
2. The orientations of the VPs in the scene depend upon the RP for which they are being rendered.

Hence the use of a shared 3D model is no longer justified. Rather, the semantic gaze configuration should be used to build different individual 3D models as appropriate for each node. These individual models are therefore animated by the semantic gaze signals in an asymmetric manner. We use a shared *semantic* space, rather than a shared Euclidean space.

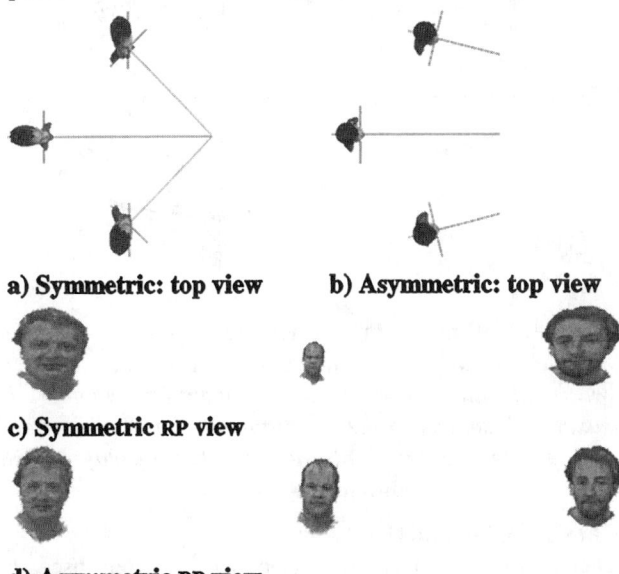

Figure 7: **VP-RP gaze and asymmetric animation.** *a,c) VPs oriented symmetrically. b,d) VPs oriented asymmetrically as shown in Figure 6a. Note that the views c) and d) are very similar. The point is that the VPs have to be orientated differently to achieve the same apparent gaze. The relative scale is again better for the asymmetric case.*

THE iCON SYSTEM

In this section, we describe our implementation of a videophone system that communicates gaze using different, yet semantically consistent, 3D models at each node. We call the system iCon (*eye-con*tact).

Optimal VP Layout

We first consider the nature of these individual 3D models. Given that we have shown that they are necessarily different for each node, we now consider how they can be constructed in an optimal way, tailored for each RP.

Until now, we have assumed that the VPs lie on $N-1$ points of the regular N-sided polygon (Figure 4-7). However, this was a direct result of the symmetry constraint on the shared virtual world, which is no longer useful. Instead, we note two design criteria.

1. In order to make switches of attention as clear as possible, we would like to maximize the minimum apparent turn that any VP has to undergo to switch attention from one participant (VP or RP) to another.
2. We would like the projections of the VPs to be evenly spaced out across the display to avoid occlusions and large amounts of unused display space.

Maximizing (1) subject to the constraint (2) leads to optimal arrangements of VPs, some of which are shown in Figure 8.

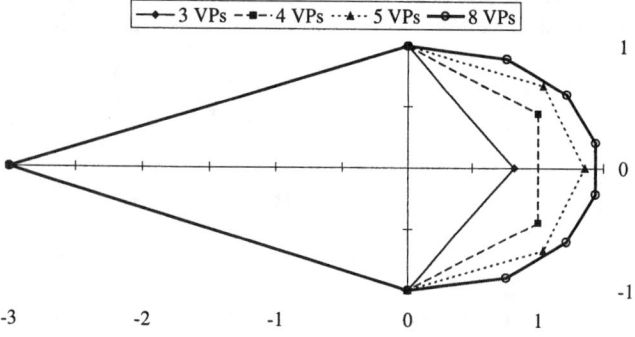

Figure 8: **Optimal VP layout:** *VPs are arranged so that the minimum apparent turn that any VP has to undergo to switch attention from one participant to another is maximized, subject to the constraint that their projections are evenly spaced out across the display. Such arrangements are shown for 3,4,5 and 8 VPs.*

Gaze Tracking and Head Animation

The handset of a normal telephone is often used in preference to a "hands-free" modality. This is usually because the remote party does not want their conversation to be heard by spurious unknown local parties. This is especially true in open–plan offices. The same applies to videophone calls, and we therefore argue that the use of a headset is often desirable, as it supports privacy and hands-free operation.

a)   b)

Figure 9: **Gaze tracking and face animation.** *a) The iCon system tracks three markers on a normal audio headset, which can be used to estimate the gaze of the RP. b) The face region in the video is segmented out and rendered on a generic 3D model of a face.*

Hence, Figure 9a shows a user wearing a headset from the videophone camera. We track RP gaze using markers. In this case, there are three: one on each earpiece, and one on the microphone. We use a standard reliable marker-based tracking algorithm [6]. It would be straightforward to use a marker-less face tracking technique [8,9] based on natural features such as eyes, but they tend to be less robust. The estimated gaze position is displayed as feedback to the RP. The resulting gaze information is used to animate the VPs' heads as described above.

Face Animation

We use a simple texture mapping technique. Using colour, background models and the positions of the markers, we segment the raw video of the face region. The face video is rendered as a video-texture onto a generic 3D model of a head (Figure 9b) to form the VP.

EVALUATION OF SYSTEM PERFORMANCE

We set out to compare the gaze perception accuracy obtained using the asymmetric models generated by the iCon system against that obtained using the symmetric models. To do this, we generated a set of five gaze configurations for 3, 4 and 5 VPs – 15 different gaze configurations in all.

The asymmetric VPs were arranged as described in the previous section (Figure 8). For each gaze configuration, we generated a view for both the symmetric and asymmetric model (shown in Figure 10 for 4 VPs). Users were asked to identify at whom each of the VPs were looking, in each configuration. We collected data for 16 users and measured the gaze miss-classification rate. The results are presented in Table 1.

a) **Symmetric**

b) **Asymmetric**

Figure 10: **Gaze perception test.** *An example of a 4 VP configuration. Users were asked to estimate at whom each VP is looking for 15 such configurations. a) The view rendered from a symmetric viewpoint of a shared space. b) The view rendered from a realistic viewpoint of an asymmetric space. The VPs are evenly spaced across the display.*

Discussion

We see that the mean gaze error rate of the asymmetric model is consistently and significantly smaller than that of the symmetric model.[1] These results therefore confirm our hypothesis that the asymmetric model supports more effective gaze communication.

As we would expect from Figure 7c,d, there were an insignificant number of cases for either model where users thought they were not being looked at when, in fact, they were. Users easily spotted a frontal view of a face, shown on the far right of both Figures 10a and Figure 10b.

---

[1] Using a Student's t-test for significantly different means, we obtain a tiny probability ($<10^{-4}$) that the two population distributions have the same mean, on the basis of these samples.

However, as expected, situations like that shown in Figure 5a and the far left VP in Figure 10a caused many users to believe they were being looked at when in fact they were not. We believe that this particular error mode is likely to have a significantly detrimental effect upon group conversation.

For both models, the error rate jumps once the number of VPs reaches five. This is caused by VP-VP gaze suddenly becoming more difficult to estimate as the display gets more crowded – compare Figure 5 with Figure 10. However, we note that in a live videoconference, motion and audio cues will make the gaze estimation problem easier.

VPs	Symmetric Model	Asymmetric Model
3	44 (14) %	11 (13) %
4	46 (13) %	8 (10) %
5	68 (11) %	31 (10) %

Table 1: **Gaze classification error rates.** *Users were asked to classify the VP gaze in a series of configurations. The table shows the mean error rate (standard deviation) against the number of VPs in the scene and the type of model used. As expected, the asymmetric model has a significantly lower mean error rate.*

Future experiments will set out to confirm Vertegaal's claim [20] that this improved gaze communication leads on to better group conversation, in the particular context of our iCon system.

CONCLUSIONS

When using traditional computer displays, this paper has demonstrated that poor gaze communication results from rendering views from the viewpoints of a avatars arranged symmetrically in a shared 3D world. We have explained why it is necessary to render the scene from the viewpoint of the real participant (RP). We then showed that it is necessary to animate the VPs differently at each node for gaze to be communicated effectively. We described a design process for optimally arranging the avatars at each node in order to communicate effectively the semantic significance of gaze. Finally, we demonstrated the improved effectiveness of gaze communication using the new asymmetric scheme with a user study. We believe that communicating gaze using these semantically consistent spaces will form an important component of future videophone systems.

REFERENCES

1. Argyle, M., Ingham, R. and McCallin, M. The different functions of gaze. *Semiotica*, 7:10-32, 1979.
2. Fields, C.I. Virtual space teleconference system. *US Patent* No 4,400,724, 1981.
3. Finn, K.E., Sellen, A.J. and Wilbur, S.B. (Editors). *Video-mediated communication*. Lawrence Erlbaum Associates, 1997.
4. Gibbs, S.J., Arapis, C. and Breiteneder, C.J., TELEPORT – Towards immersive copresence. *ACM Multimedia Systems* 7(3): pages 214-221, May 1999.
5. Heath, C. and Luff, P. Disembodied conduct: Communication through video in a multi-media office environment. In *Proceedings of CHI'91*. New Orleans, ACM, 1991.
6. Heuring, J.J. and Murray, D.W. Visual head tracking and slaving for visual telepresence. In *IEEE Intl. Conf. on Robotics and Automation*, 1996.
7. Issacs, E.I. and Tang, J.C. What video can and cannot do for collaboration: a case study. *Multimedia Systems* 2, pages 63-73, 1994.
8. La Cascia, M. and Sclaroff, S. Fast, Reilable Head Tracking under Varying Illumination. *IEEE CVPR* pages 604-610, June 1999.
9. Liu, J. Determination of the point of fixation in a head-fixed coordinate system. In *14th Inter. Conf. on Pattern Recognition*, August 1998.
10. Nakanishi, H. FreeWalk: A 3D Virtual Space for Casual Meetings. *IEEE Multimedia*, pages 20-28, 6(2), 1999.
11. O'Connaill, B., Whittaker, S. and Wilbur, S. Conversations over Video Conferences: An Evaluation of the Spoken Aspects of Video-Mediated Communication. *Human-Compter Interaction* 8, pages 389-428, 1993.
12. Ohm, J, Grüneberg, K., Hendriks, E., Izquierdo, E., Kalivas, D., Karl, M., Papadimatos, D. and Redert, A. A Realtime Hardware System for Stereoscopic Videoconferencing with Viewpoint Adaptation. In *International Workshop on Synthetic-Natural Hybrid Coding and Three Dimensional Imaging*, pages 147-150, Rhodes 1997.
13. Ohya, J., Kitamura, Y., Kishino, F. and Terashima, N. Virtual space teleconferencing: real-time reproduction of 3D human images. *Journal of Visual Communication and Image Representation*, 6(1):1-25, March 1995.
14. Okada, K., Maeda, F., Ichikawaa, Y. and Matsushita, Y. Multiparty Videoconferencing at Virtual Social Distance: MAJIC Design. In *Proceedings of CSCW*, pages 385-393, 1994.
15. Okada, K., Tanaka, S. and Matsushita, Y. MAJIC and Desktop MAJIC Conferencing System. In *Video Program, Proceedings of CSCW,* 1996.
16. Raskar, R., Welch, G., Cutts. M, Lake, A., Stesin, L. and Fuchs, H. The office of the future: a unified approach to image-based modeling and spatially immersive displays. In *SIGGRAPH*, pages 179-188, 1998.
17. Sellen, A.J. Speech Patterns in Video-Mediated Conversations. In *Proceedings CHI* , pages 49-59, 1992.
18. Sellen, A.J. Remote Conversations: The Effects of Mediating Talk With Technology. *Human-Computer Interaction,* volume 10, pages 401-444, 1995.
19. Valente, S. and Dugelay, J. A Multi-Site Tele-conferencing System Using VR Paradigms. In *ECMAST*, Milano, Italy, pages 359-374, 1997.
20. Vertegaal, R. The GAZE Groupware System: Mediating Joint Attention in Multiparty Communication and Collaboration. In *Proceedings CHI* , pages 294-301, May 1999.

Eye-Hand Co-ordination with Force Feedback

Roland Arsenault and Colin Ware

Faculty of Computer Science
University of New Brunswick
Fredericton, New Brunswick
Canada E3B 5A3

ABSTRACT

The term Eye-hand co-ordination refers to hand movements controlled with visual feedback and reinforced by hand contact with objects. A correct perspective view of a virtual environment enables normal eye-hand co-ordination skills to be applied. But is it necessary for rapid interaction with 3D objects? A study of rapid hand movements is reported using an apparatus designed so that the user can touch a virtual object in the same place where he or she sees it. A Fitts tapping task is used to assess the effect of both contact with virtual objects and real-time update of the centre of perspective based on the user's actual eye position. A Polhemus tracker is used to measure the user's head position and from this estimate their eye position. In half of the conditions, head tracked perspective is employed so that visual feedback is accurate while in the other half a fixed eye-position is assumed. A Phantom force feedback device is used to make it possible to touch the targets in selected conditions. Subjects were required to change their viewing position periodically to assess the importance of correct perspective and of touching the targets in maintaining eye-hand co-ordination, The results show that accurate perspective improves performance by an average of 9% and contact improves it a further 12%. A more detailed analysis shows the advantages of head tracking to be greater for whole arm movements in comparison with movements from the elbow.

Keywords

3d interfaces, haptics, interaction techniques, force feedback, virtual reality.

INTRODUCTION

One of the key arguments for virtual reality systems is that if artificial environments can be constructed that are like the real physical world, then we will be able to apply our everyday life skills in manipulating objects. Thus we will be able to learn to use computer software more rapidly and effectively. Applications that could benefit include 3D CAD, animated figure design for the entertainment industry, and interactive visualisation of 3D data spaces. In the present paper we report a study that investigates the value of eye hand co-ordination and simulated object contact in a limited, but high fidelity virtual workspaces.

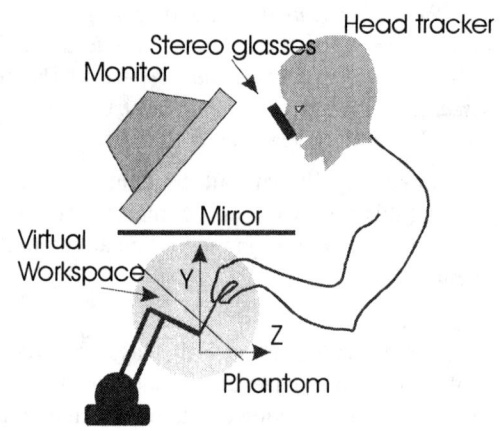

Figure 1. The apparatus used to create a small, high quality virtual environment that can be touched as well as seen.

Fish tank VR is a non-immersive type of virtual reality where a 3D virtual environment (VE) is created using a monitor display [3,18]. In order to create a correct stereoscopic view of a small virtual environment, the user's head position is tracked, from this their eye positions are calculated, and using this information a correct stereoscopic image can be displayed and continuously updated. In essence this involves making the centre-of-perspective for the computer graphics coincide with the actual viewpoint for each eye. Using this technique it is possible to create a small, high quality VR environment located just behind and just in front of the monitor screen. With the addition of a mirror to reflect the monitor, as shown in Figure 1, the user's hand can be placed in the same location as objects in the VE.

One of the thorniest problems in VR is the fact that although visual information and sound information can be simulated with reasonable fidelity, providing good touch information remains a problem. Recently, force feedback devices have become available that can provide a limited, but reasonable precise sense of touch, but only within a small working volume. The PHANToM, by Sensable Technologies, mechanically measures the position of a fingertip in 3D space and also applies a force vector to the finger tip [16]. This allows the haptic simulation of solid objects and various force-related effects, such as springs and inertia.

Because of the similarity in the working volume, fish tank VR and Phantom force feedback would seem to be complementary technologies making it possible to combine visual and haptic images. Thus we place a Phantom Force feedback device as shown in Figure 1 to create a local high fidelity VE that can be both seen and touched.

Our goal in the research presented here has been to determine the value of providing real-time head-coupled perspective and of simulated object contact for a simple task. We first review some of the perceptual issues and results from the human factors literature that are relevant to this task

Adaptation

In perception research a number of studies have investigated how eye-hand co-ordination changes when there is a mismatch between feedback from the visual sense and the proprioceptive sense of body position. A typical experiment involves subjects pointing at targets while wearing prisms that displace the visual image relative to the proprioceptive information from their muscles and joints [7,8,14]. Subjects adapt quite rapidly to the prism displacement and point accurately. Also, after they remove the prisms (having worn them for an extended period) subjects make large errors pointing at targets before recovering. The usual explanation for this is that the mapping between eye and hand has become recalibrated in the brain (although there is much debate as to the exactly where and how this takes place). Recent work by Rosetti et al. [14] suggests that there may be two mechanisms at work in prism adaptation, a long-term slow acting mechanism that is capable of spatially remapping mis-aligned systems, and a short-term mechanism that is designed to quickly optimise accuracy in situations involving temporary misalignments. There is also evidence that certain misalignments are readily compensated for, whereas others are not. Subjects seem to rapidly adapt to small lateral displacements of the visual field, but other distortions, such as inversion of the visual field can take months to adapt to, and adaptation may never be complete [7]. In any case, it is clear that active movement of the hand - or whole body - in the environment is necessary for adaptation to take place.

Adaptation experiments, such as those described above are relevant to the present study because we are interested in how useful virtual reality techniques are in making it easier for people to perform certain tasks. If it is possible to adapt quickly and completely to mismatches between hand position and visual information, then the case for VR seems much weaker. For example, if objects in small monitor-based virtual environments can be adequately manipulated using the hand placed off to the side, and viewed from a point that is not the centre of perspective, then the required equipment will be cheaper and easier to configure. On the other hand, if placing the hand in the same location as a virtual object improves performance then a stronger case can be made that 3D design systems should use VR technologies.

Perspective Distortions

For every perspective picture there is a point, called the *centre of perspective*. Viewed from this position the picture mimics the pattern of light from a scene. When an image is viewed from a point that is different from the correct centre of perspective, the laws of geometry suggest that distortions should occur as shown in Figure 2. However, although people report seeing some distortions when looking a moving pictures from the wrong point they rapidly become unaware of these distortions. Kubovy [9] called this the *robustness of linear perspective*. One of the mechanisms that can account for this lack of perceived distortion may be based on a built-in perceptual assumption that objects in the world are rigid. If the object shown in Figure 2 were to appear to change shape when the viewpoint was changed, then it would be perceived as elastic and non-rigid. A perceptual rigidity assumption may account for the fact that we perceive stable rigid 3D virtual environments under a wide range of incorrect viewpoints.

Nevertheless, even though the brain appears to compensate for an incorrect viewpoint, there will still be a discrepancy between the visual image and the haptic image if an apparatus such as that shown in Figure 1 is used. As shown in Figure 2, if the displayed object is behind the virtual picture plane, the hand must reach to a different position to be coincident with a virtual object when the viewpoint is not correct. However, a 3D cursor used to make the selection will also be distorted in the same way and this may reduce the ill effects because the *relative* position between the cursor and the object will only be distorted by a small amount. But the extent to which off-axis stereo viewing of a 3D target disrupts target selection has not, prior to the present study, been experimentally investigated.

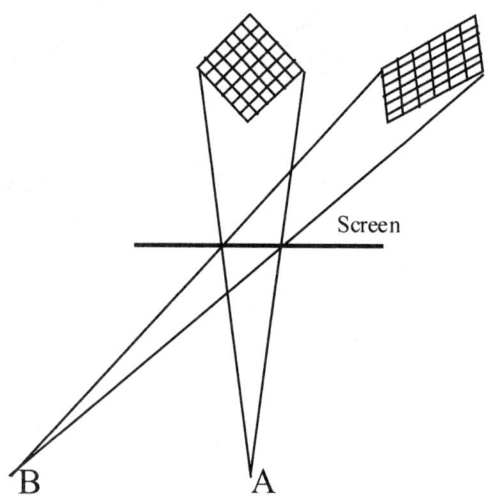

Figure 2. If an image computed to be viewed from position A is actually viewed from position B distortions occur as shown.

Previous results from VR research

There have been a number of studies reported in the human factors and virtual reality literature that bear on the importance of correct viewpoint and haptic feedback.

Ware and Franck showed that accurate perspective based on head position tracking assisted in the tasks of tracing paths in complex 3D networks [19]. However, they also showed that this is more likely to be a product of motion parallax information than correct perspective; hand linked motion of the virtual scene improved performance as much as providing head- coupled perspective. Recently, Pausch et al showed that using natural head movement to perform a visual search of an immersive environment can result in more rapid searches [13] under certain conditions. Also, head coupled perspective gives a strong sense of the three dimensionality of the virtual space [1,18].

We may be quite insensitive to translation mismatches between visual and proprioceptive information. In fact the normal practice of placing the mouse at the side of the computer is evidence for this. But there may be a significant advantage to placing the hand in the virtual workspace for object rotations. Ware and Rose [17] found that placing the subject's hand in a virtual workspace improved performance for object rotation, compared to having the subject's hand held to the side of the body.

System lag is likely to be a critical variable in how quickly people adapt to situations in which there is a mismatch between visual and haptic imagery. Held [8] found that the ability to adapt declined rapidly as lag increased beyond about 100 ms.

Simulated touch in object manipulation tasks can improve performance on a number of tasks [15]. Hannaford et al [6] showed that force feedback reduced errors substantially in the task of placing pegs in holes, and Meek et al [10] showed that the ability to grasp and lift breakable objects was markedly improved with force feedback.

The prior work that comes closest to our present study is an experiment by Boritz and Booth [2] who evaluated a reaching task for targets with and without stereo viewing and with and without head tracked perspective. They found that stereoscopic viewing did improve performance but found no effect for head tracking. However, in this experiment the default head position of the subjects appears to have been close to the correct centre of perspective, thus there may have been little difference between the head-tracked condition and the non head-tracked condition. In addition, the fact that their subjects took several seconds to carry out a simple positioning task suggests that fluid interaction was not possible in their system, perhaps due to system lag.

Although fish tank VR, as described, can provide an accurate correct perspective view calculated from the user's actual viewpoint, this is not always possible or desirable. Head tracking is expensive and requires extra apparatus. Users are generally much more accepting of interfaces where they are unencumbered. On the other hand, when an artist is working on a sculpture or a mechanic is working on an engine they may often change head position to get a better view of what they are working on. Enabling this kind of viewpoint control may be useful and an added benefit to any improvement in eye-hand co-ordination.

In addition, there is the interesting question of whether, simulated contact with virtual objects may make the ability to adapt to an incorrect viewpoint more rapid or complete.

EXPERIMENT

In order to investigate the effects of accurately estimated eye position, and simulated contact we chose a task that could be performed rapidly. In this way we hoped to understand more about skilled fluid performance. The task chosen was the classic Fitts [4] tapping task whereby subjects tap back and forth between two targets. Fitts found that each reciprocal movement, from one target to the next, could be accomplished in less than half a second. However, we did not vary target width and target separation, as in a typical Fitts' law experiment since we were more interested in varying other task parameters. Although this task is highly artificial, it requires a skill that might be used to rapidly press buttons in a 3D environment. This may become common if VR systems evolve like desktop systems.

The experiment described here has two primary objectives. The first is to determine if head tracking is advantageous when performing rapid, visually guided hand movements.

More precisely, does the distortion caused by off-axis viewing of a projected image degrade eye-hand co-ordination? A second issue is whether feedback from physical contact with a target improves performance on the same task.

Method

Apparatus. A virtual environment with a coinciding haptic and visual display was constructed for this experiment. A Phantom 1.0 from Sensable Technologies was used to provide a haptic workspace of 5" x 7" x 10" (12.7 cm x 17.8 cm x 25.4 cm). The Phantom consists of a mechanical arm, which tracks the position of a hand-held stylus and applies a force vector to the tip of the stylus [16]. A frame was built above the Phantom to support an upside-down video monitor tilted 45° towards the user to provide an image which was reflected on a mirror placed horizontally between the virtual workspace and the video monitor. The result, when viewed through the mirror is a video display tilted 45° away from the user. This virtual image coincides with the PHANToM's workspace as shown in Figure 1.

Stereoscopic vision, using LCD shutter glasses, is used throughout. Head tracking is achieved by attaching a sensor from a Polhemus 3Space Isotrack to the stereo shutter glasses. By tracking the position and orientation of the shutter glasses, the estimated position of each eye is calculated and used to provide a correct perspective image to each eye.

The co-ordinate system used to place objects originates from the centre of the workspace with the X-axis increasing towards the right, the Y-axis increasing in the up direction and the Z-axis increasing towards the user as shown in Figure 1. The units of measure used are centimetres. The screen of the visual display can be seen as an oblique plane having a normal vector perpendicular to the X-axis and 45° to the Y and Z-axis. A simple 45° rotation in software around the X-axis aligns the visual and haptic workspaces.

Calibration of the virtual workspace is verified by replacing the mirror with a pane of glass. It is now possible to place a physical object it the workspace and have a virtual object of similar dimensions superimposed on the physical object. When properly calibrated, the virtual and physical objects remain at the same position when the head is moved.

Task. Subjects are asked to alternately tap the tops of two cylindrical targets. The targets are cylinders oriented such that the flat faces are parallel to a checkerboard ground plane as illustrated in Figure 3b. The cylinders can be seen visually and felt with haptic feedback. The cylinders have a radius of 1 cm each and are separated horizontally by 6.75 cm. Two sets of positions for the cylinders are used for this experiment as illustrated in Figure 3a. For right-handed subjects, with position 1, tapping can be accomplished mainly by arm rotations of the forearm about the elbow. Whereas in position 2 subjects move their entire arm from the shoulder in order to tap back and forth. The overall location of both targets is randomly changed on successive trials by up to 1.0 cm on each axis, but the relative position of the two cylinders to each other is unchanged.

In all conditions subjects held the Phantom stylus in their right hand and used it to tap back and forth touching to tops of the targets in succession.

Subjects are required to change their viewpoint between trials. A 10-cm wide obstacle placed on the mirror, which prevents the subject from viewing the targets from a central position. In order to view the target objects the subject moves his or her head approximately 18 cm left or right of the centre. Since the subjects eye point is typically about 55 cm from the target area this results in a line-of-sight about 18 degrees off-axis. A signal in the form of sphere on the upper left or upper right portion of the display appears to indicate from which side of the obstacle the subject should look at the targets. The side is changed after every trial of 12 taps and three trials per side are run for every condition.

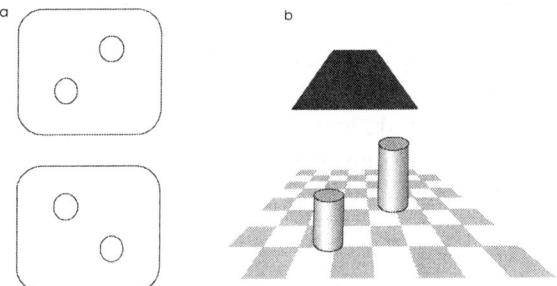

Figure 3, (a) Two sets of target positions are used in a right oblique and left oblique configuration. (b) A physical barrier, placed horizontally on the mirror above the virtual targets forces the subject to look from one side or the other.

Conditions

There are three major independent variables.

Head-tracked vs. non head-tracked. In the head tracked condition, the centre of perspective is based on the users eye position (computed based on their measured head position). In the non head tracked condition a default centre of perspective is used for each eye. This is at the midpoint of the normal range of head movement.

Touch vs. no touch. In the force condition, force feedback is provided by the Phantom to provide a sense of contact with a hard surface. In the no-force conditions visual feedback for contact with the target is provided by making the target flash to a higher colour intensity for a single frame of animation at the moment of contact.

Target position. The two sets of positions for the target

cylinders are as illustrated in Figure 3a and described above.

All combinations of the three independent variables are tested giving a total of 8 conditions.

A trial consists of twelve successive taps back and forth between the tops of the two targets, six taps on each. On alternate trials subject change their head position alternately looking at the target from the right or left of the barrier. A trial block consists of six successive trials, that are the same with respect to head tracking (or not) and virtual contact (or not). A run consists of all possible trial blocks occurring in random order. The experiment consists of two runs per subject in one sitting for a total of 96 trials.

The subjects are allowed to try the task before measurements were made to familiarise themselves with the virtual environment. Once ready, the subjects are instructed to tap the targets, always starting with the one on the same side as the green one (target 0). They are instructed to tap as fast as possible back and forth until a beep is heard. At that point, the subjects are asked to move their head position to view from the other side of the obstacle, as indicated by a red sphere appearing in the top of the workspace. At this point, before the targets where touched again, the user can take a small break to rest if desired.

Subjects. 13 subjects were chosen from within and outside the university population. 2 subjects had previous experience with the virtual environment. All subjects were right handed.

	No Force Feedback	Force Feedback	Average
No Headtracking	612	537 (-12%)	574
Headtracking	557(-9%)	491 (-20%)	524 (-9%)
Average	584	514 (-12%)	549

Table 1 - Average time (ms) for various conditions.

RESULTS

Table 1 shows the mean inter-tap interval averaged across all subject and all trials for the two main conditions. The overall mean interval was 549ms. Using head tracking to compute the correct viewpoint resulted in a reduction of 9% in the mean inter-tap interval. Using force feedback resulted in a reduction of 12% in the inter-tap interval. Both of these differences are highly significant (p<0.01). There was no significant interaction between them.

Each trial actually consisted of 12 taps giving 11 inter-tap intervals. Figure 4 shows a time series of inter-taps intervals averaged across all subjects and other conditions for head tracked and non head tracked conditions. Figure 5 shows the same series comparing performance both with and without force feedback. As can be seen over the course of each series the inter-tap interval decreased over the first four taps and then levelled off, but against our expectations there is no closing of the gap that might be expected from a rapidly acting eye-hand re-calibration.

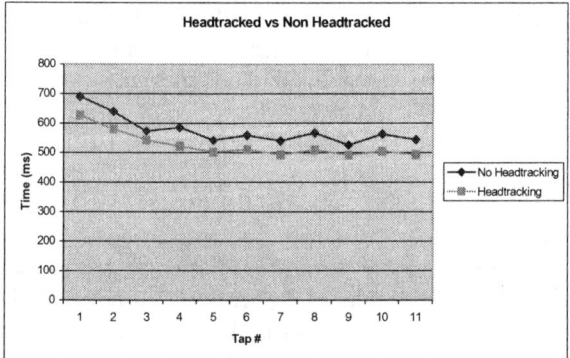

Figure 4. The average time series of inter-tap intervals is given both with and without head-tracked perspective.

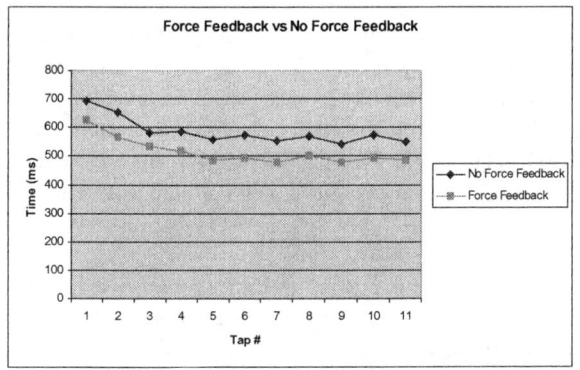

Figure 5. The average time series of inter-tap intervals is given both with and without force feedback.

For each condition there were six trial blocks divided into two runs and over the course of the experiment subjects speeded up from a mean inter-tap interval of about 605 ms. to about 535 ms.

Head tracking improved performance more for target positions 2 than for target positions 1. Figures 6 and 7 show the results both with and without head tracking with targets in positions 1 and 2 respectively. As can be seen there was approximately a 25 ms benefit for head tracking with the targets in position 1 and an 80 ms benefit in position 2. All of the subjects were right handed and position 2 required whole arm movements from the shoulder, whereas position 1 only required movements of the forearm.

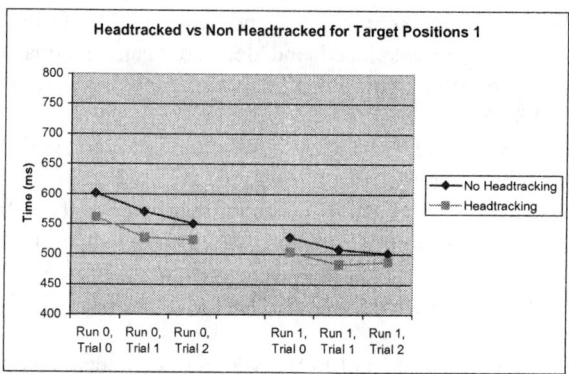

Figure 6. The results are plotted over the time course of the experiment for targets in position 1.

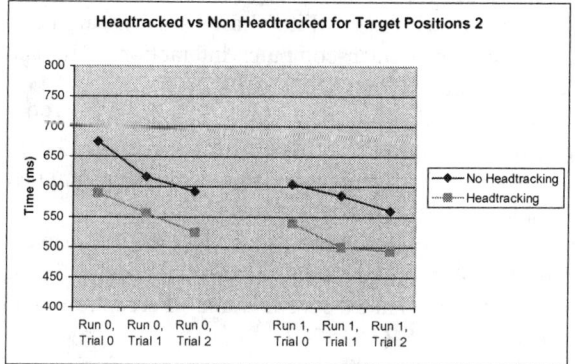

Figure 7. The results are plotted over the time course of the experiment for targets in position 2.

Errors occurred when a subject failed to make contact with a target, yet kept on tapping. This resulted in inter-tap intervals approximately 3 times as long as the norm as two extra movement were required before the "next" target was registered (i.e. the one that had been missed). We devised the following post processing strategy to deal with these occurrences. If the individual time was greater than 2.25 times the average, this measurement was treated as an error and corrected by dividing by 3.

Table 2 shows the errors broken down by the major conditions. The largest effect was that there were fewer errors with force feedback than without force feedback. This difference was significant ($p<0.05$).

	No Force Feedback	Force Feedback	Total
No Headtracking	3.29%	2.74%	3.02%
Headtracking	3.50%	2.80%	3.15%
Total	3.39%	2.77%	3.08%

Table 2 - Errors detected

There is a method, originally developed by Welford, whereby error rates can be combined with tapping times to create a single unified metric of performance [12]. When this method is applied to the force data it shows an additional 2.8% advantage to using force feedback. Thus we get an almost 15% overall benefit.

CONCLUSIONS

The ultimate goal of virtual reality systems is to allow people to work naturally and efficiently at a variety of tasks. The contribution of this paper has been to show that for a rapid tapping task, having a perspective view computed for the observers actual eye position can speed-up performance, although by a relatively small amount. In addition, making simulated contact with the targets also improves performance.

Our results differ from those of Boritz and Booth [2] in that we found an effect of head-tracked perspective, whereas they did not. One likely reason is mentioned in our introduction; they did not require their subjects to make head movements. Since the viewpoint in their non-head tracked condition was presumably quite close to the correct centre of perspective, there may have been very little difference between what the subjects saw in their head tracked and non-head tracked conditions. Thus their result cannot be taken as evidence that viewing a perspective image from an incorrect viewpoint has no ill effects. We forced head movements in the task that we devised and found a clear effect.

We also found that simulating contact using the Phantom improved performance. We are grateful to Christine MacKenzie (personal communication [11]) for pointing out to us that the tapping task with force feedback engaged is actually a rather different task to the task without force feedback. In the no-feedback mode, subjects actually made the cursor move *through* the disc shaped target region and back in order to register a target hit; this required less effort than moving and bringing the cursor to a halt in the target centre. Conversely, in the force enabled condition. The cursor could be bounced off a target actually speeding its progress back to the other target. The constraints provided by the physical environment alter the characteristics of many real-world tasks, often making them easier. Exploiting such synergies may be the most compelling reason for introducing force feedback into virtual environments.

Our results only show quite small benefits to providing a correct perspective view and force feedback, and thus might not seem to warrant the considerable technology involved. However, skilled designers can take advantage of excellent tools. Taken together, including both head tracking and force feedback, improved tapping performance by 20% and, in addition, reduced errors. If the goal is to achieve fluid and highly responsive environment, this advantage may be accrued in every interaction; such small gains can easily make the difference between an environment that is a pleasure to use and one that is barely acceptable. In our experience the combination of these technologies provides a compelling localised virtual reality experience. We are confident as costs drop and the systems improve, this kind of apparatus may provide effective support for designing virtual sculpting systems and 3D CAD operations.

ACKNOWLEDGEMENTS

We are grateful to William Knight, for help with the statistical analysis and Christine Mackenzie for useful suggestions.

REFERENCES

1. Arthur, K. W., Booth, K. S., and Ware, C. (1993). Evaluating 3D Task Performance for Fish Tank virtual Worlds. *ACM Transactions on Information Systems,* 11(3) 239-265.
2. Boritz, James and Booth, Kellogg S., A Study of Interactive 3D Point Location in a Computer Simulated Virtual Environment, In *Proceedings of ACM Symposium on Virtual Reality Software and Technology '97,* Lausanne, Switzerland, Sept., pp. 181-187.
3. Deering, M. (1992). High Resolution Virtual Reality. *Proceedings of ACM SIGGRAPH '92, Computer Graphics,* **26:2**, 195-202.
4. Fitts, P.M. (1954) The information capacity of the human motor system in controlling the amplitude of movement. Journal of Expeimental Psychology, 47(6) 381-391.
5. Graham, E.D. and MacKenzie, C.L. (1996) Physical versus virtual pointing. Proceedings of ACM CHI'96. 292-299.
6. Hannaford, B. Wood, L. Guggisberg, b., McAffee, D. and Zack, H. (1989) Performance evaluation of a six-axis universal force-reflecting hand controller. In Proceedings of the 19th IEEE Conference on Decision and Control. Albuquerque, NM, Dec, 1197-1205.
7. Harris, C. S. (1965). Perceptual adaptation to inverted , reversed, and displaced vision. *Psychological Review,* **72**, 419-444.
8. Held, R., Estanthiou, A. and Green, M. (1966) Adaptation to displaced and delayed visual feedback from the hand. Journal of Experimental Psychology, 72, 887-891.
9. Kubovy, M. (1986) The psychology of perspective and renaissance art. Cambridge University Press.
10. Meek, S.G., Jacobson, S.C., and Goulding, P.P. (1989) Extended physiologic taction, design and evaluation of a proportional force feedback system. Journal of Rehabilitation Research and Development, 2693) 53-62.
11. MacKenzie, C.L. (1992) Making contact: target surfaces and pointing implements for 3-D kinematics of humans performing a Fitts' task. Society for Neuroscience Abstracts, 18, 515
12. MacKenzie, I.S. (1992) Fitts' Law as a Research and Design tool in human-computer interaction. Human-Computer Interaction, 7(1) 91-139.
13. Pausch, R., Proffitt, D., & Williams, G (1997) Quantifying immersion in virtual reality, ACM SIGGRAPH`97 Conference Proceedings, Computer Graphics.
14. Rossetti, Y., Koga, K., and Mano, T., (1993) Prismatic displacement of vision induces transient changes in the timing of eye-hand coordination. Perception and Psychophysics, 54(3) 355-364.
15. Sheridan, T.B. (1992) Telerobotics, Automation and Human Supervisory Control, MIT Press: Cambridge, Mass.
16. Tan, H. Z., Srinivasan, M. A., Eberman, B., Cheng, B. (1994). Human Factors for the Design of Force-Reflecting Haptic Interfaces. *Dynamic Systems and Control,* 55:1
17. Ware, C. and Rose, J. (1999) Rotating Virtual Objects with Real Handles, ACM Transactions on Computer-Human Interaction, 6(2) 162-180.
18. Ware, C. Arthur, K., and Booth, K.S. (1993) Fish tank virtual reality. In Proceedings of ACM INTERCHI'93, 37-42.
19. Ware, C., and Franck, G., Evaluating Stereo and Motion Cues for Visualizing Information Nets in Three-Dimension. ACM Transactions on Graphics, 15(2) 121-140.

Putting the Feel in 'Look and Feel'

Ian Oakley, Marilyn Rose McGee, Stephen Brewster and Philip Gray

Glasgow Interactive Systems Group
Department of Computing Science
University of Glasgow, Glasgow, G12 8QQ, UK
Email: {io, mcgeemr, stephen, pdg}@dcs.gla.ac.uk   Web: www.dcs.gla.ac.uk/~stephen

ABSTRACT

Haptic devices are now commercially available and thus touch has become a potentially realistic solution to a variety of interaction design challenges. We report on an investigation of the use of touch as a way of reducing visual overload in the conventional desktop. In a two-phase study, we investigated the use of the PHANToM haptic device as a means of interacting with a conventional graphical user interface. The first experiment compared the effects of four different haptic augmentations on usability in a simple targeting task. The second experiment involved a more ecologically-oriented searching and scrolling task. Results indicated that the haptic effects did not improve users performance in terms of task completion time. However, the number of errors made was significantly reduced. Subjective workload measures showed that participants perceived many aspects of workload as significantly less with haptics. The results are described and the implications for the use of haptics in user interface design are discussed.

Keywords

Haptics, force feedback, multimodal interaction.

INTRODUCTION

Desktop interfaces are becoming increasingly complex, and with this added complexity, problems are beginning to emerge. One such problem is information overload, where so much information is presented graphically that it becomes difficult to attend to all relevant parts [4]. Presenting information in other sensory modalities has the potential to lessen this problem. Attempts have been made to overcome information overload using non-speech sound during interactions such as button clicking and scrolling [3, 5] but there have been no convincing empirical attempts to reduce overload by using haptic (or force feedback) technology. This new technology allows users to *feel* their interfaces and has the potential to radically change the way we use computers in the future. We will be able to use our powerful sense of touch as an alternative mechanism to send and receive information in computer interfaces.

Augmenting graphical user interfaces (GUIs) with haptic feedback is not a new idea. In 1994 Akamatsu and Sate [1] developed a haptic mouse with the ability to produce what they termed 'tactile feedback', the ability to vibrate a user's fingertip, and 'force feedback', a simple software controllable friction effect. Using this device they showed significantly decreased completion times in a targeting task offset by slightly increased error rates. Engel *et al.* [7] found improved speed and error rates in a generalised targeting task using a modified trackball with directional two degrees of freedom force feedback.

The devices used in these early studies have now been superseded. More advanced devices such as the Pantograph (Haptic Technologies Inc.), the FEELit mouse (Immersion Corp.), and the PHANToM (SensAble Technologies Inc.) have been developed. These devices have all been used to augment desktop interfaces. Ramstein *et al.* [11] used the Pantograph to demonstrate performance increases in desktop interactions but provided little empirical evidence to support their claims. The FEELit mouse is a commercial product that offers users a haptically-enhanced desktop but there has been little evaluation of this device published [14]. Finally, the PHANToM has been used to create a haptically enhanced XWindows desktop [10]. No formal evaluation of this enhancement can be found in the literature.

The pace of technological advancement in this field is rapid, both in terms of the hardware produced and the software developed. Current projects to 'haptify' the desktop are not constrained to use the haptic effects described by Akamutsu and Engel. However, as technology has advanced there has been no corresponding progress in its evaluation. This disparity has led to a situation where there are no formal guidelines regarding what feedback is appropriate in different situations. This, along with evidence that shows arbitrary combinations of information presented to different senses is ineffective [12, 13], leads to the conclusion that empirical evaluation of modern haptic augmentations of the desktop is urgently required if much time and effort is not to be wasted. We might even end up with haptically-enhanced interfaces that are in fact harder to use than standard ones and haptics may become just a gimmick, rather than the key improvement in interaction technology that we believe it to be.

Haptic Terminology

Many different terms with many different definitions are used throughout the literature to describe haptic interaction. One reason for this is that the area is in its infancy. To rectify this problem we propose a set of haptic definitions that should prove useful for further research in this area.

The word 'haptic' has grown in popularity with the advent of touch in computing. We define the human haptic system to consist of the entire sensory, motor and cognitive components of the body-brain system. It is therefore closest to our understood meaning of proprioceptive (see Table 1). We define haptics therefore to be anything relating to the sense of touch. Under this umbrella term, however, fall several significant distinctions. Most important of these is the division between cutaneous and kinesthetic information (see Table 1). There is some overlap between these two categories; critically both can convey the sensation of contact with an object. The distinction becomes important however when we attempt to describe the emerging technology. In brief, a haptic device provides position input like a mouse but also stimulates the sense of touch by applying output to the user in the form of forces. Tactile devices affect the skin surface by stretching it or pulling it, for example. Force feedback devices affect the finger, hand, or body position and movement. Using these definitions (summarised in Table 1), devices can be categorised and understood by the sensory system that they primarily affect.

Term	Definition
Haptic	Relating to the sense of touch.
Proprioceptive	Relating to sensory information about the state of the body (including cutaneous, kinesthetic, and vestibular sensations).
Vestibular	Pertaining to the perception of head position, acceleration, and deceleration.
Kinesthetic	Meaning the feeling of motion. Relating to sensations originating in muscles, tendons and joints.
Cutaneous	Pertaining to the skin itself or the skin as a sense organ. Includes sensation of pressure, temperature, and pain.
Tactile	Pertaining to the cutaneous sense but more specifically the sensation of pressure rather than temperature or pain.
Force Feedback	Relating to the mechanical production of information sensed by the human kinesthetic system.

Table 1: Definitions of Terminology.

EXPERIMENTAL OVERVIEW

This paper describes two experiments that empirically test the use of haptics to augment targeting in the standard GUI. It is force feedback, and not tactile feedback that is evaluated in this work. Experiment 1 compared user performance with haptically-enhanced buttons using four different haptic effects in a simple targeting task. Experiment 2 involved a more ecologically oriented task in which participants searched for and selected targets using haptic scrolling. We hypothesise that in both experiments haptics will have a positive effect on performance.

Neither of the experiments described is concerned with the influence of haptic distracters; both investigate haptic augmentation when there is guaranteed to be a clear path to target. The decision to adopt this approach reflects the preliminary nature of empirical research in this field.

Device and Software

The device used in both experiments is the PHANToM 1.0 (see Figure 1). It is a force feedback device (provides kinesthetic information as defined in Table 1) which, in the experiments, acted as a cursor control device in place of the traditional mouse.

Optical sensors detect changes in the configuration of the PHANToM. The device uses mechanical actuators to apply forces back to the user calculated from this positional information and the stored algorithmic models of the objects with which the user is currently interacting. To operate the device users hold a stylus.

The graphical interface was generated using standard (MFC) widgets and these performed in exactly the same way as standard widgets. The workspace was a box 160 mm wide x 160 mm high x 2 mm deep. The haptic effects were present only on the back wall of the workspace.

Haptic Effects

Four haptic effects were used in the experiments. These built on and added to the effects used in previous studies. The effects were all aimed at improving targeting and reducing problems of mis-hitting or slipping off interface widgets. The effects used were:

Texture: Texturing a button in a texture-less, flat workspace is a potential way of haptically signifying that the cursor is positioned over an interesting object. The texture implemented here formed a set of concentric circles 7.5 mm apart and centred around the middle of the target. The

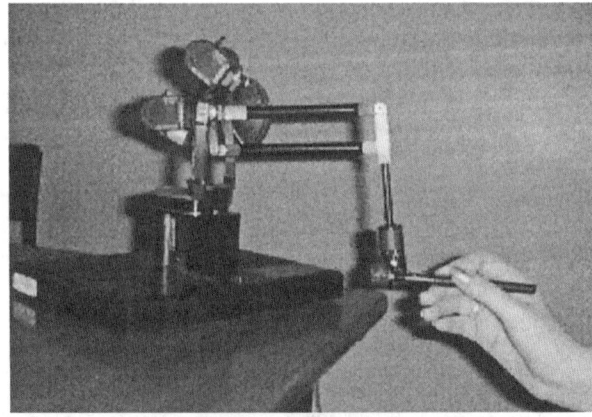

Figure 1: The Phantom 3D force feedback device from SensAble Technologies. The stylus shown has a button that can be used for performing the mouse clicks in the experiments reported.

texture was created by vector rotation (force perturbation) [15] and the maximum rotation applied was 12°. A visual representation is shown in Figure 2. This texture pattern was used because it was felt that it would maximise the possibility that users would encounter ridges irrespective of the direction they began from or travelled in.

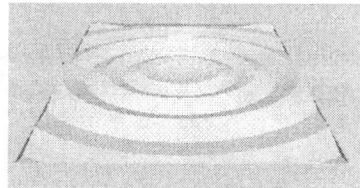

Figure 2: Diagram of the geometry of haptic texture effect.

Friction: The friction effect damped a user's velocity. Haptically-enhanced interfaces that use a friction effect are common in previous literature. This is partly because they can be produced with simple hardware – for instance with an electromagnet placed in the base of a mouse [1, 2] – and partly because it seems advantageous to provide feedback that causes a user to stop when over an interesting target. The friction effect used here was realistically modelled with both a static and a dynamic component. The static component restricted users to a point until they attained an escape velocity. The dynamic component attempted to slow them whilst they were in motion.

Figure 3: Diagram of the geometry of haptic recess effect.

Recess: The recess effect was a hole in the back of the workspace, with a depth of 2 mm and edges sloped at 45°. This effect also features strongly in previous literature [10, 11]. A diagram of the geometry of a recess is presented in Figure 3. A recess could potentially provide useful feedback by the simple fact that to leave it, the wall at the edge must be climbed. This may make it harder to accidentally slip-off a button (a problem noted by Brewster *et al.* [5]).

Gravity Well: The gravity well was a 'snap-to' effect. When users moved over a button a constant force of 0.5 N was applied that pushed them towards the button's centre. This force tapered off around the very centre so that the user could rest in the centre. The gravity well promised the same benefits as the recess – a reduction in errors through the simple mechanism of preventing a user from accidentally slipping off a button.

General Measures Used in the Experiments

In order to get a full range of quantitative and qualitative results, time, error rates, and subjective workload measures were used in both of the experiments. The subjective workload measurement was a modified version of the NASA Task Load Index (TLX) [8]. NASA reduce workload to six factors: mental demand, physical demand, time pressure, effort expended, performance level achieved, and frustration experienced. We added a seventh factor: fatigue. One potential problem with force feedback devices is the physical strain placed on the user. By adding this factor it would be possible to find out if haptic effects caused any additional perceived fatigue. Participants filled-in workload charts after each condition in both experiments.

EXPERIMENT 1

In the first experiment the haptic effects were compared to investigate which was the most effective. To do this we added each of the haptic effects to standard graphical buttons. This allowed us to investigate targeting (moving the cursor to the button) and mis-hitting errors (slipping-off the button when trying to press it).

Hypotheses

Experiment 1 was an exploratory experiment – we wanted to investigate the differences between the different haptic effects and a control condition. Therefore, the experimental hypotheses were that differences would occur in task completion time, number of errors and in the subjective data gathered. We predicted that the gravity well and recess would provide the largest reduction in errors, time and workload as they provided feedback that was highly appropriate to a simple targeting task.

Participants

There were sixteen participants. Four were female and twelve were male. All were between the ages of eighteen and thirty. Most were computing students from the University of Glasgow. All were regular and fluent computer users. Three users were left-handed and one was dyslexic. None had anything more than trivial previous exposure to the PHANToM.

Design

The experiment followed a within-subjects repeated-measures design. Each participant underwent each of the four haptic conditions, each encompassing one of the effects described above, and a control condition. The control condition used the PHANToM device but no haptic effects were applied – in essence the device worked like a normal mouse. The order of the presentation of the conditions was counterbalanced to evenly distribute the effects of practice and fatigue. Participants were randomly allocated to conditions. Training was given in each condition in a session immediately prior to the experiment. Each condition in the training session constituted 60 button presses and in the experimental session 120 presses. The experiment's duration was typically 45 minutes.

Task

A simple button pressing/targeting task was used. This task was chosen because it featured prominently in the previous literature [1, 2, 7] and also because it is a very elementary operation – it is both simple to perform and also perhaps the most fundamental cursor operation.

Two factors were engineered into the task to make it more suitable for haptic augmentation. Firstly, it was felt that participants should experience some visual distraction. This is not an unlikely circumstance in the typical operation of a GUI, particularly in the case of expert users. They concentrate on some central task and interact with graphical widgets in the periphery of their attention [4]. Secondly, in this atmosphere of visual distraction, we assert that the haptic feedback will only really prove useful if the task encompasses some repetitive motion. Without such motions the haptic task would rapidly dissolve into exhaustively searching the entire workspace for some haptically distinct area. This is clearly an inefficient strategy when compared to visually scanning the screen. Repetitive motions are also common in desktop interactions (moving to menu bars, clicking buttons, etc.).

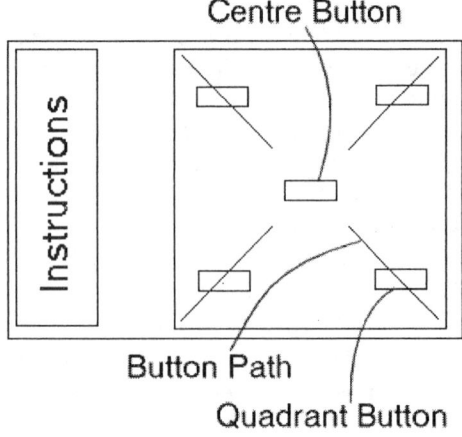

Figure 4: The interface used in Experiment 1.

To encompass these two factors two windows were placed on the screen at all times (see Figure 4). One, the instruction window, occupied the left-hand side of the screen and contained instructions as to the next target to seek. The other, larger, window occupied the centre and right-hand side of the screen and contained the targets in the form of five buttons. One button was always positioned in the centre, the other four were positioned one in each quadrant of the window, on the diagonals of the window. The position along the diagonals was changed in the course of the experiment, but each button remained in a single quadrant of the window throughout. This meant that each button remained in the same direction relative to the centre of the window at all times. The buttons moved along the diagonals to prevent users employing a purely mechanical repetition. To ensure users moved along only a few trajectories to reach each of the buttons, every second button press was the centre button. The buttons were labelled in accordance with their positions on screen, for instance "top right" or "bottom left". The instruction window indicated the next target button, on successfully pressing the named button, a new name was presented.

Measures

Data were gathered from all button presses in the experiment. The performance measures were (a) mean time per trial (secs.), (b) mean number of errors, and (c) subjective workload ratings. Times were measured at four stages: time to find target button; time to move onto target button; time to press target button; and time to move off target button. Errors were measured as when a participant moved over a button but failed to press it. There were two categories: the first was where the user simply slid over the button, arguably as a part of the normal targeting process. The second, more serious error is known as a 'slip-off' [4]. This occurs when a user presses the mouse down over a button but moves off it before releasing the mouse, thus not selecting it. The feedback for this is the same as for a successful mouse click. An error of this type can go unnoticed for some time and cause considerable confusion.

Results from Experiment 1

The error data are presented in Figures 5 and 6. Results were analysed using ANOVA tests. Significant effects were found when comparing the mean scores for each haptic effect for both slide over ($F_{4,15} = 48.487$, $p<0.001$) and slip off ($F_{4,15} = 20.81$, $p<0.001$) errors. Order effects for both slide over ($F_{4,15} = 0.152$, $p=0.961$) and slip off ($F_{4,15} = 0.123$, $p=0.974$) errors were not found.

	Gravity	Recess	Friction	Texture	Control
Gravity	---------	Not sig	p<0.001	p<0.001	p<0.002
Recess	---------	---------	p<0.01	p<0.003	Not sig
Friction	---------	---------	---------	p<0.04	Not sig
Texture	---------	---------	---------	---------	p<0.016

Table 2: Analysis of slip-off errors in Experiment 1.

	Gravity	Recess	Friction	Texture	Control
Gravity	---------	Not sig	p<0.01	p<0.01	p<0.01
Recess	---------	---------	p<0.01	p<0.01	p<0.01
Friction	---------	---------	---------	p<0.01	Not sig
Texture	---------	---------	---------	---------	p<0.01

Table 3: Analysis of slide-over errors in Experiment 1.

A summary of the results revealed by *post-hoc* analysis of the means (using Bonferroni confidence interval adjustments) is shown in Tables 2 and 3. The most dramatic results were that participants in the gravity condition made significantly fewer errors of both sorts than in the control and that the converse was true of the texture condition – it caused significantly more errors than the control.

Analysis of the temporal data was less conclusive; the total time taken to complete a trial was strongly biased by the number of errors made in each condition. It was felt that this invalidated it as a measure – it would merely be a reflection of the number of errors in each condition.

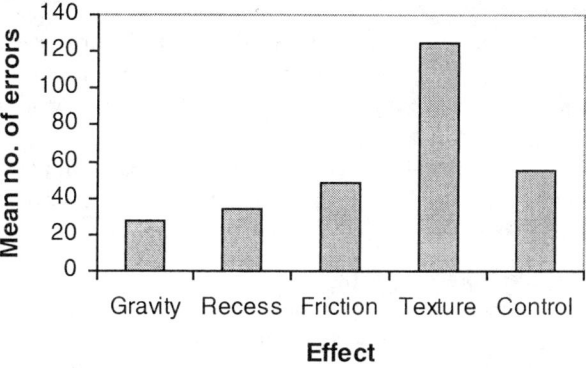

Figure 5: Slide over errors in Experiment 1.

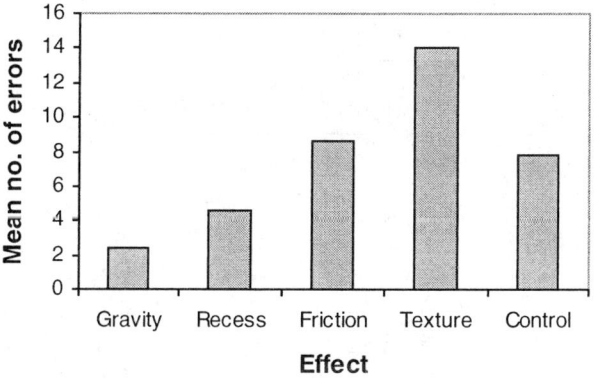

Figure 6: Slip-off errors in Experiment 1.

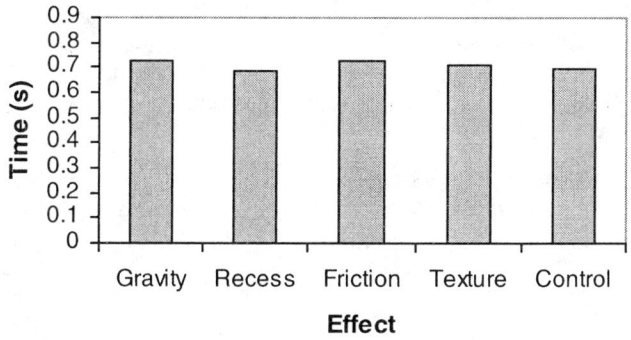

Figure 7. Total time on button in Experiment 1.

Instead, the total time on a button during a successful trial was analysed (see Figure 7). An ANOVA revealed significant differences between effects. Subsequent pair-wise comparisons (using Bonferroni adjustments) revealed that gravity was significantly slower than recess ($p<0.05$). It is also worth noting that the difference between the best and worse performing effects was only 42 ms, a very short time. No order effects were found in this temporal analysis ($F_{4,15} = 0.913$, $p=0.462$).

To validate analysing time and errors separately we ran a Pearson correlation. The timing results did not correlate with the slide over ($r=0.0$, $p<1.0$) or slip off ($r=0.019$,

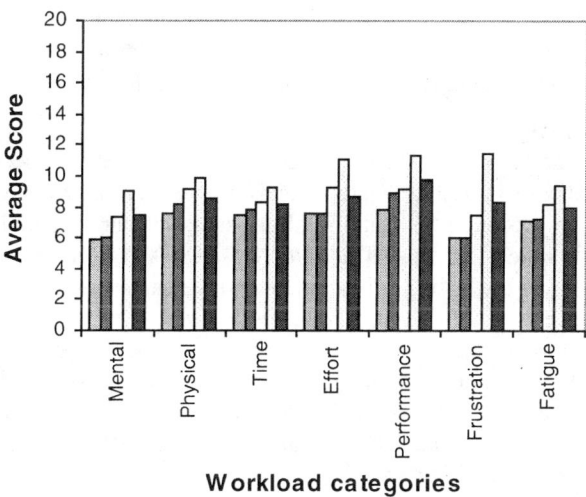

Figure 8: Workload results from Experiment 1.

$p<0.976$) errors. The two error results strongly correlated with one another ($r=0.938$, $p<0.018$).

Figure 8 shows the TLX workload scores (scored out of 20). The texture condition was significantly worse than the control across the whole board of measures. The gravity condition consistently reduced workload and, in particular, achieved a significantly better score than the control in the performance level achieved category ($p<0.018$).

EXPERIMENT 2

This experiment simulated a more realistic task where reading was accompanied by scrolling through a document, selecting from the document, and returning to the scroll bar whilst still visually attending to the material being read. When users are required to scroll through a document it is the material in it that is of interest and not the scroll bar. Users want to concentrate on reading the material but often find themselves forced to move their visual attention to the scroll bar to ensure that the cursor is positioned appropriately to operate it. The time taken to make these frequent shifts in visual attention, and the frustration experienced by the need to do so, reduce the usability of the scroll bar. Problems associated with scrolling have been addressed previously [e.g. 4, 16]. Reducing these problems using force feedback technology has not yet been empirically evaluated.

Hypotheses

It was hypothesised that when the scroll bar was haptically-enhanced, the participants would (a) take significantly less time to complete the task; (b) move on and off the scroll bar significantly less; and (c) perceive the workload during the task as significantly less.

Participants

Twenty new participants were used: one was female and the remaining nineteen male. All were between the ages of seventeen and twenty-seven. Most participants were first-year computing science students from the University of Glasgow. All were regular and fluent computer users. All users were right-handed. Participants had nothing more than trivial previous exposure to the PHANToM.

Design

The experiment again used a within-subjects repeated-measures design. Each participant underwent both a visual-only condition (visual) and a visual and haptic condition (haptic). The visual condition used a standard graphical scroll bar only. In the haptic condition, this scroll bar was overlaid with haptic effects (recess and gravity well were chosen as these were the most effective in Experiment 1). The up and down arrow buttons used gravity wells. These acted as a haptic indication that the user was in the appropriate place to press the button successfully. The rest of the scrolling area used a recess effect that allowed the user to 'fall into' the slider area. Therefore, the haptic feedback allowed the user to reserve his/her visual attention for the primary task, as being over the widget was indicated through touch. The order of the presentation of the conditions was counterbalanced to evenly distribute the effects of practice and fatigue. Training was given to each participant in each condition prior to the experiment.

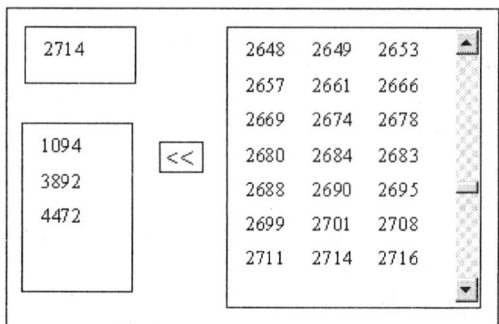

Figure 9: The interface used in Experiment 2. The top left window is the instruction window, the bottom left is the target window, the large window to the right is the data window and in the centre is the send button.

Procedure

Figure 9 shows the interface to the task. Participants had to read a four-digit numerical code from the instruction window. They then had to scroll vertically through a large file of codes (presented in the data window) to find the target code, highlight the code (either by double clicking on it or dragging across it), and press a button to send this code to the target window. The widgets operated as in standard desktop applications. The data window contained the same list of 2000 randomly generated but numerically ordered codes in each condition. Forty codes had to be entered in each condition. The list was formatted such that there were three columns of codes, simulating a standard document read from left to right and from top to bottom. The highlight operation was included to force the user off the scroll bar. This ensured repeated targeting of the scroll bar. The experiment's duration was typically 40 minutes.

Measures

The performance measures were (a) mean time per trial (secs.), (b) mean number of movements on/off scroll bar (including all required movements), and (c) workload ratings. Time was measured from when the user activated the send button at the end of the previous trial until the send button was activated at the end of the current trial. Subjective ratings were collected as before.

Results from Experiment 2

Timing results: Table 4 shows the timing and movement on/off scroll bar results. Paired T-tests established that haptic feedback did not significantly reduce the average trial time as predicted ($T_{19} = 0.46$, $p < 0.32$).

Mean Trial Time (secs.)		No. times on/off scroll bar	
Visual	Haptic	Visual	Haptic
11.7251	11.9668	107	97
SD=2.77	SD=2.84	SD=25	SD=22

Table 4: Timing and movement results from Experiment 2.

Movement on/off scroll bar: Paired T-tests showed that participants in the haptic condition moved on and off the scroll bar area significantly less than in the visual condition ($T_{19} = 2.37$, $p < 0.05$).

Workload Results: Figure 10 shows the workload scores. Paired T-tests were carried out on the visual versus haptic conditions for each of the categories. Mental demand was not significantly less in the haptic condition as expected. Both the effort and frustration ratings were significantly reduced in the haptic condition (Effort: $T_{19} = 2.80$, $p < 0.01$, Frustration: $T_{19} = 2.04$, $p < 0.05$). There was no significant difference in fatigue experienced. The hypothesis that the haptic condition would reduce workload is therefore confirmed in part.

GENERAL DISCUSSION

The timing results from the two studies indicate that the haptic effects added to the buttons and scroll bar did not

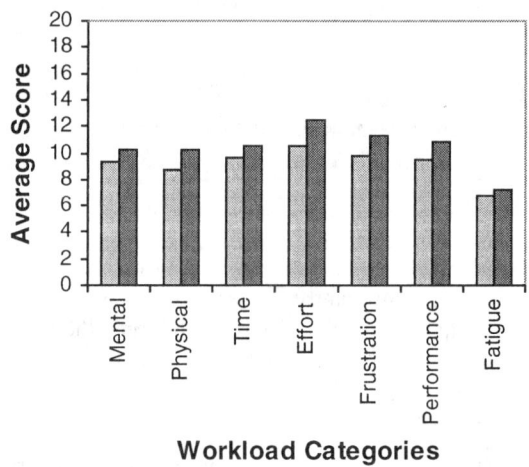

Figure 10: Workload results from Experiment 2.

reduce the time taken for either task, as hypothesised. There were also no real differences between the effects – only 42 ms between the best and worst effects (recess and gravity) in Experiment 1. The explicit separation of the error data from the timing data is no doubt a contributing factor to the lack of temporal variations across conditions. However, we suggest that one potential reason for the lack of time reduction is that, in all of the effects used, participants had to exert more force to overcome the haptic effects. In the control condition they could just slide over the interface with no obstacles, in the haptic conditions they had to climb out of recesses, overcome gravity forces applied, etc. For participants to produce the forces required to do this could have taken them more time.

Further work is needed on the haptic effects themselves and the types of desktop tasks that would benefit most from them. It may have been that the haptic effects chosen were inappropriate either for reducing time or for the tasks chosen for these experiments. Other previous work has claimed a significant reduction in performance times [10, 14]. The present work suggests that things are no so clear-cut and care must be taken when using haptics to try to reduce performance times.

The error results were more conclusive. Experiment 1 showed a significant reduction in the number of errors produced across the different haptic conditions (where gravity and recess caused the fewest errors and texture the most). Gravity and recess were the most effective for targeting tasks (which are important for using many standard GUI widgets, for example hitting a button, selecting a menu item or dragging the scrollbar thumb) in the sense that they made it very hard to slip off a target once on it; participants could not just knock the pointer off the target, they had to make an explicit movement to leave. Texture only indicated that the cursor was over a target, and did not constrain users to the target, which was one of the reasons it was less effective in this case. Texture also had the problem that it could potentially perturb users' movements, making it hard for them to stay on target. This resulted from the kinesthetic force feedback device used here. We use cutaneous stimulation to feel much of the richness of fine-grained texture in the real world [9]. A kinesthetic device can only simulate gross textures, requiring larger forces, which then make it harder for users to move precisely. Texture is much more suitable to production by tactile devices such as the *Tractile* from Campbell *et al.* [6]. The PHANToM, on the other hand, is very effective at simulating gravity and recess effects as these require movement and so are kinesthetic tasks. There are no devices, as yet, which combine both tactile and kinesthetic force feedback.

Haptic devices are now reaching the desktop. For example, the FEELit Mouse [14] adds low cost haptic effects to the standard graphical interface. Our results show that interface designers must be aware of the facilities of the devices they are using in order to generate haptic effects that will improve usability. This might seem obvious, but this area is in its infancy and new devices are appearing all the time, each having different functionality to the last.

The movement results from Experiment 2 showed a significant reduction in the number of times a participant moved on/off the scroll bar in the haptic condition. This showed that the haptic recess aided participants in remaining on target, demonstrating that haptics can provide a significant practical benefit for interaction. The haptic groove placed over the scroll bar allowed users to scroll up and down without slipping off. They could do this without looking at the bar as once the cursor was in the groove it would stay there. To move out of the recess they had to lift off the scroll bar and it was difficult to do this by mistake as it required a conscious effort.

The subjective workload measures taken across both experiments are important. Papers concerning other haptically-enhanced desktops have not presented any such data. In developing multimodal interfaces (ones that use multiple sensory modalities) it is very important to consider what effects they have on users' workload. Users may perform tasks well and quickly and yet find them frustrating and requiring more effort to complete than they would expect. This dissociation between behavioral measures and subjective experience has been addressed in studies of workload. Hart and Wickens [8] suggest that cognitive resources are required for a task and there is a finite amount of these. As a task becomes more difficult, the same level of performance can only be achieved by the investment of more resources. Just measuring time or error rates does not give the whole picture of the usability of a haptic device. Workload is particularly important in this area as we know little yet of the effects on cognitive/attentional resources of using such devices.

Experiment 1 showed that the different effects had markedly different levels of workload. Gravity well and recess came out best, indicating that they were effective at reducing error rates and decreasing workload. This suggests that they are very robust and can be successfully used in haptic interfaces of the type described here. Texture came out the worst in terms of workload, suggesting that, in general, it is hard to do effectively with the device used here. Experiment 2 showed the effect of haptics in a more realistic situation. In this case there was a significant reduction in effort and frustration – the fact that it was easy to stay on the scroll bar due to the recess effect made the task much less effortful (the reduction in the number of movements on/off the scroll bar confirms this). We had expected that this might also lead to reductions in other categories (e.g. mental demand) but these showed no significant reductions. This suggests that we need further studies of workload to learn more about the affect of haptics in desktop interactions.

One other area that we investigated was fatigue. Using a device that requires the user to apply force could cause fatigue. It is important to investigate this if force feedback devices are to be used in desktop situations (where people might use the interfaces for long periods of time). Results from Experiment 1 showed that gravity and recess effects did not cause any more fatigue than the control condition. On the other hand, texture caused significantly more fatigue than the control. This is likely to be for the reasons as discussed above – to simulate texture with a kinesthetic device required larger forces to be applied and these, in turn, required the users to exert larger forces to overcome them. Experiment 2 again showed no increase in fatigue with the use of gravity well and recess effects. This research shows that appropriate haptic effects used correctly may have no impact on fatigue, but used incorrectly may significantly increase it. This is only a first step in investigating this problem and further work is needed to ensure that we can design haptic interfaces to avoid fatigue

CONCLUSIONS

Our research has shown that haptics may have some benefits in graphical user interfaces. Reductions in the number of errors made and subjective workload experienced can be gained. We have also shown that the haptic effects used must be matched to the capabilities of the device – trying to simulate effects not supported by the device in use can have serious negative effects on all aspects of usability. As technology progresses it is easy to focus on what benefits new equipment may afford whilst forgetting to measure the benefits actually produced. Recent work on haptically-enhanced desktops has been firmly orientated towards implementation and the experiments described here begin to redress the balance. Our empirical findings provide a firm foundation for future researchers to build on and some basic principles for developers to use.

ACKNOWLEDGEMENTS

This research was supported under EPSRC project GR/L79212 and EPSRC studentship 98700418. Thanks must also go to the SHEFC REVELATION Project, SensAble Technologies and Virtual Presence Ltd.

REFERENCES

1. Akamatsu, M. & Sate, S. (1994). A multi-modal mouse with tactile and force feedback. *International Journal of Human-Computer Studies* (40), 443-453.

2. Akamatsu, M. & MacKenzie, S. (1996). Movement Characteristics using a Mouse with Tactile and Force Feedback. *International Journal of Human Computer Studies* (45), 483-493.

3. Beaudouin-Lafon, M. & Conversy, S. (1996). Auditory illusions for audio feedback. In *ACM CHI'96 Conference Companion* (Vancouver, Canada) ACM Press, Addison-Wesley, 299-300.

4. Brewster, S.A. (1997). Using Non-Speech Sound to Overcome Information Overload. *Displays*, 17,179-189.

5. Brewster, S.A. (1998). The design of sonically-enhanced widgets. *Interacting with Computers*, 11(2), 211-235.

6. Campbell, C.S., Zhai, S., May, K.W. & Maglo, P. (1999). What you feel must be what you see: adding tactile feedback to the trackpoint. In *IFIP Interact'99*, (Edinburgh, UK), IOS Press, 383-390.

7. Engel, F.L., Goossens, P. & Haakma, R. (1994). Improved Efficiency through I- and E-Feedback: A Trackball with Contextual Force Feedback. *International Journal of Human-Computer Studies*, 41(6), 949–974.

8. Hart, S.G. & Wickens, C. (1990). Workload assessment and prediction. MANPRINT, an approach to systems integration, 257-296, Van Nostrand Reinhold.

9. Lederman, S.J., Summers, C. & Klatzky, R.L. (1996). Cognitive salience of haptic object properties: Role of modality-encoding bias. *Perception*, 25, 983-998.

10. Miller, T. & Zeleznik, R. (1998). An Insidious Haptic Invasion: Adding Force Feedback to the X Desktop. In *ACM UIST'98*, (San Francisco, CA) ACM Press, 59-64.

11. Ramstein, C. (1995). A Multimodal User Interface System with Force Feedback and Physical Models. In *IFIP Interact'95*, (Lillehammer, Norway) Chapman & Hall, 157-162.

12. Ramstein, C. & Hayward, V. (1994). The Pantograph: A Large Workspace Haptic Device for Multi-Modal Human-Computer Interaction. In *Summary Proceedings of ACM CHI'94*, (Boston, MA) ACM Press, Addison-Wesley, 57-58.

13. Ramstein, C., Martial, O., Dufresne, A., Carignan, M., Chassé, P. & Mabilleau, P. (1996). Touching and hearing GUIs: Design issues for the PC-access system. In *Proceedings of ACM Assets'96*, (Vancouver, Canada), ACM Press, 2-10.

14. Rosenberg, L.B. (1997). FEELit mouse: Adding a realistic sense of FEEL to the computing experience. http://www.force-feedback.com/feelit

15. Srinivasan, M.A. & Basdogan, C. (1997). Haptics in Virtual Environments: Taxonomy, Research Status, and Challenges. *Computers and Graphics*, 21(4), 393-404.

16. Zhai, S., Smith, B.A., & Selker, T. (1997). Improving Browsing Performance: A study of Four Input Devices for Scrolling and Pointing Tasks. In *Proceedings of Interact 97*, (Sydney, Aus) Chapman & Hall, 286-293.

Force-Feedback Improves Performance For Steering and Combined Steering-Targeting Tasks

Jack Tigh Dennerlein
Harvard University
665 Huntington Ave
Boston, MA 02115
+1 617-432-2028
jax@hsph.harvard.edu

David B. Martin
Harvard University &
Dartmouth College
Hanover, NH 03755
David.B.Martin@dartmouth.edu

Christopher Hasser
Stanford University &
Immersion Corporation
2158 Paragon Drive
San Jose, CA 95131
c.hasser@ieee.org

ABSTRACT

The introduction of a force-feedback mouse, which provides high fidelity tactile cues via force output, may represent a long-awaited technological breakthrough in pointing device designs. However, there have been few studies examining the benefits of force-feedback for the desktop computer human interface. Ten adults performed eighty steering tasks, where the participants moved the cursor through a small tunnel with varying indices of difficulty using a conventional and force-feedback mouse. For the force-feedback condition, the mouse displayed force that pulled the cursor to the center of the tunnel. The tasks required both horizontal and vertical screen movements of the cursor. Movement times were on average 52 percent faster during the force-feedback condition when compared to the conventional mouse. Furthermore, for the conventional mouse vertical movements required more time to complete than horizontal screen movements. Another ten adults completed a combined steering and targeting task, where the participants navigated through a tunnel and then clicked a small box at the end of the tunnel. Again, force-feedback improved times to complete the task. Although movement times were slower than the pure steering task, the steering index of difficulty dominated the steering-targeting relationship. These results further support that human computer interfaces benefit from the additional sensory input of tactile cues to the human user.

Keywords
Mouse, pointing device, force-feedback, haptic, steering task, targeting task, Fitts' Law, index of difficulty

Permission to make digital or hard copies of all or part of this work for personal or classroom use is granted without fee provided that copies are not made or distributed for profit or commercial advantage and that copies bear this notice and the full citation on the first page. To copy otherwise, to republish, to post on servers or to redistribute to lists, requires prior specific permission and/or a fee.
CHI '2000 The Hague, Amsterdam
Copyright ACM 2000 1-58113-216-6/00/04...$5.00

INTRODUCTION

When the computer mouse was developed some 30 years ago, it ushered in a new era in computer human interfaces. The advantage of a virtual finger to point and interact with a computer graphic user interface was immediately recognizable, clearly contributing to the mouse's rapid public acceptance. Now 30 to 80 percent of computer work involves the mouse [9]. Since then, engineers and inventors have abandoned the task of building a better mousetrap to build a better mouse. That challenge, however, has proven to be as tricky. Most results to date have focused on industrial design of the mouse more so than producing a new technology for the pointing device.

Derived from telerobotic applications [12] the use of force-feedback technology in computer input device designs has long been a hot topic in the development of computer gaming, computer-aided design, surgical training, and other simulated environments [11]. Yet little has been established regarding its utility in everyday window-type computer desktop environments, an area made still more interesting in light of the ever-increasing numbers of computer-related musculoskeletal disorders [4]. Dennerlein and Yang [5] suggest that the addition of force-feedback might reduce the musculoskeletal loading during computer mouse use, a possible risk factor for chronics musculoskeletal disorders of the upper extremity.

The more general human computer interface (HCI) problem of finding a quantitative means with which to measure the performance of human motor control during completion of simple tasks is much older and better understood. Fitts [8] arrived at a quantitative predictor for movement time in peg-in-hole (targeting) type tasks. He presented the following relationship, known as Fitts' Law, for estimating movement time (*MT*) needed for successful completion of these targeting type tasks:

$$MT = a + b\log_2\left(\frac{A}{W_T} + c\right) \quad (1)$$

where A is the distance from the starting position to the target, W_T is the width of the target, and a, b, and c are empirically determined constants. The logarithmic portion of the equation is defined as the index of difficulty for a targeting task (ID_T):

$$ID_T = \log_2\left(\frac{A}{W_T}\right) \quad (2)$$

For HCI design Fitts' Law provides a practical method for comparing the performance of two different pointing devices during an identical targeting-type tasks, specifically the point-and-click operation. A set of computer programmed tasks can be chosen based on their respective indices of difficulty. Then, a human subject can be asked to perform this set of tasks twice, once using each of two different interfaces. Assuming that all other experimental conditions remain constant for the two tests, Fitts' Law thus provides a hard measure with which to gauge the performance of the two interfaces against each other. Hasser et al. [8] used these means to show that the display of tactile cues through a force-feedback mouse for pointing tasks improved movement times when compared with a standard mouse.

Modern computer interfaces, however, are more than just point-and-click targeting tasks. What can be said of more complicated HCI tasks? Accot and Zhai [1] offered the first quantitative tool for analyzing and predicting the difficulty of HCI steering tasks. A steering task (in 2D terms) requires one to move the computer's pointer a certain distance along one axis without varying more than an arbitrary amount along the opposite axis. In other words, navigating through a tunnel. Everyday examples include steering a menu and its submenus, tracing a shape in a drawing program, and moving a scroll bar down a word processor or web page. Accot and Zhai [1] proposed that the index of difficulty for an arbitrary steering task (ID_S) did not include any logarithmic element. It was simply:

$$ID_S = \frac{A}{W_S} \quad (3)$$

where A represents tunnel length, and W_S represents tunnel width. Thus, for steering tasks, the predicted movement time (MT) is:

$$MT = a + b \cdot ID_S \quad (4)$$

where a and b are empirically determined constants. As in the case of Fitts' Law, from which Accot and Zhai were able to derive this steering law, we can use this relationship as a quantitative measure for determining the performance of new interface designs.

Following Hasser's et al. [8] lead, we decided to test the effects of force-feedback on steering-type tasks. As Hasser and Goldenberg showed experimentally, the addition of a force-feedback attractive basin that pulls the mouse to the center of the desired target substantially decreased the movement time for point-and-click targeting tasks. We hypothesized that similar benefits would be observed experimentally during steering tasks.

Pilot studies indicated that movement time for steering tasks depended on the direction of the movement, perhaps due to the overall joint kinematics required for the movement. Therefore, we also hypothesized that a steering performance difference exists between horizontal and vertical movements in the virtual desktop.

Another task frequently encountered within the windows-type environment is the combined steering and targeting task. For example, within a menu driven interface, the user is required to steer down a menu *and then click on a target*. This combined steering-targeting type task is not limited to menus. Selecting text usually means clicking, then steering the cursor to the desired selection, and clicking again. Graphic design programs also make liberal use of this selection method, and spreadsheet applications. We hypothesized that the movement time is affected equally by both the steering task difficulty and the targeting task difficulty.

METHODS AND MATERIALS

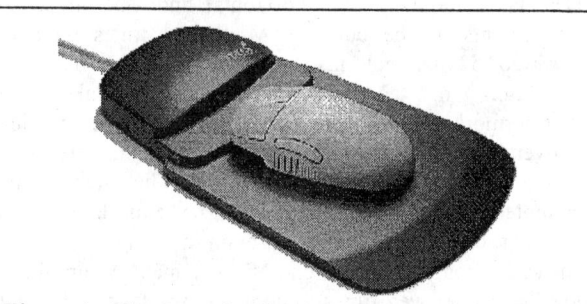

Figure 1: The force-feedback mouse (*FEELit* Mouse, Immersion Co. San Jose, CA).

Force-feedback mouse

A development force-feedback mouse (*FEELIt* Mouse, Immersion Corporation, San Jose, CA) was used throughout these experiments (Figure 1). The device consists of an ordinary two-button mouse, which is attached underneath to two motors through a bar linkage. The physical connection limits the workspace to approximately one square inch and does not allow for indexing of the mouse by picking it up and replacing it on the pad. The motors in the base can be programmed to move the mouse with up to three ounces (0.84 N) of force along any x-y vector parallel with the surface of the mouse pad. By keeping continuous track of the correspondence between the mouse's virtual location on the Windows 98 ® desktop and its location within the physical workspace, the mouse can be programmed to play force-feedback effects. For example, an attractive basin programmed around an OK button on the computer desktop produces a force that physically attracts the mouse and the hand of the user operating the mouse in the direction of the button. The end result is that the user feels as if their hand is pulled into the button itself.

Steering task and software

The testing software for the experiments completed three objectives. First the program presented the test subject with steering or combined steering-targeting tasks. Second, the program recorded their performance as measured through time to complete the task and accumulated errors during an individual successful task. Finally, the software, when requested, drove the motors of the force-feedback mouse based on the force algorithm in Figure 3.

The standard steering task consisted of a field with arbitrary length and width (Figure 2). These can readily be used, as shown previously, to calculate a steering index of difficulty. Using this index of difficulty, twenty fields of varying length and width were formulated (Table 1). The movements were also assigned directions within the visual display such that, five were in a top-to-bottom direction, five were in a bottom-to-top direction, five were in a left-to-right direction, and five were in a right-to-left direction. Hence, there were 10 vertical (top-to-bottom or bottom-to-top) movements and 10 horizontal (right-to-left and left-to-right) movements. This allowed for suitable testing of hypothesis two. The order of presentation of these twenty fields was randomized.

The test subject was presented with one of the twenty fields in random order. When a field was presented, the test subject enters the cursor within the zone, between the two solid lines from one of the open ends and moves to the other end in the direction of the arrow (Figure 2). Once the cursor enters the zone from the correct direction the timer was started. Exiting the field from the opposite side stopped the timer, and the results (including error count, field dimension, and field direction) were then recorded in an output file. When the subject exited the field through any other location the error count was incremented, the timer reset, and the user has to begin the field again. After successful completion, a new field is presented to the subject. This continued until the test was completed. Each of the twenty fields was presented four times, for a total of eighty test fields.

When force-feedback was turned on, the mouse rendered haptic walls that coincided with the graphic walls of the testing field, the solid black lines. The force fields of the haptic walls acted to repel the mouse and the operators' hand away from the solid lines of the testing field towards the center of the tunnel. A sufficient amount of effort by the human operator could still overcome the maximum force from the force fields (about three ounces), escaping the tunnel and failing the trial. The magnitude and direction of the force field is presented in Figure 3.

Table 1: Tunnel Widths and Indices of difficulty (R = Right, L = Left, D = Down, U = Up)

Tunnel Width (W_S) (Pixels)	Tunnel Length (A) (Pixels)	ID_S	Direction	Tunnel Width (W_S) (Pixels)	Tunnel Length (A) (Pixels)	ID_S	Direction
50	250	5.0	D	20	250	12.5	R
25	150	6.0	L	40	550	13.8	D
40	250	6.3	U	20	300	15.0	R
40	300	7.5	L	25	400	16.0	U
25	200	8.0	U	30	500	16.7	L
30	250	8.3	R	15	250	16.7	D
40	400	10.0	R	30	550	18.3	R
50	500	10.0	L	25	500	20.0	L
50	550	11.0	U	20	500	25.0	D
40	500	12.5	U	20	550	27.5	D

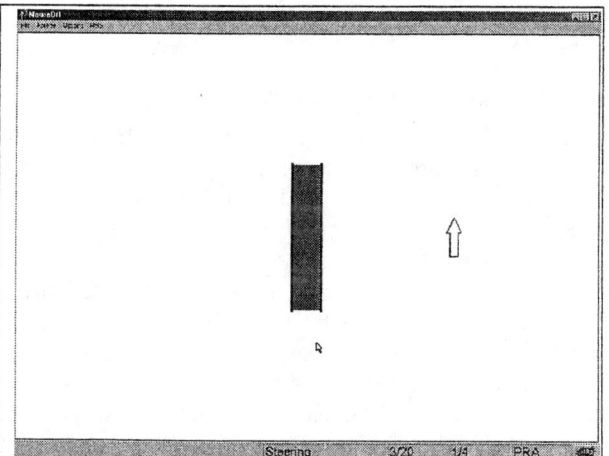

Figure 2: The steering task window. Operators had to navigate the cursor between the two solid lines in the direction denoted by the arrow on the right.

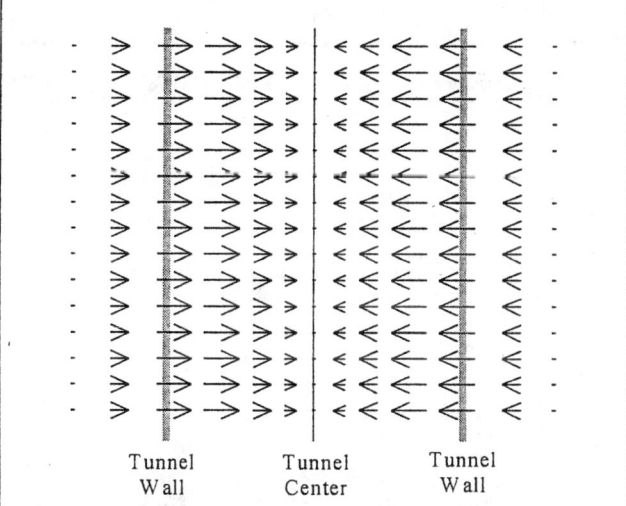

Figure 3: The steering task force field. The force magnitude was proportional to the distance from the center of the tunnel with maximum value of 0.8 N.

Combination task

For the combined steering-targeting experiment, a modified version of the steering program was used (Figure 2). Similar to the steering task, a set of ten fields with a variety of indices of steering and targeting difficulty were created. To eliminate the directional variation experienced in the steering tasks, all ten fields required left-to-right movements.

As before, field presentation was randomized. The procedure for successfully navigating a field was the same as in the steering experiment, except that the subject would exit the tunnel into an endzone area. The subject was then required to click in the endzone area. If the subject left the endzone area before clicking in it, the trial was counted as an error and the subject repeated the trial. Also as before, each of ten fields was presented four times, for a total of forty trials. The subjects completed forty trials with and without force-feedback support.

When force-feedback is turned on, the mouse rendered haptic walls around the solid lines of the steering tunnel as discussed above. In addition, the endzone area was made into an attractive enclosure; the user's mouse was physically drawn into the endzone

With our knowledge of steering and targeting, we hypothesized a hybrid steering-targeting law: that the difficulty of a combined steering and targeting task would depend in equal parts upon the difficulty of the steering and targeting components to the task.

$$MT = a + b \cdot ID_S + c \cdot ID_T \qquad (5)$$

where a, b, and c are empirically determinable constants; ID_S and ID_T are as stated in equation 2 and 3 respectively.

Subject protocol

A total of ten human subjects participated in the steering experiment and another ten participated in the combination task experiment. All subjects read and signed a consent form. The Harvard School of Public Health's Committee on Human Subjects approved the consent form and protocol. In the steering experiment, subject age ranged between 22 and 52 years. The average age was 33.8 and the median was 33. Three subjects were female and seven subjects were male. All subjects used their right hand for the experiment, which, in every case, was also the subject's dominant hand. For the combined steering-targeting experiment, subject age ranged from 22 to 47. The average age was 28.8 and the median age was 27. Three subjects were female and seven subjects were male. All used their dominant hand, which, except for one subject, was the right hand. The chair, mouse table, and monitor height were adjusted for each subject in accordance with ANSI-HFES Standard [3].

For both experiments the subjects completed the series twice, once with force-feedback and once without. The order of presentation of the force-feedback was randomized. Without force-feedback, the task takes about five minutes to complete, thus minimizing the effect of subject fatigue on the results of the later testing blocks. Each subject was also required to perform two blocks (forty fields) of practice before beginning the test.

RESULTS

Steering task and movement directions

The mean value of movement time (MT) for each of the twenty fields was calculated by first averaging the four trials within a subject and then averaging across the ten subjects.

Times to complete vertical movements were larger than the movement times for the horizontal movements for both force-feedback conditions (Figure 5). Two-sample t-test showed significance for the *without-force-feedback* condition ($p = 0.039$), but the difference for the *with-force-feedback* condition was less distinguishable ($p = 0.079$). For both cases the movement times followed a linear relationship with the steering index of difficulty and were highly correlated ($R^2 > 0.9$).

The addition of force-feedback had two effects. First, the movement times for both the vertical and

Table 2: Combined task fields.

A	W_S	W_T	ID_S	ID_T
Tunnel Length	Tunnel Width	Endzone Length	A/W_S	$ID_T = -\log_2(W_T/2A)$
160	40	40	4.00	3.00
170	40	30	4.25	3.50
180	40	20	4.50	4.17
180	30	20	6.00	4.17
160	30	40	5.33	3.00
170	30	30	5.67	3.50
360	20	40	18.00	4.17
370	20	30	18.50	4.62
380	15	20	25.33	5.25
350	15	50	23.33	3.81

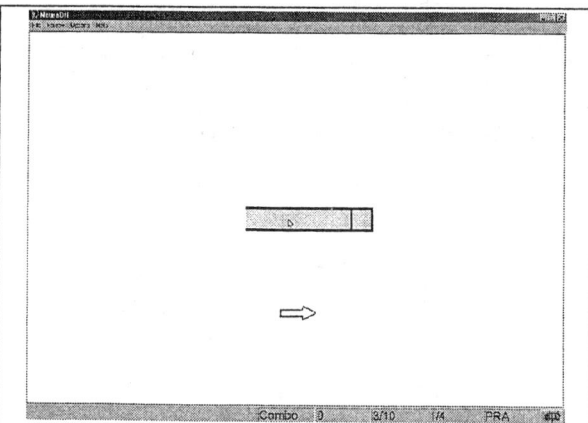

Figure 4: The Combination task. The subject navigate the cursor from the open end of the field staying within the solid lines and then stopping and clicking in the rectangle at the other end of the steering tunnel.

horizontal movements improved significantly, and were, on average, 52 percent faster (range 38 to 68 percent faster). Using the values averaged within subjects, the Student paired *t*-test for the movement times across indices of difficulty showed significance for all cases (p ranged from 0.000 to 0.015), except the smallest index of difficulty for horizontal movements ($p = 0.067$). Second, the force-feedback appeared to decrease the difference between the two directions of the movements (Figure 5).

Combined steering and targeting task

For the combined steering and targeting task, the steering index of difficulty appears to dominate the targeting index of difficulty. We used the Variance Accounted For (VAF) [10], the correlation coefficient (r^2), and the root mean square error as quantitative measures for the contribution of the steering and targeting indices of difficulty and linear combination (Equation 5) to the measured movement times. The combined estimate, based on both ID_S and ID_T (Equation 5) provided the best predictor of movement times followed closely by the steering estimate, which is based solely on the steering index of difficulty (Equation 4). The prediction based on just the targeting index of difficulty (Equation 1) correlated poorly with the movement times and had the largest error of on average 30%. The VAF, r^2, and root mean square errors (MSE) values for the three estimates are presented in Table 3.

The addition of force-feedback again improved performance for the combined task. Movement times significantly improved from 15 to 35 percent for the ten movements (Pair student *t*-test, $p < 0.023$)

DISCUSSION

All the data support the stated hypotheses, except for the combined steering and targeting task. The first hypothesis, the addition of a force-field that tends to pull the mouse to the center of the steering tunnel improves the time to complete the task, is evident in Figure 5. This hypothesis is also true for the combined steering task, although the improvement is smaller, 25% compared to the 52% improvement for the steering task alone. The design of the force field provides a physical valley or groove for the human operator to move the cursor along providing assistance for keeping the cursor in the middle of the tunnel. As a result the human operator spends little effort keeping the cursor aligned within the tunnel. Rather they rely on the force-feedback algorithm and mouse to provide the necessary physical guidance. This result was expected based on the evidence that force-feedback improves performance for targeting tasks [5, 6, 8]. Furthermore, the data also support the relationship between performance and task difficulty as proposed by Acott and Zhai [1].

The second hypothesis, that the time to complete movements in the vertical direction of the video display is larger than the times to complete horizontal movements, is supported by the same data presented in Figure 5. The difference, however, is much less evident with the application of force-feedback. The movement direction differences may be explained by the differences in the joint kinematics required for each of the movements. One of primary means of producing horizontal mouse movement is with radial and ulnar deviation of the wrist,

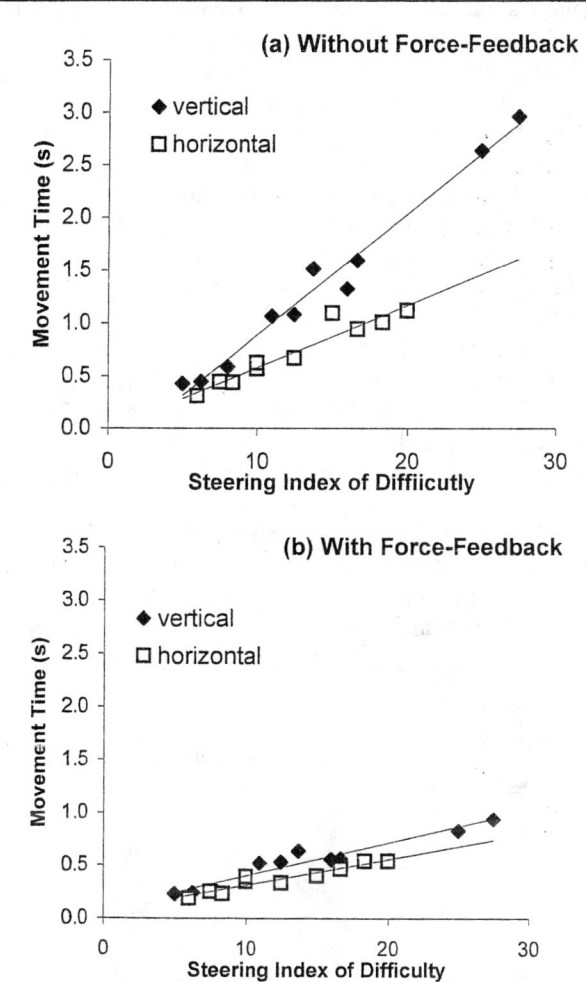

Figure 5 a and b: Steering performance comparing vertical and horizontal movements for without force feedback (a) and with force-feedback (b). For without force-feedback, the movement times for vertical movements are larger than the horizontal movements. With force-feedback (b), movement times improve for both type of movements and the difference between the two directions is less distinguishable.

Table 3: Model Fit Measures

Condition		Combined	Steering	Target
Without Force	r^2	0.98	0.98	0.52
Without Force	VAF	100%	100%	94%
Without Force	MSE	0.05 s	0.06 s	0.31 s
With Force	r^2	0.99	0.98	0.55
With Force	VAF	100%	100%	96%
With Force	MSE	0.03 s	0.04 s	0.18 s

a movement with a relatively low degree of freedom. Conversely, vertical movements require the hand to be moved in and away from the body. For small distances this movement can be achieved with some amount of wrist flexion and extension. But as the movement distances increase other joints and hence other muscles are recruited to move the mouse. For example a movement away from the body requires extension of both the shoulder and the elbow joint. Therefore, the vertical motion requires movement of greater inertia and multi-joint coordination -- a higher level of difficulty. The addition of force-feedback, which aids coordination, diminished the differences greatly. Hence the difference is more likely related to the multi-joint coordination issue. Note that the users were allowed to rest part of their arm on the tabletop, but that the chairs were armless without elbow support.

The third hypothesis, that both the steering and the targeting indices-of-difficulty equally affect the performance of a combined steering and targeting task, was not evident within these tests. Rather steering dominated, being highly correlated with the movement times whereas the targeting steering difficulties were less so. Intuitively, it is expected that the movement times for a combined task would be longer than the pure steering task, and this is the case. Comparing the data for similar steering IDs for the steering task and the combined task, the pure steering task is faster than the combined tasks on average by 37 and 54 percent for the *without-force-feedback* and *with-force-feedback* conditions, respectively. The addition of having to stop within the end zone thus affects the movement times, but the movement times are more related to the steering task difficulty and not the targeting difficulty. While we did take care to cover a large range of targeting and steering task difficulties (Table 2), our test cases for the combination are limited. For example, as the tunnel width increases, but the endzone remains constant one would expect the targeting difficulty to take over as the dominating index; however, the data here do not test that hypothesis.

Error count data showed no significant correlation with any other test condition. This may have been due to the wide variety of approaches each subject took to completing the test. While all subjects were instructed and urged during the experiment to complete the fields as quickly as possible despite the risk of making an error, many subjects still chose to navigate the fields more carefully. With force-feedback, all fields averaged to less than one error. Without force-feedback, 17 out of 20 fields averaged to less than one error. The highest mean error was 2.35, and was from the field with an ID_S of 25.

The force-feedback algorithms presented here provide the user with assistive types of tactile cues that guide the user to complete the task. The algorithms relieve the human operator of some of the difficult motor

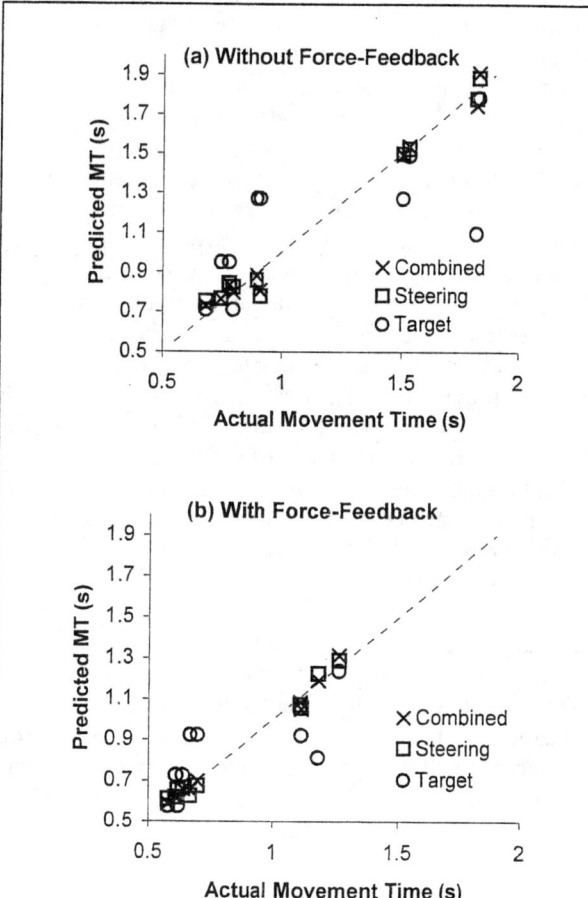

Figure 7: Performance of the combined steering and targeting task for without force feedback (a) and with force-feedback (b). The predicted values for the Combined (solid diamond), Steering (open square) and Target (open circle) were calculated from equations 5, 4 and 1, respectively. While there is little difference between the Combined and Steering predictions, there is a large difference between them and the prediction based purely on the target difficulty. Again the addition of force-feedback (b) improves movement times.

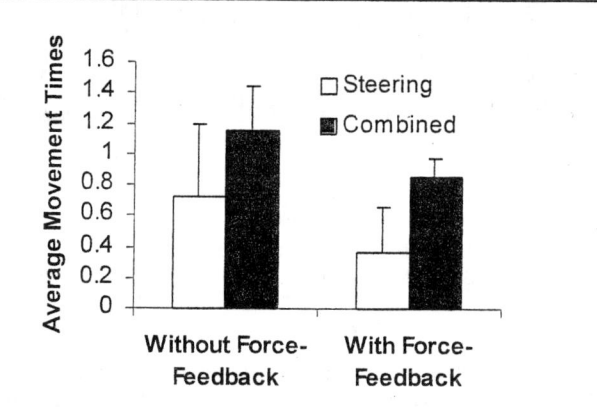

Figure 7: Differences between pure steering and combined steering and targeting task. The combined tasks take longer.

control necessary to complete a targeting or steering task. Similar to the assistive attractive basins around pointing targets [5, 8] performance for an attractive valley around a tunnel improves the steering task. Akamatsu et al [2] developed a multimodal mouse that provided simple tactile cues, such as vibrations for event detection and friction type of force when moving across certain fields. While these modes provided tactile cues, they did not assist the motor control in completing the task. As a result Akamatsu et al [2] observed small performance enhancements. Therefore, designers of force-feedback should consider the type of tactile cues and their assistive or resistive nature in order to maximize performance enhancements.

One downside of this technology is the possible display of force that does not match with the intent of the user. For example, if one gets caught in a tunnel when he or she did not intend to be inside it, the user may become frustrated with overcoming the force of the wall in order to exit the tunnel intentionally. Designers must be aware of these conflicts of interests between an intended movement and the proposed implementation of a force feedback algorithm.

The limitations of the conclusions are quite normal for most laboratory-based studies. First and foremost are that these results are for simulated tasks in a heavily controlled environment. The performance enhancements during real tasks may be affected by other factors not examined here. For the combined task, there also may be other targeting distracters that can limit or even hinder performance enhancements. The combined task also had a limited number of task difficulties tested. Examination of the extremes would provide a more complete picture of the interaction between steering and targeting.

The human computer interface was enhanced by the addition of force-feedback systems. The system adds more sensory feedback to the human pertaining to the computer environment in which they interact. It provides yet another channel for information to be exchanged.

CONCLUSIONS

For steering task completed within the virtual environment of the computer interface, the addition of tactile cues through a force-feedback device improves performance. For our configuration, steering movements in the vertical screen dimension have more difficulty. Furthermore, the combined steering and targeting task require more time to complete than a pure steering task, but are more strongly correlated with steering than targeting index of difficulty. The implementation of these results indicates a strong potential for the use of force-feedback technology for the desktop and computer aided design regimes that heavily rely on the mouse pointing device as a primary computer interface.

ACKNOWLEDGEMENTS

The authors thank Immersion Corporation for the donation of the *FEELit* Mouse and their technical support.

REFERENCES

1. Accot J, Zhai S. Beyond Fitts' Law: Models for Trajectory-Based, *Proceedings of Conference on Human Factors in Computing Systems*, CHI, 1997.
2. Akamatsu M, Sigeru S, MacKenzie IS. Multimodal Mouse: A mouse-type device with tactile and force display. *Presence*, 3(1): 73-80, 1994.
3. American National Standards Institute (ANSI) for Human Factors Engineering of Visual Display Terminal Workstations. Standard No. 100-1988. *Human Factors Society*, Santa Monica, California, 1988.
4. Armstrong TJ, Martin BJ, Franzblau A, Rempel DM, Johnson PW: Mouse input devices and work-related upper limb disorders. *Working With Display Units 94*, Elsevier Science, 375-380, 1995
5. Dennerlein JT, Yang M, Perceived Musculoskeletal Loading during Use of A Force-Feedback Computer Mouse, *Proceedings of the Human Factors and Ergonomics Conference*, Houston, 1999.
6. Eberhardt S, Neverov M, West T, Sanders C. Force Reflection for WIMPs: A Button Acquisition Experiment, *Sixth Annual Symposium on Haptic Interfaces, International Mechanical Engineering Congress and Exposition*, Dallas Texas, 1997.
7. Fitts PM. The Information Capacity of Human Motor Systems in Controlling the Amplitude of a Movement, *Journal of Experimental Psychology*, 47: 381-391, 1954.
8. Hasser C, Goldenberg A, Martin K, Rosenberg L. User performance in a GUI pointing task with a low-cost force-feedback computer mouse. *Seventh Annual Symposium on Haptic Interfaces, International Mechanical Engineering Congress and Exposition*, Anaheim, CA, 1998.
9. Johnson PW, Dropkin J, Hewes J, Rempel D: Office ergonomics: motion analysis of computer mouse usage. In: *Proceedings of the American Industrial Hygiene Conference and Exposition*, Fairfax, VA: AIHA 12-13, 1993.
10. Kearney RE, Stein RB Parameswaran L: Identification of Intrinsic and Reflex Contributions to Human Ankle Stiffness Dynamics, *IEEE Trans. Biomed. Eng.*, 44(6): 493-504, 1997.
11. Rosenberg L. *Virtual Fixtures*, Ph.D. Dissertation Stanford University, 1994.
12. Sheridan TB. *Telerobotics, automation and human supervisory control*. MIT Press, Cambridge, MA, 1992.

Power Browser: Efficient Web Browsing for PDAs

Orkut Buyukkokten, Hector Garcia-Molina, Andreas Paepcke, Terry Winograd

Digital Libraries Lab (InfoLab), Stanford University, Stanford, CA, 94305
{orkut, hector, paepcke, winograd}@cs.stanford.edu,

ABSTRACT

We have designed and implemented new Web browsing facilities to support effective navigation on Personal Digital Assistants (PDAs) with limited capabilities: low bandwidth, small display, and slow CPU. The implementation supports wireless browsing from 3Com's Palm Pilot. An HTTP proxy fetches web pages on the client's behalf and dynamically generates summary views to be transmitted to the client. These summaries represent both the link structure and contents of a set of web pages, using information about link importance. We discuss the architecture, user interface facilities, and the results of comparative performance evaluations. We measured a 45% gain in browsing speed, and a 42% reduction in required pen movements.

Keywords

Web, browser, PDA (Personal Digital Assistant), PalmPilot, wireless, HTTP, proxy

INTRODUCTION

Mobile access to information is a key to individual productivity. Small handheld computers are becoming more crucial in our daily lives. A handheld device equipped with a browser and a wireless connection provides an opportunity to connect to the Internet at any time from anywhere. Such capabilities will increase the utility of PDAs tremendously by providing access to numerous information services, like travel guides, entertainment advice, latest news, flight schedules, even driving directions.

Unfortunately, small screen size, slow text input facilities, low bandwidth, small storage capacity, limited battery life, and slow CPU speed are serious obstacles to the successful realization of that vision.

Screen size limitations in particular require special attention, because they most directly affect the user's experience. For example, a recent study [12] on the effect of screen size on completing browsing-related tasks shows that users with small screens follow links less frequently than their counterparts who were furnished with larger displays, and that their success rate was lower. The study hypothesizes that this lower success rate is in part caused by the more conservative link exploration behavior of small screen users. The study thus calls for improvements in navigation facilities for small screens.

We expect that palm-sized devices on the Internet will be used primarily for extracting particular bits of information relevant to a current task. Usually, this process begins somewhere near the correct answer, but involves some amount of navigation to home in on the information target. Navigation thus requires particularly strong support for this 'final approach' phase of information access.

One solution is to use browsers that present Web pages in their full form on small screens. Even with compression, this approach can be problematic because of bandwidth and battery limitations. More importantly, the resulting scrolling requirements in both dimensions tend to be excessive. Horizontal scrolling can be avoided on pages that use ordinary HTML text by formatting to a narrow width. Still, this will increase the page height and force the user to scroll up and down excessively.

Another solution is to provide web pages specifically for use on PDAs [22]. Two closely related examples are the Wireless Markup Language (WML) [20] and the subset of HTML that is used with Palm VII PDAs [21]. This limits the user to the subset of providers who have prepared material for the PDA (and often charge for that service). This approach also bears the danger of creating two parallel World-Wide Webs. Such duplicate effort could seriously tax human and machine resources.

Instead, our *Power Browser* for palm sized PDAs reflects a complete rethinking of small screen navigation clients. We imposed on ourselves the constraint that no server-side content adjustments are assumed. We analyze and display link structure of browsed pages dynamically, and provide specialized pen-based navigation facilities for exploring that structure. A prototype has been implemented on 3Com's popular Palm Pilot device.

POWER BROWSING

We begin our description with an overview of the architecture in which the Power Browser operates. This is followed by a description of the user's search process while using the tool. After describing several relevant user

---

Permission to make digital or hard copies of all or part of this work for personal or classroom use is granted without fee provided that copies are not made or distributed for profit or commercial advantage and that copies bear this notice and the full citation on the first page. To copy otherwise, to republish, to post on servers or to redistribute to lists, requires prior specific permission and/or a fee.
CHI '2000 The Hague, Amsterdam
Copyright ACM 2000 1-58113-216-6/00/04...$5.00

interface features, we then offer a discussion, and pointers to related work.

Architecture

The size of the Palm Screen is 160x160 pixels in a 6x6 cm area. Because the area is so small, only a small fraction of the data on a typical web page can be displayed at a time. A preprocessing stage is required to select portions of the data to show. The PDA has a processor with the power of a desktop machine in the mid-1980s. Therefore computation-intensive display processing should be performed outside the client as much as possible. More importantly, it is obviously wasteful to download full pages, only to then summarize them at the client, since the data transfer rate is generally low. Instead of downloading pages directly, our client sends requests to a proxy server (see Figure 1).

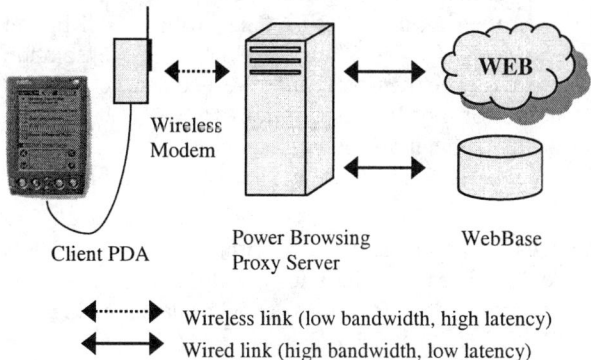

Figure 1: The Power Browser Architecture

The connection between the PDA and the Proxy Server is established through a wireless modem[1]. The server has a wired link to the web, and therefore downloads the pages faster. It processes the data and sends only a small fraction of it to the client at a time. In our experiments, our algorithm for summarization produced a reduction of almost two orders of magnitude (an average 77-fold reduction) in the number of bytes.

The proxy server uses local tools, such as an HTML parser, and an incremental crawler, which can fetch Web pages in the 'neighborhood' of the user's current page. One particularly notable tool makes use of our WebBase facility. WebBase uses an intelligent crawler [6] to collect and store Web pages. It also ranks each page by a 'Page Rank' algorithm [3]. The algorithm ranks a page high if many other pages link to that page. Given a URL for a page that the crawler has already encountered, WebBase provides its rank. When summarizing information for display on the PDA screen, the proxy server can use page rank, if available. See the section on link ordering for more detail.

---

[1] Currently we use the Metricom Modem. The Palm VII with its built in radio modem was not yet available during the research.

Each user that connects to the Power Browsing Service is identified with his/her user account. This session startup happens transparently to the user. No login process is required. The server opens a new session for each client and maintains browsing information about the user's activities for the duration of the session. The session terminates when the connection between the PDA and the proxy is closed. Whenever a client sends a request, the server returns the cached version of the result, if available.

USER INTERACTION

Starting the Browsing Process

Conventional web browsing is initiated through one of three facilities. The user may manually enter a Uniform Resource Locator (URL). Instead, the user might have a bookmark or find a URL by using a search engine In order to minimize user interaction, the Power Browser presents all of these options in a single initial display (Figure 2).

A URL can be entered on the top lines in the display, using the pen in conjunction with shortcut facilities (the *Cuts* pull-down list) described later. The user's personal bookmarks are displayed in a scrollable list at the bottom of the display. Tapping on one of the entries causes the top lines to be filled in with the corresponding URL. Once a URL is in place, the user taps the Browse button to initiate

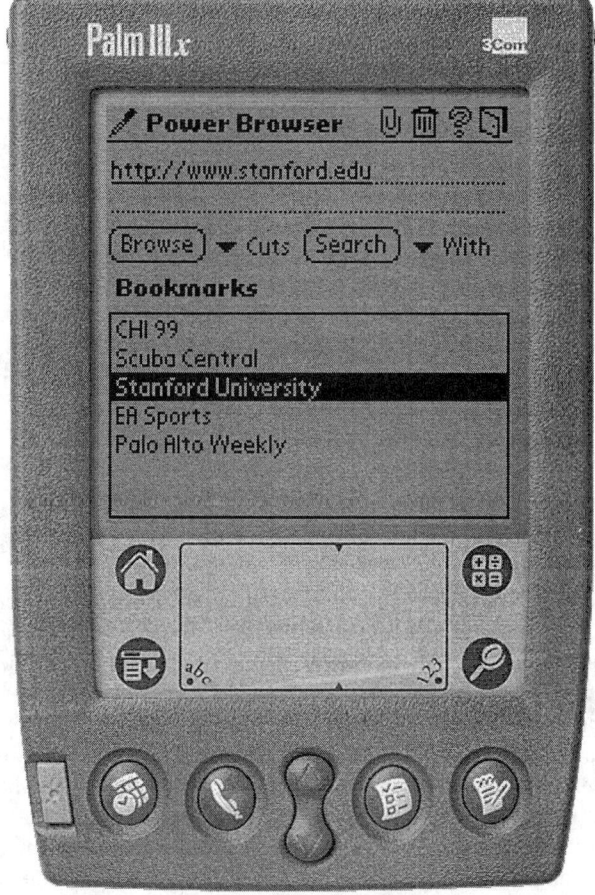

Figure 2: Initial Display for Navigation or Search

navigation.

Searches are initiated by entering keywords on the top lines, selecting a search engine from the *With* pull-down list, and tapping the Search button. The keywords are submitted to the search engine via the Power Browsing proxy server. The resulting page is then displayed on the PDA. This feature could be expanded to include results from specialized services, such as stock quotes, driving directions, restaurant, city information or a news service. The search engines that are currently supported by our proxy are Google, AltaVista, Lycos, Infoseek, Yahoo, Northern Light, and Excite.

At the top of the screen is a toolbar with buttons for (from left to right) entering a new bookmark to the bookmark list, deleting a bookmark from the list, getting help, and quitting the application. The bookmarks are stored on the PDA as PalmOS databases (persistent memory objects).

Navigating Through Pages

After the user has tapped the Browse or Search button to initiate browsing or searching, the PDA client sends an appropriate request to the Power Browsing proxy server. Depending on the request, the server either downloads the respective page, or uses a search engine to obtain the search results. The proxy caches all documents received during a user's session. Before delivery to the PDA, the proxy transforms the data into a format appropriate for the PDA's small screen.

Our technique takes advantage of the implicit structure of web pages. Web pages consist of text and multimedia elements, along with embedded links containing URLs for linked pages or files. During the user's final approach navigation phase, our Power Browser displays consists of a set of "link descriptions" which we generate heuristically from anchor text, URL structure, or ALT tags, as appropriate to the link. This structure includes not only the links on a single page, but a hierarchical structure of links on linked pages as well. The user can directly retrieve a page from any link description visible on the screen.

Figure 3 shows how the display uses minimal screen real estate to represent several levels of page structure. Each line of text in Figure 3 is a link description. The descriptions are organized in a tree, similar to the way that file browsers represent nested folders. To conserve screen space, the indentation level is marked by vertical lines rather than by folder or character icons. All text lines in one indentation block represent the links emanating from one page. For example, the page 'Database Group' contains links to pages 'DB', 'Projects', 'Members', and others not shown in Figure 3. In turn, the 'Members' page contains links to 'Andreas Paepcke', 'Andy Kacsmar', etc. Users may expand and collapse the tree through direct manipulation.

One challenge with this approach is the choice of good descriptions for each link. We make this choice heuristically as follows. If the link is associated with text that is, a regular browser would show underlined text to indicate the presence of the link, then we check whether the text is one of a few popular link descriptions that are useless for our purposes. One such 'stop description' is "Click here". If we find that the description is not one of this stop set, then we capitalize the description and use it. This kind of description is sometimes called the 'link anchor'. If we do detect the use of a stop description, we instead turn to the URL associated with the link. If it points to a directory, we use only the right-most element of the URL, and capitalize it. If the URL ends in a file name, we remove the extension, and use the capitalized name.

We experimented with using the titles of pages pointed to by links as the link descriptions. This worked well in that the titles were often good descriptions, but the solution proved to be too expensive. Since we need to generate an entire page worth of link descriptions for each display, the proxy needs to fetch all the corresponding pages from the Web. The consequent increase in latency was too high.

Sometimes, links are associated with images, rather than with text. In this case we look for the alternative text that is sometimes provided for links under images using the HTML "ALT tag". If such alternative text is available for the image link, we use it. Otherwise, we use the URL method described above.

The top of the screen in Figure 3 consists of the command toolbar. The buttons (from left to right) are used for seeing the browse history, making a link the display root, adding a bookmark, and jumping back to the initial screen. The browse history provides an overview of the user's moves

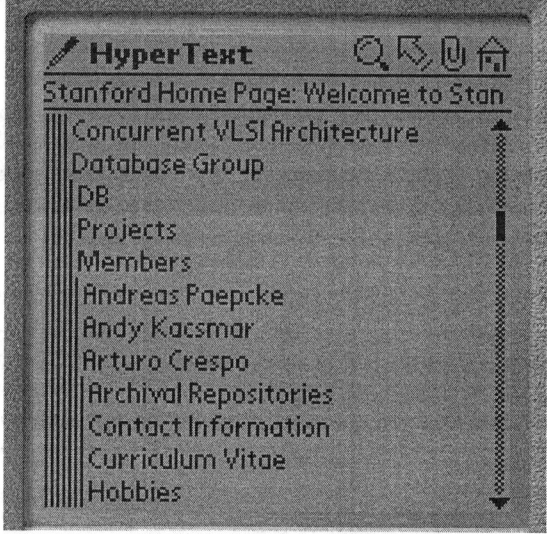

Figure 3: Tree Control Shows Link Structure

through the linked pages (described below). Turning a link into the display root is a tool for cleaning up the display if the user is confident that no backward browsing will be necessary. For example, in Figure 3, if the user did not care to return to any page further back than the 'Members' (of the Database Group) page, then the user could tap on the make-root arrow, and then tap the 'Members' node. Members would move to the top left corner of the screen, removing the indentations on the left.

Underneath the toolbar, there is the title of the root page. This is the initial page that was extracted when the browsing process started. This heading changes only when the user enters a new URL, or uses the 'make root' operation.

Expanding a node of the tree results in a request to the Power Browser proxy. Users accomplish node expansion through a left-to-right pen gesture over the item to be expanded. Such interactions are described further below.

Figure 4: Text Display

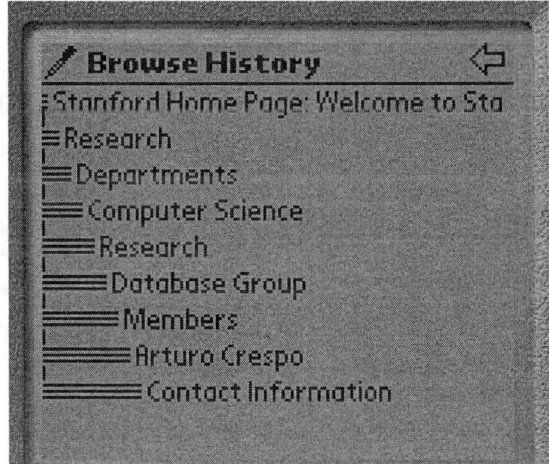

Figure 5: Browsing History

The proxy retrieves the corresponding page from the web, summarizes it by extracting the link descriptions, and returns the result to the PDA for display within the tree control.

This frequent communication could be reduced if we transmitted multiple levels of the tree at once. However, we have found that the 1-2 second delay required for the round-trip communication was worth the decreased bandwidth and PDA storage requirement. The linked structure of pages is transmitted to the PDA gradually, and conservatively, directed by the user's interests as revealed by the requests for expansion.

Viewing Pages

Once the desired page is found, users can view some of its contents. Since the display is low-resolution monochrome, it is usually not worthwhile to display images. Therefore by default, images are ignored. The ALT tag provided in the IMG and MAP environments lets an alternative text string be displayed instead of the image. The ALT text usually gives a reasonable description for the image, however it cannot carry the semantic and visual effect of the image itself. One possible solution to this problem is to display selected images on user demand. In order to display the image on the PDA, a refinement step (i.e., scaling image, decreasing color depth) should be carried out on the server. See [8] for an example of how this can be done.

The Power Browser avoids white space as much as possible. Sequences of paragraphs or line-breaks are collapsed. Many additional text attributes that are visible on standard browsers (color, size, font, alignment, etc.) are ignored as well. On the other hand, structural elements such as line breaks, paragraphs, and tables are used to format the text within the view. Lists are re-formatted into simple text blocks with breaks between successive items. Table rows and columns are folded into text blocks as well. A sample screen shot is shown in Figure 4. The buttons (upper right) let the user move to the top of the text, bottom of the text, or back to the link view, respectively. Underneath the toolbar is the title of the page being displayed. The rest consists of the text itself and a scrollbar.

Example

For explanatory purposes, we present an example here to illustrate how the browser works. Suppose our task is to find contact information for someone named Arturo who is a research member of the Database Group at Stanford University. We call this the *"Arturo task"* for reference. The URL for "Stanford University" can either be entered directly or can be found through a search engine (Figure 2). Once the page at http://www.stanford.edu is displayed on the Power Browser screen, we can expand the tree display repeatedly, thereby following links to *Research*, *Departments*, *Computer Science*, *Research*, *Database Group*, *Members*, and *Arturo Crespo*. This is the state seen in Figure 3. Finally, as we examine the links emanating from Crespo's page, we see a link to *Contact Information* (Figure 3). That is exactly what we are interested in. So instead of expanding the tree further, we look at the page text by tapping on *Contact Information*. This action takes us to the view shown in Figure 4.

The browser also provides an option for viewing the browse history. Notice in Figure 3 that while we can see the title of the root page ("Stanford Home Page: Welcome to Stan"), we cannot see any information about the pages between the root and the Database Group page. Scrolling would reveal this information, but we found that a more powerful display device for browsing history was required. Figure 5 shows the path we followed from the Stanford home page to the desired contact information in compact form. The magnifying glass icon in the tool bar at the top of Figure 3 invokes this overview display.

Finding this information using a traditional full-text browser would have required us to look at all the pages at each level, read through a large amount of text, and try to locate the hypertext that leads to our destination. In this example, instead of looking at eight different web pages, the user can easily navigate through the link structure from the initial page and reach the target. Of course, this approach works for goal-directed tasks, rather than for recreational browsing. It also works best on sites that use sets of links, rather than extensive text and images. For those sites, the browser's strengths come less into play.

Link Ordering

One of the enhancements of the system compared to a traditional browser is the option of ordering the links. A standard browser displays the links in the sequence they appear in the document. Since only a small number of links can be displayed at a time on the PDA, it becomes important to display them in an efficient order. The Power Browser allows users to specify three sorting schemes: original, alphabetical, and page ranked. For instance if we are looking for a person's first name among a large list that is ordered by last name, we would prefer to have the names re-ordered alphabetically according to their first name. This feature was used in our previous example to get the group members ordered alphabetically according to first name.

In other cases, when we see a number of similar links, we might like to visit the ones that are most popular first, to avoid browsing through a large number of pages. This feature is supported using a quality measure for web pages, called Page Rank [3]. Page Rank was first used by the Google search engine to return better search results. A mathematical analysis, calculated on more than a billion hyperlinks on the web is used to estimate the quality or importance of web pages. A page like www.stanford.edu has high importance if, recursively:

1. Many other pages point to www.stanford.edu
2. These pages in turn have high importance.

Other ranking algorithms could, of course, be used for sorting links.

NAVIGATION FACILITIES

Summarization is only one requirement for effective browsing from PDAs. We also need to consider task-specific support facilities that make user tasks easier. We describe the Power Browser's use of the pen as an input device, the use of gestures, animation, and link reduction.

Shortcuts

One of the major differences between a handheld device and a laptop or desktop computer are the input modalities. The PalmPilot does not have a keyboard and most text entry is done using a pen with the Graffiti character set. This makes text entry more difficult, so PDA applications such as the Power Browser try to minimize manual text entry. However, text entry cannot be avoided completely, so the Power Browser provides text shortcuts too. For instance while entering a URL in the initial display, a pull-down menu is provided to insert common prefixes and suffixes (http://, ftp://, www., com., etc.). Commands for cut, copy and paste are supported in the standard manner as well.

Gestures

Another key difference between PDAs and desktop machines is the PDA's lack of a mouse. The pen carries the functionality not only of the keyboard, but also of the mouse. The pen, while only as versatile in selecting as a one-button mouse, does allow us to introduce gestures into the user interface.

Gestures are limited size and duration pen-tip trajectories (strokes) of distinguishable shapes. We can save screen real estate by using gestures in place of buttons to invoke actions. The use of gestures can potentially interfere with text recognition, but this problem is avoided on the Palm Pilot, by performing text recognition only in a dedicated portion of the screen, below the information display. The display area itself is therefore available for gestures.

The tree control used for displaying the link structure (Figure 3) is operated using gestures. A node is expanded with a left-to-right gesture over the link to be expanded, and collapsed with a right-to-left gesture. Up-down and down-up gestures operate scrolling. The text of a node's associated page is displayed by a single pen-tap on the link description.

Both position-dependent and independent gestures are used in the text view (Figure 4). Here, a right-to-left gesture carries out the same action as a back button, returning the user to the previous view. Any other pen trajectory results in selecting a region of the text to enable copying/pasting operations.

Animation

Because of the limited screen size, it is essential to move the tree control up and down as the structure is modified to keep the most relevant area in view. When a tree node is expanded, the selected node moves to the top to make it possible to view as many new nodes as possible. After a node collapses, the bottom lines of the view may become empty, the tree structure is moved down.

We initially recalculated the screen and displayed the result immediately. When users tested the system, the sudden redisplay proved confusing. They found it difficult to get reoriented when the screen was redrawn, because nodes in the tree might move up or down on the display, or new nodes may be introduced. We therefore added animation. Instead of moving nodes abruptly, nodes that change their positions are scrolled to their new location. The scrolling speed is a key parameter in the animation. An animation that is too slow increases response time and is unpleasant. On the other hand if the speed is too fast, the eyes can't trace the rapid movement and the movement can result in flickering on the screen. We determined an acceptable speed experimentally. The addition of animation had a large positive impact on usability.

Seeing the Forest AND the Trees

Establishing both overviews and a notion of location throughout navigation is important for successful browsing. Conventional browsers provide buttons to move one step forward or backward, and a linear menu display of the link path leading to the current page. This makes it difficult for the user to establish a sense of location since the link structure of the web is a graph that can be traversed in arbitrary order.

For instance, while following links, the user may hit upon the same page multiple times, along varying paths. In this case, backtracking from that same page will return to a different page each time. The Power Browser's tree arrangement of links displays both the user's location and the neighboring navigation environment at all times. One positive consequence is that users can jump to a sibling page with a single action. (Siblings are pages that are pointed to by the same parent page). In conventional browsing, the user has to backtrack and move forward again to reach siblings.

Link Redundancy Reduction

Another method for navigation support is hiding; restricting the navigation space by hiding links to pages. Most web sites have many duplicate links within their link structure. For instance, some pages provide a navigation bar support that gives direct access to the most critical web pages within the site. This structure is often repeated on each page. Also, many pages provide a link to their parent page or all the way back to the root page of the site. There can be duplicate links within the same page as well. Some pages provide alternative options for the same links (e.g., a page accessible through both the anchor and an image). The proxy server removes any duplicate links that it finds. This assures that each link on the tree control is unique. If the same link has multiple descriptions, a heuristic is used to choose the best one. Link descriptions are preferred to alternative text. If there are multiple link descriptions, the longest one is chosen. When we tested the browser with duplicate removal, we observed that navigation became easier since it reduced the cognitive load for the user to recognize duplicate links. Restricting the navigation space compacted the link structure and reduced the complexity of the unrestricted space.

TESTING

We performed three kinds of experiments to test the validity of our approach. The first measured system performance, the second measured the best possible user level performance for the *Arturo task*, and the third measured actual user performance for the *Arturo task* and five others.

System Performance

One major design decision was the use of a proxy. Proxy-based designs carry intrinsic disadvantages: a proxy must be available at all times; for reasons of scalability, proxies need to be replicated when large numbers of clients are to be served, and they add an additional hop on the network. We decided on this solution nevertheless, in order to conserve bandwidth and CPU/battery activity on the PDA. Careful design was invested in the communication protocol between PDA and proxy. The size of data packets sent between the client and the proxy server was minimized.

First, there is the proxy-side reduction from the full Web page to a list of link descriptions and their associated URLs. Measured for a relatively small list of random Web pages, this yielded a factor of about 20 in byte size reduction. This number obviously varies with the composition of the pages. Rather than sending the link descriptions and URLs to the client, the proxy assigns an object identifier (OID) to each URL. These OIDs are much shorter than the URLs themselves, and added another factor of 3.5 in byte size reduction. What is transmitted to the PDA is the resulting list of link descriptions, and the associated OIDs. The overall savings is a roughly 70-fold reduction in the number of bytes. When the user performs an action on the tree that requires proxy activity, only the affected OID and the action are transmitted to the proxy.

Lower Bound for User Performance

In order to find and compare the best possible user-level performance, we measured the minimal amount of pen activity necessary for users to perform the *Arturo task* once they knew exactly what pages and information is required, and the best possible time to run through the task. We performed this analysis on the Power Browser and on three other PDA-based browsers: ProxyWeb [17], PalmScape [14], and HandWeb [18]. All of these other browsers attempt to display Web pages as similarly as possible to what a full-sized browser would show.

Table 1 summarizes the results. The total number of pen moves is the sum of pen taps needed for scrolling, the taps necessary for selection, and gestures. The time for task completion was the result of running through the task as quickly as possible, once the necessary link sequence was known to the operator. The completion time excludes the time for entering the initial URL, but it includes all connection time through the wireless network, including the initial connection setup. We ensured that no pages had been previously cached. The tests were run with a Palm IIIx, connected with a Metricom Ricochet modem with nominal speed of about 19kb/s.

	Pen-taps for Scrolling	Pen-taps for Selection	Pen Gestures	Pen Moves	Total Time (secs)
Power Browser	3	1	8	12	80
ProxiWeb	21	8	N/a	29	170
PalmScape	13	8	N/a	21	234
HandWeb	22	16	N/a	38	254

Table 1: Comparing optimally possible performance for the *Arturo task*

As we can see from the table, there was a reduction by an average factor of 2.74 of task completion time and a factor of 2.4 in the number of required user interactions on the Power Browser. For the slower browsers the numbers are even stronger: a factor of more than 3 in completion time reduction. With slower links, the reduction would be even more significant. Similar differences in performance were obtained for the tasks that are described next.

Actual User Performance

In order to measure actual user performance, we had 10 users perform a total of six tasks. Each user performed three tasks on the Power Browser, and the other three tasks on ProxiWeb. We varied the sequence in which subjects were exposed to the two browsers. We chose ProxiWeb for the comparison, because it had the best performance bound as per Table 1. The tasks were as follows:

Task 1: Beginning at http://www.mit.edu, find the date of the first day of classes in the fall semester of the '99/'00 academic year.
Task 2: Use the result of a Google search for "CHI 2000" to find the panel co-chairs for the CHI 2000 conference.
Task 3: Beginning with Google search "New York Public Library", find the Manhattan branch's opening hours.
Task 4: The *Arturo task*.
Task 5: Beginning with Google search "Metronome ballroom", find the price of group dances for *Gold International Style*.
Task 6: Beginning with http://www.usatoday.com, find the NFL league TV schedule.

All subjects were Stanford Physics, Psychology, and Computer Science graduate students who use computers at least 3 hrs/day, and perform at least one Web search in a normal working day.

Chart 1 shows the completion times for the tasks. The average time savings across all tasks and subjects was 45%. Chart 2 shows the number of pen actions performed by the subjects for each task, and the subset of these moves that were required for scrolling. The 'others' category comprises pen taps and gestures. The average pen action savings afforded by the Power Browser was 42%.

When using the Power Browser for the *Arturo task*, subjects took about double the best possible attainable time (Table 1), and they performed about twice as many pen moves as the absolute minimum. These same ratios held for task 6 for which we generated the lower bound measures as well.

RELATED WORK

Browsing the WWW from PDAs has been demonstrated in

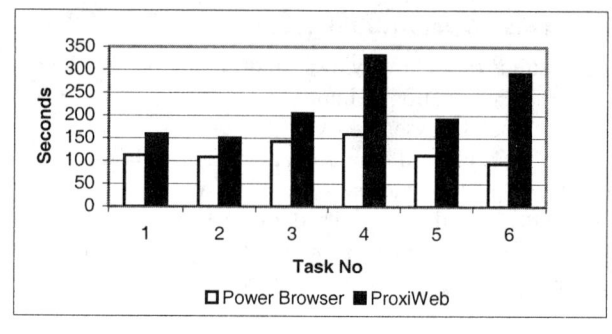

Chart 1: Average task completion times

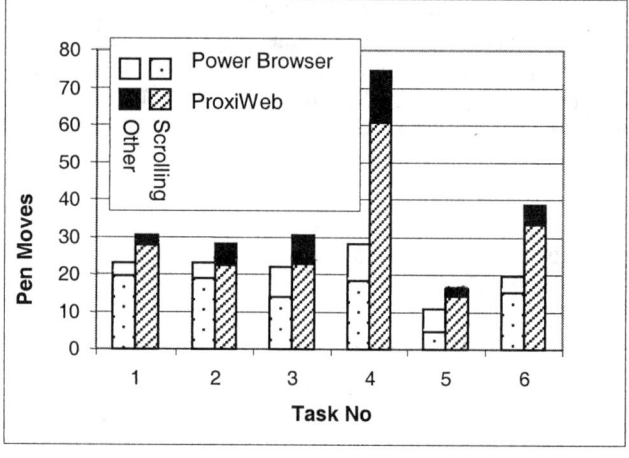

Chart 2: Average pen taps/gestures, and scrolls

research projects and in commercial applications [2, 10, 9, 17, 18, 14]. One related product is the Pocket Internet Explorer by Microsoft for Windows CE devices. Browsers for the PalmPilot include Topgun Wingman/ProxiWeb [9, 17], HandWeb [18] and PalmScape [14]. These browsers attempt to render content as fully as possible, and do not provide any additional features to assist navigation. Wingman/ProxiWeb uses a proxy server whereas HandWeb and PalmScape download the pages directly.

Much research has been done on using the hypertext structure of the web to improve navigation and to build useful applications. Bachiochi [1] added navigation buttons to a browser that enable maneuvering within hierarchical Web site structures, based on one's current position. The ParaSite system [19] exploits the link information on the Web to find moved pages and un-indexed information. Scratchpad [16] proposes a set of mechanisms based on breath-first traversal of web pages. Nif-T-nav [11] provides a hierarchical navigator and shows the state of the navigation using a tree structure. Similarly, WebToc [15] visualizes the contents of a Web site with a hierarchical table of contents. Brusilovsky [4] describes Adaptive Navigation Support (ANS) methods for Web-based systems, like link ordering. WebMap [7] creates a graphical map showing the navigation history. Cha-cha [5] uses a pre-computed tree arrangement to display results of site searches. A tree based approach has also been employed in WebTwig [13].

CONCLUSION

In summary, we have described our Power Browser for PDAs. It can significantly reduce browsing times for directed tasks. The browser achieves its gains by intelligently presenting the link structure of Web pages, and by providing effective and natural commands for navigating this structure.

REFERENCES

1. Bachiochi, D., Berstene, M., Chouinard, E., Conlan, N., Danchak, M., Furey, T., Neligon, C. and Way, D., Usability Studies And Designing Navigational Aids For The World Wide Web, in the Proceedings of the 6th WWW Conference, 511-517, 1997.
2. Bickmore, W., T., Schilit, N., B., Digestor: Device-independent Access to the WWW, in the Proceedings of the 6th WWW Conference, 655-663, 1997.
3. Brin, S. and Lawrence P., The anatomy of a large-scale Web search engine, in Proceedings of 7th International WWW Conference, 101-117, 1999.
4. Brusilovsky, P., Eklund, J., Schwarz, E., Adaptive Navigation Support in Educational Hypermedia on the World Wide Web, in the Proceedings of INTERACT, 278-185, 1997.
5. Chen, M., Hearst, A., Marti A., Hong, J., and Lin J., Cha-Cha: A System for Organizing Intranet Results, in Proceedings of the 2nd USENIX Symposium on Internet Technologies and Systems. Boulder, CO. Oct. 1999. To Appear.
6. Cho, J., Garcia-Molina H., Page L., Efficient Crawling Through URL Ordering, in the Proceedings of the 7th WWW Conference, 1998.
7. Domel, P., WebMap: A Graphical Hypertext Navigation Tool, in the Proceedings of the 2nd WWW Conference, 1994.
8. Fox, A., Brewer, A., E., Reducing WWW latency and bandwidth requirements by real-time distillation, in the Proceedings of the 5th WWW Conference, 1996.
9. Fox, A., Goldberg I., Gribble, S. D., Lee, D. C., Polito, A. and Brewe, E. A. Brewer, Experience With Top Gun Wingman: A Proxy-Based Graphical Web Browser for the 3Com PalmPilot in the Conference Reports of Middleware, 1998.
10. Gessler, S. and Kotulla, A. PDAs as mobile WWW browsers, in the Proceedings of the 2th WWW Conference, 53-59, 1994.
11. Jones, L., Krisen, nif-T-nav: A hierarchical navigator for WWW pages, in the Proceedings of the 5th WWW Conference, 1996.
12. Jones, M., Marsden, G., Mohd-Nasir, N., Boone, K. and Buchanan, G. Improving Web interaction on Small Displays, in Proceedings of 8th International WWW Conference, 51-59, 1999.
13. Jones, M., Marsden, G., Mohd-Nasir, N. and Buchanan, G. A Site-based outliner for Small Screen Web Access, Poster at 8th International WWW Conference, 1999.
14. Kazuho, O., PalmScape: http://palmscape.ilinx.co.jp/.
15. Nation, A. D., Plaisant, C., Marchionini, G. and Komlodi, A., Visualizing websites using a hierarchical table of contents browser: WebToc, in the Proceedings of the 3rd Conference on Human Factors and the Web, 1997.
16. Newfield, D., Sethi, S., B. and Ryall, K., Scratchpad: Mechanisms for Better Navigation in Directed Web Searching, in the Proceedings of UIST, 1998.
17. ProxiNet, Inc., ProxiWeb: http://www.proxinet.com/.
18. Smartcode Software, HandWeb: http://www.smartcodesoft.com/.
19. Spertus, E., Parasite: Mining Structural Information on the Web, in the Proceedings of the 6th WWW Conference, 201-211, 1997.
20. WAF (Wireless Applications Forum), WAP WAE Specification (Wireless Application Protocol, Wireless Application Environment Specification), April, 1998: http://www.wapforum.org.
21. Web Clipping Development: http://www.palm.com/devzone/webclipping/.
22. W3C, HTML 4.0 Guidelines for Mobile Access, http://www.w3.org/TR/NOTE-html40-mobile/.

A Diary Study of Information Capture in Working Life

Barry A. T. Brown, Abigail J. Sellen, Kenton P. O'Hara

Appliance Design Group
Hewlett-Packard Labs
Filton Road, Stoke Gifford
Bristol BS34 8QZ, UK
+44 117 922 9520

Barbro@hplabs.hp.com, Abisel@hplabs.hp.com , Kenoha@hplabs.hp.com

ABSTRACT

Despite the increasing number of new devices entering the market allowing the capture or recording of information (whether it be marks on paper, scene, sound or moving images), there has been little study of when and why people want to do these kinds of activities. In an effort to systematically explore design requirements for new kinds of information capture devices, we devised a diary study of 22 individuals in a range of different jobs. The data were used to construct a taxonomy as a framework for design and analysis. Design implications are drawn from the framework and applied to the design of digital cameras and hand held scanners.

KEYWORDS

Information capture, appliances, digital cameras, voice recorders, scanners, diary study, PDAs, document use

INTRODUCTION

Advances in technology are increasingly enabling devices which allow for the capture of a wide variety of information or media, whether it be for capture of paper documents, scenes, sound, or moving images. The boom in digital cameras is perhaps the most salient example, but we are also seeing gadgets hit the market in the form of voice recorders, digital video cameras and document scanners. Not only are such devices becoming more ubiquitous, they are also becoming more personal: they are becoming smaller, more portable, and often more specialised for particular kinds of information capture. In addition, we are seeing more diverse and sophisticated software for delivering, manipulating, storing and viewing captured information. These capabilities, coupled with huge advances in computer storage capacity mean that new technology is greatly enhancing our ability to record and keep images of everyday life events.

Permission to make digital or hard copies of all or part of this work for personal or classroom use is granted without fee provided that copies are not made or distributed for profit or commercial advantage and that copies bear this notice and the full citation on the first page. To copy otherwise, to republish, to post on servers or to redistribute to lists, requires prior specific permission and/or a fee.
CHI '2000 The Hague, Amsterdam
Copyright ACM 2000 1-58113-216-6/00/04...$5.00

These advances unleash great potential for technological innovation. For example, cameras or voice recording capabilities can be easily added to existing devices such as desktop computers, laptops, and palmtops. Cheaper, better capture technologies are also leading to new classes of device for the consumer market (e.g., the Sony Voice Balloon, the CoolPix camera) as well as the business market (e.g., the Crosspad, the Cpen). Since these are new classes of device, it is often not immediately obvious how these products might eventually be used by consumers. IT companies face important decisions about how to design such devices for optimal use, and how to market them.

In our own research group, we were confronted with these very issues in relation to a new kind of capture device, a handheld document scanner known as the "Capshare 910" (Figure 1). Marketing research had resulted in good positioning for the product in the mobile professional market. We wanted to understand whether other markets, and whether other kinds of activities might represent future opportunities for this kind of technology. We were also interested in changes to the hardware and software which could result in new types of Capshare type devices.

Fig 1. The Hewlett-Packard Capshare 910

To do this, we needed to first understand the range of situations in which people want to capture paper documents. We also wanted to understand the subsequent ways in which they then want to use that information. Another related question was whether or not document capture could be enhanced by considering extending the kinds of information that could be captured (such as adding sound or scene capture). We felt that without a deeper understanding of these issues, we could not systematically map out the space of design possibilities.

The Literature

Unfortunately, the existing literature is not particularly illuminating on this topic. As it stands, it has little to say about the range of ways in which people capture information or the reasons why they do it. Rather, research has tended to focus on the design and use of specific systems which make use of capture technologies. So, for example, there are descriptions of systems based on camera input designed to capture paper documents on the desktop [e.g., 6]. There are other systems designed to capture office conversations and meeting events by using combinations of audio, video and scribble information [e.g., 2, 5, 11]. Other research reports have covered the use of active badges to automatically capture activities such as people's movements within a building [4]. In addition there have been many papers written over the years describing systems which make interesting use of the capture of moving images (including hand gestures) and still images or scene capture [e.g., 3].

From a design perspective such as ours, information capture has not been recognised as a topic in its own right. Perhaps because of this we found little which could offer guidance on what sorts of devices people might find useful. The literature as it stands does not discuss the circumstances under which people might *want* to capture information if they only had the means. Also, it does not give clues as to what kinds of captured media might be usefully combined (for example audio plus scene), and it does not point to the kind of software, services or supporting infrastructure that might be valuable to help people deal with captured information.

In response to this, we designed a study to analyse people's information capture activities. The paper starts by discussing the study, and the *taxonomy of capture* which we developed using the results. This taxonomy unpacks the diversity of capture activities leading to a number of design implications. In particular, we focus on implications for the hand-held scanner we have described, but we also extend the findings to other kinds of potential capture devices.

Our Approach

Since Capshare was marketed as a business product, one of our primary interests was to understand the range of capture activities that people do in very different kinds of jobs. The focus in this initial research was therefore on understanding information capture in workplace activities. We had no specific vertical market in mind, wanting instead a diverse cross-section of people.

These considerations led us to choose a diary method with a mixture of subjects from a range of occupations. Diary methods are a popular data collection technique in sociology, but are still relatively rare in technology studies.

This is a method we have used in the past with some success in the areas of reading [1], paper use [9], and the research behaviour of library users [7].

Finally, we wanted to design a study which would not only provide data on people's paper document activities (to inform future directions for Capshare) but also to look at information capture activities more generally. Thus, for half of the subjects in the study we focussed on their paper document capture activities; for the other half we were interested in any kind of information capture they carried out or wished to carry out.

METHOD

Choice of Subjects

A mix of subjects from both within and outside our organisation was recruited – 13 from within HP, and 9 from outside HP. All were required to be PC and email users.

Eleven of the HP subjects formed the multimedia capture group (Multimedia Grp). For this group we were interested in all of their information capture activities (not just documents). To recruit these participants, an email was sent out to the Hewlett-Packard Bristol site of 2,000 non-research staff. The subjects were then chosen on the basis of their interest and occupation. Here the aim was simply to recruit a diverse mix of both professional and administrative staff. The resulting sample included a marketing manager, a sales executive, an administrative assistant, a security manager and a logistics specialist.

Eleven subjects (2 from HP and 9 external) were recruited for the paper document capture group (Paper Grp.) This was done using a list of occupations which were chosen as highly "information intensive". In particular, because of our interest in hand scanners, we chose people whose jobs were highly reliant on the collection and capture of paper documents. The resulting pool of 11 subjects in the Paper Grp. covered a wide range of occupations including a lawyer, journalist, stockbroker, office administrator, financial administrator and teacher.

While 22 subjects constitute a relatively small sample size, it is appropriate for a diary study in which each person contributes in-depth data over the course of several days. Moreover, our aim was to broaden our understanding of what information capture means rather than to produce a set of data which is statistically generalisable. This mandated the in-depth study of a wide group of different occupations.

Procedure

This study used a diary methodology which was a modification of previous methods we have used. Instead of asking subjects to keep written notes on their activities, we asked them to take photographs of the events we were interested in. We equipped subjects with digital cameras to use over the course of 7 consecutive days (covering on average 5 working days and 2 days at home). Subjects in both groups were asked to use the camera whenever during the course of each day they felt the need to "capture" some information either at work or at home. It was emphasised that they should use the camera as a *diary tool* rather than as a conventional camera. They were told to take a picture whenever they *actually* captured some information in the course of their day, or whenever they would have *liked* to have captured information but did not have the means.

The procedure for both the Multimedia Grp and the Paper Grp was the same with one exception. Subjects in the Multimedia Grp were told we were interested in opportunities to capture any kind of information they came across – be it spoken or ambient sound, document-based information (paper or electronic), moving image, or scenes. Subjects in the Paper Grp however were told that we were only interested in information captured from paper documents. These paper documents could take on various forms such as books, post-it notes, magazines, newspapers, as well as more conventional business documents.

For both groups, the pictures themselves were used later as illustrations and as memory joggers in semi-structured interviews intended to unpack the context surrounding each capture event. Subjects were interviewed three times over the week, and asked in detail about each photograph they had taken. For each photograph, they were asked a number of different questions including:

- What did they and how ideally would they have captured the information?
- How did they or how would they have used the information they captured?
- Did they share or did they want to share the captured information?
- Did they or would they have wanted to keep the captured information?

Each of these interviews lasted anywhere from thirty minutes to an hour. The interviews were tape recorded and fully transcribed, producing a large corpus of information. The methodology we used has similarities with photo-elicitation, a technique used in anthropology and sociology [8], with the exception that the study participants – rather than the researcher – actually takes the photographs.

RESULTS AND DISCUSSION

The study generated data on 381 "capture opportunities": 219 photographs from the Multimedia Grp and 162 from the Paper Grp. A capture opportunity is defined as an occasion on which they used the camera to record either an actual capture event or a potential capture event.

What The Multimedia Group Captured

As we hoped, the capture opportunities in the Multimedia Grp reflected a wide range of information media. As Figure 2 shows, subjects in this group indicated that they wanted to capture moving images, audio (both voice and ambient sound), visual scenes, and information from a variety of different kinds of paper documents (containing for example, to-do items, hand-written notes, financial data, etc). There were also some kinds of information which were harder to classify. Thus the "Other" category includes opportunities for capture of electronic information (such as Web pages), projected information (such as slides), and other events where the "type" of information is not easy to classify (such as the desire to capture information off a car license plate).

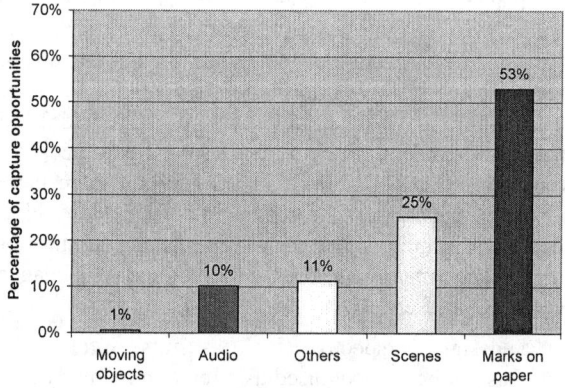

Fig 2. Kinds of information "captured" by the Multimedia Grp

This confirmed to us that subjects were able to think of the camera not necessarily as simply a camera but also as a recording tool. So for example, one subject took a picture of a colleague and told us that he had wanted to capture what that person had said; another took a picture of a meeting and told us it was a video snippet she had wanted to capture.

Perhaps because most of the diary data came from work situations, marks on paper constituted the most frequent kind of information subjects said they wanted to capture. Examples included capturing handwritten notes and photocopying paper documents. However, also prominent amongst these data were examples of subjects using the camera *as a camera* to capture scenes. This perhaps reflects a bias in use of the camera as a diary tool in this methodology. Nonetheless, what was interesting was the ways in which subjects found scene capture to be useful in different work situations. For example, one participant (we will call Phil) worked as the site security manager. Due to a large number of thefts from the site, Phil was attempting to have a security camera fitted, however he met with considerable resistance from staff who had hesitations about being "spied upon". To address these concerns Phil climbed up on a ladder and took a photograph from where he wanted to have the security camera fitted. Phil explained that he then not only used this photo in a report to management, but also showed it amongst staff to alleviate their fears (Figure 3). As he put it when interviewed: "people can say they know, but when they see the photo they say ahh! Now I see!".

Fig 3. View from a potential security camera

In addition to this example, there were many others in which scene capture provided the most effective medium by which information could be captured and used. Photos were not only shared on the camera itself, but were also captured for inclusion in documents and for emailing.

Sound capture (mainly of voice) represented another frequent kind of information capture opportunity reported. Some of the subjects stood out as sound capture enthusiasts. For example, one subject (Steve), despite worries about ethics, discussed several occasions where he would like to have recorded ongoing telephone conversations. Steve worked on technical support for medical knowledge management systems. Throughout his day he often had long unexpected calls from clients asking for help with software problems. Steve would need to calm the client and ask the right questions while taking detailed notes. By tape recording his phone conversations, he felt he might be able to better concentrate on discussing the problem with the client, reassured that he could go over the tape recorded call and take notes later.

This analysis implies that there is indeed potential for devices in the workplace that capture both sound and scene. For instance, it suggests that there may be an important new market for digital cameras designed as workplace tools rather than as consumer devices. Perhaps not surprisingly, the analysis also points to the overriding importance of devices which can effectively capture paper-based information.

What The Paper Group Captured

An analysis of what the Paper Grp captured gives us further insight into the different kinds of paper documents that people extract information from (Figure 4).

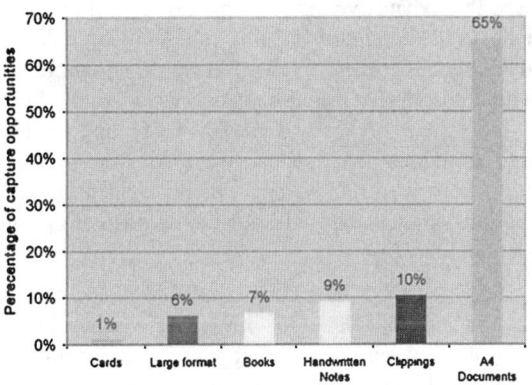

Fig 4. Kinds of paper-based information "captured" by the Paper Grp.

The documents took many different forms, from more conventional business documents (A4), to newspaper and magazine clippings, to books, large format documents (mainly flipcharts), and cards. By far the most common to be photographed, however, was regular letter-sized (A4) paper documents. Interestingly, over 62% of these were single sheets of paper, and less than 6% were documents of more than 3 pages. The second most frequent kind of paper document our subjects wanted to capture was that of clippings. These were called clippings as they represented occasions when subjects said they wanted selective pieces of text or graphics from newspapers, magazines, and other documents. Almost as frequent was the desire to capture handwritten notes and scribbles, books, and also large format documents such as flipcharts.

This kind of breakdown has implications for the kinds of document capture devices one might provide for users. These data suggest that although A4 documents are the most important type of captured document, 35% of documents that are captured are not A4. This suggests that a document capture device should have the ability to capture as wide range of different media as possible, or that one needs to think in terms of specialised devices.

Capture Taxonomy

More interesting than what people captured or wanted to capture, is *why* or *for what purpose* the capture activity took place. Capture of information is goal-oriented. That is, we do not capture information for its own sake (or at least not in most cases), but rather we do it to use that information in some other way. A key aim of the study was to elaborate on these goals since, as we will show, this significantly impacts the nature of the devices designed to support these goals.

To highlight these differences we sought to construct a taxonomy of the different capture opportunities we had collected. We did not want to develop an independently verifiable taxonomy, or a categorisation scheme which was necessarily exhaustive and comprehensive. Further, we were not constructing this in order to test any hypotheses or conduct any experimental tests. Rather we sought to come up with a descriptive framework: one which would help us (and others) to understand the diversity and range of capture activities that people want to carry out; and perhaps more importantly, one which would make useful distinctions amongst activities for the purpose of design of new technologies.

The categories were thus derived as a result of extensive discussion of the data to draw out similarities and differences between the different capture opportunities we had collected. The capture situations were categorised according to the reasons behind the capture as expressed by the subjects in each of the events in our corpus. This resulted in a ten category taxonomy which exhaustively categorised all the capture events. For some of the capture events there were multiple reasons for information capture, so the capture categories are not mutually exclusive. The categories of capture can be briefly described as follows:

1. **Capture to discuss**
Items which have been captured in order to have an interactive, synchronous conversation around the captured information or document. In this case the captured item is a part of a spoken conversation. This can be done with those physically co-located, or those who are distant, e.g. over the phone

2. **Capture to distribute**
Items which have been captured to be distributed or sent to someone. Typical methods include hand delivery, fax, email. This is different from the first category in that the document can be sent without the need to have a conversation around it. The intended receiver(s) may be remote or co-located.

3. **Capture to post in a common information space**
Items which are captured to put them into a common space, such as a shared desk, a notice board, a shared whiteboard or a fridge door. These items are often positioned in prominent places so as to allow 'incidental viewing' – noticing the item without specifically looking for it.

4. **Capture to archive**
Items which are captured for long or medium term storage. They are typically items which are captured with no specific short term purpose in mind, but are kept because there is a feeling that they might be useful 'just in case'. Materials produced in meetings often fall into this category.

5. **Capture to collect and collate**
Items which are captured to add to a collection of similar items. This might include collecting papers for a personal library, clippings for a magazine, interesting URLs, and so on. The important aspect here is that the collection unites objects with some sort of common theme, and it is the collection as a whole which is of worth.

6. **Capture to read and reflect**
Items which are captured to be read or reflected upon at a later time when more convenient. This is often the reason for "collecting" documents so they can be taken back to base and read later.

7. **Capture for task management**
Items which are captured in order to help remember things that you have to do, or have to be organised in the future. Examples include task lists, post it notes and so on. Many recent inventions such as PDAs, attempt to cater for this use.

8. **Capture to refer to**
Items which are deliberately captured to be referred to in some later activity. This includes situations such as capturing a phone number for later use, taking notes during a presentation to refer to when composing a new document, or capturing an advertisement to order a product. This does not include situations where information is captured so that it can be incorporated into a document verbatim (see next category).

9. **Capture to re-use**
Items which are captured in order that they can be re-used in the production of something else (usually a document). Examples include capturing a slide for use in composing a new presentation or capturing text or figures to incorporate into a new document. The emphasis here is on the *verbatim* reproduction of what was captured.

10. **Capture to a living document**
Items which are captured and used as "living documents", with the document being written on, annotated and edited through the day. "Living documents" are typically modified in an ongoing way, and act as structured information repositories. A project plan is a good example of such a document.

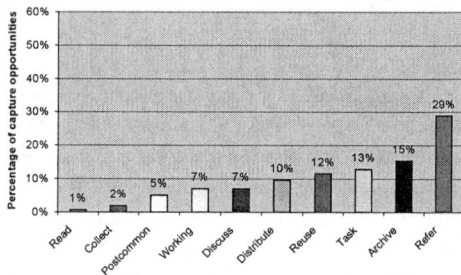

Fig 5. Frequency of capture by category (Multimedia Grp.)

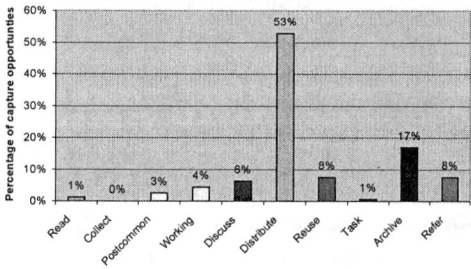

Fig 6. Frequency of capture by category (Paper Grp.)

Using this taxonomy we were then able to analyse the frequency with which subjects in both groups carried out these different kinds of capture activities. Figures 5 and 6 show differences in the most frequent kinds of capture activities the two groups carried out. The most common form of capture for the Multimedia Grp was capturing documents, scenes and even audio snippets for use as a reference in a later activity (*"capture to refer to"*). The Paper Grp, on the other hand, mainly wanted to capture documents for the purpose of sending or distributing the information to others (*"capture to distribute"*).

Rather than attempting to generalise from these somewhat small and diverse samples, the important point to take from both the taxonomy and the graphs is to recognise the very different kinds of goals people have in mind when they capture information. Further, the most frequent kind of capture activity that any one person does is likely to be highly dependent on who that person is and the kind of job that they do. Some subjects, for example, acted as something of an information hub – they would gather information from diverse sources and redistribute it to others. Accordingly, these subjects mainly *captured to distribute*. An example of this was the administrative assistant we studied. Other subjects tended to collect information for future reference and use when creating new documents – *capture to collect, capture to refer to* and *capture to re-use*. The journalist we studied was an example of this in that he often collected newspaper clippings to use when he was writing new articles.

APPLICATION OF THE TAXONOMY

We turn now to why we believe this kind of analytic framework is both useful and interesting. The first and perhaps most obvious point is that such a taxonomy is didactic – it provides a framework and a language with which to talk about information capture. In doing so, one can better understand how any particular capture event, or the capture activities of any one person fit within a broader picture.

The second point, however, is more practical. Each of the capture categories has implications for designing technology which would support those activities. To illustrate this, we will consider four of these categories and the particular implications these have for two types of capture technologies - Capshare, and digital cameras.

Capture to discuss

When people used captured information for the purpose of discussion, being able to jointly view the captured information was key to the activity. Thus, using currently available technologies, people in meetings displayed information by means of projectors, large display screens, and more old-fashioned technologies such as paper.

Of interest was that subjects also used the LCD screen on the back of the camera for the purpose of discussing captured information with others. For example, one subject used a photograph of a faulty power supply to convince a supplier that it was incorrectly fitted. The picture became an important part of the discussion, and the ability to look at it together enabled this to happen.

Once we were sensitised to these kinds of activities, we began to realise the extent to which people were using the Capshare for discussion around shared images. In doing so, it pointed to the fact that the LCD screen was sub-optimal for this purpose. This was not surprising, as the screen on Capshare was primarily intended for users to confirm what had been scanned in, and for flicking through the set of scanned images.

This finding emphasised to us that capture devices should better support the discussion of captured materials and designers should pay close attention to how information can be jointly viewed. In the case of digital cameras and Capshare their display screens – designed for capture confirmation – should be expanded and have wider viewing angles so as to better support information sharing, Moreover, capture devices could benefit from supporting the sending of information to projectors, monitors, printers and other display surfaces.

Capture to distribute

If the category "capture to discuss" puts the emphasis on information display technologies, "capture to distribute" points to the need for capture devices which are optimised to connect and communicate with other devices and infrastructures. For many capture devices, whether they be portable or desktop devices, users are required to go through several time-consuming steps in order to transmit captured information to other people.

For example, digital cameras are often designed with little support for easily sending the captured images to others. Part of the problem is uploading the contents of the camera to a networked environment such as a PC. Ease of doing this to some extent explains the success of the Sony Mavica camera on the market – this camera features a floppy disk drive built into the camera, facilitating copying photos from the camera to a PC. However, the taxonomy highlights that this does not go far enough. The problem is completing the *whole sending action*. This suggests that allowing send actions to be completed on the device itself would greatly improve usability. To investigate this further we have prototyped addressing emails and faxes directly on the Capshare. These are then sent when Capshare is put into a special "cradle" which allows quick connection and uploading of images to a networked PC. We have also prototyped sending emails and faxes directly from Capshare using a mobile phone, cutting out the need to use a PC entirely. Such solutions only arise from looking at the capture events within the context of their goals rather than focussing solely on the capture event in isolation.

Capture to archive

Capturing for the purpose of archiving is another interesting category because it has again very different implications for technology. Here, the capture opportunities we observed highlighted the need for easy connections to storage facilities, and for flexible search and retrieval mechanisms to operate on the captured information.

In the study there were many cases of subjects wanting to capture the output of meetings. This included flipcharts, whiteboard scribblings, slides and notes. Often the reason for capturing them was rather unspecific, but rather it was done to add to the group archives just in case the information was ever needed. These examples and others showed that archived information can take many different forms, that it may need to be flexibly searched, and that the ability to add information to the material may be important in providing context at the time of retrieval.

The software provided with most digital cameras provide an example of how *not* to support archiving of materials. Folders of photos are often assigned arbitrary labels or numbers which provide no useful information when later searching or browsing. Keyword facilities are often cumbersome or time consuming to use, and there are no facilities to arrange by subject matter, or to flexibly organise and store the photos. Interesting ideas in this respect seem confined mainly to the research lab at this point [3].

Early versions of Capshare nominally met the requirements of archiving. As originally designed, Capshare placed its files into a standard Windows folder, but this did not allow documents to be easily browsed. By considering the kinds of archiving activities our subjects carried out, we suggested that the PC software needed to be expanded to allow documents to be kept ready to hand in one place, and to be easily browsed using thumbnails when searching. To support archiving, the software also needed to support documents being quickly moved into archive folders. At this point we were able to experiment with different kinds of off-the-shelf software to see how users reacted during search and retrieval tasks, and again, we could make a set of recommendations for improving archiving tasks.

Capture for task management
As a final example, we will consider capturing for the purpose of task management. This class refers to the capture of information for the purpose of organising one's tasks, and reminding oneself of future actions. Most of us have many ways in which we do this, such as the use of paper diaries, ToDo lists, and PDAs. In effect, such devices support "prospective memory" or memory for things we need to do in the future.

The diary data suggested that there are two important aspects to effective support of task management activities. First is the ability to allow users to capture information in an *ad hoc* manner. This is because people remember things they need to do at unexpected times and places. In fact, research shows that people are more likely to remember ToDo or action items when they are physically mobile [10].

Second is the ability to deliver appropriate reminders at the right time and in the right place. Our subjects showed that they recognised the importance of this. For example, Post-it notes and documents often were used as reminders, and these were placed in strategic places such as on computer screens or chairs so that they could not be missed. Most subjects carried reminding systems with them such as paper calendars or PDAs. Subjects also made a point of reviewing their lists and calendars on a regular basis so as not to miss important events.

The implications here are that support for task management is well suited to mobile technologies which allow ad hoc capture of information about ToDo items, and which can also deliver important reminders anyplace, anytime. The category of technology this points toward is PDAs, and these are obviously already successfully supporting capture for task management.

What was interesting, however, was the implications from this study for how PDA technology might be extended. One subject used the digital camera as a sort of visual ToDo list. She would regularly scan through her pictures on the back of the camera in order to remind herself of things she needed to do. For example, she took a picture of the photocopying room to remind herself to pick up some documents later. One reason she found this effective was that taking a picture was quicker than writing a note for herself. By using a digital camera as input, this quick action provided rich visual output as a reminder.

In addition to digital cameras, devices like Capshare also suggest new kinds of task management tools which could offer more than conventional PDAs. For example, one of our subjects photographed a letter he had received inviting him to a career evening at a local school. In taking the photo he said he wanted to be able to associate the document with his appointment in his electronic diary, so he would have the directions to the school handy when he was reminded of the event. This led us to think about ways in which scanned in documents could be usefully linked to appointment calendars.

Summary
The above discussion provides a demonstration of how the capture taxonomy we constructed can be used to impact the process of design. Simply by considering it we make a move to understanding the *goal* of capture. As we hope we have shown, many capture devices such as digital cameras have not been designed to optimise for goals; more typically they are optimised mainly for the capture task itself. The design of these devices may be improved by considering how well they current support the different actions in the taxonomy.

Considering the capture goals discussed in the taxonomy, it is striking how a set of activities in a single category tend to point in the same general direction, and toward particular classes of technology. Some categories point to emphasising better display technologies, some point to better communications, whilst others guide the designer toward storage and archiving, better software, or PDA-like devices.

Of course, this is not to say that capture devices of necessity must be specialised for one kind of goal to the exclusion of others. For example, some of the modifications that we have suggested for Capshare have shown that easier connection to a PC and well-designed software can provide an interface to allow users to easily email, archive, or print. In other words, some solutions better support a range of capture activities.

CONCLUSION

The primary aim of this paper was to unpack the goals behind capture activity so that we have a framework with which to understand how capture devices may be used, and thus to look for opportunities to invent new ones. To do so we used an innovative camera and diary methodology to naturalistically record both descriptive and quantitative data on capture activity in the context of people's working lives. The results show that people engage in a wide spectrum of different kinds of capture tasks. Further, for each different kind of capture "goal" that people may have, there are very different kinds of design implications. We illustrated this by showing how the framework can be used to inform the design of digital cameras as well as the Capshare document scanner.

Aside from these main findings, another point to make in closing is to highlight the success of this new diary technique. By using digital cameras to support diary-keeping, one can collect naturalistic data without the large overhead of observational approaches. Since the photographs are taken at the site of action, and interviews about the photos are carried out within a few days, subjects showed few recall problems when prompted by photos even on the relatively low resolution screen used on current digital cameras. This method also reduced the demand on the subjects themselves as taking pictures was easier for them than writing notes. This is not to say that this method is comparable to techniques such as ethnography, since as with all diary methods one is relying on the participation of the subjects. However, this technique is a useful addition to existing diary methods. We are enthusiastic about the further utility of this methodology as we move on to prototype and test some of our new ideas for mobile information appliances.

ACKNOWLEDGMENTS

We thank the study participants both at Hewlett-Packard and elsewhere, for generously giving of their time and effort in this research. We are also grateful to the Appliance Design Group team who built the prototype devices and software referred to in this paper.

BIBLIOGRAPHY

1. Adler, A., Gujar, A., Harrison, B., O'Hara, K., & Sellen, A.J. (1998). A diary study of work-related reading: Design implications for digital reading devices *Proceedings of CHI '98*, ACM, New York, 241-248.

2. Hindus, D. & Schmandt, C. (1992). Ubiquitous audio: Capturing spontaneous collaboration. *Proceedings of CSCW '92,* ACM, New York, 210-217.

3. Kuchinsky, A., Pering, C., Creech, M., Freeze, D., Serra, B., and Gwizda, J. (1999). FotoFile: A consumer multimedia organisation and retrieval system. *Proceedings of CHI '99*, ACM, New York, 496-503.

4. Lamming, M. & Newman W. (1992). Activity-based information retrieval. *Personal Computers and Intelligent Systems: Information Processing '92*, 68-81.

5. Minneman, S., Harrison, S., Janssen, B., Kurtenbach, G., Moran, T., Smith, I., & van Melle, W. (1995). A confederation of tools for capturing and accessing collaborative activity. *Proceedings of Multimedia '95*.

6. Newman, W. & Wellner, P. (1992). A desk supporting computer-based interaction with paper documents. *Proceedings of CHI '92*, ACM, New York, 587-592.

7. O'Hara, K., Smith, F., Newman, W., & Sellen, A. (1998). Student readers' use of library documents: Implications for digital library technologies. *Proceedings of CHI '98*, ACM, New York, 233-240.

8. Prosser, J. (ed.) (1998) *Image based research*. Falmer, London.

9. Sellen, A.J., & Harper, R.H.R. (1997). Paper as an analytic resource for the design of new technologies. *Proceedings of CHI '97* (Atlanta, GA.) New York: ACM Press, 319-326.

10. Sellen, A.J., Louie, G., Harris, J. E., & Wilkins, A.J. (1997). What brings intentions to mind? An in situ study of prospective memory. *Memory, Vol. 5, No. 4*, 483-507.

11. Whittaker, S., Hyland, P. & Wiley, M. (1994). Filochat: Handwritten notes provide access to recorded conversations. *Proceedings of CHI '94*, ACM, New York, 271-277.

Instrumental Interaction: An Interaction Model for Designing Post-WIMP User Interfaces

Michel Beaudouin-Lafon
Dept of Computer Science
University of Aarhus
Aabogade 34
DK-8200 Aarhus N - Denmark
mbl@daimi.au.dk

ABSTRACT
This article introduces a new interaction model called Instrumental Interaction that extends and generalizes the principles of direct manipulation. It covers existing interaction styles, including traditional WIMP interfaces, as well as new interaction styles such as two-handed input and augmented reality. It defines a design space for new interaction techniques and a set of properties for comparing them. Instrumental Interaction describes graphical user interfaces in terms of *domain objects* and *interaction instruments*. Interaction between users and domain objects is mediated by interaction instruments, similar to the tools and instruments we use in the real world to interact with physical objects. The article presents the model, applies it to describe and compare a number of interaction techniques, and shows how it was used to create a new interface for searching and replacing text.

Keywords
Interaction model, WIMP interfaces, direct manipulation, post-WIMP interfaces, instrumental interaction

INTRODUCTION
In the early eighties, the Xerox Star user interface [27] and the principles of direct manipulation [26] led to a powerful graphical user interface model, referred to as WIMP (Windows, Icons, Menus and Pointing). WIMP interfaces revolutionized computing, making computers accessible to a broad audience for a variety of applications.

In the last decade, HCI researchers have introduced numerous new interaction techniques, such as toolglasses [5] and zoomable user interfaces [3]. Although some have been shown to be more efficient than traditional techniques, e.g., marking menus [19], few have been incorporated into commercial systems. A likely reason is that integrating new interaction techniques into an interface is challenging for both designers and developers. Designers find it faster and easier to stick with a small set of well-understood techniques. Similarly, developers find it more efficient to take advantage of the extensive support for WIMP interaction provided by current development tools.

The leap from WIMP to newer "post-WIMP" graphical interfaces, which take advantage of novel interaction techniques, requires both new interaction models and corresponding tools to facilitate development. This paper focuses on the first issue by introducing a new interaction model, called *Instrumental Interaction,* that extends and generalizes the principles of direct manipulation to also encompass a wide range of graphical interaction techniques. The Instrumental Interaction model has the following goals:

- cover the state-of-the-art in graphical interaction techniques;
- provide qualitative and quantitative ways to compare interaction techniques, to give designers the basis for an informed choice when selecting a given technique to address a particular interface problem;
- define a design space in which unexplored areas can be identified and lead to new interaction techniques; and
- open the way to a new generation of user interface development tools that make it easy to integrate the latest interaction techniques in interactive applications.

After a review of related work, this paper analyzes the limits of current WIMP interfaces. The Instrumental Interaction model is introduced and applied to several existing interaction techniques as well as to the design of a new interface for searching and replacing text. Finally the paper concludes with suggestions for future work.

RELATED WORK
In this paper, an *interaction model* is defined as follows:

> An interaction model is a set of principles, rules and properties that guide the design of an interface. It describes how to combine interaction techniques in a meaningful and consistent way and defines the "look and feel" of the interaction from the user's perspective. Properties of the interaction model can be used to evaluate specific interaction designs.

Direct Manipulation [26] is a generic interaction model, while style guides, e.g., Apple's guidelines [2], describe more precise and specific models. Took introduced a model called Surface Interaction [29] and Holland & Oppenheim a model called Direct Combination [15].

An interaction model differs from the *architectural model* of an interface, which describes the functional elements in the implementation of the interface and their relationships (see

[12] for a review). User interface development environments have generated a variety of *implementation* models for developing interfaces (see [24] for a review), e.g. the widget model of the X/Motif toolkit [14] or Garnet's Interactors model [23]. MVC [18] is a well-known model that was created to support the Xerox Star user interface and has influenced many other architectural and implementation models. Whereas architectural models are aimed at interface *development*, an interaction model is aimed at interface *design*.

The *model-based* approach and its associated tools [28] help bridge the gap between interaction and architectural models by offering a higher-level approach to the design of interactive systems.

Device-level models such as logical input devices [11] or Card et al.'s taxonomy [7] operate at a lower level of abstraction than interaction models. Understanding the role of the physical devices in interaction tasks is a critical component of the definition of the Instrumental Interaction model.

At the theoretical level, Activity Theory [6] provides a relevant framework for analyzing interaction as a mediation process between users and objects of interest.

Finally, Instrumental Interaction is grounded in the large (and growing) number of graphical interaction techniques that have been developed in recent years, some of which are referenced in the rest of this article.

FROM WIMP TO POST-WIMP INTERFACES

The WIMP interaction model can be outlined as follows:

- application objects are displayed in document windows;
- objects can be selected and sometimes dragged and dropped between different windows; and
- commands are invoked through menus or toolbars, often bringing up a dialog box that must be filled in before the command's effect on the object is visible.

This section uses Shneiderman's principles of direct manipulation [26] to analyze WIMP interfaces:

1. Continuous Representation of objects of interest

Objects of interest are central to direct manipulation. They are the objects that the user is interested in to achieve a given task, such as the text and drawings of a document or the formulae and values in a spreadsheet. Principle 1 asserts that objects of interest should be present at all times. Since objects of interest are often larger than the screen or window in which they are displayed, WIMP interfaces makes them *accessible* at all times through scrolling, panning or zooming. This accessibility is hindered by the growing number of interface objects that are *not* objects of interest such as toolbars, floating palettes and menu bars. These use increasing amounts of screen real-estate, forcing the user to shrink the windows displaying objects of interest. Dialog boxes also often occlude significant parts of the screen, making the rest of the interface inaccessible to the user.

Finally, there are more objects of interest than meet the eye: in many applications users must manipulate secondary objects to achieve their tasks, such as style sheets in Microsoft Word, graphical layers in Adobe Photoshop or Deneba Canvas, or paint brushes in MetaCreations Painter. Once the user is familiar with the application, these objects become part of his or her mental model and may acquire the status of object of interest. Unfortunately, these are rarely implemented as first-class objects. Thus, for example, Word's styles are editable only via transient dialog boxes that must be closed before returning to the text editing task.

2. Physical actions on objects vs. complex syntax

Most computers have only a mouse and keyboard as input devices limiting the set of user actions to: typing text or "special" keys (e.g. function keys, keyboard shortcuts and modifiers), pointing, clicking, and dragging. Given the mismatch between this small vocabulary of actions and the large vocabulary of commands, WIMP interfaces must rely on additional interface elements, usually menus and dialog boxes, to specify commands. The typical sequence of actions to carry out a command is:

- select the objet of interest by clicking it;
- select a command from a menu or keyboard shortcut;
- fill in the fields of a dialog box; and
- click the OK button to see the result.

This is conceptually no different from typing a command in a command-line interface: The user must type a command name, file name (the object of interest), arguments (the fields in the dialog box) and return key (the OK button). In both cases the syntax is complex and cannot be considered direct manipulation of the objects of interest. In fact, WIMP interfaces directly violate principle 2 and often use *indirect* manipulation of the objects of interest, through (direct) manipulation of interface elements such as menus and dialog boxes.

3. Fast, incremental and reversible operations with an immediately-apparent effect on the objects of interest

The heavy graphical syntax imposed on the user results in commands that are neither fast nor incremental. Specifying a command is not fast because of the amount of time used for non-semantic actions such as displacing windows and flipping through tabs in a tabbed dialog. Inputting parameter values for a command is often inefficient because of the small set of interactors, such as when numeric values are entered as text. Finally, the specification is not incremental: users must explicitly commit to a command that uses a dialog box before seeing the result. If the result does not match the user's expectations, the whole cycle of command activation must be started over again. This is especially cumbersome when trial-and-error is an integral part of the task, as when a graphics designer selects a font size: specifying the point size numerically is annoying when the goal is to see the visual result on the page.

4. Layered or spiral approach to learning

The small number of interaction techniques used by WIMP interfaces makes it easy to learn the basics of any new application. However, interaction shortcuts, such as combining keyboard modifiers with mouse buttons to activate the frequent commands, are concealed and inconsistent across applications and make the transition from novice to power user more difficult.

Towards a new interaction model

Some commercial applications, especially those dedicated to creative tasks such as painting, graphic design or music, extend the basic WIMP model to address some of the shortcomings identified above. For example, some painting programs make brushes first class objects that can be edited and saved into files. Some text editors have inspector windows that display the state of the current selection and update it when the user enters relevant values. Techniques such as the HotBox [21] were designed to access larger numbers of commands.

These new interaction techniques illustrate the transition from WIMP to post-WIMP interaction: Windows are not used in zoomable interfaces, icons and text are replaced by richer representations in interactive visualization, menus are complemented by more powerful interaction widgets such as toolglasses, pointing and dragging are superseded by bimanual input and gesture input. Designing Post-WIMP interfaces that are more faithful to the principles of direct manipulation and that take advantage of novel interaction techniques requires new interaction models. To guide interface designers, these models should be:

- *descriptive*, incorporating both existing and new applications;
- *comparative*, providing metrics for comparing alternative designs (as opposed to *prescriptive*, deciding a priori what is good and what is bad); and
- *generative*, facilitating creation of new interaction techniques.

INSTRUMENTAL INTERACTION

As shown in the previous analysis, WIMP interfaces do not follow the principles of direct manipulation. Instead, they introduce interface elements such as menus, dialog boxes and scrollbars that act as *mediators* between users and the objects of interest. Users have a (limited) sense of engagement, as advocated by direct manipulation, because they manipulate these intermediate objects directly. This matches our experiences in the physical world: We rarely fingerpaint, but often use pens and pencils to write. We cook with pots and pans, hang pictures with hammers and power drills, open doors with handles and turn off lights with switches. Our interaction with the physical world is governed by our use of *tools*. Direct manipulation of *physical* objects of interest occurs when we bring them into our current context of operation, before we manipulate them with the appropriate tools, usually with two hands [13]. The Instrumental Interaction model is based on how we naturally use tools (or instruments) to manipulate objects of interest in the physical world. Objects of interest are called *domain objects*, and are manipulated with computer artifacts called *interaction instruments*.

Domain objects

In computer systems, applications operate on data that represent phenomena or objects. For computer users, this data is the primary focus of their actions. For example, when creating a text document, the focus of the user is on the text of the document. Everything else on the screen is there to support the user's task of editing the text document.

Domain objects form the set of potential objects of interest for the user of a given application. Domain objects have *attributes* that describe their characteristics. Attributes can be simple values or more complex objects. For example, in a 3D modeller, the position and size of a sphere are simple values (integer or real numbers), while the material of the sphere is a complex entity (color, texture, transparency, etc.). The user may shift the object of interest, concentrating on the material as the focus of the interaction. Similarly, text styles that describe the formatting attributes of text also may also obtain the status of objects of interest. Materials and styles are therefore also domain objects in their respective interfaces.

In summary, domain objects form the basis of the interaction as well as its purpose: Users operate on domain objects by editing their attributes. They also manipulate them as a whole, e.g. to create, move and delete them.

Interaction instruments

An *interaction instrument* is a mediator or two-way transducer between the user and domain objects. The user acts on the instrument, which transforms the user's *actions* into *commands* affecting relevant target domain objects. Instruments have *reactions* enabling users to control their actions on the instrument, and provide *feedback* as the command is carried out on target objects (Figure 1).

A scrollbar is a good example of an interaction instrument. It operates on a whole document by changing the part that is currently visible. When the user clicks on one of the arrows of the scrollbar, the scrollbar sends the document a scrolling command. Note that the transduction here consists of sending scrolling commands as long as the user presses the arrow. The reaction of the scrollbar consists of highlighting the arrow being pressed. The feedback consists of updating the thumb to reflect the new position of the document. In addition, the object also responds to the instrument by updating its view in the window.

Another example is an instrument that creates rectangles in a drawing editor. As the user clicks and drags the mouse, the instrument provides a reaction in the form of a rubberband rectangle. When the user releases the button, the creation operation is actually carried out and a new domain object is created. The feedback of this operation consists in displaying the new object.

An instrument decomposes interaction into two layers: the interaction between the user and the instrument, defined as the physical *action* of the user on the instrument and the *reaction* of the instrument and the interaction between the instrument and the domain object, defined as the *command* sent to the object and the *response* of the object, which the instrument may transform into *feedback* to the user. The instrument is composed of a physical part, the input device, and a logical part, the representation of the instrument in software and on the screen.

Activating instruments

At any one time, an interface provides a potentially large number of instruments. However the user can manipulate only a few of them at the same time, usually only one, because of the limited number of input devices. In the most

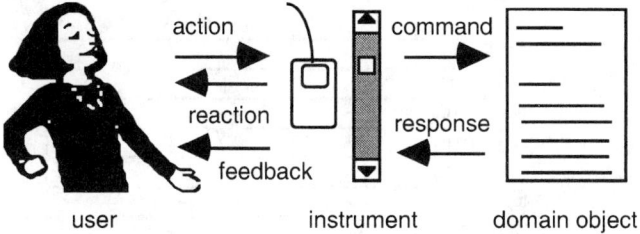

Figure 1: Interaction instrument mediating the interaction between a user and a domain object

common case of a keyboard and mouse, a single input device (the mouse) must be multiplexed between a potentially large number of instruments, i.e. a single physical part may be associated with different logical parts.

An instrument is said to be *activated* when it is under the user's control, i.e. when the physical part has been associated with the logical part. In the case of the scrollbar, the user activates the instrument by pointing at it and it remains active as long as the pointer is within the scrollbar. When creating a rectangle, the user activates the instrument by clicking a button in a tool palette and it remains active until another instrument is activated.

These two types of activation are quite different. The activation of the scrollbar is *spatial* because it is caused by moving the mouse (and cursor) inside the area of the scrollbar. The activation of the rectangle creation instrument is *temporal* because it is caused by a former action and remains in effect until the activation of another instrument. (This is traditionally called a mode). Each type of activation has an associated cost: Spatial activation requires the instrument to be visible on the screen, taking up screen real-estate and requiring the user to point at it and potentially dividing the user's attention. Temporal activation requires an explicit action to trigger the activation, making it slower and less direct.

Interface designers often face a design trade-off between temporal and spatial multiplexing of instruments because the activation costs become significant when the user must frequently change instruments. Using extra input devices can reduce these costs. For example, the thumbwheel on Microsoft's Intellimouse is a scrolling instrument that is always active. An extreme example is an audio mixing *console*, which may contain several hundred potentiometers and switches, each corresponding to a single function. This permits very fast access to all functions, which is crucial for sound engineers working in real-time and cannot afford the cost of activating each function indirectly. A large design space lies between a single mouse and hundreds of potentiometers, posing design challenges to maximally exploit physical devices and reduce activation costs.

Reification and Meta-instruments

Reification is a process for turning concepts into objects. In user interfaces, the resulting objects can be represented explicitly on the screen and operated upon. For example, a style in a text editor is the reification of a collection of text attributes; the notion of material in a 3D modeller is the reification of a set of rendering properties. This type of reification generates new domain objects such as styles and materials that complement the "primary" domain objects of the application domain.

Instrumental Interaction introduces a second type of reification: an interaction instrument is the reification of one or more commands. For example, a scrollbar is the reification of the command that scrolls a document. This link between the traditional notion of command and the notion of instrument makes it easy to analyze existing interfaces with the Instrumental Interaction model. It is also a useful guideline to identify instruments when designing a new interface. In the last part of this paper, this rule is used to reify the traditional search-and-replace command of a text editor into a search instrument.

The result of this reification rule is that instruments are themselves potential objects of interest. This is indeed the case in real life, when the focus of attention shifts from the object being manipulated to the tool used to manipulate it. For example a pencil is a writing instrument and the domain object is the text being written. When the lead breaks, the focus shifts to a new instrument, a pencil sharpener, which operates on the shifted domain object, the pencil lead. The focus may even shift to the pen sharpener, if we need a screwdriver to fix it. Such "meta-instruments" (instruments that operate on instruments) are not only useful for "fixing" instruments, but can also be used to organize instruments in the workspace, e.g. a toolbox, or to tailor instruments to particular tasks, e.g. turning a power-drill into a power-saw. In graphical user interfaces, common meta-instruments include menus and tool palettes used to select commands and tools, i.e. to activate instruments.

Properties of Instruments

An important role of an interaction model is to provide properties to evaluate and compare alternative designs. This can help interface designers who face difficult choices when selecting the interaction techniques for a particular application. The goal of defining properties of instruments is not to decide which instruments are good and which are bad, but to evaluate them so that designers can make an informed choice and so that researchers can identify and explore areas of the design space that are not mapped by existing instruments.

The literature on user interface evaluation techniques is considerable. Here, we use a particular type of evaluation based on properties. This is a common approach in software engineering and has also proved valid and useful for evaluating interactive systems [12]. The rest of this section introduces three properties of interaction instruments.

Degree of indirection

The degree of indirection is a 2D measure of the spatial and temporal offsets generated by an instrument. The *spatial offset* is the distance on the screen between the logical part of the instrument and the object it operates on. Some instruments, such as the selection handles used in graphical editors, have a very small spatial offset since they are next to or on top of the object they control. Other instruments, such as dialog boxes, can be arbitrarily far away from the object they operate on and therefore have a large spatial offset. A large spatial offset is not necessarily undesirable.

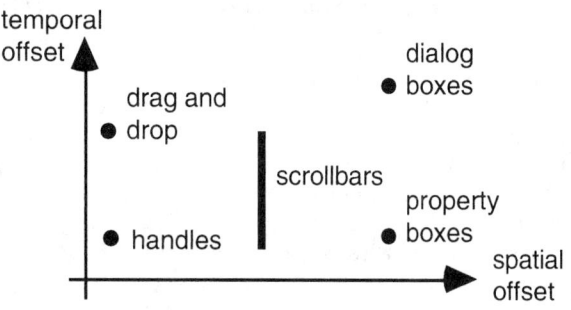

Figure 2: Degree of indirection

	Indirection	Integration	Compatibility
Menus	--	+/-	--
Toolbars	-	+	+
Dialog boxes	--	+/-	-
Property boxes	+	+/-	
Handles	++	++	++
Window titles	++	+	+
Scrollbars	+	-	-
Keyboard shortcuts	+	+	-
Drag & drop	++	++	++

Table 1: Comparing WIMP interaction techniques

For example, placing a light switch far from the light bulb it controls makes it easier to turn on the light. Similar examples can be found in user interfaces.

The *temporal offset* is the time difference between the physical action on the instrument and the response of the object. In some cases, the object responds to the user's action in real-time. For example, clicking an arrow in a scrollbar scrolls the document while the mouse button is depressed. In other cases, the object responds to the user's action only when the action reaches closure. For example, the arguments specified in a dialog box are taken into account only when the OK or Apply button is activated. In general, short temporal offsets are desirable because they exploit the human perception-action loop and give a sense of causality [22].

Figure 2 shows the degree of indirection of various WIMP instruments on a 2D chart. Scrollbars occupy a range in the diagram. For example, some scrollbars provide immediate response when the thumb is moved while others only scroll the document when the mouse button is released. The figure shows that the degree of indirection describes a continuum between direct manipulation (lower-left corner) and indirect manipulation (upper-right corner).

Degree of integration

The degree of integration measures the ratio between the number of degrees of freedom (DOF) provided by the logical part of the instrument and the number of DOFs captured by the input device. This term comes from the notion of *integral tasks* [17]: some tasks are performed more efficiently when the various DOFs are controlled simultaneously with a single device. A scrollbar is a 1D instrument controlled by a 2D mouse, therefore its degree of integration is 1/2. The degree of integration can be larger than 1: controlling 3 rotation angles with a 2D mouse [16] has a degree of integration of 3/2. This property can be used to compare instruments that perform similar operations. For example, panning over a document can be achieved with two scrollbars or a 2D panner. The latter has a degree of integration of 1 and is therefore more efficient than two scrollbars, which have a degree of integration of 1/2 and incur additional activation costs.

Degree of compatibility

The degree of compatibility measures the similarity between the physical actions of the users on the instrument and the response of the object. Dragging an object has a high degree of compatibility since the object follows the movements of the mouse. Scrolling with a scrollbar has a low degree of compatibility because moving the thumb downwards moves the document upwards. Using text input fields to specify numerical values in a dialog box, e.g. the point size of a font, has a very low degree of compatibility because the input data type is different from the output data type. Similarly, specifying the margin in a text document by entering a number in a text field has a lower degree of compatibility than dragging a tab in a ruler.

APPLYING THE MODEL

This section uses the Instrumental Interaction model to analyze existing interaction techniques, both from WIMP interfaces and from more recent research. It demonstrates the descriptive power of the model. The generative power of the model is then illustrated by the design of a new instrument for searching and replacing text.

Analyzing WIMP Interfaces

The primary components of WIMP interfaces can be easily mapped to instruments and compared (Table 1):

Menus and *toolbars* are meta-instruments used to select the command or tool to activate. This use of meta-instruments slows down interaction and generates shifts of attention between the object of interest, the meta-instrument and the instrument. Contextual menus have a small spatial offset and are therefore more efficient than toolbars and menu bars. Toolbars, which can be moved next to their context of use, have a better spatial offset than menu bars.

Dialog boxes are used for complex commands. They have a high degree of spatial and temporal indirection. They often use a small set of standard interactors such as text fields for numeric values, resulting in a low degree of compatibility.

Inspectors and *property boxes* are an alternative to dialog boxes that have a lower degree of temporal indirection. Since they can stay open, they can be activated with pointing (positional activation) rather than selection in a menu (temporal activation).

Handles are used for graphical editing and provide a very direct interaction: low degree of indirection, high degree of compatibility and good degree of integration.

Window titles and borders are instruments activated positionally to manipulate the window (move, resize, iconify, zoom, close). *Scrollbars* control the content of the window. Because of their low degree of integration, they are not optimal, especially for panning documents in 2D. Also, their spatial offset generates a division of attention, especially since they are activated positionally: the user must be sure to point at the right part of the scrollbar.

Keyboard shortcuts and *accelerator keys* are meta-instruments, used to quickly switch between instruments and save the activation costs of menus and toolbars. Some accelerator keys affect the way the current instrument works. For example, on the Macintosh, the Shift key constrains the move tool to horizontal and vertical moves and the resize tool to maintain the current aspect ratio.

Drag and drop is a generic instrument for transferring or copying information. Compared to traditional cut/copy/paste commands that use a hidden clipboard, it has a smaller degree of indirection. There is no spatial offset because the objects are manipulated directly and the temporal offset is low because there is feedback about potential drop-zones as the user drags the object.

Over the past few years, interaction techniques such as inspectors, property boxes, drag and drop and contextual menus have become more common in commercial applications. The above analysis explains why these techniques are more efficient than their WIMP counterparts, demonstrating a useful contribution of the Instrumental Interaction model.

Analyzing Post-WIMP Interaction Techniques

Table 2 summarizes the comparison of several post-WIMP interaction techniques. Interactive visualization helps users explore large quantities of visual data and make sense of it through filtering and displaying it to exhibit patterns [8]. These systems use two categories of instruments:

- navigation instruments specify which part of the data to visualize and how; and
- filtering instruments specify queries and display results.

A key aspect of these systems is a strong coupling between user actions and system response. In other words, these instruments must have a small temporal offset. For example, in the Information Visualizer [9], the instruments used to control Cone Trees and Perspective Walls provide immediate responses and use smooth animations to display changes in visualization parameters. In Dynamic Queries [1], double sliders are used to specify the range of query parameters; any change in a slider updates the display of filtered data.

Both navigation and filtering are usually multi-dimensional tasks: the user wants to control several dimensions simultaneously to navigate along arbitrary trajectories. This calls for the ability to manipulate several instruments simultaneously (which requires additional input devices) and/or for instruments with a high degree of integration. Current systems do not address this well. For example, Dynamic Queries permit only one side of a slider to be manipulated at a time, forcing the user to navigate along rectangular trajectories in the parameter space.

Zoomable user interfaces such as Pad++ [3] are based on the display of an infinite flat surface that can be viewed at any resolution. Exploring this surface requires navigation instruments to pan and zoom until the desired objects are in sight. Pad++ navigation instruments are activated by mouse buttons or modifier keys. This temporal activation is fast

	Indirection	Integration	Compatibility
Dynamic Queries	+	--	-
Pad++ navigation	++	+	+
Droppable tools	+	++	+
Toolglasses	++	++	++
Graspable interfaces	++	++	++

Table 2: Comparing post-WIMP interaction techniques

and provides access to navigation anywhere on the surface, unlike a scrollbar which requires positional activation. It also has high degrees of compatibility and integration. Editing the objects on the surface has led to Dropable Tools [4]: tools can be dropped anywhere on the surface and grabbed later. Activating these instruments is more direct than with a traditional toolbar because it does not involve a meta-instrument and the associated switch of attention.

A number of recent interaction techniques rely on new or additional input devices. This reduces activation costs by allowing several instruments to be active simultaneously. For example, the thumb wheel of the Intellimouse is always attached to a scrolling instrument. ToolGlasses [5] are semi transparent palettes operated with a track-ball in the left (or non-dominant) hand. The right hand is used to click through the palette onto a domain object, therefore specifying both the action to perform and the object to operate on. Here the toolglass is a meta-instrument under the control of the left hand, while the instruments it contains are activated by the right hand. In the TTT prototype [20], a combination of three instruments can be active simultaneously: the toolglass itself, an instrument in the toolglass and a navigation instrument to pan and zoom the drawing surface. This makes it possible, for example, to pan and zoom while creating an object. The design exploits the trackball and mouse input devices to minimize activation costs, to reduce the degree of indirection and to increase the degree of integration.

Graspable interfaces [10] use physical objects as input devices to manipulate virtual objects. In effect, they transfer most of the characteristics usually found in the logical part of the instrument into the physical part. This approach was pioneered by Augmented Reality [30], which explores ways to reconcile the physical and computer world by embedding computational facilities into physical objects. Here, the domain objects, in addition to the instruments, have a strong physical component. This increases the degrees of compatibility and integration since interaction occurs in the real world.

Designing a Text Search Instrument

In most text editors, searching and replacing text uses a dialog box where the user specifies the string to be searched and the string to replace it with. The operation is controlled with buttons to find the next or previous occurrence and replace it or not. Undoing the command usually means restoring the search string everywhere it has been replaced. This results in a sequential form of interaction where the system prompts the user and forces him or her to decide what to do with each occurrence, generating a very large temporal offset.

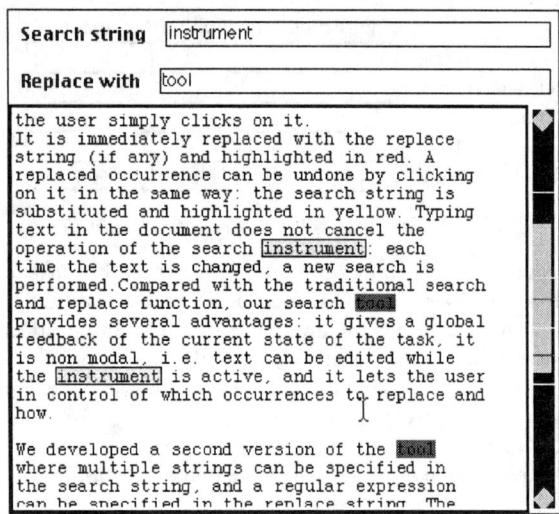

Figure 3: Search and replace instrument

An instrumental approach to search and replace has led to the following design, implemented in a Tcl/Tk prototype (Figure 3). The top part of the window is the logical part of the instrument, used to specify the search and replace strings. No buttons are necessary: as the user types a search string, the occurrences highlight in yellow both in the text window and in the scrollbar. To replace an occurrence, the user simply clicks on it, immediately replacing it with the replace string (if any) highlighted in red. Editing the replace string changes all the replaced occurrences. A replaced occurrence can be undone by clicking on it in the same way: the search string is substituted and highlighted in yellow. Typing text in the document does not cancel the operation of the search instrument: each time the text is changed, a new search is performed.

The scrollbar was modified to facilitate browsing. First, it includes tick marks representing the occurrences and giving an overview of the search. Second the arrow buttons were changed so that clicking on an arrow scrolls the document at a variable speed according to the distance between the cursor and the arrow: clicking on the up arrow scrolls the document slowly downwards (as in traditional scrollbars); moving the cursor up speeds up scrolling; moving the cursor down slows it down; moving the cursor further down, scrolling stops, then reverts. This reduces the division of attention that occurs when operating the various parts of a scrollbar while focusing on the document. It is similar in effect to the thumbwheel described earlier but does not require a separate input device.

The instrument provides feedback about the current state of the search/replace operation by highlighting *all* the occurrences in the text, as in the Document Lens [25], and in the scrollbar. This design results in a low degree of indirection: The spatial offset is reduced by getting rid of the traditional *Next*, *Previous* and *Replace* buttons; The temporal offset is reduced by displaying the occurrences as the user types the search string, by showing the effect of replacing an occurrence immediately and by allowing the user to undo any change, not necessarily in the order they were made.

The design of the scrollbar improves the degree of integration since it uses the vertical position of the mouse to control the speed and direction of scrolling (1 output DOF / 2 input DOFs) while traditional scrollbars only use the fact that the user clicked in the arrow (0 output DOF / 2 input DOFs). This design also improves the degree of compatibility since the mouse now works as a joystick. In effect, the scrollbar thumb and arrows are functionally equivalent: the thumb provides positional control while the arrows provide rate control of the visible part of the document.

Two variants of the search instrument were also developed. The first variant allows several search instruments to be active simultaneously, each independent of the others and using a different pair of colors to highlight occurrences. In the second variant, multiple strings can be specified in the search string and a regular expression can be specified in the replace string. The instrument highlights all the occurrences of all search strings at once. This variant was used to build the index of a book: a list of words to index was entered as a set of search strings. The replace string added the proper markup to include the occurrence in the index. Indexing the book became simply a matter of picking which occurrences to include in the index, taking advantage of the display of all occurrences to avoid putting in the index occurrences of the same word that are close together in the text. At any time it was possible to change the list of words in the index, the content of the text and the occurrences to index.

CONCLUSION AND FUTURE WORK

This article has introduced the Instrumental Interaction model, which generalizes and operationalizes Direct Manipulation. The model has been used to analyze WIMP interfaces as well as more recent interaction techniques and to design a new interface for searching and replacing text. This demonstrates the descriptive, comparative and generative power of the Instrumental Interaction model.

We are currently testing the model as we design a new graphical editor for Colored Petri Nets. The model is used both as a design guide and an evaluation tool to integrate existing interaction techniques and create new ones.

However further work is needed to develop the model in more detail and assess its limits. This requires a more thorough analysis of graphical interfaces and interaction techniques, the definition and evaluation of new properties, a taxonomy of interaction instruments, and an exploration of the design space defined by the model. Design principles are also needed to help designers create instruments and integrate them effectively into a user interface.

The other important area for future work is to make Instrumental Interaction useful not only to user interface designers but also to user interface developers by developing a user interface toolkit based on the model. This would support the adoption of novel interaction techniques in a wide range of applications and allow a shift from WIMP to post-WIMP interfaces.

ACKNOWLEDGMENTS

Many thanks to Wendy Mackay for numerous discussions about the model and its implications and for her help with the English. Thanks to Peter Bøgh Andersen, Susanne Bødker, Yves Guiard and Austin Henderson for discussions and feedback on earlier versions of this paper.

REFERENCES

1. Ahlberg, C., Williamson, C., Shneiderman, B. Dynamic Queries for Information Exploration: An Implementation and Evaluation. In *Proc. ACM Human Factors in Computing Systems, CHI'92*, ACM Press, p.619-626, 1992.

2. Apple Computer. *Macintosh Human Interface Guidelines*, Addison-Wesley, 1992.

3. Bederson, B. and Hollan, J. Pad++: A Zooming Graphical Interface for Exploring Alternate Interface Physics. In *Proc. ACM Symposium on User Interface Software and Technology, UIST'94*, ACM Press, p.17-26, 1994

4. Bederson, B., Hollan, J., Druin, A., Stewart, J., Rogers, D., Proft, D. Local Tools : an Alternative to Tool Palettes. In *Proc. ACM Symposium on User Interface Software and Technology, UIST'94*, ACM Press, p.169-170, 1994.

5. Bier, E., Stone, M., Pier, K., Buxton, W., De Rose, T. Toolglass and Magic Lenses : the See-Through Interface. In *Proc. ACM SIGGRAPH*, p.73-80, 1993.

6. Bødker, S. *Through the Interface. A Human Activity Approach to User Interface Design.* Lawrence Erlbaum Associates, 1991.

7. Card, S.K., Mackinlay, J.D., Robertson, G.G. A Morphological Analysis of the Design Space of Input Devices. *ACM Trans. Information Systems*, 9(2):99-122, 1991.

8. Card, S., Mackinlay, J., Shneiderman, B. *Readings in Information Visualization: Using Vision to Think.* Morgan Kaufmann Publishers, 1998.

9. Card, S., Robertson, G., Mackinlay, J. The Information Visualizer, an Information Workspace. In *Proc. ACM Human Factors in Computing Systems, CHI'91*, ACM Press, p.181-187, 1991.

10. Fitzmaurice, G., Ishii, H., Buxton, W. Laying the Foundations for Graspable User Interfaces. In *Proc. ACM Human Factors in Computing Systems, CHI'95*, ACM Press, p.442-449, 1995.

11. Foley, J., Wallace, V., Chan, P. The Human Factors of Computer Graphics Interaction Techniques. *Computer Graphics and Applications*, 4(11):13-48, 1984.

12. Gram, C. & Cockton, G. *Design Principles for Interactive Software*, Chapman & Hall, 1996.

13. Guiard, Y. Asymmetric division of labor in human skilled bimanual action: The kinematic chain as a model. *Journal of Motor Behavior*, 19:486-517, 1987.

14. Heller, D., Ferguson, P.M., Brennan, D. *Motif Programming Manual*, O'Reilly & Associates, 1994.

15. Holland, S. & Oppenheim, D. Direct Combination. *Proc. ACM Human Factors in Computing Systems, CHI'99*, ACM Press, p.262-269, 1999.

16. Jacob, I. & Oliver, J. Evaluation of Techniques for Specifying 3D Rotations with a 2D Input Device. *Proc. HCI'95 Conference, People and Computers X*, p.63-76, 1995.

17. Jacob, R., Sibert, L., McFarlane, D., Preston Mullen, M. Integrability and Separability of Input Devices. *ACM Trans. Human Computer Interaction*, 1(1), p.3-26, 1994.

18. Krasner, G.E. & Pope, S.T. A Description of the Model-View-Controller User Interface Paradigm in the Smalltalk80 System. *J. Object Oriented Programming* 1(3):26-49, 1988.

19. Kurtenbach, G. & Buxton, W. User Learning and Performance with Marking Menus. *Proc. ACM Human Factors in Computing Systems, CHI'94*, ACM Press, p.258-264, 1994.

20. Kurtenbach, G., Fitzmaurice, G., Baudel, T., Buxton. W. The Design of a GUI Paradigm based on Tablets, Two-hands, and Transparency. In *Proc. ACM Human Factors in Computing Systems, CHI'97*, ACM Press, p.35-42, 1997.

21. Kurtenbach, G., Fitzmaurice, G.W., Owen, R.N., Baudel, T. The Hotbox: efficient access to a large number of menu-items. *Proc. ACM Human Factors in Computing Systems, CHI'99*, ACM Press, p.231-237, 1999.

22. Michotte, A. *La perception de la causalité.* Publications Universitaires de Louvain, 1946.

23. Myers, B.A. A New Model for Handling Input. *ACM Trans. Information Systems*, 8(3):289-320, 1990.

24. Myers, B.A. User Interface Software Tools. *ACM Trans. Computer-Human Interaction*, 2(1):64-103, 1995.

25. Robertson, G.G. & Mackinlay, J.D. The Document Lens Visualizing Information. *Proc. ACM Symposium on User Interface Software and Technology*, p.101-108, 1993.

26. Shneiderman, B. Direct Manipulation : a Step Beyond Programming Languages. *IEEE Computer*, 16(8), pp 57-69, 1983.

27. Smith, D., Irby, C., Kimball, R., Verplank, B., Harslem E. Designing the Star User Interface. *Byte*, 7(4), p.242:282, 1982.

28. Szekely, P., Luo, P., Neches, R. Beyond Interface Builders: Model-Based Interface Tools Model-Based. *Proc. ACM Human Factors in Computing Systems, INTERCHI'93*, ACM Press, p.383-390, 1993.

29. Took, R. Surface Interaction: A Paradigm and Model for Separating Application and Interface. *Proc. ACM Human Factors in Computing Systems, CHI'90*, ACM Press, p.35-42, 1990.

30. Wellner, P., Mackay, M., Gold R. Computer-Augmented Environments. *Special Issue of Communications of the ACM*, 23(7).

ANCHORED CONVERSATIONS: CHATTING IN THE CONTEXT OF A DOCUMENT

Elizabeth F. Churchill, Jonathan Trevor, Sara Bly, Les Nelson, Davor Cubranic[†]

FX Palo Alto Laboratory Inc.
3400 Hillview Avenue
Palo Alto, CA 94304, USA
{churchill, nelson, trevor}@pal.xerox.com

Sara Bly Consulting
24511 NW Moreland Road
North Plains, OR 97133, USA
sara_bly@acm.org

ABSTRACT
This paper describes an application-independent tool called Anchored Conversations that brings together text-based conversations and documents. The design of Anchored Conversations is based on our observations of the use of documents and text chats in collaborative settings. We observed that chat spaces support work conversations, but they do not allow the close integration of conversations with work documents that can be seen when people are working together face-to-face. Anchored Conversations directly addresses this problem by allowing text chats to be anchored into documents. Anchored Conversations also facilitates document sharing; accepting an invitation to an anchored conversation results in the document being automatically uploaded. In addition, Anchored Conversations provides support for review, catch-up and asynchronous communications through a database. In this paper we describe motivating fieldwork, the design of Anchored Conversations, a scenario of use, and some preliminary results from a user study.

Keywords
Text-based chat, sticky chats, collaboration, conversations, CSCW, shared documents, synchronous communication, asynchronous communication

INTRODUCTION
The past few years have seen the development of a number of computationally lightweight text-chat systems that support synchronous and asynchronous communications between individuals and groups. Examples include Internet Relay Chat (IRC) and AOL Instant Messenger [1]. Once chat software of this kind is installed, initiating contact with others is easy, taking only a few keystrokes. By creating 'buddy lists' of regular contacts, starting a chat is even simpler – selecting a user-name initiates contact. Therefore, ongoing and frequent informal interactions are easily supported. It is also possible to have multiple simultaneous conversations with different individuals and groups by opening more than one chat window [3,5].

In many chat spaces, conversations are ephemeral, lasting only while the current conversation or session is alive. However, some applications and systems allow text-based conversations to be recorded so that they have persistence. Such text logs enable asynchronous messaging between collaborators who are not online at the same time, and can also be used for catch-up and review [2, 3, 5, 6, 16]. Much recent work has also focused on providing more social activity cues in the interfaces to such chat environments [6, 16].

The success of these chat tools is not entirely surprising – people like to talk and maintain contact with others and chat systems support a form of quick, lightweight, informal communication. These are not just important for recreational activities - in the workplace, informal quick conversations and exchanges are important for maintaining the social fabric as well as for specific information exchange. Undoubtedly, the success of messenger applications and chat spaces is also due to their being computationally and cognitively lightweight: they take little time and effort to set up, little processing power to run, and little maintenance to keep going. Yet they support rich, informal, ongoing contact whenever users require. Most people don't even have to stray from their offices – everything happens on their desktops.

Problems With Chats At Work
Although chat tools are proving very useful in the workplace for maintaining contacts and exchanging information, they do not support the close integration of documents with ongoing conversations. Many work collaborations involve the sharing of documents: Web pages, presentations, spreadsheets and word processed documents. However, text conversations tend to occur in windows that are separate from, and have no relationship to, work-related resources. Figure 1 illustrates this point. In

---
[†]Current address: Dept. of Computer Science, University of British Columbia, 20-2366 Main Mall, Vancouver, BC, Canada. Email: cubranic@cs.ubc.ca

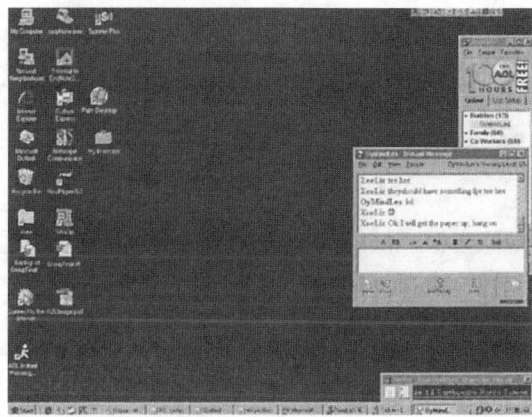 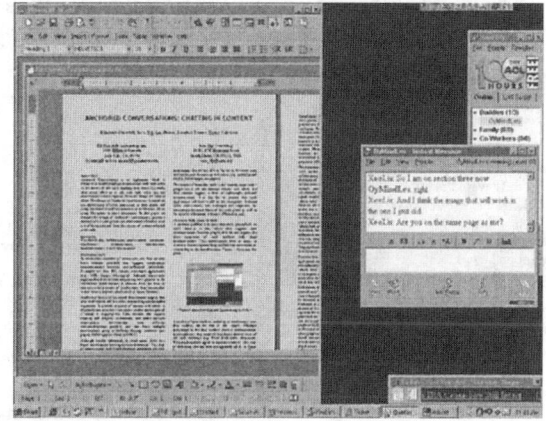

Figure 1. Chatting in messenger windows.
On the left a text chat is ongoing; on the right the word processor document that is being discussed has been opened on the desktop. Conversations and documents have no relationship

the first image a chat window is open on the desktop. In the second image, the chat window is shown alongside a word processor document that constitutes the subject of the conversation.

There are several problems with this arrangement. First sharing or copying the document that is to be discussed requires that users carry out extra steps using other applications (e.g. ftp the file, email an attachment, browse a shared repository, set up a shared window, etc). Secondly, when documents are sent or retrieved, issues arise around ensuring that all collaborators have the *same* document. Thirdly, once all collaborators have copies of the same document, establishing shared reference by navigating to the same portion of the document can be time consuming. Discussions about the content of the document require that navigation statements be typed into the chat window ("Look at section three, paragraph 2") or that the relevant materials from the document be copied and pasted into the text-chat window. In the case of shared windows, navigation cues in the form of shared pointers may be available, but shared windows tend to result in restrictions on what users can do with the content of the file itself and do not support asynchronous communications.

What we know about many collaboration situations is at odds with the technology affordances in this instance. Collaborators who are collocated and working in a "tightly coupled" way over those documents often converse about the details of documents: lines of text, figures or cells in a spreadsheet. Conversations in such collaborations have been characterized as "object laden" [7]; there is a high level of focus on the object or artifact that is under discussion and/or that is being co-constructed. Discussions are facilitated by (1) knowing which section is currently under discussion, (2) seeing the broader context in which the section is located, and (3) negotiating a shared visions of what is required and who will carry it out. Such conversations, therefore, crucially depend on shared context, i.e. that collaborators all have visual access to the artifact or object under construction or discussion. Here, "the object leads and language follows" [7].

FIELD WORK: CONVERSATIONS OVER WORK DOCUMENTS

Our intuitions about work collaborations were further developed in a series of field observations and interviews [3]. Our analyses focused on tightly-coupled collaborative work between non-collocated colleagues in a number of domains including software research, nuclear fusion experiments, geology field studies and after-sales software support. In each of these domains we noted issues arising around the sharing of artifacts.

In the first domain, software research, collaborators were using a text-based MUD to keep in touch and work together. Here, the tendency was to share files through email or by consulting shared file repositories. Specific details were shared by pasting text into the chat window and giving navigation cues like "Section 3, line 4" [5]. As well as being somewhat unwieldy and involving many steps, this copy-and-paste practice had the side-effect of taking the pasted-in material out of its local context. In the second domain, nuclear fusion experiments, problems arose around the sharing of experimental results in graphical and textual form. In the third domain, geology field experiments, we noted the need to support discussion around numerical data in a spreadsheet and graphs. In the final domain, after-sales software support, we noted the need to support collaborative problem solving over on-line software manuals.

In each of these cases problems arose around the establishment of shared context for conversations. The communication media (e.g. email, telephone and text-chat) were separate from documents under discussion and explicit navigation cues were required to literally "get everyone on the same page".

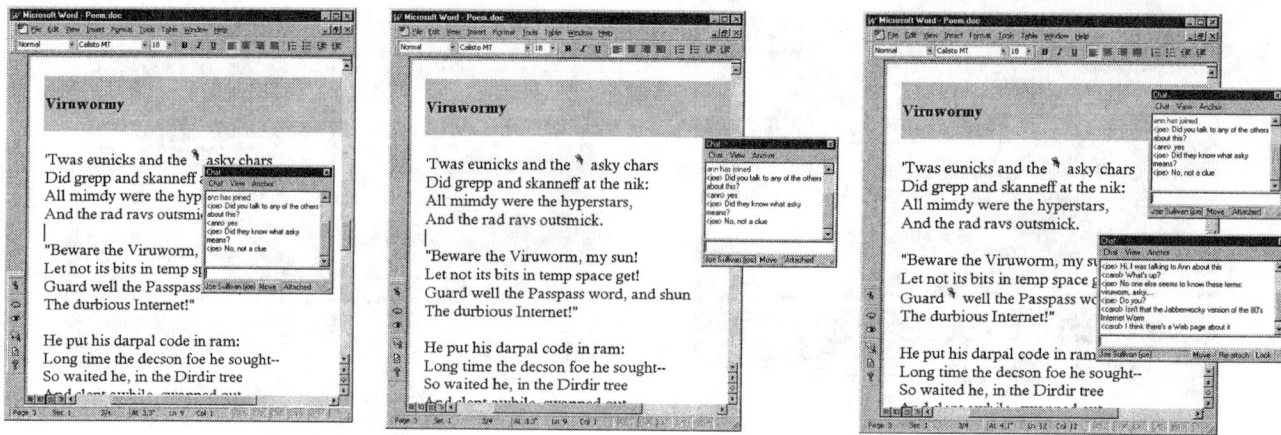

Figure 2: Chatting over a document using Anchored Conversations
Chat windows are anchored into word processor documents. Anchor points are represented as pushpins. There are no restrictions on the number of chat windows that can be anchored into a document.

ANCHORED CONVERSATIONS

The Anchored Conversations tool directly addresses the issue of coupling conversations and work artifacts when working remotely. Like chat spaces, the Anchored Conversations tool is computationally and cognitively lightweight, supporting synchronous and asynchronous communications. However, Anchored Conversations also allows text-based conversations to be coupled with the documents that provide the context for work discussions (e.g. word processor documents, presentation files, spreadsheet files, graphical simulations, etc). The user model is that of "sticky chats" - adhesive chat windows that can be stuck to documents for in-context conversations, much as a sticky note can be affixed to a printed document.

Although our examples in the following sections are all grounded in word processor documents, Anchored Conversations is application independent. Sticky chat windows can be placed into any application document, and can be moved between different application documents, just as a sticky note may be moved from one type of printed document to another.

Below we detail the features of Anchored Conversations. First we describe the "sticky chat" windows in which conversations take place. Then we describe how the Anchored Conversations tool supports document transport and sharing. Here we also describe how Anchored Conversations solves the problem of establishing and maintaining shared context.

Chat windows in documents

In Figure 2 we show the use of Anchored Conversations to support synchronous discussions over a text document. In the first image at the left of Figure 2, a text editor is shown with a document displayed. In the document is an anchored chat space, a "sticky chat" window. This is a standard chat space having a space for typing text, a window for viewing the ongoing conversation and a scroll bar at the right for viewing things that have already been said. The window can be resized easily. Again, as is standard, people's names are shown in the angled brackets at the left of the chat window.

The small pushpin icon to the top left of the chat window before the word "asky" indicates the location at which the chat is anchored - or, to put it another way, the context for the conversation. The chat space is literally *anchored* to the point where the pushpin is inserted. If the user scrolls the document, the chat window scrolls with it. It may even go off screen. On scrolling back, the sticky chat window will reappear, still attached to its anchor point. Similarly, if the user drags the application window around the desktop, the sticky chat window will stick to the application window at that location and move around with it.

Sticky chat windows may be stuck to the anchor as described above, stuck near the anchor or detached from the anchor. Thus, a sticky chat window's spatial location in relation to the anchor can be altered to prevent occlusion of the material in the body of the document.

In the second image in the center of Figure 2, the sticky chat window is shown having been moved to the right of its anchor point in the document in order to uncover the previously occluded text. The sticky chat window has been "locked" into this new location. The chat window is still attached to its anchor and will remain in this spatial relation to the anchor unless moved and again "locked" to a new location. So, if the user scrolls the document or moves the application window, the sticky chat will move as before.

Sticky chat windows can also be detached; in the third image at the right of Figure 2, two sticky chat windows are shown. The lower window in the image at the right of Figure 2 is detached. The toggle at the bottom of the chat window indicates this, as the user has the option to "reattach". When a window is detached, if the user scrolls the document or moves the application window, the chat window will not move with it. However, the sticky chat does not "forget" its "home" or context anchor point. By simply clicking on the 'Reattach' toggle at the bottom of the sticky chat window, the chat window reattaches to its "home" anchor location.

As shown at the right of Figure 2, users can have as many sticky chat windows anchored in the document as they wish. These sticky chats could be separate conversations between different groups of people or could be conversations that are separated due to theme or topic.

It is also possible to move a sticky chat window's anchor location by placing the cursor, and then selecting and clicking on the "Move" toggle at the bottom of the chat window. The new anchor location is set and the anchor is moved to a new place in the document. The sticky chat window is now anchored at this new location. When sticky chats are moved in one document, this change propagates: if I move a sticky chat to a new location, you will see the change and your chat will move too. Sticky chats can also be moved *between* documents. This is discussed below.

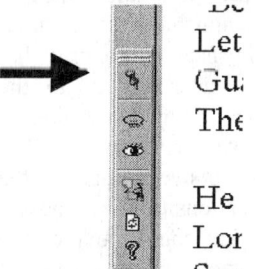

Figure 3 Pushpin Anchor Menu Item The Anchored Conversations tool adds a menu down the left side of applications. The pushpin menu item allows sticky chats to be attached to documents.

Starting conversations and sharing documents
The Anchored Conversations tool user interface consists of four parts: (1) the sticky chat window described above; (2) the pushpin that represents the point at which a sticky chat is anchored; (3) a menu bar that is added to applications and that appears down the left side of application windows (shown in Figure 3); and (4) a conversation coordinator window which interfaces to the Anchored Conversations database (shown in Figure 4).

Starting an anchored conversation requires few steps. By selecting the push-pin from the menu (shown in Figure 3) and clicking, an anchor is placed into the document at the current cursor point. Clicking on the push-pin in the document results in a sticky chat window appearing. Once the window is open, names can be selected from a "buddy" list in the menu bar. Selections result in invitations being sent automatically to invitees' desktops. This model is the same as for most messenger services. Also in accord with messenger services, once invitations are accepted, the invitees join the chat.

However, Anchored Conversations differs from other chat applications in that, on accepting an invitation to converse over a document, that document is *automatically* transported to the desktop of the invitee. The document itself opens automatically and is scrolled to the location where the sticky chat window is anchored. The sticky chat window is open and IRC available. Each person has his/her own copy of the document at this point; the sticky chat window is shared.

Anchored Conversations can also act as a simple chat application. If a chat window is not anchored within a document when created using the Conversation Coordinator, invitations result in a conversation client being opened on the desktop much as a standard chat client would. However, a chat can then be attached to a document and that document will be sent to others in the conversation.

Logging conversations and contexts
All conversations are logged in the sticky chat database (called the Conversation Database), and are available for review at any time – irrespective of whether the associated document is open. As all conversations are logged as text, searching and displaying by theme or person is trivial.

Conversation contexts are also logged. Sticky chats are not simply embedded chat objects. The "anchors" store information about the local context for the conversation. This context may be words that are nearest to the anchor, or cells in a spreadsheet – the specifics of context depend on the application in which the sticky chat is anchored. Context information of this kind is stored in the database along with information about the creation time and date of the anchor. All previous context locations in which the anchor has been inserted are also still available.

Users may query the database to see, for example, if an anchored chat has ever been located in a different place. All actions and conversations that take place within an anchored chat window are retained in the database, and are available for search, review and reuse. Even if the document in which the sticky chat was located has been deleted, it is still possible to review the conversations that were associated with the file – the filename and local context that the anchor was originally in are also preserved. Thus users can recreate conversation contexts if decisions need to be reconsidered. There is evidence that people review in this way occasionally [3].

In Figure 4 we show our current interface to the database. Information about different active chats is shown (2 chats and their contexts are shown on the right of the window) as well as a log of all the current chats is shown on the left. By selecting the chats on the right, the user can navigate to a chat window in a document. We thus support our fourth goal, to provide review facilities for recreating conversations in context.

Scenario of use
In this section we present a short scenario to illustrate the use of Anchored Conversations in a distributed, collaborative work context.

Carol, Ann and Joe are colleagues working closely together on the collection and analysis of field data. They are currently preparing an executive summary of their recent work to distribute to the company management. Ann and

Joe work in the same geographic location but Carol works 4000 miles away with a 6-hour time difference.

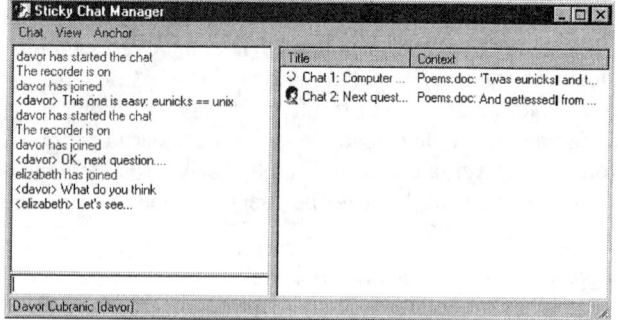

Figure 4. Interface to database. The area on the left displays the text chat and the area on the right lists the sticky chats that are currently open, their names and their current contexts. The icon beside the sticky chat name shows which chats are currently active.

Last week they had a telephone conference call so that they would all have a chance to agree on the general direction of their executive summary. In addition, they have been emailing messages back and forth daily. Today, Carol is working to incorporate her analysis into the draft document that Joe sent yesterday. She likes the graph showing the data points that they've chosen to use but thinks there is an additional interpretation that should be included based on the original data. However, it will require a substantial change to the summary graph and she's not sure how to best to represent her concern. She is anxious to interact with Ann and Joe about the data and the analysis. She has already made several modifications to the document itself. She has tried to summarize her concerns about the data in an email message. However, she wants to have a conversation about it. Although she knows she could call, Joe typically works at his home in the mornings while Ann will be in the company office. They would have to arrange for a conference phone call and still might have trouble knowing exactly what point in the original data was troubling her. She decides to use the Anchored Conversations tool. In the paragraph explaining the data, she inserts a sticky chat window and invites Ann and Joe. She knows they'll be getting into work soon and they'll see the invitation to chat with her. She then continues to edit other sections of the document, inserting and carefully positioning a second anchored conversation to explain why she reorganized the Related Work section.

As soon as Carol sees activity (she can see this both in the conversation coordinator window and in the sticky chat window itself), she returns to that section of the document. Joe is there and asks why she isn't happy with the data as presented. Carol immediately tells Joe she'll show him the issue with the original data and moves the chat window to the spreadsheet with all the original data. She anchors the conversation to the graph of data points that she thinks contradict the summary data representation. As she moves the sticky chat window, Joe's document automatically scrolls to the location Carol has selected and the sticky chat window in his document reanchored.

At this point, Ann joins in as well. She reviews the conversations that have taken place – seeing what was said and where it was said. She agrees with Carol's point and starts another sticky chat in a document where she had been trying various data representations. Neither Joe nor Carol has seen this document but copies open up for each of them on their respective desktops, and all the chats automatically follow. They continue to move around the data, the report document and their conversation, arguing and discussing the possibilities for their summary.

When Carol leaves for the day, she knows that the next morning she will have another go at the document -- and that Joe and Ann will have left conversations in critical places in the document to share their thinking with her as they continued to edit the report.

Anchored Conversations allows Joe, Ann, and Carol to work together on a shared document despite their differences in time and place. Carol was able to express her concerns about the data with a clear reference to the particular place in the data at issue. They were able to converse at the same time and to leave notes for each other at different times. Placing multiple conversations in the documents allowed different conversations to point directly to different topics. Also a conversation could move as the group discussion continued; it was not held to document boundaries or type. The anchored conversation serves as a means of annotation, of chat, of pointer in both real-time and at different times.

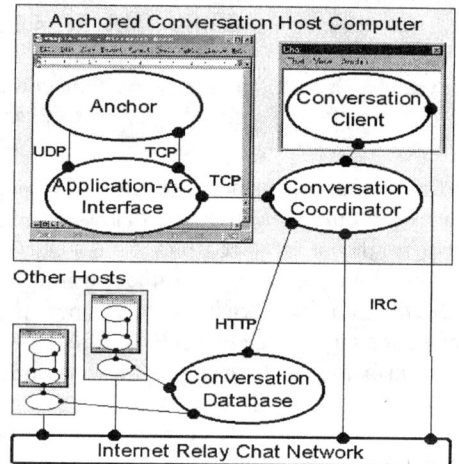

Figure 5 Architecture for Anchored Conversations

Implementation

The Anchored Conversations software design (Figure 5) consists of several components: (1) the anchor that locates a conversation within a document context; (2) the conversation client that provides the interface for one conversation; (3) the conversation coordinator that handles all conversations occurring on a host computer; (4) the

Application-Anchored Conversation interface that embeds the necessary functionality in a application required to enable conversations; (5) the Conversation Database that maintains access control and conversation history; and (6) the communication channel.

To facilitate document artifacts being accessed within their native applications (e.g. Word, PowerPoint, Photoshop) on a variety of host computers, the Anchored Conversations components are designed to be independently implemented and to communicate with each other through standard protocols (e.g. UDP, TCP/IP, HTTP). These protocols are selected for throughput and functionality needed for the different messaging requirements of the component interfaces.

The main conversational elements (Coordinator, Client, and Database) for the current prototype Anchored Conversations implementation are written in Java. The communication channel is IRC, which has been ported to a number of different systems. The Conversation Client is based on a slimmed down IRC client for text chat. Anchors are constructed using Microsoft's ActiveX technology, which enables embedding of anchors in a wide range of Microsoft Windows applications. In our study (described below), we used Microsoft Word 97, SR-2, running on Windows NT 4.0 and using different Pentium-type processors. The Word Application-AC Interface is implemented as a set of Word macros (Visual Basic for Applications, Version 5.0).

The detailed description of the software and flow of conversation and related document and control information starts with the anchors. Each anchor consists of an instance of an Anchored Conversations ActiveX control wrapped in a Word object (of type *InlineShape*). The control represents a specific point of conversation (by the pushpin graphic shown in Figure 2) and can determine and export its screen condition (location, focus, exposure etc.). Inserting an anchor thus causes a new Anchored Conversations control to be inserted within a context of the document. The context used in the current Word-based prototype to determine the conversation location consists of a range of text surrounding the anchor. Whenever the anchor object changes in visibility (by scrolling in the document) or moves (by window movement or by the user moving the anchor itself), UDP messages communicate these events. The Coordinator distributes anchor change events locally to the Client chat windows. The windows are positioned near to, and automatically follow their anchoring points within a document.

Significant events like anchor creation or deletion are distributed to other hosts using the Conversation Database. Each Coordinator maintains a local copy of the relevant parts of the database that gets updated as the system state changes. The chat client provides a user interface by which a conversation about some aspect of the artifact may be conducted. The text conversation messages are distributed from the chat clients to corresponding clients using the IRC network. IRC provides the basic machinery for invitation to a conversation. The Coordinator then copies the documents to the local host when an invitation is accepted and opens the document to the context of the inviting anchor.

There are a number of ways the software architecture permits the anchored conversation approach to be generalized beyond the current prototype including:

- The reference frame for anchor location may vary depending on the type of application and artifact (e.g., character position within a text file or [x,y] coordinate in a window, a range of text within a Word file).

- Anchor representation may include a bitmap image, text, or some combination of text and graphics. The representation may be static or may change in value based on information obtained directly from user input, from the Client, or other external sources.

- Clients may be associated with several anchors (multiple topic conversation) located in one or more artifacts. In the last case, one anchor can be considered its primary visual attachment to which the conversation homes.

Currently, we have implemented an open model of sharing: people who are invited to chat in any window associated with a document are able to access all other chats. However, more conservative, selective models of sharing can also be implemented. We have also sidestepped the issue of version control by providing users with a means of explicit resynching document versions. Prompts are given when anchor contexts across the open clients are out of synch and if users wish to synch up, they can click on a menu button to automatically update their documents to be in synch. Currently all documents synch to that of the original inviter.

RELATED WORK

The Anchored Conversations tool fits within the genre of lightweight communications technologies like chat spaces and messenger applications [1,8,14]. Unlike many of these applications Anchored Conversations retains logs of conversations and the contexts in which they took place. Further, chat windows in most applications cannot be associated with places (i.e. cannot be anchored into documents) and do not retain a history of their locations and usage in an integrated management tool. They cannot therefore be used for artifact/document navigation.

Anchored Conversations provide some of the functionality of collaborative document production tools like Quilt [9] that support collaborative document work. In addition, the storage of anchor context information is similar to that described for annotations in Quilt (e.g. creator, creation time, activity logs, notification and triggering mechanisms). In Quilt, discussion ideas in the form of annotations can be "attached directly to the relevant points in documents rather than indicated by indirect reference to those points". The authors argue that this "direct manipulation allows discussion threads to be followed more easily". However,

Quilt does not support synchronous communications in the context of the text being discussed. Anchored Conversations is primarily a conversation tool rather than a shared annotation tool.

Also similar is Microsoft's Office 2000 facility in Word which supports in document threaded conversations. Whilst this is close to our notion of in-context conversation, within document threaded discussions cannot be viewed independently of the document. By contrast, sticky chats can be accessed via the database as well as via the document in which they are anchored. Further, Anchored Conversations is application independent – sticky chats can be moved from text documents to graphical documents to spreadsheets.

There are other means of sharing contexts for conversations (e.g. 2D graphical MUDs [14] and 3D virtual worlds). Shared context is established by re-representing artifacts in virtual worlds. This process is time consuming and represents a barrier to quick and easy contact, communications and collaborations; task related spaces must be created before virtual meetings can take place. At the level of the design metaphor, Anchored Conversations inverts this model. In the Anchored Conversation model, instead of taking artifacts into the virtual place, the conversations are taken into the pre-existing work artifacts that are the focus of conversations. The pre-existing artifact becomes the environment, backdrop or periphery for the conversations.

Document sharing and conversations are also supported in work environment systems (e.g.[10,11,12,15]). Here, applications which are relevant to a work project are all "corralled" into a Web page or window-based workspace. Thus, artifacts of relevance and the chat window are all in the same desktop window space. The association that occurs, however, is at the application level – the chat window is associated with the work artifact applications *because* it is in the same workspace. These tools do not integrate the conversation or chat spaces with the text of the documents; sharing is at the level of files and folders.

USER STUDY

An informal user study was carried out using our Anchored Conversations prototype. The goal of the user study was to evaluate the basic features of Anchored Conversations, to begin to observe the ways in which people might appropriate the functionality, and to obtain feedback on possible applications of the software in people's working lives.

Six teams of three participants were asked to share a document and to answer questions regarding various parts of the document. The questions directed the conversation to specific pages, paragraphs, individual words, and picture content. One person was elected "team leader" and took responsibility for inviting the others to help with the task. This person took responsibility for recording answers to questions, and her/his document was considered to be the 'master' document. The task was intended to be one that would engage the group in the content of the document rather than focusing on prototype features and that would require use of the anchored conversation space in joint activity. All groups were given a demonstration and training instructions, then 5-10 minutes practicing with the system and then 30 minutes to answer the questions. After training, group members were separated into three different rooms so that their only communication was by using sticky chats. Their instructions were to answer each question in the conversation space, reaching agreement among the three of them. For most questions, there was no single right or best answer. A follow-up discussion with group members included questions about their overall experience using the prototype and whether or not they could imagine using such an application in their own work. All the participants said they enjoyed the exercise and all groups answered at least 3-5 questions. The "sticky chat" metaphor proved easy to grasp. A preliminary look at the data suggests that most groups chose to use multiple conversation spaces, that the participants were highly interactive and that the basic functionality was used without problems. In follow-up discussions, participants were enthusiastic about their experiences and provided numerous ideas for future design and use. One of our participants stated "This would be really good for online critiquing and discussion, like in the magazine Websites I use". More detailed analysis of the study will be forthcoming.

SUMMARY AND FUTURE WORK

In this paper we have described a tool, Anchored Conversations, which supports synchronous and asynchronous communications over documents. This application allows conversation places in the form of chat spaces to be associated with specific locations within work documents. The Anchored Conversations application and the document applications within which conversations can be anchored are associated by means of the preserved context information. Anchored Conversations windows, or sticky chat windows, are not embedded *within* documents – they are associated with locations in documents. Thus conversations may be accessed from within documents *and* from the Anchored Conversation coordinator window.

Anchored Conversations is both cognitively and computationally lightweight in the following ways:

- By taking conversations to documents, Anchored Conversations does not require re-representation in a virtual world before shared context can be established. Rather, Anchored Conversations provides a means of placing conversations within the already existent work contexts - documents.

- The Anchored Conversation tool set up shared contexts for interaction over documents automatically; users do not need to invoke other applications. Navigation to shared contexts is also automatic. Further, application control is not taken away and each collaborator has access to the application's full capabilities. There is also an easy mechanism for resynching documents if necessary.

- Anchored Conversations is easy to learn to use. As said, user's existing expertise and knowledge of how to use the document application remains relevant. Users are not placed into unfamiliar contexts to do work. By contrast, tools which re-render work artifacts require that users learn how to interact with those artifacts in new ways, and often familiar application types must be re-implemented.

- The Anchored Conversation tool is also low overhead as its use is uniform across different applications. Thus, inserting a sticky chat window into a spreadsheet requires the same actions as inserting one into a word processor document. Similarly, holding a discussion in a sticky chat window where an image provides the background (e.g. in Adobe PhotoShop) requires the same user actions as holding a discussion in a spreadsheet.

- Users can move between different documents and applications following conversation links. Following conversation links in this way results in the background documents being opened automatically, and circumvents the need for the user to search for the relevant documents to open. The Conversation Coordinator, not the user, handles the navigation to locations in documents in different applications and not the user.

There are a number of issues we are currently addressing in our work on Anchored Conversations. These include issues about sticky chat privacy and accessibility, implementation across different document types, database expansion, and shared editing of underlying documents. Our current model of document sharing is that, once invited to a single chat in a document, all other chats are available. A more conservative model may be appropriate with certain documents where security and privacy of conversations is of import.

We have successfully implemented a prototype that works with MS Word™ and are now facing the challenge of supporting sticky chat windows across different application types. Preserving context will then require appropriate representations in the database. We are also concerned to make the database more dynamic; we would like to support access to documents as well as chats via the database. Finally, Anchored Conversations could benefit from being hosted in underlying document applications that themselves provide shared editing and collaborative use.

ACKNOWLEDGEMENTS

The poem shown in Figure 2 is Virmuwormy by Ed Combs. The authors would like to thank all our study participants, and also Joe Sullivan, Cathy Marshall and Bill Schilit for their input to this project.

REFERENCES

1. AOL Instant messenger -- http://www.aol.com/aim/
2. Bradner, E., Kellog, W.A. and Erickson, T. The adoption and use of BABBLE: A field study of chat in the workplace. In Proceedings of ECSCW '99, September 1999.
3. Bly, S. and Churchill, E.F. Design Through Matchmaking: Technology is Search of Users. *interactions*, March–April 1999, 23-31
4. Churchill, E.F. and Bly, S. Virtual Environments at Work: Ongoing Use of MUDs in the Workplace. *Proceedings of WACC'99*, San Francisco, CA, USA. ACM Press, 1999.
5. Churchill, E.F. and Bly, S. It's all in the words: Supporting work activities with lightweight tools. To appear in Proceedings of Group'99, November 1999.
6. Erickson, T., Smith, D.N., Kellogg, W.A., Laff, M.R., Richards, J.T. and Bradner, E. Socially translucent systems: Social proxies, persistent conversation, and the design of 'BABBLE'. *Proceedings of CHI '99*, Pittsburgh, PA, USA, ACM, New York, 1999.
7. Fleming, D. Design talk: Constructing the Object in Studio Conversations. In Design Issues, Volume 14, Number 2, Summer, 1998.
8. ICQ -- http://www.mirabilis.com/
9. Leland, M.D.P., Fish, R.S. and Kraut, R.E. (1988) Collaborative Document Production Using Quilt. Proceedings of CSCW88, Portland, Oregon, USA, September 26-28, 1988.
10. Lotus Domino -- http://www.software.ibm.com/
11. Roseman, M. and Greenberg, S. TeamRooms: Network Places for Collaboration. In Proceedings of CSCW'96, Cambridge, MA, USA. ACM Publications.
12. Sohlenkamp, M. and Chwelos, G. Integrating Communication, Cooperaton and Awareness: The DIVA Virtual Office Environment. Proceedings of CSCW'94, Chapel Hill, NC, USA, 1994.
13. Stefik, M., Bobrow, D.G., Foster, G., Lanning, S. and Tatar, D. WYSIWIS revised: early experiences with multiuser interfaces. ACM Transactions on Information Systems, Vol. 5, No. 2 (April 1987), 147-167, 1987.
14. The Palace -- www.thepalace.com
15. TeamWave (http://www.teamwave.com/)
16. Viegas, F.B. and Donath, J.S. Chat Circles. Proceedings of CHI'99, Pittsburgh, PA, USA. ACM Press, 9-16, 1999.

The Social Life of Small Graphical Chat Spaces

Marc A. Smith, Shelly D. Farnham, and Steven M. Drucker
Microsoft Research
One Microsoft Way,
Redmond, WA 98052
+1 425 882 8080
{masmith, shellyf, sdrucker}@microsoft.com

ABSTRACT

This paper provides a unique quantitative analysis of the social dynamics of three chat rooms in the Microsoft V-Chat graphical chat system. Survey and behavioral data were used to study user experience and activity. 150 V-Chat participants completed a web-based survey, and data logs were collected from three V-Chat rooms over the course of 119 days. This data illustrates the usage patterns of graphical chat systems, and highlights the ways physical proxemics are translated into social interactions in online environments. V-Chat participants actively used gestures, avatars, and movement as part of their social interactions. Analyses of clustering patterns and movement data show that avatars were used to provide nonverbal cues similar to those found in face-to-face interactions. However, use of some graphical features, in particular gestures, declined as users became more experienced with the system. These findings have implications for the design and study of online interactive environments.

Keywords

Avatars, computer mediated communication, empirical analysis, graphical chat, log file analysis, online community, proxemics, social cyberspace, social interfaces, and virtual community.

INTRODUCTION

Text chats lack nonverbal cues that facilitate face-to-face conversations, such as gestures, physical distance, and direction of eye gaze. Graphical chats attempt to address these limitations by introducing surrogate representations for physical bodies and spaces [9, 7]. While a number of graphical chat systems have been created, little is known about the nature of social interaction in publicly accessible spaces [8, 10, 13].

What do people do in graphical chat spaces? Do they cluster together in patterns approximating those seen in face-to-face interaction? How are the graphical features used in concert with textual modes of interaction? Broadly, we want to investigate whether these spaces are *sociopetal*, drawing people together into interaction, or *sociofugal*, driving them apart and away from interaction with one another [4]. To address these questions we report the results of survey research and analyses of more than three months of log files gathered from within three rooms (Lobby, Lodge, and Red Den) in the Microsoft V-Chat graphical chat system [14].

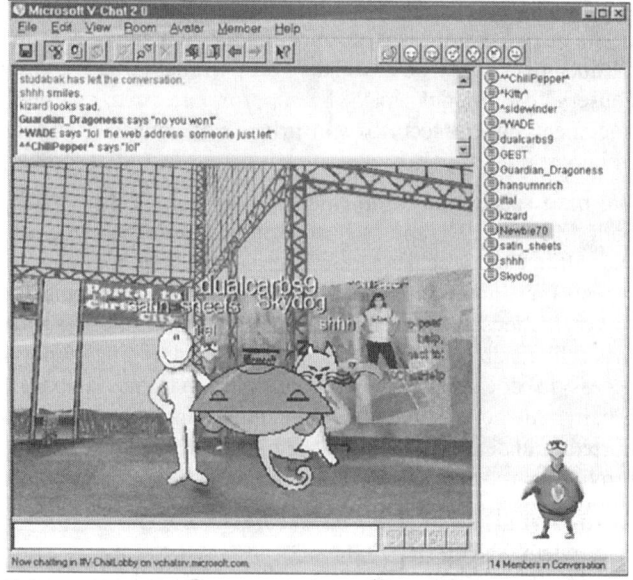

Figure 1. V-Chat interface includes a chat text box, chat history window, 3D space containing other avatars, room occupancy list, and an image of one's own avatar.

V-Chat clients connect to Internet Relay Chat (IRC) channels for communication transport. IRC is used to carry text chat as well as information about graphical events including avatar location and gestures. V-Chat provides a representation of each room as a 3D space, linked to a text chat window (Figure 1). Each space can contain up to 25 simultaneous Internet users. V-Chat allows users to puppet a graphical representation of themselves, an "avatar", in the 3D space. All users within the same room can see each other's messages (with the exception of "whispers" which are private point-to-point messages), irrespective of the distances between avatars. All avatars could also potentially see every other avatar depending on their line of sight. Traditional IRC users lack an avatar in the space, but appear in the user list and text box. People are able to select a standard avatar provided by the program, an avatar created by another user, or to create a custom avatar of their own. V-Chat avatars are represented by sprites, which have twenty frames, allowing them to communicate both direction of view in the 3D space and a series of gestures.

While V-Chat lacks object persistence, interactive objects, or user extensibility of the environment, it does implement many of the core features found in a broad range of graphical interaction tools. As such, an investigation of actual user behavior in V-Chat can shed significant light on the nature of social interaction in 3D virtual spaces.

Our investigation provides a longitudinal study of user behavior as well as analyses of user behaviors overall. These results lead back to central design and system management issues related to the development of 3D graphical environments for social interaction.

Our work follows the studies of physical social spaces pioneered by William H. Whyte [11, 12]. Whyte's studies highlighted the ways people moved through and came to rest in parks and plazas and how social interactions, from the casual to the intense, were shaped by design choices and the structure of the space.

We examined user behavior focusing on three issues: 1) general usage patterns of the chat room participants, 2) use of 3D features of V-Chat, and 3) contrasts between text only users and users of the 3D features of V-Chat.

METHODS

We address these issues by using both survey data and quantitative analyses of user behavior. While the survey data provides insight into the user's subjective experience, quantitative analyses provide a more objective representation of chat behavior. Such quantitative analyses are distinct from ethnographic studies, which take the form of direct observation of participant behavior and activity in the virtual space. While ethnographic studies provide valuable information about the content and meaning of social relationships, they have significant limitations [1,7]. Direct observation is labor intensive, misses many forms of interaction and patterns that are difficult to observe from a first person view, is subject to the biases of the observer, and often lacks broad context or duration.

Quantitative analyses of log file data provide a useful complement to such ethnographic studies. Collected logs of user activity can be used to produce a broad range of measures of the social structure and dynamics of interaction in the world. Combined with qualitative data, these measures can provide a broad backdrop for a multi-layered and complex picture of what really goes on in these graphical spaces. On their own, quantitative measures at least provide a possible basis for future comparison between varieties of graphical interaction systems.

For the present study we gathered data from three of the more popular V-Chat spaces, the "Lobby", "Lodge", and "Red Den", using a logbot. The data we report was gathered from 10/22/98 at 12:38:38 PST until 1/16/99 at 17:47:07, a total of 119 days. The bot had no avatar in the space but did show up in the user list (as "LogBert"). A sign was placed in every room being logged announcing the data collection and pointing to documents that described the project. These rooms were selected because they were the most active of all the rooms available from the public Microsoft V-Chat servers. The system did not require users to enter the any of these rooms in order to access others. Nonetheless, the "Lobby" was listed as a default choice in the V-Chat user interface.

The bot received the same information as all of the V-Chat clients; it added a time stamp and wrote the data to a set of files. Private communication between users provided by the whisper command was invisible to the logs we collected. Logs contained the following information for each V-Chat event:

TIME, DATE, NAME, ACTION, ARGUMENTS, X, Y, Z, Rotation

These logs were analyzed to generate a series of reports and graphs that profiled users, user sessions, and avatars. Log files were aggregated on the basis of the events and other world states to produce a range of behavioral measures.

We found that the data files were fairly noisy. The logbot was often disconnected from the server, introducing data dropouts and skewing login counts when it automatically logged back into the spaces. We found that the data sent to clients was noisy. Many users appeared without login events. Position data was fairly low resolution, providing coordinates of avatars in motion only once per second. The pattern of jumpy motion in the data is an artifact and does not reflect the user's experience of their own motion, but it does accurately reflect the way other user's motion was presented. Additional issues raised by the nature of the data are discussed below.

Survey data were collected from a self-selected sample of 150 V-Chat users. Respondents were recruited from within the V-Chat rooms using signs placed in the space with URL's pointing to the web-based survey. The survey asked for a broad rage of information, including demographic background, V-Chat usage patterns, and ratings of satisfaction with the V-Chat experience. These results offered a supplement to the log data.

RESULTS

General V-Chat Usage

35024 unique user names appeared in the three V-Chat rooms in the span of 119 days, averaging 5 chat sessions each. The average session length, the span of time beginning when the person arrived in a room and ending when the person left the room, was 6.6 minutes. 44% of the users logged in only once. Those who logged in more than once had an average of 8 sessions in the 119 days. Their session lengths averaged 6.4 minutes. 23.1% of the people were traditional IRC users, and 76.9% were V-Chat users.

Users were only identified by self-selected and non-persistent "handles" or user names. No email address, IP number or physical demographic data was available through the system. However, our survey data provides a picture of the basic demographic characteristics of the self-responding population. The average user was 29 years old, 72% male, and 28% female. 68% of all users had at least some college education. 45% of the users were single, 55% were not. Most of the users were from the United States or Canada (70%), and many of the remaining users were from Europe (17%).

An examination of the chat sessions shows that people tended to visit the rooms in the afternoon, from 2pm to 8pm, PST (or from 5pm to 11pm, EST) (Figure 2). While we were unable to determine the user's local time, most users are from the United States so they fall within the range of PST to EST. Afternoon use peaked sharply on Thursday afternoons, and dropped on Saturday afternoons.

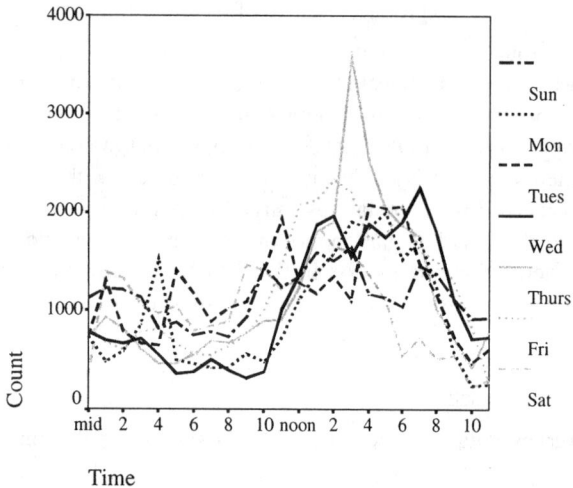

Figure 2. Count of chat sessions depending on time of day and day of week, Pacific Standard Time.

During each session, people posted an average of 3.4 messages. However, an unexpectedly large percentage of the people, 61.3%, posted no messages, observing others without participation. Session lengths were much shorter when users did not post any messages (3.1 minutes) than when they posted at least one (8.4 minutes). When people did speak, their utterances were fairly short, averaging 23 characters, or approximately 5 words.

Conversational openings were the most common form of exchange; an analysis of a subset of the data shows that out of 31,529 messages posted, 23% had some form of greeting in the text (e.g., "hello, hiya, what's up") and 4% had some form of goodbye in the text (e.g., "bye, brb"). 14% of the messages included the names of one of the others in the room.

Use of 3D V-Chat Features

Do people use the 3D features of graphical chats? If so, was that use sustained? It is important to consider the possibility that people might not use the 3D features at all, focusing for the most part on the text chat component of the program, or that people might use the 3D features initially for the sake of novelty, but use them less so as the novelty wore off. How were 3D features actually used as a component of social interactions? People might play with gestures and move around the 3D spaces without incorporating gestures and movement into their social interactions.

# sessions	# users	# gestures per minute	# positions per minute	% custom avatar
1	9165	0.57	5.9	21%
2 to 5	11105	0.53	5.2	25%
6 to 15	4548	0.37	4.6	41%
16 to 40	1517	0.35	3.3	62%
> 40	601	0.13	2.0	76%
Total	26936	0.49	5.2	31%

Table 1: Usage of 3D features by V-Chat users, broken down by user's number of sessions in 119 days

V-Chat users reported using both the text windows to chat with others, and the 3D features of V-Chat. In the survey, 76% of the people reported paying equal attention to both the text window and the graphic window, 14% mostly paid attention to the text, and 10% mostly looked at the graphics. However such self-report data provided to the V-Chat providers tends to be biased by both sampling concerns (perhaps only avid V-Chat users bothered to answer the questionnaire) and demand characteristics, where the respondents felt compelled to report using the 3D features out of a desire to be good subjects. We examined the log data to determine whether people used the 3D features, and whether they were used as a part of social interactions.

The three most prominent 3D features are the customizability of the avatars, the avatar gestures, and the position and orientation of the avatars. The following sections of the paper examine each of these features.

Avatars

People were able to either use one of 20 standard avatars provided by the V-Chat system, create one themselves, or use one created and made publicly available by another V-Chat user. A total of 1979 unique avatars were used, 99% of them custom made. V-Chat users wore a custom avatar for 45% of all the V-Chat sessions. Custom avatars ranged from simple, square photographs to complex cartoon-like characters. Overall, about 31% of the users wore a custom avatar at least once. According to the survey data, people reported using custom avatars to express their individuality (42%), stand out (24%), because they did not like the common avatars (23%) and for the challenge (11%). Two thirds of the people claimed they had avatars that represented their true gender.

Frequent users were much more likely than infrequent users to have used a custom avatar at least once (Table 1). People did not tend to change avatars during sessions. For 74% of all sessions, only one avatar was used. People used an average of 1.8 unique avatars, and each avatar was used for an average of 3.6 sessions.

Gestures

People were able to make their avatars perform one of seven gestures, representing angry, flirts, sad, shrugs, silly, smiles, and waves. As can be seen from Table 1, V-Chat users were on average using the avatar gestures .49 times per minute, or once every two minutes. Frequent users, or those who had visited the V-Chat rooms more than 15 times in 119 days, used fewer gestures: one every four to ten minutes. Given that the average session was less than 8 minutes, gestures do not appear to be a vital, sustained aspect of social interactions for the advanced users. As can be seen from Figure 3, the most common gestures were silly and waves, followed by flirts and smiles. It is important to note that when people make custom avatars, they can associate any image with the gesture buttons. The images they associate with the gestures are somewhat constrained, however, because the word appears in the chat window when the gesture button is clicked.

People may have used the silly gesture more frequently because there were three different randomly chosen sequences that represented silly, so silly provided a humorous surprise for both the user and the observer. Friendly and positive gestures (silly, smiles, waves, flirt) far outweigh (81%) conflictual or non-committal gestures (shrug, sad, angry).

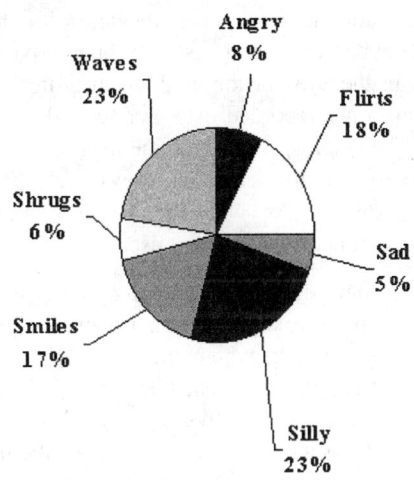

Figure 3: Breakdown of gestures used by V-chat users.

Positioning

Proxemics is the study of animal territoriality [4]. All animals, including humans, exhibit some form of territoriality. Some engage in direct physical contact with many others. Others, like humans, are predominantly non-contact species. Many people make an effort to ensure a certain space and distance is maintained around them.

Can the same proxemics be observed in graphical virtual environments as in physical spaces? That is, do people cluster together when interacting in graphical space much as they would in face-to-face interactions? Or is the graphical component ignored? How much do people orient to one another face-to-face? Do they maintain territorial buffers around themselves? If so, how does it compare in size to those seen in physical relationships?

An overhead perspective of the 3D graphical space provides a means for visualizing the proxemics of social interactions. We plotted the location of users as they moved through the V-Chat space (Figure 4). An arrow indicated the direction of each avatar's gaze. Reviewing these highlighted the movement of users into orientations that resembled conversation circles.

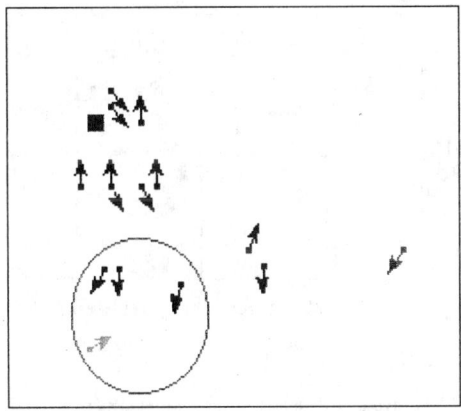

Figure 4. Top down view of the proximity and orientation of V-Chat users.

People were able to move their avatars with the use of either the keyboard or a mouse. While movement was continuous in the eyes of the user, changes in the avatar's position were only recorded once per second. As can be seen from Table 1, people had an average of 5.2 new positions every minute, indicating they spent about 8% of their time moving. As with the gestures, the rate of positioning is reduced for frequent users.

It is possible that people were moving simply to get from one end of the room to another, rather than to approach and look at the people with whom they are talking. To test whether or not people approached and looked at the people with whom they conversed, we needed to know who the target of their message was. We determined the target of a message by examining the content of the message for the name of the other users in the room. A subset of the log files from the main lobby from 12/15/1998 to 12/19/1998 was analyzed for the text content of the messages. In this period 1481 V-Chat users visited the lobby. For each person, there were an average of 20 other people co-present in the room. Messages were classified as being targeted or not targeted, depending on whether or not they contained the name of one of the other people in the room. A surprisingly large number of messages were targeted (13.8%).

For each person we calculated his or her average distance and orientation toward both targeted others and randomly selected others (selected from all of the people in the room at the time the targeted messages was produced). We calculated distances and angles of orientation using the position data provided by the logbot at the time of the message.

As can be seen from Figure 5, people were standing closer to their target than to a randomly selected other ($t(497) = 6.57, p < .001$). Nonetheless avatars kept some distance from targeted others, suggesting the maintenance of personal territories.

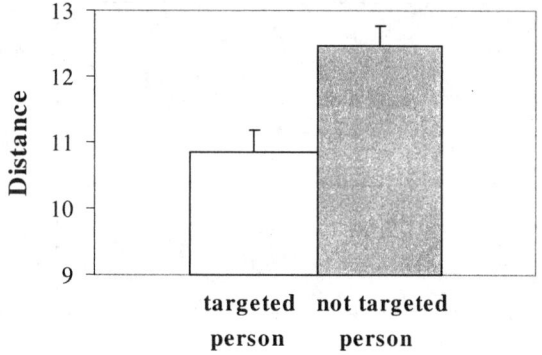

Figure 5: The average distance toward targeted persons and randomly selected non-targeted persons. Distances were measured adopting a map view of the V-Chat room using a 40 X 40 grid. People were standing closer to the people they were talking to. Note error bars represent standard errors.

Orientation toward others was calculated as the difference in angle between the vector defined by the line between the first person and second person, and the vector of the first person's gaze. As such, if a person was looking directly at another, the angle of orientation would be 0°, if the person were looking sideways relative to the other, the angle would be 90°, and if the person were looking in the opposite direction, the angle would be 180°. An examination of histograms of angle of orientation shows that people were generally not looking at randomly selected others, but rather sideways relative to randomly selected others (see Figure 6). Few people had their back turned to randomly selected others. However, people were prone towards looking toward the targets of their messages.

Randomly Selected Other **Targeted Other**

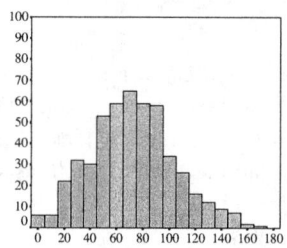 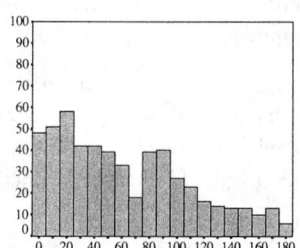

Figure 6: Histograms of people's angle of orientation relative to a randomly selected other, or relative to the target of a message. A person looking directly at another would have an angle of 0°, a person looking directly away from another would have an angle of 180°.

On average, people were more oriented toward targeted others than non targeted others (Ms = 63° and 72°, respectively, $t(496) = 4.17, p < .001$).

Just as people tended to be looking more toward a targeted other than a randomly selected other, targets were more prone to look back than were randomly selected others (Ms = 68° and 75°, $t(496) = 3.05, p < .005$).

In addition to testing whether people approached and looked at others in the 3D space, we wanted to test whether people moved their avatars during the course of their conversations, or only before and after their conversations. In other words, did people interleave chat messages and avatar movements? To measure the interleaving of chat and avatar movement we counted the frequency with which people moved their avatars in between any two messages. We found that on average, people moved their avatars in between 46% of their messages. Perhaps more importantly, the number of messages posted in a session did not affect this proportion. People moved in between messages as much for long conversations as short conversations.

These results suggest that people do appear to be using their avatars to do more than move from one end of the room to the other. They use their avatars to stand closer to people to whom they are talking, they look towards people to whom they are talking, and they frequently reposition their avatars during the course of their conversations.

Overall, V-Chat users appear to be using the 3D features of the program to reproduce the social conventions of physical proxemics.

People continued to use the 3D features over time, however the rate of gesture and positioning declined for frequent users. The reduction in the use of gestures and movement suggests that some initial use was due to the novelty, which then wore off. All users were prone to change their avatar on average once per session, and frequent users were more likely to have used a custom avatar at least once.

Contrasting Text-Only and Graphical Users

Some indication of the impact of the 3D features on social interactions is provided by the survey data. When asked in an open-ended question what they liked best about V-Chat, a full 20% of users said they liked making and seeing avatars the most. Only 4% liked gestures the most, and only 6% mentioned the ability to move around. People generally thought that V-Chat was a good place to make friends and meet people of the opposite sex. However, the survey data does not provide an objective indication of the impact that the 3D features had on people's interactions.

of participation level and type of user are significant ($\beta = 1.22, p < .0001$, and $\beta = 1.70, p < .0001$, respectively).

Although V-Chat users were more likely to return to the V-Chat space than IRC users, they did not spend more time on each session (Figure 7). For active chatters, V-Chat users spent 1.9 minutes less per session than IRC users. This difference is significant, ($t(19298) = 3.03, p, .001$).

Although V-Chat users spent slightly less time online per session than IRC users, they tended to return to the space more frequently. Over the period studied, V-Chat users frequented the space many more times than did IRC users ($t(34199) = 19.67, p < .001$), especially if they were active participants (the type of user by participation level interaction is significant, $t(34198) = 14.10, p < .001$). See Figure 8.

A comparison of traditional IRC users and V-Chat users indicates that V-Chat users were more likely to return to the V-Chat space than IRC users, and visited the space a greater number of times than the V-Chat users. However, the average duration of the V-Chat users sessions was almost two minutes less than that of the IRC users. It can be argued that return rates, number of sessions, and duration of sessions provide an indirect measure of quality of social interaction. However, IRC users may not be returning to the V-Chat space for reasons other than that of the quality

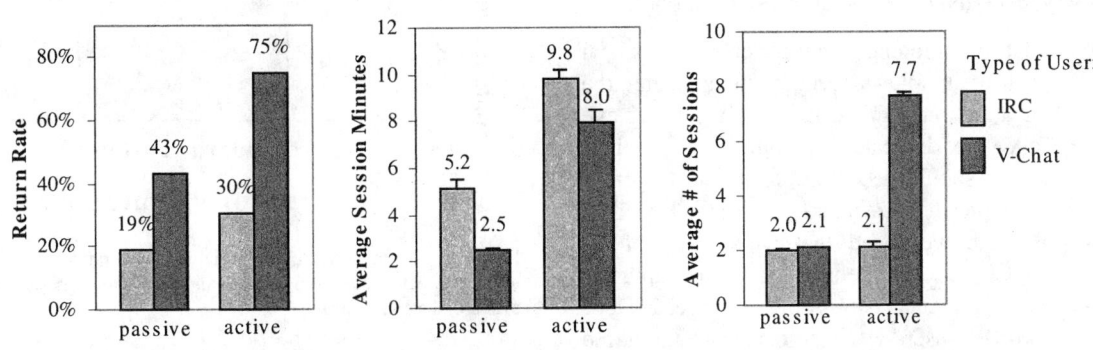

Figure 7: Rate of returns, average session length, and average # of sessions, depending on type of user and participation level.

One m only systems is the differential rate of return, length of stay and number of sessions. An important further contrast is that between active participants, who spoke at least once, and passive participants, who never spoke at all.

As mentioned earlier, a surprising number of people merely observe the space, visiting without ever saying anything (61.3%).

As can be seen from Figure 7, V-Chat users were much more likely to return to the space than conventional IRC users, especially if they actively participated in the conversation. A logistic regression with the interaction entered as a cross-product term shows that the main effects

ce. For example, they may simply feel like outsiders when they realize that many of the other users have bodies while they do not, and thus feel less inclined to return. Another possible measure of quality of social interaction might be provided by *quantity* of social interaction.

An examination of the number of messages per minute indicates that active IRC users tend to speak more than active V-Chat users (Table 2). (We focused on active V-Chat users because use of 3D features will not affect the quality of social interactions for people who only observe the space.)

Type of User	Messages per Minute	
	Mean	SD
IRC	3.37	8.12
V-Chat	0.78	1.41

Table 2: Means and standard deviations for messages posted, broken down by type of user. Only active users were included in the calculations

These results suggest that IRC users have a greater quantity of social interaction than V-Chat users. However, we were interested in whether the use of the 3D features directly affected the quantity of social interactions. As can be seen from Table 3, V-Chat people who used the 3D features at a greater rate posted more messages per minute. The rate of movement and the rate of avatar changes had the most substantial correlation with messages posted per minute.

Thus, while IRC users tend to exhibit more chat behaviors overall, V-Chat users who use the 3D features at a greater rate show higher levels of chat behaviors as well. However, given that these data are correlational in nature, we cannot make strong causal inferences. The use of 3D features may be increasing the quantity of messages, however the quantity of messages may in some way be increasing the usage of 3D features, or some third variable, such as general activity level, may be causing increases in both.

We argued that positioning would enhance social interactions because it allows people to indicate the direction of their attention. If V-Chat users are using eye gaze and distance to indicate the direction of their messages, then they should need to address the target of their message by name less frequently than standard IRC users. As predicted, we found that while 14% of all messages from V-Chat users were targeted by including the name of someone in the chat room in the message, 26% of all messages from IRC users were targeted with the name of someone in the chat room. A logistic regression indicates this difference is significant ($b = .79$, $p < .001$).

Use of 3D features	Messages per Minute
Gestures per minute	**0.22**
Positions per minute	**0.50**
Avatars per minute	**0.51**

Table 3: Correlations between use of 3D features and the messages posted for active V-Chat users. Correlation coefficients vary from –1 to 1, the greater the magnitude of the value the greater the correlation. All correlations are significant at the $p < .005$ level.

We also argued that avatars would enhance social interactions because people would be able to communicate information about themselves more effectively if they were able to represent themselves visually. Users reported feeling that they stood out more and were able to express themselves better if they had a custom avatar. If people are standing out more and expressing a richer presence if they have a custom avatar, then people should be looking at them more than if they do not have a custom avatar.

An examination of Figure 8 illustrates that randomly selected others were more likely to be looking at a person if he or she was wearing a custom avatar than if he or she was wearing a standard avatar. A within-subjects analysis shows the difference in others' orientation is highly significant ($t(727) = 7.99$, $p < .001$). That the same person receives more attention when he or she is wearing a custom avatar than when he or she is wearing a standard avatar suggests that the use of custom avatars significantly impacts the quality of people's social interactions.

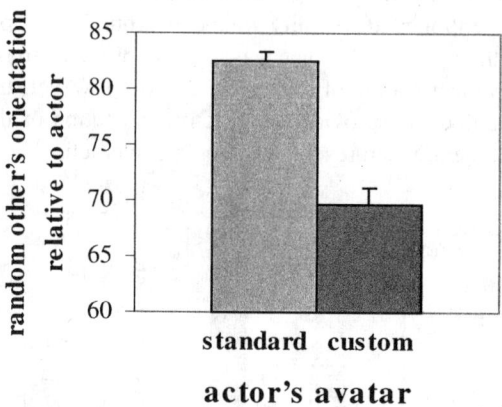

Figure 8: The average relative orientation of randomly selected others toward an actor, depending on whether the actor was wearing a custom avatar or a standard avatar.

CONCLUSIONS and DISCUSSION

Log file analysis of user behavior can illustrate the dynamics and structure of social cyberspaces. These spaces are novel environments for interaction that host familiar social norms and processes. The present research shows that people use the 3D features of graphical chat, however use of such 3D features tends to be reduced among frequent users. Spatial management of interaction occurs in a manner very similar to that in physical interactions, suggesting that proximity and orientation information are valuable additions to network interaction media. People tended to be standing near and looking toward those with whom they spoke. At the same time they maintained some personal space. A comparison of V-Chat users to IRC users showed that V-Chat users were more likely to return to the V-Chat space, returned more frequently, but did not stay as long. Traditional IRC users posted many more messages than V-Chat users. However, among V-Chat users, the use of 3D features correlated positively with the quantity of messages posted. V-Chat users tended to have fewer targeted messaged than traditional IRC users, suggesting that avatar positioning provided a nonverbal indication of attention similar to that found in face-to-face interactions. An examination of avatar usage indicates that people used about two distinct avatars across their sessions, that frequent users were more likely to have used custom avatars, and that when people used custom avatars, others were more likely to be looking at them.

The present research has several limitations. Many of the findings presented here are correlational. Further experimental studies that allow for tighter control of user conditions are necessary to draw any causal conclusions. The possibility that different people used the same names in different sessions is a very real one, as is the possibility that individuals used multiple user names in the same or different sessions. The invisibility of private interactions in the form of whispers resolved an ethical concern in the research but reduced our ability to gauge the volume of interaction and reduced the indicators of interaction ties between users. The present research compares traditional IRC users to V-Chat users, however the IRC users studied were those present in the V-Chat space. It would have been better to compare V-Chat users to IRC users who did not interact with V-Chat users. Future work should focus on contrasts between various graphical systems to explore the ways design decisions effect social interaction.

Despite these limitations, the present research does suggest that people use the 3D features of V-Chat and that the use of such features enhances social interactions. While 43% of the people who visited the V-Chat spaces did so only once, this rate is not out of line with the retention rates of many online systems. In addition, although frequent users were less likely to use some of the 3D features, even expert users continued to make use of proximity and orientation features to enhance their interactions in the space. V-Chat users did post significantly fewer messages than traditional IRC users, which may indicate that they found proxemics modes of communication sufficient to convey their intent to one another. Graphical representations, therefore, are used and may enhance social interaction in online spaces in many ways.

This research suggested important directions for future work. Producing the data set and analysis tools used in creating this research highlighted another important concept: many of the issues we were concerned with are of interest and value to the end user while in the midst of interaction. We came to think of this work and the data we generated as a form of a "social accounting" system. This system could track the number of sessions users have had in each space and how often they interacted with others. Future work will explore the effects of presenting such data in the user interfaces of such spaces in real time. We believe that social accounting data will add an important layer of context and history to online interaction environments that will improve their capacity to generate social cohesion.

ACKNOWLEDGMENTS
We thank Elizabeth Reid Steere and the Microsoft Research Virtual Worlds Group for their support of this research.

REFERENCES
1. Becker, Barbara and Gloria Mark, Social Conventions in Collaborative Virtual Environments, *Proceedings of Collaborative Virtual Environments 1998 (CVE'98)*, Manchester, UK, 17-19th June 1998.
2. Erickson, Thomas, et al. Social translucent systems: Social proxies, persistent conversation, and the design of babble. CHI 99.
3. Goffman, Erving Relations in Public
4. Hall, Edward Twitchell, 1990 The Hidden Dimension, New York: Anchor Books
5. Hill, Will and James D. Hollan. History Enriched Data Objects: Prototypes and Policy Issues, The Information Society, Volume 10, pp. 139-145.
6. Hill, Will and Loren Terveen, "Using Frequency-of-mention in public conversations for social filtering", unpublished manuscript. 1996. http://weblab.research.att.com//phoaks.cscw96.ps
7. Jeffrey, Phillip Personal Space in a Virtual Community, CHI 98, pp.347-348.
8. The Palace. http://www.thepalace.com
9. Viegas, Fernanda and Judith Donath, Chat Circles, Proceedings of CHI 99.
10. Virtual Places. http://www.vplaces.net
11. Whyte, William H. 1971 The Social Life of Small Urban Spaces, New York: Anchor Books
12. Whyte, William H., 1971 City: Rediscovering the Center, New York: Anchor Books
13. WorldsChat. http://www.worlds.net
14. Microsoft V-Chat. http://vchat.microsoft.com

The Effect of Communication Modality on Cooperation in Online Environments

Carlos Jensen[1]
carlosj@cc.gatech.edu

Shelly D. Farnham[2]
shellyf@microsoft.com

Steven M. Drucker[2]
sdrucker@microsoft.com

Peter Kollock[3]
kollock@ucla.edu

[1] College of Computing
Georgia Institute of Technology
Atlanta, GA 30332-0280 USA
+1-404-894-3152

[2] Microsoft Research
Microsoft Corporation
Redmond, WA 98052 USA
+1-425-882-8080

[3] Department of Sociology
University of California at Los Angeles
Los Angeles, CA 90095-1551 USA
+1-310-825-1313

ABSTRACT

One of the most robust findings in the sociological literature is the positive effect of communication on cooperation and trust. When individuals are able to communicate, cooperation increases significantly. How does the choice of communication modality influence this effect? We adapt the social dilemma research paradigm to quantitatively analyze different modes of communication. Using this method, we compare four forms of communication: no communication, text-chat, text-to-speech, and voice. We found statistically significant differences between different forms of communication, with the voice condition resulting in the highest levels of cooperation. Our results highlight the importance of striving towards the use of more immediate forms of communication in online environments, especially where trust and cooperation are essential. In addition, our research demonstrates the applicability of the social dilemma paradigm in testing the extent to which communication modalities promote the development of trust and cooperation.

Keywords

Computer Mediated Communication, Online Interaction, Social Interfaces, Collaboration, Social Dilemma, CSCW.

INTRODUCTION

One of the most consistent and robust findings in the sociological literature is the positive effect communication has on cooperation and trust [9]. When individuals are able to communicate, cooperation increases significantly. There are however many open questions. How does the choice of communication modality influence this effect, and how significant are the differences between different forms of communication? These are important questions for designers of collaborative online environments.

The research area we draw upon is the multi-disciplinary work on *social dilemmas*. Social dilemmas are situations in which a reasonable decision on an individual level leads to collective disaster, that is, a situation in which everyone is worse off than they might have been otherwise. Models of social dilemmas capture this tension between individual and collective outcomes, and can therefore be used as a very powerful and broadly applicable probe to assess the level of cooperation and trust in a group. Since the 1950s, a large research literature has developed in this area (for reviews, see [9, 10, 11]).

There has been surprisingly little work that has applied social dilemma models to online interaction. One recent exception is a paper by Rocco [12] which showed that face-to-face interactions increased cooperation in subsequent online interactions. Further, despite knowledge of the positive effects of communication on levels of cooperation, little is known about which aspects of communication and its medium are responsible for these effects.

There is a large literature of work examining several modes of communication and their relative effects on task performance [4, 5, 15, 16], especially in the CSCW and video-conferencing domains (for an overview see [6]). However, these studies have not specifically examined the effect of media on the development of trust and cooperation.

In most studies, communication has been used as a general term, without attempting to distinguish the various effects of different modalities. A recent survey [14] argued that the beneficial effects of communication are largely limited to verbal discussions. Written communication was not found to have a significant and consistent effect on

---

Permission to make digital or hard copies of all or part of this work for personal or classroom use is granted without fee provided that copies are not made or distributed for profit or commercial advantage and that copies bear this notice and the full citation on the first page. To copy otherwise, to republish, to post on servers or to redistribute to lists, requires prior specific permission and/or a fee.
CHI '2000 The Hague, Amsterdam
Copyright ACM 2000 1-58113-216-6/00/04...$5.00

cooperation levels in groups. However, the evidence in this review was indirect, and none of the studies discussed attempted to actually compare different communication modalities against each other.

Our hypothesis was that more immediate forms of communication (forms of communication producing a heightened sense of social presence, for instance face-to-face or voice) would prove more effective in promoting cooperation than less immediate forms such as text chat. The sociological literature supports this hypothesis, as do casual observations.

Technology coupled with bandwidth limitations have created a multitude of communication forms that are hard to evaluate. Is low quality video more conducive to cooperation than high-quality still images, and how do either compare to the use of 2D or 3D avatars? If any of the above were available with text-chat or text-to-speech (TTS), would the combination be more effective than voice communication alone?

Given a specific domain or application, the most appropriate form of communication can often be determined through traditional user testing, looking at such variables as task performance or user preference. Choosing the right communication modality is crucial, not only out of technical considerations such as bandwidth, but also to encourage and support the desired activities.

ICQ[*] and NetMeeting[†] are both examples of popular Internet communication tools. These two tools are designed for different audiences, and the choice of communication modality is an integral part of that design decision. ICQ supports lightweight, informal communication through a more anonymous and non-invasive text channel. NetMeeting on the other hand supports a more intimate form of communication through the use of voice and video.

The present research's examination of communication modality may also be applicable to the domain of electronic commerce which relies on establishing a trusting relationship between strangers. How can we make negotiation between prospective trading partners simpler and more transparent? Which forms of communication lend themselves well to the development of trust, and perhaps equally important, which do not? As consumers, it is also important for us to know which forms of communication make us overconfident, or vulnerable to deception.

The unique contribution of this study is to examine four computer mediated communication modalities in a carefully controlled experimental setting, and use quantitative models of cooperation and competition that are based on several decades of research. We examined the most common forms of online interaction: no communication, text-chat and voice. We also examined TTS communication to try to gain some insight into the differences between text and voice communication.

The results of this study are important both theoretically and practically. Theoretically, this investigation helps to explain the fundamental dynamics of cooperation in online settings. Practically, it may lead to concrete recommendations on the use of different communication modalities, and how limited bandwidth could be best used in order to encourage trust and cooperation.

METHODOLOGY

System Design

For this study we used a continuous, iterated, dyadic (two-person) Prisoner's Dilemma. The continuous nature of the game allowed for degrees of cooperation or defection to take place. By making the game iterative, relationships were allowed to evolve over time, negotiations to take place, and trends to develop.

Pre-existing relationships, no matter how brief, greatly influence future interactions, as demonstrated by the work done by Rocco [12]. Therefore, we decided to focus our efforts on studying the development of relationships, rather than how they change with the use of different mediums. We decided to pursue dyadic trials as opposed to larger groups for logistic reasons, organizing trials with two subjects is much simpler than doing so with four or five.

The Prisoner's Dilemma is the most commonly used 2-person social dilemma. The game captures the key tensions between individual and collective outcomes: There is a strong temptation to behave selfishly and exploit the partner, but both persons are hurt if they behave selfishly. What defines the Prisoner's Dilemma is the relative value of the four outcomes (see Table 1). The best possible outcome for an individual is defecting while the other player cooperates (termed DC). The next best outcome for the players as a group is mutual cooperation (CC) followed by mutual defection (DD). The worst outcome is the case when one cooperates while the other player defects (CD). Thus in a Prisoner's Dilemma: DC>CC>DD>CD.

		II C	II D
I	C	**2** , 2	**0** , 3
I	D	**3** , 0	**1** , 1

Table 1: Classic Prisoner's Dilemma

I and II designate Player I and II. C and D designate cooperation and defection respectively. Player I's outcomes are in bold. In our design, degrees of cooperation is possible.

---

[*] Available at http://www.icq.com

[†] Available at http://www.microsoft.com/windows/netmeeting/

During the course of the game, subjects were allowed to communicate with each other using one of four forms of communication; no communication, text-chat, TTS and voice (via speakerphone). All other factors were kept constant across the four cases. The game itself was built on top of an IRC-like communication channel. The other player's contribution (points given) was kept hidden until both finished whereupon both players were presented with the outcome for the current turn. A controller-bot kept track of the running score for each player, which was kept hidden until the end of the game. Communication between the players was not filtered in any way. In the text chat condition, the client was divided into two parts, one housing the game itself, and the other housing a standard chat interface, including a history of the subject's conversation.

TTS was implemented through the publicly available Microsoft Speech API 4.0[‡]. The voice used in all cases was the default, slightly feminine voice. While gender has been shown to produce biases, it was judged to be the most understandable voice by pre-test subjects. The interface for the TTS case was similar to that of the text case, except that the text messages and the history were hidden from the user. This was done to force the user to rely exclusively on the TTS technology. The subjects had their own messages read back to them in order to inspire greater confidence in the system. For the voice case, we had a speakerphone system in place and used only the game portion of the interface.

Game Rules

The particular type of Prisoner's Dilemma we used was a continuous version of the game in which *degrees* of cooperation were possible: Each turn the two players were allocated 10 points and given a choice of how many points (from 0-10) they wished to contribute to their partner. Any contributed points were doubled and given to the partner.

Therefore the situation has the structure of a Prisoner's Dilemma: The greatest possible return comes from keeping all of one's points while the partner contributes all 10 points (DC=30 points - the 10 original points plus the 20 points from the partner's doubled contribution). However, if both actors followed this strategy each will end up with only 10 points (DD - having contributed none to each other) rather than the 20 points each would receive if they both contributed all their points (CC).

In order to promote a high level of motivation (and risk), the subject's compensation was tied to their final score. Those attaining near perfect cooperation or those who consistently convinced the other player to contribute more than they did themselves, earned a piece of software of their choice. Lower scores resulted in less valuable prizes, culminating at the lowest score levels with a Frisbee or pen. Before the game, the subjects were informed that their performance was linked to their final reward. The subjects were not told how many points were needed to reach the different prize tiers.

The two players could see the amount each person had contributed on the previous round, as well as that rounds' outcome for both players. To minimize the effect of end-game conditions the subjects were told that the game would last for approximately 120 rounds. All the games ended after 96 rounds, giving the players a warning to this effect on turn 95. This was done in order to test for the presence of an end-game condition without tainting the rest of the sequence.

Subjects

The subjects were paired randomly and then randomly assigned to one of the four communication categories. They did not meet each other before, during, or after the experiment. Great care was taken to present the other subject in as neutral a language as possible, avoiding all terms such as partner or opponent. This was done in order to preserve the tension between the players without pitting them against each other. For the same reasons, we did not place any restrictions on topics of conversation. Subjects were allowed to negotiate freely or discuss their private life if they so chose.

A total of 90 adult subjects played the game using one of the four communication modalities. Our subject population was very diverse, the subjects ranged from 19 to 58 years of age, averaging at 40. The occupations of the subjects ranged from being retired, to engineers, police officers or students. Prior to running the experiments, the subjects completed a tutorial explaining the rules and how to use the program. As part of this tutorial they were asked to complete 4 questions in order to demonstrate their understanding of the game.

After the experiment, subjects filled out a questionnaire with standard questions to rate among other things their understanding of the rules and motivation. This was done in order to exclude the subjects who did not understand the game and those who did not take it seriously enough.

Two dyads were excluded because one member reported that they did not have a clear understanding of the game. Three other dyads were excluded when it became apparent from their communication that they had fundamentally misunderstood the game. All 3 of these pairs believed the equation CD+DC>CC+CC to be true. Every other turn they would alternate defection and cooperation, believing the average of this strategy to be better than continued mutual cooperation. This elaborate strategy required tremendous coordination between the two subjects, as well as demonstrating full cooperation and trust. Unfortunately, this strategy resulted in a less than optimal score, which

---

[‡] Available for from http://www.microsoft.com/iit/

would have unduly skewed the average scores if not removed.

Four dyads were excluded because one member reported that they were not motivated to earn as many points as possible. Three additional dyads were excluded from the analysis because of their refusal to use the communication modality offered to them. All told, 12 dyads were excluded from the study.

The present study had an unexpectedly high exclusion rate. While we expected to encounter both lack of motivation, and lack of understanding of the rules, we were surprised by the high percentage of our participants who did not understand the rules. Most Prisoner's Dilemma studies use college students for their study population, a fairly homogenous and intelligent segment of the population. As stated earlier, our population was much more diverse in terms of age and background. For many of our participants, the task was too abstract. They were simply intimidated and confused by the mathematical nature of the task. They could complete the pre-game quiz with the support of the tutorial, but were unable to use this knowledge in the game.

After excluding the invalid data, we were left with a total of 66 subjects, or 33 dyads: 9 dyads in the non-communication case, 9 dyads in the text chat case, 7 dyads in the TTS case, and 8 dyads in the voice case.

RESULTS

We expected that the form of communication would have a significant effect on the level of cooperation between the dyadic pairs. The more immediate forms of communication were expected to be more conducive to the development of trust and cooperation. A dyadic pair exhibited the highest degree of cooperation (CC) by exchanging all 10 of their points every round. Table 2 and Figure 1 illustrate that form of communication does affect average contributions as expected. An omnibus test of condition, using a between-subjects analysis of variance (ANOVA), shows that type of communication had a statistically significant effect ($F(3, 29) = 3.42, p < .04$).

Mode of Communication	Dyad N	Mean Contribution	Standard Deviation
A. None	9	5.3^{cd}	4.2
B. Text Chat	9	6.4^{d}	3.5
C. Text-to-Speech	7	8.4^{ad}	1.4
D. Voice	8	9.4^{abc}	0.2

Table 2: Mean dyadic contribution as a function of the mode of communication. A mean of 10 would indicate perfect cooperation between the dyadic pairs. Mean subscripts (abcd) indicate which other means are at least marginally significantly different ($p < .09$).

Dyads in the voice condition showed the greatest levels of cooperation. Most dyads using voice communication exhibited almost perfect cooperation. Planned comparisons show that dyads in the voice condition on average contributed significantly more than dyads in the text condition ($F(1, 15) = 5.82, p < .03$) and the no communication condition ($F(1, 15) = 7.69, p < .02$). The difference in average contribution between the voice condition and the TTS condition was marginally significant ($F(1, 13) = 3.50, p < .09$).

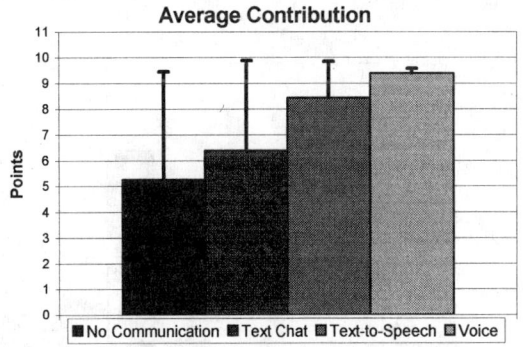

Figure 1: Mean dyadic contribution as a function of mode of communication. A mean of 10 would indicate perfect cooperation between the dyadic pairs. Error bars represent standard deviations.

Dyads also tended to be more cooperative in the TTS condition than in the text and the no communication condition. Dyads in the TTS condition contributed marginally significantly more than the dyads with no communication ($F(1, 14) = 3.66, p < .08$). The difference in contributions between TTS and text conditions was not significant ($F(1, 14) = 2.13, p < .17$), however the trend is in the predicted direction and could reach statistical significance with a larger number of dyads. That people showed greater levels of cooperation in the TTS condition than in the text conditions indicates that voice affects cooperation for reasons other than the differences in the semantic content of text versus speech, and for reasons other than the nonverbal information communicated through personal voice, such as intonation and gender.

While dyads in the text condition on average contributed more than dyads with no communication, the difference was not significant ($F(1, 16) = .37, ns$). Dyads showed a much higher variability in the level of cooperation in the text condition and the no communication conditions than in the TTS or the voice conditions. A better understanding of the nature of the greater variability in the no communication and the text chat condition can be gained through an examination of histograms of the scores as seen in Figure 2. In the no communication condition, people tended to fall into either a pattern of no cooperation, or complete cooperation. People in the chat condition showed a range of levels of cooperation. In the TTS condition and the voice condition people tended to be at least somewhat cooperative.

In the no communication condition, the bi-modal distribution was often a result of the players' first few turns in the game. Players that trusted each other from the beginning (giving large sums from the start) generally continued cooperating throughout the game. However, when players held back in the beginning, their level of cooperation deteriorated quickly. While players tested each other across all cases by holding back occasionally, this behavior tended to deteriorate much more quickly in the no communication condition.

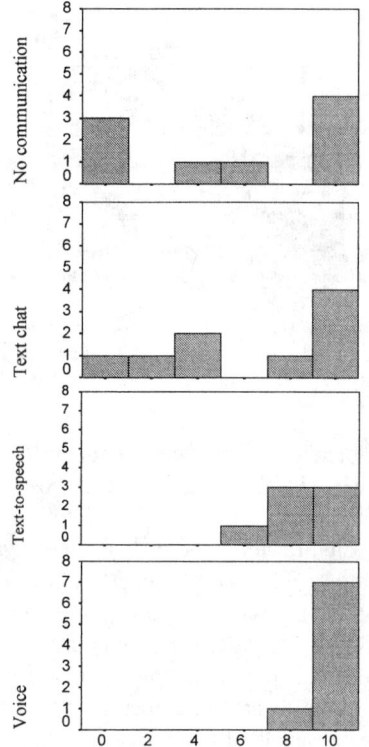

Figure 2: Distribution of average contributions for each mode of communication.

An analysis of contributions over time (see Figures 3 and 4) suggests that subjects did not initially cooperate fully with each other, but in conditions that allowed communication achieved a degree of cooperation that was greater than their initial level. A multivariate analysis of variance on 19 averaged blocks (see Figure 4), shows that there was a significant quadratic trend ($F(1, 29) = 5.14$, $p < .03$). Figure 4 shows that the quadratic trend has the shape of an inverted U, and an analysis of the trends shows that the shape of the trend was not affected by condition ($F(3, 29) = .46$, ns). This indicates that subjects tend to build trust early in the game, but that said trust tends to deteriorate over time, across all conditions. Again, this is consistent with what is found in the literature [2, 9, 12]. An examination of Figure 3 suggests that there was an end-game effect, replicating past research [9] that shows that cooperation will drop at the end of the game in a selfish attempt to maximize the final score. However, a comparison of the last round of points and the preceding rounds of points shows that this difference is not statistically significant.

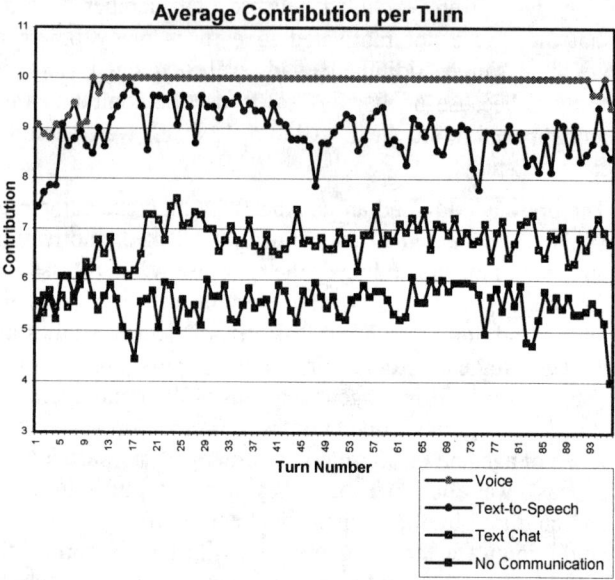

Figure 3: Average contribution across game rounds broken down by mode of communication.

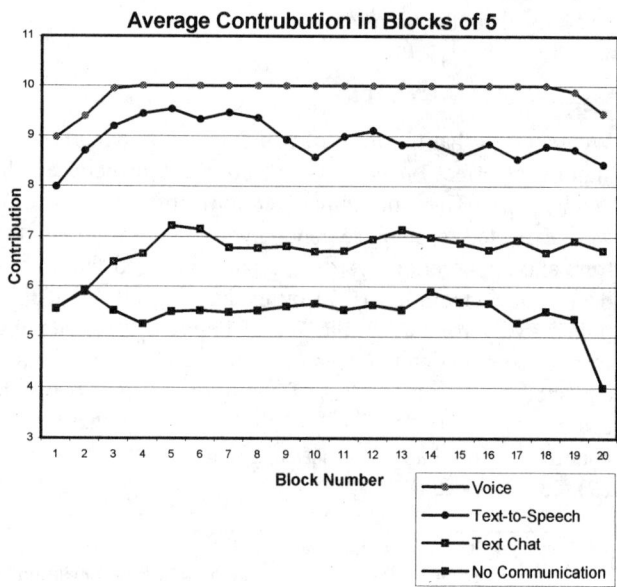

Figure 4: Average contribution across game rounds in blocks of five, broken down by mode of communication.

During the course of the game the subjects tended to vary their contributions across rounds (see Figure 3). The mean standard deviation for each dyad was 1.13, indicating that dyads tended to vary their contributions across rounds by adding or subtracting a point or two from their average.

We argued that people achieved greater levels of cooperation in the voice and text conditions because they were able to achieve greater levels of trust. To address whether people subjectively felt more trusting of the other across conditions, subjects were asked a series of questions in a post-experiment questionnaire. Subjects were asked to rate the other player on characteristics related to likeability (likable, kind, friendly, and warm), trustworthiness (honest, fair, trustworthy, and sincere) and intelligence.

As seen in Figure 5, people felt they liked and trusted the other player more if they were able to communicate with that person. An ANOVA shows a marginally significant effect of condition on ratings of the other player's likeability ($F(3, 59) = 2.44$, $p < .08$) and trustworthiness ($F(3, 59) = 2.54$, $p < .07$). Only in the voice condition did people tend to rate the other player as more intelligent. There was a significant effect of communication modality on ratings of the other player's intelligence ($F(3, 61) = 2.79$, $p < .05$)[§].

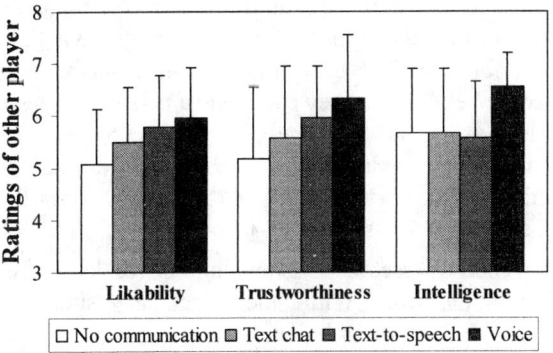

Figure 5: Evaluation of interaction partner

DISCUSSION

Consistent with sociological literature for off-line experiments, voice communication was found to have an extremely powerful effect on people's tendency to trust and cooperate with each other. We found statistically significant differences between the voice communication condition and both the text-chat and the no-communication condition in predicting cooperation. We found marginally significant differences between voice and TTS, and between TTS and no-communication.

An examination of how people evaluated their partners suggest that they had a more positive image (likable and trustworthy) of those with whom they could communicate. In addition, people felt that their partners were more intelligent when they could communicate with them by voice. This effect could be caused by the differing amounts of time that it took for people to communicate in each of the modalities. Past studies have shown a direct relationship between delays in communication and perceived levels of intelligence [13].

One possible explanation for why we did not find statistically significant differences between TTS and text-chat is that our sample size was too small. A power analysis suggests that the effect could reach significance with approximately double the number of dyads (15 in each group as opposed to 8). Also, consistent with previous findings in the sociological literature, we did not find statistically significant differences in cooperation between non-communication and text-communication modes. The lack of an effect may be due to the high variance in the two conditions. An examination of histograms of scores for each condition suggests that people tended to lapse into either full cooperation or full defection in the no communication condition, which greatly increases the variance of the condition.

Though our results are consistent with previous findings, we were nevertheless surprised by the lack of statistically significant differences between the text-chat and non-communication conditions. Strategic discussion was a primary topic of conversation in a number of the trials (approximately 80% of the dyads who could communicate discussed strategy), but while discussion was followed by robust cooperation in the voice condition, it was not as readily followed by cooperation in the text-chat condition.

We were also surprised by the relative difference between TTS communication and text-chat. Initially we expected no differences between TTS and text, because TTS contains similar semantic content to the text case, but little of the para-verbal information found in voice. In fact, since TTS might be artificially lacking cues that people use in predicting the veracity of utterances, we felt that there might be a negative effect associated with using TTS. There are several possible explanations for the higher level of cooperation in the TTS condition over the text-chat condition. Perhaps the experience of hearing a voice enhanced people's perceptions of social proximity. Another possibility is that without an available history such as found in text-chat; subjects were forced to pay greater attention to the other player, hence increasing their sense of social presence.

The fact that the TTS condition gave positive results suggests that it could be used in other situations. Perhaps we could improve on voice or video interaction by modifying the voice or aspects of the appearance to match the preferences or biases of the other participant. For instance, instead of always using a gender-neutral voice, we could appeal to a user's sense of group identity by using an appropriately gendered TTS voice. This of course quickly raises a number of ethical questions and concerns.

One of the problems with the design of our experiment was that the voice condition quickly reached a ceiling effect of perfect cooperation. This lack of resolution prevented us

---

[§] This analysis was performed on ratings of the other player at the individual level rather than at the dyadic level. Dyadic effects were controlled for by entering the dyad as a covariate in the analysis.

from getting a clearer contrast between voice and the other conditions. If more immediate forms of communications are to be examined, we will need to create a situation with a higher sense of risk and temptation to defect. We discuss possible ways to do this in the future work section.

Finally, while the voice condition did achieve fairly universal cooperation, a number of subjects showed some reluctance to using this mode of communication. These subjects reported being uncomfortable with the level of intimacy associated with voice communication, and said that they would have been more comfortable with text-chat. For these subjects, a less intimate mode of communication might well have produced better results than voice.

FUTURE WORK

In the future, we would like to run a larger number of subjects through these tests. With a larger population it could be possible to show the benefit of text over no communication, as well as see a clearer distinction between the other forms of communication. We would like to examine a number of additional modalities, including video at various resolutions and frame-rates, the use of both 2D and 3D avatars and asynchronous communication modalities such as voice mail and email.

The Prisoner's Dilemma can easily be extended to include a larger number of participants in every game. Not only do these N-person social dilemmas model an important class of interactions (group dynamics) not captured in the 2-person situation, they also create a significantly more competitive environment. This would help remove the ceiling effect of total cooperation we observed in the voice condition, and perhaps allow for a difference between voice and more personal forms of communication (e.g. video and face-to-face) to be observed.

The social dilemmas are very sensitive to individual quirks and personality traits. To truly validate the results, a large number of subjects is needed. By using the World Wide Web, large online experiments could become significantly easier to do. Such an approach would also be valuable for doing cross-national experiments, helping to determine which effects are related to culture. We are currently working on the design of such a system together with sociologists from a number of universities around the world.

Despite the simplicity of the task, a number of subjects were unable to determine what the optimal game strategy was (CC). These subjects cooperated to the best of their abilities, but did so in a sub-optimal fashion (most of these believing DC+CD>CC+CC). Some subjects were thrown by the simplicity of the task, believing that there had to be a "trick", a secret strategy or hidden motivation behind their task. Many subjects also remarked on how boring the game was, some stopped cooperating in an attempt to make things more interesting, and one pair went as far as to improvise a game of tick-tack-toe using the numeric keypad. We might be able to solve both problems by transforming the game from an abstract problem solving exercise into something more tangible, for instance some form of a card or video game.

Further experiments need to be done on the sensitivity of the technique for comparing different communication modalities. We found that this method was very sensitive to factors such as wording and the perceived level of risk among the participants. If the stakes are low, people have a tendency to take higher risks and be less selfish. As the stakes get higher, the inclination towards selfish behavior increases. An interesting question is how to encourage a high sense of risk in a WWW version of the game.

We would like to start looking at factors other than cooperation, for instance the development of empathy, or how the medium affects the ability to determine identity. From our current results it seems that people might be prone to making assumptions for which they have little basis or no for in the TTS case. In face-to-face interactions, many people believe that they can judge a person's honesty by simply looking into their eyes. Are people more prone to making such assumptions with some forms of communication than with others, and if so, how does the medium affect which assumptions are made?

A final avenue of research is extending our results to pre-existing relationships. While there are many situations where we deal with relative strangers, the existing relationship addresses an important set of interactions. Pre-existing relationships have a very significant effect on future interactions, even in high-risk situations [12]. By studying how the medium affects existing relationships, our results could be more easily applied to CSCW and groupware applications.

CONCLUSION

We have demonstrated a technique for the quantitative assessment and comparison of the effect of different forms of communication on the development of trust and cooperation. Consistent with the sociological literature, voice communication was found to have an extremely powerful effect in fostering trust and cooperation. Also consistent with previous results, text chat was not found to have a statistically significant effect beyond that of no communication. In our experiments, this may be caused by the high variance of both the text and the non-communication cases.

As expected, more immediate forms of communication showed a greater impact on the development of cooperation. The biggest surprise of this study was the difference between TTS and text-chat. While there was no statistically significant difference, the average contribution was 2 points higher in the TTS case. This indicates that

TTS technology could have the potential to positively influence computer-mediated communication.

The social dilemma methodology is a robust and widely used tool in the social sciences. By leveraging this technique we are better able to analyze and evaluate the many factors affecting social interactions.

We believe that our methodology can be adapted to analyze a wider array of social factors. By going beyond trust and cooperation, our method could become a valuable general-purpose tool in the study of computer-mediated communication.

ACKNOWLEDGEMENTS

We wish to thank Motoki Watabe for his assistance in the data analysis, as well as the Virtual Worlds group at Microsoft Research for their support.

REFERENCES

1. Abela, A., and Sacconaghi, A., Value Exchange: The Secret of Building Customer Relationships On Line. *The McKinsey Quarterly 1997*, Number 2

2. Axelrod, R. *The Evolution of Cooperation.* Basic Books, New York, 1984

3. Chapanis, A., Ochsman, R. B., Parrish, R. N. & Weeks, G. D. Studies in interactive communication: The behavior of teams during co-operative problem solving. *Human Factors, 14.* (1972). 487-509.

4. Doherty-Sneedon, G., Anderson, A., O'Malley, C., Langton, S., Garrod, S., & Bruce, V. Face-to-Face and video mediated communication: A comparison of dialog structure and task performance, *Journal of Experimental Psychology: Applied, 3, 2,* (1997). 105-123.

5. Edigo, C. Teleconferencing as a technology to support co-operative work: Its possibilities and limitations. In J. Gallegher, R. E. Kraut, & C. Edigo (Eds.) *Intellectual teamwork: Social and technological foundations of cooperative work* (pp. 351-371). Hillsdale, NJ. Erlbaum Associates

6. Finn, K., Sellen, A., Wilbur, S. (Eds.) *Video-Mediated Communication.* (1997) Hillsdale, NJ. Erlbaum Associates

7. Kelly, H. *Interpersonal Relations: A Theory of Interdependence.* Wiley and Sons, New York, 1978

8. Kiesler, S., Sproull, L. and Waters, K. A Prisoner's Dilemma Experiment on Cooperation With People and Human-Like Computers. *Journal of Personality and Social Psychology 1996*. Vol. 70, No 1, 47-65

9. Kollock, P. Social Dilemmas: The Anatomy of Cooperation. *Annual Review of Sociology 1998*. 24:183-214

10. Ledyard, J. Public Goods: A Survey of Experimental Research. In Kagel, J. and Roth, A. (Eds.), *The Handbook of Experimental Economics.* Princeton University Press, Princeton, NJ, 1995

11. Messick, D and Brewer, M. Solving Social Dilemmas. In Wheeler, L. and Shaver, P. (Eds.), *Review of Personality and Social Psychology.* Sage, Beverly Hills, CA, 1983

12. Rocco, E. Trust Breaks Down in Electronic Contexts but Can Be Repaired by Some Initial Face-to-Face Contact. *Proceedings of CHI '98* (Los Angeles CA, April 1998), ACM Press, 496-502.

13. Ruhleder, K., Jorden, B. Meaning-making across remote sites: how delays in transmission affect interaction. *Proceedings of 1999 European Conference on CSCW.*

14. Sally, D. Conversation and Cooperation in Social Dilemmas: A Meta-analysis of Experiments from 1958 to 1992. *Rationality and Society*, 7,58-92, 1995

15. Veinott, E. S., Olson, J. S., Olson, G. M., Fu, X. Video Helps Remote Work: Speakers Who Need to Negotiate Common Ground Benefit from Seeing Eachother. *Proceedings of CHI '99* (Pittsburgh PA, May 1999), ACM Press, 302-309.

16. Veinott, E. S., Olson, J. S., Olson, G. M., Fu, X. Video Matters! When communication is stressed video Helps. *Extended Abstracts of CHI '97* (Atlanta GA, April 1997) ACM Press, 315-316.

Using a Large Projection Screen as an Alternative to Head-Mounted Displays for Virtual Environments

Emilee Patrick,[1] Dennis Cosgrove, Aleksandra Slavkovic,
Jennifer Ann Rode, Thom Verratti, Greg Chiselko

Human Computer Interaction Institute, Carnegie Mellon University
5000 Forbes Avenue, Pittsburgh, PA 15213-3890 USA

ABSTRACT

Head-mounted displays for virtual environments facilitate an immersive experience that seems more real than an experience provided by a desk-top monitor [18]; however, the cost of head-mounted displays can prohibit their use. An empirical study was conducted investigating differences in spatial knowledge learned for a virtual environment presented in three viewing conditions: head-mounted display, large projection screen, and desk-top monitor. Participants in each condition were asked to reproduce their cognitive map of a virtual environment, which had been developed during individual exploration of the environment along a predetermined course. Error scores were calculated, indicating the degree to which each participant's map differed from the actual layout of the virtual environment. No statistically significant difference was found between the head-mounted display and large projection screen conditions. An implication of this result is that a large projection screen may be an effective, inexpensive substitute for a head-mounted display.

Keywords
Experiment, virtual reality, spatial knowledge, field of view, cognitive map, head-mounted display, projection screen, monitor.

INTRODUCTION

The acquisition of spatial knowledge for an unfamiliar physical environment progresses through three stages. In the first stage, people learn the locations of landmarks. This is referred to as *landmark knowledge*. As they learn to navigate from place to place in the environment following familiar paths, people gain *route knowledge*. Finally, *survey knowledge* for the environment is achieved—knowing the way around well enough to have a mental (or cognitive) map of the environment [6].

Virtual reality (VR) provides an opportunity for people to gain spatial knowledge for an environment other than the one in which they are physically located, and therefore has the potential to be an invaluable educational and training tool [1,15]. Acquisition of spatial knowledge for a virtual environment has been shown to follow the same three stages as for a physical environment: landmark knowledge, route knowledge, and survey knowledge [18]. However, previous research suggests that in order for a VR experience to seem the most realistic, an immersive experience in a head-mounted display is necessary [8].

Scientists from various disciplines have investigated differences in the accuracy of spatial knowledge acquired from still images projected onto a surface compared with spatial knowledge of the real world [7]. Variations in spatial knowledge between head-mounted displays and monitors for viewing virtual environments have also been studied [18]. This experiment was designed to augment prior work by investigating the perception of physical relationships between landmarks in a virtual environment, and the acquisition of survey knowledge under three viewing conditions: head-mounted display, large projection screen, and desk-top monitor.

Spatial Cognition

Cognitive maps, or internal, mental representations of spatial environments, are a component of spatial knowledge [9]. This internal representation is the basis for human interaction with the world, guiding people's decisions and interactions [6]. Spatial problem-solving activities such as wayfinding and navigation rely heavily on cognitive maps, which act as internal conceptualizations of the problem to be solved. Cognitive maps for real-world environments are more accurate when they are formed by viewing a paper map of the environment than from cursory navigation through the environment. However, repeated navigation in the environment results in a cognitive map that is as accurate as if it was learned from a paper map[20].

People develop a cognitive map for a virtual environment in a similar manner to the way they do a real-world

---

[1] Contact author: Emilee Patrick, User Centered Research, Motorola Labs, 1301 E. Algonquin Rd., Schaumburg IL 60196. 847-538-6886, emilee.patrick@motorola.com

environment. One way to measure differences that occur in cognitive maps arising from experiencing virtual versus real environments is by asking people to estimate distances between landmarks. Previous research examining the accuracy of distance judgments in the real world indicates that they are not perfect—they are generally 87-91% of actual distances. Interestingly, people are significantly even less accurate at estimating distances when viewing a virtual environment [22,11].

Field of view, measured in degrees, indicates how much of the world can be seen at a given time. For example, someone looking through a window towards the outdoors has a more restricted field of view than someone who is actually standing outdoors because the edges of the window make the visual field smaller. Field of view has a large impact on the underestimation of distances, both in the real world and in a virtual environment [2,3]. A smaller field of view results in compression of distance judgments—people think things are closer than they actually are. Hagen (1978) hypothesized that this is because people underestimate the unseen foreground distance between themselves and what they are viewing.

Immersion and Presence

An immersive experience can be described as one in which a person is enveloped in a feeling of isolation from the real world. One can feel immersed in movies where interaction is not possible, as well as in video games, which allow a high degree of interaction. In a virtual environment, having a task to perform increases the feeling of immersion.

A different, but related aspect of a virtual experience is presence: the extent to which a person's cognitive and perceptual systems are tricked into believing they are somewhere other than their physical location [22]. Display devices that evoke a great sense of presence often cause simulator sickness (a variant of motion sickness); symptoms include paleness, dizziness, nausea, and vomiting [12].

Head-mounted displays, which can produce an experience high in both immersion and presence, have been shown to be the most effective way to gain accurate spatial knowledge for a virtual environment. Cognitive maps formed in a head-mounted display perform significantly better than cognitive maps learned by viewing a virtual environment on a desk-top monitor [18]. Ruddle, Payne & Jones (1998) found that participants navigating a virtual building using a head-mounted display were able to do so significantly (approximately 12%) faster than participants using a desk-top monitor. Estimations of straight-line distances by participants using a head-mounted display were also significantly more accurate. In addition, participants using a head-mounted display were found to have spent a significantly greater amount of time looking around than those using a desk-top monitor. Researchers hypothesized that these significant differences were due to additional perceptual cues provided by peripheral vision and the ability to look around in the head-mounted display [18]. Peripheral vision plays a critical role in learning the spatial layout of an environment [18,19]. It is important for educational and military training applications of VR that spatial knowledge learned is as accurate as possible so that the information learned will transfer to a real environment [21]. However, the equipment required can be prohibitively expensive and uninviting to use. In addition, increased simulator sickness can result from exposure to a head-mounted display [15].

This paper discusses differences in spatial knowledge that occurred when study participants traveled through a virtual environment viewed in a head-mounted display, on a 3.35 m wide x 2.30 m tall projection screen, or on a desk-top monitor. Data consisting of judgments of the relative position of landmarks in the virtual environment was gathered after participants had experienced it. The data was then analyzed to determine its accuracy or inaccuracy when compared with the actual layout of the virtual environment. [5,17]. Participants' survey knowledge was expected to be more accurate when the environment was viewed in a head-mounted display than projected onto a screen, and least accurate when the environment was displayed on a desk-top monitor.

METHOD
Participants

Students and staff were solicited from the University of Pittsburgh and Carnegie Mellon University in Pittsburgh, Pennsylvania. Potential participants completed a questionnaire to determine their eligibility to participate in the study. Participants were excluded from the study if they reported any of the following characteristics:

- training or professional experience in architecture, mechanical or civil engineering, or industrial design
- vision that was not correctable to 20/20
- played any first-person navigation-based video games (e.g. Quake) more than 5 hours per week or 20 hours per month
- reported wearing a head-mounted display more than twice per year

Eligible participants fell within the age range of 18 to 33 years old and were randomly assigned to one of three conditions: Screen, Monitor, or Head-Mounted Display (HMD). 67 participants completed a pre-test of their spatial perceptual abilities. Data from 19 participants scoring more than one standard deviation from the mean in either direction was excluded from the statistical analysis of the experimental results. The remaining 48 participants were balanced for gender and age in each condition. The experiment took 40 - 55 minutes to complete, and each participant was paid $10.00.

Apparatus

Hardware and Software

All participants were asked to navigate through two virtual environments created using Alice, a freely available VR authoring and playback tool [14]. A Windows 95 Pentium II 300 MHz computer equipped with 128MB of RAM and two video cards was used to run the Alice software during the experiment. The computer's main video card (a Nvidia RIVA 128) was plugged into a standard desk-top monitor so the experimenter could start Alice and load the virtual environments. Meanwhile, Alice sent the output of the virtual environment to the second video card (a Diamond Monster 3D II), used to drive one of the display devices (head-mounted display, projection screen, or desk-top monitor).

Display Devices

The HMD condition used a Visette Pro head-mounted display with Ascention SpacePad tracking system (640x480 resolution, 16 bit color; see Figure 1). This brand was chosen because it offered a wide field of view, and cost less than $10,000. Tracking devices for the head-mounted display were mounted to a cardboard square and suspended from the ceiling approximately .5 m from the participant's head when seated. The head-mounted display was placed on the participant's head by an experimenter with experience in fitting these devices, and time was taken to ensure that the equipment did not strain the participant's neck or bind too tightly. Field of view for this device was 60° horizontal x 46.8° vertical. The display resolution and field of view for both the Screen and Monitor conditions was matched to the capabilities of the head-mounted

Screen condition used a rear-projection screen apparatus, consisting of a Toshiba TLP511A projector, a flat, rectangular mirror to increase the projector's throw distance, and a 3.35 m wide x 2.30 m tall screen (material custom-manufactured by Gerriets International of Revue). When mounted in the experiment room, this screen spanned floor to ceiling.

Navigation Device

Previous spatial cognition research has taken one of two approaches to the issue of navigation in a virtual environment: either participants are allowed to freely explore or they view a scripted presentation of a virtual environment. This study combined those approaches by allowing participants the freedom to navigate; yet the experimenter verbally led them through a scripted sequence of actions. A steering wheel was used to control navigation, rather than a mouse or joystick. This decision was motivated by a desire to provide a method of interactivity that would be easy to learn, and thus enhance the reality of the virtual experience [8]. Participants used a Thrustmaster Grand Prix steering wheel game controller, which allowed car-like steering (see Figure 2). The wheel had two levers that could be grasped by the fingertips of each hand while steering and used to propel the participant forward or backward in the virtual environment at a constant speed.

Procedure

The experiment consisted of a standard spatial ability pretest, exploration of two virtual environments, and a posttest to discover what participants could remember about locations of landmarks in the second, experimental virtual environment.

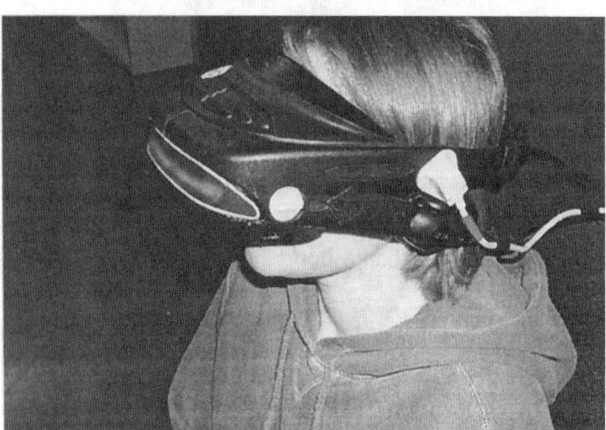

Figure 1: head-mounted display

Figure 2: steering wheel

display. Participants in the Monitor condition were seated 0.69 meters from the desk-top monitor, and 2.66 meters away from the projection screen. None of the display conditions used stereo.

The Monitor condition used a standard 21" (53 cm) computer monitor (model Iiyama Vision Master 500) raised to eye level and positioned closely on the table in front of the participant at the appropriate height. The

Pretest

The experiment took place on the Carnegie Mellon University campus. Upon their arrival, participants were asked to complete the Educational Testing Service "Surface Development Test—VZ3," an instrument to measure ability for mental manipulation of 2-dimensional objects into 3 dimensions [4]. Participants were shown several line drawings (see Figure 3) and were required to visualize how

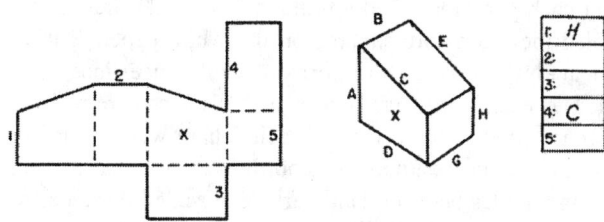

Figure 3: practice problem from the ETS "Surface Development Test—VZ3"

the items depicted might be folded to form a 3-dimensional shape. This particular test was chosen because the posttest (described below) draws upon similar cognitive abilities [3]. Data from participants scoring more than one standard deviation from the mean in either direction was excluded from the statistical analysis of the experimental results, in an effort to control for variability in the sample.

Exploration

After completion of the pre-test, participants were led to the location of the equipment for displaying the virtual environments. For the Screen condition, this room contained the previously described screen apparatus at a preset distance from the chair where participants sat. For the Monitor condition, a computer monitor was placed on a table in front of the chair. For the HMD condition, participants were shown a head-mounted display, and its function was explained. The experimenter fit the head-mounted display device to a participant's head, and instructed participants to turn their head and look around the virtual environment. Because nausea was a concern, and because higher temperatures can promote simulator sickness while viewing a virtual environment [13], two box fans were used to blow air on participants in all three conditions. Navigation through the virtual environments was restricted to ground-level navigation; that is, participants were not able to fly.

Practice Environment

Figure 4: Entrance booth

Participants first experienced a practice virtual environment, so that they could learn how to use the steering wheel for navigation. A second purpose for the practice environment was to ensure that participants were able to recognize a landmark they would later see in the experimental environment, called an entrance booth (see Figure 4). As explained below, the entrance booth was an integral part of the task to be performed in the experimental environment. The practice environment consisted of an intersection between two streets in an urban setting. Aside from the entrance booth, it bore little resemblance to the experimental environment; however, the method of interaction was identical. Participants were instructed in the use of the steering wheel, placed in the practice environment, and encouraged to freely explore. The practice session continued until the participant indicated that he or she was comfortable with the wheel and levers, approximately 3-5 minutes.

Experimental Environment

The experimental virtual environment consisted of a virtual amusement park, created for this study. The layout of the park bore no similarity to any real-world amusement park. The park contained a total of 10 rides and attractions. Attractions in the amusement park were arranged to approximate the appearance of a real amusement park (see Figure 5). Care was taken to ensure that the virtual environment had sufficient complexity to avoid a ceiling effect [17]. Participants were asked to imagine that they were the groundskeepers of the park, and were responsible for driving through it on a golf cart each morning to turn on the rides. This was accomplished by navigating into close proximity to an entrance booth, which was present in front of each attraction. When a participant moved close enough to a ride's entrance booth to activate the ride, a particular sound was played and the participant received visual feedback that the ride was activated.

To ensure that participants would recognize and understand the names used for the rides in the amusement park, a color printout consisting of images of all ten rides along with their names was shown to participants. Prior to participants' interaction with the experimental environment, the experimenter pointed to each image on the printout and spoke the name of the ride aloud. Finally, participants were instructed to pay close attention to the location and orientation of the entrance booths, because they would be asked to recall them later.

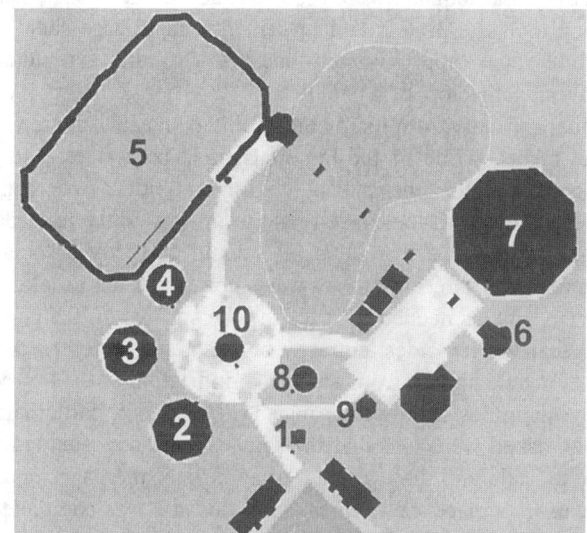

Figure 5: Layout of the Virtual Amusement Park
1 - Park Entrance; 2 - Teacup Ride; 3 - Octopus Ride;
4 - Swings; 5- Roller Coaster; 6 - Lion's Head Skyway;
7 - Haunted House; 8 - Fountain; 9 -Double Ferris Wheel; 10 - Carousel

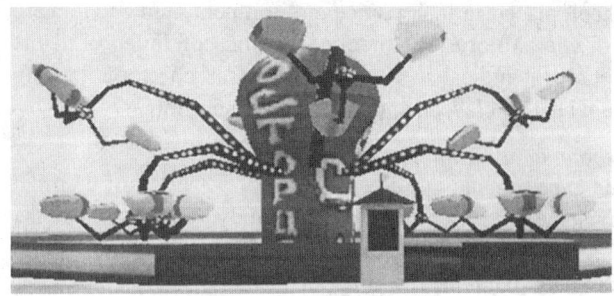

Figure 6: Octopus Ride

Participants then began interacting with the virtual amusement park. The experimenter proceeded to read aloud step-by-step instructions regarding which ride to turn on next. These instructions provided enough information to allow most participants to find the next ride with little trouble. However, in the few cases where participants appeared to be struggling or asked for help (e.g. they forgot what the Octopus Ride looked like), the experimenter assisted them by describing the ride's appearance or instructing them to "turn left" or "turn right." The experimenter was careful not to give any verbal associations between landmarks that might affect the formation of participants' cognitive maps [9]. For example, the experimenter might say, "Look to your right to see the Octopus Ride," rather than "The Octopus Ride is just to the right of the Teacup Ride." The final step in the instructions allowed participants to turn around and look at the park one more time before exiting it. While there was no strict time limit in which to complete the instructions, most participants spent 7 to 10 minutes in the virtual amusement park.

Posttest

Finally, participants were escorted back to the room where they took the pretest and presented with a large sheet of white paper (approximately one meter square, completely covering the top of a small table) and ten 3 cm x 3 cm squares (made from foam core). Printed on each square was the name of one of the ten amusement park rides, and it represented the entrance booth for that ride. Participants were allowed to look briefly at the images of the ten rides again, to ensure that he or she could correctly associate names with the rides. They were then instructed to take as much time as needed to place the ten squares on the paper so that the ten rides' entrance booths were represented as they would appear from above. No indication of the correct orientation or scale was provided on the paper; participants who asked were told that the squares were not intended to be to scale and that as much or as little of the paper could be used to place the squares. After this task was completed, participants were asked to fill out a questionnaire regarding any illness or discomfort they might have experienced while navigating in the virtual environment. Finally, they were given $10.00 for participating.

After each participant had left, the experimenter traced and labeled the foam-core squares on the white paper. During later analysis, vertical and horizontal reference lines were added to each participant's map and distances from these reference lines to the center of each square were recorded. This raw data resulted in coordinates for each ride, allowing angles between landmarks and scaled distances to be recorded for each participant.

RESULTS

The hypothesis that the HMD condition would show better performance than Screen or Monitor conditions was not supported. Results indicate no significant difference between HMD and Screen conditions or HMD and Monitor conditions. The Screen and Monitor conditions were significantly different.

Data Preparation

Distance is a conceptualization of the physical relationship between objects, used as a standard measurement in spatial cognition research. The knowledge of distances between objects is the foundation for understanding the structure of the physical world [6]. For the posttest of this experiment, participants were instructed to re-create the layout of the amusement park as it would appear if viewed from above, paying close attention to location of the rides, or landmarks, in relation to each other. We chose to provide for participants a blank piece of paper with no indication of orientation or scale, to avoid influencing the interpretation of distances between landmarks in participants' cognitive maps [6].

Because no scale or orientation information was provided, distances between landmarks in participants' reported maps could not be compared directly to the actual distances between landmarks in the virtual environment. Reported landmark relationships, encompassing both distances and angles between landmarks, were compared to their true relationships in the virtual environment to determine placement error. Simply comparing the angle between a set of three landmarks accounts for error in relative orientation, however it fails to sufficiently account for error in relative position, or distance.

It is possible to normalize for orientation and scale by transforming each reported map to most appropriately match the virtual environment, and then measure errors in relative orientation and position. One approach would be to provide participants with the location of two landmark from the virtual environment, upon which to base the rest of their reported map. While this single given relationship would provide an orientation and scale, it would bias every error calculation by exerting an external influence on participants' cognitive maps. This is undesirable. So, for each landmark pair ($_{10}C_2 = 45$), we oriented and scaled the entire reported map until the pair matched its analog in the virtual environment. Distance error (in meters) was then calculated for the remaining eight transformed landmarks.

The cumulative placement error score (360 distance measurements per participant) evenly weights every landmark relationship.

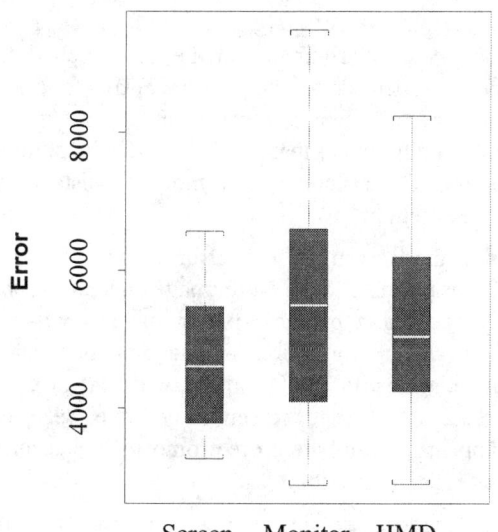

Figure 7: placement error scores across conditions

Pretest and Simulator Sickness
Pretest score was found to be a significant predictor of posttest score for all 69 participants, validating the choice of pretest for this experiment. Scores for the 48 participants scoring within one standard deviation of the mean on the pretest were used in later statistical analysis. Mean sickness scores were highest for the HMD condition, but did not affect performance on the posttest. Mean scores were about equal in Screen and Monitor conditions (see Table 1). A Wilcox sign rank test was performed on the simulator sickness data. This statistical method is more robust than others that are based on the assumption that data is normally distributed. Results showed that participants felt significantly more ill when using the head-mounted display than in the other two conditions (see Table 2).

Placement Error
Due to equipment failure and a high level of discomfort in the head-mounted display, one participant was dropped from the HMD condition after the experiment had been conducted. Table 3 shows means and variance for the normalized placement error scores. The mean placement error score was lowest for the Screen condition, and highest for the Monitor condition, meaning that Screen performed better than Monitor on average (see Figure 7).

After completing descriptive statistics, Bartlett's Test for homogeneity of variance was performed. Because results for Bartlett's Test were nearly significant (p = 0.060), indicating that the variances for the three conditions were too different for an ANOVA to yield useable results, the decision was made to perform more robust pairwise unpooled t-tests. Instead of averaging, or pooling, the variances for the two conditions compared by the t-test, each variance was used independently in computing the t-statistic.

P-values for the unpooled t-tests show that the difference between the Screen and Monitor conditions is close to significant (see Table 4). Because pairwise t-tests do not take into consideration the spatial pretest as a covariant, a regression analysis was performed, using condition as an indicator variable. A negative regression coefficient indicates that the placement error scores for the condition being compared with the base case are lower, meaning that participants performed better (see Table 5).

After accounting for the influence of pretest scores on the outcome of the posttest, t-values were calculated a second time. The Screen condition was found to outperform the Monitor condition (p=0.0497). There were no other significant differences. A complete report of the analysis can be found in Table 5.

Discussion
Results indicate that there is little difference in survey knowledge for a virtual environment when viewed through a head-mounted display or projected onto a large screen. Even with a head-mounted display's increased peripheral vision and capability to allow a participant to freely look around the virtual environment, this study found that the large screen still leaves participants with comparable spatial knowledge. While contrary to the original hypothesis, this result is consistent with Johnson (1999)

	n	Mean	Variance
Screen	16	2.625	6.65
Monitor	16	2.8125	7.09
HMD	16	5.6	10.54

Table 1: mean and variance for sickness scores

	n	z-value	p-value
Screen x Monitor	32	0.4219	0.6731
Screen x HMD	31	2.5826	0.0098
Monitor x HMD	31	2.811	0.0049

Table 2: Wilcox sign rank test on sickness scores

	n	Mean	Variance
Screen	16	4666.43	986814
Monitor	16	5640.16	3136277
HMD	15	4981.45	1274196

Table 3: mean and variance for normalized error scores

	n	t-value	p-value
Screen x Monitor	32	-1.92	0.068
Screen x HMD	31	-0.82	0.42
Monitor x HMD	31	-0.23	0.23

Table 4: results of unpooled t-tests on error scores

Screen vs. Monitor

	Coefficient	Std. Error	t-value	p value
(Intercept)	7493.9277	1422.0577	5.2698	0.0000
Pretest	-39.4142	29.2824	-1.3460	0.1887
Condition[a]	-1029.619	502.6221	-2.0485	0.0497

Monitor vs. HMD

	Coefficient	Std. Error	t-value	p value
(Intercept)	6853.9857	1576.8487	4.3466	0.0002
Pretest	-39.1196	32.7696	-1.1938	0.2429
Condition[b]	626.087	550.6973	1.1369	0.2656

Screen vs. HMD

	Coefficient	Std. Error	t-value	p value
(Intercept)	6940.5775	997.1337	6.9605	0.0000
Pretest	-40.9806	20.6173	-1.9877	0.0571
Condition[c]	-404.816	372.9730	-1.0854	0.2874

a base case monitor, coefficient 0   *b* base case HMD, coefficient 0
c base case HMD, coefficient 0

Table 5: regression and t-test results

who found no significant difference between a head-mounted display and a projection screen used to train soldiers to navigate an unfamiliar environment.

The unexpected absence of a significant difference between the HMD and Monitor conditions also contradicts the original hypothesis. The belief that the HMD condition would perform significantly better than either of the fixed-display conditions (screen, monitor) was based on the head-mounted display's additional display capabilities. By allowing the participant to turn their head, he or she could gain a greater sense of presence and potentially take in more information from the environment. However, it was consistently observed that participants in the HMD condition did not turn their head very much, as previously reported by Pausch (1996). This unintended reduction of the head mounted display to a fixed display could account for the lack of significant difference. Also, negative aspects of the head-mounted display such as weight, low acuity, and a higher degree of simulator sickness could possibly account for the higher mean error score in the HMD condition as compared with the Screen condition.

While variances in the Screen and HMD conditions were similar, the variance in the Monitor condition was found to be significantly different. A possible explanation for the greater variance observed in the Monitor condition lies in the relationship between field of view and judgments of distance. Participants' seated position when viewing the monitor was not artificially fixed, and it is possible that small movements movement forward or backward from the display might have had altered their field of view, causing differences in their interpretation of the spatial relationships between landmarks in the experimental environment.

The difference in error scores between the Screen and Monitor conditions was statistically significant. In addition, based on regression of Screen vs. HMD, the negative coefficient value, while not a significant difference, might be an indication that participants in the Screen condition tended to perform better than those in the HMD condition. Also, the smaller variance and lower mean placement error for the Screen condition compared with the HMD condition may indicate that the Screen is a more consistent and reliable display method.

What caused the screen to outperform the other two conditions? It is possible that a large image engenders more presence by tricking a person's perceptual systems into thinking they are really there, a phenomenon that is normally associated with HMD but not with flat displays. Images projected onto the screen may have been big enough to appear real, and therefore promote more accurate judgments of relative position.

This finding suggests an intriguing conclusion; that the low-cost projection screen might be as effective as a head-mounted display for educational or training exercises involving spatial cognition. The screen cost only $400 to build, while the head mounted display equipment used has a purchase price of approximately $6000. While the projector used in the Screen condition was quite expensive, it was possible to use it with a minimal amount of effort. Head-mounted display equipment is much more labor-intensive to install, as well as being invasive and uninviting technology to use. These advantages, combined with the lower incidence of discomfort due to simulator sickness, make the use of a large projection screen an attractive alternative to head-mounted displays.

Additionally, this study opens up many interesting avenues for future work. While viewing a virtual environment in a head-mounted display is a single-user experience, using a large projection screen has the potential to facilitate multi-user experiences. It is unknown at this time whether multiple participants simultaneously viewing the virtual environment would have gained the same degree of survey knowledge as the single participant who was driving. While there is no quantitative data to support the observation that participants generally did not look around when in the head-mounted display, a new study examining the impact of spatial cognitive ability and high vs. low head motion on survey knowledge for a virtual environment could produce interesting results. Finally, if indeed a large projection screen is a suitable substitute for head-mounted displays, it will be important to discover to what extent spatial knowledge learned from a virtual environment projected onto a screen is accurate in the real world.

ACKNOWLEDGMENTS

We would like to thank Dan Siewiorek, Jane Siegel, and Bonnie John for their help and support in the design and execution of this experiment. We would also like to thank Sean Householder, Bethany Rader, and Daniel van Boxel for their invaluable assistance. Finally, we owe our thanks to Asim Smailagic and the spring 1999 Rapid Prototyping class at Carnegie Mellon University for their work in developing the projection screen display.

REFERENCES

1. Ainge, D.J. Upper primary students constructing and exploring three-dimensional shapes: a comparison of virtual reality with card nets. *Journal of Educational Computing Research 14*, 4 (1996), 345-369.

2. Alfano, P.L., and Michel, G.F. Restricting the field of view: Perceptual and performance effects. *Perceptual and Motor Skills 70*, (1990), 35-45.

3. Barfield, W., and Kim, Y. Effect of Geometric Parameters of Perspective on Judgements of Spatial Information. *Perceptual and Motor Skills 73*, (1991), 619-623.

4. Educational Testing Service. Surface Development Test—VZ3. *Kit of Reference Tests for Cognitive Factors*, (1976).

5. Golledge, R.G., Dougherty, V., and Bell, S. Acquiring Spatial Knowledge: Survey Versus Route-Based Knowledge in Unfamiliar Environments. *Annals of the Association of American Geographers 85*, 1 (1995), 134-158.

6. Golledge, R.G. Cognition of Physical and Built Environments. In Garling, G. and Evans, G.W. (eds.), *Environment, Cognition and Action: An Integrated Approach*, (1991), 35-62. NY: Oxford University Press.

7. Hagen, M., Jones, R., and Reed, E. On a neglected variable in theories of pictorial perception: Truncation of the visual field. *Perception & Psychophysics 23*, 4 (1978), 326-330.

8. Hendrix, C. and Barfield, W. Presence within Virtual Environments as a Function of Visual Display Parameters. *Presence: Teleoperators and Virtual Environments 5*, 3 (1996), 274-289

9. Jackson, P.G. In search of better route guidance instructions. *Ergonomics 41*, 7 (1998), 1000-1013.

10. Johnson, D.M., and Stewart, J.E. Use of Virtual Environments for the Acquisition of Spatial Knowledge: Comparison Among Different Visual Displays. *Military Psychology 11*, 2 (1999), 129-148.

11. Neale, D. Factors influencing spatial awareness and orientation in desktop virtual environments. *Proceedings of the Human Factors and Ergonomics Society 41st Annual Meeting* (1997), 1278-1282.

12. Nichols, S. Physical ergonomics of virtual environment use. *Applied Ergonomics 30*, 1 (1999), 79-90.

13. Pausch, R., Snoddy, J., Taylor, R., Watson, S. and Haseltine, E. Disney's Aladdin: first steps toward storytelling in virtual reality. *Proceedings of the 23rd Annual Conference on Computer Graphics*, (1996), 193-203.

14. Pausch, R., Burnette, T., Capeheart, A. C., Conway, M., Cosgrove, D., DeLine, R., Durbin, J., Gossweiler, R., Koga, S., and White, J. Alice: Rapid Prototyping System for Virtual Reality. *IEEE Computer Graphics and Applications*, (1995).

15. Pausch, R., Crea, T., and Conway, M. A Literature Survey for Virtual Environments: Military Flight Simulator Visual Systems and Simulator Sickness. *Presence: Teleoperators and Virtual Environments 1*, 3 (1993).

16. Proffitt, D. A desktop display is not a window on the world. Work in progress, (1999).

17. Rossano, M. J., and Moak, J. Spatial representations acquired from computer models: Cognitive load, orientation specificity and the acquisition of survey knowledge. *British Journal of Psychology 89*, 3 (1998), 481-497.

18. Ruddle, R., Payne, S., and Jones, D. Navigating large-scale virtual environments: What differences occur between helmet-mounted and desk-top displays. *Presence: Teleoperators and Virtual Environments 8*, 2 (1999), 157-168.

19. Ruddle, R., Payne, S., and Jones, D. Navigating Buildings in "Desk-top" Virtual Environments: Experimental Investigations Using Extended Navigational Experience. *Journal of Experimental Psychology: Applied 3*, 2 (1997), 143-159.

20. Thorndyke, P. W., and Hayes-Roth, B. Differences in Spatial Knowledge Acquired from Maps and Navigation. *Cognitive Psychology 14*, (1982), 560-589.

21. Waller, D., Hunt, E., and Knapp, D. The Transfer of Spatial Knowledge in Virtual Environment Training. *Presence: Teleoperators and Virtual Environments 7*, 2 (1998), 129.

22. Witmer, B.; and Kline, P. Judging Perceived and Traversed Distance in Virtual Environments. *Presence: Teleoperators and Virtual Environments 7*, 2 (1998), 144-167.

23. Witmer, B., and Singer, M. Measuring Presence in Virtual Environments: A Presence Questionnaire. *Presence: Teleoperators and Virtual Environments 7*, 3 (1998), 225-240.

Alice:
Lessons Learned from Building a 3D System For Novices

Matthew Conway[1], Steve Audia[2], Tommy Burnette[2], Dennis Cosgrove[3], Kevin Christiansen[3], Rob Deline[3],
Jim Durbin[2], Rich Gossweiler[2], Shuichi Koga[2], Chris Long[2], Beth Mallory[2], Steve Miale[2], Kristen Monkaitis[2], James Patten[2],
Jeff Pierce[3], Joe Shochet[2], David Staack[2], Brian Stearns[3], Richard Stoakley[2], Chris Sturgill[3], John Viega[2], Jeff White[2], George Williams[2],
Randy Pausch[3]

(1) Work done at the University of Virginia
mconway@microsoft.com

(2) University of Virginia

(3) Carnegie Mellon University
http://www.alice.org

ABSTRACT

We present lessons learned from developing Alice, a 3D graphics programming environment designed for undergraduates with no 3D graphics or programming experience. Alice is a Windows 95/NT tool for describing the time-based and interactive *behavior* of 3D objects, not a CAD tool for creating object geometry. Our observations and conclusions come from formal and informal observations of hundreds of users. Primary results include the use of LOGO-style egocentric coordinate systems, the use of arbitrary objects as lightweight coordinate systems, the launching of implicit threads of execution, extensive function overloading for a small set of commands, the careful choice of command names, and the ubiquitous use of animation and undo.

Keywords

Interactive 3D graphics, animation authoring tools

INTRODUCTION

Realtime 3D graphics is becoming mainstream: most PCs shipped in 1999 will ship with some sort of 3D graphics accelerator. We see this as an opportunity to approach 3D graphics research not as a question of rendering speed, but as one of *authoring and pedagogy*. Our goal is to engineer authoring systems for interactive 3D graphics that will allow a broader audience of end-users to create 3D interactive content without specialized 3D graphics training. Implicit in this line of research are a few assumptions:

The New Audience Assumption: we believe that a larger and more diverse audience will be interested in creating interactive 3D content. It is critical to realize that this new audience will not necessarily have the mathematical or programming background that current graphics programmers have; this shapes the nature of the tools that we must provide to this audience.

Figure 1 Observing Real Subjects. Two subjects learn the Alice 3D authoring environment while an observer silently sits behind them and takes notes. The major findings of this paper were derived from observations of 100 subjects taken over several months.

The Programming Assumption: Interesting interactive 3D graphics authoring will still involve some level of logic specification/programming, at least in the near term. This is true in part because of conditional behavior, which implies the need for some sort of "if-then" construct. We have focused on scripting in this work; future systems will probably use a combination of techniques including keyframing, programming-by-demonstration, and visual programming, as well as scripting.

We began the Alice research project with the goal of creating new authoring tools that would make 3D graphics accessible to this wider audience, something that current 3D tools would not do. Our basic design principles are:

- Choose a target audience and keep their needs in mind, in our case, non-science/engineering undergraduates.

- Avoid math and cryptic notation in the API (e.g. vectors, matrices) wherever possible and introduce new terminology only when needed.

- Iteratively test our designs with real users, improving both learnability and usability of the system in the process.

We state our findings as empirical research results, not as opinions; they are supported by formal observations of 100 users and hundreds of informal observation sessions over a four year period.

THE ALICE WORKFLOW

Authoring in Alice consists of two phases: *creating an opening scene* and *scripting*. This same two-phase workflow is seen in some commercial tools, including Raydream Studio and WorldUp.

Creating an Opening Scene

Users select objects from an object gallery displayed by clicking the *add object* button (figure 2, A). Alice's library contains hundreds of low-polygon models whose high fidelity comes from carefully hand-painted texture maps [15]. Note that while Alice users can import objects in several popular file formats, Alice itself is not a CAD tool for creating object geometry: Alice is a tool for describing the time-based and interactive *behavior* of 3D objects.

Objects are placed in a PHIGS-like tree of nested objects [2][23] (figure 2, B), displayed along the left edge of the authoring window. Navigation tools (figure 2, C) provide a simple walking metaphor for moving the camera. Alice can support multiple, simultaneous windows/cameras.

Context Menus

Holding down the right hand mouse button over an object displays a *context menu* of common operations, including

- Point the camera at the object
  (Point Camera At)
- Move the camera to a position that is above and off to one side of the object
  (Get a Good Look At)
- Spin the object in place for simple examination
  (Turn Around Once)
- Show a list of methods that this object responds to
  (Show Me What You Can Do)

The Alice Command Box

Alice programmers are encouraged to explore the command set via the Alice Command Box (figure 2, E) which is used for trying individual lines of script code. If we wanted to move an object, a bunny for example, a precise distance (as opposed to using the mouse), we could type:

bunny.move (up, 1)

and press the GO button (or hit the Enter key) to make the bunny move up by one meter over a period of one second.

Commands are Always Animated

All commands in Alice animate over a period of one second with an Ease-In/Ease-Out interpolation [8] whenever it is semantically reasonable to do so. Programmers can still specify an explicit duration (including zero duration) if so desired. This is not just a flashy trick but is a *critically* important design decision. Not only does animation support the percept of object constancy [17], but it can also aid in the debugging process by providing information about how a bug unfolded.

In a system without animation, a user can easily make the mistake of using the *Move* command with a distance that takes the object off the screen. An instantaneous move effectively "teleports" the object out of sight, a mistake that is visibly indistinguishable from a delete command. By animating the move command, we give the user a chance to see the command unfold over time. In this case, the user would see the failure in progress as the object slides off the edge of the screen.

Likewise, Alice provides an animated infinite undo mechanism (figure 2, D). This mechanism always takes one second to undo an operation, regardless of the duration of the original command being undone.

Controlled Exposure of Power

Alice commands are highly overloaded, supporting several different calling patterns through a single command name. For example, *Move* can be called in all these ways:

    obj.move(forward, 1)
    obj.move(forward, 1, duration=3)
    obj.move(forward, 1, speed=4)
    obj.move(forward, speed=2)

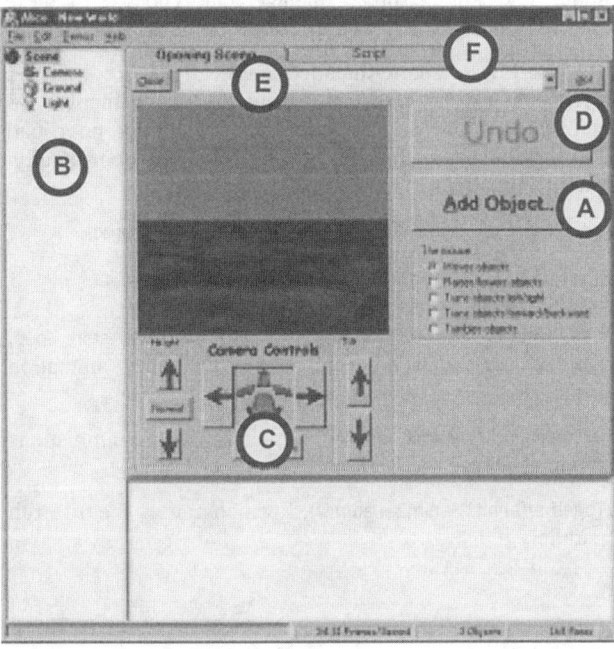

Figure 2: The Alice Authoring Environment (opening scene tab). **(A)** The Add Object button presents a gallery of 3D objects. **(B)** The Object Tree, a PHIGS-like tree of hierarchical objects **(C)** Camera controls allow the user to drive around the scene. **(D)** The Undo button provides infinite animated undo. **(E)** The Alice Command Box for evaluating single lines of Alice script. **(F)** The Script tab reveals a simple text editor where the user writes scripts that control the objects in the scene.

```
# change of coordinate system
obj.move(forward, 1, AsSeenBy=camera)
# different interpolation function
obj.move(forward, 1, style=abruptly)
```

Using overloaded methods with optional keyword parameters allows us to provide a *controlled exposure of power* to the Alice user. This characteristic of the API allows novice users to become expert users by incrementally adding to what they know, rather than forcing them to learn entirely new commands or API constructs. As one Alice user said, "you can get as complicated as you want, but not more than you need to."

Scripting
Once the objects are placed and the camera is in position, this initial state is saved into a *world file*. This file also contains the name of each object so that it can be referenced in the script that will control the object's movements.

When the opening scene is ready, the user then presses the scripting tab (figure 2, F), which reveals a text editor and a *Run Script* button. The user iteratively edits the script and runs it, with the script always starting its execution from the saved opening scene.

THE IMPLEMENTATION

For several years, we followed a "time machine" approach to Alice, doing early implementation on high-end SGIs in anticipation of low cost commodity graphics. Alice now exists solely on the PC platform, running on MS Windows 95, 98, and NT with the overall structure shown in figure 4. The layers are described below.

Rendering
The rendering software is Microsoft's Direct 3D Retained Mode (D3DRM). This layer manages the 3D database of objects, their attributes, and their texture maps; illuminates the scene; and maintains the hierarchical tree of coordinate systems.

Python
Alice uses a general purpose, interpreted language called *Python* [24]. We chose this language for its technical characteristics. Python is:

- a modern language with a rich set of built in data types (maps, strings, lists) and operators for those types.
- freely distributable and available without royalty.
- extensible in C/C++. For example, Alice uses the *Coriolis* collision detection library.

Although we resisted changing the Python implementation, our user testing data forced us to make two changes. First, we modified Python's integer division, so users could type 1/2 and have it evaluate to 0.5, rather than zero. Second, we made Python case insensitive. Over 85% of our users made case errors and many *continued* to do so even after learning that case was significant. Most novice-oriented systems (e.g. Hypercard, Pascal, LOGO) are designed to be case insensitive, a lesson we saw being ignored in the proposed standard for VRMLScript [10].

The Animation Engine
The Alice animation engine interpolates data values from a starting-state to a target-state over time, with a default duration of one second. When two or more animations run in parallel, the Alice scheduler interleaves the interpolations in round-robin fashion. This allows a user to evaluate a command while another command continues to animate, without any explicit thread management. The use of these *implicit threads* is a major contribution of the Alice system.

Alice itself is a single-threaded application. We built an experimental Alice prototype using native Windows 95 system threads, one per animation, but it exhibited poor load balancing between threads, giving rise to poor-quality, lurching animations.

Animated Alice commands return an animation object:

scoot = bob.move(forward, 1)

These objects respond to several methods (*stop, start, loop, stoplooping*) and can be composed with other animation objects:

DoInOrder(anim1, anim2,..animN), which causes the animations to run in sequence.

DoTogether(anim1, anim2,...animN), which causes them to run in parallel.

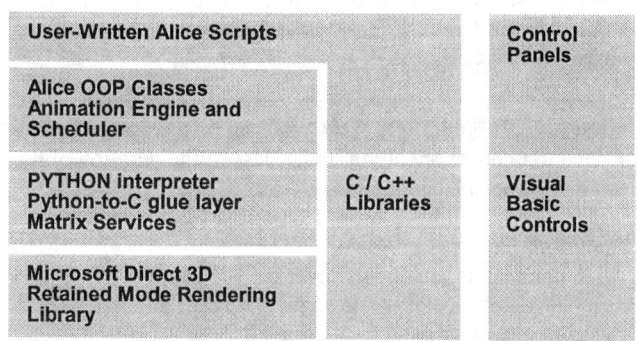

Figure 4: The Alice Software Architecture. User scripts are written in Python as is much of the Animation Engine Layer beneath it. The layers below this are written in C. Alice includes a separate Control Panels facility for creating Visual Basic GUI components directly through the Python scripting language.

Of course, DoInOrder and DoTogether animations can also be composed, giving rise to more interesting animations. For example, given a world with an object called **Bunny**, we could make the bunny beat his drum by writing:

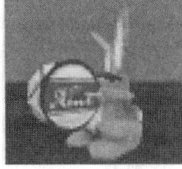

```
ArmsOut = DoTogether(
        Bunny.Body.LeftArm.Turn(Left, 1/8),
        Bunny.Body.RightArm.Turn(Right, 1/8) )
ArmsIn = DoTogether(
        Bunny.Body.LeftArm.Turn(Right, 1/8),
        Bunny.Body.RightArm.Turn(Left, 1/8) )
BangTheDrumSlowly = DoInOrder(
            ArmsOut,
            ArmsIn,
            Bunny.PlaySound('bang') )
BangTheDrumSlowly.Loop()
```

TESTING AND REAL USAGE

The table below summarizes the subjects in our tutorial observation sessions.

Number of Subjects	100	
Ages	High:	41
	Low:	18
	Mean:	22
	Std Deviation:	7.4
Sex	Female	58%
	Male	42%
Computer Experience	90% self-described as using email, some www, some word processor, no programming.	

Subjects were tested using a two-person talk-aloud protocol [12]. During a 30 minute session, pairs of users worked through the Alice tutorial. The tutorial walked the users through the steps necessary for creating a simple Alice world. Users would load in an object, practice moving it with the mouse and with Alice commands. They then learned about how to manipulate the parts of objects and other object attributes. Finally they learned how to name animations and build simple composite animations.

We used pairs because pairs of people naturally talk to each other about what is happening, what is confusing, and what they expect at each step. As they talked, we took notes.

Only under the most dire circumstances (e.g. system crash) did we assist users, a rule that sometimes requires a great deal of discipline, especially if the observer is also one of the system developers. While there was often strong temptation to show the subjects the "right way" out of a problem, we were strict about letting users find their own way, only interceding if the error was so great that it jeopardized the subjects' ability to finish the session.

Often these sessions were sobering. Encouraging everyone on the development staff to observe real users is an excellent way of sensitizing an entire team to the needs of one's target audience.

Other Sources of Observation Data

In addition to the formal usability sessions, we gathered observations from several other sources:

- suggestions from about 20 graduate students in a graduate-level graphics class.
- longer term observations of three in-house users using a critical-incident debriefing technique [4].
- deploying Alice at a magnet school in the Lynchburg, Virginia public school system and gathering on-site data from their students.
- exchanging detailed email with many members of our Alice user community, including some very valuable exchanges with home schooled grade-school children.

In short, these findings are neither opinions or the results of a marketing-style (*did you like it?*) questionnaire. The data comes from the observed behavior of real members of our target audience during use of the system, focusing on discovering which parts of the API were understood and which were not. We actively used this data to drive our design: we would build part of the system, and then user test to see where users consistently had difficulty. We would then redesign those parts, build new ones, and user test again.

FINDINGS

The Death of XYZ

Perhaps Alice's most distinguishing API feature is that it allows people to create behavior for three-dimensional objects without using the traditional mathematical names for the coordinate axes: X, Y and Z. Instead, Alice uses LOGO-style [16], object-centric direction names: Forward/Back, Left/Right, and Up/Down. We made this design decision after using XYZ for two years where we routinely observed users, even expert ones, saying things like:

> "I want to move the truck forward one unit, and that's positive X to Alice, so I will type **move(X, 1)**."

Our users already had a vocabulary for moving objects in space, but our early system was not using it. While it is true that *some* objects do not have an intrinsic forward direction, this is at least an improvement, because there are *no objects that have an intrinsic X direction*.

This seemingly tiny, cosmetic change is probably Alice's biggest contribution to making a usable API for 3D graphics. By using direction names in lieu of XYZ, we relieved the user of a cognitive mapping step that may occur thousands of times while developing a 3D program. Removing XYZ also reduced the need to talk about negative numbers to an audience that naturally shies away from mathematics.

First Class Objects and Parts

Like many 3D graphics systems, Alice uses a tree-like structure of nodes and children to organize the objects in a 3D scene.

Take, for example, the case of a vase sitting on a table. If we wanted the vase to move when the table moved, we might reasonably model the vase as a child of the table.

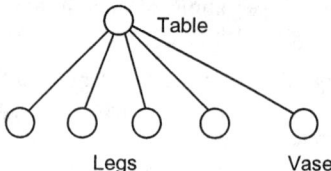

But which are parts of the table, and which are independent objects merely resting on the table? Said another way: how would we ensure that

Table.setColor(Red)

turns the table red without also turning the vase red? Intuitively, the vase is different than the table legs, but in an undifferentiated PHIGS tree, all objects look alike.

Alice marks objects loaded from disk as "first class objects" (the table and the vase) and marks the objects *defined inside those files* as parts (the legs of the table).

This allows some operations (e.g. SetTexture, SetColor, Destroy) to tell the difference between an object's parts and nodes that are attached for some other reason; the operation stops at other first class objects. First-class objects thus act like "firewalls" inside the object tree. Programmers can use

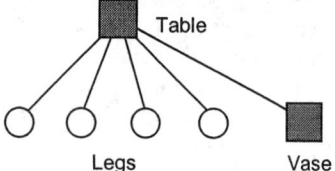

the optional parameters ObjectOnly, ObjectAndParts, ObjectAndChildren to override this behavior when they need more control.

The first-class attribute of an object is also used to control picking into the scene and other event-dispatching within Alice. This is very much like the pick-bit that some PHIGS systems have used in the past.

Objects Are Coordinate Systems (AsSeenBy)

Although 3D applications can perform all geometric operations in a global coordinate system, most 3D applications perform coordinate transformations to make geometric operations easier to compute or reason [5]. The Alice API already eases the burden of coordinate transformation to some extent by making the **move** and **turn** commands operate from the object's local coordinate system, rather than from a global frame of reference. For more general coordinate system transforms, we designed a system [7] that allowed programmers to perform any geometric operation within any other object's coordinate system. This capability is invoked by adding the coordinate system object with the optional AsSeenBy keyword parameter, as in:

bunny.move(forward, 1, AsSeenBy=chair)

Users can also use this mechanism to work in a global coordinate system by using the Scene as the reference object. Alice provides a default Scene object for every world.

Inspired by Alice, Disney Imagineering's *Player* system [15] and Microsoft's Direct 3D graphics libraries have both added this mechanism to their systems. Both are excellent examples of improving a system for experts based on lessons learned from a system for novices.

Beyond Translate, Rotate, and Scale

The Alice system provides commands that go beyond the classic translation, rotation, and scale operations:

Place – this command allows one object to be placed OnTopOf, InBackOf, ToRightOf etc, some other object: cup.place(OnTopOf, Table). This command was developed independently of and simultaneously with the similar but somewhat more expressive *Put* system, developed at SGI [3].

PointAt and **StandUp** are underconstrained rotation operations that use a global "up" direction to resolve the ambiguous nature of the command. A.PointAt(B) will make A's up vector parallel to the global up direction, while it rotates A around its other two axes so that A's forward direction passes through B. Fred.StandUp() will rotate Fred so that Fred's up direction runs parallel to the global Up direction, minimizing the rotation around the other two axes.

Nudge is similar to the *Move* command, but translates objects by a percentage of their size, not by an absolute distance. obj.nudge(forward, 0.5) will move an object forward a distance equal to half its front-to-back length. This allows people to create animations that are more portable across a wider range of object sizes.

AlignWith makes one object point in the same direction as another. This simple command is very handy and would be fairly difficult for novice users to implement given just a turn (rotate) command.

Pan rotates objects left or right around an axis that is parallel to the global up vector. This allows a camera that is tilted down toward the ground to remain pointed at the ground as it turns in place.

Resize

One of our first user observation sessions gave rise to vocal complaints about the strange side-effect that scale had on distances. Calling

bunny.scale(3)

resulted in a bunny that was 3 times as large, but it also had the effect that a subsequent call to

bunny.move(forward, 1)

Moved the bunny forward 3 meters, not 1. Our users correctly surmised that making the bunny larger also scaled its local space.

This side effect comes from our implementation: Alice uses a four dimensional homogeneous coordinate transformation matrix to represent object position, rotation and scale, an implementation that by design scales the space of objects [5].

To provide more useful semantics to this command, we use a second 4x4 matrix to keep track of scale. Alice's *Resize* command now changes the scale of an object via this second scale matrix without changing the object's position and rotation matrix; this allows us to resize the object's geometry without scaling its space. By propogating the effects of this scale matrix to the object's children we can similarly resize the geometry and offset relative to their parent of the object's children without scaling their space. As a result, a meter in Alice is always a meter regardless of whether or not an object has been resized.

Resize also takes an optional parameter, LikeRubber

> bob.resize(FrontToBack, 1/3, LikeRubber)

which scales an object in a way that preserves volume, an important technique for many animation effects [8].

Vocabulary Issues
Novice users are strongly influenced by surface issues, and seemingly inconsequential name choices can often make the difference between a clear API and a confusing one. Some notable examples:

Resize, not Scale: Scale is usually regarded as a noun, not a verb, and has strong connotations of weight, not size.

Move, not Translate: Translation is understood by our target audience to be the process by which French is converted into German and has little to do with movement.

Speed, not Rate: Alice commands can specify how fast something happens, as in bob.move(left, speed=1). Users were observed to have a few problems with *Rate* in that it seemed to have percentage or interest rate connotations, while *Speed* never caused confusion.

FrontToBack, not Depth: Previous versions of Alice used the words Depth, Width and Height to denote the dimensions of an object. We found that these terms were sufficiently ambiguous to users that we resorted to the clearer, but somewhat more cumbersome FrontToBack, LeftToRight, TopToBottom. While these terms are somewhat contrived, they at least have the advantage of clarity and are formed out of terms that a novice Alice user already knows.

AsSeenBy, not CoordSys: this name change was almost aesthetic in nature, and allowed script-writers to read scripts more naturally.

Color Names, not RGB Triples: Alice uses popular crayon color names like Red, Green, Peach, and Periwinkle to specify colors, not a numeric color model like RGB or HSV.

No individual name choice is pivotal to Alice's success, but the aggregate effect of getting these names right is quite powerful. Almost all Alice scripts use only the following commands:

Geometric Manipulation: Move, MoveTo, Turn, TurnTo, Nudge, Place, Pan, PointAt, AlignWith, StandUp, SetPointOfView, SetSize, SetScale, MoveToInPicturePlane, MoveInPicturePlane

Property and State Query: GetPosition, DistanceTo, GetAngles, GetPointOfView, GetBoundingBox, IsHidden, IsCastingShadow, BoundingBoxIsShowing, GetScale, GetTexture, GetTransparentTextureColor

Shadows: CastShadow, StopCastingShadow

Textures: SetTexture, SetTransparentTextureColor

Coloring and Rendering: Get/SetColor, Get/SetVisibility, Get/SetShininess, Get/SetHighlightColor, Get/SetEmissiveColor, Get/SetShadingStyle, Get/SetLightingStyle, GetFillingStyle

Vertex Manipulation: GetVertexPosition, SetVertices, GetVertices, GetFaces, GetVertexCount, GetFaceCount

Miscellaneous: Show, Hide, ShowBoundingBox, GetFilename, Destroy, Store, AttachCamera, MakeTransparentToInput, ShowFrustum

ROTATION RATE: ROTATIONS, NOT DEGREES

Alice's *Turn* command originally allowed programmers to specify angular amounts in degrees and the animation time in seconds, so it seemed natural that rotational *speed* be specified in degrees-per-second. Informal observation suggested that this unit was confusing.

After our test subjects had seen the first Alice tutorial and were familiar with the Alice *Turn* command, we posed the following question:

To turn objects in Alice, you specify a direction to turn (left, for example) and an amount (90 degrees, for example). Suppose you did not know an exact amount, but you wanted to make the bunny turn around and around without stopping? How would you want to describe the speed that the bunny turns?

A breakdown of the answers appears below:

Turns/Second	22
RPM	9
Unitless 1-10	7
Fast/Medium/Slow	6
Degrees/Second	3
Seconds/Turn	2
Radians/Second	1
TOTAL	**50**

Notice that turns-per-second is a clear favorite and that degrees per second, *the units we, the engineers had chosen, came in fifth*. In reaction to this, we now specify rotational speed in turns-per-second, and angular amounts in turns. In retrospect, it seems very natural to express a "quarter turn" by typing bunny.turn(left, 1/4).

Other Observations About Novices

Typing is Hard – Most of our users were non-typists and appreciated any help we could give them (mouse control, dialogs, etc.) that would keep them from having to use the keyboard. We are currently working on addressing this issue in Alice.

Problems in 3D Perception – A small percentage (~5%) of our subjects were confused about the depth of objects on screen, sometimes mistakenly seeing objects as approaching or receding when in fact they were being resized. Shadows or other depth cues might help reduce these problems.

High Expectations – Our subjects often expected collision detection and gravity and were surprised when objects passed through each other or hovered in mid-air.

The Importance of 0 and 1 – When faced with a new Alice command that required a numeric parameter, we saw many users try using a "1" to see what would happen for a wide variety of data types (distance, color, time). Partly due to this, we adopted a convention that all bounded scalar parameters to Alice calls would range between 0.0 and 1.0. "Magic ranges" like 0..255 and 0..32767 do not hold much appeal for novices.

RELATED WORK

LOGO [16], Bolio [25], and the Alternate Reality Kit (ARK) [19] and the animated Self programming environment [22] were all strong influences in the Alice project.

Smalltalk [6] and HyperCard [13] both demonstrated that programming-in-the-small was feasible by nonprogrammers.

The Simple User Interface Toolkit (SUIT) [14] used a two-user protocol to test an API for novice GUI programmers.

BAGS (Brown Animation Generation System) [20], was one of the first interactive 3D systems to use an interpreted language to describe the static layout and dynamic behavior of a 3D scene.

Like Alice, Obliq 3D [11] uses an interpreted scripting language for 3D graphics, but unlike Alice, is designed for experts.

Superscape [21] and WorldUp [18] include advanced geometric modeling capabilities and scripting languages. WorldUp shares some ease-of-use goals with Alice, but has a very different model for the distribution of scripts and the timing of animations.

FUTURE WORK

Decomposing complex animations (e.g. walking) into Alice animation primitives is still too hard. A richer set of animation primitives might help.

Writing a serial sequence of code (A then B) requires too much syntax, due to Alice's implicit threads.

We need to find ways of exposing the number and order of parameters to a function to ease the burden of typing.

There are times when programming declaratively (e.g. constraints) is more natural than expressing solutions procedurally, and vice versa. Finding the correct mix of programming styles, while folding in the advantages of some of the other animation paradigms (e.g. keyframing) remains an open problem.

SUMMARY OF LESSONS LEARNED

- Forward/Left/Up is an improvement over XYZ.
- Coordinate transformations can be made easier by allowing other objects to act as the frame of reference in which other operations happen.
- Function overloading and optional keyword parameters in a programming language can be used to support the *controlled exposure of power*, masking API complexity until the user is motivated to use it.
- Matrices appear nowhere in the Alice API.
- APIs can and should be tested against real users from one's target audience.
- Marking some objects as *first class objects* is a powerful technique for segmenting one object from another in the object tree.
- All commands should animate by default, including Undo.
- Implicit threads make it possible for novices to control surprisingly complex animations.
- Object resize and the scaling of space are both useful, but should be presented to the user as two distinct operations.
- Surface characteristics of programming languages matter to novices, especially case sensitivity and careful name choices.
- All bounded, scalar parameters should have a valid range of 0.0 to 1.0.

CONCLUSION

Alice represents the culmination of many independent design decisions based on hundreds of observations of novices. These decisions combine to form a 3D graphics API that allows 3D script writing with minimal distraction by "unrelated" issues. As one researcher in the field kindly noted, current tools inflict the "death of a thousand cuts" compared Alice's "joy of a thousand tickles." Although originally designed for undergraduates, we have observed that many middle and high school students are capable of using Alice to build interactive 3D graphics programs. Alice is available for free from http://www.alice.org. We have currently distributed over 50,000 copies of Alice.

REFERENCES

1. Card, S. K. Robertson, G., and Mackinlay, J. The Information Visualizer, an Information Workspace. *ACM SIGCHI 91 Conference Proceedings*, 1991, pp. 181-188.

2. Clarke, J. H. Hierarchical Geometric Models for Visible Surface Algorithms. *Communications of the ACM*, 19(10), October 1976, pp. 547-554.

3. Clay, S. R., and Wilhelms, J. Put: Language-Based Interactive Manipulation of Objects. *IEEE Computer Graphics and Applications*, March 1996. Vol 16, Number 2, pp. 31-39.

4. Fitts, P. M., and Jones, R. E. Pychological Aspects of Instrument Display: Analysis of 270 "Pilot Error" Experiences in Reading and Interpreting Aircraft Instrument. *Memorandum Report TSEAA-694-12A*, Aero Medical Labaroatory, Air Materiel Command, Wright Patterson Air Force Base, Dayton, Ohio, October 1, 1947, pp. 47.

5. Foley, J. D., van Dam, A., Feiner, S. K., and Hughes, J. F. *Fundamentals of Interactive Computer Graphic*, Addison-Wesley Reading, MA 1990.

6. Goldberg, A., and Robson, D. *Smalltalk80: The Language*, Addison-Wesley, Reading, MA, 1989.

7. Gossweiler, R., Long, C., Koga, S., and Pausch, R. DIVER: A Distributed Virtual Environment Research Platform. *IEEE Symposium on Research Frontiers in Virtual Reality*, October 25-26, 1993, San Jose, CA, pp. 10-15.

8. Lasseter, J. Principles of Traditional Animation Applied to 3D Computer Animation. *SIGGRAPH 87 Conference Proceedings*, pp. 35-44.

9. Mackinlay, J. D., Card, S. K., and Robertson, G. G. Rapid Controlled Movement Through a 3D Virtual Workspace. *ACM SIGGRAPH 1990, Conference Proceedings*, pp 171-179.

10. Marrin, C., and Kent, J. Proposal for a VRML Script Node Authoring Interface, *VRMLScript Reference*, Silicon Graphics, Inc.October 6, 1996.

11. Najork, M. Obiq-3D Tutorial and Reference Manual. *DEC SRC Research Report #129*, December 1, 1994.

12. Nielsen, J. *Usability Engineering*, Academic Press, Boston, 1993.

13. Nielsen, J., Frehr, I., and Nymand, H. O. The learnability of HyperCard as an object-oriented programming system. *Behaviour & Information Technology* 10, 2 (March-April), 111-120.

14. Pausch, R., Conway, M., and DeLine, R. Lessons Learned from SUIT, the Simple User Interface Toolkit. *ACM Transactions on Office Information Systems* October 1992, 10:4, pp. 320-344.

15. Pausch, R., Snoddy, J., Taylor, R., Watson, S., and Haseltine, E. Disney's Aladdin: First Steps Toward Storytelling in Virtual Reality. *ACM SIGGRAPH 96 Conference Proceedings*, August 1996.

16. Papert, S. *MindStorms: Children, Computers, and Powerful Ideas*, Basic Books, New York, 1980.

17. Robertson, G. G., Card, S. K., and Mackinlay, J. D. The Cognitive Coprocessor Architecture For Interactive User Interfaces. *ACM Symposium on User Interface Software and Technology*, 1989, pp. 10-18.

18. Sense8 Corporation: http://www.sense8.com.

19. Smith, R. B. The Alternate Reality Kit: An Animated Environment for the Creation of Interactive Simulations. *Proceedings of the 1986 IEEE Computer Society Workshop on Visual Languages*, 1986, 99-106.

20. Strauss, P. BAGS: The Brown Animation Generation System. *Technical Report No. CS-88-22*, Brown University, May 1988.

21. Superscape: http://www.superscape.com.

22. Ungar, D., and Smith, R. SELF: The Power of Simplicity. *OOPSLA 87, Conference Proceedings*, published as SIGPLAN Notices, Volume 22, Number 12, 1987, pp. 227-241.

23. van Dam, A., et. al. PHIGS+ Functional Description Revision 3.0, *Computer Graphics* 22, 3, (July 1988), 124-218.

24. van Rossum, G., and de Boer, J. Interactively Testing Remote Servers Using the Python Programming Language. *CWI Quarterly*, Volume 4, Issue 4 (December 1991), Amsterdam, pp 283-303. For more information on Python, see http://www.python.org.

25. Zeltzer, D., Pieper, S., and Sturman, D. J. An Integrated Graphical Simulation Platform, *Graphics Interface 89 Conference Proceedings*, pp. 266-274.

The Task Gallery:
A 3D Window Manager

George Robertson, Maarten van Dantzich, Daniel Robbins, Mary Czerwinski,
Ken Hinckley, Kirsten Risden, David Thiel, and Vadim Gorokhovsky

Microsoft Research
One Microsoft Way
Redmond, WA 98052, USA
Tel: 1-425-703-1527
E-mail: ggr@microsoft.com

ABSTRACT

The Task Gallery is a window manager that uses interactive 3D graphics to provide direct support for task management and document comparison, lacking from many systems implementing the desktop metaphor. User tasks appear as artwork hung on the walls of a virtual art gallery, with the selected task on a stage. Multiple documents can be selected and displayed side-by-side using 3D space to provide uniform and intuitive scaling. The Task Gallery hosts any Windows application, using a novel *redirection mechanism* that routes input and output between the 3D environment and unmodified 2D Windows applications. User studies suggest that the Task Gallery helps with task management, is enjoyable to use, and that the 3D metaphor evokes spatial memory and cognition.

Keywords: Window managers, 3D user interfaces, spatial cognition, spatial memory.

INTRODUCTION

Management of multiple user tasks is an activity that, if made easier, could help enrich users' computing experience. A task is a collection of documents and applications organized around a particular user activity. For example, a user may rapidly switch between working on finances, writing a paper, and managing correspondence. Each of these may involve many applications. Task management has several components: creating, locating, and bringing tasks into focus, and window management within a task. Within a task, users need to manage placement and size of windows and quickly shift focus of attention from one window to another. Users also need to be able to bring relevant information to bear on the task being performed. In some cases, this requires bringing two or more windows into a useful view simultaneously.

Permission to make digital or hard copies of all or part of this work for personal or classroom use is granted without fee provided that copies are not made or distributed for profit or commercial advantage and that copies bear this notice and the full citation on the first page. To copy otherwise, to republish, to post on servers or to redistribute to lists, requires prior specific permission and/or a fee.
CHI '2000 The Hague, Amsterdam
Copyright ACM 2000 1-58113-216-6/00/04...$5.00

Figure 1. The Task Gallery.

The Task Gallery is designed to meet the goals of task management, while providing other features available in a window manager. It is a 3D environment designed so users can be productive using familiar, existing applications. Our design premise is that 3D virtual environments can more effectively engage spatial cognition and perception. Almost all new personal computers are now delivered with 3D graphics acceleration hardware. Although this innovation has been driven by the computer game industry, it could usher in a whole new class of productivity applications with 3D interfaces.

In the Task Gallery (Figure 1), the current task is displayed on a stage at the end of a virtual art gallery. It contains opened windows for that task. Other tasks are placed on the walls, floor, and ceiling of the gallery. The user switches to a new task by clicking on it, which moves it to the stage. Viewing multiple windows simultaneously is done with a button click, using automatic layout and movement in the 3D space to provide uniform and intuitive scaling. Applications and frequently used documents are kept in a Start Palette (Figure 6, described later) carried in the user's virtual left hand. Studies suggest that users are enthusiastic about the Task Gallery, that it is easy to navigate the space, and that it is easy to find tasks and switch between them. The Task Gallery uses a novel redirection mechanism for routing input events and graphics output between the 3D environment and existing, unmodified Windows applications.

PREVIOUS WORK

The window manager is the part of the computer user interface that manages display and input device resources [7]. It allows the user to bring up windows, menus, and dialogue boxes associated with running applications, manipulate windows, and minimize them. It takes mouse and keyboard input and directs it to the appropriate applications. The window manager also determines the look and feel of much of the user interface.

Window management systems have been a fundamental part of computer user interfaces for the last 25 years. From the mid-1970's to the mid-1980's, there was much research on window systems [2, 12, 13, 20, 21, 22]. By the mid-1980's, Unix and MacOS window management had converged on the desktop metaphor with overlapped windows [18]. This metaphor has served the computer industry well for 15 years, making it possible for many new users to use computers effectively.

The desktop metaphor has changed little since it was created. However, the way computers are used has changed significantly. The growing range of applications and online services have made computers applicable to many more real-world activities. People often engage in a number of tasks and need to switch between them frequently and quickly [1, 9]. In the desktop metaphor, switching between tasks can involve dozens of operations (iconifying, opening, moving and resizing windows). Users often need to see multiple documents simultaneously [10]. Again, this can take many steps (opening windows, moving and resizing them, and scrolling). The desktop metaphor has inadequate support for task switching, leading to wasted effort and frustration on the part of the user [4, 9, 10, 16, 17]. In the Task Gallery, switching between tasks and viewing multiple windows simultaneously are simple actions. In addition, the Task Gallery provides a strong spatial framework for encoding location information and front to back relationships, thereby engaging the user's spatial memory to help retrieve tasks and services.

Rooms [9] was created to deal with problems that early PC users had in managing their tasks [1]. Users switch between tasks frequently and there is strong locality of window reference based on task: particular windows are associated with particular tasks. This can be exploited by creating a visible representation of a task, and allowing the user to easily switch between tasks (which combines opening, sizing, and placement of multiple windows into one act). The Task Gallery takes advantage of the user's spatial memory for task management, while Rooms lays out tasks in a linear alphabetic order. The Task Gallery currently lacks one feature of Rooms, which is the ability to share a window so that it appears in multiple tasks; however, we plan to implement this in the future.

The Andrew window system [12] explored a space-filling tiled window layout, where windows are resized automatically (when one window grows others shrink by cropping to keep the space filled). Users found it confusing and the approach was abandoned.

Elastic Windows [10] also uses a space-filling tiled layout, with tasks replaced by hierarchical user roles. The lowest level role is similar to the Rooms notion of task. The Task Gallery returns to tasks as the basic unit, and uses spatial layout of tasks for task management, instead of a role hierarchy. Elastic Windows addresses the problem of simultaneous display of multiple windows by allowing the user to create a container into which multiple windows can be dragged. The Task Gallery has similar functional advantages, but no special container is needed and only a single button click is required to select each window. The Task Gallery also maintains spatial continuity whereas Elastic Windows can do significant window repositioning.

3D Rooms [16] was built as an information workspace that used 3D virtual environments to extend the ideas of Rooms. This was not actually a window manager; abstract information visualizations replaced windows. The basic motivation was to engage human spatial cognition and perception in order to make task management easier. The Task Gallery shares that motivation, but manages windows associated with existing applications in order to bring the advantages of human spatial cognition and perception to our current set of computer applications.

Web Forager [4] and Data Mountain [17] are virtual environments designed for managing documents. They each use a 3D virtual environment to more fully engage human spatial cognition and memory. Studies of the Data Mountain [5, 17] demonstrate that placing documents in a 3D space helps the user remember where the documents are during later retrievals. The Task Gallery also seeks to use spatial memory to help the user remember where tasks are placed in the gallery.

TASK GALLERY DESIGN

The choice of a navigable spatial metaphor was partly motivated by a desire to leverage human spatial memory [17]. An art gallery was chosen because of its familiarity. To increase ease of retrieval, the Task Gallery includes the images of documents and tasks in the space in addition to their spatial location and title cues. Classical mnemonic research has documented that mental cues in the form of visual images are an excellent way to enhance memory for items [14]. Our previous studies have shown the strong influence of snapshot/thumbnail cues to aid spatial memory during the storage and retrieval of web pages [5].

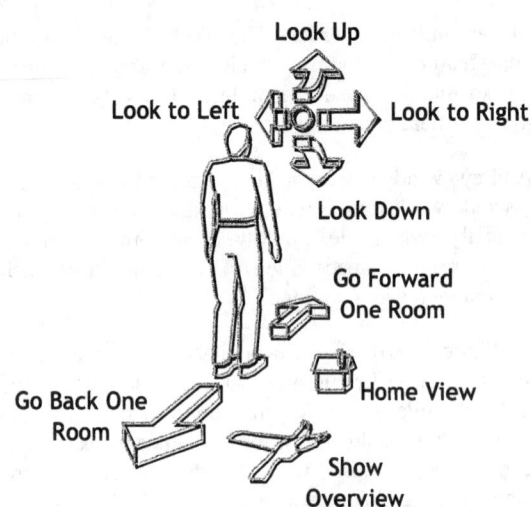

Figure 2. On-screen 3D navigation controls appear in the lower left corner of the screen.

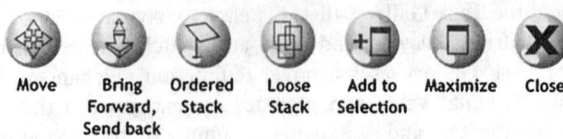

Figure 3. Window manipulation controls appear over a window banner when the user points to it.

Animation is used to reinforce the spatial metaphor. For example, when a user switches to a task by clicking on it, a one-second animation starts. The current task is closed by creating a snapshot which is moved back to the task's frame in the gallery. The newly selected task is then moved from its frame to the stage. When it arrives at the stage, it is transformed from artwork into live windows. A "ghosted" view of the task remains in the gallery, to mark the spot that it came from.

The initial and primary working view is a close-up of the stage (Figure 4), showing the current task and its live windows. To view other tasks, the user backs up to see more of the gallery, as in Figure 1. The gallery is composed of a sequence of rooms, with only one closed end; more rooms are revealed without limit as the user moves back. This provides a simple way of managing the user's attention. As the user backs away, attention is widened. Moving to the stage focuses attention on the current task.

The user can move tasks wherever desired with a dragging movement. Tasks are constrained to remain on walls, floor, or ceiling, but can be moved between these surfaces in a way inspired by Object Associations [3]. The transition from wall to floor, for example, causes the task to shift to the appropriate orientation on the floor. Task frames are tilted outward so that they are more legible from a distance. Task frames on walls are mounted on a stand to make the metaphor more obvious and to ground them visually in depth. Segmentation of the gallery into separate rooms, grouping of task windows into mounted pieces of artwork, and using distinctive backgrounds all provide landmark and spatial cues that act as memory aids.

Users (especially non-gamers) tend to get lost in many 3D systems that require them to navigate. We avoid this problem by keeping the space simple (a linear hallway), by choosing a metaphor appropriate for the context (viewing art in a gallery), and by constraining the navigation. Thus, we provide a few simple controls rather than a general egocentric navigation mechanism. Figure 2 shows these on-screen controls, which allow the user to "jump" backward, forward, home (primary view), and to a bird's eye view showing all the tasks in the Task Gallery. Each jump control starts a one-second camera animation from the current position to the desired target. Our studies showed that users did not become disoriented in the 3D space when using these controls, and that they could easily find their desired tasks.

New tasks can be created by picking the "new task" item on a menu or on the Start Palette (described later). A background image is chosen by the system to distinguish the new task from existing tasks. The user's desired location of the new task is not yet known, so it is placed on the floor in front of the stage. Other tasks on the floor are moved back away from the stage to make room for the new task. This is done with a three-step animation: move the camera back to make the action visible, move the tasks on the floor back and place the new task on the floor, and finally do a task switch as described earlier. The three-step animation was implemented as a result of user testing, and greatly improved the usability of task creation. It is assumed that the user will move the task to a more appropriate location in the gallery later.

Window Management

The current task on the stage is composed of several components, including a loose stack, an ordered stack, and a selected windows set. The loose stack is used for overlapped windows in much the same way as the current desktop metaphor. These windows are mounted on stands to visually ground them to the stage. Clicking on one of these windows will bring it forward to a selected window position, replacing the current selected window. The window manipulation controls shown in Figure 3 are used for moving windows around and placing them on various stacks. These controls appear over the window banner when the user points to the banner. Windows in the loose stack can be directly moved anywhere on the stage, using a technique similar to Point of Interest object movement [11]; mouse movement controls movement in the plane perpendicular to the line of sight, and the shift and control keys control movement toward or away from the user.

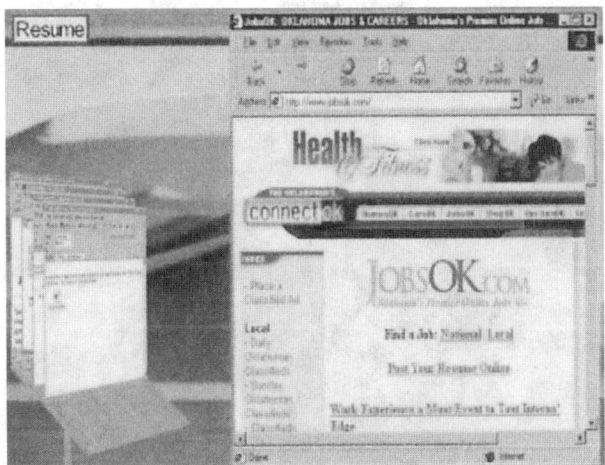

Figure 4. Ordered stack, one selected window.

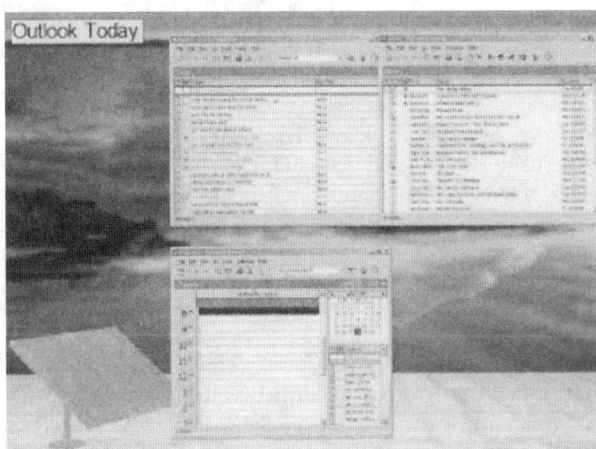

Figure 5. Multiple selected windows.

The ordered stack appears to the left of the stage, as shown in Figure 4. Users choose to place windows in the ordered stack to keep currently unused windows organized (e.g., open email messages). If one of the windows on the podium is moved, the stack is tidied to have a fixed distance between each window. Clicking on a page in the ordered stack moves it to the selected window region.

When windows are selected, the system moves them closer to the user for greater legibility. Multiple windows can be selected using the "Add to Selection" icon in Figure 3. Each time a window is added, an automatic layout moves the windows so they are all visible side by side (Figure 5). Unlike tiled window managers that crop windows and may force users to scroll, this operation does not affect what is visible in selected windows. Thus we use distance in 3D to provide uniform scaling in an intuitive way.

Toolspaces

The existing Windows desktop metaphor uses menus (especially the Start Menu) and toolbars to give the user access to commonly used tools and documents. To better fit the metaphor of moving through a hallway, we designed the Task Gallery so that the user carries tools and documents associated with the virtual body, using an adaptation of Glances and Toolspaces [15]. Glances are a lightweight, ephemeral way of looking around in a virtual environment without moving the virtual body. Toolspaces are placed around the user, and hold various tools or objects, keeping them associated with the virtual body as it moves through the virtual environment.

The Task Gallery has toolspaces left of, right of, above, and below the user. Hands and feet are shown to make the scale of the objects in the toolspaces more obvious and to suggest that these tools stay with the user as the user navigates through the environment. In the Task Gallery, glances are initiated with the controls shown in Figure 2. Glances remain in effect until the user selects something in the toolspace or glances elsewhere.

The Left toolspace contains the "Start Palette", which is a Data Mountain [17] with the appearance of an artist's palette (Figure 6). The Data Mountain was originally a tilted plane in 3D holding favorite web pages. The objects on the Start Palette are icons and snapshots for applications, favorite documents, or web pages. The behavior of the Start Palette is similar to a Data Mountain, including object movement and occlusion avoidance. The only difference is that selecting an object from the Start Palette causes an application to launch with its window(s) in the current task. When an application is launched, the glance is terminated. Our user testing demonstrated that participants learned to add applications and documents to their tasks easily using the Start Palette. Earlier studies of the Data Mountain [17] suggest that users should be able to find icons on the Start Palette much faster than in the traditional Start Menu.

TASK GALLERY USER TESTING

During the design and implementation of the Task Gallery, we gathered empirical evidence to support our design decisions. Our first three studies examined task management before and after various usability issues were resolved. The third study took place several months after the first two, and included evaluation of features added in response to the first two studies (e.g., icon identification). In addition, we were interested in how spatial cognition pertains to 3D environments like the Task Gallery, and whether or not aspects of real world spatial location memory transfer to electronic environments.

We were interested specifically in how well users could create and modify tasks and arrange the overall task space. In addition, detailed information about organizing and retrieval strategies was collected, to support those strategies in future designs. We wanted to know whether organizing strategies were based on frequency, size, type of content or time. While the art gallery metaphor suggests use of the walls over the floor and ceiling, previous research suggests that certain bodily axes are considered primary in the real world [8, 19]. We wanted to know if participants' organizing strategies and subsequent retrieval performance and representation of the space related to properties of the metaphor or to up-down, front-back and left-right axes relative to the user's orientation in the space.

We also wanted to know how subjectively satisfying participants considered the 3D user interface for task management. A benchmark comparison between the Task Gallery and Windows is planned after further design iteration.

Experiments 1 and 2—Prototype System
Method
Participants
Eleven participants (5 female) between the ages of 16-65 participated in two iterations of the same study. All were intermediate to expert Windows users. Five participants evaluated the first iteration prototype, and six participants evaluated the second iteration prototype.

Materials
Materials included two prototype versions of the Task Gallery, which used "snapshots" of documents instead of live applications. The prototype environments were fully interactive except that the applications were not live. Eight tasks and their contents were created prior to the study, based on common computer tasks collected during actual Windows' user home visits. Tasks typically contained 2 to 5 documents (like Word documents, Excel spreadsheets, web pages, and email). Images of documents comprising these tasks were saved onto the Start Palette in a default arrangement used for each subject. During the study, participants used the Start Palette for items to add to tasks and to create new tasks. The Start Palette had 33 items.

The study was run on a 300Mhz Gateway Pentium computer with a (1024x768 resolution) 15" NEC Mutisync LCD flat panel monitor. Participants interacted with the software using a standard Microsoft serial mouse. No audio was included in this prototype.

Procedure
Participants carried out 6 tutorial trials, and 20 experimental trials. In the tutorial trials, users were introduced to the concepts of navigating, selecting, and arranging documents and tasks in the environment. Once it was determined that the participants could perform all of the tutorial trials easily, the experimental trials were begun. During the experimental trials, users created tasks, organized the tasks in a way that was meaningful to them, retrieved eight tasks and their specific content items, and finally carried out various Windows operations. After the first experimental trial, we asked users to draw what the hallway looked like to them, and what location and orientation they had within the hallway. At the end of the session, users drew their information layout in the hallway in as detailed a manner as they were able. In addition, they filled out a user satisfaction questionnaire.

Figure 6. Start Palette - A Data Mountain held in the user's left-hand toolspace.

Between the first and second study, several changes were made to the prototype in response to observed user problems. We changed the manner in which tasks were created, named, and labeled when selected.

Results
Trial Times
Trial times for each subject were averaged across trials for each trial type in the experimental phase of the study. Overall trial times improved after changes were made to the user interface by about 7 seconds (range 25.4 to 9 seconds), on average. None of the performance improvements reached statistical significance due to the small number of participants and the large individual differences observed.

Organizational Strategies
Participants placed significantly more tasks on the left and right walls of the gallery than the ceiling or floor. On average 4.18 of the tasks were placed on the walls while 2.18 were placed on the ceiling and floor ($t(10) = 2.54$, $p < .05$). This tendency to conform to the way space is typically used in a real world gallery suggests that participants were using the metaphor to guide interaction. Legibility was the same on walls, floor, and ceiling in these two studies.

Participants' organization of tasks involved spatially grouping related tasks. Tasks that "went together" were placed close together on the same surface. A variety of organizational strategies were observed including ordering by frequency of use, location of use (i.e., home versus work), semantic relations, and alphabetical. Furthermore, most participants used more than one organizing strategy.

Spatial Representations
Most participants thought of the hallway as a square, rectangle, or quadrilateral in shape. All participants correctly identified their face forward orientation. The fact that participants chose a canonical gallery shape suggests that the metaphor and 3D cues were sufficient for them to perceive a 3D space.

Participants' drawings of the Task Gallery were scored for correct recall of tasks and correct placement of recalled tasks in their depictions of the space. Sixty-four per cent of the tasks were correctly recalled, and forty-six percent were correctly recalled and also drawn in the same position that they were placed in the Task Gallery. These are acceptable figures given the complex nature of the environment, and the fact that participants were not told they would be given a memory test. There were no reliable differences between tasks that were "pre-arranged" in the space and those that the participants created themselves. This suggests that an initial default Task Gallery layout could be provided for users, to simplify the work of laying out tasks.

Eighteen percent of the tasks were recalled but placed incorrectly. Analysis of those errors showed that it was more difficult to remember whether a task had been placed on the left or right wall than to remember its depth order (i.e., was it closest to the stage, next closest, and so on). Ninety-two per cent of the placement errors were due to drawing tasks on the wrong wall. Only eight percent of these errors were due to drawing tasks in the wrong relative depth order ($t(5) = 2.74$, $p < .05$). This is consistent with the literature on memory for spatial arrays [8, 19], which finds that front-back relations are easier to represent than left-right relations. This supports our design by showing that users leverage the front-back relations afforded by the use of 3D to represent and recall task location.

User Satisfaction Ratings
Overall, user satisfaction ratings were positive, given that this is the first evaluation of the prototype. Average satisfaction ratings were 4.9 for both the first and second iterations, using a 7 point scale, with 7=positive.

Experiment 3—Live Task Gallery
Method
Participants
Nine participants (3 female) between the ages of 16-52 participated in this iteration of testing with a version of the system including live Windows applications. All were intermediate to expert Windows users.

Materials
For this study, eight tasks and their contents were identified and created prior to the study. Tasks typically contained between 5 and 11 documents (like Word documents, Excel spreadsheets, web pages, and email). Note that this iteration of testing included many more documents in tasks than the previous two iterations, as we were interested in how the Task Gallery might scale up to larger numbers of documents. Therefore, we did not attempt any quantitative comparisons to the previous two iterations. As before, thumbnails of documents comprising these tasks were saved onto the Start Palette in a default arrangement that was used for each subject. During the study, participants went to the Start Palette to get items to add to tasks and to create new tasks. There were 27 items on the Start Palette.

The study was run on a 400Mhz Gateway Pentium computer with a (1024x768 resolution) 15" NEC Mutisync LCD flat panel monitor. Participants interacted with the software using a standard Microsoft serial mouse. Fully spatialized audio was used in this iteration of testing, which had not been available in the earlier iterations.

Procedure
Participants carried out icon identification and purpose-matching on the windows controls shown in Figure 3. These identification and matching tasks were carried out on paper, requiring subjects to label and match pictures of the icons to their actual functions without ever using the Task Gallery or seeing hover text titles. Next, participants ran through 2 tutorial trials introducing them to the navigation and window controls, and 12 experimental trials. Five tasks were pre-arranged in the Task Gallery, and these tasks were used in the early phase of the experiment in order to give users a reason to move about the environment (i.e., familiarizing themselves with the predetermined layout). Users were introduced to the notion of arranging various documents into tasks, which could be saved away to a permanent spatial location (the floor, either wall, or the ceiling). Next, the users created three new tasks and saved them during the experimental trials. During the experimental trials, users organized all the tasks in a way that was meaningful to them.

Results
Icon Identification and Matching
On average, users identified the icons 44% of the time and matched the icons correctly 48% of the time. Given the users had not seen the Task Gallery nor did they know what could be done in the environment at the time of the icon evaluation, this is not a surprising result. After using the system for under 10 minutes, all users understood how the novel 3D windows controls operated, and what their unique functions were.

User Satisfaction Ratings
Satisfaction ratings were even higher with this iteration. The overall average satisfaction ratings were 5.3 (standard deviation=1.4) using a 7 point scale, with 7=positive. The only satisfaction question which received a lower than average rating was "I always knew what to do in this software", with an average of 3.1 (1.1 standard deviation). Given the highly novel nature of this environment for most users, we believe a lower rating here is acceptable for a first session, but may suggest an area to focus on for improvement. On average, users rated the Task Gallery as preferable to their current Windows software (average =5.0, 7=prefer Task Gallery).

Spatial Representation
We asked participants where they had laid out their tasks at the end of the session, and why they chose those spatial locations. The majority of the participants felt that placing tasks on the ceiling or floor would violate the Task Gallery

metaphor. Some participants simply did not like the idea of tasks lying on the floor. Two participants, however, mentioned that tasks on the ceiling and floor were more difficult to read, due to the angle at which they are placed. This was not true in the prototype tested in experiments 1 and 2. Legibility problems were introduced in the final version of the system as a result of addressing some serious texture management issues. We are currently exploring alternative layouts to make tasks on the ceiling and floor easier to read.

Usability Issues

We identified several usability issues in this iteration of testing. Some of the novel icons and controls have already been changed based on feedback. In particular, many of the navigation and control icons were confusable or not grouped properly by function. In addition, it was clear that some participants had trouble differentiating glances from hallway navigation. Some users wanted to multi-select items from the palette for addition into a new task. These issues will be addressed in a revised design.

IMPLEMENTATION ISSUES

The key technical challenge in building a 3D window manager is to get existing applications to work in the 3D environment without changing or recompiling them. This requires both output and input redirection facilities in the operating system. Output redirection causes applications to render to off-screen bitmaps instead of the screen, gain access to those bitmaps so that they can be used as textures in the 3D environment, and receive notification whenever an application has updated its visual display. Input redirection causes mouse and keyboard events to be received by an application rather than the 3D environment's main window, but with mouse coordinates translated from 3D to 2D.

The details of our implementation unfortunately fall beyond the scope of the current paper, but will be published in a separate paper. Although these details are specific to the Windows 2000 operating system, the components needed will be similar in other operating systems. For example, similar changes are possible for any OS that uses the X Window System, as long as the X server runs on the same machine as the client applications and the window manager so that bitmap sharing is efficient. For example, Feiner [6] modified an X server to put 2D windows into a 3D augmented reality.

The Task Gallery runs on current high-end PCs with a modified version of Windows 2000 and a standard 3D graphics accelerator (NVidia TNT2). All Task Gallery code was implemented in C++, using Win32 and Direct3D APIs.

Task Persistence

One key problem which Rooms [9] faced was the capture of information necessary to persist the state of tasks with running applications, so that on restart all of those applications are re-launched and the user sees exactly the same layout last seen in each task. The Task Gallery faces the same problem. The best we can currently do is to record the information used to launch the application. Unfortunately, that is far from ideal. Applications allow the user to change what files are open, and some even provide a form of window management within the application. Without some standard way of getting the state of open files and sub-windows within an application, it is extremely difficult to solve the general problem. Some Windows applications allow inspection of their open documents through COM interfaces. We are exploring what can be done by tracking file opens and closes and window creation, but this approach is difficult without modifying existing applications.

DISCUSSION AND FUTURE WORK

The Task Gallery is an exploration of the use of 3D virtual environments for window and task management. It is motivated by the desire to leverage human spatial cognition and perception and to take advantage of the coming ubiquity of 3D graphics hardware for more than computer games. Early user tests suggest that the Task Gallery does help with task management and is enjoyable to use. But we have only scratched the surface.

In our usability studies we observed users exhibiting many of the same principles of spatial cognition as are exhibited in the real world. Users had a strong sense of front to back ordering of their tasks, rarely confusing that ordering in memory. We will continue to explore metaphors leveraging users' real world knowledge in our future 3D environments.

There are a number of things that we plan to do as we continue to evolve the Task Gallery. Better landmarks could make a significant difference in helping users remember on which wall they placed tasks. The Data Mountain occlusion avoidance algorithm can be used to help avoid occlusion problems while moving task frames. As discussed earlier, the task persistence mechanism may benefit from application-level changes, although we hope to avoid those. These changes will make it possible to effectively use the Task Gallery as a replacement for the current desktop on a day-to-day basis. Once these necessary changes are made, we intend to do a benchmark study comparing the Windows desktop shell with the Task Gallery. Beyond that, we plan to explore integration of novel uses of 3D visualizations living side-by-side with existing Windows applications.

Our goal was to design a 3D window manager that solves two problems with the current desktop metaphor: task management and comparison of multiple windows. The Task Gallery is a first-generation system that addresses these issues, and is built on a general-purpose application redirection technology which will allow us to explore alternative user interfaces for application environments.

ACKNOWLEDGMENTS
We thank Matt Conway, Susan Dumais, George Furnas, and Dennis Proffitt for valuable assistance during the design of the Task Gallery. The Windows 2000 USER and GDI teams (primarily one of the authors, VG) were instrumental in architecting appropriate low-level support for this work; we owe great thanks in particular to Corneliu Lupu and Andrew Goossen.

REFERENCES
1. Bannon, L., Cypher, A., Greenspan, S., Monty, M. L., Evaluation and analysis of users' activity organization, *CHI'83*, ACM, New York, NY, (1983), pp. 54-57.

2. Bly, S., Rosenberg, J., A comparison of tiled and overlapping windows, *CHI '86*, ACM, New York, NY, (1986), pp. 101-106.

3. Bukowski, R., and Sequin, C., Object associations: a simple and practical approach to virtual 3D manipulation, in *Proceedings of Symposium on Interactive 3D Graphics*, 1995, pp. 131-138.

4. Card, S., Robertson, G., York, W., The WebBook and the Web Forager: An information workspace for the World-Wide Web, *CHI'96*, ACM, New York, NY, (1996), pp. 111-117.

5. Czerwinski, M., van Dantzich, M., Robertson, G., and Hoffman, H., The contribution of thumbnail image, mouse-over text and spatial location memory to web page retrieval in 3D, *Proceedings of Interact '99*.

6. Feiner, S., MacIntyre, B., Haupt, M., and Solomon, E., Windows on the world: 2D windows for 3D augmented reality, *UIST '93, ACM*, November 1993, pp. 145-155.

7. Foley, J., van Dam, A., Feiner, S., and Hughes, J.F., *Computer graphics, principles and practice.* Addison Wesley, 1996.

8. Franklin, N., & Tversky, B. Searching imagined environments. Journal of Experimental Psychology: General, 199, (1990), pp. 63-76.

9. Henderson, A., Card, S. K., Rooms: The use of multiple virtual workspaces to reduce space contention in a window-based graphical user interface, *ACM Transactions on Graphics 5*, 3, (1986), pp. 211-243.

10. Kandegon, E., and Shneiderman, B., Elastic Windows: Evaluation of Multi-Window Operations, *CHI'97*, ACM, New York, NY, (1997), pp. 250-257.

11. Mackinlay, J., Card, S., and Robertson, G., Rapid controlled movement through a virtual 3D workspace, *SIGGRAPH '90*, ACM, 1990, pp. 171-176.

12. Morris, J., Satyanarayanan, M., Conner, M., Howard, J., Rosenthal, D., and Smith, F., Andrew: a distributed personal computing environment, *CACM*, 29(3), March 1986, 184-201.

13. Myers, B., Window interfaces: A taxonomy of window manager user interfaces, *IEEE Computer Graphics and Applications 8*, 5 (September 1988), pp. 65-84.

14. Patten, B.M. (1990). The history of memory arts. *Neurology, 40*, 346-352.

15. Pierce, J., Conway, M., van Dantzich, M., and Robertson, G., Toolspaces and glances: storing, accessing, and retrieving objects in 3D desktop applications, in *Proceedings of Symposium on Interactive 3D Graphics*, April 1999, pp. 163-168.

16. Robertson, G., Card, S., and Mackinlay, J., Information visualization using 3D interactive animation, *CACM*, 36, 4, (1993), pp. 57-71.

17. Robertson, G., Czerwinski, M., Larson, K., Robbins, D., Thiel, D. & van Dantzich, M.. Data Mountain: Using Spatial Memory for Document Management, *UIST '98*, ACM, November 1998, pp. 153-162.

18. Scheifler, R.W., and Gettys, J., The X window system, *ACM TOG*, 5(2), April 1986, 79-109.

19. Siegel, A., & White, S. The development of spatial representations of large-scale environments. In H. Reese (Ed.), Advances in child development and behavior, 10, (1975), pp. 9-55. NY: Academic Press.

20. Smith, D.C., Irby, C., Kimball, R., and Harslem, E. The Star user interface: An overview. In *Proceedings of the National Computer Conference*, 1982.

21. Teitelman, W., Ten years of window systems – a retrospective view, in Hopgood, F.R.A., et al., eds., *Methodology of window management*, Springer-Verlag, New York, 1986, pp. 35-46.

22. Tesler, L., The Smalltalk environment, *Byte*, 6, 8, August 1981, pp. 90-147.

A Comparison of Tools for Building GOMS models

Lynn K. Baumeister
HCI Institute
Carnegie Mellon University
Pittsburgh, PA 15213
lkb@cs.cmu.edu

Bonnie E. John
HCI Institute
Carnegie Mellon University
Pittsburgh, PA 15213
bej@cs.cmu.edu

Michael D. Byrne
Department of Psychology
Rice University
Houston, TX 77005
byrne@acm.org

ABSTRACT

We compare three tools for creating GOMS models, QGOMS [2], CATHCI [17] and GLEAN3 [12], along several dimensions. We examine the representation and available constructs in each tool, the qualitative and quantitative design information provided, the support for building cognitively plausible models, and pragmatics about using each tool (e.g., how easy it is to modify a model). While each tool has its strengths, they all leave something to be desired as a practical UI design tool.

Keywords: Tool support for evaluation, GOMS.

INTRODUCTION

The utility of GOMS models [3] has precipitated their use in industry (e.g., [2]; Nielsen's sidebar in [5]; and 11 cases in [8]). However, because of the large amount of detail involved in describing the users' *G*oals, *O*perators, *M*ethods, and *S*election rules, GOMS has often been viewed as extremely time- and labor-intensive, a "Cadillac" of techniques to be used only when cost is no object (e.g. [15; 14]). Although we have argued elsewhere that this may be a misconception [8], there is no doubt that appropriate tool-support would ease the burden and allow the method to become more widely used.

Currently, GOMS models are usually built by hand, or with generic tools (e.g., spreadsheets). Occasionally, researchers implement GOMS models in sophisticated computational cognitive architectures like those reviewed in [16], e.g. a GOMS model of browsing implemented in Soar [9], or GOMS models of a telephone operator implemented in EPIC [11]. Both generic tools and computational cognitive architectures do not directly support building GOMS models: the former are not powerful enough, the latter are too complex and time-consuming to be practical for practitioners.

There are a few tools specifically designed to support building GOMS models, but they have not yet seen wide acceptance in any modeling or development community. This paper reviews the three tools that are returned when the GOMS bibliography (http://www.usabilityfirst.com/ goms/gomsbib.html) is searched with the terms "tool" or "editor": QGOMS [2], CATHCI [17] and GLEAN3 [12]. We hope to give UI designers enough information to decide whether they want to use one of these tools in its present form. In addition, by describing weaknesses with the current systems we hope to inspire tool-designers to build superior GOMS tools.

PROCEDURE

To compare these three tools, we built a GOMS model in each tool for the same set of tasks. We first represented the procedures in the task suite using Hierarchical Task Analysis (HTA, [13]) as a baseline from which to build the three GOMS models. HTA is a generic representation of a task analysis that includes all the constructs found in GOMS and is easy for people to understand. After building GOMS models in each tool, we compared them quantitatively on a specific instance of a specific task.

We modeled Atropos (built by the last author), the system used by Student Volunteers at CHI'98 to sign up for work time slots. Atropos is simple enough to model in a tractable amount of time, while still having a goal hierarchy, many operators, and several methods for accomplishing the same goals. Although Atropos is a simple system it exercises all the constructs of GOMS and thus represents a minimal benchmark for tool evaluation. Atropos has also been modeled previously with the Keystroke-Level Model (KLM) version of GOMS, both by hand and by an automatic model-generating tool [4]. Thus, we have additional models to compare to beyond those built with these three tools. The task suite consisted of logging in, adding a work time slot, viewing slots, deleting a slot, and quitting. Our HTA representation of these tasks is shown in Figure 1.

Mapping HTA to GOMS

The HTA maps to a GOMS model using the following conversions. Root nodes (rectangular blocks) in the HTA are equivalent to GOMS goals; they represent a task that requires further sub-division in order for the goal to be completed. Leaf nodes (roundtangles) in the HTA are the equivalent of operators in a keystroke-level GOMS model; they represent the actions (e.g. typing, clicking, pointing) that the user must exercise to accomplish the goal. The plan for each level of the HTA is equivalent to GOMS methods; it captures the different ways the goal may be accomplished. For instance, Plan 1.2 specifies a keyboard method (step 1)

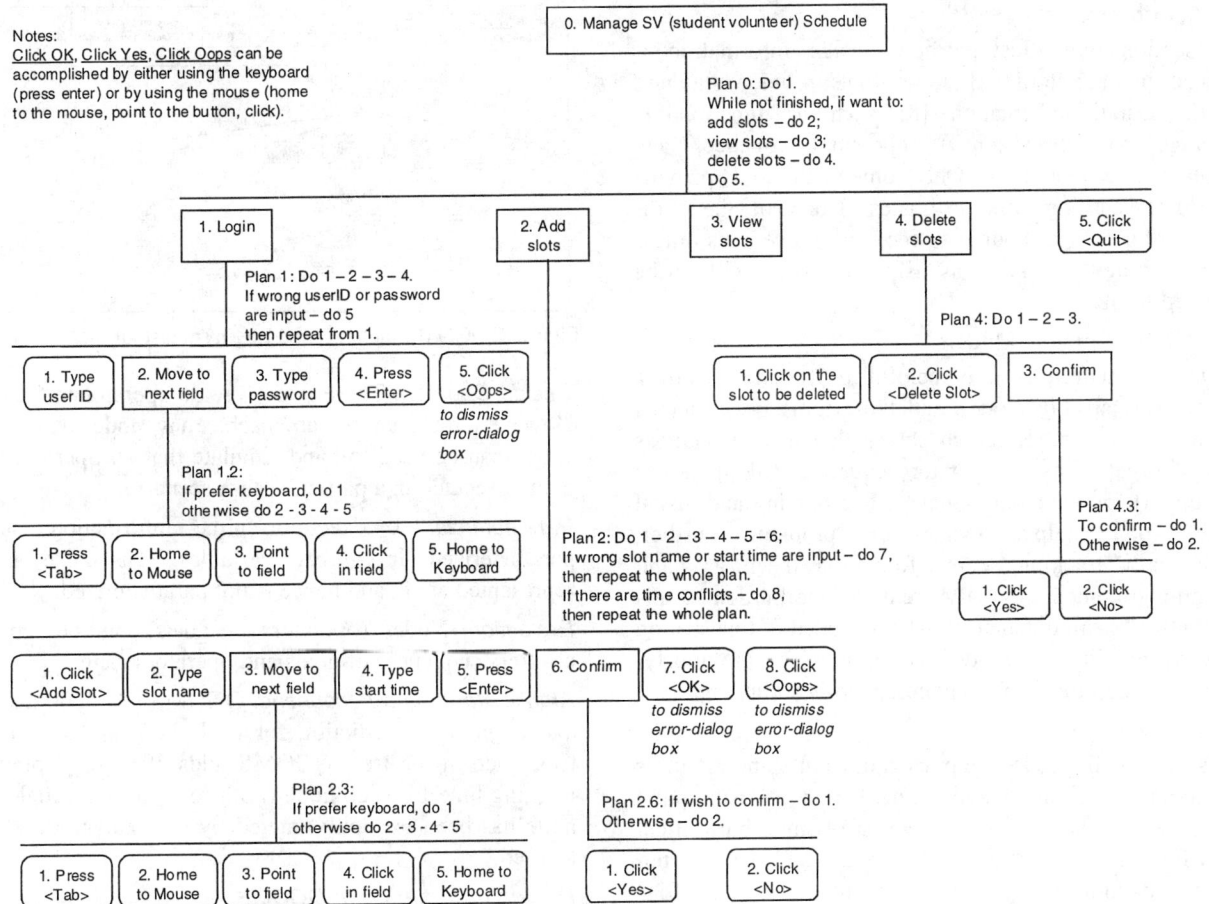

Figure 1. The HTA for Atropos. We deliberately left off checking the location of the user's hand before typing, pointing or clicking operations to illustrate how an analysis can appear complete without all the details. GOMS tools should aid in remembering such details.

and a mouse method (steps 2 - 5) to move to another field. Conditionals in the HTA correspond to GOMS selection rules (in Plan 1.2, "if prefer keyboard") or decide operators (in Plan 1, "if wrong userID or password are input").

DESCRIPTION OF COMPARISON POINTS
Representation and Available Constructs

Basic Representation. This dimension captures how each tool represents the basic constructs of GOMS: goals, operators, methods and selection rules. It also discusses the overall representation of the model, for example, graphically in a tree diagram, or textually in a program.

Decide Operator. The original formulation of GOMS made a distinction between selecting which method to use to complete a goal when several well-known methods exist, and making a decision about task-state as an operation within a method. If there is no Decide operator, there is no ability to have an in-line contingency in a method and to build a complete model every contingency must spawn a new method. This is subtlety of GOMS models that may or may not be embodied in a tool.

Parameterization. There are several possibilities for parameterization in GOMS models. For instance, parameterizing operator execution times allows sensitivity analyses to be performed easily. Parameterizing state information allows multiple benchmark tasks to be easily evaluated with a single model.

Repetition. Since repetition of a subgoal or operator is common, it is helpful to be able to represent loops in a model. For tools that don't provide this capability, the analyst must do the multiplication by hand.

Architectural Time Accounting. There are different ways to include the architectural overhead for execution times. For instance, [3] included execution time only for operators, not for goal manipulation, but [10] included time to push and pop goals. Both accounting schemes have been shown to make good matches to human performance, and the research has not yet been done to discriminate between the two (and pragmatically, it may not matter!). Tools may also include architectural accounting for learning time.

Design Information

This section will illustrate how design information is obtained in each tool. All these tools provide standard GOMS design information [8] such as functionality (coverage and consistency), operator sequence, and execution time predictions. Only some of the tools provide procedural learning time estimates. None of the tools provide information about error-recovery procedures unless they are explicitly included as tasks; this aspect will not be discussed further.

Cognitively Plausible Models

Design information is only helpful if it comes from a cognitively plausible model, and these tools give varying degrees of support. Research shows that novice analysts tend to forget steps in operator sequences [6], therefore tool-support for remembering steps like homing and visual search would help alleviate this problem. Further, automatically tracking state information (such as hand position and contents of LTM) removes the burden of this "accounting" from the analyst. Also included in this section is the availability of pre-defined timings for previously-researched operators such as pointing, typing, clicking, etc.

Using the Tool

This section will discuss the practicalities of using the tools for building a model, changing a model, and re-using parts of a model. It also covers a collection of small but critical issues for use in the real-world: copying and pasting into another document, printing, crashing, saving files reliability, and documentation.

COMPARISION OF THE TOOLS

All three tools were built using [3] and/or [10] as their primary inspiration, thus there should be many similarities in their expressiveness and results. All the previously described dimensions will be summarized here.

QGOMS

QGOMS (Quick and Dirty GOMS) was developed at UNC-Chapel Hill CS and Radiology and at ISU College of Business [2]. As the name suggests, it embodies a simplified version of GOMS. QGOMS displays the model in a single-window tree-like form much like the HTA, where root nodes are goals and leaf nodes are operators (Figure 2). The analyst primarily uses the mouse to place nodes when building the tree, and the keyboard to enter execution-time parameters for each node.

Representation and Constructs Available in QGOMS

Basic Representation. Goals in QGOMS are the root nodes and operators are leaf nodes of the tree. Methods are not explicit; they are captured only in the overall tree structure. Instead of symbolic selection rules triggered by characteristics of the users' preferences or tasks, QGOMS allows the analyst to attach probabilities to sub-trees to indicate the frequency with which each sub-tree will be executed.

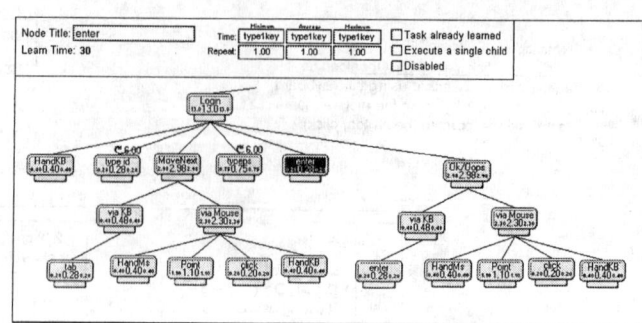

Figure 2. QGOMS model of the Login subtask of the Atropos HTA.

Decide Operator. There is no Decide operator in QGOMS. However, the analyst can disable any node at any time, which can be used to hand-simulate that an operator is not being executed in a particular task scenario.

Parameterization. Execution time information can be parameterized in global variables. Task-state is not represented at all, and hence is not parameterized.

Repetition. Nodes (operators or goals) can be repeated; arbitrary sequences of goals and operators cannot.

Architectural Time Accounting. QGOMS adds 200 ms to its execution time prediction for each subgoal, i.e. for each root node in the tree. QGOMS adds 30 s to its predicted learning time for each node (both root and leaf), unless the node has been explicitly tagged by the analyst as "already learned."

Design Information in QGOMS

Functionality. A model's coverage of functionality can only be found through visual inspection of a QGOMS model. The analyst must ensure that there is a path through the model for every task the user would want to perform. Evaluating the consistency of the interface is also the responsibility of the analyst, and requires careful comparison to determine if sub-trees for the same task are identical. For example, sub-trees for the same sub-task should be identical (e.g. all sub-trees for responding to an error-notification dialog box in Atropos should be the same).

Operator Sequence. QGOMS presents only a static representation of the GOMS model, without a "run" or "step-through" capability. Therefore, in determining operator sequence the analyst must pay close attention to which nodes in the tree would be executed in any particular task scenario.

Execution time. QGOMS automatically calculates the execution time for each node as the model is built. Thus, at all times the analyst can see the execution times for the entire task as well as for each sub-task. The analyst tests different scenarios by changing values on the global variables, changing probabilities on nodes, and/or disabling nodes (including whole sub-trees).

Procedural learning time. QGOMS automatically sums procedural learning time for all nodes (30 s per node).

Nodes for sub-tasks can be marked as an 'already learned task' in which case their learning times are not added to the total. The analyst must do this marking deliberately.

QGOMS's Support for Cognitively Plausible Models

There is no real support for building a cognitively plausible model in QGOMS. It does not come with a library of pre-defined operators with previously-researched execution times. It gives no support for remembering steps in a method, e.g. it possible to build a model in which the user points to an icon without moving her hand from the keyboard to the mouse. Since QGOMS does not "run" the analyst must also carefully track state information such as hand position and contents of LTM when deciding which nodes to enable in any particular task scenario.

Our Experience Using QGOMS

A QGOMS model is built by creating and connecting nodes using the mouse. For each leaf node, timing information must be entered from the keyboard. Constructing a model with common sub-tasks (e.g. selecting and dragging on a menu, responding to a yes/no dialog) can be quick because it is easy to copy and paste entire sub-trees. Nodes, including whole sub-trees, can be disconnected and re-connected arbitrarily, so it is easy to re-structure the tree. It is possible to construct a library of sub-trees for common tasks which the analyst can re-use in future models.

The analyst can view a QGOMS model as a tree on the screen, or in an outline-like format that comes from printing the model. To include part of the tree in another document the analyst must take a screenshot. QGOMS allows complete control over the position of the nodes, and which nodes are expanded or collapsed, so the tree can be finely rendered in preparation for a screen shot. It is not possible to include a section of the outline-like paper format in another document without resorting to scissors and glue.

QGOMS crashed often once the model extended beyond the bounds of the initial window. We also experienced unexpected and unexplainable behavior when scrolling the window. The most disturbing problem, however, was that several of our models did not re-open after being saved.

The documentation comes packaged with QGOMS on-line. It was short but adequate to build models. Included with the package were some working models. However, the sources of GOMS theory that were the basis for the tool were not consistently referenced in the documentation.

CATHCI

CATHCI (Cognitive Analysis Tool for Human Computer Interfaces [17]) was built at Virginia Tech as an aid for expert tacticians to develop models for training scenarios. Like QGOMS, CATHCI provides a tree-like visualization of the model (Figure 3). However, branches in a CATHCI tree are used to denote different methods, not subgoals; subgoals are displayed in separate windows. A wizard-like guidance mode is provided for creating new models.

Representation and Constructs Available in CATHCI

Basic Representation. A goal is represented as a blue node which, when clicked, opens another window displaying the methods to accomplish the goal. Operators are green nodes which do not expand further. Methods are encoded as a sequence of nodes with an explicit red label-node at the top. Multiple methods to accomplish a single goal are represented as branches beneath the blue goal-node. Selection rules, which are symbolic conditionals that request analyst input during execution, are represented by magenta nodes at the top of each method.

Decide Operator. Decide operators are the same color as selection rules (magenta) but appear in-line with the other steps in the method rather than at the top of a method.

Parameterization. CATHCI contains an extensive collection of primitive operators with pre-researched execution times. Although the analyst cannot make changes to the existing operators, she can add to the collection. Task state cannot be parameterized.

Repetition. Only a sequence of operators within the same sub-tree can be repeated. The documentation includes a GOTO operator that could be the action of a Decide operator, making a more powerful repeat function programmable, but we were unable to get that operator to work.

Architectural Time Accounting. According to the documentation, both pushing and popping a goal takes 2000

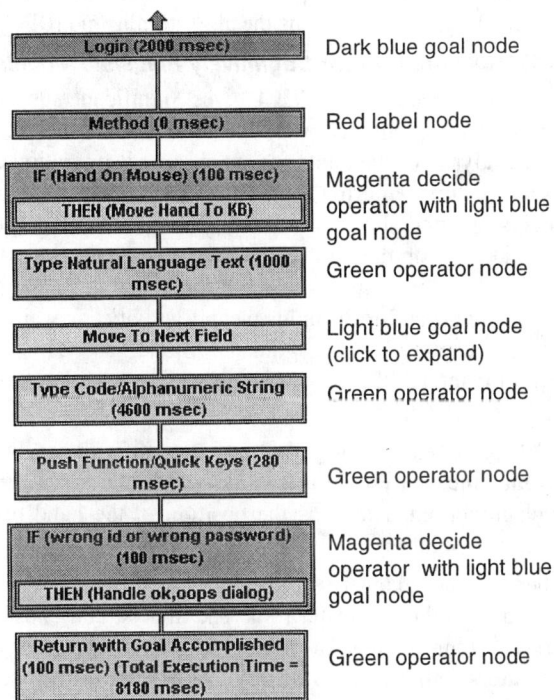

Figure 3. CATHCI model of the Login subtask in the Atropos HTA. Note that this subtask has only one method to accomplish its goal, therefore multiple methods and selection rules do not appear in this figure.

ms, for a total of 4 seconds each time there is a subgoal. (CATHCI actually incurs the entire cost at the beginning of each subgoal and charges an additional 100ms at the end of a subgoal for the Return_from_goal_accomplished step). If their conditions match, Decide operators take 100 ms in addition to any time needed to actually do the action in their THEN or ELSE clauses.

Design Information in CATHCI

Functionality. Like QGOMS, coverage and consistency are only extracted by the analyst visually inspecting the model.

Operator Sequence. Operator sequence is available by the analyst following the operator nodes along an execution path. CATHCI does not record these operators, so if the analyst wants a permanent record of an operator sequence, she must write them down herself.

Execution time. To obtain execution times from CATHCI the analyst executes the model. As the model is executed, the analyst is queried at each Selection rule and Decision step for the status of the conditional. Testing different scenarios is accomplished by providing difference answers for the conditionals on different executions. The execution time prediction appears on the screen. CATHCI does not provide any permanent record of the answers the analyst gave at each decision point (which encode the specific task scenario being executed) nor the final execution time prediction. If the analyst wants a permanent record of these things, she must write them down as she goes along.

Procedural learning time. At this writing, CATHCI does not provide tool support for procedural learning time. However, the documentation explains the algorithm from [10].

CATHCI's Support for Cognitively Plausible Models

Guidance mode in CATHCI gives significant support for building a cognitively plausible model. CATHCI reminds the analyst to account for unobserved operations. For example when entering a pointing operation not only is the analyst queried about the hand position prior to the operator (is it already on the mouse?) but the analyst is also asked to indicate the event that triggered the pointing (prior event, visual input, auditory input, mental activity, or waiting for system response). After guidance mode has been used to build the model, the analyst is free to fine-tune the model by adding or deleting operators.

CATHCI does not formally track state information but will prompt the analyst during model construction to consider such important details as the position of the hand prior to typing or pointing.

There is an extensive collection of primitive operators included in the CATHCI tool. The analyst can chose from several long menus of perceptual, cognitive and motor operators, all with estimates of execution times. The documentation provides the references to research papers that support these execution time estimates.

Our Experience Using CATHCI

There are two ways to build a CATHCI model, freeplay edit mode and guidance mode. In guidance mode CATHCI steps the analyst through the model in breadth-first order, prompting for the specifics of each node. In freeplay edit mode the analyst can indicate where nodes should be added and deleted but guidance mode takes over when the details of the node must be supplied. Important dialog boxes that arose in guidance mode seem to be inaccessible from freeplay edit mode. Moreover there are dialog boxes that appeared once in guidance mode and could not be retrieved once dismissed. Thus, building a model is easy if the analyst is ready to build the model in the breadth first order required by the guidance mode. If the analyst wants to build the model in pieces, or in an alternative order (e.g., depth first), requiring the use of freeplay edit mode, then building the model becomes more difficult.

On the positive side, because methods can be named in CATHCI, common sub-tasks can be referenced (much like a procedure calls) from different sections of the model.

Once the model is built, making structural changes is remarkably difficult. Sub-trees cannot be copy and pasted nor can they be moved around by any other mechanism. By the same token, assembling pieces of different pre-existing models to construct a new model is also impossible.

Copying and pasting into another document can only be achieved by taking a screenshot and including the screenshot in the other document. When the model is saved CATHCI writes a program-like text file, which can be opened in any text editor. However, the text view includes artifacts of the underlying programming language which are not documented and not immediately understandable. When the model is printed it is logically paginated; generally each subgoal prints on its own page.

CATHCI crashed often while building our benchmark model. The system also frequently got stuck in an unrecoverable error state and we had to restart from scratch.

The documentation is extensive with many references to the research sources used for the pre-defined operators and architectural accounting. However, there is little discussion of how to use the tool itself in the documentation. The environment provides hints to its operation in a window at the bottom of the screen, but we found them to be too general to solve every problem we encountered. CATHCI comes with a tutorial, which might more fully cover the use of the tool, but we were unable to get the tutorial to work on our machines. There were no running example models available (unless they were included in the tutorial and thereby inaccessible to us).

GLEAN3

GLEAN was developed at the University of Michigan as a tool for building generative GOMS models [12]. A model is built by writing a program in GOMSL, a programming language specifically designed to express GOMS constructs (Figure 4). GLEAN provides both run-time analyses (execution time, working-memory load) and static analyses (learning time estimates). Several versions of GLEAN have been developed, the latest being version3.

Representation and Constructs Available in GLEAN3

Basic Representation. Goals are represented by a name and at least one method to accomplish that goal. Operators are steps that are not expanded any further, but have a duration assigned to them. A method is a named series of steps (operators or goals). Selection rules are an ordered series of if statements that fire in a production-like manner (in parallel). They must be written so that they are exclusive, i.e. so that one and only one fires in a given situation.

Decide Operator. GLEAN3 provides a formal Decide operator to be used when the analyst is modeling a user's decision about which path to take.

Parameterization. Previous versions of GLEAN allowed the analyst to define new operators with parameterized execution times, but that functionality has not yet been implemented in GLEAN3. Task state can be parameterized with global variables, so it is easy to set up new task scenarios.

Repetition. Anything can be repeated using the conditional Decide operator and the GOTO operator.

Architectural Time Accounting. GLEAN3 adds 50 ms for the execution of every GOMSL statement, plus whatever extra time an operator might take (e.g., mouse click takes an additional 200 ms). Because pushing a goal takes two GOMSL statements (`Accomplish_goal <GoalName>` in the calling method and `Method_for_goal <GoalName>` at the top of the method for that subgoal), GLEAN3 adds 100 ms for pushing a goal. Likewise, GLEAN3 adds 50 ms for the `Return_with_goal_accomplished` statement that pops a goal. No matter how many selection rules there are in a set, only 50 ms is added for the entire set, because only one of those rules will match any task situation. Decide operators only cost 50 msec (plus the time to do their action) if their conditions match.

GLEAN3 counts GOMSL statements that must be learned in a model to help predict learning time. Since learning time is dependent on the type of learning environment (e.g., classroom lecture vs. computer-based mastery), GLEAN merely counts the statements and gives guidance in the documentation about how to convert those counts to learning time predictions. It uses an algorithm for determining *similar* methods which do not have to be learned again and automatically gives credit for transfer of that procedure-learning [10].

```
Method_for_goal: Login Atropos
  Step 1. Recall_LTM_item_whose Content is
            User_info and_store_under <User_info>.
  Step 2. Type_in User_ID of <User_info>.
  Step 3. Accomplish_goal: Move_to Next_field.
  Step 4. Type_in Password of <User_info>.
  Step 5. Keystroke Enter_key.
  Step 6. Decide:
            If System_response of <User_info> is
              Error_dialog,
            Then Accomplish_goal:
              Recover_from Login_Error_dialog.
  Step 7. Return_with_goal_accomplished.

Selection_rules_for_goal: Move_to Next_field
  If User_prefers_KB of <User_info> is True,
    Then Accomplish_goal: Move_via KB.
  If User_prefers_KB of <User_info> is False,
    Then Accomplish_goal: Move_via Mouse.
  Return_with_goal_accomplished.

Method_for_goal: Move_via KB
  Step 1. Keystroke Tab_key.
  Step 2. Return_with_goal_accomplished.

Method_for_goal: Move_via Mouse
  Step 1. Point_to Password_field.
  Step 2. Click Mouse_button.
  Step 3. Return_with_goal_accomplished.
```

Figure 4. GLEAN3 model of the Login subtask in the Atropos HTA.

Design Information in GLEAN3

Functionality. GLEAN3 forces every goal declared with an `Accomplish_goal` step to have a method; a model with no method will compile, but not run. Thus, GLEAN3 gives explicit feedback when all known user-goals are not covered by methods. Information about consistency is provided in the Learning Analysis table. Each method is shown with its count of new statements to learn. If two methods are similar, the one that appears later in the table will have a very small count because knowledge from the previously-learned method transfers. The table lists which method contributed the transfer knowledge, explicitly displaying this measure of consistency.

Operator Sequence. When a GLEAN3 model runs, it records the operator sequence to a text file.

Execution time. When a GLEAN3 model runs, the text file that records the operator sequence also shows the cumulative execution time prediction.

Procedural learning time. Learning Analysis is a static analysis done on a GLEAN3 model. As described above, it provides a count of the GOMSL statements to be learned in the model, accounting for transfer of knowledge between similar methods. This analysis can be displayed on the screen or saved to a file.

GLEAN3's Support for Cognitively Plausible Models

GLEAN3 gives quite a bit of support for constructing cognitively plausible models. Unlike CATHCI, GLEAN3's feedback comes upon execution of a model instead of during the construction of a model. For instance, it tracks the position of the hands and automatically inserts a homing operator if the model attempts to point when its hands are on the keyboard. GLEAN3 also generates an error if the code attempts to use a variable whose value has not been retrieved either from Long Term Memory or from the Task definition. Previous versions of GLEAN could be used to simulate the interface as well as the user's state, so the state of the world could also be tracked and used by the model (e.g., whether a button was highlighted). GLEAN3, however, cannot do this modeling within itself, but can be integrated with a running C++ program that implements the interface.

GLEAN3 comes with an extensive set of predetermined primitive operators with associated execution-time estimates. In previous versions of GLEAN the analyst could define new operators, but this has not yet been implemented in GLEAN3. The documentation includes extensive references to validated models and parameters and several examples to follow. The system comes with several runnable example models.

Our Experience Using GLEAN3

Using GLEAN3 is like programming in a classic programming langauge where the editor is separate from the compiler, with all the associated benefits and difficulties of such an environment. It is a purely textual experience, with no graphic depictions of the model or execution.

Since a GLEAN3 model is a text file, it can be created and edited in any text editor. Thus, re-using components from previous models is simply a matter of copying and pasting the relevant lines of GOMSL code.

Copying and pasting portions of the model or an execution trace into another document is straightforward. Printing is similarly straightforward.

GLEAN3 proved remarkably stable. We experienced no crashing or unusual behavior while building the benchmark models. Documentation is extensive, both for the theory and the environment, and it comes with several worked examples of models.

COMPARING EXECUTION TIME PREDICTIONS

To investigate the quantitative aspects of these three tools, we examined the timings predicted for the Login subtask (Figure 5). Time predictions from previously built KLMs [4] are also included as an additional comparison point. All these models used the following assumptions.

- The user's hand did not start on the keyboard.
- The user transfers between fields using the TAB key (one of two possible methods in the models).
- The user made no errors typing her UserID or password.
- There were six characters in both the UserID and the password.
- 280 ms per keystroke for the UserID because it was the user's last name plus her first initial (a relatively familiar thing to type for a skilled non-secretary typist, [3])
- 750 ms per keystroke for password because these were randomly assigned alphanumeric codes that the users had never seen before [3].

Model Type	Predicted Time for Logging in (sec)
QGOMS	7.94
CATHCI	16.68
GLEAN3	9.60
Hand-done KLM [4]	9.54*
Automatically-generated KLM [4]	9.54*

Figure 5. Predicted times for the Login subtask. *These values were calculated from the KLMs displayed in [4], for just the login portion of the models, using a 6-character UserID.

The most striking difference between these predictions is that CATHCI predicts almost double the time of any of the other models. We attribute this to the 4 second architectural overhead of having two subgoals in the login procedure (move-hand-to-mouse and move-to-next-field). The GLEAN3 model also uses two subgoals to implement these actions, but incurs a much lower cost. The CATHCI documentation cites [1] as the sources of this estimate for goal-manipulation overhead, whereas the GLEAN documentation bases its overhead cost on EPIC. It is our belief that CATHCI's sources are estimates for much higher-level goals (like driving to the airport as the first subgoal in the process of getting to CHI'00) than are present in this model. This difference emphasizes the importance of thoroughly understanding the implications of the tool's dependence on cognitive theory. CATHCI's use of the subgoal concept does not seem to scale down to small movements.

The smaller difference between QGOMS and the other three models (GLEAN3 and the two KLMs) is within the 20% accuracy often expected of GOMS models. These differences are similar to those in [7]'s comparison of four different versions of GOMS. Comparing the content and operation of the models reveals that the difference between GLEAN3 and QGOMS can be attributed to QGOMS not charging an architectural time for each leaf node like GLEAN3 does (50 ms), but using only the operator-specific time (e.g., 280 ms per keystroke).

CONCLUSIONS AND FUTURE WORK

This preliminary comparison revealed that even on a task as simple as Atropos, where the analysts were skilled in GOMS and started with a detailed HTA, none of these tools can be considered the perfect GOMS tool. Each of them has strengths that should be preserved in future versions and each has weaknesses that should be overcome, especially if it is to be used in real design practice.

For basic representation, an ideal GOMS tool would be capable of expressing all the concepts in the original GOMS models (not doing so is a weakness of QGOMS). We found the graphic representation in QGOMS to be a nice way to visualize the goal hierarchy and hope that some 2-dimensional representation would be available in future GOMS tools. For running what-if scenarios or running higher-level analyses, the ability to parameterize both operator duration and task state, and save both the parameter values and their resulting predictions is invaluable.

Support for creating cognitively plausible models should be part of an ideal GOMS tool. First, complete documentation that references all sources of theoretical contributions is a necessity (as CATHCI and GLEAN3 have attempted to provide), but also extensive documentation about the implications of the theoretical choices for the modeling style must be included. Such insight will develop over time and use, but tool developers should begin the process with the first "snapshot" of understanding and have a mechanism for sharing future revelations with tool-users.

To come into widespread use, all the pragmatics of tool-use need to be handled: crashing, saving, reusing models, copying to other documents, etc. In addition, although this feature did not appear in any of the tools we compared and is emphasized only by its omission, we have argued elsewhere [4] that to fit well with the development process, an ideal GOMS tool should be integrated with a prototyping tool and other usability evaluation aids. We place these challenges before the modeling and tool-building communities as the next area for progress in UI evaluation methods.

AKNOWLEDGMENTS

We would like to thank the members of the "Cognitive Modeling for HCI" Graduate Seminar at Carnegie Mellon University, Spring 1999, for their effort creating the HTA and initial models in all the tools and their insights and help finalizing the models. We also thank the tool authors for supplying the most recent version of their tools and documentation as of January 1999. This research was sponsored in part by NSF Award #IRI-9457628 and NIMH fellowship 2732-MH19102. The views and conclusions contained herein are those of the authors and should not be interpreted as representing the official policies or endorsements, either expressed or implied, of the NSF, the NIMH, or the U.S. Government.

REFERENCES

1. Anderson, J. R. (1993). Rules of the mind. Hillsdale, NJ: Lawrence Erlbaum Associates.Lawrence.
2. Beard, David V., Smith, Dana K. & Denelsbeck, Kevin M. (1996). "Quick and Dirty GOMS: A Case Study of Computed Tomography", *Human-Computer Interaction, 11* (2), 157-180.
3. Card, S. K., Moran, T.P. & Newell, A. (1983). *The Psychology of Human-Computer Interaction.* Hillsdale, NJ: Lawrence Erlbaum Associates.
4. Hudson, S. E., John, B. E., Knudsen, K., & Byrne, M. D. (1999). A tool for creating predictive performance models from user interface demonstrations. UIST'99: *Proceedings of the ACM Symposium on User Interface Software and Technology.*
5. John, B. E. (1995), "Why GOMS?", *interactions,* v.2(4), 80-89.
6. John, B. E. (1994). "Toward a deeper comparison of methods: A reaction to Nielsen & Phillips and new data", *Proceedings Companion of CHI, 1994,* pp. 285-286. New York: ACM Press.
7. John, B. E. & Kieras, D. E. (1996a). "The GOMS family of user interface analysis techniques: Comparison and Contrast", *ACM Transactions on Computer-Human Interaction,* v.3(4), 320-351. New York: ACM Press.
8. John, B. E. & Kieras, D. E. (1996b) "Using GOMS for user interface design and evaluation: Which technique?", *ACM Transactions on Computer-Human Interaction,* v.3(4), 287-319. New York: ACM Press.
9. Peck, V. A. & John, B. E. (1992) Browser-Soar: A cognitive model of a highly interactive task. In Proceedings of CHI, 1992 (Monterey, CA, May 3-May 7, 1992) ACM, New York, 1992. pp. 165-172.
10. Kieras, D. E. (1988). Towards a practical GOMS model methodology for user interface design. In M. Helander (Ed.), *The handbook of human-computer interaction.* (pp. 135-158). Amsterdam: North-Holland.
11. Kieras, D. E., Wood, S. D., Meyer, D. E. (1997) Predictive Engineering Models Based on the EPIC Architecture for a Multimodal High-Performance Human-Computer Interaction Task. *ACM Transactions on Computer-Human Interaction, 4*(3) 230-275.
12. Kieras, D. E., Wood, S. D., Abotel, K., & Hornof, A. (1995) GLEAN: A Computer-Based Tool for Rapid GOMS Model Usability Evaluation of User Interface. *Proceedings of the ACM Symposium on User Interface Software and Technology* 1995 p.91-100
13. Kirwan, B., & Ainsworth, L. K. (1992). *A Guide to Task Analysis.* London: Taylor and Francis.
14. Landauer, T. K. (1995). The Trouble with Computers: Usefulness, Usability, and Productivity, Cambridge: MIT Press.
15. Lewis, C. & Rieman, J. (1994). *Task-centered User Interface Design: A Practical Iintroduction.* Shareware book at ftp.cs.colorado.edu/pub/cs/distribs/clewis/HCI-Design-Book.
16. Pew, R. W., & Mavor, A. S. (1998), Modeling human and organizational behavior: Application to military simulations. Washington, DC: National Academy Press.
17. Williams, K. E. (1993) Automating the cognitive task modeling process: An extension to GOMS for HCI. In *Proceedings of the Fifth International Conference on Human-Computer Interaction Poster Sessions: Abridged Proceedings* (vol 3. p. 182).

DENIM: Finding a Tighter Fit Between Tools and Practice for Web Site Design

James Lin, Mark W. Newman, Jason I. Hong, James A. Landay
Group for User Interface Research, Computer Science Division
University of California, Berkeley
Berkeley, CA 94720-1776 USA
+1 510 643-3043
{jimlin, newman, jasonh, landay} @cs.berkeley.edu

ABSTRACT

Through a study of web site design practice, we observed that web site designers design sites at different levels of refinement—site map, storyboard, and individual page—and that designers sketch at all levels during the early stages of design. However, existing web design tools do not support these tasks very well. Informed by these observations, we created DENIM, a system that helps web site designers in the early stages of design. DENIM supports sketching input, allows design at different refinement levels, and unifies the levels through zooming. We performed an informal evaluation with seven professional designers and found that they reacted positively to the concept and were interested in using such a system in their work.

Keywords

Web design, Zooming User Interface (ZUI), Sketching, Informal, Pen-based Computers, Rapid Prototyping

INTRODUCTION

Web site design has much in common with other types of design, such as graphic design and "traditional" graphical user interface design, but it is also emerging as its own discipline with its own practices and its own set of problems. We have taken a fresh look at web site design in order to determine what kinds of tools would be helpful to support designers. In this paper, we describe some of our observations of web site design practice and introduce a system named DENIM that is aimed at supporting the early phases of the web site design process.

We conducted an ethnographic study in which we observed and interviewed several professional web designers. This study showed that the process of designing a web site involves an iterative progression from less detailed to more detailed representations of the site. For example, designers often create *site maps* early in the process, which are high-level representations of a site in which each page or set of pages is depicted as a label. They then proceed to create *storyboards* of interaction sequences, which employ minimal page-level detail and focus instead on the navigational elements required to get from one page to another. Later still, designers create *schematics* and *mock-ups*, which are different representations of individual pages.

The design process often includes rapid exploration early on, with designers creating many low-fidelity sketches on paper. These sketches are considered crucial to the process. Designers can quickly sketch the overall look and feel of a web site without having to deal with unnecessary low-level details and without having to commit a large amount of time and effort to a single idea. Furthermore, sketches are important for communicating ideas with other team members and gaining valuable feedback early in the design process. These uses of sketches are similar to what has been previously reported for GUI design [12, 26].

Yet, there is a gulf between the needs of web designers during early design phases and the tools available to them. Most web design tools focus only on the creation of production web sites. The high-fidelity nature of these tools tends to force premature formalization of ideas and require undue attention to low-level details.

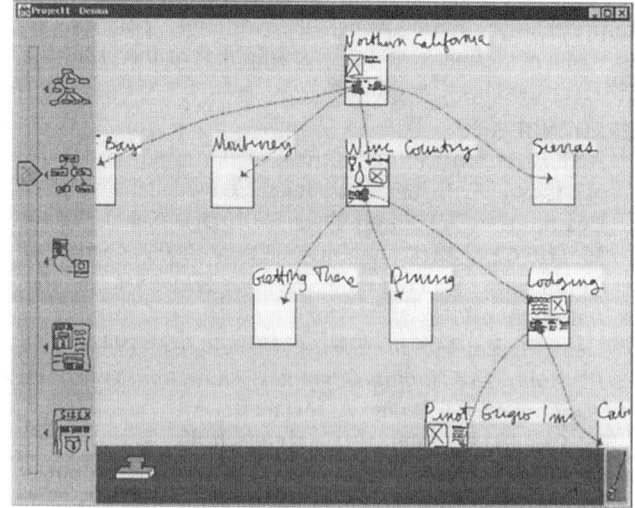

Figure 1. The DENIM interface in *site map* view. This is the sample web site used in the evaluation.

Permission to make digital or hard copies of all or part of this work for personal or classroom use is granted without fee provided that copies are not made or distributed for profit or commercial advantage and that copies bear this notice and the full citation on the first page. To copy otherwise, to republish, to post on servers or to redistribute to lists, requires prior specific permission and/or a fee.
CHI '2000 The Hague, Amsterdam
Copyright ACM 2000 1-58113-216-6/00/04...$5.00

These were the primary observations that led to the design and implementation of DENIM, a system to assist web designers in the early stages of information, navigation, and interaction design. DENIM (see Figure 1) is an informal pen-based system [10] that allows designers to quickly sketch web pages, create links among them, and interact with them in a run mode. The different ways of viewing a web site, from site map to storyboard to individual pages, are integrated through the use of zooming. An informal evaluation of this system has yielded positive comments, subjectively rating high on usefulness and fair on usability.

AN INVESTIGATION INTO WEB SITE DESIGN

We conducted a series of ethnographic interviews with designers about how they work when designing web sites. In total, eleven designers from five different companies were interviewed, representing a range of backgrounds, experience levels, and roles with respect to web site design. During each interview, the designer was asked to choose a recent project that was completed or nearly completed, and walk the interviewer through the entire project, explaining what happened at each phase. We asked the designer to show examples of documents (including sketches) that he or she produced during each phase and explain the meaning of the document with respect to the process as a whole. In many cases, we were able to obtain copies of these documents for later analysis. Projects discussed ranged from a university site to a municipal aquarium site to sub-sites of a large Internet portal. A more complete description of the study can be found in [18].

Progressive Refinement

The designers we studied generally followed a process of progressive refinement of their designs from less detail to greater detail, and simultaneously from coarse granularity to fine granularity. By this we mean that there was a tendency to think about the larger picture, such as the overall site architecture, early on in the process, and then progressively focus on finer and finer details, such as the appearance of specific page elements, typefaces, and colors. The importance of iterating through web site designs at multiple levels of detail is also discussed in [19] and [21].

During the course of our interviews, we identified several types of documents that are commonly used by web designers to represent a site design at different granularities. *Site maps* generally represent an entire web site at a coarse granularity, where the smallest unit represented is a page or a related group of pages (Figure 2). At a finer level of granularity, some designers used *storyboards* to represent specific interaction sequences, such as how a user might execute a task using a part of the site (Figure 3). Finally, designers create representations of individual pages, which can range from *thumbnails*, which are miniature representations of pages; to *roughs*, which are usually hand-drawn sketches of pages; to *schematics*, which are medium-fidelity representations of the information and navigation components on a page; to

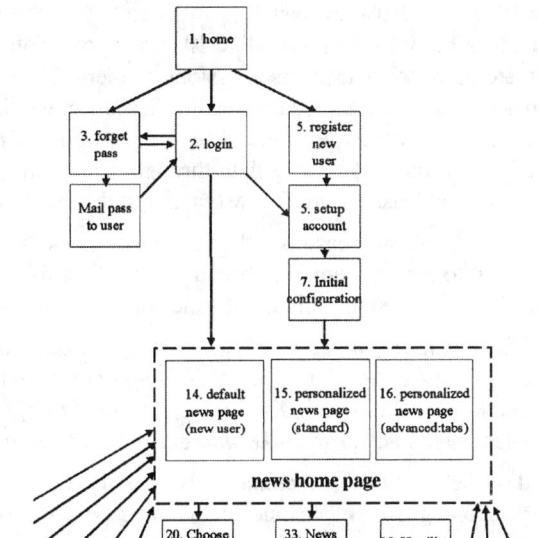

Figure 2. Part of a site map for a news web site. Site maps show entire web sites at a low level of detail.

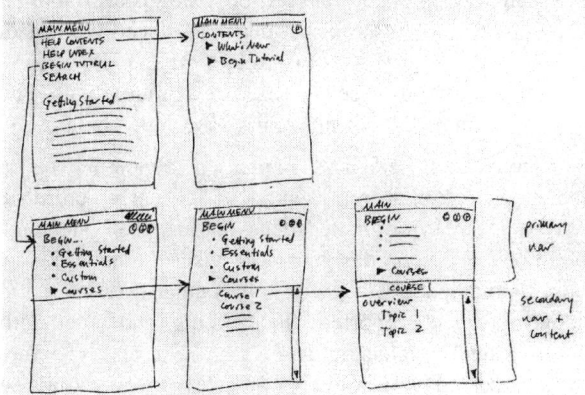

Figure 3. A storyboard showing an interaction with a tutorial system. Storyboards depict individual interaction sequences.

mock-ups, which are high-fidelity representations of the exact contents and appearance of a specific page.

The observation that designers create site visualizations at different levels of detail inspired us to offer a similar range of options in DENIM. We introduced zooming to allow multiple visualizations of a site while preserving a unified context in which to iteratively refine the site design.

Sketching

All of the designers we interviewed sketch with pen on paper as a regular part of their design process, even though eventually all of them end up using computerized tools. Some designers work for relatively long periods on paper before transferring to electronic media, while some merely make quick sketches on scrap paper before using computer-based tools to realize their ideas. It is worth noting that hand-sketched versions were observed for basically all of the document types described above, including site maps, storyboards, and individual pages.

Designers said that they sketch in order to "work through" their ideas before using tools like Illustrator or Photoshop to create more formal, precise versions of them. Several of them also said they use sketching to "try different things out," i.e., they can explore the space of possibilities more effectively through sketching than through using computer-based tools, at least during the early parts of the process.

There are several reasons why designers switch from sketching to using computer-based tools. The following quote from one designer highlights one common reason:

> *The beginning of each step I'll do on paper. As soon as I feel like I'm going to be doing any design revisions, I'll move to [an electronic tool]... because it's easier to make changes to these things.*

Besides the ability to incrementally modify documents, other advantages of electronic media over pen and paper include the ease of replication and distribution. Electronic tools also offer the ability for designers to express themselves more precisely and to a greater level of detail than sketching on paper, and this precision is desirable later in the process when the basic ideas have been worked out. Another reason for switching to more formal representations is the need to share design ideas with others outside the immediate design team, especially clients.

In many cases designers expressed concern over the tendency of formal representations of early, unfinished ideas to cause viewers to focus on inappropriate details [26]. For example, a designer may wish to obtain feedback about the navigational flow of a particular user interaction. Many designers reported that clients and even other designers tend to focus on details like color and typography when presented with a set of high-fidelity mock-ups and have trouble focusing on the larger concepts. To strike a balance between the need to present "professional" representations and the desire to constrain feedback to relevant aspects of the current state of the design, several designers use *medium-fidelity* representations like schematics to represent web pages. Such representations can be made attractive without overspecifying graphical details that can confuse and mislead viewers.

The fact that all of the designers sketch as part of their design process supports our hypothesis that they would find a sketch-based tool familiar. Several of them indicated that they find themselves switching to electronic media earlier than they would like. This indicates that a sketch-based tool could meet a need that currently exists. A tool to support web site design should support the need of designers to design and view sites at multiple granularities and levels of detail. It should also support representations at multiple levels of formality. As we describe in the rest of this paper, DENIM provides both an informal, sketch-based interface and the ability to view sites at several levels through zooming. Currently DENIM does not support the generation of representations at different levels of formality, though we plan to explore this area in the future.

RELATED WORK

Sketching and writing are natural activities used by many designers as part of the design process. DENIM captures this activity with an informal ink-based interface [10]. Using an informal interface is a key aspect of DENIM, as it allows designers to defer the details until later and focus on their task without having to worry about precision. Many research systems have taken this direction in recent years, either by not processing the ink [4, 23, 24] or by processing the ink internally while displaying the unprocessed ink [8, 12, 17, 22].

DENIM is most closely related to SILK [12, 13], a sketch-based user interface prototyping tool. Using SILK, individual screens can be drawn, with certain sketches recognized as interface widgets. These screens can be linked to form storyboards [14], which can be tested in a run mode. DENIM takes many of these ideas and extends them to the domain of web site design. However, DENIM de-emphasizes the screen layout aspects of SILK, focusing instead on the creation of whole web sites. Furthermore, instead of the separate screen and storyboard views in SILK, all of the views are integrated through zooming. Also, SILK attempts to recognize the user's sketches and display its interpretation as soon as possible. DENIM intentionally avoids doing much recognition in order to support more free-form sketching.

DENIM's use of storyboarding for behaviors is similar to SILK. Other systems that use storyboarding include Anecdote [9] and PatchWork [24].

WebStyler [10] is another sketch-based tool for prototyping individual web pages. However, DENIM addresses more aspects of web site design, including designing the site structure and being able to interact with the sketches.

Others have noted that designers often sketch basic designs [25, 26]. Sketching has many advantages over traditional user-interface design tools that focus on creation of high-fidelity prototypes. Sketches are inherently ambiguous, which allows the designer to focus on basic structural issues instead of unimportant details [2]. The ambiguity also allows multiple interpretations of the sketch, which can lead to more design ideas. Sketching is quick, so designers can rapidly explore different ideas, which leads to a more thorough exploration of the design space [6]. Rapid sketching also encourages iteration, which is widely considered to be a valuable technique for designing interfaces [7].

There is a lack of early-stage prototyping tools for the web. Our ethnographic study showed us that web designers use other tools to fill this gap. Macromedia Director is often used to assemble storyboards, while Visio is used for modeling the high-level information architecture of a web site. However, Director is a multimedia authoring tool, and

Visio is a general purpose diagramming tool. This makes using them for such high-level web site design awkward at best, since they are not designed for those tasks.

Currently, the most popular tools for creating web sites include Microsoft FrontPage, Adobe GoLive, Macromedia Dreamweaver, and NetObjects Fusion. These tools focus on designing page layout rather than the site architecture. Admittedly, each of them has a "site structure view" of a web site. However, this view often constrains any edits so that the tree structure remains intact. Furthermore, the site structure view and the page layout view are usually distinct and not unified. Most importantly, these tools focus on producing high-fidelity representations, which is inappropriate in the early stages of design. These are all important issues that we chose to address in DENIM.

THE DENIM SYSTEM

Informed by our study, we designed and implemented a prototyping tool to assist designers in the early stages of web site design. Intended to be more informal than SILK, we named our system DENIM, which also conveniently stands for *Design Environment for Navigation and Information Models*. DENIM is intended for prototyping in the early stages of design, but not for the creation of finished web sites. For example, it does not output finished HTML pages. We built DENIM in Java 2, on top of SATIN, a toolkit for supporting informal pen-based interaction [11].

The DENIM interface is shown in Figures 1 and 4. The center area is a canvas where the user can create web pages, sketch the contents of those pages, and draw arrows between pages to represent their relationship to one another. On the left is a slider that reflects the current zoom level and allows the level to be set. The bottom area is a toolbox that will hold tools for inserting reusable components, such as templates. However, this part is not currently implemented.

Zooming

To change the zoom level, the user either drags the slider's elevator or clicks directly on one of the icons. Changing the zoom level initiates an animated transition from the current zoom level to the desired zoom level. The center point for a zoom operation can be set by tapping on the background of the canvas. Such a tap causes crosshairs to be displayed at the point tapped, and any subsequent zoom operation will center on that point. Alternatively, if any objects are selected, the center of the selected object or objects is used as the zoom target.

There are five main zoom levels in DENIM, which are identified on the zoom slider with icons representing the type of view available at that level (see Figure 5). There is also an intermediate zoom level in between each main level. Three zoom levels—the site map, storyboard, and sketch levels—map directly to the most common representations of web site designs that we observed during our ethnographic study. The *site map* level (Figure 1) gives a view of the site as connected labels with attached thumbnails of individual pages. The *storyboard* level (Figure 4a) allows the user to view several pages simultaneously and more clearly see the navigational relationships between the pages. The *sketch* level (Figure 4b) displays pages at "100%" scale, and is intended to allow users to sketch the page contents. In addition to these levels, there are two major levels at the extreme ends of the scale, with the *overview* level providing a more abstract, higher-level representation of the entire site, and the *detail* level providing a more fine-grained view of individual pages, for more precise sketching.

Creating Pages

In DENIM, web pages are accompanied by a label that represents the name or description of the page. The labels remain the same size throughout all the zoom levels, so that they can always be read.

There are two ways to create a new web page in DENIM. One way is to simply write some words on the canvas while in site map view. A blank page is created with those words as its label. The other way is to draw a rectangle, which is converted to a page with an empty label.

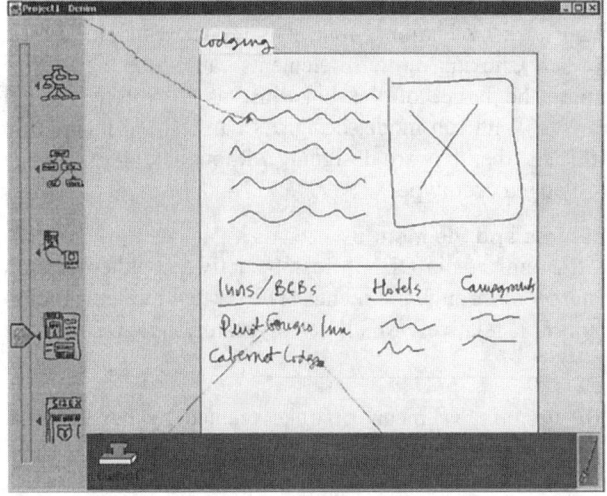

Figure 4. The *storyboard (a)* and *sketch (b)* zoom levels

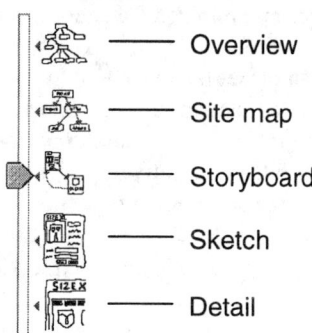

Figure 5. The zoom slider

Links

An arrow between two pages represents a link between those pages. We provide *navigational* and *organizational* links. Navigational links are links in the HTML sense: they represent the reference from an item on one page (e.g., a word or image) to another page. Organizational links are used to represent a conceptual link between two pages; that is, the designer eventually wants to make a navigational link from one page to another, but does not want to fill in the details at this time.

To create a link, the user draws a stroke between two pages. The system checks if the stroke is a link. Organizational links start on one page and end in another. This creates a *gray* arrow from the source to the destination. Navigational links start on a specific object on one page and end in some other page. This creates a *green* arrow from the source to the destination. When creating a navigational link, any organizational links from the source page to the destination page are removed. As additional feedback, the source of the navigational link becomes blue.

Run Mode

After a number of pages have been sketched and navigational links drawn between them, it is possible to preview the interaction by entering run mode. In run mode, a simplified "browser" window appears on the screen. The browser displays one page at a time, like a real web browser, except the pages displayed are the sketches that the designer has created. If an element inside a page is the source of a navigational link, it is rendered in blue in the browser. Clicking on these elements causes the browser to display the target of the link, just as in a conventional browser. With run mode, designers can test the interaction with a site that they are designing without having to create a full-fledged prototype.

Gestures and Pie Menus

Most commands in DENIM can be activated either through gestures[1] or through pie menus. The current implementation supports a relatively small set of gestures, as we are still experimenting with how to best map the functions of DENIM to a set of gestures. To activate a gesture, the user presses the button on the barrel of the pen and makes a stroke. Using a modified version of gdt [16] and Rubine's recognizer [20], we implemented gestures for panning, undo, redo, group select (select everything enclosed by a circular gesture), cut, copy, and paste. Tapping on an object without depressing the barrel button selects or deselects that object. Tapping on the canvas, outside of any web page, clears the selected objects and sets the zoom-center target, denoted by crosshairs. The selected object can also be dragged, moving it to a new location.

We use a form of semantic zooming [1] in which the interaction with objects changes with zoom. In the two broadest views, the overview and site map views, gestures work *shallowly:* you can select, move, or edit web pages, but not anything inside of a web page. Since these views focus on whole pages and the relationships between them, it follows that editing commands should operate on entire pages. In the two narrowest views, the sketch and detail views, gestures work *deeply:* you can select, move, or edit individual ink objects inside a web page, but not web pages themselves. These views focus on the contents of individual pages, so operations work on the page contents. The middle zoom view, the storyboard view, supports operations at both levels of detail. For example, the user taps the page's label to select the page but can tap any object inside a page to select that object.

Pie menus [3] are used to provide access to functions not easily mapped to gestures, as well as providing redundant access to certain commands. The user activates the pie menu by tapping the screen with the barrel button depressed. Keyboard shortcuts are available for several commands, including cut, copy, paste, delete, undo, redo, pan, and zoom.

DENIM does not attempt to recognize most of what users write or sketch. The exceptions are the small set of gestures described above, words written directly on the canvas in site map view that are interpreted as new page labels, and lines drawn between two pages that are treated as links between those pages.

EVALUATION

We conducted an informal evaluation of DENIM in order to gain feedback about the usefulness of the functionality of the tool and the usability of the basic interactions, such as creating pages, creating links between pages, zooming, panning, and interacting with a design in run mode. Seven professional designers participated in the study, one of whom had participated in the initial investigation. Five of the participants said that web site design projects constituted at least half of their current workload. The remaining two participants were a user interface designer working on non-web related projects and a manager of a usability group for a large software company.

---

[1] By gesture, we mean a stroke created by the pen that activates a command.

The system that we used for the evaluation consisted of an IBM 560Z ThinkPad (300MHz Pentium II) laptop running Windows NT 4.0, and an ITI VisionMaker Sketch 14 display tablet (see Figure 6). The participants interacted primarily with the display tablet, although they could also use the keyboard for shortcuts.

One evaluation session was conducted per participant, and each evaluation session consisted of three parts. First, the participant was asked to add a few elements to a drawing in Microsoft Paint to become familiar with using the display tablet and pen. The second task was to get the participant used to interacting with DENIM. We loaded a previously-created web site design (shown in Figure) and asked the user to create a new page, link the page to the site, and then run through the site using run mode starting from the home page and ending at the page they just created.

The final part was a large design task, intended to be difficult to complete in the time allotted. We were interested in seeing how participants approached a realistic design task and how they used DENIM to help them. To help motivate the participants to create the best design they could, we offered US$250 to the best design. The participant was asked to develop a web site for a fictitious start-up company. The web site was to help renters find places to rent and to help landlords find tenants. We provided an analysis of a competitor's web site, market research on what renters and landlords said they wanted, and a description of what the client company required and desired. The participant had 45 to 60 minutes to come up with a preliminary site design, and then he or she presented the designs to us as if we were the rest of the design team.

While the participants performed the tasks, we recorded what types of actions they did (e.g., panning, drawing, and creating new pages) and at what zoom levels they performed those actions. This was to give us a sense as to what features of DENIM they used and how well zooming supported the different design activities. We also recorded any critical incidents that occurred and their general comments and reactions.

After the participants were finished with the tasks, they filled out a questionnaire. We asked what they thought of DENIM in terms of usefulness, ease of use, and how they thought using it would affect their design process. The questionnaire also asked for basic demographics, primary job responsibilities, what tools they normally used, and how much web design experience they had.

Observations

Users made substantial use of different zoom levels, with usage concentrated primarily in the middle three levels (site map, storyboard, and sketch). Several users verbally expressed that they liked the concept of the different zoom levels and liked the ability to maintain a unified representation of the site, while interacting with it at different levels of detail. It appears that users felt that the integrated view would help them iterate more quickly through different design ideas. One user highlighted the advantages of the integrated view by observing:

> It's not like 'OK, that's one idea,' then open a new file and work on a new [idea]. You don't need to do that. The iteration goes on within this [tool] and I can see the relationships.

Another user described how she thought DENIM would improve her current process by remarking:

> I usually [create site maps] in PowerPoint, then I go back to the navigational flow, then I go back to PowerPoint... And here it would be so easy to do that iterative kind of thing.

However, the current integration of these views through zooming sometimes proved to be problematic. Several of the users became frustrated navigating around their site designs and found that they often had to zoom out to a higher level in order to find their desired target and then zoom back in on that target.

Likewise, users had trouble creating navigational links between pages that they had initially drawn far apart on the canvas. It was difficult to find a view of the site that would include both the source and the target, yet have enough detail to be able to find the specific object on the source page that they wished to serve as the link source.

In response to these issues, we have made two changes to DENIM. We introduced auto-panning, which pans the screen when the user draws a line towards a side of the screen. This makes it easier to link two pages that are not visible at the same time. A user can start drawing from one page and draw until he or she sees the desired page. We also changed the display of pages at the site map level so that only their labels appear. This encourages users to draw their initial site maps more densely, since the total size of each page is less, and the density of pages makes it more likely that the source and target page will be visible on the screen at the same time in the storyboard view. These features have been implemented but still need to be evaluated.

We also plan to explore focus+context techniques [5] to address the navigation and linking problems we observed.

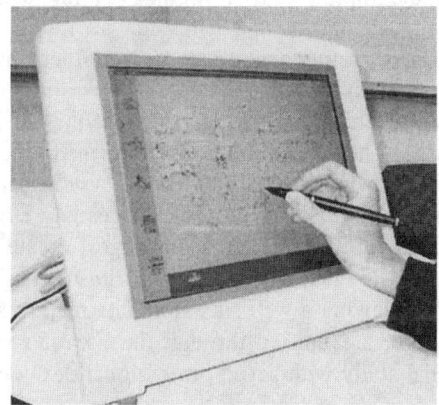

Figure 6. The display tablet used in the evaluation

Being able to see more of the site in the periphery while zoomed in to a particular portion of the site could help reduce the difficulty of finding one's place in the site. Similarly, being able to compress the distance between a source and target page while maintaining a high level of detail in the source page would help relieve the problem of linking pages that were originally drawn far apart from each other in the site map.

Users appreciated the informal mode of interaction provided by DENIM. One user compared the interaction to other tools with the comment:

> You draw a box in Illustrator or Freehand or Quark, and it's got attributes *that have to be dealt with, and it interrupts the thought process....* It's nice to be able to get rid of all the business with the pictures and all the definite object attributes. That is such a hassle.

At the same time, the free-form sketching interface provided some stumbling blocks. For example, handwriting on the screen was difficult, given the average performance of the application, the simple stroking algorithm used, and the lack of feedback from writing on a smooth screen. Two users experienced difficulty reading their own page labels. Another user wanted to type her page labels. Other users said that they like to handwrite while brainstorming, but would like the ability to replace handwritten labels with typed labels as their ideas become solidified. We plan to address these concerns by improving handwriting input, as well as supporting the progressive refinement of text objects by allowing their replacement with typed text.

Feedback

The responses to the post-test questionnaire, though informal, were instructive in several ways. Opinions about DENIM's perceived effect on the respondent's work practices were sharply divided based on the amount of the respondent's workload that consisted of web design projects. The two individuals not involved in web design ranked DENIM relatively low on factors such as "the perceived benefit using the tool would have on their ability to communicate with team members" and on "DENIM's overall usefulness" to them. The five web designers, on the other hand, had generally positive opinions of DENIM along these lines.

First, while the web designers ranked the ease-of-use just above average (6.4 out of 10), they ranked the usefulness fairly high (9.0 out of 10). This seems to indicate that, despite the shortcomings of the current implementation in terms of performance and fluid interaction, users felt that the basic concepts were on target.

Also, the web designers gave very high rankings when asked to rate DENIM according to its perceived ability to communicate with others involved in the design process. Those users rated DENIM better than 8.5 out of 10 in terms of ability to communicate with design team members (8.6), internal managers (8.8), and usability engineers and testers (8.8). They also gave similarly high marks to DENIM's improvement in their ability to express their ideas (9.0), iterate quickly through versions of a design (8.6), and overall efficiency (8.6). All users gave DENIM relatively low marks in terms of ability to communicate with clients (6.14 out of 10 overall), which we attribute largely to DENIM's inability to produce "cleaned-up" versions of sketches that would be acceptable to show to clients.

FUTURE DIRECTIONS

Our initial evaluation of DENIM was informal; we intend to follow up with more formal evaluations on later versions of the system, including field tests in which DENIM is used to design real web sites. We hope that such evaluations will tell us whether long-term use of DENIM can help designers work more efficiently and produce better web sites.

We are also looking into ways to support the generation of medium-fidelity prototypes from low-fidelity sketches. As noted before, such a feature could allow designers to give more "professional" presentations, while staying with sketching longer.

We would like DENIM to work with existing web design tools to fit more naturally into the entire web design cycle. This includes generating HTML and other artifacts that can be imported by other tools. DENIM should also be able to import files from other tools, so that designers can smoothly move back and forth in the design process.

In order to make DENIM scale for large web sites, we plan to explore additional visualizations and interactions that operate on higher levels of abstraction. For example, it would be desirable to allow the designer to identify sub-sites and collapse and expand their representation. The overview zoom level in particular could be used to support interactions with the overall site structure.

We have devised extensions to the storyboarding mechanism to support the design of more sophisticated web interfaces as well as traditional GUIs [15], including methods to allow designers to specify their own reusable components. These components can be as simple as a new kind of widget or as complex as a template for a web page.

CONCLUSION

Our ethnographic study showed us that in the early stages of design, web designers go through an iterative process of progressive refinement, that each refinement focuses on finer levels of granularity and an increasing level of detail, and that sketching is used throughout the early part of this process. These observations informed the design of DENIM, an informal sketch-based system supporting web designers in the early stages of design. DENIM allows designers to quickly sketch out pages, create links among them, and interact with them in a run mode. The different ways of viewing a web site, from site map to storyboard to web page, are unified through the use of zooming. In an informal study with seven professional designers, we found that they were enthusiastic about DENIM's concepts and would like to use such a system in their work.

ACKNOWLEDGEMENTS
We would like to thank the designers that participated in all of the phases of this research and NEC USA for its support.

REFERENCES
1. Bederson, B.B. and J.D. Hollan. Pad++: A Zooming Graphical Interface for Exploring Alternative Interface Physics. In Proceedings of *the ACM Symposium on User Interface Software and Technology: UIST '94*. Marina del Rey, CA. pp. 17-26, November 2–4 1994.
2. Black, A., Visible Planning on Paper and on Screen: The Impact of Working Medium on Decision-making by Novice Graphic Designers. *Behaviour & Information Technology*, 1990. **9**(4): p. 283-296.
3. Callahan, J., D. Hopkins, M. Weiser, and B. Shneiderman. An Empirical Comparison of Pie vs. Linear Menus. In Proceedings of *Human Factors in Computing Systems*. pp. 95-100 1988.
4. Davis, R.C., *et al.* NotePals: Lightweight Note Sharing by the Group, for the Group. In Proceedings of *Human Factors in Computing Systems: CHI '99*. Pittsburgh, PA. pp. 338-345, May 15-20 1999.
5. Furnas, G.W. Generalized Fisheye Views. In Proceedings of *Human Factors in Computing Systems: CHI '86*. Boston, MA. pp. 16-23 1986.
6. Goel, V., *Sketches of Thought*. Cambridge, MA: The MIT Press. 279, 1995.
7. Gould, J.D. and C. Lewis, Designing for Usability: Key Principles and What Designers Think. *Communications of the ACM*, 1985. **28**(3): p. 300-311.
8. Gross, M.D. and E.Y.-L. Do. Ambiguous Intentions: A Paper-like Interface for Creative Design. In Proceedings of *ACM Symposium on User Interface Software and Technology: UIST '96*. Seattle, WA. pp. 183-192, November 6–8 1996.
9. Harada, K., E. Tanaka, R. Ogawa, and Y. Hara. *Anecdote*: A Multimedia Storyboarding System with Seamless Authoring Support. In Proceedings of *ACM International Multimedia Conference 96*. Boston, MA. pp. 341-351, November 18-22 1996.
10. Hearst, M.A., M.D. Gross, J.A. Landay, and T.E. Stahovich, Sketching Intelligent Systems. *IEEE Intelligent Systems*, 1998. **13**(3): p. 10-19.
11. Hong, J.I. and J.A. Landay, *A Toolkit Supporting Pen-Based Interfaces*. Technical Report UCB//CSD-99-1058, University of California, Berkeley, Computer Science Division, Berkeley, CA 1999.
12. Landay, J.A., *Interactive Sketching for the Early Stages of User Interface Design*. Technical Report CMU-CS-96-201, Carnegie Mellon University, Pittsburgh, PA 1996.
13. Landay, J.A. and B.A. Myers. Interactive Sketching for the Early Stages of User Interface Design. In Proceedings of *Human Factors in Computing Systems: CHI '95*. Denver, CO. pp. 43-50, May 7–11 1995.
14. Landay, J.A. and B.A. Myers. Sketching Storyboards to Illustrate Interface Behavior. In Proceedings of *Human Factors in Computing Systems: CHI '96*. Vancouver, Canada. pp. 193-194, April 13–18 1996.
15. Lin, J. A Visual Language for a Sketch-Based UI Prototyping Tool. In Proceedings of *Human Factors in Computing Systems: CHI '99 Extended Abstracts*. Pittsburgh, PA. pp. 298-299, May 15-20 1999.
16. Long, A.C., Jr., J.A. Landay, and L.A. Rowe. Implications For a Gesture Design Tool. In Proceedings of *Human Factors in Computing Systems: CHI '99*. Pittsburgh, PA. pp. 40-47, May 15-20 1999.
17. Moran, T.P., P. Chiu, and W.v. Melle. Pen-Based Interaction Techniques For Organizing Material on an Electronic Whiteboard. In Proceedings of *the ACM Symposium on User Interface Software and Technology: UIST '97*. Banff, Alberta, Canada. pp. 45-54, October 14-17 1997.
18. Newman, M. and J.A. Landay, *Sitemaps, Storyboards, and Specifications: A Sketch of Web Site Design Practice as Manifested Through Artifacts*. Technical Report UCB//CSD-99-1062, University of California, Berkeley, Computer Science Division, Berkeley, CA 1999.
19. Rosenfeld, L. and P. Morville, *Information Architecture for the World Wide Web*. Sebastopol, CA: O'Reilly, 1998.
20. Rubine, D., Specifying Gestures by Example. *Computer Graphics: ACM SIGGRAPH*, 1991: p. 329-337.
21. Sano, D., *Designing Large-Scale Web Sites: A Visual Design Methodology*. New York, NY: John Wiley & Sons, 1996.
22. Saund, E. and T.P. Moran. A Perceptually-Supported Sketch Editor. In Proceedings of *the ACM Symposium on User Interface Software and Technology: UIST '94*. Marina del Rey, CA. pp. 175-184, November 2–4 1994.
23. Schilit, B.N., G. Golovchinksy, and M.N. Price. Beyond Paper: Supporting Active Reading with Free Form Digital Ink Annotations. In Proceedings of *Human Factors in Computing Systems: CHI '98*. Los Angeles, CA. pp. 249-256, April 18-23 1998.
24. van de Kant, M., S. Wilson, M. Bekker, H. Johnson, and P. Johnson. PatchWork: A Software Tool for Early Design. In Proceedings of *Human Factors in Computing Systems: CHI '98*. Los Angeles, CA. pp. 221-222, April 18-23 1998.
25. Wagner, A., Prototyping: A Day in the Life of an Interface Designer, in *The Art of Human-Computer Interface Design*, B. Laurel, Editor. Addison-Wesley: Reading, MA. p. 79-84, 1990.
26. Wong, Y.Y. Rough and Ready Prototypes: Lessons From Graphic Design. In Proceedings of *Human Factors in Computing Systems: CHI '92*. Monterey, CA. pp. 83-84, May 3–7 1992.

Tool Support for Cooperative Object-Oriented Design: Gesture Based Modeling on an Electronic Whiteboard

Christian Heide Damm, Klaus Marius Hansen, Michael Thomsen
Department of Computer Science, University of Aarhus
Aabogade 34, 8200 Aarhus N, Denmark
{damm, marius, miksen}@daimi.au.dk

ABSTRACT

Modeling is important in object-oriented software development. Although a number of Computer Aided Software Engineering (CASE) tools are available, and even though some are technically advanced, few developers use them. This paper describes our attempt to examine the requirements needed to provide tool support for the development process, and describes and evaluates a tool, Knight, which has been developed based on these requirements. The tool is based on a direct, whiteboard-like interaction achieved using gesture input on a large electronic whiteboard. So far the evaluations have been successful and the tool shows the potential of greatly enhancing current support for object-oriented modeling.

KEYWORDS

Gesture input, electronic whiteboards, cooperative design, object-oriented modeling, user study, CASE tools.

INTRODUCTION

Software developers use models to develop object-oriented software. In the early stages of a software development project, developers focus on understanding the part of the world that the computer system should support. Throughout the project, they represent their understanding in the form of models. The models are not only used in order to understand and discuss the world, but are also implemented in code and thus form an important part of the final software system.

A variety of Computer Aided Software Engineering (CASE) tools have been created to support the developers' work throughout the development process [12]. However, in practice these tools are supplemented with whiteboards, especially in creative phases of development.

The most appealing aspects of whiteboards are their ease of use and their flexibility. Whiteboards require no special skills, they do not hamper the creativity of the user, and they can be used for a variety of tasks. Their many advantages aside, for most development projects whiteboards are not enough. Capturing diagrams electronically with CASE tools facilitates code generation, documentation, and allows developers much more flexibility in editing and changing the diagrams. The conflicting advantages and disadvantages of whiteboards and CASE tools can lead to frustrating and time consuming switches between the two technologies. Our goal is to design a tool that offers the best of both worlds.

Paper Structure

The next section presents the motivation for our design. We then discuss two user studies from which we derive a set of design implications. We describe the Knight tool we developed based on these observations, and present an evaluation of the tool. Finally, we discuss directions for future research and draw our conclusions.

BACKGROUND

Use of whiteboards has been studied in various contexts including meeting rooms [9][15], classrooms [1], and personal offices [17]. Whiteboards are very simple to use and many activities are ideally suited for this simple interaction. Computational augmentation [24] can potentially solve problems with whiteboards such as lack of space and efficient editing facilities. We are concerned with the use of whiteboards in a specific work practice, cooperative object-oriented modeling, and the potential use of augmentation in that setting.

CASE tools seek to support software development techniques such as diagramming, code generation, and documentation. Nevertheless, adoption of CASE tools in organizations is slow and partial [5][8]. A main reason is that current CASE tools are targeted at technically-oriented methods rather than user-oriented processes. In particular, CASE tools are weak in supporting creativity, idea generation, and problem solving [7].

The *Tivoli* system [20][16] inspired us and was a starting point for our work. It is designed to support small, informal meetings on an electronic whiteboard. The similarity of user interaction to that on ordinary whiteboards is stressed. In order to be able to support specific meeting practices, Tivoli introduced domain objects that allow customizations of the tool to support, e.g., brainstorming sessions and decision-making meetings. We focus on the creation and manipulation of a certain kind of domain objects and the integration of these into a computational environment.

OBSERVING DESIGN IN PRACTICE

We conducted two field studies of software developers using CASE tools and whiteboards in order to understand the current practice of object-oriented modeling. In both studies, we observed a group of developers with mixed competencies and then interviewed them. The developers used the Unified Modeling Language (UML [22]), which is a formal graphical notation containing several different diagram types. We concentrate on UML class diagrams, which are used to model central concepts and relationships found in the real world (or an imagined world).

Each study focused on three aspects of the design activity: *cooperation, action,* and *use*. *Cooperation* includes the communicative, coordinative, and collaborative aspects of design. *Action* and *use* are akin to the categories Bly et al. used in their observations of shared drawings [3]. *Action* involves the physical interaction of the designers with tools. *Use* involves the semantics of the result of actions.

User Study 1: Building a New System

Research setting. COT [29] is a technology transfer project between Danish universities and private companies. As part of COT, a university research group and developers from a private company cooperatively designed an object-oriented control system for a flow meter.

Participants. The developers from the private company had no previous knowledge of object-oriented development, whereas the university research group consisted of experienced object-oriented developers. The developers from the private company acted as domain experts in initial phases of development, while the university researchers were object-oriented software designers and, to some extent, mentors for the developers from the private company. During the two-week period in which the project was studied, the number of people attending design meetings varied. Typical sessions involved 2-4 developers from the private company and 2-4 university people.

Procedure. The sessions took place in meeting rooms equipped with multiple ordinary whiteboards, an overhead projector, and a computer. We observed three sessions in detail. In each of these an observer took notes.

Observation 1: Alternating Between Tools

Tom, Mike and Peter[1] are about to discuss a new area of the problem domain. To brainstorm an initial design, Tom and Mike draw initial models on a whiteboard. Just before lunch they run out of whiteboard space and Peter captures the work so far using a digital camera. After lunch Tom and Mike continue on a freshly-wiped whiteboard, while Peter redraws the diagrams from before lunch in a CASE tool using the photos as a reference.

Developers alternated between whiteboards and CASE tools. A typical work sequence involved sketching a model and then transferring it to a CASE tool, which could then use the formal model to generate code.

The next morning, Mike uses the CASE tool to generate diagrams from existing code. These illustrate details he and Tom discussed the previous day. Mike places printouts of the diagrams on the overhead projector (Figure 1) and Tom uses whiteboard markers to make amendments.

Work on an area of the problem domain involved several cycles of drawing and redrawing both on whiteboards and in CASE tools.

Figure 1. Using transparencies on a whiteboard

Observation 2: Working with a Formal Notation

Tom explains how a flow meter works conceptually. As he explains, he tries to model this using a UML diagram containing classes and relationships. He stops several times in order to ask Mike and Peter how to draw UML elements correctly. Moreover, to understand the semantics of the drawings, Mike and Peter often interrupt Tom.

The formal UML notation was hard to learn for the inexperienced developers. These syntactic problems caused a number of interruptions and breakdowns during modeling.

Peter and Tom are modeling how a number of objects interact with each other in the system. They want to show how these are synchronized. However, the UML is incapable of describing these aspects. Mike suggests a new notational element for this. Peter and Tom pick this up and use it in subsequent design problems.

The semantics of the notation was extended in order to make it support the work process and to add expressive

---

[1] The names of participating designers have been changed throughout this paper.

power. In this way, the developers effectively extended the UML notation on the fly.

Observation 3: Combining Informal and Formal Drawings
Tom sketches the physical appearance of a flow meter on the whiteboard. He uses this drawing while explaining a diagram of the flow meter's electrical circuits. Following this, Mark models the interface to the circuits. He connects the elements to Tom's sketches.

The domain experts used illustrations, in connection with diagram elements, to explain important problem domain concepts. New ideas were often sketched informally, just before the introduction of UML notation.

Peter photographs the sketches of the circuits and flow meter before he continues to elaborate on Mark's diagram. Later, he realizes that he needs the sketches as a reminder. He consults the digital photo and redraws a part of the circuits before he continues modeling.

The informal drawings were temporary and were usually erased after the corresponding formal diagram was drawn. Central drawings were, however, redrawn on paper or photographed to make them persistent.

User Study 2: Restructuring an Existing System
Research settings. Mjølner Informatics [30] is a small company that makes compilers and other software development tools for the object-oriented language BETA [13]. A design meeting was held to design a new tool that integrated several separately-developed tools.

Participants. Six developers attended the meeting. They had varying experience in object-orientation and varying knowledge of the separate tools. Four of the developers had been previously responsible for a separate tool each. The last two developers had limited knowledge of the tools.

Procedure. We videotaped the ongoing discussion. In addition, we took notes with special emphasis on what was drawn on the blackboard.

Observation 1: Filtering of UML Drawings
John and Michael have each drawn a model of the tool they have developed. They now focus on how to integrate the two tools. Michael erases the parts of the tools that are irrelevant for the integration. He groups several associations into one in order to show a relationship even though he has erased the intermediate classes.

Often, the developers filtered information to handle large models (Figure 2). The developers idealized the model when it improved the understandability, reduced the interfaces of classes whenever full classes were too detailed, and kept transitive relations between classes to a minimum.

Observation 2: Editing Diagrams
Eric draws a model of a tool. To explain this, he adds details to the classes John and Michael have drawn. After a while this clutters the diagram. Eric erases some of the extended classes and redraws them further apart. For most of the "moved" classes he only redraws the box and name.

Participants frequently changed the diagrams. Such changes were time-consuming and annoyed the developers.

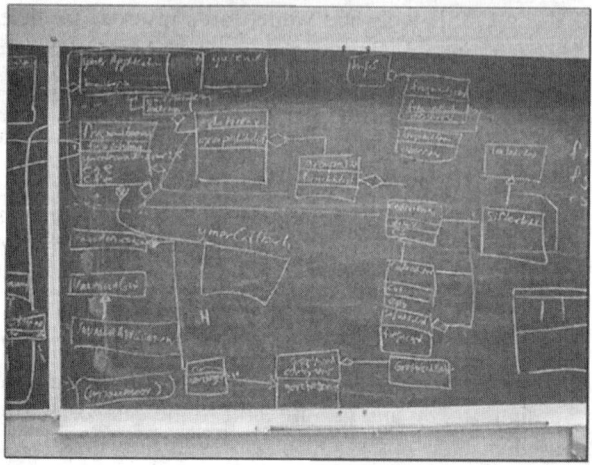

Figure 2. Blackboard snapshot

Observation 3: Drawing Informal and Incomplete elements
John and Eric are modeling a part of the integration. In order to show the lack of knowledge of this area, they only draw a partial diagram. Several times they draw relationships that are only connected to one class.

Approximately 25% of the meeting was spent on actual drawing on the blackboard. Of this, about 80% was spent drawing formal UML diagrams, and the remaining 20% on informal and incomplete drawings (or about 5% of the total meeting time). UML elements were drawn in incomplete variants, such as classes without names, incomplete inheritance trees, or associations connected to only one class.

Observation 4: Cooperation Between Developers
John starts to draw a model of a tool on the blackboard. The other developers sit at a table and ask questions. Following this, Eric starts to draw the tool that he has developed. Soon he discovers relations between the two tools and asks John to elaborate on his drawing. After John finishes, Eric continues on his drawing.

Figure 3. Turn-taking at the blackboard

The developers frequently cooperated by taking turns at the blackboard. Figure 3 shows two developers standing at the blackboard, taking turns adding, deleting or changing the model. Abrupt interruptions were rare and only one developer drew on the model at any time. However, other discussion often took place while another person was drawing.

Design Implications

The two user studies highlighted the effectiveness of ordinary whiteboards as tools for cooperative design. They support a direct interaction that is easy to understand, and they never force the developer to focus on the interaction itself. Whiteboards allow several developers to work simultaneously and thus facilitate cooperation. They do not require a special notation and thus support both formal and informal drawings. Notational conventions can easily be changed and extended.

Whiteboards, however, miss several desirable features of CASE tools. Without the computational power of CASE tools, making changes to the drawings is laborious, the fixed space provided by the board is too limited, and there is no distinction between formal and informal elements. There is also no support for saving and loading drawings.

These observations led to the following design criteria for a tool to support object-oriented modeling:

- *Provide a direct and fluid interaction.* A low threshold of initial use is needed and the tool should never force the developer to focus on the interaction itself. The whiteboard style of interaction is ideally suited for this.

- *Support cooperative work.* Several developers must be able to work with the tool cooperatively. Informal cooperative work with domain experts as well as software developers must be supported.

- *Integrate formal, informal, and incomplete elements.* Besides support for formal UML elements, there must be support for incomplete UML elements and informal freehand elements. Also, the support for formal UML elements must be extensible, to allow for the introduction of new formal elements.

- *Integrate with development environment.* Integration with traditional CASE and other tools is needed. Diagrams must be saved and restored, and code must be generated and reverse engineered.

- *Support large models.* A large workspace is needed. In addition, there must be support for filtering out information that is not needed at a given time.

DESIGN OF THE KNIGHT TOOL

Based on the user studies, we have designed and implemented the *Knight* tool. The Knight tool uses an electronic whiteboard, currently a SMART Board [26], as its input medium. The SMART Board is a 72-inch touch-sensitive computer screen mounted in a cabinet to resemble a traditional whiteboard. Users can draw on the surface using a number of pens (or just using their fingers). In contrast to other electronic whiteboards, such as, e.g., the Xerox Liveboard [20], the SMART Board unfortunately only allows input from one pen at a time.

The prototype is implemented in Tcl/Tk [19] with the [incr Tcl] extension [14], runs on the Windows and Unix platforms and is available for download from the Knight homepage [25]. We kept the interface very simple (Figure 4). A large workspace, resembling an ordinary whiteboard, is the central part of the user interface. The interaction is based on pen-strokes made directly on the workspace.

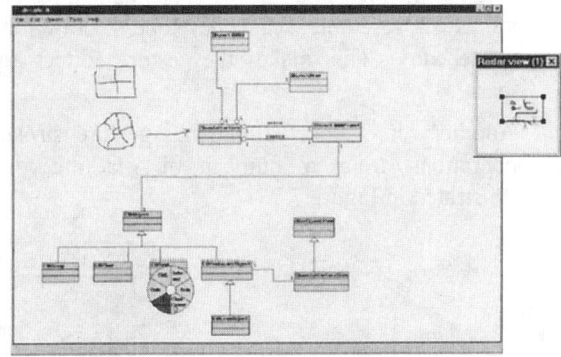

Figure 4. Knight user interface

Formality, Informality, and Directness

We wanted the tool to support a continuum of drawing formality, ranging from informal sketching elements over incomplete UML elements to formal UML elements. To allow this, the tool currently operates in one of two modes: Freehand mode or UML mode. We recognize that the use of modes is potentially problematic. However, our studies indicate that users already naturally operate in these two different modes, in different phases of the design. Freehand mode supports idea generation and UML mode supports design formulation. Two different background colors indicate the different modes.

In freehand mode, the pen strokes are not interpreted. Instead, they are simply transferred directly to the drawing surface. This allows the users to make arbitrary sketches and annotations just as on an ordinary whiteboard. Unlike on whiteboards, these can be moved around or hidden. Each freehand session creates a connected drawing element that can be manipulated as a single whole.

In UML mode, pen strokes are interpreted as gestural commands that create UML elements. If, e.g., a user draws a box, the tool will immediately interpret this as the gesture for a UML class and replace the pen stroke by a UML class symbol (Figure 5).

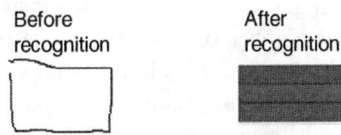

Figure 5. Recognition of the gesture for a UML class

The diagrams need not adhere to the UML semantics completely, in that incomplete diagram elements are allowed. Figure 6 shows how the user can input a relationship between two classes with only one of the two classes specified. The relationship can later be fully specified.

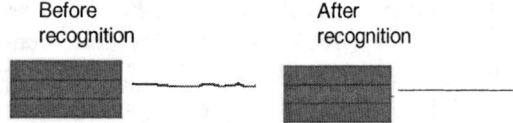

Figure 6. A relationship with only one class specified

The gestures for creating UML elements have been chosen so as to resemble what developers draw on ordinary whiteboards. This makes the gestures direct and easy to learn.

Another set of short directional gesture strokes chooses operations from a number of marking menus [10] illustrated in Figure 7.

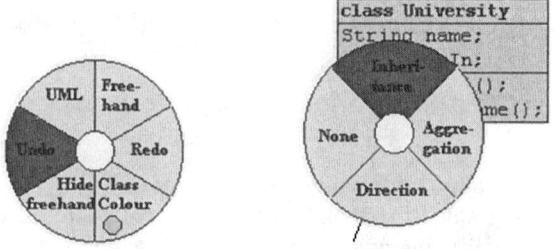

Figure 7. Context-dependent pie menus

For example, in order to undo or redo, the user may either make a short left or right stroke, or press and hold the pen and choose the corresponding field in a popup pie menu. The marking menus are also used to switch between UML and freehand mode. Marking menus support the interaction on a large surface well, because they are always ready at hand, unlike usual buttons and menus [20]. Apart from supporting a transition from initial to expert use, the marking menus also conveniently provide an alternative way of creating certain diagram elements (Figure 7 right). The marking menus are context-dependent. Depending on the immediate context in which a stroke or press was made, it will be determined whether it should be interpreted as a gesture command or as a marking menu shortcut. In the latter case a context-specific menu will be shown.

Use of Gestures

We use Rubine's algorithm [21] to recognize the gestures. This algorithm has the advantage of being relatively easy to train: To add a new gesture command, one simply draws a number of gesture examples. Potential problems with the algorithm, and gesture recognition in general, include that only a limited number of gestures can be recognized and that no feedback is given while gestures are drawn. To address these problems, we use *compound gestures* [11] and *eager recognition* [21], respectively.

Compound gestures combine gestures that are either close in time or space to a diagram element. For example, a user can change an association relationship (represented by an undecorated line) to a unidirectional association (represented by a line with an arrowhead) by drawing an arrowhead at the appropriate end. In this way, users can gradually build up a diagram, refining it step-by-step.

With eager recognition, the tool continuously tries to classify gestures as they are being drawn. Whenever the gesture can be classified with a high confidence, feedback is given to show that the gesture was recognized, and the rest of the gesture is used as parameters to the recognized gesture's command. For example, when a move gesture is recognized, the elements located at the starting point of the gesture will follow the pen while it is pressed down.

Support for Large Models

The workspace is potentially infinite, allowing users to draw very large models. It also supports zooming, as in zoomable interfaces [2]. In order to provide overview and context awareness, one or more floating "radar" windows can be opened (Figure 8; see also Figure 4 upper right).

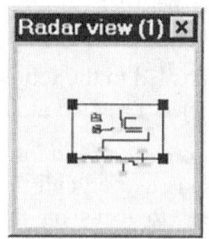

Figure 8. Radar windows provide context awareness

These radar windows show the whole drawing workspace, with a small rectangle indicating the part currently visible. Clicking and dragging the rectangle pans while dragging the handles of the rectangle zooms. By opening more radar windows, multiple users can have convenient access to pan and zoom, even though the physical drawing space is large.

Filtering is in a preliminary stage. Currently, it is possible to suppress details of the formal UML model and toggle the visibility of informal elements.

Tool Integration

The Knight tool must be integrated with existing CASE tools, to facilitate code generation from the models. Although the Knight tool is currently only able to exchange data with the WithClass CASE tool [27], we are currently working on making it a plug-in front-end to a variety of tools. In this way it is possible to use the CASE tool capabilities to create or edit models outside a cooperative modeling situation.

EVALUATION OF THE KNIGHT TOOL

We evaluated the current design of the Knight tool in two sessions. Both sessions were actual design sessions in which Knight was the primary tool. The purpose was to evaluate the usability of the tool in a realistic work setting.

First, a facilitator introduced the Knight tool to the participating designers and taught them the basic use of the tool. During the evaluation, he also helped if the designers had problems and asked for help.

We videotaped the sessions and took notes. As in our user studies, we focused on three aspects of design: cooperation, action, and use. Following the design sessions, we conducted qualitative interviews.

Both sessions were encouraging. Each lasted approximately one and a half hours, and the participants were able to maintain focus on their job, rather than on the tool or the evaluation.

Next we discuss the results of the evaluations with respect to the design criteria identified and summarized in "Design Implications" above.

Evaluation 1: Designing a New System Using Knight
Research setting. The CPN2000 project [6][31] is concerned with developing and experimenting with new interaction techniques for a Petri Net editor with a complex graphical user interface. The original user interface is a traditional window-icon-menu-pointer interface, whereas the new interface will use interaction styles such as tool-glasses, marking menus, and gestures. As part of the design, three object-oriented models for the handling of events and for the implementation of certain interface elements had been constructed. We observed the meeting in which these three models were integrated into one model using the Knight tool.

Participants. Three designers participated in the meeting. One of these had modeled the event handling and was knowledgeable of UML and traditional CASE tools. The other two modeled the interaction styles and had little knowledge of UML.

Results
The resulting diagram is shown in Figure 9. This rather large model was constructed with few problems and mishaps.

Provision for a direct and fluid interaction. After the short introduction, the participants were able to use the tool for long periods without help: the tool had a low threshold for initial use.

The use of gestures was mostly unproblematic. However, some participants had trouble drawing certain gestures. This may be due in part to too little training, but the gesture set can also be improved. For example, people draw differently with respect to size, orientation, and speed, and the gesture examples used to train the recognizer must encompass such variations.

Support for cooperative work. The electronic whiteboard worked well as a center of cooperation. The only problem participants reported was that only one person could draw

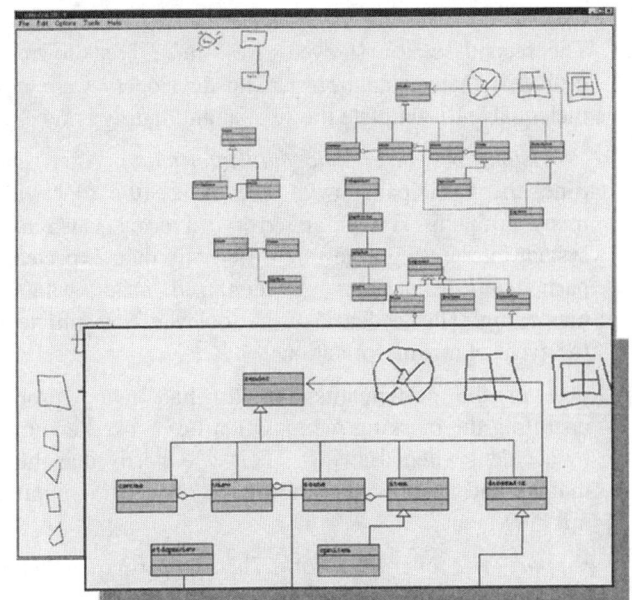

Figure 9. Diagram produced in the first session (with a blow-up of the upper right part)

at a time. Nevertheless, each developer was able to hold his or her own pen, and they all coordinated their actions when necessary.

Integration of formal, informal and incomplete elements. Freehand drawings were widely used and appreciated. The low resolution of the actual electronic whiteboard meant that the freehand drawings were relatively coarse-grained. In addition, response from the electronic whiteboards was delayed when a user drew quickly. This meant that freehand text was hard to do both legibly and quickly.

Support for large models. As the size of the model grew, the radar window was used to pan and zoom efficiently. However, the fact that the radar window was the only way to move around the workspace was problematic. In order to move a class from one corner of the large diagram to another, one participant had to move, pan, and then move again. This suggests the need for some other means of scrolling.

Evaluation 2: Restructuring a System Using Knight
Research setting. The DESARTE project [32] is concerned with designing an electronic support environment for architects. As part of this environment, a 3D replacement of the workstation desktop is implemented. A conceptual model had previously been designed and was to be restructured during this meeting using the Knight tool.

Participants. Two designers attended the meeting: The designer responsible for implementing the 3D desktop and a user involvement expert with an understanding of architectural work practice. Both had a good knowledge of object-oriented modeling.

Results

The second session showed a few more breakdowns and problems than evaluation 1. The developers were nevertheless able to complete the session and their work.

Provision for a direct and fluid interaction. After a short time, the participants were able to use the tool without many problems. When an error did occur, such as the system interpreting a gesture differently than expected, the participants sometimes got confused about what was happening: The feedback of the tool was not sufficient in the event of misinterpretations.

One of the participants initially had many problems operating the marking menu: Often he invoked commands by accident when drawing. This was partly due the fact that he had no previous knowledge of gestural input and marking menus.

Support for cooperative work. The two participants had no trouble cooperating around the tool.

Integration of formal, informal, and incomplete elements. The participants often made freehand drawings to illustrate the user interface of the designed environment. A minor problem in this case, was that that informal and formal elements could only be rudimentary connected, and there was little support for advanced grouping. Incomplete UML elements were considered useful, but were not widely used.

Support for large models. In this evaluation, the focus was on restructuring an existing diagram of moderate size, and the radar window was mostly used for zooming.

DESIGN IMPLICATIONS & FUTURE WORK

The observations and subsequent interviews showed that the Knight tool is a valuable tool for modeling in practice. However, as the above results point out, improvements are needed.

A number of physical problems with the actual electronic whiteboard hindered cooperation. Only one person at a time could draw on the whiteboard and informal drawing was only slowly rendered. The latter problem may be handled in part by the Knight tool, whereas the former is intrinsic to the specific electronic whiteboard. However, the lack of support for synchronous drawings was not construed as a major problem in the two evaluations. Since design processes are becoming increasingly distributed, we are currently investigating distributed cooperative design using the Knight tool. Integration with a mediaspace [4] may provide a non-intrusive way of supporting distributed communication in relation to this.

Problems with gestures caused a number of breakdowns. Several possibilities exist for alleviating this. First, more appropriate feedback can be given when a gesture has been drawn. Second, personalized gestures may be necessary, e.g., in the form of a *personal pen*, as in Tivoli [20]. For each personal pen there could be a separate gesture set, a separate mode, separate colors, or other personal settings.

The integration of formal, informal, and incomplete elements is not complete. It is not, e.g., possible to connect formal and informal elements. A dynamic extension of the formal notation is a step towards this integration. The environment, with gesture recognition based on examples and an interpreted programming environment, makes such extensions technically feasible. *Flatland* [18] defines non-overlapping *segments* with different behaviors. Such segments may be used to group formal and informal elements separately. A notion of overlapping groups may be used to link these different segments.

Many of our observations of object-oriented modeling seem to be true for other types of formal modeling such as task modeling. A natural step would be to implement support for these as well, especially if combined with the idea of overlapping groups. This would facilitate combination of informal elements with formal elements, as well as handling types of formal elements together.

Filtering should also be considered in depth. Especially in evaluation 1, after the diagram had reached a certain size, navigation in the workspace became time-consuming. It should be possible to selectively hide parts of a model and give drawing elements temporality so elements may exist only for a certain period of time.

An important future area of research is the use of the Knight tool as a plug-in interface for different tools, which we are currently working on. This will involve longitudinal studies of the use of the Knight tool in development projects.

CONCLUSION

We have developed a tool for object-oriented development: Knight. The design of the tool is based on user studies of software developers creating object-oriented models. These show that important design criteria for a usable tool are (1) a direct and fluid interaction, (2) support for collaborative work, (3) an integration of both formal and informal drawing elements, (4) support for modeling in the large and (5) integration with existing development tools.

The Knight tool was designed to meet these criteria by using a large electronic whiteboard as input medium and by using an interaction style similar to that of traditional whiteboards. Input is done using gestures that resemble what is drawn on whiteboards. Both formal and more informal elements are supported and several developers can easily cooperate at the electronic whiteboard. Knight thus maintains the advantages of whiteboards and additionally adds features only possible in a computer based tool: Models can be easily modified, diagrams can be exported and imported to and from CASE tools, elements can be hidden and later restored, and a much larger workspace is provided.

ACKNOWLEDGEMENTS

We thank Michael Tyrsted who participates in the Knight project. We also thank Wendy Mackay for many discussions and for critique and help in writing this paper. Furthermore, we thank Michel Beaudouin-Lafon as well as the people from Danfoss Instruments, Mjølner Informatics, the DESARTE project, and the CPN/2000 project.

The Knight Project is carried out in the Centre for Object Technology that has been partially funded by the Danish National Centre for IT Research [28].

REFERENCES

1. Abowd, G., Atkeson, C., Feinstein, A., Hmelo, C., Kooper, R., Long, S., Sawhney, N., Tani, M.: Teaching and Learning as Multimedia: The Classroom 2000 Project. *Proceedings of Multimedia '96*, 1996, 187-198.

2. Bederson, B.B., Hollan, J.D. Pad++: A Zooming Graphical Interface for Exploring Alternate Interface Physics. *Proceedings of UIST*, 1994, 17-26.

3. Bly, S.A, Minneman, S.L. Commune: A Shared Drawing Surface. *Proceedings of the Conference on Office Information Systems*, 1990, 184-192

4. Bly, S.A., Harrison, S.R., Irwin, S. Mediaspaces: Bringing people together in a video, audio and computing environment. *Communications of the ACM*, 36(1), January 1993.

5. Iivari, J. Why Are CASE Tools Not Used? In *Communications of the ACM*, 39 (10), 1996.

6. Janecek, P., Ratzer, A.V., Mackay, W.E. Redesigning Design/CPN: Integrating Interaction and Petri Nets in Use. *Proceedings of the Second Workshop on Practical Use of Coloured Petri Nets and Design/CPN*, 1990, 119-133.

7. Jarzabek, S., and Huang, R. The Case for User-Centered CASE Tools. *Communications of the ACM*, 41 (8), 1998.

8. Kemerer, C.F. Now the learning curve affects CASE tool adoption. In *IEEE Software*, 9 (3), 1992.

9. Kraut, R., Fish, R., Root, R., Chalfonte, B. Informal Communication in Organizations: Form, Function and Technology. *Groupware and Computer-Supported Cooperative Work.* 1993, 287-314.

10. Kurtenbach, G. *The Design and Evaluation of Marking Menus*. Ph.D. Thesis, University of Toronto, 1993.

11. Landay, J.A., and Myers, B.A. Interactive Sketching for the Early Stages of User Interface Design. *Proceedings of CHI'95*, 45-50.

12. Lyytinen, K., Tahvanainen; V.-P. *Next Generation CASE Tools*. IOS Press, 1992.

13. Madsen, O.L., Møller-Pedersen, B., Nygaard, K. *Object-Oriented Programming in the BETA Programming Language*. ACM Press, Addison Wesley, 1993.

14. McLennan, M.J. [incr Tcl]: Object-Oriented Programming. In *Proceedings of the Tcl/Tk Workshop*, University of California at Berkeley, June 10-11, 1993.

15. Moran, T.P., Chiu, P., Harrison, S., Kurtenbach, G., Minneman, S., van Melle, W. Evolutionary Engagement in an Ongoing Collaborative Work Process: A Case Study. *Proceedings of CSCW'96*, 150-159.

16. Moran, T.P., van Melle, W., and Chiu, P. Tailorable Domain Objects as Meeting Tools for an Electronic Whiteboard. *Proceedings of CSCW'98*, 295-304.

17. Mynatt, E.D. The Writing on the Wall. *Proceedings of INTERACT'99*, 1999, 196-204.

18. Mynatt, E.D., Igarashi, T., Edwards, W.K., and LaMarca, A. Flatland: New Dimensions in Office Whiteboards. *Proceedings of CHI'99*, 346-353.

19. Ousterhout, J.K. *Tcl and the Tk Toolkit*. Addison-Wesley, 1994.

20. Pedersen, E.R., McCall, K., Moran, T.P., and Halasz, F.G. Tivoli: An Electronic Whiteboard for Informal Workgroup Meetings. *INTERCHI'93*, 391-398.

21. Rubine, D. Specifying gestures by example. *Proceedings of SIGGRAPH'91*, 329-337.

22. Rumbaugh, J., Jacobson, I., Booch, G. *The Unified Modeling Language Reference Manual*. Addison-Wesley, 1999.

23. Russell, F. The case for CASE. *Software Engineering: A European Perspective.* Thayer, R., McGettrick, A. (Eds.) IEEE Computer Society Press, 1993, 531-547.

24. Wellner, P., Mackay, W., Gold, R.: Guest Editors' Introduction to the Special Issue on Computer-Augmented Environments: Back to the Real World. In *Communications of the ACM*, 36(7), 1993.

ONLINE REFERENCES

25. http://www.daimi.au.dk/~knight
26. http://www.smarttech.com
27. http://www.microgold.com
28. http://www.cit.dk
29. http://www.cit.dk/COT
30. http://www.mjolner.com
31. http://www.daimi.au.dk/CPnets/CPN2000
32. http://desarte.tuwien.ac.at/

The Cubic Mouse
A New Device for Three-Dimensional Input

Bernd Fröhlich **John Plate**
GMD/IMK.VE
D-53754 Sankt Augustin, Germany
{Bernd.Froehlich, John.Plate}@gmd.de

ABSTRACT
We have developed a new input device that allows users to intuitively specify three-dimensional coordinates in graphics applications. The device consists of a cube-shaped box with three perpendicular rods passing through the center and buttons on the top for additional control. The rods represent the X, Y, and Z axes of a given coordinate system. Pushing and pulling the rods specifies constrained motion along the corresponding axes. Embedded within the device is a six degree of freedom tracking sensor, which allows the rods to be continually aligned with a coordinate system located in a virtual world. We have integrated the device into two visualization prototypes for crash engineers and geologists from oil and gas companies. In these systems the Cubic Mouse controls the position and orientation of a virtual model and the rods move three orthogonal cutting or slicing planes through the model. We have evaluated the device with experts from these domains, who were enthusiastic about its ease of use.

Keywords
User interface hardware, two-handed interaction, virtual reality

INTRODUCTION
Many graphics applications require the input of three-dimensional coordinates to position objects in a virtual world. Desktop applications typically use a mouse, trackball or a more exotic device like a dial box for such input. This works reasonably well as long as the coordinate system of the world is aligned with the screen. However, once the virtual world is rotated the mapping of mouse movements is often no longer intuitive, e.g. it might happen that the mouse is moved to the right, but the virtual object moves to the left. This is often quite confusing.

We present a new input device that allows intuitive input of three-dimensional coordinates in virtual environment applications. The device (Figure 1), which we named the Cubic Mouse, consists of a cube-shaped case, three rods, and control buttons. Each rod passes approximately through the center of two parallel faces of the case. The rods are perpendicular to each other and movable. They represent the X, Y, and Z axes of a coordinate system. There is also a six degree of freedom (6DOF) tracker embedded in the cube-shaped case, which we use to orient and position the virtual world in three-space relative to the observer. In this way the rods stay aligned with the coordinate system axes. By pushing and pulling the rods we specify motions of virtual objects constrained along the X, Y, and Z axes. Typically users hold the device in their non-dominant hand to position and orient the world, while the dominant hand operates the rods and the control buttons.

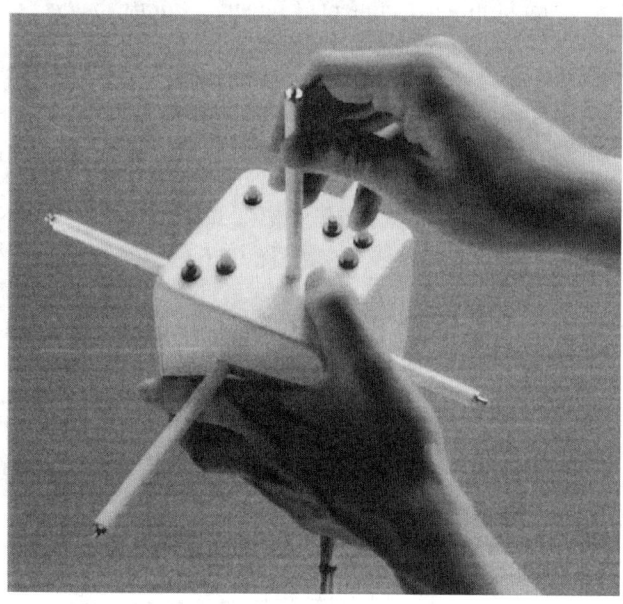

Figure 1: The Cubic Mouse device

For the development of the Cubic Mouse we had two driving applications. Within a consortium of car manufacturers and crash software vendors we are developing an application prototype for steering and visualizing car crash simulations in virtual environments. Stereoscopic virtual environments like Caves [1], Responsive Workbenches [6,7] or large screen projections facilitate the understanding of complex three-dimensional deformations occurring during a car crash. The main tool for investigating simulated car crash results are cutting planes and chair cuts. We use the Cubic Mouse for navigating around the car model, for positioning cutting planes inside the model (Figure 2), and for performing chair cuts.

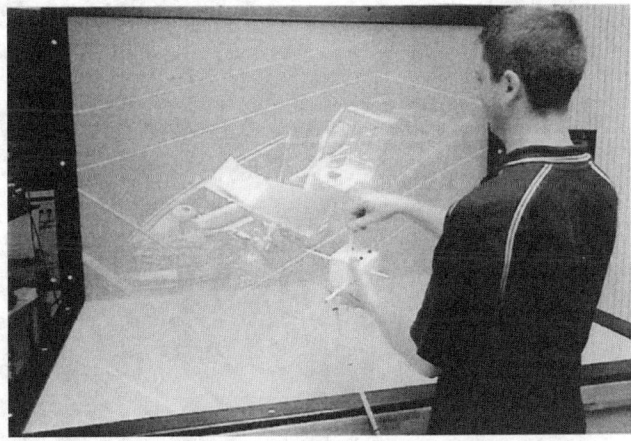

Figure 2: A car crash visualization on the Responsive Workbench. The Cubic Mouse is aligned with the principal axes of the car model. The user moves one of the three orthogonal cutting planes from the top into the model using the appropriate rod.

Our second driving application stems from the geo-scientific domain. We work with geologists and geo-physicists from oil and gas companies to evaluate virtual environment technology for reservoir discovery and characterization. Their data is mostly based on seismic measurements and information acquired from the actual drilling of wells. This data is three-dimensional in nature and stereoscopic virtual environments allow users to explore and understand complex subsurface structures in three dimensions. One traditional method of exploring the seismic data is by moving three orthogonal slices through a 3D seismic volume. In our system these slices can be positioned intuitively with the three rods. We also experimented with the positioning of orthogonal slices in volumetric data sets from CT and MRI scans in a similar way. In all cases users were able to use the basic features of our device with only two or three sentences of introduction and they were immediately enthusiastic about its ease of use.

Our main contribution is the development of an intuitive device for the input of three-dimensional coordinates in interactive graphics applications. We describe the realization of the device, introduce a set of possible application domains, and discuss some experiences of experts from these domains and results from an initial user study. We also present some variations on the basic design of our device.

RELATED WORK

In most of our application scenarios, the Cubic Mouse serves as a coordinate system prop. From this perspective, the closest relative is the head prop in [5] for neurosurgical visualization. In their system, users hold a small rubber sphere or a doll's head with an embedded tracker in one hand. This head prop is used to control the orientation of a head model on the screen. The other hand holds a second prop which, for example, is used to position a cutting plane relative to the head prop. This is in contrast to our system, where the dominant hand is used to manipulate controls located on the Cubic Mouse held in the non-dominant hand.

A variety of systems use two-handed interaction techniques based on hand-held widgets, e. g. in [8] users hold a virtual widget in one hand and operate it with the other. In [9] users hold a miniature model of the virtual world in one hand and manipulate objects in the miniature with the other. These systems do not employ real world props other than tracked wands or data gloves.

The TouchCube by ITU Research presented at the Siggraph'98 exhibition is a cube-shaped input device with touch sensitive faces. By applying certain gestures to the touch sensitive surfaces, objects are moved in a three-dimensional world. The version presented was not tracked. A tracking sensor could be easily added, but the device relies largely on being mounted on a stand. Otherwise it would be difficult to apply two finger gestures simultaneously to two opposite surfaces as they are suggested in the patent (US patent US-A-5 729 249). The key concept of the Cubic Mouse is that moving a rod into a certain direction results in the movement of a virtual object into exactly the same direction thus heavily relying on tracking.

At the University of North Carolina there was a system developed that used pairs of X, Y, and Z sliders mounted in the corresponding directions in a fixed world coordinate system to specify pairs of orthogonal clipping planes [Prof. Brooks, UNC, personal communication]. This setup is very similar to the Cubic Mouse, except that the Cubic Mouse can be used to specify clipping planes relative to an arbitrarily oriented object.

From another point of view, the Cubic Mouse is related to devices which allow the separate input of X, Y, and Z coordinates. For example, the dial box uses separate dials to specify constraint motions along the X, Y, and Z axis, but there is no intuitive connection between a given dial and the corresponding axis.

HARDWARE AND SOFTWARE IMPLEMENTATION

The Cubic Mouse was built using off the shelf parts. The cube-shaped case's edge length is 9 centimeters (3.5 inches) and was determined by the size of readily available potentiometers. The total weight of the device plus the 1.3 meters of cable is around 300 grams (2/3 of a pound). The case and the cables contribute most of the weight. They could be built with much lighter materials, which would reduce the weight considerably. The latest prototype has buttons mounted at both ends of each rod plus six application programmable buttons mounted on one of the Cubic Mouse's faces as shown in Figure 1. The two cables come off the opposite face. Most of our applications have a natural up and down direction, so the face with the buttons becomes typically the "up" face and the cables come off the "down" face. Linear potentiometers are used to measure the positions of the rods. A Polhemus Fastrak sensor provides the spatial position and orientation information for the Cubic Mouse.

We built the device within a few days and integrated it into the Avocado graphics framework [11] without any difficulties. Avocado is based on SGI's Performer toolkit and OpenGL. The system supported already an analog/digital (A/D) converter. After plugging the Cubic Mouse into the A/D box the corresponding button states and potentiometer values were immediately available in Avocado. Most of the programming was done in Scheme, Avocado's scripting language. Avocado supports a variety of output devices, so we were able to experiment with the Cubic Mouse in a CAVE, on a two-sided Responsive Workbench system (a Responsive Workbench with an additional vertical screen at the back), and in a monitor based environment.

APPLICATION SCENARIOS

We experimented with the Cubic Mouse in four different application domains. Most of our experiences were collected from an application prototype developed for the oil and gas industry, where we use the Cubic Mouse for positioning three orthogonal slices in a seismic data set. We used the Cubic Mouse in a similar way for the visualization of CT and MRI data. Clipping planes and chair cuts are important tools in engineering visualization systems. We specify these cuts with the Cubic Mouse and presented our system to a crash engineer from a large automotive company.

Data Exploration for the Oil and Gas Industry

The oil and gas industry acquires enormous amounts of seismic data for the exploration of potential new reservoirs. This data has to be sighted by geologists and geo-physicists to discover the precious oil and gas containing subsurface structures. The raw seismic data is processed into regular three-dimensional arrays - the seismic cubes. These seismic cubes along with information obtained from drilled wells are the basis for building a subsurface model. Geo-engineers roam through the seismic cubes to find areas of interest and model subsurface structures within these areas. The whole process is much more complicated than this, but these two phases are very important and they are repeated many times.

Often geo-engineers visualize their data sets by rendering subsurface structures as polygonal models and by representing the seismic volume as three orthogonal slices. A typical data set is shown in Figure 3. The Cubic Mouse allows geo-engineers to look at the data set from different directions while moving the slices through the seismic volume. This is realized in the following way: The seismic cube's orientation is always kept in sync with the Cubic Mouse's orientation and the rods move the seismic slices.

This way the rods stay always perpendicular to the slicing planes, which makes it very easy and intuitive to find the desired rod without looking at the device.

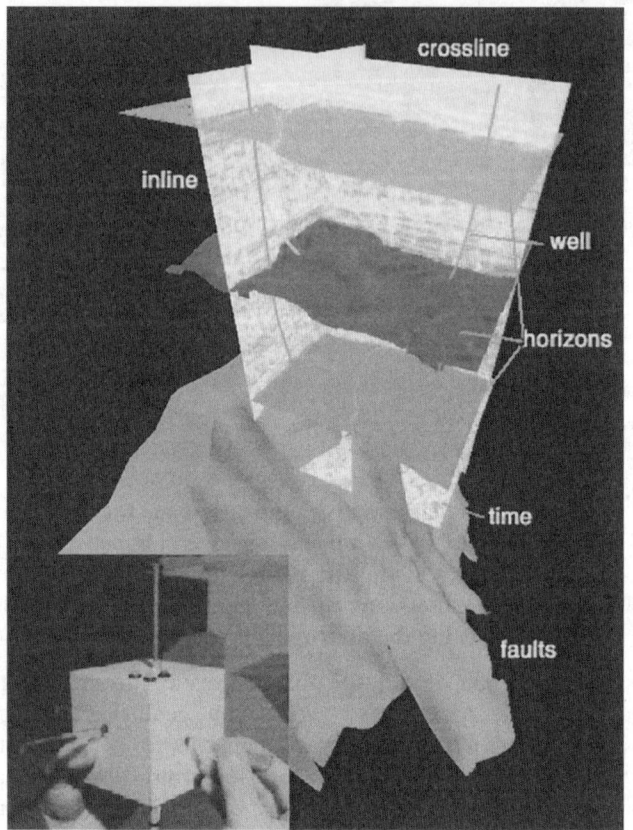

Figure 3: A typical oil exploration data set containing subsurface structures, wells, and seismic slices. The subsurface model consists of two main structures: horizons and faults. Horizons separate two earth layers, and faults are breaks in the rocks, where one side is moved relative to the other. Horizons are typically horizontal structures and faults are typically vertical structures. Three orthogonal slicing planes are used to visualize the seismic volume. Typically, one slice, the so called inline-slice, is oriented perpendicular to the main fault direction. The time-slice is oriented horizontally and the crossline-slice is perpendicular to both.

Two buttons are used for scaling the model up and down. A third button serves as a clutch, which allows users to freeze the model in the current orientation and lay down the Cubic Mouse. In our system, which was originally developed for the two-sided Responsive Workbench, the Cubic Mouse's translation was mapped 1:1 to the translation of the model. Releasing the clutch attaches the model to the Cubic Mouse's current location. This allows users for example to pan the model on the workbench by releasing the clutch on the left side, moving the Cubic Mouse and therefore the model to the right side, pressing the clutch and moving the hand back to the left side, releasing the clutch again, and so on. This worked quite well on the workbench, since most of the visible parts of the model are within an arms length organization. None of the subjects had ever used the Cubic Mouse before. The subjects were presented with the medical scenario at the two-sided Responsive Workbench shown in Figure 4. After handing over the Cubic Mouse to a subject we gave a short introduction of a few sentences explaining how to rotate the data set, describing the functionality of the rods, the two zoom buttons, and the clutch button. The subjects had to perform various tasks like finding a cross section through the nose and showing it from the side, finding a cross section through the eyes and showing it from the top, investigating the brain, finding the brain stem, and so on. After the subjects had performed these tasks, they received a questionnaire with twenty questions to assess the overall reaction to the input device, learnability, and the reaction to specific device characteristics. The questions were answered on a scale from 0 to 7.

On the whole, users found the Cubic Mouse natural and easy to manipulate (see video). They became proficient with the device in a few seconds and were able to fulfill our requests very quickly. Particular results from our questionnaire are:

- The test took 4 to 8 minutes for each person, one person being 4 minutes, one person being 8 minutes, the rest between 5 and 7 minutes. There was no correlation between the time used and the experience people had with virtual environments.
- Users felt very much in control of the application.
- Users with small hands found the device slightly too big and the rods slightly too long. The device can be comfortably held in a large hand, but should be smaller for small hands.
- Nobody complained strongly about the weight and surprisingly the cables were mostly not an issue, but our subjects used the device only for a short time.

Even though we did not tell our subjects how to hold the Cubic Mouse, it turned out that in general they held the device in their non-dominant hand and operated the rods with the dominant hand. These observations correspond closely with Guiard's findings [3] on how humans use their hands in asymmetric bimanual tasks. Both hands are used symmetrically to perform rotations of the model that can not be achieved by just twisting the wrist of the non-dominant hand.

The most surprising and convincing observation is that almost all users did not look at the device for switching rods from the first use. Instead they focus on the task and on the model shown on the screen. We observed that operating the buttons required sometimes a quick glance at the Cubic Mouse. These observations are also backed by observations done at the Siggraph 1999 exhibition, where the Cubic Mouse was used by a few hundred people.

An important reason for the intuitive operation of the Cubic Mouse can be found in the strong proprioceptive cues provided. The cubic shape of the device in the non-dominant hand provides a strong cue for the dominant hand for finding the rods, which is additionally supported by visual cues from the application.

Experiences with Experts

When developing the prototype for the oil and gas visualization, we first implemented a virtual tools based approach similar to the one described in [10], [2]. Users had to pick up different tools for each task, e.g. a zoom tool, a rotation tool, and so on. To drag around a seismic slice, the user had to pick up a drag tool with a tracked wand, point to the slice, press the button on the wand, and move the slice by moving the wand. Sometimes the slices were hard to find since they were hidden behind faults or horizons. Among other demos we worked with one geologist for three days, did a one day evaluation session with three geologists, and presented the system for four hours to 20 experts from different oil companies. We showed both versions of our system and our users found that with the Cubic Mouse, the most common tasks were immediately available and easy to perform. They had to resort to the tool based interface only occasionally.

A few month ago a crash engineer from a large automotive company visited us to review our progress in a crash visualization project our group is involved in. We presented to him the simple cutting plane and chair cut scenario shown in Figure 5 on a Responsive Workbench system with his crash test data. Even though the demonstration ran at a low frame rate due to the complex finite element model, he stated that the Cubic Mouse is all he ever needed. He added that the intuitiveness of the device was "hard to beat". His single complaint was that reversing the direction of an individual clipping plane was non-intuitive, because there were no buttons mounted at the ends of the rods at that time, which has been fixed with our latest version of the Cubic Mouse.

The medical scenario has not yet been presented to medical doctors, but based on our experience with the oil and gas experts and the automotive engineers, we are highly motivated to explore this application domain in the near future.

CONCLUSIONS AND FUTURE WORK

Coordinate systems are an essential building block for most computer graphics applications. The Cubic Mouse literally puts an application or object coordinate system into the user's hand. The rods provide a strong affordance and make the device obvious to use. The Cubic Mouse's two-handed operation allows one hand to rest against the other which reduces fatigue and allows extended use.

When presenting the Cubic Mouse, people are often inspired to think about further enhancements of the device. In particular everybody wants to be able to turn the rods and we have just completed the development of a first prototype. These additional three degrees of freedom lead to an intuitive six degree of freedom input device. Turning reach. When we started using the application in a CAVE, the model was typically displayed at a much larger scale. To compensate we used a 1:3 or 1:4 hand to model movement ratio.

Another issue is how to define the center of the rotate and zoom operations applied to the model. At the start of our application, we display the model at a scale such that the entire model is visible. The origin for the zoom and rotate operations is defined as the origin of the model's bounding sphere. After zooming and panning the model, occasionally our users want to change the origin to a new location or feature within the model. Usually they are already investigating this area by moving the slicing planes back and forth through the feature. Thus, by simultaneously pushing the two zoom buttons, we let them define the new origin of rotation as the intersection point of the three slicing planes.

Visualizing Volumetric Medical Data

CT, MRI and PET scanners generate volumetric data sets similar to the seismic data sets used in our geological scenario. Traditionally, medical visualization uses three orthogonal slicing planes (Figure 4) to view human cross sections. We use the Cubic Mouse for this scenario in exactly the same way as for the geo-scientific visualization. This gives users the skull in their hand, similar to the system described in [4], and allows them to position the slicing planes using the rods.

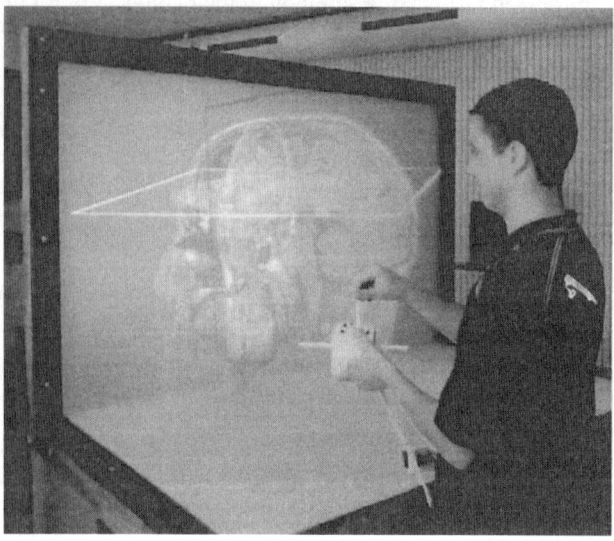

Figure 4: Three cross sections through a human head: The transversal plane is oriented horizontally, the frontal plane is parallel to the front, and the sagittal plane is perpendicular to both and moves from ear to ear.

Cutting Planes and Chair Cuts

For engineering applications, cross sections are an important tool for viewing and understanding the structure of three-dimensional models. Often two or three orthogonal cutting planes are applied at once or a so called chair cut is performed as shown in Figure 5. A chair cut removes one octant of the model, whereas three clipping planes cut away seven octants. These operations are complementary to each other and we allow users to switch between them. The definition of the octant being cut away or kept depends on the orientation of the clipping planes, which can be toggled using the buttons mounted on the two ends of each rod.

Since clipping planes in OpenGL can only be used to create convex cuts, we use a three pass rendering algorithm similar to [12] to create the chair cut. The first pass renders one half of the model using one clipping plane. The second pass renders an additional quarter of the model using a second clipping plane and the first clipping plane reversed. The third pass renders the last octant and uses a third clipping plane and the first and second clipping plane reversed. This approach sends the model three times down the graphics pipeline, which increases rendering times typically by a factor of two to three. Our system is based on SGI's Performer graphics tool kit, which uses a three process pipeline to render graphical objects. The application process positions objects in the virtual world, the culling stage removes polygons outside the viewing frustum, and the rendering stage feeds the remaining polygons into the graphics hardware. We mostly avoid the slow down for the chair cut by considering clipping planes already during the Performer culling stage similar to the already employed view frustum culling. This removes invisible polygons before they are handed over to the graphics hardware.

Figure 5: This chair cut clips away the left, upper, rear octant of a car model to permit a better view of the interior.

EXPERIENCES AND RESULTS

We observed people using the Cubic Mouse during a planned user observation session, while giving presentations to our consortium partners, and during a variety of demos.

User Observation

For a user observation session we recruited 12 subjects (8 male, 4 female) from staff and students of our research one of the rods specifies a rotation of a virtual object around the appropriate axis of a given reference frame represented by the Cubic Mouse. In contrast to a standard six degree of freedom tracker we have here the possibility to specify the six degrees of freedom independently from each other. This is often very desirable, since it allows a much more precise input much like a dial box, just in a more intuitive way. Another use for these additional degrees of freedom would be the fine adjustment of X, Y, and Z coordinates during a positioning task.

Currently the device is mostly used as an absolute positioning system. Moving a rod from one stop to the other moves the corresponding virtual object a given distance. This requires the frequent use of the clutch when moving objects over a larger distance. With a spring mechanism to reset the rods to a centered rest position the device turns into a well designed relative positioning device, without range limitations.

Our first prototype tracks the rods' movement with potentiometers, which is really a primitive approach. Obviously, optical tracking could be used to overcome the static friction problem of potentiometers and it would have higher resolution. Even more interesting is the use of step motors attached to the rods. In addition to tracking the rods' movements this would allow haptic feedback and turns the device into an *active* real world prop.

We presented an input device, which intuitively combines constrained translation and potentially rotation input with a 6DOF free-motion device. There is still more work to do to explore the full potential of this setup. However some geologists predicted already from their first experience that the Cubic Mouse will become a standard for their application domain within the next years.

ACKNOWLEDGMENTS

This work was partially supported by the VRGeo and SimVR consortia. We thank the members of the consortia for their valuable feedback during our meetings. Special thanks to Ernst Kruijff for help with the user testing, and to David Tonnesen for a thorough critique of an earlier version of the paper. Thanks also to Jakob Beetz, Hartmut Seichter, Jan Springer, Michael Tirtasana, Henrik Tramberend, and Jürgen Wind for help with the paper, and to the rest of the VE group at GMD for their support.

The first author is grateful to Pat Hanrahan and the Stanford Graphics Group as well as Frank Crow and Interval Research for providing a unique research environment at Stanford, where many of the foundations for this research were laid.

REFERENCES

1. Cruz-Neira, C., Sandin, D.J., and DeFanti, T.A. Surround-screen Projection-based Virtual Reality: The Design and Implementation of the CAVE. *Proceedings of SIGGRAPH '93,* 135-142, 1993.

2. Cutler, L. D., Fröhlich, B., and Hanrahan, P. Two-Handed Direct Manipulation on the Responsive Workbench. *1997 Symposium on Interactive 3D Graphics,* 107-114, 1997.

3. Guiard, Y. Symmetric division of labor in human skilled bimanual action: the kinematic chain as a model. *The Journal of Motor Behaviour,* 19(4):486–517, 1987.

4. Hinckley, K., Pausch, R., Goble, J.C., and Kassell, N.F. A Survey of Design Issues in Spatial Input. *In Proceedings of the ACM Symposium on User Interface Software and Technology,* pages 213–222, 1994.

5. Hinckley, K., Pausch, R., Goble, J.C., and Kassell, N.F. Passive real-world interface props for neurosurgical visualization. *In Proceedings of ACM CHI'94 Conference on Human Factors in Computing Systems,* pages 452–458, 1994

6. Krüger, W., Bohn, C.-A., Fröhlich, B., Schüth, H., Strauss, W., and Wesche, G. The Responsive Workbench. *IEEE Computer,* 42-48, July 1995

7. Krüger, W., and Fröhlich B. The Responsive Workbench. *IEEE Computer Graphics and Applications,* 12-15, May 1994

8. Mine, M.R., Brooks F. P. Jr., and H. Sequin, C. H. Moving Objects in Space: Exploiting Proprioception In Virtual-Environment Interaction. *In SIGGRAPH 97 Conference Proceedings,* pages 19–26. August 1997

9. Pausch, R., Burnette, T., Brockway, D., and Weiblen, M.E.. Navigation and locomotion in virtual worlds via flight into Hand-Held miniatures. *In SIGGRAPH 95 Conference Proceedings,* pages 399–400, August 1995.

10. Poston, T., and Serra, L. The Virtual Workbench: Dextrous VR. *In Virtual Reality Software and Technology (Proceedings of VRST'94, August 23-26, 1994, Singapore),* pages 111–122, August 1994.

11. Tramberend, H. Avocado: A Distributed Virtual Reality Framework. *Proceedings of VR'99 Conference, Houston,* Texas, 14-21, March 1999.

12. Viega, J., Conway, M. J., Williams G., and Pausch, R. 3D Magic Lenses. *In Proceedings of the ACM Symposium on User Interface Software and Technology, Papers:* Information Visualization, pages 51–58, 1996.

The Role of Contextual Haptic and Visual Constraints on Object Manipulation in Virtual Environments

Yanqing Wang and Christine L. MacKenzie
School of Kinesiology
Simon Fraser University
Burnaby, BC V5A 1S6
Canada
+1 604 291 5794
{wangy, cmackenz}@move.kines.sfu.ca

ABSTRACT
An experiment was conducted to investigate the role of surrounding haptic and visual information on object manipulation in a virtual environment. The contextual haptic constraints were implemented with a physical table and the contextual visual constraints included a checkerboard background ("virtual table"). It was found that the contextual haptic constraints (the physical table surface) dramatically increased object manipulation speed, but slightly reduced spatial accuracy, compared to free space. The contextual visual constraints (presence of the checkerboard) actually showed detrimental effects on both object manipulation speed and accuracy. Implications of these findings for human-computer interaction design are discussed.

KEYWORDS
Human performance, virtual reality, visual information, haptic information, 3D, docking, controls and displays, task context, force feedback, graphic interface, degrees of freedom, augmented environment.

INTRODUCTION
Virtual environments provide domain constraints and contextual constraints for human object manipulation. Domain constraints include intrinsic properties of the object being manipulated such as controller and cursor size and shape [14]. Contextual constraints are the surrounding information and environment for object manipulation. The goal of this experiment is to explore the role of contextual haptic and visual constraints on multidimensional object manipulation in virtual environments.

Permission to make digital or hard copies of all or part of this work for personal or classroom use is granted without fee provided that copies are not made or distributed for profit or commercial advantage and that copies bear this notice and the full citation on the first page. To copy otherwise, to republish, to post on servers or to redistribute to lists, requires prior specific permission and/or a fee.
CHI '2000 The Hague, Amsterdam
Copyright ACM 2000 1-58113-216-6/00/04...$5.00

Contextual haptic constraints
There are two kinds of contextual haptic constraints in virtual environments: active and passive haptic constraints. Active haptic constraints are provided with force feedback devices. Passive haptic constraints are implemented with the real object in augmented environments, for example, where graphic cues are augmented with surrounding physical cues. This study deals with the effects of passive haptic constraints on object manipulation. Recent research shows that such passive haptic feedback not only provides realism in virtual environments [7], but also enhances human performance [9]. Lindeman, Sibert and Hahn compared human performance on docking a graphic object to a "floating" graphic panel with docking when the panel was augmented with a physical paddle [9]. They found that the passive haptic feedback provided by the augmented paddle resulted in a 44% decrease in the movement time and a 38% increase in accuracy. Zhai reported a study on a six degrees of freedom elastic controller for object manipulation tasks [19]. The elastic constraint can be considered as a kind of haptic constraint on the controller. They found that human performance was better when the elastic device was used, compared with isometric devices, and suggested that the elastic property of the controller provided more sensitivity for position control [19].

Contextual haptic constraints on object manipulation are essentially the problem of degrees of freedom (DOF) for movement control. Previous research has not explicitly examined the effect of contextual haptic constraints in terms of degrees of freedom of movements. Furthermore, in an augmented environment, contextual haptic constraints are usually spatially related to contextual visual constraints. We are unaware of research that addresses contextual haptic constraints in relation to contextual visual constraints. These aspects of contextual haptic constraints are investigated in this study.

Contextual visual constraints

Contextual visual constraints such as graphic checkerboard or groundplane background have been a standard technique to provide depth cues in graphic interfaces [2]. The depth cues provided by contextual visual constraints can facilitate human performance in virtual environments [1] [12]. Robertson, Czerwinski and Larson suggested that the spatial graphic background improves the user's spatial memory for information visualization [11]. The role of contextual visual constraints on object manipulation in virtual environments needs further investigation.

Research hypotheses

In this experiment, we used a physical table to provide the contextual haptic constraints for object manipulation. We compared human performance on object manipulation on the table surface with that in free space. The movement on the table surface had fewer degrees of freedom than in free space. A graphic checkerboard, a "virtual table", served as the contextual visual constraint. The physical table surface was overlaid with the checkerboard. We tested two hypotheses:

1. Contextual haptic constraints enhance human performance on object transportation and orientation;
2. Contextual visual constraints facilitate human performance on object transportation and orientation.

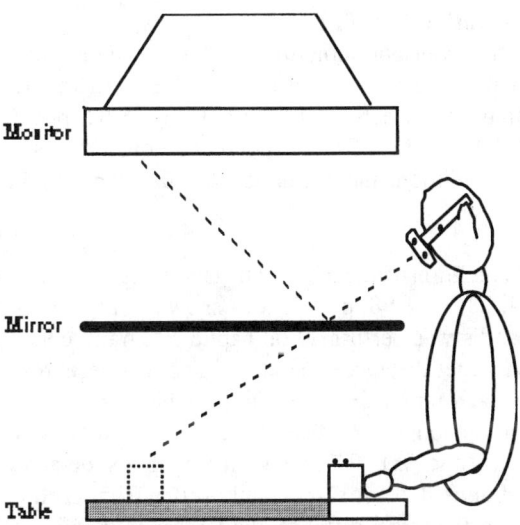

Figure 1. The Virtual Hand Laboratory setup for this experiment. The stippled part of the table surface is removable. When this part is removed, the subject manipulates the wooden cube in free space. A graphic checkerboard, or "virtual table" appears as if on the table surface. The graphic target cube (dashed line) is drawn to appear on the table surface. The wooden cube (solid line) is the controller. Markers on the goggles and wooden cube drive the stereoscopic, head-coupled graphics display.

METHOD

Subjects

Eight university student volunteers were each paid $20 for participating in one, two-hour experimental session. All subjects were right-handed, and had normal or corrected-to-normal vision. All subjects had experience using a computer. Informed consent was provided before the experimental session.

Experimental setup

This experiment was conducted in the Virtual Hand Laboratory (VHL) [14], shown in Figure 1. A SGI monitor was placed screen down on a specially constructed cart. A half-silvered mirror was placed parallel to and between the computer screen and the table surface. The image on the screen was reflected by the mirror, and was perceived by the subject as if it were in a workspace on the table surface. The images for the left and right eye were alternatively displayed and were synchronized with goggles to provide the subject with a stereoscopic view. Three infrared (IRED) markers were place on the side of goggles. An OPTOTRAK 3D motion system monitored 3D position information from these markers to provide a head-coupled view.

The physical object, a wooden cube, was the controller, serving as the input device. Three IREDs were placed on top of the controller cube. The 3D position information from these three IREDs was used to drive a red six-degree-of-freedom wireframe graphic object, the cursor cube (not shown in Figure 1). The information from these IREDs on the top of the wooden cube was also recorded for data analysis. The transportation data were collected with the marker on the top center of the wooden cube. The orientation data were derived with two markers on the top of the wooden cube.

The target was a red wireframe graphic cube generated on the monitor, appearing on the table surface for the subject looking into the mirror, as shown in Figure 1. The graphic target was placed along the horizontal center axis on display, which was aligned with the subject's body midline. The controller, cursor and target cubes had the same size of 30 mm. The graphic target was located 70, 140 or 210 mm away from the starting position, with an angle rotated 22.5 or 45 degrees clockwise about a vertical axis.

As shown in Figure 1, one part of the table surface was removable. The other part of the table with the same height was used to support the controller at its start position. The graphic target was perceived by the subject through the mirror, as if on the table surface. When the table surface was present, the subject could slide the controller on the table surface; when the table surface was removed, the subject had to move the controller to the target in the air, without the table as a supporting surface. The part of the table surface at the start location of the controller was always present so that the controller was supported at the beginning of the movement.

This setup provided three contextual haptic conditions: table-slide, table-lift and no-table. In the table-slide condition, the physical table was present and the subject was instructed to slide the controller on the table surface. The table-lift condition was when the physical table was present, but the subject was instructed to slightly lift the controller from the table surface and land the controller on the table surface at the final position. In the no-table condition, the table was removed and the subject had to move the controller in the air to its final position. In all cases, the controller started in a constant, supported location and the subject's task was to align the cursor cube with the target cube.

In the table-slide condition, the wooden cube (controller) and the cursor cube were constrained to three degrees of freedom, two for translation and one for rotation. In the table-lift condition, the wooden cube and cursor cube were constrained to three degrees of freedom only at the start and the end of the movement; the controller and cursor had six degrees of freedom for free motion between the start and the end. In the no-table condition, the wooden cube and the cursor cube had six degrees of freedom after it left the start position.

A black and white checkerboard was displayed with a block size of 13.5 by 13.5 mm. The checkerboard was superimposed on the planar table surface. When the checkerboard was not present, object manipulation was performed on a black background.

Procedure

System calibration was first performed [13]. The workspace on the table surface, including the checkerboard, was calibrated so that the checkerboard was aligned to the tabletop. The cursor cube was registered to superimpose with the wooden cube. The individual subject eye positions were also calibrated to obtain a customized, stereoscopic, head-coupled view.

During the experiment, subjects saw only the graphic cursor and target, with no vision of the hand and the wooden cube. The task was to match the location and angle of the cursor cube to those of the graphic target as fast and accurately as possible. Trials were blocked on contextual haptic constraint and visual constraint conditions. At the beginning of each block, subjects were given 20 trials for practice. The order of target distances and angles were randomly generated over trials within each block. Ten trials were repeated in each experimental condition.

Data analysis

Independent variables for this experiment were contextual haptic constraints, contextual visual constraints, target distances and target angles. Object transportation and orientation data collected from the IRED markers on the wooden cube were analyzed separately. Temporal dependent measures were: total task completion time (CT), object transportation time (TT) and object orientation time (OT). CT was the time between the start of the cube movement to the end of the cube movement, either cube translation or orientation. TT was the cube translation time and OT was the cube orientation time. Spatial error measures were: constant errors of distance (CED), variable errors of distance (VED), constant errors of angle (CEA), and variable errors of angle (VEA). The constant error was the average distance or angle off the target for all trials under an experiment condition, and the variable error was the standard deviation of trials for that condition. ANOVAs were performed on the balanced design of *3 haptic constraints × 2 visual constraints × 3 target distances × 2 target angles* with repeated measures within subjects.

RESULTS

Time Measures

Object transportation and orientation showed a parallel structure, that is, the transportation process completely contained the orientation process [17]. Noting that the results of transportation time (TT) were almost the same as task completion time (CT), only TT and OT are reported in detail here.

Transportation time (TT)

The average transportation time (TT) was 906 ms, taking up 97% of the task completion time (CT) of 933 ms. Haptic constraints had effects on TT, $F(2, 14) = 78.62$, $p < .001$. TT was 749 ms in the table-slide condition, 778 ms in the table-lift condition, and increased to 1192 ms in the no-table condition.

Haptic conditions interacted with target distance to affect TT, $F(4, 28) = 16.98$, $p < .001$, as shown in Figure 2. Post hoc analysis was performed on haptic constraint conditions for each target distance separately. Post hoc tests revealed that for each target distance, the no-table condition had a significantly longer TT that the table-slide and table lift conditions ($p < .05$). TTs between the table-slide and table lift conditions did not significantly differ from each other. The longer the target distance, the larger the no-table effect on TT.

Visual constraints significantly interacted with the target angle, $F(1, 7) = 6.20$, $p < .05$. The checkerboard actually increased TT, but it appeared that at 22.5 degrees, the presence of the checkerboard resulted in a larger increase in TT than at 45 degrees, as shown in Figure 3.

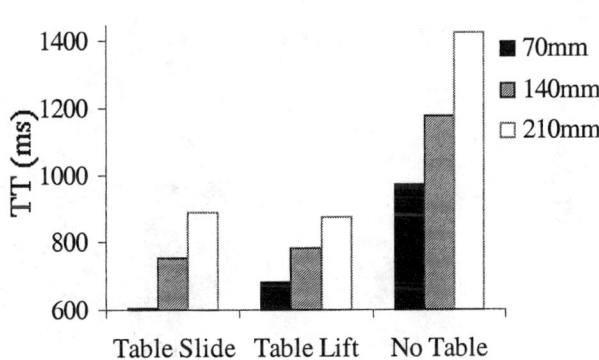

Figure 2. Transportation time with haptic constraints and target distances.

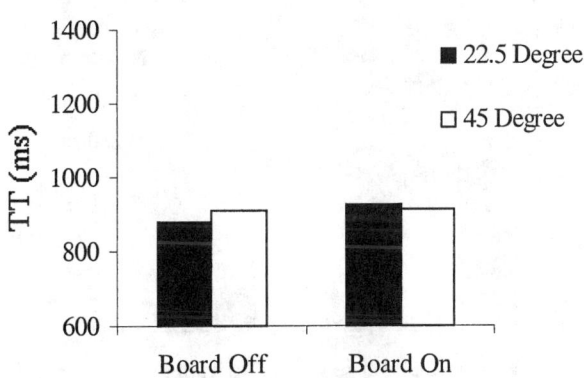

Figure 3. Transportation time with visual constraints and target angles. Board Off = Black background; Board On = Checkerboard background.

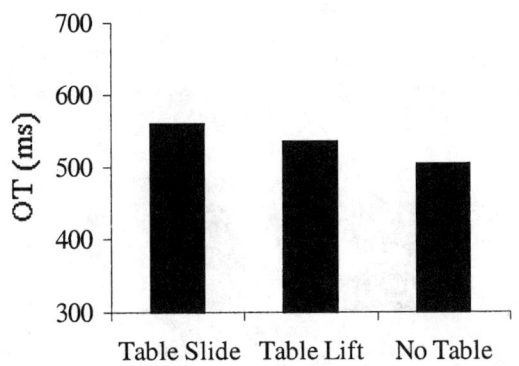

Figure 4. Orientation time with haptic constraints.

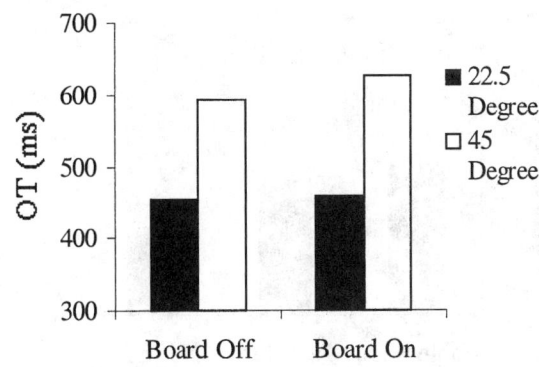

Figure 5. Orientation time with visual constraints and target angles.

Orientation time (OT)

The orientation time (OT) was 533 ms on average, 57% of the task completion time (CT). The effect of haptic constraints on OT was significant, $F(2, 14) = 4.38$, $p < .05$, and showed a trend opposite to TT, as demonstrated in Figure 4. The longest OT occurred in the table-slide condition, 560 ms, and reduced to 536 ms in the table-lift condition, and then further decreased to 504 ms in the no table condition. Post hoc tests revealed that OT in the no-table condition was significantly different from that in the table-slide condition ($p < .05$). No significant difference in OT was found between the no-table and table-lift conditions nor between the table-lift and table-slide conditions.

There was a significant interaction on OT between visual constraints and target angles, $F(1, 7) = 10.18$, $p < .05$, as shown in Figure 5. It appeared that the checkerboard had more impact on OT at 45 degrees than at 22.5 degrees of target angle.

Spatial Errors

Overall, the spatial errors were very small. The average value of constant errors of distance (CED) was 0.04 mm, not significantly different from the target location. The average value of constant errors of angle (CEA) was 0.8 degree, not significantly off the target angle. Constant errors reflect system errors and subjects' individual bias while

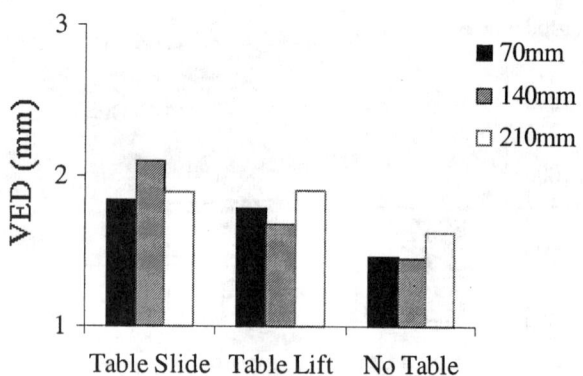

Figure 6. Variable errors of distance with haptic constraints and target distances.

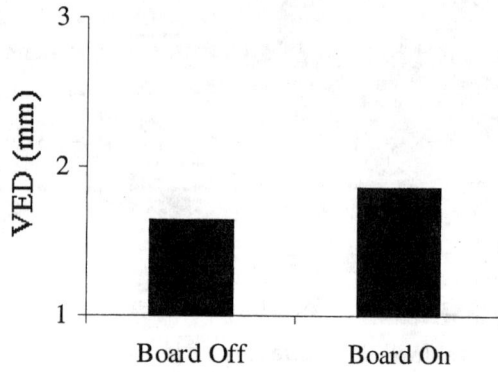

Figure 7. Variable errors of distance with visual constraints

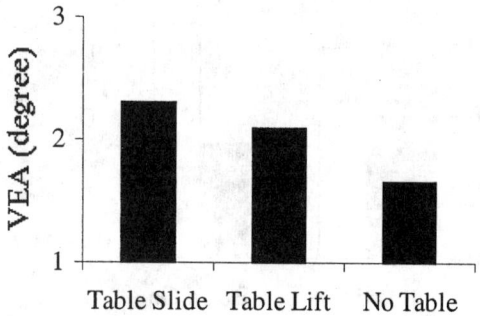

Figure 8. Variable errors of angle with haptic constraints.

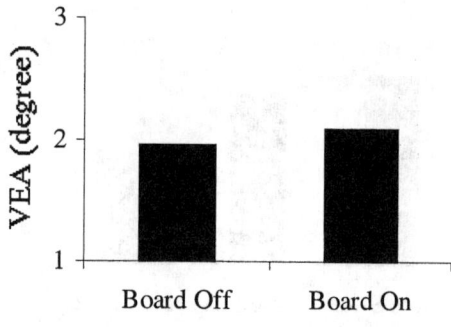

Figure 9. Variable errors of angle with visual constraints.

variable errors indicate human performance consistency under a certain interface system. Only the results of variable errors are presented as follows.

Variable errors of distance (VED)
Haptic constraints, visual constraints, and target angles had main effects on the variable errors of distance (VED). VED increased a small amount with changing haptic constraints (e.g., 1.5 mm in the no-table condition, 1.8 mm in the table-lift condition, and 1.9 mm in the table-slide condition), $F(2, 14) = 5.12$, $p < .05$. Post hoc analysis revealed a significant difference in VED between the table-slide and no-table conditions ($p < .05$), but there were no significant differences between the table-slide and table-lift conditions nor between the no-table and table-lift conditions.

There was an interaction between the haptic constraint and the target distance, $F(4, 28) = 3.67$, $p < .05$, as shown in Figure 6. Post hoc analysis revealed that at the target distance of 140 mm, VED in the table-slide condition was significantly larger than that seen in the table-lift condition or in the no-table condition, but VEDs between the table-lift condition and the no-table condition did not differ from each other. The differences in VED among haptic constraints were not significant at target distances of 70 mm and 210 mm.

The presence of the checkerboard background increased VED from 1.6 mm to 1.9 mm ($F(1, 7) = 16.54$, $p < .01$), as shown in Figure 7. This was a small and unexpected result.

There was a three-way interaction on VED among the haptic constraint, visual constraint and target distance, $F(4, 28) = 4.33$, $p < .01$. VEDs consistently increased in the presence of the checkerboard at the target distance 70 mm across haptic constraints. The checkerboard appeared not to make a difference in VED in the table-lift at 210 mm or in the no-table at 140 mm.

Variable errors of angle (VEA)
The variable errors of angle (VEA) had an average value of 2.0 degrees. The haptic constraint had significant effects on VEA, $F(2, 14) = 12.27$, $p < .001$. VEA was 1.7 degrees in no-table, 2.1 degrees in table-lift, and 2.3 degrees in table-slide, as shown in Figure 8. Post hoc revealed that VEA in the no-table condition was significantly smaller than that both in the table-lift and table-slide conditions ($p < .05$). There was no difference in VEA between the table-lift and table-slide conditions.

The checkerboard led to a small increase in VEA, $F(1, 7) = 6.56$, $p < .05$, from 2.0 to 2.1 degrees (Figure 9). It was unexpected that the presence of the checkerboard would have detrimental effects on both VED and VEA.

There was a three-way interaction on VEA among the haptic constraint, visual constraint and target distance, $F(4, 28) = 4.35$, $p < .01$. VEAs were similar or generally increased with the presence of the checkerboard with the exception in the table-lift condition at the target distance of 70 mm. In this particular condition, it appeared that the presence of the checkerboard reduced VEA.

DISCUSSION
The role of contextual haptic constraints
Haptic constraints had profound effects on human performance in object transportation and orientation. The task completion time (transportation time) was reduced dramatically with the tabletop, compared to when no supporting surface was present. This was consistent with the speed findings by Lindeman et al. [9]. This result also supports our suggestions in previous research that haptic information has more impact on object manipulation speed than visual information [15] [16]. The fact that the contextual haptic constraint was imposed in the control space indicates the importance of human motor control systems in object manipulation.

In this experiment, the task required three degrees of freedom for control. When the controller was moved in free space, it had six degrees of freedom. The degrees of freedom were reduced to three when the controller slid on the tabletop, and the controller actually became a three-degree-of-freedom input device. Jacob, Sibert, McFarlane, and Mullen found that object manipulation speed increased when the structure (dimensions) of tasks was matched with the structure of input devices [8]. Evidence from this experiment supports their theory. Hinckley, Tullio, Pausch, Proffitt and Kassell conducted an experiment to compare two DOF input with three DOF input for three DOF rotation tasks [6]. They found in an orientation matching task, that users completed the task up to 36% faster when using three DOF input than two DOF input devices, without significant loss of accuracy.

No significant difference was found in task completion time between table slide and table lift conditions. At the end of movements, the controller was constrained to three degrees of freedom for both table slide and table lift. This suggests that the contextual haptic constraint on the end of the movement or the target is more critical than during the course of the movement. Similar results were found by MacKenzie for a 3D pointing task where subjects pointed to a solid target faster than to a hole [10]. MacKenzie suggested that the solid target (haptic constraint) helps to stop the movement, compared to pointing to a hole where subjects have to take extra time to decelerate the pointing device.

At the same time, however, the haptic constraint, the table surface, consistently increased the spatial variable errors in task performance, although the increase was quite small. This finding is counter-intuitive and contrary to Lindeman et al.'s results [9]. Hinckley et al. did not find that the constraints on degrees of freedom for object orientation had effects on spatial errors [6]. It is not clear what factors in these experiments cause such inconsistency. This needs further investigation.

The role of contextual visual constraints
It was originally predicted that the contextual visual constraint would be used as guidance for object manipulation. It is surprising that the visual constraint, the checkerboard background, generally deteriorated human performance in both times and spatial errors. Even though the effect of visual constraint was small, it is theoretically and practically important.

It appeared that the visual constraint, the checkerboard, interfered with the manipulation task rather than provided extra visual cues to enhance object manipulation performance. One interpretation is that the checkerboard background distracted the subject's attention from the target. Other factors such as the pattern and color of the checkerboard may also cause the interference.

If the same effect is replicated with various depth cues such as groundplanes or stereoscopic view, it poses an important question. Can depth cues of graphics, supposed to help in visualization, benefit interaction in general? The results from previous research are not conclusive [1] [3] [4] [11] [18]. Recent research by Boritz shows that the depth cues provided by head-coupled view are generally detrimental for object docking in virtual environments [3]. Cao, MacKenzie and Payandeh found that the depth cues from stereoscopic view help for certain tasks, but not for others [4]. Our results show that the cues provided by the checkerboard actually hindered object manipulation, yet the checkerboard-like background has been widely used in human-computer interaction design [2].

There is strong evidence showing that humans have two visual systems, one for perception and another for prehension [5]. The theory of two visual pathways suggests that the depth cues that facilitate perception may not necessarily benefit action (object manipulation). We hypothesize that contextual visual constraints play different roles in object visualization than object manipulation. Future work is definitely needed to test this hypothesis.

Implications for human-computer interaction design
Virtual environment design should take advantage of passive haptic constraints. The significant benefit gained in the task completion time from the haptic constraint can be weighed against the relatively small reduction in accuracy control. Passive haptic constraints such as table surfaces, walls or paddles are cheap and reliable, and could be easily implemented in virtual environments. For example, a graphic 2D command menu in virtual environments could be augmented with a physical plate whenever possible.

Recently, force feedback devices have been implemented in virtual environments to enhance the realism of interaction. It is generally believed that force feedback devices improve human performance in accuracy control, but lack of empirical evidence. We found that the significant benefit of passive haptic constraints is for speed control rather than accuracy control. Thus, more attention should be paid to utilization of passive haptic feedback for applications where speed is a major concern. Force feedback devices should be designed to simulate passive haptic feedback as well as active haptic feedback.

We issue a caveat about the role of contextual visual constraints in virtual environment design. This experiment suggests that the background depth cues may benefit object perception, but degrade object manipulation. For example, in medical virtual reality applications, a checkerboard background may be used for diagnosis, but not for surgery. We cannot assume that graphic cues beneficial for perception will always be beneficial for interaction. The effect of contextual visual constraints should be carefully evaluated before implementation.

Contextual haptic and visual constraints have significant effects on human object manipulation in virtual environments. Research in this area is very limited, compared to research on domain constraints of object manipulation. Contextual constraints are ubiquitous in virtual environments, and human-computer interaction design should be guided by understanding contextual constraints as well as domain constraints.

CONCLUSIONS
It is concluded from this experiment:
1. Contextual haptic constraints, such as a table surface, improve human performance in the task completion time, but slightly reduce the spatial accuracy.
2. Contextual visual constraints, such as a checkerboard background, degrade human performance on object manipulation for both speed and accuracy.

ACKNOWLEDGEMENTS
This research was supported by the National Science and Engineering Research Council of Canada (NSERC) Strategic Project program. We would like to thank Evan D. Graham, Valerie A. Summers and Kellogg S. Booth for their help in software design for the Virtual Hand Laboratory.

REFERENCES
1. Arthur, K.W., Booth, K.S. and Ware, C. (1993). Evaluating 3D task performance for fish tank virtual world. *ACM Transactions On Information Systems*, 11 (3), 239-265.

2. Balakrishnan, R. and Kurtenbach, G. (1999). Exploring bimanual camera control and object manipulaiton in 3D graphics interfaces. *Proceedings of the Conference on Human Factors in Computing Systems CHI '99 /ACM*, 56-63.

3. Boritz, J. (1998). *The Effectiveness of Three-dimensional Interaction*. Ph.D. Thesis, Dept. of Computer Science, University of British Columbia, Vancouver, B.C., Canada.

4. Cao, C. G. L., MacKenzie, C. L. and Payandeh, S. (1996). Task and motion analyses in endoscopic surgery. *Proceedings of the ASME Dynamic Systems and Control Division*, 58, 583-590.

5. Goodale, M.A., Jakobson, L.S and Servos, P. (1996). The visual pathways mediating perception and prehension. In Wing, A. M., Haggard, P. and Flanagan, J. R. (Eds.), *Hand and Brain*, 15-31, New York: Academic Press.

6. Hinckley, K., Tullio, J., Pausch, R., Proffitt, D. and Kassell, N. (1997). Usability analysis of 3D rotation techniques. *Proceedings of UIST '97*, 1-10.

7. Hoffman, H.G. (1998). Physical touching virtual objects using tactile augmentation enhances the realism of virtual environments. *Proceedings of IEEE VRAIS*, 59-63.

8. Jacob, R. J. K., Sibert, L. E., McFarlane, D. C. and Mullen, M. P. Jr. (1994). Integrality and separability of input devices. *ACM Transactions on Computer-Human Interaction*, 1 (1), 1-26.

9. Lindeman, R.W., Sibert, J.L. and Hahn, J.K. (1999). Towards usable VR: An empirical study of user interfaces for immersive virtual environments. *Proceedings of the Conference on Human Factors in Computing Systems CHI '99/ACM*, 64-71, Pittsburgh, PA.

10. MacKenzie, C.L. (1992). Making contact: Target surfaces and pointing implements for 3D kinematics of humans performing a Fitts' task. *Society for Neuroscience Abstracts*, 18, 515.

11. Robertson, G., Czerwinski, M. and Larson, K. (1998). Data mountain: Using spatial memory for document management. *Proceedings of UIST '98/ACM*, 153-162.

12. Servos, P., Goodale, M.A., and Jakobsen, S. C. (1992). The role of binocular vision in prehension: A kinematic analysis. *Vision Research*, 32 (80), 1513-1521.

13. Summers, V.A., Booth, K.S., Calvert, T., Graham, E. and MacKenzie, C.L. (1999). Calibration for augmented reality experimental testbeds. *ACM Symposium on Interactive 3D Graphics*, 155-162.

14. Wang, Y. (1999). *Object Transportation and Orientation in Virtual Environments*. Ph.D. Thesis, School of Kinesiology, Simon Fraser University, Burnaby, BC, Canada.

15. Wang, Y. and MacKenzie, C.L. (1999a). Object manipulation in virtual environments: Relative size matters. *Proceedings of the Conference on Human Factors in Computing Systems CHI '99/ACM*, 48-55.

16. Wang, Y. and MacKenzie, C.L. (1999b). Effects of orientation disparity between haptic and graphic displays of objects in virtual environments. *INTERACT '99*, 391-398.

17. Wang, Y., MacKenzie, C.L., Summers, V. and Booth, K.S. (1998). The structure of object transportation and orientation in human-computer interaction. *Proceedings of the Conference on Human Factors in Computing Systems CHI '98/ACM*, 312-319.

18. Wickens, C.D. (1992). *Engineering Psychology and Human Performance*. Columbus, OH: Harper Collins.

19. Zhai, S. (1995). *Human Performance in Six Degree of Freedom Input Control*. Ph.D. Thesis, Dept. of Computer Science, University of Toronto, Toronto, Ontario, Canada.

Non-Isomorphic 3D Rotational Techniques

Ivan Poupyrev [1], Suzanne Weghorst [2], Sidney Fels [3]

[1] ATR MIC Research Laboratories
2-2 Hikaridai, Seika, Soraku-gun
Kyoto 619-02, Japan
(774) 95-1432
poup@mic.atr.co.jp

[2] HIT Lab, University of Washington
Box 352142
Seattle, WA 98195, USA
(206) 685-3215
weghorst@hitl.washington.edu

[3] Department of ECE,
University of British Columbia
Vancouver, BC Canada, V6T 1Z4
(604) 822-5338
ssfels@ece.ubc.ca

ABSTRACT

This paper demonstrates how non-isomorphic rotational mappings and interaction techniques can be designed and used to build effective spatial 3D user interfaces. In this paper, we develop a mathematical framework allowing us to design non-isomorphic 3D rotational mappings and techniques, investigate their usability properties, and evaluate their user performance characteristics. The results suggest that non-isomorphic rotational mappings can be an effective tool in building high-quality manipulation dialogs in 3D interfaces, allowing our subjects to accomplish experimental tasks 13% faster without a statistically detectable loss in accuracy. The current paper will help interface designers to use non-isomorphic rotational mappings effectively.

Keywords: 6DOF input devices, interactive 3D rotations, 3D user interfaces, interaction techniques, motor control.

INTRODUCTION

Three-dimensional (3D) computer graphics has advanced from a subject of research curiosity to an indispensable tool in many areas of human activities. While visual quality and rendering efficiency have been rapidly improving, the design of efficient interfaces for 3D applications remains a practical concern for application developers and a vexing problem for researchers in industry and academia [12].

Direct manual control has a very special place in 3D user interfaces design and research: human hands remain the dominant channel of interaction not only for 3D interfaces, but also for traditional 2D graphical user interfaces (GUI) as well for our everyday interaction with the physical world. The quality of interface components that enable users to manipulate objects and scenes in virtual environments has a profound effect on the quality of the whole interface – if the user cannot manipulate effectively, many specific application tasks simply cannot be performed. Consequently, a large amount of research has already addressed various issues in multidimensional manipulation, e.g., designing and evaluating multiple degrees-of-freedom (DOF) input devices, innovating new manipulation interaction techniques, investigating implications of human motor skills on 3D interface design, and many others [e.g. 1, 8, 12, 22, 25].

The design of 3D mappings and interaction techniques, which translate user-operated device motions into object movements in virtual environments, is certainly one of the core issues in designing manipulation interfaces. The challenge was very well defined by Sheridan (cited from [24]): "How do the geometrical mappings of body and environmental objects, both within the virtual environment and the true one, and relative to each other, contribute to the sense of presence, training, and performance? ... In some cases there may be a need to deviate significantly from strict geometric isomorphism because of hardware limits, or constraints of the human body. At present we do not have design/operating principles for knowing what mapping ... is permissible, and which degrades performance."

Designing and investigating non-isomorphic mappings for 3D spatial user interfaces has recently been an area of active research [e.g. 2, 8, 12, 13, 14, 15], and the current paper adds to this body of work. In particular, we explore how non-isomorphic rotational mappings can be designed and used to enhance 3D rotations of objects and scenes in virtual worlds. The paper attempts to close a current gap in the literature on multidimensional interaction, where 3D mappings and interaction techniques have been used only with the translation components of multiple DOF input. When it comes to 3D rotations, most researchers, as well as producers of commercial devices and software, have used only the simplest one-to-one (*isomorphic*) mapping between rotations of the multiple DOF controller and virtual objects. In fact, even the basic equations of control-display (C-D) gain for 3D rotations and their properties have not been reported.[1] In comparison, C-D mappings for translation tasks have been used and studied since the early 1940s.

This paper demonstrates how non-isomorphic 3D rotational mappings can be constructed and effectively used to design 3D interfaces. First, we introduce a basic mathematical framework that allows design of both linear and non-linear C-D mappings between device rotations and rotations in 3D interface space. This framework is based on the idea of extrapolating the orientation of a multiple DOF device on a quaternion sphere in four dimensions. Second, we identify basic idiosyncratic properties of rotational mappings, such as relations between the mappings and device form-factor, and discuss issues of interaction techniques design. Finally, we report experiments which have shown that by using our technique, subjects could accomplish an experimental task 13% faster without any significant loss in accuracy.

BACKGROUND AND RELATED WORK

Any interface between humans and machines that uses continuous manual control includes three basic components:

---

[1] We reported preliminary results in the CHI'99 late-breaking paper [16].

1) input devices, which capture user actions, 2) display devices, which present the effect of these actions back to the user, and 3) transfer functions, often referred to as control-display mappings, which map the movements of the device into the movements of controlled elements of the system or interface [11, 24] (Figure 1). The goal is to design input devices, displays and transfer functions that facilitate high user performance and comfort, while diminishing the impact from human and hardware limitations [11].

The design of mapping functions for manual control and studies of their impact on operator performance stretch back to the 1940s [10]. It has also been an active research area in 3D user interfaces where two philosophies have emerged [24]: The *isomorphic* view suggests a strict geometrical isomorphism (i.e. one-to-one mapping) between motions in the physical and virtual worlds, on the grounds that it is the most natural and therefore is better for users. The results of early human factor studies indicated that while isomorphism is, indeed, often more natural [overview in 11], it also has important shortcomings. First, isomorphic mappings are often impractical because of constraints in the input technologies, e.g., the limited tracking range of input devices. Second, isomorphism is often ineffective due to the limitations of human operators, e.g., anatomical constraints. Finally, it has been argued that 3D interfaces can be more effective, intuitive and richer if, instead of imitating the physical reality, we create mappings and interaction techniques that are specifically tailored to virtual environments, providing in some sense a "better" reality [e.g. 21].

Hence, the *non-isomorphic* approach suggests that manipulation mappings and techniques can significantly deviate from strict realism, providing users with "magic" virtual tools, e.g., laser rays, rubber arms, Voodoo Dolls [13, 14] and others. These non-isomorphic mappings and techniques allow users to manipulate objects quite differently than in the physical world, yet rather effectively [2, 15]. In fact, the majority of 3D direct manipulation techniques today are non-isomorphic techniques.

We should note, however, that non-isomorphic mappings are not an entirely new idea; they have been used for decades in a variety of everyday controls, e.g., dials, pedals, handlers, and wheels, where our input is scaled, shifted or integrated using different mapping or *transfer functions* [11]. Traditionally, the human factors literature categorizes transfer functions by a number of integrations, applied to the user input [11, 24]. Thus, in *zero-order* mappings, displacement of the input device results in displacement of the controlled element, while in *first-order* mappings, it results in a change of its velocity. Consequently, they are often referred to as position and rate control, respectively.

The simplest example of zero-order mapping is a linear control-display gain function which scales the user input:

$$D_d = kD_c, \qquad (1)$$

where D_c and D_d are displacements of the controller and displayed elements, respectively, and k is a ratio of scaling. The zero-order control should not necessarily be linear; for example, various dials in consumer electronic devices often use non-linear mappings. Non-linear position control has

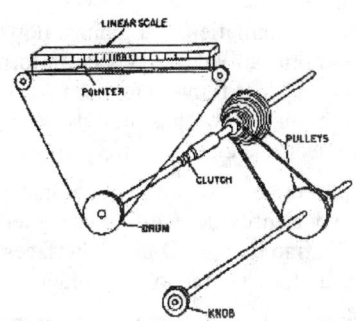

Figure 1: Basic components of any direct manipulation system: input device, output device, and transfer function (in this figure: knob, pointer, and pulleys respectively) [10].

also been used in VR interaction techniques for object manipulation [14] and navigation [20]. A good example of first-order control is the steering wheel of a car, where the displacement of the steering wheel results in the change of the car's angular velocity.

The design of non-isomorphic mappings cannot be accomplished without considering the properties of input devices. The most important device property is the number of the degrees of freedom: early research on 3D user interfaces was often concerned with the design and evaluation of techniques for performing 3D tasks with 2D input devices, e.g., ARCBALL or Virtual Trackball techniques, which use a mouse to rotate 3D objects [9, 18]. In multiple DOF input, additional device properties have to be considered. For example, studies by Zhai [24, chapter 2] have shown that isometric devices, such as force-resistant joysticks, allow for better rate control performance, while isotonic devices, such as free-moving magnetic trackers, are preferable for position control. Given that the same device permits a large variety of mappings, device-mappings compatibility is an important and interesting research direction.

The non-isomorphic mappings and interaction techniques have been designed, until now, only for translation components in multiple DOF input. When it comes to 3D rotations, most researchers, as well as producers of commercial devices and software, use only the simplest one-to-one mapping between the 3D rotation of the device and virtual objects. In fact, even the basic equations of mappings that would linearly amplify the device rotations, i.e., linear C-D gain, have not been reported.

What advantages can we gain by using non-isomorphic 3D rotational mappings? Indeed, it can be argued that, unlike translations, the rotation space is limited to 360 degrees, so any desired orientation can be easily achieved. This issue, however, may well be moot. First, the entire 360 degrees of rotations cannot always be tracked; for example, in computer vision-based tracking, the range of rotations that can be reliably measured is often less then 180 degrees [16]. The non-isomorphic mappings would allow a more effective use of this limited tracking range.

Second, the effective range of rotations in manual control is naturally constrained by human anatomy: our joints can only rotate up to a certain angle. Hence, controlling the large range of rotations is difficult and requires *clutching*, i.e., releasing a virtual object, re-adjusting the hand, and continuing the manipulation. Clutching, however, is frustrating and can noticeably degrade user performance. While an appropriate device form-factor can reduce clutching [25], it cannot eliminate it. Thus, non-isomorphic mappings can be used to decrease clutching in 3D rotations.

Finally, the introduction of non-isomorphic mappings for 3D rotations would provide interface designers with an additional tool for fine-tuning 3D user interfaces and creating new mappings and 3D interaction techniques.

CONTROL-DISPLAY MAPPINGS IN 3D ROTATIONS

In this section we introduce a basic mathematical framework that allows design of both linear and non-linear C-D mappings between device rotations and rotations in 3D interface space. The design of these mappings is not obvious and requires a consideration of the fundamental mathematical properties of rotations in space. The resulting framework is based on the idea of extrapolating multiple DOF device orientations on a quaternion sphere in 4 dimensions.

Rotations in space

Rotations in 3D space are significantly more confusing then they appear, since they do not follow familiar laws of Euclidean geometry. For example, rotate an object in some direction and it would eventually return to its initial starting orientation, something which cannot happen in a vector space. This is because the space of rotations is not a vector space but a closed and curved surface, a manifold, in four dimensions, which can also be represented as a 4D sphere.

The connection between spatial rotations and spherical geometry is quite natural and can be illustrated using a simple physical example. Imagine rotating a rigid physical object, e.g., a pencil, about a fixed point. Apparently, the tip of the pencil would travel on the surface of a sphere and each *orientation* of the pencil can be identified as a *point* on this sphere. Furthermore, a pencil *rotation* around an axis would draw an *arc* and if the pencil has unit length, then the length of this arc equals the rotation angle. Thus, the orientation of the body can be conveniently represented as a point on a unit sphere, while rotation can be represented as an arc on a sphere, connecting the starting and final body orientations.

This example is illustrative, albeit not quite correct: a point on a 3D sphere specifies a family of rotations, since twisting the pencil along the longest axis would not draw any arcs. Since a sphere in 3D specifies only two degrees of rotational freedom, we need to move into a higher, fourth dimension to specify all three degrees of rotations. This is exactly what *unit quaternions* allow us to do.

Quaternions

Quaternions were discovered by Hamilton in 1843 [17]. Since then, they have been widely used in robotics, avionics and any other application field that requires an efficient way to describe and operate 3D rotations. Introduced into computer graphics and interface design by Shoemake [17, 18], today quaternions are a standard tool in the arsenal of the interactive computer graphics professional.

Quaternion q is a four-dimensional vector often represented as a pair $(\mathbf{v}, w)$, where w is a real number and $\mathbf{v}$ is a 3D vector. Given quaternion q, we can compute its length $|q|$ and inverse q^{-1}; given quaternion q', we can compute their multiplication qq' and a dot product $q \cdot q'$. A quaternion of unit length can be used to represent a single rotation about unit axis $\hat{\mathbf{u}}$ by angle ϑ in two equal forms as follows:

$$q = (\sin\frac{\vartheta}{2}\hat{\mathbf{u}}, \cos\frac{\vartheta}{2}) = e^{\frac{\vartheta}{2}\hat{\mathbf{u}}}.$$

Rotating a vector $\mathbf{v}$ about axis $\hat{\mathbf{u}}$ by angle ϑ can be computed as the double quaternion multiplication $\mathbf{v}' = q\mathbf{v}q^{-1}$. A sequence of rotations $q_1, q_2 \ldots q_n$ can be easily computed as the multiplication $q_n \ldots q_2 q_1$ (notice the reversed order; see Appendix for operation definitions).

The set of all unit quaternions forms a unit sphere in four dimensions and each point on its surface represents an orientation of a rigid body. It was proven by Euler that a combination of any number of rotations can be represented as a single rotation from an reference orientation. A unit quaternion represents this single rotation as a *great arc* connecting the reference and current body orientations on quaternion sphere. The length of this arc equals ½ of the rotation angle. Thus, just as we use vectors to represent translations, we also can use spherical arcs to represent 3D rotations. If the reference orientation is not explicitly specified, a quaternion defines the rotation from the identity quaternion $\mathbf{1} = (\bar{\mathbf{0}}, 1)$, which has a special meaning as a zero orientation or no rotation – an equivalent to the origin in a vector space.

Linear zero-order C-D gain for 3D rotations

Given an orientation of the multiple DOF input device, what mapping allows us to amplify or scale this orientation in a manner similar to scaling translations of the device?

Amplifying rotation means changing the amplitude while preserving the direction of rotation. Let q_c be the orientation of a multiple DOF input device:

$$q_c = (\sin\frac{\vartheta_c}{2}\hat{\mathbf{u}}_c, \cos\frac{\vartheta_c}{2}) = e^{\frac{\vartheta_c}{2}\hat{\mathbf{u}}_c},$$

where $\hat{\mathbf{u}}_c$ is the axis of rotation and ϑ_c is the angle. The zero-order C-D gain should amplify the angle of rotation ϑ_c by coefficient k while leaving axis $\hat{\mathbf{u}}_c$ intact:

$$q_d = (\sin\frac{k\vartheta_c}{2}\hat{\mathbf{u}}_c, \cos\frac{k\vartheta_c}{2}) = e^{k\frac{\vartheta_c}{2}\hat{\mathbf{u}}_c} = q_c^k.$$

Therefore, the basic equation for the zero-order linear C-D gain for spatial rotations is a *power* function of the form:

$$q_d = q_c^k, \qquad (2)$$

where q_c is the device rotation, q_d is the displayed orientation, and k is the C-D gain coefficient.

Quaternion q_c in Equation 2 specifies device orientation as a rotation from an unspecified initial orientation designated by identity quaternion $\mathbf{1}$. However, it is often important to amplify rotation relative to some explicitly specified reference orientation q_0. This can be done by calculating the rotation that connects q_0 and q_c, amplifying it, and combining it with reference orientation q_0:

$$q_d = (q_c q_0^{-1})^k q_0. \qquad (3)$$

Notice that Equation 3 is identical to the *slerp* function introduced by Shoemake for rotation interpolation [17]. This should not come as a surprise; indeed, while Shoemake

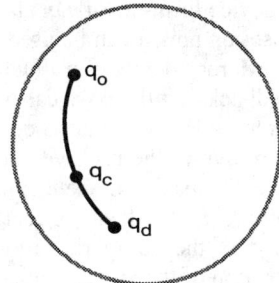

Figure 2: Extrapolating device orientation q_c on a quaternion sphere; q_0 and q_d are initial and displayed orientations.

interpolates quaternions using a great arc on a quaternion sphere, we *extrapolate* the device orientation using a great arc connecting q_0 and q_c (Figure 2)[2]. Therefore, we can use an equivalent formula that is easier to apply [17]:

$$q_d = q_0 \frac{\sin((1-k)\Omega)}{\sin(\Omega)} + q_c \frac{\sin(k\Omega)}{\sin(\Omega)}$$

where Ω can be obtained from $\cos\Omega = q_c \cdot q_d$.

Equations 2 and 3 are fundamental equations of rotational C-D gain. They are fundamental in the sense that they represent a basic form of zero-order C-D mappings between rotations of the device and rotations in a 3D interface space and provide a generic method for constructing a variety of rotational techniques suitable for particular application.

These mappings are *linear*, since the rate of amplification does not change no matter how far the user rotates the device. The non-linear mappings, which might be useful, are not discussed here; we refer the interested reader to [16].

INTERACTION TECHNIQUES: DESIGN GUIDLINES

In the previous section, we derived the basic equations of mapping that allow us to linearly amplify rotations of multiple DOF input devices. This section discusses how these equations can be used to design non-isomorphic techniques for rotating objects in VEs. At the center of this discussion are important and non-intuitive differences between absolute and relative mapping schemes in 3D rotations.

Absolute and relative mappings: it makes a difference

Typical isotonic multiple DOF devices, such as magnetic trackers, are absolute devices, i.e., they measure and return the absolute displacement of the device relative to the initial, zero orientation [4]. Hence, the easiest method for implementing non-isomorphic techniques is to map the absolute orientation of device q_{c_i}, measured on *i*-th cycle of the simulation loop, using Equations 2 or 3:

$$q_{d_i} = q_{c_i}^k,$$

and apply the resulting absolute orientation q_{d_i} to virtual objects, scenes, and virtual viewpoints.

An alternative way to implement non-isomorphic techniques using the same equations is to amplify only relative changes in the device orientation, i.e. on *i*-th cycle of the simulation loop, we calculate the relative rotation of device rotation from its orientation on the *i-1*th cycle and amplify it. The orientation of virtual object q_{d_i} is then calculated by combining this amplified relative rotation with the orientation of virtual object on the *i*-1 step of the simulation loop:

$$q_{d_i} = (q_{c_i} q_{c_{i-1}}^{-1})^k q_{d_{i-1}}. \tag{4}$$

Hence, the difference between these two mapping schemes is that in the first one we amplify the absolute orientation of the device, while in the second one we amplify its relative rotations. Consequently, we will refer to them as *absolute* and *relative* non-isomorphic rotation mappings.

Differentiating between absolute and relative mappings in spatial rotations is important for two reasons. First, they are different from a mathematical point of view: *given the same rotation path of the device, these two mappings produce different rotation paths of the displayed object* [3]. This might be unexpected; indeed, in the case of translations, relative and absolute mappings would obviously yield the same trajectory of movement. This, however, is yet another example of the peculiar nature of curved rotational space.

Second, *absolute and relative mappings are very different from the usability point of view*. The "feel" of the manipulation largely depends on the choice between relative and absolute mappings. The next section compares and contrasts the usability characteristics of relative and absolute mappings and their implications for 3D interface design.

Usability properties of absolute and relative techniques

Our ability to self-regulate motor movements, e.g., object manipulation, depends on spatial and temporal correspondence between a large variety of sensory feedbacks: visual, tactile, kinesthetic, proprioceptive and others. If the computer response, e.g., visual feedback, conflicts with kinesthetic or proprioceptive feedback produced by the human motor system, then the user performance degrades [19]. Therefore, the effectiveness of manipulation techniques depends on whether they preserve *compliances* between the user motor movements and the sensory feedback s/he receives, i.e., a stimulus-response (S-R) compatibility [6, 19].

In this section, we examine whether absolute and relative rotational mappings preserve two particular compliances which are important for effective direct manipulation: 1) compliance between the rotation directions of the input device and virtual object, i.e., *directional compliance* and 2) compliance between initial orientations of the object and input device, which we refer to as a *nulling compliance*.

Directional compliance

Directional compliance in spatial rotations simply means that as the user rotates the multiple DOF input device, the virtual object rotates in the same direction, i.e., around the same axis. Directional compliance ensures correspondence between visual, kinesthetic, proprioceptive and other feedbacks of motor movement [7, 19]. Britton [3] introduced

---

[2] Spherical arcs have also been used in Arcball [18]. However, using spherical arcs to represent 3D rotations is a standard practice while the purpose and realization of Arcball are different from the present work.

[3] To prove this, we need to show that if for a sequence of *n* incremental rotations $q_n q_{n-1} ... q_1 = q$ then generally $q_n^k q_{n-1}^k ... q_1^k \neq q^k$ (*). Although an analytical proof is beyond the scope of the paper, it can be easily tested empirically: for *n*=3, *k*=2, q_1=(0.8,0.6,0,0), q_2=(0.8,0,0.6,0) and q_3=(0.64,-0.48,-0.48,-0.36), the left part of equation (*) is (0.7, 0.3,-0.5,-0.2), while the right part is (0,0,0,1).

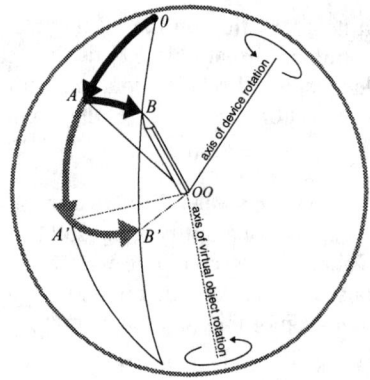

Figure 3: Directional *in*compliance in absolute mapping: (black – actual rotations; gray – amplified rotations).

directional compliance to computer graphics as a principle of kinesthetic correspondence. We can show that:

1) Relative non-isomorphic mappings always maintain directional compliance between rotations in physical and virtual spaces. As shown in Equation 4, in each cycle of the simulation loop, the virtual object is incrementally rotated in the same direction as the device, though with a different amplitude. *2) Absolute non-isomorphic mappings generally do not preserve directional compliance* between rotations of the input device and virtual objects. To illustrate this let us consider a simple physical example. Suppose our input device is a pencil fixed in point *00* and it rotates from the initial orientation *0* to *A* and then to *B* (Figure 3). The absolute orientation of the pencil can be represented as an arc connecting initial orientation *0* and the tip of the pencil. The absolute mapping would always scale this arc *relative to the initial orientation 0,* and then apply the resulting amplified orientation to the virtual pencil, which would rotate from *0* to *A'* and then to *B'*. It is obvious from Figure 3 that rotation *AB* of the physical pencil and rotation *A'B'* of the virtual pencil will happen around different axes.

Nulling compliance

Nulling compliance ensures that nulling the device, i.e., rotating it into an initial, zero orientation [4], would also rotate the controlled virtual object into a zero orientation. Nulling compliance preserves the consistent correspondence between the origins of the coordinate systems in physical and virtual spaces. We can show that:

1) Absolute non-isomorphic mappings strictly preserve nulling compliance. This follows directly from Equation 2; indeed, raising the identity quaternion, i.e., zero rotation, to a power will always yield the identity quaternion. *2) The relative mappings do not generally preserve nulling compliance.* Nulling the device would not necessarily rotate the virtual object into the expected initial state, but rather into some unpredictable orientation. This follows directly from the discussion in footnote 2.

How important is nulling compliance for direct manipulation in 3D interfaces? The answer depends on the other design variable – the form-factor of the input device.

Non-isomorphic mappings and device form-factor

The shape of a multiple DOF input device can make its manipulation easier or harder, by involving different muscle groups [25]. The device form-factor can also provide cognitive clues to the user on how it can be used [8]. In addition, we observe that different device form-factors provide different sensory feedbacks on the physical orientation of the device. For example, a device mounted on the user's hand, e.g., a data glove, provides the user with strong kinesthetic and proprioceptive feedback on the device's orientation. The user inherently *feels* the device orientation, hence the inconsistency between the zero orientation of the device and the zero orientation of the virtual object, will be noticed and may degrade user performance.

On the other hand, devices that are not worn on the body but are freely rotated in the user's fingers do not inherently provide any kinesthetic or proprioceptive feedback on their orientation. In cases where the device's geometrical shape can be recognized through tactile feedback, the inconsistencies between orientations of the device and virtual objects can still be perceived. However, a homogeneous form, such as a sphere, provides very little sensory information about its actual physical orientation. For such a device, all orientations are equivalent and it is difficult, if impossible, for the user to discriminate between them. Hence, the nulling compliance becomes unnecessary.

To conclude, multiple DOF devices, which have a spherical form and can be manipulated in the fingers, such as Zhai's Finger Ball [25], are not subjected to nulling compliance constraints. Such devices can be effectively used with non-isomorphic relative mappings, as experimental studies reported later in the paper will demonstrate.

Design trade-off in 3D rotational techniques

The difference between using relative and absolute C-D mappings in designing rotational techniques is a question of a trade-off between the directional and the nulling consistency: we cannot have both (Table 1).

Absolute non-isomorphic mappings can only have a limited use since they do not always preserve the directional com-

	Absolute mapping	*Relative mapping*
Directional compliance	**No,** virtual object does *not* always rotate in the same direction as the device	**Yes,** virtual object *always* rotates in the same direction as the device
Nulling compliance	**Yes,** nulling device *always* returns virtual object to initial orientation	**No,** nulling device does *not* necessarily return virtual object to initial orientation
Device form factor	Worn on body or easy recognizable shape	Homogeneous, sphere

Table 1: Design trade-off in 3D rotational techniques

pliance, and therefore do not allow the user to consistently predict the response of the virtual object on the device rotations. These mappings, however, can be useful when the device rotations do not change the axis much. For example, we have used them with satisfactory results for viewpoint control using head rotations tracked by a camera [16].

Relative non-isomorphic mappings can be very efficient in manual control tasks if the multiple DOF input device provides little tactile and kinesthetic feedback on its actual orientation and can be freely rotated in the fingers, such as in the case of a Finger Ball and to a certain degree the Polhemus Space Ball. We will support this observation by presenting the results of the following experimental studies.

EXPERIMENTAL USABILITY STUDY

An experimental usability evaluation was conducted to investigate the performance characteristics of a relative non-isomorphic rotational technique compared with conventional one-to-one mapping in a 3D object rotation task. Based on the results of pilot studies, the following preliminary hypotheses were formulated prior to the experiments.

H_1: A relative amplification of multiple DOF input device rotations will allow subjects to accomplish a rotation task faster than with traditional isomorphic mappings when a large range of rotations is required. A non-isomorphic mapping will not have a significant effect on subject performance for a small range of rotations.

A large range of rotations usually requires clutching or involves larger muscles of the arms and shoulders, and this decreases user performance [25]. We suggest that a non-isomorphic interaction technique with moderate amplification of rotations will allow subjects to use their fingers more effectively, reduce the need for clutching, and therefore result in faster task completion. We hypothesize that non-isomorphic mapping would not result in better performance for small rotations, and we were interested in whether there would be a decrease in user performance.

H_2: Non-isomorphic techniques with moderate amplification of rotations will decrease rotational accuracy.

The higher the sensitivity of a device, the more difficult it is to rotate the device precisely into the required orientation. While it seems logical to assume that accuracy will suffer, we were interested in how significant the decreases in rotational accuracy would be.

Finally, we were interested in estimating subjects' preferences for rotation techniques. The rotation task has an inherently limited range – 360 degrees – and can be accomplished with or without amplification. Strong subject preferences for some of the techniques would indicate that the choice of mapping does make a difference and therefore should be considered in the design of spatial interfaces.

Subjects and apparatus

Twenty unpaid subjects, eighteen male and two female, all right handed, age range from 19 to 35, were recruited from the laboratory subjects pool. None of the subjects had previous experience with 6DOF input devices.

The experiments were conducted in desktop environments using the SGI O2 workstation with a 17" 1280x1024 pixels true color monitor. The update rate was controlled between 19 and 25 Hz. The Polhemus SpaceBall 6DOF magnetic sensor was selected as the input device, and a mouse was used as the trigger device. The coefficient of amplification in non-isomorphic technique was chosen empirically at 1.8.

Experimental task

The experimental task design followed the design of orientation matching experiments used by Chen [5] and Hinckley [8]. Participants were instructed to rotate a solid shaded 3D model of a house from a randomly generated orientation into a requested, a priori-specified target orientation (Figure 4). The target orientation was a front of the house, indicated by the front door, facing the user. The house model was made to provide maximum clues to understanding its orientation, e.g., asymmetric location of chimney and windows.

The user picked up and released a house by pressing and releasing the mouse button with the non-dominant hand. The user could rotate the house iteratively using clutching – pick, rotate, release, re-adjust the hand, and re-pick the house as many times as necessary to orient it within the threshold of the specified accuracy. When the error of orientation fell below the threshold, which was approximately 18 degrees, the house would disappear, cueing subjects that the task had been accomplished successfully. The next trial was presented after a three-second delay.

This task design differed from Chen's in two respects. First, Chen required subjects to rotate a house from a fixed initial orientation into a randomly generated one. We slightly simplified the task by reversing it: the user rotated the house from the initial random orientation into a known one. Second, Chen rated and scored the user's completion accuracy after each task as "Excellent," "Good match," and so on. We used the accuracy threshold instead, because it allowed us to implicitly control the difficulty of the task as well as to provide participants with clear criteria of task completion.

Experiment design and procedure

The repeated measures-within subject experimental design was used. The independent variables were *interaction technique* (one-to-one and non-isomorphic mapping) and *amplitude* of rotation, defined as the shortest rotation required to rotate the house into the target orientation. The amplitude variable had two levels: *small* (a random angle from 20 to 60) and *large* (a random angle from 70 to 180).

The dependent variables were *completion time* and *orientation error*. The completion time was measured from the

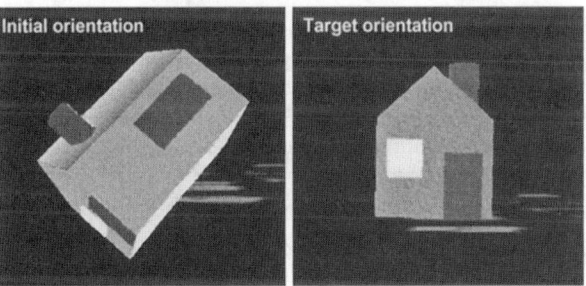

Figure 4: Task required users to rotate house model from randomly generated initial orientation (left) to target orientation, with front of the house facing user (right).

moment the user picked up a house until the moment the house was oriented with the required accuracy. Error was measured as the angular difference between the final orientation of the house and target orientation.

The experiments started with an explanation of the techniques, experimental task and procedure, followed by a 15 to 20 minute training to stabilize the manipulation performance and ensure understanding of the task and techniques. The training was followed by the experimental session consisting of two blocks of trials, one with one-to-one and the other with non-isomorphic mapping. Each block consisted of ten trials: five with large and five with small amplitude of rotation, randomized. All subjects matched the same randomly generated orientations. To control for order effect, half of the subjects started with one-to-one mapping while the other half started with the non-isomorphic technique.

In a questionnaire administered after completion of the experiments, subjects were asked to rate the techniques on a scale from 0 to 4 (0 = very bad, 1 = bad, 2 = OK, 3 = good, and 4 = excellent) and explain their choices. The experiments took from 45 minutes to 1 hour for each subject.

Results

A repeated-measures two-way analysis of variance (ANOVA) was performed for each of the dependent variables with *interaction techniques* and *amplitude* as independent variables. Data for completion time was transformed using a natural logarithm, since analysis revealed that the data was skewed away from a normal distribution.

Table 2 outlines the main effects of independent variables as well as their interaction for each dependent variable. Both technique and amplitude significantly affected the completion time. The interaction technique, however, was not a significant factor for the orientation error. A significant interaction between technique and amplitude for completion time suggests that the effect of the interaction technique depends on rotational amplitudes.

A separate comparison of techniques for small and large rotations shows that the non-isomorphic mapping was on average 13.4 percent faster when a large amplitude was required ($F_{1,19} = 7.3$, $p < 0.01$, Figure 5), while no significant difference was found for small rotations ($F_{1,19} = 0.03$, $p < 0.87$). This finding supports the first hypothesis. Both techniques resulted in almost the same orientation error: the average was 6.9 and 6.6 degrees for non-isomorphic and one-to-one mappings, respectively (Figure 5). This difference is insignificant both statistically and qualitively.

Subjects preferences

In the questionnaire 18 subjects (95%) preferred non-isomorphic to one-to-one mapping; on average they were rated

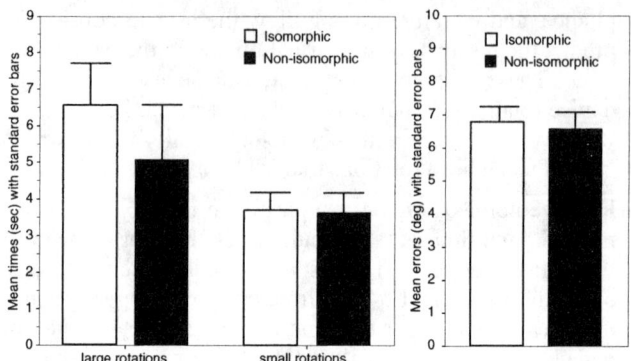

Figure 5: Left: mean completion times; right: mean errors of orientation, collapsed for all amplitudes

3.15 and 2.3, respectively, on a scale from 0 to 4. A paired *t*-test confirmed that this difference was statistically significant: $t_{19} = 3.9$, $p < 0.001$. Subjects noted that non-isomorphic mappings allowed them to rotate objects faster, with little re-adjustment of hand or device and less physical effort when a large range of rotations was required. Three subjects specifically commented that amplified rotations allowed them to use their fingers over a larger range of rotations, which they found was more efficient.

Many subjects reported that it was slightly more difficult to precisely control device rotations with non-isomorphic mapping. However, many of them noted that it was a question of experience and practice. Two subjects suggested that it would be useful to be able to control the sensitivity of mapping and use slower rotations, especially when accuracy was important. The cable of the SpaceBall tracker was found to be the most disturbing factor in the experiments.

Discussion

The experiments demonstrated that a non-isomorphic interaction technique, which linearly amplifies rotations of a multiple DOF input device, allowed the subjects to accomplish the experimental task 13% faster compared with one-to-one mapping for a large range of rotations. The performance for a small range of rotations was the same. The subjects' strong preferences for the non-isomorphic techniques also suggest that rotational mappings are an important design variable in constructing 3D user interfaces.

Furthermore, mappings had no effect on the accuracy of rotation, or at least none that could be detected with 20 subjects. If we compare our results with Hinkley's experiments [8], who used a similar experimental design, our subjects averaged 6.8 degrees of error, while Hinkley's 24 subjects averaged 6.7 degrees of error. Although our experimental design emphasized speed, our results closely replicated Hinkley's, even though his experiments emphasized accuracy instead. This supports Hinkley's observation that the accuracy of rotation might be less affected by the manipulation capabilities of the interface then by the difficulties subjects had in perceiving and adjusting the rotation error. Furthermore, the experiments of Ware and Rose [23] demonstrated that even when subjects rotated ordinary physical objects in real world, there was a natural limit in accuracy that averaged 4.64 degrees. In 3D interfaces, the accuracy can deteriorate further due to insufficient depth cues or lag.

	technique	amplitude	interaction t•a
Time	$F_{1,19} = 5.52$, $p < 0.03$	$F_{1,19} = 113.932$, $p < 0.0001$	$F_{1,19} = 5.68$, $p < 0.028$
Error	$F_{1,19} = 2.29$, $p < 0.15$	$F_{1,19} = 15.7$, $p < 0.001$	$F_{1,19} = 0.026$, $p < 0.874$

Table 2: The main effects and interaction

For completion time, Hinkley's subjects averaged 17.8 seconds while our subjects averaged 5.15 seconds for one-to-one mapping. This difference can be explained, first, by the emphasis of accuracy in Hinkley's experiments, i.e., his subjects spent more time trying to match orientation; second, in the training level, i.e., his subjects were given as little instructions as possible, while we tested the stabilized manipulation performance. The experiments of Ware and Rose, on the other hand, resulted in quite comparable subject performance: their 4.96 versus our 5.15 seconds. Thus, we believe that our experiments produced quite accurate estimates of user rotational performance and accuracy.

CONCLUSIONS

This paper demonstrates how non-isomorphic 3D rotational interaction techniques can be constructed and used to design effective spatial user interfaces. We attempted to provide a thorough treatment of this subject, by designing the mathematical foundations of rotational mappings, investigating their usability properties, and evaluating their user performance characteristics. Our results suggest that non-isomorphic rotational mappings are an effective tool in building high-quality manipulation dialogs in 3D interfaces. This paper will help designers to use them effectively.

ACKNOWLEDGMENTS

Research reported in this paper has been partially conducted as part of the first author's Ph. D. work at Hiroshima University. We are thankful to many people for their suggestions, discussion and help. The basic idea for this work emerged from a discussion between the first author and Jock McKinlay. Michael Svinin and Horst Kraemer have provided invaluable help in discussing the subtle aspects of the mathematics involved in 3D rotations. Takeo Igorashi, Mark Billinghurst, Michael Kowalski and Prof. Ichikawa provided crucial feedback that helped us present these materials in the best possible way. We are also thankful to all subjects who participated in the experiments as well as anonymous reviewers for their comments.

REFERENCES

1. Boritz, J., Booth, K., A study of interactive 3D point location in a computer simulated virtual environment. *Proceedings of VRST'97*. 1997. ACM. pp. 181-187.
2. Bowman, D., Hodges, L., An evaluation of techniques for grabbing and manipulating remote objects in immersive virtual environments. *Proceedings of I3DCG*. 1997. ACM. pp. 35-38.
3. Britton, E., Lipscomb, J., Pique, M., Making nested rotations convenient for the user. *Proceedings of SIGGRAPH'78*. 1978. ACM. pp. 222-227.
4. Buxton, W., There's more to interaction then meets the eye: some issues in manual input. In *User Centered System Design: New Perspectives on Human-Computer Interaction*, D. Norman and S. Draper, Editors. 1986, Lowrence Erlbaum, pp. 319-337.
5. Chen, M., Mountford, S., Sellen, A., A study in interactive 3D rotation using 2-D control devices. *Proceedings of SIGGRAPH'88*. 1988. ACM. pp. 121-129.
6. Fitts, P., Jones, R., Compatibility: spatial characteristics of stimulus and response codes. *Journal. of Experimental Psychology*, 1953 (46).
7. Gould, J., Smith, K., Angular displacement of the visual feedback in motion. *Science*, 1962(137): pp. 619-620.
8. Hinckley, K., Tullio, J., Pausch, R., et al., Usability analysis of 3D rotation techniques. *Proc. of ACM UIST'97*. 1997. pp. 1-10.
9. Hultquits, J., A Vrtual Trackball. In *Graphics Gems I*. 1990, Academic Press. pp. 462-463.
10. Jenkins, W., Connor, M., Some design factors in making settings on a linear scale. *Journal of Applied Psychology*, 1949. 33(4): pp. 395-409.
11. Knight, J., Manual control and tracking. In *Handbook of human factors*, Salvendy, Ed. 1987, John Wiley&S. pp. 182-218.
12. Mine, M., Brooks, F., Sequin, C., Moving objects in space: exploiting proprioception in virtual-environment interaction. *Proceedings of SIGGRAPH'97*. 1997. ACM. pp. 19-26.
13. Pierce, J., Stearns, B., Pausch, R., Voodoo Dolls: Seamless interaction at the multiple scales in virtual environments. *Proceedings of I3DCG'99*. 1999. ACM. pp. 141-145.
14. Poupyrev, I., Billinghurst, M., Weghorst, S., Ichikawa, T., Go-Go Interaction Technique: Non-Linear Mapping for Direct Manipulation in VR. *Proc. of UIST'96*. 1996. ACM. pp. 79-80.
15. Poupyrev, I., Weghorst, S., Billinghurst, M., Ichikawa, T., Egocentric object manipulation in virtual environments: empirical evaluation of interaction techniques. *Computer Graphics Forum, EUROGRAPHICS'98 issue*, 1998. 17(3): pp. 41-52.
16. Poupyrev, I., Weghorst, S., Otsuka, T., Ichikawa, T., Amplifying rotations in 3D interfaces. *Proc. of CHI'99 Conference Abstracts, Late Breaking Results*. 1999. ACM. pp. 256-257.
17. Shoemake, K., Animating rotations with quaternion curves. *Proceedings of SIGGRAPH'85*. 1985. ACM. pp. 245-254.
18. Shoemake, K., ARCBALL: a user interface for specifying three-dimensional orientation using a mouse. *Proceedings of Graphics Interface'92*. 1992. pp. 151-156.
19. Smith, T., Smith, K., Feedback-control mechanisms of human behavior. In *Handbook of human factors*, G. Salvendy, Editor. 1987, John Wiley and Sons. pp. 251-293.
20. Song, D., Norman, M., Nonlinear interactive motion control techniques for virtual space navigation. *Proceedings of VRAIS'93*. 1993. IEEE. pp. 111-117.
21. Stoakley, R., Conway, M., Pausch, R., Virtual reality on a WIM: interactive worlds in miniature. *Proceedings of CHI'95*. 1995. pp. 265-272.
22. Wang, Y., MacKenzie, L., Object manipulation in virtual environments: relative size matters. *Proceedings of CHI'99*. 1999. ACM. pp. 48-55.
23. Ware, C., Rose, J., Rotating virtual objects with real handles. ACM *Transactions on Computer-Human Interaction*, 1999. 6(2): pp. 162-180.
24. Zhai, S., Human performance in six degrees of freedom input control, Ph.D. Thesis, *Department of Industrial Engineering*. 1995, University of Toronto, Canada.
25. Zhai, S., Milgram, P., Buxton, W., The influence of muscle groups on performance of multiple degree-of-freedom input. *Proceedings of CHI'96*. 1996. ACM. pp. 308-315.

APPENDIX: QUATERNIONS

The definitions of the quaternion operations used in this paper are as follows:

$$q^* = (-\mathbf{v}, w); \quad |q| = \sqrt{x^2 + y^2 + z^2 + w^2}; \quad q^{-1} = \frac{q^*}{|q|}$$

$$qq' = \mathbf{v} \times \mathbf{v}' + w\mathbf{v}' + w'\mathbf{v}, ww' - \mathbf{v} \cdot \mathbf{v}'; \quad q \cdot q' = \mathbf{v} \cdot \mathbf{v}' + w \cdot w'$$

More information on quaternions as well code samples can be found at http://www.hitl.washington.edu/people/poup/ or http://www.mic.atr.co.jp/~poup/

Joking, storytelling, artsharing, expressing affection: A field trial of how children and their social network communicate with digital images in leisure time

Anu Mäkelä

Helsinki University of Technology
P.O. Box 5400
02015 HUT, Finland
+358 9 451 5041
ahmakela@cc.hut.fi

Verena Giller   Manfred Tscheligi   Reinhard Sefelin

CURE & University of Vienna
Lenaugasse 2/8
A-1080 Vienna, Austria
+43 1 4277 38452
cure@cure.at

ABSTRACT

Increasing use of mobile phones in leisure and communication with digital images are important and current issues in the field of telecommunications. However, little is known about how images would be used in leisure related communication. According to our experience field trials are the best way of studying it. In this paper, we describe a field trial case study of leisure related communication with digital images. Moreover, we discuss the advantages of conducting field trials as part of product concept design process.

Keywords

Product concept design, field trial, prototypes, wireless communication, digital images, children, family, leisure.

INTRODUCTION

People have to send and receive a huge number of messages and information to handle daily life duties like picking up children from school but also to communicate emotions and feeling to loved ones and relatives. How are these goals performed by sending digital images? What kinds of changes will occur in people's communication behavior if still images carry the information of their children's, friends' etc. current activities? How is the perception of images changing? What kinds of recommendations can be given for designing mobile applications for communicating with digital images in leisure? Questions like these have been approached by two field trials with portable prototypes that enabled users to send and receive digital images. The focus of the field trials in Vienna (Austria) and Helsinki (Finland) was on the potential of communication with digital images over a wireless network in the everyday life of children and their social network. The main goal of these trails was to discover typical use situations of digital images in communication among an extended family, and a peer group.

Major findings and results of these field trials presented in this paper were:

- Images were mainly used for joking, expressing emotions and creating art
- Users' perception of images changed from memory support to the expression of current activities and feelings
- Users wanted and used large sets of picture editing possibilities
- The usage of images for communication purposes depended on the user's willingness to develop a picture language with the receiver
- Images were not enough for functional communication and users did not want to carry several mobile communication devices with them. Therefore, images should provide possibility for annotation with text or audio
- Further design recommendations for further development of devices for sending of digital images could be given

In this paper our field trials will be described in detail. Secondly, the importance of conducting field trials as a part of user-centered product concept design process will be pointed out.

The field trials were a part of product concept design done in Maypole. Maypole was a two-year European Project (1997-1999) within the European Initiative on Intelligent Information Interfaces (I3). The other Maypole partners were IDEO Europe, Meru Research, Netherlands Design Institute and Nokia Research Center (see more [16]).

Permission to make digital or hard copies of all or part of this work for personal or classroom use is granted without fee provided that copies are not made or distributed for profit or commercial advantage and that copies bear this notice and the full citation on the first page. To copy otherwise, to republish, to post on servers or to redistribute to lists, requires prior specific permission and/or a fee.
CHI '2000 The Hague, Amsterdam
Copyright ACM 2000 1-58113-216-6/00/04...$5.00

Today's Communication with Digital Images

There is not much knowledge on designing applications for leisure related or informal communication. Most of the research on communication in the HCI community has concentrated on video-mediated communication at work (e.g., [17], [13]). The focus has been on task-related communication, even though there is evidence [1] that task-unrelated "goofing around" happens and is needed at adults' workplaces, too. The interest in task-unrelated aspects of interaction is increasing. An example is the workshop, "Towards a Framework of Interaction and Experience As It Relates to Product Design" [10], at UPA'99.

There is also not much knowledge on the use of still images in communication. However, researchers and designers are more and more interested in digital cameras in consumer use. For example, Frohlich and Tallyn reported [11] a field trial of a digital camera with audio capture. They wanted to understand why and how people would record sound with their digital cameras.

In our own studies done before prototype development we found out that leisure related communication with images is done more and more by publishing digital pictures on WWW. Analyzing those web-sites helped to generate ideas that were used in the following design phase of the prototypes.

Field Trials

A huge number of products are brought to the market with very little idea of how people will use them or whether they will use them at all. Of course, often usability tests with user involvement are made beforehand but they are, nevertheless, not useful to find answers to questions like, "How, how long, in which kind of situations etc. will users use the product and which kind of improvements seem to be reasonable for their purposes?"

Etnographic user research methods, such as Contextual Inquiry [2] and others [see e.g. 18], used at the beginning of the development process can give ideas for new concepts by describing how current technologies are used and what kind of values users have. They can not, however, forecast what completely new technology would mean to the users.

The designer cannot really understand the role and the meaning of a new product concept in the user's life until she/he sees how the user uses the product concept in his/her own environment. For example, when mobile phones became popular in Finland developers and designers were asking whether Short Message Service (SMS) is needed at all, and who would use it. They expected that if it was needed it would be suitable mainly for professional use. Today, the service is very popular among many user groups including teenagers. It is used both during the office hours and leisure.

In work related environments some essential findings about so called Group Support Systems (GSS) could be discovered by field studies [1,8,12]. However, also those studies were carried out at a stage in the design process where redesign and essential changes of functions would initiate much costs and effort.

Unfortunately there seems to be rather little enthusiasm to carry out field trials to test products for leisure time communication in the field. The research is mainly limited to the investigation of network communities or "electronic villages", like Blacksburg [4, 5], and to studies on the products used by people with special needs [15]. Also some exploitation of new technologies (like digital TV, wireless LAN etc.) has been done but the focus has mainly been on the technical realization.

Product managers often try to avoid field testing of prototypes because they worry about increasing production costs and loss of time.

In the following chapters we will describe a case study where "medium-fidelity" prototypes were tested in the field. The experiences we gained is hoped to be an impulse for researchers, designers, and managers working in the field of product concept design not to wait until it is too late but to start as soon as possible with the realization of tests with real users in their own environments.

METHODS

Development of the Concept of the Prototypes

The goal of the project was the development of communication devices for children and their social networks. During the project we conducted several user studies both in Austria and Finland. Our main user group was children but members of their social network were also included in the studies. We started with focus groups and interviews that were held in the user's own environments. Also photo diaries and field trials of antecedent products were used to understand everyday communication of children and their social network. Then, several different product concepts were generated and evaluated by users in laboratory settings. The concepts were presented to the users with blank models, screen demos and in some cases with storyboards (see more [14]).

Our main discoveries from initial user studies and user evaluations done before starting the prototypes were that:

- Socializing was the most important issue in leisure related communication among children. Socializing refers to a vital social activity that is not necessarily task-related. Social contacts to other individuums can be searched for emotional reasons as well [9].

- Parents' need to communicate with their children was very functional on the one hand (who picks up kids, where) but has on the other hand very much to do with trust between children and their parents. Most parents were very interested to know what their children were doing without giving them the feeling of being watched all the time.

- There was a need for an application that supports group communication rather than one to one communication.
- Communication with images could enable socializing better than audio or text based communication.

On the basis of these findings one of the generated concepts was chosen for prototyping which enabled users to send, receive and edit digital images.

Tested Prototypes

The prototypes (see Figure 1) which were tested during the field trials offered mainly three functions to the users:

- Taking and saving of digital images.
- Editing of images; Users could change the appearance of the taken and received messages.

    The prototype included functions for making short, maximum 5 pictures long series (Figure 3 and 5) which were enhanced with transition and sound effects. Pictures also could be manipulated by blending two pictures into one, and changing the colors of the new picture.

- Sending and receiving of digital pictures (jpg-files) wirelessly (over GSM): Images could be sent to all participants of the field studies, and every participant was able to receive pictures from the other participants.

The device was a prototype and had two parts: the interface module and laptop in a rucksack to guarantee maximum portability of the prototypes (Figure 1 and 2).

The functions of the prototypes were limited to pictures to ensure that the research focus keeps on communication with digital images. The possibility to use text or audio would have biased the results. It would not have been possible for researchers to fully understand the communicative potential of digital images by comparison with other media like mobile phones, pagers etc.

Every sent picture was saved in a log-file on a server, which gave us the possibility to observe participants' communication real time. This also enabled us to detect technical problems immediately, and to ask more concrete and contextual questions during the interviews.

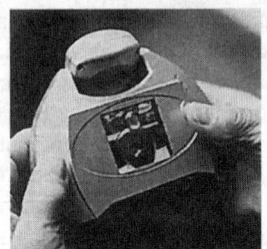

Figure 1: The prototype

Subjects

The field trials were conducted at two different test sites in Helsinki and Vienna. Since there was only four weeks time to conduct the trial in both sites it was thought that it is better to have test users who were familiar with each other before the trial, and who were motivated to explore the prototypes. Two very different groups participated in the trial: a group of four chummy, Finnish boys (12 years old) and a family in Vienna. The Finnish boys lived in the same neighborhood. They were classmates, and had common hobbies. The Viennese participants were two parents with four children (two boys and two girls, 8 - 15 years old) living together at the same place and their 70 year old grandmother living in a very different area of Vienna.

Figure 2: Finnish Boy using the prototype

The prototypes were handed out to the users at both test sites for four weeks which was enough to test the usage in typical communication situations.

Goal of the Field Trials

As mentioned before, the aim of the field trials was to understand how digital images would be used by children and their social network in leisure related communication.

The research questions of the field trials were the following:

1. How is the perception of taking pictures changing?
2. What is the key motivator to use the device?
3. In which communication situations do users use images?
4. What is the special advantage for a user to have the possibility of sending images instead of text or audio messages?

5. What do users want to add to their images, what do they want to remove or hide?
6. Which kind of pictorial communication is needed in a family and in a peer group, and what are the differences between generations?

User Research Methods
The users were selected according to two criteria. At least some of them should use (have used) mobile communication devices before and they should be interested in figuring out new possibilities of communication. Those criteria should guarantee that subjects are capable of making subjective comparisons to mobile phones and similar devices and that they do not loose interest if there are any kind of technical problems. To ascertain if those requirements were met and to explore current communication behavior of test persons before starting the field trials intensive interviews were carried out with all participants of the field studies. After the prototypes were handed out and a usage introduction had been given to the participants, users were interviewed weekly in their own environment. These interviews were carried out partially as group and partially as single interviews whereas especially children had to be separated from their parents to guarantee valid and true answers.

The interviews during the trials were based on natural user experiences, and therefore they were very concrete and contextual. That means that users could freely tell their experiences and that they were encouraged to make their own recommendations about functions they wanted to be included and which they felt were superfluous. Additionally they were motivated to report situations of usage and to detected changes in their communication behavior.

The log kept from sent images in the server during the trials turned out to be a good source of user data. Researchers could understand the content of the images much better after going through them with the users. Typical use situations, and specifically advantages and drawbacks of the usage of images for the typical daily life communication were reported.

MAIN RESULTS
Users' Subjective Attitudes towards Prototypes
The user interface and the appearance of the prototypes were designed mainly for children older than 10 years. The design was successful in the sense that the children at that age liked the prototype and were enthusiastic to explore its possibilities. However, adults and younger children had difficulties in learning to use the prototype, and the adults thought that the design was childish.

Furthermore, especially during the first days of field trials we recognised that prototypes were seen as expensive equipment that might get broken easily. This problem often arises and researchers have to consider these anxieties very carefully and develop strategies to avoid them. In our case letters guaranteeing users that they will not be held responsible for any damages and that any hardware troubles will be fixed immediately helped a lot.

Older and younger (children younger than 10 years old) users definitely needed more help during the first days. So researchers had to make sure that all features and functions of the prototypes were understood by those users and again that their anxieties of making mistakes or of initiating any damages were relieved.

During field trials all users had fun using the prototypes and also the number of images sent during the field trials did not decrease after some weeks. In both cases (Vienna and Helsinki) the number increased over the first two weeks and then kept stable until the end of field trials.

Meaning of Photographs Changed
Most of the children thought that photographing and photo albums were for older people, which is consistent with the results of Csikszentmihalyi and Rochberg-Halton [6]. In their study on the most cherished objects in the home for the youngest generation photographs were the sixteenth category in order of frequency, for grandparents they were the first. Photographs were the primary tools for preserving the memory of one's close relations among older respondents of their study.

Our field trials showed that the meaning of photographs can change among both children and older people at least when taking pictures is cheap and instant.

Children started to use digital images in a completely different way than they had used traditional photographs. In general, the content of their images seemed to be pretty much like what they used to draw by themselves when not having a device for taking and editing images easily. The camera was not any more used only for capturing a special event but for illustrating everyday items and people in a funny way and for creating stories.

Also the participating grandmother discovered new possibilities in photographing during the field trial in Vienna. First, she started to make series of images she had taken from her everyday items at home but later she worked mainly with the blend mode, which enabled the user to combine two pictures and created artistic images. For her the experience that images are still editable after the trigger was pressed was completely new. She spent much time (sometimes more than two hours) with editing her images to guarantee that they met her own artistic requirements.

Key Motivator for Different User Groups
The motivation to use the prototype differed very much between the user groups participating in our field studies. Whereas for children having fun before and after school with the device itself and the potential usage of it for further games was most important, for the grandmother the key motivator was the possibility of sharing feelings.

For parents who did not send as much images as the other user groups receiving of pictures that told them about current activities of their relatives was the most substantial part of the prototype's potential.

During the field studies two major variables emerged as good predictors for using digital images for intra family communication:

- Availability of time
- Family orientation of communication

The following Table 1 shows how those two variables were distributed among our four user groups.

	Time	Family oriented communication
Grandparents	YES	YES
Parents	NO	YES
Younger Children (< 10 years)	YES	YES
Older Children (> 10 years)	YES	NO

Table 1. Variables predicting the intra family usage of the tested prototype for our four user groups.

Note that also participants who did not meet both of those requirements still were very interested in at least receiving images for sharing other family members' life and moods.

The number of messages the children older than ten years sent to their parents was smaller than younger ones sent. This does not mean that the older children were less interested in using digital images in communication. The studies with the Finnish boy group showed that children also older than ten years sent a huge number of images to peer groups using their own image languages and codes. Their parents who often did not have the knowledge about the context of certain images would not have understood many of those images.

For the parents who suffered from lack of time the possibility of receiving images and getting information about what their children currently were doing was very important. However, often they refused to use the prototype because composing meaningful images just was much more time consuming than using the mobile phone for making appointments or to make sure that everything is all right with their children.

Use Situations

Situation 1: creating stories
The children loved to create stories with series of images. The stories were fictional. They were used to joke or illustrate movie-like scenes. Creating a story was very interactive in a way that the children set up the scene for a story by themselves. They created the settings (e.g. blood or fried chicken on a table) and acted the situations by themselves (e.g. the murder scene in Figure 3). Most of the images used in storytelling were not edited afterwards but the Finnish boys wished that they could have had more editing possibilities, such as changing the background. Manipulation of the background would have allowed them to create easily new contexts for a story.

The image stories required a lot of work, and the children were proud of their masterpieces. They shared the stories with others by showing them directly from the screen or by sending them over the network.

Figure 3. A murder scene created by two of the Finnish boys. The small picture on the left corner is the symbol of the sender.

Situation 2: expressing spirituality
The possibility of image editing was a new and exiting experience for the grandmother. She could create art (Figure 4) with the prototype and shared her creations with her grandchildren. "Now I understand my grandchildren playing with computer toys all the time", she said. She wished even more ways of image editing.

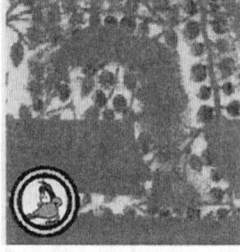 

Figure 4. Images created by the grandmother using the implemented modes for picture editing

Situation 3: expressing affection
For the boys in Finland pets but also girls they liked from their school (Figure 5) were one of their main photographing motives. Those pictures are a good example for images that can only be understood with knowledge about their special context. For us as neutral observers they were just pictures from dogs or girls. However, for the receivers they contained a big message.

Situation 4: increasing or maintaining group cohesion
One of the purposes of sending images seemed to be also to maintain attractiveness of a group or attraction between group members. For example, a boy who was not an active member of the Finnish boy group when the trial

started sent a lot of pictures of himself and his dog at home to the others. He also received images from other boys, and could in that way participate more in their activities. At the end of the field trial the boy had become much closer with the other boys.

Also the grandmother of the family in Vienna sent images for participating in the family's everyday life remotely. For her the possibility of sending artistic images and images from her dog to her children and grandchildren gave her the possibility to share her life with her family without being obtrusive.

Situation 5: supporting conversation
Toward the end of the field trial the boys in Helsinki additionally invented more utility uses for the device. For example, a boy needed to describe a feature of a computer game to his friend on the phone. It was difficult, and therefore, he sent a picture of the computer screen and used it as a tool of collaboration.

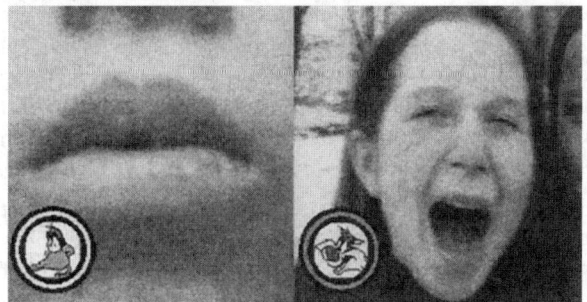

Figure 5. Expressing affection with a series of images

Images vs. Text or Voice Messages

Users used images mainly for expressing emotions, moods and humor, and for telling stories and sharing self-made art. There was hardly any usage of images for functional reasons like the arrangement of meeting points or activities etc. The reason for this is rather simple: The composition of an image which would be understood undoubtedly needed much more time than an ordinary phone call or SMS. The special language, which was used by children for communicating activities and moods, is not precise enough to guarantee clear and understandable messages. Moreover, in interviews children also said that most of their appointments are made at school and via face to face communication. Therefore their need for remote functional communication among each other is fairly small.

Field studies clearly showed that images are only useful if the receiver has a clear understanding of the sender's daily life and the context in which the image was sent. To be sure: An image can contain much more information than a voice or text message. Nevertheless, each message must be translated by the receiver according to certain rules. In the case of text or voice messages the receivers' "linguistic competence" [7] allows them to give meaning to other's "linguistic performance". In the case of image messages senders and receivers have to produce rules of their own. However, that also means that people not belonging to the same peer group have no possibility to understand image messages or to send ones.

Note that the necessity to develop a language of one's own also makes image communication interesting and playful. So for boys a picture of a certain girl is a clear message whereas for their parents it means almost nothing. This dealing with a secret language also encouraged the grandmother to send her artistical masterpieces and encouraged her grandchildren to uncover the mystery included and made them call her to find out more about those images.

Additionally we should not forget that linguistic conventions are developing rapidly. Some years ago ":-)" was not understood by anybody. Nowadays it is a generally agreed convention to mark nonserious sentences. The future progress of communication with digital images in leisure time will show if such a commonly accepted convention will also arise for the sending and receiving of images.

For the development of a digital image communication product, nevertheless, the need to provide the possibility of annotation with a text or voice message seems unavoidable. Since images definitely are not good for the transmission of functional messages and since users (also children) are not willing to carry more than one wireless communication device with them the additional functions for text and voice messages will be essential.

Communication with Digital Images in Family Settings

In typical family settings the main incentive for communication using communication devices are functional reasons. However, also the family participating in our field trials reported typical socializing (non-functional communication) activities before the field studies started.

However, there are rather big differences between family members concerning their needs for communication and for socializing. Sharing things with family was more important for the grandmother, the parents and the children under ten years than for the children older than ten years who were more interested in peer group values and rules. Note that for the parents, who tend to be very busy, receiving images was much more important than sending.

Especially for the grandmother and the parents received images were a big chance for coming a step nearer to their children and for overcoming the barriers between generations. So e.g. during one interview the mother pointed out that she often do not have the chance to join one of her children at certain events like sport competitions etc. and that she would be extremely happy to see how her child is doing at the moment. "It just would be great to see that he/she is happy", she said.

However the problem of how to encourage children to send images to their parents respectively grandparents still seems to be unsolved. Especially this problem could occur after some time of usage and habituation when the device isn't that cool and new anymore.

The parents' competence of solving this problem will depend on their flexibility to adopt to their children's image language or at least to a part of it. Nevertheless, experiences made during the field trials show that children's willingness to communicate with their parents e.g. by sending pictures of their current activities is much bigger than their willingness to call them. Firstly, using images it is much easier to conceal to whom one is communicating with. (For children older than ten sometimes it is painful to contact parents during peer group activities). Secondly, children can control the information they are sending better and without having the feeling that they are cheating. Annoying questions like "Who is with you?", "Are you wearing your pullover" etc. are almost impossible to ask with images.

To be sure, communication with digital images never will replace other kinds of communication means completely. It just can be a kind of completion. During the field trials typical functional communication like the request to be picked up was never done by using images, because images are too ambiguous, and the effort of production and translation too big.

Design Recommendations
One of the major goals of the field trials was to provide recommendations for further design of devices for communication with digital images in leisure related activities. According to the users who participated in the field trials the devices should enable users to:

- Create series of images.
- Edit images in various ways. The challenge is to design the interaction so that the user can edit pictures on a relative small screen as directly as possible.
- Annotate image message with text or/and audio.
- Store and print images.
- Send and receive images also with fixed applications at home or office. Currently, web-based applications are preferred.
- Share images both via wireless network and from screen. Sharing from a screen requires that the screen can be viewed from several angles.

CONCLUSIONS
Field trials were useful
The field trials helped us to understand the possibilities of the product concept much better than any other user research we did before. On the basis of previous user studies we knew that digital images would possibly support socialization, and children enjoyed editing pictures but we did not know in which kind of situations the children and their social network would use digital images in communication.

Before starting to build the prototypes the product concept was tested in the laboratory settings both in Vienna and Helsinki. The concept was presented to a pair of users with help of a blank model and screen demo. The laboratory tests gave some support for the concept but did not really give understanding on how the users would use it in their everyday life.

In field trials the concept is embedded in the user's behavioral patterns or it creates new patterns, which is something that blank models, storyboards, and demos can only try to simulate. In our case the fact that users refused to send functional messages for the handling of typical daily life duties exemplifies that field trials are irreplaceable. Only the observation during users' typical daily life actions showed that their answers given during user studies beforehand are not trustworthy since they just couldn't imagine the actual time pressure during such situations.

Therefore, field trials are an important part of product concept design process. However, more research is needed on how long trials should last in order to get rid of the "honey-moon effect" of introducing new technology into the users' life. Moreover, more case studies are needed on conducting field studies as part of product concept development processes in industry.

Potentials of Communication with Digital Images
The field trial results suggest that the concept of wireless communication with digital images has great possibilities in consumer markets. Already during four weeks of trial the users discovered their own ways of using the prototypes. Moreover, the usage was not occasional but happened as part of everyday patterns (such as socializing with friends before and after school), or it created new patterns (such as creating art).

The meaning of photographing changed during the trials. Digital images were not used as memories of past events or relationships but as tools for creating playful stories, expressing affection, and creating art. The users clearly wanted to have possibility to create stories with series of images. They also liked to edit and annotate images before sending them. Sometimes the images were not sent but shared with others on the screen, which requires special qualities from the product. Some of the images became so precious that users wanted to have them stored or printed because they were piece of art or picture of someone special. Moreover, not always the mobile device was the best application to use in family settings.

Therefore, the device should be interoperable with other everyday applications.

These field trials focused on leisure. The potentials of using digital images in work environment would be interesting to study, too. The future research should not focus only on task-based activities at work but also on "goofing around" [1] at work.

ACKNOWLEDGMENTS

We thank Maypole partners: IDEO Europe, Meru Research, Netherlands Design Institute and Nokia Research Center. We are also grateful for Katja Battarbee, Thomas Grill, Juha Huuhtanen, Kristiina Karvonen, Pia Kurimo, Gerhard Leitner, and Aapo Puskala who participated in conducting the user studies with us. Moreover, many thanks for Professor Kari Kuutti for giving constructive feedback on our text for this paper.

REFERENCES

1. Abramis, D.J. (1990). Play in Work. Childish Hedonism or Adult Enthusiasm? American Behavioral Scientist, 33 (3), 353-373
2. Beyer, H., and Holtzblatt, K. (1998). Contextual Design. Defining Customer-Centred Systems. Morgan Kaufmann Publishers, Inc.
3. Bly, S. (1997). Field Work: Is it product work? Interactions, 4 (1), 25 – 30
4. Carroll, J.M, and Rosson M. B. (1996). Developing the Blacksburg electronic village. Communications of the ACM, 39(12), 69 – 74
5. Carroll, J.M. et al. (1995) Building a History of the Blacksburg Village. In Proceedings of Designing Interactive Systems: Processes, Practices, Methods, & Techniques , 1995, 1
6. Chikszentmihalyi, M., and Rochberg-Halton, E. (1981). The meaning of things. Domestic symbols and the self. Cambridge University Press
7. Chomsky, N. (1957). Syntactic Structures. Mouton, The Hague
8. De Vrede, G, and van Wijk, W. (1997). A field study into the organizational Application of Group Support Systems. In Proceedings of Computer Personnel Research '97, ACM Press
9. Dunbar, R. (1996). Grooming, Gossiping and the Evolution of Language. Faber & Faber
10. Ford, S., and Forlizzi, J. (1999). Towards a Framework of Interaction and Experience As It Relates to Product Design. Workshop in Usability Professional's Association Conference 1999. URL: http://www.goodgestreet.com/UPAweb/home.html
11. Frohlich, D., and Tallyn, E. (1999). AUDIOPHOTOGRAPHY: Practice and prospects. In Extended Abstracts of CHI 99, ACM Press
12. Hindus, D. et al. (1996). Thunderwire: A field study of an audio-only media space. In Proceedings of Computer supported Cooperative Work '96, ACM Press
13. Kristoffersen, S., and Ljungberg, F. (1999). An Empirical Study of How People Establish Interaction: Implications for CSCW Session Management Models. In Proceedings of CHI 99, ACM Press
14. Mäkelä, A., and Battarbee, K. (1999). Applying Usability Methods to Concept development of a Future Wireless Communication Device – Case in Maypole. In Proceedings of 17th International Symposium on Human Factors in Telecommunication Copenhagen, Denmark, May 4-7, 1999, 291-298
15. Poulson, D., and Richardson, S. (1994). Developing adaptable smarter homes for elderly and visually impaired people. In Proceedings of International Ergonomics Association, vol.4: Ergonomics and Design. Ergonomics Association
16. Giller, V. et al. (1999). Image makers. Interactions, 6 (6), 12 - 15
17. Veinott, E.S. et al. (1999). Video Helps Remote Work: Speakers Who Need to Negotiate Common Ground Benefit from Seeing Each Other. In Proceedings of CHI 99, ACM Press
18. Wixon, D., and Ramey, J. (1996). Field Methods Casebook for Software Design. John Wiley & Sons, Inc.

Designing Storytelling Technologies to Encourage Collaboration Between Young Children

Steve Benford[1], Benjamin B. Bederson[3,4], Karl-Petter Åkesson[3], Victor Bayon[1],
Allison Druin[2,4], Pär Hansson[3], Juan Pablo Hourcade[3,4], Rob Ingram[1], Helen Neale[1],
Claire O'Malley[1], Kristian T. Simsarian[3], Danaë Stanton[1], Yngve Sundblad[2], Gustav Taxén[2]

[1]The University of Nottingham, University Park Nottingham, UK +44 115 9515151 sdb@cs.nott.ac.uk	[2]The Royal Institute of Technology, KTH Stockholm, Sweden +46 8 790 7147 yngve@nada.kth.se	[3]The Swedish Institute of Computer Science, SICS Stockholm, Sweden +46 8 752 1586 kristian@sics.se	[4]The University of Maryland, College Park, MD 20742, USA +1 301 405 2764 bederson@cs.umd.edu

ABSTRACT

We describe the iterative design of two collaborative storytelling technologies for young children, KidPad and the Klump. We focus on the idea of designing interfaces to subtly encourage collaboration so that children are invited to discover the added benefits of working together. This idea has been motivated by our experiences of using early versions of our technologies in schools in Sweden and the UK. We compare the approach of encouraging collaboration with other approaches to synchronizing shared interfaces. We describe how we have revised the technologies to encourage collaboration and to reflect design suggestions made by the children themselves.

Keywords

Children, Single Display Groupware (SDG), Computer Supported Cooperative Work (CSCW), Education, Computer Supported Collaborative Learning (CSCL).

INTRODUCTION

Collaboration is an important skill for young children to learn. Educational research has found that working in pairs or small groups can have beneficial effects on learning and development, particularly in early years and primary education [14, 19, 20]. Technology offers an opportunity to support and facilitate collaborative learning in many respects [1, 13]. The computer can provide a common frame of reference and can be used to support the development of ideas between children. However, neither learning nor collaboration will occur simply because two children share the same computer [13]. Numerous factors must be addressed, not least of which is the learner-machine interface. Today's technology is designed to support either one individual at one computer, or one individual collaborating with another individual at a different computer. However, much if not most, classroom computer use involves pairs or small groups sharing the same computer, especially in primary or elementary schools. What we have come to call *shoulder-to-shoulder collaboration*, as distinct from distributed collaboration, is not well supported with today's interfaces.

In this paper, we explore the design of storytelling technologies to help develop collaboration skills in children aged 5-7 years. This is a particularly interesting group to work with because previous research has shown significant changes in the ability to collaborate effectively within this age range [21]. Young children find it difficult to collaborate effectively. Informal observation of behavior in our project has found that the youngest children (aged 4 and 5) have the most difficulty in working collaboratively and cannot work effectively at all in groups greater than 2.

We introduce an approach to the design of shared interfaces that involves subtly *encouraging* children to explore the possibilities of collaborating, without forcing them to do so. The aim is to provide opportunities for children to discover the positive benefits of working together. This is achieved through the approach of 'tool mixing' where interface tools can be combined to give new effects. In one example, using two different colored crayon tools in close proximity on the screen produces a shaded area, filled with a mixture of the two colors.

Encouraging collaboration is more proactive than only *enabling* collaboration. Something new is gained by choosing to work together, although the children may work independently if they wish. On the other hand, it is not as rigid as *enforcing* collaboration, for example by demanding that two children have to synchronize their actions in order to succeed, an approach that has been tried before with some positive gains in terms of individual development [5]. The approach of encouraging collaboration is intended to combine the educational goal of learning collaboration skills with our design philosophy of giving children control as much as possible. We also suspect that long-term educational gains might be made when children discover collaboration for themselves.

From an HCI point of view, the terms encouraging, enabling and enforcing collaboration can be related to previous approaches to the design of shared interfaces. Early approaches such as "What You See is What I See" (WYSIWIS) enforced strict synchronization of different users' views onto a shared workspace [16]. Subsequent approaches such as relaxed-WYSIWIS [15], coupled with techniques for promoting multi-user awareness [11] and concurrency control mechanisms for interleaving users' actions [10] have focussed on enabling the possibility of collaboration while retaining a high degree of individual autonomy. The approach of encouraging collaboration lies somewhere between these two and so offers a new variant on approaches to designing shared interfaces.

The research described here has been carried out within the KidStory project, a collaboration between researchers, classroom teachers, and children (5-7 years old) from England, Sweden, and the United States. The goal of the project is to develop collaborative storytelling technologies for young children. The KidStory technologies are based on the approach of Single Display Groupware (SDG), where several children interact with a single display using multiple input devices, for example, two independent mice [6,4,12,18,17]. In its first phase, KidStory has worked with two pre-existing technologies, a shared drawing tool called KidPad [8] and a shared 3D environment called the Klump (an application of the DIVE collaborative virtual environment system [9]), both initially with one mouse and later with multiple mice. KidStory has used the methods of cooperative inquiry [7], to involve children as technology design partners in an intergenerational and interdisciplinary design team. To accomplish this, a year-long series of technology design sessions were conducted in two schools in England and Sweden involving more than 100 children.

The following section describes the initial KidStory technologies. We then introduce the approach of designing interfaces to encourage collaboration and describe its use in the redesign of KidPad and the Klump.

THE INITIAL VERSIONS OF KIDPAD and THE KLUMP

We have been working with two collaborative storytelling technologies, KidPad and the Klump. Both enable two or more children to create and tell stories together, but differ in style, KidPad being derived from drawing and the Klump from sculpting or modeling. In the following we describe them as they were at the start of this research, before being extended to encourage collaboration.

KidPad

KidPad is a shared 2D drawing tool that incorporates a zooming interface. Children can bring their stories to life by zooming between drawing elements (see Figure 1). Zooming and spatial structure lie at the heart of KidPad, since they enable children to add narrative structure to their stories by dynamically moving between different parts of a drawing. The creation of a story in KidPad, which involves creating links and zooming between picture/scenes or zooming deeper into the scene, is intended to allow the development of non-linear, complex structured stories. These story representations might make salient the links between scenes and the overall structure of the story. We anticipate that the focus of the children's attention on these features of the story structure will provide new opportunities for learning, in a different and complementary way to the creation of a story using more traditional drawing or word-processing packages.

The KidPad interface is designed around a series of graphical 'local tools' that children pick up and apply using

Figure 1: A sequence of views in KidPad as we zoom into a simple story (from left to right, and then top to bottom)

a mouse [3]. The tools are:

Crayons – different coloured crayons can be used to create drawing elements.

Arrow – a selection tool that can pick up and move objects.

Eraser – can be used to delete drawing elements.

Magic wand – can be used to create zooms between different drawing elements. The child selects the drawing element to be the start of the zoom followed by the destination element and sees an arrow linking the two.

Hand – can be used to activate zooms when the story is being told. Selecting the start point of the zoom initiates an animated zoom to the end point.

Turn alive – this tool animates a story element by causing its outline to ripple, making it appear to be alive.

Bulletin Board – this tool enables children to save stories to a bulletin board.

Toolbox – this special tool is used to organize the other tools, and can be opened or closed.

KidPad is a Single Display Groupware system, which means that it supports several mice plugged into a single computer. Two or more children can independently grab and use different tools at the same time using their own mice. Any free tool can be picked up and the children see each other's cursors. As a result, this initial version of KidPad could be said to *enable* collaboration – the children can choose to work together or individually. Figure 2 shows an example of the KidPad interface.

Figure 2: The initial version of KidPad showing all the toolboxes open at once with four simultaneous users.

KidPad is built on the Jazz[1] [2] and MID[2] open source Java toolkits. Jazz supports Zoomable User Interfaces by creating a hierarchical scenegraph for 2D graphics and MID supports multiple input devices for Java.

The Klump

In contrast to the drawing based approach of KidPad, our second storytelling tool, the Klump is based on a modeling approach. The Klump is a collaborative tool based around an amorphous 3D object (in fact, a textured deformable 3D polygon mesh) that can be stretched, textured and coloured and that makes sounds as it changes and is manipulated. Figure 3 shows an image of the Klump after it has been stretched and textured.

Figure 3: the Klump, a deformable 3D modeling object

As with KidPad, two or more children can manipulate the Klump at the same time. The Klump is intended to be an improvisational tool to help generate ideas in the early stages of story development. In our experience, the real-time exploration of the properties of the Klump provides a starting point for inspiring story characters and objects in a way that a blank page sometimes may not – it is an aid to creativity. The flexible and amorphous nature of the Klump is intended to inspire a wide range of stories. Again, by supporting synchronous multi-user access and by displaying the children's cursors to one another, the Klump enables collaboration. The initial version of the Klump can be manipulated in the following ways:

Stretching – a point on the surface of the Klump can be grabbed using the mouse and can be pulled to deform its shape. There is an option to switch between pulling a single vertex and a group of vertices, thereby changing the kind of deformation that occurs. The single vertex option pulls out a thin volume of the Klump, whereas the group of vertices pulls out a thick volume. There is also a button to return the Klump back to its original spherical shape.

Texturing – a variety of pre-defined textures may be applied to the surface of the Klump by selecting buttons on the interface. These textures allow different facial expressions to be added to the front side of the Klump,

---

[1] Jazz is available at http://www.cs.umd.edu/hcil/jazz

[2] MID is available at http://www.cs.umd.edu/hcil/mid

giving it a sense of character, and enable its background colors to be changed.

Rotating – the texture on the surface of the Klump can be grasped and rotated around to a new position.

Finally, the Klump makes a variety of sounds to reflect these different manipulations.

INTERFACES TO ENCOURAGE COLLABORATION

The core technical innovation of this paper is the idea of designing interfaces to encourage or invite children to collaborate. This has been motivated by our experiences of using the initial versions of KidPad and the Klump in two schools, one in Sweden and one in England, during the 1998-1999 school year as part of a program of activities that included:

- **contextual inquiry** – sessions to observe how children work with existing storytelling technologies (e.g., crayons and paper) and how they collaborate.
- **participatory design** – initial sessions to establish the children in the role of design partners and co-inventors of technology, followed by sessions with KidPad and The Klump aimed at eliciting specific design suggestions. These are reflected in the redesign of these technologies described later on.
- **evaluation of the technologies** – observations of how the children used the initial versions of KidPad and the Klump.

Over the course of the year, the combination of these activities has resulted in more than fifty sessions in schools involving more than one hundred five and seven year olds. At the peak of this activity, there were weekly participatory design and contextual inquiry sessions.

Children were observed with respect to collaborative behavior and their ability to use the technology to tell stories. Children and teachers were encouraged to provide feedback on these technologies that would instigate changes in design. Although after a few months, small-group and whole-class collaborative storytelling activities were being performed using these technologies, it was evident that some children found collaborating difficult. We frequently saw children competing or working alone instead of collaborating. They would not share their ideas to create a joint story, they would physically stop other children from using their input device, and would ignore or even delete each other's efforts.

Interfaces that encourage collaboration were proposed as a way of addressing this problem. Such interfaces should provide opportunities for children to discover the positive benefits of working together. Ideally, this should be achieved in as subtle and natural a way as possible, avoiding forced solutions. As noted in the introduction, encouraging collaboration is more proactive than only *enabling* it as was the case with the initial versions of KidPad and the Klump described previously. On the other hand it is not as extreme as strictly requiring collaboration, for example, demanding that two children have to press a button together to achieve an action, the approach that we described as 'enforcing collaboration'.

In its strictest interpretation, the approach of encouraging collaboration without enforcing it would require that a single child could achieve on their own any action that two children could achieve together, but that the two would do so in an easier, more efficient or more fun way. However, a more relaxed interpretation, is that a single child can carry out all of the major classes of action supported by the tool, but that by working together, two children can achieve subtle extensions to and variations on these actions. For example, a single child or two children working independently can create a functioning drawing in KidPad, but two children collaborating can create an enhanced one. This more relaxed approach is the one that we have adopted in revising KidPad and the Klump and several examples are given later in the paper. However, we first briefly digress to explore the more general relationship between the approach of encouraging collaboration and previous work on the design of shared interfaces.

Relationship to previous work on shared interfaces

Up to now, we have introduced the idea of interfaces that encourage collaboration within the context of educational applications. We now consider its broader relationship to CSCW technologies, especially how it compares to other approaches to synchronizing shared interfaces

How to synchronize shared interfaces has been a major concern for CSCW research. This has predominantly focused on distributed groupware where multiple users share a common workspace, for example a shared document, 2-D sketch tool or 3-D virtual world, using separate displays connected over a computer network. In such cases, the problem of synchronization can be broadly broken down into two parts.

How to synchronize what different users see? One of the first approaches was WYSIWIS (What You See Is What I See) where different users at different displays were forced to see the same part of a virtual workspace [16]. Experience with WYSIWIS led to less strictly coupled approach called relaxed WYSIWIS where different user's views could diverge [15]. Systems adopting this approach typically introduce additional functionality to support users in being aware of where others are looking and what they are doing. This may take the form of various awareness widgets, such as 'radar views' in 2D workspaces [11] or visible user embodiments ('avatars') in 3D systems [9].

How to synchronize object manipulations? Many CSCW systems allow users to collaboratively manipulate objects, changing their state. Examples include jointly editing a shared document or grasping and moving objects in a virtual world. This raises the problem of how to prevent conflicting updates. The most common solution is some form of locking, including simple turn-taking protocols, optimistic locking, non-optimistic locking and serialization protocols that allow participants to interleave their actions at various granularities [10]. Another option is social locking where given sufficient mutual awareness, user's may be able to negotiate mutual access with minimal system intervention.

We suggest that these various strategies can be located along a 'collaboration continuum' according to the extent to which they constrain individual autonomy and demand collaboration or leave users free to act independently. One extreme of this continuum involves what we have called *enforcing collaboration*, where the users are locked in step with one another. WYSIWIS and strict turn-taking can be found here. So can the work of Light, Foot and Colbourn, who modified the input of a standard computer so that two students had to enter information at the same time to succeed at a task [5]. A kind of dual key control was used. It was found that this enforcement of collaboration improved individual cognitive development. At the other extreme is what we have called *enabling collaboration*, where the users can act independently, are mutually aware and are free to coordinate their actions if they wish. Relaxed-WYSIWIS and social locking can be found here.

Our approach of *encouraging collaboration* lies somewhere between the two. It is not so strict as to require users to work together, but it provides some explicit motivation for them to do so in terms of added benefit. As noted earlier, encouraging collaboration can be interpreted in different ways. The case where a single user could achieve any action, but multiple users can achieve it in a way that is easier or more fun lies towards the enabling end of the continuum. The case where a single user can carry out each general class of action, but where multiple users can achieve enhanced actions lies towards the enforcing end.

It should be noted that a single CSCW system can use different approaches for different actions. For example, collaborative virtual environments often enable collaboration for viewpoint control (each user steers their own viewpoint, but is made aware of others' viewpoints through their embodiments), but enforce it for object manipulation (there is a turn-taking or coarse locking protocol regarding who can grab a virtual object).

This discussion raises the question of how the approach of encouraging collaboration might be applied in areas other than educational applications. One possible application area is in entertainment and games applications where participants might choose to collaborate, pooling abilities and resources to mutual benefit. Another more subtle approach might be in situations where participants can benefit by sharing costs. People increasingly have to pay for the use of network resources, for example in video and audio streaming. Users who agree to collaborate, for example to receive or manipulate the same information might be rewarded by sharing the costs between them.

REDESIGNING KIDPAD AND THE KLUMP TO ENCOURAGE COLLABORATION

We now describe how KidPad and the Klump were redesigned according to the lessons learned from the various schools sessions. Our overall strategy was to introduce design changes that satisfied two criteria:

- first they should encourage collaborative activity, reflecting the project's educational agenda and reacting to the observations noted previously.

- second, they should be based on the children's own design suggestions, emerging from the cooperative inquiry process.

Our general approach has been to use the more frequently occurring of the children's ideas as the basis for deciding on new functionality, but to realize this functionality through the approach of encouraging collaboration.

Redesign of KidPad

The basic approach that we followed in redesigning KidPad to encourage collaboration was to support 'tool mixing'. By this, we mean that when two (or sometimes more) children each use mixable tools at about the same time and place, the tools give enhanced functionality.

As a concrete example of this approach, consider the operation of the crayons in KidPad. The initial version provided three colors. A frequent design suggestion from the children was to provide more colors. We immediately added three more crayons, but that wasn't enough. Our final solution is to enable children to collaborate and combine their crayons to produce new colors. If two children draw with two crayons close together, then the result is a filled area between the two crayons whose color is the mix of the two. In this case, the children are not prevented from drawing as individuals, but they can gain additional benefit (new colors and filled areas) by working together. Applying our approach involves examining combinations of tools to look for interesting benefits and effects. We can consider all tools combined with themselves, for example, what happens when two selection tools are used together in KidPad? We also consider how tools combine with other tools, for example, what might happen if one child rotates the Klump while another stretches it? In each case, we look for effects that are natural and useful rather than contrived.

As described above, crayons in KidPad now work this way by drawing a filled in area between the two crayons using a color that mixes the two crayon's colors. By introducing collaborative color mixing, we added 15 mixed colors with the six crayons, and filled areas while encouraging collaboration and without adding any new tools. (see figure 4). Also, we added a special 'duplicating tool' that makes copies of other tools so several children could use the same tool type simultaneously. Figure 4 shows the redesigned interface with two children using mixed crayons.

Figure 4: Redesigned KidPad interface with mixed crayons being used. Note that inactive tools are faded. There are three active crayons, and two are currently being used to create a "mixed" area.

We built in mixing capability for multiple uses of all tools, except the magic wand and toolboxes. In every case, we tried to add a special behavior that acts as if it is a natural extension from the behavior with a single user. We felt this design ideal to be important in order to make it as easy as possible for children to anticipate what the mixed behavior might be. The mixing behavior we added is:

Crayons – As described above.

Arrow – Two or more children can squash and stretch selected drawing objects.

Eraser – One user can erase bits of a drawing object, but two children can erase an entire drawing object at once.

Hand – Two or more children can zoom in and out by moving their hands apart, or closer together, respectively.

Turn Alive – Two or more children can control the animation properties of a wiggling object by moving the turn alive tools closer together or further apart.

Redesign of the Klump

In redesigning the Klump to encourage collaboration, we have focused on combining the actions of stretching and texturing with themselves.

Stretching – the initial version of the Klump enabled toggling between two modes of stretching, pulling out a single vertex and pulling out a group of vertices. The revised version enables a single child to pull out only a single vertex on their own. However, if two children synchronously pull out two vertices that are close together on the Klump's surface, the result is to pull out a whole group of vertices. Thus, the added benefit of collaborating is to be able to make a different shaped deformation.

Texturing – our redesigned version of the Klump enables the children to apply a limited number of textures to its surface by pressing buttons. The textures represent happy and sad faces as well as background textures for the three primary colors. These may be applied independently so as to combine each of the two faces with the three background colors. However, by pressing some buttons together, the children may arrive at new combined textures. Three new faces become possible: laughing (pressing happy and happy), a kind of surprised expression (pressing happy and sad) and crying (pressing sad and sad). In addition, the background colors can be selected together to make new combined colors (similar to combining the crayons in the revised KidPad). A single user can also select the combined textures by selecting one button and then another a short time after (while the first is seen to rotate), but it requires speed and skill.

We have also extended the sounds made by the Klump to provide feedback as to when collaborative effects are being triggered, for example, by saying "cool" and "yippee".

Figure 5 shows the revised Klump interface. In the center we see the Klump, currently with its laughing face on a red background. To its left are the two buttons that are used to apply happy and sad face textures. To its rights are the three buttons for applying the colors. Above the Klump are two buttons that toggle between using a mouse for stretching and using it for rotating. The red button at the bottom returns the Klump to its original shape.

Figure 6 shows the difference between single-user and collaborative stretching. On the left we see the results of a single user stretching the Klump, pulling out a single vertex. On the right we see a collaborative stretch that pulls out a group of vertices, making a larger deformation.

Figure 7 shows the different facial expressions that can be obtained using the two buttons at the left of the interface. Faces 1 (happy) and 2 (sad) are obtained by a single user pressing the button. Faces 3 (laughing), 4 (surprised) and 5 (crying) are obtained when two users select combinations of the buttons at once (happy and happy gives laughing, happy and sad gives surprised, sad and sad gives crying).

Initial reflections on the revised interfaces

Although no formal program of evaluation has yet been carried out, the revised versions of KidPad and the Klump have been tested with a few groups of children.

The revised version of KidPad was introduced to our school in Nottingham. Pairs of children were given the common goal of recreating a well-known nursery rhyme. The children appeared to collaborate effectively, working on separate parts of the story and then joining together to use the collaborative tools to color in their picture.

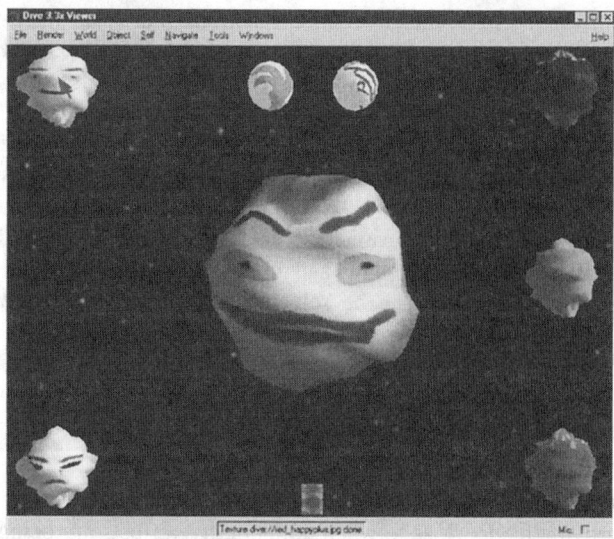

Figure 5: the revised Klump interface

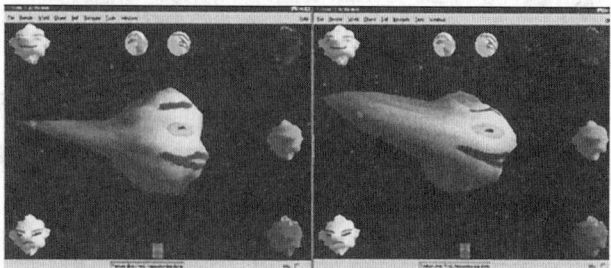

Figure 6: single user and collaborative stretching

Figure 7 : facial expressions for the Klump

Two children from the UK tested the re-designed version of the Klump. While the children explored features of the Klump, including the collaborative features, they did not show much interest in working together. This may in part, have been the result of them having no explicit 'shared goal'. This session, however, did raise an issue that should be considered when developing tools to encourage children's collaboration. When two young children carry out a collaborative action, the resulting effect has to be obvious and noticeably different from the effect displayed when the children carry out the action independently.

The revised versions of both KidPad and the Klump were also informally tested with a small group of children that are design partners at the University of Maryland's Human-Computer Interaction Lab. This formative evaluation showed that it took considerable experience with KidPad and the Klump for children to make use of the collaborative tools. For example, in a one-hour session where two boys (ages 10 and 8) used the Klump, it took almost 25 minutes for the children to discover the collaborative features. (These children on a previous occasion had used a less collaborative version of the Klump for a twenty minute session). They were then shown the collaborative features by an adult. In their comments afterwards said that they had enjoyed changing the faces and mixing colors.

Another formative study was carried out with six children (4 boys/2 girls; ages 7-10) using KidPad. For an hour and a half session, the three children who had previously worked with KidPad (a single-mouse version) showed strong differences in their use of collaborative tools, than the three other children who had never seen KidPad before. The children formed two teams, and each team worked on a computer with three mice. The children that already had used KidPad formed one group, and the children that hadn't used KidPad formed another group. After introducing KidPad and the new collaborative tools to the group, the children freely explored the tools for 20 minutes. Then, the children were asked to create a story with at least three 'scenes' to zoom to and from. The experienced children had little trouble creating a story. They collaborated throughout the process, making extensive use of the collaborative tools before starting the story, trying out the different possibilities. However, interestingly enough, they did not use the collaborative tool behaviors in the actual story creation.

The children that used KidPad for the first time had a harder time collaborating to create a story. They tended to experiment with the tools, including the collaborative tool behaviors. Most of what they did however was scribbling. This group found it hard to identify each other's cursors and to negotiate collaboration.

These early observations suggest that young children are able to use some of the collaborative features of KidPad and the Klump and that they can enjoy doing so. On the other hand, the way these features work has to be made more obvious in some cases. Furthermore, discovering them in the first place is a problem and they had to be pointed out by an adult on several occasions. On reflection, we realize that our designs only showed the results of collaborating, but did not highlight in advance when the possibility existed. We have therefore begun to revise KidPad and the Klump to more explicitly show the potential to collaborate. An example of this can be seen in figure 4 that is actually taken from the most recent version of KidPad. The two dots above the crayons are eyes that only appear when the crayons are close enough for the color mixing and filling to happen. We hope that indicating the potential to collaborate in advance will help the children discover collaborative features for themselves.

SUMMARY AND FUTURE WORK

In summary, we have proposed a new approach to designing shared interfaces that is intended to support children in learning to collaborate. The approach, called encouraging collaboration, allows children to work as individuals, but gives added benefits if they choose to work together. We have demonstrated this approach applied to the design of two storytelling technologies within the more general framework of cooperative inquiry within UK and Swedish schools. We have compared our approach with other user interface mechanisms from CSCW.

Future work will involve further design changes to KidPad and the Klump to reflect our early experiences. We will then undertake a more rigorous programme of evaluation including the development of a more intricate coding system, focusing on verbal and non-verbal collaborative behaviors, tracked from video recordings of the children and computer tracking of the children's interactions.

ACKNOWLEDGEMENTS

KidStory is funded under the ESPRIT i[3] Experimental Schools Environment initiative. We are deeply grateful to our partners at the Albany Infant School in Nottingham, England and at Rågsvedsskolan in Stockholm, Sweden. We would like to thank our summer evaluation team of children at the University of Maryland's HCIL.

REFERENCES

1. Barfurth, M.A. (1995) Understanding the Collaborative learning process in a technology rich environment: The case of children's disagreements. *Proc CSCL 1995.*

2. Bederson, B. B., & McAlister, B. (1999). Jazz: An Extensible 2D+Zooming Graphics Toolkit in Java., *University of Maryland Computer Science Tech Report #CS-TR-4015.*

3. Bederson, B. B., Hollan, J. D., Druin, A., Stewart, J., Rogers, D., & Proft, D. (1996). Local Tools: An Alternative to Tool Palettes. *UIST 96*, pp.169-170.

4. Bier, E. A., & Freeman, S. (1991). MMM: A User Interface Architecture for Shared Editors on a Single Screen. *UIST 91*, pp. 79-86.

5. Light, P., Foot, T., and Colbourn, C., (1997) Collaborative interactions at the microcomputer keyboard. *Educational Psychology*, 7, 1, 13-21.

6. Buxton, W., & Myers, B. A. (1986). A Study in Two-Handed Input. *CHI 86*, pp. 321-326.

7. Druin, A., (1999) Cooperative Inquiry: Developing New Technologies for Children with Children. *CHI'99*, pp. 223-230.

8. Druin, A., Stewart, J., Proft, D., Bederson, B.., & Hollan, J. D. (1997). KidPad: A Design Collaboration Between Children, Technologists, and Educators. *CHI 97*, pp. 463-470.

9. Fahlén, L. E., Brown C. G., Stahl, O., Carlsson, C., (1993) A Space Based Model for User Interaction in Shared Synthetic Environments, *InterCHI'93.*

10. Greenberg, S. & Marwood, D. (1994) Real Time groupware as a Distributed System: Concurrency Control and its Effect on the Interface, *CSCW'94.*

11. Gutwin, C. & Greenberg, S., (1998) Design for individuals, Design for Groups: Tradeoffs betwen Power and Workspace Awareness, *CSCW'98*, 207-217.

12. Inkpen, K., Booth, K. S., Klawe, M., & McGrenere, J. (1997). The Effect of Turn-Taking Protocols on Children's Learning in Mouse-Driven Collaborative Environments. *In Proc Graphics Interface (GI 97)* Canadian Information Processing Society, pp. 138-145.

13. O'Malley, C (1992) Designing Computer Systems to support peer learning, in *European Journal of Psychology of Education*, Vol. VII, No. 4, 339-352.

14. Rogoff, T. (1990) *Apprenticeship in Thinking: Cognitive development in social context.* Oxford University Press, Oxford.

15. Stefik, M., Bobrow, D., Foster, G. Lanning, S., & Tatar, D., (1997) WYSIWIS Revised: Early experiences with multi-user interfaces, *ACM TOIS*, 5(2), 147-167.

16. Stefik, M., Foster, G., Bobrow, D., Kahn, K., Lanning, S. & Suchman, L., (1987) Beyond the Chalkboard: Computer Support for Collaboration and problem Solving in Meeting, CACM, 30(1), 32-47.

17. Stewart, J., Bederson, B. B., & Druin, A. (1999). Single Display Groupware: A Model for Co-Present Collaboration. *CHI 99*, pp. 286-293.

18. Stewart, J., Raybourn, E., Bederson, B. B., & Druin, A. (1998). When Two Hands Are Better Than One: Enhancing Collaboration Using Single Display Groupware. *CHI'98 Extended Abstracts*, pp. 287-288.

19. Topping, K. (1992) Cooperative learning and peer tutoring: An overview. *The Psychologist*, 5(4), 151-157

20. Wood, D. & O'Malley, C., (1996) Collaborative learning between peers: An overview. *Educational Psychology in Practice*, 11(4), 4-9.

21. Wood, D., Wood, H., Ainsworth, S. & O'Malley, C., (1995) On becoming a tutor: Toward an ontogenetic model. *Cognition and Instruction*, 13(4), 565-581.

Storytelling with Digital Photographs

Marko Balabanović
marko@rsv.ricoh.com

Lonny L. Chu
lonny@ccrma.stanford.edu

Gregory J. Wolff
wolff@rsv.ricoh.com

Ricoh Silicon Valley
2882 Sand Hill Road, Suite 115
Menlo Park, CA 94025 USA
+1 650 496 5700

ABSTRACT
Photographs play a central role in many types of informal storytelling. This paper describes an easy-to-use device that enables digital photos to be used in a manner similar to print photos for sharing personal stories. A portable form factor combined with a novel interface supports local sharing like a conventional photo album as well as recording of stories that can be sent to distant friends and relatives. User tests validate the design and reveal that people alternate between "photo-driven" and "story-driven" strategies when telling stories about their photos.

Keywords
Digital storytelling, multimedia organization, digital photography, browsing.

INTRODUCTION
One of the most common and enjoyable uses for photographs is to share stories about experiences, travels, friends and family [2]. Almost everyone has experience with this form of storytelling, which ranges from the exchange of personal reminiscences to family and cultural histories. The World-Wide Web can facilitate the sharing of such stories in digital form, and has inspired a movement towards "digital storytelling." For instance, *Bubbe's Back Porch* [3] is a collection of family stories expressed as Web pages containing text and photographs, captured during a series of workshops—a grandmother's conversation as she makes soup, a grandfather's tale of how he met his wife. On a grander scale, national institutions such as the US Library of Congress store oral histories from, for instance, migrant farm workers in 1940s California [9].

The goal of our project is to support the sharing of digital photographs and associated stories. The design objectives were motivated by both formal research [2,7] and informal observations and interviews regarding photo usage. This background research revealed the need for a means of relating to digital photos as one typically relates to print photos. For example, all of us have had the experience of handing around a photo album while the photographer tells us the story behind the pictures, or receiving a couple of photographs in the mail with a short note from a friend or family member. In developing our prototype "StoryTrack" device, we imagined two specific scenarios which guided the design and embody both the imposed constraints and assumptions about user needs and priorities.

- Ben asks Fred about his recent trip to Alaska. Fred fetches his box of pictures and pulls out the most recent batch of photos. He flips through the pictures one by one, and relates interesting anecdotes associated with some of the photos. Occasionally Ben points at an image and asks about it. At one point Fred gets sidetracked and launches into a story about last year's trip to Canada. He looks back through his other photos to find a picture that illustrates the point.

- Fred's children Amy and Johnny want to send some pictures to Grandma. They pick out some pictures of themselves playing from the recent photos. While doing so they come across a funny picture of Dad in Alaska and add that in as well. In addition to creating a brief message to go along with their photos, they spend a lot of time recalling (and arguing) about what happened in each picture.

These two scenarios illustrate the two basic modes of interaction we address:

1. Sharing of stories locally, that is, with at least one other person present and viewing the same photos;
2. Sharing of stories remotely, that is, by sending someone a set of photos along with some commentary.

Existing tools do not naturally support these activities. As explained below, current form factors and user interfaces present barriers to sharing digital photographs. To overcome these limitations, we set out to build a prototype device that is easy to hold and pass around like a regular photo album, with an interface that requires no training to select, display and comment on a sequence of photos.

Current Approaches
With printed photographs, people spontaneously generate stories as they view pictures and interact with one another around the kitchen table, living room couch, or other

---
Permission to make digital or hard copies of all or part of this work for personal or classroom use is granted without fee provided that copies are not made or distributed for profit or commercial advantage and that copies bear this notice and the full citation on the first page. To copy otherwise, to republish, to post on servers or to redistribute to lists, requires prior specific permission and/or a fee.
CHI '2000 The Hague, Amsterdam
Copyright ACM 2000 1-58113-216-6/00/04...$5.00

communal space. However, digital photographs are generally viewed on personal computers that do not facilitate shared interactions. People can only create digital stories with special-purpose software requiring computer skills. (For our purposes, a story is an ordered set of photos with an accompanying audio narration.) Most digital photo album software (e.g., [10]) explicitly distinguishes between authoring and viewing stories. Hypermedia composition tools such as Isis [6] or MediaDesc [1] also support story creation, but require the user to focus on structures such as temporal constraints and hyperlinks rather than just telling a story. In contrast, our design does not distinguish between authoring and viewing modes thereby allowing people to view existing stories while simultaneously creating a new one.

Our work also differs in some respects from research on image retrieval and organization [5]. Digital photos have an advantage over print photos in that users can search for and retrieve them both by their content (e.g., features such as color and texture) and by their metadata (e.g., user-supplied text annotations). Much previous work has focused on such searching as a key aspect of working with digital photos. For example, the *FotoFile* system [7] provides a unified interface for annotation and search, using categories such as *people*, *places* and *events* that are commonly used for labeling photographs. With prints, however, people are generally limited to very simple search strategies—scanning through chronologically ordered images in an album or batch of photos. Our prototype derives great simplicity by operating on very similar principles without any sophisticated retrieval mechanisms.

Stories as an Organizational Metaphor

Stories form an intriguing organizational metaphor, blending aspects of chronological orderings and user-created groupings such as folders or directories in a file system. There are two kinds of stories represented in the StoryTrack:

1. Imported stories are the "rolls of film" with which a user starts. In the case of scanned prints, they might correspond to literal rolls of film. However, in the case of digital photos, they correspond to a set of photos downloaded from the camera in one session. Within an imported story, photos are ordered by time of creation.

2. Authored stories are selections of photos that have been grouped and ordered by the user.

The sets of imported and authored stories can themselves be ordered according to the time of creation of the stories. At any time, the story under construction has special status. When complete, it joins the set of authored stories. A photo appears exactly once in the set of imported stories, and can appear in zero or more authored stories.

Objects from other domains could equally be organized in this way. Imagine a Web browser that represents a user's browsing history as one or more imported stories, consisting of chronologically ordered Web pages. The authored stories then correspond to folders of bookmarks that a user creates, representing interests that change over time.

The remainder of this paper discusses the application of the story metaphor to digital photos. The following section describes our prototype device and explains how it attempts to meet the stated requirements. Next, the results of informal usability tests are presented, demonstrating the differences between anticipated and actual use of the device. The conclusion offers several suggestions for developers of digital photo albums and story sharing devices.

DESIGN OF THE DEVICE

Our design can be broken down into two components: the physical form of the device and users' interactions with it. We consider each in turn, and follow with a brief discussion of the prototype implementation.

Form Factor

Sharing photos in a natural setting requires a portable device that can be used in different locations throughout one's home. The device should also be large enough to show photos at a size similar to regular prints, viewable by more than one person. Initial investigation with mocked-up interfaces led to a design for two-handed usage that allows the device to be held easily, rested on one's lap or a table, or shared with another person. Figure 1 shows our current prototype being shared by two people.

Note that a typical personal computer cannot easily be shared. One user typically controls both the mouse and keyboard and has the best view of the monitor. In contrast, sharing photos requires that control pass easily from one user to another.

These design constraints led us to avoid using a touchscreen or pointing device since it is easier for two

Figure 1: The device in use

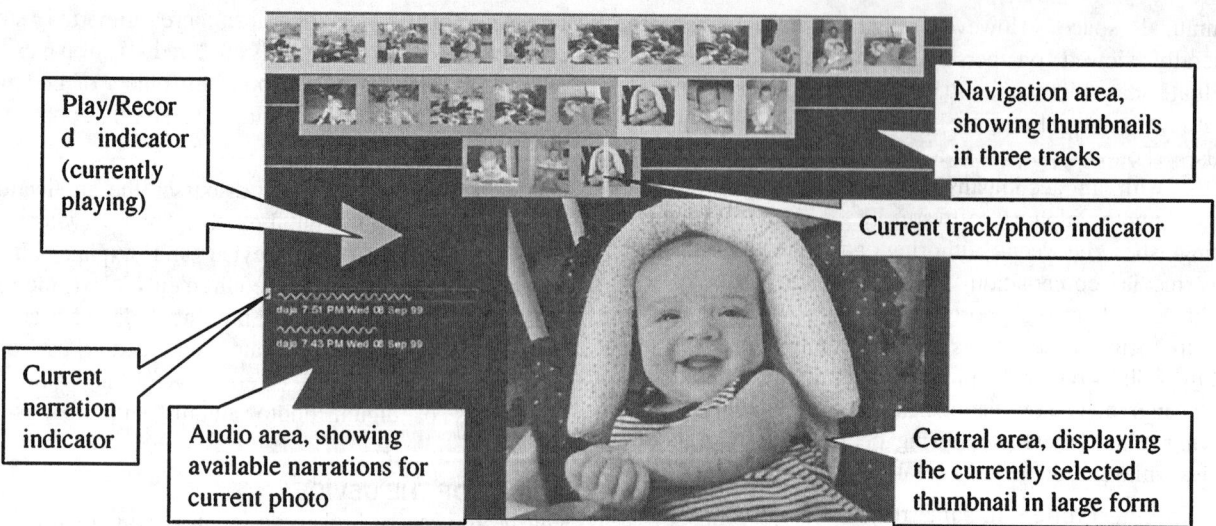

Figure 2: Screen layout

users to share a device when there is not a single locus of control. Furthermore, people point at pictures when talking about them. Using the same gestures to control the device might be confusing and produce unexpected behavior. Instead, all input controls are mounted on the edges of the device. As seen in Figure 1, this enables control to pass more fluidly between two users.

Interaction Design

The imagined usage scenarios presented quite a design challenge. At any point, a user may switch from recording an anecdote to viewing another set of images, may hand the device off to somebody else, or may simply browse through the latest photos. To accommodate such usage, we developed a novel interface based on the story metaphor.

Figure 2 shows the layout of the display. It is divided into three main regions. The large *central area* of the screen displays the currently highlighted thumbnail. The *audio area* shows available audio narrations for the highlighted photo. The *navigation area* at the top of the screen is a graphical representation for browsing and navigating through photos. Each of the three horizontal tracks of photo thumbnails serves a distinct purpose:

1. The top track shows the imported stories: all existing photos currently stored in the device, ordered chronologically[1]. Photos from digital cameras are ordered according to when they were taken, while images scanned from print photos are ordered by scanning time. Alternating background colors distinguish separate stories (corresponding to different rolls of film or download batches).

2. The middle track shows the authored stories: photos that have been grouped into stories by the user. Each story appears as a sequence of thumbnails. Stories are ordered according to their time of creation; the display again visually distinguishes separate stories using colored backgrounds.

3. The bottom track represents a story in progress: the "working set" of photos selected during the current session. A photo will appear in the bottom track if it has been added to the working set by pressing either of the + (add) or record buttons, as detailed below.

The permanent display of all three tracks enables an essentially modeless interface where a user can simultaneously view stories, see their original unedited set of photos, and see the story they are creating. The display also provides helpful context for viewing the current photo.

In a typical interaction, a user comes across a photo that is interesting and adds it to the working set, possibly also recording a related voice annotation. At the end of the session, the bottom track is grouped into a single story that is then appended to the middle track. We now explain the interaction in detail.

Navigation

Figure 3 labels the control buttons. The top two buttons on each side, easily accessible to the thumbs, scroll the current track either to the left or the right. A bright yellow vertical line (labeled in Figure 2) indicates the current track and the center thumbnail of this track. This selected photo is always also displayed in the large central area. When one of the scroll buttons is pressed, the current track shifts to the left or right. As a different thumbnail moves under the indicator, the corresponding photo is displayed.

Variable-speed scrolling allows the user to quickly traverse the photos on a track. At slow speeds, the display appears as shown in Figure 2. Faster scrolling speeds are enabled

---

[1] Since meaningful timestamps for the photos used in the user tests were not available, the prototype as illustrated does not display them.

Figure 3: Prototype device showing functions of buttons

by rendering only low-quality versions of the thumbnails (quickly accessible from a separate index), and by not rendering the central or audio areas at all. An earlier prototype used a center-weighted dial to control scrolling speed. The left and right buttons are better suited to the way the device is held, but cannot control scrolling as easily. At present, a single "click" of a scroll button moves the track by exactly one thumbnail. Holding down a scroll button for a longer time causes scrolling at increasing speed.

There are two remaining navigation buttons. The track selection button moves the indicator between the three different tracks. As illustrated in Figure 4, the expand/collapse button controls which of two different views of a track is shown:

- The expanded view, shown by default, where every photo is shown in thumbnail form.
- The collapsed view, where each story is represented by a single image, allowing faster navigation. The thumbnail of the first photo in a story is used to represent the story.

Story Playback
The cluster of buttons at the bottom left controls the creation and playback of stories. The play button starts playback from the currently selected thumbnail. If an audio narration exists, it is played through the built-in loudspeaker. Subsequently, or after a short pause if there is no audio narration, the currently selected track automatically scrolls forward to the next photo (we assume left-to-right as a default viewing and storytelling direction) and continues playing until the stop button is pushed or the end of story is reached. If the user navigates to a different photo while playing, playback will continue from that point.

Story Creation
Pressing the + (add) button appends a copy of the currently selected thumbnail onto the working set (the bottom track). The − (remove) button, conversely, removes the selected thumbnail from the working set.

Pressing the record button begins recording of an audio clip associated with the currently selected photo in the working set. If the photo is not already in the working set, it is appended before recording begins (as though the + button were pushed first). If the selected photo is already on the working track, the new recording overwrites any previous recording associated with it in the story under construction. The recording can be stopped with the stop button. If the user scrolls to a new photo while recording, the recording for the first photo terminates and a new audio clip is started that is associated with the new photo. Our aim is to make recording a story as similar as possible to viewing a story. However, to prevent accidental erasures, scrolling backwards automatically halts the recording.

We hypothesized that some users would compose a story by first selecting a working set of photos using the +/− buttons and then annotating each photo in order, a "select then narrate" strategy. We believed other users users would continuously record annotations while navigating and selecting photos. This "select while narrating" strategy is also supported: when recording is active, each new photo that a user views for longer than a short time interval is automatically added to the working set along with any narration.

Hyperlinks
The same photo may appear in many stories, with a corresponding audio narration for each. Whenever a photo is displayed, all of its associated narrations are displayed in the audio area. Figure 2 shows two available narrations for the current photo. Each is marked with the time of

 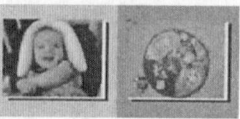

Figure 4: Expanded and collapsed views of two stories

recording and the name of the recording user. The length of the wavy lines is proportional to the duration of the audio. The narration associated with the currently selected story is listed first and is played by default when the play button is pressed. Pressing the play button multiple times in quick succession selects one of the alternate audio clips and playback "jumps" to the corresponding story, providing a kind of automatic hyperlinking between stories.

Saving Stories
The final group of buttons, at the bottom right of Figure 3, controls story operations. The save button "saves" the current working story (on the bottom track) by moving it to the end of the middle track. Note that the current state of the device is also saved to disk at that time, matching users' expectations about "save" and persistent storage.

The user would also have the option of electronically sending a completed story to another user for viewing on a similar device or on a regular PC via a media player application or standard Web browser. However, the implementation of this feature was not part of the user testing reported here.

Design Summary
This interaction design achieves our primary goals for modeless sharing of digital photos and stories. A user can seamlessly switch between browsing, viewing, authoring, and showing. In addition, by removing the complexities of dragging, menu selection and other artifacts of windows-based interfaces, we arrived at something that we hoped would be intuitive and simple to use in conjunction with our hardware controls.

Implementation
The prototype described here is based on a portable tablet computer (a Fujitsu Stylistic 2300) augmented with several specially built control buttons. Although fairly heavy at 3.9lbs (1.8kg), we expect future prototypes to be lighter as hardware technology advances. The screen viewing angle permits up to two users seated side by side, and the machine provides sufficient battery life and computational resources for testing our design. Photos displayed in the central area are 4.5x3" (11.4x7.6cm), about 60% of the size of standard 4x6" photo prints.

The software is implemented using Java 2 on Windows 98, with additional native libraries for audio I/O. The control buttons are simple pushbutton switches connected to the tablet's serial port via a BASIC Stamp 2 microcontroller. Relative to other input devices, these buttons are very easy to use, robust, and provide sufficient functionality for the desired interaction. For sound recording, a microphone is attached to the external body of the tablet, while the built-in speakers are used for playback.

Stories and metadata about photos are stored on the tablet's hard drive in Extensible Markup Language (XML)[2]. This allows for easy translation to Hypertext Markup Language (HTML) or Synchronized Multimedia Integration Language (SMIL) so stories can be shared with others who do not have a StoryTrack. The SMIL format is especially appropriate as it allows the synchronization of audio with a series of images, exactly matching the structure of our stories.

Digital photos can be loaded onto the device from digital cameras by inserting a flash memory card, or may be downloaded through a wireless network connection.

OBSERVATIONS OF USAGE
To evaluate the efficacy of this design for sharing digital photos, we conducted an informal usage test [4,8]. Subjects were presented with the device in a natural setting and encouraged to think aloud while using it. Observation of the user interaction helped identify features of the interface that did or did not work well. Moreover, it provided valuable insight into the ways that people use photos to tell stories.

Usage Study Description
We observed nine sessions of subjects. In seven of these sessions, we observed pairs of subjects consisting of one primary (**P**) and one secondary user (**S**). The remaining two sessions involved a single primary user. For each session, the StoryTrack was preloaded with 2–3 recent sets of photos provided by **P**. We intended that **P** treat the device as its owner. Eight subjects provided prints that we scanned, one subject provided photos taken with a digital camera. The total number of photos provided by each subject ranged from 45 to 234, with half of the subjects falling in the 45-50 range (the user with a digital camera provided the 234 photos). Note that on initial introduction to the device the first track included all of **P**'s photos; the second and third tracks were blank.

Each session was divided into three stages: initial exploration, sharing, and feedback.

- In the initial exploratory stage, **P** was given a brief introduction to the device without explanation of the

---

[2] XML, HTML and SMIL are all standards of the World-Wide Web Consortium, available via http://www.w3.org

interface. **P** then played with the device for 10–15 minutes without any further input from the experimenters. At the conclusion of this stage, a written instruction sheet was provided detailing the functions of each of the control buttons as well as an overview of the interface. At this point subjects were welcome to ask any questions regarding the basic functionality of the device, but were not provided answers regarding usage strategies. For example, we would explain that the + button added an image to the current working track, but would not explain how to use the + button along with the recording capabilities to create audio-annotated stories. Once **P** felt comfortable with the device, **S** was brought into the room and seated adjacently. If desired, the device was reset, erasing any stories created during the first stage.

- In the second stage, lasting 20–30 minutes, **P** was assigned two tasks. First, to share photos locally with **S**. Second, to create a story that could be sent electronically to any specified recipient (although note that for these tests no stories were actually sent). In the two single-subject sessions, the local sharing task was skipped.

- The third stage consisted of a series of follow-up questions directed at both subjects, broken down into three discussion topics. First, comments on the interface (e.g., describe usage difficulties, confusing aspects of the interface, overall likes and dislikes about StoryTrack). Second, comparisons of the StoryTrack experience with the users' current handling of print or digital photos (e.g., methods of organization, methods of sharing). Finally, feedback regarding the storytelling aspects of StoryTrack.

Two of the tests were conducted in subjects' homes, the remainder in an office library. With comfortable seats and no office furniture, the library somewhat resembles a home environment. All sessions were videotaped.

Usage Study Results

Overall we were pleased to find that subjects could operate the prototype with little or no instruction. Eight of the nine primary users were able to create sequences of photos and "save" them onto the second track during the initial exploratory stage. At the end of the first stage subjects asked few questions and spent very little time, less than 30 seconds on average, reviewing the instruction sheet. Primary users had no difficulty demonstrating the device to the secondary users without experimenter help. Everyone intuitively understood the use of the three tracks although they did not necessarily label the groupings as "stories." Subjects saved 2–8 stories each. Some of these were unintentional, occurring during the exploration stage, and some did not include audio recording. Most of the stories ranged in length from 3–7 photos. We examined these saved stories along with unsaved browsing behavior in our analysis.

Physical Interaction and Form Factor

The primary complaints about the physical nature of the device referred to its weight and bulkiness. Our prediction during the design of StoryTrack was that subjects would show photos to a local audience by holding the tablet in both hands while pressing the buttons. However, the weight of the tablet and size of the BASIC Stamp assembly (a box attached to the underside of the tablet) made it difficult for subjects to lift and maneuver the device comfortably during usage. All of the primary subjects dealt with these problems by resting the tablet on their laps and rotating it as needed to show the other subject the photos.

By holding the tablet in this position, the subjects maintained easy access to the button controls. Many subjects mentioned that the buttons were conveniently placed so that it was easy to simultaneously hold and control the device. In several of the multiple-subject sessions, one manipulated the controls along the right edge of the device while the other manipulated the controls along the left edge. Occasionally, the device was handed over to the other subject. Subjects frequently used one hand to point to images on the screen. This is similar to a way of browsing through print photo albums—one person holds one side of the album, another person holds the other side and both of them can point to photos.

Subjects had no difficulty operating the control buttons although very few subjects discovered and used the fast scroll capability. Rather than hold down the scroll button, subjects would press it multiple times quickly to advance through a track. Since many subjects asked for a fast method of advancing through the tracks (especially returning to the beginning of a track) we suspect that an alternative input device, such as a thumb wheel or pressure-sensitive pad, may be more effective at facilitating variable-speed scrolling. The buttons used in this study give a satisfying tactile click when pressed, and so create the expectation of a discrete rather than a continuous action.

Software Interaction

As already explained, subjects had few difficulties with basic navigation and the saving of stories. The experimenters did explain the save function to one subject and the operation of the recording function to two subjects. All other subjects learned to operate the basic functions of the device during the exploratory phase without experimenter help. The expand/collapse feature (Figure 4) occasionally caused confusion, due in part to the alternating background colors. Upon first use, some subjects thought most of their photos had disappeared. However, the meaning usually became clear after pressing the button again (to expand the photos) and several subjects achieved faster navigation using the expand/collapse function.

The record function also caused confusion for several subjects. A number were surprised to hear their own voices while experimenting with the play button after having inadvertently recorded some narration. Subjects uniformly favored the "select then narrate" strategy, preferring to add photos using the + button and explicitly start and stop audio recording. In general, few subjects used audio recording except when creating a story to be sent.

Several subjects requested additional functions for editing stories. Users wanted to insert photos into the middle of stories, and to bring saved stories back down to the bottom track for editing.

Sharing and Storytelling

During the two-person tests many stories about photos were told. As observed in [2], it is socially inappropriate to be silent while showing your photos to someone. There appeared to be two different styles of storytelling:

1. *Photo-driven*—the subject explains every photo in turn, the story prompted by the existing sequence of pictures. Narration often comprises a sequence of sentences of the form "This is my wife," "This is my parents at home," etc. This corresponds to the well-documented use of picture-taking to preserve memory and aid recall [2].

2. *Story-driven*—the subject has a particular story in mind (e.g., "Zachary's first camping trip"), then gathers the appropriate photos and recounts the story.

Rather than sticking to one or another of these styles, subjects would segue from one to the other. A familiar photo would remind them of a particular story (shifting from photo-driven to story-driven), or in the midst of telling a story an unexpected photo would come up (shifting from story-driven to photo-driven). For example, one subject was creating a story about a camping trip until she came across a Thanksgiving photo. This photo received a brief commentary before she moved on to creating a new story about a musical performance.

Table 1 summarizes the characteristics of each strategy. All subjects started with a photo-driven style when showing photos to a local audience and were more likely to use story-driven strategies when assembling photos to send to a remote audience. Note that there was often not a clear distinction between photo driven and story driven usage. People normally expect photos to be in chronological order and often explain the sequence of events as they go through each photo. At least one subject created a story for local sharing that was simply a re-ordering of the photos in correct chronological order.

When sending photos to a remote friend or family member, subjects first selected a set of photos, using a combination

	Local sharing	Remote sharing
Photo-driven	Use mainly top track Scroll forward and comment on each photo Lots of conversation but little or no audio recording Working set used to reorder photos chronologically	Use mainly top track Scroll forward and add interesting photos to working set Record annotation after all photos selected
Story-driven	Usually prompted by a particular photo Search for other photos to illustrate anecdote Related photos added to working set	Search for particular/related photos and add to working set Narrate story after selecting all photos

Table 1: Observed interaction types

of photo-driven and story-driven browsing, then recorded an accompanying audio narration.

These observations shed some light on the relative merits of browsing and search. If subjects cannot think of a story to tell (or a photo to search for) until prompted by another photo, then browsing has to precede search. Furthermore, since subjects often move between photo-driven and story-driven styles, it is important to support both without context switching in the interface. Whether these styles were the preferred strategies or due to characteristics of the prototype remains to be determined. However, subjects seemed to enjoy using the device and did not ask for or mention the lack of search or other retrieval tools.

Audio clearly plays a big role in sharing photos. All subjects talked a great deal while showing photos to their partners in the study. Subjects did not record these conversations, nor did they record audio annotations for their own use. However, they did record narrations for sequences of photos to be sent to a friend or family member ("Hi Laura, here are some photos…"). It seems quite likely that the use of audio may change with experience as users become accustomed to multimodal albums. Indeed, our youngest tester (age 6) had a very different way of composing a story. Favorite pictures were added multiple times, and the voice annotation consisted of sound effects such as "Splash!" and "Neee-haw!"

Comparison to Current Behaviors

Subjects' strategies for managing their current photo collections influenced their use and perception of StoryTrack. All subjects reported using one or more of three different organizational tools:

- The "shoebox:" A disorganized container for all sets of photos, possibly in approximate chronological order. Apart from actual shoeboxes, closets and desk drawers were commonly cited containers.
- The album: A carefully selected and ordered set of photos, presented in an album.
- The Web site: Subjects with digital cameras or scanners select a small number of photos at regular intervals to post on a personal Web site, in order to share them with friends and relatives.

Approximately one third of the subjects use the shoebox exclusively, one third use a combination of the shoebox and the album, and one third use a combination of the album for print photos and the Web site for digital photos.

In many cases, the albums and Web sites include short text annotations describing the photos. This is the conventional way to record stories with personal photos. It may also help explain why subjects showed a preference for "select then narrate" over "select while narrating," since the process of annotating a print album or Web site typically occurs after the photos have been selected.

Different advantages were cited for the StoryTrack depending on the subject's current organizational methods:

- "Shoebox" subjects liked the idea that all of their photos would be easily browsable and all in one place;
- Album creators liked the fact that a photo could be in more than one story;
- Subjects with digital photos and Web sites liked the fact that now they could share these pictures without having to sit around the computer screen, which was not seen as a sociable activity.

CONCLUSIONS

The StoryTrack device demonstrates that digital photos can be used to support some of the same kinds of story sharing that people enjoy with print photos. It also provides a convenient way of recording stories and sending them to family and friends, much more easily than is possible with conventional albums or tools.

The novel "three track" interface enables a very clean design that was easy to use for all of our test subjects (including the 6-year-old). In less than 15 minutes of using the device, people very naturally mixed browsing, composition, and annotation of photos while seamlessly switching between "photo driven" and "story driven" strategies. It remains to be seen whether these simple navigational tools and story metaphors will suffice for very large collections of photos. Looking ahead, we are curious whether access to this kind of device would alter the quantity or types of photos people take.

The lessons learned from the design and testing of the StoryTrack prototype are summed up with the following observations for developers of digital photo albums and story sharing devices:

- A shareable device with shareable controls facilitates spontaneous interpersonal interaction;
- Browsing linear structures allows easily understood controls and interaction, providing sufficient functionality for at least hundreds of images;
- Users constantly mix creating, viewing and telling stories. Modeless interfaces that simultaneously support these activities should be preferred.

ACKNOWLEDGMENTS

We thank all of the testers, and also Zachary Phillips (age 6 months), whose likeness is featured in all of the figures.

REFERENCES

1. Caloini, A., Taguchi, D., Yanoo, K. and Tanaka, E. Script-free scenario authoring in MediaDesc, in *Proceedings of ACM Multimedia '98* (Bristol, UK, 1998), ACM Press, 273-278.
2. Chalfen, R. Snapshot versions of life. Bowling Green State University Press, Bowling Green OH, 1987.
3. Don, A. Bubbe's Back Porch. Available at http://www.bubbe.com
4. Gomoll, K. Some techniques for observing users. In Laurel, B., editor, *The Art of Human-Computer Interface Design*, Addison-Wesley, Reading MA, 1990, 85-90.
5. Idris, F. and Panchanathan, S. Review of image and video indexing techniques. *Journal of Visual Communication and Image Representation 8*, June 1997, 146–166.
6. Kim, M.Y. Creative multimedia for children: Isis story builder, in *Proceedings of CHI '95* (Denver CO, May 1995), ACM Press, 37-38.
7. Kuchinsky, A., Pering, C., Creech, M.L., Freeze, D., Serra, B. and Gwizdka, J. FotoFile: A consumer multimedia organization and retrieval system, in *Proceedings of CHI '99* (Pittsburgh PA, May 1999), ACM Press, 496-503
8. Nielsen, J. Guerilla HCI: Using discount usability engineering to penetrate the intimidation barrier. In Bias, R.G. and Mayhew, D.J., editors, Cost-Justifying Usability, Academic Press, Boston MA, 1994, 245-272
9. Todd, C.L., and Sonkin, R. Voices from the dust bowl. Available at http://memory.loc.gov/ammem/afctshtml/tshome.html
10. Yagawa, Y., Iwai, N., Yanagi, K., Kojima, K. and Matsumoto, K. The Digital Album: A Personal File-tainment System. In *Proceedings of MULTIMEDIA '96* (Hiroshima, Japan, June 1996), IEEE Computer Society Press, 433-439.

Video Figures

The following papers of the *CHI 2000 Conference Proceedings* have video figures which can be found at the end of the *CHI 2000 Video Program*:

HandSCAPE: A Vectorizing Tape Measure for On-Site Measuring Applications (p137)
Jay Lee, Victor Su, Sandia Ren, Hiroshi Ishii, *MIT Media Lab*

The Task Gallery: A 3D Window Manager (p 494)
George Robertson, Maarten van Dantzich, Mary Czerwinski, Ken Hinckley, David Thiel, *Microsoft Research*
Daniel Robbins, Kirsten Risden, Vadim Gorokhovsky, *Microsoft*

The Cubic Mouse: A New Device for Three-Dimensional Input (p 526)
Bernd Fröhlich, John Plate, *GMD*

Designing Storytelling Technologies to Encourage Collaboration Between Young Children (p 556)
Steve Benford, Victor Bayon, Rob Ingram, Helen Neale, Claire O'Malley, Danaë Stanton, *University of Nottingham*
Benjamin Bederson, Allison Druin, Juan Pablo Hourcade, *University of Maryland*
Karl-Petter Åkesson, Pär Hansson, Kristian Simsarian, *SICS*
Yngve Sundblad, Gustav Taxén, *Royal Institute of Technology*

CHI 2000 Video Program

Copies of the *CHI 2000 Video Program* are available in VHS (PAL or NTSC) and may be ordered prepaid from:

ACM Member Services
1515 Broadway
New York, NY 10036, USA
Telephone: +1 212 626 0500
Fax: +1 212 944 1318
Email: acmorder@acm.org

NTSC Video
ACM Order Number: 608004
ISBN Number: 1-58113-246-8

Pal Video
ACM Order Number: 608003
ISBN Number: 1-58113-247-6

Author Index

Abowd, Gregory 368
Åkesson, Karl-Petter 556
Anderson, John 273
Arsenault, Roland 408
Audia, Steve 486
Avery, Brian 313
Bälter, Olle 105
Balabanović, Marko 564
Balakrishnan, Ravin 33
Baumeister, Lynn 502
Bayon, Victor 556
Beaudouin-Lafon, Michel . . 446
Bederson, Benjamin 556
Benford, Steve 233, 556
Bhatti, Nina 297
Bly, Sara A. 454
Boreczky, John 185
Bouch, Anna 297
Bouwhuis, Don 376
Brereton, Margot 217
Brewster, Stephen 415
Brown, Barry 438
Burmester, Michael 201
Burnette, Tommy 486
Buyukkokten, Orkut 430
Byrne, Mike 502
Cadiz, Jonathan 392
Campbell, Christopher S. . . 241
Card, Stuart K. 153
Chang, Bay-Wei 249
Chen, Hao 145
Cheverst, Keith 17
Chi, Ed H. 153, 161
Chiselko, Greg 478
Christian, Andrew 313
Christiansen, Kevin 486
Chu, Lonny 564
Churchill, Elizabeth 454
Considine, Michael 321
Conway, Matthew 486
Corbett, Albert 97
Cosgrove, Dennis . . . 478, 486
Cubranic, Davor 454
Czerwinski, Mary 494
Damm, Christian Heide . . . 518
Dantzich, Maarten van . . . 494
Davies, Nigel 17
Davis, Richard 89
Deline, Rob 486
Dennerlein, Jack Tigh 423
Donath, Judith 81
Drucker, Steven 462, 470
Druin, Allison 556
Dumais, Susan 145
Durbin, Jim 486
Efstratiou, Christos 17
Eggen, Berry 376
Ehrlich, Sheryl M. 193
Espinosa, Alberto 392
Faltings, Boi 289
Farnham, Shelly 462, 470
Fels, Sidney 540
Frei, Phil 129
Friday, Adrian 17
Friedman, Alinda 113
Fröhlich, Bernd 526
Frøkjær, Erik 345
Garcia-Molina, Hector . . . 430
Gaver, Bill 209
Giller, Verena 548
Girgensohn, Andreas 185
Golovchinsky, Gene 185
Gorokhovsky, Vadim 494
Gossweiler, Rich 153, 486
Gray, Philip 415
Greenhalgh, Chris 233
Grudin, Jonathan 177, 384
Gupta, Anoop . . 169, 177, 384
Hansen, Klaus Marius 518
Hansson, Pär 556
Hassenzahl, Marc 201
Hasser, Christopher 423
He, Liwei 169, 177
Hertzum, Morten 345
Hinckley, Ken 33, 494
Hirschberg, Julia 89
Hong, Jason 510
Hornbæk, Kasper 345
Hourcade, Juan Pablo 556
Hudson, Scott 368
Humburg, Judee 337
Ingram, Rob 556
Isbister, Katherine 57
Ishida, Toru 57
Ishii, Hiroshi 129, 137
Jacob, Robert 265, 281
Jancke, Gavin 384
Jensen, Carlos 470
John, Bonnie 502
Kim, Jinwoo 257, 305
Kobayashi, Motoki 121
Kobayashi, Yoshinori 121
Koga, Shuichi 486
Koike, Hideki 121
Koleva, Boriana 233
Kollock, Peter 470
Korhonen, Panu 9
Kraut, Robert 392
Kuchinsky, Allan 297
Lai, Jennifer 321
Landay, James 360, 510
Lautenbacher, Glenn 392
Lee, Jay 137
Lee, Jungwon 305
Lehner, Katrin 201
Levin, Golan 225
Li, Francis 169
Lie, Kin Pou 41
Lieberman, Henry 65
Lin, James 510
Long, Chris 360, 486
MacKenzie, Christine L. . . 532
MacKenzie, I. Scott 9
MacLean, Karon 225
Mackinlay, Jock 249
Maglio, Paul 241
Mäkelä, Anu 548
Mallory, Beth 486
Mankoff, Jennifer 368
Martin, David 423
Martin, Heather 209
Masliah, Maurice 25
McClard, Anne 1
McGaffey, Aaron 113
McGarry, Ben 217
McGee, Marilyn Rose 415
Miale, Steve 486
Michiels, Joseph 360
Mikhak, Bakhtiar 129
Milgram, Paul 25
Min, Kwan 329
Mitchell, Keith 17
Monkaitis, Kristen 486
Moon, Jae Yun 305
Muller, Urs 89
Myers, Brad 41
Nakanishi, Hideyuki 57
Nass, Clifford 57, 329
Neale, Helen 556
Nelson, Lester 454
Newman, Mark 510
Nonnecke, Blair 73
O'Hara, Kenton 438
O'Malley, Claire 556
Oakley, Ian 415
Paepcke, Andreas 430
Park, Joonah 257
Patrick, Emilee 478
Patten, James 486
Pausch, Randy 486
Pauws, Steffen 376
Pierce, Jeff 486
Pirolli, Peter 161
Pitkow, James 153, 161
Plate, John 526
Platz, Axel 201
Poupyrev, Ivan 540
Preece, Jenny 73
Pu, Pearl 289
Rahardja, Krishnawan 193
Reeves, Byron 49
Regli, Susan Harkness . . . 249
Ren, Sandia 137
Rickenberg, Raoul 49
Rico-Gutierrez, Luis 392
Risden, Kirsten 494
Robbins, Daniel 494
Robertson, George 494
Rode, Jennifer Ann 478
Rodenstein, Roy 81
Rohn, Janice Anne 337
Rosenbaum, Stephanie . . . 337
Rowe, Lawrence 360
Rowe, Simon 400
Rui, Yong 169
Salvucci, Dario 273
Sanocki, Elizabeth . . . 169, 177
Sato, Yoichi 121
Scherlis, William 392
Schiano, Diane J. 193
Schnädelbach, Holger 233
Sefelin, Reinhard 548
Sellen, Abigail 438
Sheridan, Kyle 193
Shochet, Joe 486
Sibert, Linda 281
Silfverberg, Miika 9
Simsarian, Kristian 556
Slavkovic, Aleksandra 478
Smith, Marc 462
Snibbe, Scott 225
Somers, Patricia 1
Spencer, Richard 353
Staack, David 486
Stanton, Danaë 556
Stearns, Brian 486
Stoakley, Richard 486
Sturgill, Chris 486
Su, Victor 129, 137
Sundblad, Yngve 556
Tanriverdi, Vildan 265
Taxén, Gustav 556
Taylor, Michael 400
Thiel, David 494
Thomsen, Michael 518
Tobita, Hiroaki 121
Trask, Holly 97
Trevor, Jonathan 454
Tscheligi, Manfred 548
Uchihashi, Shingo 185
Verratti, Thom 478
Viega, John 486
Vivacqua, Adriana 65
Wang, Yanqing 532
Ware, Colin 408
Watson, Benjamin 113
Weghorst, Suzanne 540
White, Jeff 486
Whittaker, Steve 89
Williams, George 486
Winograd, Terry 430
Wolff, Gregory 564
Wood, David 321
Woodruff, Allison 153
Yang, Bo-Chieh 41
Zellweger, Polle 249

Keyword Index

3D 532
- 3D book 153
- 3D interfaces 408, 494, 540
- interactive 3D graphics 486
- interactive 3D rotations 540

6
- 6 degree-of-freedom control 25
- 6 degree-of-freedom input devices .. 540

abstraction
- video abstraction 177

access
- speech access 89

activation
- spreading activation 153

actors
- computers as social actors 329

adaptive hypermedia 17

added
- gaze-added interfaces 273

affect 193
- facial affect 193

affective computing 193

agents 65
- social agents 49
- social interface agents 57

allocation of control 25

ambiguous input 368

analysis
- empirical analysis 462
- log file analysis 462

animated characters 49

animation 400
- animation authoring tools 486

annotation 89

appliances 438
- information appliances 209
- internet appliances 1

applications
- on-site applications 137

architecture
- client-server architecture 289

assistants
- personal digital assistants 41, 430

asynchronous
- asynchronous communication .. 89, 454
- asynchronous work 392

attraction
- similarity-attraction effect 329

audio 81

augmented
- augmented environments 532
- augmented reality 121, 217, 233

authoring
- animation authoring tools 486

avatars 400, 462
- talking avatars 313

aware
- orientation-aware 137

awareness
- awareness devices 392
- context-awareness 17
- task awareness 392

BBS 73

bibliometrics 153

book
- 3D book 153

browser 430

browsing 257, 564
- video browsing 169, 177, 185

cameras
- digital cameras 438

capture
- information capture 438

CASA 329

CASE tools 518

categorization 145

CE
- Windows CE 41

change
- organizational change 337

characters
- animated characters 49

chat
- graphical chat 462
- sticky chat 454
- text-based chat 454

children 129, 548, 556

classification 145

client-server architecture 289

cognition
- spatial cognition 494

cognitive
- cognitive map 478
- cognitive models 217
- cognitive walkthrough 353

collaboration 454, 470

collaborative
- computer supported collaborative learning 556

commerce
- electronic commerce 305

communication
- asynchronous communication .. 89, 454
- computer-mediated communication 81, 392
- cross-cultural communication 57
- nonverbal communication 193
- synchronous communication 454
- wireless communication 548

community
- online community 462
- virtual community 462

components
- hedonic components 201

comprehension 321

compression
- time compression 169

computer
- computer-mediated communication 81, 392
- computer supported collaborative learning 556
- computer supported cooperative work 556
- computer supported learning 121
- computer vision 121

computers
- computers as social actors 329
- hand-held computers 1, 41
- pen-based computers 510

computing
- affective computing 193
- mobile computing 17
- pen-based computing 1
- ubiquitous computing 41

concept
- product concept design 548

conceptual design 209

constraint solver 289

container 225

context
- context-awareness 17
- context information 257
- focus + context 249
- task context 532

continuous 225

control
- 6 degree-of-freedom control 25
- allocation of control 25
- locus of control 49
- motor control 25, 540

controls and displays 532

conversations 454

cooperative
- computer supported cooperative work 556
- cooperative design 518

coordination 25

corporate planning 337

cost
- transaction cost 305

cross-cultural communication 57

CSCL 556

CSCW 454, 470, 556

cultural
- cross-cultural communication 57

customer
- customer interface 305
- customer loyalty 305

cyberspace
- social cyberspace 462

data
- data mining 161
- speech as data 89

degree
- 6 degree-of-freedom control 25
- 6 degree-of-freedom input devices .. 540
- degree of interest 153

degrees of freedom 532

demographic 73

design
- conceptual design 209
- cooperative design 518
- design process 225

Keyword Index

design research 209
design thinking 217
instructional interface design 97
interaction design 81, 217
interface design 376
product concept design 548
user interface design 17, 241, 313
web design 510
devices
   6 degree-of-freedom input devices .. 540
   awareness devices 392
   input devices 25, 137
   pointing devices 423
diagram
   Venn diagram 121
diary study 438
difficulty
   index of difficulty 423
digital
   digital cameras 438
   digital images 548
   digital photography 564
   digital video 169
   digital video library 177
   personal digital assistants 41, 430
digraph frequencies 9
dilemma
   social dilemma 470
dimensional
   multi-dimensional scaling 360
direct manipulation 446
discrete 225
discussion list 73
display
   single display groupware 556
displays
   controls and displays 532
   head-mounted displays 478
   information displays 313
distributed work 392
docking 532
   virtual docking task 25
document use 438
documents
   fluid documents 249
   shared documents 454
dome tree 161
drawing 81
dual-task tradeoffs 241
dynamic
   studies of dynamic user interfaces .. 249
eCommerce 289
education 121, 129, 556
effect
   similarity-attraction effect 329
effectiveness 345
efficiency 345
electronic
   electronic commerce 305

electronic whiteboards 518
email 73, 105
embodiment
   tele-embodiment 233
emotion 193
   facial expression of emotion 193
emotional usability 201
empirical
   empirical analysis 462
   empirical evaluation 89
entry
   text entry 9
environments
   augmented environments 532
   smart environments 41
   virtual environments 233, 265
ergonomics 1
errors
   recognition errors 368
ethnography 1
evaluation 17
   empirical evaluation 89
   evaluation methods 25
   tool support for evaluation 502
   user evaluation 376
experiment 478
expertise location 65
expression
   facial expression of emotion 193
extraction
   key-frame extraction 185
eye
   eye movements 265, 273, 281
   eye tracking 249, 265, 281
face 193
facial
   facial affect 193
   facial expression of emotion 193
facilitation
   social facilitation 49
family 548
feedback
   force feedback .. 225, 408, 415, 423, 532
field
   field measurement tool 137
   field of view 478
   field trial 548
file
   log file analysis 462
finger/hand recognition 121
Fitts' law 9, 423
fluid
   fluid documents 249
   fluid user interfaces 249
focus + context 249
foraging
   information foraging 161
force feedback 225, 408, 415, 423, 532
frame
   key-frame extraction 185

freedom
   6 degree-of-freedom control 25
   6 degree-of-freedom input devices .. 540
   degrees of freedom 532
frequencies
   digraph frequencies 9
gaze 400
   gaze-added interfaces 273
   gaze-based interfaces 273
generation
   next-generation video playback interfaces 169
gesture input 518
gestures 81
   pen gestures 360
GOMS 502
graphic interfaces 532
graphical chat 462
graphics
   interactive 3D graphics 486
groupware
   single display groupware 556
Guiard theory 33
hand
   finger/hand recognition 121
   hand-held computers 1, 41
handed
   two-handed input 33, 41
   two-handed interaction 526
haptic information 532
haptics 225, 408, 415, 423
hardware
   user interface hardware 526
HCI professionals 337
head-mounted displays 478
health-support 73
hedonic components 201
held
   hand-held computers 1, 41
help systems 65
home 209
HTTP 430
human
   human-human interaction 57
   human performance 532
   human performance modeling 9
   human vision 113
hypermedia
   adaptive hypermedia 17
hypertext 257
   hypertext navigation 249
image quality 113
images
   digital images 548
index of difficulty 423
indexing
   video indexing 169
informal 510
information
   context information 257
   haptic information 532

575

Keyword Index

information appliances 209
information capture 438
information displays 313
information foraging 161
information retrieval 121, 345
information scent 161
information visualization 153, 161
peripheral information 241
visual information 532
input 33
   6 degree-of-freedom input devices .. 540
   ambiguous input 368
   gesture input 518
   input devices 25, 137
   input models 368
   keypad input 9
   two-handed input 33, 41
inspection
   usability inspection 353
instructional interface design 97
instrumental interaction 446
intelligent
   intelligent interfaces 273
   intelligent tutoring systems 97
interaction
   human-human interaction 57
   instrumental interaction 446
   interaction design 81, 217
   interaction model 446
   interaction techniques 25, 33, 265,
                        281, 368, 408, 540
   multimodal interaction 376, 415
   nonvisual interaction 376
   online interaction 470
   physical interaction 137
   symmetric interaction 33
   two-handed interaction 526
interactive
   interactive 3D graphics 486
   interactive 3D rotations 540
   interactive music systems 376
interest
   degree of interest 153
interface
   customer interface 305
   instructional interface design 97
   interface design 376
   social interface agents 57
   user interface design 17, 241, 313
   user interface hardware 526
interfaces
   3D interfaces 408, 494, 540
   fluid user interfaces 249
   gaze-added interfaces 273
   gaze-based interfaces 273
   graphic interfaces 532
   intelligent interfaces 273
   multimodal interfaces 81
   next-generation video playback
     interfaces 169

pen-based interfaces 360, 368
post-WIMP interfaces 446
recognition-based interfaces 368
social interfaces 462, 470
speech user interfaces 329
studies of dynamic user interfaces .. 249
tangible interfaces 129, 137
user interfaces 145, 281
WIMP interfaces 446
zooming user interfaces 510
internet 297
   internet appliances 1
   internet shopping 305
involvement 305
Java 65
joy of use 201
key-frame extraction 185
keypad input 9
kiosk
   public kiosk 313
knowledge
   spatial knowledge 478
law
   Fitts' law 9, 423
layout
   usage-based layout 161
learning 129
   computer supported
     collaborative learning 556
   computer supported learning 121
leisure 548
library
   digital video library 177
line
   on-line reading 249
   on-line travel planning systems .. 289
list
   discussion list 73
location
   expertise location 65
locus of control 49
log file analysis 462
longest repeated subsequences 161
loyalty
   customer loyalty 305
lurker 73
lurking 73
M
   the M-metric 25
machine
   machine vision 313
   support vector machine 145
managers
   window managers 494
manipulation
   direct manipulation 446
map
   cognitive map 478
matchmaking 65

measurement
   field measurement tool 137
measures
   usability measures 345
media
   media space 81
   streaming media 384
   tangible media 217
mediated
   computer-mediated 81, 392
     communication
meeting
   virtual meeting 400
   virtual meeting place 57
membership 73
memory
   spatial memory 494
messages
   organization of messages 105
methodology 337
methods
   evaluation methods 25
metric
   the M-metric 25
metrics
   simplification metrics 113
mining
   data mining 161
mixed reality 233
mobile
   mobile computing 17
   mobile phones 9
   mobile systems 9
model 105
   interaction model 446
   model simplification 113
modeling
   human performance modeling 9
   object-oriented modeling 518
   student modeling 97
models
   cognitive models 217
   input models 368
   user models 217, 273
monitor 478
motor control 25, 540
mounted
   head-mounted displays 478
mouse 423
movements
   eye movements 265, 273, 281
multi-dimensional scaling 360
multicast 81
multimedia 177
   multimedia organization 564
multimodal
   multimodal interaction 376, 415
   multimodal interfaces 81
music
   interactive music systems 376

Keyword Index

naming time 113
navigation 257
   hypertext navigation 249
   social navigation 81
newsgroup 73
next-generation video playback
   interfaces 169
nonverbal communication 193
nonvisual interaction 376
note-taking 89
object
   object-oriented modeling 518
   tagged object 225
online
   online community 462
   online interaction 470
organization
   multimedia organization 564
   organization of messages 105
organizational change 337
orientation-aware 137
overview
   visual overview 289
Palm Pilot 41, 430
pause removal 169
PDAs 41, 430, 438
pebble 41
pen
   pen-based computers 510
   pen-based computing 1
   pen-based interfaces ... 360, 368
   pen gestures 360
perceived software quality 201
perception 360
   user perception 297
performance
   human performance 532
   human performance modeling ... 9
peripheral information 241
personal digital assistants .. 41, 430
personality 329
phones
   mobile phones 9
photography
   digital photography 564
physical interaction 137
Pilot
   Palm Pilot 41, 430
place
   virtual meeting place 57
plan scaffolding 97
planning
   corporate planning 337
   on-line travel planning systems 289
playback
   next-generation video playback
     interfaces 169
pointing devices 423
Polhemus tracker 265

post-WIMP interfaces 446
presence
   tele-presence 233
presentation
   tele-presentation 384
process
   design process 225
product concept design 548
professionals
   HCI professionals 337
projection screen 478
prototypes 548
prototyping
   rapid prototyping 510
proxemics 462
proxy 430
public kiosk 313
quality
   image quality 113
   perceived software quality .. 201
   quality of service 297
rapid prototyping   510
reading
   on-line reading 249
reality
   augmented reality 121, 217, 233
   mixed reality 233
   virtual reality ... 265, 408, 478, 526, 532
recognition
   finger/hand recognition 121
   recognition-based interfaces ... 368
   recognition errors 368
   speech recognition 313, 368
recommendations 153
recorders
   voice recorders 438
removal
   pause removal 169
repeated
   longest repeated subsequences .. 161
representation 81
research
   design research 209
retrieval
   information retrieval ... 121, 345
rotations
   interactive 3D rotations ... 540
satisfaction 345
scaffolding
   plan scaffolding 97
scaling
   multi-dimensional scaling .. 360
scanners 438
scent
   information scent 161
screen
   projection screen 478
SDG 556
search 145

server
   client-server architecture .. 289
service
   quality of service 297
shared documents 454
shopping
   internet shopping 305
similarity 360
   similarity-attraction effect .. 329
simplification
   model simplification 113
   simplification metrics 113
single display groupware 556
site
   on-site applications 137
sketching 510
skim
   video skim 177
smart environments 41
social
   computers as social actors .. 329
   social agents 49
   social cyberspace 462
   social dilemma 470
   social facilitation 49
   social interface agents 57
   social interfaces 462, 470
   social navigation 81
software
   perceived software quality .. 201
solver
   constraint solver 289
space
   media space 81
spatial
   spatial cognition 494
   spatial knowledge 478
   spatial memory 494
speech 81
   speech access 89
   speech as data 89
   speech recognition 313, 368
   speech user interfaces 329
   synthetic speech 321
   text-to-speech 321, 329
spreading activation 153
steering task 423
sticky chat 454
storytelling 548, 556, 564
strategic usability 337
streaming media 384
structure 257
student modeling 97
studies of dynamic user interfaces .. 249
study
   diary study 438
   user study ... 145, 321, 345, 518
subsequences
   longest repeated subsequences .. 161

Keyword Index

summarization
    video summarization 177, 185
support
    health-support 73
    support vector machine 145
    tool support for evaluation 502
supported
    computer supported
        collaborative learning 556
    computer supported
        cooperative work 556
    computer supported learning 121
symmetric interaction 33
synchronous communication 454
synthetic speech 321
systems
    help systems 65
    intelligent tutoring systems 97
    interactive music systems 376
    mobile systems 9
    on-line travel planning systems 289
    web systems 257
tagged object 225
taking
    note-taking 89
talking avatars 313
tangible 225
    tangible interfaces 129, 137
    tangible media 217
targeting task 423
task
    dual-task tradeoffs 241
    steering task 423
    targeting task 423
    task awareness 392
    task context 532
    virtual docking task 25
techniques
    interaction techniques 25, 33, 265,
        .... 281, 368, 408, 540
tele
    tele-embodiment 233
    tele-presence 233
    tele-presentation 384
testing
    usability testing 345
text 145
    text-based chat 454
    text entry 9
    text-to-speech 321, 329
theory
    Guiard theory 33
thinking
    design thinking 217
time
    naming time 113
    time compression 169
token 225
tool 225

field measurement tool 137
    tool support for evaluation 502
toolkits 368
tools
    animation authoring tools 486
    CASE tools 518
toys 129
tracker
    Polhemus tracker 265
tracking
    eye tracking 249, 265, 281
tradeoffs
    dual-task tradeoffs 241
traffic 73
transaction cost 305
travel
    on-line travel planning systems 289
tree
    dome tree 161
trial
    field trial 548
trust 305
TTS 329
tutoring
    intelligent tutoring systems 97
two-handed
    two-handed input 33, 41
    two-handed interaction 526
ubiquitous computing 41
usability 161, 337
    emotional usability 201
    strategic usability 337
    usability inspection 353
    usability measures 345
    usability testing 345
usage-based layout 161
use
    document use 438
    joy of use 201
user 105
    fluid user interfaces 249
    speech user interfaces 329
    studies of dynamic user interfaces .. 249
    user evaluation 376
    user interface design 17, 241, 313
    user interface hardware 526
    user interfaces 145, 281
    user models 217, 273
    user perception 297
    user study 145, 321, 345, 518
    zooming user interfaces 510
vector
    support vector machine 145
Venn diagram 121
video
    digital video 169
    digital video library 177
    next-generation video playback
        interfaces 169

    video abstraction 177
    video browsing 169, 177, 185
    video indexing 169
    video skim 177
    video summarization 177, 185
videophones 400
view
    field of view 478
virtual
    virtual community 462
    virtual docking task 25
    virtual environments 233, 265
    virtual meeting 400
    virtual meeting place 57
    virtual reality ... 265, 408, 478, 526, 532
vision
    computer vision 121
    human vision 113
    machine vision 313
visual
    visual information 532
    visual overview 289
visualization
    information visualization 153, 161
voice recorders 438
voicemail 89
walkthrough
    cognitive walkthrough 353
Web 430
    web design 510
    web systems 257
    World Wide Web 145, 161
whiteboards
    electronic whiteboards 518
WIMP
    post-WIMP interfaces 446
    WIMP interfaces 446
window managers 494
Windows CE 41
wireless 430
    wireless communication 548
work
    asynchronous work 392
    computer supported
        cooperative work 556
    distributed work 392
workgroups 392
World Wide Web 145, 161
zooming user interfaces 510
ZUI 510

Talking In Circles: Designing A Spatially-Grounded Audioconferencing Environment (Page 81)

Roy Rodenstein, Judith Donath

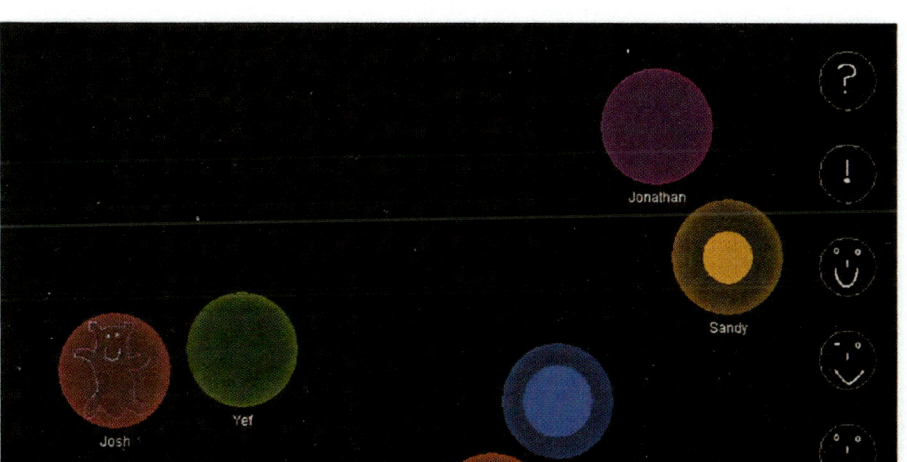

Figure 1: Three conversations in a *Talking in Circles* chat session. Al is talking with Helen and Andy, and Sandy converses with Jonathan. Meanwhile, Josh draws for Yef.

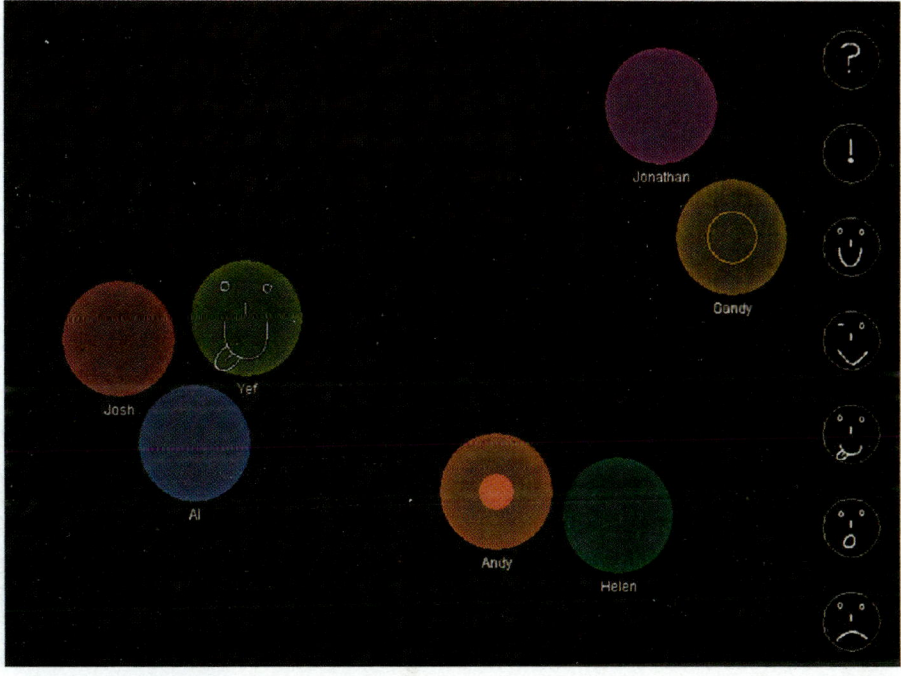

Figure 2: Screenshot from the point of view of Al, the blue circle, after he has wandered over to speak with Josh and Yef. Both Sandy and Andy are speaking, but Andy remains within Al's hearing range while Sandy is now outside it.

Enhancing a Digital Book with a Reading Recommender (Page 153)
Allison Woodruff, Rich Gossweiler, James Pitkow, Ed H. Chi, Stuart K. Card

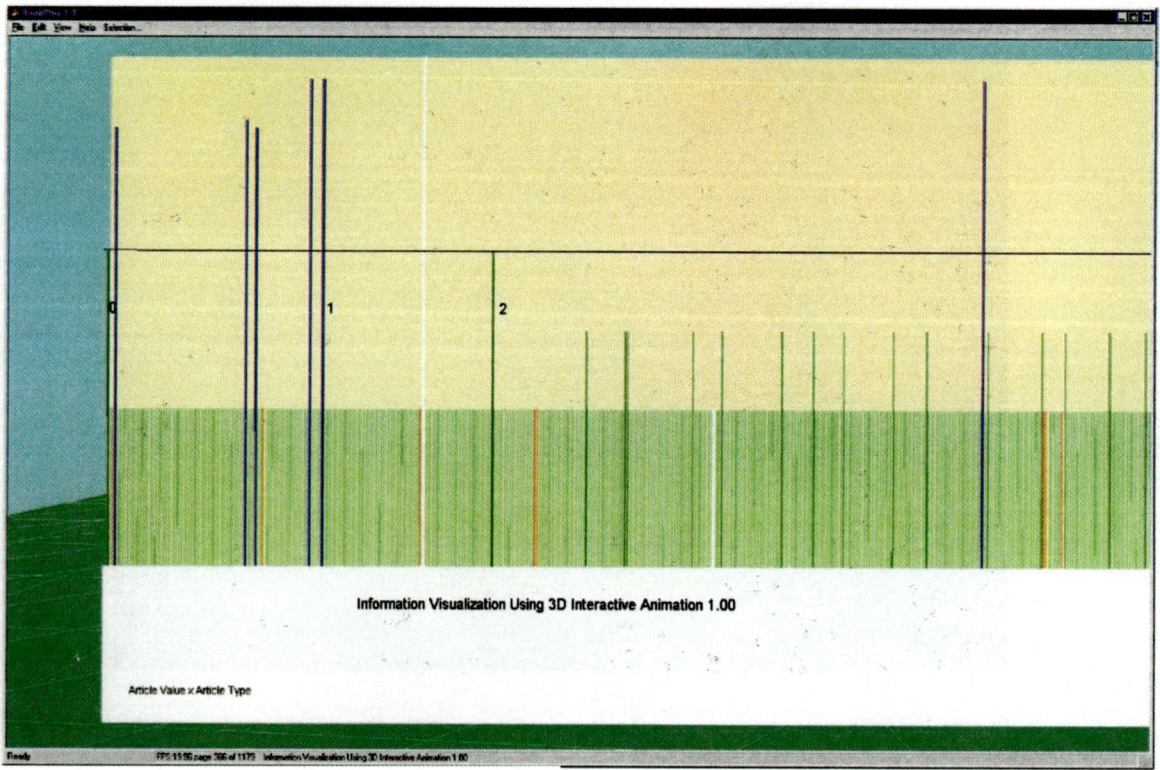

Figure 3. Close-up of a section of the Book Ruler.

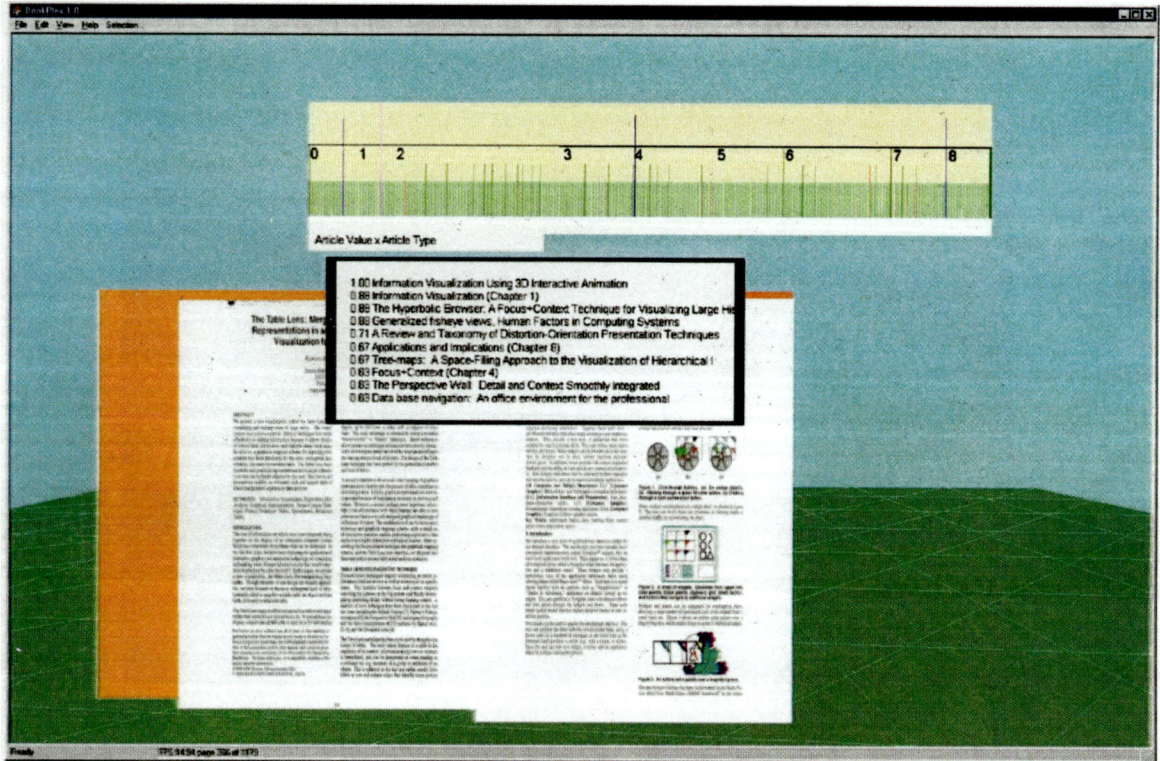

Figure 4. Book, Book Ruler, and recommendation list.

The Scent of a Site: A System for Analyzing and Predicting Information Scent, Usage, and Usability of a Web Site (Page 161, plate 1)

Ed H. Chi, Peter Pirolli, James Pitkow

Figure 3: Dome Tree with Usage-Based Layout (left) shows that links (shown in yellow) are laid along significant paths (shown by orange arrow), eliminating crossings. In comparison, the traditional Disk Tree approach (right) has many crossing yellow links (shown in enclosed orange box). White arrows point to the current document being examined (investor.html).

Figure 4: Multi-way Branching Point (investor/sitemap.htm) shown enclosed by orange lines.

Figure 5: Pass-through Point in a series of pages (marked by orange arrows and current page pointed by white arrow is annualreport/1997/market.htm)

Color Plates

The Scent of a Site: A System for Analyzing and Predicting Information Scent, Usage, and Usability of a Web Site (Page 161, plate 2)
Ed H. Chi, Peter Pirolli, James Pitko

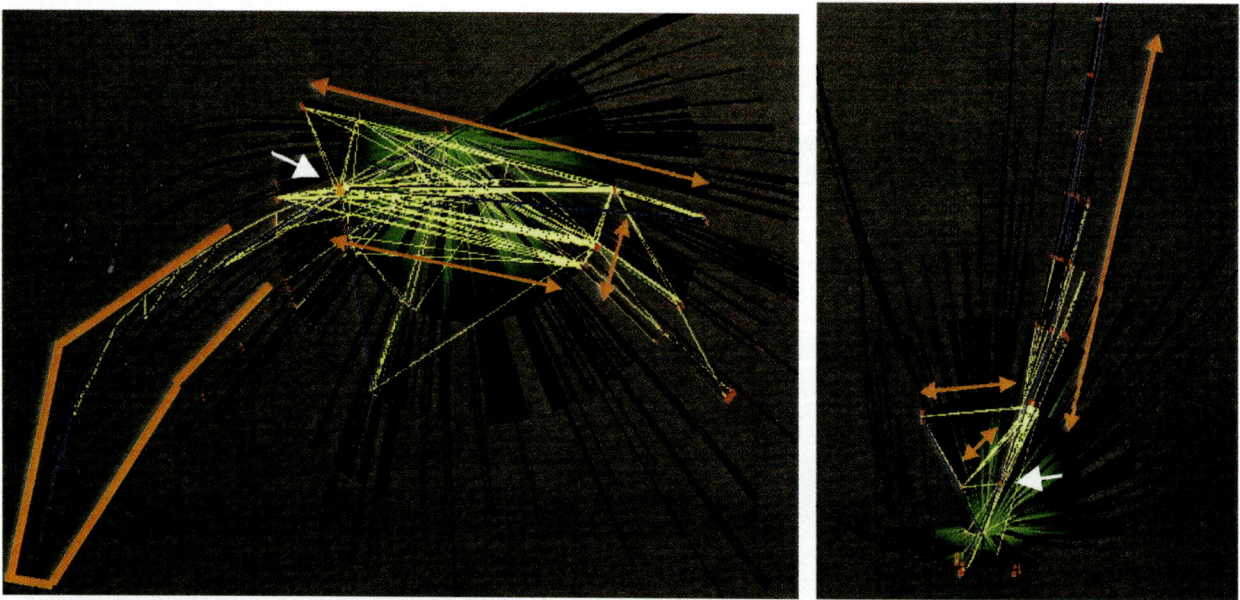

Figure 6: Well-traveled paths related to scansoft/tbpro98win/index.htm (left) and scansoft/pagis/index.htm (right), where major traffic routes are marked by orange lines and arrows.

Figure 9: Given an information need related to "Pagis", Scent Flow simulation results in good match in scansoft/pagis/index.html (left, good match points pointed by orange arrows), but poor match from products.html (right, bad match points pointed to by purple arrows).

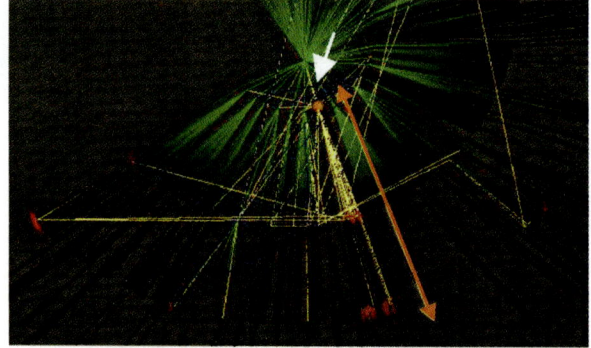

Figure 8: from xis/tbpro96win/index.htm

Color Plates

Hedonic and Ergonomic Quality Aspects Determine a Software's Appeal (Page 201)
Marc Hassenzahl, Axel Platz, Michael Burmester, Katrin Lehner

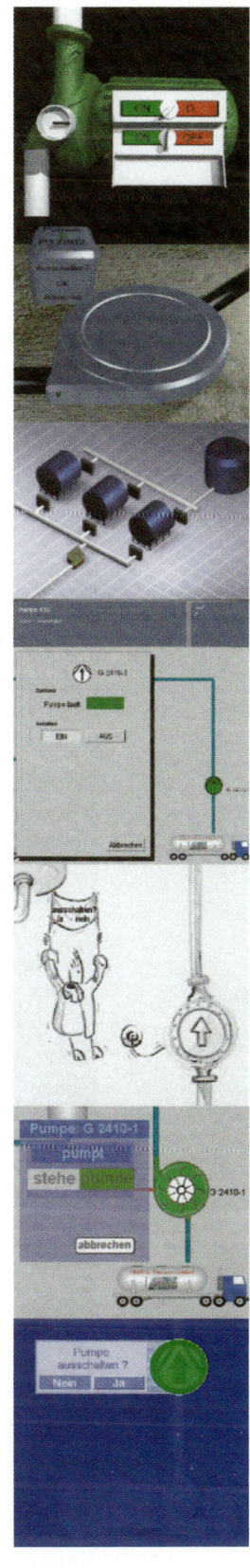

Color Plates

Traversable Interfaces Between Real and Virtual Worlds (Page 233)
Boriana Koleva, Holger Schnädelbach, Steve Benford, Chris Greenhalgh

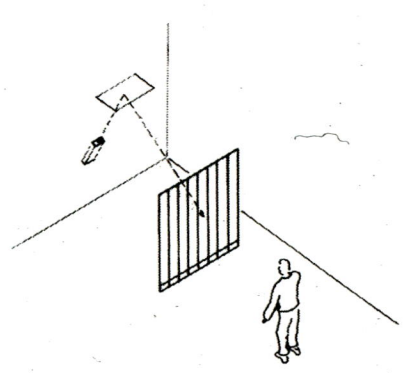
(a) Fabric curtain design

(b) Emerging through the fabric curtain

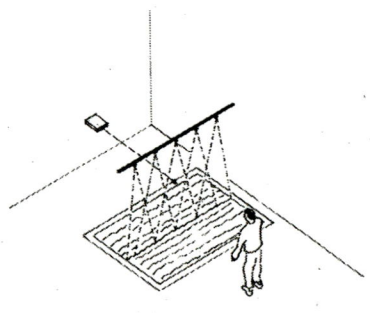
(c) Rain curtain design

(d) Rain curtain infrastructure   (e) Rain curtain in use

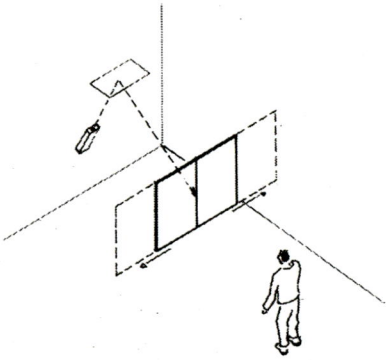
(f) Sliding door design

(g) Opening the sliding door

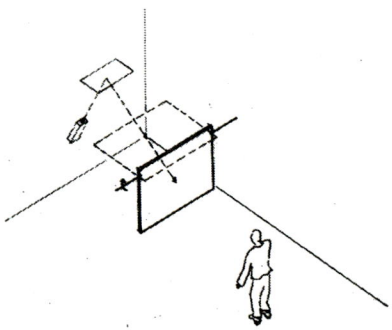
(h) Flip-up screen design

(i) Raised as ambient display   (j) Raising the screen

Figure 4: four designs for non-solid projection

The Impact of Fluid Documents on Reading and Browsing: An Observational Study (Page 249)

Polle Zellweger, Susan Harkness Regli, Jock Mackinlay, Bay-Wei Chang

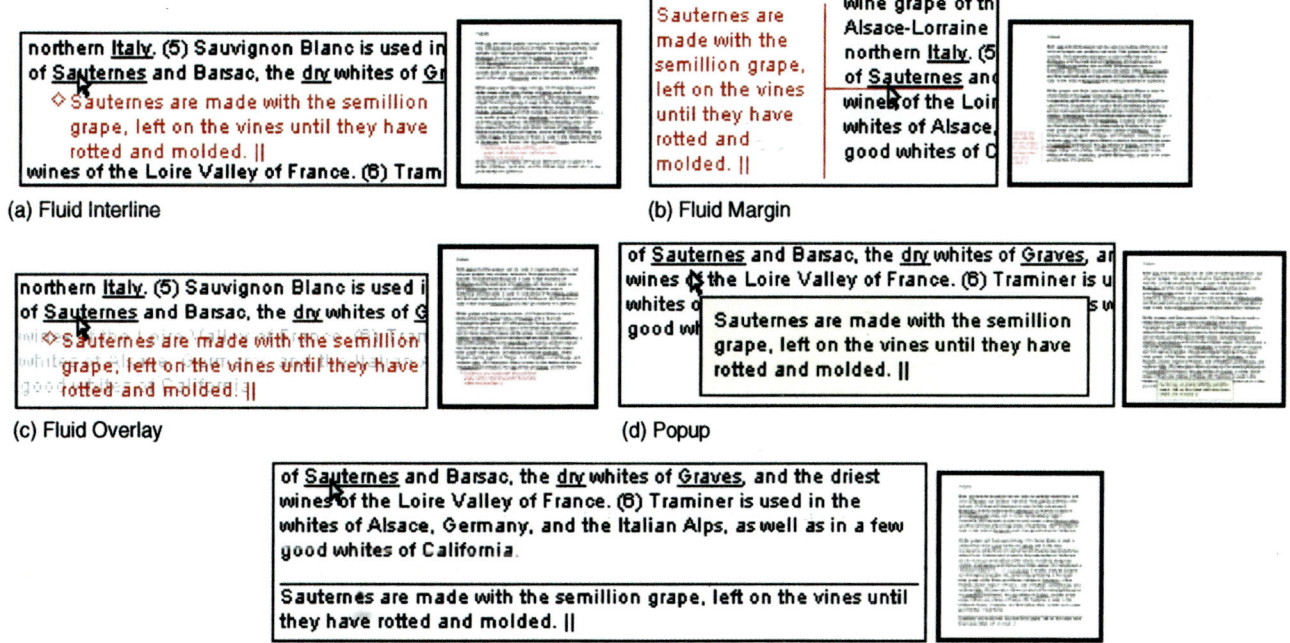

Figure 1. Fluid and conventional glosses, along with thumbnails of the pages of text from which they are excerpted. Font sizes have been greatly reduced for this figure; in the study, the primary text size was chosen to have a visual size of 12 points.

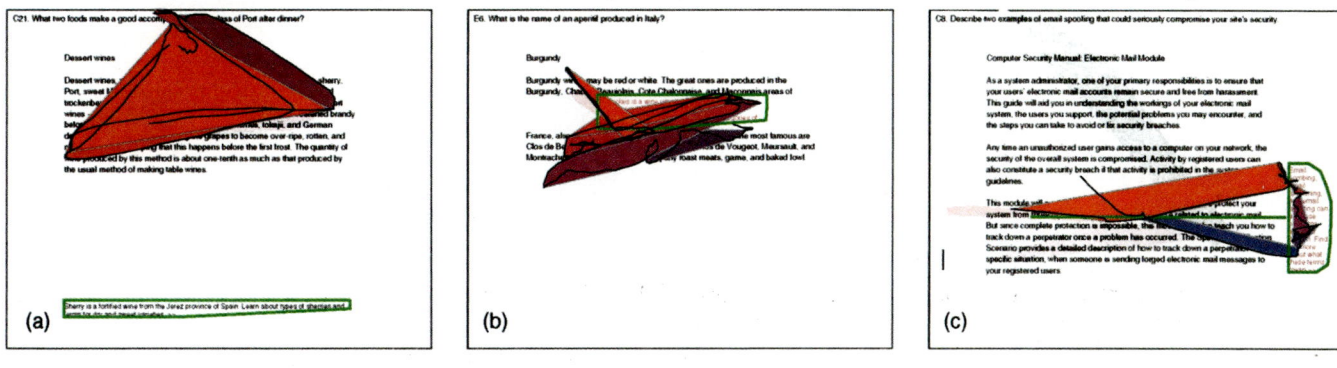

Figure 3. Gaze region visualizations of gloss events. The bold green line outlines the gloss, while the thin black line traces the eye's point of regard within the gaze region. The gaze region during the first second of each event is shown in bright red and may occlude later regions; gaze regions for successive seconds are shown in darker colors. The pale colors behind the text show gaze regions for three seconds before the event. (a) shows a Footnote event that the subject does not see. (b) shows a Fluid Interline event that the subject sees and reads. (c) shows a Fluid Margin event that the subject reads with seeming difficulty (note twists in black line).

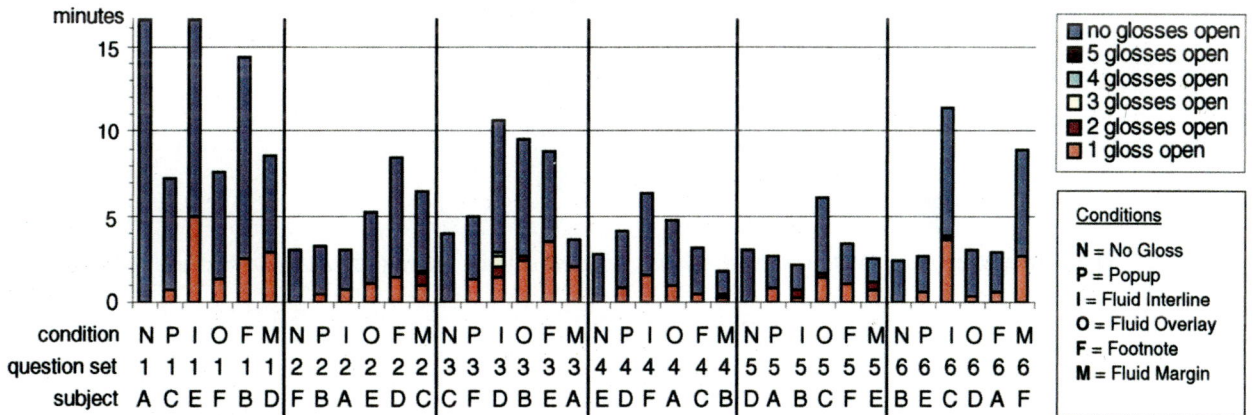

Figure 4. Gloss usage (pruned static time) per subject in each question set.

Color Plates

Interacting with Eye Movements in Virtual Environments (Page 265)
Vildan Tanriverdi, Robert Jacob

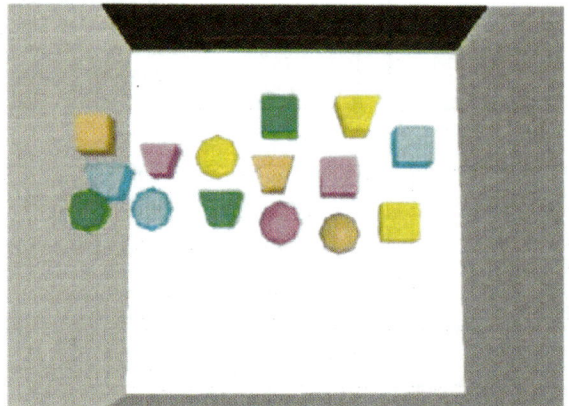

Figure 1. The entire virtual room.

Figure 2. A portion of the virtual room. None of the objects is selected in this view.

Figure 3. The purple object near the top of the display is selected, and its internal details have become visible.

Appendix A: Pilot Survey Questions (from Rosenbaum, Rohn, and Humburg paper)

A. About Your Company
1. How large is your company? a. Number of employees _____ b. Annual revenue _____ (individual consultants need not supply this)
2. How long has your company been in business? _____ years
3. Please describe the organizational structure of your company.
4. Please describe in a sentence or two what your company does.
5. In what category is your company?
   - _____ Aerospace
   - _____ Automotive
   - _____ Computer
   - _____ Education/Training
   - _____ Financial Services
   - _____ Government
   - _____ Health/Medical Services
   - _____ HCI/Usability
   - _____ Internet/E-Commerce
   - _____ Manufacturing
   - _____ Telecommunications
   - _____ Oil & Gas/Petroleum
   - _____ Publishing
   - _____ Retail/Wholesale
   - _____ Securities
   - _____ Other _____
6. What functional areas in your company determine product requirements and business direction?
7. Does your company have an R & D (advanced technology) function that isn't directly tied to product development? _____ YES _____ NO
8. If so, what impact does it have on the product development cycle?
9. Are any reward systems in your company formally tied to usability goals? If so, how? (Consultants please leave this question blank.)

B. About Your HCI/Usability Group
10. How many people are in your group and in your company, what are their respective roles, and how long has each worked in the usability field?
11A. If your group includes both interface (visual and interaction) or information designers AND usability testers, do the roles overlap in any way? YES _____ NO _____
11B. Also, do the designers evaluate, or test, their own designs? YES _____ NO _____ SOMETIMES _____
12A. To what department or functional area does your group report?
12B. What is the title of the person outside your group to whom it reports?
12C. To help us understand levels within respective organizations and reporting chains, please list the management titles between your group and the CEO of your company.
13. How are independent consultants used to complement the work of your group (check all that apply)?
    - _____ As ongoing staff resources
    - _____ For specialized expertise (Describe briefly:_____)
    - _____ For overflow work, on an 'as needed' basis
    - _____ For field studies, or some specific type of research
    - _____ When employees are on leave
    - _____ Other _____
    - _____ Don't use consultants
14. On the average, what percentage of your group's work is performed by independent consultants?
15. How centralized or decentralized are your organization's group(s) in their relationships to product development?
16. Briefly describe your product development process and how human factors activities fit into it (1 or 2 paragraphs or a bulleted list). Consultants, please describe the process you encounter most frequently.
17. How is your group funded (for example, annual budget, bill-back by project)? Consultants, please describe the process you encounter most frequently.
18. What are the top two *obstacles* you face in creating greater strategic impact for usability engineering/HCI within your organization?

C. What Techniques Have You Tried and Were They Effective in Creating a More Strategic Impact for Your Efforts?
19A. Please indicate with a check mark the *organizational approaches* you have tried and indicate with a number rating how effective each was in creating or improving your strategic impact within your company. Effectiveness Rating Scale:
1=EXTREMELY Effective in creating or improving my (group's) strategic impact
2=SOMEWHAT Effective 3=Neutral 4=LESS Effective 5=NOT AT ALL Effective

ORGANIZATIONAL APPROACHES
Have Tried? Rate How Effective [Approaches are listed in Table 1]

19B. Please indicate with a check mark the *usability methodologies* you have tried and indicate with a number rating how effective each was in creating or improving your strategic impact within your company. Effectiveness Rating Scale:
1=EXTREMELY Effective in creating or improving my (group's) strategic impact
2=SOMEWHAT Effective 3=Neutral 4=LESS Effective 5=NOT AT ALL Effective

USABILITY METHODOLOGIES
Have Tried? Rate How Effective [Approaches are listed in Table 1]

D. Attributes for the 'Ideal' Company Environment Where Usability has a Strategic Role
20. Choose a metaphor that best describes how you might conceptualize the 'Ideal' Company Environment wherein usability efforts play a strategic role in setting business and product direction.
    - _____ Garden
    - _____ Zoo
    - _____ Train and Train Station
    - _____ Circus
    - Choose one of your own! _____

Next, use phrases or keywords in the context of the metaphor to identify the TOP 5 characteristics or attributes that describe this environment, or culture, wherein usability contributes at the strategic level of setting business and product direction.

A Toolkit for Strategic Usability: Results from Workshops, Panels, and Surveys (Page 337, plate 2)

Stephanie Rosenbaum, Janice Anne Rohn, Judee Humburg

Appendix B: CHI and UPA Survey Questions (from Rosenbaum, Rohn, and Humburg paper)

1. How large is your company?
   - ____ sole practitioner
   - ____ 2-5 employees
   - ____ 6-10 employees
   - ____ 11-25 employees
   - ____ 26-50 employees
   - ____ 51-100 employees
   - ____ 101-250 employees
   - ____ 251-500 employees
   - ____ 501-1,000 employees
   - ____ 1,001-5,000 employees
   - ____ 5,001-10,000 employees
   - ____ over 10,000 employees

2. How long has your company been in business?
   - ____ less than 1 year
   - ____ 1-2 years
   - ____ 3-5 years
   - ____ 6-10 years
   - ____ 11-15 years
   - ____ 16-20 years
   - ____ 21-30 years
   - ____ 31-40 years
   - ____ 41-50 years
   - ____ over 50 years

3. Please describe in a sentence or two what your company does.

4. In what category is your company?
   - ____ Aerospace
   - ____ Automotive
   - ____ Computer
   - ____ Education/Training
   - ____ Financial Services
   - ____ Government
   - ____ Health/Medical Services
   - ____ HCI/Usability Consulting
   - ____ Internet/E-Commerce
   - ____ Manufacturing
   - ____ Telecommunications
   - ____ Oil & Gas/Petroleum
   - ____ Publishing
   - ____ Retail/Wholesale
   - ____ Securities
   - ____ Other_____

5. How many HCI/usability people are in your company? _____

6. How is your group funded (for example, annual budget, bill-back by project)? Consultants, please describe the process you encounter most frequently.

7. What are the top two OBSTACLES you face in creating greater strategic impact for usability engineering/HCI within your organization?

8. Please indicate the ORGANIZATIONAL APPROACHES you have tried and rate how effective each was in creating or improving strategic usability within your company. See above definition of strategic usability. Only assign ratings to approaches you personally have used.

   Effectiveness Rating Scale:
   1=EXTREMELY Effective in creating or improving my (group's) strategic impact
   2=SOMEWHAT Effective
   3=Neutral
   4=LESS Effective
   5=NOT AT ALL Effective

   RATE HOW EFFECTIVE
   - _____ Organizational Audits (UCD Analysis of Org.)
   - _____ High-Level/Founder Support
   - _____ UI Group Reports to UI, not Development
   - _____ Leveraging Related Initiatives
   - _____ Fit into Current Engineering Processes
   - _____ High-Profile Projects
   - _____ User Interface Committees
   - _____ Usability Open Houses
   - _____ Usability Advocates/Champions
   - _____ Other (Please name & describe BRIEFLY)
   - _____ Organizational Usability Planning
   - _____ Partnering/Collaborating with Marketing on Projects
   - _____ UI Staff Members Co-located with Engineering
   - _____ Corporate Mandates/Usability Objectives
   - _____ Internal Task Forces
   - _____ Communities of Practice—Alliances with Academia/Industry
   - _____ Coach/Support Grass Roots Efforts
   - _____ Design Review Boards
   - _____ Educate/Train Other Functional Groups (e.g., Marketing, Development and/or Documentation)

9. Please indicate the USABILITY METHODOLOGIES you have tried and rate how effective each was in creating or improving strategic usability within your company. See above definition of strategic usability. Only assign ratings to approaches you personally have used.

   Effectiveness Rating Scale:
   1=EXTREMELY Effective in creating or improving my (group's) strategic impact
   2=SOMEWHAT Effective
   3=Neutral
   4=LESS Effective
   5=NOT AT ALL Effective

   RATE HOW EFFECTIVE
   - _____ Contextual Inquiry
   - _____ Task Analysis
   - _____ Participatory Design
   - _____ Surveys
   - _____ Lab Usability Testing
   - _____ Usability Testing with Portable Lab Equipment
   - _____ Field Studies other than CI
   - _____ Usage Scenarios
   - _____ Focus Groups
   - _____ Heuristic Evaluation
   - _____ Usability Testing Outside of a Lab Facility
   - _____ Other (Please name & describe BRIEFLY)

10. How successful overall is strategic usability in your organization (if you're a consultant, how successful overall is strategic usability in the client organizations where you consult)? See above definition of strategic usability.
    ____ 1 Very successful ____ 2 Quite successful ____ 3 Somewhat successful ____ 4 Neutral ____ 5 Somewhat unsuccessful ____ 6 Quite unsuccessful ____ 7 Very unsuccessful

NOTES

NOTES

NOTES

NOTES

NOTES

NOTES